Principles of Physical Chemistry

Principles of Physical Chemistry

Understanding Molecules, Molecular Assemblies, Supramolecular Machines

Hans Kuhn
*Max Planck Institute for Biophysical Chemistry,
Göttingen, Germany*

and

Horst-Dieter Försterling
Philipps Universität Marburg, Germany

JOHN WILEY & SONS, LTD
Chichester • New York • Weinheim • Brisbane • Singapore • Toronto

Copyright © 2000 by John Wiley & Sons Ltd,
Baffins Lane, Chichester,
West Sussex PO19 IUD, England

National 01243 779777
International (+44) 1243 779777
e-mail (for orders and customer service enquiries): cs-books@wiley.co.uk
Visit our Home Page on http://www.wiley.co.uk
or http://www.wiley.com

All Rights Reserved. No part of this publication may be reproduced, stored in a retrieval system, or transmitted, in any form or by any means, electronic, mechanical, photocopying, recording, scanning or otherwise, except under the terms of the Copyright, Designs and Patents Act 1988 or under the terms of a licence issued by the Copyright Licensing Agency, 90 Tottenham Court Road, London UK WIP 9HE, without the permission in writing of the publisher.

Other Wiley Editorial Offices

John Wiley & Sons, Inc., 605 Third Avenue,
New York, NY 10158-0012, USA

WILEY-VCH Verlag GmbH, Pappelallee 3,
D-69469 Weinheim, Germany

Jacaranda Wiley Ltd, 33 Park Road, Milton,
Queensland 4064, Australia

John Wiley & Sons (Asia) Pte Ltd, 2 Clementi Loop #02-01,
Jin Xing Distripark, Singapore 129809

John Wiley & Sons (Canada) Ltd, 22 Worcester Road,
Rexdale, Ontario M9W 1L1, Canada

Library of Congress Cataloging-in-Publication Data

Kuhn, H. (Hans)
   Principles of physical chemistry / H. Kuhn and H.D. Försterling.
     p. cm.
   Includes bibliographical references and index.
   ISBN 0-471-95902-2 (hb). − ISBN 0-471-96541-3 (pb)
   1. Chemistry, Physical and theoretical.   I. Försterling, H. D.
II. Title
QD453.2.K84   1999
541.3 − dc21   98-48542
                                                                                   CIP

British Library Cataloguing in Publication Data

A catalogue record for this book is available from the British Library

ISBN 0 471 95902 2 (HB)   0 471 96541 3 (PB)

Typeset in 10/12pt Times by Laser Words, Madras, India
Printed and bound in Great Britain by Bookcraft (Bath) Ltd, Midsomer Norton, Avon
This book is printed on acid-free paper responsibly manufactured from sustainable forestry, in which at least two trees are planted for each one used for paper production.

Contents

Acknowledgements			xxiii
Preface and Guided Tour			xxv
1	**Wave–Particle Duality**		**1**
	1.1 Light		1
		1.1.1 The Particle Nature of Light: Photoelectric Effect	1
		1.1.2 Wave Nature of Light: Diffraction	3
		1.1.3 Interpretation of the Experiments	7
	1.2 Electrons		10
		1.2.1 The Particle Nature of Electrons	10
		1.2.2 Wave Nature of Electrons	11
		1.2.3 Interpretation of the Experiments	13
	1.3 Questions Arising About Wave–Particle Duality		13
		1.3.1 Single Event: Probability Statement; Collective Behavior: Definite Statement	13
		1.3.2 Wave–Particle Duality: The Need To Abandon Familiar Ways of Thinking	15
	Problems		17
		Problem 1.1 – Double-slit Experiment: Distribution of $|\psi|$ on the Screen	17
		Problem 1.2 – Diffraction of Photons, Electrons, and Neutrons	17
2	**Basic Features of Bonding**		**19**
	2.1 Distinct Energy States		20
		2.1.1 Atomic Spectra	20
		2.1.2 The Franck–Hertz Experiment	20
	2.2 Standing Waves		22
		2.2.1 A Particle Between Parallel Walls	22
		2.2.2 The Heisenberg Uncertainty Relation	25
		2.2.3 Meaningless Questions	26
		2.2.4 Electron in a Box	26

	2.3	H Atom in the Ground State	27
		2.3.1 Box Model for H Atom	28
		2.3.2 Variation Principle	28
	2.4	H_2^+ Molecule Ion in the Ground State	31
		2.4.1 Forming H_2^+ from an H Atom and a Proton	32
		2.4.2 Box Model for H_2^+	32
	2.5	He^+ Ion, He Atom, and Similar Systems	33
		2.5.1 He^+ Ion and He Atom	33
		2.5.2 He-like Systems: H^-, Li^+, Be^{2+}	34
		2.5.3 H_2 Molecule	35
	2.6	Pauli Exclusion Principle	36
		2.6.1 Nonexistence of He_2	36
		2.6.2 Li^+H^- (Ionic Crystal)	36
	Problems		37
		Problem 2.1 – Franck–Hertz Experiment	37
		Problem 2.2 – Standing Waves Formed by Overlap of Traveling Waves	37
3	**Schrödinger Equation and Variation Principle**		**39**
	3.1	Wave Equation and Schrödinger Equation	39
		3.1.1 Wave Equation	39
		3.1.2 Schrödinger Equation (One-dimensional)	42
		3.1.3 Schrödinger Equation (Three-dimensional)	44
	3.2	Normalization of the Wavefunctions	47
	3.3	Orthogonality of the Wavefunctions	50
	3.4	Time-dependent Schrödinger Equation	52
	3.5	H Atom in the Ground State	53
		3.5.1 Wavefunction	53
		3.5.2 Energy	54
		3.5.3 Radial Probability Distribution of Electron	54
		3.5.4 Average Potential Energy and Virial Theorem	57
		3.5.5 Most Probable and Average Distance of Electron from Nucleus	58
	3.6	H Atom in Excited States	58
		3.6.1 Energies and Wavefunctions	58
		3.6.2 Emission Spectra	61
	3.7	Variation Principle	65
		3.7.1 Justification of the Variation Principle	65
	3.8	Bohr Atomic Model and Correspondence Principle	69
		3.8.1 Bohr Atomic Model	69
		3.8.2 Correspondence Principle	70
	Problems		71
		Problem 3.1 – H Atom: Ground State Derived from the Schrödinger Equation	71
		Problem 3.2 – H Atom: Normalization of Wavefunction	72
		Problem 3.3 – Probability Density ρ	72
		Problem 3.4 – H Atom: Average Distance Between Electron and Nucleus	72
		Problem 3.5 – H Atom: Particle in a Box Trial Function	73
		Problem 3.6 – He^+: Schrödinger Equation	73

CONTENTS

   Problem 3.7 – H Atom: Comovement of the Nucleus   74
   Problem 3.8 – Ground State Energy of H Atom by Variation
    Principle   75
  Foundation 3.1 – Electron in a Potential Trough of Finite Depth   77
  Foundation 3.2 – Relation Between Time-independent and Time-dependent
   Schrödinger Equation   79
  Foundation 3.3 – H-Atom: Solution of the Schrödinger Equation   80
  Foundation 3.4 – Proof of Variation Principle   82

4 Chemical Bonding and the Pauli Principle   85
 4.1 H_2^+ Molecule Ion   85
  4.1.1 Electron Described by Exact Wavefunction   85
  4.1.2 Electron Described by Box Wavefunctions   86
  4.1.3 Electron Density and Chemical Bond   89
  4.1.4 Virial Theorem   91
  4.1.5 Nature of the Chemical Bond   91
  4.1.6 Electron Described by Combination of Atomic Orbitals   91
 4.2 He Atom and Similar Systems   96
  4.2.1 Ground State of He   96
  4.2.2 Excited States of He   98
 4.3 Antisymmetry of Wavefunctions   99
  4.3.1 Energy Splitting in a Magnetic Field   99
  4.3.2 Spin Variables   102
  4.3.3 Pauli Exclusion Principle as Antisymmetry Postulate   102
  4.3.4 Singlet–Triplet Splitting Caused by Coulomb Forces   103
 4.4 Quantum Mechanical Tunneling   104
 Problems   104
  Problem 4.1 – Kinetic Energy in Box Model   104
  Problem 4.2 – Repulsion of the Two Electrons in He   104
  Problem 4.3 – Energy Separation of EPR Lines   105
 Foundation 4.1 – H_2^+ Ion: Exact Wave Function and Energy. Virial
  Theorem   106
 Foundation 4.2 – Evaluation of LCAO Integrals in H_2^+ 108
 Foundation 4.3 – He Atom: Energy in the Grounds State   111
 Foundation 4.4 – Oscillation of Electron between Protons at Distance d
  (Tunneling)   112

5 The Periodic Table and Simple Molecules   117
 5.1 Periodic Table of the Elements   117
  5.1.1 Basic Principles   117
  5.1.2 Hydrogen and Helium ($Z = 1$ and $Z = 2$)   118
  5.1.3 Lithium ($Z = 3$)   119
  5.1.4 Aufbau Principle and Periodic Table   120
 5.2 The Structure of Simple Molecules   124
  5.2.1 Simple Bond Models   124
  5.2.2 The Polarity of Bonds and Electronegativity   126
  5.2.3 Bond Lengths and Bond Angles   129
  5.2.4 Stretching and Bending Force Constants   133

	Problems	137
	Problem 5.1 – Partial Charges and Dipole Moment	137
	Problem 5.2 – Stretching Force Constant of H_2^+ Ion	137
	Problem 5.3 – Bond Angle of H_2O	139

6 Bonding Described By Hybrid and Molecular Orbitals — 140

- 6.1 Degeneracy of Energy Levels — 140
  - 6.1.1 Hybrid Functions — 142
  - 6.1.2 Hybridization of H Atom Functions — 144
- 6.2 Localized Electrons: Hybrid Atomic Orbitals — 147
  - 6.2.1 Li Atom (No Hybridization) — 147
  - 6.2.2 BeH_2 (Linear (sp) Hybridization: Two $s^{1/2}p^{1/2}$ Hybrid Orbitals) — 148
  - 6.2.3 H_2S and H_2O (Two Orbitals Between p and $s^{1/4}p^{3/4}$ Hybrid) — 148
  - 6.2.4 CH_4 (Tetrahedral (sp^3) Hybridization: Four $s^{1/4}p^{3/4}$ Hybrid Orbitals) — 150
- 6.3 Properties of Electron Pair Bonds — 151
  - 6.3.1 Formal and Effective Charges — 152
  - 6.3.2 Polymerization of BeH_2 — 154
- 6.4 Delocalized Electrons: Molecular Orbitals — 154
  - 6.4.1 Box Wavefunctions — 154
  - 6.4.2 LCAO Wavefunctions — 155
  - 6.4.3 Improvement of Trial Function — 159
- Problems — 159
  - Problem 6.1 – Linear Combinations of Wave Functions — 159
  - Problem 6.2 – Degeneracy and Hybridization of Box Functions — 159
  - Problem 6.3 – Orthogonality and Normalization of Hybrid Functions — 160
  - Problem 6.4 – Symmetry Properties of Linear Hybrid Functions — 160
  - Problem 6.5 – Energy of Hybrid States — 160
  - Problem 6.6 – Structure of $B(CH_3)_3$ and Hg_2Cl_2 — 161
  - Problem 6.7 – Dative Bond — 161
  - Problem 6.8 – Three-center Bond (CH_5^+) — 161
  - Problem 6.9 – LCAO Model for O_2 — 162
  - Problem 6.10 – Molecular Orbitals of Some Diatomic Molecules — 162

7 Molecules with π Electron Systems — 163

- 7.1 Bonding Properties of π Electrons — 163
- 7.2 Free-electron Model — 165
  - 7.2.1 Linear π Electron Systems — 165
  - 7.2.2 Cyclic π Electron Systems — 171
  - 7.2.3 Charge Density dQ/ds — 173
  - 7.2.4 Resonance — 175
  - 7.2.5 Branched Molecules — 177
- 7.3 HMO Model — 181
  - 7.3.1 Wavefunctions and Energies — 181
  - 7.3.2 Charge Density $dQ/d\tau$ — 184
- 7.4 Bond Lengths, Dipole Moments — 184
  - 7.4.1 Bond Length and Charge Density — 184
  - 7.4.2 Bond Alternation in Polyenes and Fullerenes — 187
  - 7.4.3 Dipole Moment — 187

	Problems	189
	Problem 7.1 – HMO Method: Ethene	189
	Problem 7.2 – HMO Method: Butadiene	190
	Problem 7.3 – HMO Method: Fulvene	191
	Problem 7.4 – Resonance of Hückel $(4n+2)$ Rings	192
	Problem 7.5 – Bond Lengths from Bond Orders	192
	Problem 7.6 – Cyclobutadiene: Bond Lengths	192
	Problem 7.7 – Dipole Moment of Fulvene	193
	Foundation 7.1 – Free Electron Model	195
	Foundation 7.2 – HMO Model	198
	Foundation 7.3 – Self-consistency in Bond Alternation	200
8	**Absorption and Emission of Light**	**204**
8.1	Basic Experimental Facts	204
	8.1.1 Transmittance and Absorbance	204
	8.1.2 Polyenes and Cyanines	207
8.2	Absorption Maxima of Dyes	209
	8.2.1 Band Broadening	210
	8.2.2 Single-molecule Absorption	210
	8.2.3 Cyanine Dyes	211
8.3	Strength and Polarization of Absorption Bands	213
	8.3.1 Oscillator Strength	213
	8.3.2 Polarization of Absorption Bands	215
8.4	Heteroatoms as Probes for Electron Distribution	216
8.5	HOMO–LUMO Gap by Bond Alternation	219
8.6	Dyes with Cyclic Electron Cloud: Phthalocyanine	223
8.7	Coupling of π Electrons	226
8.8	Light Absorption of Biomolecules	228
	8.8.1 β-Carotene	228
	8.8.2 Retinal	230
	8.8.3 Vitamin B_{12}	231
	8.8.4 Chlorophyll, Bacteriochlorophyll	232
8.9	Spontaneous Emission	233
	8.9.1 Fluorescence and Phosphorescence	233
	8.9.2 Single Molecule Emission	235
	8.9.3 Singlet and Triplet States	236
	8.9.4 Shift of Fluorescence and Phosphorescence Relative to Absorption	238
	8.9.5 Absorption from Excited States	240
	8.9.6 Quenching of Fluorescence	242
8.10	Stimulated Emission	242
	8.10.1 Inversion of Population	242
	8.10.2 Dye Laser	243
	8.10.3 Excimer Laser	246
8.11	Optical Activity	247
	8.11.1 Rotatory Dispersion	247
	8.11.2 Ellipticity	248
	8.11.3 Circular Dichroism	249

		8.11.4 Circular Dichroism of Spirobisanthracene	250
		8.11.5 Circular Dichroism of Chiral Cyanine Dye	253
	Problems		253
		Problem 8.1 – Lone Electron Pair at the Nitrogen	253
		Problem 8.2 – Light Absorption of Different Classes of Dyes	254
		Problem 8.3 – Energy Shift in Azacyanines	254
		Problem 8.4 – Shift of Energy Levels by Bond Alternation	256
		Problem 8.5 – Light Absorption of Phthalocyanine and Porphyrin	256
		Problem 8.6 – Oscillatory Strength in Phthalocyanine and Porphyrin	257
		Problem 8.7 – Cis-Peak in β-Carotene	258
		Problem 8.8 – Splitting of Absorption Band	258
		Problem 8.9 – Circular Dichroism of Cyanine Dye	259
	Foundation 8.1 – Integrated Absorption: Classical Oscillator		261
	Foundation 8.2 – Oscillator Strength: Quantum mechanical Treatment		265
	Foundation 8.3 – Coupling Transitions with Parallel Transition Moment		267
	Foundation 8.4 – Normal modes of Coupled Oscillators		270
	Foundation 8.5 – Fluorescence Life Time		274
	Foundation 8.6 – Proof of Relation for g (Anisotropy Factor)		276
9	**Nuclei: Particle and Wave Properties**		**278**
	9.1	Quantum Mechanical Rotator	278
		9.1.1 Exact Solution	279
		9.1.2 Simplified Model	280
	9.2	Rotational Spectra	282
	9.3	Quantum Mechanical Oscillator	287
		9.3.1 Exact Solution	288
		9.3.2 Box Model for Oscillator	291
		9.3.3 Comparison of a Quantum Mechanical Oscillator with a Classical Oscillator	293
	9.4	Vibrational–Rotational Spectra	294
		9.4.1 Diatomic Molecules	294
		9.4.2 Polyatomic Molecules	297
	9.5	Raman Spectra	303
		9.5.1 Rayleigh Scattering	303
		9.5.2 Rotational Raman Spectra	305
		9.5.3 Vibrational–Rotational Raman Spectra of Diatomic Molecules	307
		9.5.4 Raman Spectra of Polyatomics	310
	9.6	Vibrational Structure of Electronic Spectra	312
		9.6.1 Diatomic Molecules	312
		9.6.2 Photoelectron Spectroscopy	312
		9.6.3 Polyatomic Molecules	317
	9.7	Nuclear Spin (Orthohydrogen and Parahydrogen)	318
		9.7.1 Spin of Protons in H_2	319
		9.7.2 Nuclear Wavefunctions	319
		9.7.3 Antisymmetry Postulate in H_2	320

9.8	Nuclear Magnetic Resonance	320
	9.8.1 Fundamentals	320
	9.8.2 Chemical Shift	321
	9.8.3 Fine Structure of NMR Signals	323
Problems		324
	Problem 9.1 – Wavefunction of an Harmonic Oscillator	324
	Problem 9.2 – Oscillator: Box Wavefunctions	325
	Problem 9.3 – Classical Oscillator: Probability Density $\rho(x)$	326
	Problem 9.4 – Isotope Effect in Vibrational–Rotational IR Spectrum	326
	Problem 9.5 – Bond Length of HCl from Raman Spectrum	327
	Problem 9.6 – Bond Length of CO from Infrared Spectrum	328
	Problem 9.7 – Force Constant of CO from IR Spectrum	328
Foundation 9.1 – Rotator: Solution of the Schrödinger Equation		330
Foundation 9.2 – Vibrational Structure of Electronic Absorption Bands		331

10 Intermolecular Forces and Aggregates — 333

10.1	Forces in Ionic Crystals	333
	10.1.1 Attracting and Repelling Forces	333
	10.1.2 Lattice Types	334
10.2	Forces in Metals	340
	10.2.1 Coulomb Energy: Electrons Considered as Being Localized at Lattice Points	341
	10.2.2 Kinetic Energy: Electrons Considered as Being Delocalized Over the Lattice	341
	10.2.3 Lattice Energy	344
10.3	Dipole Forces	344
10.4	Hydrogen Bonds	347
10.5	Induction Forces	348
10.6	Dispersion Forces	349
10.7	Molecular Crystals	353
Problems		353
	Problem 10.1 – The Energy of an Ion Pair	353
	Problem 10.2 – Dipole–Dipole Attraction	354
	Problem 10.3 – Polarizability of a Conducting Plate	354

11 Thermal Motion of Molecules — 356

11.1	Kinetic Gas Theory and Temperature	356
	11.1.1 Thermal Motion and Pressure	357
	11.1.2 Avogadro's Law	362
	11.1.3 Temperature Equilibration and Heat	364
	11.1.4 Ideal Gas Law: Definition of Absolute Temperature	364
11.2	Speed of Molecules in a Gas	369
	11.2.1 Mean Speed	369
	11.2.2 Effusion	370
11.3	Mean Free Path and Collision Frequency	370
	11.3.1 Mean Free Path	371
	11.3.2 Collision Frequency	374

11.4 Diffusion ... 374
   11.4.1 Mean Displacement ... 374
   11.4.2 Equation of Einstein and Smoluchowski ... 378
   11.4.3 Fick's Law ... 379
   11.4.4 Gravity Competing with Thermal Motion ... 381
11.5 Viscosity Arising from Collisions of Molecules ... 383
   11.5.1 Viscous Flow ... 383
   11.5.2 Calculating Viscosity ... 384
   11.5.3 Dependence of Viscosity on Pressure ... 386
   11.5.4 Increasing η with Increasing Temperature ... 386
   11.5.5 Calculating Collision Diameters ... 386
11.6 Thermal Motion in Liquids ... 387
   11.6.1 Collisions in Liquids ... 387
   11.6.2 Diffusion Coefficient D of a Liquid ... 388
   11.6.3 Viscosity of a Liquid ... 389
   11.6.4 Stokes–Einstein Equation: Diffusion, Assembling, Interlocking ... 389
11.7 Molecular Motion and Phases ... 393
   11.7.1 Melting Point ... 394
   11.7.2 Boiling Point ... 395
   11.7.3 Critical Point ... 395
   11.7.4 Phases ... 396
   11.7.5 Clusters and Liquid Crystals ... 396
Problems ... 397
   Problem 11.1 – Elastic Collisions of Spheres ... 397
   Problem 11.2 – Gas Bubbles Rising from the Bottom of a Sea ... 398
   Problem 11.3 – Pressure Change on Cooling in a Refrigerator ... 398
   Problem 11.4 – Gas Thermometer ... 398
   Problem 11.5 – Separation of Isotopes by Effusion ... 399
   Problem 11.6 – Separation of Isotopes by Using an Ultracentrifuge ... 399
   Problem 11.7 – Diffusion Path ... 400
   Problem 11.8 – Distribution Function for Diffusion Path ... 401
   Problem 11.9 – Evaporation of Water at Room Temperature ... 404
Foundation 11.1 – Averaging the Free Path λ ... 405
Foundation 11.2 – Intermolecular Forces Affecting the Mean Free Path ... 405

12 Energy Distribution in Molecular Assemblies ... 408
12.1 The Boltzmann Distribution Law ... 408
   12.1.1 System Consisting of Two Quantum States ... 408
   12.1.2 Systems Consisting of Many Quantum States ... 411
12.2 Distribution of Vibrational Energy ... 412
   12.2.1 Population Number N_n ... 413
   12.2.2 Total Vibrational Energy U_{vib} ... 414
12.3 Distribution of Rotational Energy ... 417
   12.3.1 Population Numbers N_n ... 418
   12.3.2 Total Rotational Energy U_{rot} ... 418
12.4 Translational Energy ... 420
   12.4.1 Average Translational Energy According to Quantum Mechanics ... 420

	12.4.2 Maxwell–Boltzmann Distribution	422
	12.4.3 Deriving the Maxwell–Boltzmann Distribution	424
	12.4.4 Number of Translational Quantum States Available per Molecule	427
12.5	Distribution of Electronic Energy	428
12.6	Proving the Boltzmann Distribution	429
	12.6.1 Distinguishable Particles	429
	12.6.2 Indistinguishable Particles	434
Problems		435
	Problem 12.1 – Population of Quantum States of a Rotator at Temperature T	435
	Problem 12.2 – Maxwell–Boltzmann Distribution: v_P, $\bar{v}$, $\overline{v^2}$ and E_{trans}	435
	Problem 12.3 – Number of Quantum States in a One-dimensional Gas	436
	Problem 12.4 – Population Probability	437
	Problem 12.5 – Boltzmann Distribution	437
	Problems 12.6 and 12.7 – Superhelix as a Frozen Boltzmann Distribution	438
	Problem 12.8 – Internal Energy for Particles Rotating on a Circle	441
Foundation 12.1 – Oscillator: Population Numbers N_n and Total Energy U_{vib}		442
Foundation 12.2 – Rotator: Population Number N_n and Total Energy U_{rot}		444
Foundation 12.3 – Boltzmann Distribution		446

13 Internal Energy U, Heat q, and Work w — 450

13.1	Change of State at Constant Volume	450
	13.1.1 Change of Internal Energy ΔU; Heat q	450
	13.1.2 Heat Capacity C_V	452
13.2	Temperature Dependence of C_V	453
	13.2.1 Rotational and Vibrational Contribution to C_V	453
	13.2.2 ortho- and para-H_2: Fascinating Quantum Effects on C_V	456
	13.2.3 Electronic Contribution to C_V ($C_{V,\text{electr}}$)	457
	13.2.4 C_V of Solids	457
	13.2.5 Characteristic Temperature	459
13.3	Change of State at Constant Pressure	460
	13.3.1 Change of Internal Energy ΔU; Heat q and work w	460
	13.3.2 Heat Capacity C_P	461
13.4	System and Surroundings; State Variables	462
	13.4.1 Defining System and Surroundings	462
	13.4.2 Defining State and State Variables	462
	13.4.3 Closed, Isolated, and Open Systems	464
	13.4.4 Cyclic Processes	465
Problems		466
	Problem 13.1 – Isothermal Expansion of an Ideal Gas	466
	Problem 13.2 – Adiabatic Expansion of an Ideal Gas	468

14 Principle of Entropy Increase — 469

14.1	Irreversible and Reversible Changes of State	469
	14.1.1 Irreversible Changes	469
	14.1.2 Reversible Changes	470

14.2	Distribution Possibilities	472
	14.2.1 Mixing of Two Gases	473
	14.2.2 Expansion of a Gas	474
14.3	Counting the Number of Configurations Ω	477
	14.3.1 Probability P of Reversion of an Irreversible Process	477
	14.3.2 Particles Each in One of Three Energy States	477
	14.3.3 Number of Configurations Ω of an Atomic Gas	477
14.4	Entropy of a System: $S = k \cdot \ln \Omega$	480
	14.4.1 Entropy of Atomic Gases	480
	14.4.2 Entropy of Diatomic Gases	481
14.5	Entropy Change ΔS	482
	14.5.1 Temperature Equilibration	482
	14.5.2 Mixing of Two Gases	483
	14.5.3 Entropy Increase in an Irreversible Process in an Isolated System	483
	14.5.4 Entropy of Subsystems	484
	14.5.5 Entropy Change in Non-isolated Systems	486
Problems		486
	Problem 14.1 – Increase in the Number of Configurations with Temperature Equilibration	486
	Problem 14.2 – Mixing Entropy	488
Foundation 14.1 – Entropy for Rotation and Vibration		490

15 Entropy S and Heat q_{rev} — 492

15.1	Heat and Change of Entropy in Processes with Ideal Gases	493
	15.1.1 Expansion at Constant Temperature	493
	15.1.2 Thermal Equilibration	494
	15.1.3 Cyclic Processes and Processes in Isolated Systems	498
	15.1.4 Reversible Heat Engine with Ideal Gases (Carnot Cycle)	498
15.2	Heat and Change of Entropy in Arbitrary Processes	501
15.3	Entropies of Substances	507
15.4	Thermodynamic Temperature Scale and Cooling	511
15.5	Laws of Thermodynamics	511
Problems		513
	Problem 15.1 – Reversible Adiabatic Expansion	513
	Problem 15.2 – Entropy of Mixing	514
	Problem 15.3 – Efficiency of Different Heat Engines	515
	Problem 15.4 – Efficiency of a Power Station	515
	Problem 15.5 – Heat Pump	515
	Problem 15.6 – Air Conditioning	515
	Problem 15.7 – Entropy and Configurations of Water	516

16 Criteria for Chemical Reactions — 517

16.1	Heat Exchange	518
	16.1.1 Reaction at Constant Volume: $q = \Delta U$	518
	16.1.2 Reaction at Constant Pressure: $q = \Delta H$	520

16.2	Change of Internal Energy and Enthalpy	522
	16.2.1 Temperature Dependence of ΔU and ΔH	522
	16.2.2 Molar Enthalpies of Formation from Elements $\Delta_f H^\ominus$	524
	16.2.3 Molar Enthalpy of Reaction $\Delta_r H^\ominus$	525
16.3	Conditions for Spontaneous Reactions	528
	16.3.1 Helmholtz Energy and Gibbs Energy	528
	16.3.2 Reversible Work w_{rev}	529
16.4	Change of Gibbs Energy	530
	16.4.1 Molar Gibbs Energy of Formation from Elements $\Delta_f G^\ominus$	530
	16.4.2 Molar Gibbs Energy of Reaction $\Delta_r G^\ominus$	531
	16.4.3 Temperature Dependence of $\Delta G^\ominus$	531
	16.4.4 Pressure Dependence of ΔG	533
Problems		535
	Problem 16.1 – Bunsen Burner Fed with Methane	535
	Problem 16.2 – $\Delta_r H$ from Standard Enthalpies of Formation	535
	Problem 16.3 – Temperature Regulation in the Human Body	536
	Problem 16.4 – Spontaneous Reactions	537
	Problem 16.5 – Burning Limestone on Mount Everest	537
Foundation 16.1 – How to Calculate $S_{T_1}^{T_2} \Delta C_{P,m}^\ominus \cdot dT$		539
Foundation 16.2 – How to Calculate ΔG_{T_2} from ΔG_{T_1}		542

17 Chemical Equilibrium — 545

17.1	ΔG for Reactions in Gas Mixtures	546
	17.1.1 Mass Action Law and Equilibrium Constant K	546
	17.1.2 Equilibrium Constant K from $\Delta_r G^\ominus = -RT \cdot \ln K$	549
	17.1.3 Reactions Involving Gases and Immiscible Condensed Species	553
	17.1.4 Van't Hoff Equation	554
	17.1.5 Statistical Interpretation of K	555
	17.1.6 Estimation of $K = f(T)$	557
	17.1.7 ΔH and ΔS from Measured K	558
	17.1.8 Vapor Pressure	559
17.2	ΔG for Reactions in Dilute Solution	560
	17.2.1 Osmotic Pressure and Concentration	560
	17.2.2 Concentration and Molality	562
	17.2.3 Depression of Vapor Pressure	562
	17.2.4 Reversible Change of Concentration	565
	17.2.5 Mass Action Law: Solutions of Neutral Particles	565
	17.2.6 Mass Action Law: Solutions of Charged Particles	567
	17.2.7 Gibbs Energy of Formation in Aqueous Solution	570
	17.2.8 Part of Reactants or Products in Condensed or Gaseous State	573
Problems		575
	Problem 17.1 – Equilibrium Constant Calculated from ΔH and ΔS	575
	Problem 17.2 – Dissociation of I_2	575
	Problem 17.3 – Hydrogen–Iodine Equilibrium	576
	Problem 17.4 – Formation of NH_3 from Its Elements	576
	Problem 17.5 – Boiling Point of Water on Mount Everest	577
	Problem 17.6 – Reverse Osmosis	577

	Problem 17.7 – Elevation of Boiling Point	577
	Problem 17.8 – Depression of Melting Point	578

18 Reactions in Aqueous Solution and in Biosystems — 579

- 18.1 Proton Transfer Reactions: Dissociation of Weak Acids — 579
  - 18.1.1 Henderson–Hasselbalch Equation — 579
  - 18.1.2 Degree of Dissociation — 580
  - 18.1.3 Acid in a Buffer — 581
  - 18.1.4 Titration Curve of a Weak Acid — 582
  - 18.1.5 Stepwise Proton Transfer: Amino Acids — 583
- 18.2 Electron Transfer Reactions — 584
  - 18.2.1 Electron Transfer from Metal to Proton: Dissolution of Metals in Acid — 584
  - 18.2.2 Electron Transfer from Metal 1 to Metal 2 Ion: Coupled Redox Reactions — 585
  - 18.2.3 Electron Transfer Coupled with Proton Transfer — 586
  - 18.2.4 Electron Transfer to Proton at pH 7: $\Delta G^{\ominus\prime}$ — 587
- 18.3 Group Transfer Reactions in Biochemistry — 588
  - 18.3.1 Group Transfer Potential — 589
  - 18.3.2 Coupling of Reactions by Enzymes — 590
- 18.4 Bioenergetics — 591
  - 18.4.1 Synthesis of Glucose — 591
  - 18.4.2 Combustion of Glucose — 592
  - 18.4.3 Energy Balance of Formation and Degradation of Glucose — 592
- Problems — 594
  - Problem 18.1 – pH of Weak Acid for Different Total Concentrations — 594
  - Problem 18.2 – Buffer Solutions — 595
  - Problem 18.3 – pH of Amino Acids — 596
  - Problem 18.4 – Absorption Maximum of NADH — 596
- Foundation 18.1 – Titration of Acetic Acid by NaOH — 597

19 Chemical Reactions in Electrochemical Cells — 600

- 19.1 ΔG and Potential E of an Electrochemical Cell Reaction — 600
- 19.2 Concentration Cells — 604
  - 19.2.1 Metal Electrodes — 604
  - 19.2.2 Gas Electrodes — 607
- 19.3 Standard Potential $E^{\ominus}$ — 609
  - 19.3.1 Nernst Equation — 609
  - 19.3.2 Practical Determination of $E^{\ominus}$ — 610
- 19.4 Redox Reactions — 611
  - 19.4.1 Fe^{3+}/Fe^{2+} Electrode — 611
  - 19.4.2 Quinone/Hydroquinone Electrode — 611
  - 19.4.3 $NAD^+/NADH$ Electrode — 612
  - 19.4.4 Oxygen Electrode — 613
- 19.5 Applications of Electrochemical Cells — 613
  - 19.5.1 Reference Electrodes — 613
  - 19.5.2 Glass Electrodes — 615

	19.5.3 Galvanic Elements	616
	19.5.4 Fuel Cells	618
	19.5.5 Electrolysis	619
19.6	Conductivity of Electrolyte Solutions	620
	19.6.1 Mobility of Ions	620
	19.6.2 Generalization	622
Problems		624
	Problem 19.1 – Acceleration of Na^+ Ions	624
	Problem 19.2 – Thermodynamic and Electrochemical Data	625
	Problem 19.3 – Complex Formation	625
	Problem 19.4 – Calculation of $\Delta G^{\ominus\prime}$ and $E^{\ominus\prime}$	626

20 Real Systems — 627

20.1	Phase Equilibria	627
20.2	Equation of State for Real Gases	629
	20.2.1 van der Waals Equation	630
	20.2.2 Critical Point and van der Waals Constants	633
	20.2.3 Virial Coefficients	634
20.3	Change of State of Real Gases	636
	20.3.1 Isothermal Compression of a van der Waals Gas	636
	20.3.2 Fugacity and Equilibrium Constant	636
	20.3.3 Adiabatic Expansion into Vacuum	640
	20.3.4 Joule–Thomson Effect	642
20.4	Change of State of Real Solutions	646
	20.4.1 Partial Molar Volume	646
	20.4.2 Chemical Potential	648
	20.4.3 Activities and Equilibrium Constants in Solutions with Ions	649
Problems		653
	Problem 20.1 – Freezing Point of Mercury	653
	Problem 20.2 – Determination of Fugacity Coefficient ϕ	653
	Problem 20.3 – Charge Distribution $\rho(r)$	654
	Problem 20.4 – Activity Coefficient $\gamma_{\pm}$ for HCl	655
Foundation 20.1 – Distribution of Ions at a Charged Plate		657

21 Kinetics of Chemical Reactions — 663

21.1	Collision Theory for Gas Reactions	663
	21.1.1 Counting the Number of Collisions	663
	21.1.2 Activation	665
	21.1.3 Accumulation of Kinetic Energy	666
21.2	Rate Equation for Gas Reactions	669
	21.2.1 Rate Constant and Frequency Factor	669
	21.2.2 Reactants A and B are Identical Molecules	672
21.3	Rate Equation for Reactions in Solution	672
	21.3.1 Reaction Through Activated Complex	672
	21.3.2 Diffusion Controlled Reaction	674
21.4	Transition State Theory	675
	21.4.1 Eyring Equation for Bimolecular Reaction	675
	21.4.2 Activation Enthalpy and Activation Entropy	676
	21.4.3 Decay Reaction	677

21.5	Treatment of Experimental Data	677
	21.5.1 Rate Constants	677
	21.5.2 Activation Energy and Frequency Factor	682
	21.5.3 Activation Enthalpy and Activation Entropy	685
	21.5.4 Tunneling of Proton and Deuteron	687
21.6	Complex Reactions	689
	21.6.1 Reactions Leading to Equilibrium	689
	21.6.2 Parallel Reactions	692
	21.6.3 Consecutive Reactions	693
	21.6.4 Chain Reactions	697
	21.6.5 Branching Chain Reactions	699
	21.6.6 Enzyme Reactions (Michaelis–Menten Mechanism)	700
	21.6.7 Autocatalytic Reactions	703
	21.6.8 Inhibition of Autocatalysis	707
	21.6.9 Bistability	709
	21.6.10 Oscillating Reactions (Belousov–Zhabotinsky Reaction)	710
	21.6.11 Chemical Waves	714
21.7	Experimental Methods	719
	21.7.1 Flow Methods	720
	21.7.2 Flash Photolysis	721
	21.7.3 Relaxation Method	722
Problems	726
	Problem 21.1 – Second-order Reaction	726
	Problem 21.2 – Reaction Leading to Equilibrium	727
	Problem 21.3 – Yield of Main Reaction	727
	Problem 21.4 – Consecutive Reaction	728
	Problem 21.5 – Autocatalytic Reaction	729
	Problem 21.6 – Unimolecular Reactions	730

22 Organized Molecular Assemblies   732

22.1	Interfaces	732
	22.1.1 Surface Tension and Interfacial Tension	732
	22.1.2 Surface Films	737
	22.1.3 Insoluble Monolayers	738
	22.1.4 Solid Surfaces	741
	22.1.5 Micelles	742
22.2	Liquid Crystals	743
	22.2.1 Birefringence	746
	22.2.2 Selective Reflection	746
	22.2.3 Electro-optical Effects	747
22.3	Membranes	750
	22.3.1 Soap Lamella	750
	22.3.2 Black Lipid Membranes	751
	22.3.3 Liposomes	752
	22.3.4 Biomembranes	752
22.4	Macromolecules	755
	22.4.1 Random Coil: A Chain of Statistical Chain Elements	755
	22.4.2 Length of Statistical Chain Element from Light Scattering	760

	22.4.3 Length of Statistical Chain Element from Hydrodynamic Properties	762
	22.4.4 Refined Theory: Macroscopic Models	765
	22.4.5 Uncoiling Coil	769
	22.4.6 Restoring Coil	774
	22.4.7 Motion Through Entangled Polymer Chains	776
	22.4.8 Rubber Elasticity	781
22.5	Supramolecular Structures	786
Problems		787
	Problem 22.1 – Contact angles	787
	Problem 22.2 – Transfer of Charge e from Water into a Membrane	787
	Problem 22.3 – Light Scattering (diameter of molecules $\ll \lambda$)	788
	Problem 22.4 – Light Scattering (diameter of molecules $< \lambda/4$)	788
	Problem 22.5 – Special Case of Scattering Relation (22.29)	789
	Problem 22.6 – Molar Mass from Diffusion and Sedimentation	789
	Problem 22.7 – Extension of an unraveled coil	790
Foundation 22.1 – Mobility of DNA in Meshwork		792

23 Supramolecular Machines 794

23.1	Energy Transfer Illustrating the Idea of a Supramolecular Machine	794
23.2	Programmed Interlocking Molecules	795
23.3	Manipulating Photon Motion	800
	23.3.1 Energy Transfer Between Dye Molecules	800
	23.3.2 Functional Unit by Coupling Dye Molecules	803
	23.3.3 Dye Aggregate as Energy Harvesting Device	805
	23.3.4 Solar Energy Harvesting in Biosystems	810
	23.3.5 Manipulating Luminescence Lifetime by Programming Echo Radiation Field	812
	23.3.6 Nonlinear Optical Phenomena	815
23.4	Manipulating Electron Motion	818
	23.4.1 Photoinduced Electron Transfer in Designed Monolayer Assemblies	818
	23.4.2 Switching by Photoinduced Electron Transfer	819
	23.4.3 Monolayer Assemblies for Elucidating the Nature of Photographic Sensitization	821
	23.4.4 Conducting Molecular Wires	823
	23.4.5 Solar Energy Conversion: The Electron Pump of Plants and Bacteria	826
	23.4.6 Artificial Photoinduced Electron Pumping	830
	23.4.7 Tunneling Current Through Monolayer	831
	23.4.8 Electron Transfer Through Proteins	833
23.5	Manipulating Nuclear Motion	833
	23.5.1 Light-induced Change of Monolayer Properties	833
	23.5.2 Solar Energy Conversion in Halobacteria	836
	23.5.3 Photoinduced Sequence of Amplification Steps: The Visual System	844
	23.5.4 Mechanical Switching Devices	845

23.6	Molecular Recognition and Replica Formation	847
	23.6.1 Multisite Recognition of a Molecule at a Surface Layer	847
	23.6.2 Catalytic Reaction in Solution	848
	23.6.3 Catalytic Reaction in Organized Media	848
	23.6.4 Molecular Replica Formation	850
	23.6.5 Biosensors	852
23.7	Addressing and Positioning Molecules	853
	23.7.1 STM (scanning tunneling microscopy)	853
	23.7.2 AFM (atomic force microscopy)	854
	23.7.3 SNOM (scanning near-field optical microscopy)	856
Problems		858
	Problem 23.1 – Derivation of Equation for ε in Foundation 23.4	858
	Problem 23.2 – Dielectric Medium as a Tunneling Barrier	858
	Problem 23.3 – Induced Dipole Moment	859
	Problem 23.4 – Proton Pump: Field of Charged Amino Acids at Chromophore	860
	Problem 23.5 – Circular Dichroism and Structural Features of Chlorosomes	861
Foundation 23.1 – Energy Transfer		862
Foundation 23.2 – Energy transfer from Exciton to Acceptor		865
Foundation 23.3 – Radiation Echo field		867
Foundation 23.4 – Electron Transfer Between π-Electron Systems		871
Foundation 23.5 – Electron Transfer in Soft Medium		877

24 Origin of Life — 880

24.1	Investigation of Complex Systems	880
	24.1.1 Need for Simplifying Models	880
	24.1.2 Increasing Simplification with Increasing Stages of Complexity	881
24.2	Can Life Emerge by Physicochemical Processes?	881
	24.2.1 Bioevolution as Process of Learning	881
	24.2.2 Model Case for the Learning Mechanism	882
	24.2.3 Modeling the Emergence of the Genetic Apparatus	885
	24.2.4 Later Evolutionary Steps: Emergence of an Eye With Lens	894
24.3	General Aspects of Life	895
	24.3.1 Information and Knowledge	895
	24.3.2 Processing Information and Genesis of Information and Knowledge	895
	24.3.3 Limits of Physicochemical Ways of Thinking	899
Problems		903
	Problem 24.1 – Aggregation of Folded Strands	903
	Problem 24.2 – Replication Error Rate of a Bacterium and a Human	904
	Problem 24.3 – Time Needed to Evolve a Bacterium	904
	Problem 24.4 – Maximum Genetic Information Carried by DNA	905
Foundation 24.1 – The Emergence of a Simple Genetic Apparatus Viewed as a Supramolecular Engineering Problem. A thought Experiment		906

Foundation 24.2 – Attempts to Model the Origin of Life 913
  Foundation 24.3 – Maxwell's Demon: Production of Entropy 919

Glossary 922

Appendices 928

Further Reading 937

Index 955

Acknowledgements

We are greatly indebted to Martin Röthlisberger from John Wiley & Sons, Ltd for his encouragement and advice during all stages of the development of this project.

A number of colleagues have read some chapters and have given us important advice. We deeply appreciate comments from David Beratan on the introductory chapters, from Hugo Franzen on the chapters on thermodynamics and from Jürg Waser and Peter Waser on the final chapters. Of particular importance was the advice from Edgar Heilbronner which strongly influenced the writing of the chapters on chemical equilibria. We wish to express our gratitude to Dieter Oesterhelt and to Wolfgang Zinth for reading the section on proton and electron pumps and for important suggestions and expert advice.

We are grateful to Pierre-Gilles de Gennes and Erich Sackmann for drawing our attention to recent work on uncoiling of polymers, and Albert Eschenmoser for reading the chapter on the origin of life and for very helpful and stimulating discussions. Rolf Landauer's thought-provoking comments on the Maxwell demon are gratefully acknowledged as is useful advice from Jerry Swalen, Tom Fields and Helmut Schreiber.

We are particularly grateful to Ian McNaught and Christoph Kuhn for reading the manuscript and for their important advice and suggestions. We wish to thank Bob Golden for his careful editing are most indebted to the people at John Wiley & Sons, Ltd, especially Martin Tribe for his helpfulness and cooperativity during the production of this book, and to Andy Slade and Debbie Scott for their care.

Preface and Guided Tour

Preface

This textbook for an undergraduate course in physical chemistry is addressed equally to chemists, polymer chemists, biologists, chemical engineers, material scientists, and physicists.

Chemical structures and processes can be interpreted as an interplay of atoms and molecules. Matter can be regarded as a collective interaction of these atoms and molecules, with macroscopic properties based on a few fundamental principles governing chemistry and related sciences. To emphasize this situation, we have adopted in this book a presentation where physical chemistry is first introduced by the treatment of atoms and chemical bonds, followed by a discussion of how these lead to molecules and, subsequently, to more and more complex representations of matter. By this structure in the sequence of chapters, the student is encouraged to focus on a systematic understanding of the subject. Simple theoretical models will illustrate many of these concepts.

Although the presentation is logical, the course can proceed in almost any order because the chapters are relatively self-contained (see "Guided Tour" below). The concepts determining the processes at the atomic level are based on the postulates of quantum mechanics. Many of these concepts seem to be counterintuitive and strange at a first glance. The book starts with a simple introduction to quantum mechanics in a way that should be easy to grasp.

We feel that an early confrontation with quantum mechanics is advantageous to the student. Thinking in terms of quantum mechanics is needed in all parts of physical chemistry and is crucial in understanding the nature of the chemical bond. Students in chemistry and biochemistry are trained to think in terms of chemical bonds and processes at the atomic level; it is important for them to see immediately the interlinkage of physical chemistry with inorganic and organic chemistry. Obviously, this aim can be much more easily reached by emphasizing the atomic approach instead of treating thermodynamics without reference to molecular interplay.

Inventing idealized theoretical descriptions of real situations is crucial in understanding increasingly complex organized systems, and ingenious experiments are of particular importance in modern fields of physical chemistry. Studying typical examples is an excellent opportunity for learning how to design simple experiments and theoretical models. With this goal in mind, we do not attempt to cover as many topics as possible. Experimental methods must stay in the background. Selected cases are emphasized in order to focus on fundamental aspects, for example, π electron systems are discussed in more detail than other subjects because they illustrate particularly well the use of simple theoretical models.

The treatment of organized systems of molecules, including supramolecular machines, is a fast-developing subject in physical chemistry. We attempt to introduce the student to this fascinating and dramatically growing field by discussing selected examples illustrating the new approaches. Of course, an overview of the broad variety of interdisciplinary topics cannot be given. The final chapter should lead to an understanding of the basic processes in the origin of life; it demonstrates the usefulness of typical physicochemical modeling in the development of exciting new areas.

A Guided Tour Through Physical Chemistry

The sequence of chapters in this book – starting with quantum mechanics and its application to simple atoms and molecules and then proceeding stepwise to increasingly complex systems – should emphasize the fact that chemical phenomena can be considered as processes among molecules and that the properties of molecules can be deduced from the concepts of quantum mechanics. This straightforward organization of the book retains the logical structure of the subject of physical chemistry.

Various approaches to using this textbook can lead toward the goal that the postulates of quantum mechanics are to be recognized as the conceptual basis for rationalizing the great variety of chemical phenomena. Following the given sequence of chapters is one possible pathway. It requires that we start with quantum theory, which is unfamiliar to the beginner, but is ultimately needed for understanding chemical bonds and the world of structures based on chemical bonding.

The interplay of atoms and molecules, however, can be approximated by using the familiar laws of classical mechanics. Consequently, as an alternative, the student may begin with Chapter 11, which describes the thermal motion of molecules in the familiar classical way of thinking. This leads to a lively picture of the molecular interplay, and the student learns the concepts of temperature and heat.

After reading Chapter 11 it is advisable to continue with Chapters 1 and 2 in order to grasp the fundamental importance of quantum theory and chemical bonding. The student may then proceed with Chapters 3–10, which focus on molecular structure and intermolecular forces. He or she would subsequently read Chapters 12–24, which are concerned with criteria for chemical reactions, equilibria, rate of chemical reactions, and the properties of increasingly complex systems.

Alternatively, reading the chapters in the sequence 11, 1, 2, 12–24, 3–10 is also acceptable. In both cases, it is important that the student is exposed to thinking in terms of quantum mechanics by reading Chapters 1 and 2 at an early stage. The content of each chapter is summarized below.

Chapters 1 and 2. Here we describe experiments (e.g., diffraction of electrons) that focus on the essence of quantum mechanics. At first glance, these considerations seem to be remote from physical chemistry. However, they show essential aspects of atoms and molecules. At the same time, these experiments demonstrate how developments in one field in scientific research can revolutionize other scientific areas.

Experiments demonstrating dramatic consequences of the quantum nature of matter are discussed and the simplest atom and the simplest molecule, the hydrogen atom and

PREFACE   xxvii

the hydrogen molecule ion, are considered. The *variation principle*, which is crucial in understanding the nature of the chemical bond, is introduced as a postulate. Helium-like systems and the hydrogen molecule are discussed.

Chapters 3 and 4. In Chapter 3 a deeper understanding of quantum mechanics is given by introducing the Schrödinger equation as a postulate and deducing the variation principle as a theorem. On this basis the stationary states of the H atom are discussed.

In Chapter 4 the treatment is applied to H_2^+ and extended to the hydrogen molecule and to systems containing more than one electron. The *Pauli principle* is introduced as an additional postulate.

Beginners might restrict themselves to looking at the energy levels and wavefunctions of the excited states of the H atom and to the elementary form of the Pauli principle, proceed to Chapter 5, and then come back to Chapters 3 and 4 at a later stage.

Chapters 5–10. In Chapters 5 and 6 different types of chemical bonds, their occurrence, and their specific properties are discussed. Chapters 7 and 8 are devoted to π electrons. These chapters exemplify the use of simple theoretical models which are crucial to understanding complex molecular assemblies, the heart of today's physical chemistry. Phenomena such as light absorption and laser action establish the importance of quantum mechanics. Chapter 9 is concerned with the quantum mechanical view of nuclear motion. Chapter 10 deals with intermolecular bonding and the occurrence of molecular assemblies.

Chapters 11–16. Chapter 11 introduces the concept of temperature as a measure of the motion of molecules. The classical aspects of thermal motion are emphasized and the phase transitions gas-liquid-solid are discussed at a molecular level. In Chapter 12 the intramolecular distribution of thermal energy is considered; here the quantum mechanical aspects are crucial. Chapters 13–15 are concerned with quantities and theorems important for an understanding of macroscopic behavior of matter: internal energy, heat, work, and entropy. Chapter 16 introduces the quantities useful in discussing the criteria for chemical reactions besides temperature and entropy: enthalpy and Gibbs energy.

Chapters 17–20. These chapters deal with applications of the basic concepts developed in Chapters 11–16: chemical equilibrium, bioenergetics and electrochemistry.

Chapters 21–24. Chapter 21 is concerned with the kinetics of chemical reactions. Chapters 22–24 consider increasingly complex systems: interfaces, membranes, polymers, supramolecular machines, and the mechanisms involved in the origin of life.

A Note on Introducing Thermodynamics (Chapters 13–16). Historically, the macroscopic properties of matter were investigated without taking the atomic structure into account. Classical thermodynamics emerged in connection with the development of the steam engine, which drove the industrial revolution in the nineteenth century. The classical picture of molecules developed in the second half of the nineteenth century led to the molecular view of temperature and heat. The quantum theoretical approach driving toward an understanding of the particular properties of chemical species originated in the first half of the twentieth century. This traditional sequence is still kept in most introductory courses in physical chemistry, which start with the laws of thermodynamics. When proceeding to classical and then to quantum mechanical statistical thermodynamics, it is confusing for the student and difficult to accept that the laws of thermodynamics are first set as postulates, then understood as properties of collections of molecules, and finally deduced from the postulates of quantum mechanics.

This difficulty is avoided in our approach by first viewing the quantum mechanical nature of matter and then deducing the macroscopic properties as a consequence.

The significance of entropy is easily seen when introducing it as a measure of the number of possible configurations of a macroscopic system in a given state: this number does not decrease in an isolated system, so the entropy of an isolated system cannot decrease (Chapter 14). We then demonstrate that the entropy defined by the number of configurations is closely connected with the heat transfer in reversible processes in classical thermodynamics (Chapter 15).

1 Wave–Particle Duality

Matter is made up of atoms bound together to form molecules and assembled to form larger structures. What are the basic principles leading to an understanding of chemical bonds? How can we rationalize the huge variety of molecules and their properties? An understanding of these fascinating features requires us to abandon some of our familiar ways of thinking. We must get used to thinking in terms of *quantum mechanics* in order to understand the most astonishing behavior of small particles. This is achieved by considering fundamental experiments with light and with electrons. These experiments lead us to the laws of quantum mechanics, which are basic to an understanding of the behavior of atoms, molecules, and molecular assemblies.

1.1 Light

Light is described in classical physics as a wave phenomenon. It can also be described as a stream of particles. Which view is correct? The answer depends on the experiment. Some experiments (such as investigations into the photoelectric effect or the photochemistry on a photographic plate) demonstrate the particle nature of light; interference experiments (such as the diffraction of light from an edge or by a slit) demonstrate the wave nature of light. In Section 1.1.3 we will discuss the theory which combines these two types of properties.

1.1.1 The Particle Nature of Light: Photoelectric Effect

Electrons can be knocked out of the surface of a metal by light (Figure 1.1). This phenomenon is called the *photoelectric effect*. The speed v of the electrons leaving the surface depends on the color of the light: it is higher for violet light and lower for red light. Increasing the intensity of the light will cause more electrons to leave the surface but will not increase their speed.

Figure 1.2(a) shows the experimental setup. Two metal electrodes are placed in an evacuated glass flask and connected to an external circuit. One of the electrodes is a metal such as Cs, which emits electrons easily. A voltage V is placed across the electrodes, with the polarity such that electrons are decelerated in traveling from the Cs electrode to the opposite electrode. When monochromatic light is shone on the Cs electrode, a current

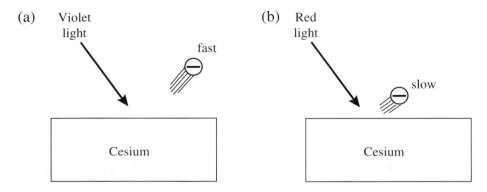

Figure 1.1 Ejection of electrons from a metal by light: (a) violet light, electron fast; (b) red light, electron slow.

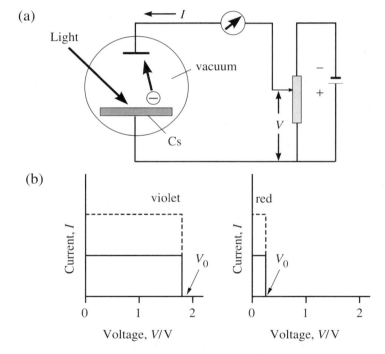

Figure 1.2 Experimental arrangement for the photoelectric effect. (a) Construction of the photocell with decelerating voltage. (b) Current I measured by the ammeter as a function of the decelerating voltage V (greatly simplified; because of the thermal energy of the electrons, the current does not fall off so suddenly at the cutoff voltage, and V_0 must be determined by extrapolation). The dashed lines are for experiments with twice the light intensity.

is observed in the circuit as long as V is low enough. As V is increased from zero with a constant light source, the current remains constant until a certain value, V_0, is reached where the current drops abruptly to zero (Figure 1.2(b)). It is observed that the current depends on the intensity of the light, while the cutoff voltage V_0 depends on the color of the light.

1.1 LIGHT

To analyze the results, we note that an electron traveling through a voltage V changes its energy by the amount eV, where $-e$ is the electron charge. When an electron (mass m_e) leaves the metal surface with speed v, it has kinetic energy $\frac{1}{2}m_e v^2$. It will reach the opposite electrode if the energy eV lost to the electric field is less than its starting energy, $\frac{1}{2}m_e v^2$. At the cutoff voltage V_0, the energy loss is just equal to the initial kinetic energy:

$$E_{\text{kin}} = \tfrac{1}{2}m_e v^2 = eV_0 \tag{1.1}$$

For violet light and Cs metal the cutoff voltage is 1.80 V, so with $e = 1.60 \times 10^{-19}$ C we obtain,

$$E_{\text{kin}} = 1.60 \times 10^{-19}\,\text{C} \times 1.80\,\text{V} = 2.88 \times 10^{-19}\,\text{J} = 1.80\,\text{eV}$$

It is known from other experiments that it takes 2.08×10^{-19} J (1.30 eV) to pull an electron out of the metallic bonding of the Cs metal (work function). So the violet light supplies each ejected electron with $(2.88+2.08) \times 10^{-19}\,\text{J} = 4.96 \times 10^{-19}\,\text{J}$ of energy. By repeating the photoelectric effect experiment with light of other colors, we find that yellow light transfers 3.31×10^{-19} J of energy to each electron, and red light transfers 2.48×10^{-19} J of energy to each electron.

We conclude from these experiments that light interacts with electrons in the form of energy packets (energy packets of 4.96×10^{-19} J for violet light and of 2.48×10^{-19} J for red light). These energy packets are called *light quanta* or *photons*. The energy of each photon depends on the color of the light.

1.1.2 Wave Nature of Light: Diffraction

The properties of waves are most easily seen in water waves. Figure 1.3 shows water waves passing through a small hole in a barrier. The hole acts like a source, sending new waves into the region beyond the barrier. This phenomenon is called *diffraction*. If two

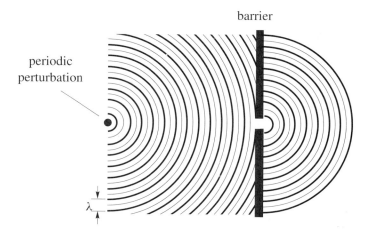

Figure 1.3 Formation of secondary waves from a small hole (water waves; the surface of the water is forced to a wave motion by periodically dipping a stick into the water). Thick lines represent maxima of the waves; thin lines represent minima. Reflection of the primary waves and disturbance of the secondary waves by the barrier are neglected. The distance between two subsequent thick lines corresponds to the wavelength λ.

holes are opened in the barrier, then waves originate from each hole (Figure 1.4). These waves overlap; there are places where they reinforce each other, but there are also places where they cancel each other. This phenomenon is called *interference*.

The surface of the water at a selected point P oscillates between the distance ψ above the average height and the distance $-\psi$ below the average height at the point P. Generally speaking, ψ is the maximum displacement, and in our case the displacement is the distance of the water surface above or below the average height (ψ positive or negative, respectively). Figure 1.5 shows $|\psi|$ for each point along a line on the water surface when the surface is disturbed by the interference of two waves.

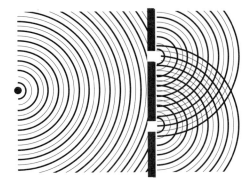

Figure 1.4 Formation of secondary waves from two small holes. The distance between two subsequent thick lines corresponds to the wavelength λ.

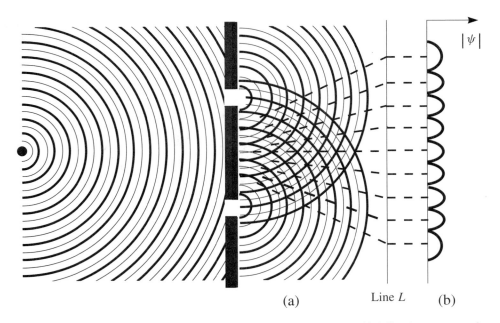

Figure 1.5 Interference of two water waves. (a) Amplitude maxima (thick lines) propagate along the dashed lines. The distance between two subsequent thick lines corresponds to the wavelength λ. (b) Plot of the maximum displacement $|\psi|$ along the line L.

1.1 LIGHT

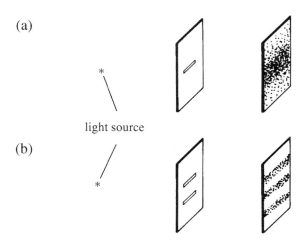

Figure 1.6 (a) Diffraction of light by a single slit. (b) Interference of light from a double slit. The light source emits monochromatic light.

Now we explore the wave-like characteristics of light. If light travels through a single slit or a double slit, a pattern of light can be observed on the screen behind the slits, as in Figure 1.6. We can interpret this experimental fact as evidence that light has properties of a wave. By measuring the distance between successive minima (places where the intensity is zero) or maxima on the screen in Figure 1.6(b), we can determine the wavelength λ (Box 1.1). We find $\lambda = 400$ nm for violet light, $\lambda = 600$ nm for yellow light, and $\lambda = 800$ nm for red light.

Box 1.1 – Wavelength from Double Slit Experiment

We consider two slits (separation d) and a screen at distance a from the slits (Figure 1.7). At point P_1 the distances r_1 and r_2 between both slits and the screen are the same. Then the phases of the light waves arriving at P_1 are the same and, on the time average, we obtain an intensity maximum at P_1.

At point P_2 we find the next intensity maximum: the distances r_1 and r_2 differ by λ, thus the phases of the two waves arriving at P_2 are the same again. From Figure 1.7 we calculate the distance x between both intensity maxima:

$$\frac{x+d/2}{a} = \tan\alpha, \quad \frac{\lambda}{d} = \sin\alpha$$

Because of

$$\tan\alpha = \frac{\sin\alpha}{\cos\alpha} = \frac{\sin\alpha}{\sqrt{1-\sin^2\alpha}}$$

we obtain

$$\frac{x+d/2}{a} = \frac{\lambda/d}{\sqrt{1-(\lambda/d)^2}}$$

Continued on page 6

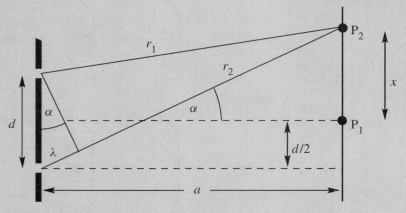

Figure 1.7 Diffraction on a double slit (slits separated by d). The intensity of the light arriving on the screen at distance a is measured at points P_1 and P_2.

and

$$x = \frac{a\lambda}{d} \cdot \frac{1}{\sqrt{1 - \left(\frac{\lambda}{d}\right)^2}} - \frac{d}{2}$$

Solving for λ we obtain

$$\lambda = \frac{(x + d/2) \cdot d}{a} \cdot \frac{1}{\sqrt{1 + \left(\frac{x + d/2}{a}\right)^2}}$$

Example

We illuminate a double slit with width $d = 1.7\,\mu\text{m}$ with the light of a red HeNe laser. On a screen at a distance of $a = 1\,\text{m}$ we measure $x = 40.1\,\text{cm}$; then we obtain $\lambda = 633\,\text{nm}$. Using the light of a green HeNe laser, we measure $x = 33.6\,\text{cm}$ and obtain $\lambda = 542\,\text{nm}$. A slit width of $d = 1.7\,\mu\text{m}$ corresponds to the distance of two adjacent grooves on a typical grating used in UV/VIS spectroscopy. The relation derived here for a double slit is also valid for such a grating.

For $x \ll a$ and $d/2 \ll x$ this relation can be simplified as

$$\lambda = \frac{x \cdot d}{a}$$

Note that in the example the condition $d/2 \ll a$ is fulfilled, but x is on the same order as a. Thus using the simplified relation for λ is not appropriate in this case.

1.1 LIGHT

An easy way to determine the wavelength of light is the diffraction of light from an edge (Figure 1.8a). There is no sharp shadow on the screen but stripes from interference allowing to determine λ. Once the wavelength has been determined, it is possible to calculate the displacement ψ at each point on the screen for the diffraction and interference experiments. On the other hand, the intensity distribution of the light on the screen can be measured directly (see Figure 1.8(b)). The brightness at a given point P on the screen measures the intensity I of the light at point P:

$$I = \frac{\text{power of light on area of screen}}{\text{cross-sectional area}} \tag{1.2}$$

(power = energy per time. The area A at point P is sufficiently small to be uniform in brightness. The cross-sectional area equals A if the screen is perpendicular to the direction of the light beam. Otherwise the cross-sectional area is $A \cdot \cos \alpha$, where α is the angle between the direction of the light beam and the direction of the normal on the screen.)

Comparison of calculated displacement ψ and measured intensity I shows that the intensity is proportional to ψ^2. Figure 1.8(c) shows the comparison of theory and experiment for diffraction from an edge. Three theoretical curves are calculated: $|\psi|, \psi^2$, and ψ^4 versus distance x. It is evident that the experimental curve of intensity versus distance is proportional to the calculated curve of ψ^2 versus distance x.

1.1.3 Interpretation of the Experiments

We can describe the observed phenomena on the basis of the following postulates.

> *Postulate.* Light behaves as a wave with wavelength λ traveling from a light source.
>
> *Postulate.* The probability that a light quantum is detected in a small area at a position x, y, z is proportional to this area and proportional to ψ^2, where ψ is the displacement of the postulated wave at the position x, y, z.

The first postulate does not say that light is a wave phenomenon (which would be in contradiction to the results of the photoelectric effect experiments). It only makes the weaker assertion that light does exactly what we would expect a wave to do in certain experiments.

The second postulate connects the wave and particle descriptions of light. We saw in Section 1.1.1 that the number of photons interacting with matter is proportional to the intensity of light falling on the object. In Section 1.1.2, we found that the intensity of light is proportional to the square of the displacement ψ in the wave picture. Thus the number of photons in a given small area on the screen is proportional to ψ^2 and proportional to the area; this means also the probability of detecting one photon by its interaction with matter is proportional to ψ^2 and this area. In other words, the probability that a light quantum is detected in a small area is proportional to this area and proportional to ψ^2.

For a given color of light, the energy E per photon can be measured in photoelectric effect experiments. The wavelength λ of the same color of light can be measured in diffraction experiments. Table 1.1 lists the corresponding energies (per photon) and wavelengths for several colors of light.

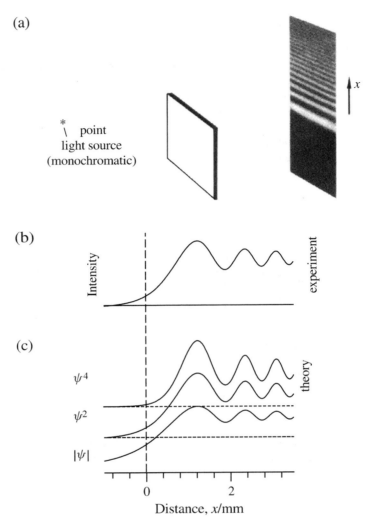

Figure 1.8 Diffraction of light from an edge. (a) Experimental arrangement and interference bands on the screen. (b) Measured intensity pattern on the screen as a function of distance x from the geometrically constructed shadow boundary. The intensity can be measured by replacing the screen by a photographic plate and measuring the density of the photographic image at each point on the screen. (c) Calculated dependencies of $|\psi|$, ψ^2, and ψ^4 on the distance x.

Table 1.1 Energies (per photon) and wavelengths for various colors of light, as measured by photoelectric effect and diffraction experiments

Color	$E/10^{-19}$ J	λ/nm	$E\lambda/10^{-26}$ J m
Violet	4.96	400	19.8
Yellow	3.31	600	19.9
Red	2.48	800	19.8

1.1 LIGHT

The fourth column of Table 1.1 gives the product of E and λ for each case. We see that it is the same for each entry. We can guess that there is a law of nature which makes the product of E and λ the same for each color of light. When the experiment is repeated carefully for different colors of light, the same relationship is always found:

$$E\lambda = 19.86 \times 10^{-26}\, \text{J m} \tag{1.3}$$

From this relationship we can assign the wavelength λ to the photons whose energies were determined in the photoelectric effect experiments. We can write equation (1.3) in a different form by relating the wavelength λ to the frequency ν. For waves, it is generally true that

$$\nu\lambda = c_0 \tag{1.4}$$

where c_0 is the propagation speed of the wave (see Figure 1.9). For light we have $c_0 = 2.998 \times 10^8\, \text{m s}^{-1}$, so for violet light we obtain

$$\nu = \frac{c_0}{\lambda} = \frac{2.998 \times 10^8\, \text{m s}^{-1}}{400 \times 10^{-9}\, \text{m}} = 0.749 \times 10^{15}\, \text{Hz}$$

From equations (1.3) and (1.4) it follows that

$$E = \frac{19.86 \times 10^{-26}\, \text{J m}}{\lambda}$$
$$= \frac{19.86 \times 10^{-26}\, \text{J m}}{c_0} \times \nu = (6.62 \times 10^{-34}\, \text{J s}) \times \nu = h\nu \tag{1.5}$$

The constant h is called *Planck's constant*. Exact measurements establish the value $h = 6.6260755 \times 10^{-34}\, \text{J s}$. Thus our law of nature is written most simply as

$$\boxed{E = h\nu} \tag{1.6}$$

This is a fundamental relationship between frequency and energy of a light quantum.

We will see that another formulation of the same experimental fact is of interest. From relativity theory, a particle of mass m has an equivalent energy

$$E = mc_0^2 \tag{1.7}$$

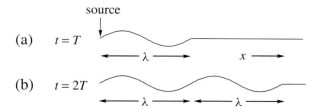

Figure 1.9 Traveling wave. A string is shaken periodically; the disturbance propagates to the right. (a) at time $t = T$, (b) at time $t = 2T$. After the time T (T = period length) the wavefront has traveled the distance $\lambda = c_0 T$. Since $T = 1/\nu$ (ν = frequency) we obtain $\lambda = c_0/\nu$ or $\lambda\nu = c_0$.

For example, a mass of 1 g is equivalent to the energy $E = 1.0\,\text{g} \times (3 \times 10^8\,\text{m s}^{-1})^2 = 9 \times 10^{13}$ J. For comparison, the annual energy consumption of the world is on the order of 10^{20} J, about the amount contained in 1 m^3 of water.

In this sense, the photon can be assigned a mass

$$m_{\text{photon}} = E/c_0^2 \tag{1.8}$$

Combining equations (1.6) and (1.7), we obtain

$$E = m_{\text{photon}} c_0^2 = h\nu = \frac{hc_0}{\lambda} \tag{1.9}$$

or

$$\lambda = \frac{h}{m_{\text{photon}} c_0} \tag{1.10}$$

Equations (1.3), (1.6), and (1.10) are different forms of the same relationship. In equation (1.6), λ is replaced by the frequency defined by equation (1.4), while in equation (1.10), m_{photon} is introduced instead of E.

From the above discussion it is clear that the properties of light cannot be described by the wave picture alone or by the particle picture alone. We need both pictures in totality to describe the observed phenomena (*wave–particle duality*).

1.2 Electrons

1.2.1 The Particle Nature of Electrons

The particle nature of electrons is evident in many situations, such as the Wilson cloud chamber (Figure 1.10), Geiger counters, and scintillation on a fluorescent screen. From the deflection of the electron's trajectory in an electric field and from the deflection in a magnetic field, the charge of the electron is found to be $-e = -1.60217733 \times 10^{-19}$ C (e = elementary charge) and its mass is found to be $m_e = 9.1093897 \times 10^{-31}$ kg.

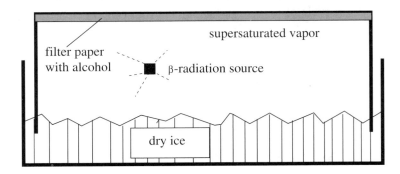

Figure 1.10 The Wilson cloud chamber. Supersaturated vapor condenses preferentially in places where there are small particles produced by electrons from a β-radiation source (nucleation of ions obtained by binding electrons to the air): along the path of an electron, a trail of small alcohol drops appears (similar to the condensation trail of an airplane).

1.2.2 Wave Nature of Electrons

The wave nature of electrons can be demonstrated by diffraction experiments analogous to those for light. Instead of monochromatic light, electron diffraction experiments use a beam of electrons of a definite speed v. This can be produced from a heated filament in an evacuated tube with an accelerating voltage (Figure 1.11). Electrons leaving the heated filament acquire kinetic energy eV_a in traveling through the accelerating voltage V_a. Thus we can write

$$E_{kin} = \tfrac{1}{2}m_e v^2 = eV_a \tag{1.11}$$

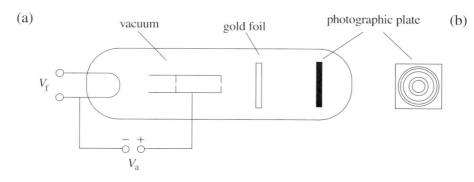

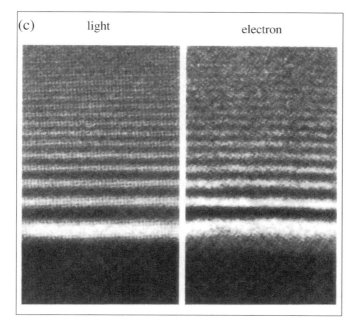

Figure 1.11 Diffraction of electrons. (a) Production of an electron beam (V_f: voltage for heating filament, V_a: accelerating voltage) and diffraction from a gold foil. (b) Diffraction rings on a photographic plate (schematic). (c) Interference bands from diffraction by an edge in the case of monochromatic light ($\lambda = 633$ nm, left, see also Figure 1.8(a)) compared to the case of electrons of distinct speed ($v = 1 \times 10^6$ m s^{-1}, right). The enlargement of the photographs is chosen so that the distance between the first and second maximum is the same in each case. This distance is related to the wavelength λ and the distance between edge and screen, see Figure 1.8.

From this we calculate the speed as

$$v = \sqrt{\frac{2eV_a}{m_e}} \qquad (1.12)$$

If $V_a = 100\,\text{V}$, for example, then

$$v = \sqrt{\frac{2 \times 1.602 \times 10^{-19}\,\text{C} \times 100\,\text{V}}{9.109 \times 10^{-31}\,\text{kg}}} = 5.93 \times 10^6\,\text{m s}^{-1}$$

If an electron beam passes through a double slit, then a diffraction pattern as a pattern of bright and dark bands (as in Figure 1.6(b)) can be observed on a fluorescent screen. The wavelength can be calculated from the distance between bands, as in the case for light. We call the electron wavelength Λ, to distinguish it from the wavelength λ of light.

The double-slit method for electrons requires special techniques, since the wavelength of electrons in practical experiments is very small. The slits must be placed extremely close together to produce an interference pattern. Experimentally, it is easier to use a thin piece of gold foil or an edge. The crystal lattice of the gold foil acts as a diffraction grating for electrons, in the same way as for X-rays. The wavelength Λ can be calculated from the spacing of the diffraction rings, as in Figure 1.11(b). Diffraction from an edge is also feasible. Figure 1.11(c) shows the edge diffraction patterns for light and for electron waves.

Table 1.2 lists the wavelengths Λ of electrons for different accelerating voltages V_a. The last column gives the product of wavelength Λ and speed v. We see that in this case the same value $v\Lambda = 7.2 \times 10^{-4}\,\text{m}^2\,\text{s}^{-1}$ is obtained for both cases. If we try the experiment with still more speeds, we find the same relationship every time. It seems reasonable, then, to see if the constant $7.2 \times 10^{-4}\,\text{m}^2\,\text{s}^{-1}$ can be expressed in terms of the constant we found from the photoelectric effect experiments. For light, $c_0\lambda = h/m_{\text{photon}}$, so we try

$$v\Lambda = \frac{h}{m_e} = \frac{6.63 \times 10^{-34}\,\text{J s}}{9.109 \times 10^{-31}\,\text{kg}} = 7.27 \times 10^{-4}\,\text{m}^2\,\text{s}^{-1}$$

So within the bounds of experimental error, the new constant $v\Lambda$ really can be written as h/m_e! We can formulate a new law of nature for electrons as

$$\boxed{\Lambda = \frac{h}{m_e v}} \qquad (1.13)$$

This is called the *de Broglie relationship*. De Broglie first postulated the equation in analogy with equation (1.10). The electron diffraction experiments which confirmed equation (1.13) were carried out later.

Table 1.2 Speeds and wavelengths Λ for electrons accelerated through different voltages V_a

V_a/V	$v/\text{m s}^{-1}$	Λ/m	$v\Lambda/\text{m}^2\,\text{s}^{-1}$
35	3.51×10^6	2.05×10^{-10}	7.20×10^{-4}
100	5.93×10^6	1.22×10^{-10}	7.24×10^{-4}

The de Broglie relationship is the fundamental expression relating the wave and particle properties of the electron. With increasing speed v of the electron viewed as a particle its wavelength Λ decreases; v and Λ are inversely proportional.

1.2.3 Interpretation of the Experiments

The postulates which we developed for light can be rewritten for electrons.

> *Postulate. Electrons behave as a wave with wavelength Λ traveling from an electron source.*
>
> *Postulate. The probability that an electron is detected in a small area at a position x, y, z is proportional to this area and proportional to ψ^2, where ψ is the displacement of the postulated wave at the position x, y, z.*

In spite of the formal similarities between equations (1.10) and (1.13), there are fundamental differences between photons and electrons.

Photons. The speed c_0 is a constant of nature, and the mass m_{photon} depends on the wavelength λ. We can only speak of a mass of the photon in a sense of imagining the photon as moving with the speed of light; the rest mass is zero. When a photon interacts with matter, its entire quantum of energy can be transferred to a bound electron, in which case the photon ceases to exist.

Electrons. The mass m_e is a constant of nature, and the wavelength Λ depends on the speed v. (We have assumed that the mass of the electron is independent of its speed v. That is not completely correct. The energy, and therefore the mass given by relativity theory, increases with increasing speed. The effect is not significant, however, until the electron's speed v approaches the speed of light c_0. We neglect relativistic effects here.)

1.3 Questions Arising About Wave–Particle Duality

1.3.1 Single Event: Probability Statement; Collective Behavior: Definite Statement

Light or moving free electrons behave, on the one hand, like particles, and on the other hand, like waves of wavelength

$$\lambda = \frac{h}{m_{\text{photon}} c_0} \text{ (light)} \quad \text{or} \quad \Lambda = \frac{h}{m_e v} \text{ (electrons)}$$

The probability that a photon or an electron appears in a given small range at the position x, y, z is proportional to $\psi^2(x, y, z)$.

Experiments with very many photons or electrons lead to interference patterns like the ones in Figure 1.12. If we carry out the corresponding experiments with only a few particles, however, then the patterns can look very different. Figure 1.13 shows a computer simulation with (a) 26, (b) 182, (c) 1900 particles. In case (a) no pattern is seen, in case (b) the pattern is vaguely seen, in case (c) it is clearly seen. With increasing particle

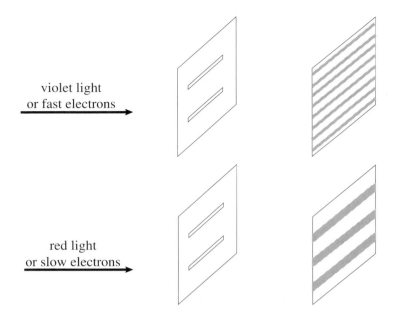

Figure 1.12 Diffraction of light and of electrons.

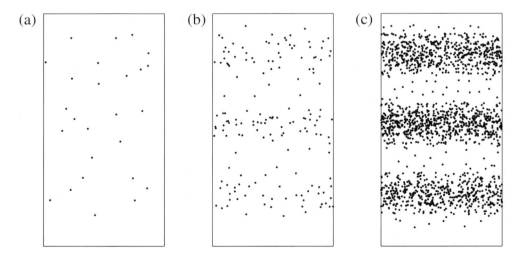

Figure 1.13 Interference pictures: computer simulation with (a) 26, (b) 182, and (c) 1900 particles.

number, it becomes more and more improbable that the interference pattern actually observed deviates from the pattern which should be produced by an infinite number of particles.

We can assume for simplicity that a particle appears in the area of each of the three interference bands with probability

$$P = \tfrac{1}{3} \tag{1.14}$$

1.3 QUESTIONS ARISING ABOUT WAVE–PARTICLE DUALITY

Then for N particles, the probability that all of them appear in the top band, for example, is

$$P = \left(\tfrac{1}{3}\right)^N \quad (1.15)$$

That is, 1/9 for $N = 2$, 1/81 for $N = 4$, and 2×10^{-48} for $N = 100$.

Only probabilistic statements can be made about single events ("where will the next particle appear on the screen?"). Definite statements can be given for collective behavior; the interference pattern is reproducible with very large numbers of particles. In other words, a causal relationship is possible only for the collective behavior of large numbers of particles but not for single events.

We now consider what happens when the wavelength is more and more decreased. For the water waves of Figure 1.3, the diffraction becomes less and less noticeable. The diffraction patterns of both light waves and electron waves are transformed gradually into shadow patterns. Thus wave optics goes over to ray optics and wave mechanics goes over to classical mechanics. So with energetic photons or very fast electrons, we can say practically with certainty that the particle will appear on the screen within the geometrically constructed shadow boundary; thus in this limiting case definite statements about single events are possible.

1.3.2 Wave–Particle Duality: The Need To Abandon Familiar Ways of Thinking

The experiments considered here force us to abandon old ways of thinking. It is therefore of fundamental significance to reflect on the way in which we obtain new knowledge, how conclusions are drawn from experiments, and how this leads to proposing new experiments.

You Cannot Trace the Path of a Photon or an Electron. We can detect single photons or electrons as particles by their effect on the screen in Figure 1.12. However, it is not possible to trace a path, to say that the particle has traveled through either one slit or the other. If one slit or the other is closed, something completely new is observed on the screen. Photons or electrons in the single-slit case appear at places where they were never observed before, that is, they appear at the positions where the waves cancel through interference in the double-slit case. It is thus senseless to ask about the trajectory of the electron. It is just as senseless to look for a medium in which the waves move.

Intuitive Aids Assist You to Grasp the Mathematical Formalism Describing Experiments with Photons or Electrons. Water waves were introduced in Section 1.1.2 as an intuitive aid, an instrument to express the mathematical formulas describing the phenomena (Figure 1.6). Particles and waves are mental pictures to describe concretely the complete physical phenomena. These aids should then be thrown away, and we should use only the mathematical formalism to describe events (Figure 1.14).

Reflect on the Development of Our Everyday View to Understand the Strangeness of the Behavior of Photons or Electrons. The goal of the physical sciences is to reduce the multitude of data received by our senses to a simple order. Throughout the history of the physical sciences this order has been developed and refined. Numerous models are utilized to classify our experience as simply and completely as possible, in an attempt to present a logical picture from the totality of data based on sensory perception.

Figure 1.14 Intuitive aids useful in the beginning are thrown away.

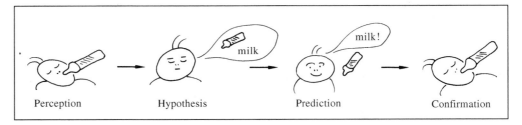

Figure 1.15 Recognition of facts. Hypothesis confirmed (this means it is not falsified).

This process is not different from our own experience of growing up. Infants perceive a multitude of events and they start to make connections between them. They start to become aware of underlying facts, and they try to generalize these facts in an attempt to create a picture of reality (Figure 1.15). These pictures are constantly tested and used to predict facts. With time these pictures become a model of reality, which widens increasingly. More and more new models (or hypotheses) are established and predictions are made. According to the outcome of these predictions the hypotheses are accepted, modified, or rejected.

An important step in the development of a model of reality is reached when perceptions appear as objects in space, changing their location and form. This step leads to an enormous simplification: the growing infant experiences events in space and time. This model succeeds in describing everyday experiences and is therefore deeply rooted in human beings. In the case of the experiments we have discussed in this chapter, however, we are forced to change our hypothesis of reality to which we are all so accustomed.

The unusual model of reality taught us by the experiments in this chapter must be carefully considered to avoid future pitfalls. Reflect on how you arrive at the given

Problems

Problem 1.1 – Double-slit Experiment: Distribution of $|\psi|$ on the Screen

Calculate the distribution of $|\psi|$ displayed in Figure 1.5.

Solution. We generalize the treatment in Box 1.1 (Figure 1.7) for arbitrary distances r_1 and r_2 between the slits and the point P. From the wave displacements φ_1 and φ_2 we calculate the resulting displacement φ at point P.

$$\varphi = \varphi_1 + \varphi_2 = A \sin \frac{2\pi(vt - r_1)}{\lambda} + A \sin \frac{2\pi(vt - r_2)}{\lambda}$$

With the relation

$$\sin \alpha + \sin \beta = 2 \sin \frac{\alpha + \beta}{2} \cos \frac{\alpha - \beta}{2}$$

we get

$$\varphi = 2A \sin \frac{\pi(2vt - r_1 - r_2)}{\lambda} \cos \frac{\pi(r_2 - r_1)}{\lambda}$$

This corresponds to a wave with the maximum displacement

$$\psi = 2A \cos \frac{\pi(r_2 - r_1)}{\lambda}$$

Following the consideration in Box 1.1 (where the special case $r_2 - r_1 = \lambda$ was considered) we obtain

$$r_2 - r_1 = \frac{d}{a} x \quad \text{for } x \ll a \text{ and } d/2 \ll x$$

Then we obtain

$$\psi = 2A \cos \frac{\pi d}{\lambda a} x$$

The maxima of $|\psi|$ are at the positions

$$x = 0, \quad x = \lambda \frac{a}{d}, \quad \text{and} \quad x = \lambda \frac{2a}{d}$$

Problem 1.2 – Diffraction of Photons, Electrons, and Neutrons

Given an aperture with diameter d, let d be 1 cm or 10^{-7} cm.

(a) We direct a light beam onto the aperture; how high must the frequency be so that diffraction can be observed?
(b) We direct an electron beam onto the aperture; how large must the speed of the electrons be so that diffraction can be observed?
(c) We assume that the de Broglie relationship holds not only for electrons, but also for any particle. How large must the speed of neutrons be for the aperture to diffract a neutron beam?

Do not be disturbed if the answers to these exercises are not experimentally feasible. The problems should help clarify the content of equations (1.10) and (1.13).

Solution. In all cases the wavelength must be in the order of d.

(a) $v = c_0/\lambda = c_0/d$. For $d = 1$ cm: $v = 3 \times 10^{10}$ Hz (this is in the microwave region). For $d = 10^{-7}$ cm: $v = 3 \times 10^{17}$ Hz.

(b) $v = h/(m_e \Lambda) = h/(m_e d)(m_e = 9.1 \times 10^{-31}$ kg). For $d = 1$ cm: $v = 7.3 \times 10^{-2}$ cm s^{-1}. For $d = 10^{-7}$ cm: $v = 7.3 \times 10^{5}$ cm s^{-1}.

(c) $v = h/(m_n \Lambda) = h/(m_n d)(m_n = 1.67 \times 10^{-27}$ kg). For $d = 1$ cm: $v = 4 \times 10^{-5}$ cm s^{-1}. For $d = 10^{-7}$ cm: $v = 400$ cm s^{-1}.

2 Basic Features of Bonding

The behavior of macroscopic bodies can be described by the laws of classical mechanics. According to these laws, their speed (and thus their energy) can change continuously. In contrast, an electron in an atom or molecule can assume only certain energies. These energy states are also called *stationary states*. Normally an atom or molecule is found in the lowest energy state (*ground state*); by absorption of energy it can make a transition to a higher energy state (*excited state*). This most astonishing fact, that atoms and molecules can exist only in distinct energy states, can be shown experimentally, for example, by measurement of light absorption (atomic and molecular spectroscopy) and in collision experiments with electrons (the Franck–Hertz experiment).

How is this to be understood? We will show that the existence of distinct energy states is a consequence of our previous considerations in Chapter 1. We will picture, in a thought experiment, electrons confined by rigid walls. We will see that such electrons can only exist in certain discrete states. This fact is related to the wave–particle duality of electrons. The situation is similar to that of a vibrating string, which can form standing waves of certain frequencies.

In the ground state of the H atom the electron cloud around the nucleus has a size of about 100 pm. How can we understand this? Due to the Coulomb attraction of electron and nucleus the cloud should be as small as possible in order to obtain a potential energy as low as possible; then the de Broglie wavelength of the electron would be very small and its kinetic energy very large, according to quantum mechanics. The actual size of the electron cloud corresponds to the energy minimum; it results as a compromise between both tendencies (*variation principle*).

The H_2^+ molecule ion is formed from an H atom and a proton. This is the simplest case showing the formation of a covalent chemical bond. Again, according to the variation principle, the actual size and shape of the electron cloud corresponds to the energy minimum. The same kind of reasoning leads to an understanding of more complex molecules and allows one to grasp the nature of the chemical bond.

For understanding atoms and molecules with more than two electrons an additional postulate must be introduced: at the most two electrons can occupy the same orbital. (Electronic wavefunctions, in the simplified picture considered here, are called orbitals.) This is the Pauli exclusion principle.

2.1 Distinct Energy States

2.1.1 Atomic Spectra

If an electron in an atom is transferred from the ground state to an excited state by a beam of visible or UV light (absorption of light) or by collisions in a gas discharge, then it can return to the ground state by giving off light. This process is described as emission of light (Figure 2.1). In the absorption and emission spectra there are sharp lines which can only be explained if we assume discrete energy levels in the atom. The energy of the outgoing light quantum must be the same as the energy difference ΔE between the higher energy state and the lower energy state. Because of

$$\Delta E = h\nu = \frac{hc_0}{\lambda} \tag{2.1}$$

we find that the emitted light has the wavelength

$$\lambda = \frac{hc_0}{\Delta E} \tag{2.2}$$

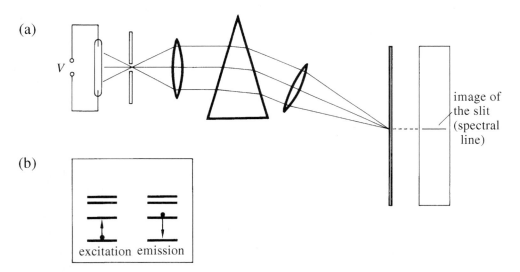

Figure 2.1 (a) Excitation of Na vapor in an evacuated tube by application of a voltage V (glow discharge) and measurement of the wavelength of emitted light. (b) Energy diagram of the Na atoms: energy absorption (excitation from the glow discharge) and energy emission (emission of light with wavelength $\lambda = hc_0/\Delta E$).

2.1.2 The Franck–Hertz Experiment

Consider the setup in Figure 2.2 (the Franck–Hertz experiment). In a tube filled with Na vapor, electrons are emitted from the heated cathode and accelerated from the cathode to the grid by the applied voltage V_{grid}. The electrons which fly through the grid have the kinetic energy

$$E_{\text{kin}} = eV_{\text{grid}} \tag{2.3}$$

2.1 DISTINCT ENERGY STATES

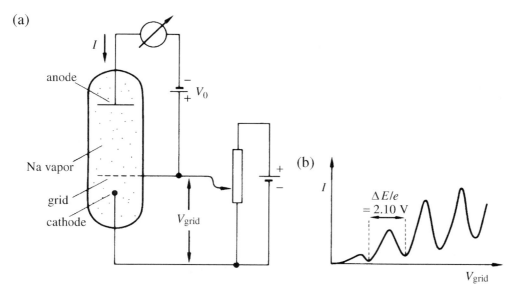

Figure 2.2 Franck–Hertz experiment. (a) Experimental arrangement. (b) Anode current I as a function of grid voltage V_{grid}; ΔE is the excitation energy of Na.

They reach the anode if V_{grid} is greater than the opposing voltage V_0 (about 0.5 V) between grid and anode. On the way to the anode the electrons can collide with Na atoms. If E_{kin} is smaller than the excitation energy of the Na atom then the collisions are elastic, and the electrons reach the anode as if Na atoms are absent. If in fact

$$E_{kin} = \Delta E \tag{2.4}$$

then the colliding electron gives up its kinetic energy completely to the Na atom and comes to rest. Since its kinetic energy is now zero, it can no longer travel against the opposing voltage V_0 and contribute to the current I. The observable effect is that the current, upon reaching the voltage

$$V_{grid} = \frac{\Delta E}{e} \tag{2.5}$$

suddenly decreases (Figure 2.2(b)). With further increase of V_{grid}, the electrons do not come to rest after giving up the energy ΔE, since they still have the kinetic energy $(eV_{grid} - \Delta E)$. Thus they can still reach the anode, and the current I increases further. This increase lasts until

$$V_{grid} = \frac{2\Delta E}{e} \tag{2.6}$$

and then an electron can transfer the energy ΔE to each of two Na atoms by consecutive collisions. We obtain for I a curve with distinct minima. The distances between the minima correspond exactly to the excitation energy ΔE of the Na atoms ($\Delta E = 3.37 \times 10^{-19}$ J).

The Na atoms give up the energy ΔE which they gained from the collisions by radiation of light (the yellow Na line, wavelength $\lambda = 589$ nm, corresponding to $\Delta E = 3.37 \times 10^{-19}$ J $= 2.10$ eV, see Figure 2.1).

2.2 Standing Waves

2.2.1 A Particle Between Parallel Walls

Electrons in atoms and molecules are confined to a small region around the nuclei by Coulomb forces, so they behave approximately as if they were contained in a box of molecular dimensions. In this section we ask the following question: What happens to an electron which finds itself between two parallel walls, moving with velocity v perpendicular to the walls (Figure 2.3)? This is a simplification to focus on the essential features of an electron in a box used later in Sections 2.3 and 2.4 to model atoms and molecules.

According to equation (1.13) we can describe the electron as a planar de Broglie wave of wavelength

$$\Lambda = \frac{h}{m_e v} \qquad (2.7)$$

This wave is reflected back and forth from the walls, so that the original wave interferes with the reflected waves. The waves cancel each other out under most conditions. Cancellation of the waves means, however, that ψ is zero, so the electron cannot be detected anywhere. ψ is not zero only for certain values of Λ, which allow amplification of the waves to occur. We can find these values from the analogous case of a vibrating string which is fixed at both ends (Figure 2.4). Here standing waves appear (i.e., nodes and

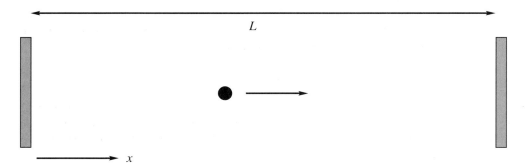

Figure 2.3 Electron reflected back and forth between two parallel walls a distance L apart.

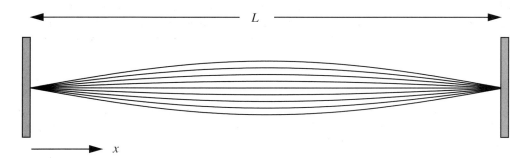

Figure 2.4 Displacement $\varphi(x, t)$ and maximum displacement $\psi(x)$ of a vibrating string. Each curve displays the displacement at a particular time (fundamental vibration: $\Lambda = 2L$).

2.2 STANDING WAVES

maxima of the waves (antinodes) remain at the same places) if the condition

$$\frac{\Lambda}{2} = L \tag{2.8}$$

is fulfilled. The displacement $\varphi(x, t)$ of the standing wave in Figure 2.4 is given by

$$\varphi(x, t) = A \cdot \cos \omega t \cdot \sin \frac{\pi x}{L} \tag{2.9}$$

where A is a constant and $\omega = 2\pi \nu$. The maximum displacement $\psi(x)$ is obtained from equation (2.9) if we set

$$\cos \omega t = \pm 1 \tag{2.10}$$

(this is the case for $\omega t = 0, \pi, 2\pi, \ldots$):

$$\psi(x) = A \cdot \sin \frac{\pi x}{L} \tag{2.11}$$

ψ is called the *wavefunction*. For the electron between two parallel walls, we can calculate the speed, and thus the kinetic energy T_1, in the lowest electron state from the de Broglie relationship (equation (2.7)), using $\Lambda = 2L$:

$$T_1 = \frac{1}{2}m_e v^2 = \frac{1}{2}m_e \left(\frac{h}{m_e \Lambda}\right)^2 = \frac{1}{2}m_e \left(\frac{h}{m_e 2L}\right)^2 = \frac{h^2}{8m_e L^2} \tag{2.12}$$

We can generalize these results if we consider other possible vibrational states of the vibrating string (Figure 2.5) and transfer the results to standing waves of electrons:

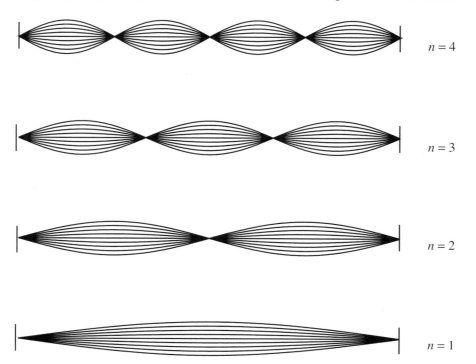

Figure 2.5 Various vibrational states of a string.

$$\frac{\Lambda}{2} = \frac{L}{n}, \quad n = 1, 2, 3, \ldots \tag{2.13}$$

$$\boxed{\psi_n(x) = A \cdot \sin\frac{n\pi x}{L}} \tag{2.14}$$

$$\boxed{T_n = \frac{h^2}{8m_e L^2} n^2} \tag{2.15}$$

The energy E of the electrons is given by

$$E = T + V \tag{2.16}$$

The potential energy V of the electrons in our case does not depend on its position between the walls. We set $V = 0$ arbitrarily and thus obtain the energy diagram in Figure 2.6 for an electron between two parallel walls.

An essential result of this analysis is that an electron has kinetic energy even in the lowest energy state ($n = 1$); that is, it cannot be at rest. The case $n = 0$ is not possible since cancellation of waves would take place. In the lowest state, according to equation (2.12), the energy amounts to $E_1 = 6.0 \times 10^{-18}$ J for $L = 100$ pm and $E_1 = 6.0 \times 10^{-34}$ J for $L = 1.0$ cm.

According to Section 1.2, the probability of finding a particle in a region is proportional to ψ^2. In Figure 2.7, ψ^2 for an electron between two parallel walls is shown for $n = 1$, $n = 2$, and $n = 20$. While it is expected classically that the particle should be found with

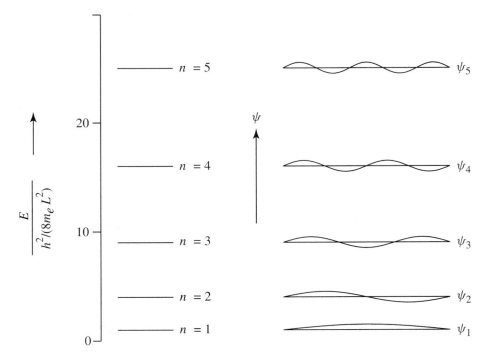

Figure 2.6 Energy scheme and functions ψ of an electron between two parallel walls.

2.2 STANDING WAVES

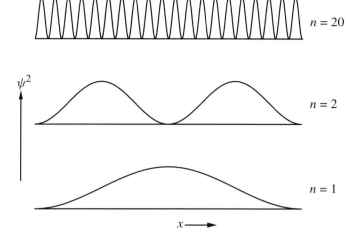

Figure 2.7 ψ^2 (proportional to the probability of finding an electron) as a function of x for an electron between two parallel walls a distance L apart.

equal probability anywhere between the walls (since the speed is the same everywhere), we see obvious maxima and minima in the wave mechanics treatment.

2.2.2 The Heisenberg Uncertainty Relation

According to equations (2.7) and (2.8), an electron between two walls, in its lowest quantum state, must have the speed

$$v = \frac{h}{m_e \Lambda} = \frac{h}{2m_e L} \tag{2.17}$$

Since it may be traveling toward the left or the right wall, the direction of its motion is uncertain. Therefore, the uncertainty in velocity is

$$\Delta \text{velocity} = v - (-v) = 2v = \frac{h}{m_e L} \tag{2.18}$$

and thus the uncertainty in its momentum ($p = m_e \cdot$ velocity) is equal to

$$\Delta p = \frac{h}{L} \tag{2.19}$$

Since the electron can be found somewhere between the two walls with a probability proportional to ψ^2, the uncertainty in its position (the uncertainty in giving the x coordinate) is

$$\Delta x = L \tag{2.20}$$

The product of these uncertainties is

$$\boxed{\Delta p \cdot \Delta x = h} \tag{2.21}$$

This is the *Heisenberg uncertainty principle*.

In contrast to a particle obeying classical physics, the position and the momentum of an electron cannot be given exactly, but within a certain range Δp and Δx, respectively. An accurate determination of the position (Δx small) is given on the expense of accuracy in momentum (Δp large).

As an example, for an electron with $\Delta x = 100\,\text{pm}$ we find

$$\Delta \text{velocity} = \frac{h}{m_e \cdot \Delta x} = \frac{6.63 \times 10^{-34}\,\text{J s}}{9.11 \times 10^{-31}\,\text{kg} \times 100 \times 10^{-12}\,\text{m}} = 7.3 \times 10^6\,\text{m s}^{-1}$$

This means that the speed attributed to an electron which is confined to 100 pm (this is the size of an atom) is on the order of $10^7\,\text{m s}^{-1}$ (this is 1/30 of the speed of light). For a proton, the mass is larger by a factor of 1800, and thus $\Delta \text{velocity} = 4 \times 10^3\,\text{m s}^{-1}$.

2.2.3 Meaningless Questions

In the case of a particle in the quantum state $n = 2$, a question often arises as to how the electron travels from one area of large probability to the other, since there is a location with $\psi^2 = 0$ between them. The postulates of quantum theory allow us to describe the electron in a certain sense as a wave and in another sense as a particle. However, one cannot say that the electron *is* a particle, and therefore the question of how these particles travel from one area to another is nonsensical in the context of the theory. Heisenberg's uncertainty principle elucidates this feature.

The postulates of quantum theory are the basis of describing phenomena in nature; questions which extend beyond what the postulates express are not answerable and make no sense.

Note: The standing wave along the string illustrates the mathematical formalism. However, it is crucial to see that this illustration is no more than an aid for the beginner to grasp some features of the basic postulate expressed by the mathematical formalism (remember Figure 1.14).

2.2.4 Electron in a Box

Our considerations on the electron between two walls at distance L can be extended to the three-dimensional case (this will be shown in detail in Chapter 3). We restrict our present discussion to the lowest quantum state of an electron in a box.

For an electron between two walls the kinetic energy in the lowest energy state ($n = 1$) is given by equation (2.12). For an electron in a box (edge length L) the kinetic energy is the sum of the contributions in the x-, y- and z-directions:

$$T_1 = \frac{h^2}{8m_e L^2} + \frac{h^2}{8m_e L^2} + \frac{h^2}{8m_e L^2} = \frac{h^2}{8m_e L^2} \cdot 3 \quad (2.22)$$

The corresponding wavefunction can be written as the product

$$\psi(x, y, z) = A \cdot \sin\frac{\pi x}{L} \cdot \sin\frac{\pi y}{L} \cdot \sin\frac{\pi z}{L} \quad (2.23)$$

where A is a constant. The wavefunction equals zero at the walls (that is, for $x = 0$ and $x = L$, for $y = 0$ and $y = L$, and for $z = 0$ and $z = L$); it has a maximum at the center

2.3 H ATOM IN THE GROUND STATE

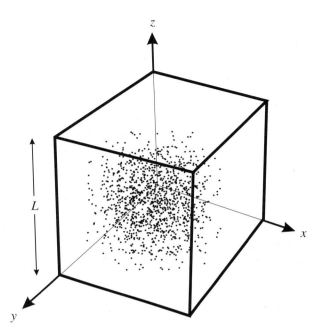

Figure 2.8 Electron in a cubic box (side length L). Electron cloud in the ground state; antinode in the center (computer simulation with 1000 random points).

of the cube ($x = y = z = L/2$). According to Chapter 1, the probability of finding the electron at a given point is proportional to $\psi^2(x, y, z)$. Therefore this probability has a maximum at the center of the cube and it decreases toward the walls. The probability distribution (electron cloud) is illustrated in Figure 2.8.

2.3 H Atom in the Ground State

The hydrogen atom consists of a proton (mass = 1.672623×10^{-27} kg, charge = $+e$) and an electron. Then the Coulomb attractive force

$$f = \frac{1}{4\pi\varepsilon_0} \cdot \frac{e^2}{r^2} \tag{2.24}$$

acts on the electron. Here r is the distance from the electron to the nucleus and ε_0 is a constant called the *permittivity of vacuum*; $\varepsilon_0 = 8.854187816 \times 10^{-12}\, \text{C}^2\,\text{J}^{-1}\,\text{m}^{-1}$. The potential energy $V(r)$ of the electron in the field of the nucleus is

$$V = \int_\infty^r f \cdot dr = -\int_r^\infty f \cdot dr = -\frac{1}{4\pi\varepsilon_0} \cdot \frac{e^2}{r} \tag{2.25}$$

The H atom is the simplest of all atoms, and therefore its behavior is the easiest to understand theoretically. The results obtained from the H atom, as we will see later, can be transferred in a straightforward manner to more complicated electronic systems. For this reason we look at the H atom in a simplified manner which is particularly helpful in paving the way for an understanding of the more complex systems. A rigid treatment of the H atom is given in Chapter 3.

2.3.1 Box Model for H Atom

How can we describe the standing wave of an electron which is attracted by a positively charged atomic nucleus? We start with the approximation that an electron in the Coulomb field of an atomic nucleus behaves similarly to an electron confined to a cubic box of appropriate side length L.

The Coulomb field of the nucleus keeps the electron in its proximity, comparable to the walls of a cubic box. We imagine the walls of the box removed and the positive nuclear charge placed in the center of the cube. We assume in the model that the wavefunction remains unchanged and that the average kinetic energy $\overline{T}$ of the electron is still given by equation (2.22). Thus we take care of the fact that the speed of the electron is determined by its de Broglie wavelength.

Now we view the electron distributed over the cloud in Figure 2.8 as an electron in the Coulomb field of the proton instead of being enclosed in the box. Then its average potential energy $\overline{V}$ can be roughly estimated from the size of the cloud. Since $L/2$ is about the maximum distance of the electron from the center of the box in Figure 2.8, it is reasonable to approximate the average distance $\overline{r}$ by

$$\overline{r} = \tfrac{1}{4}L \qquad (2.26)$$

By inserting this distance in equation (2.25) we obtain

$$\overline{V} = -\frac{1}{4\pi\varepsilon_0} \cdot \frac{4e^2}{L} \qquad (2.27)$$

In a more rigorous treatment $\tfrac{1}{4}L$ has to be replaced by $\tfrac{1}{4.18}L$ (see Section 3.8.2). Then the total energy E is

$$E = \overline{V} + \overline{T} = -\frac{1}{4\pi\varepsilon_0} \cdot \frac{4.18 \cdot e^2}{L} + \frac{3h^2}{8m_e L^2} \qquad (2.28)$$

2.3.2 Variation Principle

What is the appropriate size of the box that takes account of the Coulomb field keeping the electron in the vicinity of the nucleus? To answer this question we consider relation (2.28). For sufficiently large L the total energy E becomes zero (proton and electron at infinite distance). For sufficiently small L the total energy becomes positive (the term L^2 for the kinetic energy overcomes the term L in the expression for the potential energy). For medium-sized L the total energy becomes negative (Figure 2.9).

Intuitively you would say that the most appropriate size of the box corresponds to the energy minimum and this is what quantum mechanics tells us.

The actual size and shape of the electron cloud in the ground state corresponds to the minimum of the total energy E assigned to any imaginable electron cloud (*variation principle*). Among a given set of trial wavefunctions (in our case these are box wavefunctions) the wavefunction corresponding to the energy minimum is the best approximation. The variation principle can be considered as a postulate in addition to the postulates given in Chapter 1. It will be discussed in detail in Chapter 3.

Decreasing the size of the electron cloud results in a decrease of the potential energy $\overline{V}$ (the electron is more strongly attracted by the nucleus); however, at the same time the kinetic energy $\overline{T}$ increases (the electron is confined to a smaller space). The actual size of the electron cloud follows from a compromise between the energy corresponding to the

2.3 H ATOM IN THE GROUND STATE

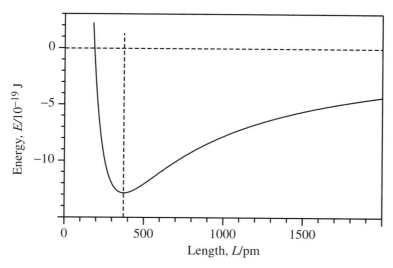

Figure 2.9 Box model of the H atom. Energy E versus box side length L according to (2.28). The energy minimum corresponds to the best approximation (variation principle).

Coulomb attraction of nucleus and electron and kinetic energy, such that the total energy reaches its minimum.

$$\boxed{\text{Postulate: } E = \overline{T} + \overline{V} = \text{minimum} \quad \text{(variation principle)}} \quad (2.29)$$

According to Figure 2.9, and as outlined in Box 2.1, the minimum of E in the box model is reached as

$$E = -\overline{T} = \frac{1}{2}\overline{V} = -\frac{1}{4\pi\varepsilon_0} \cdot \frac{4.18 \cdot e^2}{2 \cdot L} = -12.9 \times 10^{-19}\,\text{J} \quad (2.30)$$

with

$$L = 4\pi\varepsilon_0 \cdot \frac{3h^2}{16.7 \cdot m_e e^2} = 376\,\text{pm} \quad (2.31)$$

The electron is linked to the nucleus by Coulomb forces. But, on the average, half of the Coulomb energy gained in bonding is converted into kinetic energy. So the net bonding energy is $-\overline{V}/2$.

The result of the exact calculation (see Chapter 3) is

$$E = -\overline{T} = \frac{1}{2}\overline{V} = -\frac{1}{4\pi\varepsilon_0} \cdot \frac{e^2}{2 \cdot a_0} = -21.8 \times 10^{-19}\,\text{J} \quad (2.32)$$

where

$$a_0 = 4\pi\varepsilon_0 \frac{h^2}{4\pi^2 m_e e^2} = 52.9177249\,\text{pm} \quad (2.33)$$

is called the *Bohr radius* (see Section 3.8.1).

This is in essential agreement with the result of the model calculation, although the estimated value $\overline{r} = \frac{1}{4}L = 94\,\text{pm}$ is too high by a factor of two.

Box 2.1 – H Atom: Energy for Box Trial Function

For the energy E in equation (2.28) depending on box side length L

$$E = \frac{3h^2}{8m_e L^2} - \frac{1}{4\pi\varepsilon_0} \frac{4.18 \cdot e^2}{L}$$

we seek the minimum:

$$\frac{dE}{dL} = -\frac{6h^2}{8m_e L^3} + \frac{1}{4\pi\varepsilon_0} \frac{4.18 \cdot e^2}{L^2} = 0$$

Solving for L and calculation of energies gives

$$L = 4\pi\varepsilon_0 \frac{h^2}{m_e e^2} \cdot \frac{3}{4 \times 4.18} = 376 \, \text{pm}$$

$$\overline{V} = -\frac{1}{4\pi\varepsilon_0} \cdot \frac{4.18 \cdot e^2}{L}$$

$$\overline{T} = \frac{3h^2}{8m_e L^2} = \frac{3h^2}{8m_e L} \cdot \frac{1}{L}$$

We replace L in the second term:

$$\overline{T} = \frac{3h^2}{8m_e L} \cdot \frac{1}{4\pi\varepsilon_0} \frac{m_e e^2}{h^2} \cdot \frac{4 \times 4.18}{3}$$

$$= \frac{1}{4\pi\varepsilon_0} \cdot \frac{4.18 \cdot e^2}{2L} = -\frac{1}{2}\overline{V}$$

$$E = \overline{V} + \overline{T} = -\frac{1}{4\pi\varepsilon_0} \cdot \frac{4.18 \cdot e^2}{2L}$$

According to equation (2.32) $-E = 21.8 \times 10^{-19}$ J $= 13.6$ eV. This value agrees with the experimentally determined *ionization energy* of the H atom (i.e., the energy necessary to remove the electron to an infinite distance from the atomic nucleus). This means that an electron bouncing on an H atom must have been accelerated by an applied voltage of at least 13.6 eV in order to remove the electron bound to the nucleus.

The prediction which results from the postulates of quantum theory thus agrees with experiment. The kinetic energy $\overline{T}$ of the electron in the H atom is

$$\overline{T} = \frac{m_e}{2}\overline{v^2} = -E = 21.80 \times 10^{-19} \, \text{J} = 13.6 \, \text{eV} \quad (2.34)$$

This corresponds to an average speed of $2 \times 10^6 \, \mathrm{m\,s^{-1}}$. The time needed for the electron to travel a distance $s = 100 \, \mathrm{pm}$ is $t = s/v = 100 \, \mathrm{pm}/(2 \times 10^6 \, \mathrm{m\,s^{-1}}) = 5 \times 10^{-17} \, \mathrm{s}$. Note that we can only speak of a speed or traveled distance in the sense that the electron behaves as if it were a moving particle.

The speed $2 \times 10^6 \, \mathrm{m\,s^{-1}}$ is large compared with the speed with which atoms and molecules move ($\approx 10^3 \, \mathrm{m\,s^{-1}}$ at room temperature). The time t required for two H atoms to approach each other by a distance $s = 10 \, \mathrm{pm}$ is $t = s/v = 10 \, \mathrm{pm}/(10^3 \, \mathrm{m\,s^{-1}}) = 10^{-14} \, \mathrm{s}$; during this time the electrons behave as if each electron would traverse the region of its nucleus several hundred times.

Thus the electrons act like a smeared-out cloud of charge $-e$ (an electron shell). Two colliding H atoms repel each other through the Coulomb repulsive force as soon as the electron shells begin to overlap. Therefore we can regard the H atoms approximately as hard spheres with a radius of about $2a_0$. This value is close to the value of 128 pm (the collision radius of the H atom) obtained from experiment (see Table 11.1). (Under the experimental conditions no H_2 is formed: the process would require a third collision partner taking up the energy appearing in forming the bond.)

In conclusion, the actual electron cloud of an atom or molecule, among all conceivable electron clouds, is the one attributed to the lowest energy. The best approximation among a proposed set of electron clouds is obtained by searching for the cloud assigned to the lowest energy.

The energy minimum is reached as a compromise between the average potential energy $\overline{V}$ ($\overline{V}$ decreases with decreasing size of the electron clouds) and the average kinetic energy $\overline{T}$ ($\overline{T}$ increases with decreasing size). The search for this compromise is the key to an understanding of the occurrence and properties of atoms and molecules. In the following sections this is illustrated by the examples of H_2^+, He^+, He, H^-, Li^+, Be^{2+}, and H_2.

2.4 H_2^+ Molecule Ion in the Ground State

We consider the simplest molecule, the H_2^+ ion, which is composed of two protons and one electron. H_2^+ is produced by the bombardment of H_2 molecules with electrons:

$$H_2 + 1 \text{ electron} \longrightarrow H_2^+ + 2 \text{ electrons} \tag{2.35}$$

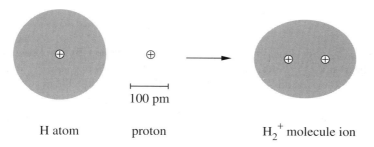

Figure 2.10 Formation of an H_2^+ molecule from an H atom and a proton. Within the gray zone, the charge density is greater than 1% of its value at the nucleus of the H atom.

2.4.1 Forming H_2^+ from an H Atom and a Proton

For our analysis, we consider H_2^+ to be made from an H atom and a proton (Figure 2.10). The energy minimum is reached when the nuclei are a certain distance apart (106 pm), with an electron cloud around both nuclei. Compared to the electron cloud in the H atom, this electron cloud is longitudinally expanded (the extension of the electron cloud in direction of the bond line is increased); on the other hand, the electron cloud is laterally compressed. Perpendicular to the bond line the electron is in the much stronger field of both nuclei.

The average kinetic energy increases in forming the bond because the effect of lateral compression outweighs the effect due to the longitudinal expansion. Exactly one-half of the Coulomb energy gained in forming the bond is lost by increasing the kinetic energy, as you have seen in the case of forming an H atom from a proton and an electron.

We must imagine that the two positive nuclei in the negative cloud of the electron are pulled in until the force which pulls the nuclei into the electron cloud is balanced with the repulsive force between the nuclei. The fact that the chemical bond is based on the Coulomb attraction of the electron cloud and the nuclei must be strongly emphasized.

2.4.2 Box Model for H_2^+

In analogy to the H atom we describe the H_2^+ molecule ion by an electron confined to a rectangular box (short axes $L_y = L_z = L$, long axis $L_x = bL$) around the two protons at distance d in its center (Figure 2.11). The mean kinetic energy of the electron in its ground state ($n_x = n_y = n_z = 1$) is

$$\overline{T} = \frac{h^2}{8m_e L_x^2} + \frac{h^2}{8m_e L_y^2} + \frac{h^2}{8m_e L_z^2} = \frac{h^2}{8m_e}\left(\frac{1}{L_x^2} + \frac{1}{L_y^2} + \frac{1}{L_z^2}\right) \quad (2.36)$$

$$= \frac{h^2}{8m_e L^2}\left(\frac{1}{b^2} + 2\right)$$

corresponding to the contributions of the three directions of movement in space. The potential energy V depends on the attraction of the electron by the two positive charges at distances r_1 and r_2 (negative contributions) and the mutual repulsion of the positive charges at distance d (positive contribution):

$$V = \left(-\frac{1}{4\pi\varepsilon_0}\frac{e^2}{r_1} - \frac{1}{4\pi\varepsilon_0}\frac{e^2}{r_2}\right) + \frac{1}{4\pi\varepsilon_0}\frac{e^2}{d} \quad (2.37)$$

In this case L, b, and d are parameters which are obtained by applying the variation principle. The result (for details see Chapter 4) is $L_x = 409$ pm, and $L_y = L_z = 301$ pm (L_x is longer, while L_y and L_z are shorter than the box length 376 pm for the H atom). The energy is $E = -\overline{T} = \frac{1}{2}\overline{V} = -16.9 \times 10^{-19}$ J. E is more negative than for the corresponding value -12.9×10^{-19} J obtained in the box model for the H atom: this means that a stable bond is formed.

The bond energy is

$$E_{\text{bond}} = -(E_{H_2^+} - E_H) = (16.9 - 12.9) \times 10^{-19}\,\text{J} = 4.0 \times 10^{-19}\,\text{J}$$

2.5 HE⁺ ION, HE ATOM, AND SIMILAR SYSTEMS

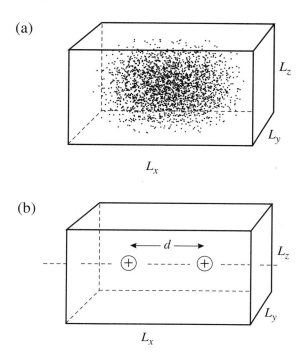

Figure 2.11 Box model of H_2^+. (a) Electron in a rectangular box with side lengths L_x and $L_y = L_z$. (b) Protons placed at distance d as indicated, electron cloud considered unchanged.

The result of the exact treatment (see Chapters 3 and 4) is

$$E_{\text{bond}} = -(E_{H_2^+} - E_H) = (26.2 - 21.8) \times 10^{-19}\,\text{J} = 4.4 \times 10^{-19}\,\text{J}$$

in agreement with the experiment. In a rigorous treatment it must be taken into account that the energy of an H_2^+ ion, as a result of the vibrational motion of the nuclei, lies about 0.2×10^{-19} J higher (see Section 9.2); the energy which is necessary to pull the bond apart must be smaller by this amount. Indeed, the experimental value is $E_{\text{bond}} = 4.2 \times 10^{-19}$ J. The bond energy of H_2^+ is about one-third of the bond energy of N_2, one of the strongest chemical bonds. Comparison with equation (2.32) shows that the energy binding a proton to an H atom is 20% of the energy of binding an electron to a proton.

2.5 He⁺ Ion, He Atom, and Similar Systems

2.5.1 He⁺ Ion and He Atom

The He⁺ ion differs from the H atom only in that the He nucleus has twice the positive charge (Figure 2.12). For this reason the electron is more strongly attracted to the nucleus, so we expect to find the electron closer to the nucleus on the average; that is, the charge cloud should be compressed compared with the charge cloud in the H atom.

The He atom consists of a twofold positively charged nucleus and two electrons (Figure 2.13). If we disregard the repulsion of the two electrons then the cloud of the electrons would be the same as in the He⁺ ion; the probability density ψ^2 should be twice the value of He⁺ and the same should be true for the energy E.

Actually, we have to include the repulsion of the two electrons. On average, the electrons are at a greater distance from each other, and the energy minimum is reached at a higher energy and a less compact electron cloud. If the charge of the second electron were localized at the nucleus (this means the first electron is in the field of only one positive charge), we would get the same electron cloud as for the H atom. Actually we obtain an electron cloud which is smaller than the electron cloud of the H atom and larger than the electron cloud of the He^+ ion (see Figure 2.12). Experimentally, the collision radius of He is 110 pm instead of 128 pm for the H atom (see Table 11.1).

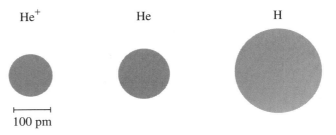

Figure 2.12 Charge clouds of He^+, He, and H. Within the gray zone, the electron density is greater than 1% of the density at the nucleus.

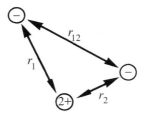

Figure 2.13 He atom: twofold positively charged nucleus and two electrons.

2.5.2 He-like Systems: H^-, Li^+, Be^{2+}

In analogy to He we can understand the dimensions of the electron clouds of ions with two electrons, such as in He, but with different nuclear charges: H^-, Li^+, Be^{2+} (Figure 2.14).

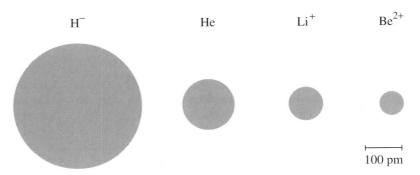

Figure 2.14 Cloud sizes for ions with two electrons similar to He.

2.5 HE+ ION, HE ATOM, AND SIMILAR SYSTEMS

The electron cloud is less condensed in H$^-$ and more condensed in Li$^+$ (nuclear charge $+3e$) and Be^{2+} (nuclear charge $+4e$) as a result of the different nuclear charges. Note that the extension of the electron cloud for H$^-$ is extremely large. This is due to the shielding effect of the electrons on the nucleus, which plays an important role for a nucleus consisting of only one proton.

2.5.3 H$_2$ Molecule

The H$_2$ molecule consists of two protons and two electrons. The cloud of these two electrons is similar to that of the H$_2{}^+$ ion. The internuclear distance of 74 pm is, however, smaller than that of H$_2{}^+$ since the nuclei are sucked in by a cloud of two electrons. Because of this, the electron cloud should be strongly condensed; however, by the repulsion of the two electrons the electron cloud should be extended. The two effects compensate each other, so that finally the electron clouds in H$_2{}^+$ and H$_2$ are very similar (Figure 2.15).

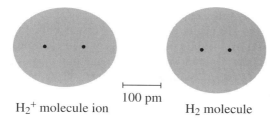

Figure 2.15 H$_2{}^+$ ion and H$_2$ molecule. Within the gray zone, the electron density is greater than 1% of the density at the nucleus.

If the two electrons did not repel and if the nuclear distance were exactly the same as in H$_2{}^+$, then the bond energy would have to be twice as great in H$_2$ as in the H$_2{}^+$ ion; because of the repulsion of the two electrons it is somewhat less than double: for H$_2{}^+$ it is 4.2×10^{-19} J, while for H$_2$ it is 7.2×10^{-19} J. The bond energy calculated on the basis of the variation principle, taking into account the electrons' evasion effect, agrees very well with the experimental value. (As in the case of H$_2{}^+$, we have to consider the vibrational energy of H$_2$ in the ground state.) This result also shows how well questions about chemical bonds can be treated by the postulates of wave mechanics.

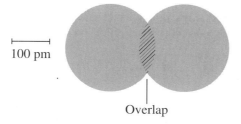

Figure 2.16 Electron cloud of the H$_2$ molecule, roughly represented by overlapping electron clouds of two H atoms (striped area: overlap region). Note the neglect of the compression of the electron cloud in forming the bond.

The electron cloud in the H_2 molecule can be approximated by an overlap of the electron clouds of the two H atoms (Figure 2.16). The region where the electron clouds overlap is called the *overlap region*; in this region the electrons experience the strong Coulomb attractive force of the two atomic nuclei. In atomic bonds (covalent bonds), the greater the overlap region, the greater the bond energy.

2.6 Pauli Exclusion Principle

2.6.1 Nonexistence of He_2

According to the foregoing analysis, it would have to be possible for two He nuclei and four electrons to form an He_2 molecule similar to H_2^+ and H_2, but the bond energy from four electrons would be much stronger than that of H_2. Such a molecule has not been experimentally observed (although two He atoms can weakly bind because of dispersion forces, see Chapter 10).

The observations on He cannot be understood on the basis of the postulates which have proven useful in all cases considered up to now. A further postulate is needed to take account of these and many other experimental facts.

Let us assume that two He nuclei are fixed at a distance of, say, 100 pm and we add four electrons compensating for the nuclear charges. Two electrons, as in H_2, occupy an electron cloud approximated by the lowest energy state in the box model (they occupy the so-called bonding orbital with an antinode in the center of the box); now we assume, as a new postulate, that the next two electrons, according to this new postulate, are prevented from occupying the same orbital, they occupy an electron cloud approximated by the next higher energy state (they occupy the antibonding orbital with a nodal plane in the center of the box). Generally we refer to one-electron wavefunctions as orbitals.

In general terms, our new postulate can be expressed as follows

> *Postulate. In an atom or molecule, at most two electrons can occupy the same orbital.*

This postulate is called the *Pauli exclusion principle*. While the first two electrons in the bonding orbital hold the nuclei together, the other two electrons in the antibonding orbital effectively push the atoms apart. The two effects compensate, so that no bond appears. If there are two electrons in the bonding orbital and only one in the antibonding orbital, then the attraction dominates, and He_2^+ is known as a molecule. The bond energy is about the same as in H_2^+; this is plausible, since the binding effect of the first two electrons is about half compensated by the repulsion of the third.

2.6.2 Li^+H^- (Ionic Crystal)

Li^+ and H^- contain two electrons in their clouds, as does the He atom, so according to the Pauli exclusion principle no bond should be possible between the two ions. The two ions carry opposite charge, however, and thus attract each other according to Coulomb's law. The two ions approach each other until the attractive force is balanced by the force originating from shell repulsion (Figure 2.17(a)); the equilibrium distance lies approximately

PROBLEM 2.1

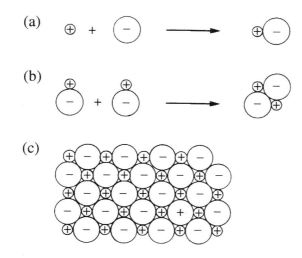

Figure 2.17 Coulomb attraction of Li$^+$ and H$^-$ and formation of an ionic crystal. (a) Aggregate formed from two ions. (b) Aggregate formed from four ions. (c) Crystal lattice, formed by further packing of ions.

where the probability density ρ is about 1% of its value at the nucleus. From the pairs of ions further aggregations can be made with release of potential energy (Figure 2.17(b)). The greatest imaginable release of Coulomb energy results from the formation of a crystal with the same crystal lattice as NaCl (Figure 2.17(c)).

Experimentally, LiH is actually found to be a stable ionic crystal. The ionic radii (Figure 2.14) are: Li$^+$, 58 pm; H$^-$, 154 pm. The ion spacing in the crystal should be about 212 pm; experimentally it is found to be 204 pm. We have thus understood the simplest case of ionic bonds on the grounds of our postulates.

Problems

Problem 2.1 – Franck–Hertz Experiment

The separation of two consecutive minima in the Franck–Hertz experiment for sodium is 2.10 V. Calculate the wavelength for the yellow sodium emission line.

Solution. The voltage $V = 2.10$ V corresponds to the excitation energy $\Delta E = 2.10$ eV $= 1.602 \times 10^{-19} \times 2.10$ C V $= 3.36 \times 10^{-19}$ J and to the wavelength $\lambda = hc_0/\Delta E = 590$ nm.

Problem 2.2 – Standing Waves Formed by Overlap of Traveling Waves

Show that a standing wave with nodes at the positions $x = 0$ and $x = L$ is formed from the overlap of two traveling waves of the same wavelength traveling in opposite directions.

Solution. A wave traveling from left to right with speed v can be described by

$$\varphi_+ = A \sin \frac{2\pi(vt - x)}{\Lambda}$$

This wave is reflected at the right boundary. Then the back traveling wave is described by

$$\varphi_- = -A \sin \frac{2\pi(vt + x)}{\Lambda}$$

Both waves interfere, and thus the displacement φ at a point x is

$$\varphi = \varphi_+ + \varphi_- = A\left[\sin\frac{2\pi(vt-x)}{\Lambda} - \sin\frac{2\pi(vt+x)}{\Lambda}\right]$$

Using the theorem

$$\sin\alpha - \sin\beta = 2\cos\frac{\alpha+\beta}{2}\sin\frac{\alpha-\beta}{2}$$

we get

$$\varphi = 2A\cos\frac{2\pi vt}{\Lambda}\cdot\sin\frac{2\pi x}{\Lambda}$$

or

$$\varphi = \cos\frac{2\pi vt}{\Lambda}\cdot\psi \quad \text{with} \quad \psi = 2A\cdot\sin\frac{2\pi x}{\Lambda}$$

where ψ is the equation for a standing wave. With $\Lambda = 2L/n$ we obtain

$$\psi = 2A\cdot\sin\frac{\pi x n}{L}, \quad n = 1, 2, \ldots$$

3 Schrödinger Equation and Variation Principle

In Chapters 1 and 2 a simple quantum mechanical picture of small atoms and molecules was developed; after reading these chapters, one should be able to understand the basic concepts and to grasp the essentials of chemical bonding.

In this chapter the formalism of quantum theory is introduced, allowing a quantitative treatment. The Schrödinger equation is discussed as a postulate generalizing and replacing the first postulate in Chapter 1. This equation is applied to a quantitative description of the stationary states of the H atom.

Furthermore, the variation principle, introduced as a postulate in Chapter 2, is derived as a theorem resulting from the Schrödinger equation.

Beginners who are less interested in quantum mechanical formalism are advised to take a look at the energy levels and wavefunctions of the excited states of the H atom. They should proceed to Chapter 5 and come back to Chapters 3 and 4 at a later stage.

3.1 Wave Equation and Schrödinger Equation

3.1.1 Wave Equation

In the preceding chapters we have discussed the fundamental experiments which lead to the postulates of quantum mechanics, and we have discussed standing electron waves in simple examples. For a deeper understanding of atoms and molecules, we want to become familiar with a more general prescription for calculating the wavefunctions and the energies of electronic states.

It is useful to consider the wave equation of a stretched string as an aid that has to be thrown away afterwards (remember Figure 1.14). The wave equation is a differential equation. Differential equations govern guitar string vibration, heat flow in metals, evolution of populations, planetary motion, and diffusion. It will be helpful to give a short look at the wave equation for readers who are not familiar with it.

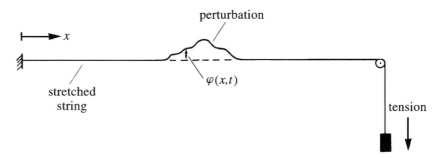

Figure 3.1 Spreading of a disturbance along a stretched string.

Vibrating string

We start out with the vibrating string (Figure 3.1). If the string is struck anywhere, waves move to the left and to the right with speed v; v depends on the tension and on the mass–length ratio.

If we create an arbitrary displacement $\varphi(x, t = 0)$ at time $t = 0$ and afterwards leave the string to itself, then the displacement φ at any position x will be a complicated function of time t. $\varphi(x, t)$ can be described by the wave equation

$$\frac{\partial^2 \varphi}{\partial t^2} = v^2 \cdot \frac{\partial^2 \varphi}{\partial x^2} \tag{3.1}$$

Standing waves on a string

We are interested in the special case of an initial displacement that leads to a standing wave. In this case $\varphi(x, t)$ consists of one factor dependent only on x and one dependent only on t. We propose a solution

$$\varphi(x, t) = \psi(x) \cdot \cos \omega t \tag{3.2}$$

Then we find

$$\frac{\partial \varphi}{\partial t} = -\psi(x) \cdot \omega \sin \omega t, \quad \frac{\partial^2 \varphi}{\partial t^2} = -\psi(x) \cdot \omega^2 \cos \omega t, \quad \frac{\partial^2 \varphi}{\partial x^2} = \frac{d^2 \psi}{dx^2} \cdot \cos \omega t \tag{3.3}$$

and by substituting into the wave equation (3.1) we have

$$-\psi \cdot \omega^2 = v^2 \cdot \frac{d^2 \psi}{dx^2} \quad \text{or} \quad \frac{d^2 \psi}{dx^2} = -\left(\frac{\omega}{v}\right)^2 \cdot \psi \tag{3.4}$$

Because of $\omega = 2\pi \nu$ and $\Lambda \nu = v$ we can relate ω/v to the wavelength Λ. Thus

$$\frac{\omega}{v} = \frac{2\pi \nu}{\Lambda \nu} = \frac{2\pi}{\Lambda} \tag{3.5}$$

and

$$\frac{d^2 \psi}{dx^2} = -\frac{4\pi^2}{\Lambda^2} \cdot \psi \tag{3.6}$$

3.1 WAVE EQUATION AND SCHRÖDINGER EQUATION

Example – check of equation (3.6)

Let us check equation (3.6) for standing waves along a string fixed at the ends. According to Chapter 2, the condition for standing waves is

$$n\frac{\Lambda}{2} = L, \quad n = 1, 2, 3, \ldots \tag{3.7}$$

and the maximum displacement ψ is

$$\psi = A \cdot \sin\frac{n\pi x}{L} \tag{3.8}$$

Is this expression really a solution of equation (3.6)? By differentiation of equation (3.8) we obtain

$$\frac{d^2\psi}{dx^2} = -A\left(\frac{n\pi}{L}\right)^2 \cdot \sin\frac{n\pi x}{L} = -\left(\frac{n\pi}{L}\right)^2 \cdot \psi \tag{3.9}$$

Then because of condition (3.7) we obtain

$$\frac{d^2\psi}{dx^2} = -\left(\frac{2\pi}{\Lambda}\right)^2 \cdot \psi \tag{3.10}$$

in agreement with equation (3.6). This shows that expression (3.8) for standing waves is a solution of the differential equation (3.6) obeying the boundary conditions

$$\psi = 0 \quad \text{for} \quad x = 0 \text{ and } x = L \tag{3.11}$$

Electron between two walls

Now we consider an electron. In this case we can make use of the de Broglie relation

$$\Lambda = \frac{h}{m_e v} \tag{3.12}$$

to express Λ in terms of the energy E:

$$E = T + V = \frac{1}{2}m_e v^2 + V = \frac{1}{2}m_e\left(\frac{h}{m_e \Lambda}\right)^2 + V \tag{3.13}$$

$$\Lambda^2 = \frac{h^2}{2m_e(E - V)} \tag{3.14}$$

By substituting Λ in equation (3.6) we obtain

$$-\frac{h^2}{8\pi^2 m_e}\frac{d^2\psi}{dx^2} + V\psi = E\psi \tag{3.15}$$

In this equation the potential energy V is a constant. This equation represents the de Broglie relation in a way useful to understand how the Schrödinger equation (see Section 3.1.2) was introduced. Although equation (3.15) has formally the same structure as the Schrödinger equation (3.16), by no means this consideration can be understood as a derivation of the Schrödinger equation.

3.1.2 Schrödinger Equation (One-dimensional)

Schrödinger assumed on a trial basis that equation (3.15) is also true for the case that $V = V(x)$ is a function of x:

$$\text{Postulate:} \quad -\frac{h^2}{8\pi^2 m_e}\frac{d^2\psi}{dx^2} + V(x)\psi = E\psi \qquad (3.16)$$

with the boundary condition

$$\lim_{x \to \pm\infty} = 0 \qquad (3.17)$$

This generalization is a postulate, replacing the first postulate in Section 1.2.3. It was freely invented, so it cannot be derived from anything else. Whether or not this postulate is reasonable can be checked by comparing its consequences with experiment. We will see that equation (3.16) is the basis for understanding the entire field of chemistry.

In the case of an electron between two parallel walls, the potential energy V is infinitely large outside the walls ($x < 0$ and $x > L$), so ψ must be zero everywhere outside the

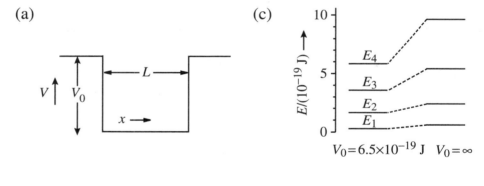

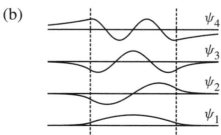

Figure 3.2 Electron in a potential trough of finite depth. (a) Potential energy V (potential trough). (b) Wavefunctions of the lowest four quantum states; the dashed lines mark the positions of the walls. (c) Energies of the four quantum states (left) compared to the energies of the four lowest energy states of a particle in box with infinitely high walls (right). The distance between the walls is $L = 1000\,\text{pm}$, and the height of the walls is $V_0 = 10 \times h^2/(8m_e L^2) = 6.5 \times 10^{-19}\,\text{J}$.

3.1 WAVE EQUATION AND SCHRÖDINGER EQUATION

region $0 < x < L$. In this special case the boundary condition (3.17) is satisfied by the wavefunctions discussed in Chapter 2.

In all cases when the potential energy V is not infinitely large outside the region $0 < x < L$, the wavefunction ψ vanishes at infinity. Figure 3.2 shows the wavefunctions ψ and the energies E ($E < V_0$) for the given potential obtained by solving Equation (3.16) with the boundary condition (3.17), as outlined in Foundation 3.1. The energy levels are lower than the levels of the corresponding states with $n = 1, 2, 3, 4$ for $V_0 = \infty$.

Classically, in the considered case $V_0 > E$ the electron would be confined to the region inside the potential trough. According to the solution of the Schrödinger equation the probability of observing the electron in the region outside the trough decreases exponentially with the distance; it decreases more strongly with increasing difference $V_0 - E$ (see Figure 3.2(b) and Box 3.1).

This can be easily seen by rewriting the Schrödinger equation (3.16) in the form

$$\frac{d^2\psi}{dx^2} = \frac{8\pi^2 m_e}{h^2}(V - E) \cdot \psi \qquad (3.18)$$

In a region where the potential energy is larger than E

$$V = V_0, \quad \text{where } V_0 > E \qquad (3.19)$$

the wavefunction ψ has the property that its second derivative is equal in sign to ψ. This means that the solution of the Schrödinger equation in this region is not periodic, but exponential:

$$\psi = A \cdot e^{-bx}, \quad \frac{d^2\psi}{dx^2} = Ab^2 \cdot e^{-bx} \qquad (3.20)$$

Then we find by inserting in equation (3.18)

$$b = \frac{2\pi}{h} \cdot \sqrt{2m_e(V_0 - E)} \qquad (3.21)$$

Box 3.1 – Probability of Finding an Electron Outside a Potential Trough

According to equations (3.20) and (3.21), in the region with $V = V_0$ (Figure 3.2(a)) the wavefunction ψ decays exponentially:

$$\psi = A \cdot e^{-bx}, \quad b = \frac{2\pi\sqrt{2m_e}}{h}\sqrt{V_0 - E}$$

$V_0 - E$ can be considered as an energy barrier E_{barrier} which must be surmounted by an electron in order to escape the potential trough:

Continued on page 44

> *Continued from page 43*
>
> $$E_{\text{barrier}} = V_0 - E$$
>
> Then the probability dP of finding the electron between x and $x + dx$ is
>
> $$dP = \psi^2 \cdot dx = A^2 \cdot e^{-2bx} \cdot dx$$
>
> This relation is important in the treatment of electron tunneling processes (see Chapters 4 and 23).
>
> We compare dP at point $L + \Delta x$ with dP at $x = L$ (i.e., at the position of the right wall). Then
>
> $$dP(L + \Delta x) = A^2 \cdot e^{-2b(L+\Delta x)} \cdot dx$$
> $$= A^2 \cdot e^{-2bL} \cdot e^{-2b\Delta x} \cdot dx$$
>
> and
>
> $$\frac{dP(L + \Delta x)}{dP(L)} = e^{-2b\Delta x}$$
>
> For $E_{\text{barrier}} = 2 \times 10^{-19}$ J we obtain $b = 6 \times 10^9$ m^{-1}. Then for $\Delta x = 1$ nm
>
> $$\frac{dP(L + \Delta x)}{dP(L)} = e^{-12} \approx 6 \times 10^{-6}$$

It is important at this point to make clear how we progress in science: we try to fit a large number of phenomena to a theory, that is, to derive them from a small number of basic assumptions. We gain confidence in a theory when we succeed in fitting more and more areas of facts to the theory. If we find contradictions and if the facts are more successfully interpreted by a new theory the new theory will replace the older one. In this way greater and greater areas of facts are described by increasingly general theories. The basic assumptions are established by the success with which they interpret the widest possible area of observed phenomena and predict new phenomena.

In the present case, it will be fascinating to see that the theory allows us to understand an immense amount of experimental facts.

3.1.3 Schrödinger Equation (Three-dimensional)

Equation (3.16) is true for a one-dimensional wave, where ψ depends only on x. In a molecule, however, ψ depends on all three spatial coordinates x, y, and z. To go from the one-dimensional to the three-dimensional problem, the wave equation must be generalized as

$$\boxed{\text{Postulate:} \quad -\frac{h^2}{8\pi^2 m_e}\left(\frac{\partial^2}{\partial x^2} + \frac{\partial^2}{\partial y^2} + \frac{\partial^2}{\partial z^2}\right)\psi + V\psi = E\psi} \quad (3.22)$$

3.1 WAVE EQUATION AND SCHRÖDINGER EQUATION

with the boundary condition

$$\lim_{x,y,z \to \pm\infty} \psi = 0 \qquad (3.23)$$

Relation (3.22) is called the *Schrödinger equation*. The abbreviation

$$\nabla^2 = \frac{\partial^2}{\partial x^2} + \frac{\partial^2}{\partial y^2} + \frac{\partial^2}{\partial z^2} \qquad (3.24)$$

is often used. It is denoted as the *Laplacian operator*. So we can write the Schrödinger equation in the form

$$\left(-\frac{h^2}{8\pi^2 m_e}\nabla^2 + V\right)\psi = E\psi \qquad (3.25)$$

A further useful abbreviation is

$$\mathcal{H} = -\frac{h^2}{8\pi^2 m_e}\nabla^2 + V \qquad (3.26)$$

$\mathcal{H}$ is called the *Hamiltonian operator*. Thus equation (3.22) can be written in the short form

$$\mathcal{H}\psi = E\psi \qquad (3.27)$$

Note: The quantum mechanical wavefunction ψ should not be confused with the wavefunction φ in equation (3.1) that describes the displacement of a string.

Particle in a box

As a simple example for the application of the Schrödinger equation we consider a particle in a box (Figure 3.3); we assume the particle can move freely within this box; this means that $V(x, y, z)$ is constant. We try to determine if the wavefunction

$$\psi(x, y, z) = A \cdot \sin\frac{n_x \pi x}{L_x} \cdot \sin\frac{n_y \pi y}{L_y} \cdot \sin\frac{n_z \pi z}{L_z} \qquad (3.28)$$

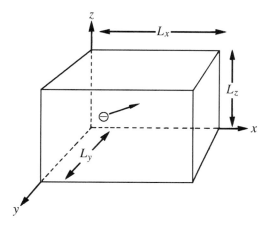

Figure 3.3 Electron in a box with side lengths L_x, L_y, and L_z.

where A is a constant, is a solution of the Schrödinger equation. Equation (3.28) is a generalization of the wavefunction (2.23) applied to a cubic box. This function is zero at the walls of the box, thus fulfilling the boundary condition for this problem. We insert the wavefunction into the Schrödinger equation; first we calculate the derivatives $\partial \psi / \partial x$ and $\partial^2 \psi / \partial x^2$:

$$\frac{\partial \psi}{\partial x} = A \cdot \frac{n_x \pi}{L_x} \cdot \cos \frac{n_x \pi x}{L_x} \cdot \sin \frac{n_y \pi y}{L_y} \cdot \sin \frac{n_z \pi z}{L_z}$$

$$\frac{\partial^2 \psi}{\partial x^2} = -A \cdot \left(\frac{n_x \pi}{L_x}\right)^2 \sin \frac{n_x \pi x}{L_x} \cdot \sin \frac{n_y \pi y}{L_y} \cdot \sin \frac{n_z \pi z}{L_z} = -\left(\frac{n_x \pi}{L_x}\right)^2 \cdot \psi$$

where A is a constant. Correspondingly, we obtain the derivatives $\partial^2 \psi / \partial y^2$ and $\partial^2 \psi / \partial z^2$. We insert into the differential equation (3.22) with $V = 0$ (the zero point of the energy is arbitrary):

$$\frac{h^2}{8\pi^2 m_e} \left[\left(\frac{n_x \pi}{L_x}\right)^2 + \left(\frac{n_y \pi}{L_y}\right)^2 + \left(\frac{n_z \pi}{L_z}\right)^2\right] \cdot \psi = E \cdot \psi$$

This solution fulfills the Schrödinger equation for

$$E = \frac{h^2}{8 m_e} \left(\frac{n_x^2}{L_x^2} + \frac{n_y^2}{L_y^2} + \frac{n_z^2}{L_z^2}\right) \tag{3.29}$$

Figure 3.4 shows the nodal planes of some wavefunctions according to equation (3.28). In Figure 3.5(b) we display the wavefunctions and the energies of a particle in a cubic box ($L_x = L_y = L_z = L$). For comparison, in Figure 3.5(a) the energies of a particle between two parallel walls are shown.

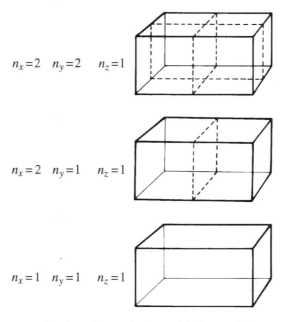

Figure 3.4 Standing waves in a box; here only the nodal planes of the waves are displayed.

3.2 NORMALIZATION OF THE WAVEFUNCTIONS

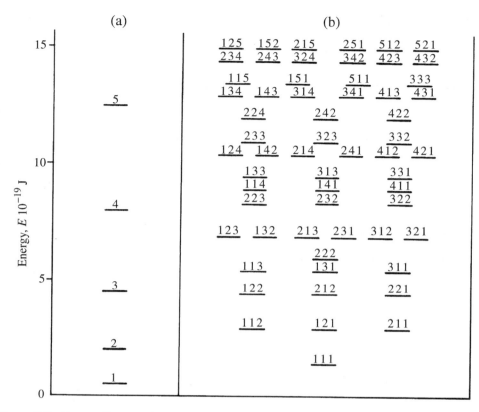

Figure 3.5 Energy diagrams for stationary states of an electron (a) Electron between two parallel walls at distance L ($L = 1000$ pm): $E = \dfrac{h^2}{8m_e L^2} \cdot n^2$ (b) Electron in a cubic box with side length L: $E = \dfrac{h^2}{8m_e L^2} \cdot (n_x^2 + n_y^2 + n_z^2)$ The numbers above the energy levels correspond to the quantum numbers n_x, n_y, and n_z.

3.2 Normalization of the Wavefunctions

According to the postulates in Chapter 1, the probability dP that a particle appears in a volume element $d\tau = dx\,dy\,dz$ at a location x, y, z in space is proportional to $\psi^2(x, y, z)$ and $d\tau$:

$$dP = C \cdot \psi^2(x, y, z) \cdot d\tau \tag{3.30}$$

where C is a constant.

The wavefunction ψ contains an arbitrary proportionality constant (e.g., the constant A in equation (3.28)). We can choose this proportionality constant such that we can simplify equation (3.30) as

$$dP = \psi^2(x, y, z) \cdot d\tau \tag{3.31}$$

Wavefunctions with accordingly fixed proportionality constants are called *normalized*. Using normalized wavefunctions, we define the probability density ρ as

$$\rho = \frac{dP}{d\tau} = \psi^2(x, y, z) \tag{3.32}$$

How can we normalize a wavefunction? The probability P is understood as the ratio

$$P = \frac{\text{number of favorable cases}}{\text{number of possible cases}} \tag{3.33}$$

Therefore, the probability P_{total} of finding an electron anywhere in space, is

$$P_{\text{total}} = 1 \tag{3.34}$$

Thus we obtain the following as a normalization condition

$$P_{\text{total}} = \int_{\text{space}} dP = 1 \tag{3.35}$$

Example – electron between two parallel walls

In this case (Figure 3.6) we have

$$dP = \psi^2(x) \cdot dx$$

and

$$P_{\text{total}} = \int_0^L \psi^2 \cdot dx = A^2 \int_0^L \sin^2 \frac{n\pi x}{L} \cdot dx = 1 \tag{3.36}$$

Because of

$$\int_0^L \sin^2 \frac{n\pi x}{L} \cdot dx = \frac{L}{n\pi} \int_0^{n\pi} \sin^2 u \cdot du = \frac{L}{n\pi} \frac{n\pi}{2} = \frac{L}{2} \tag{3.37}$$

we obtain

$$A = \sqrt{\frac{2}{L}} \tag{3.38}$$

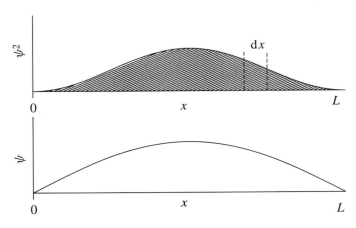

Figure 3.6 Wave function $\psi(x)$ and the square $\psi^2(x)$ for an electron between two walls a distance L apart. The dashed area corresponds to the integral $\int \psi^2 \cdot dx$.

3.2 NORMALIZATION OF THE WAVEFUNCTIONS

Example – electron in a rectangular box

In this case we have
$$dP = \psi^2(x, y, z) \cdot dx\, dy\, dz \tag{3.39}$$

and with (3.28) we obtain

$$P_{\text{total}} = \int_{x=-\infty}^{+\infty} \int_{y=-\infty}^{+\infty} \int_{z=-\infty}^{+\infty} \psi^2(x, y, z) \cdot dx\, dy\, dz$$

$$= A^2 \cdot \int_{x=0}^{L_x} \sin^2 \frac{n_x \pi x}{L_x} \cdot dx \cdot \int_{y=0}^{L_y} \sin^2 \frac{n_y \pi y}{L_y} \cdot dy \cdot \int_{z=0}^{L_z} \sin^2 \frac{n_z \pi z}{L_z} \cdot dz = 1$$

Thus
$$A^2 \cdot \frac{L_x}{2} \cdot \frac{L_y}{2} \cdot \frac{L_z}{2} = 1 \tag{3.40}$$

and
$$A = \sqrt{\frac{8}{L_x L_y L_z}} \tag{3.41}$$

In the following we will only use normalized wavefunctions. ψ is then determined up to the sign. The sign can be chosen freely; for example, for an electron between two walls, in quantum state $n = 1$, ψ can be set to

$$\psi = \sqrt{\frac{2}{L}} \cdot \sin \frac{\pi x}{L} \quad \text{or to} \quad \psi = -\sqrt{\frac{2}{L}} \cdot \sin \frac{\pi x}{L}$$

For either choice, the quantity
$$\psi^2 = \frac{2}{L} \cdot \sin^2 \frac{\pi x}{L}$$

is the same in both cases.

ψ^2 must be single-valued because it corresponds to a measurable quantity, namely the probability of observing an electron at a particular location. The amplitude ψ is not directly observable, so the sign is arbitrary.

It is customary to write the normalization condition (3.35) for simplicity as

$$\boxed{\int \psi^2 \cdot d\tau = 1} \tag{3.42}$$

The integral sign does not signify an indefinite integral, as elsewhere, but does signify an integral which extends over all space.

The probability density $\rho = \psi^2$ can be displayed as a cloud (Figure 3.7). For this the spatial locations of high probability are drawn with a high density of points, and the locations of low probability are drawn with a low density of points.

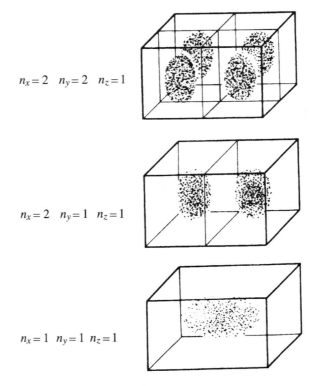

Figure 3.7 Electron in a box. Cloud pictures for the probability density ρ. The quantum numbers are chosen as in Figure 3.4.

The integral (3.42) can only obtain a finite value if ψ is zero at infinity. So in accordance with equation (3.23) a normalizable wavefunction must have the property

$$\lim_{x,y,z \to \pm\infty} \psi = 0$$

3.3 Orthogonality of the Wavefunctions

An important property of electron wavefunctions is shared by the displacement functions of the vibrational states of a string. Take, for example, the wavefunctions

$$\psi_1 = A \cdot \sin \frac{\pi x}{L}, \quad \psi_2 = A \cdot \sin \frac{2\pi x}{L} \qquad (3.43)$$

and picture the product function $\psi_1 \psi_2$ (Figure 3.8(a)). The areas of the two striped regions are equal on grounds of symmetry, so we have

$$\int_0^L \psi_1 \psi_2 \cdot dx = 0 \qquad (3.44)$$

If we replace the wavefunction ψ_2 with the wavefunction

$$\psi_3 = A \cdot \sin \frac{3\pi x}{L} \qquad (3.45)$$

3.3 ORTHOGONALITY OF THE WAVEFUNCTIONS

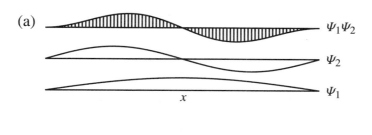

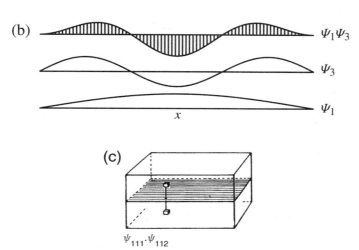

Figure 3.8 Orthogonality of the wavefunctions. (a) and (b) Functions for a particle between two parallel walls. Striped area above x axis equals striped area below x axis. (c) Function $\psi_{111} \cdot \psi_{112}$ for a particle in a rectangular box. Values in the two elements are equal, but opposite in sign.

then from Figure 3.8(b) we can suspect that

$$\int_0^L \psi_1 \psi_3 \cdot dx = 0 \tag{3.46}$$

This is really the case, as we can see by evaluating the integral analytically. With $u = \pi x/L$ we obtain

$$\int_0^L \sin\frac{\pi x}{L} \cdot \sin\frac{3\pi x}{L} \cdot dx = \frac{L}{\pi}\int_0^\pi \sin u \cdot \sin 3u \cdot du$$

Because of $\sin u \cdot \sin 3u = \frac{1}{2}(\cos 2u - \cos 4u)$ we have

$$\int_0^L \sin\frac{\pi x}{L} \cdot \sin\frac{3\pi x}{L} \cdot dx = \frac{L}{2\pi}\int_0^\pi \cos 2u \cdot du - \frac{L}{2\pi}\int_0^\pi \cos 4u \cdot du = 0$$

A corresponding relation also holds for the wavefunctions for a box. We consider the wavefunctions ψ_{111} and ψ_{112} in Figure 3.8(c). Each volume element under the striped plane is the mirror image of an element over the plane. In the integral

$$\int_{x=-\infty}^{+\infty}\int_{y=-\infty}^{+\infty}\int_{z=-\infty}^{+\infty} \psi_{111}\psi_{112} \cdot dx\, dy\, dz$$

the two elements make the same contribution with opposite signs, as one can see immediately by symmetry. Thus the integral has the value zero.

It can be shown that the relation found for these wavefunctions is generally valid for the solutions of the Schrödinger equation. It is called the *orthogonality relation*. The orthogonality relation will be crucial for the understanding of the electronic structure of atoms and molecules with many electrons.

A particular situation occurs in the case of wavefunctions corresponding to the same energy level. Such a level is said to be n-fold degenerate if a set of n orthogonal wavefunctions correspond to the level (e.g., three-fold degenerate state in Figure 3.5b corresponding to quantum numbers 112, 121, and 211). In this case any superposition of wavefunctions of the orthogonal set is a solution of the Schrödinger equation (e.g., $\psi = a\psi_{112} + b\psi_{121} + c\psi_{211}$, where a, b, and c are constants) which is not in general orthogonal to the wavefunctions of the orthogonal set. In this way an infinite number of different orthogonal sets can be constructed by superposition. See also the discussion in Chapter 6.

3.4 Time-dependent Schrödinger Equation

The Schrödinger equation (3.22, 3.27) is valid for the special case that $\rho = \psi^2$ does not depend on time: $\rho = \rho(x, y, z)$. This equation is sufficient for almost all problems discussed in this book. Schrödinger generalized his postulate for the case that ρ depends on time. This extension is important for a deeper understanding, but it requires practice with complex numbers. In this generalization we have

$$\rho = \rho(x, y, z, t) = \Psi\Psi^* \tag{3.47}$$

where the wavefunctions Ψ and Ψ^* are complex conjugates; Ψ is defined by the time-dependent Schrödinger equation

$$\boxed{\text{Postulate: } \mathcal{H}\Psi = i\hbar\frac{\partial \Psi}{\partial t}} \tag{3.48}$$

($i = \sqrt{-1}$, $\hbar = h/(2\pi)$). We show in Foundation 3.2, that the time-dependent Schrödinger equation goes over into the time-independent equation (3.27), if $\Psi\Psi^*$ does not depend on time.

Later on we will discuss applications of the time-dependent Schrödinger equation. In equation (3.48) we find the term $\partial\Psi/\partial t$, this is the first time derivative of the wavefunction. If Ψ is known at time $t = 0$, then it is possible to calculate Ψ at any time thereafter.

This time evolution will be of particular interest when discussing the time required for the electron transfer from one molecule (the donor) to another molecule (the acceptor) (see Chapter 4).

3.5 H Atom in the Ground State

3.5.1 Wavefunction

We ask next what the standing waves look like for an electron which is attracted by a positively charged atomic nucleus. As outlined in Chapter 2, the potential energy $V(r)$ of the electron in the field of the nucleus is

$$V = -\frac{1}{4\pi\varepsilon_0} \cdot \frac{e^2}{r} \tag{3.49}$$

where $r = \sqrt{x^2 + y^2 + z^2}$ is the distance from the electron to the nucleus. This expression is substituted into the Schrödinger equation (3.22):

$$-\frac{h^2}{8\pi^2 m_e}\left(\frac{\partial^2}{\partial x^2} + \frac{\partial^2}{\partial y^2} + \frac{\partial^2}{\partial z^2}\right)\psi - \frac{1}{4\pi\varepsilon_0} \cdot \frac{e^2}{r}\psi = E\psi \tag{3.50}$$

The solution to this Schrödinger equation, for the lowest energy state of the electron, is a wavefunction which depends only on r (Problem 3.1):

$$\boxed{\psi(r) = \frac{1}{\sqrt{\pi a_0^3}} \cdot e^{-r/a_0}} \tag{3.51}$$

with

$$\boxed{a_0 = 4\pi\varepsilon_0 \frac{h^2}{4\pi^2 m_e e^2} = 52.92 \, \text{pm}} \tag{3.52}$$

For reasons explained in Section 3.8, a_0 is called the *Bohr radius*.

The wavefunction ψ in equation (3.51) is displayed in Figure 3.9 (ψ along the x-axis) and in Figure 3.10(a) (ψ versus r). Figure 3.10(b) shows the graph of the probability density

$$\boxed{\rho = \psi^2 = \frac{1}{\pi a_0^3} \cdot e^{-2r/a_0}} \tag{3.53}$$

as a function of r. A "cloud" picture of ρ is shown in Figure 3.10(c). According to equation (3.53) we have for ρ at the location $r = a_0/2 = 26.5 \, \text{pm}$

$$\rho = \frac{1}{\pi a_0^3} \cdot e^{-1} = \frac{1}{\pi a_0^3} \cdot 0.368$$

So at $r = 26.5 \, \text{pm}$, ρ is less than half as great as at the location $r = 0$; at a distance $r = 122 \, \text{pm}$ from the nucleus, ρ has already fallen to 1% of its value at $r = 0$.

These considerations on size and shape of the electron cloud will be important for understanding the interplay between atoms and molecules in later chapters.

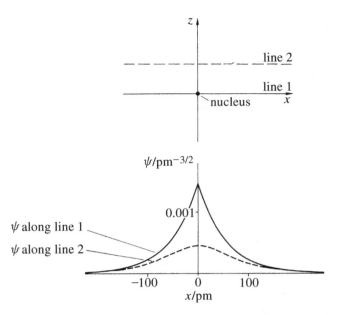

Figure 3.9 Plot of the wavefunction of the lowest energy state of an H atom (1s state) along the x axis (Line 1) and along a line 62.5 pm above the x axis (Line 2). The x axis goes through the nucleus.

3.5.2 Energy

In Problem 3.1 it is further shown that the energy E for the lowest energy state of the H atom is given by the expression

$$E = -\frac{1}{4\pi\varepsilon_0} \cdot \frac{e^2}{2a_0} = -21.80 \times 10^{-19}\, \text{J} \tag{3.54}$$

as already mentioned in Section 2.3.2. The experimental value is -21.80×10^{-19} J. (This value follows from spectroscopic data as the limit of the Lyman series (equation (3.66)).) The prediction from the postulates of quantum theory is thus in excellent agreement with experiment.

3.5.3 Radial Probability Distribution of Electron

Of further interest is the probability that an electron is found in a spherical shell of radius r and thickness dr. This probability (Figure 3.11) equals

$$\rho 4\pi r^2 \cdot dr = \frac{4}{a_0^3} \cdot r^2 e^{-2r/a_0} \cdot dr \tag{3.55}$$

The radial probability distribution $\rho 4\pi r^2$ is displayed in Figure 3.12.

Although the probability density ρ is greatest at the nucleus and then falls off exponentially, the probability of finding an electron in a sphere directly at the nucleus

3.5 H ATOM IN THE GROUND STATE

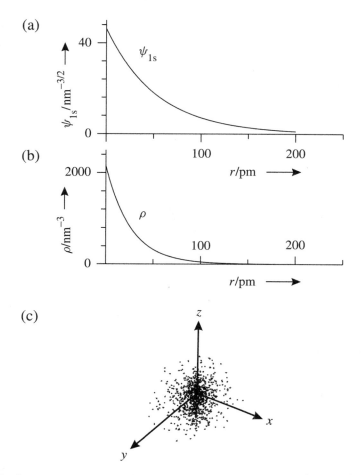

Figure 3.10 Lowest energy state of the electron in a hydrogen atom (1s state): (a) wavefunction $\psi(r)$; the index 1s denotes the ground state, see Section 3.6.1) (b) probability density ρ versus distance r; (c) cloud picture of ρ (computer simulation with 1000 random points).

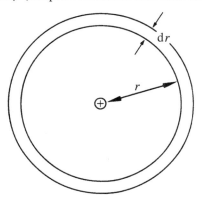

Figure 3.11 Section through the center of a sphere of radius r with a spherical shell of thickness dr (spherical surface: $4\pi r^2$); nucleus is the center of the sphere.

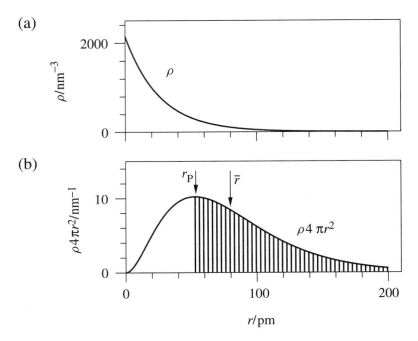

Figure 3.12 Probability densities for the electron in the H atom ground state: (a) $\rho = \psi^2$ versus radius r; (b) $\rho \cdot 4\pi r^2$ versus radius r. The most probable distance r_P and the average distance $\bar{r}$ are indicated. $\rho 4\pi r^2 \, dr$ is the probability of finding the electron in a spherical shell of radius r and thickness dr. The striped area is larger than the blank area, thus $\bar{r} > r_P$.

($r = 0$) approaches zero, since the volume element approaches zero. $\rho 4\pi r^2$ runs through a maximum: on the one hand, ρ decreases exponentially, and on the other hand, the volume of the volume element $4\pi r^2 \, dr$ increases quadratically with r. At large values of r, $\rho 4\pi r^2$ approaches zero: the exponential decrease of r outweighs the quadratic increase of the volume element.

The probability of finding an electron within a spherical shell of radius r and thickness dr, namely $\rho 4\pi r^2 \cdot dr$, is thus to be distinguished from the probability $\rho \cdot d\tau$ that an electron is found in a fixed, given volume element $d\tau$, which can be located at various distances from the nucleus. In the latter case $d\tau$ is held constant, while in the former case only dr is held constant, so that the volume element under consideration grows with r.

The probability of finding the electron inside a sphere with radius r

$$\int_0^r \rho 4\pi r^2 \cdot dr = 1 - e^{-2r/a_0} \cdot \left[2\left(\frac{r}{a_0}\right)^2 + 2\frac{r}{a_0} + 1 \right]$$

is 32% for $r = a_0$, 76% for $r = 2a_0$, and 85% for $r = 2.36 \cdot a_0 = 125$ pm; thus the probability of finding the electron outside a sphere of $r = 125$ pm is 15%, although ρ at $r = 125$ pm is only 1% of the value at $r = 0$. Remember that the collision radius of H is 128 pm (Section 2.3.2).

3.5 H ATOM IN THE GROUND STATE

3.5.4 Average Potential Energy and Virial Theorem

The average potential energy $\overline{V}$ is calculated in (Box 3.2) as

$$\overline{V} = \int_0^\infty V \cdot d\tau = -\frac{1}{4\pi\varepsilon_0} \cdot \frac{e^2}{a_0} \tag{3.56}$$

Box 3.2 – Averaging the Potential Energy V

We calculate $\overline{V}$ as

$$\overline{V} = \int V \cdot \rho \cdot d\tau$$

with

$$V = \frac{-e^2}{4\pi\varepsilon_0 r}, \quad \rho = \psi^2 = \frac{1}{\pi a_0^3} e^{-2r/a_0}, \quad \text{and} \quad d\tau = 4\pi r^2 \cdot dr$$

(volume element of a sphere). Then

$$\overline{V} = \frac{-e^2}{4\pi\varepsilon_0} \cdot \frac{1}{\pi a_0^3} 4\pi \int_0^\infty r \cdot e^{-2r/a_0} \cdot dr$$

$$= \frac{-e^2}{4\pi\varepsilon_0} \cdot \frac{1}{\pi a_0^3} 4\pi \cdot \frac{a_0^2}{4} = -\frac{1}{4\pi\varepsilon_0} \cdot \frac{e^2}{a_0}$$

From $\overline{V}$ and E we calculate the average kinetic energy $\overline{T}$:

$$\overline{T} = E - \overline{V} = -\frac{1}{4\pi\varepsilon_0} \cdot \frac{e^2}{2a_0} + \frac{1}{4\pi\varepsilon_0} \cdot \frac{e^2}{a_0} = \frac{1}{4\pi\varepsilon_0} \cdot \frac{e^2}{2a_0} \tag{3.57}$$

The magnitudes of $\overline{T}$ and E are equal; only the signs are opposite:

$$\boxed{E = -\overline{T} = \tfrac{1}{2}\overline{V}} \tag{3.58}$$

This relation is valid for all systems consisting of nuclei and electrons (in this case the attraction and repulsion forces are proportional to $1/r^2$) at the minimum of energy; equation (3.58) is called the *virial theorem* (see Figure 3.13).

This is of crucial importance for understanding the stability of atoms and molecules: half of the energy gain by Coulomb forces when forming the electron cloud from separated electrons and nuclei is used up on the average to speed up the electrons. This was pointed out already in Section 2.3.2.

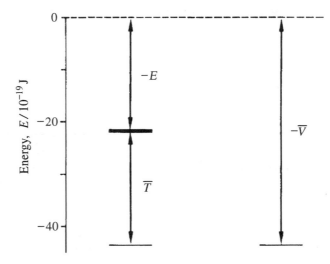

Figure 3.13 Virial theorem for the H atom: $E = -\overline{T}$, $\overline{V} = -2\overline{T}$.

3.5.5 Most Probable and Average Distance of Electron from Nucleus

We can easily calculate the *most probable distance* r_P of the electron from the nucleus; for this distance, $\rho 4\pi r^2$ is maximum, so the derivative with respect to r is zero:

$$\frac{d(\rho 4\pi r^2)}{dr} = 4\pi \cdot \frac{1}{\pi a_0^3}\left(-\frac{2}{a_0}r^2 e^{-2r/a_0} + 2r e^{-2r/a_0}\right) = 0 \qquad (3.59)$$

From this follows

$$r_P = a_0 \qquad (3.60)$$

As derived in Problem 3.4, the average distance of the electron from its nucleus is $\overline{r} = \frac{3}{2}a_0$. $\overline{r}$ is greater than r_P because the electron is more often at distances $r > r_P$ (striped area in Figure 3.12) than at distances $r < r_P$. Thus large values of r contribute more in forming the average than do small values.

3.6 H Atom in Excited States

We have already established that atoms supplied with energy (e.g., through collisions) can be transformed to excited states. They can return to the ground state by emitting light. The energies of the excited states can be determined experimentally from the positions of spectral lines.

3.6.1 Energies and Wavefunctions

We now want to calculate the energies and wavefunctions of excited states of the H atom in order to see whether the hydrogen spectrum can be interpreted using the Schrödinger equation. In general, the energies for the H atom from the Schrödinger equation, see Foundation 3.3, are

$$E_n = -\frac{1}{4\pi\varepsilon_0} \cdot \frac{e^2}{2a_0} \cdot \frac{1}{n^2}, \qquad n = 1, 2, 3, \ldots \qquad (3.61)$$

3.6 H ATOM IN EXCITED STATES

The case $n = 1$ corresponds to the ground state with spherically symmetric wavefunction denoted by index 1s, which we have already discussed. For the quantum number $n = 2$, in addition to one spherically symmetric wavefunction (Figure 3.14), three more wavefunctions with the same energy E are found (Figure 3.15). To distinguish these wavefunctions, the spherically symmetric solution is denoted by the index 2s, and the other solutions are labeled by the indices $2p_x$, $2p_y$, and $2p_z$. The four wavefunctions for the second energy level are

2s	$\psi_{2s} = \dfrac{1}{4\sqrt{2\pi a_0^3}} \left(2 - \dfrac{r}{a_0}\right) \cdot e^{-r/(2a_0)}$	
$2p_x$	$\psi_{2p_x} = \dfrac{1}{4\sqrt{2\pi a_0^3}} \cdot \dfrac{x}{a_0} \cdot e^{-r/(2a_0)}$	(3.62)
$2p_y$	$\psi_{2p_y} = \dfrac{1}{4\sqrt{2\pi a_0^3}} \cdot \dfrac{y}{a_0} \cdot e^{-r/(2a_0)}$	
$2p_z$	$\psi_{2p_z} = \dfrac{1}{4\sqrt{2\pi a_0^3}} \cdot \dfrac{z}{a_0} \cdot e^{-r/(2a_0)}$	

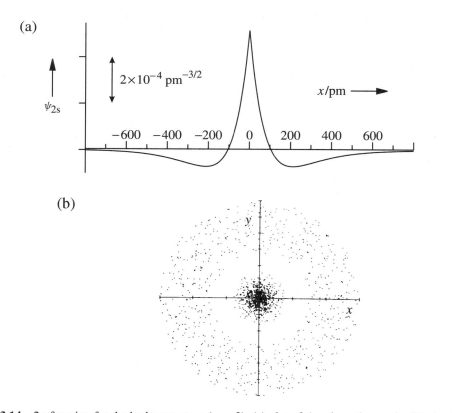

Figure 3.14 2 s function for the hydrogen atom ($n = 2$): (a) plot of ψ_{2s} along the x axis; (b) cloud display of $\rho = \psi^2$.

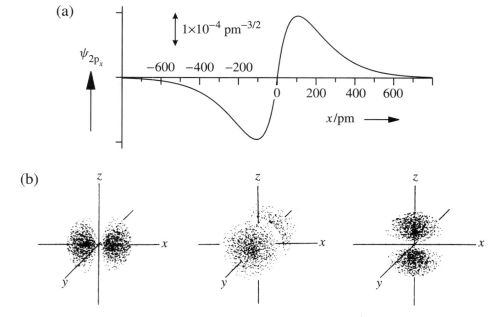

Figure 3.15 2p functions for the hydrogen atom ($n = 2$). (a) ψ_{2p_x} along the x axis. (b) Cloud representation of ψ_{2p_x} (left), ψ_{2p_y} (center), and ψ_{2p_z} (right).

This is a fourfold degenerate energy level, since a set of four orthogonal wavefunctions corresponds to the level. See the final paragraph of Section 3.3 and Chapter 6 for properties of wavefunctions corresponding to degenerate energy levels.

For the quantum number $n = 3$, there are nine wavefunctions which solve the Schrödinger equation: one spherically symmetric (3s), three axially symmetric (3p$_x$, 3p$_y$, 3p$_z$), and five others (3d$_{xy}$, 3d$_{xz}$, 3d$_{yz}$, 3d$_{z^2}$, 3d$_{x^2-y^2}$):

3s	$\psi_{3s} = \dfrac{1}{81\sqrt{3\pi a_0^3}} \left(27 - 18\dfrac{r}{a_0} + 2\dfrac{r^2}{a_0^2} \right) \cdot e^{-r/(3a_0)}$
3p$_x$	$\psi_{3p_x} = \dfrac{\sqrt{2}}{81\sqrt{\pi a_0^3}} \left(6 - \dfrac{r}{a_0} \right) \cdot \dfrac{x}{a_0} \cdot e^{-r/(3a_0)}$
3p$_y$	$\psi_{3p_y} = \dfrac{\sqrt{2}}{81\sqrt{\pi a_0^3}} \left(6 - \dfrac{r}{a_0} \right) \cdot \dfrac{y}{a_0} \cdot e^{-r/(3a_0)}$
3p$_z$	$\psi_{3p_z} = \dfrac{\sqrt{2}}{81\sqrt{\pi a_0^3}} \left(6 - \dfrac{r}{a_0} \right) \cdot \dfrac{z}{a_0} \cdot e^{-r/(3a_0)}$

3.6 H ATOM IN EXCITED STATES

$$3d_{xy} \quad \psi_{3d_{xy}} = \frac{\sqrt{2}}{81\sqrt{\pi a_0^3}} \cdot \frac{x}{a_0} \cdot \frac{y}{a_0} \cdot e^{-r/(3a_0)}$$

$$3d_{yz} \quad \psi_{3d_{yz}} = \frac{\sqrt{2}}{81\sqrt{\pi a_0^3}} \cdot \frac{y}{a_0} \cdot \frac{z}{a_0} \cdot e^{-r/(3a_0)}$$

$$3d_{xz} \quad \psi_{3d_{xz}} = \frac{\sqrt{2}}{81\sqrt{\pi a_0^3}} \cdot \frac{x}{a_0} \cdot \frac{z}{a_0} \cdot e^{-r/(3a_0)} \quad (3.63)$$

$$3d_{x^2-y^2} \quad \psi_{3d_{x^2-y^2}} = \frac{1}{81\sqrt{2\pi a_0^3}} \left[\left(\frac{x}{a_0}\right)^2 - \left(\frac{y}{a_0}\right)^2 \right] \cdot e^{-r/(3a_0)}$$

$$3d_{z^2} \quad \psi_{3d_{z^2}} = \frac{1}{81\sqrt{6\pi a_0^3}} \left[3\left(\frac{z}{a_0}\right)^2 - \left(\frac{r}{a_0}\right)^2 \right] \cdot e^{-r/(3a_0)}$$

All wavefunctions with the quantum numbers $n = 1, 2, 3$ are displayed in Figure 3.16. In general, n^2 different wavefunctions belong to a state with quantum number n. The energy diagram of an electron in an H atom is displayed in Figure 3.17, illustrating both the number and spacing of states for $n = 1$ to 4.

3.6.2 Emission Spectra

In a glow discharge, the electron in an H atom can be excited into a high-energy state; with emission of light it transforms to a lower state. Dispersion of this light in a spectrophotometer shows that it contains contributions only from certain wavelengths. Different series of spectral lines are obtained from the set of emissions corresponding to transitions down to each energy state (Figure 3.18).

We can calculate the wavelength of each spectral line from equation (3.61) using the relation

$$\lambda = \frac{hc_0}{\Delta E} \quad (3.64)$$

For example, the transition from a higher energy level to the level with $n = 2$ corresponds to an energy difference of

$$\Delta E = E_n - E_2 = \frac{1}{4\pi\varepsilon_0} \cdot \frac{e^2}{2a_0} \cdot \left(\frac{1}{4} - \frac{1}{n^2}\right), \quad n = 3, 4, 5, \ldots \quad (3.65)$$

This is the Balmer series. In Table 3.1, the calculated and measured wavelengths for the Balmer series are compared; we establish that (after the application of two minor corrections) there is excellent agreement. We count this as a further indication that the Schrödinger equation is a useful postulate.

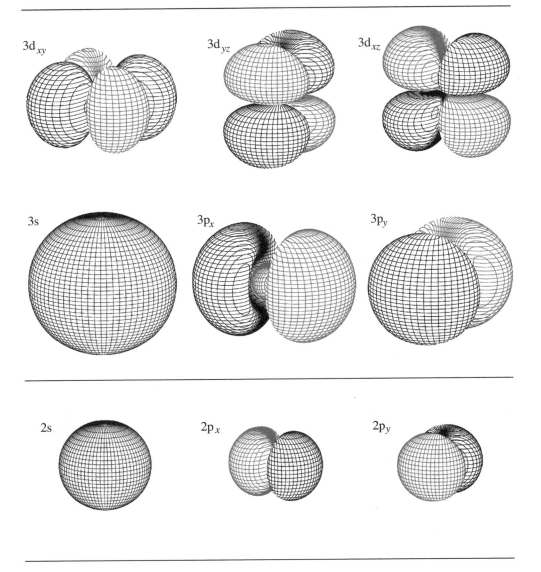

Figure 3.16 Three-dimensional diagrams of the hydrogen atom wavefunctions with the quantum numbers 1 (1s), 2 (2s, 2p), and 3 (3s, 3p, 3d). The individual picture elements correspond to the positions in space in which the functional values of the normalized functions amount to $\pm 1 \times 10^{-5}$ pm$^{-3/2}$.
Black lines: positive functional values; red lines: negative functional values
In the coordinate system the distances from the origin to the corresponding tic are 500 pm.

3.6 H ATOM IN EXCITED STATES

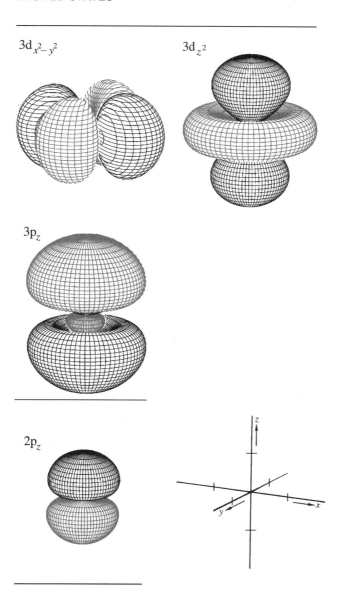

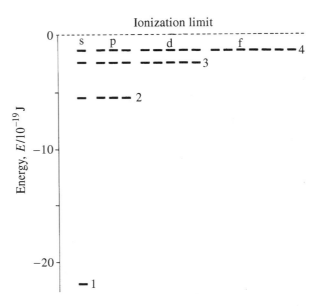

Figure 3.17 Energy diagram of the electron in the H atom with the quantum numbers 1, 2, 3, and 4. To the quantum number $n = 4$ belong one s function, three p functions, five d functions, and seven f functions (see Foundation 3.3). In the ground state the electron occupies the 1 s state.

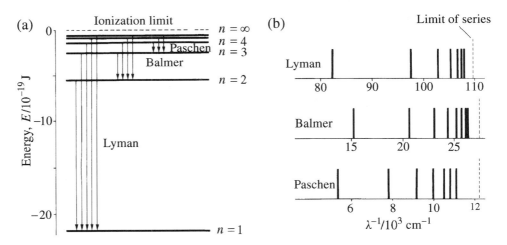

Figure 3.18 Series of spectral lines for the H atom. (a) Energy diagram with transitions. (b) Display of the corresponding spectral lines: Lyman Series (in the ultraviolet), Balmer Series (visible), and Paschen Series (in the infrared). λ^{-1} is the wavenumber.

From the Lyman series (transitions to the level with $n = 1$), the ionization energy E_{ion} can be determined. In this case we have

$$\Delta E = \frac{1}{4\pi\varepsilon_0} \cdot \frac{e^2}{2a_0} \cdot \left(1 - \frac{1}{n^2}\right), \quad n = 2, 3, 4, \ldots \quad (3.66)$$

We find for $n \to \infty$ that $E_{\text{ion}} = 21.80 \times 10^{-19}$ J.

3.7 Variation Principle

Table 3.1 Wavelengths λ (in nm) for transitions in the H atom from $n = 3, 4, 5,$ and 6 to $n = 2$ (Balmer series) – experiment and theory

n	Experiment	Theory	
		Uncorrected	Corrected
3	656.466	656.112	656.466
4	486.271	486.009	486.272
5	434.170	433.937	434.171
6	410.290	410.070	410.291

The experimental values are for vacuum conditions. The theoretical values in column 3 (uncorrected) are calculated from equations (3.64) and (3.65). Two minor corrections must be applied. The values must be corrected for the comovement of the nucleus (proton mass m_p) in the H atom, (leading to an increase of wavelength), and for the effect of relativity – The electron mass increases with velocity, this leads to a decrease of wavelength.

3.7 Variation Principle

In the last section, the energies and wavefunctions of the electron in the hydrogen atom were obtained by direct solution of the Schrödinger equation. Since an exact solution of the Schrödinger equation can be found only in a few cases (intractable mathematical difficulties already appear in the H$_2$ molecule), in practically all cases we have to rely on an approximate solution.

3.7.1 Justification of the Variation Principle

In general we can calculate the energy E, that belongs to a given wavefunction with the help of the Schrödinger equation

$$\mathcal{H}\psi = E\psi \qquad (3.67)$$

by the following method. We multiply both sides by $\psi \cdot d\tau$ ($d\tau = dx\,dy\,dz$) and integrate from $x, y, z = -\infty$ to $x, y, z = +\infty$

$$\int (\mathcal{H}\psi)\psi \cdot d\tau = E \cdot \int \psi^2 \cdot d\tau \qquad (3.68)$$

E can be pulled out of the integral as a constant (the energy of the state does not depend on the coordinates x, y, z). With the normalization condition

$$\int \psi^2 \cdot d\tau = 1$$

we obtain

$$E = \int \psi \mathcal{H}\psi \cdot d\tau \qquad (3.69)$$

Equation (3.69) gives a procedure for calculating the energy belonging to a known wavefunction.

For example, for the wavefunction ψ_1 of the lowest quantum state of the H atom, we obtain

$$E_1 = \int \psi_1 \mathcal{H} \psi_1 \cdot d\tau \tag{3.70}$$

Trial function for the ground state

Now as a trial, we consider another function ϕ_1 (trial function) which also satisfies the normalization condition $\left(\int \phi^2 \cdot d\tau = 1\right)$. We determine the energy ε_1 assigned to this trial function defined by:

$$\varepsilon_1 = \int \phi_1 \mathcal{H} \phi_1 \cdot d\tau \tag{3.71}$$

It is demonstrated in Foundation 3.4 that ε_1 for an arbitrary function ϕ_1 must be greater than the true energy E_1 for the lowest quantum state. Thus

$$\varepsilon_1 \geq E_1 \tag{3.72}$$

The limiting value $\varepsilon_1 = E_1$ will be reached if ϕ_1 is the exact wavefunction of the lowest quantum state. In nature, then, the electron distribution corresponding to the smallest possible value of $\varepsilon_1 (\varepsilon_1 = E_1)$ is realized. The lower the energy ε_1 of the trial function ϕ_1, the more accurately ϕ_1 approximates the exact wavefunction of the ground state (Figure 3.19).

The best trial function taken into consideration is obtained by minimizing ε_1:

$$\boxed{\varepsilon_1 = \int \phi_1 \mathcal{H} \phi_1 \cdot d\tau = \text{minimum}} \tag{3.73}$$

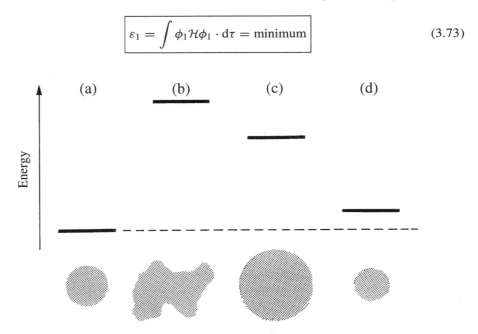

Figure 3.19 Variation method explained in the case of an H atom. (a) Exact energy E by solution of the Schrödinger equation (exact wavefunction ψ). (b)–(d) Energies ε assigned to various trial functions ϕ (displayed by schematic electron clouds); note that $\varepsilon \geq E$.

3.7 VARIATION PRINCIPLE

In practice, one proceeds by starting off with an arbitrary function ϕ_1 and varying this function until a minimum of ε_1 is reached. This procedure is therefore called the *variation principle*.

Example – trial function (3.74)

We consider the electron in the H atom and choose as trial function for the ground state

$$\phi_1 = \frac{1}{\sqrt{\pi C^3}} \cdot e^{-r/C} \tag{3.74}$$

where C is an adjustable constant. If C is large, then the exponential function falls off gently; if C is small, then the decrease is steep. For different values of C we obtain different diameters of the electron cloud corresponding to the trial function (Figure 3.20).
According to Problem 3.8 the energy

$$\varepsilon = \overline{T} + \overline{V} = \frac{h^2}{8\pi^2 m_e} \frac{1}{C^2} - \frac{1}{4\pi\varepsilon_0} \frac{e^2}{C} \tag{3.75}$$

depends on C; if we plot ε as a function of C, we obtain the curve displayed in Figure 3.21. There is a minimum at $C = a_0 = 52.92$ pm and $\varepsilon = -21.8 \times 10^{-19}$ J.
The minimum thus appears when C is just equal to a_0. The smaller the electron cloud (the smaller C), the lower the potential energy and the greater the kinetic energy.

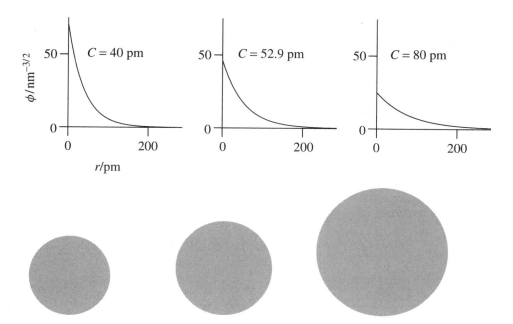

Figure 3.20 H atom. Graphs of the functions and electron cloud pictures (schematic) for the trial function $\phi_1 = \frac{1}{\sqrt{\pi C^3}} \cdot e^{-r/C}$ for various values of C. At the edge of the gray zone, ϕ^2 is 1% of its maximum value.

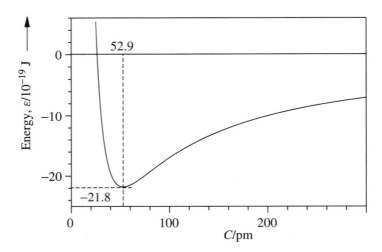

Figure 3.21 Energy ε for the trial function $\phi_1 = \frac{1}{\sqrt{\pi C^3}} \cdot e^{-r/C}$ for various values of C.

The minimum of the energy ε corresponds to a compromise between the two opposite tendencies.

Trial functions for excited states

For the wavefunction ψ_2 and the energy E_2 of the next higher state, according to equation (3.69), we obtain

$$E_2 = \int \psi_2 \mathcal{H} \psi_2 \cdot d\tau \tag{3.76}$$

At the same time the orthogonality condition

$$\int \psi_1 \psi_2 \cdot d\tau = 0 \tag{3.77}$$

must be fulfilled (see Section 3.3).

We now consider a trial function ϕ_2, which is normalized and orthogonal to ϕ_1, which thus satisfies the conditions

$$\int \phi_2^2 \cdot d\tau = 1, \quad \int \phi_1 \phi_2 \cdot d\tau = 0 \tag{3.78}$$

Foundation 3.4 shows then that

$$\varepsilon_2 = \int \phi_2 \mathcal{H} \phi_2 \cdot d\tau \tag{3.79}$$

cannot be smaller than E_2, if ϕ_1 agrees with ψ_1.

An approximation for E_2 can be found by variation of ϕ_2 by the same procedure used for the lowest quantum state. An approximation for ψ_3 and E_3 is obtained in an analogous

3.8 Bohr Atomic Model and Correspondence Principle

3.8.1 Bohr Atomic Model

Based on the postulates of wave mechanics we can understand the characteristics of simple atoms. Before the introduction of wave mechanics, Bohr tried to set up an atomic model. He started out with the idea of an electron in the H atom as a localizable particle revolving around the atomic nucleus at a distance r (Figure 3.22); the electrostatic attraction supplies the centripetal force which holds the electron in a circle. If v is the speed of the electron, then we obtain by equalizing centrifugal and centripetal force

$$\frac{m_e v^2}{r} = \frac{1}{4\pi\varepsilon_0} \cdot \frac{e^2}{r^2} \tag{3.80}$$

To determine v, Bohr postulated intuitively that the angular momentum of the electron equals $\hbar = h/(2\pi)$, thus

$$m_e r^2 \frac{v}{r} = m_e v r = \hbar \tag{3.81}$$

From equations (3.80) and (3.81) we obtain the following for v and r

$$v = \frac{1}{4\pi\varepsilon_0} \cdot \frac{e^2}{\hbar}, \quad r = 4\pi\varepsilon_0 \frac{\hbar^2}{m_e e^2} = a_0 \tag{3.82}$$

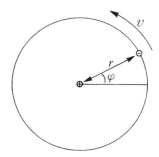

Figure 3.22 Bohr atom model; the electron circles at a distance r with the speed v around the nucleus.

For the radius we obtain the same expression as for the most probable distance a_0; so a_0 is called the *Bohr radius*. We can calculate the kinetic and potential energy from v and r. The results are

$$T = \frac{1}{2} m_e v^2 = \frac{1}{2} m_e \left(\frac{1}{4\pi\varepsilon_0} \cdot \frac{e^2}{\hbar} \right)^2 = \frac{1}{4\pi\varepsilon_0} \cdot \frac{e^2}{2a_0}$$

$$V = -\frac{1}{4\pi\varepsilon_0} \cdot \frac{e^2}{a_0} \tag{3.83}$$

as in the case of the wave mechanical analysis.

In spite of this agreement, the Bohr theory leads to a false prediction; a nucleus and circling electron together make up a rotating dipole, and according to the theory of electromagnetism this dipole would have to emit electromagnetic radiation of frequency

$$\nu = \frac{v}{2\pi r} = \frac{1}{4\pi\varepsilon_0} \cdot \frac{2\pi e^2}{2\pi h a_0} = \frac{1}{4\pi\varepsilon_0} \cdot \frac{e^2}{h a_0} = 6 \times 10^{15} \text{ s}^{-1}$$

Apparently, this is not the case. But the precise agreement of equation (3.83) with equations (3.56) and (3.57) and of the Bohr radius with the most probable distance shows the close relation between classical and quantum mechanical considerations.

3.8.2 Correspondence Principle

In the Bohr model the electron moves in a distinct path, which contradicts the results of the electron interference experiments (Chapter 1). The Bohr model is, in spite of these contradictions, of great historical significance as a forerunner of quantum mechanics and by its impact to recognize that the ideas of classical physics lead to sensible predictions, even though they are fundamentally wrong (*Bohr correspondence principle*). This correspondence is based on the fact that quantum mechanics and classical mechanics correspond to each other in their conceptual structure.

This is seen if we replace the Schrödinger equation as our postulate by another postulate from which the Schrödinger equation is then derived. The energy of a particle of mass m_e, which has the energy E and whose potential energy $V(x, y, z)$ depends on its position in space is described in classical mechanics by the equation

$$H = \frac{1}{2m_e}(p_x^2 + p_y^2 + p_z^2) + V(x, y, z) = E \tag{3.84}$$

where p_x, p_y, and p_z are the components of the momentum in the three spatial directions. The first term is thus the kinetic energy. H is called the *Hamiltonian function*.

We postulate a recipe to convert classical mechanics into quantum mechanical expressions. This postulate converts the classical equation

$$H = E \tag{3.85}$$

into the Schrödinger equation

$$\mathcal{H}\Psi = i\hbar \frac{\partial \Psi}{\partial t} \tag{3.86}$$

PROBLEM 3.1

The Hamiltonian function H is replaced by the Hamiltonian operator $\mathcal{H}$ applied to the wavefunction Ψ.

$\mathcal{H}$ is obtained as follows. We postulate that the operator $-i\hbar \cdot \partial/\partial x$ replaces p_x, and $i\hbar \cdot \partial/\partial t$ replaces the energy E ($i = \sqrt{-1}$). Instead of the term p_x^2 we obtain the operator

$$-i\hbar \frac{\partial}{\partial x}\left(-i\hbar\frac{\partial}{\partial x}\right) = i^2\hbar^2 \frac{\partial^2}{\partial x^2} = -\hbar^2 \cdot \frac{\partial^2}{\partial x^2} \qquad (3.87)$$

while p_y and p_z are replaced by the corresponding operators. V is identical with the operator V, and with these the Hamiltonian operator is obtained:

$$\mathcal{H} = -\frac{\hbar^2}{2m_e}\left(\frac{\partial^2}{\partial x^2} + \frac{\partial^2}{\partial y^2} + \frac{\partial^2}{\partial z^2}\right) + V(x, y, z) \qquad (3.88)$$

which we have already considered. Of course, the Schrödinger equation is not justified by this procedure; we have only replaced one postulate (namely, the Schrödinger equation) by another one (namely, the recipe leading to (3.87)), which lets us better recognize certain structural relationships.

The idea of the correspondence between the classical and quantum mechanical pictures will be extremely useful in many cases in subsequent chapters. There we replace the quantum mechanical approach by an equivalent, but more transparent classical model (see Sections 8.7, 8.8, 8.9, 8.11, 22.2, 22.4, and 23.3).

Problems

Problem 3.1 – H Atom: Ground State Derived from the Schrödinger Equation

Show that equation (3.51) is a solution of the Schrödinger equation for the H atom. Calculate the energy of the ground state.

Solution. From equation (3.51) we obtain with $r = \sqrt{x^2 + y^2 + z^2}$ and $A = 1/\sqrt{\pi a_0^3}$

$$\frac{\partial \psi}{\partial x} = -A \cdot \frac{1}{a_0} x \cdot \frac{1}{\sqrt{x^2+y^2+z^2}} \cdot e^{-\sqrt{x^2+y^2+z^2}/a_0}$$

$$\frac{\partial^2 \psi}{\partial x^2} = -A \cdot \frac{1}{a_0} \cdot \frac{r - x^2/r - x^2/a_0}{x^2+y^2+z^2} \cdot e^{-\sqrt{x^2+y^2+z^2}/a_0}$$

Similar expressions are obtained for the derivatives according to y and z (x^2 in the numerator has to be replaced by y^2 or z^2). Then

$$\frac{\partial^2 \psi}{\partial x^2} + \frac{\partial^2 \psi}{\partial y^2} + \frac{\partial^2 \psi}{\partial z^2} = -A \cdot \frac{1}{a_0} \cdot \left(\frac{2}{r} - \frac{1}{a_0}\right) \cdot e^{-r/a_0}$$

We insert this expression into the Schrödinger equation for the H atom:

$$\frac{h^2}{8\pi^2 m_e} \cdot A \cdot \frac{1}{a_0} \cdot \left(\frac{2}{r} - \frac{1}{a_0}\right) \cdot e^{-r/a_0} - \frac{1}{4\pi\varepsilon_0} \cdot \frac{e^2}{r} \cdot \frac{1}{\sqrt{\pi a_0^3}} \cdot e^{-r/a_0} = E \cdot A \cdot e^{-r/a_0}$$

With $a_0 = 4\pi\varepsilon_0 h^2/(4\pi^2 m_e e^2)$ we obtain

$$E = \frac{h^2}{8\pi^2 m_e} \cdot \frac{2}{a_0 r} - \frac{h^2}{8\pi^2 m_e} \cdot \frac{1}{a_0^2} - \frac{1}{4\pi\varepsilon_0} \cdot \frac{e^2}{r} = -\frac{1}{4\pi\varepsilon_0} \cdot \frac{e^2}{2a_0}$$

Problem 3.2 – H Atom: Normalization of Wavefunction

Calculate the normalization constant for the wavefunction of the H atom in its ground state.

Solution. As volume elements for the integration we choose spherical shells of thickness dr. According to Figure 3.11 and equation (3.51) we get

$$\int_0^\infty \psi^2 \cdot 4\pi r^2 \cdot dr = A^2 \int_0^\infty e^{-2r/a_0} \cdot 4\pi r^2 \cdot dr = 1$$

By substituting $u = 2r/a_0$ we transform the integral into

$$\int_0^\infty e^{-2r/a_0} \cdot 4\pi r^2 \cdot dr = 4\pi \frac{a_0^3}{8} \int_0^\infty e^{-u} \cdot u^2 \cdot du = 4\pi \frac{a_0^3}{8} \cdot 2 = \pi a_0^3$$

and it follows $A = 1/\sqrt{\pi a_0^3}$.

Problem 3.3 – Probability Density ρ

Calculate the distance r for the hydrogen atom when ρ is decreased to 1/1000 of its maximum value.

Solution: ρ is proportional to e^{-2r/a_0}, and ρ has its maximum at $r = 0$. Then we find r for $\rho = \rho_{\max}/1000$ by solving the equation $e^{-2r/a_0} = 0.001$: $r = 3.5 a_0$.

Problem 3.4 – H Atom: Average Distance Between Electron and Nucleus

Calculate the average distance of the electron from the nucleus in the H atom in its ground state.

Solution.

$$\bar{r} = \int_{\text{space}} r \cdot \rho \cdot d\tau = \int_0^\infty r \cdot \rho 4\pi r^2 \cdot dr = \frac{1}{\pi a_0^3} \int_0^\infty r \cdot e^{-2r/a_0} 4\pi r^2 \cdot dr$$

From this we obtain (with the substitution $\xi = 2r/a_0$)

$$\bar{r} = \frac{4}{a_0^3} \left(\frac{a_0}{2}\right)^3 \frac{a_0}{2} \int_0^\infty e^{-\xi} \xi^3 \cdot d\xi = \frac{3}{2} a_0$$

This calculation shows that the average distance $\bar{r}$ is indeed greater than the most probable distance r_P (Figure 3.12).

Problem 3.5 – H Atom: Particle in a Box Trial Function

Calculate the average potential energy $\overline{V}$ of the hydrogen atom using the box trial function.
Solution. We calculate $\overline{V}$ as

$$\overline{V} = \int_{\text{box}} V \cdot \rho \cdot dx\, dy\, dz = -\frac{e^2}{4\pi\varepsilon_0} \int_{\text{box}} \frac{1}{r} \cdot \phi^2 \cdot dx\, dy\, dz$$

with $r = \sqrt{x^2 + y^2 + z^2}$, using the box trial function discussed in Chapter 2 with the zero point shifted from a corner of the box to its center:

$$\phi = \sqrt{\frac{8}{L^3}} \cdot \cos\frac{\pi x}{L} \cdot \cos\frac{\pi y}{L} \cdot \cos\frac{\pi z}{L}$$

Then

$$\overline{V} = -\frac{8}{L^3} \frac{e^2}{4\pi\varepsilon_0} \int_{\text{box}} \frac{1}{r} \cos^2\frac{\pi x}{L} \cos^2\frac{\pi y}{L} \cos^2\frac{\pi z}{L} \cdot dx\, dy\, dz$$

where x, y, and z extend from $-L/2$ to $+L/2$. The integral can be simplified by the substitutions $u = x/L$, $v = y/L$, and $w = z/L$. u, v, and w extend from $-1/2$ to $+1/2$. In this way we obtain

$$\overline{V} = -\frac{e^2}{4\pi\varepsilon_0} \frac{1}{L} \cdot \alpha$$

where

$$\alpha = 8 \int_{u=-1/2}^{1/2} \int_{v=-1/2}^{1/2} \int_{w=-1/2}^{1/2} \frac{\cos^2 \pi u \cos^2 \pi v \cos^2 \pi w}{\sqrt{u^2 + v^2 + w^2}} \cdot du\, dv\, dw$$

The integral is calculated numerically by replacing the integration by a summation; equal integration intervals $\Delta u = \Delta v = \Delta w$ are used. The result for different values of the integration interval is

Δu	Number of intervals	α
1/2	$2^3 = 8$	2.31
1/4	$4^3 = 64$	3.75
1/20	$20^3 = 8000$	4.16
1/160	$160^3 = 4\,096\,000$	4.18

α is converging to the value 4.18.

Problem 3.6 – He$^+$: Schrödinger Equation

Derive the expression for the energy E for the ground state of He$^+$. Generalize for nuclei with charge number Z.
Solution. We consider an electron attracted by the nuclear charge Ze:

$$-\frac{h^2}{8\pi^2 m_e} \nabla^2 \psi - \frac{1}{4\pi\varepsilon_0} \frac{Ze \cdot e}{r} \psi = E\psi$$

74   SCHRÖDINGER EQUATION AND VARIATION PRINCIPLE

With
$$\psi = A e^{-Zr/a_0}$$
we find in analogy to Problem 3.1
$$\nabla^2 \psi = -A \frac{Z}{a_0} \cdot e^{-Zr/a_0} \left(\frac{2}{r} - \frac{Z}{a_0} \right)$$

Inserting ψ and $\nabla^2 \psi$ into the Schrödinger equation, we find with $a_0 = 4\pi\varepsilon_0 h^2/(4\pi^2 m_e e^2)$

$$E = \frac{h^2}{8\pi^2 m_e} \frac{Z}{a_0} \left(\frac{2}{r} - \frac{Z}{a_0} \right) - \frac{1}{4\pi\varepsilon_0} \frac{Ze \cdot e}{r} = -\frac{1}{2} \frac{1}{4\pi\varepsilon_0} \frac{Z^2 e^2}{a_0}$$

In the case of He$^+$ we have $Z = 2$:

$$\psi = A e^{-2r/a_0}, \quad E = -\frac{1}{4\pi\varepsilon_0} \frac{2e^2}{a_0}$$

The experimentally obtained ionization energy (87.184×10^{-19} J) is indeed four times as great for He$^+$ as for the H atom. Actually 87.184×10^{-19} J $= 4.0016 \times 21.787 \times 10^{-19}$ J; the ionization energy of He$^+$ is thus somewhat greater than four times the ionization energy of the H atom. The difference lies in the fact that the He nucleus is heavier, so the comovement of the He nucleus is less important.

Calculation of the average distance (with $A = \sqrt{8/(\pi a_0^3)}$) yields

$$\bar{r} = \frac{8}{\pi a_0^3} \int_0^\infty e^{-4r/a_0} r \cdot 4\pi r^2 \cdot dr = \frac{3}{4} a_0$$

$\bar{r} = \frac{3}{4} a_0$ is half the value $\bar{r} = \frac{3}{2} a_0$ found for the H atom. This is because the nuclear charge in He$^+$ is twice the value in the H atom.

Problem 3.7 – H Atom: Comovement of the Nucleus

The energy of the H atom was calculated with the assumption that the nucleus is at rest; actually, it is the center of mass which is stationary. Calculate the energy regarding the comovement of the nucleus (Figure 3.23).

Solution. The electron (mass m_e) and the proton (mass m_p) are rotating around the center of mass (distance r_1 and speed v_1 for the electron, distance r_2 and speed v_2 for the

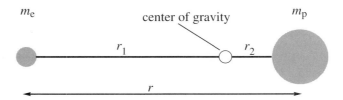

Figure 3.23 Calculation of reduced mass m_{red} in the case of the H atom (electron mass m_e and proton mass m_p).

nucleus). Thus, instead of writing equation (3.80), we write with

$$v_1 = r_1 \cdot \frac{d\varphi}{dt}$$

$$\frac{m_e v_1^2}{r_1} = m_e r_1 \left(\frac{d\varphi}{dt}\right)^2 = \frac{1}{4\pi\varepsilon_0} \cdot \frac{e^2}{r^2}$$

With $r_1 + r_2 = r$ and $m_e r_1 = m_p r_2$ we find

$$m_e r_1 = \frac{m_e m_p}{m_e + m_p} r = m_{\text{red}} \cdot r \quad \text{or} \quad m_{\text{red}} \cdot \left(r\frac{d\varphi}{dt}\right)^2 = \frac{1}{4\pi\varepsilon_0} \frac{e^2}{r}$$

Correspondingly, instead of writing equation (3.81), we write

$$m_e r_1^2 \frac{d\varphi}{dt} + m_p r_2^2 \frac{d\varphi}{dt} = \hbar \quad \text{or} \quad m_{\text{red}} \cdot \left(r\frac{d\varphi}{dt}\right) = \frac{\hbar}{r}$$

Thus we obtain two equations with the unknowns r and $(r \cdot d\varphi/dt)$. We solve for these unknowns:

$$\left(r\frac{d\varphi}{dt}\right) = \frac{e^2}{4\pi\varepsilon_0 \hbar} \quad \text{and} \quad r = \frac{4\pi\varepsilon_0 \hbar}{m_{\text{red}} e^2} = a_0 \cdot \frac{m_e + m_p}{m_p}$$

The potential energy is

$$V = -\frac{1}{4\pi\varepsilon_0} \cdot \frac{e^2}{r} = -\frac{1}{4\pi\varepsilon_0} \cdot \frac{e^2}{a_0} \cdot \frac{m_p}{m_e + m_p}$$

and for the kinetic energy we obtain

$$T = \tfrac{1}{2} m_{\text{red}} \cdot \left(r\frac{d\varphi}{dt}\right)^2$$

$$= \frac{1}{2} \frac{1}{4\pi\varepsilon_0} \cdot \frac{e^2}{r} = \frac{1}{4\pi\varepsilon_0} \cdot \frac{e^2}{2a_0} \cdot \frac{m_p}{m_e + m_p}$$

For the H atom, we obtain $m_{\text{red}} = m_e \times 0.99946$; the corresponding expression for the He atom is $m_{\text{red}} = m_e \times 0.99987$.

The tiny shift of the positions of the emission lines by this effect (correction in Table 3.1) is large compared with the accuracy of spectroscopic measurements and therefore easily observed.

Problem 3.8 – Ground State Energy of H Atom by Variation Principle

Calculate the energy of the ground state of the H atom by using the trial function

$$\phi = \frac{1}{\pi C^3} \cdot e^{-r/C}$$

Solution.

$$\varepsilon = \int \phi \mathcal{H} \phi \cdot d\tau = \frac{1}{\pi C^3} \int e^{-r/C} \mathcal{H} e^{-r/C} \cdot d\tau$$

$$= -\frac{h^2}{8\pi^2 m_e} \frac{1}{\pi C^3} \cdot \int e^{-r/C} \left(\frac{\partial^2}{\partial x^2} + \frac{\partial^2}{\partial y^2} + \frac{\partial^2}{\partial z^2} \right) e^{-r/C} \cdot d\tau$$

$$- \frac{e^2}{4\pi\varepsilon_0} \frac{1}{\pi C^3} \cdot \int e^{-r/C} \frac{1}{r} e^{-r/C} \cdot d\tau$$

The second term corresponds to the average potential energy $\overline{V}$; by replacing a_0 in Box 3.2 by C we obtain

$$\overline{V} = -\frac{1}{4\pi\varepsilon_0} \frac{e^2}{C}$$

The first term corresponds to the average kinetic energy $\overline{T}$. By replacing a_0 in Problem 3.1 we find

$$\left(\frac{\partial^2}{\partial x^2} + \frac{\partial^2}{\partial y^2} + \frac{\partial^2}{\partial z^2} \right) e^{-r/C} = -\frac{1}{C} \left(\frac{2}{r} - \frac{1}{C} \right) \cdot e^{-r/C}$$

Then we obtain with $d\tau = 4\pi r^2 \cdot dr$

$$\overline{T} = \frac{h^2}{8\pi^2 m_e} \frac{1}{\pi C^3} \frac{4\pi}{C} \int_0^\infty \left(\frac{2}{r} - \frac{1}{C} \right) r^2 \cdot e^{-r/C} \cdot dr$$

$$= \frac{h^2}{8\pi^2 m_e} \frac{1}{C^2}$$

We determine the position of the minimum of $\varepsilon = \overline{T} + \overline{V}$ analytically, by setting

$$\frac{d\varepsilon}{dC} = -\frac{h^2}{8\pi^2 m_e} \cdot \frac{2}{C^3} + \frac{1}{4\pi\varepsilon_0} \cdot \frac{e^2}{C^2} = 0$$

From this we obtain

$$C_{\min} = 4\pi\varepsilon_0 \frac{h^2}{4\pi^2 m_e e^2} = a_0$$

$$\varepsilon_{\min} = \frac{h^2}{8\pi^2 m_e} \cdot \frac{4\pi^2 m_e e^2}{4\pi\varepsilon_0 h^2} \cdot \frac{1}{a_0} - \frac{1}{4\pi\varepsilon_0} \cdot \frac{e^2}{a_0} = -\frac{1}{4\pi\varepsilon_0} \cdot \frac{e^2}{2a_0}$$

Foundation 3.1 – Electron in a Potential Trough of Finite Depth

We consider an electron in a one-dimensional box with finitely high walls (Figure 3.2):

$$V(x) = 0 \quad \text{for} \quad -\frac{L}{2} \leq x \leq +\frac{L}{2}$$

$$V(x) = V_0 \quad \text{for} \quad x < -\frac{L}{2} \text{ and } x > +\frac{L}{2}$$

Note that, for convenience, the origin of the coordinate system is in the center between the walls. V_0 is assumed to be larger than the energy of the electron. We start with the differential equations

$$-\frac{h^2}{8\pi^2 m_e}\frac{d^2\psi_a}{dx^2} = E\psi_a \quad \text{for} \quad -\frac{L}{2} \leq x \leq +\frac{L}{2}$$

$$-\frac{h^2}{8\pi^2 m_e}\frac{d^2\psi_b}{dx^2} + V_0\psi_b = E\psi_b \quad \text{for} \quad x < -\frac{L}{2} \text{ and } x > +\frac{L}{2}$$

We try to solve these equations with $\psi_a = A\cos ax$ and $\psi_b = B \cdot e^{-bx}$. The second expression fulfills the boundary condition $\psi = 0$ for $x \to \pm\infty$. Inserting into the differential equations gives

$$E = \frac{h^2}{8\pi^2 m_e}a^2 = -\frac{h^2}{8\pi^2 m_e}b^2 + V_0$$

For the calculation of the constants a, b, A, and B we have the conditions

$$\psi_a = \psi_b, \quad \frac{\partial \psi_a}{\partial x} = \frac{\partial \psi_b}{\partial x}$$

at $x = -L/2$ and $x = +L/2$. In the following we show that for $V_0 = 10h^2/(8m_e L^2)$ we obtain the wavefunctions and energies displayed in Figure 3.2.

Evaluation of the data for Figure 3.2

From the conditions

$$\psi_a = \psi_b \quad \text{and} \quad \frac{d\psi_a}{dx} = \frac{d\psi_b}{dx} \quad \text{at } x = L/2$$

it follows

$$A\cos\frac{aL}{2} = Be^{-bL/2} \quad \text{and} \quad -Aa\sin\frac{aL}{2} = -Bbe^{-bL/2}$$

Dividing the second equation by the first one results in

$$b = a \cdot \tan\frac{aL}{2}$$

Continued on page 78

Continued from page 77

From the equation for the energy E we obtain

$$a^2 = \frac{8\pi^2 m_e}{h^2} V_0 - b^2$$

By combining these two equations we get

$$a^2 + a^2 \tan^2 \frac{aL}{2} - V_0 \frac{8\pi^2 m_e}{h^2} = 0$$

We set

$$V_0 = \frac{h^2}{8 m_e L^2} k$$

This means V_0 is the k-fold of the energy of an electron in a one-dimensional box in its ground state. Furthermore, we use the abbreviation $w = aL$. Then we obtain

$$w^2 + w^2 \cdot \tan^2 \frac{w}{2} - \pi^2 k = 0$$

$$E = \frac{h^2}{8\pi^2 m_e} a^2 = \frac{h^2}{8 m_e L^2} \left(\frac{w}{\pi}\right)^2$$

The equation for w is solved numerically. As an example, we choose $k = 10$. Then the solutions for w are $w = 2.61$ and $w = 7.66$. Furthermore, $bL = w \cdot \tan(w/2)$ and $B/A = \cos(w/2) \cdot e^{bL/2}$.

These are solutions for symmetric wavefunctions (cosine functions, maximum at $x = 0$, states 1 and 3).

The corresponding antisymmetric solutions follow from $\psi = A \cdot \sin ax$; in this case we get w from the equation

$$w^2 + w^2 \Big/ \left(\tan^2 \frac{w}{2}\right) - \pi^2 k = 0$$

The numerical solutions for this equation are $w = 5.18$ and $w = 9.78$. Furthermore, $bL = -w/\tan(w/2)$ and $B/A = \sin(w/2) \cdot e^{bL/2}$. The final results for $L = 1000$ pm are as follows:

State	w	$a = \frac{w}{L}$	bL	b	$\frac{B}{A}$	$E \cdot \frac{8 m_e L^2}{h^2}$
1	2.61	0.00261 pm^{-1}	9.58	0.00958 pm^{-1}	31.6	0.69
2	5.18	0.00518 pm^{-1}	8.42	0.00842 pm^{-1}	35.3	2.72
3	7.66	0.00766 pm^{-1}	6.30	0.00630 pm^{-1}	−18.0	5.95
4	9.78	0.00978 pm^{-1}	1.76	0.00176 pm^{-1}	−2.4	9.69

From these data the wavefunctions and energies displayed in Figures 3.2(b) and (c) are calculated.

Foundation 3.2 – Relation Between Time-independent and Time-dependent Schrödinger Equations

We consider an electron in the quantum state with quantum number n; it has the energy E_n and the probability density distribution $\rho_n = \psi_n^2$, where

$$\mathcal{H}\psi_n = E_n \psi_n$$

We shall show that the time-independent Schrödinger equation describes a special case of the time-dependent Schrödinger equation

$$\mathcal{H}\Psi = i\hbar \frac{\partial \Psi}{\partial t}$$

We consider $\mathcal{H}$ as time-independent – that is, the electron moves in a time-independent potential field. We assume that the function

$$\Psi = \psi_n \cdot e^{iat}$$

describes the same stationary state as ψ_n if constant a is appropriately chosen.

Is this assumption really true? If it is true, then the product $\Psi\Psi^*$ must be identical with ψ_n^2. Indeed, this is the case:

$$\Psi\Psi^* = \psi_n \cdot e^{iat} \cdot \psi_n \cdot e^{-iat} = \psi_n^2$$

Furthermore, Ψ must be a solution of the time-dependent Schrödinger equation:

$$\mathcal{H}\psi_n e^{iat} = i\hbar \frac{\partial \psi_n e^{iat}}{\partial t}$$

Thus

$$e^{iat}\mathcal{H}\psi_n = i\hbar \psi_n \frac{d(e^{iat})}{dt} = i\hbar \psi_n \cdot ia \cdot e^{iat}$$

and therefore (with $i^2 = -1$)

$$\mathcal{H}\psi_n = -\hbar a \cdot \psi_n$$

Comparison with the time-independent Schrödinger equation

$$\mathcal{H}\psi_n = E_n \cdot \psi_n$$

shows that Ψ is indeed a solution of the time-dependent Schrödinger equation and that a is related to the energy E_n:

$$a = -\frac{E_n}{\hbar}$$

As a general solution of the time-dependent Schrödinger equation for time-independent $\mathcal{H}$ we can write

Continued on page 80

Continued from page 79

$$\Psi = \sum_n c_n \psi_n e^{-(iE_n/\hbar)\cdot t}$$

This is true because each term in the sum obeys the time-dependent Schrödinger equation as we have just shown. The coefficients c_n are determined by the initial condition; then Ψ describes the temporal evolution of $\rho = \Psi\Psi^*$.

This relation is important for describing the electron transfer process from a donor to an acceptor molecule. In this case, the solution of the time-dependent Schrödinger equation of the system of donor and acceptor at $t = 0$ (electron at donor) can be expressed in terms of the wavefunctions ψ_n of the system:

$$\Psi_{t=0} = \sum_n c_n \psi_n$$

Then the temporal evolution of Ψ and thus of ρ can be obtained.

Foundation 3.3 – H Atom: Solution of the Schrödinger Equation

The Schrödinger equation for the H atom is

$$-\frac{h^2}{8\pi^2 m_e}\left(\frac{\partial^2 \psi}{\partial x^2} + \frac{\partial^2 \psi}{\partial y^2} + \frac{\partial^2 \psi}{\partial z^2}\right)\psi - \frac{1}{4\pi\varepsilon_0}\cdot\frac{e^2}{r}\psi = E\psi$$

In contrast to the Schrödinger equation of the particle in a box, we cannot describe the wavefunction ψ as a product of functions depending on x, y, or z. This is because of the spherical symmetry of the potential energy V. Thus we transform the Cartesian coordinates x, y, and z into spherical coordinates r, ϑ, and φ (Figure 3.24).

Rewriting the Schrödinger equation in these coordinates we solve the equation with

$$\psi(r, \vartheta, \varphi) = R(r)\cdot\Theta(\vartheta)\cdot\Phi(\varphi)$$

This way the Schrödinger equation can be split into the following three ordinary differential equations:

$$\frac{d^2\Phi}{d\varphi^2} = -k^2\cdot\Phi$$

$$\frac{1}{\Theta}\cdot\frac{1}{\sin\vartheta}\frac{d}{d\vartheta}\left(\sin\vartheta\frac{d\Theta}{d\vartheta}\right) - \frac{k^2}{\sin^2\vartheta} = -l(l+1)$$

Continued on page 81

Continued from page 80

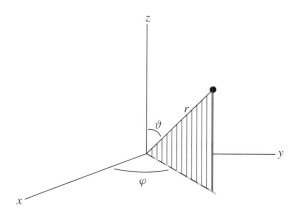

Figure 3.24 Spherical coordinates r, ϑ, and φ.

$$-\frac{h^2}{8\pi^2 m_e r^2}\left[\frac{1}{R}\frac{d}{dr}\left(r^2\frac{dR}{dr}\right) - l(l+1)\right] - \frac{1}{4\pi\varepsilon_0}\cdot\frac{e^2}{r} = E$$

where k and l are constants which are given by the boundary conditions. For Φ, Θ, and R the unnormalized solutions are

$\Phi_0 = 1$	$k = 0$
$\Phi_{1a} = \cos\varphi$	$k = 1$
$\Phi_{1b} = \sin\varphi$	$k = 1$
$\Phi_{2a} = \cos 2\varphi$	$k = 2$
$\Phi_{2b} = \sin 2\varphi$	$k = 2$
$\Phi_{3a} = \cos 3\varphi$	$k = 3$
$\Phi_{3b} = \sin 3\varphi$	$k = 3$

$\Theta_{00} = 1$	$l = 0$	$k = 0$
$\Theta_{10} = \cos\vartheta$	$l = 1$	$k = 0$
$\Theta_{11} = \sin\vartheta$	$l = 1$	$k = 1$
$\Theta_{20} = 3\cos^2\vartheta - 1$	$l = 2$	$k = 0$
$\Theta_{21} = \sin\vartheta\cos\vartheta$	$l = 2$	$k = 1$
$\Theta_{22} = \sin^2\vartheta$	$l = 2$	$k = 2$

$R_{10} = e^{-r/a_0}$	$n = 1$	$l = 0$
$R_{20} = \left(2 - \dfrac{r}{a_0}\right)\cdot e^{-r/(2a_0)}$	$n = 2$	$l = 0$
$R_{21} = \dfrac{r}{a_0}\cdot e^{-r/(2a_0)}$	$n = 2$	$l = 1$
$R_{30} = \left[27 - 18\dfrac{r}{a_0} + 2\left(\dfrac{r}{a_0}\right)^2\right]\cdot e^{-r/(3a_0)}$	$n = 3$	$l = 0$
$R_{31} = r\left(6 - \dfrac{r}{a_0}\right)\cdot e^{-r/(3a_0)}$	$n = 3$	$l = 1$
$R_{32} = \left(\dfrac{r}{a_0}\right)^2\cdot e^{-r/(3a_0)}$	$n = 3$	$l = 2$

Continued on page 82

Continued from page 81

The energy E is determined by the quantum number n:

$$E = -\frac{1}{4\pi\varepsilon_0} \cdot \frac{e^2}{2a_0} \cdot \frac{1}{n^2}, \quad n = 1, 2, 3, \ldots$$

where

$$a_0 = 4\pi\varepsilon_0 \frac{h^2}{4\pi^2 m_e e^2}$$

It is easy to check the validity of these solutions by inserting them into the corresponding differential equations.

The wavefunctions of the H atom are obtained by combining all the individual functions for the quantum numbers n, l, and k with the restriction $k \leq l < n$.

ψ_{1s}	$n = 1$	$l = 0$	$k = 0$	$1 = 1^2$ functions
ψ_{2s}	$n = 2$	$l = 0$	$k = 0$	
ψ_{2p_z}		$l = 1$	$k = 0$	$4 = 2^2$ functions
ψ_{2p_x}, ψ_{2p_y}		$l = 1$	$k = 1$	
ψ_{3s}	$n = 3$	$l = 0$	$k = 0$	
ψ_{3p_z}		$l = 1$	$k = 0$	
ψ_{3p_x}, ψ_{3p_y}		$l = 1$	$k = 1$	$9 = 3^2$ functions
$\psi_{3d_{z^2}}$		$l = 2$	$k = 0$	
$\psi_{3d_{yz}}, \psi_{3d_{xz}}$		$l = 2$	$k = 1$	
$\psi_{3d_{xy}}, \psi_{3d_{x^2-y^2}}$		$l = 2$	$k = 2$	

After normalization we obtain the wavefunctions displayed in equation (3.51), (3.62), and (3.63); for better visualization of the directions of the different orbitals the spherical coordinates were replaced by Cartesian coordinates (except for the exponential factors).

Foundation 3.4 – Proof of Variation Principle

We consider a molecule described by the normalized trial function ϕ_1 and the energy ε_1; we want to show that

$$\varepsilon_1 = \int \phi_1 \mathcal{H} \phi_1 \cdot d\tau \geq E_1$$

where E_1 is the energy of the molecule for the exact solution of the Schrödinger equation. We describe the trial function ϕ_1 by a linear combination of

Continued on page 83

Continued from page 82

the normalized wavefunctions $\psi_1, \psi_2, \ldots$ belonging to the exact energies $E_1, E_2, \ldots$. E_1 is the exact energy of the lowest energy state.

$$\phi_1 = c_1\psi_1 + c_2\psi_2 + \cdots = \sum_i c_i\psi_i$$

We insert this expression into the integral for ε_1

$$\varepsilon_1 = \int \left(\sum_i c_i\psi_i\right) \mathcal{H} \left(\sum_i c_i\psi_i\right) \cdot d\tau$$

$$= \sum_i c_i^2 \int \psi_i \mathcal{H} \psi_i \cdot d\tau + \sum_{j \neq i} c_i c_j \int \psi_i \mathcal{H} \psi_j \cdot d\tau$$

For the exact wavefunctions $\mathcal{H}\psi_i = E_i\psi_i$, thus:

$$\int \psi_i \mathcal{H} \psi_i \cdot d\tau = E_i \int \psi_i^2 \cdot d\tau = E_i, \quad \int \psi_i \mathcal{H} \psi_j \cdot d\tau = E_j \int \psi_i \psi_j \cdot d\tau = 0$$

Then we obtain

$$\varepsilon_1 = \sum_i c_i^2 E_i = \sum_i \left[c_i^2 E_1 + c_i^2(E_i - E_1)\right]$$

$$= E_1 \sum_i c_i^2 + \sum_i c_i^2(E_i - E_1)$$

The trial function ϕ_1 is normalized:

$$\int \phi_1^2 \cdot d\tau = \sum_i c_i^2 \int \psi_i^2 \cdot d\tau + \sum_{j \neq i} c_i c_j \int \psi_i \psi_j \cdot d\tau$$

$$= \sum_i c_i^2 = c_1^2 + c_2^2 + \cdots = 1$$

Thus we have

$$\varepsilon_1 = E_1 + \sum_i c_i^2(E_i - E_1)$$

Because of $E_i \geq E_1$ the second term is greater than zero and thus we obtain $\varepsilon_1 \geq E_1$. If ϕ_1 is identical with ψ_1, then all coefficients except c_1 are zero and we have $\varepsilon_1 = E_1$.

In analogy, for the first excited state we use a trial function ϕ_2 which is orthogonal to ψ_1. This means that $c_1 = 0$ and it can be shown in the same way that $\varepsilon_2 \geq E_2$. The same procedure can be applied for $\varepsilon_3, \varepsilon_4, \ldots$.

In practical cases ψ_1 is unknown, and one chooses functions ϕ_2 which are orthogonal to ϕ_1. In this case they are no longer automatically orthogonal to

Continued on page 84

> *Continued from page 83*
>
> ψ_1 and we cannot claim that $\varepsilon_2 \geq E_2$. The better ϕ_1 approaches ψ_1, the better the orthogonality to ψ_1 will be fulfilled. The variation method leads to values of $\varepsilon_1, \varepsilon_2, \ldots$ which become near to the values of $E_1, E_2, \ldots$ when the process is improved.

4 Chemical Bonding and the Pauli Principle

In this chapter the quantitative quantum chemical treatment of matter is extended to the H_2^+ molecule ion. The ground-state properties of H_2^+ are obtained by solving the Schrödinger equation exactly and by using simple trial functions (box model and linear combination of H atom wavefunctions).

Furthermore, atoms and molecules with two electrons, such as He and H_2, are considered. In these cases, due to the mutual repulsion of the electrons, the electron cloud is expanded, in accordance with the variation principle.

Two new strange properties of electrons in a many-electron system are considered. Firstly, electrons are indistinguishable, that is, the total wavefunction must be such that exchanging coordinates leads to the same density distribution. This leads to the same wavefunction, or to the same wavefunction but with opposite sign: the wavefunction must be symmetric or antisymmetric. Secondly, the total wavefunction of a molecule must be antisymmetric when exchanging the coordinates of two electrons. This is the Pauli exclusion principle in its generalized form: only the antisymmetric alternative is realized by nature.

Finally, the phenomenon of quantum mechanical tunneling is discussed.

4.1 H_2^+ Molecule Ion

For our analysis, we consider H_2^+ to be made from an H atom and a proton:

$$H + \text{proton} \longrightarrow H_2^+ \tag{4.1}$$

(see Figure 2.10).

4.1.1 Electron Described by Exact Wavefunction

In analogy to the electron in the H atom, the Schrödinger equation for the H_2^+ ion is

$$-\frac{h^2}{8\pi^2 m_e}\left(\frac{\partial^2}{\partial x^2} + \frac{\partial^2}{\partial y^2} + \frac{\partial^2}{\partial z^2}\right)\psi + \frac{e^2}{4\pi\varepsilon_0}\left(-\frac{1}{r_1} - \frac{1}{r_2} + \frac{1}{d}\right)\psi = E\psi \tag{4.2}$$

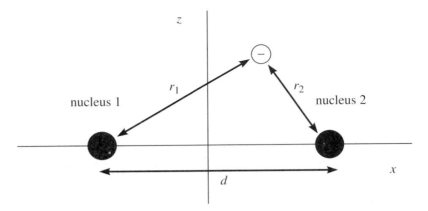

Figure 4.1 H_2^+ molecule ion: electron charge in the field of two positive charges separated by a distance d. r_1 and r_2 are the distances between the electron and both positive charges.

where d is the distance between the nuclei and r_1 and r_2 are the distances between the electron and the nuclei 1 and 2 (Figure 4.1). The wavefunction ψ and the energy E of the H_2^+ ion depend on the distance d. In Foundation 4.1 it is demonstrated how E and ψ can be calculated for various distances d. The lowest energy E is obtained for $d = 106\,\text{pm}$:

$$E = -\frac{1}{4\pi\varepsilon_0} \cdot \frac{e^2}{2a_0} \times 1.203 = -26.2 \times 10^{-19}\,\text{J} \qquad (4.3)$$

This energy is lower than the energy $E = -21.8 \times 10^{-19}\,\text{J}$ of the electron in an H atom obtained in Chapter 3. For this reason, the reaction of an H atom with a proton releases energy and the H_2^+ ion is a stable molecule. In order to pull the molecule apart, the energy $(26.2 - 21.8) \times 10^{-19}\,\text{J} = 4.4 \times 10^{-19}\,\text{J}$ must be supplied; this is in excellent agreement with the experiment (see Section 2.4.2).

The discussion was based on the exact energies and wavefunctions of H_2^+ obtained by solving the Schrödinger equation analytically. Nevertheless, it is extremely helpful to consider, in addition, simple descriptions based on the variation principle. Such descriptions lead to a deeper insight into the electronic structure of complex systems, such as those considered in Chapters 7 and 8. In Chapter 2 we used box wavefunctions as trial functions for H_2^+ on a qualitative level. Now we proceed with a more detailed discussion.

4.1.2 Electron Described by Box Wavefunctions

As the simplest trial function for the ground state we use, similar to the case of the H atom, the wavefunction of an electron in a box of dimensions L_x, L_y, and L_z. L_y and L_z must be equal on grounds of symmetry; as outlined in Chapter 2 we set $L_y = L_z = L$ and $L_x = bL$ (Figure 4.2) and use the ground-state quantum numbers $n_x = n_y = n_z = 1$.

If we place the origin of the coordinate system in the center of the box, then we can write the following for the wavefunction ϕ

$$\phi = \sqrt{\frac{1}{b}\left(\frac{2}{L}\right)^3} \cdot \cos\frac{\pi x}{bL} \cdot \cos\frac{\pi y}{L} \cdot \cos\frac{\pi z}{L} \qquad (4.4)$$

4.1 H₂⁺ MOLECULE ION

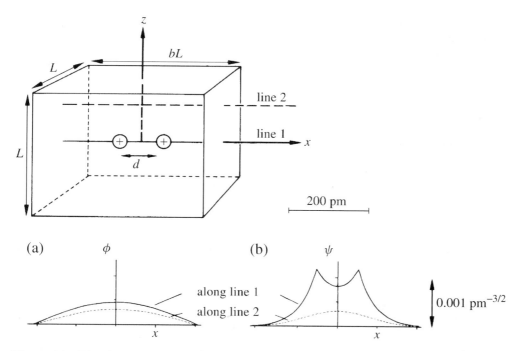

Figure 4.2 Wavefunction of the H$_2^+$ ion. Graph of the wavefunction of the lowest energy state along the x axis (line 1) and along a line a distance 75 pm above (line 2). (a) Trial function ϕ (box model). (b) Exact wavefunction ψ (see Foundation 4.1).

where x extends from $-bL/2$ to $+bL/2$ and y and z extend from $-L/2$ to $+L/2$. For the average kinetic energy of the electron, we obtain the following equation according to Chapters 2 and 3:

$$\overline{T} = \frac{h^2}{8m_e}\left[\frac{1}{(bL)^2} + \frac{1}{L^2} + \frac{1}{L^2}\right] = \frac{h^2}{8m_e L^2}\left(\frac{1}{b^2} + 2\right) \tag{4.5}$$

The two atomic nuclei are placed a distance d apart symmetrically to the origin of the coordinate system on the abscissa (Figure 4.2). The potential energy of the system consists of the repulsive energy of the two atomic nuclei, the attractive energy between the electron and nucleus 1, and the attractive energy between the electron and nucleus 2. Thus the average potential energy $\overline{V}$ is given by

$$\overline{V} = \frac{1}{4\pi\varepsilon_0}\cdot\frac{e^2}{d} - \frac{1}{4\pi\varepsilon_0}\int\left[\frac{e^2}{r_1} + \frac{e^2}{r_2}\right]\phi^2\cdot dx\,dy\,dz \tag{4.6}$$

where r_1 and r_2 are the distances between the electron and the nuclei 1 and 2:

$$r_1 = \sqrt{\left(x+\frac{d}{2}\right)^2 + y^2 + z^2} \quad \text{and} \quad r_2 = \sqrt{\left(x-\frac{d}{2}\right)^2 + y^2 + z^2} \tag{4.7}$$

The total energy

$$\varepsilon = \overline{T} + \overline{V} \tag{4.8}$$

depends on the nuclear distance d and on the parameters L and b; in contrast to the corresponding problem for the H atom, we must vary three parameters instead of only one. This variation is carried out numerically; L and b are varied for a series of nuclear distances d until the lowest energy state is found (Table 4.1). In Figure 4.3 the energies $\overline{T}$, $\overline{V}$, and ε are shown as a function of nuclear distance d. From Table 4.1 and Figure 4.3

Table 4.1 Energy ε of H_2^+, calculated according to the variation principle (box wavefunction, $L = L_y = L_z$) for various nuclear distances d

L/pm	$d = 140$ pm		L/pm	$d = 150$ pm		L/pm	$d = 160$ pm	
	b	$\varepsilon/(10^{-19}$ J)		b	$\varepsilon/(10^{-19}$ J)		b	$\varepsilon/(10^{-19}$ J)
280	1.3	−16.795	290	1.3	−16.828	290	1.3	−16.627
	1.4	−16.809		1.4	−16.886		1.4	−16.809
	1.5	−16.622		1.5	−16.768		1.5	−16.783
290	1.2	−16.681	300	1.2	−16.696	300	1.3	−16.798
	1.3	**−16.858**		1.3	**−16.914**		1.4	**−16.900**
	1.4	−16.801		1.4	−16.888		1.5	−16.806
300	1.2	−16.763	310	1.2	−16.768	310	1.3	−16.866
	1.3	−16.842		1.3	−16.839		1.4	−16.881
	1.4	−16.696		1.4	−16.810		1.5	−16.739

The integral for the potential energy was calculated numerically in a similar way as demonstrated in Chapter 3 for the H atom: discretization of the box into $60^3 = 216\,000$ points. The energy minimum is reached for $d = 150$ pm, $L = 300$ pm, and $b = 1.3$. Further refinement of the method results in $d = 152$ pm, $L = 301$ pm, and $b = 1.36$.

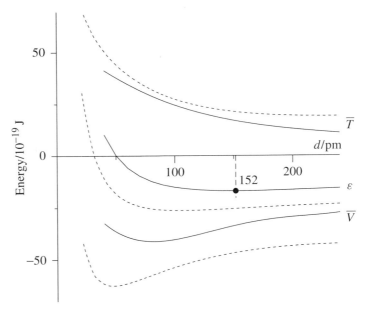

Figure 4.3 Energy ε, average kinetic energy $\overline{T}$, and average potential energy $\overline{V}$ of the H_2^+ ion as a function of nuclear distance d. Solid lines: according to the variation principle with the box trial function (4.4). Dashed lines: exact solution. The dot indicates the minimum of the energy ε.

4.1 H_2^+ MOLECULE ION

we discover that the lowest energy state is reached at a nuclear distance $d = d_0 = 152$ pm and with box parameters $L = 301$ pm and $b = 1.36$.

The best cloud shape is thus given by a box with the dimensions $L_x = 409$ pm and $L_y = L_z = 301$ pm. The electron cloud in the H_2^+ ion has a greater long axis and a smaller minor axis than the electron cloud in the H atom ($L = 376$ pm). In this way we obtain $\varepsilon = -16.9 \times 10^{-19}$ J $= -10.5$ eV. This energy is lower than the value $\varepsilon = -12.9 \times 10^{-19}$ J calculated in Chapter 2 for the energy of the electron in an H atom as described by a box wavefunction. In order to pull the molecule apart, the energy $(16.9 - 12.9) \times 10^{-19}$ J $= 4.0 \times 10^{-19}$ J must be supplied; this is very near to the value obtained in the exact calculation (4.4×10^{-19} J). The corresponding distance $d = 152$ pm agrees reasonably with the experimental value 106 pm.

The box model will be very useful in subsequent chapters for understanding the electronic structure of molecules. For that reason a careful analysis of the model in the simple case of H_2^+ is important for judging its power and its limits. In this case, an exact treatment is possible, in contrast to more complex electron systems.

4.1.3 Electron Density and Chemical Bond

The average electron density in the vicinity of the nuclei increases when we proceed from an H atom and a proton to H_2^+. For an electron in the H atom the probability density is $\rho_{r_1=0} = 1/(\pi a_0^3) = 0.32 a_0^{-3}$ and for a proton $\rho_{r_2=0} = 0$. On the average, the probability density is

$$\rho_{\text{average}} = \tfrac{1}{2}\left(\rho_{r_1=0} + \rho_{r_2=0}\right) = \tfrac{1}{2}(0.32 + 0)a_0^{-3} = 0.16 a_0^{-3} \tag{4.9}$$

For the H_2^+ molecule ion, according to the exact calculation, we find

$$\rho_{r_1=0} = \rho_{r_2=0} = 0.21 a_0^{-3} \tag{4.10}$$

Thus the electron density at the nuclei in H_2^+ is larger than the average electron density at the nuclei of an H atom and a proton. This increase of the electron density is due to the compression of the electron cloud perpendicular to the bond line.

For this reason the electron density is also increased on the line connecting the nuclei: halfway between the nuclei it is 47% of the maximum value at the nuclei. The electron is thus more frequently to be found between the nuclei; through this the nuclei are held together by Coulomb forces.

As already mentioned in Chapter 2, the bond energy of H_2^+ compared to the H atom is only half the value of the difference between the average potential energies of the two systems (Figure 4.4). Initially this is surprising, since the electron cloud stretches in the bond direction (Figure 4.5) and thus the de Broglie wavelength increases, so the kinetic energy should decrease. Since at the same time, however, the wavelength perpendicular to the bond line is smaller, this contribution to the kinetic energy increases; this increase outweighs the decrease on grounds of the length stretching, so that the kinetic energy in the transition from H to H_2^+ increases overall.

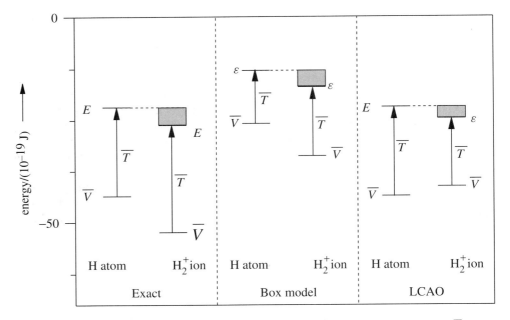

Figure 4.4 H atom and H_2^+ ion; average potential energy $\overline{V}$, average kinetic energy $\overline{T}$, energy E according to the exact treatment ($d = 106$ pm), and energy ε according to the box model ($d = 152$ pm) and according to the LCAO model ($d = 132$ pm). The energy is taken to be zero when the charges are separated (nucleus and electron removed infinitely far apart). In each case the bond energy is indicated by the height of the shaded area. Note that in the LCAO model the virial theorem is not fulfilled ($\overline{V}$ increases and $\overline{T}$ decreases in forming the bond).

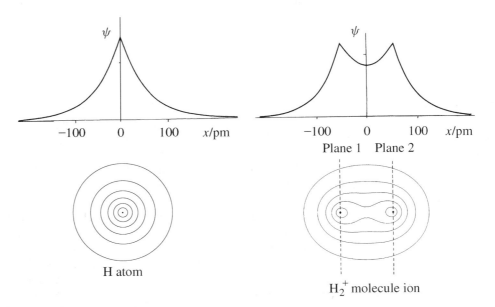

Figure 4.5 H atom and H_2^+ ion. Graph of the exact wavefunction along the x axis as in Figures 4.2 and 3.9. Contour lines in the x, y plane ($z = 0$). The outermost contour line has the value $\psi = 2 \times 10^{-4}$ pm$^{-3/2}$, the next one has a value of 4×10^{-4} pm$^{-3/2}$, and so on.

4.1 H_2^+ MOLECULE ION

4.1.4 Virial Theorem

The exact calculation for the H_2^+ molecule ion shows that the virial theorem

$$E = -\overline{T} = \tfrac{1}{2}\overline{V} \tag{4.11}$$

is fulfilled (Foundation 4.1). It is also fulfilled for our box trial function. With the values obtained above for L and b, the result according to equation (4.5) is $\overline{T} = 16.9 \times 10^{-19}$ J. Then $\varepsilon = -16.9 \times 10^{-19}$ J $= -1.00 \cdot \overline{T}$.

4.1.5 Nature of the Chemical Bond

In order to see as clearly as possible how the two nuclei are held together by the electron cloud, we imagine the nuclei at a distance somewhat larger than the equilibrium separation. The two nuclei are pulled toward each other along the bond line by the Coulomb attraction of the charge elements between planes 1 and 2 (Figure 4.5). This attractive force overcomes the forces acting in the opposite direction: (a) Coulomb repulsion of the nuclei and (b) Coulomb attraction of the nuclei toward the charge elements outside the two planes 1 and 2.

The nuclei approach each other, until these forces balance each other at the equilibrium distance. With this nuclear approach a certain compression of the electron cloud takes place due to the increased field of the nuclei, and with it to an increase in the kinetic energy of the electron. Through this, the bond energy is smaller as one would expect only from taking into account the Coulomb energy.

The fact that the chemical bond is based on Coulomb attraction of the electron cloud and the nuclei must be strongly pointed out.

4.1.6 Electron Described by Combination of Atomic Orbitals

Approximating the electron system of a molecule by a linear combination of atomic orbitals is the most popular approach to understanding chemical bonds. Therefore, using this approach for H_2^+ is of particular importance.

LCAO trial function

By use of the variation principle on the H_2^+ ion we can try out other functions in place of the box trial functions. A very common procedure for molecules is to imagine that the molecular wavefunctions are built from the wavefunctions of the constituent atoms. These components are called *atomic orbitals*. In the case of the H_2^+ ion we overlap the 1s functions

$$\varphi_a = \frac{1}{\sqrt{\pi a_0^3}} \cdot e^{-r_a/a_0}, \quad \varphi_b = \frac{1}{\sqrt{\pi a_0^3}} \cdot e^{-r_b/a_0} \tag{4.12}$$

of the two hydrogen atoms (Figure 4.6) in the trial function

$$\phi = c \cdot (\varphi_a + \varphi_b) \tag{4.13}$$

where c is a constant. The method of proposing a trial function from a combination of atom functions is called the *LCAO method* (*linear combination of atomic orbitals*).

CHEMICAL BONDING AND THE PAULI PRINCIPLE

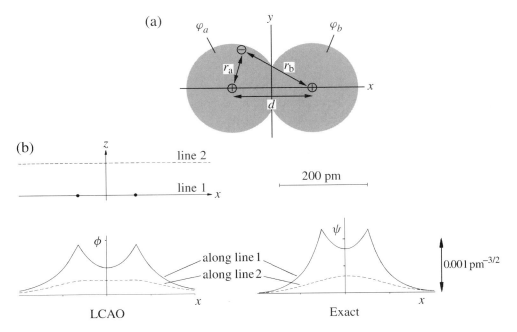

Figure 4.6 H_2^+ ion. (a) Atom functions φ_a and φ_b of two hydrogen atoms a and b, which are located at the equilibrium distance d from each other. (b) Graph of the wavefunction along the nuclear bond line (line 1) and along line 2 at a distance of 75 pm. Left curve: LCAO method (wavefunction ϕ; $d = 132$ pm). Right curve: exact (wavefunction ψ; $d = 106$ pm).

According to the variation principle we have to seek for the minimum of

$$\varepsilon = \int \phi \mathcal{H} \phi \cdot d\tau \tag{4.14}$$

Since not only the energy of the electrons but also the repulsive energy of the two nuclei contribute to the energy of the molecule, the expression

$$V = \frac{e^2}{4\pi\varepsilon_0} \cdot \left(-\frac{1}{r_a} - \frac{1}{r_b} + \frac{1}{d} \right)$$

must be inserted in the Hamiltonian $\mathcal{H}$, as in equation (4.6).

Calculation of the energy

The constant c can be evaluated by using the normalization condition

$$\int \phi^2 \cdot d\tau = c^2 \cdot \left(\int \varphi_a^2 \cdot d\tau + 2\int \varphi_a \varphi_b \cdot d\tau + \int \varphi_b^2 \cdot d\tau \right) = 1 \tag{4.15}$$

Because of the normalization of the atomic functions we have $\int \varphi_a^2 \cdot d\tau = \int \varphi_b^2 \cdot d\tau = 1$; if we shorten the integral $\int \varphi_a \varphi_b \cdot d\tau$ by

$$S_{ab} = \int \varphi_a \varphi_b \cdot d\tau \tag{4.16}$$

4.1 H₂⁺ MOLECULE ION

then

$$c = \frac{1}{\sqrt{2(1+S_{ab})}} \tag{4.17}$$

Now we insert the trial function (4.13) into the energy equation (4.14):

$$\varepsilon = \int \phi \mathcal{H} \phi \cdot d\tau = c^2 \int (\varphi_a + \varphi_b) \mathcal{H} (\varphi_a + \varphi_b) \cdot d\tau$$

$$= c^2 \left(\int \varphi_a \mathcal{H} \varphi_a \cdot d\tau + \int \varphi_a \mathcal{H} \varphi_b \cdot d\tau + \int \varphi_b \mathcal{H} \varphi_a \cdot d\tau + \int \varphi_b \mathcal{H} \varphi_b \cdot d\tau \right) \tag{4.18}$$

Because the atomic functions φ_a and φ_b are identical we have

$$\int \varphi_a \mathcal{H} \varphi_b \cdot d\tau = \int \varphi_b \mathcal{H} \varphi_a \cdot d\tau$$

and

$$\int \varphi_a \mathcal{H} \varphi_a \cdot d\tau = \int \varphi_b \mathcal{H} \varphi_b \cdot d\tau$$

We shorten the remaining integrals by

$$\int \varphi_a \mathcal{H} \varphi_a \cdot d\tau = \int \varphi_b \mathcal{H} \varphi_b \cdot d\tau = H_{aa}, \quad \int \varphi_a \mathcal{H} \varphi_b \cdot d\tau = \int \varphi_b \mathcal{H} \varphi_a \cdot d\tau = H_{ab} \tag{4.19}$$

Finally we obtain

$$\boxed{\varepsilon = \frac{H_{aa} + H_{ab}}{1 + S_{ab}}} \tag{4.20}$$

The energy ε depends on the bond distance d, because H_{aa}, H_{ab}, and S_{ab} are complicated functions of d. In Foundation 4.2 it is shown how ε depends on d; as a result, $\overline{T}$, $\overline{V}$, and ε are displayed in Figure 4.7 in dependence of d. For the energy minimum we deduce

$$\overline{V} = -41.3 \times 10^{-19} \, \text{J}, \overline{T} = 16.7 \times 10^{-19} \, \text{J}, \varepsilon = -24.6 \times 10^{-19} \, \text{J},$$

and $d = d_0 = 132\,\text{pm}$. Compared to the H atom, ε decreases by $(24.6 - 21.8) \times 10^{-19}\,\text{J} = 2.8 \times 10^{-19}\,\text{J} = 1.7\,\text{eV}$; in the box model, as mentioned above, we find $4.0 \times 10^{-19}\,\text{J} = 2.5\,\text{eV}$.

Virial theorem

A closer inspection shows that for the LCAO trial function (4.13) the virial theorem is not fulfilled ($\varepsilon = -1.47 \times \overline{T}$). Additionally, near the equilibrium distance the curve for $\overline{T}$ goes through a minimum, although we expect from Figure 4.3 that T increases monotonically with decreasing d. On the other hand, $\overline{V}$ increases with decreasing d, instead of passing through a minimum.

Thus, there is a serious problem in the LCAO approximation. The model predicts erroneously that charge is removed from the regions near the nuclei and brought into the overlap region. Then the charge density near the nuclei decreases in the model, although actually an increase occurs due to the contraction of the electron cloud perpendicular to

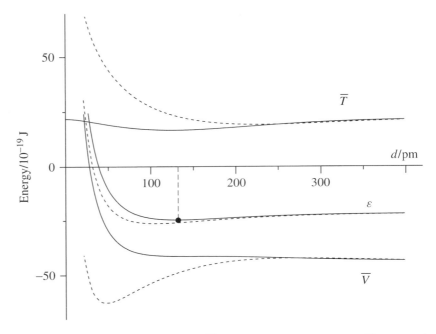

Figure 4.7 Energy ε, average kinetic energy $\overline{T}$, and average potential energy $\overline{V}$ of the H_2^+ ion as a function of nuclear distance d. Solid lines: according to the LCAO method. Dashed lines: according to the exact treatment. The dot indicates the minimum of the energy ε.

the bond line. The model predicts a decrease of kinetic energy of the electron in forming the bond; in reality, however, the kinetic energy increases when proceeding from H to H_2^+, and the potential energy decreases.

A refined treatment using modified atomic orbitals is discussed in Box 4.1.

Box 4.1 – H_2^+: Linear Combination of Compressed Atomic Orbitals as Trial Functions

The discrepancy in the LCAO model of H_2^+ (virial theorem not fulfilled) can be avoided by using modified trial functions. As an example, the trial functions

$$\phi = A \cdot \left[\exp(-\eta r_a/a_o) + \exp(-\eta r_b/a_o)\right]$$

can be used, where η is treated as a variation parameter. Thus the extension of the constituent orbitals can change with internuclear distance. In the simple LCAO treatment of H_2^+ the value $\eta = 1$ is used (the constituent atomic orbitals are hydrogen atom orbitals).

To obtain the energy ε we proceed as in Foundation 4.2. Then ε is calculated for a given value of d and for different values of η. The value of η corresponding to the minimum of ε is selected. This is shown in Figure 4.8

Continued on page 95

4.1 H_2^+ MOLECULE ION

Continued from page 94

for the experimental equilibrium distance $d = 106\,\text{pm}$. In Figure 4.8(a) the energy ε, and in Figure 4.8(b) the ratio $-\varepsilon/\overline{T}$ are plotted versus η; at the energy minimum ($\eta = 1.24$) we find $-\varepsilon/\overline{T} = 1$, in accordance with the virial theorem.

The value of η at distance d corresponding to the minimum of ε extends from $\eta = 1$ for large d (separated atomic orbitals of the H atom) to $\eta = 2$ for small d (this is the limiting case for $d = 0$: contraction of the electron cloud as in He^+). η increases with decreasing d; this indicates that the electron cloud becomes more and more compressed perpendicular to the bond line (Figure 4.9(a)).

Finally, using these η values the energies ε, $\overline{V}$, and $\overline{T}$ are obtained (Figure 4.9(b)). These energy curves are very close to the result of the exact calculation (see Figure 4.7, dashed lines). Particularly, the energy minimum at $d = 106\,\text{pm}$ is identical with the experimental value, and the kinetic energy increases with decreasing distance d, contrary to the prediction of the simple LCAO model.

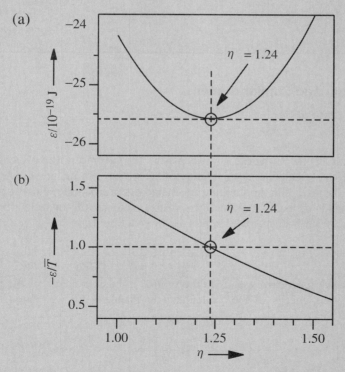

Figure 4.8 H_2^+ molecule ion, linear combination of compressed atomic orbitals. Determination of parameter η using the variation principle for $d = 106\,\text{pm}$ (experimental equilibrium distance). (a) Energy ε versus parameter η. (b) Ratio $-\varepsilon/\overline{T}$ versus η. At $\eta = 1.24$ the virial theorem is fulfilled.

Continued on page 96

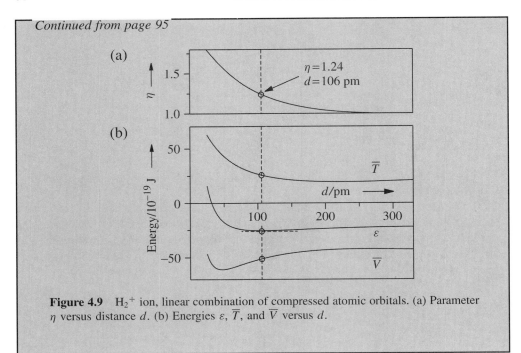

Figure 4.9 H_2^+ ion, linear combination of compressed atomic orbitals. (a) Parameter η versus distance d. (b) Energies ε, $\overline{T}$, and $\overline{V}$ versus d.

4.2 He Atom and Similar Systems

4.2.1 Ground State of He

Energy

As outlined in Chapter 2, the He atom consists of a twofold positively charged nucleus and two electrons. In calculating the energy of He we have to include the repulsion of the two electrons. Using a simple quantum mechanical perturbation method (which includes the electron repulsion but disregards the evasion effect of the two electrons) we obtain for the energy of He in the ground state (see Foundation 4.3)

$$\varepsilon_{He} = -\frac{1}{4\pi\varepsilon_0} \cdot \frac{4e^2}{a_0} + \frac{1}{4\pi\varepsilon_0} \cdot \frac{5}{4}\frac{e^2}{a_0} = -\frac{1}{4\pi\varepsilon_0} \cdot \frac{11}{4} \cdot \frac{e^2}{a_0} \qquad (4.21)$$

where the positive term corresponds to the mutual repulsion energy of the two electrons. By subtracting the energy of He^+ calculated in Problem 3.6 we obtain the ionization energy E_{ion} of He

$$E_{ion} = E_{He^+} - E_{He} = \frac{1}{4\pi\varepsilon_0} \cdot \left(\frac{11}{4} \cdot \frac{e^2}{a_0} - 2\frac{e^2}{a_0}\right) = \frac{1}{4\pi\varepsilon_0} \cdot \frac{3}{4} \cdot \frac{e^2}{a_0} = 32.7 \times 10^{-19} \text{ J} \qquad (4.22)$$

Using more sophisticated methods in finding a good trial function (taking into account that the electron cloud in He is larger than that in He^+ and that small distances between electrons 1 and 2 are avoided because of the mutual repulsion: evasion effect) the result is

$$E_{ion} = 39.39 \times 10^{-19} \text{ J}$$

(Figure 4.10). This value agrees well with the experiment (39.391×10^{-19} J).

4.2 HE ATOM AND SIMILAR SYSTEMS

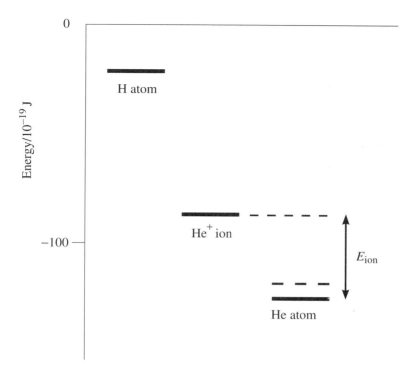

Figure 4.10 Energies of the H atom, the He$^+$ ion, and the He atom (thick lines) and ionization energy E_{ion} of He. The dashed line for the He atom corresponds to the result (equation (4.21)) of the simple perturbation theory. The ionization energy of He$^+$ is four times the ionization energy of H, that is $4 \times 21.8 \times 10^{-19}$ J $= 87.2 \times 10^{-19}$ J.

In the following we will neglect the evasion effect completely. We attribute each electron its own wavefunction (*one electron approximation*); these one-electron wavefunctions are called *orbitals* (*atomic* or *molecular orbitals*).

Two-electron Wavefunction

In the exact description both electrons in the system are indistinguishable and it is impossible to attribute separate wavefunctions to individual electrons. This is only possible for the system of both electrons 1 and 2 as a whole:

$$\psi(x_1, y_1, z_1, x_2, y_2, z_2) \tag{4.23}$$

If we neglect the evasion effect of the electrons, we can approximate ψ by a trial function ϕ which is the product of two wavefunctions depending on one-electron coordinates each:

$$\phi_1 = \varphi_{1s}(x_1, y_1, z_1) \cdot \varphi_{1s}(x_2, y_2, z_2) \tag{4.24}$$

The two electrons cannot be distinguished, this means that the function ψ^2 (which is proportional to the probability ρ) does not change if the coordinates of electrons 1 and 2 are exchanged. It is evident that this symmetry condition is fulfilled by the trial function ϕ_1.

4.2.2 Excited States of He

We have considered the indistinguishability of electrons in a very simple case (two electrons occupying the same orbital). How can we account for the fact that electrons are indistinguishable in cases where two or more orbitals are involved? This question is of general importance for understanding multi-electron systems. We consider the excited state of He as the simplest example to study this question.

Similar to an H atom, a He atom can be excited in a gas discharge. An electron can be transferred from the ground state to the lowest excited state approximated by a 2s wavefunction. We use as a trial function

$$\phi_2(1, 2) = \varphi_{1s}(1) \cdot \varphi_{2s}(2) \tag{4.25}$$

The symbols 1 and 2 are used for the coordinates x_1, y_1, z_1 and x_2, y_2, z_2, respectively. This function fulfills the Schrödinger equation if the repulsion term is neglected (see treatment of the ground state in Foundation 4.3). Apparently this approximation does not fulfill the condition that the two electrons are indistinguishable; if we exchange coordinates 1 and 2 we obtain a completely new function:

$$\phi_2(2, 1) = \varphi_{1s}(2) \cdot \varphi_{2s}(1) \tag{4.26}$$

To fulfill the indistinguishability condition we use the linear combinations

$$\phi_{2,\text{sym}} = A \cdot \left[\varphi_{1s}(1) \cdot \varphi_{2s}(2) + \varphi_{1s}(2) \cdot \varphi_{2s}(1)\right] \tag{4.27}$$

$$\phi_{2,\text{anti}} = A \cdot \left[\varphi_{1s}(1) \cdot \varphi_{2s}(2) - \varphi_{1s}(2) \cdot \varphi_{2s}(1)\right] \tag{4.28}$$

where A is a constant. In the first case the function remains unchanged when the two electrons are exchanged (*symmetric function*), and in the second case the sign of the function changes (*antisymmetric function*). In both cases the probability density $\rho = \phi^2$ remains unchanged. Regarding the normalization condition, we find for the constant $A = 1/\sqrt{2}$. In the first case the Coulomb repulsion of the electrons is stronger than in the second case: according to equation (4.28) the wavefunction $\phi_{2,\text{anti}}$ is zero when the coordinates of electrons 1 and 2 are equal, i.e. when both electrons are close together, in contrast to the wavefunction $\phi_{2,\text{sym}}$ described by equation (4.27); therefore the two states differ in energy:

$$E_{2,\text{sym}} - E_1 = 32.92 \times 10^{-19}\,\text{J} = 20.55\,\text{eV}$$

$$E_{2,\text{anti}} - E_1 = 31.67 \times 10^{-19}\,\text{J} = 19.77\,\text{eV}$$

The excitation energy from the ground state E_1 into the state with symmetric wavefunction (singlet state, see Section 4.3.1) is slightly higher than the energy for exciting to the state with antisymmetric wavefunction (triplet state, see Section 4.3.1).

These theoretical results are justified in spectroscopic experiments as well as in the Franck–Hertz experiment. In the latter case (tube filled with He instead of Na vapor, see Figure 2.2) minima of the current are observed which are 19.8 and 20.6 eV apart, corresponding to the excitation of He from its ground state (energy E_1) to the antisymmetric (4.28) and symmetric (4.27) state, respectively.

In the spectroscopic experiment (light emission from a gas discharge of He) many spectral lines appear corresponding to transitions from higher excited states into the states (4.27) and (4.28).

4.3 Antisymmetry of Wavefunctions

4.3.1 Energy Splitting in a Magnetic Field

If an H atom is placed in a magnetic field, a splitting of the energy level as displayed in Figure 4.11 is found. We cannot understand this on the basis of our ideas up to this point. This difficulty is connected with the fact that our description is not yet complete, although it is sufficient for a fundamental understanding of the aspects given in this introduction.

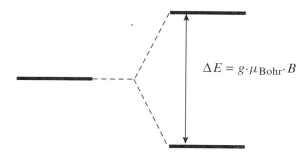

Figure 4.11 Splitting of the energy level of the H atom in a magnetic field.

To understand the magnetic properties, we limit ourselves to a correspondence consideration. The electron behaves as if it were a charged spinning top. Accordingly it produces a magnetic field, and it behaves like a small magnet itself (Box 4.2). In electron pairs the spins are opposite, so the magnetic effects of the two electrons cancel each other. In a single unpaired electron the top can be considered to behave like a magnet with magnetic moment

$$\boxed{\mu_{\text{Bohr}} = \frac{e^2 \hbar}{2m_e} = 9.274 \times 10^{-24}\,\text{J}\,\text{T}^{-1}} \quad (4.29)$$

Box 4.2 – Current Loop in a Magnetic Field (Bohr Magneton)

Consider a current I flowing through a current loop (Figure 4.12) when applying a magnetic field of flux density B. A force f acts on a section Δy of the wire in the direction perpendicular to the direction of the current I and perpendicular to the direction of the magnetic flux density B

$$f = IB \cdot \Delta y$$

Continued on page 100

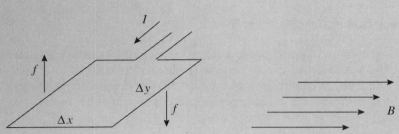

Figure 4.12 Current loop to calculate the magnetic moment μ.

Thus, the moment acting on the loop is

$$M = 2f \cdot \frac{\Delta x}{2} = 2IB \cdot \Delta y \cdot \frac{\Delta x}{2} = IBA = \mu B$$

where $A = \Delta x \cdot \Delta y$ is the area enclosed by the loop.

$\mu = IA$ is called the *magnetic moment*. It is directed perpendicular to the plane of the loop. The energy ΔE needed to rotate μ by $180°$ (e.g., from being directed parallel to B to being directed in the opposite direction) is

$$\Delta E = 2f \cdot \Delta x = 2IB \cdot \Delta x \cdot \Delta y = 2\mu B.$$

In the Bohr model the electron moves with a speed

$$v = \frac{1}{4\pi\varepsilon_0} \cdot \frac{e^2}{\hbar}$$

on a circle of radius

$$a_0 = 4\pi\varepsilon_0 \frac{\hbar^2}{m_e e^2}$$

(see Chapter 3). Then its frequency is

$$\nu = \frac{v}{2\pi a_0}$$

and with $A = \pi a_0^2$ we obtain for the corresponding magnetic moment (Bohr magneton)

$$\mu_{\text{Bohr}} = I \cdot A = \nu e \cdot \pi a_0^2 = \frac{1}{2} e \cdot v a_0 = \frac{e\hbar}{2m_e}$$

directed parallel or antiparallel to the magnetic field (*Bohr magneton*). (1 T = 1 tesla is the unit of magnetic flux density.) In a magnetic field each energy level splits into two levels. The energy difference in a magnetic field of flux density B is (see Box 4.2)

$$\boxed{\Delta E = g \cdot \mu_{\text{Bohr}} \cdot B} \qquad (4.30)$$

4.3 ANTISYMMETRY OF WAVEFUNCTIONS

with $g = 2$. This proposal is not completely correct. The experiment shows that this splitting indeed occurs, but with $g = 2.0023$ for a free electron.

In a magnetic field with $B = 1\,\text{T}$ the energy is $\Delta E = 2.0023 \times 9.274 \times 10^{-24}\,\text{J}\,\text{T}^{-1} \times 1\,\text{T} = 18.54 \times 10^{-24}\,\text{J}$. (For comparison, the magnetic flux density of the magnetic field of the earth is $B = 4 \times 10^{-5}\,\text{T}$; $B = 10\,\text{T}$ can be reached with the strongest magnets.) This means that the electron can be excited by irradiation with a frequency

$$\nu = \frac{\Delta E}{h} = 2.8 \times 10^{10}\,\text{s}^{-1} = 28\,\text{GHz}$$

corresponding to microwave radiation. This effect is the basis of the *electron paramagnetic resonance* (EPR) method. This method is useful in chemistry to study free radicals or other molecular species with unpaired electrons: organic radicals (Figure 4.13), transition metal complexes, triplet states (see Chapter 8).

For two unpaired electrons (e.g., in the excited state of a He atom, see Section 4.2) the electron spins can compensate (singlet excited state: spins antiparallel) or add (triplet excited state: spins parallel). In the latter case the axis of the top can be parallel or antiparallel to the applied field, or it can precess in the plane perpendicular to the field. The energy level then splits into three levels in a magnetic field (thus the name *"triplet"* as opposed to *"singlet"*). In systems with more than two electrons, an excited state can split into more levels (multiplet splitting). The distinct orientations of the axis of the top is a quantum feature similar to the distinct energies of an electron confined in space, e.g. an electron in a box.

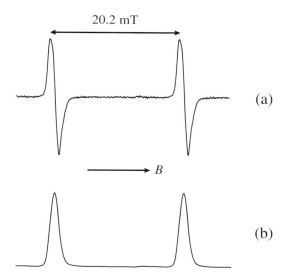

Figure 4.13 EPR spectrum of the malonyl radical HOOC–ĊH–COOH (absorbed microwave power versus magnetic flux density B). In this case we expect a doublet because of coupling of the magnetic moment of the electron with the nuclear magnetic moment of the proton next to the excited electron. The resonance frequency of the spectrometer is $\nu = 9.63\,\text{GHz}$, and the magnetic flux density at the center of the spectrum is $B = 0.34\,\text{T}$. (a) First derivative of the signal as usually measured in the EPR spectrometer; therefore for each transition we obtain a maximum and a minimum of the signal. (b) After integration of the signal the two absorption bands occur.

4.3.2 Spin Variables

In quantum mechanics the behavior of the electron spin is described symbolically by introducing a spin variable ζ, which gives the orientation of the spin of the electron in the magnetic field. The two orientations of the spin of a single unpaired electron in a magnetic field are described by the spin wavefunctions $\alpha(\zeta)$ and $\beta(\zeta)$. For example, the electron in the H atom placed in a magnetic field has the wavefunctions

$$\varphi_{1s}(x, y, z) \cdot \alpha(\zeta) \quad \text{or} \quad \varphi_{1s}(x, y, z) \cdot \beta(\zeta) \tag{4.31}$$

For an He atom in the ground state, the wavefunction for antiparallel spins is written as

$$\varphi_{1s}(x_1, y_1, z_1) \cdot \varphi_{1s}(x_2, y_2, z_2) \cdot \alpha(\zeta_1)\beta(\zeta_2)$$

or

$$\varphi_{1s}(x_1, y_1, z_1) \cdot \varphi_{1s}(x_2, y_2, z_2) \cdot \alpha(\zeta_2)\beta(\zeta_1)$$

where ζ_1 and ζ_2 denote the spin variables of electrons 1 and 2, respectively. Usually these wavefunctions are written in the abbreviated form

$$\varphi_{1s}(1)\varphi_{1s}(2) \cdot \alpha(1)\beta(2) \quad \text{or} \quad \varphi_{1s}(1)\varphi_{1s}(2) \cdot \alpha(2)\beta(1) \tag{4.32}$$

These expressions are not sufficient for the condition of indistinguishability of the two electrons. This condition is fulfilled, however, for the linear combinations

$$\varphi_{1s}(1)\varphi_{1s}(2) \cdot \{\alpha(1)\beta(2) + \alpha(2)\beta(1)\} \tag{4.33}$$

$$\varphi_{1s}(1)\varphi_{1s}(2) \cdot \{\alpha(1)\beta(2) - \alpha(2)\beta(1)\} \tag{4.34}$$

4.3.3 Pauli Exclusion Principle as Antisymmetry Postulate

The Pauli exclusion principle makes a further limitation. It states that the only wavefunctions allowed are those in which the exchange of the coordinates of two electrons changes the sign of the wavefunction. This is equivalent to the statement

> **Postulate.** *Pauli exclusion principle: Wavefunctions must be antisymmetric with respect to the exchange of electrons.*

Thus combination (4.34) is allowed, while combination (4.33) is forbidden. In the case considered in Section 4.2 for the excited states of the He atom, the combination

$$\{\varphi_{1s}(1)\varphi_{2s}(2) + \varphi_{1s}(2)\varphi_{2s}(1)\} \cdot \{\alpha(1)\beta(2) - \alpha(2)\beta(1)\} \tag{4.35}$$

satisfies the conditions for indistinguishable particles and the Pauli exclusion principle (symmetric spatial wavefunction, antisymmetric spin wavefunction). The same is true for the combinations

$$\{\varphi_{1s}(1)\varphi_{2s}(2) - \varphi_{1s}(2)\varphi_{2s}(1)\} \cdot \{\alpha(1)\alpha(2)\}$$

$$\{\varphi_{1s}(1)\varphi_{2s}(2) - \varphi_{1s}(2)\varphi_{2s}(1)\} \cdot \{\beta(1)\beta(2)\} \tag{4.36}$$

$$\{\varphi_{1s}(1)\varphi_{2s}(2) - \varphi_{1s}(2)\varphi_{2s}(1)\} \cdot \{\alpha(1)\beta(2) + \alpha(2)\beta(1)\}$$

(antisymmetric spatial function, symmetric spin function). For simplicity, wavefunctions (4.33–4.36) are not normalized.

4.3 ANTISYMMETRY OF WAVEFUNCTIONS

Combination (4.35) corresponds to the singlet excited state (which does not split in a magnetic field), while the combinations (4.36) correspond to the triplet excited state (which splits into three states in a magnetic field).

The antisymmetry postulate holds for other elementary particles besides electrons. It is certainly a very strange requirement in the context of our daily experience. However, experimental facts force us to accept this strangeness. Otherwise, the important interplay between molecules cannot be understood in many respects.

4.3.4 Singlet–Triplet Splitting Caused by Coulomb Forces

The energy difference between the singlet and triplet excited states is practically exclusively caused by the different Coulomb interaction effects (see Section 4.2), and only very slightly by the magnetic interaction (see Box 4.3) between the two electrons (antiparallel spins in the first case, parallel spins in the second case).

Box 4.3 – Magnetic Interactions of Electrons

We consider two magnets of magnetic moment μ_{Bohr} at a distance of a_0 (Bohr radius). The magnets are placed parallel to each other and in the direction of their bond line. One magnet is then rotated to the antiparallel configuration (180°). The energy expended in this operation is $2\mu_{Bohr} \cdot B$ according to equation (4.30). A magnetic dipole produces a magnetic field with flux density

$$B = \frac{2}{a_0^3} \cdot \frac{\mu_{Bohr}}{4\pi\varepsilon_0 c_0^2} \qquad (4.37)$$

along its axis at a distance a_0 (point dipole approximation). The energy is thus

$$E_{magn} = \frac{4}{a_0^3} \cdot \frac{\mu_{Bohr}^2}{4\pi\varepsilon_0 c_0^2} \qquad (4.38)$$

and we obtain (with $\varepsilon_0 c_0^2 = 1/\mu_0$)

$$E_{magn} = \frac{4\mu_{Bohr}^2 \mu_0}{4\pi a_0^3} = 1.1 \times 10^{-22} \, J$$

where μ_0 is the permeability of vacuum ($\mu_0 = 1.257 \times 10^{-6} \, C^{-2} \, J \, s^2 \, m^{-1}$). The energy of 10^{-22} J is negligibly small compared to the energy difference between the singlet and triplet excited states (e.g., 0.78 eV = 1.25×10^{-19} J in the case of helium). On the same grounds, magnetic forces play no role in comparison to Coulomb forces in the formation of the chemical bond.

Note that an expression corresponding to (4.38) results in the case of an electric dipole (see Chapter 10).

4.4 Quantum Mechanical Tunneling

We consider a hydrogen atom (nucleus a, electronic wavefunction φ_a) and a proton b at a distance d ($d \gg a_0$, $a_0 =$ Bohr radius) at time $t = 0$. What is the probability of finding the electron at time t at atom b (wavefunction φ_b)? We demonstrate in Foundation 4.4 that we can describe the electron as oscillating between the nuclei a and b; that is, it effectively tunnels back and forth through an energy barrier of 13.6 eV (ionization energy of the H atom).

The period of oscillation depends strongly on the distance between the two nuclei: it is 1 ns for $d = 1$ nm and 1 s for $d = 2$ nm.

Tunneling is in contrast to what is possible for a particle that follows the laws of classical physics. In classical physics the particle surmounts the barrier when it has sufficient energy, but it cannot overcome the barrier when its energy is smaller than that required for surmounting. Electron tunneling is detected in examples with fixed potential troughs. Its principle is seen in the present case of an electron in the potential trough of a proton when a second proton is close by.

The tunneling effect is the basis of understanding important electron transfer reactions, such as the primary process in photosynthesis and the process on which scanning tunneling microscopy (see Chapter 23) is based. In these cases, in contrast to the example discussed above, the electron after tunneling through a potential barrier, is shifted further instead of oscillating back and forth.

Problems

Problem 4.1 – Kinetic Energy in Box Model

The mean kinetic energy of the H_2^+ molecular ion in the box model was calculated as $T = 16.9 \times 10^{-19}$ J. Calculate $\overline{T}$ for a hypothetical H_2^+ ion with the same box length L_x, but with box length $L_y = L_z = 376$ pm (box length for the H atom, see Section 4.2.1).

Solution. From Table 4.1 we obtain for the energy minimum $L_x = 1.36 \times 301$ pm $= 409$ pm. Then in analogy to (4.5) we obtain

$$\overline{T} = \frac{h^2}{8m_e}\left(\frac{1}{L_x^2} + \frac{1}{L_y^2} + \frac{1}{L_z^2}\right) = 12.1 \times 10^{-19}\,\text{J}$$

This is less than the mean kinetic energy of the electron in the H atom ($\overline{T} = 12.9 \times 10^{-19}$ J). When forming the bond, the resulting box lengths $L_y = L_z = 301$ pm are smaller than in the H atom ($L_x = L_y = L_z = 376$ pm), and, as a consequence, the kinetic energy rises to $\overline{T} = 16.9 \times 10^{-19}$ J, higher than the kinetic energy of the electron in the H atom.

Note that in the LCAO model of H_2^+ the kinetic energy of the molecule in its equilibrium distance is smaller than that of the H atom (Figure 4.7), because the extension of the electron cloud perpendicularly to the bond line remains unchanged.

Problem 4.2 – Repulsion of the Two Electrons in He

In Section 4.2.1 the mutual repulsion energy of the two electrons in He was roughly calculated as

$$\varepsilon_{\text{repulsion}} = \frac{1}{4\pi\varepsilon_0}\frac{5}{4}\frac{e^2}{a_0}$$

Estimate the average distance $\bar{r}$ between the two electrons.

Solution. The repulsion energy of two negative charges e is

$$\varepsilon_{\text{repulsion}} = \frac{1}{4\pi\varepsilon_0} \frac{e^2}{\bar{r}}$$

From comparison of both equations it follows that

$$\bar{r} = \tfrac{4}{5}a_0 = 42\,\text{pm}$$

Problem 4.3 – Energy Separation of EPR Lines

In Figure 4.13 the two EPR lines of the malonyl radical are apart by 20.2 mT. What is the energy difference between the corresponding transitions? How do you explain the splitting, considering the fact that the magnetic moment of a proton is $\mu_N = \frac{1}{2000} \cdot \mu_{\text{Bohr}}$ as outlined in Section 9.8.12?

Solution. According to (equation 4.30) the energy difference is

$$\Delta E_2 - \Delta E_1 = g \cdot \mu_{\text{Bohr}} \cdot (B_2 - B_1)$$
$$= 2.0023 \cdot 9.274 \times 10^{-24} \cdot 20.2 \times 10^{-3}\,\text{J} = 3.75 \times 10^{-25}\,\text{J}$$

The shift in energy of the electron by the next proton, according to Box 4.3, is on the order of

$$\pm \frac{4\mu_{\text{Bohr}}\mu_N\mu_0}{4\pi a_0^3} = \pm \frac{1}{2000} \times 1.1 \times 10^{-22}\,\text{J} = \pm 0.55 \times 10^{-25}\,\text{J}.$$

Thus the splitting of the EPR signal (twice the shift) should be on the order of 10^{-25} J, in agreement with the experiment.

Foundation 4.1 – H_2^+ Ion: Exact Wavefunction and Energy; Virial Theorem

The Schrödinger equation for the H_2^+ ion is:

$$-\frac{h^2}{8\pi^2 m_e}\left(\frac{\partial^2 \psi}{\partial x^2}+\frac{\partial^2 \psi}{\partial y^2}+\frac{\partial^2 \psi}{\partial z^2}\right)+\frac{1}{4\pi\varepsilon_0}\left(-\frac{e^2}{r_1}-\frac{e^2}{r_2}+\frac{e^2}{d}\right)\psi = E\psi$$

where r_1 and r_2 are the distances of the electron from the nuclei 1 and 2 (Figure 4.1). For the separation of this differential equation we use elliptic coordinates φ (angle between xz plane and plane of r_1 and r_2), λ, and μ. For λ and μ we have

$$\mu = \frac{r_1 - r_2}{d}, \quad \lambda = \frac{r_1 + r_2}{d}$$

μ can assume values from -1 to $+1$ and λ can assume values from 1 to ∞. We can easily see the meaning of μ and λ, if we discuss points on the bond line. Between both nuclei there is $r_1 + r_2 = d$, thus $\lambda = 1$; to the left of nucleus 1 we have $r_1 - r_2 = -d$, thus $\mu = -1$; to the right of nucleus 2 we have $\mu = +1$. In the center of the bond, $r_1 = r_2 = d/2$; thus $\mu = 0$. λ increases with increasing distance from the nuclei.

Energy and wavefunction

In the following we consider the ground state of the H_2^+ ion. In this case ψ has rotation symmetry according to the bond line, and it is sufficient to calculate values of the wavefunction in the xz plane. We solve the differential equation with the wavefunction

$$\psi(\mu, \lambda) = M(\mu) \cdot \Lambda(\lambda)$$

M and Λ are solutions of the ordinary differential equations

$$\frac{d^2 M}{d\mu^2} + \frac{1}{\mu^2 - 1}\left[2\mu\frac{dM}{d\mu} + (A - \gamma\mu^2)M\right] = 0$$

$$\frac{d^2 \Lambda}{d\lambda^2} + \frac{1}{\lambda^2 - 1}\left[2\lambda\frac{d\Lambda}{d\lambda} + \left(A + 2\frac{d}{a_0}\lambda - \gamma\lambda^2\right)\Lambda\right] = 0$$

a_0 is the Bohr radius, and A and γ are parameters which are determined by the boundary condition $\psi = 0$ for $\lambda \to \infty$ and by the condition that M and Λ must remain finite at the points $\mu^2 = 1$ and $\lambda^2 = 1$. γ is related to E according to

$$E = \frac{1}{4\pi\varepsilon_0}\left(\frac{e^2}{d} - \frac{2\gamma a_0 e^2}{d^2}\right) = \frac{1}{4\pi\varepsilon_0}\frac{e^2}{d}\left(1 - \frac{2\gamma a_0}{d}\right)$$

Continued on page 107

Continued from page 106

Both ordinary differential equations are solved numerically. For example, for $d = 2a_0 = 106\,\text{pm}$ we obtain $A = 0.8109$ and $\gamma = 2.203$. Then

$$E = \frac{1}{4\pi\varepsilon_0} \cdot \frac{e^2}{2a_0} \cdot (1 - 2.203) = -26.20 \times 10^{-19}\,\text{J}$$

The corresponding wavefunction is displayed in Figure 4.2(b). In Figure 4.14 the energy E is displayed as a function of the nuclear distance d together with the average kinetic energy $\overline{T}$ and the average potential energy $\overline{V}$. At $d = 106\,\text{pm}$ there is a minimum of the energy E.

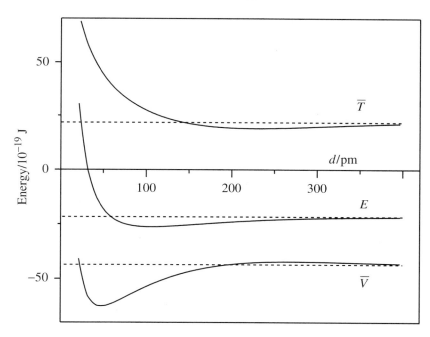

Figure 4.14 Energy E, average kinetic energy $\overline{T}$, and average potential energy $\overline{V}$ of the H_2^+ ion as a function of the bond distance d. The dashed lines correspond to the energies for an H atom and a proton at $d \to \infty$. (Calculated from: D. R. Bates, K. Ledsham, and A. L. Stewart, Wavefunctions of the hydrogen molecule ion, *Phil. Trans. R. Soc. London Ser.*, A, 1953, **246**, 215).

Virial theorem for molecules

From Figure 4.14 we find for the H_2^+ molecule ion in the equilibrium distance $d = 106\,\text{pm}$ the following: $E = -26.20 \times 10^{-19}\,\text{J}$, $\overline{V} = -52.0 \times 10^{-19}\,\text{J}$, and $\overline{T} = E - \overline{V} = 26.20 \times 10^{-19}\,\text{J}$. Apparently, we have

$$\overline{T} = -E = -\tfrac{1}{2}\overline{V}$$

Continued on page 108

Continued from page 107

as expected from the virial theorem discussed in Chapter 3. However, it must be pointed out that this simple relation is restricted to the equilibrium distance (in this case, $d = 106\,\text{pm}$). It is not valid for non-equilibrium distances, as can be seen in Figure 4.14.

Foundation 4.2 – Evaluation of LCAO Integrals in H_2^+

According to equation (4.20), the energy ε in the LCAO treatment of H_2^+ is

$$\varepsilon = \overline{T} + \overline{V} = \frac{H_{aa} + H_{ab}}{1 + S_{ab}}$$

The corresponding expressions for the mean kinetic energy $\overline{T}$ and the mean potential energy $\overline{V}$ are

$$\overline{T} = \frac{T_{aa} + T_{ab}}{1 + S_{ab}}, \quad \overline{V} = \frac{V_{aa} + V_{ab}}{1 + S_{ab}}$$

with

$$T_{aa} = \int \varphi_a \mathcal{T} \varphi_a \cdot d\tau, \quad T_{ab} = \int \varphi_a \mathcal{T} \varphi_b \cdot d\tau$$

and

$$V_{aa} = \int \varphi_a \mathcal{V} \varphi_a \cdot d\tau, \quad V_{ab} = \int \varphi_a \mathcal{V} \varphi_b \cdot d\tau$$

$\mathcal{T}$ and $\mathcal{V}$ are the operators of the kinetic and the potential energy, respectively:

$$\mathcal{T} = -\frac{h^2}{8\pi^2 m_e}\left(\frac{\partial^2}{\partial x^2} + \frac{\partial^2}{\partial y^2} + \frac{\partial^2}{\partial z^2}\right), \quad \mathcal{V} = \frac{e^2}{4\pi\varepsilon_0}\cdot\left(-\frac{1}{r_a} - \frac{1}{r_b} + \frac{1}{d}\right)$$

The first term in $\mathcal{V}$ corresponds to the attraction of the electron by nucleus a ($-1/r_a$), the second one corresponds to the attraction of the electron by nucleus b ($-1/r_b$), and the third term corresponds to the repulsion of the two nuclei ($+1/d$). The wavefunctions φ_a and φ_b are chosen as

$$\varphi_a = \sqrt{\frac{\eta^3}{\pi a_0^3}} \cdot e^{-\eta r_a/a_0}, \quad \varphi_b = \sqrt{\frac{\eta^3}{\pi a_0^3}} \cdot e^{-\eta r_b/a_0}$$

Continued on page 109

FOUNDATION 4.2

Continued from page 108

where $\eta = 1$ in the simple LCAO method; in the method of compressed atomic orbitals η is a parameter determined by the variation principle. As outlined in Box 4.1 η is in the range between 1 (H atom) and 2 (He$^+$ ion).

T_{aa} is simply the kinetic energy of an isolated atom with nuclear charge e:

$$T_{aa} = \frac{e^2}{4\pi\varepsilon_0} \cdot \frac{\eta^2}{2a_0}$$

$\eta = 1$ corresponds to the mean kinetic energy of the electron in an H atom, and $\eta = 2$ corresponds to the mean kinetic energy of the electron in He$^+$.

V_{aa} can be written as

$$V_{aa} = -\frac{e^2}{4\pi\varepsilon_0} \cdot \int \varphi_a \frac{1}{r_a} \varphi_a \cdot d\tau - \frac{e^2}{4\pi\varepsilon_0} \cdot \int \varphi_a \frac{1}{r_b} \varphi_a \cdot d\tau + \frac{e^2}{4\pi\varepsilon_0} \cdot \int \varphi_a \frac{1}{d} \varphi_a \cdot d\tau$$

The first term corresponds to the mean potential energy of an isolated atom with nuclear charge e:

$$-\frac{e^2}{4\pi\varepsilon_0} \cdot \int \varphi_a \frac{1}{r_a} \varphi_a \cdot d\tau = -\frac{e^2}{4\pi\varepsilon_0} \cdot \frac{\eta}{a_0}$$

$\eta = 1$ corresponds to the mean potential energy of the electron in an H atom and $\eta = 2$ corresponds to the mean potential energy of an electron distributed as in a He$^+$ ion, but attracted by nuclear charge $+e$ only.

The evaluation of the second term (Coulomb energy of nucleus b in the field of electron in wavefunction φ_a) is more involved, see below.

The third term in V_{aa} can be simplified as

$$\frac{e^2}{4\pi\varepsilon_0} \cdot \int \varphi_a \frac{1}{d} \varphi_a \cdot d\tau = \frac{e^2}{4\pi\varepsilon_0} \cdot \frac{1}{d} \cdot \int \varphi_a \varphi_a \cdot d\tau = \frac{e^2}{4\pi\varepsilon_0} \cdot \frac{1}{d}$$

For V_{ab} we obtain

$$V_{ab} = -\frac{e^2}{4\pi\varepsilon_0} \cdot \int \varphi_a \frac{1}{r_a} \varphi_b \cdot d\tau - \frac{e^2}{4\pi\varepsilon_0} \cdot \int \varphi_a \frac{1}{r_b} \varphi_b \cdot d\tau + \frac{e^2}{4\pi\varepsilon_0} \cdot \int \varphi_a \frac{1}{d} \varphi_b \cdot d\tau$$

The last term can be simplified as

$$\frac{e^2}{4\pi\varepsilon_0} \cdot \int \varphi_a \frac{1}{d} \varphi_b \cdot d\tau = \frac{e^2}{4\pi\varepsilon_0} \cdot \frac{1}{d} \cdot \int \varphi_a \varphi_b \cdot d\tau = \frac{e^2}{4\pi\varepsilon_0} \cdot \frac{1}{d} \cdot S_{ab}$$

where S_{ab} is the overlap integral.

The remaining integrals have been summarized by Roothaan as a function of d; taking his results we obtain:

$$S_{ab} = \left(1 + \eta \frac{d}{a_0} + \frac{1}{3}\eta^2 \frac{d^2}{a_0^2}\right) \cdot e^{-\eta d/a_0}$$

Continued on page 110

Continued from page 109

$$T_{ab} = \frac{e^2}{4\pi\varepsilon_0} \cdot \frac{\eta^2}{2a_0} \cdot \left(2\sqrt{2}S'_{ab} - S_{ab}\right), \text{ where } S'_{ab} = \frac{1}{\sqrt{2}}\left(1 + \eta\frac{d}{a_0}\right) \cdot e^{-\eta d/a_0}$$

$$V_{aa} = -\frac{e^2}{4\pi\varepsilon_0} \cdot \left\{+\frac{\eta}{a_0} + \frac{1}{d}\left[1 - \left(1 + \eta\frac{d}{a_0}\right) \cdot e^{-2\eta d/a_0}\right]\right\} + \frac{e^2}{4\pi\varepsilon_0} \cdot \frac{1}{d}$$

$$V_{ab} = -\frac{e^2}{4\pi\varepsilon_0} \cdot \frac{2\sqrt{2}\cdot\eta\cdot S'_{ab}}{a_0} + \frac{e^2}{4\pi\varepsilon_0} \cdot \frac{1}{d} \cdot S_{ab}$$

(See C. C. J. Roothaan, A study of Two-center integrals useful in calculations on molecular structure, I, *J. Chem. Phys.*, 1951, **19**, 1445.)

Special case d = 0

In this case we obtain $S_{ab} = 1$ and $S'_{ab} = 1/\sqrt{2}$. Then it follows immediately that

$$T_{ab} = T_{aa} = \frac{e^2}{4\pi\varepsilon_0} \cdot \frac{\eta^2}{2a_0}$$

and

$$\overline{T} = \frac{e^2}{4\pi\varepsilon_0} \cdot \frac{\eta^2}{2a_0}$$

Correspondingly, for the electronic contributions to V_{aa} and V_{ab} we obtain

$$V_{aa} - \frac{e^2}{4\pi\varepsilon_0} \cdot \frac{1}{d} = -\frac{e^2}{4\pi\varepsilon_0} \cdot \left\{+\frac{\eta}{a_0} + \frac{\eta}{a_0}\right\} = -\frac{e^2}{4\pi\varepsilon_0} \cdot \frac{2\eta}{a_0}$$

$$V_{ab} - \frac{e^2}{4\pi\varepsilon_0} \cdot \frac{1}{d} \cdot S_{ab} = -\frac{e^2}{4\pi\varepsilon_0} \cdot \frac{2\eta}{a_0}$$

and

$$\overline{V} - \frac{e^2}{4\pi\varepsilon_0} \cdot \frac{1}{d} = -\frac{e^2}{4\pi\varepsilon_0} \cdot \frac{2\eta}{a_0}$$

For $\eta = 2$ the expressions for $\overline{T}$ and $\overline{V}$ can be understood as the mean potential energy of the electron in a He$^+$ ion (nuclear charge $+2e$).

Calculation of $\overline{T}$, $\overline{V}$, and $\varepsilon = \overline{T} + \overline{V}$ at the Energy Minimum ($\eta = 1.24$) in comparison with the simple LCAO ($\eta = 1$):

η	d_0 pm	S_{ab}	T_{aa}	T_{ab}	V_{aa}	V_{ab}	H_{aa}	H_{ab}	$\overline{T}$	$\overline{V}$	ε
							10^{-19} J				
1.24	106	0.462	33.5	4.0	−53.5	−21.4	−20.0	−17.4	25.6	−51.2	−25.6
1.00	132	0.460	21.8	2.6	−43.2	−17.1	−21.4	−14.5	16.7	−41.3	−24.6

Foundation 4.3 – He Atom: Energy in the Ground State

For a system with two electrons the coordinates of both electrons appear in the Hamiltonian operator:

$$\mathcal{H} = -\frac{h^2}{8\pi^2 m_e}\left[\left(\frac{\partial^2}{\partial x_1^2}+\frac{\partial^2}{\partial y_1^2}+\frac{\partial^2}{\partial z_1^2}\right) + \left(\frac{\partial^2}{\partial x_2^2}+\frac{\partial^2}{\partial y_2^2}+\frac{\partial^2}{\partial z_2^2}\right)\right]$$

$$+\frac{1}{4\pi\varepsilon_0}\cdot\left(-\frac{2e^2}{r_1}-\frac{2e^2}{r_2}+\frac{e^2}{r_{12}}\right)$$

$$= -\frac{h^2}{8\pi^2 m_e}\left(\nabla_1^2+\nabla_2^2\right) + V_1 + V_2 + \frac{1}{4\pi\varepsilon_0}\frac{e^2}{r_{12}}$$

Because of the term

$$\frac{1}{4\pi\varepsilon_0}\frac{e^2}{r_{12}}$$

which is caused by the repulsion of both electrons it is not possible to separate the Schrödinger equation into differential equations which depend on coordinate 1 or coordinate 2 only. For this reason we consider a simplified Schrödinger equation without the repulsion term:

$$\mathcal{H}^0\psi^0 = E^0\psi^0$$

$$\mathcal{H}^0 = -\frac{h^2}{8\pi^2 m_e}\left(\nabla_1^2+\nabla_2^2\right) + V_1 + V_2$$

This equation is solved by the wavefunction

$$\psi^0(r_1, r_2) = \varphi_{1s}(r_1)\cdot\varphi_{1s}(r_2) = \frac{8}{\pi a_0^3}e^{-2r_1/a_0}\cdot e^{-2r_2/a_0}$$

By inserting this wavefunction into the simplified Schrödinger equation we obtain the energy

$$E^0 = E_1^0 + E_2^0 = -\frac{1}{4\pi\varepsilon_0}\frac{4e^2}{2a_0} - \frac{1}{4\pi\varepsilon_0}\frac{4e^2}{2a_0} = -\frac{1}{4\pi\varepsilon_0}\frac{4e^2}{a_0}$$

To take account of the mutual repulsion of the two electrons we calculate the average repulsion energy $\overline{V}_e$ of these electrons. Since $\rho_1 e \cdot d\tau_1$ is the charge of electron 1 in the volume element $d\tau_1$ and $\rho_2 e \cdot d\tau_2$ is the charge of electron 2 in the volume element $d\tau_2$ we obtain for $\overline{V}_e$

$$\overline{V}_e = \frac{1}{4\pi\varepsilon_0}e^2\int\rho_1\frac{1}{r_{12}}\rho_2 d\tau_1\, d\tau_2$$

Continued on page 112

Continued from page 111

With
$$\rho_1 = \frac{8}{\pi a_0^2} e^{-4r_1/a_0} \quad \text{and} \quad \rho_2 = \frac{8}{\pi a_0^2} e^{-4r_2/a_0}$$

we obtain (see W. Kauzman, Quantum Chemistry, Academic, New York 1957; D.A. McQuarrie, J.D. Simon, Physical Chemistry, Univ.Science Books, Sausalito 1997, Problems 7–30, 7–31, page 270)

$$\overline{V}_e = \frac{1}{4\pi\varepsilon_0} \frac{5}{4} \frac{e^2}{a_0}$$

Thus the result for the energy of the He atom is

$$\varepsilon = E^0 + \overline{V}_e = \frac{1}{4\pi\varepsilon_0} \left(-4\frac{e^2}{a_0} + \frac{5}{4}\frac{e^2}{a_0} \right) = -\frac{1}{4\pi\varepsilon_0} \frac{11}{4} \frac{e^2}{a_0} = -119.9 \times 10^{-19} \, \text{J}$$

In this calculation the evasion effect of the electrons is neglected; this effect can be taken into account by using trial functions which depend on the mutual distance r_{12}.

In the Schrödinger equation both electrons are indistinguishable. Thus we cannot attribute separate wavefunctions to each of the electrons, but only one wavefunction to both electrons as a whole. The indistinguishability of both electrons means that the function ψ^2 remains unchanged when the electrons are exchanged. This symmetry condition is fulfilled in our example.

Foundation 4.4 – Oscillation of Electron between Protons at Distance d (Tunneling)

Calculating ε_1 and ε_2 in H_2^+ for large distance d

We consider a system of an electron and two protons (H_2^+ ion); we assume that the distance d of the protons is large compared to a_0 ($d \gg a_0$). Then as in equations (4.13) to (4.20) the wavefunction ϕ_1 and the energy ε_1 in the ground state are

$$\phi_1 = \frac{1}{\sqrt{2(1+S_{ab})}} \cdot (\varphi_a + \varphi_b) \approx \frac{1}{\sqrt{2}}(\varphi_a + \varphi_b)$$

$$\varepsilon_1 = \frac{H_{aa} + H_{ab}}{1 + S_{ab}} \approx (H_{aa} + H_{ab}) \cdot (1 - S_{ab})$$

if we treat S_{ab} as a small quantity ($S_{ab} \ll 1$).

Continued on page 113

FOUNDATION 4.4

Continued from page 112

Correspondingly, the wavefunction ϕ_2 in the first excited state is antisymmetric

$$\phi_2 = c_2 (\varphi_a - \varphi_b)$$

Following the derivation in equations (4.15) to (4.17) we obtain

$$c_2 = \frac{1}{\sqrt{2(1 - S_{ab})}}$$

Calculation of the energy $\varepsilon_2 = \int \phi_2 \mathcal{H} \phi_2 \cdot d\tau$ as in equations (4.18) to (4.20) results in

$$\varepsilon_2 = \frac{H_{aa} - H_{ab}}{1 - S_{ab}}$$

Then, again treating S_{ab} as a small quantity, we obtain

$$\phi_2 = \frac{1}{\sqrt{2}} (\varphi_a - \varphi_b)$$

and

$$\varepsilon_2 = (H_{aa} - H_{ab}) \cdot (1 + S_{ab})$$

Thus we find for the energy difference

$$\varepsilon_2 - \varepsilon_1 = 2(H_{aa} S_{ab} - H_{ab})$$

Calculating $\varepsilon_2 - \varepsilon_1$

The quantities H_{aa}, S_{ab}, and H_{ab} depend on the distance d. From the relations displayed in Foundation 4.2 we find the following equations for $d \gg a_0$

$$H_{aa} = T_{aa} + V_{aa} = -E_1 + 2E_1 = E_1$$

$$H_{ab} = T_{ab} + V_{ab} = -2E_1 \left(\frac{d}{a_0} e^{-d/a_0} - \frac{1}{2} S_{ab} \right) + 2E_1 \left(2\frac{d}{a_0} e^{-d/a_0} - \frac{a_0}{d} S_{ab} \right)$$

$$= 2E_1 \cdot \left[\left(\frac{1}{2} - \frac{a_0}{d} \right) \cdot S_{ab} + \frac{d}{a_0} \cdot e^{-d/a_0} \right]$$

$$S_{ab} = \frac{1}{3} \left(\frac{d}{a_0} \right)^2 \cdot e^{-d/a_0}$$

with

$$E_1 = -\frac{1}{4\pi\varepsilon_0} \cdot \frac{e^2}{2a_0}$$

where E_1 is the energy of an H atom in its ground state. Then we obtain

$$\varepsilon_2 - \varepsilon_1 = -E_1 \frac{8}{3} \cdot \frac{d}{a_0} e^{-d/a_0}$$

Continued on page 114

Continued from page 113

It is useful to replace e^2 in the above expression for E_1 by the Bohr radius a_0:

$$a_0 = \frac{4\pi\varepsilon_0 \hbar^2}{m_e e^2}, \quad e^2 = \frac{\varepsilon_0 h^2}{\pi m_e a_0}$$

Then we obtain

$$E_1 = -\frac{1}{4\pi\varepsilon_0} \cdot \frac{e^2}{2a_0} = -\frac{1}{4\pi\varepsilon_0} \cdot \frac{\varepsilon_0 h^2}{2\pi m_e a_0^2} = -\frac{h^2}{8\pi^2 m_e a_0^2}$$

and

$$\frac{1}{a_0} = \frac{\sqrt{-E_1 \cdot 8\pi^2 m_e}}{h^2} = \frac{2\pi}{h} \cdot \sqrt{-2E_1 m_e}$$

Thus for $\varepsilon_2 - \varepsilon_1$ we obtain

$$\varepsilon_2 - \varepsilon_1 = -E_1 \frac{8}{3} \cdot \frac{d}{a_o} \exp\left(-\frac{2\pi}{h} \cdot \sqrt{-2E_1 m_e} \cdot d\right)$$

$-E_1$ is the ionization energy of H and this is essentially equal to the energy barrier between the H atom and a proton at distance $d \gg a_0$. Thus

$$\varepsilon_2 - \varepsilon_1 = E_{\text{barrier}} \frac{8}{3} \cdot \frac{d}{a_0} \cdot \exp\left(-\frac{2\pi}{h} \cdot \sqrt{2E_{\text{barrier}} m_e} \cdot d\right)$$

Calculating the time-dependent probability density ρ

We assume that at time $t = 0$ the electron is at atom a. Then its wavefunction is

$$\Psi(t = 0) = \varphi_a = \frac{1}{\sqrt{2}}(\phi_1 + \phi_2)$$

Now we describe the behavior of the system at $t > 0$ by the time-dependent Schrödinger equation

$$\mathcal{H}\Psi = i\frac{h}{2\pi} \cdot \frac{\partial \Psi}{\partial t}$$

as discussed in Chapter 3. We start with the general solution (see Foundation 3.2)

$$\Psi = \sum_n c_n \psi_n \exp\left(-i\frac{E_n}{\hbar} \cdot t\right)$$

and approximate Ψ by the first two terms

$$\Psi = \frac{1}{\sqrt{2}}\left[\phi_1 \exp\left(-i\frac{\varepsilon_1}{\hbar}t\right) + \phi_2 \exp\left(-i\frac{\varepsilon_2}{\hbar}t\right)\right]$$

($c_1 = c_2 = 1/\sqrt{2}$ to meet the condition for $t = 0$). Then

$$\rho = \Psi\Psi^* = \frac{1}{2}\left(\phi_1 e^{-i\varepsilon_1 t/\hbar} + \phi_2 e^{-i\varepsilon_2 t/\hbar}\right) \cdot \left(\phi_1 e^{i\varepsilon_1 t/\hbar} + \phi_2 e^{i\varepsilon_2 t/\hbar}\right)$$

Continued on page 115

FOUNDATION 4.4

Continued from page 114

$$= \frac{1}{2}\left[\phi_1^2 + \phi_2^2 + \phi_1\phi_2\left(e^{i(\varepsilon_2-\varepsilon_1)t/\hbar} + e^{-i(\varepsilon_2-\varepsilon_1)t/\hbar}\right)\right]$$

$$= \frac{1}{2}\left[\phi_1^2 + \phi_2^2 + 2\phi_1\phi_2 \cos\frac{\varepsilon_2-\varepsilon_1}{\hbar}t\right]$$

Inserting the wavefunctions ϕ_1 and ϕ_2 we get

$$\rho = \frac{1}{2}\left[\varphi_a^2 + \varphi_b^2 + (\varphi_a^2 - \varphi_b^2)\cos\frac{\varepsilon_2-\varepsilon_1}{\hbar}t\right]$$

We rewrite this expression by using the relation $\cos 2x = 1 - 2\sin^2 x$:

$$\rho = \varphi_a^2 + (\varphi_b^2 - \varphi_a^2)\sin^2\frac{\varepsilon_2-\varepsilon_1}{2\hbar}t$$

Result: the electron oscillates between $\rho = \varphi_a^2$ and $\rho = \varphi_b^2$. In the first case the electron is at nucleus a, e.g., at time $t = 0$; in the second case the electron is at nucleus b, e.g., at time

$$t = \frac{h}{2(\varepsilon_2 - \varepsilon_1)}$$

Thus the probability P_b of finding the electron at nucleus b is

$$P_b = \sin^2\frac{\varepsilon_2-\varepsilon_1}{2\hbar}t$$

Period of oscillation

The oscillation period T is given by

$$\frac{\varepsilon_2-\varepsilon_1}{2\hbar} = \frac{\pi}{T}, \quad \text{or} \quad T = \frac{\hbar \cdot 2\pi}{\varepsilon_2-\varepsilon_1} = \frac{h}{\varepsilon_2-\varepsilon_1}$$

With $E_1 = -21.79 \times 10^{-19}$ J we calculate for different distances d:

d/a_0	$(\varepsilon_2 - \varepsilon_1)/$J	$T/$s
10	2.6×10^{-21}	3×10^{-13}
20	2.4×10^{-25}	3×10^{-9}
30	1.6×10^{-29}	4×10^{-5}
40	1.0×10^{-33}	7×10^{-1}
50	5.6×10^{-38}	1×10^{4}
100	2.2×10^{-59}	3×10^{25}

Continued on page 116

> *Continued from page 115*
>
> *Tunneling probability*
>
> In Chapter 23 (Foundation 23.4) the expression
>
> $$\varepsilon_2 - \varepsilon_1 = E_{\text{barrier}} \frac{8}{3} \frac{d}{a_0} \exp\left(-\frac{2\pi}{h}\sqrt{2E_{\text{barrier}} m_e} \cdot d\right)$$
>
> will be used in a modified form.
> With $\varepsilon = \frac{1}{2}(\varepsilon_2 - \varepsilon_1)$ we obtain
>
> $$\varepsilon = \frac{4}{3} E_{\text{barrier}} \frac{d}{a_0} \cdot e^{-\alpha d} \quad \text{with} \quad \alpha = \frac{1}{\hbar}\sqrt{2 m_e E_{\text{barrier}}}$$
>
> and for the probability P_b we obtain
>
> $$P_b = \sin^2 \frac{\varepsilon}{\hbar} t$$

5 The Periodic Table and Simple Molecules

Can we understand the particularities of the periodic table from the electronic structure of the atoms? The electrons are described by standing waves similar to the electron waves in the H atom (ground state and excited states wavefunctions). H atom-like orbitals are filled according to increasing energy, taking account of the Pauli principle. The shape of the electron cloud is determined by the variation principle. With increasing nuclear charge the orbitals are increasingly contracted and the volume of the cloud is periodically extended when proceeding from one shell to the next in adding electrons.

The bonding in simple molecules is described by overlap of atomic orbitals and pairing of electrons. Conclusions are drawn concerning polarity of bonds, bond lengths, bond angles, bond deformabilities, and steric constraints, depending on the number of electrons and the charge of the nuclei.

The architecture of a molecule, its charge distribution, the stability of each bond, and the elastic properties of the molecular skeleton are crucial for understanding the interplay between molecules and the resulting chemical processes.

5.1 Periodic Table of the Elements

5.1.1 Basic Principles

Elements consist of atoms with a definite atomic number Z (the number of protons in the nucleus). In Chapters 3 and 4 we proceeded from hydrogen ($Z = 1$) to helium ($Z = 2$). In the same way we can understand the building principles of heavier elements (Table 5.1).

Table 5.1 Atomic numbers Z for some elements

Z	1	2	3	4	5	6	7	8	9	10
Element	H	He	Li	Be	B	C	N	O	F	Ne

As outlined in Chapters 2, 3, and 4, the basic features for understanding the electronic structure of atoms are the following.

(i) The energy level scheme and the corresponding wavefunctions of the H atom (Figures 3.16 and 3.17).
(ii) The Pauli exclusion principle in its simple form stating that in an atom or molecule at most two electrons can occupy the same orbital (Section 2.6).
(iii) The variation principle (Chapters 2 and 3) stating that, among all imaginable wavefunctions, the wavefunction in the ground state corresponds to the minimum of the total energy. In a multielectron atom this minimum is reached at a balance of the attracting Coulomb forces between electrons and nucleus and the repelling Coulomb forces between electrons.

If we neglect the repulsion forces between the electrons, we should expect, in principle, the same energy scheme as that of the H atom for all atoms; merely the distances between energy levels with different quantum numbers n should increase with increasing Z. As we have seen in the case of the He atom, the repulsion forces between the electrons cause significant changes in the shape of the electron cloud and in the energy of the atom. Thus neglecting the repulsion forces between the electrons would lead to false predictions. When the repulsion effect is taken into account even sophisticated mathematical approaches give only approximate results.

We will limit our discussion to a simple treatment. We neglect the antisymmetry of the total wavefunction, that is, we assume that each electron is in its distinct orbital in an effective field (nuclear field shielded by all other electrons). The orbitals with the lowest energies are occupied by two electrons each, according to the Pauli principle.

5.1.2 Hydrogen and Helium (Z = 1 and Z = 2)

Let us first look at the ionization energies of the atoms and ions that we have investigated so far. For the ionization of a hydrogen atom from its ground state

$$H \longrightarrow H^+ + e^- \tag{5.1}$$

an energy of

$$E_{\text{ion}} = \frac{1}{4\pi\varepsilon_0} \cdot \frac{e^2}{2a_0} = 21.80 \times 10^{-19} \, \text{J} \tag{5.2}$$

is needed, according to Chapters 2 and 3 (E_{ion} is called the *ionization energy*). The ionization energy from the 2s orbital would be $\frac{1}{4} \cdot 21.80 \times 10^{-19}$ J $= 5.45 \times 10^{-19}$ J only.

For the He atom we have to distinguish between the ionization of the first electron (ionization energy $E_{\text{ion},1}$) and of the second one (ionization energy $E_{\text{ion},2}$).

$$\text{He} \xrightarrow{E_{\text{ion},1}} \text{He}^+ + e^- \xrightarrow{E_{\text{ion},2}} \text{He}^{2+} + 2e^- \tag{5.3}$$

In Chapters 3 and 4 (see Problem 3.6 and Section 4.2.1) we have calculated

$$E_{\text{ion},1} = 39.39 \times 10^{-19} \, \text{J} \quad \text{and} \quad E_{\text{ion},2} = 87.18 \times 10^{-19} \, \text{J} \tag{5.4}$$

Both values are in good agreement with the experiment.

Even though both electrons occupy the same 1s orbital, the energy needed to remove the first electron is much less than for the second one. The reason for this is that while the first electron is being removed, the second one remains in the vicinity of the nucleus, that is, much less than the total nuclear charge is acting on the first electron. At large distances

5.1 PERIODIC TABLE OF THE ELEMENTS

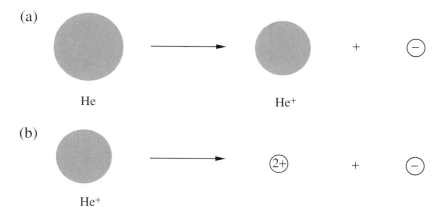

Figure 5.1 Removal (a) of the first and (b) of the second electron from the He atom. Because of the repulsion of the two electrons, the electron cloud of He is more spread than that of He⁺.

from the nucleus, only one positive charge acts upon the first electron. The charge of the nucleus is diminished by the charge of the second electron (Figure 5.1).

5.1.3 Lithium (Z = 3)

Now we consider the Li atom. Li consists of a threefold positively charged nucleus and three electrons; two of them share a 1s orbital. Because of the Pauli principle the third electron has to occupy another orbital; it can occupy either a 2s or a 2p orbital.

$$\text{Li} \xrightarrow{E_{ion,1}} \text{Li}^+ + e^- \xrightarrow{E_{ion,2}} \text{Li}^{2+} + 2e^- \xrightarrow{E_{ion,3}} \text{Li}^{3+} + 3e^- \tag{5.5}$$

We expect three different ionization energies for Li. The experimental data are

$$E_{ion,1} = 8.64 \times 10^{-19}\,\text{J}, \quad E_{ion,2} = 121.2 \times 10^{-19}\,\text{J}, \quad E_{ion,3} = 196.2 \times 10^{-19}\,\text{J} \tag{5.6}$$

$E_{ion,3}$ corresponds to the removal of the remaining 1s electron from a threefold positively charged nucleus, so we expect

$$E_{ion,3} = 3^2 \cdot 21.80 \times 10^{-19}\,\text{J} = 196.2 \times 10^{-19}\,\text{J} \tag{5.7}$$

in agreement with the experiment.

$E_{ion,2}$ is smaller than $E_{ion,3}$ because the nuclear charge is shielded by the other electron; however, the effective nuclear charge is somewhat larger than the charge $+3e - e = +2e$ in the case of complete shielding. Therefore, $E_{ion,2}$ of Li is somewhat larger than $E_{ion,2}$ of He (nuclear charge $+2e$).

$E_{ion,1}$ of Li is much smaller than $E_{ion,1}$ of He. This is because the average distance of an electron from the nucleus in a 2s or 2p orbital is larger than for an electron in a 1s orbital. The energy of ionization from a 2s or 2p state of a H atom is 5.45×10^{-19} J. $E_{ion,1}$ for Li should be similar; however, the charge of the nucleus is only partially shielded by the other two electrons, and for this reason the ionization energy is somewhat larger.

Now, we still have to decide whether the third electron of the Li atom occupies a 2s or a 2p orbital, or whether both possibilities are equally feasible. There is an important

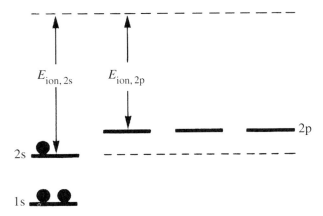

Figure 5.2 Population scheme for the Li atom in the ground state. The ionization energy from the 2s state ($E_{ion,2s}$) is larger than the corresponding energy from the 2p state ($E_{ion,2p}$), because a 2p electron is more strongly shielded by the 1s electrons than a 2s electron. Therefore the third electron of the Li atom occupies a 2s orbital and not a 2p orbital.

difference between the two orbitals. The electron density of the 2s orbital is nonzero close to the nucleus, whereas for the 2p orbital it is zero at this position (Figures 3.14, 3.15 and 3.16).

Thus, when compared to a 2s electron, a 2p electron is less often in the region upon which the unshielded field of the nucleus acts. Therefore a 2p electron is better shielded by the remaining 1s electrons than a 2s electron. As a result, the energy level of a 2p electron is higher than for a 2s electron. Because of this shielding effect, these levels are not equal as in the case of the hydrogen atom. Thus in the ground state, according to the variation theorem, the third electron of the Li atom must occupy the 2s orbital (Figure 5.2).

5.1.4 Aufbau Principle and Periodic Table

We can predict the electron configuration of the elements up to boron following the method described above (Table 5.2; left-hand side). In the boron atom, the last electron can, of course, also occupy a $2p_y$ or $2p_z$ orbital, since these are completely equivalent.

With the next element, carbon with six electrons, a new problem arises: the sixth electron can occupy either the $2p_x$ orbital (together with the fifth electron, Figure 5.3(a))

Table 5.2 Electron configurations of elements

Element	Configuration	Element	Configuration
H	1s	C	$1s^2 2s^2 2p_x 2p_y$
He	$1s^2$	N	$1s^2 2s^2 2p_x 2p_y 2p_z$
Li	$1s^2 2s$	O	$1s^2 2s^2 2p_x^2 2p_y 2p_z$
Be	$1s^2 2s^2$	F	$1s^2 2s^2 2p_x^2 2p_y^2 2p_z$
B	$1s^2 2s^2 2p_x$	Ne	$1s^2 2s^2 2p_x^2 2p_y^2 2p_z^2$

The notation $1s^2$ means that two electrons are in the 1s orbital, etc.

5.1 PERIODIC TABLE OF THE ELEMENTS

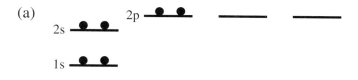

Figure 5.3 Schemes for the distribution of the six electrons in the C atom. (a) Two electrons in the $2p_x$ orbital; (b) one electron in the $2p_x$ orbital and one electron in the $2p_y$ orbital in accordance with Hund's rule.

or one of the two remaining 2p orbitals ($2p_y$ or $2p_z$, Figure 5.3(b)). The electrons in the first case would be closer together, on the average, than those in the second case; this means that in the second case the resulting repulsion energy of the two electrons would be smaller. Thus the electron configurations on the right-hand side of Table 5.2 will be realized for the subsequent elements. This procedure is in accordance with *Hund's rule*:

Hund's rule. If several electrons are to occupy different orbitals of equal energy, they will occupy as many orbitals as possible.

In neon, the charge distribution around the nucleus is spherical. The 1s and 2s electrons are spherically symmetrically distributed, and the 2p electrons together form a spherical cloud, since according to equation (3.62) we have

$$\psi^2_{2p_x} + \psi^2_{2p_y} + \psi^2_{2p_z} = \frac{1}{32\pi a_0^5} \cdot e^{-r/a_0} \cdot (x^2 + y^2 + z^2) \tag{5.8}$$

and the last term in this expression is equal to r^2.

The consideration about the shielding of 2s and 2p orbitals applies for the 3s, 3p, and 3d and the 4s, 4p, 4d, and 4f orbitals as well:

$$\text{orbital energy:} \quad s < p < d < f$$

In Figure 5.4 the elements are arranged in a scheme according to the rules discussed above. This scheme is called the *periodic table of the elements*. An approximate description of atomic orbitals by Slater functions is given in Box 5.1.

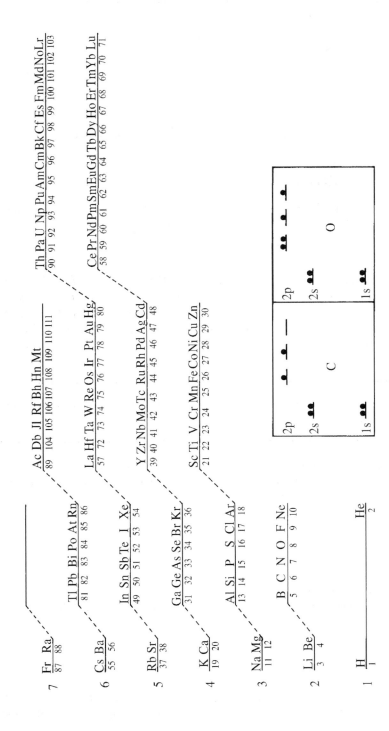

Figure 5.4 The periodic table of the elements. The elements are arranged in a scheme according to increasing atomic number Z; the scheme results from the energy scheme of the H atom accounting for the orbital contraction increasing with increasing nuclear charge Ze and for the shielding of the nucleus when filling the p, d, and f levels. Hund's rule has to be considered (examples for C, and O are shown in the boxes). **Note**: The periodic table given here is inverted relative to the usual representation to indicate the occupation of hydrogen atom-like orbitals according to the Aufbau principle.

Box 5.1 – Slater Wavefunctions

In order to take the shielding problem into account, Slater suggested that we describe the atoms by effective atomic numbers Z_{eff}:

$$Z_{eff} = Z - \sigma$$

where σ is called the *shielding constant*; it is calculated from contributions of each of the electrons in the groups 1s; 2s, 2p; 3s, 3p, 3d; 4s, 4p; 4d, 4f, ... as shown in Figure 5.5.

Selected electron	Number of shielding electrons	Contribution to σ per electron
4s	1 (4s)	0.35
3p, 3s	8 (3s, 3p)	0.85
2p, 2s	8 (2s, 2p)	1.0
1s	2 (1s)	1.0

Figure 5.5 Calculation of the shielding constant σ from contributions of different electrons.

The remaining electrons of the same group as the selected electron contribute 0.35 to σ (exception: 0.30 in the 1s group). The groups d and f contribute 1.0 each, the s and p groups 0.85 each, when they are just below the group of the selected electron; if they are in a lower group they contribute 1.0.

In the scheme in Figure 5.5 a 4s electron (Ca atom) is selected. We obtain $\sigma = 2 + 8 + 6.8 + 0.35 = 17.15$. This yields $Z_{eff} = 20 - 17.15 = 2.85$. In the following table, Z_{eff} is displayed for the outer-shell electrons of several elements.

For example, for Li the effective atomic number is $Z_{eff} = 1.3$, thus $E_{ion,1} = \frac{1}{4} \times 21.8 \times 1.3^2 \times 10^{-19}$ J $= 9.2 \times 10^{-19}$ J. The experimental value is 8.64×10^{-19} J. For Li$^+$, $Z_{eff} = 3 - 0.3 = 2.7$; thus $E_{ion,2} = 21.8 \times 2.7^2 \times 10^{-19}$ J $= 159 \times 10^{-19}$ J. The experimental values are $E_{ion,2} = 121 \times 10^{-19}$ J, and $E_{ion,3} = 196 \times 10^{-19}$ J; thus the Slater value corresponds to the average of $E_{ion,2}$ and $E_{ion,3}$, namely 159×10^{-19} J.

Continued on page 124

Continued from page 123

Atom	σ	Z_{eff}
He	0.30	$2 - 0.30 = 1.70$
Li	$2 \cdot 0.85 = 1.70$	$3 - 1.70 = 1.30$
Be	$0.35 + 2 \cdot 0.85 = 2.05$	$4 - 2.05 = 1.95$
B	$2 \cdot 0.35 + 2 \cdot 0.85 = 2.40$	$5 - 2.40 = 2.60$
C	$3 \cdot 0.35 + 2 \cdot 0.85 = 2.75$	$6 - 2.75 = 3.25$
N	$4 \cdot 0.35 + 2 \cdot 0.85 = 3.10$	$7 - 3.10 = 3.90$
O	$5 \cdot 0.35 + 2 \cdot 0.85 = 3.45$	$8 - 3.45 = 4.55$
F	$6 \cdot 0.35 + 2 \cdot 0.85 = 3.80$	$9 - 3.80 = 5.20$
Ne	$7 \cdot 0.35 + 2 \cdot 0.85 = 4.15$	$10 - 4.15 = 5.85$
Na	$8 \cdot 0.85 + 2 \cdot 1.00 = 8.80$	$11 - 8.80 = 2.20$
Mg	$1 \cdot 0.35 + 8 \cdot 0.85 + 2 \cdot 1.00 = 9.15$	$12 - 9.15 = 2.85$

Using these assumptions we obtain the same functions as discussed in Chapter 3 with the only difference that Z is substituted by Z_{eff}. As a further simplification Slater suggested that we use a simple power r^{n-1} instead of the polynomial of r that appears in the radial term of the wavefunctions ("*Slater functions*"), for example,

$$\psi_{2s,\text{Slater}} = A \cdot r \cdot e^{-rZ_{\text{eff}}/(2a_0)}$$

$$\psi_{3s,\text{Slater}} = A \cdot r^2 \cdot e^{-rZ_{\text{eff}}/(3a_0)}$$

instead of

$$\psi_{2s} = A \cdot \left(r - \frac{2a_0}{Z_{\text{eff}}}\right) \cdot e^{-rZ_{\text{eff}}/(2a_0)}$$

$$\psi_{3s} = A \cdot \left(r^2 - 9\frac{a_0}{Z_{\text{eff}}}r + \frac{27}{2}\left(\frac{a_0}{Z_{\text{eff}}}\right)^2\right) \cdot e^{-rZ_{\text{eff}}/(3a_0)}$$

where A is given by the normalization condition. Only for ψ_{1s} and ψ_{2p} are the Slater functions hydrogen-like:

$$\psi_{1s} = A \cdot e^{-rZ_{\text{eff}}/(a_0)}$$

$$\psi_{2p} = A \cdot x \cdot e^{-rZ_{\text{eff}}/(2a_0)}$$

5.2 The Structure of Simple Molecules

5.2.1 Simple Bond Models

The bond in the H_2 molecule can be viewed as being formed by the overlapping of the 1s orbitals of two H atoms. Since each atom contributes one electron to the bond this means that two electrons share the bond as a pair; this type of bonding is called *electron pair bonding*. Other orbitals can overlap in a similar way, e.g., p orbitals.

5.2 THE STRUCTURE OF SIMPLE MOLECULES

Example – F₂ molecule

The electron configuration of the F atom is $1s^2 2s^2 2p_y^2 2p_z^2 2p_x$. The $2p_x$ orbital is occupied by an unpaired electron. The choice of p_x is arbitrary; of course, we can choose p_y or p_z as well (see Table 5.2). The s orbitals and the $2p_y$ and $2p_z$ orbitals are occupied by two electrons each. According to the Pauli principle, they cannot contribute to bonding. The $2p_x$ orbitals overlap forming the F_2 molecule (Figure 5.6). Thus the fluorine atom can form one bond only.

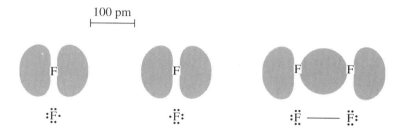

Figure 5.6 Electron pair bonding of F_2 (overlap of $2p_x$ orbitals). In the chemical formula representation a dot denotes a single electron; only the electrons of the outer shell are displayed. Inside the shaded area the electron density is larger than $e \cdot 0.02 \times 10^{-30}$ m^{-3}. Electron clouds are depicted according to A. C. Wahl, Molecular orbital densities: Pictorial approach, *Science*, 1966, **151**, 961; and J. R. Wazar and I. Absar, *Electron Densities in Molecules and Molecular Orbitals*, Academic Press, New York, 1975.

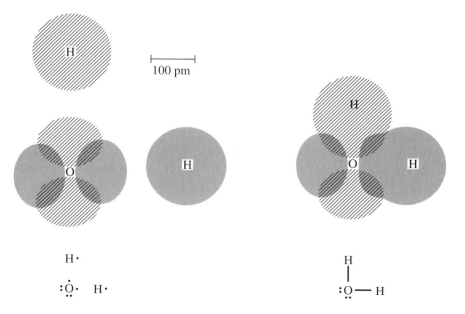

Figure 5.7 Formation of an electron pair bond in a H_2O molecule. The electron density in the shaded area is larger than $e \cdot 0.02 \times 10^{-30}$ m^{-3}.

The electron distribution in molecules can be schematically represented by dots, as indicated in the structural formula in Figure 5.6. A double dot corresponds to an electron pair, and a dot represents an unpaired electron.

In H_2O a p_z and p_x orbital of the oxygen atom overlap with the 1s orbitals of the hydrogen atoms (Figure 5.7). Accordingly the oxygen forms two bonds, and the H_2O molecule exists in a bent form.

In the N atom there are three unpaired electrons, and three bonds can be formed. The N_2 molecule is described by a formal triple bond; in the NH_3 molecule the three bonds are perpendicular to each other (Figure 5.8). This is a very rough description of the chemical bond; in fact the H—O—H bond angle is 105° and the H—N—H bond angle is 108° instead of 90° for reasons discussed in Chapter 6.

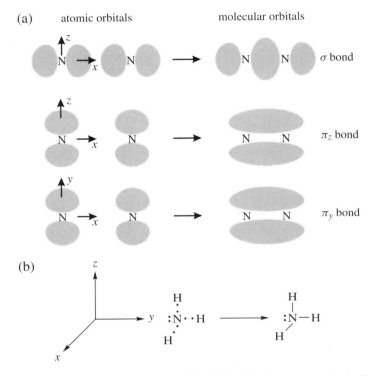

Figure 5.8 Electron pair bonds (a) in N_2 and (b) in NH_3. The electron density in the shaded area is larger than $e \cdot 0.02 \times 10^{-30}$ m^{-3}.

5.2.2 The Polarity of Bonds and Electronegativity

The electrons in a molecule built up of two atoms of the same element (e.g., F_2) are equally distributed over both atoms. If, however, the molecule is composed of different atoms this is no longer true.

As an example, we consider an F—Cl molecule; it differs from the F_2 molecule in that one F atom is replaced by a Cl atom, which is one row higher in the periodic table (see Figure 5.4). The electron pair bond is actually the same as for F_2, but in the Cl atom the bonding electron occupies a 3p orbital; thus the average distance of the electron from the

5.2 THE STRUCTURE OF SIMPLE MOLECULES

nucleus is larger than that for the electron of the F atom, and the electron is less strongly attracted by the nucleus. As a result, the electron cloud in F—Cl is distorted toward the fluorine (Figure 5.9). As a consequence, F—Cl has a dipole moment

$$\mu = Q \cdot d \tag{5.9}$$

The center of gravity of the negative charge $-Q$ is in the fluorine atom, and the center of gravity of the positive charge $+Q$ is at a distance d in the chlorine atom.

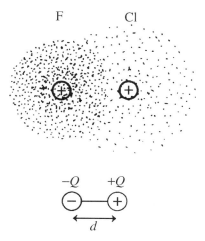

Figure 5.9 Polarity of the electron pair bond in F—Cl. The dipole moment is described by a positive and a negative point charge at distance d: $\mu = Q \cdot d$.

The tendency of an atom to attract a bonding electron is quantified by the value of the *electronegativity*. The electronegativity increases along a line of the periodic table

Li Be B C N O F

because the nuclear charge increases from left to right.

The electronegativity decreases from bottom to top of Figure 5.4, e.g., from F to Cl to Br to I to At or from O to S to Se to Te to Po because the average distance of the electron from the nucleus increases. From this it follows that F is the most electronegative element (Table 5.3).

Table 5.3 Electronegativity in the periodic table

→	→	→	→	→	→	→	
Li	Be	B	C	N	O	F	↑
			Si	P	S	Cl	↑
				As	Se	Br	↑

Increasing electronegativity from left to right and from bottom to top.

The polarity of a bond increases with increasing difference in electronegativity of the bonding atoms, for example,

$$
\begin{array}{ll}
\text{F–F} & \text{homopolar bonding} \\
\text{F–Cl} & \downarrow \\
\text{F}^-\text{Li}^+ & \text{ionic bonding}
\end{array}
$$

For the quantitative description of the electronegativity a scale has been derived from theoretical considerations (see Box 5.2), which is shown in Table 5.4.

Table 5.4 Electronegativities of the main group elements from H to Br, calculated as described in Box 5.2

			H 2.79			
Li 1.00	Be 1.48	B 1.84	C 2.35	N 3.16	O 3.52	F 4.00
Na 0.89	Mg 1.24	Al 1.40	Si 1.64	P 2.11	S 2.52	Cl 2.84
K 0.73	Ca 0.96	Ga 1.54	Ge 1.69	As 1.99	Se 2.40	Br 2.52

Box 5.2 – Electronegativity scale

The energies of hydrides HA of element A are calculated according to the variation principle using the best possible trial functions. These trial functions are set up from three terms: two atomic terms φ_H and φ_A with the center of gravity at H or A, respectively, and a bond term φ_{HA} with the center of gravity on the line connecting the nuclei in a distance δ from the H atom (Figure 5.10). The latter contribution is relevant for the electron distribution in the bonding region. The trial function is

$$\phi = \alpha \varphi_H + \beta \varphi_{HA} + \gamma \varphi_A$$

The constants α, β, and γ as well as the distance δ are varied until the energy minimum is achieved. If both atoms H and A have the same electronegativity, the center of gravity of the bond term will be localized halfway between the nuclei, that is, according to Figure 5.10 we have $\delta = d/2$.

When the electronegativity is larger or smaller, δ will be larger or smaller than $d/2$. It is therefore reasonable to use the quantity $\delta/d - \frac{1}{2}$ as a measure of the electronegativity difference between H and A. An electronegativity scale was obtained by assuming proportionality between $(x_A - x_H)$ and $(\frac{\delta}{d} - \frac{1}{2})$,

Continued on page 129

5.2 THE STRUCTURE OF SIMPLE MOLECULES

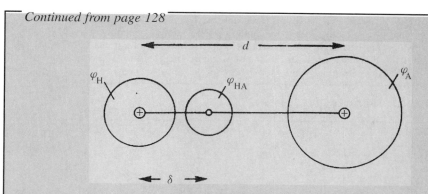

Figure 5.10 Trial function for the bonding electrons in hydrides HA.

where x_A and x_H are the electronegativities of A and H, respectively:

$$x_A - x_H = C \cdot \left(\frac{\delta}{d} - \frac{1}{2}\right)$$

The electronegativities of Li and F are arbitrarily chosen as $x_{Li} = 1.0$ and $x_F = 4.0$. Then, x_A is obtained from δ/d of A, Li, and F (see Table 5.4).

5.2.3 Bond Lengths and Bond Angles

Bond lengths

The length of a bond depends on the electron distribution within the bond, and therefore on the type of bonding partners. For polyatomic molecules the bond length is almost independent of the nature of the substituents (Table 5.5).

The bond length decreases from N–H to O–H, because the nuclear charge increases from nitrogen to oxygen. The bond length increases from HF to HI, because iodine uses a 5p orbital for bonding instead of a 2p orbital used by fluorine. Since the bond length depends upon the extension of the atomic orbitals, we can estimate the length of any chosen bond from atomic contributions. We define one-half of the length of a bond between equal atoms as covalent radius of the atom. We assume that the length of the bond between different atoms is the sum of the covalent radii (Figure 5.11). For HCl we obtain

$$d_{HCl} = r_H + r_{Cl} = 37\,\text{pm} + 99\,\text{pm} = 136\,\text{pm}$$

This distance is in reasonable agreement with the experimental value (127 pm). The covalent radii shown in Table 5.6 are obtained accordingly.

Covalent and van der Waals radii

From the space requirements of an I_2 molecule in a crystal, one finds a distance of 400 pm between the nuclei of two adjacent I_2 molecules (Figure 5.12), whereas the distance

Table 5.5 Bond lengths d and force constants k_f and bond energies ε_B for diatomic and polyatomic molecules. (Bond energies in polyatomic molecules are extracted from the energies of dissociation into atoms. It is postulated that these energies are the sum of the energies for each bond. In this way, appropriate values for bond energies are obtained.)

Diatomic Molecules	$\dfrac{d}{\text{pm}}$	$\dfrac{k_f}{\text{N m}^{-1}}$	$\dfrac{\varepsilon_B}{\text{kJ mol}^{-1}}$
H_2^+	106	160	252
H_2	74	574	436
O_2	121	1177	498
N_2	110	2296	945
F_2	142	471	159
Cl_2	199	328	243
Br_2	228	246	193
I_2	267	173	151
HF	92	966	570
HCl	127	516	432
HBr	141	412	366
HI	161	314	298
CO	113	1903	1077
NO	115	1600	631

Polyatomic Molecules		$\dfrac{d}{\text{pm}}$	$\dfrac{k_f}{\text{N m}^{-1}}$	$\dfrac{\varepsilon_B}{\text{kJ mol}^{-1}}$
CH_4	C–H	109	550	438
C_2H_4	C–H	107		444
C_2H_2	C–H	106	640	552
$CH_3–CH_3$	C–C	154	434	376
diamond	C–C	154	2500	
$CH_2=CH_2$	C=C	133	1080	720
CH≡CH	C≡C	120	1490	962
NH_3		101	635	449
H_2O		96	780	498
H_2S		134	390	381
SiH_4		148	280	
CO_2		116	1500	532
NO_2		119	1100	305
F_2O		142	400	
SO_2		143	956	552

Bond lengths: *Tables of Interatomic Distances and Configuration in Molecules and Ions.* Spec. Pub. **11** (1958); **18** (1965), The Chemical Society, London.
Force constants: Landolt–Börnstein: *Zahlenwerte und Funktionen*, 1. Bd., 2. Teil, Springer, Berlin 1950. A. Fadini, Molekülkraftkonstanten, Steinkopff, Darmstadt 1976.
Bond energies: D. R. Lide (Ed.), *CRC Handbook of Chemistry and Physics*, CRC Press 1992.

between the nuclei in one molecule is much smaller (267 pm, Table 5.5). Bonding orbitals overlap and bind; however, doubly occupied orbitals push off when touching because of the Coulomb repulsion of the electron clouds. Therefore covalent bonding radii and atomic radii must be distinguished; the latter are called *van der Waals radii*. Both radii are compared in Table 5.6.

5.2 THE STRUCTURE OF SIMPLE MOLECULES

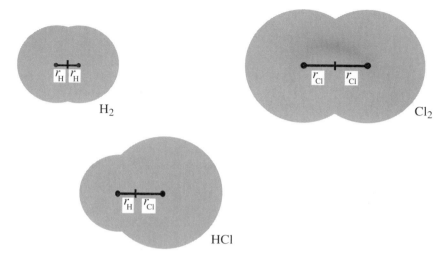

Figure 5.11 Calculating bond lengths from bond radii.

Table 5.6 Covalent radii for some single bonds (pm; calculated from Table 5.5) and van der Waals radii (pm) for some atoms

Atom	covalent	van der Waals
F	72	135
Cl	99	180
Br	114	195
I	133	215
C	77	160
H	37	120

Atom models can be constructed from the covalent radii and the van der Waals radii. One cuts a cap from a sphere of van der Waals radius: when two such models are put together, the correct internuclear distance is obtained. Such a "space-filling model" (after applying a chosen enlargement factor of about 1×10^8) is shown in Figure 5.13.

Bond angles

Molecules with more than two atoms are characterized, not only by the bond lengths, but by the bond angles as well. As mentioned above, we expect that molecules such as H_2S, H_2O, and F_2O, which are formed by the overlap of two free p orbitals of oxygen or sulfur, will have a bond angle of 90°. As can be seen in Figure 5.14, the experimental values are somewhat larger, mainly because of the electrostatic repulsion of the two H or F atoms (Problem 5.3).

The molecular orbitals of single bonds have a rotational symmetry along the line connecting the nuclei (bond axis). Rotation around this axis does not change the overlap of the bonding orbitals, and the molecule can rotate essentially freely around this axis (Figure 5.15).

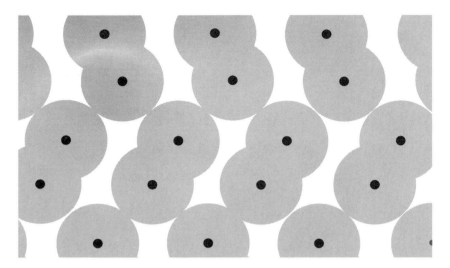

Figure 5.12 The arrangement of I_2 molecules in a crystal. The smallest distance between the nuclei of different molecules is much larger than the internuclear distance inside a molecule.

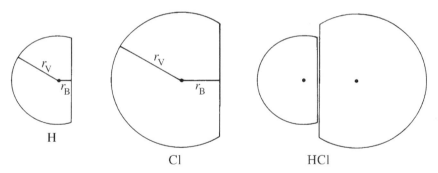

Figure 5.13 Space-filling models of H_2 and Cl_2 (r_B is the covalent bond radius, r_V is the van der Waals radius).

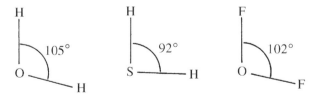

Figure 5.14 Bond angles of H_2O, H_2S, and F_2O.

5.2 THE STRUCTURE OF SIMPLE MOLECULES

Figure 5.15 Free rotation of the C−O bond in CH_3OH.

5.2.4 Stretching and Bending Force Constants

Stretching (Hooke's law)

An increase or decrease in the bond lengths and bond angles causes an increase in the energy of the molecule, which can be easily explained. A decrease in the bond length causes an increase in the repulsion between the electronic shells and between the nuclei; an increase of the distance decreases the overlap in the bonding region. Figures 4.3, 4.7 and 4.14 show the energy E of a H_2^+ molecule as a function of the distance between the nuclei; E is a complicated function of the distance d between the nuclei. For small deflections of the nuclei we can assume, however, that the force f needed for an elongation x of the bond is proportional to x:

$$f = k_f x = k_f \cdot (d - d_{eq}) \tag{5.10}$$

This assumption is in accordance with Hooke's Law for the deflection of a spring (k_f is the stretching force constant). From equation (5.10) we obtain the energy $\Delta\varepsilon$ that has to be supplied to elongate the bond by x

$$\Delta\varepsilon = \varepsilon - \varepsilon_{eq} = \int_0^x f \cdot dx = \int_0^x k_f x \cdot dx = \frac{1}{2} k_f \cdot x^2 \tag{5.11}$$

(ε is the energy at a distance d between the nuclei, ε_{eq} is the energy at the equilibrium distance d_{eq}). This relation looks qualitatively plausible, because the energy curves show a parabolic shape for small deflections (Figure 5.16). Using the data in Figure 5.16 we obtain the following values of k_f for the H_2^+ ion (see Problem 5.2): $k_f = 169\,\text{N m}^{-1}$ (exact wavefunction), $k_f = 65\,\text{N m}^{-1}$ (box model), $k_f = 102\,\text{N m}^{-1}$ (LCAO model). The experimental value is $k_f = 160\,\text{N m}^{-1}$.

In Table 5.5 values of the force constant k_f are displayed. For the C−C bond of ethane, $k_f = 434\,\text{N m}^{-1}$; a force of $f = 10^{-9}\,\text{N}$ then causes an elongation of the bond by

$$x = \frac{f}{k_f} = \frac{10^{-9}\,\text{N}}{434\,\text{N m}^{-1}} = 2 \times 10^{-12}\,\text{m} = 2\,\text{pm}$$

(Figure 5.17). This corresponds to an elongation of 1%.

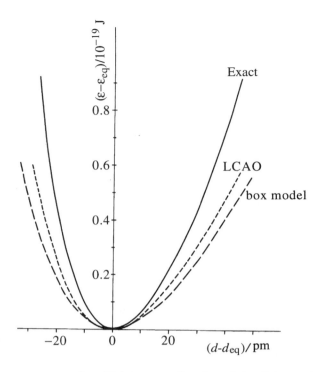

Figure 5.16 Energy $\varepsilon - \varepsilon_{eq}$ of an H_2^+ ion as a function of the distance d of the nuclei (see Problem 5.2) at small elongations (ε_{eq} is the energy at equilibrium distance).

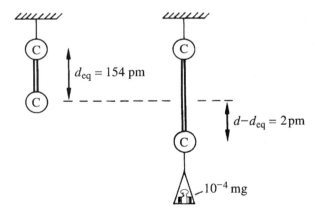

Figure 5.17 Elongation of the C–C bond in an ethane molecule by applying a force of 10^{-9} N (this is a weight of 10^{-4} mg or the weight of 10 lead cubes of 0.01 mm edge length).

Stretching (Morse function)

Hooke's law does not hold for larger forces; the energy curve is steeper for bond shortening than for bond lengthening (Figure 5.16). At large distances the energy ε becomes constant and equals the energy of the separate bond partners (e.g., H_2^+ dissociates into an H atom

5.2 THE STRUCTURE OF SIMPLE MOLECULES

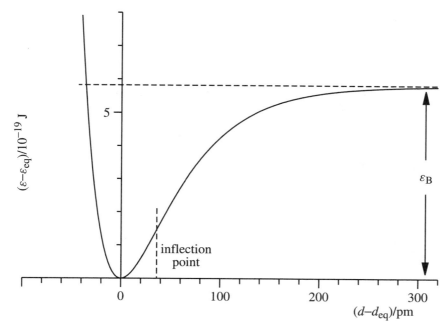

Figure 5.18 Energy of a diatomic molecule (Morse function) according to equation (5.12); the values used are $\varepsilon_B = 5.8 \times 10^{-19}$ J and $k_f = 434$ N m^{-1} (corresponding to the C–C bond in ethane).

and a proton). Then $(\varepsilon - \varepsilon_{eq})$ corresponds to the bond energy ε_B. The energy curve can be approximately described by

$$\varepsilon - \varepsilon_{eq} = \varepsilon_B \cdot (1 - e^{-\sqrt{k_f/(2\varepsilon_B)} \cdot (d - d_{eq})})^2 \tag{5.12}$$

(*Morse function*, Figure 5.18). The energy minimum is at $d = d_{eq}$, and for $d \to \infty$ the energy becomes equal to the bond energy ε_B; close to the minimum (at small enough values of $x = d - d_{eq}$) equation (5.12) becomes identical with equation (5.11). The Morse function has an inflection point at

$$x = \ln 2 \cdot \sqrt{\frac{2\varepsilon_B}{k_f}} \tag{5.13}$$

At this distance the force acting between the bonded atoms reaches a maximum

$$f_{max} = \frac{1}{2}\sqrt{\frac{k_f \varepsilon_B}{2}} \tag{5.14}$$

For an H$_2^+$ molecule $\varepsilon_B = 4.2 \times 10^{-19}$ J (see Chapter 3) and $k_f = 160$ N m^{-1}. Then $f_{max} = 2.9 \times 10^{-9}$ N at $x = 50$ pm. For a C–C bond with $k_f = 434$ N m^{-1} and $\varepsilon_B = 5.8 \times 10^{-19}$ J we obtain $f_{max} = 5.6 \times 10^{-9}$ N at $x = 36$ pm. The force needed to break a C–C bond is about 6 times larger than that for stretching the bond by 1%. The bond breaks at an elongation of 36 pm, that is, at an elongation of 20% above the equilibrium bond length. Indeed, a single polysaccharide chain attached to the tip and the substrate (glass) of an atomic force microscope (Figure 22.41) breaks at $f_{max} = 2 \cdot 10^{-9}$ N due to the rupture of a silicon–carbon bond (M. Grandbois, M. Beyer, M. Rief, H. Clausen-Schaumann, H.E. Gaub, *Science*, 283, 1727 (1999)).

Bending

In addition to the stretching of a bond, a deformation of the bond angles occurs in molecules with three or more atoms. Since there is only a small change of the overlap of the bonding orbitals when changing the bond angles, the force needed for bending a bond is much smaller than for stretching. A change $\Delta\alpha$ of the bond angle α (Figure 5.19) causes a shift of the atom by

$$x = d_{eq} \cdot \Delta\alpha \tag{5.15}$$

The force required for this shift is

$$f = k'_f \cdot x \tag{5.16}$$

(k'_f = bending force constant). The experimental value for the bending force constant of the C–C–C bond angle in propane is $k'_f = 36\,\text{N}\,\text{m}^{-1}$. According to Figure 5.19, this corresponds to a shift of one C atom by

$$x = d_{eq} \cdot \Delta\alpha = \frac{f}{k'_f} = \frac{10^{-9}\,\text{N}}{36\,\text{N}\,\text{m}^{-1}} = 30\,\text{pm}$$

at a force of $10^{-9}\,\text{N}$. This shift is larger by a factor of about 15 than for stretching the bond. This bending force can be directly measured by an atomic force microscope by fixing the ends of a single chain molecule, spanning it and measuring the stretching force (see Section 22.4.5).

In a similar way, we can estimate the energy barrier for the rotation of a methyl group around the C–C axis in ethane. In this case already a force of $2 \times 10^{-10}\,\text{N}$ is sufficient for overcoming the energy barrier against rotation of the methyl group (Figure 5.20).

We have explained the structure of the periodic table by considering the electronic structure of the elements depending on their nuclear charges. Furthermore, we have discussed the very basic features for understanding molecules and their mutual interplay: bond lengths, bond angles, polarity, and the elastic forces against changes of the molecular skeleton by stretching and compressing bonds and changing bond angles.

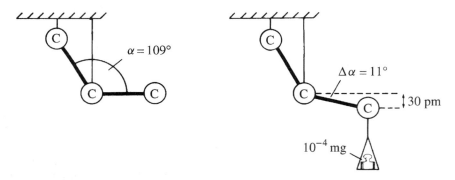

Figure 5.19 Bending a C–C–C bond.

PROBLEM 5.1

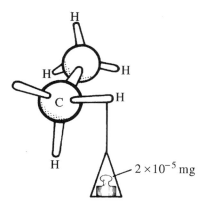

Figure 5.20 Free rotation (C–C bond in ethane).

Problems

Problem 5.1 – Partial Charges and Dipole Moment

The dipole moment of an HCl molecule is 3.4×10^{-30} C m = 1.03 debye. What are the partial charges on the H and Cl atoms when we assume that the center of gravity of these charges coincides with the nuclei? The distance between the nuclei is 127 pm.

Solution. According to equation (5.9) the partial charge is

$$Q = \frac{\mu}{d} = \frac{3.4 \times 10^{-30} \,\text{C m}}{127 \,\text{pm}} = 0.27 \times 10^{-19}\,\text{C} = 0.17 \cdot e$$

This is about 20% of the charge of an electron.

Problem 5.2 – Stretching Force Constant of H_2^+ Ion

Values of the energy ε of the H_2^+ ion as a function of the distance d of the nuclei are given below (d_{eq} is the equilibrium distance, ε_{eq} is the energy at the equilibrium distance).

$(d - d_{eq})$/pm	$(\varepsilon - \varepsilon_{eq})/10^{-19}$ J		
	Exact	Box model	LCAO model
−20	0.451	0.15	0.259
−15	0.237	0.08	0.136
−10	0.093	0.03	0.056
−5	0.022	0.008	0.013
0	0	0	0
5	0.019	0.008	0.011
10	0.071	0.03	0.043
15	0.147	0.07	0.091
20	0.240	0.12	0.151

Calculate the stretching force constant of the H–H bond in the H_2^+ ion from these data.

Solution. The data in the table can be well described by the function

$$\varepsilon - \varepsilon_{eq} = \tfrac{1}{2} k_f x^2 + b x^3$$

where $x = d - d_{eq}$ and the term bx^3 accounts for the fact that the curve is steeper on the left-hand side than on the right-hand side or by

$$y = \frac{\varepsilon - \varepsilon_{eq}}{x^2} = \frac{1}{2} k_f + bx$$

When plotting y as a function of x (Figure 5.21) we expect a straight line with slope b and intercept $k_f/2$. For the three models we obtain the following results:

Model	k_f/N m^{-1}
Exact	163
Box model	61
LCAO model	97

The slopes of the solid lines in Figure 5.21 are larger for negative elongations of the bond than those for positive ones; this indicates that the power series describing $(\varepsilon - \varepsilon_{eq})$ with only two terms (x^2 and x^3) is not completely adequate. For this reason the force constant obtained for the exact calculation is slightly larger than the value $k_f = 160$ N m^{-1} shown in Table 5.5.

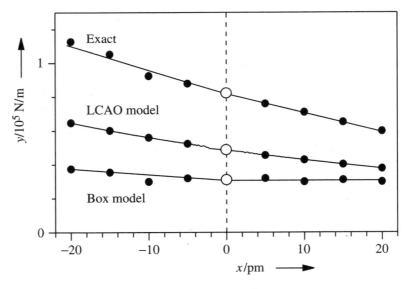

Figure 5.21 Evaluation of the force constant k_f of the H_2^+ ion (exact calculation, box model, and LCAO model). The quantity $y = (\varepsilon - \varepsilon_{eq})/x^2$ is plotted versus $x = (d - d_{eq})$. The intercept at point $x = 0$ equals one half of the force constant. Full circles: data points, solid lines: extrapolation of the data points to $x = 0$ (open circles).

Problem 5.3 – Bond Angle of H₂O

Discuss whether the electrostatic repulsion of the H atoms in the H_2O molecule is sufficient to cause a widening of the bond angle from 90° to 105°. The dipole moment of H_2O is $\mu = 6.16 \times 10^{-30}$ C m $= 1.85$ debye, the bending force constant k'_f is 68 N m^{-1}, and the bond length d is 96 pm.

Solution. Consider Figure 5.22. The distance a between the O atom and the center of gravity of the two H atoms is $a = d_1/\sqrt{2} = 67.9$ pm. In analogy to Problem 5.1 we obtain for the partial charge $Q = \mu/a = 0.9 \times 10^{-19}$ C. The force acting between the two hydrogen atoms is

$$f = \frac{1}{4\pi\varepsilon_0} \cdot \frac{Q^2}{(2a)^2} = 4 \times 10^{-9} \text{ N}$$

We compare this force with the force required to increase the bond angle by $\Delta\alpha = 15°$ ($x = d \cdot \Delta\alpha = 25$ pm):

$$f = 68 \text{ N m}^{-1} \times 25 \times 10^{-12} \text{ m} = 2 \times 10^{-9} \text{ N}$$

This is of the same order as the Coulomb repulsion force. For more details see Chapter 6.

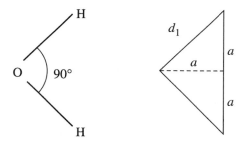

Figure 5.22 Geometry of an H_2O molecule with a bond angle of 90°; a is the distance between the O atom and the center of gravity of the two H atoms.

6 Bonding Described By Hybrid and Molecular Orbitals

How can we understand the occurrence of a molecule such as methane with tetrahedral bonds? The bond between two atoms can be described by the overlap of atomic orbitals forming electron pair bonds. The assignment of electron pairs to distinct bonds is a very useful approximation for an understanding of the directionality of bonds. In reality, we cannot relate distinct electrons to distinct bonds, since electrons are indistinguishable, and therefore individual electrons cannot be allocated.

It is equally legitimate to describe the electron cloud of a molecule using orbitals that are spread all over the molecule. This procedure (the molecular orbital method) will be very useful in Section 6.4 and in Chapter 7. Both methods are different approximations of the same system. We will use the molecular orbital method to investigate O_2 with its unusual electronic structure and particular properties.

6.1 Degeneracy of Energy Levels

First we investigate simple box wavefunctions, limiting ourselves to two-dimensional examples. The energies and wavefunctions of an electron moving between four walls with the boundaries L_x and L_y are analogous to those of a particle in a box (see Chapters 2 and 3)

$$E_{n_x,n_y} = \frac{h^2}{8m} \cdot \left(\frac{n_x^2}{L_x^2} + \frac{n_y^2}{L_y^2} \right) \tag{6.1}$$

$$\psi(x, y) = \sqrt{\frac{4}{L_x L_y}} \cdot \sin \frac{n_x \pi x}{L_x} \cdot \sin \frac{n_y \pi y}{L_y} \tag{6.2}$$

We investigate the special case where $L_x = L_y = L$

$$E_{n_x,n_y} = \frac{h^2}{8mL^2} \cdot (n_x^2 + n_y^2) \tag{6.3}$$

6.1 DEGENERACY OF ENERGY LEVELS

$$\psi(x, y) = \frac{2}{L} \cdot \sin \frac{n_x \pi x}{L} \cdot \sin \frac{n_y \pi y}{L} \qquad (6.4)$$

In this case quantum states which are obtained by exchanging n_x and n_y ($n_x \neq n_y$) are equal in energy (equation (6.3)). This is the case, for example, for $n_x = 1$, $n_y = 2$, and $n_x = 2$, $n_y = 1$ (Figure 6.1):

$$E_{1,2} = E_{2,1} = \frac{h^2}{8mL^2}(1^2 + 2^2) = \frac{h^2}{8mL^2} \cdot 5$$

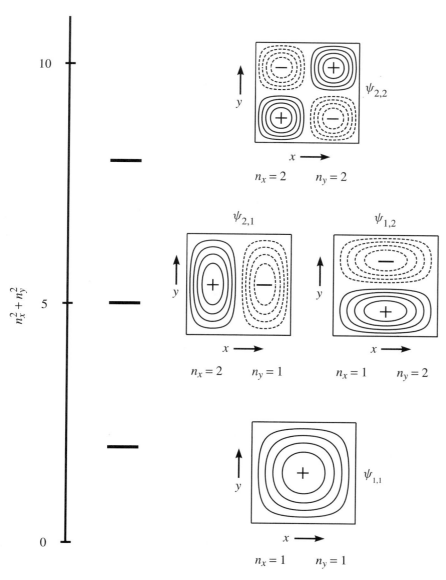

Figure 6.1 Degeneracy of electronic states in a two-dimensional box. Energy scheme and wavefunctions displayed as contour lines. Positive values: solid lines; negative values: dashed lines.

$$\psi_{1,2} = \frac{2}{L} \cdot \sin\frac{\pi x}{L} \cdot \sin\frac{2\pi y}{L} \quad \text{and} \quad \psi_{2,1} = \frac{2}{L} \cdot \sin\frac{2\pi x}{L} \cdot \sin\frac{\pi y}{L} \tag{6.5}$$

Such electronic states are called *degenerate*; in our case the state is twofold degenerate.

6.1.1 Hybrid Functions

Any linear combination of wavefunctions (6.5) belonging to the same energy

$$\psi = a\psi_{1,2} + b\psi_{2,1} \tag{6.6}$$

is also a solution of the Schrödinger equation (see Problem 6.1). Such linear combinations are called *hybrid functions*.

As outlined in Section 3.3, wavefunctions ψ_1 and ψ_2 – which are solutions of the Schrödinger equation and which belong to different energy states – are orthogonal to each other; that is,

$$\int \psi_1 \psi_2 \cdot d\tau = 0 \tag{6.7}$$

Furthermore it was shown in Section 3.2 that it is advantageous to use normalized wavefunctions, that is, wavefunctions satisfying the condition

$$\int \psi^2 \cdot d\tau = 1 \tag{6.8}$$

From this normalization condition we obtain the following equations for the hybrid function (6.6)

$$a^2 + b^2 = 1 \tag{6.9}$$

and

$$\psi = a \cdot \psi_{1,2} + \sqrt{1-a^2} \cdot \psi_{2,1} \tag{6.10}$$

In the present case of twofold degeneracy, each function

$$\psi_I = a \cdot \psi_{1,2} + \sqrt{1-a^2} \cdot \psi_{2,1} \tag{6.11}$$

has an orthogonal companion

$$\psi_{II} = \sqrt{1-a^2} \cdot \psi_{1,2} - a \cdot \psi_{2,1} \tag{6.12}$$

This can be easily seen:

$$\int \psi_I \psi_{II} \cdot d\tau = a\sqrt{1-a^2} \cdot \int \psi_{1,2}^2 \cdot d\tau - a^2 \cdot \int \psi_{1,2}\psi_{2,1} \cdot d\tau$$

$$+ (1-a^2) \cdot \int \psi_{2,1}\psi_{1,2} \cdot d\tau - a\sqrt{1-a^2} \cdot \int \psi_{2,1}^2 \cdot d\tau$$

$$= a\sqrt{1-a^2} - 0 + 0 - a\sqrt{1-a^2} = 0$$

The two functions form an orthogonal set. Pairs of normalized functions with $a = 0$, $\frac{1}{2}$, $\frac{1}{2}\sqrt{2}$, $\frac{1}{2}\sqrt{3}$, and 1 are shown in Figure 6.2. For $a = 0$ and $a = 1$ we obtain functions

6.1 DEGENERACY OF ENERGY LEVELS

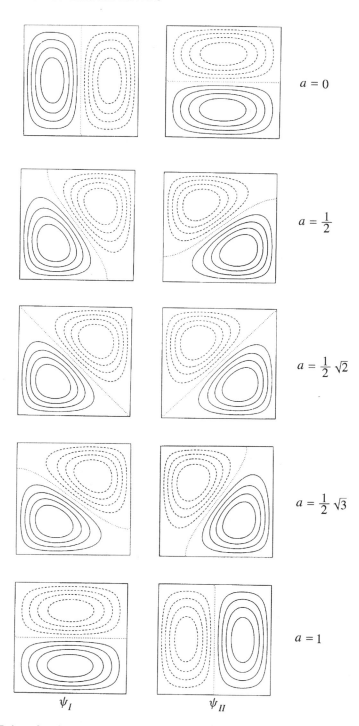

Figure 6.2 Pairs of orthogonal hybrid functions ψ_I and ψ_{II} according to functions (6.11) and (6.12). The parameter a is in the range from 0 to 1. For $a = 0$ we obtain $\psi_I = \psi_{2,1}$ and $\psi_{II} = \psi_{1,2}$ (see Figure 6.1). Positive values: solid lines; negative values: dashed lines; nodal lines: dotted lines.

with horizontal and vertical nodal lines, respectively; for $a = 1/\sqrt{2}$ the nodal lines are diagonal. All the other cases fall in between. Accordingly, in an n-fold degenerate state, n functions form an orthogonal set.

6.1.2 Hybridization of H Atom Functions

In the hydrogen atom the functions ψ_{2s}, ψ_{2p_x}, ψ_{2p_y}, and ψ_{2p_z} form a fourfold degenerate set of orthogonal functions. We can choose any linear combination

$$\psi = a\psi_{2s} + b\psi_{2p_x} + c\psi_{2p_y} + d\psi_{2p_z} \tag{6.13}$$

to describe the state with quantum number $n = 2$. Out of the infinite number of possibilities to construct orthogonal sets of four hybrid functions ψ_I, ψ_{II}, ψ_{III} and ψ_{IV} we discuss some special cases:

Possibility 1. No hybridization

$$\psi_I = \psi_{2s}, \quad \psi_{II} = \psi_{2p_x}, \quad \psi_{III} = \psi_{2p_y}, \quad \psi_{IV} = \psi_{2p_z} \tag{6.14}$$

Possibility 2. Two linear hybrid functions (sp). Each of the two orbitals consists of $\frac{1}{2}$s and $\frac{1}{2}$p ($s^{1/2}p^{1/2}$ orbitals).

$$\psi_I = \sqrt{\tfrac{1}{2}} \cdot (\psi_{2s} + \psi_{2p_x}), \quad \psi_{II} = \sqrt{\tfrac{1}{2}} \cdot (\psi_{2s} - \psi_{2p_x})$$

$$\psi_{III} = \psi_{2p_y}, \quad \psi_{IV} = \psi_{2p_z} \tag{6.15}$$

In this case we combine the functions ψ_{2s} and ψ_{2p_x} in analogy to equations (6.11) and (6.12) with $a = b = 1/\sqrt{2}$; the functions ψ_{2p_y} and ψ_{2p_z} remain unchanged. In Figure 6.3 the two linear combinations resulting from ψ_{2s} and ψ_{2p_x} are illustrated. These functions are called *linear hybrid functions*.

Possibility 3. Three trigonal planar hybrid functions (sp^2). Each of the three orbitals consists of $\frac{1}{3}$s, $\frac{2}{3}$p ($s^{1/3}p^{2/3}$ orbitals).

$$\psi_I = \sqrt{\tfrac{1}{3}} \cdot (\psi_{2s} + \sqrt{2}\psi_{2p_y}) \tag{6.16}$$

$$\psi_{II} = \sqrt{\tfrac{1}{3}} \cdot \left(\psi_{2s} + \sqrt{\tfrac{3}{2}}\psi_{2p_x} - \sqrt{\tfrac{1}{2}}\psi_{2p_y}\right)$$

$$\psi_{III} = \sqrt{\tfrac{1}{3}} \cdot \left(\psi_{2s} - \sqrt{\tfrac{3}{2}}\psi_{2p_x} - \sqrt{\tfrac{1}{2}}\psi_{2p_y}\right)$$

$$\psi_{IV} = \psi_{2p_z}$$

The combinations ψ_I, ψ_{II}, and ψ_{III} are illustrated in Figure 6.4. The symmetry axes of these functions are shifted by 120°, but all of them are in the xy plane. Therefore these functions are called *trigonal planar hybrid functions*.

6.1 DEGENERACY OF ENERGY LEVELS

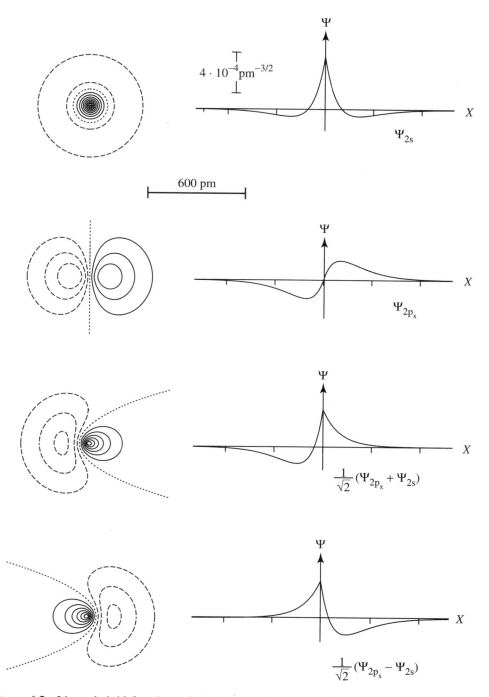

Figure 6.3 Linear hybrid functions of the H atom. Formation of the combinations ψ_I and ψ_{II} in functions (6.15) from ψ_{2s} and ψ_{2p_x}. On the right: wavefunctions along the x axis. On the left: wavefunctions in the xz plane displayed by contour lines. Positive values: solid lines; negative values: dashed lines; nodal lines: dotted lines. Spacing of contour lines: $0.5 \times 10^{-4}\,\text{pm}^{-3/2}$.

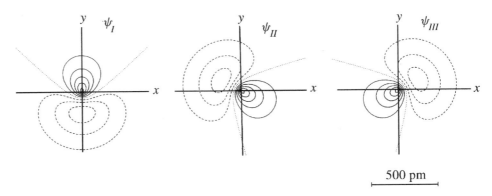

Figure 6.4 Trigonal planar hybrid function of the H atom according to functions (6.16), displayed by contour lines. Positive values: solid lines; negative values: dashed lines; nodal lines: dotted. Spacing of contour lines: 0.5×10^{-4} pm$^{-3/2}$.

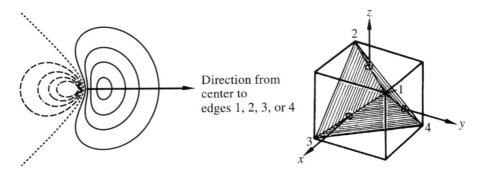

Figure 6.5 Tetrahedral hybrid function of the H atom according to functions (6.17). Spacing of contour lines 0.5×10^{-4} pm$^{-3/2}$. Positive values: solid lines, negative values: dashed lines, nodal lines: dotted. Left: the axis of the hybrid function is indicated by an arrow. Right: direction of axis of hybrid function ψ_I to corner 1 of tetrahedron, axes of functions ψ_{II}, ψ_{III}, and ψ_{IV} accordingly to corners 2, 3 and 4 respectively.

Possibility 4. Four tetrahedral hybrid functions (sp^3). Each of the four orbitals consists of $\frac{1}{4}$s, $\frac{1}{4}$p$_x$, $\frac{1}{4}$p$_y$, and $\frac{1}{4}$p$_z$ (s$^{1/4}$p$^{3/4}$ orbitals).

$$\psi_I = \tfrac{1}{2} \cdot (-\psi_{2s} + \psi_{2p_x} + \psi_{2p_y} + \psi_{2p_z}) \quad (6.17)$$

$$\psi_{II} = \tfrac{1}{2} \cdot (-\psi_{2s} - \psi_{2p_x} - \psi_{2p_y} + \psi_{2p_z})$$

$$\psi_{III} = \tfrac{1}{2} \cdot (-\psi_{2s} + \psi_{2p_x} - \psi_{2p_y} - \psi_{2p_z})$$

$$\psi_{IV} = \tfrac{1}{2} \cdot (-\psi_{2s} - \psi_{2p_x} + \psi_{2p_y} - \psi_{2p_z})$$

These combinations are illustrated in Figure 6.5. The symmetry axes of these functions are arranged tetrahedrally; therefore these functions are called *tetrahedral hybrid functions*.

These four closely considered hybrid types possess high symmetry and therefore they are important limiting cases.

6.2 Localized Electrons: Hybrid Atomic Orbitals

For the description of all atoms and molecules, besides H and H_2^+, we have to search for suitable trial wavefunctions and energies according to the variation principle. In the previously considered simple cases it was reasonable to restrict the trial functions to H atom-like wavefunctions, that is, wavefunctions corresponding to appropriate effective nuclear charges, as discussed in Chapter 5. This treatment can be significantly improved by including H atom-like hybrid functions as trial functions.

6.2.1 Li Atom (No Hybridization)

As a first example, we consider the Li atom in the ground state. The potential energy of an electron in the 2s orbital of the Li atom is lower than that in the 2p orbital, because a 2p electron is shielded by the 1s electrons more strongly than a 2s electron:

$$E_{2p_x} = E_{2s} + \Delta E \tag{6.18}$$

where ΔE is the energy to promote the electron from 2s to 2p. As a result of that, the third electron occupies the 2s orbital (Figure 6.6(a)). An Li atom in which the third electron occupies a 2p orbital (Figure 6.6(b)) must have higher energy and therefore cannot correspond to the lowest energy state. Now let us imagine that the third electron of an Li atom is described by the trial function

$$\phi = \frac{1}{\sqrt{2}} \cdot (\psi_{2s} + \psi_{2p_x}) \tag{6.19}$$

(a linear H atom like hybrid function). Then

$$\varepsilon_\phi = \tfrac{1}{2}(E_{2s} + E_{2p_x}) = \tfrac{1}{2}(E_{2s} + E_{2s} + \Delta E) = E_{2s} + \tfrac{1}{2}\Delta E \tag{6.20}$$

Thus the energy corresponding to the linear hybrid trial function is higher in energy than that corresponding to the 2s orbital (Figure 6.6(c)); as a consequence, according to the variation principle (Chapters 2 and 3), isolated atoms in the ground state cannot be described by hybrid functions.

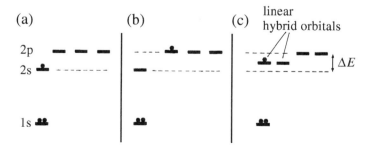

Figure 6.6 Li atom: discussion of different possibilities for the electron configuration. The third electron occupies (a) a 2s orbital, (b) a 2p orbital, (c) a linear hybrid orbital. The lowest energy is assigned to case (a) according to the variation principle.

6.2.2 BeH₂ (Linear (sp) Hybridization: Two $s^{1/2}p^{1/2}$ Hybrid Orbitals)

In the BeH₂ molecule four electrons are available for the bonds (the outer electrons of Be and the electrons of each of the H atoms). Two electrons can be attributed to wavefunctions that are obtained from the hybrid function (6.19) and the 1s function of one of the H atoms. The other two electrons are attributed, respectively, to wavefunctions that are derived from the hybrid function

$$\phi = \frac{1}{\sqrt{2}}(\psi_{2s} - \psi_{2p_x}) \tag{6.21}$$

which is orthogonal to function (6.19) and the 1s function of the second H atom (Figure 6.7). BeH₂ should be linear.

We used this example for illustration, even though BeH₂ is not known as a simple molecule – a polymeric form has lower energy. This polymeric form appears in the solid state, where each Be is surrounded by four H atoms (for details see Figure 6.15 in Section 6.3.2).

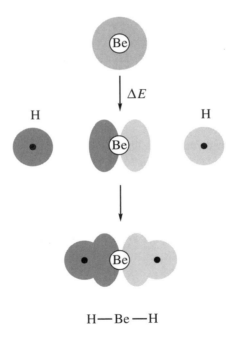

Figure 6.7 Formation of BeH₂: Description by localized hybrid orbitals.

6.2.3 H₂S and H₂O (Two Orbitals Between p and $s^{1/4}p^{3/4}$ Hybrid)

Now let us check if hybrid functions can be meaningful in describing the bonding situation in H₂S. The two hydrogens, for example, can bind to the sulfur atom using the $3p_x$ or $3p_y$ orbitals of the sulfur (Figure 6.8(a)), or linear hybrid orbitals (Figure 6.8(b)), or tetrahedral orbitals (Figure 6.8(c)). Which of these possibilities corresponds to the lowest energy?

6.2 LOCALIZED ELECTRONS: HYBRID ATOMIC ORBITALS

We take into account that energy is released when the bond with the H atom is formed. Tetrahedral hybrid functions overlap better with the hydrogen function than 3p functions; consequently, in the case of Figure 6.8(c) a larger amount of overlap energy is released than in the case of Figure 6.8(a). However, this difference is still not sufficient to compensate for the energy expense of

$$2 \cdot \tfrac{3}{4}\Delta E - 4 \cdot \tfrac{1}{4}\Delta E = \tfrac{1}{2}\Delta E \tag{6.22}$$

(promotion energy) corresponding to the formation of tetrahedral hybrid orbitals. Also it makes no sense to use linear hybrid orbitals, because the promotion energy $2 \cdot \tfrac{1}{2}\Delta E = \Delta E$ is even larger. Thus a bond according to Figure 6.8(a) should be realized (using pure p orbitals of sulfur for bonding).

In the H_2O molecule, however, the hydrogen atoms are positively charged, because of their low electronegativity; therefore they repel each other strongly. The bond angle increases and the Coulomb repulsion energy decreases when we proceed from bonding by p orbitals to bonding by hybrid orbitals. This energy gain can compensate for the expense of energy in forming the hybrid orbitals. Correspondingly, the bond angle of the H_2O molecule (105°) is considerably larger than that of the H_2S molecule (92°). In H_2S, the hydrogen atoms are less positively charged and nearly pure p orbitals are used for bonding; in H_2O, approximately $s^{1/8}p^{7/8}$ hybrid orbitals are involved.

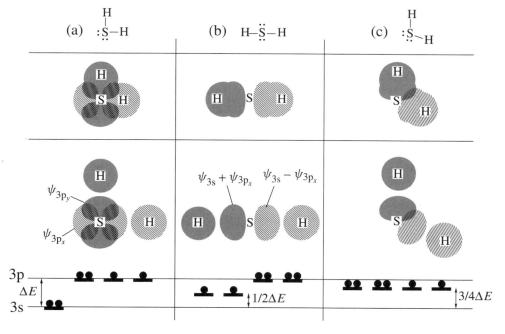

Figure 6.8 H_2S molecule: discussion of different possibilities for the electron configuration. Electrons with the quantum number $n = 3$ occupy (a) one 3s and three 3p orbitals of S (analog to functions (6.14)), (b) two linear hybrid orbitals (analog to functions (6.15)), and (c) four tetrahedral hybrid orbitals (analog to functions (6.17)). Only those orbitals of S are displayed which are considered to form bonds with the H atoms. Possibility (a) is favored according to the variation principle, because no promotion energy is needed.

6.2.4 CH₄ (Tetrahedral (sp³) Hybridization: Four $s^{1/4}p^{3/4}$ Hybrid Orbitals)

According to Figure 6.9, three possibilities are imaginable for the bonding of the H atoms to the C atoms.

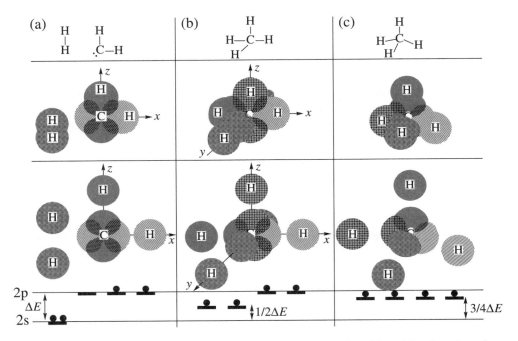

Figure 6.9 Four H atoms and one C atom. (a) Two H atoms form H_2, while the others form CH_2 (bonding through 2p orbitals of C). (b) and (c) Four H atoms form CH_4; (b) bonding through linear hybrid orbitals of C (x axis) and through 2p orbitals (y and z axes); (c) bonding through tetrahedral hybrid orbitals of C. Possibility (c) is favored. The promotion energy in (b) and (c) are equal but the overlap with the 1s orbitals of the H atoms is better.

Case a. Bonding of two H atoms to a C atom using 2p orbitals; no promotion energy has to be supplied, but only two bonds are formed.

Case b. Bonding of four H atoms to a C atom using two linear hybrid and to two 2p orbitals; the energy for the promotion of hybrid orbitals is

$$E_H = 2 \cdot \tfrac{1}{2}\Delta E = \Delta E \quad \text{(linear hybrid orbital)} \tag{6.23}$$

In this case also the energy ΔE has to be supplied, but four bonds are formed.

Case c. Bonding of four H atoms to a C atom using four tetrahedral hybrid orbitals; in this case the promotion energy is

$$E_H = 2 \cdot \tfrac{3}{4}\Delta E - 2 \cdot \tfrac{1}{4}\Delta E = \Delta E \quad \text{(tetrahedral hybrid orbital)} \tag{6.24}$$

The overlap through hybrid orbitals in case c is better than that in case b. Therefore tetrahedral orbitals should be preferred for the formation of the bonds. Indeed, a tetrahedral arrangement of the H atoms around the C atom is found experimentally. In methane derivatives the bonding angle is very close to the tetrahedral angle (109°28′); similar relations are found for silicon compounds.

6.3 Properties of Electron Pair Bonds

We summarize the conditions for the formation of an electron pair bond.

- Both bond partners provide one suitable orbital each (the corresponding energy levels are sufficiently low) (Figures 6.10 and 6.11).
- Two electrons are needed for the bond; each partner can contribute one electron as discussed above. Alternatively, both electrons can be contributed by one partner (dative bond). Examples are NH_4^+ and $\overset{\oplus}{N}H_3-\overset{\ominus}{B}(CH_3)_3$ (Figure 6.12). The partner which contributes to the bonding electrons is called the *donor* (here the NH_3 molecule), the other one is the *acceptor*.

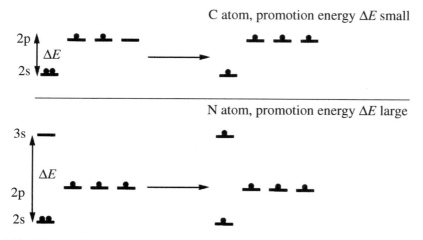

Figure 6.10 Electron structure of N in comparison with C. In the C atom the promotion energy ΔE is small enough, and C can form four bonds. To do this in the case of an N atom, a 2s electron must be raised to a 3s level; the promotion energy ΔE is too large, and N cannot form four bonds. On the other hand, N^+ has the same number of electrons as carbon, and NH_4^+ is a stable ion.

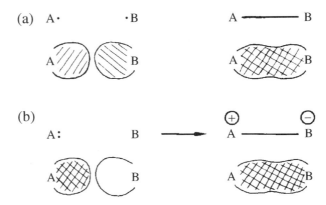

Figure 6.11 Formation of an electron pair bond. (a) Each of the two partners contributes one electron. (b) Bond partner A donates both electrons (donor), and bond partner B has an empty orbital (acceptor).

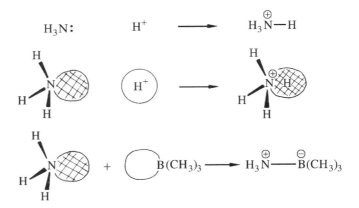

Figure 6.12 Examples of dative bonds.

6.3.1 Formal and Effective Charges

Formally we may associate one electron of an electron pair bond with one partner and associate the second electron with the other partner. Then in the case of a dative bond the donor carries a charge $+e$ while the acceptor carries a charge $-e$ (Figure 6.13 and Figure 6.11(b)). These formal charges must be distinguished from the effective charge. The bonding electrons are not uniformly spread over the bond partners, but are shifted toward the more electron-attracting atom.

Then the formal positive charge on the N atom in NH_4^+ decreases, because on the average the bonding electrons spend more time close to the N than to the H atoms. The effective charges are nearly equally distributed over the N atom and the H atoms (Figure 6.14).

In order to clarify what type of bond exists in a dative bond, one has to start with the formal charge of the considered atom. Formally the N atom in NH_4^+ has four electrons that are distributed, like in the neutral C atom in CH_4, over the 2s and 2p states. Thus we conclude that the N atom in NH_4^+ has to be tetrahedral like the C atom in CH_4.

6.3 PROPERTIES OF ELECTRON PAIR BONDS

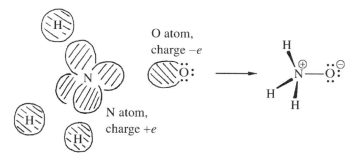

Figure 6.13 An example of formal charges.

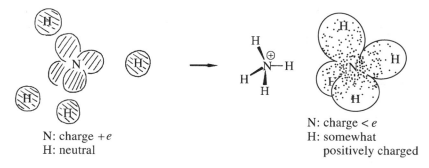

Figure 6.14 Formal and effective charge in NH_4^+. The effective charge is almost equally distributed over H and N.

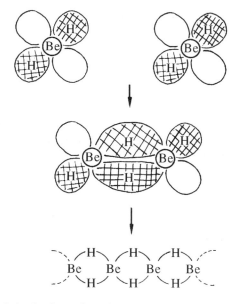

Figure 6.15 Dative bonds in the formation of $(BeH_2)_n$.

In $\overset{\oplus}{\text{NH}}_3 - \overset{\ominus}{\text{B}}(\text{CH}_3)_3$ the B atom has a formal single negative charge; that is, it has the same electron arrangement as the C atom, along with a tetrahedral hybridization, contrary to the trigonal planar BH_3 molecule.

6.3.2 Polymerization of BeH$_2$

$(BeH_2)_n$ constitutes a special case. Both bonding electron pairs in BeH_2 share the two empty orbitals of another BeH_2 molecule to form electron pair bonds between Be atoms. The two protons are pulled into the charge cloud of both electron pairs and thus stabilize the system. The first BeH_2 is a donor while the second one is an acceptor, but at the same time the second one is the donor for the first one, and a third BeH_2 binds to the second one, accordingly, and so on. In an infinite chain each Be atom forms four equal bonds using tetrahedral hybrid orbitals (Figure 6.15). This type of bond is called a *three-center two-electron bond* since a proton is found in the cloud of each electron pair.

6.4 Delocalized Electrons: Molecular Orbitals

6.4.1 Box Wavefunctions

For the description of bonding in BeH_2, H_2O, and CH_4 we have used the concept of hybridization. We can, however, describe the bonds in a totally different way, using orbitals extending over the whole molecule (*molecular orbitals*).

Both approaches lead to essentially the same result. Electrons are indistinguishable, only the total electron cloud can be observed. It is irrelevant whether this cloud is considered as being composed of localized or delocalized contributions. First we use the box wavefunction as a simple version of a molecular orbital.

Box model for BeH$_2$

We imagine a box which contains the Be nucleus and both H atom nuclei (Figure 6.16(a)). Furthermore, we imagine the two 1s electrons of Be as localized at the Be nucleus (they do not contribute to the bond). The other four electrons in the box occupy the two wavefunctions ψ_1 and ψ_2 in Figure 6.16(b). Figure 6.16(c) shows the charge cloud of BeH_2.

The following consideration shows that both models, namely, the description by localized (Figure 6.7) or delocalized orbitals (Figure 6.16), are practically equivalent: the difference in energy attributed to the two molecular orbitals ψ_1 and ψ_2 is small. If we neglect this energy difference we can form two hybrid functions orthogonal to each other

$$\psi = \psi_1 + \psi_2, \quad \psi' = \psi_1 - \psi_2 \tag{6.25}$$

These hybrid functions correspond to the bonding orbitals in Figure 6.7.

Box model for O$_2$

With a proper box model we can also describe the behavior of the valence electrons of other small molecules like O_2. Again we assume that the 1s electrons are localized on the nuclei. Then in O_2 a total of 12 valence electrons is to be distributed. When we proceed as for BeH_2, we obtain the eight box wavefunctions and energy levels illustrated in

6.4 DELOCALIZED ELECTRONS: MOLECULAR ORBITALS

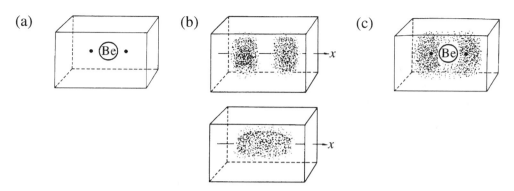

Figure 6.16 Bonding in the BeH$_2$ molecule. (a) Box model (replacing the coulomb field of Be^{2+} and the two protons by a box which is filled with the four valence electrons. (b) Charge cloud representation of the two box wavefunctions with the lowest energies. (c) Total charge cloud representation.

Figure 6.17, when we choose a box of dimensions $L_x = 450$ pm and $L_y = L_z = 300$ pm. In O$_2$ these orbitals are occupied consecutively by two electrons each. Adding up the squares of the wavefunctions of the occupied orbitals, we obtain the charge cloud in Figure 6.18.

Molecular orbitals with antinodes in the line connecting the nuclei are called σ *orbitals* (Figure 6.17, left); orbitals in which the line connecting the nuclei is in a nodal plane are called π *orbitals* (Figure 6.17, right).

The σ orbitals 2, and 4 and the π orbitals 3 and 4 have a node in the center of the bond and therefore are antibonding, whereas the σ orbitals 1 and 3 and the π orbitals 1 and 2 have an antinode in the center of the bond and therefore are bonding.

There are, then, four bonding and three occupied nonbonding orbitals. Two electrons have to be distributed over the degenerate π orbitals 3 and 4. As a result, O$_2$ has two unpaired electrons. This explains why O$_2$ is paramagnetic.

6.4.2 LCAO Wavefunctions

As outlined in Chapter 4, the wavefunction of the H$_2^+$ molecule ion can be described not only by box wavefunctions, but also by a linear combination of atomic orbitals (LCAO functions):

$$\phi = \tfrac{1}{2}\sqrt{2} \cdot (\varphi_a + \varphi_b) \qquad (6.26)$$

where φ_a and φ_b are 1s atomic orbitals located at the positions of the two protons a and b.

LCAO model for O$_2$

Correspondingly, the O$_2$ molecule wavefunctions of the valence electrons can be described by linear combinations of 2s and 2p atomic orbitals of the oxygen atom.

The two lowest molecular orbitals are obtained from the linear combination of two 2s atomic functions, a bonding and an antibonding one. The same is true for the combination of the 2p atomic orbitals. In this way we obtain the scheme in Figures 6.19 and 6.20 corresponding to the scheme in Figure 6.17.

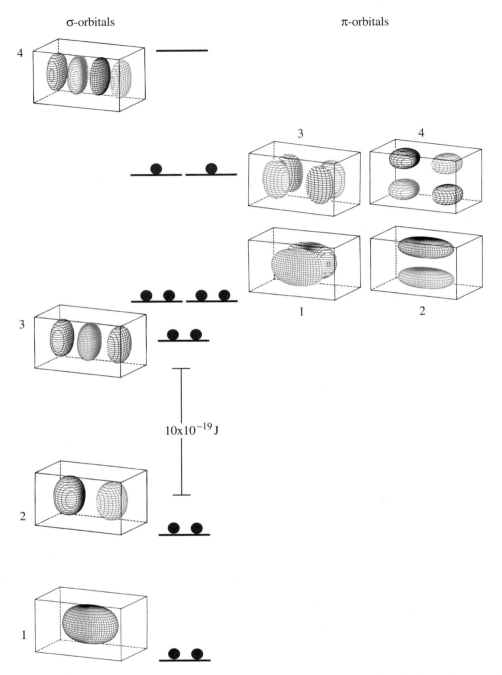

Figure 6.17 Box model for O$_2$ ($L_x = 450$ pm, $L_y = L_z = 300$ pm). Population scheme and illustration of the eight lowest box wavefunctions. Left: σ orbitals, right: π orbitals. The contour lines connect points in space where the absolute value of the wavefunction is 2.5×10^{-4} pm$^{-3/2}$. Black lines: positive values, red lines: negative values.

6.4 DELOCALIZED ELECTRONS: MOLECULAR ORBITALS

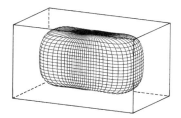

Figure 6.18 Charge cloud of the O_2 molecule, calculated from the wavefunctions in Figure 6.17. The contour lines connect points in space which have a charge density of $e \cdot 0.075 \cdot a_0^{-3}$.

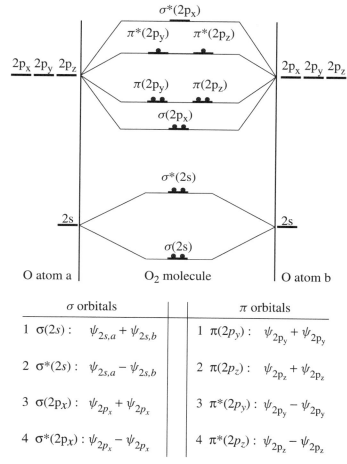

Figure 6.19 LCAO model for O_2. The eight lowest LCAO wavefunctions. The relations hold for the assumption that the atomic orbitals φ_a and φ_b have the same sign in the overlap region. Therefore, the molecular σ orbitals 1 and 3 and the π orbitals 1 and 2 have an antinode in the center of the molecule (bonding orbitals), whereas the molecular σ orbitals 2 and 4 and the π orbitals 3 and 4 have a node (antibonding orbitals, marked by an asterisk).

158 BONDING DESCRIBED BY HYBRID AND MOLECULAR ORBITALS

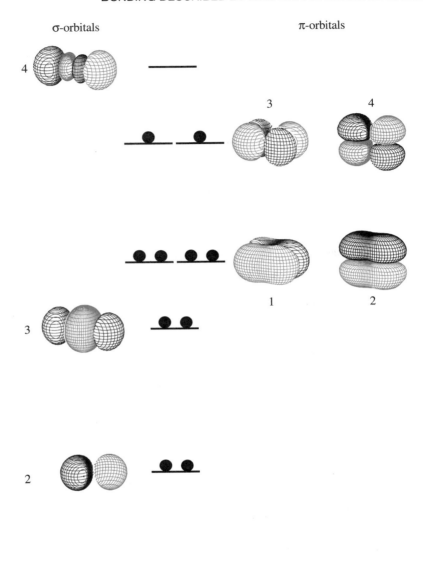

Figure 6.20 LCAO model for O_2 (Slater functions with effective charge number 4.55, internuclear distance 121 pm). Population scheme and illustration of the eight lowest orbitals. Left: σ orbitals, right: π orbitals. The contour lines connect points in space where the absolute value of the wavefunction is 2.5×10^{-4} pm$^{-3/2}$. Black lines: positive values, red lines: negative values.

6.4.3 Improvement of Trial Function

In this chapter we have discussed two methods to describe bonding electrons; (a) the description by localized hybrid orbitals and (b) the description by molecular orbitals (using box and LCAO functions). Hybrid orbitals provide a clear picture of σ bonds, whereas molecular orbitals are very useful in describing π bonds. It should be pointed out that in this chapter the discussion on methods for describing bonds is based on very simplified models; therefore, we should not expect that the obtained results will agree with experimental data quantitatively.

We must always keep in mind that electrons are indistinguishable. Improvements are obtained by using many-electron wavefunctions taking account of indistinguishability and antisymmetry (see Section 4.3.3). In spite of the neglect of indistinguishability, the two simplifications considered in this chapter are useful to describe the electron structure of a molecule and to understand specific molecular properties. Both methods will be applied in Chapters 7, 8, and 23.

Problems

Problem 6.1 – Linear Combinations of Wave Functions

Show, by insertion of equation (6.6) in the Schrödinger equation, that any linear combination of wavefunctions corresponding to the same energy is a solution of the Schrödinger equation. Show that this does not hold for a linear combination of wavefunctions belonging to different energies.

Solution. The Schrödinger equation is fulfilled if

$$\mathcal{H}(a\psi_{12} + b\psi_{21}) = const \cdot (a\psi_{12} + b\psi_{21})$$

In general, because of

$$\mathcal{H}\psi_{12} = E_{12}\psi_{12}, \quad \mathcal{H}\psi_{21} = E_{21}\psi_{21}$$

we obtain

$$\mathcal{H}(a\psi_{12} + b\psi_{21}) = aE_{12}\psi_{12} + bE_{21}\psi_{21}$$

This means that in the general case ($E_{21} \neq E_{12}$) the trial function does not satisfy the Schrödinger equation. Only in the special case that the energy levels are twofold degenerate ($E_{21} = E_{12} = E$) we find

$$\mathcal{H}(a\psi_{12} + b\psi_{21}) = E(a\psi_{12} + b\psi_{21})$$

in accordance with the requirement of the Schrödinger equation.

Problem 6.2 – Degeneracy and Hybridization of Box Functions

Discuss the degeneracy in the case of an electron in a cubic box.
Solution. For example, the state with $n_x = 1$, $n_y = 2$, and $n_z = 3$ is sixfold degenerate.

Problem 6.3 – Orthogonality and Normalization of Hybrid Functions

Show that the wavefunctions (6.15) are normalized and orthogonal to each other and that the same is true for wavefunction (6.16).

Solution. Linear combination of wavefunctions (6.15):

$$\int \psi_I^2 \cdot d\tau = \frac{1}{2} \cdot \left(\int \psi_{2s}^2 \cdot d\tau + 2 \int \psi_{2s}\psi_{2p_x} \cdot d\tau + \int \psi_{2p_x}^2 \cdot d\tau \right) = 1$$

$$\int \psi_{II}^2 \cdot d\tau = \frac{1}{2} \cdot \left(\int \psi_{2s}^2 \cdot d\tau - 2 \int \psi_{2s}\psi_{2p_x} \cdot d\tau + \int \psi_{2p_x}^2 \cdot d\tau \right) = 1$$

$$\int \psi_I \psi_{II} \cdot d\tau = \frac{1}{2} \cdot \left(\int \psi_{2s}^2 \cdot d\tau - \int \psi_{2s}\psi_{2p_x} \cdot d\tau + \int \psi_{2s}\psi_{2p_x} \cdot d\tau - \int \psi_{2p_x}^2 \cdot d\tau \right) = 0$$

Linear combination of wavefunctions (6.16):

$$\int \psi_I^2 \cdot d\tau = \frac{1}{3} \cdot (1 + 2 \cdot 1) = 1$$

$$\int \psi_{II}^2 \cdot d\tau = \frac{1}{3} \cdot \left(1 + \frac{3}{2} + \frac{1}{2} \right) = 1$$

$$\int \psi_I \psi_{II} \cdot d\tau = \frac{1}{3} \cdot (1 - 1) = 0$$

Problem 6.4 – Symmetry Properties of Linear Hybrid Functions

Show that the second hybrid function in (6.15) can be obtained from the first one by rotation of 180° around the z axis.

Solution. Rotation around the z axis: ψ_{2s} remains unchanged and ψ_{2p_x} changes its sign; thus from $\psi_I = a\psi_{2s} + b\psi_{2p_x}$ we obtain $\psi_{II} = a\psi_{2s} - b\psi_{2p_x}$.

Problem 6.5 – Energy of Hybrid States

Show that for a hybrid function

$$\phi = a\psi_{2s} + b\psi_{2p_x} + c\psi_{2p_y} + d\psi_{2p_z}$$

the expression

$$\varepsilon_{\text{hybrid}} = E_{2s} + (b^2 + c^2 + d^2) \cdot \Delta E$$

is obtained; calculate $\varepsilon_{\text{hybrid}}$ for the tetrahedral hybrid state.

Solution.

$$\varepsilon_{\text{hybrid}} = \int \phi \mathcal{H} \phi \cdot d\tau = a^2 E_{2s} + b^2 E_{2p_x} + c^2 E_{2p_y} + d^2 E_{2p_z}$$

With $E_{2p_x} = E_{2p_y} = E_{2p_z} = E_{2s} + \Delta E$ we obtain

$$\varepsilon_{\text{hybrid}} = E_{2s} \cdot (a^2 + b^2 + c^2 + d^2) + (b^2 + c^2 + d^2) \cdot \Delta E$$

Because of normalization of ϕ we obtain $a^2 + b^2 + c^2 + d^2 = 1$. Then for the tetrahedral hybrid state ($a^2 = b^2 = c^2 = d^2 = \frac{1}{4}$) we obtain

$$\varepsilon_{\text{hybrid}} = E_{2s} + \tfrac{3}{4} \cdot \Delta E$$

Problem 6.6 – Structure of B(CH$_3$)$_3$ and Hg$_2$Cl$_2$

Show that the B(CH$_3$)$_3$ molecule is expected to be trigonal planar. Consider the formation of Hg$_2$Cl$_2$, Hg(phenyl)$_2$, and O(phenyl)$_2$. In the case of Hg it is assumed that a hybrid is formed from 6s and 6p orbitals.

Solution. In B, one electron is promoted from the 2s into the 2p orbital: formation of a trigonal planar hybrid orbital. Then each orbital of B and each CH$_3$ provide one electron to form three electron pair bonds. Tetrahedral hybridization would require more promotion energy to form three $s^{1/4}p^{3/4}$ hybrid orbitals.

In Hg, one electron is promoted from the 6s into the 6p state: formation of a linear hybrid orbital. Thus Hg$_2$Cl$_2$ and Hg(phenyl)$_2$ should be linear. In the case of O, no empty 2p orbitals are available for hybridization, and O(phenyl)$_2$ should have a bent structure like H$_2$O.

Problem 6.7 – Dative Bond

Show that in the formation of amine oxide $(H_3)\overset{\oplus}{N}-\overset{-}{O}|^{\ominus}$ the NH$_3$ molecule is the donor and the O atom is the acceptor.

Solution. One electron of the lone electron pair in NH$_3$ (donor) is transferred into an unoccupied 2p orbital of O (acceptor), see Figure 6.13. Then N has a formal positive charge, the O has a formal negative charge. Thereafter, each bonding partner contributes one electron to form the electron pair bond.

Problem 6.8 – Three-center Bond (CH$_5^+$)

In strongly acidic solution the carbonium ion CH$_5^+$ and the carbenium ion CH$_3^+$ are formed according to

$$CH_4 + H^+ \longrightarrow CH_5^+ \longrightarrow CH_3^+ + H_2$$

Discuss the structure of these ions.

Solution. CH$_3^+$ can be formed from C$^+$ and three H atoms; in C$^+$ a trigonal planar hybrid orbital offers the best bonding possibility, and thus CH$_3^+$ should be trigonal planar. CH$_5^+$ can be formed by addition of a proton to methane; one of the tetrahedral hybrid orbitals is shared by two protons (two-electron three-center bond similar to the case of polymer BeH$_2$).

Problem 6.9 – LCAO Model for O₂

Why are the energy levels of the linear combinations of the $2p_x$ orbitals of the two atoms (orbitals $\sigma(2p_x)$ and $\sigma^*(2p_x)$ in Figures 6.19, 6.20) more strongly separated than those of the linear combinations of the $2p_y$ and $2p_z$ orbitals (orbitals $\pi(2p_y)$, $\pi(2p_z)$ and $\pi^*(2p_y)$, $\pi^*(2p_y)$, respectively?

Solution. In σ orbital 3 there are antinodes of the wavefunction along the bond line; in π orbitals 1 and 2 the antinodes are at some distance from the bond line. Thus the electrons in σ orbital 3 are more strongly attracted by the nuclei than those in π orbitals 1 and 2; thus σ orbital 3 has a lower energy. In σ orbital 4 and in π orbitals 6 and 7 we have nodal planes perpendicular to the bond line in the center of the bond, thus the energies of these orbitals must be higher than those in the isolated atoms. In σ orbital 4 the electrons are closer together and strongly repel each other, so the corresponding energy level is higher than that for π orbitals 3 and 4.

Problem 6.10 – Molecular Orbitals of Some Diatomic Molecules

Set up the schemes for the molecular orbitals of Li₂, Be₂, B₂, C₂, N₂, and F₂.

Solution: In analogy to O₂, the following schemes are obtained (Figure 6.21).

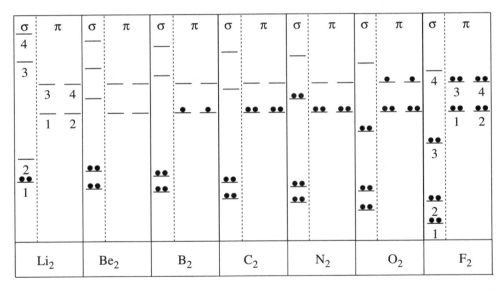

Figure 6.21 Molecular orbital scheme for Li₂ to F₂ in the box model approximation (Figure 6.17 in case of O₂). Note that σ level 3, when proceeding from Li₂ to F₂ is continuously decreasing in height relative to π levels 1 and 2, since the electrons in σ level 3 are less shielded from nuclei and therefore more strongly stabilized by increasing nuclear charge.

7 Molecules with π Electron Systems

How can we understand the extraordinary properties of molecules with double or triple bonds? These molecules can be viewed as systems where π electrons move relatively freely and are extending over the entire molecular skeleton formed by σ electrons. The charge distribution of the π electrons (given by the position of the antinodes of the wavefunctions) disturbs the molecular skeleton and vice versa. The result of this interaction depends on details of the geometry of the molecular skeleton.

Bonds of equal lengths result in benzene, and alternating short and long bonds occur in polyenes. Tiny changes in the electronic structure have important consequences on the spectroscopic properties, which will be discussed in Chapter 8. It is fascinating to see these basic features in the simple picture of de Broglie waves.

7.1 Bonding Properties of π Electrons

So far we have mainly examined molecules with single bonds (one dash in the structural formula) – for example,

$$\begin{array}{cccc} \mathrm{H} & & & \\ | & \mathrm{H_3C} \quad \mathrm{CH_3} & & \\ \mathrm{H-C-H} & \mathrm{B} & \mathrm{H-H} & \mathrm{H-Cl} \\ | & | & & \\ \mathrm{H} & \mathrm{CH_3} & & \end{array}$$

Now we consider molecules with double or triple bonds (two or three dashes in the structural formula) – for example,

Ethene   Butadiene   Benzene

H—C≡C—H   H—C≡C—C≡C—H   H—C≡N
   Ethyne   Diethyne   Hydrogen cyanide

As a first example we consider ethene. Each C atom has three neighbors, so the best suitable bonding orbital is a trigonal planar hybrid orbital (Figure 7.1(a)). The C atoms are kept together by trigonal planar σ bonds and a π bond (Figure 7.2). It is obvious that the energy needed for breaking the C–C bond in ethene (the bond energy) must then be larger than for ethane; furthermore, the force constant is larger and the bond length is shorter (Table 7.1). Rotation of the two parts of the ethene molecule around the C–C axis causes a strong decrease in the overlap of the π orbitals (Figure 7.3): the molecule is brought into a very unfavorable energetic state. This means that unrestricted rotation is not possible, unlike in the ethane molecule.

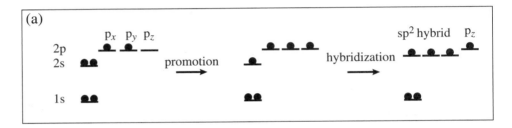

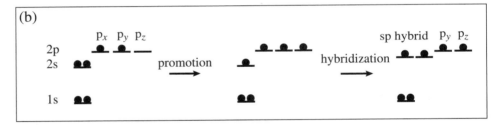

Figure 7.1 Formation of (a) a trigonal planar and (b) a linear hybrid orbital for a carbon atom.

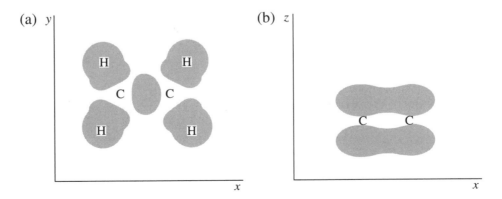

Figure 7.2 Ethene molecule. (a) Trigonal planar hybrid orbitals (xy plane: view from above); (b) $2p_z$ orbitals (xz plane: side view). Inside the shaded area the charge density is larger than $e \cdot 0.02 \times 10^{30}$ m^{-3}.

7.2 FREE-ELECTRON MODEL

Table 7.1 Bond energies, force constants, and bond lengths of C—C bonds (H_2^+ ion and H_2 molecule for comparison)

	Bond energy/10^{-19} J	Force constant/N m^{-1}	Bond length/pm
Ethane	4.1	434	154
Ethene	7.0	1080	133
Ethyne	8.6	1490	120
$(H-H)^+$	4.2	160	106
H—H	7.2	574	74

Bond energies in polyatomic molecules are extracted from the energies of dissociation into atoms. It is postulated that these energies are the sum of the energies for each bond. In this way, appropriate values for bond energies are obtained.

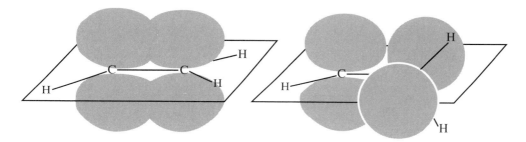

Figure 7.3 Decrease of the overlap of the π orbitals at a rotation of the ethene molecule.

In the ethyne molecule, each C atom has two neighbors. The most suitable σ bonding orbital is, therefore, a linear hybrid orbital (Figure 7.1(b)). In this case, two singly occupied π orbitals are available ($2p_y$ and $2p_z$), both of which can form a π bond (Figure 7.4). Therefore, the bond energy and the force constant of the C≡C bond in ethyne are even larger than for the C=C bond, and the bond is shorter (Table 7.1).

7.2 Free-electron Model

7.2.1 Linear π Electron Systems

Three-dimensional box model

We have already used the box model successfully for describing the chemical bond in the H_2^+ ion (Chapters 2 and 4). Now we apply this model to the ethene molecule. The cores (nucleus and two 1s electrons) of both C atoms and the nuclei of the four hydrogens (protons) are arranged according to Figure 7.5 in a box of dimensions $L_x = 500$ pm, $L_y = 400$ pm, and $L_z = 300$ pm. These dimensions are chosen such that the box extends the nuclear skeleton in the x, y, and z directions by about 150 pm. This corresponds to our finding in Section 4.1.2 that the box dimensions for H_2^+ are $L_x = 400$ pm, $L_y = L_z = 300$ pm. We assume that the 12 bonding electrons (four valence electrons of each of the two C atoms and four electrons of the H atoms) are described by box wavefunctions. The 12 electrons are distributed according to Figure 7.6(a) (lowest six energy levels); the orbital energies resulting from this box model agree well with experimental findings (Figure 7.6(b)).

The corresponding orbitals for ethene are shown in Figure 7.7. The five lowest orbitals have an antinode in the plane of the molecule ($n_z = 1$); the ten electrons in this group are

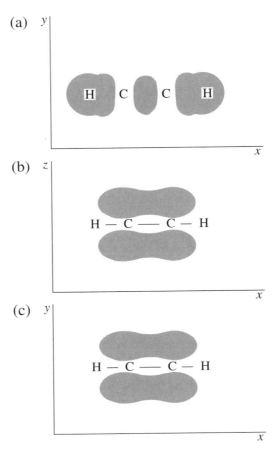

Figure 7.4 Ethyne molecule. (a) Linear hybrid orbitals (xy plane); (b) and (c) $2p_y$, $2p_z$ orbitals (xz, xy plane). Inside the shaded area the charge density is larger than $e \cdot 0.02 \times 10^{30}$ m^{-3}.

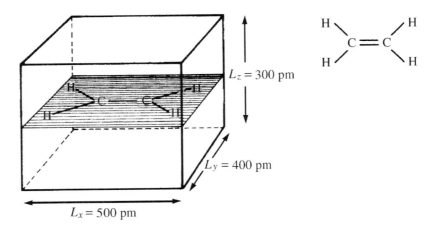

Figure 7.5 Box model for the ethene molecule.

7.2 FREE-ELECTRON MODEL

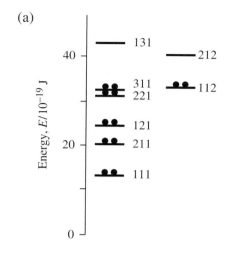

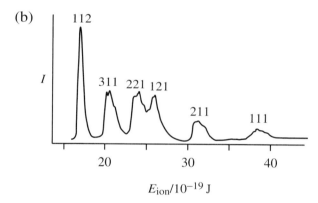

Figure 7.6 (a) Energy scheme for the box model of ethene ($L_x = 500\,\text{pm}$, $L_y = 400\,\text{pm}$, $L_z = 300\,\text{pm}$); the quantum numbers are in the order n_x, n_y, n_z. (b) Photoelectron spectrum of ethene (for details on photoelectron spectra see Section 9.6.2). The peaks correspond to the ionization of electrons from the occupied molecular orbitals as indicated. They appear in the same sequence as the levels in the box model; also the energy spacing agrees well with the spacing expected from the box model (part (a)). For example, the spacing between orbitals 111 and 112 is $22 \times 10^{-19}\,\text{J}$, while it is $20 \times 10^{-19}\,\text{J}$ in the box model. (Photoelectron spectrum of ethene see C. R. Brundle, M. B. Robin, H. Basch, M. Pinsky, and A. Bond, *J. Am. Chem. Soc.*, 1970, **92**, 3863 (Figure 1)).

called σ electrons. The highest occupied orbital has a node in the plane of the molecule ($n_z = 2$); the two electrons in this group are called π electrons.

The σ electrons represent the C—C bond and the four C—H bonds which we have described in Section 7.1 by bonds made of sp³ hybrid orbitals of the two C atoms and the 1s orbitals at the four H atoms. The two π electrons are responsible for the second bond in C=C.

Figures 7.8(a)–(c) display the π orbitals for the ethene molecule obtained in the box model. In Figure 7.8(a) the orbitals are represented by an electron cloud, in Figure 7.8(b) a contour line representation is given, and in Figure 7.8(c) a one-dimensional representation shows the wavefunction along a line parallel to the bond line a distance of 75 pm above the plane of the molecule.

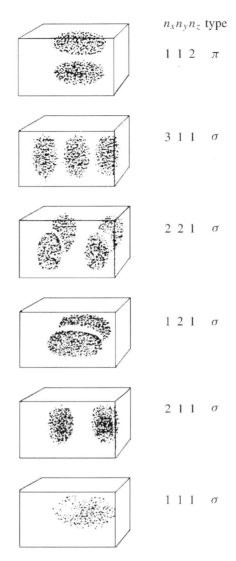

Figure 7.7 Electron cloud representation of the charge distribution in the six lowest occupied molecular orbitals in the ethene molecule ($L_x = 500$ pm, $L_y = 400$ pm, $L_z = 300$ pm). Electrons with an antinode in the plane of the molecule are called σ electrons. Electrons with an node in the plane of the molecule are called π electrons. The numbers denote the quantum numbers n_x, n_y, and n_z.

In butadiene there are 22 valence electrons (16 electrons of the four C atoms and 6 electrons of the H atoms). They are considered as electrons in the box in Figure 7.8(d), which corresponds to the *trans* form of the molecule. 18 electrons occupy the lowest nine σ orbitals representing the three C–C and six C–H bonds. The four remaining electrons occupy two π orbitals.

In the following we focus on the π subsystem of orbitals with one node in the plane of the molecule. Each C atom in butadiene contributes one π electron; the lowest two orbitals are occupied by two electrons each, according to the Aufbau principle. Figure 7.8(e) shows

7.2 FREE-ELECTRON MODEL

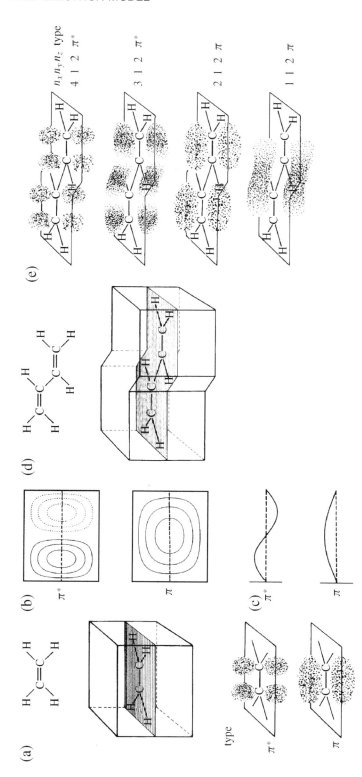

Figure 7.8 Box model for ethene and butadiene. Only π states are shown. Bonding orbitals are denoted by π, antibonding orbitals are denoted by π^*. Ethene: (a) cloud representation of the charge distribution; (b) contour line representation of the wavefunction (in a plane $L_z/4 = 75$ pm above the molecular plane, distance of the contour lines 0.1×10^{15} m$^{-3/2}$); (c) representation by one dimensional wavefunction (course along the dashed line in part (b)); (d) and (e) Butadiene: cloud representation of the charge distribution.

the lowest four π electron orbitals which differ in the number of nodes and antinodes along the bond line. The lowest π orbital has one antinode along the bond line, the next orbitals have two, three, and four antinodes.

One-dimensional free-electron model of π electrons

Thus it is reasonable to approximate the π orbitals by one-dimensional waves along the bond line (length L) with de Broglie wavelengths (see Chapter 2) of

$$\Lambda = \frac{2L}{n}, \quad n = 1, 2, 3 \ldots \tag{7.1}$$

This simplified method is called the *free-electron model*. In this model the wavefunctions are given by

$$\psi_n = \sqrt{\frac{2}{L}} \sin \frac{\pi n s}{L}, \quad n = 1, 2, 3 \ldots \tag{7.2}$$

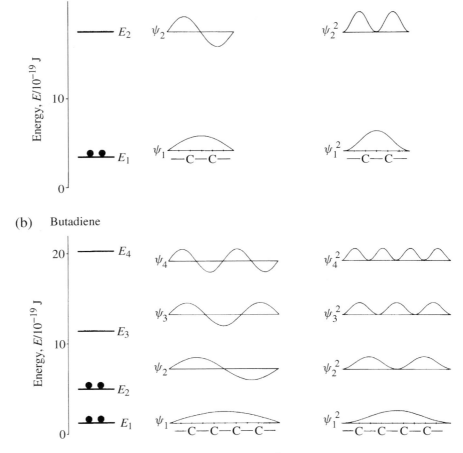

Figure 7.9 Energy E, wavefunction ψ, and square ψ^2 in the free-electron model. (a) Ethene ($L = 3 \cdot 140\,\text{pm} = 420\,\text{pm}$). (b) Butadiene ($L = 5 \cdot 140\,\text{pm} = 700\,\text{pm}$).

7.2 FREE-ELECTRON MODEL

where s is the coordinate along the chain of carbon atoms. The corresponding energies E are

$$E = \frac{1}{2}m_e v^2 = \frac{h^2}{2m_e \Lambda^2} = \frac{h^2}{8m_e L^2} \cdot n^2, \quad n = 1, 2, 3 \ldots \quad (7.3)$$

When considering the box length for a box including all valence electrons we assumed that the box extends 150 pm beyond the outermost H atom in the molecular skeleton (see Figure 7.5). Correspondingly, a box for π electrons extends about 150 pm beyond the outermost carbon atoms. Therefore, we determine the box length L for π electrons by the length of the bond line including one bond length beyond each of the outermost carbon atoms (see Figure 7.8).

The bond lengths in conjugated systems are in the range between 148 pm and 133 pm. As an average, we use $d_0 = 140$ pm. The energies and wavefunctions for ethene ($L = 3d_0 = 420$ pm) and for butadiene ($L = 5d_0 = 700$ pm) are shown in Figure 7.9.

An interesting case, in comparison with butadiene, is the amidinium ion

$$H_2 \overset{\oplus}{N}=CH-\overset{..}{N}H_2$$

($L = 4d_0 = 560$ pm) which has the same number of π electrons occupying orbitals with $n = 1$ and $n = 2$. However, the bonding features are quite different in the two cases, as we will discuss in Section 7.2.3.

7.2.2 Cyclic π Electron Systems

As an example, we consider the benzene molecule. In the free-electron model (Figure 7.10(a)) we obtain the individual π electron orbitals by investigating the stationary waves along a cyclic closed string.

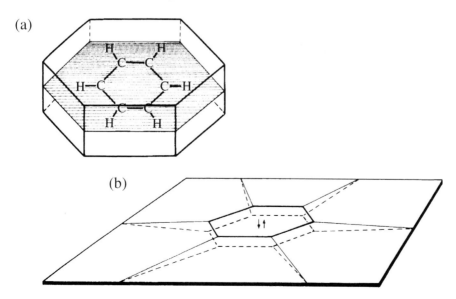

Figure 7.10 Free-electron model of benzene. (a) Box model. (b) One-dimensional string model (ground-state vibration).

In its lowest vibration state the string is vibrating up and down as a whole (Figure 7.10(b)); the vibration amplitude is then constant along the string. The next vibrational mode is twofold degenerate (e.g., ψ_{1a} and ψ_{1b}, Figure 7.11). We obtain the energies and wavefunctions of this cyclic system from the condition

$$\Lambda = \frac{L_c}{n}, \quad n = 0, 1, 2, \ldots \tag{7.4}$$

where L_c is the circumference (L_c = sum of the bond lengths) and Λ is the de Broglie wavelength. There is an integer number of wavelengths along the circumference. The case $n = 0$ corresponds to a constant value of the wavefunction. All other energy levels ($n = 1, 2, 3, \ldots$) are twofold degenerate and the wavefunctions have $2n$ nodes. In each orthogonal set the nodes and antinodes are exchanged. If we denote the coordinate along the bond line with s, we obtain

$$\psi_0 = \frac{1}{\sqrt{L_c}} \quad \text{for } n = 0$$

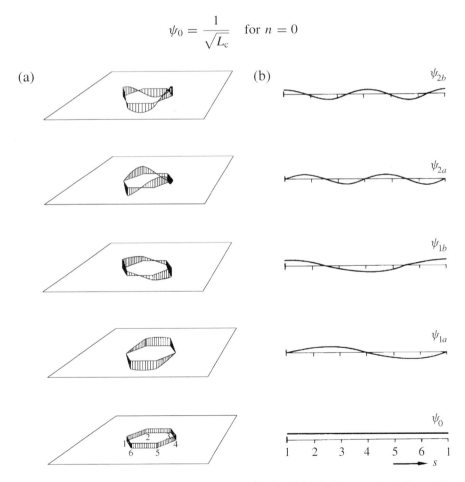

Figure 7.11 Vibrational states of a closed hexagonal string. (a) Maximum and minimum displacement of the string. (b) Displacement along the coordinate s. These vibrations illustrate the standing de Broglie waves of the π electrons and should not be confused with vibrations of the molecular skeleton.

7.2 FREE-ELECTRON MODEL

$$\psi_{1a} = \sqrt{\frac{2}{L_c}} \cdot \sin \frac{2\pi s}{L_c}, \quad \psi_{1b} = \sqrt{\frac{2}{L_c}} \cdot \cos \frac{2\pi s}{L_c} \quad \text{for } n = 1 \quad (7.5)$$

$$\psi_{2a} = \sqrt{\frac{2}{L_c}} \cdot \sin \frac{4\pi s}{L_c}, \quad \psi_{2b} = \sqrt{\frac{2}{L_c}} \cdot \cos \frac{4\pi s}{L_c} \quad \text{for } n = 2$$

We calculate the energy E from equation (7.3) by introducing condition (7.4) instead of condition (7.1):

$$E = \frac{h^2}{2m_e \Lambda^2} = \frac{h^2}{2m_e L_c^2} \cdot n^2, \quad n = 0, 1, 2, \ldots \quad (7.6)$$

For benzene we obtain $L_c = 6 \cdot d_0 = 840 \, \text{pm}$. The electron cloud and the energies of benzene are illustrated in Figure 7.12.

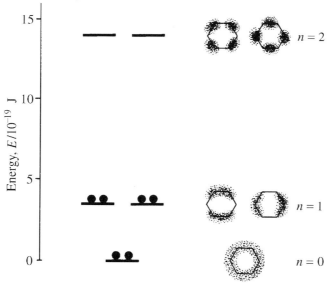

Figure 7.12 Energy E and cloud representation of the π electron states of benzene in the free-electron model.

7.2.3 Charge Density dQ/ds

What is the electron distribution in a π electron system? To answer this question we calculate the electron charge dQ between s and $s + ds$ from the wavefunctions by summation over all occupied orbitals

$$dQ = -e \sum_k N_k \psi_k^2(s) \cdot ds \quad (7.7)$$

where N_k is the number of π electrons in orbital k. For ethene, for example, we obtain

$$\frac{dQ}{ds} = -2e \cdot \frac{2}{L} \cdot \sin^2 \frac{\pi s}{L} \quad (L = 420 \, \text{pm}) \quad (7.8)$$

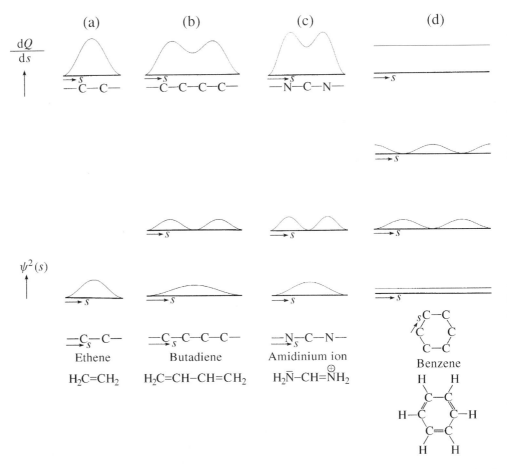

Figure 7.13 Probability density $\psi^2(s)$ of occupied π-electron orbitals and charge density dQ/ds along the bond line: (a) ethene, (b) butadiene, (c) amidinium cation, and (d) benzene.

and for butadiene we obtain

$$\frac{dQ}{ds} = -2e \cdot \frac{2}{L} \cdot \left(\sin^2\frac{\pi s}{L} + \sin^2\frac{2\pi s}{L}\right) \quad (L = 700 \text{ pm}) \quad (7.9)$$

In Figure 7.13, dQ/ds is displayed as a function of s for ethene, butadiene, amidinium ion, and benzene. For ethene we find a maximum of dQ/ds at the center of the bond. For butadiene dQ/ds is larger in the centers of the formal double bonds than in the center of the formal single bond; note that dQ/ds of the single bond is by no means zero (as would be in the case of a "real" single bond, namely a pure σ bond). dQ/ds is equally large at both bonds in the amidinum ion and it is constant along the ring in benzene.

The larger dQ/ds, the stronger and shorter is the corresponding bond. Therefore, we expect that in butadiene the C—C bond is longer than the C=C bond. The bond in ethene should be slightly shorter than the double bond in butadiene; this is really the case (Table 7.2). The uneven distribution of the electron charge causes an attraction of the nuclei to the positions where the electron cloud is accumulated (Figure 7.14). This shows

7.2 FREE-ELECTRON MODEL

Table 7.2 Charge density dQ/ds in the center of the bond and corresponding measured bond length for some π bonds

Molecule	Bond	$\dfrac{(dQ/ds)}{e}$ / nm^{-1}	Bond length/pm
Ethene	C=C	9.52	133
Butadiene	C–C	5.71	148
	C=C	8.91	134
Amidinium	C⋯N	9.67	140
Guanidinium	C⋯N	8.83	140
Benzene	C⋯C	7.14	140

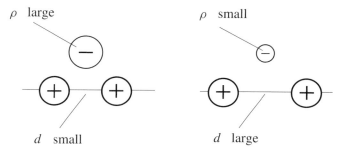

Figure 7.14 Attraction of a π electron between two positive nuclear charges. Comparison of π electron density ρ and nuclear distance d.

that you can directly probe the charge density along the chain by looking at measured bond lengths: the higher the charge density, the shorter the bond.

7.2.4 Resonance

Resonance energy

In benzene the charge density

$$\frac{dQ}{ds} = -2e \cdot \frac{1}{L_c} \cdot \left(1 + 2 \cdot \sin^2 \frac{2\pi s}{L_c} + 2 \cdot \cos^2 \frac{2\pi s}{L_c}\right) = -\frac{6e}{L_c} \qquad (7.10)$$

is constant along the ring (Figure 7.13(d)). Because of $L_c = 6 \times 140\,\text{pm} = 6 \times 0.140\,\text{nm}$ we obtain

$$\frac{dQ}{ds} = \frac{-6e}{6 \times 140\,\text{pm}} = -e \cdot 7.14\,\text{nm}^{-1}$$

Thus the electron charge is uniformly distributed along the bond line. The measured bond length is 140 pm (Table 7.2) (this is in between the lengths of a double and a single bond). For the amidinium ion dQ/ds is equal at the centers of both C–N bonds (Figure 7.13(c)). Thus we conclude that bars symbolizing the bonds in the structural formula representation do not correspond to the real electron distribution in the amidinium cation and in benzene. The π electron distribution is represented by limiting structures – for example,

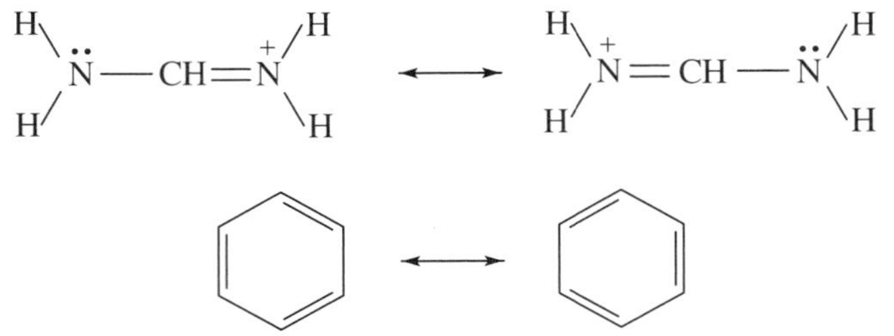

$$H_2N-CH=NH_2^+ \longleftrightarrow H_2N^+=CH-NH_2$$

(benzene resonance structures)

Molecules like benzene have lower energy than comparable systems with the same number of isolated double bonds. To demonstrate this we consider the energies of hydrogenation of benzene and cyclohexene (Figure 7.15). From calorimetric measurements (see Chapter 16) we obtain $\Delta E_{benzene} = -3.42 \times 10^{-19}$ J and $\Delta E_{cyclohexene} = -1.93 \times 10^{-19}$ J. There are three formal double bonds in benzene and one double bond in cyclohexene; thus for isolated double bonds in benzene we expect that the hydrogenation energy of benzene would be three times the value of cyclohexene. Actually, it is smaller (Figure 7.16). For the energy difference we obtain

$$\varepsilon_R = \Delta E_{benzene} - 3 \times \Delta E_{cyclohexene} = 2.36 \times 10^{-19} \text{ J} \quad (7.11)$$

(corresponding to 142 kJ mol^{-1}). This means that the π electron energy of benzene is smaller than would be expected from its number of formal single and double bonds.

The energy difference ε_R is called *resonance energy*. ε_R becomes large when the real electron distribution in a conjugated system differs significantly from the formal single double bond structure (localized single and double bonds). In particular, benzene is a very stable system because the orbitals with $n = 1$ are totally occupied by four electrons (see Figure 7.12). This situation is similar to the closed-shell configuration of rare gases. This holds not only for benzene, but for any cyclic system with six π electrons – for example,

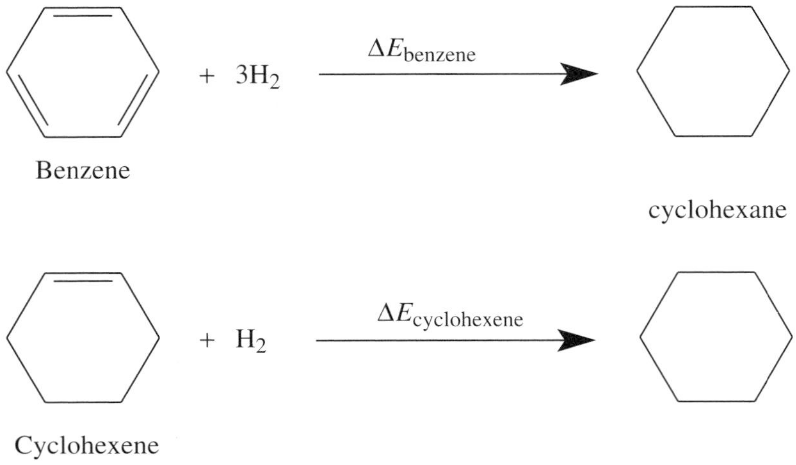

Benzene + 3H$_2$ $\xrightarrow{\Delta E_{benzene}}$ cyclohexane

Cyclohexene + H$_2$ $\xrightarrow{\Delta E_{cyclohexene}}$

Figure 7.15 Hydrogenation of benzene and cyclohexene.

7.2 FREE-ELECTRON MODEL

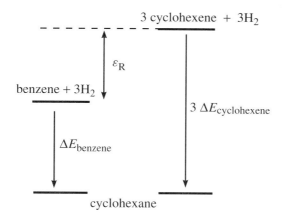

Figure 7.16 Energy change for the hydrogenation of benzene and cyclohexene; for the hydrogenation of three cyclohexene molecules (i.e., three double bonds) we find $3 \times \Delta E_{\text{cyclohexene}} = -5.78 \times 10^{-19}$ J. For the hydrogenation of benzene (i.e., three formal double bonds) we find $\Delta E_{\text{benzene}} = -3.42 \times 10^{-19}$ J; thus the energy of benzene is smaller by $\varepsilon_R = 2.36 \times 10^{-19}$ J than would be expected from its structural formula.

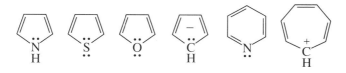

where the lone pairs below the heteroatoms are those which do not contribute to the π electron system.

Hückel $4n + 2$ rule

A closed-shell configuration is also present in cyclic π systems with $4 \times 2 + 2 = 10$, $4 \times 3 + 2 = 14$ (i.e., $4 \times n + 2$) electrons. Therefore, coplanar rings with $4n + 2$ electrons should have especially low energy (*Hückel $4n + 2$ rule*). This is an important criterion to judge the stability of ring compounds. Note that larger rings require special consideration (see Section 7.4.2).

7.2.5 Branched Molecules

As an example of a branched π electron cloud, we consider the guanidinium cation (six π electrons indicated in the structural formula by the double bond and the lone pairs at two N atoms) with the following three limiting structures:

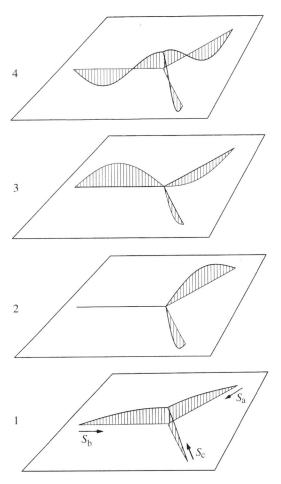

Figure 7.17 Vibrations of a branched string illustrating the wavefunctions of the guanidinium cation.

The orbitals of these π electrons can be described by the wavefunctions of an electron in a box extending over the π electron system. The wavefunctions along the bond line can be approximated by the amplitudes of the standing waves of a branched string (Figures 7.17, 7.18, and 7.19).

Energies

We consider the de Broglie wavelength Λ in terms of the length L, where $L/2$ is the length of each branch: $L = 4d_0$ with $d_0 = 140$ pm.

Orbital 1 ($\Lambda/2 = L$) and orbital 4 ($\Lambda/2 = L/3$) belong to non-degenerate energy states, while orbitals 2 and 3 are an orthogonal set belonging to a doubly degenerate energy state ($\Lambda/2 = L/2$), where $L = 4d_0$ with $d_0 = 140$ pm. According to equation (7.3), the energies E result from the relation

$$E = \frac{1}{2}m_e v^2 = \frac{h^2}{2m_e} \cdot \frac{1}{\Lambda^2} \tag{7.12}$$

7.2 FREE-ELECTRON MODEL

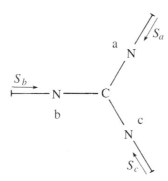

Figure 7.18 Coordinates s_a, s_b, and s_c in a branched system (guanidinium cation). The electron cloud extends by one bond length beyond the last atom.

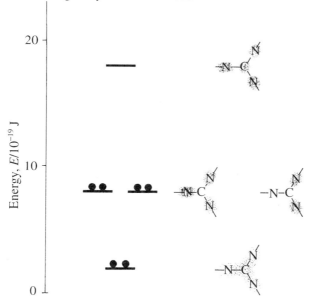

Figure 7.19 Guanidinium cation; energy scheme and charge distribution of the different states. The energies are calculated from (7.13) with $L = 4d_0 = 560$ pm.

Then the energies for the states 1 to 4 are

$$E_1 = \frac{h^2}{8m_e L^2} \cdot 1 \qquad (7.13)$$

$$E_2 = \frac{h^2}{8m_e L^2} \cdot 4$$

$$E_3 = \frac{h^2}{8m_e L^2} \cdot 4$$

$$E_4 = \frac{h^2}{8m_e L^2} \cdot 9$$

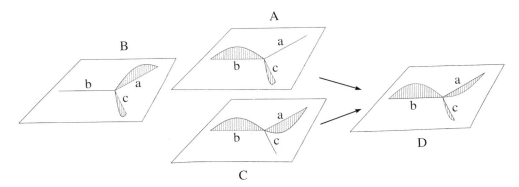

Figure 7.20 Degenerate vibration modes of a string analogous to the guanidinium ion.

Wavefunctions

Orbitals 1 and 4 have an antinode at the branching point. Orbitals 2 and 3, however, have a node at the branching point; in this case the situation is somewhat puzzling for the beginner. We can easily recognize the three equivalent standing waves A, B, and C of a branched string with a node in the branching point (Figure 7.20). In A the wave extends over branches b and c; branch a is at rest. Correspondingly, in B and C branches b and c, respectively, are at rest.

The three wavefunctions ψ_A, ψ_B, and ψ_C are not orthogonal to each other: For example, the product $\psi_A \psi_B$ is zero in branches a and b and positive in branch c, thus $\int \psi_A \psi_B \cdot ds$ is positive. When superposing ψ_A and ψ_C we obtain wavefunction ψ_D (another possibility for a standing wave of the branched string). Wavefunctions B and D form an orthogonal set. $\int \psi_B \psi_D \cdot ds = 0$, because $\psi_B \psi_D$ is positive in branch c, negative in branch a (by the same amount at corresponding points) and zero in branch b.

π electron density

The three lowest energy levels in the guanidinium cation are occupied by six π electrons. We can show that the π electron density in each branch is the same as suggested by the three equivalent limiting structures:

Orbital 1. The charge of $\frac{2}{3}$ electrons are distributed over each of the three branches.

Orbital 2. The charge of one electron is distributed over branch a, and the charge of one electron over branch c.

Orbital 3. The displacement ψ in branch b is twice the total displacement in branches a and c together; thus the electron density in branch b is four times larger. Therefore, the charge of $\frac{4}{3}$ electrons is distributed over branch b, and the charge of $\frac{1}{3}$ electrons over both branches a and c.

Summing over all six electrons we obtain

Branch a charge of $\frac{2}{3} + 1 + \frac{1}{3} = 2$ electrons

Branch b charge of $\frac{2}{3} + 0 + \frac{4}{3} = 2$ electrons

Branch c charge of $\frac{2}{3} + 1 + \frac{1}{3} = 2$ electrons

The electron charge is then equally distributed over the branches of the nuclear skeleton and the same is true for any chosen orthogonal set.

Note. The situation is the same as in benzene, where we can arbitrarily choose any point on the ring as the zero point of the coordinate s (the positions of the nodes and antinodes). This does not have any effect on the total charge distribution and is of no physical meaning in reality, since we cannot distinguish between electrons (see Section 4.3).

In the guanidinium ion there is a closed-shell configuration and a corresponding stabilization similar to benzene. The stability of the guanidinium ion relative to the neutral guanidine is reflected by the fact that guanidine is a very strong base. It easily binds a proton forming a guanidinium ion:

$$HN=C(NH_2)(NH_2) + H^+ \longrightarrow H_2\overset{+}{N}=C(\ddot{N}H_2)(\ddot{N}H_2)$$

For larger molecules, the energies and wavefunctions of branched π electron systems are obtained by numerical methods (see Foundation 7.1).

7.3 HMO Model

7.3.1 Wavefunctions and Energies

Instead of using the free-electron model, we can describe π electrons by the LCAO method, as in the case of the H_2^+ ion in Chapter 4 and in the case of O_2 in Chapter 6. For H_2^+ we used 1s hydrogen functions as atomic orbitals which are now replaced by 2p Slater atomic orbitals of the carbon atom (Chapter 5)

$$\varphi_{2p_z} = A \cdot z \cdot e^{-3.25r/(2a_0)} \qquad (7.14)$$

For ethene the wavefunction ϕ_1 in the ground state is

$$\phi_1 = c_1 \cdot (\varphi_a + \varphi_b) \qquad (7.15)$$

where φ_a and φ_b are Slater atomic orbitals at the centers a and b, respectively; c_1 is a constant determined by the normalization condition. In this case we obtain formally the same result as for the H_2^+ ion (Chapter 4):

$$\phi_1 = \frac{1}{\sqrt{2(1+S_{ab})}} \cdot (\varphi_a + \varphi_b), \quad \varepsilon_1 = \frac{H_{aa} + H_{ab}}{1+S_{ab}} \qquad (7.16)$$

The wavefunction ϕ_2 for the first excited state must, according to the variation principle considered in Chapter 4, be orthogonal to ϕ_1:

$$\phi_2 = c_2 \cdot (\varphi_a - \varphi_b) \qquad (7.17)$$

Then the energy ε_2 is, following the derivation in Section 4.1.6,

$$\varepsilon_2 = \int \phi_2 H \phi_2 \cdot d\tau = 2c_2^2 \cdot (H_{aa} - H_{ab}) \qquad (7.18)$$

By normalizing ϕ_2, we obtain

$$\phi_2 = \frac{1}{\sqrt{2(1-S_{ab})}} \cdot (\varphi_a - \varphi_b), \quad \varepsilon_2 = \frac{H_{aa} - H_{ab}}{1 - S_{ab}} \quad (7.19)$$

In order to generalize this approach to other π electron systems, drastic simplifications are introduced in the model.

1. As in the free-electron model, the π electrons are considered to be in the field of a rigid molecular skeleton.
2. The integral S_{ab} which depends on the overlap of neighboring atomic orbitals φ_a and φ_b is neglected.
3. The integrals H_{aa} (mainly determined by the atomic orbital φ_a) and H_{ab} (mainly determined by the overlap of the neighboring atomic orbitals φ_a and φ_b) are not calculated explicitly but treated as adjustable parameters. H_{ab} is considered to be zero for atoms which are not linked by a σ bond. For atoms linked by a σ bond, H_{ab} depends on the nature of the atoms a and b and on their distance.

These approximations were first introduced by Erich Hückel, and therefore this method is called the *Hückel molecular orbital method* or *HMO method*. The neglect of the overlap integral S_{ab} seems strange in light of the role of overlap in bonding considerations. But H_{ab} as well as S_{ab} depend on the overlap, and H_{ab} is the determining factor. However, the neglect has considerable consequences (Box 7.1) and this should be kept in mind in judging results of HMO calculations.

Box 7.1 – Neglect of the Overlap Integral S_{ab} in HMO

For H_2^+ in its equilibrium state ($d_0 = 132$ pm) we calculated in Foundation 4.2: $S_{ab} = 0.46$, $H_{aa} = -21.4 \times 10^{-19}$ J, and $H_{ab} = -14.6 \times 10^{-19}$ J, thus

$$\varepsilon_1 = \frac{H_{aa} + H_{ab}}{1 + S_{ab}} = -24.6 \times 10^{-19} \text{ J},$$

$$\varepsilon_2 = \frac{H_{aa} - H_{ab}}{1 - S_{ab}} = -12.6 \times 10^{-19} \text{ J}$$

Then the excitation energy is $\varepsilon_2 - \varepsilon_1 = 12.0 \times 10^{-19}$ J.
The corresponding values with $S_{ab} = 0$ are

$$\varepsilon_1 = -36.0 \times 10^{-19} \text{ J}, \quad \varepsilon_2 = -6.8 \times 10^{-19} \text{ J}, \quad \varepsilon_2 - \varepsilon_1 = 29.2 \times 10^{-19} \text{ J}$$

In the case considered, the Hückel assumption $S_{ab} = 0$ causes a strong change in the value of $\varepsilon_2 - \varepsilon_1$ (29.2×10^{-19} J instead of 12.0×10^{-19} J). In the case of ethene this mismatch is somewhat less.

For carbon atoms in a conjugated π electron system with equal bond lengths we set

$$\alpha = H_{aa}, \quad \beta = H_{ab} \quad (7.20)$$

7.3 HMO MODEL

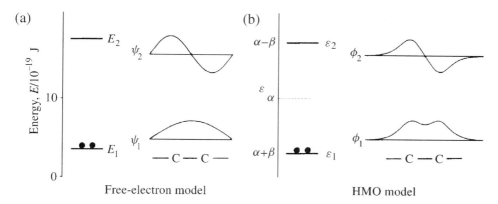

Figure 7.21 Energies and wavefunctions for ethene. (a) Free-electron model. (b) HMO model. Wavefunction along the bond line 75 pm above the molecule plane.

With these simplifications the result for the ethene molecule is

$$\phi_1 = \frac{1}{\sqrt{2}} \cdot (\varphi_a + \varphi_b), \quad \varepsilon_1 = \alpha + \beta \qquad (7.21)$$

$$\phi_2 = \frac{1}{\sqrt{2}} \cdot (\varphi_a - \varphi_b), \quad \varepsilon_2 = \alpha - \beta$$

Note that α and β are negative quantities.

These energies and wavefunctions are illustrated in Figure 7.21 and compared to those calculated according to the free-electron method. ϕ_1 is a bonding orbital (antinode in the center of the bond); ϕ_2 is an antibonding orbital (node in the center of the bond).

For larger molecules we proceed in a corresponding way (see Foundation 7.2 and Problems 7.1–7.3). The resulting energies ε_1 to ε_4 and wavefunctions ϕ_1 to ϕ_4 for butadiene are illustrated in Figure 7.22, and compared with the results of the free-electron method.

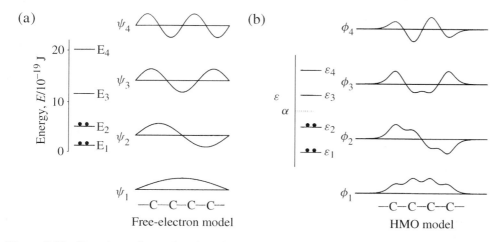

Figure 7.22 Energies and wavefunctions for butadiene. (a) Free-electron model. (b) HMO model ($\varepsilon_1 = \alpha + 1.681\beta$, $\varepsilon_2 = \alpha + 0.681\beta$, $\varepsilon_3 = \alpha - 0.681\beta$, $\varepsilon_4 = \alpha - 1.681\beta$). Wavefunction along the bond line 75 pm above the molecule plane.

7.3.2 Charge Density dQ/dτ

For the quantitative description of the electron distribution in the HMO method we proceed as in the case of the free-electron method. We assume that a volume element $d\tau$ contains the charge dQ

$$dQ = -e \cdot \sum_k N_k \phi_k^2(x, y, z) \cdot d\tau \qquad (7.22)$$

where $\phi_k^2(x, y, z)$ is the probability density of an π electron in orbital k and N_k is the number of π electrons in that orbital. According to the HMO approximation we find for ethene (orbital 1 occupied by two π electrons)

$$\frac{dQ}{d\tau} = -2e \cdot \phi_1^2 \qquad (7.23)$$

and for butadiene (orbitals 1 and 2 occupied by two π electrons each)

$$\frac{dQ}{d\tau} = -2e \cdot (\phi_1^2 + \phi_2^2) \qquad (7.24)$$

where ϕ_1 and ϕ_2 must be expressed by the corresponding $2p_z$ Slater atomic functions. In Figure 7.23 the charge density $dQ/d\tau$ of ethene and butadiene along the bond line (coordinate s) is illustrated for a plane 75 pm above the plane of the molecule. The charge density is compared with the result of the free electron method.

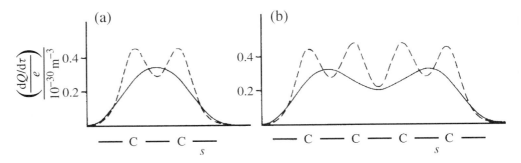

Figure 7.23 Charge density $dQ/d\tau$ of the π electrons along the bond line 75 pm above the molecule plane. Solid line: free-electron model. Dashed line: HMO model. (a) Ethene. (b) Butadiene. In the case of the free electron model we have $\frac{dQ}{d\tau} = \frac{4}{L_y L_z} \cdot \frac{dQ}{ds}$ with $L_y = 400$ pm and $L_z = 300$ pm (see Figure 7.5 and Table 7.2).

7.4 Bond Lengths, Dipole Moments

7.4.1 Bond Length and Charge Density

According to Table 7.2, the bond length d in π electron systems depends on the charge density dQ/ds (free-electron model) at the center of the bond. In order to facilitate the comparison between the bond length d and (dQ/ds) in more complicated molecules, we consider the quantity

$$P = \frac{(dQ/ds) \text{ at the center of the bond}}{(dQ/ds) \text{ at the center of the bond in ethene}} \qquad (7.25)$$

7.4 BOND LENGTHS, DIPOLE MOMENTS

Table 7.3 Quantity P describing the charge density at the center of a bond relative to ethene and experimental bond length (free-electron and HMO model)

	P, free-electron model	P, HMO model	bond length/pm
Ethene	1	1	133
Butadiene	0.94	0.89	134
	0.60	0.45	148
Benzene	0.75	0.67	140

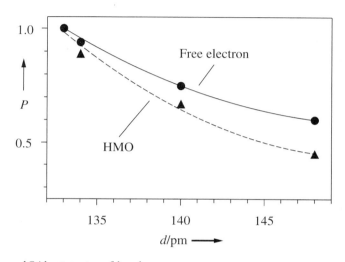

Figure 7.24 $P = \dfrac{\mathrm{d}Q/\mathrm{d}s \text{ at center of bond}}{\mathrm{d}Q/\mathrm{d}s \text{ in ethene}}$ according to the free-electron model (solid line) and to the HMO model (dashed line) for ethene, butadiene, and benzene as a function of the bond length d. For numerical data (full circles and triangles) see Table 7.3.

For ethene, butadiene, and benzene, P is calculated from the data in Table 7.2 and displayed in Table 7.3, together with experimental bond lengths.

In Figure 7.24, P is plotted as a function of the bond length d in the case of the free-electron model (solid line). This curve can be used to predict the bond lengths in arbitrary π electron systems from calculated P values. To do this we calculate P from $\mathrm{d}Q/\mathrm{d}s$ at the center of the bond and find the corresponding length d in Figure 7.24. The bond lengths calculated in this way are displayed in Table 7.4 and compared with experimental data.

The agreement between theory and experiment shows that the simple free-electron model correctly predicts details of the π electron density distribution (measured by differences in bond length), even in complicated π electron systems.

In the HMO method the quantity P for bonded atoms i and j is approximated by

$$P = \sum_k N_k c_{ik} c_{jk} \tag{7.26}$$

where c_{ik} and c_{jk} are the HMO coefficients (see Foundation 7.2) of the atoms i and j in orbital k and N_k is the number of π electrons in orbital k. For ethene ($N_1 = 2$, $N_2 = 0$) we obtain

$$P = 2 \cdot c_{11} \cdot c_{21} = 2 \cdot \frac{1}{\sqrt{2}} \cdot \frac{1}{\sqrt{2}} = 1$$

Table 7.4 Comparison of experimental and calculated bond lengths for some π-electron systems

Molecule	Bond	Bond length/pm		
		Free-electron	HMO	Experiment
Naphthalene	1–2	138	138	136
	2–3	142	142	142
	1–9	140	144	142
	9–10	142	145	142
Anthracene	1–2	137	138	137
	2–3	143	143	142
	1–11	141	144	144
	9–11	137	142	140
	11–12	143	147	143
Chrysene	1–2	138	139	138
	2–3	142	142	139
	1–14	139	143	141
	3–4	138	139	136
	4–15	140	143	143
	5–15	142	145	142
	5–6	137	137	137
	6–16	141	144	143
	13–16	139	143	140
	13–14	143	147	147
	14–15	141	144	141
Fulvene	1–2	138	137	134
	2–3	141	148	144
	3–4	135	137	135
	4–5	142	145	144
Triphenylene	1–2	139	139	138
	1–14	138	142	142
	2–3	141	141	140
	14–15	139	143	142
	13–14	145	149	145
Azulene	1–2	137	140	139
	1–9	135	142	141
	8–9	141	143	138
	7–8	141	140	140
	6–7	143	141	139
	9–10	146	150	148

7.4 BOND LENGTHS, DIPOLE MOMENTS

In the HMO model the quantity P (see Table 7.3) is called the *bond order*. Table 7.4 includes bond lengths calculated by the HMO model (obtained in a corresponding way as for the free-electron model using Fig. 7.24, dashed line).

7.4.2 Bond Alternation in Polyenes and Fullerenes

We have considered the influence of the π electron distribution on bond lengths. In this consideration we have neglected the resulting influence of bond length changes on the π electron distribution. This is a too simple procedure in the case of polyenes where the mutual interaction leads to the formation of strongly alternating bonds.

In a bond shortened due to the Coulomb attraction between the π electron cloud and the adjacent nuclei (Figure 7.14), the potential energy of the π electrons is decreased. By this effect an additional accumulation of the π electron charge at the shortened bond takes place. This effect becomes crucial in long polyene chains, where the difference between the charge density on a formal double bond and the charge density on a formal single bond decreases with increasing chain length.

The effect causes an equally strong alternation of the single and double bond lengths, as in butadiene (148 and 133 pm). This can be demonstrated by considering the alternation of the potential energy $V(s)$ along the coordinate s. Solving the Schrödinger equation for this potential energy (step-potential model) reveals quantitatively that the alternation of bond lengths is maintained for long polyenes (Foundation 7.3, Figure 7.28).

This procedure (see Foundation 7.3) leads to alternating double and single bonds in polyenes. It also leads to equal bond lengths for benzene (this is not trivial) and generally for $4n+2$ annulenes (polyene rings with $4n+2$ carbon atoms) up to $n=3$, in agreement with experiment and in accordance with the Hückel rule. Alternating bonds are obtained for $n>3$ in agreement with experiment and in contradiction to the Hückel rule.

In C_{60} fullerene the calculated single and double bond lengths agree with the observation (145 and 140 pm).

We have considered the amidinium cation in Section 7.2.4. It possesses four π electrons like butadiene but has equal bond lengths in contrast to butadiene. Similarly, in the cyanine cations

$$H_2\overset{..}{N}-(-CH=CH-)_n-CH=\overset{+}{N}H_2 \longleftrightarrow H_2\overset{+}{N}=(=CH-CH=)_n=CH-\overset{..}{N}H_2$$

(where amidinium ion is the compound with $n=0$) there are equal bonds along the chain in contrast to the polyenes. The fundamental difference between the cyanine cations and the polyenes will become obvious when discussing the light absorption in Chapter 8.

7.4.3 Dipole Moment

If the centers of gravity of the π electron charges do not coincide with the positive charges of the atom cores, a dipole moment results. If we calculate the excess charge ΔQ_i on each atom resulting from the charge of the core ($+e$ for a C atom) and the electron charge Q_i in the region around the atom i, we obtain the x component of the dipole moment

$$\mu_x = \sum_{i=1}^{N} \Delta Q_i \cdot x_i \tag{7.27}$$

where x_i is the x coordinate of the participating atom i, and the summation has to include all the N atoms of the molecule. We calculate the charge Q_i by summation of the charges around the atom i (from center to center of the adjacent bonds).

In the case of the free electron method the charges Q_i are obtained as follows: Q_i is proportional to

$$\left(\frac{dQ}{ds}\right)_{\text{at atom } i}$$

and the proportionality constant is given by the condition that $\sum_{i=1}^{N} Q_i$ is the charge of all π electrons, namely, $-eN_{\text{electr}}$, where N_{electr} is the number of π electrons:

$$Q_i = A \cdot \left(\frac{dQ}{ds}\right)_{\text{at atom } i} \qquad \sum_{i=1}^{N} Q_i = -eN_{\text{electr}} \qquad (7.28)$$

As an example, we consider fulvene. We choose the coordinate system such that atom 2 has the coordinates $x = 0$ and $y = 0$:

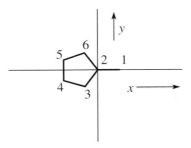

Using the data in Foundation 7.1 we obtain for the charge Q_1 at atom 1: $Q_1 = -e \cdot 2(0.192^2 + 0.567^2 + 0.000^2) = -e \cdot 0.717$ where the factor of two takes into account that each orbital is occupied by two electrons. The charges $Q_2, \ldots, Q_6$ are obtained accordingly (column 2 in Table 7.5). Then the excess charges are calculated as

$$\Delta Q_i = Q_i + e \qquad (7.29)$$

(column 3 in Table 7.5). We see that the excess charges are positive outside the five-membered ring and at the branching point and negative inside the ring (Figure 7.25). Note that the sum over the excess charges equals zero; this means that the molecule, as a whole, is neutral. Then, using the coordinates x_i in column 4 of Table 7.5, the contributions $\Delta Q_i x_i$ to the dipole moment are calculated (column 5). Summing up over these contributions, according to equation (7.27), we obtain $\mu_x = 1.9 \times 10^{-29}$ C m = 5.7 debye where 1 debye = 3.33×10^{-30} C m. In the case of the HMO method the coefficients c_i are used instead of the values C_i to calculate the charges Q_i. As demonstrated in Problem 7.7 the dipole moment $\mu_x = 1.6 \times 10^{-29}$ C m = 4.8 debye is obtained.

Although the direction of the calculated dipole moment is in agreement with experiment, the absolute values are much larger than those in the experiment ($\mu_x = 0.7 \times 10^{-29}$ C m = 2.1 debye). This is caused, in both methods, by oversimplification; in a refined consideration the polarization of the σ electrons in the field of the dipole has to be accounted for. This leads to a partial neutralization of the dipole charge.

Table 7.5 Calculation of dipole moment of fulvene according to equation (7.27)

i	$\dfrac{Q_i}{e}$	$\dfrac{\Delta Q_i}{e}$	x_i/pm	$\dfrac{\Delta Q_i}{e} \cdot x_i$/pm
1	−0.717	+0.283	−140	−39.6
2	−0.779	+0.221	0	0
3	−1.110	−0.110	82.3	−9.1
4	−1.141	−0.141	215.4	−30.4
5	−1.141	−0.141	215.4	−30.4
6	−1.110	−0.110	82.3	−9.1

Q_i is the electron charge at atom i and ΔQ_i is the excess charge.

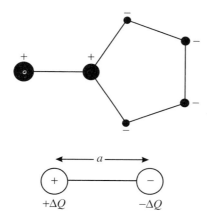

Figure 7.25 Dipole moment of fulvene and equivalent dipole (excess charges $-\Delta Q$ and $+\Delta Q$ at distance a).

We have seen in this chapter that π electron systems are goods examples to test quantum mechanical models in detail by comparison with experimental data such as bond lengths. It is fascinating to experience that π electrons modeled by standing de Broglie waves give a realistic picture of the actual situation.

The stability of aromatic systems, such as benzene, is related to the presence of a closed-shell configuration. The model shows what is meant when chemists deal with limiting structures to describe resonance hybrids.

Problems

Problem 7.1 – HMO Method: Ethene

Perform the derivation carried out for the HMO model in Foundation 7.2 in the special case of the wavefunction

$$\phi = c_1 \varphi_1 + c_2 \varphi_2$$

(ethene molecule).

Solution. Using the abbreviations α and β we obtain from the third equation in Foundation 7.2

$$\varepsilon = c_1^2 \alpha + c_2^2 \alpha + 2 \cdot c_1 c_2 \beta$$

We multiply ε by $c_1^2+c_2^2$ (because of the normalization condition $c_1^2+c_2^2 = 1$, the left-hand side of the equation remains unchanged by this operation):

$$\varepsilon \cdot (c_1^2 + c_2^2) = \alpha(c_1^2 + c_2^2) + 2 \cdot c_1 c_2 \beta$$

and differentiate with respect to c_1:

$$\frac{\partial \varepsilon}{\partial c_1} \cdot (c_1^2 + c_2^2) + \varepsilon \cdot 2c_1 = 2c_1 \alpha + 2c_2 \beta$$

Setting $\partial \varepsilon / \partial c_1 = 0$:

$$c_1 \cdot (\alpha - \varepsilon) + c_2 \cdot \beta = 0$$

Correspondingly, differentiating with respect to c_2, we obtain the equations

$$c_1 \cdot (\alpha - \varepsilon) + \quad c_2 \cdot \beta \quad = 0$$
$$c_1 \cdot \beta \quad + c_2 \cdot (\alpha - \varepsilon) = 0$$

Problem 7.2 – HMO Method: Butadiene

Calculate the orbital energies ε and the wavefunction ϕ for butadiene.

Solution.

$$\overset{1}{C}H_2 = \overset{2}{C}H - \overset{3}{C}H = \overset{4}{C}H_2$$

The number of atoms is $N = 4$. Then, according to Foundation 7.2, the system of linear equations is

$$\begin{aligned} c_1 \cdot x + \quad c_2 \quad\quad\quad\quad\quad\quad &= 0 \\ c_1 \quad + c_2 \cdot x + \quad c_3 \quad\quad\quad &= 0 \\ c_2 \quad + c_3 \cdot x + \quad c_4 &= 0 \\ c_3 \quad + c_4 \cdot x &= 0 \end{aligned}$$

An analytical method to solve this system of homogeneous linear equations is to set the determinant of the coefficients equal to zero:

$$\begin{vmatrix} x & 1 & 0 & 0 \\ 1 & x & 1 & 0 \\ 0 & 1 & x & 1 \\ 0 & 0 & 1 & x \end{vmatrix} = x \cdot \begin{vmatrix} x & 1 & 0 \\ 1 & x & 1 \\ 0 & 1 & x \end{vmatrix} - \begin{vmatrix} 1 & 0 & 0 \\ 1 & x & 1 \\ 0 & 1 & x \end{vmatrix}$$

$$= x^2 \cdot \begin{vmatrix} x & 1 \\ 1 & x \end{vmatrix} - x \cdot \begin{vmatrix} 1 & 0 \\ 1 & x \end{vmatrix} - \begin{vmatrix} x & 1 \\ 1 & x \end{vmatrix} + \begin{vmatrix} 0 & 0 \\ 1 & x \end{vmatrix}$$

$$= x^4 - x^2 - x^2 - x^2 + 1 = x^4 - 3x^2 + 1 = 0$$

From this equation (the *characteristic polynomial*) we get $x_1 = -1.618$, $x_2 = -0.618$, $x_3 = 0.618$, and $x_4 = 1.618$. The coefficients c_i are obtained by inserting the solution for x into the determinant and calculating the subdeterminant corresponding to the first column:

$$c_1 = \begin{vmatrix} x & 1 & 0 \\ 1 & x & 1 \\ 0 & 1 & x \end{vmatrix} = x \cdot (x^2 - 1) - x = x \cdot (x^2 - 2)$$

$$c_2 = - \begin{vmatrix} 1 & 0 & 0 \\ 1 & x & 1 \\ 0 & 1 & x \end{vmatrix} = -x^2 + 1$$

On grounds of symmetry, $c_3 = c_2$ and $c_4 = c_1$. For example, with $x = x_1 = -1.618$ we obtain $c_1 = -1.000$ and $c_2 = -1.618$. These values must be normalized

$$c_1^2 + c_2^2 + c_3^2 + c_4^2 = 1$$

and we obtain as a final result:

Orbital	1	2	3	4
	$x_1 = -1.618$	$x_2 = -0.618$	$x_3 = 0.618$	$x_4 = 1.618$
c_1	0.372	0.602	0.602	0.372
c_2	0.602	0.372	−0.372	−0.602
c_3	0.602	−0.372	−0.372	0.602
c_4	0.372	−0.602	0.602	−0.372

Then the orbital energies are

$$\varepsilon_1 = \alpha + 1.618 \cdot \beta, \quad \varepsilon_2 = \alpha + 0.618 \cdot \beta$$
$$\varepsilon_3 = \alpha - 0.618 \cdot \beta, \quad \varepsilon_4 = \alpha - 1.618 \cdot \beta$$

and the wavefunction ϕ_1 is

$$\phi_1 = 0.372\varphi_1 + 0.602\varphi_2 + 0.602\varphi_3 + 0.372\varphi_4$$

The wavefunctions ϕ_2, ϕ_3, and ϕ_4 are obtained accordingly.

Problem 7.3 – HMO Method: Fulvene

Calculate the eigenvalues x and the coefficients c for fulvene.

Solution. If we number the atoms of fulvene as in Table 7.4, we obtain the following system of linear equations:

$$\begin{aligned}
c_1 \cdot x + c_2 &= 0 \\
c_1 + c_2 \cdot x + c_3 + c_6 &= 0 \\
c_2 + c_3 \cdot x + c_4 &= 0 \\
c_3 + c_4 \cdot x + c_5 &= 0 \\
c_4 + c_5 \cdot x + c_6 &= 0 \\
c_2 + c_5 + c_6 \cdot x &= 0
\end{aligned}$$

Numerical solution of this system of linear equations results in the following values of x and of the coefficients c_i:

Orbital	1	2	3	4	5	6
	$x_1 = -2.115$	$x_2 = -1.000$	$x_3 = -0.618$	$x_4 = 0.254$	$x_5 = 1.618$	$x_6 = 1.861$
c_1	0.247	−0.500	0	0.759	0	−0.357
c_2	0.523	−0.500	0	−0.189	0	0.664
c_3	0.429	0	0.602	−0.344	−0.372	−0.439
c_4	0.385	0.500	0.373	0.275	0.602	0.153
c_5	0.385	0.500	−0.372	0.275	−0.602	0.153
c_6	0.429	0.000	−0.602	−0.344	0.372	−0.439

Problem 7.4 – Resonance of Hückel (4n + 2) Rings

Show that cyclic π systems with $(4n + 2)$ π electrons are well stabilized by resonance, contrary to systems with $4n$ π electrons ($n = 1, 2, 3, \ldots$).

Solution. According to Figure 7.12, we find two-fold degenerate orbitals in cyclic π electron systems. In analogy to the benzene molecule, the charge is equally distributed and the bond lengths are equal if each of the degenerate orbitals is occupied by two electrons. This is the case if there are $4n + 2$ electrons in the molecule ($n = 1$ for benzene, $n = 2$ for cyclodecapentaene).

For cyclic molecules with $4n$ electrons the charge cannot be equally distributed. For example, in cyclobutadiene two electrons are in the orbital with quantum number $n = 0$, and the two degenerate orbitals with $n = 1$ can be occupied by one electron each.

However, the energy of the system is lower if two electrons occupy one of the degenerate orbitals and the second one remains empty. The antinodes of the wavefunction are then only at the locations of the formal double bonds (decreasing the energy because of the stronger attraction of the electrons by the positively charged nuclei at a double bond). This effect leads to an alternation of the bonds, as for the polyenes.

Problem 7.5 – Bond Lengths from Bond Orders

Calculate the bond lengths of fulvene from the bond orders using the HMO coefficients listed in Problem 7.3 and the bond lengths curve in Figure 7.24.

Solution. For the bond order P_{12} of the bond 1–2 we obtain $P_{12} = 2 \cdot 0.247 \cdot 0.523 + 2 \cdot 0.500 \cdot 0.500 + 2 \cdot 0.000 \cdot 0.000 = 0.758$. Similarly, we obtain $P_{23} = P_{26} = 0.449$, $P_{34} = P_{56} = 0.778$, and $P_{45} = 0.520$. With these values we obtain, from Figure 7.24, $d_{12} = 137$ pm, $d_{23} = d_{26} = 148$ pm, and $d_{34} = d_{56} = 137$ pm, and $d_{45} = 145$ pm. We see that strong bond alternation is reached.

Problem 7.6 – Cyclobutadiene: Bond Lengths

Calculate the energies, wavefunctions, and charge densities in the centers of the bonds for cyclobutadiene. Discuss a square and a rectangular shape of the molecule.

Solution. In analogy to benzene, the energies and the wavefunctions are, assuming a square structure:

$$E = \frac{h^2}{2m_e L_c^2} \cdot n^2$$

PROBLEM 7.7

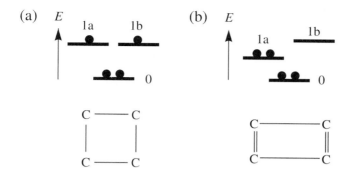

Figure 7.26 Occupied orbitals in cyclobutadiene. (a) Equal bond lengths. (b) Bond length alternation.

$$\psi_0 = \frac{1}{\sqrt{L_c}}, \quad \psi_{1a} = \sqrt{\frac{2}{L_c}} \cdot \sin\frac{2\pi s}{L_c}, \quad \psi_{1b} = \sqrt{\frac{2}{L_c}} \cdot \cos\frac{2\pi s}{L_c}$$

Two electrons occupy the orbital with $n = 0$, and the remaining two electrons occupy the degenerate orbitals with $n = 1$ (Figure 7.26(a)). Thus cyclobutadiene should be a radical.

If we assume, however, that there are two single and two double bonds in cyclobutadiene (rectangular structure, Figure 7.26(b)), then the energies of the orbitals $1a$ and $1b$ are different. If the double bonds are at the place of the antinodes of the wavefunction $1a$, then $E_{1a} < E_{1b}$ and both electrons share the orbital $1a$.

The rectangular structure is much more reasonable than the square structure. Let us start with a square-shaped cyclobutadiene; if, by chance, one of the bonds is somewhat decreased, then the electrons will be more strongly attracted by this bond. Consequently, more negative charge will accumulate in this bond, and the bond will be automatically shortened further. This process continues until double and single bonds have been established. This can be demonstrated in a corresponding calculation as shown for dodecahexaene (Foundation 7.3).

In a more detailed consideration, the energy to change the σ bonds has to be taken into account, and the energy decrease of the π electrons is partially compensated by the energy increase of the σ electrons. (See W. T. Borden, E. R. Davidson, and P. Hart, The potential surfaces for the lowest singlet and triplet states of cyclobutadiene, *J. Am. Chem. Soc.*, 1978, **100**, 388; also see *Angew. Chem.*, 1991, **103**, 1048; 1988, **100**, 317.)

Problem 7.7 – Dipole Moment of Fulvene

Calculate the dipole moment of fulvene by using the π electron charge distribution obtained from the HMO method.

Solution. First we calculate the electron charges Q_i from the data in Problem 7.3. For example, for atom 1 we get $Q_1/e = 2 \cdot (0.247^2 + 0.500^2 + 0.000^2) = 0.622$. Thus we obtain the values in the second column in the following table:

i	$\dfrac{Q_i}{e}$	$\dfrac{\Delta Q_i}{e}$	x_i/pm	$\dfrac{\Delta Q_i}{e} \cdot x_i \Big/ \text{pm}$
1	−0.622	+0.387	−140	−52.9
2	−1.047	−0.047	0	0
3	−1.092	−0.092	82.3	−7.6
4	−1.073	−0.073	215.4	−15.7
5	−1.073	−0.073	215.4	−15.7
6	−1.092	−0.092	82.3	−7.6

In the third column the excess charge on the different atoms is calculated. With the x coordinates in column 4 we obtain the terms in column 5, which must be summed up to result in the dipole moment. We obtain $\sum \Delta Q_i \cdot x_i = -e \cdot 99.5\,\text{pm}$ and, according to equation (7.27), $\mu_x = -99.5 \cdot e\,\text{pm} = -1.6 \times 10^{-29}\,\text{C m} = 4.8$ debye. The negative sign means that the negative part of the equivalent dipole is inside the five-membered ring.

In this calculation we assumed that the zero point of the x axis is at atom 2. What will happen if we shift the coordinate system by a constant value c? Then the new coordinates are $x'_i = x_i + c$, and the dipole moment becomes

$$\mu' = \sum_i (e - Q_i) \cdot (x_i + c) = \sum_i (e - Q_i) \cdot x_i + c \cdot \sum_i (e - Q_i)$$
$$= \mu + c \cdot \sum_i (e - Q_i)$$

Because the molecule as a whole is neutral, the sum of the excess charges on the different atoms must be zero; thus $\mu' = \mu$. This means that the value of the dipole moment does not depend on the arbitrary choice of the coordinate system.

Foundation 7.1 – Free-electron model

Here we present a formal method which can be applied to both unbranched and branched π electron systems; additionally, this method is well suited for solution by means of a computer as it is for application to the HMO method. We start with fulvene as an example. First we consider the wavefunction in branch a (Figure 7.27):

$$\psi_a(s_a) = A \cdot \sin\left(\frac{2\pi}{\Lambda} s_a + \alpha\right)$$

where α denotes the phase of the sinus function

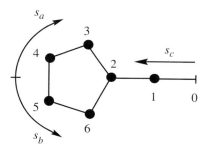

Figure 7.27 Coordinates s_a, s_b, and s_c for fulvene.

We proceed from a point with coordinate s_a to a point $s_a + d$, where d denotes the bond length:

$$\psi_a(s_a + d) = A \cdot \sin\left(\frac{2\pi}{\Lambda} s_a + \alpha + \frac{2\pi}{\Lambda} d\right)$$

$$= A \cdot \sin\left(\frac{2\pi}{\Lambda} s_a + \alpha\right) \cdot \cos\frac{2\pi}{\Lambda} d + A \cdot \cos\left(\frac{2\pi}{\Lambda} s_a + \alpha\right) \cdot \sin\frac{2\pi}{\Lambda} d$$

Because of

$$\frac{d\psi_a}{ds_a} = A \cdot \frac{2\pi}{\Lambda} \cdot \cos\left(\frac{2\pi}{\Lambda} s_a + \alpha\right)$$

we obtain

$$\psi_a(s_a + d) = \psi_a(s_a) \cdot \cos\frac{2\pi}{\Lambda} d + \frac{\Lambda}{2\pi} \cdot \frac{d\psi_a}{ds_a} \cdot \sin\frac{2\pi}{\Lambda} d$$

Accordingly, for a point $s_a - d$ we have

$$\psi_a(s_a - d) = \psi_a(s_a) \cdot \cos\frac{2\pi}{\Lambda} d - \frac{\Lambda}{2\pi} \cdot \frac{d\psi_a}{ds_a} \cdot \sin\frac{2\pi}{\Lambda} d$$

Now we consider the values of the wavefunction at the coordinates of the atoms. For example, ψ_a at $s_a = 3d/2$ is the value of the wavefunction at atom

Continued on page 196

Continued from page 195

3 (Figure 7.27). We connect this value with the values at the adjacent atoms 2 and 4. With C_i = value of wavefunction at atom i we find

$$C_2 = C_3 \cdot \cos \frac{2\pi}{\Lambda} d + \frac{\Lambda}{2\pi} \cdot \left(\frac{d\psi_a}{ds_a}\right)_3 \cdot \sin \frac{2\pi}{\Lambda} d$$

$$C_4 = C_3 \cdot \cos \frac{2\pi}{\Lambda} d - \frac{\Lambda}{2\pi} \cdot \left(\frac{d\psi_a}{ds_a}\right)_3 \cdot \sin \frac{2\pi}{\Lambda} d$$

We add up:

$$\boxed{C_2 + C_4 = 2 \cdot C_3 \cdot \cos \frac{2\pi}{\Lambda} d}$$

Correspondingly, by connecting C_2 and C_0 with C_1 (with C_0 = value of the wavefunction at the end point $s_c = 0$: $C_0 = 0$), we obtain

$$\boxed{C_2 + C_0 = C_2 = 2 \cdot C_1 \cdot \cos \frac{2\pi}{\Lambda} d}$$

Finally we consider atoms 1, 3, and 6, which are adjacent to the branching point 2:

$$C_1 = C_2 \cdot \cos \frac{2\pi}{\Lambda} d - \frac{\Lambda}{2\pi} \cdot \left(\frac{d\psi_c}{ds_c}\right)_2 \cdot \sin \frac{2\pi}{\Lambda} d$$

$$C_3 = C_2 \cdot \cos \frac{2\pi}{\Lambda} d - \frac{\Lambda}{2\pi} \cdot \left(\frac{d\psi_a}{ds_a}\right)_2 \cdot \sin \frac{2\pi}{\Lambda} d$$

$$C_6 = C_2 \cdot \cos \frac{2\pi}{\Lambda} d - \frac{\Lambda}{2\pi} \cdot \left(\frac{d\psi_b}{ds_b}\right)_2 \cdot \sin \frac{2\pi}{\Lambda} d$$

We add up these three equations:

$$C_1 + C_3 + C_6 = 3 \cdot C_2 \cdot \cos \frac{2\pi}{\Lambda} d$$
$$- \frac{\Lambda}{2\pi} \cdot \left[\left(\frac{d\psi_c}{ds_c}\right)_2 + \left(\frac{d\psi_a}{ds_a}\right)_2 + \left(\frac{d\psi_b}{ds_b}\right)_2\right] \cdot \sin \frac{2\pi}{\Lambda} d$$

For the slopes at the branching point the condition

$$\left(\frac{d\psi_c}{ds_c}\right)_2 + \left(\frac{d\psi_a}{ds_a}\right)_2 + \left(\frac{d\psi_b}{ds_b}\right)_2 = 0$$

must be fulfilled, as can be seen for the displacement of the vibrating string. Then the expression in the brackets equals zero, and we obtain

$$\boxed{C_1 + C_3 + C_6 = 3 \cdot C_2 \cdot \cos \frac{2\pi}{\Lambda} d}$$

Continued on page 197

FOUNDATION 7.1

Continued from page 196

Correspondingly, we proceed for atoms 4, 5, and 6. Then with the abbreviation

$$x = -2 \cdot \cos \frac{2\pi}{\Lambda} d$$

we obtain the following system of linear equations:

$$xC_1 + C_2 = 0$$
$$C_1 + \tfrac{3}{2}xC_2 + C_3 + C_6 = 0$$
$$C_2 + xC_3 + C_4 = 0$$
$$C_3 + xC_4 + C_5 = 0$$
$$C_4 + xC_5 + C_6 = 0$$
$$C_2 + C_5 + xC_6 = 0$$

This system of homogeneous linear equations corresponds formally to the system of linear equations obtained in the HMO model (see Foundation 7.2); the only difference is that at the branching point x is replaced by $\tfrac{3}{2}x$. Of course, the physical meaning of x is different in the two methods.

By solving this system of linear equations by standard numerical methods, we obtain the following values C_i of the wavefunction at atom i for different states (x_1 to x_6) for the fulvene molecule:

Orbital	1	2	3	4	5	6
x	−1.925	−0.888	−0.618	0.250	1.563	1.618
C_1	0.192	−0.567	0	0.759	−0.364	0
C_2	0.369	−0.504	0	−0.189	0.569	0
C_3	0.437	−0.052	0.602	−0.344	−0.485	−0.372
C_4	0.472	0.458	0.372	0.275	0.189	0.602
C_5	0.472	0.548	−0.372	0.275	0.189	−0.602
C_6	0.437	−0.052	−0.602	−0.344	−0.485	0.372

From these x values we obtain the de Broglie wavelengths Λ and energies E (with $d = 140\,\text{pm}$):

$$\Lambda = \frac{2\pi d}{\arccos\left(-\frac{x}{2}\right)}, \quad E = \frac{h^2}{8 m_e \Lambda^2}$$

Orbital	x	Λ/pm	$E/10^{-19}\,\text{J}$
1	−1.925	3200	0.23
2	−0.888	791	3.85
3	−0.618	700	4.92
4	0.250	518	8.96
5	1.563	356	18.97
6	1.618	350	19.68

Foundation 7.2 – HMO Model

For N atoms, the trial function is

$$\phi = \sum_{i=1}^{N} c_i \varphi_i$$

We calculate the energy ε according to

$$\varepsilon = \int \phi \mathcal{H} \phi \cdot d\tau = \int \left(\sum_{i=1}^{N} c_i \varphi_i \right) \mathcal{H} \left(\sum_{j=1}^{N} c_j \varphi_j \right) \cdot d\tau$$

$$= \sum_{i=1}^{N} c_i^2 \int \varphi_i \mathcal{H} \varphi_i \cdot d\tau + 2 \sum_{j>i} c_i c_j \int \varphi_i \mathcal{H} \varphi_j \cdot d\tau$$

With $\alpha = \int \varphi_i \mathcal{H} \varphi_i \cdot d\tau$ (if only carbon atoms are in the π system and) $\beta_{ij} = \int \varphi_i \mathcal{H} \varphi_j \cdot d\tau$ (atoms i,j are linked by a σ bond).
Then we obtain

$$\varepsilon = \sum_{i=1}^{N} c_i^2 \alpha + 2 \sum_{i>j} c_i c_j \beta_{ij}$$

Lagrange method to find the minimum of ε

Because of the normalization of the trial function and the neglect of the overlap integrals we find

$$\int \left(\sum_{i=1}^{N} c_i \varphi_i \right)^2 \cdot d\tau = \sum_{i=1}^{N} c_i^2 = 1$$

We multiply ε with $\sum_{i=1}^{N} c_i^2$ (this sum has the value 1). Then:

$$\varepsilon \cdot \sum_{i=1}^{N} c_i^2 = \sum_{i=1}^{N} c_i^2 \cdot \alpha + 2 \sum_{j>i} c_i c_j \cdot \beta_{ij}$$

To calculate the energy minimum according to the variation principle, we differentiate with respect to the coefficients c_i. For example, the result of differentiation with respect to c_1 is

$$\frac{\partial \varepsilon}{\partial c_1} \sum_{i=1}^{N} c_i^2 + \varepsilon \cdot 2c_1 = 2c_1 \alpha + 2 \sum_{j>1} c_j \cdot \beta_{1j}$$

Continued on page 199

FOUNDATION 7.2

Continued from page 198

The condition for the energy minimum is $\partial \varepsilon / \partial c_1 = 0$; thus

$$\varepsilon \cdot 2c_1 = 2c_1 \alpha + 2 \sum_{j>1}^{N} c_j \cdot \beta_{1j}$$

or

$$c_1(\alpha - \varepsilon) + c_2 \beta_{12} + c_3 \beta_{13} + \cdots + c_N \beta_{1N} = 0$$

Including the equivalent equations which are obtained by differentiating with respect to $c_2, c_3, \ldots, c_N$, we obtain the following system of linear equations:

$$\begin{aligned}
c_1(\alpha - \varepsilon) + c_2 \beta_{12} + c_3 \beta_{13} + \cdots + c_N \beta_{1N} &= 0 \\
c_1 \beta_{21} + c_2(\alpha - \varepsilon) + c_3 \beta_{23} + \cdots + c_N \beta_{2N} &= 0 \\
c_1 \beta_{31} + c_2 \beta_{32} + c_3(\alpha - \varepsilon) + \cdots + c_N \beta_{3N} &= 0 \\
\cdots + \cdots + \cdots + \cdots + \cdots &= 0 \\
c_1 \beta_{N1} + c_2 \beta_{N2} + c_3 \beta_{N3} + \cdots + c_N(\alpha - \varepsilon) &= 0
\end{aligned}$$

If atoms i and j are direct neighbors, we set $\beta_{ij} = \beta$, otherwise we set $\beta_{ij} = 0$. These linear equations can be solved by standard numerical methods.

Example – ethene $\overset{1}{C}H_2 = \overset{2}{C}H_2$

The number of atoms is $N = 2$:

$$\begin{aligned}
c_1 \cdot (\alpha - \varepsilon) + c_2 \beta &= 0 \\
c_1 \beta + c_2 \cdot (\alpha - \varepsilon) &= 0
\end{aligned}$$

For convenience, we use the abbreviation

$$x = \frac{\alpha - \varepsilon}{\beta}$$

Then we obtain

$$\begin{aligned}
c_1 \cdot x + c_2 &= 0 \\
c_1 + c_2 \cdot x &= 0
\end{aligned}$$

Solving these equations leads to $x_1 = -1$, $c_1 = +c_2$, $x_2 = +1$, $c_1 = -c_2$. Then $\varepsilon_1 = \alpha + \beta$, $\varepsilon_2 = \alpha - \beta$. Normalization of the wavefunctions $c_1^2 + c_2^2 = 1$ results in $\phi_1 = \frac{1}{\sqrt{2}} \cdot (\varphi_1 + \varphi_2)$, $\phi_2 = \frac{1}{\sqrt{2}} \cdot (\varphi_1 - \varphi_2)$

This result corresponds to equation (7.21).

For additional examples see Problems 7.1 and 7.2.

Foundation 7.3 – Self-consistency in Bond Alternation

We show that the bonds in dodecahexaene, when starting with equal bond lengths, converge to self-consistent alternating bond lengths.

$$CH_2=CH-CH=CH-CH=CH-CH=CH-CH=CH-CH=CH_2$$

First we assume an equally strong bond length alternation in dodecahexaene as in butadiene. For simplicity, the potential $V(s)$ is approximated by a step potential with step height $2b$ (Figure 7.28). As shown below $2b = 6.45 \times 10^{-19}$ J. The Schrödinger equation is solved and the resulting charge density dQ/ds (Figure 7.28) shows an equally strong alternation as in butadiene indicating selfconsistancy between bond lengths and π-electron density in the bonds.

Estimate of Constant b

We estimate the constant b by calculating the potential energy of an electron in the center of a bond in the electric field of the adjacent carbon atoms. We assume that the electron is at a distance $\delta = 75$ pm above the molecular plane (Figure 7.29). The adjacent atoms are assumed to act as point charges $e \cdot (Z_{eff} - 1)$: the Slater charge is approximately shielded by one electron. The remaining atoms are considered as completely shielded. According to Figure 7.28 the quantity $2b$ is the difference in the Coulomb attraction energy of the electron in the center of a double bond or in the center of a single bond,

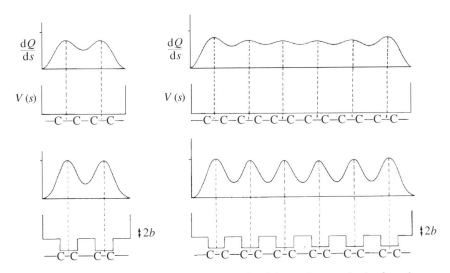

Figure 7.28 Charge density distribution dQ/ds of the π electrons in the free electron model (top) and in the step potential model with amplitude b (bottom). Comparison of butadiene (left) and dodecahexaene (right).

Continued on page 201

respectively:

$$2b = \frac{2 \cdot (Z_{\text{eff}} - 1) \cdot e^2}{4\pi\varepsilon_0} \cdot \left(\frac{1}{r_{C=C}} - \frac{1}{r_{C-C}}\right)$$

(a)

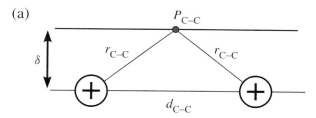

(b)

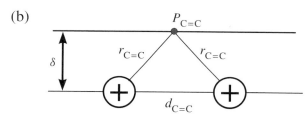

Figure 7.29 Estimating constant b in Figure 7.28. The electron is considered to be in points P_{C-C} and $P_{C=C}$, respectively, exposed to the effective charges of the adjacent C atoms. The difference in Coulomb energy in the two cases corresponds to $2b$.

The distances $r_{C=C}$ and r_{C-C} are, according to Figure 7.29,

$$r_{C=C} = \sqrt{\left(\frac{d_{C=C}}{2}\right)^2 + \delta^2} \qquad r_{C-C} = \sqrt{\left(\frac{d_{C-C}}{2}\right)^2 + \delta^2}$$

where $d_{C=C}$ and d_{C-C} are the bond lengths of a double bond and a single bond. With $d_{C=C} = 133$ pm, $d_{C-C} = 148$ pm, $\delta = 75$ pm, and $Z_{\text{eff}} = 3.25$ (Slater effective charge of carbon) we obtain $r_{C=C} = 100.4$ pm, $r_{C-C} = 105.3$ pm and $b = 2.4 \times 10^{-19}$ J.

Step Potential Model

We calculate the potential energy $V(s)$ as $V(s) = V_i$ at bond i directly from the self-consistent quantity P_i defined in (7.25) at each bond. By calibrating V_i and P_i on the experimental bond lengths of ethene, butadiene, and benzene we find the following relations between bond length d_i, bond potential V_i and π electron density P_i:

$$\frac{V_i}{10^{-19} \text{ J}} = 0.43 \cdot \frac{d_i - 140 \text{ pm}}{\text{pm}}$$

Continued from page 201

$$P_i = 1 - 0.0464 \cdot \frac{d_i - 133\,\text{pm}}{\text{pm}}$$

where 140 pm is the bond length in benzene, and 133 pm is the bond length in ethene. Then the potential V_i for $d_i = 140$ pm equals zero, and the potential difference between double bond and single bond is

$$\Delta V = V_{C-C} - V_{C=C}$$
$$= 0.43 \cdot (148 - 133) \times 10^{-19}\,\text{J} = 6.45 \times 10^{-19}\,\text{J}$$

This is slightly larger than our estimated value $2b = 4.8 \times 10^{-19}$ J. Finally, for $d_i = 133$ pm we obtain the π density $P_i = 1$, in accordance with Table 7.3.

Bond Lengths in Octatetraen

By starting the calculation for octatetraene with equal bond lengths we obtain in the first step for the different bonds:

$$\mathrm{CH_2 \overset{1}{=} CH \overset{2}{-} CH \overset{3}{=} CH \overset{4}{-} CH \overset{5}{=} CH \overset{6}{-} CH \overset{7}{=} CH_2}$$

bond	1	2	3	4	5	6	7
P_i	0.81	0.57	0.74	0.59	0.74	0.57	0.81
d_i/pm	137.1	142.3	138.6	141.8	138.7	142.4	137.1
$V_i/10^{-19}\,\text{J}$	−1.23	+1.04	−0.58	+0.83	−0.58	+1.04	−1.23

Then the calculation is repeated starting with a step potential with height V_i, and new values P_i, d_i, and V_i are obtained. The procedure is repeated until self-consistency is reached. This means that all bond lengths d_i assumed in the calculation agree with the bond lengths resulting from the calculated P_i's (introducing the P_i's in the above equation relating P_i with bond distance d_i). This is the case for:

bond	1	2	3	4	5	6	7
P_i	0.943	0.397	0.895	0.428	0.895	0.397	0.943
d_i/pm	134.2	146.0	135.3	145.3	135.3	146.1	134.2
$V_i/10^{-19}\,\text{J}$	−2.58	+2.43	−2.13	+2.14	−2.13	+2.43	−2.58

Obviously, in the case of the single bonds represented by high potential steps, P_i is much smaller than according to the simple free electron and HMO models. A linear relationship between P_i and d_i is obtained in contrast to the free-electron model and HMO (Figure 7.30). This is remarkable; such a simple relation is expected when considering the π electrons in the elastic skeleton

Continued on page 203

Continued from page 202

of the σ bonds. The charges of the nuclei connected by a bond are attracted by the π electron cloud in between, its density being proportional to P, so the compression of the bond is proportional to P.

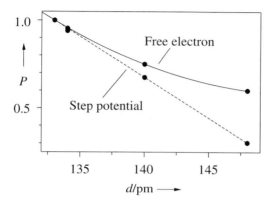

Figure 7.30 $P = \dfrac{\mathrm{d}Q/\mathrm{d}s \text{ at center of bond}}{\mathrm{d}Q/\mathrm{d}s \text{ in ethene}}$ according to the free-electron model (solid line) and to the step-potential model (dashed line) for ethene, butadiene, and benzene (full circles) as a function of the bond length d.

8 Absorption and Emission of Light

We discuss important relationships between the electronic structure and properties of π electron systems. We elucidate how tiny changes in the electronic structure affect the color of dyes. These considerations are of fundamental importance for an understanding of the role of light-absorbing biomolecules.

Using color changes as indicators we can directly probe the electron density distribution in molecules. High-resolution emission spectroscopy allows us to detect single molecules.

A fascinating effect is the stimulated emission of light in dye lasers, which is closely connected with the structure and the properties of π electron systems.

By measuring the optical rotation, information about chiral molecules can be obtained.

The picture of the de Broglie waves allows an easy understanding of the interaction of π-electron systems with light. Coupling effects requiring a more careful study are discussed in the foundations, which are addressed to specially interested and more advanced readers.

8.1 Basic Experimental Facts

8.1.1 Transmittance and Absorbance

We excite π electrons by shining light in the visible or ultraviolet spectral region on organic dye molecules (Figure 8.1(a)). A prism or a grating subsequently analyzes the light, which is then observed on a screen. Without the dye molecules in the light path, the intensity distribution of the different wavelengths on the screen corresponds to the intensity distribution of the light emitted by the lamp. With the dye molecules in the light path, the light intensity in a certain range of wavelengths is decreased (see double arrow in Figure 8.1(a)). The ratio

$$T = \frac{I}{I_0} \tag{8.1}$$

(I light intensity with dye, I_0 light intensity without dye) is called *transmittance*. At the wavelength where the dye absorbs the light most strongly ($\lambda = \lambda_{max}$), the transmittance is minimal. I and I_0 can be measured quantitatively with a photoelectric cell or with a photomultiplier. Then we can plot I/I_0 as a function of λ (Figure 8.1(b)). From the

8.1 BASIC EXPERIMENTAL FACTS

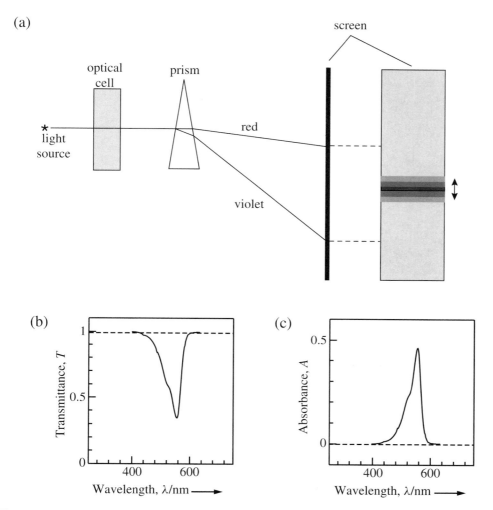

Figure 8.1 Light absorption of a solution of a dye. (a) Experimental setup. The double arrow marks the range of light absorbed by the dye molecules. (b) Transmittance $T = I/I_0$, (c) absorbance $A = -\log(I/I_0)$ versus wavelength λ for branched dye cation structure (8.4) with $k = 1$. The concentration is 6×10^{-6} mol L^{-1} (dye dissolved in methanol), and the optical path length is 1 cm.

transmittance the absorbance

$$A = -\log \frac{I}{I_0} \tag{8.2}$$

can be calculated (Figure 8.1(c)); the absorbance is proportional to the concentration c of the dye molecules in the optical cell

$$\boxed{A = \varepsilon c l} \tag{8.3}$$

(*law of Lambert and Beer*, see Box 8.1), where l is the length of the path through the cell (Fig. 8.1(a)) and ε is the molar decadic absorption coefficient (or extinction coefficient). ε is a quantity depending on the nature of the absorbing molecules and on the wavelength.

Box 8.1 – Law of Lambert and Beer

We consider a cell with absorbing molecules (Figure 8.2). At point x the incident light intensity is I, and at point $x + dx$ it is $I + dI$. Then

$$dI = -\alpha \cdot c \cdot I \cdot dx$$

where c is the concentration, α is a constant. Integration:

$$\int_{I_0}^{I} \frac{dI}{I} = -\alpha c \int_{0}^{x} dx$$

Then

$$\ln \frac{I}{I_0} = -\alpha \cdot c \cdot x \quad \text{or} \quad I = I_0 \cdot e^{-\alpha \cdot c \cdot x}$$

We rewrite for the decadic logarithm:

$$\log \frac{I}{I_0} = \frac{1}{\log e} \cdot \ln \frac{I}{I_0} = -\frac{\alpha}{\log e} \cdot c \cdot x$$

Then, with $\varepsilon = \alpha / \log e$ we obtain

$$\log \frac{I}{I_0} = -\varepsilon \cdot c \cdot x$$

ε is the molar decadic absorption coefficient and A is the absorbance. This is the *law of Lambert and Beer* of light absorption.

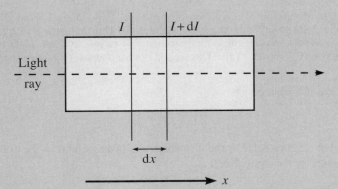

Figure 8.2 Derivation of the law of Lambert and Beer. Note that I decreases with increasing x, thus dI is negative.

8.1 BASIC EXPERIMENTAL FACTS

In Figure 8.3 the absorption spectra of the dye cations

$$\text{(structure 8.4)} \tag{8.4}$$

with $k = 0, 1, 2,$ and 3 are shown. In this case the absorption is due to the excitation of the π electrons in the dye molecule.

8.1.2 Polyenes and Cyanines

The position of the absorption maximum λ_{max} and the strength of the absorption band depend sensitively on the electronic structure. The wavelength λ_{max} increases with the length of the chain of conjugated carbon atoms. The color of solutions of dyes structure (8.4) changes from yellow ($k = 0$) through red ($k = 1$) and blue ($k = 2$) to blueish green ($k = 3$), as can be seen in Figure 8.3 and Table 8.1.

This is quite different with polyenes

$$CH_2=(=CH-CH=)_k=CH_2, \quad k = 0, 1, 2, \ldots \tag{8.5}$$

where the compounds with $k = 0$ ($CH_2=CH_2$, ethene) and $k = 1$ ($CH_2=CH-CH=CH_2$, butadiene) are colorless, and only compounds with $k > 8$ are colored.

By exchanging the CH_2 groups in the polyenes by the groups $(CH_3)_2\ddot{N}-CH=$ and $=\overset{\oplus}{N}(CH_3)_2$ we proceed to the class of symmetric cyanine dye cations, which can be

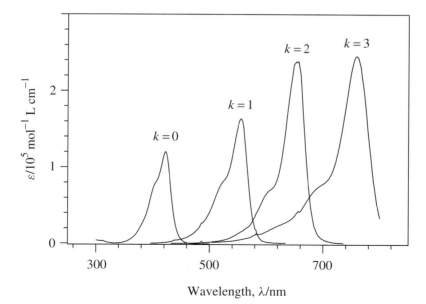

Figure 8.3 Absorption spectra (extinction coefficient ε versus wavelength) for dye cations (8.4) with $k = 0, 1, 2,$ and 3. It is customary to measure ε in $mol^{-1} L\, cm^{-1}$, since the usual units for c and l are $mol \cdot L^{-1}$ and cm, respectively. Shift of the absorption maximum with increasing chain length (i.e., with increasing k). Dyes dissolved in methanol.

Table 8.1 Absorption of dye cations (structure (8.4)) with different values of k

k	λ_{max}/nm	Color of absorbed light	Color of solution
0	420	Blue	Yellow
1	540	Green	Red
2	640	Orange	Blue
3	740	Red	Blueish green

written in two equivalent limiting structures:

$$(CH_3)_2\ddot{N}-(-CH=CH-)_k-CH=\overset{\oplus}{N}(CH_3)_2, \qquad (8.6)$$

and

$$(CH_3)_2\overset{\oplus}{N}=CH-(-CH=CH-)_k-\ddot{N}(CH_3)_2, \qquad (8.7)$$

As already discussed in Chapter 7 for $k = 0$ (amidinium cation), the symmetric cyanines can be considered as resonance hybrids between the structures (8.6) and (8.7).

In Figure 8.4(a) the positions of the absorption maxima of the symmetric cyanine cations are compared with those of the corresponding polyenes. The cyanine cations absorb at much longer wavelengths than do the corresponding polyenes.

There is another remarkable phenomenon, namely, that with increasing chain length the absorption maxima of the polyenes approach a limiting value, whereas the maxima of the cyanine cations increase by about 100 nm when the chain length increases by one double bond.

The shift of the absorption maximum of a cyanine dye cation compared with a polyene depends on the type of the end group. In the case of the cyanine cations considered there is a $(CH_3)_2\overset{\oplus}{N}=$ group on one side (electron acceptor) and a $(CH_3)_2\ddot{N}-$ group on the other side (electron donor). In the two possible limiting structures these two groups are simply exchanged; this is why these dye cations are called *symmetric cyanine dyes*. The end groups, however, can also be asymmetric; for example, merocyanines have one amino group and one carbonyl group

$$(CH_3)_2\ddot{N}-CH=CH-CH=\underset{..}{\overset{..}{O}}:$$

Here nitrogen is an electron donor and oxygen is an electron acceptor; in the second limiting structure

$$(CH_3)_2\overset{\oplus}{N}=CH-CH=CH-\underset{..}{\overset{..}{O}}\overset{\ominus}{:}$$

the situation is reversed. Both limiting structures are different. Generally, a dye can be characterized by the scheme

$$\text{donor}-CH=CH-CH=CH-CH=\text{acceptor} \qquad (8.8)$$

There is a surprising observation when we compare the absorption maxima of cyanine cations with those of azacyanine cations

$$(CH_3)_2\ddot{N}-CH=\underset{..}{N}-CH=\overset{\oplus}{N}(CH_3)_2 \qquad (8.9)$$

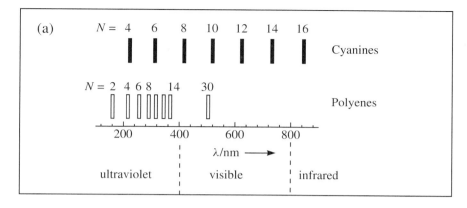

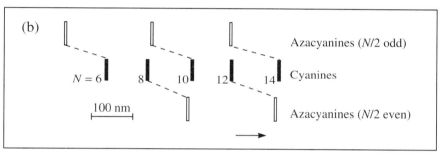

Figure 8.4 Absorption maxima λ_{max} of π electron systems for different numbers N of π electrons. (a) Comparison of cyanine cations with polyenes. (b) Comparison of cyanine cations with azacyanines (schematic). For numerical data see Tables 8.2, and 8.3.

and

$$(CH_3)_2\ddot{N}-CH=CH-\underset{..}{N}=CH-CH=\overset{\oplus}{N}(CH_3)_2 \tag{8.10}$$

In an azacyanine cation the central CH group of a cyanine cation is replaced by an N atom. This causes a shift of λ_{max} for compounds with an even number of conjugated double bonds by about 100 nm to shorter wavelengths. Compounds with an odd number of conjugated double bonds show a shift of about 100 nm to longer wavelengths (Figure 8.4(b)).

8.2 Absorption Maxima of Dyes

How can we understand these astonishing features of light absorption? The molecules are transferred from the electronic ground state to an excited state. Our task is to find the excitation energy ΔE in each case. Light is absorbed only when the condition (2.1)

$$h\nu = \Delta E \tag{8.11}$$

is fulfilled, where ν is the frequency of light and ΔE is the excitation energy. From ν we obtain the wavelength λ of the absorption maximum

$$\lambda = \frac{c_0}{\nu} = \frac{hc_0}{\Delta E} \tag{8.12}$$

where c_0 is the speed of light.

8.2.1 Band Broadening

Since ΔE has a distinct value, we expect sharp absorption lines in the absorption spectra as we found for the spectra of atoms (see Chapter 3). What is the reason for the broad absorption bands shown in Figures 8.1 and 8.3? The position of the absorption line of a selected molecule depends on its particular surroundings; the absorption band can be understood as a superposition of a multitude of absorption lines of each molecule in the population.

8.2.2 Single-molecule Absorption

A sample with dye molecules embedded in a polymer is exposed to monochromatic light with a high intensity; as a consequence, the dye molecules absorbing within the frequency range of the light beam are bleached. A narrow gap in the absorption band of the dye is observed (Figure 8.5). This shows that the molecules that were destroyed by light had an absorption line just in the narrow range around the wavelength of the light that caused their bleaching. This method is called *spectral hole burning*. Such experiments can be done with sufficiently small samples where only one molecule is present that has its absorption line at the wavelength of the incident light. By thermal motion the environment of the individual molecules changes; this results in a shift of the individual absorption lines and to a broadening and disappearance of the gap.

The absorption line of a molecule can be directly observed. Occasional shifts of this line by fluctuations in the individual environment appear. Either one or no (or occasionally

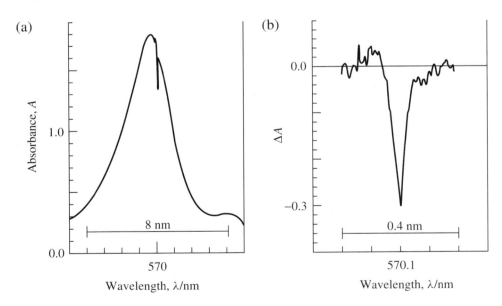

Figure 8.5 Spectral hole burning. A solution of the dye pseudoisocyanine bromide in polyvinyl alcohol was cooled to the temperature of liquid helium (4.2 K). The dye was bleached by exposure to an intense light beam at a wavelength of 570.1 nm. After bleaching, an absorption spectrum was taken (a) showing a spectral hole at this wavelength. The half width of this hole (b) is about 0.03 nm. K. Misawa, T. Kobayashi, Hierarchical Structure in Oriented J-aggregates, in T. Kobayashi (Ed.), J-Aggregates, World Scientific 1996, p. 50.

8.2 ABSORPTION MAXIMA OF DYES

more than one) molecules absorbing in the given spectral range can be present. The position of the absorption peak can be shifted by applying an electric field, because the field changes the relative position of ground state and excited state depending on the electron distribution. Thus single-molecule spectroscopy allows direct access to molecular parameters. (*Single Molecule Optical Detection, Imaging and Spectroscopy*, Eds. T. Basche, W. E. Moerner, M. Orrit, V. P. Wild, Wiley-VCH Weinheim 1998).

8.2.3 Cyanine Dyes

First we look at a symmetric cyanine dye cation. All C—C bonds of these molecules are equal in length; as a result the potential energy $V(s)$ of a π electron is the same for all bonds. For the cyanine cation

$$(CH_3)_2\ddot{N}-CH=CH-CH=CH-CH=CH-CH=\overset{\oplus}{N}(CH_3)_2 \qquad (8.13)$$

we obtain, therefore, according to the free-electron model, the wavefunctions and energies shown in Figure 8.6. The energies are

$$E = \frac{h^2}{8m_e L^2} \cdot n^2, \quad n = 1, 2, 3, \ldots \qquad (8.14)$$

where L is the length of the π-electron system.

There are four conjugated double bonds and one free π electron pair at the nitrogen; that is altogether there are $10\,\pi$ electrons. We distribute these π electrons over the lowest five energy levels. Level 5 is then the highest occupied level (*HOMO = highest occupied molecular orbital*), and level 6 is the lowest unoccupied level (*LUMO = lowest unoccupied molecular orbital*). When the dye cation is irradiated with light, the molecule can

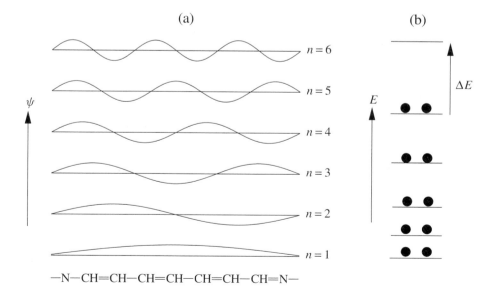

Figure 8.6 Wavefunctions (a) and energies (b) of the symmetrical cyanine structure (8.13) according to the free-electron model.

take up energy from the electromagnetic field of the light source:

$$\text{molecule} + \text{light quantum} \rightarrow \text{molecule*} \tag{8.15}$$

The asterisk indicates that the molecule is in an excited state. The HOMO–LUMO transition ($\Delta n = 1$) is responsible for the color of a dye; thus we focus on this transition. Other transitions occur at higher energies. In the present case a π electron is lifted from level 5 to level 6.

Now, we calculate the excitation energy of the dye according to the free-electron model, starting with equation (8.14):

$$\Delta E = \frac{h^2}{8m_e L^2} \cdot (n_{\text{LUMO}}^2 - n_{\text{HOMO}}^2) \tag{8.16}$$

n_{LUMO} is the quantum number of the first unoccupied level, and n_{HOMO} is the quantum number of the highest occupied level. These numbers are connected with the number N of π electrons. Since each level is occupied by two electrons, we have

$$n_{\text{HOMO}} = \tfrac{1}{2}N, \quad n_{\text{LUMO}} = \tfrac{1}{2}N + 1 \tag{8.17}$$

and

$$\boxed{\Delta E = \frac{h^2}{8m_e L^2} \cdot (N+1)} \tag{8.18}$$

The length L of the π system can be expressed by the number N of π electrons and the bond length d_0. For the cyanine cations (structures (8.6) and (8.7)) we obtain

$$L = N \cdot d_0 \tag{8.19}$$

For example, for structure (8.13) we have $N = 10$ and $L = 10 \cdot d_0$ assuming that the π system extends the outermost N atoms by one bond length d_0, as indicated in Figure 8.6(a). Then

$$\Delta E = \frac{h^2}{8m_e N^2 d_0^2} \cdot (N+1) = \frac{h^2}{8m_e d_0^2} \cdot \frac{N+1}{N^2} = 30.7 \times 10^{-19} \text{ J} \cdot \frac{N+1}{N^2} \tag{8.20}$$

and

$$\lambda_{\max} = \frac{hc_0}{\Delta E} = \frac{8m_e d_0^2 c_0}{h} \cdot \frac{N^2}{N+1} = 64.7 \text{ nm} \cdot \frac{N^2}{N+1} \tag{8.21}$$

with $d_0 = 140$ pm.

For cyanine cations with a large number of π electrons we have

$$\frac{N^2}{N+1} \approx N \tag{8.22}$$

Hence $\lambda_{\max}$ is approximately linearly increasing with N.

In Table 8.2 the calculated $\lambda_{\max}$ values are compared with experimental data. The experimental $\lambda_{\max}$ values really increase linearly with the length of the chain. The increase of about 100 nm per π-electron pair (i.e., per double bond) is somewhat smaller than the 2×65 nm $= 130$ nm expected from equation (8.21). Considering the very simple model used as a basis for our calculations, the agreement between theory and experiment is remarkably good.

8.3 STRENGTH AND POLARIZATION OF ABSORPTION BANDS

Table 8.2 Symmetrical cyanine cations $(CH_3)_2 \overset{\oplus}{N}=CH-(CH=CH)_k-\overset{..}{N}(CH_3)_2$ with $k = 0$ to $k = 6$ ($N = 4$ to $N = 16\pi$ electrons). Calculated and experimental absorption maxima λ_{max} and oscillator strengths f (see Section 8.3)

k	N	λ_{max}/nm		Color of solution	f	
		Calculated	Experiment		Calculated	Experiment
0	4	206	224	Colorless	0.7	
1	6	332	313	Colorless	1.0	0.9
2	8	459	416	Yellow	1.3	1.0
3	10	587	519	Red	1.6	1.2
4	12	716	625	Blue	1.7	1.5
5	14	844	735	Green	2.2	2.0
6	16	973	848	Colorless	2.4	

The color of the solution of the dye is complementary to the color of the absorbed light; e.g., if red light is absorbed the color of the solution is green. The absorption maxima are calculated according to equation (8.21), the oscillator strengths according to equation (8.29) with $\hat{n} = 1.42$ (index of refraction of the solvent dichloromethane). References see Further Reading

It is most exciting that we find such simple relations for λ_{max}. λ_{max} depends merely on the number N of π electrons, on the length L of the molecular chain, and on universal constants (h, m_e, and c_0). This is an important check for the reliability of the model. Using this model we can understand why organic dyes are colored and how they are colored.

The dyes used in everyday life have a somewhat more complicated molecular skeleton. The essential features, however, are the same as discussed here, but deeper considerations are needed to understand their colors. Such considerations will follow in the next sections.

8.3 Strength and Polarization of Absorption Bands

8.3.1 Oscillator Strength

In order to obtain the strength of an absorption band, we replace the quantum mechanical approach by a classical consideration. This can be a very useful simplification of the problem (compare the correspondence principle, Section 3.8.2). In the present case, the dye molecule can be replaced by a classical oscillator of charge e, mass m_e, and frequency $v_0 = \Delta E/h$ oscillating in the direction of the molecular chain. ΔE follows from the quantum mechanical treatment, (see equation (8.18)). The oscillator is excited in the electric field of the incident light. Its amplitude ξ is strongly increasing when the frequency v of the light is approaching the resonance frequency v_0 of the oscillator. The absorption is strongest for $v = v_0$.

The molar absorption coefficient ε of an assembly of such oscillators can be calculated (see Foundation 8.1). We obtain for the considered classical oscillators with charge e and mass m_e (Figure 8.8)

$$\left(\int_{band} \varepsilon \cdot dv \right)_{classical(e,m_e)} = \frac{N_A e^2}{4\varepsilon_0 2.303 m_e c_0} \quad (8.23)$$

The integral is taken in the range of the absorption band. The measured integrated absorption of a dye, the quantity $\int_{band} \varepsilon \cdot dv$, is frequently expressed in terms of the absorption

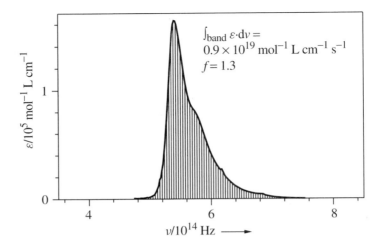

Figure 8.7 Evaluation of the oscillator strength f according to equation (8.24) for the symmetrical cyanine dye cation structure (8.4) with $k = 1$. ν is the frequency of the light.

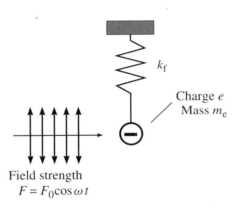

Figure 8.8 Classical oscillator with force constant k_f, mass m_e, and charge $Q = e$ excited by an electric field with field strength $F = F_0 \cos \omega t$.

strength, f, of a classical oscillator

$$f = \frac{\int_{band} \varepsilon \cdot d\nu}{\left(\int_{band} \varepsilon \cdot d\nu\right)_{classical(e,m_e)}} = \frac{4\varepsilon_0 2.303 m_e c_0}{N_A e^2} \cdot \int_{band} \varepsilon \cdot d\nu$$

$$= 1.44 \times 10^{-19} \text{ mol L}^{-1} \text{ cm s} \cdot \int_{band} \varepsilon \cdot d\nu \qquad (8.24)$$

f is called the oscillator strength. Figure 8.7 demonstrates the evaluation of f from the absorption spectrum of dye cation structure (8.4) with $k = 1$. Experimental values for f for some symmetrical cyanine dye cations are collected in Table 8.2.

8.3.2 Polarization of Absorption Bands

This simple classical model based on the idea of Bohr's correspondence principle is also useful in describing a property called *polarization of an absorption band*. We consider a cyanine dye dissolved in a polymer (e.g., polyvinyl alcohol); when the polymer is stretched, the polymer chains and the rod-shaped dye molecules are oriented in the stretching direction. Then light at wavelength λ_{max} polarized in the stretching direction (x direction) is absorbed, but light polarized perpendicularly (y direction) is not absorbed (Figures 8.9 and 8.10).

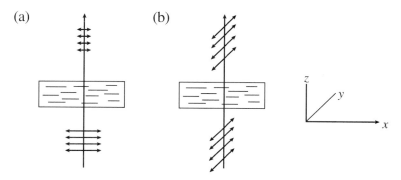

Figure 8.9 Absorption of light by a cyanine dye dissolved in a polymer. (a) Light polarized in the stretching direction of the polymer (x direction). (b) Light polarized perpendicularly (y direction).

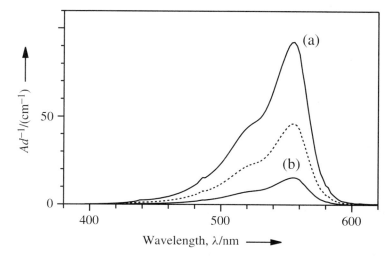

Figure 8.10 Absorption spectrum of dye cation structure (8.4) with $k = 1$ dissolved in polyvinyl alcohol. A is the absorbance and d is the thickness of the polymer foil ($d = 0.2$ mm for the unstretched polymer). Electric field strength of the polarized light: (a) in direction of stretching of polymer, (b) perpendicular to direction of stretching. In the actual experiment the molecular chains are not exactly in the stretching direction; thus there is a residual absorption in case (b).

This effect is useful to produce and analyze polarized light in the range of the absorption band. The light passing the polarizing filter is linearly polarized perpendicular to the stretching direction.

In the classical approach, the absorbing dye molecule is replaced by an oscillator with an electron charge oscillating in the direction of the molecular chain (x direction). The electric field of the light wave accelerates the oscillator when acting in the x direction; the field has no effect, however, when the oscillator is acting in the y direction.

In the quantum mechanical approach we have to apply the time-dependent Schrödinger equation including the electric field of the incident light wave. Then the calculated oscillator strength (Foundation 8.2) is

$$f = \frac{8\pi^2 m_e}{3h^2 e^2 \hat{n}} \cdot M^2 \cdot \Delta E \qquad (8.25)$$

where M is the transition moment

$$M^2 = M_x^2 + M_y^2 + M_z^2 \qquad (8.26)$$

with

$$M_x = e \cdot \int \psi_i \cdot x \cdot \psi_j \cdot d\tau \qquad (8.27)$$

(M_y and M_z can be written correspondingly) and $\hat{n}$ is the index of refraction of the solvent. The direction of the movement of the classical oscillator corresponds to the direction of the transition moment, and the oscillating charge corresponds to the value of M. In the present case (molecular chain is in x direction)

$$M_y = M_z = 0 \quad \text{and} \quad M_x = e \cdot \int \psi_n \cdot x \cdot \psi_{n+1} \cdot ds \qquad (8.28)$$

where n is the quantum number of the highest occupied orbital.

This means that only light with electrical vector oscillating in the x direction is absorbed. As we have seen in Figures 8.9 and 8.10 this is in agreement with experiment. Again, this confirms our model.

The f values for the cyanine dyes considered in Section 8.2.3 are calculated in Foundation 8.2:

$$f = \frac{2}{\pi^2 \hat{n}} \cdot (N+1) \qquad (8.29)$$

where $\hat{n}$ is the index of refraction of the solvent and N is the number of π electrons. The f values are in good agreement with experiment (Table 8.2). They are on the order of 1, as for a classical oscillator with mass m_e and charge e.

8.4 Heteroatoms as Probes for Electron Distribution

Next we compare the cyanine cation (structure (8.13))

$$(CH_3)_2\overset{..}{N}-CH=CH-CH=CH-CH=CH-CH=\overset{\oplus}{N}(CH_3)_2$$

with the azacyanine cation

$$(CH_3)_2\overset{..}{N}-CH=CH-CH=\underset{..}{N}-CH=CH-CH=\overset{\oplus}{N}(CH_3)_2 \qquad (8.30)$$

8.4 HETEROATOMS AS PROBES FOR ELECTRON DISTRIBUTION

where the central N atom is trigonally hybridized like every C atom of the chain. While in the C atom one of the hybrid orbitals is used for a C—H bond, it is occupied in the N atom by two electrons (lone electron pair). This orbital is in the molecular plane, and thus the lone pair electrons are σ electrons. Consequently, azacyanine cations have the same number of π electrons and the same wavefunctions as a corresponding cyanine cation.

However, we must pay attention to the fact that an N atom is more electronegative than a C atom. This means that the π electrons are more strongly attracted by the core of the N atom than by the core of a carbon atom.

Whereas for cyanine cations the potential energy $V(s)$ along the chain of carbon atoms can be assumed as constant (all the atoms in the chain are identical, Figure 8.11(a)), the potential energy at the center of an azacyanine cation is strongly decreased (Figure 8.11(b)). This causes the energies of all orbitals of an azacyanine cation to be, in principle, lower than those of a comparable cyanine.

The amount by which the energy of a selected orbital decreases depends upon the probability

$$dP = \psi^2 \cdot ds \tag{8.31}$$

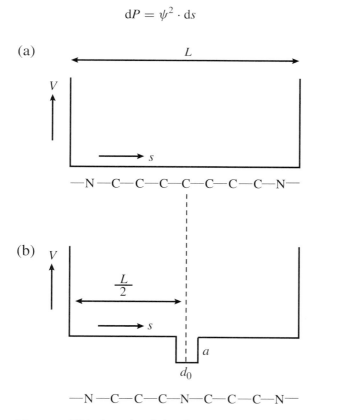

Figure 8.11 Potential energy $V(s)$ along the chain of (a) a cyanine cation and (b) an azacyanine cation. We assume a potential trough at the center of the molecule with depth a and width d_0 (the trough extends from center to center of the adjacent bonds). Note: The nitrogen atoms at the chain ends of the cyanine and azacyanine cations should also be taken into account by potential troughs. These troughs cause an equal shift of the HOMO's and LUMO's to lower energies, since these troughs are at antinodes of the wavefunction; therefore we neglect them in treating excitation energies.

of finding an electron in a region ds around the N atom. If ψ^2 is large at the position of the N atom, then the decrease in energy will be large for this orbital. If $\psi^2 = 0$, then the electron can never be found in the vicinity of the N atom, and the energy of the orbital remains unchanged.

Orbitals 5 and 6 of the cyanine cation (structure (8.13)) and azacyanine cation (structure (8.30)) are illustrated in Figure 8.12 (HOMO and LUMO). ψ^2 has an antinode at the central N atom in orbital 5 and a node in orbital 6. From this it follows that orbital 5 in the azacyanine has a lower energy than in the corresponding cyanine; orbital 6 practically does not change its energy when we proceed from a cyanine to an azacyanine. This means that the excitation energy ΔE increases, and dye structure (8.30) absorbs at shorter wavelengths than does dye structure (8.13).

In the cyanine

$$(CH_3)_2\overset{..}{N}-CH=CH-CH=CH-CH=\overset{+}{N}(CH_3)_2 \tag{8.32}$$

and azacyanine

$$(CH_3)_2\overset{..}{N}-CH=CH-\overset{..}{N}=CH-CH=\overset{+}{N}(CH_3)_2 \tag{8.33}$$

where both dyes are shorter by two carbon atoms, the situation is reversed. According to Figure 8.13 the energy of the LUMO decreases, and the energy of the HOMO remains unchanged. This aza dye then absorbs at longer wavelengths than does the corresponding cyanine.

Generally we find that λ_{max} is shifted to shorter wavelengths when the HOMO has an antinode in the center of the molecule. This is always the case for dyes with an odd number of occupied states ($N/2$ odd). For an even number of occupied states ($N/2$ even) we expect the reverse, namely, a shift to longer wavelengths.

In Problem 8.3 the aza shifts are calculated quantitatively:

$$\text{energy shift} = -a \cdot \frac{2}{N} \quad \text{for} \quad \frac{N}{2} = 2, 4, 6, \ldots \tag{8.34}$$

$$\text{energy shift} = +a \cdot \frac{2}{N} \quad \text{for} \quad \frac{N}{2} = 3, 5, 7, \ldots$$

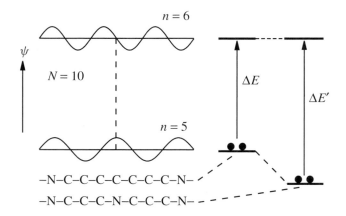

Figure 8.12 Orbitals 5 (HOMO) and 6 (LUMO) of the cyanine cation structure (8.13) and of the azacyanine cation structure (8.30); decrease of level 5 for the azacyanine: $\Delta E' > \Delta E$.

8.5 HOMO–LUMO GAP BY BOND ALTERNATION

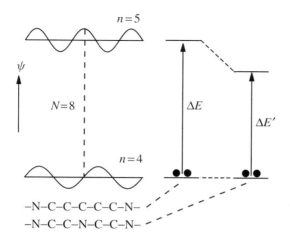

Figure 8.13 Orbitals 4 (HOMO) and 5 (LUMO) of the cyanine cation structure (8.32) and of the azacyanine cation structure (8.33) with $N = 8$; decrease of level 5 for the azacyanine: $\Delta E' < \Delta E$.

Table 8.3 Calculated and experimental aza shifts for some cyanine cations

$\frac{N}{2}$	Shift $\Delta E/10^{-19}$ J	
	Calculated	Experiment
2	−1.75	
3	+1.17	+0.79
4	−0.88	−0.93
5	+0.70	+0.53
6	−0.58	−0.59
7	+0.50	+0.62

$N/2$ = number of occupied π electron states. Calculation according to equation (8.34) with $a = 3.5 \times 10^{-19}$ J. References see Further Reading.

where a is the depth of the potential trough extending over the nitrogen atom (Figure 8.11(b)). The constant a corresponds to the energy difference of an electron in a p orbital of a carbon atom and of a nitrogen atom. From the ionization energies of C (17.2×10^{-19} J) and N (20.7×10^{-19} J), it follows that $a = 3.5 \times 10^{-19}$ J. The result of the calculation is in agreement with the experimental data displayed in Figure 8.4(b) and Table 8.3.

A dramatic shift of the absorption maximum to shorter and longer wavelengths, respectively, is observed. This experimental result supports our considerations about the positions of nodes and antinodes of the wavefunctions. It shows that the model can give guidelines for designing dyes. Such guidelines are useful for chemists aiming to synthesize dyes absorbing at certain wavelengths.

8.5 HOMO–LUMO Gap by Bond Alternation

Even though polyenes are simpler molecules than cyanine cations, the theoretical treatment causes more trouble, because of the alternating single and double bonds

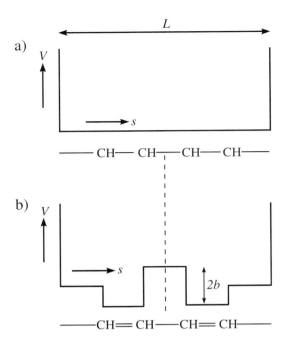

Figure 8.14 Potential energy $V(s)$ of a π electron along a chain of carbon atoms. (a) All C–C bonds equally long. (b) Alternating bond lengths (polyene).

(Figures 7.28 and 8.14). Thus the potential energy $V(s)$ along the chain of carbon atoms cannot be considered as constant as in a cyanine cation (Figure 8.14(a)). According to Figure 8.14(b), $V(s)$ is lower at a double bond and higher at a single bond, because the positive charges of the nuclei are closer together at the double bond, and the π electron is more strongly attracted.

Thus we have a "potential trough" at a double bond and a "potential hill" at a single bond. In analogy to the consideration of the azacyanine cations, we expect that the energy is decreased for orbitals with antinodes of the wavefunction at double bonds; for states with antinodes of the wavefunction at single bonds the energy is increased (Figure 8.15).

In Problem 8.4 the resulting energy shift is calculated as

$$\Delta E' = \Delta E + \frac{6}{5}b = \Delta E + 3.0 \times 10^{-19} \, \text{J} \tag{8.35}$$

where

$$\Delta E = \frac{h^2}{8m_e \cdot L^2} \cdot (N+1) = \frac{h^2}{8m_e d_0^2} \cdot \frac{1}{N+1} \tag{8.36}$$

with $L = (N+1) \cdot d_0$, where d_0 is the average of the length of C–C and C=C ($d_0 = 140$ pm); b is the depth of the potential trough (see Figure 8.14b and Foundation 7.3).

The result of the calculation is in good agreement with the experiment (Table 8.4). Whereas the excitation energy ΔE of cyanines is approaching zero with increasing chain length, $\Delta E'$ of the polyenes approaches 3.0×10^{-19} J for an infinitely long chain and λ_{max} approaches a limiting value of 660 nm. This big difference between cyanines and polyenes is merely caused by the differences in the bond lengths.

8.5 HOMO-LUMO GAP BY BOND ALTERNATION

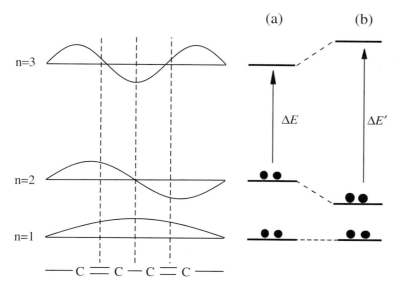

Figure 8.15 Wavefunctions and energy levels of butadiene. Because of the alternation of the bond lengths, level 2 (HOMO) is shifted downward, and level 3 (LUMO) is shifted upward compared with a corresponding π-electron system with equal bond lengths.

Table 8.4 Absorption maxima λ_{max} of polyenes with N π electrons ($N/2$ Double Bonds)

Polyene	N	$\Delta E/10^{-19}$ J	$\Delta E'/10^{-19}$ J	λ_{max}/nm	
				Calculated	Experiment
Ethene	2	10.4	13.4	150	162
Butadiene	4	6.2	9.2	220	217
Hexatriene	6	4.5	7.5	270	257
Octatetraene	8	3.5	6.5	310	290
	10	2.8	5.8	340	317
	12	2.4	5.4	370	344
	14	2.1	5.1	390	368
	16	1.8	4.8	410	386
	18	1.6	4.6	430	413
	20	1.5	4.5	440	420
(β-Carotene)	22	1.4	4.4	450	453
	24	1.2	4.2	470	461
	26	1.2	4.2	470	471
	28	1.1	4.1	480	500
	30	1.0	4.0	500	504
Polyacetylene	∞	0	3.0	660	650

ΔE is the excitation energy of a corresponding π electron system with equally long bonds according to equation (8.36) with $d_0 = 140$ pm, $\Delta E'$ is the excitation energy with correction for bond alternation according to equation (8.35). References see Further Reading.

Example – Indicator dyes

From these considerations the characteristic color changes of indicator dyes, e.g. of methylorange,

$$SO_3^- \text{—} \langle \text{—} \rangle \text{—} \ddot{N}{=}N\text{—} \langle \text{—} \rangle \text{—} \ddot{N}(CH_3)_2 \quad (460 \text{ nm})$$

$\downarrow H^+$

$$SO_3^- \text{—} \langle \text{—} \rangle \text{—} \overset{\oplus}{\underset{H}{N}}{=}N\text{—} \langle \text{—} \rangle \text{—} \ddot{N}(CH_3)_2$$

$\updownarrow$ (505 nm)

$$SO_3^- \text{—} \langle \text{—} \rangle \text{—} \underset{H}{\ddot{N}}\text{—}\ddot{N}{=}\langle \text{=} \rangle {=}\overset{\oplus}{N}(CH_3)_2$$

can be understood. Reaction of the neutral or alkaline form (absorption maximum at 460 nm) with H^+ ions leads to the acidic form with an absorption maximum at $\lambda_{max} = 505$ nm (Figure 8.16). The alkaline form has single bonds between benzene ring and adjacent nitrogen atoms, the acidic form can be described, similar to the cyanines, by two limiting structures with the positive charge at the nitrogen atom on the left and the nitrogen atom on the right, respectively. Thus the acidic form absorbs at longer wavelengths.

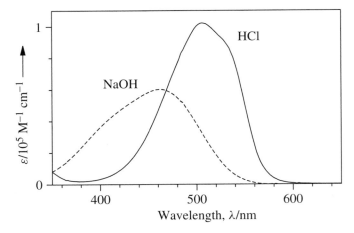

Figure 8.16 Absorption spectrum of solutions of the indicator dye methylorange. Solid line: dissolved in HCl (0.1 mol L^{-1}); dashed line: dissolved in NaOH (0.1 mol L^{-1}).

The distinction between the limiting cases of (a) symmetric cyanines (structures (8.6)) and (8.7) and of (b) polyenes (structure (8.5)) is crucial for understanding basic features of π electron systems. The important features are equal bond lengths in case (a) and bond length alternation in case (b). Bond length alternation causes a gap between the HOMO and the LUMO. Therefore, the absorption band of a polyene is strongly shifted to shorter wavelengths relative to the band of a cyanine of comparable size. A solution of the cyanine structure (8.13) with $N = 10$ is red, a solution of a polyene with $N = 10$ is colorless.

8.6 Dyes with Cyclic Electron Cloud: Phthalocyanine

Cu^{2+} phthalocyanine is an important dye with a blue color used as a coloring agent in industry; moreover, it plays a role as a catalyst in dehydrogenation reactions. The blue color is due to an absorption band in the red part of the visible spectrum ($\lambda = 675$ nm, see Figure 8.17).

The benzene rings are connected to the inner ring by single bonds, and we can consider the inner ring as an isolated π electron system. In the inner ring each C and N atom donates one π electron; in addition, two electrons are donated by the Cu atom. Altogether there are $16 + 2 = 18\,\pi$ electrons in the ring. Note that the structural formula representation

does not explain that the benzene branches are essentially isolated.

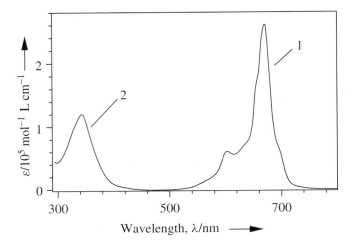

Figure 8.17 Absorption spectrum of Cu-phthalocyanine (solution of the tetrasulfonate compound in a mixture of 67% ethanol and 33% water). Absorption band 1 corresponds to transition 4b → 5b, and absorption band 2 corresponds to transition 4a → 5a (see Figure 8.20).

In analogy to benzene (Chapter 7) we obtain for the energies of Cu^{2+} phthalocyanine

$$E_n = \frac{h^2}{2m_e L_c^2} \cdot n^2, \quad n = 0, 1, 2, \ldots \tag{8.37}$$

where L_c is the circumference of the inner ring. The corresponding wavefunctions are displayed in Figure 8.18. According to this figure, the transitions 4a → 5a, 4a → 5b, 4b → 5a, and 4b → 5b are possible. All these transitions correspond to the same energy difference

$$\Delta E = \frac{h^2}{2m_e L_c^2} \cdot (25 - 16) = \frac{9h^2}{2m_e L_c^2} \tag{8.38}$$

In this treatment the potential energy along the π electron ring was assumed to be constant. Actually, we have to take into account the fact that the nitrogen atoms in the ring are more electronegative than the carbon atoms, and thus the potential energy is no longer constant along the ring (as for the azacyanine cations).

We consider the wavefunctions of the orbitals 4 and 5 (Figure 8.19). In orbital 4a the wavefunction has an antinode at each N atom and the energy of this state is decreased; orbital 4b remains unchanged (a node at the positions of the N atoms). The energies of orbitals 5a and 5b are both decreased by the same amount (Figure 8.20).

For the calculation of the energy splitting we proceed as in the case of the azacyanine cations. As shown in Problem 8.5 we obtain the result in Table 8.5, which is in good agreement with the experimental spectrum.

Essentially the electron responsible for the color is initially in an orbital with nodes of the wavefunction at each N atom while the electron of the second band in the UV is initially in an orbital with antinodes at each N atom. This is the reason for the strong separation of the two bands.

8.6 DYES WITH CYCLIC ELECTRON CLOUD: PHTHALOCYANINE

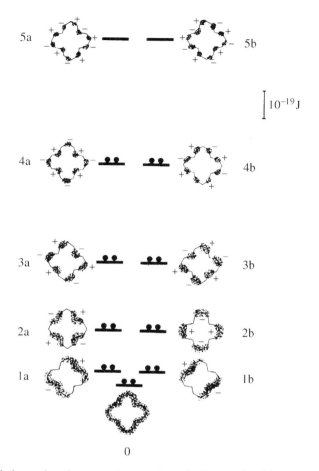

Figure 8.18 Phthalocyanine (benzene rings neglected). Charge densities of occupied and lowest unoccupied orbitals.

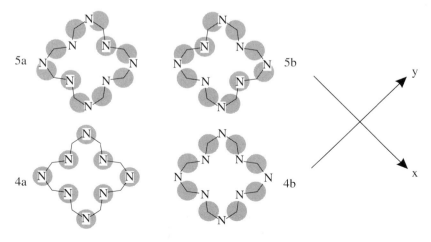

Figure 8.19 Phthalocyanine: orbitals 4a, 4b, 5a, and 5b; location of charge density accumulations (antinodes of the wavefunctions) relative to the N atoms.

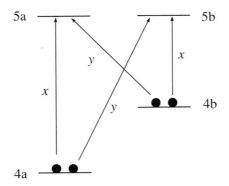

Figure 8.20 Energy shift in phthalocyanine caused by the electronegativities of the nitrogen atoms.

Table 8.5 Transitions in Cu^{2+} phthalocyanine and in Zn^{2+} porphyrin

	Transitions	Polarization	Cu^{2+} Phthalocyanine			
			λ_{max}/nm Theory	f	λ_{max}/nm Experiment	f
1	4b → 5b, 4b → 5a	x, y	772	1.2	675	0.8
2	4a → 5a, 4a → 5b	x, y	327	2.8	345	1.5
	Transitions	Polarization	Zn^{2+} Porphyrin			
			λ_{max}/nm Theory	f	λ_{max}/nm Experiment	f
1	4b → 5b, 4b → 5a	x, y	576	1.6	550	0.1
2	4a → 5a, 4a → 5b	x, y	382	2.4	423	1.6

8.7 Coupling of π Electrons

We consider porphyrin as an example:

In a good approximation, the chromophore can be described as a 16-membered ring with $18\,\pi$ electrons occupying nine orbitals (similar to the case of phthalocyanine).

The apparently small difference between porphyrin and phthalocyanine (four N atoms instead of eight N atoms in the inner ring) has a considerable consequence. The separation

8.7 COUPLING OF π ELECTRONS

of the two absorption bands 4b → 5a, 4b → 5b, and 4a → 5a, 4a → 5b is half as large in porphyrin as in phthalocyanine (Problem 8.5). This is in agreement with experiment (Table 8.5, Figure 8.21). The first transition at $\lambda = 550$ nm of porphyrin is related to the red color of blood.

However, the calculated oscillator strengths of porphyrin (Table 8.5) are completely different from the experimental values: the calculated oscillator strengths for the absorption band at long wavelengths (f_1) and for the absorption at short wavelengths (f_2) are similar ($f_1 : f_2 = 0.7$). In the experiment, however, the ratio of these oscillator strengths is only $f_1 : f_2 = 0.06$.

This is due to a coupling effect between the two transitions. This coupling effect is caused by the repulsion of the electrons responsible for the two transitions. It is shown in Foundation 8.3 that the quantum mechanical consideration can be replaced by a classical picture relating to the correspondence principle (Section 3.8).

The electrons responsible for the two transitions are replaced by two classical oscillators (charge e) of frequencies

$$\nu_1 = \frac{hc_0}{\lambda_1} \quad \text{and} \quad \nu_2 = \frac{hc_0}{\lambda_2} \tag{8.39}$$

They are coupled by a force related to the repulsion of the electrons in the actual system (Figure 8.22(a)). This coupling plays a role only when the frequencies ν_1 and ν_2 are similar. In this case we have to distinguish between the in-phase oscillation and the out-of-phase oscillation of the two oscillators (Figures 8.22(b) and (c)). The in-phase oscillation represents a strong absorption band (both charges are accelerated by the electric field of the incident light). The out-of-phase oscillation represents a weak absorption band (one charge is accelerated, while the other one is decelerated by the electric field). As a consequence, the oscillator strength of transition 1 is decreased and that of transition 2 is increased. According to Foundation 8.4, the effect of coupling can be calculated; the result $f_1/f_2 = 0.05$ is in good agreement with the experimental value (Table 8.5).

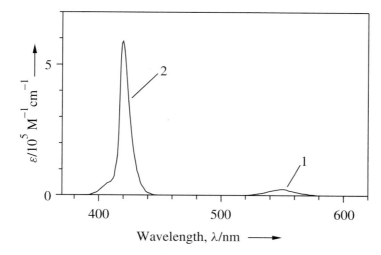

Figure 8.21 Absorption spectrum of Zn-tetraphenyl-porphyrin.

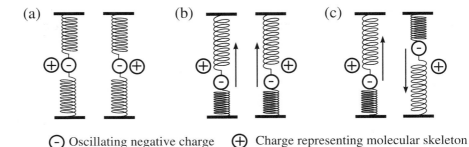

⊖ Oscillating negative charge ⊕ Charge representing molecular skeleton

Figure 8.22 Coupled oscillators: (a) oscillators at rest; (b) in-phase oscillation; (c) Out-of-phase oscillation. The positive charges represent the positively charged molecular skeleton.

In principle, a coupling effect is also observed in the case of Cu^{2+} phthalocyanine. Because of the eight nitrogen atoms in the ring, however, the energy of orbital 4a is more strongly decreased than for porphyrin. Therefore the transition energies are farther apart:

$$\Delta E_1 = 2.57 \times 10^{-19}\,\text{J}, \quad \Delta E_2 = 6.07 \times 10^{-19}\,\text{J} \qquad (8.40)$$

and the coupling effect is much weaker:

$$\frac{f_1}{f_2} = 0.5 \qquad (8.41)$$

Here we considered porphyrin as an example. Porphyrins are important biomolecules (hemoglobin, cytochromes) and it is important to understand their absorption features.

8.8 Light Absorption of Biomolecules

Important biomolecules besides porphyrin absorb in the visible region of the spectrum. This is highly relevant. Chlorophyll, bacteriochlorophyll, and β-carotene are responsible for the conversion of the energy of the sun by plants and bacteria. Retinal is the pigment for vision and for sun energy conversion in an important kind of bacteria (halobacteria). The π electron systems of these compounds are essential for their biological functions. Their role as components of biomachineries will be discussed in Chapter 23.

8.8.1 β-Carotene

β-Carotene is a polyene with 22 π electrons:

It is used as a natural color for food. According to Table 8.4, we obtain for the HOMO–LUMO transition (11 → 12) $\lambda_{max} = 450$ nm in accordance with the experimental value

8.8 LIGHT ABSORPTION OF BIOMOLECULES

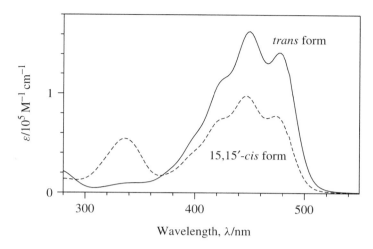

Figure 8.23 Absorption spectrum of β-carotene in n-hexane. Solid line: *trans* form; dashed line: 15,15′-*cis* form. The structure of the absorption band at long wavelengths (maxima at 425, 450, and 475 nm) is due to transitions into different vibrational levels of the excited electronic state (see Chapter 9).

$\lambda_{max} = 453$ nm (Figure 8.23). Thus we can understand that β-carotene has a yellow color. Without taking into account the bond length alternation, an absorption maximum at $\lambda_{max} = 1400$ nm (this is in the infrared region) would be expected.

β-Carotene exists in different *cis–trans* isomers. In contrast to the all-*trans* β-carotene the 15,15′-*cis* β-carotene

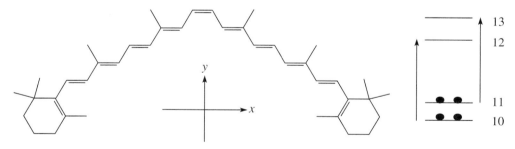

has, additionally to the absorption at 453 nm, an absorption band at 340 nm (Figure 8.23, dashed line: *cis* peak) which is polarized in the y direction, while the main absorption band (transition 11 → 12) is polarized in the x direction.

As outlined in Problem 8.7 the *cis* peak can be considered as being due to a coupling of the transitions 11 → 13 ($\lambda_{max} = 386$ nm, $f = 0.44$) and 10 → 12 ($\lambda_{max} = 390$ nm, $f = 0.47$) leading to one absorption band shifted to higher frequency (*cis* peak, $\lambda_{max} = 341$ nm, $f = 0.90$) and a second band with small oscillator strength shifted to lower frequency ($\lambda_{max} = 462$ nm, $f = 0.001$).

It is of interest to see that this band is expected to be at 462 nm, that is at longer wavelengths than the HOMO → LUMO peak. What is the reason for this unexpected behavior? The gap between HOMO and LUMO is much larger than the gap between HOMO − 1 and HOMO and the gap between LUMO and LUMO + 1. Therefore, the transition energies HOMO − 1 → LUMO and HOMO → LUMO + 1 are only slightly larger than the transition energy HOMO → LUMO; for this reason a small shift by coupling is sufficient to bring the out-of-phase coupled transition to longer wavelengths than the HOMO → LUMO transition. Experimentally, this effect can be observed in polyenes with four to six double bonds. (B. S. Hudson, B. E. Kohler, *Chem. Phys. Lett.* **14**, 299 (1972)).

8.8.2 Retinal

Retinal, the chromophore responsible for the process of vision, has its absorption band at 380 nm in solution. This is at somewhat longer wavelengths than observed for a polyene with six double bonds (344 nm, Table 8.4), as expected for the polarization effect of the C=O bond.

This effect is stronger when forming Schiff's base in the protonated form which absorbs at 440 nm in solution; it is weaker in the unprotonated form, which absorbs at 359 nm.

The protonation leads to a decreasing bond length alternation and thus to a bathochromic shift of the absorption maximum for reasons discussed at the end of Section 8.5. The shift was 60 nm in the case of Figure 8.16 as compared to 80 nm in the present case.

In the biosystem the Schiff's base of the 11,12-*cis* retinal is in a protein environment inducing a further shift to 500 nm in the rod pigments and to 580 nm in the cone pigments (the dark adapted pigment). This indicates an environmental influence discussed in Chapter 23.

The primary process of vision is a photo-induced *cis–trans* isomerization taking place in less than a picosecond. The energy barrier for the *cis–trans* isomerization process is 2.7×10^{-19} J. The energy of a photon at 580 nm is 3.4×10^{-19} J, more than enough to surmount this barrier. The photo-induced *cis–trans* isomerization triggers an excitation of a nerve cell connected with the vision center in the brain.

In the light-induced proton pump of halobacteria, the Schiff base of the photoactive all-*trans* retinal (absorption at 570 nm in the biosystem) is converted by light into the 13-*cis*

8.8 LIGHT ABSORPTION OF BIOMOLECULES

isomer (absorption at 520 nm) inducing the proton pumping (Chapter 23). Thus the shift from solution to biosystem is equally large as in pigments used in the vision process.

8.8.3 Vitamin B₁₂

Vitamin B_{12} is important in DNA biosynthesis. Its chromophore can be considered as an azacyanine with a bent electron cloud:

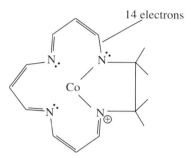

The wavefunctions (indicated by antinodes) for the HOMO–LUMO transition (7 → 8) are:

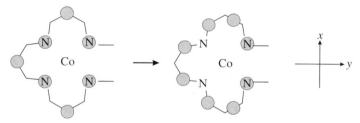

According to Section 8.4 we obtain

$$\Delta E = \frac{h^2}{8m_e L^2} \cdot (N+1) + \Delta E_{\text{nitrogen}} \qquad (8.42)$$

where $\Delta E_{\text{nitrogen}}$ is the change of ΔE due to the lowering of the potential energy at the sites of the nitrogen atoms.

The HOMO is stabilized by $4 \cdot (2/L)ad_0$, because the four nitrogen atoms are at antinodes. The LUMO is stabilized by $2 \cdot (2/L)ad_0$, because two nitrogen atoms are at antinodes. Then

$$\Delta E_{\text{nitrogen}} = 2 \cdot \frac{2}{L} ad_0 \qquad (8.43)$$

where a is the depth of the potential trough associated with a nitrogen atom (Figure 8.11).

With $L = 14d_0$ we obtain

$$\Delta E_{\text{nitrogen}} = 2 \cdot \frac{2}{14 d_0} \cdot ad_0 = \frac{2}{7} a$$

Then with $d_0 = 140\,\text{pm}$, $N = 14$, and $a = 3.5 \times 10^{-19}$ J we obtain

$$\Delta E = \frac{h^2}{8m_e L^2} \cdot (N+1) + \frac{2}{7} a$$
$$= 2.35 \times 10^{-19}\,\text{J} + 0.86 \times 10^{-19}\,\text{J} = 3.21 \times 10^{-19}\,\text{J}$$

This corresponds to $\lambda_{max} = 590$ nm, a value which is in fair agreement with the experiment (560 nm). The transition is polarized in the x direction; the transition moment is relatively small due to the bent form, leading to $f = 0.3$, in agreement with the experiment.

The transitions $7 \to 9$ ($\lambda_{max} = 420$ nm) and $6 \to 8$ ($\lambda_{max} = 340$ nm) are polarized in the y direction. The transition moments M_y are relatively large due to the bent structure, in contrast to the linear cyanine dyes, where the corresponding absorption bands have negligible transition moments. The large M_y values in the present case lead to relatively large oscillator strengths. As a consequence of the coupling effect, the 420 nm band is essentially suppressed, and the 340 nm band becomes correspondingly stronger and is slightly shifted to shorter wavelengths; this is in accordance with the experiment (band at $\lambda_{max} = 360$ nm with $f = 0.5$).

8.8.4 Chlorophyll, Bacteriochlorophyll

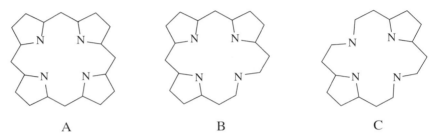

In chlorophyll (B) and bacteriochlorophyll (C) (indicated by the branches of the π-electron system) the situation is similar to porphyrin (A), but the branching of the electron cloud plays a more important role. We consider the branching and neglect bond length differences. We restrict our discussion to case C, because of the convenient symmetry of this structure. The orbitals engaged in light absorption can be easily drawn by comparison with Figure 8.18. For bacteriochlorophyll the orbitals are:

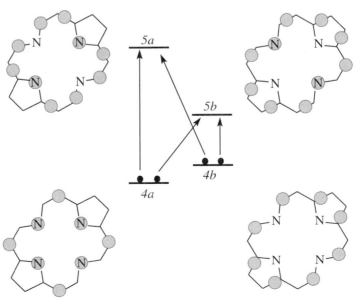

8.9 SPONTANEOUS EMISSION

They are very similar to the orbitals in phthalocyanin and porphyrin and thus we call them 4a, 4b, 5a, and 5b. Transitions 4a→5a, 4b→5b are y polarized, transitions 4a→5b, 4b→5a are x polarized. The λ_{max} values and the f values resulting from the simple model (neglecting coupling effects) are compared with experimental data:

transition	theory λ_{max}/nm	f	experiment λ_{max}/nm	f
4b→5b	790	0.7	750	0.4
4a→5a	370	1.2	357	1.0
4b→5a	400	0.6	384	0.6
4a→5b	690	0.7	525	0.2

The small discrepancies in the f values disappear when taking coupling effects into account.

The transition 4b→5b is essential in bacterial energy harvesting and electron pumping (see Chapters 23.3.4 and 23.4.5). In chlorophyll (B) the situation is between porphyrin (A, λ_{max} = 550 nm) and bacteriochlorophyll (C, λ_{max} = 750 nm) and, accordingly, the relevant transition 4b→5b is at about λ_{max} = 650 nm. This transition and its spectral position are most important for solar energy conversion in plants. The red color of blood and the green color of plants is due to the absorption bands at 550 nm and 650 nm, respectively (see Table 8.2).

8.9 Spontaneous Emission

A dye in solution is illuminated with light of wavelength λ within its absorption band. The excited dye molecules return into the ground state, either in a radiationless process (energy loss by collisions with solvent molecules) or in a radiative process. The radiative process is called *spontaneous emission*; it takes place without external influence. The spectral distribution of the emitted light is measured with the experimental setup displayed in Figure 8.24.

8.9.1 Fluorescence and Phosphorescence

We consider a solution of the cyanine cation

$$\text{[structure]} \tag{8.44}$$

The absorption spectrum (solid line, λ_{max} = 420 nm) and the corresponding emission spectrum (dashed line, λ_{max} = 470 nm, corresponding to blue light) at room temperature are shown in Figure 8.25. This emission, decaying in a short time interval of about 1×10^{-9} s after excitation, is called *fluorescence*. Its maximum is shifted to long wavelengths by 50 nm relative to the maximum of the absorption band. At low temperature (80 K) an additional emission (λ_{max} = 540 nm, corresponding to yellow light) is observed with a decay time of some seconds (dotted line in Figure 8.25). This slowly decaying emission is called *phosphorescence*. Its maximum is shifted to long wavelengths by 120 nm relative to the maximum of the absorption band and by 70 nm relative to the fluorescence band.

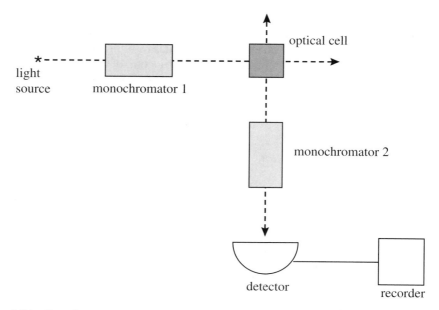

Figure 8.24 Experimental setup for the measurement of fluorescence and phosphorescence. Light from a source, selected for a wavelength $\lambda_{\text{excitation}}$ by monochromator 1 excites dye molecules in an optical cell. The light emitted from the excited dye molecules and selected for a wavelength $\lambda_{\text{emission}}$ by monochromator 2 is measured by a photomultiplier.

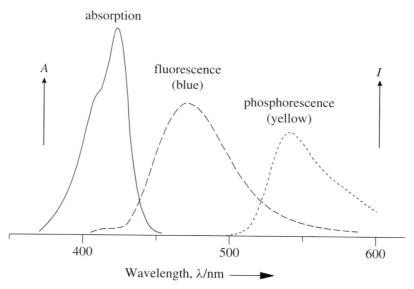

Figure 8.25 Absorption and emission spectra of the cyanine cation (8.44) dissolved in methanol. Absorbance A (solid curve), relative fluorescence intensity I (dashed curve), and phosphorescence intensity I (dotted curve). The fluorescence was measured at room temperature and the phosphorescence was measured at liquid nitrogen temperature. To separate the phosphorescence (decay time ≈ 1 s) from the fluorescence (decay time ≈ 1 ns), rotating shutters in the illuminating and the emission light beams, which can be phase shifted by 180°, are applied.

8.9 SPONTANEOUS EMISSION

How can we understand these fascinating phenomena? What is the reason for the tremendous difference in the decay time of fluorescence and phosphorescence?

We first discuss the decay of fluorescence. In the spirit of the correspondence principle we replace the excited molecule by a classical oscillator – that is, by a dipole antenna emitting light. The rate of de-excitation increases strongly with the frequency ν_0 of the oscillator and with the oscillating charge Q, which is determined by the oscillator strength f. According to Foundation 8.5 the decay time τ_0 is

$$\tau_0 = \frac{3m_e \varepsilon_0 c_0^3}{2e^2 \pi} \cdot \frac{1}{\hat{n} f \cdot \nu_0^2} \tag{8.45}$$

τ_0 depends on ν_0, f, refractive index $\hat{n}$, and on the universal constants m_e, ε_0, and e. In the present case ($\nu_0 = 6 \times 10^{14}$ Hz, calculated from $\lambda_{max} = 470$ nm $\hat{n} = 1.33$ and $f = 0.7$) we obtain $\tau_0 = 12$ ns in agreement with experiment.

The width $\Delta \nu$ of the emission band of a molecule is related to τ_0: $\Delta \nu$ becomes smaller with increasing τ_0:

$$\Delta \nu = \frac{1}{\tau_0} = \frac{1}{12 \text{ ns}} = 0.08 \text{ GHz}$$

$\Delta \nu$ is called the *natural line width*.

However, the observed fluorescence band (Figure 8.25, $\Delta \nu \approx 1 \times 10^{14}$ Hz) is by far broader than the natural band width. The reason for this discrepancy is the same as we have discussed in Section 8.2.1 for the broadening of the absorption band. The position of the fluorescence of a selected molecule depends on its surroundings. The observed fluorescence band is a superposition of the bands of all molecules in the population.

8.9.2 Single Molecule Emission

A fascinating possibility is the detection of the fluorescence of one single molecule absorbing in the narrow spectral range of a laser beam. As an example, terrylene,

in a highly diluted polyethylene glass at a temperature of 2 K. A single terrylene molecule is excited by a tuned laser. Figure 8.26(a) shows its fluorescence intensity versus the frequency. This plot, the fluorescence excitation spectrum, reflects the absorption of the single molecule. The bandwidth of 0.1 GHz is 10^6 times narrower than the width of the overall band constituting the sum of each molecular band (Note that 0.1 GHz is on the order of the natural bandwidth). The single molecule spectrum stays unchanged for about a minute and then, all of a sudden, jumps to spectrum in Figure 8.26(b). The subsequent jumping back and forth between the two spectra in the time range of minutes indicates a flip-flop of this individual molecule between two energy levels by some change in the surrounding matrix.

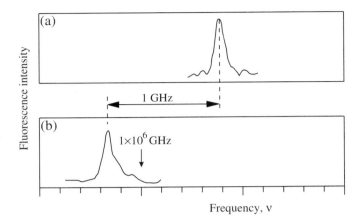

Figure 8.26 Fluorescence excitation spectra of a single terrylene molecule (highly diluted solution in a polyethylene glass at a temperature of 2 K) taken within a time span of several minutes. Occasionally the spectrum changes, all of a sudden, from (a) to (b). This shows that a terrylene molecule which is fixed to a certain surrounding (a) jumps to another lattice site with a different surrounding, thus changing the absorption (b) On the ordinate is the fluorescence intensity, on the abscissa the frequency ν of the exciting laser beam. (See B. Basche, *Ber. Bunsenges. Phys. Chem.*, **100**, 1996, 1269).

8.9.3 Singlet and Triplet States

In Section 4.2 we discussed that He as a two-electron system can be excited into a symmetric and an antisymmetric excited state. Symmetric means that the two-electron wavefunction remains unchanged when the coordinates of the two electrons are exchanged; antisymmetric means that the wavefunction exchanges its sign.

These considerations can be extended to explain the fluorescence and phosphorescence of the dye cation structure (8.44). For simplicity, we approximate the branched dye as a cyanine cation

$$(CH_3)_2 \overset{\oplus}{N}=CH-CH=CH-\overset{..}{N}(CH_3)_2 \tag{8.46}$$

with six π electrons occupying three energy levels in the ground state (Figure 8.27(a)). By irradiation with light, one electron is excited from level 3 to level 4 (Figure 8.27(b)), while

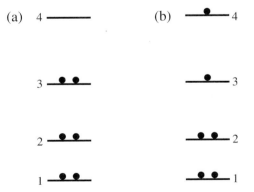

Figure 8.27 One-electron energy levels of the cyanine cation structure (8.46) in the ground state (a) and in the excited state (b).

8.9 SPONTANEOUS EMISSION

the electron distribution in levels 1 and 2 remains unchanged. Thus when considering the excitation of the dye, we restrict our discussion to the two π electrons in the highest occupied (HOMO) orbital, neglecting the interaction with the electrons in levels 1 and 2. Then in analogy to the treatment of He (Section 4.2) the two-electron wavefunction for the ground state (singlet state S^0) is

$$\psi_{S^0} = \phi_3(1) \cdot \phi_3(2) \tag{8.47}$$

where ϕ_3 is the one-electron wavefunction in orbital 3. After excitation the system can be found in the singlet state S^1

$$\psi_{S^1} = \frac{1}{\sqrt{2}} \cdot [\phi_3(1) \cdot \phi_4(2) + \phi_3(2) \cdot \phi_4(1)] \tag{8.48}$$

or in the triplet state T^1

$$\psi_{T^1} = \frac{1}{\sqrt{2}} \cdot [\phi_3(1) \cdot \phi_4(2) - \phi_3(2) \cdot \phi_4(1)] \tag{8.49}$$

In this consideration we have omitted the spin contribution (Section 4.3) since it has no effect on the final result. We approximate the one-electron wavefunctions by using the free-electron model

$$\phi_3(1) = \sqrt{\frac{2}{L}} \cdot \sin \frac{3\pi s_1}{L} \tag{8.50}$$

$$\phi_4(1) = \sqrt{\frac{2}{L}} \cdot \sin \frac{4\pi s_1}{L} \tag{8.51}$$

where s_1 is the coordinate of electron 1 along the bond line in the cyanine cation structure (8.46). The energy of the two electrons is

$$E'_{S^0} = \frac{h^2}{8m_e L^2}(3^2 + 3^2) = \frac{h^2}{8m_e L^2} \cdot 18 = 7.69 \times 10^{-19} \text{ J} \tag{8.52}$$

in the ground state and

$$E'_{S^1} = E'_{T^1} = \frac{h^2}{8m_e L^2}(3^2 + 4^2) = \frac{h^2}{8m_e L^2} \cdot 25 = 13.66 \times 10^{-19} \text{ J} \tag{8.53}$$

in the excited state; the prime in E' indicates that in calculating the energy the Coulomb repulsion of the two electrons has been neglected; we expect the same energy for the singlet and for the triplet state (Figure 8.28(a)). We add the Coulomb repulsion energies

$$\int \psi_{S^0}^2 \cdot g(1,2) \cdot ds_1 ds_2 = 4.0 \times 10^{-19} \text{ J}$$

$$\int \psi_{S^1}^2 \cdot g(1,2) \cdot ds_1 ds_2 = 4.8 \times 10^{-19} \text{ J}$$

$$\int \psi_{T^1}^2 \cdot g(1,2) \cdot ds_1 ds_2 = 3.1 \times 10^{-19} \text{ J}$$

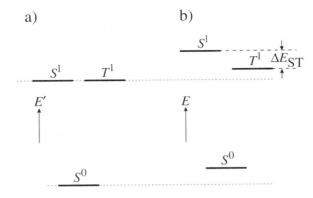

Figure 8.28 Energy scheme of a two-electron system in the ground state S^0 and in the first excited singlet state S^1 and triplet state T^1. (a) Electron repulsion neglected (energies E'). (b) Electron repulsion included (energies E).

where $g(1, 2)$ is the repulsion energy $(e^2/4\pi\varepsilon_0)/r_{12}$ of electrons 1 and 2 depending on their distances r_{12}. Then for the energies E we obtain

$$E_{S^0} = (7.69 + 4.0) \times 10^{-19}\,\text{J} = 11.7 \times 10^{-19}\,\text{J}$$

$$E_{S^1} = (13.66 + 4.8) \times 10^{-19}\,\text{J} = 18.5 \times 10^{-19}\,\text{J}$$

$$E_{T^1} = (13.66 + 3.1) \times 10^{-19}\,\text{J} = 16.8 \times 10^{-19}\,\text{J}$$

These energy levels are displayed in Figure 8.28(b). The energy difference between the singlet and triplet state is

$$\Delta E_{ST} = E_{S^1} - E_{T^1} = 1.7 \times 10^{-19}\,\text{J}.$$

The corresponding calculation for the branched cyanine cation structure (8.44) yields $\Delta E_{ST} = 1.1 \times 10^{-19}$ J. This smaller value is due to the larger extension of the π electron system in the branched cyanine resulting in a smaller Coulomb repulsion.

From the experimental values $\lambda_S = 470$ nm and $\lambda_T = 540$ nm (see Figure 8.25) we obtain $\Delta E_{ST,\text{exp}} = 4.2 \times 10^{-19}\,\text{J} - 3.7 \times 10^{-19}\,\text{J} = 0.5 \times 10^{-19}\,\text{J}$ in fair agreement with the calculation.

The transition $S^0 \to T^1$ requires an inversion of the spin of an electron, thus a transition by absorption of light is not possible, in contrast to the transition $S^0 \to S^1$. For the same reason the transition $T^1 \to S^0$ by emission of light is not allowed. However, we can observe this transition in the experiment (yellow emission) with a lifetime of about 1 second. From this very long lifetime (it is larger than for the $S^1 \to S^0$ transition by 9 powers of magnitude) we conclude that the transition is not completely forbidden but its rate is extremely small.

8.9.4 Shift of Fluorescence and Phosphorescence Relative to Absorption

Why is the fluorescence band shifted to longer wavelengths relative to the absorption band (Figure 8.25)? Light absorption is accompanied by a change of the π electron distribution.

8.9 SPONTANEOUS EMISSION

Therefore, the nuclei, after excitation by light, are no longer in their equilibrium positions and the molecular skeleton immediately starts to oscillate. The vibronic energy is transferred to the surroundings by collisions with the solvent molecules. Furthermore, the changed π electron distribution leads to a change in the interaction forces with the solvent molecules; the solvent molecules reorganize, and this process leads to an additional energy decrease of the dye molecule. The same process happens when the molecule in the excited state changes over into the ground state.

Before excitation the π system is in level 1 (ground state S^0, lowest vibronic state) in Figure 8.29. After excitation into level 4 (excited singlet state S^1, excited vibronic state) the π system changes over into level 3 (excited singlet state S^1, lowest vibronic state); this process occurs on a time scale of about 10^{-12} s, comparable to the vibration frequency of the molecular skeleton (see Chapter 9).

The system remains in level 3 for about 10^{-9} s (lifetime of the excited state). Then a transition occurs to level 2 (electronic ground state, excited vibronic state) and energy is emitted in the form of light (spontaneous emission). Finally, because of the vibrational motion, the system is de-excited to level 1 (electronic ground state, lowest vibronic state), again very rapidly.

Due to these processes, the fluorescence (F in Figure 8.29) occurs at a lower energy than the absorption A (shift to longer wavelengths).

An alternative pathway for the π system in level 3 is to cross over into an excited vibronic level of the triplet state T^1 (dotted arrow in Figure 8.29; this process is called *intersystem crossing*); then the π system goes over rapidly into level 5 (excited triplet state T^1, lowest vibronic level). From here a transition into level 2 (emission of phosphorescence light) occurs. Then it is clear that the phosphorescence P occurs at a lower energy than the fluorescence.

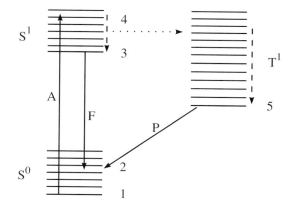

Figure 8.29 Energy states of a dye molecule in solution. Vibrational levels of electronic states S^0, S^1, T^1 (schematic) 1: Lowest vibronic level of electronic ground state S^0. 4: Vibronically excited level of electronic singlet state S^1 reached immediately after light absorption (process A). 3: Lowest vibronic level of electronic state S^1 reached by the molecule in level 4 after transfer of vibrational energy to the solvent and solvent reorganization. 2: Vibronically excited level of electronic ground state S^0 reached immediately after emission of fluorescence (process F) or later by emission of phosphorescence (process P). 5: Lowest vibronic level of electronic triplet state T^1. This level is reached by singlet triplet transition (dotted arrow) and transfer of vibrational energy of the molecular lattice to the solvent and solvent reorganization.

8.9.5 Absorption from Excited States

The energy scheme in Figure 8.29 can be extended by including higher excited singlet and triplet states (Figure 8.30). Whereas singlet states S^1 and S^2 can be directly reached by absorption of light from the ground state S^0, transitions from S^0 to triplet states are not allowed.

However, after excitation of the singlet state S^1, the triplet state T^1 can be reached in a radiationless process as discussed in Section 8.9.4 (*intersystem crossing*). Then a transition from T^1 to T^2 can be induced by a light wave with appropriate wavelength (*triplet–triplet absorption*).

As an example, we consider anthracene. In the free-electron model the orbitals and energy levels in Figure 8.31 are obtained. We induce transition 3b→4a by a flashlight

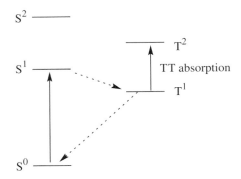

Figure 8.30 Energy scheme for a two-electron system including higher singlet (S^2) and triplet (T^2) electronic states.

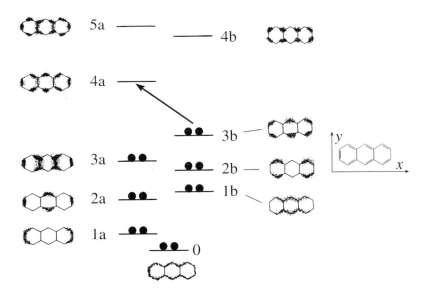

Figure 8.31 π electron energies of anthracene. Flashlight-induced transition from ground state S^0 (3b) to excited singlet state S^1 (4a).

8.9 SPONTANEOUS EMISSION

pulse. Then, by intersystem crossing, the triplet state T^1 of 4a is populated (Figure 8.32). When simultaneously illuminating the sample with light, a transition from T_1 to T_2 can be observed.

Note that the transition 4a → 4b is forbidden: The transition moments are zero: $M_x = M_y = M_z = 0$. This can be seen from the symmetry of the wavefunctions ψ_{4a} and ψ_{4b}.

The lifetime of the T^1 state is on the order of some seconds. Then the population of the T^1 state diminishes in this time interval and the absorbance corresponding to the $T^1 \to T^2$ transition decreases accordingly (Figure 8.33).

Instead of following the $T^1 \to T^2$ absorption, the $S^1 \to S^2$ absorption can be measured, if a fast laser pulse is used instead of flashlight (pulse much shorter than the lifetime of S_1 of about 10^{-9} s); for measuring the $S_1 \to S_2$ absorption a second fast laser pulse must be used (singlet–singlet absorption).

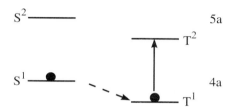

Figure 8.32 Triplet–triplet absorption in anthracene. Intersystem crossing from singlet state S^1 to triplet state T^1. The light induced transition from T^1 to T^2 is observed at $\lambda = 424$ nm. Numbering of levels 4a and 5a see Figure 8.31.

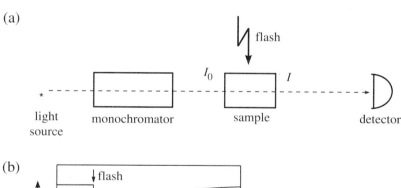
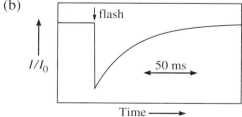

Figure 8.33 Triplet–triplet absorption of anthracene. (a) Experimental setup. (b) Intensity of light ($\lambda = 424$ nm) passing the sample. By starting the flash, T^1 is occupied and a strong $T^1 \to T^2$ absorption at 424 nm is observed. According to the de-excitation of T^1 the $T^1 \to T^2$ absorbance decreases exponentially (increase of the transmittance I/I_0) with time.

8.9.6 Quenching of Fluorescence

The decay of an excited electronic singlet state of a molecule A by spontaneous emission of light

$$A^* \longrightarrow A + h\nu \tag{8.54}$$

can be described by the natural lifetime τ_0 of fluorescence; the rate of decay is proportional to the concentration c_{A^*} of the excited molecules:

$$\frac{dc_{A^*}}{dt} = -k_{fl} \cdot c_{A^*} \tag{8.55}$$

The rate constant k_{fl} is connected with the natural lifetime τ_0 by

$$k_{fl} = \frac{1}{\tau_0} \tag{8.56}$$

In most cases radiation is not the only de-excitation process; there are many other possibilities, such as thermal deactivation or a chemical reaction of the molecule in the excited state (*quenching of fluorescence*). Then

$$\frac{dc_{A^*}}{dt} = -(k_{fl} + k') \cdot c_{A^*} \tag{8.57}$$

where k' denotes the rate constant for the competitive process. The rate constant $k_{fl} + k'$ is connected with the lifetime τ by

$$k_{fl} + k' = \frac{1}{\tau} \tag{8.58}$$

For $k' = 0$ the only de-exciting process is the fluorescence; then the maximum fluorescence intensity I_0 is observed. For $k' \gg k_{fl}$ the competitive process dominates and no fluorescence light appears. In between these extremes a fluorescence intensity I is observed which is smaller than I_0. The ratio $\phi = I/I_0$ is given by

$$\phi = \frac{I}{I_0} = \frac{k_{fl}}{k_{fl} + k'} = \frac{1}{1 + k'/k_{fl}} \tag{8.59}$$

(*Stern–Volmer equation*). ϕ is called the *quantum yield of fluorescence*. Then the lifetime τ of the excited state is

$$\tau = \frac{1}{k_{fl} + k'} = \tau_0 \cdot \phi \tag{8.60}$$

8.10 Stimulated Emission

8.10.1 Inversion of Population

In the preceding section we discussed the process of spontaneous emission. *Spontaneous* means that the transition from the excited state occurs without any external influence (Figure 8.34(a)). On the other hand, we can induce the transition by irradiating the π system with light (Figure 8.34(b)). Then the excitation energy is emitted as a photon corresponding to a light wave with the same phase and the same direction as the incident (stimulating) light wave (coherence of stimulating and emitted light wave). This

8.10 STIMULATED EMISSION

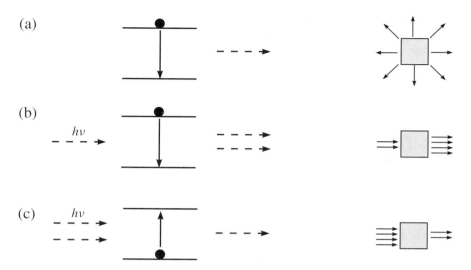

Figure 8.34 Emission and absorption of light: (a) spontaneous emission; (b) stimulated emission; (c) absorption.

process is therefore called *stimulated emission*. Stimulated emission is just the reverse of absorption; in the absorption process the intensity of the light wave decreases when passing molecules in the ground state (Figure 8.34(c)), whereas the intensity increases when passing molecules in the excited state (Figure 8.34(b)).

To observe stimulated emission in a solution of dye molecules, there must be more molecules in the excited state (Figure 8.34(b)) than in the ground state. Otherwise the reverse process would predominate, namely, the absorption of a light quantum leading to a transition from the ground state to the excited state.

The situation required for observing stimulated emission is reached by exposing the dye solution to sufficiently intense absorbing light, usually to a flash light pulse. Then, by absorption of light, a transition $1 \rightarrow 4$ (Figure 8.29) occurs, followed by a rapid jump (within 1 ps) into level 3. Then, for a short time there will be more molecules in level 3 of the excited singlet state than in level 2, the vibronically excited level of the ground state (inversion of ground-state population).

Let us consider a light quantum of frequency $v_{23} = (E_3 - E_2)/h$, where E_3 and E_2 are the energies of vibronic states 3 and 2 in Figure 8.29. We assume that this light quantum approaches a molecule within the very short time interval of 1 ns after the exciting flash. Then the light quantum will usually find the molecule in state 3 and induce the stimulated emission of a second quantum of frequency v_{23}. The probability of finding the molecule in state 2 and thus the probability of the absorption of the light quantum is much smaller. Thus in this case stimulated emission is more probable than absorption.

8.10.2 Dye Laser

Pumping

A dye laser (*laser* is an abbreviation for *light amplification by stimulated emission of radiation*) is obtained by positioning a solution of dye molecules in an optical cell between two parallel mirrors and illuminating the solution with a flash light pulse (Figure 8.35). We consider the dye cation rhodamine 6G (structure a).

a

b

Its π electron system is similar to a cyanine cation (structure b) with 12π electrons. Then, according to Table 8.1, we expect an absorption maximum at 625 nm. Due to the branched structure, rhodamine 6G absorbs at somewhat shorter wavelengths (experimental absorption maximum at 522 nm). In contrast to the cyanine cation, rhodamine 6G has a rigid structure keeping the π electrons well in a plane; as a consequence, the lifetime of the excited state is long enough to allow emission of the energy as fluorescence light (i.e., the quantum yield of fluorescence is nearly one). The emission maximum of fluorescence is at 556 nm (Figure 8.36).

Then, after flash light excitation from level 1 to level 4 (Figure 8.29), an inversion of the population of levels 2 and 3 is reached. Due to the Boltzmann distribution, only few molecules are in the vibrationally excited level 2 of the electronic ground state; after excitation many molecules accumulate in the lowest level of the excited state (level 3).

Stimulated emission

Accidentally, a light quantum of frequency ν_{23} can be spontaneously emitted by some dye molecule; this light quantum induces stimulated emission of light quanta with identical wavelength and phase when approaching other dye molecules in the optical cell. This process is illustrated in Figure 8.37.

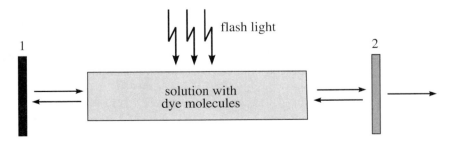

Figure 8.35 Experimental setup of a dye laser. The dye (e.g., rhodamine 6G) dissolved in methanol is in an optical cell which is placed between two parallel mirrors 1 and 2, where mirror 2 is partially transparent (1%). The dye molecules are excited by means of a flash light pulse.

8.10 STIMULATED EMISSION

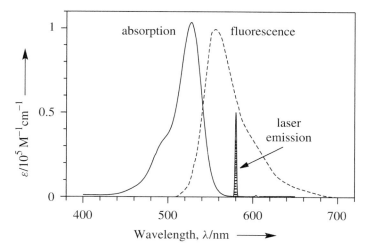

Figure 8.36 Rhodamine 6G as an example for a laser dye. Absorption spectrum (solid line, left-hand scale), fluorescence spectrum (dashed line, intensity in arbitrary units), and laser emission (concentration of dye 1×10^{-5} M, optical path length 5 cm).

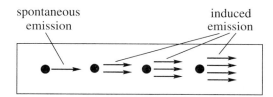

Figure 8.37 Amplification of induced emission in a laser cavity. The process is initialized by a spontaneous emission process of the molecule on the left.

Only those light waves which are accidentally moving just perpendicular to the mirrors are reflected at both mirrors again and again, inducing more and more dye molecules to stimulated emission. Thus the intensity of the narrow beam of light perpendicular to the mirrors is rapidly increasing. If one of the two mirrors (mirror 2 in Figure 8.35) is partially transparent (by about 1%), then a narrow beam of coherent light perpendicular to the mirrors leaves the cavity.

The dye laser is an example of a four-level system (Figure 8.38(a)): external excitation from level 1 to level 4, radiationless transition to level 3, laser action from level 3 to level 2, radiationless transition to level 1. During the lifetime of level 3 the population in level 3 can easily exceed the population in level 2 (inversion).

Laser action is impossible in a two-level system (Figure 8.38(b)): When 50% of the molecules have been pumped into the higher level 4, then the probability of transition $1 \rightarrow 4$ equals the probability of transition $4 \rightarrow 1$, thus it is impossible to reach inversion of the population.

Emission band narrowing

The wavelength of the emitted laser light is determined by the accidental spontaneous emission initiating the process. Level 2 in Figure 8.29 is not exactly fixed. A large number

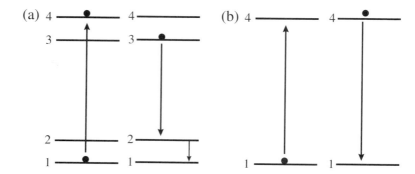

Figure 8.38 Comparison of four-level system (a) and two-level system (b). Dye laser operation is only possible in a four-level system.

of levels are close to each other, corresponding to a large number of wavelengths in spontaneous and stimulated emission. However, transitions approaching the fluorescence maximum are preferentially amplified in the laser, and therefore the laser light emission occurs in a narrow range of wavelengths near to the fluorescence maximum. Thus dye laser light has a much narrower bandwidth than does spontaneous fluorescence: 5 nm compared with 40 nm (see Figure 8.36).

Tuning

In Figure 8.36 it can be seen that the laser emission occurs at longer wavelengths relative to the maximum of the fluorescence band. What is the reason for this unexpected behavior? It is the overlap of fluorescence and absorption bands. Part of the emitted fluorescence light can be reabsorbed by surrounding molecules which are in the electronic ground state. The reabsorption is strongest at the maximum of the absorption band. But even at the fluorescence maximum the reabsorption can be strong enough to decrease the quantum yield of fluorescence significantly. Then the laser emission is more probable at longer wavelengths where the overlap of fluorescence and absorption is much smaller.

The reabsorption increases with increasing concentration of the dye molecules. For this reason the laser emission is shifted to longer wavelengths with increasing dye concentration. The laser emission at 580 nm indicated in Figure 8.36 occurs at a dye concentration of 10^{-5} mol L^{-1}; the emission wavelengths are 585 nm at 10^{-4} mol L^{-1} and 600 nm at 10^{-3} mol L^{-1}. Thus changing the dye concentration can be used to tune the laser wavelength within the range of the fluorescence band.

Moreover, one of the mirrors in Figure 8.35 can be replaced by a grating. A grating reflects only light in a very small wavelength range in the direction of the laser beam. By rotating the grating, the laser light can be tuned to any wavelength in the range of the fluorescence band (tunable dye laser) and the spectral band width can be further decreased. Using dyes with various positions of the fluorescence maximum, laser light can be produced at any wavelength between 300 and 900 nm.

8.10.3 Excimer Laser

In the UV region, laser emission is available by stimulated light emission of excimers. An excimer is a molecule that is stable only in the excited state. For example, an excited atom of a noble gas can form an excimer with a halogen molecule. The excimer emits UV light during its transition to the ground state; the ground state is unstable, and the molecule dissociates into the component atoms within a time of 1 ps. Consequently, it is easy to maintain the inversion of the ground-state population (more molecules are in the excited state than in the ground state). The wavelength range of excimer lasers extends from 193 nm (ArF laser) to 352 nm (XeF laser). The ArF laser is important in the production of high density integrated circuits (microlithography).

8.11 Optical Activity

Molecules which are not congruent with their mirror image are called *chiral*. Chiral molecules play a crucial role as components in biosystems. The highly specific interlocking of molecules is the key to build the biomachinery which is an entity of interlocking parts.

The interaction of a chiral molecule with polarized light is important to obtain structural information.

8.11.1 Rotatory Dispersion

A beam of linear polarized light travels through a solution of D-saccharose. The plane of polarization (i.e., the plane in which the electric vector of the light oscillates) rotates by an angle α (Figure 8.39). α is proportional to the path length l and proportional to the concentration c:

$$\alpha = [\alpha] \cdot l \cdot c \tag{8.61}$$

$[\alpha]$ is called *molar rotation*. The angle α is assigned to be positive if the observer sees the plane of polarization rotated clockwise (as in Figure 8.39). For a wavelength $\lambda = 589$ nm (Na-D line) at 20 °C we find $[\alpha]_D^{20} = +66.5°$ for natural saccharose.

The observed effect (*optical rotation*) is related to the fact that the molecules are chiral. For simplicity we describe the interaction of the molecule with the incident light by replacing it by two coupled oscillating dipoles (frequency ν_0). In contrast to the case considered in Foundation 8.3, the two coupled dipoles are in a right-handed screw-like arrangement (Figure 8.40(a)). The oscillators are coupled such that when one of the oscillators is displaced, a force is exerted on the second oscillator.

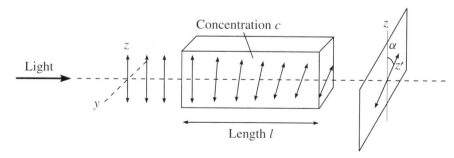

Figure 8.39 Optical activity. A light beam polarized in the z direction passes through a solution of saccharose (concentration c, path length l). The plane of polarization is changed by an angle α into the z' direction.

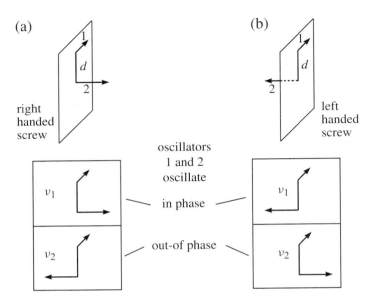

Figure 8.40 Model of chiral molecule: Two coupled oscillators 1 and 2. When displacing oscillator 1 (arrow 1) a force is exerted on oscillator 2 (arrow 2) in the direction of arrow 2. (a) Right-handed screw, (b) left-handed screw. Note: We must distinguish in both cases (a) and (b) between in-phase oscillation of the two coupled oscillators at a lower frequency (v_1) than the frequency of the uncoupled oscillators. Correspondingly, the out-of-phase oscillation occurs at higher frequency v_2.

When light travels through an assembly of such model oscillators, each oscillator acts as an antenna emitting a light wave, as shown in Foundation 8.5 for a single oscillator. The wave of the incident light and the waves scattered by each model oscillator interfere, resulting in a rotation of the plane of polarization. In a solution of artificial L-saccharose the sense of rotation is reversed compared with natural D-saccharose. Each L-saccharose molecule is replaced by model b (Figure 8.40(b)), which is the mirror image of model a. (Figure 8.40(a)).

In the considered case, the incident light has a frequency v much lower than the absorption frequency v_1 (oscillation in phase with field of incident light). When the frequency v is increased, α changes sign when passing the center of the absorption band v_1 (change from in-phase to out-of-phase oscillation; see Figure 8.41(b)).

8.11.2 Ellipticity

In the range of the absorption band, the transmitted light has a component polarized perpendicular to the plane of polarization of the light when leaving the sample ($y'z'$ plane, Figure 8.41(a)), and it is elliptically polarized:

$$F_{z'} = a \sin 2\pi v t \quad \text{(component in the } z' \text{ direction)} \tag{8.62}$$

$$F_{y'} = b \cos 2\pi v t \quad \text{(component in the } y' \text{ direction)}$$

The angle

$$\Theta = \arctan \frac{b}{a} \tag{8.63}$$

8.11 OPTICAL ACTIVITY

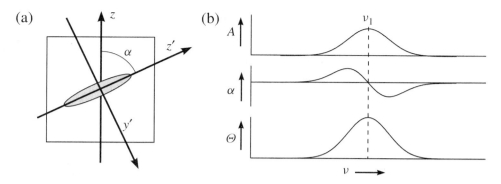

Figure 8.41 Optical activity. (a) Plane of polarization rotated by an angle α viewed by an observer facing the oncoming beam. α is assigned to be positive in the present case. Long axis of ellipse $= a$; short axis of ellipse $= b$. (b) Relation between absorption spectrum (absorbance A), rotation angle α, and ellipticity $\Theta = \arctan(b/a)$. We consider the absorption band corresponding to the in-phase oscillation of the two coupled oscillators in case (a) in Figure 8.40 (frequency ν_1). Θ is assigned to be positive in the present case (α positive for $\nu < \nu_1$).

is called *ellipticity* (Figure 8.41b). The sign convention is that Θ is positive in the case of Figure 8.41 – that is, when α is positive for $\nu < \nu_1$. Θ is proportional to l and c:

$$\Theta = [\Theta] l c \tag{8.64}$$

where $[\Theta]$ is the molar ellipticity. $[\Theta]$ depends on the molecular structure. To investigate this dependence it is useful to introduce the term "*circular dichroism*".

8.11.3 Circular Dichroism

Linear polarized light can be regarded as a superposition of two equal right (R) and left (L) circular polarized components. For right (left) circular polarized light the observer sees the electric vector rotating clockwise (counterclockwise) (Figure 8.42). The molar decadic absorption coefficients ε_R and ε_L are different for an optically active compound. The difference

$$\Delta\varepsilon = \varepsilon_L - \varepsilon_R \tag{8.65}$$

is called *circular dichroism* (commonly abbreviated as "CD"). A detailed analysis shows that the molar ellipticity $[\Theta]$ is proportional to $\Delta\varepsilon$:

$$[\Theta] = \frac{2.303}{4} \Delta\varepsilon \tag{8.66}$$

We consider the ratio

$$g = \left(\frac{\Delta\varepsilon}{\varepsilon}\right)_{\text{absorption maximum}} \tag{8.67}$$

where

$$\varepsilon = \frac{\varepsilon_L + \varepsilon_R}{2} \tag{8.68}$$

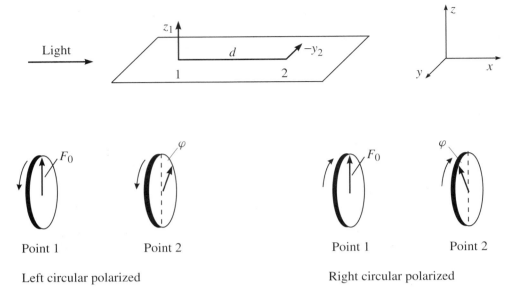

Figure 8.42 Left and right circular polarized light waves. The angle φ denotes the phase shift between points 1 and 2.

g is called the *anisotropy factor* of the given absorption band. In the simple case of Figure 8.40 (absorption bands with maxima at $\lambda_1 = c_0/\nu_1$ and $\lambda_2 = c_0/\nu_2$) we obtain (see Foundation 8.6)

case (a) right handed srew	case (b) left handed screw
$g_1 = -\frac{2\pi d}{\lambda_1}$	$g_1 = +\frac{2\pi d}{\lambda_1}$
$g_2 = +\frac{2\pi d}{\lambda_2}$	$g_2 = -\frac{2\pi d}{\lambda_2}$

8.11.4 Circular Dichroism of Spirobisanthracene

The chiral spirobisanthracene

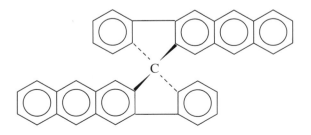

8.11 OPTICAL ACTIVITY

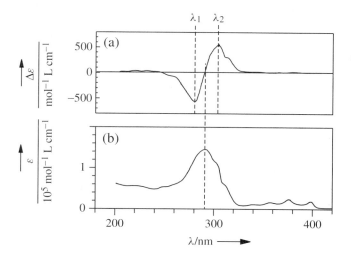

Figure 8.43 CD spectrum (a) and absorption spectrum (b) of spirobisanthracene. (N. Harada, H. Ono, T. Nishiwaki, and H. Uda, *J. Chem. Soc., Chem. Commun.*, 1991, 1753). The in-phase oscillation of the two coupled oscillators at $\nu_1 = c_0/\lambda_1$ and the out-of-phase oscillation at $\nu_2 = c_0/\lambda_2$ are only weakly separated in this case. The two bands cannot be resolved in the absorption spectrum, but clearly appear in the CD spectrum.

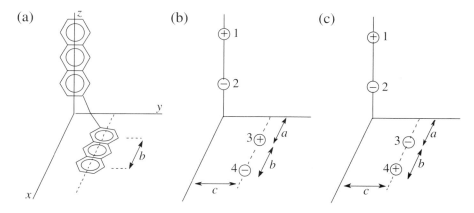

Figure 8.44 Arrangement of the anthracene units in spirobisanthracene. (a) Screw-like arrangements of the two anthracene units within the molecule. (b) and (c) coupling in the in-phase and in the out-of-phase oscillation.

has a strong absorption band at 290 nm (Figure 8.43(b)), as an isolated anthracene molecule. The transition moment is oriented parallel to the long axis of anthracene (x direction)

$$M_x = e \int \psi_{3a} x \psi_{4a} \cdot d\tau = 3 \times 10^{-29} \, \text{C m} \qquad (8.69)$$

The two anthracene units are arranged as shown in Figure 8.44(a).

Note that in Figure 8.31 we have considered the first transition $3b \to 4a$ (calculated $\lambda = 490$ nm, observed 370 nm). Now we are interested in the second transition $3a \to 4a$ (calculated $\lambda = 310$ nm, observed 290 nm).

In this case we expect a strong circular dichroism because the oscillator replacing the molecule oscillates in the x direction (Figure 8.31). In the arrangement of the two anthracene molecules in Figure 8.44(a) the oscillators are in the approximate geometry of the oscillators in Figure 8.40.

In contrast, the oscillators related to the first transition 3b → 4a oscillate in the y direction (Figure 8.31). Therefore, in the arrangement of the two anthracene molecules in Figure 8.44(a) the oscillators are parallel, and thus no circular dichroism can be expected.

Splitting of absorption band

The electrons responsible for the 3a → 4a transition in both units are coupled to each other. This results in a shift to longer wavelengths for the in-phase oscillation (Figure 8.44(b)) and a shift to shorter wavelengths for the out-of-phase oscillation (Figure 8.44(c)).

The splitting of the absorption band can be calculated by estimating the Coulomb interaction of the two oscillators (see Problem 8.8) as $4 \times J_{12} = 0.6 \times 10^{-19}$ J $= 0.4$ eV. This splitting is reflected in the circular dichroism band in Figure 8.43. The negative branch is at 278 nm ($\Delta E = 7.14 \times 10^{-19}$ J), the positive branch is at 300 nm ($\Delta E = 6.62 \times 10^{-19}$ J), and the difference is 0.52×10^{-19} J $= 0.33$ eV – in good agreement with the model.

Anisotropy factor g

For simplicity we treat the two anthracene units arranged such that the dipoles are located as indicated in Figure 8.45. This arrangement is similar to the one in Figure 8.40(a), and we obtain for the oscillation depicted in (Figures 8.44(b) and 8.45(a)), according to equation (8.69),

$$g = +\frac{2\pi d}{\lambda} \quad \text{(case a, in-phase)}$$

With $d = c = 318$ pm and $\lambda = 278$ nm, this results in $g = +7 \times 10^{-3}$, corresponding to the anisotropy factor in the maximum of the experimental CD curve:

$$g = \left(\frac{\Delta\varepsilon}{\varepsilon}\right)_{\text{maximum}} = \frac{500}{10 \times 10^4} = 5 \times 10^{-3}$$

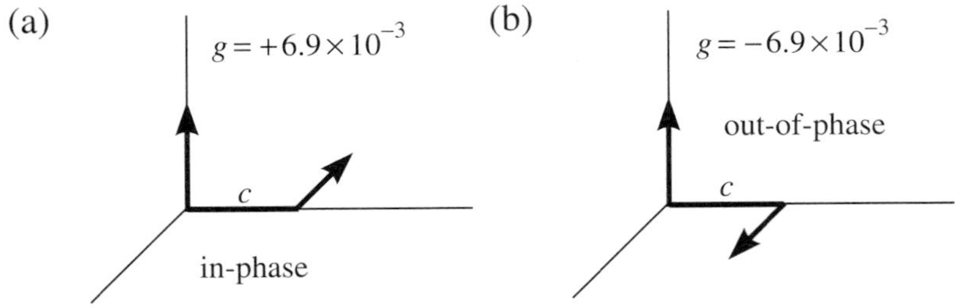

(a) $g = +6.9 \times 10^{-3}$ in-phase

(b) $g = -6.9 \times 10^{-3}$ out-of-phase

Figure 8.45 Simplified arrangement of the transition moments of the anthracene units in spirobisanthracene.

Correspondingly, for the oscillation depicted in (Figures 8.44(c) and 8.45(b)) we calculate $g = -7 \times 10^{-3}$, whereas the experimental value (minimum of the CD curve) is $g = -6.5 \times 10^{-3}$. The agreement in sign as well as in the values is remarkable.

8.11.5 Circular Dichroism of Chiral Cyanine Dye

The chiral dye cation

absorbs at $\lambda_{max} = 500$ nm; the absorption coefficient for this wavelength is $\varepsilon = 8 \times 10^4$ mol^{-1} L cm^{-1}; the circular dichroism is $\Delta\varepsilon = \varepsilon_L - \varepsilon_R = 15$ mol^{-1} L cm^{-1} (V.Buß, D. Ulbrich, Ch. Reichardt, Chiral Tri- and Pentamethinium Cyanine Dyes with 1,2,3,4-Tetrahydro-6-methylquinolyl End Groups: UV/Vis and CD Spectroscopy and Structure Correlations, *Liebigs Ann.* 1996, 1823). In this case, in contrast to the example in the previous Section the circular dichroism is not related to a splitting of the absorption band but it is caused by the right handed screw like shape of the π-electron cloud between the N atoms. Thus the sign of the dichroism does not change within the absorption band. The sign and the value of the quantity $\Delta\varepsilon/\varepsilon = 1.9 \times 10^{-4}$ can easily be explained (see Problem 8.9).

Problems

Problem 8.1 – Lone Electron Pair at the Nitrogen

In the following cases discuss whether the lone electron pairs of the nitrogen atoms contribute to the number of π electrons in the conjugated system.

Solution. In contrast to a carbon atom (four electrons in the 2s, 2p state), a nitrogen atom has five electrons (two of them in the 2s state, three in the 2p state). There are two possibilities to distribute these electrons in a trigonal planar hybrid state. 1. Three electrons in the hybrid function forming three σ bonds, as in the carbon atom; the remaining two electrons occupy a $2p_z$ orbital. 2. Four electrons in the hybrid function forming two σ bonds and a lone electron pair (which extends in the plane of the σ electrons); the remaining electron occupies a $2p_z$ orbital.

From this it follows that the lone electron pair in the nitrogen in the center of the azacyanine does not contribute to the number of π electrons; the same is true for the nitrogen in the six-membered ring. The nitrogen in the five-membered ring forms three σ bonds, and thus it contributes two electrons to the number of π electrons.

Problem 8.2 – Light Absorption of Different Classes of Dyes

Discuss the color of the dyes A to P in Figure 8.46.

Solution. A, in electronic structure, compares with case $N = 6$ in Table 8.2 (332 nm) and C with $N = 10$ (587 nm), but the polarization by the benzene rings causes a shift to longer wavelengths in both cases. B, as compared with A, and D, as compared with C, have an absorption maximum shifted to short wavelengths if we count the π electrons along the chain and neglect branches. λ_{max} of F, as compared to E, is shifted to long waves according to Section 8.4. The same is true for J as compared to I, for L as compared to K, and for N as compared to M and O. In all these cases the number of π electrons along the chain is $N = 12$. P compares with the case $N = 14$ in Table 8.2 (735 nm).

Problem 8.3 – Energy Shift in Azacyanines

Calculate the energy shift when proceeding from a cyanine cation to the corresponding azacyanine cation.

Solution. For calculating the wavelength shift we proceed as in the calculation of the mean potential energy of an electron in an H atom (Section 3.2.1); we calculate the mean potential energy of a π electron in the electric field of the charge of the core of the aza nitrogen atom

$$\overline{V} = \int_0^L V_N \psi^2 \cdot ds$$

V_N is the potential energy of a π electron in the field of the N atom (more precisely: the difference between the potential energies of N and C). We approximate V_N by assuming that the potential energy of an electron around a N atom is lower by a constant amount a compared to the constant potential energy of a chain of carbon atoms. According to Figure 8.11 with $s_1 = L/2 - d_0/2$ and $s_2 = L/2 + d_0/2$ we have $V_N = -a$ for $s_1 < s < s_2$ and $V_N = 0$ for s elsewhere. Then

$$\overline{V} = \int_{s_1}^{s_2} -a\psi^2 \cdot ds = -a \int_{s_1}^{s_2} \psi^2 \cdot ds$$

In orbitals with a node at the center of the molecule, ψ^2 is practically zero for the whole region around the N atom, so $\overline{V} = 0$. For orbitals with an antinode at the center of the molecule, ψ^2 in the region of the N atom is only slightly different from its maximal value at the center, thus

$$\overline{V} = -a \int_{x_1}^{x_2} \psi^2 \cdot dx = -a \cdot (\psi^2)_{\text{at center}} \cdot d_0 = -a \cdot \frac{2}{L} \cdot d_0$$

This amount decreases with increasing chain length L. Because of $L = N \cdot d_0$

$$\overline{V} = -a \cdot \frac{2}{L} \cdot d_0 = -a \cdot \frac{2}{N}$$

and the corrected energy is

$$\Delta E' = \Delta E - a \cdot \frac{2}{N} \quad \text{for} \quad \frac{N}{2} = 2, 4, 6, \ldots$$

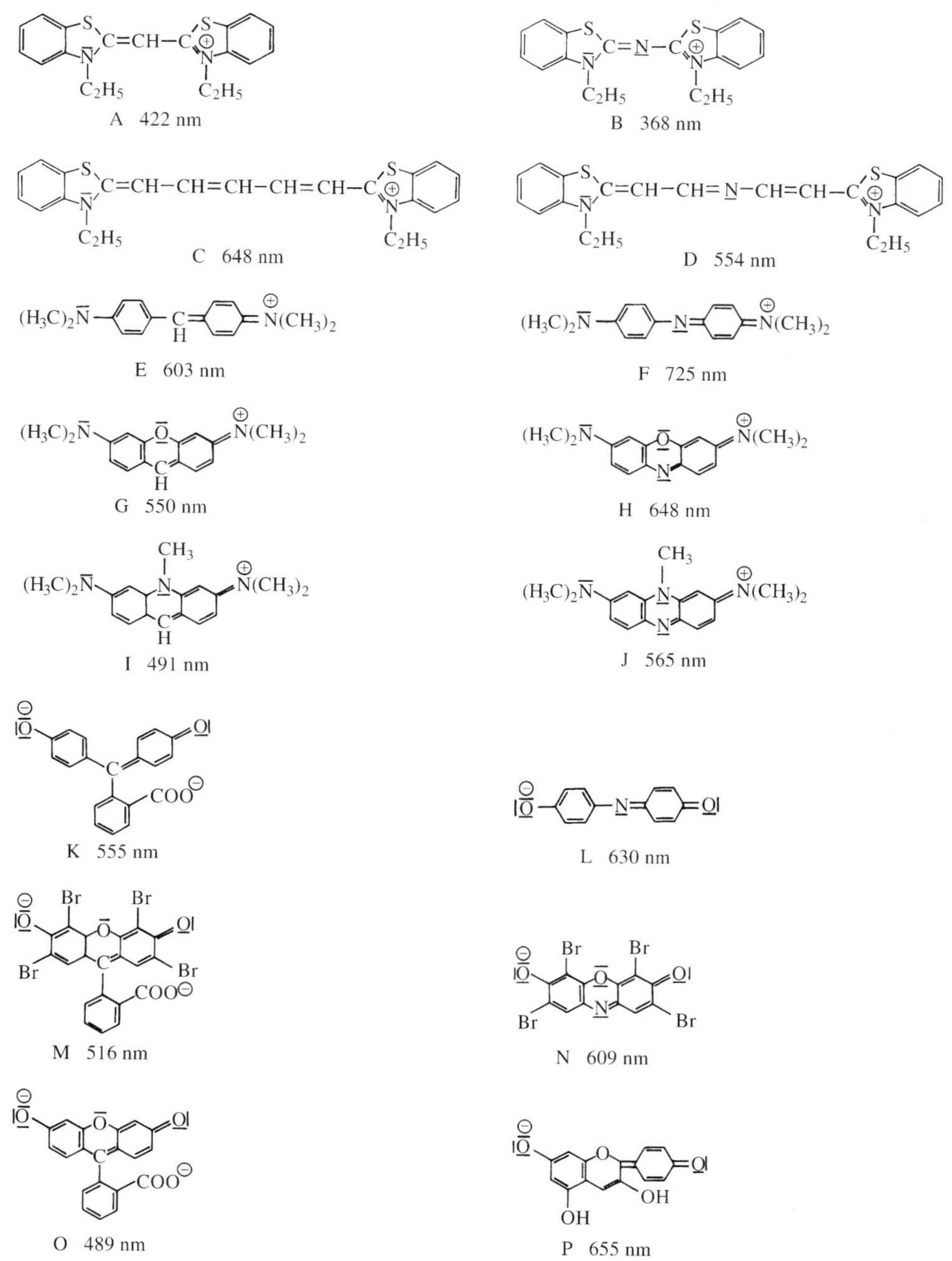

PROBLEM 8.3

Figure 8.46 Dyes from different classes of π electron system. A Thiacyanine, B Azathiacyanine, C Thiadicarbocyanine, D Azathiadicarbocyanine, E Michler's Hydrol Blue, F Bindschedler's Green, G Pyronine, H Capri Blue, I Acridine Orange, J Diazine Violet, K Phenolphthaleine, L Indophenol, M Eosin, N Iris Blue, O Fluoresceine, P Pelargonidine. The numbers refer to the absorption maximum.

$$\Delta E' = \Delta E + a \cdot \frac{2}{N} \quad \text{for} \quad \frac{N}{2} = 3, 5, 7, \ldots$$

The constant a corresponds to the energy difference of an electron in a p orbital of a carbon atom and of a nitrogen atom. From the ionization energies of C (17.2×10^{-19} J) and N (20.7×10^{-19} J), it follows that $a = 3.5 \times 10^{-19}$ J. The calculated shifts (Table 8.2) are in satisfactory agreement with the experimental data.

Problem 8.4 – Shift of Energy Levels by Bond Alternation

Calculate the energy shift when proceeding from a hypothetical butadiene with equal bond lengths to butadiene with alternating single and double bonds.

Solution. In analogy to the considerations in Problem 8.3 (see above) the energy in orbital n should be

$$E' = E - 2 \cdot b \cdot \psi_{n,C=C}^2 \cdot d_0 + b \cdot \psi_{n,C-C}^2 \cdot d_0$$

(two double bonds, one single bond in butadiene; constant b see Figure 8.14). $\psi_{n,C=C}$ is the value of the wavefunction in orbital n at the place of the double bond, and $\psi_{n,C-C}$ is the corresponding value at the place of the single bond. From Figure 8.15 we find in a good approximation for the orbital with $n = 2$

$$\psi_{2,C=C}^2 = \frac{2}{L}, \quad \psi_{2,C-C}^2 = 0$$

and for the orbital with $n = 3$

$$\psi_{3,C=C}^2 = 0, \quad \psi_{3,C-C}^2 = \frac{2}{L}$$

Then with $L = 5d_0$ we obtain

$$\Delta E' = \Delta E + 2b\frac{2}{L}d_0 + b\frac{2}{L}d_0 = \Delta E + \frac{6}{5}b$$

Proceeding in a similar manner for longer polyenes we find that the difference between $\Delta E'$ and ΔE is approximately the same: on the one hand the number of correction terms increases, but on the other hand the length L increases to the same extent.

Problem 8.5 – Light Absorption of Phthalocyanine and Porphyrin

Calculate the transition energies for the transitions 4b → 5b and 4a → 5a.

Solution. Phthalocyanine – We discuss the wavefunctions displayed in Figure 8.19. In state 4a all eight N atoms are placed at the antinodes of the wavefunction, then according to the considerations in Problem 8.3 (see above),

$$-a \int_{s_1}^{s_2} \psi^2 \cdot ds = -a \cdot (\psi^2)_{\text{at center}} \cdot d_0 = -a \cdot \frac{2}{L_c} \cdot d_0$$

and

$$E_{4a} = \frac{h^2}{2m_e L_c^2} \cdot 16 - \frac{16}{L_c} \cdot d_0 \cdot a$$

In state 4b all N atoms are at nodes of the wavefunction; thus

$$E_{4b} = \frac{h^2}{2m_e L_c^2} \cdot 16$$

In state 5a two N atoms are at an antinode, two are at a node and four are adjacent to an antinode; for the latter atoms, ψ^2 is

$$\psi^2 = \frac{2}{L_c} \cdot \cos^2 \frac{10\pi \cdot L_c/8}{L_c} = \frac{1}{L_c}$$

Thus

$$E_{5a} = \frac{h^2}{2m_e L_c^2} \cdot 25 - 2a \cdot \frac{2}{L_c} \cdot d_0 - 4a \cdot \frac{1}{L_c} \cdot d_0 = \frac{h^2}{2m_e L_c^2} \cdot 25 - \frac{8}{L_c} \cdot d_0 \cdot a$$

The same expression is obtained for state 5b.

Thus with $L_c = 16 \cdot d_0$, $d_0 = 140\,\text{pm}$, and $a = 3.5 \times 10^{-19}\,\text{J}$ (this is the same value as we used in the case of the azacyanines) we obtain

$$\Delta E_{4b \to 5b} = \Delta E_{4b \to 5a} = \frac{9h^2}{512 m_e d_0^2} - \frac{1}{2}a = (4.32 - 1.75) \times 10^{-19}\,\text{J} = 2.57 \times 10^{-19}\,\text{J}$$

$$\Delta E_{4a \to 5a} = \Delta E_{4a \to 5b} = \frac{9h^2}{512 m_e d_0^2} + \frac{1}{2}a = (4.32 + 1.75) \times 10^{-19}\,\text{J} = 6.07 \times 10^{-19}\,\text{J}$$

This corresponds to $\lambda_{\max} = 772\,\text{nm}$ and $\lambda_{\max} = 327\,\text{nm}$.

Solution. Porphyrin – In principle, in porphyrin we have the same π electron system as in phthalocyanine. However, the shift of the energy levels due to the nitrogen atoms is smaller, because the four outermost nitrogen atoms are missing. The energy shift is just one-half of the shift in phthalocyanine. Thus we obtain

$$\Delta E_{4b \to 5b} = \Delta E_{4b \to 5a} = \frac{9h^2}{512 m_e d_0^2} - \frac{1}{4}a = \left(4.32 - \frac{1.75}{2}\right) \times 10^{-19}\,\text{J} = 3.45 \times 10^{-19}\,\text{J}$$

$$\Delta E_{4a \to 5a} = \Delta E_{4a \to 5b} = \frac{9h^2}{512 m_e d_0^2} + \frac{1}{4}a = \left(4.32 + \frac{1.75}{2}\right) \times 10^{-19}\,\text{J} = 5.20 \times 10^{-19}\,\text{J}$$

This corresponds to $\lambda_{\max} = 576\,\text{nm}$ and $\lambda_{\max} = 382\,\text{nm}$.

Problem 8.6 – Oscillatory Strength in Phthalocyanine and Porphyrin

Calculate the oscillatory strengths for the transitions 4b $\to$ 5b and 4a $\to$ 5a.

Solution. We calculate the corresponding transition moments by approximating the 16-membered ring with circumference $L_c = 16 d_0$ as a circle with radius

$$r = \frac{L_c}{2\pi} = \frac{16 d_0}{2\pi}$$

Then

$$\psi_{4a} = \frac{1}{\sqrt{\pi}} \cos 4\varphi, \quad \psi_{5a} = \frac{1}{\sqrt{\pi}} \cos 5\varphi, \quad x = r \cos \varphi$$

and

$$M_{x,4a\to 5a} = e\int_0^{2\pi} \psi_{4a} x \psi_{5a} \cdot d\varphi = e\frac{16d_0}{2\pi^2}\int_0^{2\pi} \cos 4\varphi \cdot \cos\varphi \cdot \cos 5\varphi \cdot d\varphi$$

$$= e\frac{16d_0}{2\pi^2}\cdot\frac{\pi}{2} = 2.85\times 10^{-29}\,\text{C m}$$

The same value is obtained for $M_{y,4b\to 5b}$.

According to equation (8.25) the oscillatory strength for the transition of one electron is

$$f = \frac{8\pi^2 m_e}{3h^2 e^2 \hat{n}}\cdot M^2 \cdot \Delta E$$

In our case f is two times larger, because two electrons are excited, and we obtain with $\Delta E_{4b\to 5b} = 2.57\times 10^{-19}$ J, $\Delta E_{4a\to 5a} = 6.07\times 10^{-19}$ J, and $\hat{n} = 1.5$

$$f_{4b\to 5b} = 1.19, \quad f_{4a\to 5a} = 2.81$$

In the case of porphyrin, the transition energies are $\Delta E_{4b\to 5b} = 3.45\times 10^{-19}$ J, and $\Delta E_{4a\to 5a} = 5.20\times 10^{-19}$ J, and the result for the oscillatory strengths is

$$f_{4b\to 5b} = 1.60, \quad f_{4a\to 5a} = 2.40$$

Problem 8.7 – Cis-Peak in β-Carotene

We consider the light absorption of β-Carotene (see Section 8.8.1).

In the step potential model, using the value $2b = 6.00\times 10^{-19}$ J (slightly different from the value $2b = 6.45\times 10^{-19}$ J derived in Foundation 7.3) we obtain for the $11 \to 12$ transition an absorption peak at $\lambda_{max} = 453$ nm which is identical with the measured value. Then the model predicts $\lambda_{max,1} = 386$ nm, $f_1 = 0.44$ for the $11 \to 13$ transition, and $\lambda_{max,2} = 390$ nm, $f_2 = 0.47$ for the $10 \to 12$ transition; the coupling integral is $J_{12} = 0.38\times 10^{-19}$ J (see Foundations 8.3 and 8.5). Calculate the location of the *cis*-peak in β-carotene using the method of coupled oscillators.

Solution. We first simplify by using a mean value for $\lambda_{max,1}$ and $\lambda_{max,2}$: $\lambda_{max,1} = \lambda_{max,2} = 388$ nm ($\Delta E_1 = \Delta E_2 = \Delta E = 5.11\times 10^{-19}$ J). Then, according to Foundation 8.7, we obtain

$$\Delta E_{coupled} = \sqrt{\Delta E^2 \pm 4\cdot J_{12}\cdot \Delta E}$$

and therefore $\Delta E_{coupled,1} = 5.82\times 10^{-19}$ J ($\lambda_{coupled,1} = 341$ nm), $\Delta E_{coupled,2} = 4.29\times 10^{-19}$ J ($\lambda_{coupled,2} = 463$ nm). For the oscillator strength we obtain $f_{coupled,1} = f_1 + f_2 = 0.91$, and $f_{coupled,2} = 0$.

When considering the slight difference between $\lambda_{max,1}$ and $\lambda_{max,2}$ and following the procedure in Foundation 8.7 we obtain $\lambda_{coupled,1} = 341$ nm, $\lambda_{coupled,2} = 462$ nm, and $f_{coupled,1} = 0.904$, $f_{coupled,2} = 0.001$.

Problem 8.8 – Splitting of Absorption Band

Calculate the splitting of the absorption band in spirobisanthracene.

Solution. According to Foundation 8.4 the shift ΔE is $\Delta E = -2J_{12}$ (in-phase) and $\Delta E = +2J_{12}$ (out-of phase) where J_{12} is the coupling integral. (For the coupling of two electrons (each HOMO occupied with one electron) the shift is J_{12}, and for four electrons (each HOMO occupied with two electrons) it is $2J_{12}$.)

We approximate the calculation of J_{12} by considering the coupling energy between the two oscillating dipoles (dipoles 1 → 2 and 3 → 4 in Figures 8.44(b) and (c) with dipole length 480 pm (this is the distance of the centers of the outer rings in anthracene) and charge $Q = e/2$. This corresponds to the transition dipole moment of $M_x = 3 \times 10^{-29}$ C m. Then

$$J_{12} = \frac{Q^2}{4\pi\varepsilon_0\varepsilon_r}\left[-\frac{1}{r_{23}} - \frac{1}{r_{14}} + \frac{1}{r_{24}} + \frac{1}{r_{13}}\right]$$

with $\varepsilon_r = 2.5$ (r_{ij} = distance between charges i and j). From the geometry in Figure 8.44 we obtain

$$r_{23} = \sqrt{2a^2 + c^2}, \quad r_{14} = \sqrt{2(a+b)^2 + c^2}, \quad r_{24} = r_{13} = \sqrt{(a+b)^2 + a^2 + c^2}$$

Then with $a = 246$ pm, $b = 484$ pm, and $c = 318$ pm we obtain $r_{23} = 471$ pm, $r_{14} = 1080$ pm, $r_{24} = r_{13} = 833$ pm, and $J_{12} = 2.32 \times 10^{-29} \cdot (-6.48 \times 10^8)$ J $= 0.15 \times 10^{-19}$ J $= 0.1$ eV.

Problem 8.9 – Circular Dichroism of Cyanine Dye

Explain the sign and the value of the quantity $\Delta\varepsilon/\varepsilon$ for the chiral cyanine cation discussed in Section 8.11.5.

Solution. The absorption coefficient ε is proportional to

$$\tfrac{1}{3}(M_x^2 + M_y^2 + M_z^2)$$

where M_x, M_y, and M_z are the components of the transition moment in the x-, y-, and z-direction. We set the transition moments approximately proportional to the extension of the electron cloud (Figure 8.47):

$$M_x \sim d, \quad M_y \sim a, \quad M_z \sim a$$

where $d = 0.73$ nm, and $a = 0.1$ nm, thus

$$\varepsilon \sim \tfrac{1}{3}(d^2 + 2a^2).$$

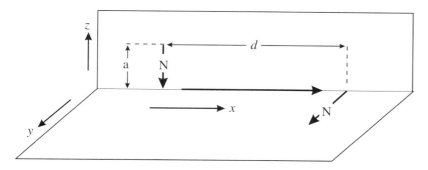

Figure 8.47 Chiral cyanine cation replaced by charge oscillating from nitrogen atom on the left to nitrogen atom on right and back along the line indicated by the three arrows.

The circular dichroism $\varepsilon_L - \varepsilon_R$ is due to the interaction of light with M_y and M_z (equations 8.67, 8.69):

$$\frac{\varepsilon_L - \varepsilon_R}{\varepsilon_{yz}} = \frac{2\pi d}{\lambda}$$

Where

$$\varepsilon_{yz} \sim \tfrac{1}{3}(M_y^2 + M_z^2) \sim \tfrac{1}{3}(a^2 + a^2)$$

Thus we obtain

$$\frac{\varepsilon_{yz}}{\varepsilon} = \frac{2a^2}{d^2 + 2a^2}$$

and therefore

$$\frac{\varepsilon_L - \varepsilon_R}{\varepsilon} = \frac{2\pi d}{\lambda} \frac{2a^2}{d^2 + 2a^2}$$

For $\lambda = 500$ nm, $a = 0.1$ nm, $d = 0.73$ nm we obtain

$$\frac{\varepsilon_L - \varepsilon_R}{\varepsilon} = \frac{2\pi \times 0.73}{500} \frac{0.02}{0.73^2 + 0.02} = 3 \times 10^{-4}$$

This result agrees in sign and approximate value with the result of the measurement

$$\frac{\varepsilon_L - \varepsilon_R}{\varepsilon} = \frac{15}{8 \times 10^4} = 2 \times 10^{-4}$$

Foundation 8.1 – Integrated Absorption: Classical Oscillator

Absorbed power and absorption coefficient

From the Law of Lambert and Beer (Box 8.1) we calculate the molar decadic absorption coefficient ε

$$\varepsilon = \frac{\alpha}{\log e} = -\frac{1}{\log e} \cdot \frac{dI}{I \cdot c \cdot dx} = -\frac{dI}{c \cdot I \cdot dx \cdot 2.303}$$

where c is the concentration of the dissolved molecules, and I is the intensity of the incident light. dI is the intensity change by the absorbing molecules in a volume element with thickness dx ($dV = A \cdot dx$, A = cross-section of the cell). The number of absorbing molecules in this volume element is (N_A is Avogadro's constant)

$$dN = \frac{N}{V} \cdot dV = \frac{nN_A}{V} \cdot dV = N_A \cdot c \cdot A \cdot dx$$

Intensity is the mean power of the light wave divided by the area A of the light beam:

$$I = \frac{\text{mean power}}{A}$$

Each molecule absorbs the energy dE in time dt from the incident electromagnetic wave, and $\overline{dE/dt} \cdot dN$ is the mean power absorbed by dN molecules. Then the intensity change dI of the light wave is

$$dI = -\frac{\overline{dE/dt} \cdot dN}{A} = -\overline{\left(\frac{dE}{dt}\right)} \cdot N_A \cdot c \cdot dx$$

On the other hand, the intensity I of the incident light can be expressed by the square of the amplitude F_0 of the electric field strength of the electromagnetic wave (c_0 = velocity of light, ε_0 = permittivity of the vacuum)

$$I = \tfrac{1}{2} \cdot c_0 \cdot \varepsilon_0 \cdot F_0^2$$

Then

$$\varepsilon = \frac{2 \cdot N_A}{2.303 \cdot c_0 \cdot \varepsilon_0} \cdot \overline{\left(\frac{dE}{dt}\right)} \cdot \frac{1}{F_0^2}$$

Calculating the absorption coefficient

First we calculate the mean power $\overline{dE/dt}$ for an isotropic damped classical oscillator of mass m and charge Q. The elongation ξ of the oscillator is obtained

Continued on page 262

Continued from page 261

by solving the differential equation

$$m \cdot \frac{d^2\xi}{dt^2} = -k_f \cdot \xi - \rho \cdot \frac{d\xi}{dt} + Q \cdot F_0 \cdot \cos \omega t$$

where ρ is the damping constant and k_f is the force constant. The solution is

$$\xi = \frac{Q}{m} F_0 \cdot \frac{1}{\sqrt{(\omega_0^2 - \omega^2)^2 + \rho^2 \omega^2/m^2}} \cdot \cos(\omega t + \alpha)$$

with $\omega = 2\pi\nu$, where ν is the frequency of the electric field of the light wave. The energy dE absorbed by the oscillator in time dt equals the product of the damping force $\rho \cdot d\xi/dt$ and the change of elongation $d\xi$; then we obtain

$$\frac{dE}{dt} = \rho \cdot \left(\frac{d\xi}{dt}\right)^2$$

We calculate $d\xi/dt$ from ξ:

$$\frac{d\xi}{dt} = -\frac{Q}{m} F_0 \cdot \frac{\omega}{\sqrt{(\omega_0^2 - \omega^2)^2 + \rho^2 \omega^2/m^2}} \cdot \sin(\omega t + \alpha)$$

Then

$$\frac{dE}{dt} = \rho \frac{Q^2}{m^2} F_0^2 \cdot \frac{\omega^2}{(\omega_0^2 - \omega^2)^2 + \rho^2 \omega^2/m^2} \cdot \sin^2(\omega t + \alpha)$$

and, after averaging $\sin^2(\omega t + \alpha)$,

$$\overline{\left(\frac{dE}{dt}\right)} = \rho \frac{Q^2}{m^2} F_0^2 \cdot \frac{1}{2} \cdot \frac{\omega^2}{(\omega_0^2 - \omega^2)^2 + \rho^2 \omega^2/m^2}$$

Thus the absorption coefficient ε is

$$\varepsilon = \frac{2N_A}{2.303 c_0 \varepsilon_0} \rho \frac{Q^2}{m^2} \cdot \frac{1}{2} \cdot \frac{\omega^2}{(\omega_0^2 - \omega^2)^2 + \rho^2 \omega^2/m^2}$$

and we can rewrite the expression for the mean power $\overline{(dE/dt)}$ as

$$\overline{\left(\frac{dE}{dt}\right)} = \frac{2.303 \cdot c_0 \varepsilon_0}{2N_A} \varepsilon \cdot F_0^2$$

Calculating the integrated absorption

In Figure 8.48 the absorption coefficient ε is displayed versus ω for different values of the damping constant ρ. The half width of the resonance curve depends on ρ, but the area under the curve is independent of ρ. This can be

Continued on page 263

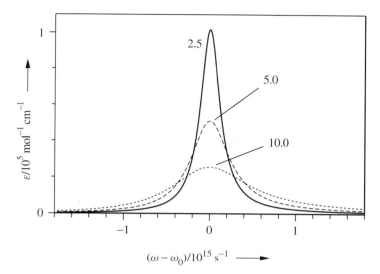

Figure 8.48 Absorption coefficient ε versus $(\omega-\omega_0)$ in the classical oscillator model. Damping constants $\rho = 2.5$, 5.0, and $10.0 \times 10^{-16}\,\text{N s m}^{-1}$; $\omega_0 = 3.77 \times 10^{15}\,\text{s}^{-1}$ (corresponding to $\lambda_{\max} = 500\,\text{nm}$).

proved by calculating the integral

$$\int_0^\infty \varepsilon \cdot d\omega = \frac{N_A}{2.303 c_0 \varepsilon_0} \rho \frac{Q^2}{m^2} \cdot \int_0^\infty \frac{\omega^2}{(\omega_0^2 - \omega^2)^2 + \rho^2 \omega^2/m^2} \cdot d\omega$$

We rearrange the remaining integral

$$J = \int_0^\infty \frac{\omega^2}{(\omega_0^2 - \omega^2)^2 + \rho^2 \omega^2/m^2} \cdot d\omega$$

$$= \int_0^\infty \frac{\omega^2}{(\omega_0 - \omega)^2 \cdot (\omega_0 + \omega)^2 + \rho^2 \omega^2/m^2} \cdot d\omega$$

The denominator reaches a minimum for $\omega_0 = \omega$, when the oscillator is in resonance with the frequency of the light wave; when ω is significantly different from ω_0, the denominator becomes very large, and the integrand does no longer contribute to the integral. For this reason we simplify the evaluation of the integral by replacing ω^2 in the nominator and in the damping term by ω_0^2 and $\omega_0 + \omega$ by $2\omega_0$. Then

$$J = \int_0^\infty \frac{1}{4(\omega_0 - \omega)^2 + \rho^2/m^2} \cdot d\omega$$

With the substitution

$$z = \omega_0 - \omega \quad \text{and} \quad dz = -d\omega$$

Continued from page 263

we obtain

$$J = \int_{-\infty}^{\infty} \frac{1}{4z^2 + \rho^2/m^2} \cdot dz = \frac{m^2}{\rho^2} \int_{-\infty}^{\infty} \frac{1}{(2mz/\rho)^2 + 1} \cdot dz$$

$$= -\frac{m}{2\rho} \cdot \left| \arctan \frac{2mz}{\rho} \right|_{-\infty}^{\infty} = \frac{m\pi}{2\rho}$$

Then

$$\int_0^{\infty} \varepsilon \cdot d\nu = \frac{1}{2\pi} \int_0^{\infty} \varepsilon \cdot d\omega = \frac{N_A}{2\pi \cdot 2.303 c_0 \varepsilon_0} \rho \frac{Q^2}{m^2} \cdot \frac{m\pi}{2\rho}$$

$$= \frac{N_A}{4 \cdot 2.303 c_0 \varepsilon_0} \frac{Q^2}{m}$$

Note that $\int_0^{\infty} \varepsilon \cdot d\nu$ does not depend on the damping constant ρ. This means that the integrated absorption is independent of the shape of the absorption band. In the special case $Q = e$ and $m = m_e$ (oscillator with charge and mass of an electron) we obtain

$$\left(\int_0^{\infty} \varepsilon \cdot d\nu \right)_{\text{classical}(e, m_e)} = \frac{N_A e^2}{4 \cdot 2.303 c_0 \varepsilon_0 m_e}$$

The ratio

$$f = \frac{\int_0^{\infty} \varepsilon \cdot d\nu}{\left(\int_0^{\infty} \varepsilon \cdot d\nu \right)_{\text{classical}(e, m_e)}} = \frac{Q^2}{m} \cdot \frac{m_e}{e^2}$$

is called the *oscillator strength* of the absorption band.

According to Section 8.3.2 oscillators are linear, not isotropic as assumed here. Then in the case of a statistical distribution of dye molecules present in the solution of the dye the absorption coefficient ε is smaller. Effectively, one third of the molecular chains are in the direction of the electric field, and two third are perpendicular. Thus we obtain

$$f = \frac{1}{3} \cdot \frac{Q^2}{m} \cdot \frac{m_e}{e^2}.$$

For more details, see Foundation 8.2.

Foundation 8.2 – Oscillator Strength: Quantum Mechanical Treatment

Quantum mechanical expression for f

The time-dependent Schrödinger Equation for an electron in the field of the molecular skeleton and in the electric field of the light wave is, according to equation (3.48),

$$\mathcal{H}\Psi = i\hbar \frac{\partial \Psi}{\partial t}$$

with

$$\mathcal{H} = \mathcal{H}_0 + exF_0 \cdot \cos \omega t$$

$\mathcal{H}_0$ relates to the Hamiltonian operator (3.26) in the time-independent Schrödinger equation (3.27). F_0 is the amplitude of the electric field of the incident light wave with frequency ν ($\omega = 2\pi\nu$), and $exF_0 \cdot \cos \omega t$ is the potential energy of an electron in this field at time t.

From this time-dependent Schrödinger Equation we obtain the time-dependent part of the dipole moment of the molecule in the radiation field. The problem of finding the time-dependent dipole moment turns out to be the same as to ask for the displacement of a classical oscillator (force constant k_f, mass m, charge Q) in the radiation field (Foundation 8.1).

Both treatments are equivalent if we set

$$\nu_0 = \frac{1}{2\pi}\sqrt{\frac{k_f}{m}} = \frac{\Delta E}{h} \quad \text{or} \quad \frac{k_f}{m} = \frac{4\pi^2(\Delta E)^2}{h^2}$$

and

$$\frac{Q^2}{m} = \frac{8\pi^2}{h^2}\Delta E \cdot M^2$$

where ΔE is the excitation energy and M is the transition moment defined in (8.26), (8.27). The transition moment of the molecule is assumed to be in the direction of the electric field of the incident light. (W. Kauzmann, *Quantum Chemistry*, Acad. Press, 1957).

According to Foundation 8.1 the oscillator strength for a classical oscillator with charge Q and mass m is

$$f = \frac{1}{3} \cdot \frac{Q^2}{m} \cdot \frac{m_e}{e^2}$$

Then inserting the quantum mechanical expression for Q^2/m it follows that

$$f = \frac{1}{3} \cdot \frac{Q^2}{m} \cdot \frac{m_e}{e^2} = \frac{8\pi^2 m_e}{3h^2 e^2} \cdot \Delta E \cdot M^2$$

Continued on page 266

Continued from page 265

This equation holds for an oscillator in vacuo. In the case of dye molecules in solution the relation

$$I = \frac{1}{2} \cdot c_0 \cdot \varepsilon_0 \cdot F_0^2$$

for the intensity of the incident light wave (see Foundation 8.1) has to be replaced by

$$I = \frac{1}{2} \cdot \frac{c_0}{\widehat{n}} \cdot (\varepsilon_0 \widehat{\varepsilon}) \cdot F_0^2$$

where $\widehat{n}$ is the refractive index and $\widehat{\varepsilon}$ is the permittivity of the medium. Because of

$$\widehat{\varepsilon} = \widehat{n}^2$$

we obtain

$$I = \frac{1}{2} \widehat{n} \cdot c_0 \cdot \varepsilon_0 \cdot F_0^2$$

Then the expression for the oscillator strength becomes

$$f = \frac{1}{3} \cdot \frac{Q^2}{m} \cdot \frac{m_e}{e^2} \cdot \frac{1}{\widehat{n}} = \frac{8\pi^2 m_e}{3h^2 e^2 \widehat{n}} \cdot \Delta E \cdot M^2$$

Oscillator strength for cyanine dyes

As an example, we calculate the transition moment for a transition from orbital i to orbital j for a linear π-electron system. The molecular chain (and thus the transition moment M) is assumed to be oriented in the x-direction:

$$M = e \int_0^L \sqrt{\frac{2}{L}} \sin \frac{\pi x \cdot i}{L} \cdot x \cdot \sqrt{\frac{2}{L}} \sin \frac{\pi x \cdot j}{L} \cdot dx$$

$$= e \frac{2}{L} \frac{L}{\pi} \frac{L}{\pi} \int_0^\pi \sin zi \cdot z \cdot \sin zj \cdot dz$$

$$= e \frac{2L}{\pi^2} \cdot \left[\frac{1}{(i-j)^2} - \frac{1}{(i+j)^2} \right]$$

for i even and j odd, or i odd and j even; otherwise the transition moment is zero.

Then for the HOMO–LUMO transition ($j = i + 1$) we obtain

$$M = e \frac{2L}{\pi^2} \cdot \left[1 - \frac{1}{(2i+1)^2} \right] \approx e \frac{2L}{\pi^2}$$

Continued on page 267

Continued from page 266

The excitation energy is (N = number of π electrons)

$$\Delta E = \frac{h^2}{8m_e L^2} \cdot (N+1)$$

and we obtain the following for the oscillator strength f:

$$f = \frac{8\pi^2 m_e}{3h^2 e^2 \widehat{n}} \cdot \frac{h^2}{8m_e L^2} \cdot (N+1) \cdot e^2 \frac{4L^2}{\pi^4} = \frac{4}{3\pi^2 \widehat{n}} \cdot (N+1)$$

This is the oscillator strength calculated for one absorbing electron. In the cyanine dyes the HOMO is occupied by two electrons, thus

$$f = \frac{8}{3\pi^2 \widehat{n}} \cdot (N+1)$$

We can easily account for the fact that the cyanine is a zigzag chain with an angle of 30° relative to the x-direction. Then the coordinate s along the chain is connected with x by

$$x = s \cdot \cos 30° = \tfrac{1}{2}\sqrt{3} \cdot s$$

and the oscillator strength is somewhat smaller:

$$f = \frac{8}{3\pi^2 \widehat{n}} \cdot \frac{3}{4} \cdot (N+1) = \frac{2}{\pi^2 \widehat{n}}(N+1)$$

Foundation 8.3 – Coupling Transitions with Parallel Transition Moments

Replacing quantum mechanical system by coupled classical oscillators

We consider two π electrons giving rise to transitions polarized in the x-direction and assume that the incident light is polarized in the same direction. As in Foundation 8.2 we consider the time-dependent Schrödinger Equation for electron 1

$$\mathcal{H}_1 \Psi_1 = i\hbar \frac{\partial \Psi_1}{\partial t}$$

$$\mathcal{H}_1 = \mathcal{H}_{10} + ex_1 F_0 \cdot \cos \omega t + \mathcal{H}_{12}$$

$$\mathcal{H}_{12} = \int g(1,2) \cdot \Psi_2 \Psi_2^* \cdot d\tau_2$$

Continued on page 268

Continued from page 267

$\mathcal{H}_{10}$ relates, as in Foundation 8.2, to the time-independent Schrödinger equation for electron 1 in the field of the molecular skeleton. Correspondingly, $ex_1 F_0 \cdot \cos \omega t$ is the potential energy of electron 1 in the radiation field.

In addition to the case considered in Foundation 8.2, we have to include the repulsion energy of electron 1 in the field of electron 2. This corresponds to the term $\mathcal{H}_{12}$, where $\Psi_2 \Psi_2^*$ is the time-dependent probability density distribution of electron 2, and $g(1,2)$ is the repulsion energy of electrons 1 and 2 depending on the coordinates of electrons 1 and 2. Then $\mathcal{H}_{12}$ is the repulsion energy for a given coordinate of electron 1, averaged over the distribution of electron 2.

A corresponding time-depending Schrödinger Equation holds for electron 2:

$$\mathcal{H}_2 \Psi_2 = i\hbar \frac{\partial \Psi_2}{\partial t}$$

$$\mathcal{H}_2 = \mathcal{H}_{20} + ex_2 F_0 \cdot \cos \omega t + \mathcal{H}_{21}$$

$$\mathcal{H}_{21} = \int g(1,2) \cdot \Psi_1 \Psi_1^* \cdot d\tau_1$$

Similarly as in Foundation 8.2, we obtain the time-depending part of the dipole moment of the molecule in the radiation field by solving the two coupled Schrödinger equations. The problem of finding this time-dependent dipole moment turns out to be the same as to ask for the displacements ξ_1 and ξ_2 of two coupled classical oscillators (force constants $k_{f,1}$ and $k_{f,2}$, mass m, charges Q_1 and Q_2, damping constants ρ_1 and ρ_2) in the radiation field:

$$m \cdot \frac{d^2 \xi_1}{dt^2} = -k_{f,1} \cdot \xi_1 - \rho_1 \cdot \frac{d\xi_1}{dt} + Q_1 \cdot F_0 \cdot \cos \omega t - k_{f,12} \xi_2$$

$$m \cdot \frac{d^2 \xi_2}{dt^2} = -k_{f,2} \cdot \xi_2 - \rho_2 \cdot \frac{d\xi_2}{dt} + Q_2 \cdot F_0 \cdot \cos \omega t - k_{f,21} \xi_1$$

The coupling between the two oscillators with the frequencies

$$\nu_{10} = \frac{1}{2\pi} \sqrt{\frac{k_{f,1}}{m}} \quad \text{and} \quad \nu_{20} = \frac{1}{2\pi} \sqrt{\frac{k_{f,2}}{m}}$$

is included in the terms $k_{f,12} \xi_2$ and $k_{f,21} \xi_1$.

A detailed consideration shows that formal identity with the quantum mechanical treatment is obtained by the following relations:

Continued on page 269

> $$\frac{k_{f,1}}{m} = \frac{4\pi^2}{h^2} \cdot \Delta E_1^2, \quad \frac{k_{f,2}}{m} = \frac{4\pi^2}{h^2} \cdot \Delta E_2^2$$
>
> $$\frac{k_{f,12}}{m} = \frac{k_{f,21}}{m} = \frac{8\pi^2}{h^2} \cdot \sqrt{\Delta E_1 \Delta E_2} \cdot J_{12}$$
>
> $$J_{12} = \int \psi_{i,1} \psi_{j,1} \cdot g(1,2) \cdot \psi_{i,2} \psi_{j,2} \cdot d\tau_1 d\tau_2$$
>
> $$\frac{Q_1^2}{m} = \frac{8\pi^2}{h^2} \Delta E_1 M_1^2, \quad \frac{Q_2^2}{m} = \frac{8\pi^2}{h^2} \Delta E_2 M_2^2$$
>
> The coupling of electrons 1 and 2 results in a shift of the resonance frequencies of the two oscillators and in a change of the oscillator strength corresponding to the transitions 1 and 2. The uncoupled oscillators would be in resonance with the electric field of the light wave of frequency ν for $\nu = \nu_{10}$ and $\nu = \nu_{20}$, respectively. The system of two coupled oscillators, however, is in resonance at somewhat different frequencies ν_1 and ν_2, where ν_1 corresponds to an in-phase oscillation and ν_2 corresponds to an out-of-phase oscillation of both oscillators (Figure 8.22).
>
> In the out-of-phase oscillation oscillator 1 is accelerated in the electric field of the light wave when oscillator 2 is retarded, and vice versa. As a consequence, the power absorbed from the electric field at ν_2 is small; the power is zero in the geometry of Figure 8.22 (two identical oscillators). On the other hand, in the in-phase oscillation both oscillator 1 and oscillator 2 are accelerated in the electric field of the light wave, and the interaction with the electric field is strong.
>
> *Doubly occupied orbitals*
>
> These equations are valid for transitions from singly occupied orbitals. For transitions from doubly occupied orbitals to empty orbitals the right side of the equations for Q_1, Q_2, and $k_{f,12}$ must be multiplied by a factor of 2:
>
> $$\frac{Q_1^2}{m} = \frac{16\pi^2}{h^2} \Delta E_1 M_1^2, \quad \frac{Q_2^2}{m} = \frac{16\pi^2}{h^2} \Delta E_2 M_2^2$$
>
> $$\frac{k_{f,12}}{m} = \frac{16\pi^2}{h^2} \cdot \sqrt{\Delta E_1 \Delta E_2} \cdot J_{12}$$
>
> (Ch. Kuhn, H. Kuhn, Considerations on correlation effects in π-electron systems, *Synthetic Metals* **68**, 173 (1995)).

Foundation 8.4 – Normal Modes of Coupled Oscillators

Resonance frequencies of coupled oscillators

We consider the general case of N coupled oscillators. In the absence of the field of a light wave and for sufficiently small damping constants ρ_k we obtain

$$m\frac{d^2\xi_k}{dt^2} + k_{f,k} \cdot \xi_k + \sum_{l \neq k} k_{f,kl} \cdot \xi_l = 0 \quad k,l = 1, 2, \ldots N$$

where $k_{f,kl} = k_{f,lk}$. With

$$\xi_k = \gamma_k \cdot \sin \omega_0 t$$

this system of differential equations reduces to a system of linear homogeneous equations:

$$\gamma_k \cdot (-m\omega_0^2 + k_{f,k}) + \sum_{l \neq k} k_{f,kl} \cdot \gamma_l = 0$$

These equations are solved by setting the determinant of the coefficients equal to zero:

$$\begin{vmatrix} (-m\omega_0^2 + k_{f,1}) & k_{f,12} & k_{f,13} & \ldots \\ k_{f,21} & (-m\omega_0^2 + k_{f,2}) & k_{f,23} & \ldots \\ k_{f,31} & k_{f,32} & (-m\omega_0^2 + k_{f,3}) & \ldots \\ \ldots & \ldots & \ldots & \ldots \end{vmatrix} = 0$$

By evaluation of this determinant we obtain the resonance frequencies ω_0 belonging to the N different normal modes.

Oscillator strength of coupled oscillators

We consider the system oscillating in a distinct normal mode (circular frequency ω_0, amplitude γ_k of oscillator k with charge Q_k and mass m). We replace the system by a single oscillator of mass m, circular frequency ω_0, and charge Q oscillating with amplitude γ. If this oscillator has the same dipole moment

$$Q\gamma = \sum_k Q_k \gamma_k$$

and the same energy

$$\gamma^2 = \sum_k \gamma_k^2$$

as the system of coupled oscillators, then it has the same oscillator strength f. According to Foundation 8.2 the oscillator strength is

Continued on page 271

Continued from page 270

$$f = \frac{1}{3}\frac{Q^2}{m}\cdot\frac{m_e}{e^2}\cdot\frac{1}{\widehat{n}} = \frac{1}{3}\frac{m_e}{e^2 m}\cdot\frac{\left(\sum_k Q_k \gamma_k\right)^2}{\sum_k \gamma_k^2}\cdot\frac{1}{\widehat{n}}$$

Coupling two oscillators

In this case the linear homogeneous equations are

$$\gamma_1 \cdot (-m\omega_0^2 + k_{f,1}) + \gamma_2 \cdot k_{f,12} = 0$$
$$\gamma_1 \cdot k_{f,12} + \gamma_2 \cdot (-m\omega_0^2 + k_{f,2}) = 0$$

and from the first equation we obtain

$$\frac{\gamma_2}{\gamma_1} = \frac{m\omega_0^2 - k_{f,1}}{k_{f,12}}$$

The corresponding determinant is

$$\begin{vmatrix} (-m\omega_0^2 + k_{f,1}) & k_{f,12} \\ k_{f,12} & (-m\omega_0^2 + k_{f,2}) \end{vmatrix} = 0$$

leading to

$$(-m\omega_0^2 + k_{f,1})(-m\omega_0^2 + k_{f,2}) = k_{f,12}^2$$

and

$$\omega_0^2 = \frac{1}{2}\frac{k_{f,1} + k_{f,2}}{m} \pm \frac{\sqrt{k_{f,12}^2 - k_{f,1}k_{f,2} + \frac{1}{4}(k_{f,1} + k_{f,2})^2}}{m}$$

$$= \frac{1}{2}\frac{k_{f,1} + k_{f,2}}{m} \pm \frac{\sqrt{k_{f,12}^2 + \frac{1}{4}(k_{f,1} - k_{f,2})^2}}{m}$$

For the oscillator strength f it follows

$$f = \frac{1}{3}\frac{m_e}{e^2 m}\frac{(Q_1\gamma_1 + Q_2\gamma_2)^2}{\gamma_1^2 + \gamma_2^2}\cdot\frac{1}{\widehat{n}} = \frac{1}{3}\frac{m_e}{e^2 m}\frac{\left(Q_1 + Q_2\frac{\gamma_2}{\gamma_1}\right)^2}{1 + \left(\frac{\gamma_2}{\gamma_1}\right)^2}\cdot\frac{1}{\widehat{n}}$$

Coupling two identical oscillators

In the special case $k_{f,1} = k_{f,2} = k_f$ and $f_1 = f_2$ we obtain

$$m\omega_0^2 = k_f \pm k_{f,12}, \quad \gamma_2/\gamma_1 = \pm 1$$

Continued on page 272

Continued from page 271

and it follows for the resonance frequencies ω_0

$$\omega_{0,\text{in-phase}}^2 = \frac{k_f + k_{f,12}}{m}, \quad \omega_{0,\text{out-of-phase}}^2 = \frac{k_f - k_{f,12}}{m}$$

We use the relations (see Foundation 8.3)

$$\frac{k_f}{m} = \frac{4\pi^2}{h^2}\Delta E^2, \quad \frac{k_{f,12}}{m} = \frac{8\pi^2}{h^2}\Delta E \cdot J_{12}, \quad \omega_0 = \frac{2\pi}{h}\Delta E_{\text{coupled}}$$

Then for the excitation energies $\Delta E_{\text{coupled}}$ and for the corresponding oscillatory strengths we obtain

$$\Delta E_{\text{coupled}} = \sqrt{\Delta E^2 \pm 2\Delta E \cdot J_{12}}$$

$$f_{\text{in-phase}} = 2 \cdot f, \quad f_{\text{out-of-phase}} = 0$$

where f is the oscillatory strength of each uncoupled oscillator. This means that only one transition occurs:

$$\Delta E_{\text{coupled,in-phase}} = \sqrt{\Delta E^2 + 2\Delta E \cdot J_{12}} \approx \Delta E + J_{12}, \quad f_{\text{in-phase}} = 2 \cdot f$$

Example – Coupling effect in cyanines

According to Table 8.2 for the cyanine cation with $N = 8$ π electrons we obtain for the HOMO $\rightarrow$ LUMO transition

$$\lambda_{\text{max}} = 459\,\text{nm}, \quad \Delta E = 4.33 \times 10^{-19}\,\text{J}, \quad f = 1.1$$

For the coupling integral we obtain $J_{12} = 0.5 \times 10^{-19}$ J, then

$$\Delta E_{\text{coupled, in-phase}} = (4.33 + 0.5) \times 10^{-19}\,\text{J} = 4.83 \times 10^{-19}\,\text{J}, \quad f = 1.1$$

The corresponding absorption wavelength is $\lambda = 411$ nm, in excellent agreement with the experimental value 416 nm.

Example – Coupling effect in porphyrin

From the calculated absorption maxima $\lambda_1 = 576$ nm and $\lambda_2 = 382$ nm of porphyrin in Table 8.5 (uncoupled transitions) we obtain

$$\Delta E_1 = 3.45 \times 10^{-19}\,\text{J}, \quad \Delta E_2 = 5.20 \times 10^{-19}\,\text{J}$$

The coupling integral for the coupling of the two electrons in the electron pairs is $J_{12} = 0.5 \times 10^{-19}$ J. Then for the transition energies of the electron pairs we obtain, as in the case of the cyanine,

Continued on page 273

FOUNDATION 8.4

Continued from page 272

$$\Delta E_{1,\text{pair}} = 3.45 \times 10^{-19}\,\text{J} + 0.5 \times 10^{-19}\,\text{J} = 3.95 \times 10^{-19}\,\text{J}$$

$$\Delta E_{2,\text{pair}} = 5.20 \times 10^{-19}\,\text{J} + 0.5 \times 10^{-19}\,\text{J} = 5.70 \times 10^{-19}\,\text{J}$$

As a second step we calculate the coupling between the two electron pairs (coupling integral $J_{12} = 0.6 \times 10^{-19}$ J), following the treatment in Foundation 8.3 for doubly occupied orbitals:

$$\frac{k_{f,1}}{m} = \frac{4\pi^2}{h^2} \Delta E_{1,\text{pair}}^2 = 1.41 \times 10^{31}\,\text{s}^{-1},$$

$$\frac{k_{f,2}}{m} = \frac{4\pi^2}{h^2} \Delta E_{2,\text{pair}}^2 = 2.94 \times 10^{31}\,\text{s}^{-1}$$

$$\frac{k_{f,12}}{m} = \frac{16\pi^2}{h^2} \sqrt{\Delta E_{1,\text{pair}} \Delta E_{2,\text{pair}}} \cdot J_{12} = 1.02 \times 10^{31}\,\text{s}^{-1}$$

Then we obtain

$$\omega_0^2 = \frac{1}{2}\frac{k_{f,1}+k_{f,2}}{m} \pm \frac{\sqrt{k_{f,12}^2 + \frac{1}{4}(k_{f,1}-k_{f,2})^2}}{m}$$

$$= (2.18 \pm 1.28) \times 10^{31}\,\text{s}^{-2}$$

and

$$\omega_{0,1} = 5.88 \times 10^{15}\,\text{s}^{-1}, \quad \omega_{0,2} = 3.00 \times 10^{15}\,\text{s}^{-1}$$

or

$$\Delta E_{\text{coupled},1} = 6.19 \times 10^{-19}\,\text{J}, \quad \Delta E_{\text{coupled},2} = 3.16 \times 10^{-19}\,\text{J}$$

corresponding to

$$\lambda_{\text{coupled},1} = 321\,\text{nm}, \quad \lambda_{\text{coupled},2} = 628\,\text{nm}$$

Now we calculate the oscillatory strengths of the coupled transitions, again following the treatment in Foundation 8.3 for doubly occupied orbitals. With $M_{x,1} = M_{x,2} = 2.85 \times 10^{-29}$ Cm we obtain

$$\frac{Q_1^2}{m} = \frac{16\pi^2}{h^2} \Delta E_1 M_{x,1}^2 = 10.01 \times 10^{-8}\,\text{C}^2\,\text{kg}^{-1}$$

$$\frac{Q_2^2}{m} = \frac{16\pi^2}{h^2} \Delta E_2 M_{x,2}^2 = 15.19 \times 10^{-8}\,\text{C}^2\,\text{kg}^{-1}$$

Furthermore, for the ratio γ_2/γ_1 we obtain

$$\frac{\gamma_2}{\gamma_1} = \frac{\omega_0^2 - k_{f,1}/m}{k_{f,12}/m},$$

$$\frac{\gamma_2}{\gamma_1} = 2.0 \text{ for normal mode 1}, \quad \frac{\gamma_2}{\gamma_1} = -0.5 \text{ for normal mode 2}$$

Continued on page 274

> *Continued from page 273*
>
> Then for the oscillatory strengths we obtain
>
> $$f = \frac{1}{3}\frac{m_e}{e^2}\frac{\left(Q_1/\sqrt{m} + Q_2/\sqrt{m}\cdot\frac{\gamma_2}{\gamma_1}\right)^2}{1 + \left(\frac{\gamma_2}{\gamma_1}\right)^2}\cdot\frac{1}{\widehat{n}}$$
>
> $$f_{\text{coupled},1} = 1.9, \quad f_{\text{coupled},2} = 0.09$$
>
> These are the oscillatory strengths for the transitions with transition moment in the x-direction. The transitions in the y-direction contribute the same value, thus the total oscillatory strengths are 3.8 and 0.18, respectively. Note that the sum of the oscillatory strengths is not affected by the coupling.
> The ratio of the oscillatory strengths is
>
> $$\frac{f_{\text{coupled},2}}{f_{\text{coupled},1}} = 0.05$$
>
> The experimental data are $\lambda_1 = 421$ nm, $\lambda_2 = 546$ nm, and $f_2/f_1 = 0.063$. The coupling effect on the intensities of the absorption bands is well represented by the model. However, the separation of the two bands is overestimated; this is due to the neglect of branching as can be seen by refining the treatment. (H. D. Försterling, H. Kuhn, Projected Electron Density Method of π-Electron Systems. II. Excited States, *Int. J. Quant. Chem.* 2, 413 (1968)).

> ### Foundation 8.5 – Fluorescence Life Time
>
> *Antenna equation of Hertz*
>
> A classical oscillator (charge Q, displacement ξ) emits electromagnetic waves. The electric field in point P in distance r (Figure 8.49(a)) has components F_a and F_b:
>
> $$F_a = \frac{2Q\cos\vartheta}{4\pi\varepsilon_0}\left\{\frac{1}{c_0 r^2}\frac{d\xi}{dt} + \frac{1}{r^3}\xi\right\}$$
>
> $$F_b = \frac{Q\sin\vartheta}{4\pi\varepsilon_0}\left\{\frac{1}{c_0^2 r}\frac{d^2\xi}{dt^2} + \frac{1}{c_0 r^2}\frac{d\xi}{dt} + \frac{1}{r^3}\xi\right\}$$
>
> *Continued on page 275*

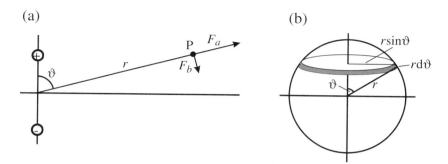

Figure 8.49 (a) Electric field strength at point P in distance r from an oscillating dipole: components F_a (in direction of r) and F_b (perpendicular to the direction of r). (b) Area element $2\pi r^2 \sin\vartheta \cdot d\vartheta$ on the surface of a sphere (shaded area).

ϑ is the angle between the directions of the dipole and r. By inserting

$$\xi = \xi_0 \cos(\omega t - \gamma), \quad \gamma = \frac{\omega r}{c_0} = \frac{2\pi r}{\lambda}$$

we obtain

$$F_a = \frac{2Q\xi_0 \cos\vartheta}{4\pi\varepsilon_0}\left\{-\frac{\omega}{c_0 r^2}\sin(\omega t - \gamma) + \frac{1}{r^3}\cos(\omega t - \gamma)\right\}$$

$$F_b = \frac{Q\xi_0 \sin\vartheta}{4\pi\varepsilon_0}\left\{-\frac{\omega}{c_0 r^2}\sin(\omega t - \gamma) + \left(\frac{1}{r^3} + \frac{\omega^2}{c_0^2 r}\right)\cos(\omega t - \gamma)\right\}$$

The energy emitted by the oscillator in time dt is obtained by considering the light intensity I on a sphere of large radius with the dipole in its center:

$$\text{Mean emitted power} = -\overline{\left(\frac{dE}{dt}\right)} = \int_{\text{surface of sphere}} I \cdot dA$$

dA is the surface area element on the sphere (Figure 8.49(b)). The intensity I can be expressed by the amplitude F_0 of the electric field (see Foundation 8.1):

$$I = \frac{c_0 \varepsilon_0}{2} F_0^2$$

The amplitude F_0 follows from the expressions for F_a and F_b at large distances (neglecting the terms with $1/r^2$ and $1/r^3$):

Continued from page 275

$$F_a = 0, \quad F_b = -\frac{Q\xi_0 \sin\vartheta}{4\pi\varepsilon_0} \cdot \frac{\omega^2}{c_0^2 r} \cdot \cos(\omega t - \gamma)$$

$$F_0 = F_{b,0} = \frac{Q\xi_0}{4\pi\varepsilon_0} \frac{\omega^2}{c_0^2 r} \sin\vartheta$$

Then for the mean emitted power we obtain

$$-\overline{\left(\frac{dE}{dt}\right)} = \frac{Q^2\xi_0^2}{(4\pi\varepsilon_0)^2} \frac{\omega^4}{c_0^4} \frac{c_0\varepsilon_0}{2} \int \frac{\sin^2\vartheta}{r^2} \cdot 2\pi r^2 \sin\vartheta \cdot d\vartheta = \frac{Q^2\xi_0^2}{4\pi\varepsilon_0} \frac{\omega^4}{4c_0^3} \cdot \frac{4}{3}$$

or

$$-\overline{\left(\frac{dE}{dt}\right)} = \frac{1}{4\pi\varepsilon_0} \frac{Q^2\omega^4}{3c_0^3} \xi_0^2$$

With

$$E = \frac{k_f}{2}\xi_0^2 \quad \text{and} \quad \omega^2 = \frac{k_f}{m}$$

(k_f is the force constant of the oscillator) we obtain

$$-\overline{\left(\frac{dE}{dt}\right)} = \frac{1}{4\pi\varepsilon_0} \frac{Q^2\omega^2}{3c_0^3} \cdot \xi_0^2 \omega^2 = \frac{1}{4\pi\varepsilon_0} \frac{2Q^2\omega^2}{3mc_0^3} \cdot E = \frac{1}{\tau_0} \cdot E$$

Life time

Correspondingly, for an excited molecule emitting a light quantum the average lifetime of the excited state equals τ_0. In this case, Q^2/m must be replaced, according to Foundation 8.2, by

$$\frac{Q^2}{m} = f \cdot \frac{3e^2\hat{n}}{m_e}$$

where f is the oscillator strength, and $\hat{n}$ is the index of refraction. τ_0 is called the *natural lifetime*.

$$\tau_0 = \frac{3m\varepsilon_0 c_0^3}{2Q^2 \pi v_0^2} \quad \text{or} \quad \tau_0 = \frac{m_e \varepsilon_0 c_0^3}{2e^2 \pi v_0^2 \hat{n}} \cdot \frac{1}{f}$$

where $2\pi v_0 = w$. Note that the natural lifeitem τ_0 is large compared with the oscillation period, i.e. the energy loss in one oscillation is small compared with the energy E of the oscillator.

Foundation 8.6 – Proof of Relation for g

We estimate ε_R and ε_L of the model in Figure 8.40(a). In the geometry of Figure 8.42 the phase of the electric field of left circular polarized light is delayed at point 2 relative to the field at point 1 by $\varphi = 2\pi d/\lambda$. According to Figure 8.42 the electric field strength driving oscillator 1 in the z direction is $F_z = F_0 \cdot \cos \omega t$ and the field strength driving oscillator 2 in the y direction is $F_y = -F_0 \cdot \sin(\omega t - \varphi)$. The displacements z_1 and $-y_2$ of the two oscillating charges Q_1 and Q_2 are equal. Thus we can replace the two oscillators by a single oscillator (charge $(Q_1 + Q_2)$, force constant $2k_f$, mass $2m$) driven by the external field.

In Foundation 8.1 we calculated the absorption coefficient for an oscillator driven by the external force

$$f_{\text{external}} = QF_0 \cos \omega t$$

It turned out that the absorption coefficient is proportional to Q^2.

In the present case we can follow the same consideration to calculate ε_L, but now the external force is

$$f_{\text{external}} = Q_1 F_z + Q_2 F_y = Q_1 F_0 \cos \omega t - Q_2 F_0 \sin(\omega t - \varphi)$$
$$= QF_0(\cos \omega t - \sin(\omega t - \varphi))$$

if both charges are equal ($Q_1 = Q_2 = Q$). It can be shown that this equation can be converted into

$$f_{\text{external}} = QF_0 \cdot \sqrt{2} \cdot \sqrt{1 + \sin \varphi} \cdot \cos\left(\omega t - \frac{\varphi}{2} + \frac{\pi}{4}\right)$$

Then, because ε_L is proportional to Q^2,

$$\varepsilon_L = C \cdot 2(1 + \sin \varphi)$$

where C is a constant. Accordingly, for right circular polarized light the forces counteract:

$$\varepsilon_R = C \cdot 2(1 - \sin \varphi)$$

Thus, taking the mean value ε from ε_L and ε_R

$$\varepsilon = \frac{\varepsilon_L + \varepsilon_R}{2} = 2C$$

we obtain

$$\frac{\Delta \varepsilon}{\varepsilon} = \frac{\varepsilon_L - \varepsilon_R}{2C} = 2\sin \varphi \approx 2\varphi = \frac{4\pi d}{\lambda}$$

In the present geometry (d in the direction of incident light: x direction) the circular dichroism is strongest, and there is no effect when d is in the y or z direction. For the statistical average we obtain

$$g = \frac{\Delta \varepsilon}{\varepsilon} = \frac{1}{2} \cdot \frac{4\pi d}{\lambda} = \frac{2\pi d}{\lambda}$$

9 Nuclei: Particle and Wave Properties

We have seen that the behavior of electrons is governed by their wave–particle duality. So far we have treated nuclei as fixed point charges because their masses are much larger than the electron mass. But nuclei, in a molecular skeleton, cannot stay at rest. They are restricted to discrete energy states because of their wave–particle duality. In classical terms they oscillate in a molecule relative to each other. Molecules can rotate as a whole. During the vibrational motion, bond lengths or bond angles change and the electron cloud is deformed accordingly.

The system of nuclei in a molecule is restricted to discrete energy states, because of the wave nature of the nuclei. The absorption of light in the infrared (IR) spectral region excites oscillations of the molecular skeleton and rotations of the molecule. This gives us detailed information on vibrational and rotational states and thus detailed knowledge on the fine structure of molecules. The possible energy states depend in a most astonishing way on the nuclear spin, which constitutes a unique probe for details in molecular structure.

Nuclear motions (translation, rotation, and vibration) are not completely independent from each other. Their treatment as independent motions, however, makes it easier to understand the most important features of nuclear motion.

Translation can be described as a motion of the center of gravity of the molecules; thus the classical description of the kinetic gas theory (Chapter 11) should be sufficient at the moment, and we restrict our analysis to the rotational and vibrational motions.

Nuclei have similar strange properties as electrons. Nuclei of the same kind are indistinguishable; nuclei have a spin and behave in a very particular way in a magnetic field. This is the basis of nuclear magnetic resonance spectroscopy, a most important technique for the determination of chemical structures.

9.1 Quantum Mechanical Rotator

As an example we consider the HCl molecule. The Cl atom is much heavier than the H atom ($m_{Cl} = 35 \cdot m_H$). In a classical picture the H atom circles around the Cl atom on the surface of a sphere of radius d (Figure 9.1(a)). Accordingly, the H atom can circle with

9.1 QUANTUM MECHANICAL ROTATOR

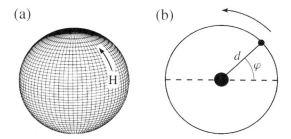

Figure 9.1 Rotational motion of an HCl molecule. The mass of the Cl atom is 35 times larger than that of H, so the Cl atom is practically at rest and the H atom circles at a distance d. (a) Rotation in space; H atom circles on the surface of a sphere. (b) Rotation in a plane.

any speed but the experiments discussed in Section 9.2 establish the astonishing fact that the system exists in discrete rotational energy states. How can we understand this?

9.1.1 Exact Solution

We have to solve the Schrödinger equation for a particle of mass m moving at constant potential energy on the surface of a sphere with radius d. For the energy we obtain (Foundation 9.1)

$$E = \frac{h^2}{8\pi^2 m d^2} \cdot n(n+1), \quad n = 0, 1, 2, \ldots \tag{9.1}$$

This relation tells us that the rotational energy can be zero and can have distinct values depending on nuclear mass m and distance d between nuclei and on the quantum number n. In the present case (HCl molecule), we have $m = m_H \approx 2000 \times m_e$. Usually, the quantum number of rotation is denoted by J. Then it can be formally distinguished from the quantum number n for electrons and the quantum number v for vibration. Instead, we use n throughout for quantum numbers.

The wavefunctions correspond to the angular parts of the wavefunctions of the hydrogen atom (angles φ and ϑ; see Foundation 9.1)

$n = 0$:
$$\psi_0 = \frac{1}{\sqrt{4\pi}} \tag{9.2}$$

$n = 1$:
$$\psi_{1a} = \sqrt{\frac{3}{4\pi}} \cdot \sin\vartheta \cdot \cos\varphi, \quad \psi_{1b} = \sqrt{\frac{3}{4\pi}} \cdot \cos\vartheta, \quad \psi_{1c} = \sqrt{\frac{3}{4\pi}} \cdot \sin\vartheta \cdot \sin\varphi$$

$n = 2$:
$$\psi_{2a} = \sqrt{\frac{15}{16\pi}} \cdot \sin^2\vartheta \cos 2\varphi, \quad \psi_{2b} = \sqrt{\frac{15}{4\pi}} \cdot \sin\vartheta \cos\vartheta \cos\varphi$$

$$\psi_{2c} = \sqrt{\frac{5}{16\pi}} \cdot (3\cos^2\vartheta - 1), \quad \psi_{2d} = \sqrt{\frac{15}{4\pi}} \cdot \sin\vartheta \cos\vartheta \sin\varphi$$

$$\psi_{2e} = \sqrt{\frac{15}{16\pi}} \cdot \sin^2\vartheta \sin 2\varphi$$

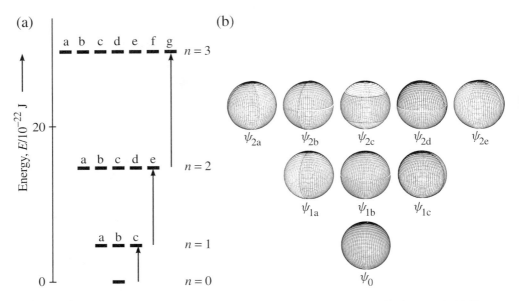

Figure 9.2 Rotational motion of the H nucleus in the HCl molecule (rotation in space). (a) Energy levels; (b) wavefunctions. Black: wavefunction positive, red: wavefunction negative. The arrows indicate the allowed transitions ($\Delta n = 1$).

The energies and wavefunctions are displayed in Figure 9.2. We obtain one wavefunction for $n = 0$, three wavefunctions for $n = 1$, and five wavefunctions for $n = 2$. The degeneracy of the energy levels is similar as for the s, p, and d states for the hydrogen atom. The angular dependence is the same for $n = 0$ and s, for $n = 1$ and p_x, p_y, p_z, and for $n = 2$ and d_{xy}, d_{yz}, d_{xz}, $d_{x^2-y^2}$, d_{z^2}. Generally, the energy levels are

$$g_n = 2n + 1 \tag{9.3}$$

fold degenerate.

9.1.2 Simplified Model

For a better understanding of the essence of the quantization of the nuclear motion we simplify the description. We assume that the H atom is moving with a speed v on the circumference of a circle with radius d (Figure 9.1(b)). Then its kinetic energy is

$$E = \tfrac{1}{2}mv^2 \tag{9.4}$$

According to Chapter 1, wave–particle duality is an important feature not only of electrons, but of nuclei as well. For this reason we describe the H nucleus by a standing wave of wavelength Λ along a circle of circumference $2\pi d$. According to the de Broglie equation

$$\Lambda = \frac{h}{mv} \tag{9.5}$$

Λ depends upon the velocity v. As in the case of an electron in a benzene molecule (Section 7.2), for Λ we obtain

9.1 QUANTUM MECHANICAL ROTATOR

$$\Lambda = \frac{2\pi d}{n}, \quad n = 0, 1, 2, \ldots \quad (9.6)$$

(an integer number of wavelengths extends along the circle). Thus we obtain for the energy E from equations (9.4), (9.5), and (9.6)

$$E = \frac{h^2}{8\pi^2 m d^2} \cdot n^2, \quad n = 0, 1, 2, \ldots \quad (9.7)$$

The corresponding wavefunctions are

$n = 0$:
$$\psi_0 = \frac{1}{\sqrt{2\pi}}$$

$n = 1$:
$$\psi_{1a} = \frac{1}{\sqrt{2\pi}} \cdot \sin\varphi, \quad \psi_{1b} = \frac{1}{\sqrt{2\pi}} \cdot \cos\varphi \quad (9.8)$$

$n = 2$:
$$\psi_{2a} = \frac{1}{\sqrt{\pi}} \cdot \sin 2\varphi, \quad \psi_{2b} = \frac{1}{\sqrt{\pi}} \cdot \cos 2\varphi$$

if we replace the coordinate s in benzene (Chapter 7) by the angle of rotation φ (Figure 9.1(b)); $n(n+1)$ in equation (9.1) is replaced by n^2 in equation (9.7). The energies and wavefunctions are shown in Figure 9.3. The simplified model focuses on the heart of the problem by viewing the de Broglie wave interference to a standing wave.

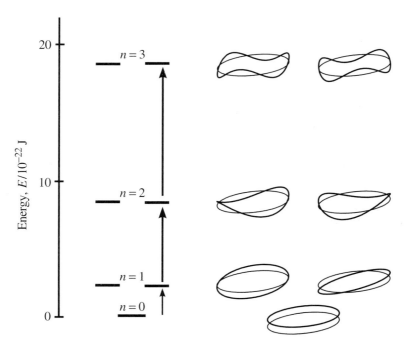

Figure 9.3 Energy levels and wavefunctions of the rotating H nucleus in the HCl molecule (simplified model: rotation in a plane). The arrows indicate the allowed transitions ($\Delta n = 1$).

9.2 Rotational Spectra

In Chapter 2 we have established discrete electronic states of atoms showing discrete spectral lines. Only light quanta are absorbed or emitted that have the appropriate energy to excite and de-excite the atom. The situation is similar in the present case of rotational energy states of a molecule; however, the required energy quanta are much smaller (microwave region).

The quantization of the rotational motion can be experimentally demonstrated. We irradiate an optical cell filled with HCl gas by electromagnetic radiation in the microwave region (far IR). The intensity of the transmitted radiation is measured and then the spectrum in Figure 9.4 is obtained. Many equidistant absorption lines appear, and we conclude that many transitions are possible.

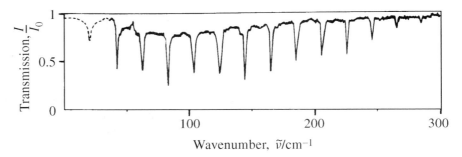

Figure 9.4 Absorption spectrum of HCl gas in the far IR region; the transmission I/I_0 is displayed as a function of the wavenumber $\tilde{v} = v/c_0$ From G.M. Barrow, Physical Chemistry, McGraw-Hill Book Company, New York 1966, P. 338.

Box 9.1 – Selection Rules for Rotation of Linear Molecules

We consider a linear molecule with the permanent dipole moment μ_{perm}. In analogy to the light absorption of an electron (Foundation 8.2), we find for the transition moment $M^2 = M_x^2 + M_y^2 + M_z^2$

$$M_x = \int \psi_i \mu_x \psi_j \cdot d\tau \quad \text{(with } M_y, M_z \text{ correspondingly)}$$

μ_x is the x component of the permanent dipole moment μ_{perm} (Figure 9.5)
$\mu_x = \mu_{\text{perm}} \cdot \cos \varphi = \mu_{\text{perm}} \cdot x/d$
($d =$ bond length). Then

$$M_x = \mu_{\text{perm}} \frac{1}{d} \int \psi_i x \psi_j \cdot d\tau$$

Continued on page 283

9.2 ROTATIONAL SPECTRA

Continued from page 282

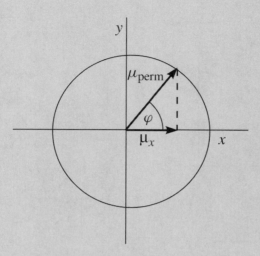

Figure 9.5 Permanent dipole moment μ and its x component μ_x.

From this relation we conclude that the rotator can only be excited by light if it possesses a permanent dipole moment μ_{perm} (e.g., HCl, NO, CO). We can easily see from the symmetry properties of the rotator wavefunctions that M_x is nonzero for

$$\Delta n = \pm 1$$

Example – rotation in a plane

In this case, ψ depends on the angle φ, thus

$$M_x = \mu_{\text{perm}} \frac{1}{d} \int \psi_i \cdot x \cdot \psi_j \cdot d\varphi$$

We consider transitions from the ground state ($n = 0$) and from the states with $n = 1$ (Figure 9.3) to higher states. From the symmetries of the corresponding wavefunctions and the symmetries of the product functions in the transition integral, we see that only transitions are allowed for $\Delta n = \pm 1$.

Particularly, we consider the transition $0 \rightarrow 1b$. The wavefunction ψ_0 is symmetric, and ψ_{1b} is antisymmetric with respect to the x coordinate. Then the product $(\psi_0 \psi_{1b}) \cdot x$ is symmetric and the integral is nonzero. Accordingly, M_x for the transition $0 \rightarrow 1a$ is zero. The same is true for transitions $0 \rightarrow 2b$ and $0 \rightarrow 2a$ (Figure 9.6).

Continued on page 284

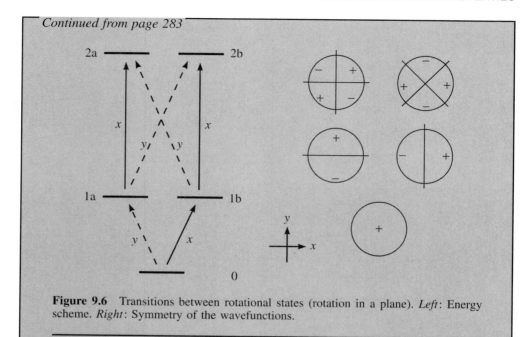

Figure 9.6 Transitions between rotational states (rotation in a plane). *Left*: Energy scheme. *Right*: Symmetry of the wavefunctions.

As shown in Box 9.1, there are strong selection rules. The transitions can only be excited by irradiation with light, if the diatomic molecules have a nonzero value of the permanent dipole moment μ_{perm}, and if the quantum number n increases by 1. Let us assume that before irradiation all the HCl molecules are in the lowest energy level ($n = 0$). Then only one transition would be possible with the excitation energy (see equation (9.1))

$$\Delta E = \frac{h^2}{8\pi^2 md^2} \cdot [1 \cdot (1+1) - 0] = \frac{2h^2}{8\pi^2 md^2} \tag{9.9}$$

and only one single line would appear in the spectrum. How can we understand the observed sequence of equidistant lines?

The spacing of the rotation levels is much smaller than that of the energy levels of bonding electrons. Because the mass of the proton is 2000 times larger than the mass of the electron, the distances between the energy levels will be smaller by a factor of about 2000. Thus the nuclei can accumulate so much energy through thermal collisions that the molecules are distributed over many rotational energy levels. By excitation in the far IR, many transitions with $\Delta n = +1$ are possible (arrows in Figure 9.2(a)).

We consider the excitation from a state with a quantum number n to a quantum number $n + 1$. Using the abbreviation

$$B = \frac{h^2}{8\pi^2 md^2} \tag{9.10}$$

(*rotational constant*), we obtain according to equation (9.1),

$$\Delta E = B \cdot [(n+1) \cdot (n+2) - n \cdot (n+1)] = 2B \cdot (n+1), \quad n = 0, 1, 2, \ldots \tag{9.11}$$

9.2 ROTATIONAL SPECTRA

ΔE increases linearly with n, in agreement with Figure 9.4; the spacing of the lines, $\Delta(\Delta E)$, is constant:

$$\Delta(\Delta E) = \Delta E_{n+1} - \Delta E_n = 2B \tag{9.12}$$

This spacing depends on the mass m and on the bond distance d.

Example

We compare equation (9.12) with the experiment; to do this we convert the wavenumbers $\tilde{\nu} = 1/\lambda$ in Figure 9.4 into excitation energies ΔE:

$$\Delta E = h\nu = hc_0 \frac{1}{\lambda} \tag{9.13}$$

According to Figure 9.4 and Table 9.1 we obtain $\Delta(\Delta E) = 40.85 \times 10^{-23}$ J as an average. Note the small decrease of $\Delta(\Delta E)$ with increasing n; this is due, in the classical picture, to slightly larger values of d at higher rotational frequencies (centrifugal force stretching bond). Then we calculate the bond distance d of HCl from equation (9.12):

$$d = \sqrt{\frac{h^2}{4\pi^2 m} \cdot \frac{1}{\Delta(\Delta E)}} = 128\,\text{pm}$$

This example demonstrates that the evaluation of rotational spectra is an important method for determining bond distances.

The strengths of the absorption lines in Figure 9.4 reflect the rotational energy distribution by thermal collisions. The translational energy of the molecule that can be converted into rotational energy in a collision is on the order of 4×10^{-21} J at room

Table 9.1 Evaluation of the absorption lines in the rotational spectrum of HCl in Figure 9.4 according to equation (9.12) Numerical data from G. Herzberg, Molecular Spectra and Molecular Structure I. Spectra of Diatomic Molecules, D. Van Nostrand, New York 1950

n	$\frac{1}{\lambda}/\text{cm}^{-1}$	$\Delta E/10^{-23}$ J	$\Delta(\Delta E)/10^{-23}$ J
0	20.8	41.3	
1	41.6	82.6	41.3
2	62.3	123.7	41.1
3	83.0	164.8	41.1
4	104.1	206.7	41.9
5	124.3	246.9	40.2
6	145.0	288.0	41.1
7	165.5	328.7	40.7
8	185.9	369.2	40.5
9	206.4	409.9	40.7
10	226.5	449.8	39.9

temperature (see Chapter 11 eq. 11.20). This corresponds to the energy in rotational state 4 in HCl:

$$E_4 = B \cdot 4 \cdot (4+1) = 20B = 4 \times 10^{-21} \text{ J}$$

This means that rotational states around $n = 4$ can be significantly populated.

In our treatment of the HCl molecule we assumed that the Cl atom is at rest. This is an approximation, of course. In a rigorous treatment (Box 9.2) we obtain the rotational constant B for a diatomic molecule A–B with arbitrary masses m_A and m_B:

$$B = \frac{h^2}{8\pi^2 \mu d^2} \quad \text{with} \quad \mu = \frac{m_A \cdot m_B}{m_A + m_B} \tag{9.14}$$

Box 9.2 – Rotator: Reduced Mass

During the rotational motion the center of gravity of the molecule stays at rest. Then according to Figure 9.7 we have $am_A = bm_B$ and $a + b = d$. Thus

$$a = d \cdot \frac{m_B}{m_A + m_B} \quad \text{and} \quad b = d \cdot \frac{m_A}{m_A + m_B}$$

In (9.7) $I = md^2$ is the moment of inertia of a point mass rotating around a fixed point in distance d. This quantity must be replaced by the moment of inertia of a diatomic molecule:

$$\begin{aligned} I &= m_A a^2 + m_B b^2 \\ &= m_A d^2 \cdot \frac{m_B^2}{(m_A + m_B)^2} + m_B d^2 \cdot \frac{m_A^2}{(m_A + m_B)^2} \\ &= d^2 \cdot \frac{m_A m_B}{m_A + m_B} = d^2 \cdot \mu \end{aligned}$$

with

$$\mu = \frac{m_A m_B}{m_A + m_B}$$

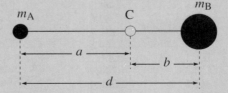

Figure 9.7 Rotation of diatomic molecule. How to find the center of gravity C.

9.3 QUANTUM MECHANICAL OSCILLATOR

instead of equation (9.10). μ is called the *reduced mass*; the quantity

$$I = \mu d^2 \tag{9.15}$$

is the *moment of inertia* of the rotator.

Example

For HCl the reduced mass

$$\mu = \frac{m_H \cdot m_{Cl}}{m_H + m_{Cl}} = \frac{35}{36} \cdot m_H = 0.972 \cdot m_H$$

differs only by 3% from the mass of the H atom. If the masses of the atoms A and B are identical ($m_A = m_B$), both nuclei circle around the center of the bond (Figure 9.8); the kinetic energy belonging to this motion is the same as for a rotation of the mass $(m_A + m_B) = 2m_B$ on a circle of radius $d/2$. The moment of inertia is then

$$I = 2m_A \left(\frac{d}{2}\right)^2 = \frac{1}{2} m_A d^2$$

in accordance with equations (9.14) and (9.15). For molecules with more than two atoms more complicated expressions for I and μ are obtained.

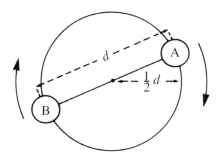

Figure 9.8 Rotational motion of a diatomic molecule with nuclei A and B of identical mass.

9.3 Quantum Mechanical Oscillator

We consider an HCl molecule in its equilibrium state (bond distance d); we imagine the bond being stretched or contracted by x (Figure 9.9). In both cases, the energy of the molecule increases because the nuclear distance and the distribution of the bonding electrons change. We limit ourselves to a very small deflection x. Then the energy is proportional to x^2 (see Problem 5.2); that is, in this approximation the vibrational motion of a molecule can be described by the motion of two point masses connected by a spring with a force constant k_f (Figure 9.9(b)). During the vibration of the HCl molecule the Cl

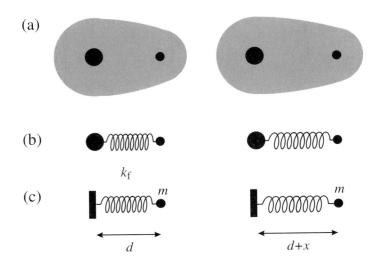

Figure 9.9 The deflection of the H atom in a HCl molecule by x from the equilibrium distance d. (a) Deformation of the electron cloud during the deflection. (b) Spring model of the HCl molecule. (c) Spring model of the HCl molecule, Cl is assumed to be at rest.

nucleus remains practically at rest, as in the case of rotation. We can therefore approximate the vibrational motion of the H atom by the vibrational motion of a spring with force constant k_f connected to a mass m: *a linear harmonic oscillator*.

In classical mechanics this oscillator can assume any energy. However, the experiments discussed in Section 9.4 show that only definite vibronic energy states exist. This is explained by quantum mechanics.

9.3.1 Exact Solution

We obtain the energies of a linear harmonic oscillator by solving the Schrödinger equation for a particle of mass m (mass of an atom) moving in the x direction with the potential energy $V(x) = \frac{1}{2}k_f x^2$:

$$-\frac{h^2}{8\pi^2 m} \cdot \frac{d^2 \psi}{dx^2} + \frac{1}{2}k_f x^2 \psi = E\psi \tag{9.16}$$

The exact solutions of the Schrödinger equation for the two lowest energies are (see Problem 9.1 for E_0 and ψ_0)

$$E_0 = \frac{1}{2}h\nu_0, \quad \psi_0 = \sqrt[4]{\frac{a}{\pi}} \cdot e^{-ax^2/2} \tag{9.17}$$

$$E_1 = \frac{3}{2}h\nu_0, \quad \psi_1 = \sqrt[4]{\frac{a}{\pi}} \sqrt{2a} \cdot x \cdot e^{-ax^2/2}$$

with

$$a = \frac{2\pi}{h}\sqrt{k_f m} \tag{9.18}$$

9.3 QUANTUM MECHANICAL OSCILLATOR

and

$$\nu_0 = \frac{1}{2\pi}\sqrt{\frac{k_\mathrm{f}}{m}} \qquad (9.19)$$

ν_0 is identical with the frequency of a classical oscillator (Box 9.3).

Box 9.3 – Classical Oscillator: Frequency ν_0

The force acting on a point mass m connected to a spring with force constant k_f and moving in the x direction is

$$m\frac{\mathrm{d}^2 x}{\mathrm{d}t^2} = -k_\mathrm{f} x$$

This is a differential equation for $x(t)$:

$$\frac{\mathrm{d}^2 x}{\mathrm{d}t^2} = -\frac{k_\mathrm{f}}{m} x$$

This equation is solved by $x = A \cdot \sin 2\pi\nu_0 t$ where A is a constant and ν_0 is the frequency. Insertion into the differential equation yields

$$-A \cdot 4\pi^2 \nu_0^2 \cdot \sin 2\pi\nu_0 t = -\frac{k_\mathrm{f}}{m} \cdot A \cdot \sin 2\pi\nu_0 t$$

Thus

$$4\pi^2 \nu_0^2 = \frac{k_\mathrm{f}}{m} \quad \text{and} \quad \nu_0 = \frac{1}{2\pi}\sqrt{\frac{k_\mathrm{f}}{m}}$$

For arbitrary quantum numbers n the exact solutions for the energy are

$$\boxed{E_n = h\nu_0 \cdot \left(\tfrac{1}{2} + n\right), \quad n = 0, 1, 2, \ldots} \qquad (9.20)$$

These energies and the corresponding wavefunctions are displayed in Figure 9.10 for $n = 0$ to $n = 5$.

In a more rigorous treatment, we have to replace the mass m by the reduced mass μ (see Box 9.4) in equations (9.18) and (9.19), as for the rotator:

$$\nu_0 = \frac{1}{2\pi}\sqrt{\frac{k_\mathrm{f}}{\mu}}, \quad a = \frac{2\pi}{h}\sqrt{k_\mathrm{f}\mu} \qquad (9.21)$$

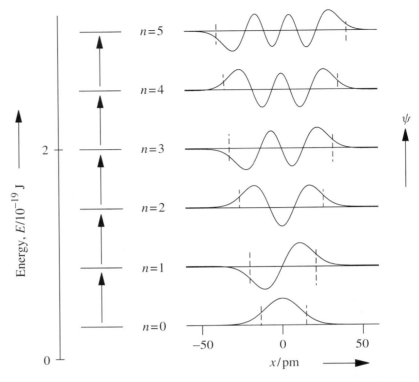

Figure 9.10 Energies and wavefunctions of a quantum mechanical oscillator for the quantum numbers $n = 0$ to $n = 5$ (H atom in the HCl molecule: $m = m_H = 1.67 \times 10^{-27}$ kg, $k_f = 480\,\text{N}\,\text{m}^{-1}$). The dashed lines mark the turning points of the corresponding classical oscillator with the same energy.

Box 9.4 – Oscillator: Reduced Mass of Diatomic Molecules

We consider the vibration as a motion of the masses m_A and m_B relative to the center of gravity (Figure 9.11). Then atom A is an oscillator with mass m_A and force constant k_{fA}. Because the force constant is inversely proportional to the length of the spring representing the bond, we have

$$k_{fA} = k_f \cdot \frac{d}{a} \quad \text{with} \quad a = d\frac{m_B}{m_A + m_B}$$

(k_f = force constant of the spring). Then the frequency of this oscillator is

$$\nu_0 = \frac{1}{2\pi} \cdot \sqrt{\frac{k_{fA}}{m_A}} = \frac{1}{2\pi} \cdot \sqrt{\frac{k_f d}{a m_A}} = \frac{1}{2\pi} \cdot \sqrt{\frac{k_f d}{m_A d \frac{m_B}{m_A + m_B}}} = \frac{1}{2\pi} \cdot \sqrt{\frac{k_f}{\mu}}$$

Continued on page 291

9.3 QUANTUM MECHANICAL OSCILLATOR

with

$$\mu = \frac{m_A m_B}{m_A + m_B}$$

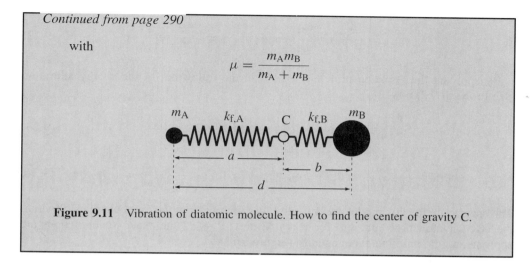

Figure 9.11 Vibration of diatomic molecule. How to find the center of gravity C.

9.3.2 Box Model for Oscillator

For a better physical insight into the problem we estimate the energies of the two lowest quantum states by approximating the wavefunctions ψ_0 and ψ_1 of the oscillator by the trial functions for a particle in a one-dimensional box (Figure 9.12). Our box trial functions are

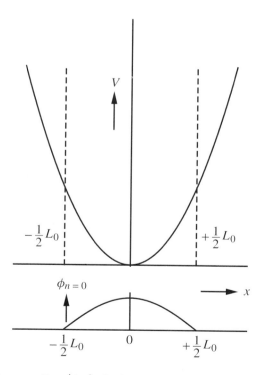

Figure 9.12 Potential energy $V = \frac{1}{2}k_f x^2$ of a harmonic oscillator and trial function $\phi_{n=0}$ (estimation of the ground level energy $\varepsilon_{n=0}$).

$$\phi_{n=0} = \sqrt{\frac{2}{L_0}} \cos \frac{\pi x}{L_0} \quad \text{and} \quad \phi_{n=1} = \sqrt{\frac{2}{L_1}} \sin \frac{2\pi x}{L_1} \quad (9.22)$$

By applying the variation principle we obtain the following for the energy minimum (see Problem 9.2)

$$L_0^2 = 2.77 \cdot \frac{h}{\sqrt{k_f m}} \quad \text{and} \quad L_1^2 = 3.76 \cdot \frac{h}{\sqrt{k_f m}} \quad (9.23)$$

$$\varepsilon_{n=0} = 0.57 \cdot \frac{h}{2\pi} \cdot \sqrt{\frac{k_f}{m}} = 0.57 \cdot h\nu_0, \quad \varepsilon_{n=1} = 1.67 \cdot \frac{h}{2\pi} \cdot \sqrt{\frac{k_f}{m}} = 1.67 \cdot h\nu_0 \quad (9.24)$$

The energies in the box model are only slightly different from the exact energies. Whereas the box wavefunctions are zero for $x > L/2$ or $x < -L/2$ the exact wavefunctions are approaching the baseline exponentially (Figure 9.13).

Note that in the lowest energy state the energy is not zero as in the case of the rotational motion; this means the oscillator has a zero-point energy because of its finite de Broglie wavelength. In contrast, the de Broglie wavelength for a rotator in the ground state is infinite (the wavefunction has the same constant value for all values of ϑ and φ).

The energy E increases proportionally with the quantum number n. This is in contrast to a particle in a box, where E increases with n^2. This is because for the oscillator, approximated by the box trial functions, the corresponding box length L increases with increasing quantum number, whereas in the case of a particle in a box the distance between the two walls is constant.

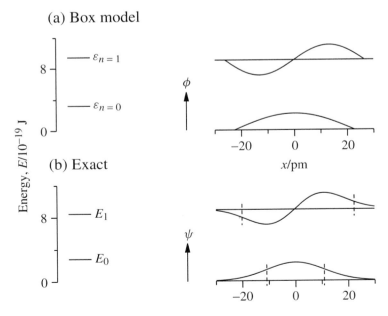

Figure 9.13 Energies and wavefunctions of a quantum mechanical oscillator (HCl molecule). (a) Box model (energies $\varepsilon_{n=0}$ and $\varepsilon_{n=1}$, wavefunctions ϕ). (b) Exact (energies E_0 and E_1, wavefunctions ψ).

9.3 QUANTUM MECHANICAL OSCILLATOR

The aim of the simplified model is to appreciate the essence in viewing nuclear de Broglie waves.

9.3.3 Comparison of a Quantum Mechanical Oscillator with a Classical Oscillator

Consider the points of inflection (coordinate x_0) of the wavefunctions in Figure 9.10. At an inflection point the second derivative of the wavefunction is zero:

$$\left(\frac{\partial^2 \psi}{\partial x^2}\right)_{x_0} = 0 \qquad (9.25)$$

Inserting this condition into the Schrödinger equation (9.16), we get

$$E = \tfrac{1}{2} k_f x_0^2 \qquad (9.26)$$

Thus for $x = x_0$ the energy E equals the potential energy V. The point of inflection x_0 corresponds to the turning point of a classical oscillator: a classical oscillator cannot reach any point beyond x_0. For the quantum mechanical oscillator, however, the wavefunction is nonzero even at $x > x_0$ or at $x < -x_0$, and we can find the oscillator at points where the classical oscillator can never be found. The coordinate

$$x_0 = \sqrt{\frac{2E}{k_f}} \qquad (9.27)$$

increases with increasing energy of the oscillator (dashed lines in Figures 9.10 and 9.13).

We calculate the probability P to find the oscillator between x and $x + dx$. This probability is

$$dP = \psi^2 \cdot dx \qquad (9.28)$$

in the quantum mechanical description. In the classical description the probability dP_{class} of finding the oscillator between x and $x + dx$ is inversely proportional to its classical speed v

$$dP_{\text{class}} \sim \frac{1}{v} \cdot dx \qquad (9.29)$$

It follows (Problem 9.3) that

$$dP_{\text{class}} = \frac{1}{\pi x_0} \cdot \frac{1}{\sqrt{1 - (x/x_0)^2}} \cdot dx \qquad (9.30)$$

In Figure 9.14 the probability densities

$$\frac{dP}{dx} = \psi^2 \qquad (9.31)$$

for the quantum numbers $n = 0$ and $n = 50$ are displayed and compared with the classical expression

$$\frac{dP}{dx} = \frac{1}{\pi x_0} \cdot \frac{1}{\sqrt{1 - (x/x_0)^2}} \qquad (9.32)$$

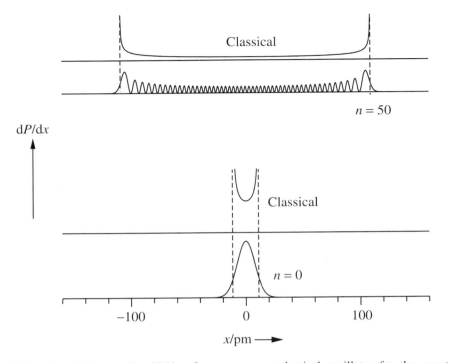

Figure 9.14 Probability density dP/dx of a quantum mechanical oscillator for the quantum numbers $n = 0$ and $n = 50$ and of a classical oscillator with the same energies ($k_f = 480\,\mathrm{N\,m^{-1}}$, $m = 1.67 \times 10^{-27}\,\mathrm{kg}$). (We consider the case $n = 50$ only as a matter of principle, to show the correlation between the classical and quantum mechanical treatment. In molecules, such high quantum numbers do not play a role).

In the classical consideration we expect a minimum of dP/dx at $x = 0$ independent of the energy (at $x = 0$ the oscillator has its maximum speed). In the quantum mechanical treatment for $n = 0$, however, we obtain a maximum at this point; this means that the classical model is not correct at low energies. On the other hand, the curve for $n = 50$ corresponds quite well with the quantum mechanical calculation. This is in accordance with Bohr's correspondence principle discussed in Section 3.8.

9.4 Vibrational–Rotational Spectra

9.4.1 Diatomic Molecules

As we have shown in Section 9.2, many thermally excited rotational energy states are occupied at room temperature (for details see Chapter 11). The energy needed to excite the molecule into a higher vibrational state is much larger than for the rotational states. Therefore, at room temperature an HCl molecule is usually in the lowest vibrational state. By irradiation with light in the near IR it can be excited (see Box 9.5) from the ground state to the next vibrational state); the excitation energy for this process is $h\nu_0$. At the same time, however, its rotational energy changes according to the selection rules discussed in Section 9.2 (Figure 9.16): the quantum number of rotational energy either increases by 1

9.4 VIBRATIONAL–ROTATIONAL SPECTRA

or decreases by 1. Because of equation (9.11), in the first case we obtain

$$\Delta E = h\nu_0 + 2B, \quad h\nu_0 + 4B, \quad h\nu_0 + 6B, \ldots \quad (9.33)$$

In the second case we have

$$\Delta E = h\nu_0 - 2B, \quad h\nu_0 - 4B, \quad h\nu_0 - 6B, \ldots \quad (9.34)$$

From this consideration we expect a system of equally spaced absorption lines around the wavenumber $\tilde{\nu} = \nu_0/c_0$. According to Table 9.1 (see Section 9.2), the lines should be spaced by $41\,\text{cm}^{-1}$. This is what we see in the absorption spectrum (Figure 9.17). The center of both line systems (indicated by the dashed line) corresponds to the energy ΔE for the pure vibrational transition which is forbidden, as mentioned above.

As in the case of the rotational spectrum, the strengths of the absorption lines reflect the population of the energy levels at room temperature.

Box 9.5 – Selection Rules for Vibration

Diatomic molecules

To calculate the transition moment we proceed as discussed in Box 9.1, but insert the oscillator wavefunctions. From the symmetry of the oscillator wavefunctions it follows that transitions $n \to (n+1)$ are allowed and that transitions $n \to (n+2)$ are forbidden. A quantitative treatment reveals that the $n \to (n+1)$ transition is the only one which is allowed. As in the case of

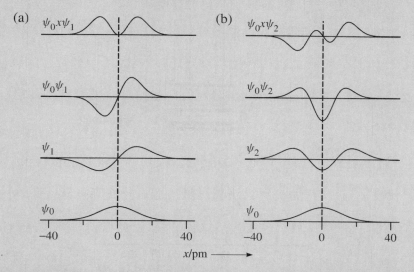

Figure 9.15 How to calculate the transition moment M_x for vibrational transitions. (a) Transition $n = 0 \to n = 1$. (b) Transition $n = 0 \to n = 2$.

Continued on page 296

> *Continued from page 295*
>
> the rotator, transitions are only allowed for a nonzero value of the permanent dipole moment.
>
> ---
>
> *Example*
>
> We consider the transition $0 \to 1$. $\psi_0 \psi_1$ is antisymmetric with respect to the x direction, then $\psi_0 \psi_1 \cdot x$ is symmetric and $M_{0\to 1}$ is nonzero (Figure 9.15). On the other hand, for the transition $0 \to 2$ the product $\psi_0 \psi_1$ is symmetric and we obtain $M_{0\to 2} = 0$.

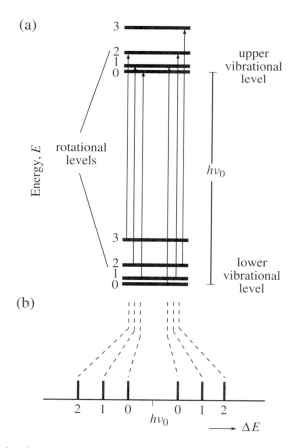

Figure 9.16 Vibrational and rotational motion of a diatomic molecule. (a) Energy scheme and transitions that can be excited by irradiating with IR light. (b) Expected absorption lines in the IR spectrum.

9.4 VIBRATIONAL–ROTATIONAL SPECTRA

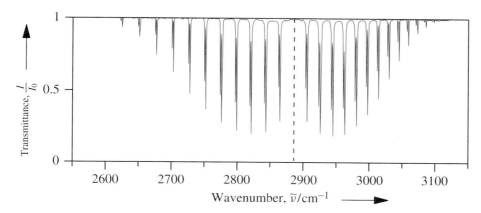

Figure 9.17 IR spectrum of HCl gas. Transmittance $T = I/I_0$ versus wavenumber $\tilde{\nu} = \nu/c_0$. The line system on the left of the dashed line is called the P branch; the line system on the right is called the R branch. Optical path length 10 cm, temperature 25 °C, HCl pressure 65 mbar (D. Schiöberg, Marburg, private communication).

Example

From the wavenumber $\tilde{\nu} = 2885\,\text{cm}^{-1}$ at the center of the line system in the HCl spectrum (Figure 9.17) we calculate $\Delta E = h\nu_0 = hc_0\tilde{\nu} = 0.5731 \times 10^{-19}\,\text{J}$ and $\nu_0 = c_0\tilde{\nu} = 8.65 \times 10^{13}\,\text{Hz}$. ΔE is much larger than that for the excitation of the first rotational level ($0.0041 \times 10^{-19}\,\text{J}$, see Table 9.1). From the frequency ν_0 we calculate the force constant k_f using equation (9.21):

$$\mu_{\text{HCl}} = \frac{m_\text{H} \cdot m_{\text{Cl}}}{m_\text{H} + m_{\text{Cl}}} = 0.972 \cdot m_\text{H} = 1.627 \times 10^{-27}\,\text{kg}$$

$$k_\text{f} = 4\pi^2 \nu_0^2 \mu_{\text{HCl}} = 4\pi^2 \cdot (8.65 \times 10^{13}\,\text{Hz})^2 \cdot 1.627 \times 10^{-27}\,\text{kg} = 480\,\text{N m}^{-1}$$

Taking into account that the vibrations of HCl are not strictly harmonic, we obtain the value $k_\text{f} = 516\,\text{N m}^{-1}$ (see Table 5.5).

The absorption lines in Figure 9.17 are not exactly spaced by $41\,\text{cm}^{-1}$ as expected for an harmonic oscillator; the distance between adjacent lines is larger than $41\,\text{cm}^{-1}$ at small wavenumbers and smaller at large wavenumbers. This is a result of the deviation of the potential curve of HCl from that of a harmonic oscillator (see Section 5.2). Furthermore, each line is accompanied by a weaker line; the stronger lines belong to the ^{35}Cl isotope and the weaker ones belong to ^{37}Cl (see Problem 9.4).

9.4.2 Polyatomic Molecules

Number of vibrational modes

In polyatomic molecules, different types of vibrations are possible (*vibrational modes*). A molecule with N_atom atoms possesses $3N_\text{atom}$ degrees of freedom for the motions of the atoms in space. Out of these are 3 degrees of freedom for the translation. If the molecule is linear, it has 2 degrees of freedom for the rotation corresponding to two axes of rotation. Then the number of vibrational modes is

$$3N_{atom} - 5 \quad \text{(linear molecule)} \tag{9.35}$$

A nonlinear polyatomic molecule has three axes of rotation, then we obtain

$$3N_{atom} - 6 \quad \text{(nonlinear molecule)} \tag{9.36}$$

vibrational modes.

Stretching and bending modes

We consider CO_2 as an example for a linear polyatomic molecule (Figure 9.18). In this case we have $3 \times 3 - 5 = 4$ vibration modes. These vibration modes are shown in Figure 9.18. They are twofold degenerate bending modes (xy and xz plane (Figure 9.18(a) and (b)), symmetric stretching mode (Figure 9.18(c)), and antisymmetric stretching mode (Figure 9.18(d)). Each of these vibrational modes occurs with a different frequency (Box 9.6):

The total vibration energy of CO_2 is the sum of the contributions of the four vibrational modes:

$$\begin{aligned} E = &\, h\nu_{0,\text{bending}} \cdot \left(\tfrac{1}{2} + n_{\text{bending},1} + \tfrac{1}{2} + n_{\text{bending},2}\right) \\ &+ h\nu_{0,\text{sym}} \cdot \left(\tfrac{1}{2} + n_{\text{sym}}\right) + h\nu_{0,\text{anti}} \cdot \left(\tfrac{1}{2} + n_{\text{anti}}\right) \end{aligned} \tag{9.37}$$

The quantum numbers n_{sym} and n_{anti} correspond to the symmetric and antisymmetric stretching modes, respectively; the bending mode is twofold degenerate (Figure 9.18) and therefore has to be characterized by two quantum numbers ($n_{\text{bending},1}$ and $n_{\text{bending},2}$). Figure 9.19 shows the energy levels corresponding to the different vibration modes. At low enough temperature the lowest energy levels are occupied. IR radiation excites the CO_2 molecules to higher energies (arrows in Figure 9.19).

Selection rules

Transitions in the IR are only possible if the vibration of the molecule is connected with a periodically changing dipole moment. For diatomic molecules this means that they must have a permanent dipole moment (like HCl and CO); in polyatomic molecules, however, a dipole moment can occur in certain vibration modes, even if the permanent dipole moment of the molecule is zero.

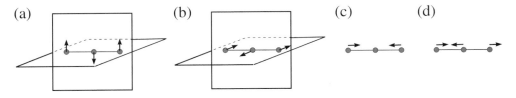

Figure 9.18 Vibrational modes of a CO_2 molecule. (a) and (b) Bending mode; this vibration mode can occur in two perpendicular planes and is therefore twofold degenerate. (c) Symmetric stretching mode. (d) Antisymmetric stretching mode.

9.4 VIBRATIONAL–ROTATIONAL SPECTRA

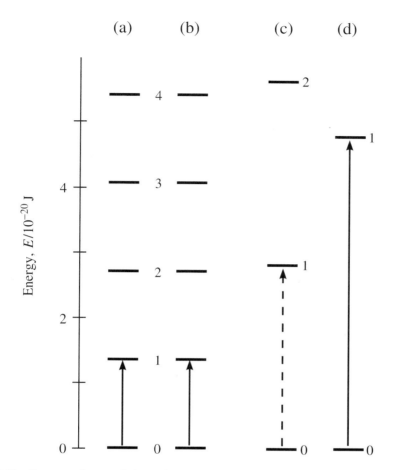

Figure 9.19 Energy scheme of the different vibrational modes of a CO_2 molecule. (a) and (b) Bending modes. (c) Symmetric stretching mode. (d) Antisymmetric stretching mode. The illustration is based on the vibrational frequencies: $\nu_{0,\text{bending}} = 0.20 \times 10^{14}$ Hz, $\nu_{0,\text{sym}} = 0.42 \times 10^{14}$ Hz, and $\nu_{0,\text{anti}} = 0.71 \times 10^{14}$ Hz. These values are obtained from the vibrational spectrum of CO_2. Solid arrows: transitions from the lowest quantum state that can be observed in the IR spectrum; dashed line: this transition does not appear in the IR spectrum but appears, instead, in the Raman spectrum.

Example – let us discuss the CO_2 molecule

CO_2 is a linear molecule, and because of its symmetry the permanent dipole moment is zero. However, a periodically changing dipole moment occurs in the bending mode and in the antisymmetric stretching mode (Figure 9.20(a)). In the symmetric stretching mode the dipole moment is always zero, and thus the corresponding transition cannot be observed in the IR spectrum but instead in the Raman spectrum (Section 9.5). The IR spectrum of CO_2 is displayed in Figure 9.20(b); force constants can be derived from the main absorption bands, as shown in Box 9.6.

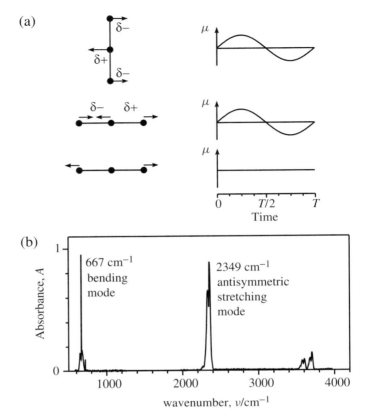

Figure 9.20 IR absorption of CO_2. (a) Dipole moment μ of a CO_2 molecule during vibration. *Top*: Bending mode (Direction of μ perpendicular to the molecular axis). *Center*: Antisymmetric stretching mode. *Bottom*: Symmetric stretching mode. (b) IR spectrum. The absorbance $A = -\log I/I_0$ is plotted versus the wavenumber $\tilde{\nu} = \nu/c_0$. The bending mode and the antisymmetric stretching mode give rise to strong absorption bands ($n_{\text{bending}}: 0 \to 1$, $n_{\text{anti}}: 0 \to 1$). Additionally, three weak absorption bands occur, which correspond to transitions where different quantum numbers change simultaneously (for more details see Box 9.6):

No.	$\tilde{\nu}/\text{cm}^{-1}$	initial state n_{bending}	n_{sym}	n_{anti}		final state n_{bending}	n_{sym}	n_{anti}
1	720	1	0	0	→	0	1	0
2	3609	0	0	0	→	2	0	1
3	3716	0	0	0	→	0	1	1

Optical path length 10 cm, temperature 25 °C, CO_2 pressure 100 mbar.

Box 9.6 – Oscillator: Reduced Mass and Force Constants of Polyatomic Molecules

We restrict our discussion to CO_2 as an example of a triatomic linear molecule.

Continued on page 301

Continued from page 300

Symmetric stretching mode (Figure 9.21(b))

The carbon atom is at the center of gravity and both oxygen atoms are moving with the same amplitude, but in opposite directions: $x_2 = 0$, $x_1 = -x_3$. Then

$$m_O \cdot \frac{d^2 x_1}{dt^2} = -k_f \cdot x_1$$

and

$$\nu_{0,\text{sym}} = \frac{1}{2\pi} \cdot \sqrt{\frac{k_f}{\mu_{\text{sym}}}}, \quad \mu_{\text{sym}} = m_O$$

Antisymmetric stretching mode (Figure 9.21(c))

In this case the oxygen atoms are again moving with the same amplitude, but in the same direction: $x_1 = x_3$. In order to keep the center of gravity at the same place, the carbon atom moves in the opposite direction

$$m_C \cdot x_2 = -2 \cdot m_O \cdot x_1, \quad x_2 = -\frac{2 m_O}{m_C} \cdot x_1$$

Then

$$m_O \cdot \frac{d^2 x_1}{dt^2} = -k_f \cdot (x_1 - x_2) = -k_f \cdot \left(1 + \frac{2 m_O}{m_C}\right) \cdot x_1$$

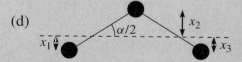

Figure 9.21 Linear triatomic molecule (carbon dioxide). How to calculate the moment of inertia from the displacements x_1, x_2, and x_3 (a). Symmetric (b) and antisymmetric (c) stretching modes; bending mode (d).

Continued on page 302

Continued from page 301

and
$$\nu_{0,\text{anti}} = \frac{1}{2\pi} \cdot \sqrt{\frac{k_f}{\mu_{\text{anti}}}}, \quad \mu_{\text{anti}} = \frac{m_C m_O}{m_C + 2m_O}$$

Bending mode (Figure 9.21(d))

Both oxygen atoms are moving in the same direction, but opposite to the direction of the carbon atom. During vibration, the bond length d is constant, and the bond angle α is changing periodically. Then for conservation of the center of gravity we obtain

$$2x_1 \cdot m_O = x_2 \cdot m_C \quad \text{or} \quad \frac{x_1}{x_2} = \frac{m_C}{2m_O}$$

Furthermore,
$$x_1 + x_2 = d \sin\frac{\alpha}{2} \approx \frac{\alpha}{2} \cdot d$$

Thus
$$x_1 = d \cdot \frac{m_C}{2(2m_O + m_C)} \cdot \alpha$$

The force acting on an oxygen atom is
$$m_O \cdot \frac{d^2 x_1}{dt^2} = -k_f' \cdot d \cdot \alpha$$

where k_f' is the bending force constant, see Chapter 5. Thus
$$\frac{d^2 \alpha}{dt^2} = -\frac{k_f'}{\mu_{\text{bending}}} \cdot \alpha$$

and
$$\nu_{0,\text{bending}} = \frac{1}{2\pi} \cdot \sqrt{\frac{k_f'}{\mu_{\text{bending}}}}, \quad \text{with} \quad \mu_{\text{bending}} = \frac{m_C m_O}{2(2m_O + m_C)}$$

Force constants

For k_f we obtain:
$$k_f = \mu(2\pi\nu_0)^2 = 4\pi^2 c^2 \cdot \mu \cdot \widetilde{\nu}^2$$

Antisymmetric stretching mode: $\mu = 7.25 \times 10^{-27}$ kg, $\widetilde{\nu} = 2349 \text{ cm}^{-1}$ (Figure 9.20(b)); thus $k_f = 1420 \text{ N m}^{-1}$.

Symmetric stretching mode: $\mu = 26.6 \times 10^{-27}$ kg, $\widetilde{\nu} = 1388 \text{ cm}^{-1}$ (Figure 9.31); thus $k_f = 1820 \text{ N m}^{-1}$.

Bending mode: $\mu = 3.62 \times 10^{-27}$ kg, $\widetilde{\nu} = 667 \text{ cm}^{-1}$ (Figure 9.20(b)); thus $k_f' = 57 \text{ N m}^{-1}$.

Continued on page 303

> Continued from page 302
>
> We see that k_f' is much smaller than the force constant k_f for the stretching modes: the deformation of bond angles is much easier than the deformation of bond lengths.
>
> It seems strange that the force constant k_f for the symmetric stretching mode is larger by 20% compared with that for the antisymmetric stretching mode. This discrepancy is due to the simple spring model used to describe bond length deformations (*harmonic oscillator*). Actually, when lengthening one bond in the CO_2 molecule, the electron density in the second bond will be affected and this results in a change of the force constant attributed to this bond. Therefore, only a rough value of the force constant in CO_2 can be estimated ($k_f \approx 1500\,\text{N\,m}^{-1}$, see Table 5.5).
>
> For a quantitative treatment of vibrations of molecules, nonlinear terms must be added in the expression (equation (9.37)) for the vibrational energy (correction for *anharmonicity*). Furthermore, in polyatomic molecules coupling can occur between different vibration modes, in a similar way as coupling of electronic transitions occurs (see Chapter 8): this type of coupling is called *Fermi resonance*. These are the reasons why the Raman line corresponding to the bending mode of CO_2 is found at $1285\,\text{cm}^{-1}$ (Figure 9.31) instead of $2 \times 667\,\text{cm}^{-1} = 1334\,\text{cm}^{-1}$, as expected from the harmonic oscillator model. The same is true for the weak transitions observed in Figures 9.20(b) and 9.31.

9.5 Raman Spectra

In a Raman spectrophotometer the molecules are irradiated with light in the visible, usually with the green light of an argon ion laser ($\lambda = 514$ nm). The laser light is scattered by the molecules and the scattered light is observed perpendicular to the direction of the laser beam after passing a monochromator (Figure 9.22).

The Raman effect gives similar information on molecular parameters as IR spectra (bond lengths, force constants, geometry of molecules), but simply visible light is used in the experiment. Furthermore, the selection rules are different from those in an IR spectrum: transitions which are inactive in the IR spectrum can be active in the Raman spectrum.

To understand the effect, we first consider light scattering using a classical picture.

9.5.1 Rayleigh Scattering

In the electric field of the incident light, a dipole moment μ_{ind} is induced in the molecule:

$$\mu_{\text{ind}} = \alpha \cdot F_0 \cdot \cos \omega_{\text{laser}} t \tag{9.38}$$

where α is the polarizability of the molecule and $\nu_{\text{laser}} = \omega_{\text{laser}}/(2\pi)$ is the frequency of the laser light (Figure 9.23).

The induced dipole oscillating with the frequency ν_{laser} can be viewed as a radiating antenna emitting light in different directions: *Rayleigh scattering*. The induced dipole moment μ_{ind} depends on the polarizability α of the molecule. This leads to an emission of light of circular frequency ω_{laser}. In the Hertz equation (see Foundation 8.5) we replace

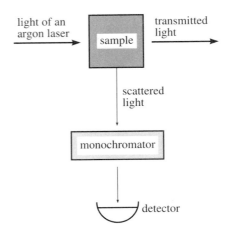

Figure 9.22 Raman effect: experimental setup.

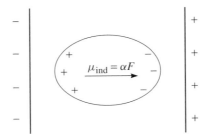

Figure 9.23 Induced dipole moment $\mu_{\text{ind}} = \alpha \cdot F$ of a molecule in an applied electric field of field strength F where α is the polarizability of the molecule.

$Q\xi_0$ by αF_0. Because of $F_0^2 = 2I/(\varepsilon_0 c_0)$, where I is the intensity of the incident light wave of wavelength $\lambda = 2\pi c_0/\omega_{\text{laser}}$ we obtain

$$\text{Power of scattered light} = \overline{\left(\frac{dE}{dt}\right)} = \frac{1}{4\pi\varepsilon_0} \frac{\alpha^2 F_0^2 \omega_{\text{laser}}^4}{3c_0^3}$$

$$= \left(\frac{1}{4\pi\varepsilon_0}\right)^2 \cdot \frac{8\pi\alpha^2}{3} \left(\frac{2\pi}{\lambda}\right)^4 \cdot I$$

Note the difference between Rayleigh scattering (molecule emitting light due its polarized atoms in the field of the incident light) and fluorescence (molecule emitting light by being in the excited state). Rayleigh scattering is discussed in more detail in Chapter 22.

Spectral satellite lines are observed on both sides of the Rayleigh line. In the case of a rotational Raman spectrum these lines are very close (514 ± 1 nm) to the Rayleigh line at 514 nm. They are due to the fact that the molecules can take up rotational energy from the incident light, or they can release rotational energy to the scattered light. Then the frequency of the scattered light can be slightly smaller or larger than the frequency of the incident light.

In the case of a vibrational–rotational Raman spectrum the molecules take up or release vibrational energy, additionally to the rotational energy. Then the satellite lines appear about ± 20 nm apart from the Rayleigh line.

9.5.2 Rotational Raman Spectra

The polarizability α of a molecule depends on its orientation to the electric vector of the incident light beam. μ_{ind} is large if the long axis of the molecule is parallel to the electric field of the laser beam, and it is small for the short axis. Then μ_{ind} becomes small and large two times during a rotation, and therefore α is changing with a frequency twice the rotation frequency $\nu_{\text{rot}} = \omega_{\text{rot}}/(2\pi)$ (Figure 9.24).

$$\alpha = \alpha_0 + \Delta\alpha \cdot \cos 2\omega_{\text{rot}} t \tag{9.39}$$

Then the induced dipole moment is

$$\mu_{\text{ind}} = (\alpha_0 + \Delta\alpha \cdot \cos 2\omega_{\text{rot}} t) \cdot F_0 \cdot \cos \omega_{\text{laser}} t$$
$$= \alpha_0 \cdot F_0 \cdot \cos \omega_{\text{laser}} t + \Delta\alpha \cdot F_0 \cdot \cos 2\omega_{\text{rot}} t) \cdot \cos \omega_{\text{laser}} t$$

This expression can be written as (see mathematical appendix):

$$\mu_{\text{ind}} = \alpha_0 \cdot F_0 \cdot \cos \omega_{\text{laser}} t$$
$$+ \tfrac{1}{2}\Delta\alpha \cdot F_0 \cdot [\cos(\omega_{\text{laser}} + 2\omega_{\text{rot}})t + \cos(\omega_{\text{laser}} - 2\omega_{\text{rot}})t] \tag{9.40}$$

This means that a molecule rotating with frequency ν_{rot} is emitting light of frequency ν_{laser}, $\nu_{\text{laser}} + 2\nu_{\text{rot}}$, and $\nu_{\text{laser}} - 2\nu_{\text{rot}}$. The rotational energy at room temperature is about

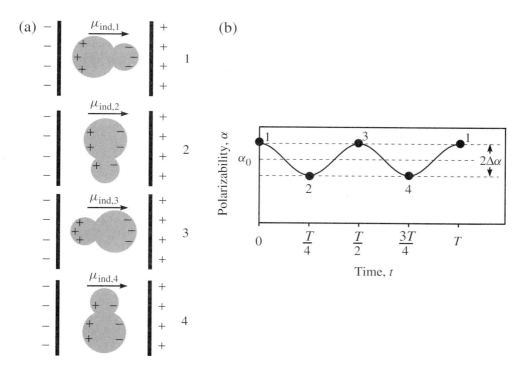

Figure 9.24 Change of the polarizability α for a rotating molecule. (a) Four stages of rotation inside a capacitor replacing the x component of the electric field strength $F_0 \cos \omega_{\text{laser}} t$ of the exciting laser beam. (b) Polarizability α versus time t during one rotation cycle (period T): $\alpha = \alpha_0 + \Delta\alpha \cos 2\omega_{\text{rot}} t$, with $\alpha_0 = \tfrac{1}{2}(\alpha_1 + \alpha_2)$, $\Delta\alpha = \tfrac{1}{2}(\alpha_1 - \alpha_2)$, and $\omega_{\text{rot}} = 2\pi/T$.

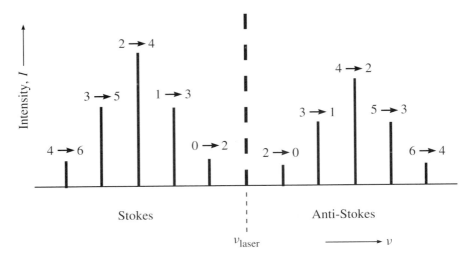

Figure 9.25 Intensity I expected in a Raman experiment with rotational excitation of diatomic molecules. ν_{laser}: Rayleigh line; the lines at lower frequencies $\nu_{\text{laser}} - 2\nu_{\text{rot}}$ relative to the Rayleigh line correspond to a change in quantum numbers $0 \to 2$, $1 \to 3$, ... (Stokes lines). The transitions $\nu_{\text{laser}} + 2\nu_{\text{rot}}$ correspond to a change in quantum numbers $2 \to 0$, $3 \to 1$, ... (anti-Stokes lines).

4×10^{-21} J, and this corresponds to a rotational frequency of about $\nu_{\text{rot}} = 10^{12}$ Hz for CO. So there should be satellite bands with maxima at $\nu_{\text{laser}} \pm 2 \times 10^{12}$ Hz or $\tilde{\nu}_{\text{laser}} \pm 60\,\text{cm}^{-1}$. Indeed, these bands are observed. They consist of lines at equal distances; this is due to the fact that the rotational energy of a molecule is restricted to discrete energy states, according to equation (9.1). The transitions $\nu_0 - 2\nu_{\text{rot}}$ correspond to the case where the rotational quantum number increases by two:

$$\Delta E_{n \to n+2} = B \cdot [(n+2)(n+3) - n(n+1)] \tag{9.41}$$

$$= 2B \cdot (2n+3), \quad n = 0, 1, 2, \ldots$$

Such transitions are shown in Figure 9.25. The lines at lower frequencies relative to the Rayleigh line correspond to a change of quantum numbers $0 \to 2$, $1 \to 3$, ... (*Stokes lines*); the lines at higher frequencies relative to the Rayleigh line correspond to changes $2 \to 0$, $3 \to 1$, ... (*anti-Stokes lines*). The spacing between two adjacent lines

$$\Delta(\Delta E) = \Delta E_{n+1 \to n+3} - \Delta E_{n \to n+2}$$

$$= 4B = \frac{h^2}{2\pi^2 \mu d^2} \tag{9.42}$$

is just twice the spacing of the lines in the IR spectrum, as can be seen by comparison with equation (9.12). As in the case of the IR spectrum, the strengths of the Raman lines reflect the population of the rotational energy levels at room temperature.

As an example, the rotational Raman spectrum of CO is shown in Figure 9.26.

9.5 RAMAN SPECTRA

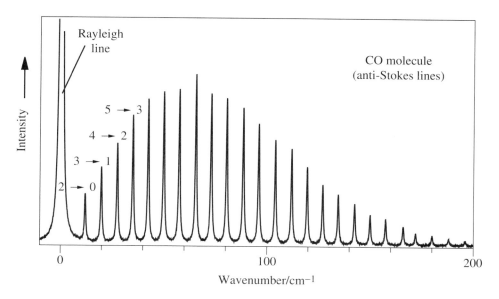

Figure 9.26 Rotational Raman spectrum of CO (anti-Stokes lines). Intensity I versus wavenumber $(\tilde{\nu} - \tilde{\nu}_{\text{laser}}) = (\nu - \nu_{\text{laser}})/c_0$. (P. Dechant, Marburg, private communication).

Example – bond length of CO from Raman spectrum

The spacing between two adjacent rotational Raman lines in Figure 9.26 is $\tilde{\nu} = 7.75\,\text{cm}^{-1}$. This corresponds to

$$\Delta(\Delta E) = \frac{hc_0}{\lambda} = hc_0\tilde{\nu} = 1.54 \times 10^{-22}\,\text{J}$$

Then it follows from equation (9.42) with

$$\mu = \frac{m_C m_O}{m_C + m_O} = 1.14 \times 10^{-26}\,\text{kg}$$

$$d = \sqrt{\frac{h^2}{2\pi^2 \mu} \cdot \frac{1}{\Delta(\Delta E)}} = 113\,\text{pm}$$

9.5.3 Vibrational–Rotational Raman Spectra of Diatomic Molecules

In the classical picture, during the vibration of a diatomic molecule, the polarizability α becomes periodically small and large with the same frequency as the vibration (Figure 9.27). Then

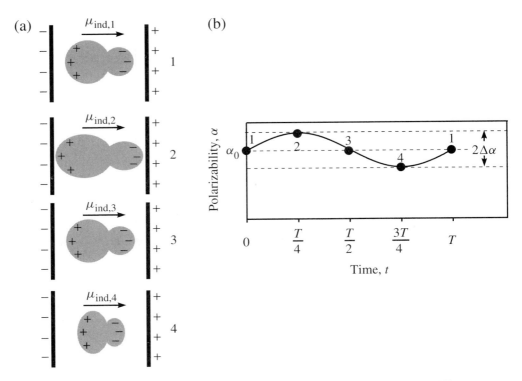

Figure 9.27 Polarizability α for vibrational motion of a diatomic molecule. (a) Different stages of vibration. (b) Polarizability α versus time t during one vibration cycle (period T).

$$\mu_{\text{ind}} = (\alpha_0 + \Delta\alpha \cdot \cos\omega_{\text{vib}}t) \cdot F_0 \cdot \cos\omega_{\text{laser}}t \tag{9.43}$$

and a similar analysis as in equation (9.40) shows that the frequencies of the light emitted by the dipole are ν_{laser}, $\nu_{\text{laser}} + \nu_{\text{vib}}$, and $\nu_{\text{laser}} - \nu_{\text{vib}}$. Thus the selection rules for vibrational Raman transitions of diatomic molecules are the same as for the corresponding vibrational transitions (Section 9.4), in contrast to the rotational Raman transitions.

As in the case of the IR spectrum, the vibrational transitions are accompanied by rotational transitions. The quantum mechanical treatment reveals that in a vibrational transition the rotational quantum number can change by $+2$, -2, or it can remain unchanged. This is shown in Figure 9.28. Because all of the transitions where the rotational quantum number remains unchanged appear at the same place in the Raman spectrum we expect a strong line corresponding to the pure vibrational transition and many weak lines at both sides of the strong line (Figure 9.28).

The vibrational–rotational Raman spectrum of CO is shown in Figure 9.29. The lines correspond to transitions from the vibrational ground state (Stokes lines). Because of the large energy difference between vibrational ground state and the first vibrational excited state, the excited state is only slightly occupied at room temperature. Therefore, the anti-Stokes lines corresponding to the vibrational transition (expected at $\tilde{\nu} - \tilde{\nu}_{\text{laser}} = -2142\,\text{cm}^{-1}$) are extremely weak.

9.5 RAMAN SPECTRA

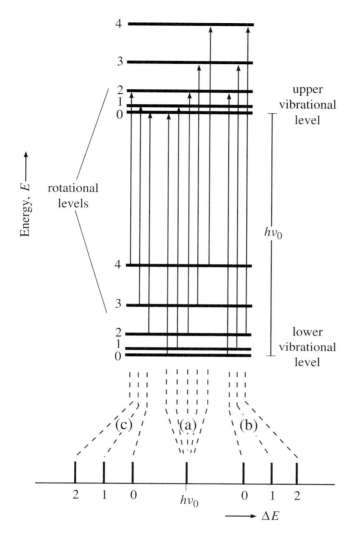

Figure 9.28 Transitions in a vibrational–rotational Raman spectrum. $h\nu_0$ is the energy for the pure vibrational transition. (a) No change of the rotational quantum number. (b) and (c) Rotation quantum number changing by $+2$ and -2, respectively.

Example – force constant of CO from Raman spectrum

According to Figure 9.29 the pure vibrational transition of CO is at $\tilde{\nu} = 2142 \text{ cm}^{-1}$. This corresponds to $\Delta E = 4.25 \times 10^{-20}$ J. Then it follows with $\mu = 1.14 \times 10^{-26}$ kg

$$k_\text{f} = 4\pi^2 \mu \cdot \left(\frac{\Delta E}{h}\right)^2 = 1850 \text{ N m}^{-1}$$

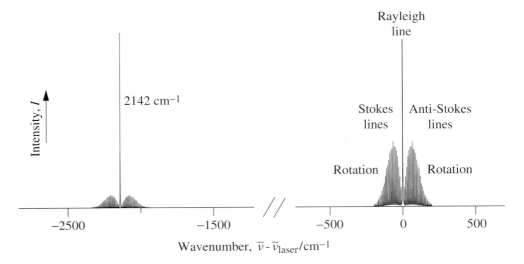

Figure 9.29 Complete Raman spectrum of CO. Intensity I versus wavenumber $\tilde{\nu} - \tilde{\nu}_{laser} = (\nu - \nu_{laser})/c_0$. (P. Dechant, Marburg, private communication) *Right*: Rotational lines (Stokes and anti-Stokes lines, see Figure 9.26 for better spectral resolution). *Left*: Vibrational–rotational lines. Note the strong line corresponding to the transitions where the rotational quantum numbers do not change.

9.5.4 Raman Spectra of Polyatomics

In the following we discuss CO_2 as an example of a polyatomic molecule. In Figure 9.30 the polarizability α during the vibration of CO_2 is displayed for the three different vibration modes. In Figure 9.30(a) α is changing with twice the vibration frequency for the bending mode: a Raman line corresponding to twice the vibration frequency is expected. In Figure 9.30(b) α is changing periodically with the same frequency as the vibration for the symmetric stretching mode: a Raman line corresponding to the vibration frequency is expected. In Figure 9.30(c) α is constant for the antisymmetric stretching mode: no Raman line is expected. Consequently, we find two strong lines in the Raman spectrum of CO_2 (Figure 9.31).

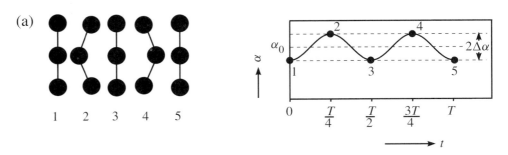

Figure 9.30 Vibrational modes for CO_2: Raman activity. Different stages of vibration (*left*) and corresponding polarizability α versus time t (*right*) during one vibration cycle. (a) Bending mode. (b) Symmetric stretching mode. (c) Antisymmetric stretching mode. The bending, compression, and expansion of atoms is strongly exaggerated.

9.5 RAMAN SPECTRA

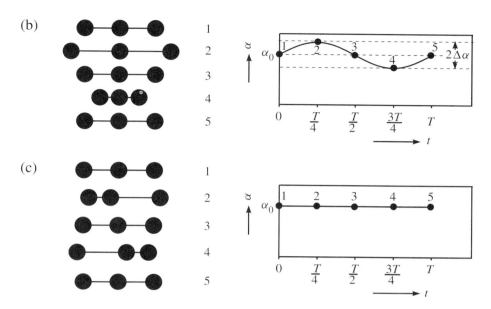

Figure 9.30 (*continued*).

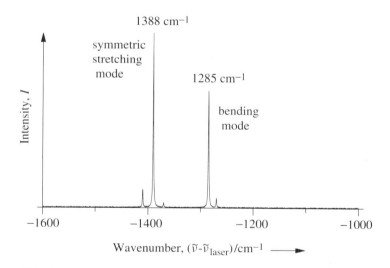

Figure 9.31 Vibrational Raman spectrum for CO_2. Intensity I versus wavenumber $(\tilde{\nu} - \tilde{\nu}_{laser}) = (\nu - \nu_{laser})/c_0$. (P. Dechant, Marburg, private communication) The strong Raman lines correspond to an excitation of the bending mode ($n_{bending}$: $0 \to 2$) and of the symmetric stretching mode (n_{sym}: $0 \to 1$). Additionally, three weak absorption bands occur, which correspond to the following transitions (for more details see Box 9.6).

No.	$(\tilde{\nu} - \tilde{\nu}_{laser})/\text{cm}^{-1}$	initial state				final state		
		$n_{bending}$	n_{sym}	n_{anti}		$n_{bending}$	n_{sym}	n_{anti}
1	1264	1	0	0	$\to$	3	0	0
2	1369	0	1	0	$\to$	2	1	0
3	1409	1	0	0	$\to$	1	1	0

9.6 Vibrational Structure of Electronic Spectra

9.6.1 Diatomic Molecules

In electronically excited molecules the equilibrium bond length d^* is different from the bond length d in the ground state due to the different electron distributions. Since the electronic motion is fast compared to the motion of the nuclei, the molecule, immediately after excitation, is in the geometry of the ground state.

As an example, we consider the diatomic molecule in Figure 9.32 where the nucleus on the left-hand side has a much larger mass than the nucleus on the right-hand side. Then the light nucleus in Figure 9.32 (mass m_r), in the classical approach, starts to oscillate with amplitude $d^* - d$. The vibration frequency ν_0 and the vibrational energy E^* are

$$\nu_0 = \frac{1}{2\pi}\sqrt{\frac{k_f}{m_r}} \quad \text{and} \quad E^* = \tfrac{1}{2}k_f \cdot (d^* - d)^2 \tag{9.44}$$

where k_f is the stretching force constant.

In this classical picture the molecule vibrates with the amplitude $d^* - d$ (vibrational energy E^*): *Franck–Condon Principle*. According to quantum mechanics any vibrational energy

$$E_n^* = h\nu_0 \cdot \left(\tfrac{1}{2} + n\right) \tag{9.45}$$

can be taken up in the electronic excitation, but the quantum state with energy E_n^* closest to E^* is reached with highest probability ($n = 4$ in the case of Figure 9.32); but also, to a less extent, states with smaller and larger values of n: the electronic absorption band has a vibrational structure as indicated in Figure 9.32. A more detailed approach is discussed in Foundation 9.2.

9.6.2 Photoelectron Spectroscopy

An interesting version of absorption spectroscopy is the ionization of molecules by UV radiation of distinct energy followed by the measurement of the kinetic energy of the emitted electrons (*photoelectron spectroscopy*). The kinetic energy E_{kin} is measured by deflecting the electrons in a magnetic or electric field (Figure 9.33).

We consider a gas of H_2 molecules. The H_2 molecules are excited by using an He discharge tube producing radiation of 58.4 nm wavelength (corresponding to an energy of 21.2 eV), and H_2^+ ions are generated. From the measured kinetic energy E_{kin} we obtain the ionization energy E_{ion} as $E_{ion} = 21.2\,\text{eV} - E_{kin}$. In Figure 9.34 the counting rate at the detector is shown as a function of E_{ion} (photoelectron spectrum).

In this ionization process H_2^+ is produced in various vibrational states ($n_{vib} = 0, 1, 2, \ldots$):

$$H_2(n_{vib} = 0) \xrightarrow{h\nu = 21.2\,\text{eV}} H_2^+(n_{vib}) + e^-(E_{kin})$$

9.6 VIBRATIONAL STRUCTURE OF ELECTRONIC SPECTRA

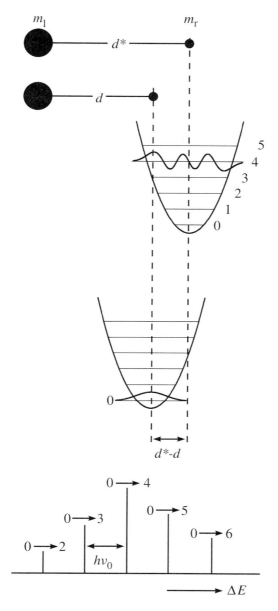

Figure 9.32 Vibrational structure of an electronic transition in a diatomic molecule according to the Franck–Condon principle. HCl is used as an example, where the Cl nucleus (large mass m_l) is practically at rest and only the H nucleus (small mass m_r) is vibrating. (a) Potential curves in the electronic ground state (equilibrium distance d, center) and in the electronically excited state (equilibrium distance d^*, top). The transition from vibrational quantum number 0 in the electronic ground state to the excited state with quantum number 4 has maximum probability because of the large overlap of the vibrational wave functions. Bottom: Corresponding intensity distribution of the absorption band.

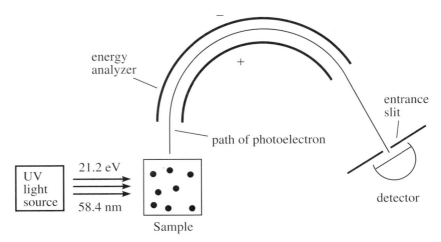

Figure 9.33 Experimental setup of a photoelectron spectrophotometer. The sample molecules are irradiated with UV light of 58.4 nm and photoelectrons are ejected from the molecules. The kinetic energy of the photoelectrons is the difference of the excitation energy and the energy required to transfer a molecule into the corresponding ion. Using a charged capacitor and an entrance slit in front of the detector photoelectrons in a small energy range are selected.

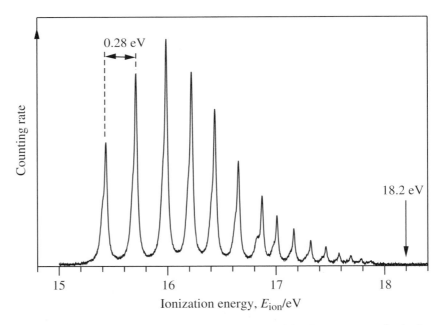

Figure 9.34 Photoelectron spectrum of H_2. Counting rate of the detector versus ionization energy E_{ion} (Redrawn from N. Knöpfel, Th. Olbricht, A. Schweig in B. Schröder, J. Rudolph (Eds.) Physikalische Methoden in der Chemie, VCH Weinheim 1985, p. 205.). The arrow denotes the dissociation limit of the H_2^+ ion at 18.2 eV. The energy difference of 0.28 eV between the first and second band corresponds to change of vibration energy of H_2^+ for the $0 \to 1$ transition.

9.6 VIBRATIONAL STRUCTURE OF ELECTRONIC SPECTRA

In the case $H_2^+(n_{vib} = 0)$ the electron arrives at the detector with the largest energy ($E_{kin} = 5.77$ eV) corresponding to the $0 \to 0$ transition with an energy difference

$$\Delta E_{0 \to 0} = (21.2 - 5.77)\,\text{eV} = 15.43\,\text{eV}$$

$\Delta E_{0 \to 0}$ is the difference between the energy of $H_2(n_{vib} = 0)$ and $H_2^+(n_{vib} = 0)$. This energy is given by the difference between the energy for the process

$$H_2 \longrightarrow H + H^+ + e^-$$

and the energy for the process

$$H_2^+ \longrightarrow H + H^+$$

$$\Delta E_{0 \to 0} = (4.5 + 13.6)\,\text{eV} - 2.6\,\text{eV}$$
$$= 15.5\,\text{eV} \qquad (9.46)$$

where 4.5 eV is the bond energy of H_2, 13.6 eV is the ionization energy of H, and 2.6 eV is the bond energy of H_2^+ (Figure 9.35).

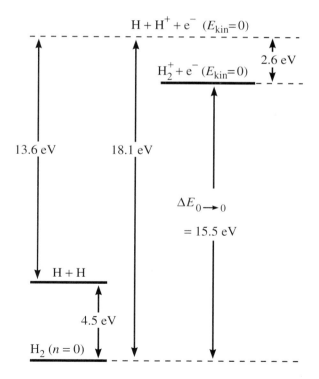

Figure 9.35 Energy scheme for H_2 and H_2^+ (photoionization of H_2); calculation of $\Delta E_{0 \to 0}$ in equation (9.46). The bond energies of H_2 and H_2^+ are 7.2×10^{-19} J = 4.5 eV and 4.2×10^{-19} J = 2.6 eV, respectively (see Chapter 2).

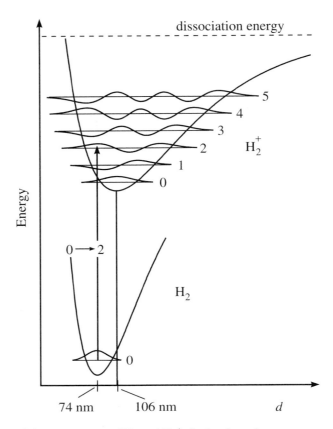

Figure 9.36 Potential energy curves of H_2 and H_2^+. In the photoelectron spectroscopy experiment transitions from the lowest vibrational level in H_2 into different vibrational levels in H_2^+ are induced. Transition $0 \rightarrow 2$ has maximum transition probability in accordance with the experimental spectrum in Figure 9.34.

Correspondingly, $\Delta E_{0 \rightarrow n_{vib}} = 15.5\,\text{eV} + n_{vib} h\nu_0$, where $21.2\,\text{eV} - (15.5\,\text{eV} + n_{vib} h\nu_0) = \nu_0$ is the distance between subsequent vibrational levels in H_2^+. Thus we obtain $E_{kin} = 5.77\,\text{eV} - n_{vib} h\nu_0$.

The difference between the bond lengths of H_2 (74 pm) and H_2^+ (106 pm) determines the intensities of the vibrational transitions (see the considerations in connection with Figure 9.32 and Figure 9.36). The most probable transition, according to the Franck–Condon Principle, is the $0 \rightarrow 2$ transition, and this is really the case (Figure 9.34).

Example – force constant of H_2^+ from photoelectron spectrum

The difference between the two lines on the left in Figure 9.34 is $\Delta E = 0.28\,\text{eV}$. This energy corresponds to the vibrational excitation energy $h\nu_0$. Then with

$$\mu = \frac{m_H m_H}{m_H + m_H} = 8.31 \times 10^{-28}\,\text{kg}$$

9.6 VIBRATIONAL STRUCTURE OF ELECTRONIC SPECTRA

we obtain

$$k_\text{f} = 4\pi^2 \mu \cdot \left(\frac{\Delta E}{h}\right)^2 = 150\,\text{N m}^{-1}$$

in good agreement with the directly determined value $160\,\text{N m}^{-1}$ (see Section 5.2.4).

9.6.3 Polyatomic Molecules

In polyatomic molecules the situation is similar as for diatomic molecules, but in a dye molecule, such as a cyanine, the electron density distribution is only slightly changed when exciting the molecule; this is because the π electron density is distributed over many atoms, and therefore only a slight change of the bond lengths can occur. Consequently, the absorption peak corresponding to a transition into a vibrationless excited state is strongest (vibrational structure according to Figure 9.37(a), observed absorption band Figure 9.37(b)).

Polyenes constitute a somewhat different case because of the strong bond alternation. This bond alternation is caused by the fact that the electrons in the highest occupied π orbital have antinodes at the double bonds (Section 7.4.2). When exciting one of these electrons into the next orbital the bond lengths are equalized, and a valence vibration

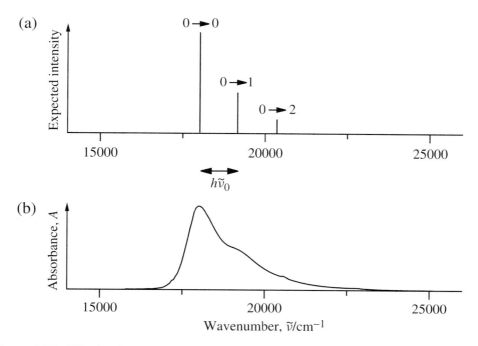

Figure 9.37 Vibrational structure of an electronic absorption band for a cyanine dye cation according to the Franck–Condon principle. (a) Expected intensity distribution for different vibrational transitions. (b) Experimental absorption spectrum of the cyanine dye cation with structure (8.4), $k = 1$.

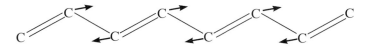

Figure 9.38 Vibrational mode in a polyene.

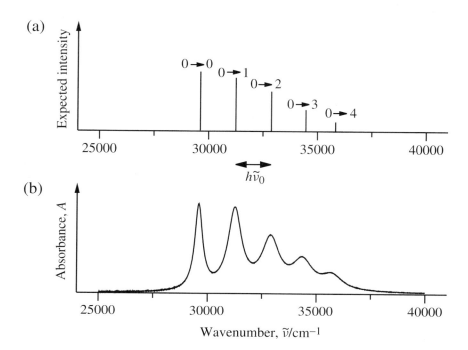

Figure 9.39 Vibrational structure of the electronic absorption band for a polyene. (a) Expected intensity distribution. (b) Experimental absorption spectrum of decapentaene: $CH_2=CH-CH=CH-CH=CH-CH=CH-CH=CH_2$ (F. Sondheimer, D. A. Ben-Efraim, R. Wolovsky, J. Amer. Chem. Soc. *83*, 1675 (1961)).

according to Figure 9.38 takes place. The frequency of this C–C vibration is $\nu_0 = 4.2 \times 10^{13}$ Hz, corresponding to $h\nu_0 = 0.17$ eV.

In terms of quantum theory, the vibrational structure according to Figure 9.39(a) is expected (strongest transition for $n_{\text{vib}} = 0$ to $n_{\text{vib}} = 1$) and observed (Figure 9.39(b)).

9.7 Nuclear Spin (Orthohydrogen and Parahydrogen)

This section is concerned with an extraordinary consequence of the antisymmetry postulate. We discuss some experiments with H_2 to emphasize the strangeness of the quantum nature of matter.

The electron can exist in states with spin function α or β (Section 4.6). States with different spin are split in an external magnetic field. As in the case of electrons, a property, called *nuclear spin*, can be attributed to nuclei.

9.7 NUCLEAR SPIN (ORTHOHYDROGEN AND PARAHYDROGEN)

9.7.1 Spin of Protons in H_2

A proton has a spin like an electron, and it can exist in two nuclear spin states. In an H_2 molecule the spins of the two protons can be parallel (ortho-H_2) or antiparallel (para-H_2), thus H_2 is a mixture of ortho- and para-H_2 in a ratio of $3:1$. This can be easily understood when symbolically describing the nuclear spin states by α and β. Then in a hydrogen molecule, similar to the case of the electron spin discussed in Chapter 4, nuclear spins combine as follows:

symmetric combinations (orthohydrogen):

$$\alpha(1)\alpha(2), \quad \beta(1)\beta(2), \quad \alpha(1)\beta(2) + \beta(1)\alpha(2) \tag{9.47}$$

Antisymmetric combination (parahydrogen):

$$\alpha(1)\beta(2) - \beta(1)\alpha(2) \tag{9.48}$$

Ortho-H_2 and para-H_2 can be separated experimentally. As will be shown below, ortho-H_2 and para-H_2 exist only in rotational states with odd and even quantum numbers n, respectively. Therefore, at very low temperature, para-H_2 ($n = 0$) is the stable form.

Para-H_2 is produced from the natural gas mixture by contacting the gas with charcoal, catalyzing the conversion of nuclear spins. Para-H_2 is metastable when heating the gas up to room temperature. The Raman lines of para-H_2 originate from rotational levels with even n ($n = 0, 2, \ldots$), while those of ortho-H_2 originate from levels with odd n ($n = 1, 3, \ldots$).

9.7.2 Nuclear Wavefunctions

How can we understand this most astonishing fact? The explanation is based on a fundamental quantum mechanical feature, and it is useful to recall our procedure.

In Chapters 1 and 2 we stressed the need to abandon familiar ways of thinking. In Chapter 4 we discussed the fact that electrons are indistinguishable and therefore, that exchange of coordinates leave wavefunctions either unchanged (*symmetric* wavefunctions) or just changed in sign (*antisymmetric* wavefunctions). Furthermore, we concluded that the electronic wavefunction of a molecule with N electrons depending on $3N$ positional coordinates and N spin coordinates is antisymmetric; that is, the electronic wavefunction changes its sign when any two-electron coordinates exchange.

In Chapters 5–8 we considered the wavefunction ψ as a product of one-electron wavefunctions. This extremely simplified picture leads to a reasonable description of the electronic structure of molecules (filling orbitals according to the aufbau principle).

The complete wavefunction of a molecule results as a solution of a Schrödinger equation of electrons and all nuclei. Making use of the fact that the mass of an electron is small compared to the mass of every nucleus, we can write the wavefunction as a product of the electronic wavefunction (depending on the electron coordinates) and a nuclear wavefunction (*Born–Oppenheimer* approximation).

The nuclear wavefunction can be written as a product of rotational, vibrational, and nuclear spin contributions:

$$\psi = \psi_{\text{electronic}} \cdot \psi_{\text{nuclear}} \tag{9.49}$$

$$\psi_{\text{nuclear}} = \psi_{\text{rotation}} \cdot \psi_{\text{vibration}} \cdot \psi_{\text{nuclear spin}} \tag{9.50}$$

9.7.3 Antisymmetry Postulate in H_2

$\psi_{electronic}$ depends on the coordinates of the electrons, but not on the coordinates of the nuclei. Thus exchanging nuclei 1 and 2 does not change $\psi_{electronic}$.

$\psi_{vibration}$ depends on the distance of the nuclei. Thus exchanging nuclei 1 and 2 does not change $\psi_{vibration}$.

$\psi_{nuclear\ spin}$ is antisymmetric for para-H_2. Thus exchanging nuclei 1 and 2 changes the sign of $\psi_{nuclear\ spin}$.

$\psi_{rotation}$ is a constant in the ground state ($n = 0$, see Figure 9.4). Thus exchanging nuclei 1 and 2 does not change $\psi_{rotation}$.

This means that ψ changes its sign when exchanging the nuclei. Consequently, according to the Pauli principle, this wavefunction is allowed for a para-H_2 molecule.

On the other hand, $\psi_{rotation}$ for the first excited rotational state ($n = 1$) changes its sign when passing the node line (Figure 9.4). Thus exchanging nuclei 1 and 2 changes the sign of $\psi_{rotation}$. Therefore ψ remains unchanged, and this wavefunction is forbidden for a para-H_2 molecule.

Correspondingly, ortho-H_2 (with symmetric nuclear spin wavefunction) cannot exist in the rotational state with $n = 0$, but in the state with $n = 1$.

This effect illustrates a strange quantum feature: it is impossible to distinguish between identical particles. This holds for protons as well as for electrons. There is no such effect in the case of the HD molecule, because in this case there is no such symmetry restriction (we can distinguish between a proton and a deuteron).

9.8 Nuclear Magnetic Resonance

This important technique is based on the nuclear magnetic moment. This moment is oriented in a magnetic field and it is turned when the nucleus is exposed to ultrashort radiowaves. The local magnetic field and therefore the resonance condition depend on the close neighborhood of the nucleus.

9.8.1 Fundamentals

We consider a proton (spin α or β) in a magnetic field with magnetic flux density B. In analogy to the case of electrons (Section 4.6), the nuclear magnetic moment is oriented in two different directions relative to the magnetic field. The energy difference between these two orientations is

$$\Delta E = g_N \cdot \mu_N \cdot B \quad (9.51)$$

where g_N is the nuclear g factor and μ_N is the nuclear magneton:

$$\mu_N = \frac{eh}{4\pi m_p} = 5.051 \times 10^{-27}\,\text{J}\,\text{T}^{-1} \quad (9.52)$$

(m_p is the mass of the proton). μ_N is smaller by a factor of 2000 compared to the Bohr magneton (Section 4.3) because of the large proton mass. The nuclear g factor for a proton is $g_N = 2.79$. The energy splitting is $\Delta E = 3 \times 10^{-26}\,\text{J}$ for $B = 1\,\text{T}$. The nucleus can then be excited by irradiation with an electromagnetic wave of frequency $\nu = \Delta E / h = 50\,\text{MHz}$. This technique to induce transitions between different nuclear spin states is called *nuclear magnetic resonance (NMR)*.

9.8 NUCLEAR MAGNETIC RESONANCE

A difficulty with this method is that ΔE is extremely small compared to the thermal energy, and the population of the ground state exceeds the population of the excited state only by 10^{-6} (see Chapter 12). The probability of absorption is almost the same as that for stimulated emission, and the overall absorption is very small.

By using very strong superconducting magnets ($B > 10$ T) and correspondingly high frequencies of the electromagnetic wave ($\nu > 500$ MHz) the energy splitting ΔE and thus the difference in the population between ground state and excited state can be substantially increased. In addition, sensitive detection methods are used. Instead of scanning the magnetic field in the region of resonance, a short high-frequency pulse of electromagnetic radiation is applied on the probe. In this way all spins in the probe can be excited simultaneously, and the relaxation of the nuclear spins is measured as a function of time.

By using fast Fourier transformation techniques the experimental data are transformed from the time domain into the frequency domain. Then the measurement of the NMR spectrum needs only some seconds. The signal-to-noise ratio can be significantly improved by the accumulation of a large number of measurements.

Besides the proton (^{1}H), many other nuclei can be excited by the NMR technique – for example, ^{13}C, $g_N = 1.405$; ^{14}N, $g_N = 0.403$; ^{19}F, $g_N = 5.257$.

9.8.2 Chemical Shift

NMR is a very powerful experimental method in chemistry, because the resonance frequency of a nucleus in a molecule depends on the chemical structure. When a magnetic field is applied to the molecule, the electrons in the molecule interact with the field in such a way that the local magnetic field at the nucleus is less than the external magnetic field:

$$B_{\text{nucleus}} = B \cdot (1 - \sigma) \tag{9.53}$$

(σ = shielding constant). σ depends on the electron distribution in the neighborhood of the nucleus and on the orientation of the molecule relative to the direction of the magnetic field. We restrict our considerations to molecules in a liquid, where the measured shielding is an average over all possible orientations.

Instead of measuring the absolute value of σ it is customary to measure the difference of σ relative to a reference. The difference

$$\delta = (\sigma_{\text{ref}} - \sigma) \times 10^6 \tag{9.54}$$

is called the *chemical shift*. Tetramethylsilane (TMS)

$$\begin{array}{c} \text{CH}_3 \\ | \\ \text{H}_3\text{C}-\text{Si}-\text{CH}_3 \\ | \\ \text{CH}_3 \end{array}$$

is used as a reference for proton NMR spectra. The resonance frequencies of the sample and of the reference are, because of equations (9.51) and (9.53), and $\Delta E = h\nu$,

$$\nu = A \cdot (1 - \sigma) \tag{9.55}$$

$$\nu_{\text{ref}} = A \cdot (1 - \sigma_{\text{ref}})$$

with
$$A = \frac{g_N \cdot \mu_N}{h} \cdot B \qquad (9.56)$$

Thus for the chemical shift we obtain
$$\delta = (\sigma_{\text{ref}} - \sigma) \times 10^6 = (\nu - \nu_{\text{ref}}) \times 10^6 \times \frac{1}{A} \qquad (9.57)$$

The shielding constant σ is very small (on the order of 10^{-6}), and then A in equation (9.55) can be approximated by
$$A = \frac{\nu_{\text{ref}}}{1 - \sigma_{\text{ref}}} \approx \nu_{\text{ref}}$$

Thus for the chemical shift we obtain
$$\boxed{\delta = \frac{\nu - \nu_{\text{ref}}}{\nu_{\text{ref}}} \times 10^6} \qquad (9.58)$$

The protons in tetramethylsilane (TMS) are more shielded than most protons in organic compounds. Using TMS as a reference in proton NMR results in positive values for δ in the range from 0 to 12.

The chemical shift increases with decreasing electron density at the proton. Thus δ increases if the neighboring atoms of the proton are more negative:

		δ
methyl proton	$-CH_3$	1
carboxyl proton	CH_3-COOH	9.5
carboxyl proton	CCl_3-COOH	12

The chemical shift is exceptionally large in aromatic compounds: $\delta = 7$ for benzene, compared to $\delta = 5$ for ethene and $\delta = 1$ in ethane. To explain this observation, we have to regard the mobility of the π electrons in a cyclic π system. Along the ring a ring current can be induced (Figure 9.40). The additional magnetic field generated by the ring current enlarges the magnetic field outside the ring – that is, at the positions of the protons. Consequently, the shielding effect becomes smaller and δ becomes larger.

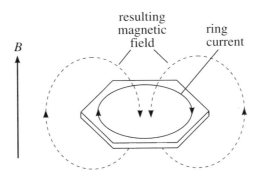

Figure 9.40 Ring current in benzene.

9.8 NUCLEAR MAGNETIC RESONANCE

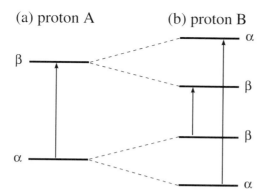

Figure 9.41 Splitting scheme for two nonequivalent H (spins α and β, respectively). (a) Two energy levels for proton A (one transition possible: singlet). (b) Proton B in the field of proton A: four energy levels for proton B (two transitions possible: doublet).

9.8.3 Fine Structure of NMR Signals

The magnetic moment of a proton can couple with another proton in its neighborhood. Let us consider two protons A and B far away from each other. Then we observe two single peaks in the NMR spectrum, provided that the protons are not equivalent (Figure 9.41(a)). In the case of coupling, the energy levels are split (Figure 9.41(b)), because it makes a difference if the nuclear spin of proton A is parallel or antiparallel to the nuclear spin of proton B. Then there are two transitions possible for proton A, where the spin quantum numbers of proton B remain unchanged. In this way we obtain a doublet for proton A and a doublet for proton B.

Example – NMR spectrum of ethanol

In ethanol

$$\overset{A}{CH_3}-\overset{B}{CH_2}-\overset{C}{OH}$$

we have three nonequivalent types of protons (A, B, and C) which give rise to three NMR peaks. Because of coupling a fine structure is observed (Figure 9.42).

Let us start with the protons A, which are coupled to protons B. There are four possibilities to couple the two spins of protons B (Figure 9.43), where two possibilities are equivalent. Thus the signal of protons A splits into a triplet with intensities $1:2:1$.

The protons B couple with the protons A. There are eight possibilities to arrange the three spins of protons A; however, three possibilities with two spins up and one spin down and three possibilities with one spin up and two spins down are equivalent. Thus the signal of protons A splits into a quadruplet with intensities $1:3:3:1$.

Furthermore, we expect a coupling of protons B with proton C, which would result in an octuplet for signal B and in a triplet for signal C. In the experiment, however, this is not the case because the OH proton is exchanging rapidly with protons in other molecules. Thus the protons B are in the average field of proton C, and the expected fine structure cannot be seen.

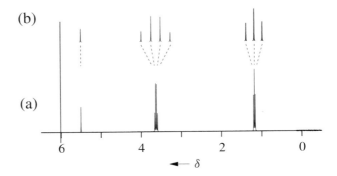

Figure 9.42 ^{1}H NMR spectrum of ethanol. (a) Peaks of protons A, B, and C (from right to left). (b) Enlargement to show the fine structure (triplet, quadruplet, singlet).

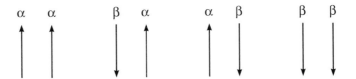

Figure 9.43 Combinations of spins α and β of protons B in the ethanol molecule.

Actually, the exchange of the OH proton is an acid-catalyzed process, and a very small concentration of acid is enough for the exchange. In acid-free ethanol the full fine structure can be observed.

Problems

Problem 9.1 – Wavefunction of an Harmonic Oscillator

Show by inserting in the Schrödinger equation (9.16), that the wavefunction ψ_0 (equation (9.17)) is an eigenfunction.

Solution.

$$\psi_0 = A \cdot e^{-ax^2/2}, \quad \frac{d\psi_0}{dx} = -A \cdot ax \cdot e^{-ax^2/2}$$

$$\frac{d^2\psi_0}{dx^2} = -A \cdot a \cdot e^{-ax^2/2} + A \cdot a^2 x^2 \cdot e^{-ax^2/2}$$

$$= -A \cdot e^{-ax^2/2} \cdot (a - a^2 x^2) = -(a - a^2 x^2) \cdot \psi_0$$

Insertion into the Schrödinger equation:

$$\frac{h^2}{8\pi^2 m} \cdot (a - a^2 x^2) + \frac{1}{2} k_f x^2 = E$$

$$\frac{ah^2}{8\pi^2 m} + x^2 \cdot \left(\frac{1}{2} k_f - \frac{a^2 h^2}{8\pi^2 m} \right) = E$$

PROBLEM 9.2

This equation can only be fulfilled for arbitrary x when the expression in parentheses equals zero. Then

$$a = \sqrt{k_f m} \cdot \frac{2\pi}{h}$$

and we obtain

$$E = \frac{ah^2}{8\pi^2 m} = \frac{1}{2} \cdot \frac{h}{2\pi} \cdot \sqrt{\frac{k_f}{m}} = \frac{1}{2} h\nu_0$$

Problem 9.2 – Oscillator: Box Wavefunctions

In the trial functions ϕ_0 and ϕ_1 in equation (9.22) the length L is an adjustable parameter. Determine L by using the variation principle.

Solution. First we calculate the energy ε_0 (length $= L_0$)

$$\varepsilon_{n=0} = \overline{T_0} + \overline{V_0} = \frac{h^2}{8mL_0^2} + \int_{-L_0/2}^{L_0/2} \frac{1}{2} k_f x^2 \phi_0^2 \cdot dx$$

$$= \frac{h^2}{8mL_0^2} + \frac{k_f}{L_0} \int_{-L_0/2}^{L_0/2} x^2 \cdot \cos^2 \frac{\pi x}{L_0} \cdot dx$$

$$= \frac{h^2}{8mL_0^2} + k_f L_0^2 \cdot a \quad \text{with} \quad a = \frac{\pi^3}{24} - \frac{\pi}{4} = 0.0163$$

According to the variation principle,

$$\frac{d\varepsilon_0}{dL_0} = 0$$

Then

$$L_0^2 = \frac{h}{\sqrt{k_f m}} \cdot \alpha \quad \text{with} \quad \alpha = \frac{1}{\sqrt{8a}} = 2.77$$

and

$$\varepsilon_{n=0} = 0.57 \cdot \frac{h}{2\pi} \cdot \sqrt{\frac{k_f}{m}}$$

With the trial function ϕ_1 we obtain in a similar way (length $= L_1$)

$$\varepsilon_{n=1} = \frac{h^2}{8mL_1^2} + k_f L_1^2 \cdot b \quad \text{with} \quad b = 0.0353$$

$$L_1^2 = \frac{h}{\sqrt{k_f m}} \cdot \beta \quad \text{with} \quad \beta = \frac{1}{\sqrt{2b}} = 3.76$$

$$\varepsilon_1 = 1.67 \cdot \frac{h}{2\pi} \cdot \sqrt{\frac{k_f}{m}}$$

Problem 9.3 – Classical Oscillator: Probability Density $\rho(x)$

Calculate the probability dW_{class}

$$dW_{class} = A \cdot \frac{1}{v} \cdot dx$$

of finding a classical oscillator moving with speed v in the x direction between x and $x + dx$.

Solution. dP is inversely proportional to the speed v. We calculate v from the displacement $x = x_0 \cdot \sin 2\pi\nu_0 t$, where x_0 is the amplitude of the oscillator. Thus

$$v = \frac{dx}{dt} = x_0 \cdot 2\pi\nu_0 \cdot \cos 2\pi\nu_0 t = x_0 \cdot 2\pi\nu_0 \cdot \sqrt{1 - \sin^2 2\pi\nu_0 t}$$
$$= x_0 \cdot 2\pi\nu_0 \cdot \sqrt{1 - (x/x_0)^2}$$

and

$$dP_{class} = A \cdot \frac{1}{v} \cdot dx = A \cdot \frac{1}{x_0 \cdot 2\pi\nu_0 \cdot \sqrt{1 - (x/x_0)^2}} \cdot dx$$

The constant A is determined by the condition that the probability of finding the oscillator between $-x_0$ and $+x_0$ equals 1. Thus we obtain $A = 2\nu_0$ and

$$\rho = \frac{dW_{class}}{dx} = \frac{1}{x_0 \cdot \pi} \cdot \frac{1}{\sqrt{1 - (x/x_0)^2}}$$

Problem 9.4 – Isotope Effect in Vibrational–Rotational IR Spectrum

Calculate the distances between the H–^{35}Cl and the H–^{37}Cl line in the IR spectrum of HCl (Figure 9.16).

Solution. The main influence between the different isotopes is a change of the pure vibrational transition energy $\Delta E^{35} = h\nu_0^{35}$ and $\Delta E^{37} = h\nu_0^{37}$. The energy difference $\Delta E' = \Delta E^{35} - \Delta E^{37}$ is

$$\Delta E' = h\nu_0^{35} - h\nu_0^{37} = \frac{h\sqrt{k_f}}{2\pi} \cdot \left(\frac{1}{\sqrt{\mu^{35}}} - \frac{1}{\sqrt{\mu^{37}}} \right)$$

Because of

$$\mu^{35} = \frac{m_H \cdot m_{^{35}Cl}}{m_H + m_{^{35}Cl}} = 1.627 \times 10^{-27} \, \text{kg},$$

$$\mu^{37} = 1.630 \times 10^{-27} \, \text{kg}$$

we obtain $\Delta E' = 5.3 \times 10^{-23}$ J corresponding to a shift of the lines by $\tilde{\nu} = 2.7 \, \text{cm}^{-1}$ in accordance with the experiment (Figure 9.16). Note that in this figure all ^{37}Cl lines are mainly equally shifted to smaller wavenumbers.

Correspondingly, the calculated shift associated with the difference in rotational energy is, according to equ. (9.11),

$$\Delta E'_{rot} = 2B^{35}(n+1) - 2B^{37}(n+1)$$
$$= 2(n+1) \cdot (B^{35} - B^{37})$$
$$= 2(n+1) \cdot \frac{h^2}{8\pi^2 d^2} \left(\frac{1}{m^{35}} - \frac{1}{m^{37}} \right)$$
$$= 6.3 \times 10^{-25} \cdot 2(n+1) \text{ J}$$

For $n = 0$ this is only 1% of the shift calculated for the vibrational transition; even for $n = 10$ it is only 10%.

Problem 9.5 – Bond Length of HCl from Raman Spectrum

Figure 9.44 shows the rotational Raman spectrum of HCl. From the location of the rotational lines calculate the bond length.

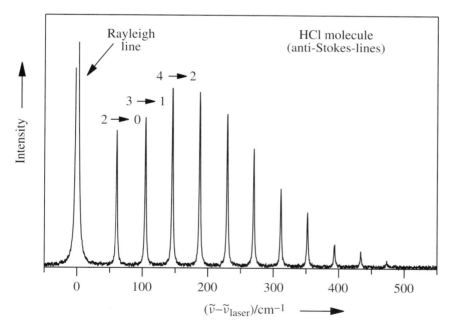

Figure 9.44 Rotational Raman spectrum of HCl (courtesy P. Dechant).

Solution. From the spectrum we read the wavenumbers corresponding to the different transitions, and we calculate the excitation energies ΔE as well as the differences $\Delta(\Delta E)$ of two subsequent lines. For $\Delta(\Delta E)$ we obtain, as an average, $\Delta(\Delta E) = 0.828 \times 10^{-21}$ J.

Then from equ. (9.42) with $\mu = 1.615 \times 10^{-27}$ kg we obtain

$$d = \sqrt{\frac{h^2}{2\pi^2 \mu \cdot \Delta(\Delta E)}} = 129 \, \text{pm}$$

Note that in the Raman spectrum the distances of two adjacent lines is larger by a factor of two compared with those in the infrared spectrum (Figure 9.6).

Problem 9.6 – Bond Length of CO from Infrared Spectrum

Figure 9.45 shows the rotational infrared spectrum of CO. From the location of the rotational lines calculate the bond length.

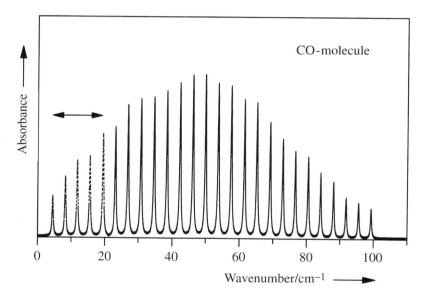

Figure 9.45 Rotational infrared spectrum of CO. The dashed lines are obtained by extrapolation (the technique used in this experiment did not allow to measure below $20 \, \text{cm}^{-1}$).

Solution. From the spectrum we read the wavenumbers corresponding to the different transitions, and we calculate the excitation energies ΔE as well as the differences $\Delta(\Delta E)$ of two subsequent lines. For $\Delta(\Delta E)$ we obtain, as an average, $\Delta(\Delta E) = 0.750 \times 10^{-22}$ J. Then from equ. (9.12) with $\mu = 1.14 \times 10^{-26}$ kg we obtain

$$d = \sqrt{\frac{h^2}{4\pi^2 \mu \cdot \Delta(\Delta E)}} = 114 \, \text{pm}$$

Problem 9.7 – Force Constant of CO from IR Spectrum

Figure 9.46 shows the vibrational-rotational infrared spectrum of CO. Calculate the force constant of CO.

PROBLEM 9.7

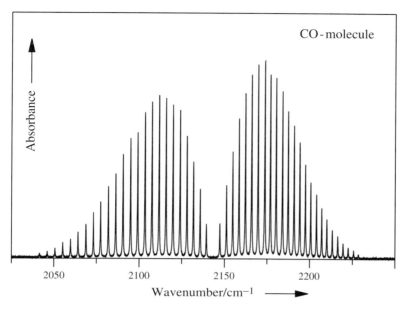

Figure 9.46 Rotational-vibrational infrared spectrum of Co.

Solution. The wavenumber $\tilde{\nu} = 2143\,\text{cm}^{-1}$ at the center between both rotational branches corresponds to the vibrational transition $0 \to 1$. Then $\Delta E_{0\to 1} = hc_0\tilde{\nu} = 4.26 \times 10^{-20}$ J. On the other hand, from equ. (9.17), (9.19), (9.21) we obtain

$$\Delta E_{0\to 1} = h\nu_0 = h\frac{1}{2\pi}\sqrt{\frac{k_f}{\mu}} \quad \text{and} \quad k_f = \frac{4\pi^2 \Delta E^2 \mu}{h^2}$$

Then with $\mu = 1.14 \times 10^{-26}$ kg we calculate $k_f = 1860\,\text{N}\,\text{m}^{-1}$.

Foundation 9.1 – Rotator: Solution of the Schrödinger Equation

The Schrödinger equation for a mass m rotating in space (like the proton in an HCl molecule) is

$$-\frac{h^2}{8\pi^2 m}\left(\frac{\partial^2 \psi}{\partial x^2} + \frac{\partial^2 \psi}{\partial y^2} + \frac{\partial^2 \psi}{\partial z^2}\right) + V(x, y, z)\cdot \psi = E\psi$$

where

$$V(x, y, z) = \frac{1}{2}k_f(r-d)^2$$

is a function of r

$$r = \sqrt{x^2 + y^2 + z^2}$$

As in the case of the Schrödinger equation for the electron in the H atom (see Chapter 3) the Cartesian coordinates x, y, and z are replaced by the spherical coordinates r, φ, and ϑ. Then the wavefunction ψ can be written as

$$\psi(\varphi, \vartheta, r) = \Phi(\varphi)\cdot \Theta(\vartheta)\cdot R(r)$$

The wavefunctions Φ and Θ are the same as listed in Box 3.4 for the H atom. However, the function $R(r)$ is very different from $R(r)$ for the electron in the H atom, because the potential energy $V(r)$ is very different. $V(r)$ increases strongly as soon as the proton in HCl deviates from the surface of the sphere with radius d.

The energy E of the rotator is obtained from the corresponding expression for the H atom (Chapter 3):

$$-\frac{h^2}{8\pi^2 mr^2}\cdot\left[\frac{1}{R}\frac{d}{dr}\left(r^2\frac{dR}{dr}\right) - l\cdot(l+1)\right] + \frac{1}{2}k_f(r-d)^2 = E$$

For simplification we assume that $r - d$, now called x, is small compared with the equilibrium distance d. Then $r \approx d$, $dr = dx$.

$$\frac{1}{R}\frac{d}{dr}\left(r^2\frac{dR}{dr}\right) = \frac{1}{R}\frac{d}{dx}\left(r^2\frac{dR}{dx}\right) \approx \frac{1}{R}r^2\frac{d}{dx}\left(\frac{dR}{dx}\right) = \frac{1}{R}r^2\frac{d^2R}{dx^2}$$

Then

$$E = \frac{h^2}{8\pi^2 md^2}\cdot l\cdot(l+1) - \frac{h^2}{8\pi^2 m}\cdot\frac{1}{R}\frac{d^2R}{dx^2} + \frac{1}{2}k_f\cdot x^2$$

Continued on page 331

FOUNDATION 9.2

Continued from page 330

The sum of the second and third term must be independent of x:

$$-\frac{h^2}{8\pi^2 m}\frac{1}{R}\cdot\frac{d^2 R}{dx^2}+\frac{1}{2}kf\cdot x^2 = const$$

or

$$-\frac{h^2}{8\pi^2 m}\cdot\frac{d^2 R}{dx^2}+\frac{1}{2}kf\cdot x^2\cdot R = const\cdot R$$

This is the same expression as the differential equation (9.16) for the harmonic oscillator. Thus the solution for $const$ is the energy of the harmonic oscillator, equation (9.20), and for E we obtain

$$E = \frac{h^2}{8\pi^2 m d^2}\cdot l\cdot(l+1)+h\nu_0\left(\frac{1}{2}+n\right)$$

where n is the quantum number of vibration and l is the quantum number of rotation. The case $n = 0$ is of particular interest. Then

$$E = \frac{h^2}{8\pi^2 m d^2}\cdot l\cdot(l+1)+\frac{1}{2}h\nu_0, \quad l = 0, 1, 2, \ldots$$

where $\frac{1}{2}h\nu_0$ is the energy of the oscillator in the ground state.

Foundation 9.2 – Vibrational Structure of Electronic Absorption Bands

The absorption intensity is related to the electronic transition moment

$$M_x = e\int \psi_i x \psi_j \cdot dxdydz$$

(M_y, M_z accordingly; x, y, z are the coordinates of the electrons, see (8.25) to (8.27)). Including the nuclear motion (nuclear coordinate ξ) the wavefunction ψ_i before the transition is

$$\psi_i = \psi_{el,i}(x, y, z)\cdot\psi_{vib,i}(\xi)$$

Continued on page 332

Continued from page 331

if we assume that $kT \ll h\nu_0$ (molecule in the vibrational ground state). The nuclear part of the wavefunction is separated from the electronic part since the nuclei are much slower than electrons (*Born–Oppenheimer approximation*).

Accordingly, for the excited state,

$$\psi_j = \psi_{el,j}(x, y, z) \cdot \psi_{vib,j}(\xi)$$

Then the transition moment is

$$M_{x,vib} = e \int \psi_{el,i} \cdot \psi_{vib,i} \cdot x \cdot \psi_{el,j} \cdot \psi_{vib,j} \cdot dx\,dy\,dz \cdot d\xi$$

$$= M_x \cdot \int \psi_{vib,i} \cdot \psi_{vib,j} \cdot d\xi$$

This means that the transition moment depends on the overlap of the vibrational wavefunctions.

10 Intermolecular Forces and Aggregates

How do molecules interact? What forces lead to aggregation? The structure of ionic crystals can be described by a picture of essentially rigid charged spheres in an arrangement corresponding to the energy minimum. Metals can be considered as metal ions embedded in a cloud of free electrons. Due to the Coulomb attraction between the metal ions and the free electrons the metal is compressed. On the other hand, the metal expands due the mutual Coulomb repulsion of the metal ions; moreover, an expansion of the metal increases the de Broglie wavelength of the free electrons and thus decreases their kinetic energy. An equilibrium between both tendencies is established.

The aggregation of uncharged molecules is governed by dispersion forces besides dipole and induction forces. The dispersion is based on the evasion effect (electrons escape each other, thus reducing their mutual Coulomb repulsion). The unique properties of water and the importance of hydrogen bonds (which are predominantly based on electrostatic forces) must be emphasized. Atomic force scanning microscopy (AFM) has great capabilities with regard to directly measuring intermolecular forces.

10.1 Forces in Ionic Crystals

10.1.1 Attracting and Repelling Forces

As already shown in Section 4.5 in the example of LiH, attracting Coulomb forces interact between two oppositely charged ions (charges $+e$ and $-e$)

$$f = \frac{1}{4\pi\varepsilon_0} \cdot \frac{e^2}{d^2} \tag{10.1}$$

(d = distance of the ions, ε_0 = permittivity of free space). As a result, the distance between two oppositely charged ions decreases until the electrostatic attracting force equals the repulsive force resulting from the penetration of the electron clouds. The energy that has to be supplied in order to bring a singly positively and a singly negatively charged ion from infinity to a distance d is

$$E_{\text{attraction}} = \int_{\infty}^{d} f \cdot dr = -\frac{1}{4\pi\varepsilon_0} \cdot \frac{e^2}{d} \tag{10.2}$$

It is difficult to describe the repulsion energy of the electron clouds theoretically, but we can use the empirical formula

$$E_{\text{repulsion}} = C \cdot \frac{1}{d^n} \tag{10.3}$$

where C and n are constants. A range for n from $n = 6$ to $n = 12$ can be estimated from the compressibility of ionic crystals. The energy of an ion pair is then

$$E = E_{\text{attraction}} + E_{\text{repulsion}} = -\frac{1}{4\pi\varepsilon_0} \cdot \frac{e^2}{d} + C \cdot \frac{1}{d^n} \tag{10.4}$$

At large distances the attracting energy dominates and E is then negative; at small distances the repulsion energy dominates and E becomes positive. For a distinct distance d_{eq} (equilibrium distance) E becomes minimal, the derivative of E equals zero:

$$\frac{1}{4\pi\varepsilon_0} \cdot \frac{e^2}{d^2} - C \cdot n \cdot \frac{1}{d^{n+1}} = 0 \tag{10.5}$$

Then

$$d_{\text{eq}}^{n-1} = \frac{nC \cdot 4\pi\varepsilon_0}{e^2} \tag{10.6}$$

By expressing the constant C by d_{eq} we obtain

$$E = -\frac{1}{4\pi\varepsilon_0} \cdot \frac{e^2}{d} \cdot \left[1 - \frac{1}{n}\left(\frac{d_{\text{eq}}}{d}\right)^{n-1}\right] \tag{10.7}$$

(Figure 10.1). At equilibrium ($d = d_{\text{eq}}$) we obtain

$$E = -\frac{1}{4\pi\varepsilon_0} \cdot \frac{e^2}{d_{\text{eq}}} \cdot \left[1 - \frac{1}{n}\right] \tag{10.8}$$

Since n takes values between 6 and 12, $1/n$ is only a small correction in comparison to 1; this means that only a minor deformation of the electron clouds takes place when an ion pair is formed. Then the energy needed for separating the ions is mainly determined by the Coulomb attracting force. Approximately, the ions can be treated as hard spheres of radius r_+ and r_-; then the equilibrium distance d_{eq} is given by $d_{\text{eq}} = r_+ + r_-$.

10.1.2 Lattice Types

Ion pairs can aggregate (Figure 2.17). An energetically favorable arrangement is reached when a three-dimensional lattice has been formed, as outlined in Section 2.62 for LiH. The situation is the same for NaCl.

Na$^+$ and Cl$^-$ ions form the lattice shown in Figure 10.2; the cations (as well as the anions) occupy the corners and the centers of the faces of cubes. Each Na$^+$ ion is surrounded by six Cl$^-$ ions as next neighbors (coordination number 6): this is the

10.1 FORCES IN IONIC CRYSTALS

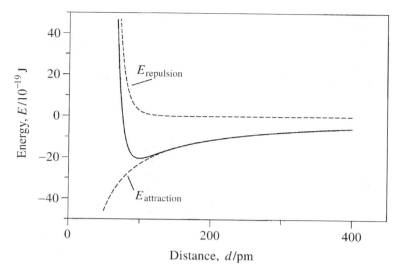

Figure 10.1 The energy of an ion pair as a function of the distance d between the ions according to equation (10.7) (with $n = 9$ and $d_{eq} = 100$ pm).

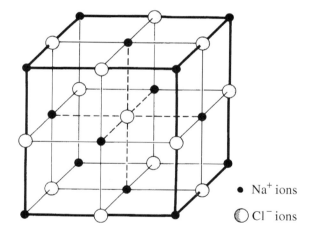

Figure 10.2 Arrangement of the Na^+ and Cl^- ions in the sodium chloride lattice.

sodium chloride lattice. The crystals grow in cubes as a result of the arrangement of Na^+ and Cl^- in the lattice, and therefore cleavage of NaCl crystals results in cubic pieces (Figure 10.3).

The lattice structure is energetically more favorable in comparison to the isolated ion pair, because in the lattice the ions of an ion pair are surrounded by ions with opposite charges. In Box 10.1, we show that the energy E' per ion pair in the NaCl crystal is 1.748 times the energy E of an isolated ion pair:

$$E' = -\frac{1}{4\pi\varepsilon_0} \cdot \frac{e^2}{d} \cdot 1.748 + C' \cdot \frac{1}{d^n} \quad (10.9)$$

Figure 10.3 The cleavage of a NaCl crystal results in cubic pieces.

Box 10.1 – Madelung Constant

We consider an infinite one-dimensional arrangement of NaCl ion pairs, where d is the distance between two adjacent ions. At one end we add a new ion pair (Figure 10.4). First we add an Na$^+$ at the end of the infinite chain. The energy required is the energy of the Na$^+$ in the electric field of the ions already present in the arrangement

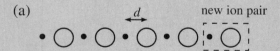

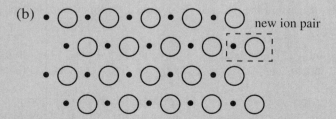

Figure 10.4 Calculation of the Madelung constant. (a) One-dimensional arrangement of positively and negatively charged ions; d is the distance between two adjacent ions. (b) Two-dimensional arrangement.

Continued on page 337

10.1 FORCES IN IONIC CRYSTALS

> Continued from page 336
>
> $$E_+ = \frac{1}{4\pi\varepsilon_0}\frac{e^2}{d} \cdot \left[-1 + \frac{1}{2} - \frac{1}{3} + \frac{1}{4} - \frac{1}{5} + \cdots\right]$$
>
> Because of (see mathematical appendix)
>
> $$\ln 2 = 1 - \frac{1}{2} + \frac{1}{3} - \frac{1}{4} + \frac{1}{5} + \cdots$$
>
> we obtain
>
> $$E_+ = -\frac{1}{4\pi\varepsilon_0}\frac{e^2}{d}\cdot \ln 2 = -\frac{1}{4\pi\varepsilon_0}\frac{e^2}{d}\cdot 0.693$$
>
> Adding the Cl^- in a second step requires the same energy. Then the energy to add one ion pair is
>
> $$E = E_+ + E_- = -\frac{1}{4\pi\varepsilon_0}\frac{e^2}{d}\cdot 1.386$$
>
> Correspondingly, we proceed for a two-dimensional and a three-dimensional arrangement. In these cases the sums are evaluated numerically. Instead of the factor 1.386 we obtain 1.614 (two-dimensional) and 1.748 (three-dimensional).

(C' = constant) and we calculate for the equilibrium distance d_{eq}

$$E'_{eq} = -\frac{1}{4\pi\varepsilon_0}\cdot \frac{e^2}{d_{eq}}\cdot 1.748 \cdot \left[1 - \frac{1}{n}\right] \tag{10.10}$$

The factor of 1.748 is called the *Madelung constant* for the NaCl lattice. The Madelung constant depends upon the geometry of the crystal. The sodium chloride lattice is especially favorable for the Na^+ and Cl^- ions because the ions just touch each other (Figure 10.5(a)).

Let us consider an ion pair with larger anions and smaller cations (e.g., BeO), again assuming that the ions (Be^{2+} and O^{2-}) are arranged in a sodium chloride lattice. Then the anions touch each other, but there is no contact between anions and cations, because the cations are too small (Figure 10.5(b)). Thus the sodium chloride lattice becomes unfavorable in this case. The energy can be decreased if the cations and anions get closer to each other. This is the case in a Zincblende *lattice*, where each cation is tetrahedrally surrounded by anions (Figure 10.6). The Madelung constant is accordingly smaller than for the sodium chloride lattice (1.639 instead of 1.748); but cations and anions get closer to each other and d_{eq} is therefore smaller. This effect overcompensates the effect due to the smaller Madelung constant.

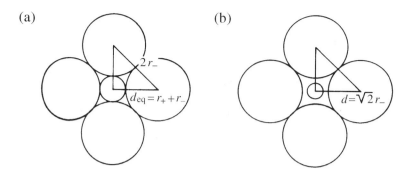

Figure 10.5 (a) The sodium chloride lattice. Anions (radius r_-) and cations (radius r_+) touch each other. Then the distance between the ions in the ion pair equals $(r_- + r_+)$; that is, $d_{eq} = r_- + r_+ = \sqrt{2} \cdot r_-$. (b) The anions touch each other, but no contact is possible between cations and anions. In this case the distance d is larger than $(r_- + r_+)$.

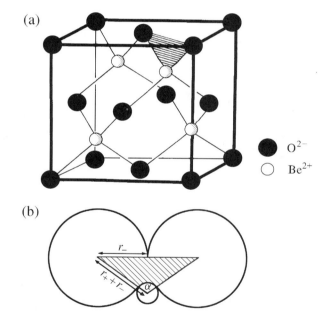

Figure 10.6 (a) Three-dimensional arrangement of Be^{2+} and O^{2-} ions in a Zincblende lattice. (b) Cross-section through the plane shown in (a) when anions and cations touch each other. The ratio of the radii is obtained from $r_-/(r_- + r_+) = \sin \alpha/2$, where $\alpha = 109.47°$ is the tetrahedral angle: $r_+/r_- = 0.225$.

Example

The ionic radii of Be and O are $r_+ = 35$ pm and $r_- = 140$ pm. In the Zincblende lattice with $d_{eq} = r_+ + r_- = 175$ pm we obtain ($Q_+ = +2e$, $Q_- = -2e$)

$$E'_{BeO} = -\frac{1}{4\pi\varepsilon_0} \cdot \frac{4e^2}{175 \text{ pm}} \cdot 1.639 \cdot \left[1 - \frac{1}{n}\right] \qquad (10.11)$$

10.1 FORCES IN IONIC CRYSTALS

For a crystal in a sodium chloride lattice, the distance d_{eq} would be, according to Figure 10.5(b), $d_{eq} = 140 \times \sqrt{2}$ pm $= 198$ pm and

$$E'_{BeO} = -\frac{1}{4\pi\varepsilon_0} \cdot \frac{4e^2}{198 \text{ pm}} \cdot 1.748 \cdot \left[1 - \frac{1}{n}\right] \quad (10.12)$$

Then the energy gain in forming the Zincblende lattice is larger by a factor of $(1.639 \times 198)/(175 \times 1.748) = 1.06$ than for the sodium chloride lattice. Indeed, BeO crystallizes in the Zincblende lattice.

According to Figure 10.5(a) a contact between anions and cations is possible for

$$r_+ + r_- \geq \sqrt{2} \cdot r_- \quad (10.13)$$

resulting in a ratio

$$\frac{r_+}{r_-} \geq \sqrt{2} - 1 = 0.414 \quad \text{(sodium chloride lattice)} \quad (10.14)$$

for the radii of cation and anion.

If this ratio is smaller than 0.414, the anions and cations cannot touch each other and the ions should crystallize in the Zincblende lattice. If the ratio is larger than 0.732 a cation can be surrounded by eight next neighbors. Then a cesium chloride lattice (Figure 10.7) is formed with a larger Madelung constant (1.763). Table 10.1 shows examples of lattice types. We can see the general trend to crystallize in the NaCl lattice when the ratio r_+/r_- is between 0.414 and 0.732; the Zincblende or the CsCl lattice is preferred when this ratio is smaller or larger than these limits, respectively. Deviations result from additional effects.

So far we have only considered crystals consisting of ions with the same charge number; the smallest electrically neutral unit is then an ion pair AB. In components with different

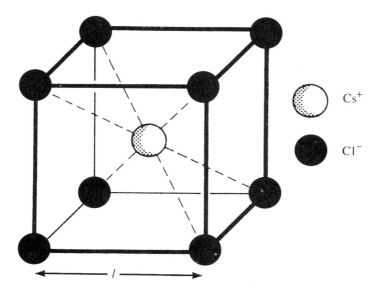

Figure 10.7 CsCl lattice. The distance between a Cs$^+$ and the Cl$^-$ is $\frac{1}{2}\sqrt{3} \cdot l$, where l is the length of the cube. $N_A \cdot l^3$ is the volume of 1 mol CsCl, thus $l = \sqrt[3]{m_{CsCl}/\rho}$, where m_{CsCl} is the mass of CsCl and ρ is the density of the crystal.

Table 10.1 Lattice types and radii ratios r_+/r_- of different crystals of type AB

ZnS		NaCl		CsCl	
ZnO	0.56	NaF	0.74	CsCl	0.91
ZnS	0.42	NaCl	0.54	CsBr	0.84
BeO	0.26	NaBr	0.50	CsI	0.75
BeS	0.20	NaI	0.44	RbCl	0.81
BeSe	0.18	KF	1.00	TlCl	0.81
MgTe	0.37	KCl	0.73	TlI	0.67
AgI	0.57	KBr	0.68		
CdS	0.56	KI	0.60		
CdSe	0.51	LiF	0.59		
		LiCl	0.43		
		LiBr	0.40		
		LiI	0.35		
		CaO	0.80		

ZnS (Zincblende or Wurtzite): coordination number = 4, Madelung constant = 1.638; NaCl: coordination number = 6, Madelung constant = 1.748; CsCl: coordination number = 8, Madelung constant = 1.763.

charge numbers (units $A_n B_m$) we have to proceed differently. The resulting structures are more complicated. The considerations become even more complicated when ions cannot be approximated as spheres (e.g., CH_3COO^-, CN^-, CO_3^{2-}). In these cases the structure is determined by the specific geometry.

10.2 Forces in Metals

Metals like lithium crystallize in simple lattices. An Li atom releases one electron, such that the crystal is composed of Li^+ ions and valence electrons (Figure 10.8). The Li^+ ions are embedded in the electron cloud of the valence electrons; they are attracted by the negative charges and thus the system is kept together by Coulomb forces.

What will happen when we apply a stress on a piece of Li metal? The Li^+ ions, after deformation, are still kept together very well by the electron cloud. Therefore, lithium is

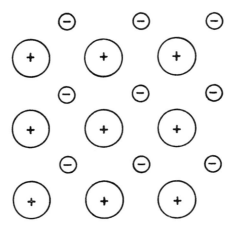

Figure 10.8 Metal lattice.

10.2 FORCES IN METALS

plastic and ductile. It does not resist external deformation as strongly as does a crystal with located bonds (e.g., a sodium chloride crystal or diamond).

Another consequence of the even distribution of the valence electrons in the electron cloud is the high conductivity of metals. When we apply an electric potential to a metal, the electrons freely move, since numerous unoccupied energy levels are available. In contrast, the electrons in an ionic crystal cannot move at all: ionic crystals are insulators. The same is true for crystals of covalently bonded elements, like diamond.

With decreasing volume of the crystal, the kinetic energy of the electron cloud increases because of the decrease of the de Broglie wavelengths of the electrons. The energy minimum is reached at a certain volume which results as a compromise between potential and kinetic energy.

Note that this tendency of the electron gas to expand the lattice is caused by the relation between the de Broglie wavelength and the electron speed. Do not confuse this phenomenon with the pressure of a gas such as air on the walls of a container which is caused by the thermal motion of the molecules (see Chapter 11).

Now let us look at the two compromising tendencies of Coulomb attraction and free electron expansion: the Coulomb energy decreases by decreasing the size, and the kinetic energy increases. At a certain size the energy reaches a minimum: this is a compromise between the two counteracting tendencies.

10.2.1 Coulomb Energy: Electrons Considered as Being Localized at Lattice Points

To estimate the lattice energy, we assume in a simplified manner that the lithium ions are arranged in a cubic lattice (actually they are arranged in a body-centered lattice). At first we imagine that an electron occupies the center of each cube like in a CsCl lattice in which the Cl^- ions are exchanged by electrons. This model illustrates how electrons and Li ions are kept together. The model gives us an estimate of the potential energy of the system of metal ions and electrons:

$$\overline{E}_{pot} = -\frac{1}{4\pi\varepsilon_0}\frac{e^2}{d_{eq}} \cdot 1.763 \tag{10.15}$$

According to Figure 10.7 we have

$$d_{eq} = \frac{\sqrt{3}}{2} \cdot l \quad \text{and} \quad l = \sqrt[3]{V_{Li}} = \sqrt[3]{\frac{m_{Li}}{\rho}} \tag{10.16}$$

where l is the side length of a lattice cell containing one Li^+ ion; V_{Li} is the volume occupied by one lithium ion in the lattice, m_{Li} is the mass of an lithium ion and ρ is the density of the Li crystal. Then

$$\overline{E}_{pot} = -A \cdot \rho^{1/3} \quad \text{with} \quad A = \frac{e^2}{4\pi\varepsilon_0} \cdot 1.763 \frac{2}{\sqrt{3}} \cdot \frac{1}{m_{Li}^{1/3}} \tag{10.17}$$

For $m_{Li} = 1.151 \times 10^{-26}$ kg and $\rho = 0.534$ g cm^{-3} we obtain $\overline{E}_{pot} = -16.8 \times 10^{-19}$ J.

10.2.2 Kinetic Energy: Electrons Considered as Being Delocalized Over the Lattice

Unfortunately, our model is incomplete with regard to one very important point. According to the wave–particle model of matter, an arrangement in which the electrons are localized

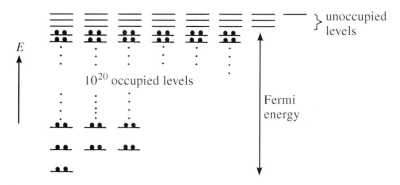

Figure 10.9 Distribution of the freely moving electrons on the different orbitals in a cube of metal (volume 1 cm³).

in a very small volume would correspond to de Broglie waves with an extremely short wavelength or to a particle with an extremely large energy

$$E_{\text{kin}} = \frac{1}{2} \cdot \frac{h^2}{m_e \Lambda^2} \tag{10.18}$$

Thus it is advantageous for the electrons to spread over the whole space inside the crystal. According to equation (3.29) and Figure 3.5, the kinetic energy of an electron in a cubic box of length L is

$$E_{\text{kin}} = \frac{h^2}{8m_e L^2}(n_x^2 + n_y^2 + n_z^2) \tag{10.19}$$

where n_x, n_y, and n_z are the quantum numbers. The N electrons in the metal are distributed in pairs over the lowest orbitals (Figure 10.9). The mean kinetic energy of these electrons (Box 10.2) is

$$\overline{E}_{\text{kin}} = B \cdot \rho^{2/3} \quad \text{with} \quad B = \frac{1}{5}\left(\frac{3}{\pi}\right)^{3/2} \frac{3h^2}{8m_e} \cdot \frac{1}{m_{\text{Li}}^{2/3}} \tag{10.20}$$

For Li we obtain $\overline{E}_{\text{kin}} = 4.5 \times 10^{-19}$ J.

Box 10.2 – Energy of Electron Gas in an Li Crystal

Calculation of E_{max}

The energies of an electron moving in the x direction between two parallel walls in distance L are

$$E_{n_x} = \frac{h^2}{8m_e L^2} \cdot n_x^2$$

Continued on page 343

10.2 FORCES IN METALS

Continued from page 342

Then the quantum number n belonging to the energy E_n is

$$n_x = \left(\frac{8m_e L^2}{h^2}\right)^{1/2} \cdot E_n^{1/2}$$

The quantum number n_x equals the number of quantum states for energies $E \leq E_{n_x}$. Thus we can approximate the number of quantum states of an electron moving in a cube with side length L in the x, y, and z directions for energies between 0 and E by

$$g = n_x n_y n_z = n_x^3 = \left(\frac{8m_e L^2}{h^2}\right)^{3/2} \cdot E^{3/2}$$

N electrons occupy $N/2$ orbitals with two electrons each, thus $g_{max} = N/2$ and

$$E_{max} = \frac{h^2}{8m_e} \cdot \frac{1}{L^2} \cdot \left(\frac{N}{2}\right)^{2/3} = \frac{h^2}{8m_e} \cdot \left(\frac{1}{2}\right)^{2/3} \cdot \left(\frac{N}{L^3}\right)^{2/3}$$

$$= \frac{h^2}{8m_e} \cdot \left(\frac{1}{2}\right)^{2/3} \cdot \left(\frac{N}{V}\right)^{2/3}$$

where E_{max} is the energy of the highest occupied energy level, and g_{max} is the corresponding number of quantum states; furthermore, $V = L^3$ is the volume of the considered cube.

We substitute the volume V of the crystal by the density ρ of the crystal and the mass m_{Li} of an Li atom. $N \cdot m_{Li}$ is the mass of all Li atoms in the volume V; thus

$$\frac{N \cdot m_{Li}}{V} = \rho$$

and we obtain

$$E_{max} = \left(\frac{1}{2}\right)^{2/3} \cdot \frac{h^2}{8m_e} \cdot \left(\frac{\rho}{m_{Li}}\right)^{2/3}$$

The result of a more rigorous treatment (see Chapter 12) is

$$g = \frac{\pi}{6} \cdot \left(\frac{8m_e L^2}{h^2}\right)^{3/2} \cdot E^{3/2} \quad \text{and} \quad E_{max} = \left(\frac{3}{\pi}\right)^{2/3} \cdot \frac{h^2}{8m_e} \cdot \left(\frac{\rho}{m_{Li}}\right)^{2/3}$$

Calculation of the mean energy $\overline{E}$

To calculate the mean energy $\overline{E}$, we first calculate the number dg of quantum states in the interval dE by differentiating the expression for g:

$$dg = \frac{\pi}{6} \left(\frac{8m_e L^2}{h^2}\right)^{3/2} \cdot \frac{3}{2} E^{1/2} \cdot dE$$

Continued on page 344

> *Continued from page 343*
>
> Then the mean energy $\overline{E}$ is
>
> $$\overline{E} = \frac{\int_0^{g_{max}} E \cdot dg}{\int_0^{g_{max}} dg} = \frac{\int_0^{E_{max}} E \cdot E^{1/2} \cdot dE}{\int_0^{E_{max}} E^{1/2} \cdot dE} = \frac{\frac{2}{5}(E_{max})^{5/2}}{\frac{2}{3}(E_{max})^{3/2}} = \frac{3}{5}E_{max}$$
>
> or
>
> $$\overline{E} = \frac{3}{5}\left(\frac{3}{\pi}\right)^{2/3} \cdot \frac{h^2}{8m_e} \cdot \left(\frac{\rho}{m_{Li}}\right)^{2/3}$$

10.2.3 Lattice Energy

The lattice energy is

$$E = \overline{E}_{pot} + \overline{E}_{kin} = -A \cdot \rho^{1/3} + B \cdot \rho^{2/3} \tag{10.21}$$

With the experimental data for Li we obtain $E = (-16.8 + 4.5) \times 10^{-19}\,\text{J} = -12.3 \times 10^{-19}\,\text{J}$. This means that the energy $12.3 \times 10^{-19}\,\text{J}$ has to be supplied to decompose an Li atom in an Li crystal into an Li$^+$ ion and an electron. The experimental value is $11 \times 10^{-19}\,\text{J}$ (the sum of the heat of sublimation, $2.32 \times 10^{-19}\,\text{J}$, and the ionization energy, $8.64 \times 10^{-19}\,\text{J}$).

Instead of using the experimental value of ρ, we can also minimize (10.21) with respect to ρ:

$$\frac{dE}{d\rho} = -A\frac{1}{3}\rho^{-2/3} + B\frac{2}{3}\rho^{-1/3} = 0$$

Then

$$\rho^{1/3} = \frac{A}{2B}$$

and

$$l = \sqrt[3]{\frac{m_{Li}}{\rho}} = m_{Li}^{1/3} \cdot \frac{2B}{A} = 150\,\text{pm}$$

This result is on the order of the experimental value, which is 280 pm.

10.3 Dipole Forces

We imagine dipoles arranged according to Figure 10.10(a) and ask about their attraction energy. This energy is the sum of all attraction and repulsion energies between the point charges Q_1 and Q_2 belonging to different dipoles

$$E = \frac{1}{4\pi\varepsilon_0} \cdot \left[\frac{Q_1Q_2}{r} + \frac{Q_1Q_2}{r} - \frac{Q_1Q_2}{(r+a)} - \frac{Q_1Q_2}{(r-a)}\right] \tag{10.22}$$

$$= \frac{1}{4\pi\varepsilon_0} \cdot Q_1Q_2 \cdot \left[\frac{2}{r} - \frac{1}{(r+a)} - \frac{1}{(r-a)}\right]$$

10.3 DIPOLE FORCES

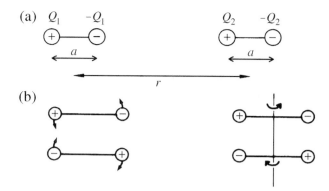

Figure 10.10 Attraction of two dipoles (charge Q, dipole length a). (a) Dipoles arranged linearly. (b) Dipoles arranged parallel; (*left*) rocking mode, (*right*) twisting mode.

For $a \ll r$ we obtain (Box 10.3)

$$E = -\frac{1}{4\pi\varepsilon_0} \cdot \frac{2\mu_1\mu_2}{r^3} \qquad (10.23)$$

where $\mu_1 = Q_1 a$ and $\mu_2 = Q_2 a$ are the dipole moments in the point dipole approximation. The energy between two dipoles depends much more sensitively upon the distance r than does the energy between two point charges. The energy to rotate one of the dipoles by 180° (from antiparallel to parallel) is $2E$. The attracting force f between the dipoles is

$$f = \frac{dE}{dr} = \frac{1}{4\pi\varepsilon_0} \cdot \frac{6\mu_1\mu_2}{r^4} \qquad (10.24)$$

Box 10.3 – Dipole and Induction Energies

Dipole energy

The expression in brackets in equation (10.22) can be rearranged:

$$\frac{2}{r} - \frac{1}{r+a} - \frac{1}{r-a} = \frac{2r^2 - 2a^2 - r^2 + ra - r^2 - ra}{r(r+a)(r-a)}$$

$$= -\frac{2a^2}{r^3} \cdot \frac{1}{1+(a/r)^2} \approx -\frac{2a^2}{r^3}$$

Then the energy E is

$$E = -\frac{1}{4\pi\varepsilon_0} \cdot \frac{2\mu_1\mu_2}{r^3} \quad \text{with} \quad \mu_1 = Q_1 a, \mu_2 = Q_2 a$$

Continued on page 346

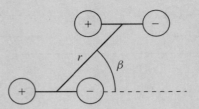

Figure 10.11 Orientation of two dipoles with arbitrary angle β.

This expression holds for angle $\beta = 0$ (Figure 10.11). For arbitrary angle β we obtain

$$E = -\frac{1}{4\pi\varepsilon_0} \cdot \frac{\mu_1\mu_2}{r^3} \cdot (3\cos^2\beta - 1)$$

For example, with $\beta = 90°$ we obtain

$$E = +\frac{1}{4\pi\varepsilon_0} \cdot \frac{\mu_1\mu_2}{r^3}$$

(a corresponding case is discussed in Problem 10.2).

Induction energy

The expression in brackets in equation (10.30) can be rearranged:

$$\frac{1}{(r-a/2)^2} - \frac{1}{(r+a/2)^2} = \frac{1}{r^2} \cdot \left(\frac{(1+a/(2r))^2 - (1-a/(2r))^2}{(1-a^2/(4r^2))^2}\right)$$

$$= \frac{1}{r^2} \cdot \frac{2a/r}{(1-a^2/(4r^2))^2}$$

$$= \frac{2a}{r^3} \cdot \frac{1}{(1-a^2/(4r^2))^2} \approx \frac{2a}{r^3}$$

Thus it follows for the electric field strength F in distance r from a dipole ($\beta = 0$)

$$F = \frac{1}{4\pi\varepsilon_0} \cdot \frac{2Qa}{r^3} = \frac{1}{4\pi\varepsilon_0} \cdot \frac{2\mu}{r^3}$$

For arbitrary angle β we obtain

$$F = \frac{1}{4\pi\varepsilon_0} \cdot \frac{\mu}{r^3} \cdot (3\cos^2\beta - 1)$$

As an example, we consider two CH_3CN molecules. Because of their large dipole moment they are arranged as a pair (Figure 10.10(b)) that can be examined in a solid argon matrix. The molecules vibrate towards each other in different vibrational modes. By irradiation with IR light, vibrational modes can be excited that are accompanied by a change in the total dipole moment. These are the two vibrational modes shown in Figure 10.10(b). Indeed, absorption bands corresponding to these vibrations can be seen at $1/\lambda = 80\,\text{cm}^{-1}$ and $130\,\text{cm}^{-1}$. They measure the dipole–dipole interaction. (E. Knözinger, D. Leutloff: Far infrared spectra of strongly polar molecules in solid solution. I. Acetonitrile, *J. Chem. Phys.* 74, 4812 (1981)).

10.4 Hydrogen Bonds

Hydrogen bonds are particularly important in processes where the formation and disruption of weak bonds is crucial. Among these reactions are almost all fundamental processes taking place in biosystems.

When polar parts of molecules are close together, a description by the point dipole approximation is no longer reasonable. This is particularly the case in molecules with hydrogen atoms bound to strongly electronegative atoms (e.g., in NH_3 and H_2O). On the one hand, the electronegative atoms have a high negative charge, and on the other hand the van der Waals radius of the hydrogen atom is small. This causes such a strong dipole interaction that the H atom of one molecule is preferentially oriented toward an unshared electron pair of a second molecule (*hydrogen bond*):

$$
\begin{array}{c}
\text{O—H----O} \overset{\displaystyle H}{} \\
|| \\
\text{H}\text{H}
\end{array}
$$

As a consequence, in the dimer $H_2O \cdots HOH$ the distance $O \cdots H$ (180 pm) is much smaller than the sum of the van der Waals radii of O and H (140 pm + 120 pm = 260 pm), but larger than the O—H bond distance (97 pm).

The energy gain in forming the hydrogen bond in this example is about 0.5×10^{-19} J. In general, the bond energy of hydrogen bonds ($\approx 1 \times 10^{-19}$ J) is less than 10% of typical covalent bond energies (see Table 7.1).

F—H forms such a strong hydrogen bond with F^- (bond energy 4×10^{-19} J) that in the energy minimum the H nucleus is centered between the F nuclei. Thus this aggregate can vibrate at frequencies comparable with frequencies of molecules with covalent bonds (Box 10.4).

Box 10.4 – Vibrations of H-Bond Aggregates

We consider the aggregate

$$(F \cdots H \cdots F)^-$$

The H nucleus can be considered as a harmonic oscillator, which can be excited by IR light. Absorption bands are found at $1/\lambda = 1364\,\text{cm}^{-1}$ (excitation in the direction of the bond line: antisymmetric vibration mode) and

Continued on page 348

> *Continued from page 347*
>
> at 1217 cm^{-1} (excitation perpendicular to the bond line: bending mode). If the hydrogen atom is exchanged by a deuterium with mass $m_D = 2m_H$, these bands are shifted toward $1/\lambda = 1364/\sqrt{2}$ cm$^{-1} = 964$ cm^{-1} and $1/\lambda = 1217/\sqrt{2}$ cm$^{-1} = 861$ cm^{-1}, respectively. The experimental values are 969 and 880 cm^{-1}. (B. S. Ault, *Acc. Chem. Res.* **15**, 103 (1982)).

Hydrogen bonds affect many properties of water: high boiling point, high heat of evaporation, loose structure (the directed interaction does not allow a dense packing) – the unique properties of water. Hydrogen bonds are essential for base pairing in DNA:

Thymine — Adenine   Cytosine — Guanine

Essentially an H_2O molecule can be modeled with charges $+0.24e$ centered at each hydrogen atom and two compensating charges of $-0.24e$ on the opposite side of the oxygen atom, representing the two unshared electron pairs. This model accounts for essential properties of water. The directional bonding and the tendency of H_2O molecules to aggregate in a tetrahedral open arrangement can be explained. The low density of ice and its high melting point are caused by the stability of the three-dimensional tetrahedral network. When ice melts, some of these bonds are broken, and a denser packing of the H_2O molecules is possible. As a result, the density of water is higher than that of ice, and it further increases after melting up to a temperature of 4 °C.

10.5 Induction Forces

Attraction forces appear not only between two permanent dipoles, but also when a dipole moment μ_{ind} is induced by a second molecule (Figure 10.12). Such forces are called *induction forces*. If F is the strength of the electric field (originating from a permanent dipole) at the position of a polarizable molecule, then

$$\mu_{ind} = \alpha F \tag{10.25}$$

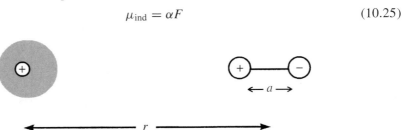

Figure 10.12 Force between a dipole and a polarizable molecule.

where α is the polarizability. The electron cloud of the molecule is deformed in an electric field. This causes a shift of the center of gravity of the negative charge relative to the nuclear charge. The effect is similar to what happens when a metal sphere of radius r is put into an electric field. In this case

$$\alpha = 4\pi\varepsilon_0 \cdot r^3 \qquad (10.26)$$

(Problem 10.3). Assuming, for comparison, that the sphere has the size of an H atom ($r = 125$ pm $= 2.4a_0$, obtained from the collision diameter, see Table 11.1), then

$$\alpha = 4\pi\varepsilon_0 \cdot (2.4a_0)^3 = 4\pi\varepsilon_0 \cdot 14a_0^3 \qquad (10.27)$$

The value found experimentally for the polarizability of H is

$$\alpha = 4\pi\varepsilon_0 \cdot 0.67 \times 10^{-24} \text{ cm}^3 = 4\pi\varepsilon_0 \cdot 4.5a_0^3 \qquad (10.28)$$

It is remarkable that the polarizability of an electron cloud is only moderately smaller than the polarizability of a piece of metal of the same size considered to maintain macroscopic behavior. An He atom is smaller than an H atom (Figure 2.12), so its polarizability is smaller; with increasing size the polarizability increases (see Table 10.2).

To calculate the energy caused by the attraction of a polarizable molecule by a permanent dipole, we consider the strength of the electric field F at a distance r from a point charge Q

$$F = \frac{1}{4\pi\varepsilon_0} \cdot \frac{Q}{r^2} \qquad (10.29)$$

Thus the field strength of the dipole in Figure 10.12 at the position of the polarizable molecule is

$$F = \frac{1}{4\pi\varepsilon_0} \cdot \left[\frac{+Q}{(r-a/2)^2} + \frac{-Q}{(r+a/2)^2} \right] \qquad (10.30)$$

For $a \ll r$ (point dipole approximation) this expression (Box 10.3) can be simplified as

$$F = \frac{1}{4\pi\varepsilon_0} \cdot \frac{2\mu}{r^3} \qquad (10.31)$$

where μ is the dipole moment of the permanent dipole.

Then with equations (10.24) and (10.25) we obtain for the attracting force f between the permanent dipole and the polarizable molecule with $\mu_1 = aQ$, $\mu_2 = \mu_{ind}$

$$f = \frac{1}{4\pi\varepsilon_0} \cdot \frac{6\mu \cdot \mu_{ind}}{r^4} = \frac{1}{(4\pi\varepsilon_0)^2} \cdot \frac{12\mu^2 \cdot \alpha}{r^7} \qquad (10.32)$$

The energy E is obtained by integration:

$$E = \int_\infty^r f \cdot dr = -\frac{1}{(4\pi\varepsilon_0)^2} \cdot \frac{2\mu^2 \cdot \alpha}{r^6} \qquad (10.33)$$

10.6 Dispersion Forces

From our considerations it is clear that two HCl molecules or an HCl and an H_2 molecule attract each other. However, experimentally it has been found that two He atoms also attract each other, even though none of these atoms have a permanent dipole moment. How can we explain this surprising fact?

In the following we show that the reason is a remarkable consequence of the quantum mechanical variation principle. Electrons behave as if they would move around the nucleus, resulting in a fuzzy cloud at each atom. They behave as if the electrons of both atoms would escape each other, forming induced dipoles in both atoms that attrack each other by Coulomb forces.

Both electrons of the He atom occupy a spherically symmetric 1s orbital, thus the centers of gravity of the positive and of the negative charges coincide, and no permanent dipole occurs. In slow processes (like the shift of nuclei by a distance of the size of an atom), the electrons behave as if they were distributed over a cloud (Section 3.2). If, however, we consider fast processes (like the shift of two electrons toward each other), the momentary position of the electrons becomes decisive.

We have seen this already in the case of an He atom (Section 4.2). Each electron is in the electric field of the nucleus and the field of the other electron. It behaves as if it would escape from the considered electron. In Section 4.2 we considered this evasion effect only qualitatively. It can be exactly described by solving the Schrödinger equation.

Here we again restrict our discussion to a simplified consideration. The essence can be seen in Figure 10.13 where atom 1 is replaced by fluctuating electrons and atom 2 by an electron cloud polarized in the electric field of these charges leading to an attraction in both cases (a) and (b). He atom 2 is in the instantaneous field of the He atom 1 – that is, in the field of the nucleus and the two localized electrons. Atom 2 is polarized in the field of this momentary dipole (dipole moment $\mu_{\text{mom},1}$), and an induced dipole is formed (dipole moment $\mu_{\text{mom},2}$). The momentary dipole moment is on the order of $\mu_{\text{mom},1} = ea_0$. Then according to equation (10.32) and with

$$\mu = \mu_{\text{mom},1} \text{ and } \alpha = \alpha_2$$

we obtain for the attractive force

$$f = \frac{1}{(4\pi\varepsilon_0)^2} \cdot \frac{12\mu_{\text{mom},1}^2 \cdot \alpha_2}{r^7} \tag{10.34}$$

It is advantageous to express $\mu_{\text{mom},1}^2 = e^2 a_0^2$ by the polarizability α_1 and the ionization energy $E_{\text{ion},1}$. As has been shown already, the polarizability of the H atom equals $4\pi\varepsilon_0 \cdot$

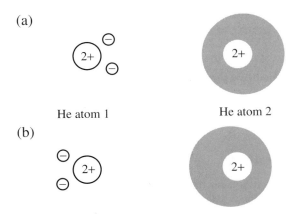

Figure 10.13 He atom 2 in different instantaneous dipole fields of He atom 1.

10.6 DISPERSION FORCES

Table 10.2 Dipole moment μ, polarizability α, ionization energy E_{ion}, and collision diameters of some atoms and molecules

	μ /10^{-30} C m	/Debye	$\alpha/(4\pi\varepsilon_0)$ /10^{-30} m^3	E_{ion} /10^{-19} J	Collision diameter/pm
H	0	0	0.67	21.8	256
He	0	0	0.21	39.2	220
Ne	0	0	0.39	41.2	260
Ar	0	0	1.63	28.0	370
Kr	0	0	2.46	23.5	420
Xe	0	0	4.0	19.5	495
H$_2$	0	0	0.81	23.2	276
HCl	3.42	1.03	2.63	21.5	450
HI	1.26	0.38	5.4	16.8	530
NH$_3$	4.97	1.5	2.24	18.7	450
H$_2$O	6.11	1.84	1.48	28.8	462

The dipole moments are given in SI units as well as in debye units (1 debye = 3.33×10^{-30} C m)

$4.5a_0^3$. Then the polarizability of a helium atom is, because of its smaller size,

$$\alpha_1 = 4\pi\varepsilon_0 \cdot a_0^3.$$

According to Section 4.2 we obtain

$$E_{\text{ion},1} = \frac{1}{4\pi\varepsilon_0} \cdot \frac{e^2}{a_0} = \frac{1}{4\pi\varepsilon_0 \cdot a_0^3} \cdot e^2 a_0^2 = \frac{1}{\alpha_1} \cdot e^2 a_0^2$$

Then

$$\mu_{\text{mom},1}^2 = e^2 a_0^2 = \alpha_1 E_{\text{ion},1} \tag{10.35}$$

The force f in equation (10.34) can be rewritten as

$$f = \frac{1}{(4\pi\varepsilon_0)^2} \cdot \frac{12\alpha_1 E_{\text{ion},1} \cdot \alpha_2}{r^7} \tag{10.36}$$

Then the energy (*dispersion energy*) is

$$E_{\text{disp}} = \int_\infty^r f \cdot dr = -\frac{1}{(4\pi\varepsilon_0)^2} \cdot \frac{2\alpha_1 \cdot \alpha_2}{r^6} E_{\text{ion},1} \tag{10.37}$$

An exact treatment results in the expression

$$E_{\text{disp}} = -\frac{1}{(4\pi\varepsilon_0)^2} \cdot \frac{3}{2} \cdot \frac{\alpha_1 \cdot \alpha_2}{r^6} \frac{E_{\text{ion},1} E_{\text{ion},2}}{E_{\text{ion},1} + E_{\text{ion},2}} \tag{10.38}$$

Table 10.2 shows the dipole moment μ, the polarizability α, and the ionization energy E_{ion} of some atoms and molecules.

Since $\alpha_1 \cdot \alpha_2$ appears in the expression for the dispersion energy, this energy is much smaller for the He atom than for H$_2$ molecules, even though the ionization energy is twice as large. Indeed He, in contrast to H$_2$, does not solidify even at the lowest possible temperatures under atmospheric pressure and it has a unique structure at low temperatures.

Figure 10.14 shows the dispersion energy of H, He, and Xe as a function of the distance r. The attraction between the molecules decreases very rapidly with increasing distance. In Figure 10.15 the contributions of dipole energy (going over to hydrogen bond energy at small distances), induction energy, and dispersion energy are shown for NH$_3$ molecules.

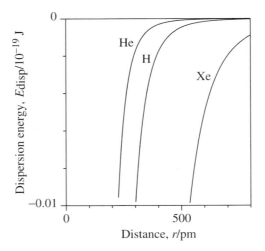

Figure 10.14 Dispersion energy E_{disp} as a function of the distance r.

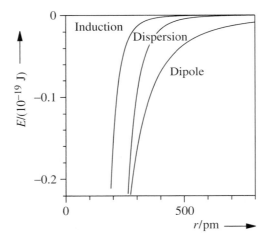

Figure 10.15 Attraction energy as a function of the distance r for NH_3 molecules; comparison of the contributions of the dipole energy, induction energy, and dispersion energy.

Even though the dipole moment of NH_3 is large, the contribution of the dispersion energy is larger than that of the induction energy and it approaches that of the dipole energy at $r < 300$ pm. At larger distances, only the dipole forces play a role.

The force to separate two NH_3 molecules, dE/dr, has its maximum at 300 pN (Figure 10.15; we assume that the slope dE/dr at 280 pm corresponds to the slope at the inflection point when considering the potential energy curve including the repulsion energy, see equation 10.3). Forces required to separate molecules can be directly measured by the atomic force microscope (see Section 22.4.5). Biotin (vitamin H) binds specifically to the protein avidin. By fixing biotin to the tip of an atomic force microscope and avidin to the surface below, the force required to break the intermolecular bond of a single biotin molecule to its specific site on the protein can be measured. This force is 200 pN (E. L. Florin, V. T. May, H. E. Gaub, Science 264, 415 (1994)).

10.7 Molecular Crystals

The energy gain in forming a molecular crystal is much smaller than that for an ionic crystal. For two NH_3 molecules brought into contact (distance $d = 300$ pm) we obtain, according to Figure 10.15,

$$E = E_{\text{dipole}} + E_{\text{ind}} + E_{\text{disp}} = -(0.17 + 0.01 + 0.10) \times 10^{-19} \text{ J} = -0.28 \times 10^{-19} \text{ J}$$

According to equation (10.7) the energy of an ion pair with the same distance $d = 300$ pm is $E = -12 \times 10^{-19}$ J; this energy is larger by a factor of about 50. Therefore, an NH_3 crystal is much less stable than a NaCl crystal. This is the reason why NH_3 is a gas at room temperature.

According to our considerations the polarizability α increases strongly with the size of the molecule; this means that large organic molecules (e.g., naphthalene, anthracene) can really form crystals at room temperature, even if they do not possess a permanent dipole moment. A complex made up of a molecule with electron donor properties and an electron acceptor molecule (e.g., quinhydrone, a molecular crystal of quinone and hydroquinone) can be additionally stabilized by partial charge transfer (charge transfer complex).

Problems

Problem 10.1 – The Energy of an Ion Pair

Estimate how much energy has to be supplied in order to increase or decrease the distance d of an ion pair by 1% and 10% ($d = 100$ pm, $n = 10$).

Solution. According to equation (10.7)

$$E = A \cdot \left[\frac{1}{1+x} - \frac{1}{n} \cdot \frac{1}{(1+x)^n} \right]$$

with

$$A = -\frac{1}{4\pi\varepsilon_0} \cdot \frac{e^2}{d_{\text{eq}}} \quad \text{and} \quad d = d_{\text{eq}} \cdot (1+x)$$

for $x \ll 1$ we find

$$E = A \cdot \left[1 - x + x^2 - \cdots - \frac{1}{n}\left(1 - nx - \frac{n(n-1)}{2}x^2 + n^2x^2 + \cdots\right) \right]$$

$$= A \cdot \left[1 - \frac{1}{n} + x^2 - x^2 \cdot \left(-\frac{n}{2} + \frac{1}{2} + n\right) \right]$$

$$= A \cdot \left[1 - \frac{1}{n} - x^2 \cdot \left(\frac{n}{2} - \frac{1}{2}\right) \right] = A \cdot \left[1 - \frac{1}{n} \right] \cdot \left[1 - \frac{nx^2}{2} \right]$$

$$= E_{\text{eq}} \cdot \left[1 - \frac{nx^2}{2} \right]$$

Note that E_{eq} is negative, thus E is increasing proportional to $nx^2/2$. For $x = 1\%$ and $n = 10$ the change of the energy E is 0.05%; it is 5% for $x = 10\%$.

Problem 10.2 – Dipole–Dipole Attraction

Calculate the attraction energy when two dipoles are arranged in the following ways (Figure 10.16).
Solution.

Figure 10.16(a): $E = \dfrac{Q^2}{4\pi\varepsilon_0} \cdot \left[-\dfrac{2}{r} + \dfrac{2}{\sqrt{r^2+a^2}} \right] \approx -\dfrac{Q^2}{4\pi\varepsilon_0} \cdot \dfrac{a^2}{r^3} = -\dfrac{1}{4\pi\varepsilon_0}\dfrac{\mu^2}{r^3}$

Figure 10.16(b): $E = \dfrac{Q^2}{4\pi\varepsilon_0} \cdot \left[+\dfrac{2}{r} - \dfrac{1}{r-a} - \dfrac{1}{r+a} \right] \approx -\dfrac{1}{4\pi\varepsilon_0} \cdot \dfrac{2\mu^2}{r^3}$

(The approximations are valid for $a \ll r$.)

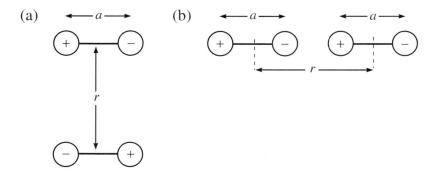

Figure 10.16 Attraction of two dipoles in two different arrangements (a) and (b).

Problem 10.3 – Polarizability of a Conducting Plate

Calculate the polarizability of an electrically conducting plate. Imagine a plate in a homogeneous electric field and calculate the induced dipole moment of the plate. Compare with the polarizability of a sphere of the same volume.

Solution. We consider a capacitor with area A and plate distance l (Figure 10.17). Its capacitance is

$$C = \frac{Q}{V} = \varepsilon_0 \frac{A}{l}$$

where Q is the charge on the plates, and V is the voltage. Now we place a metallic plate of the same area (thickness d) inside the capacitor. Then the charge Q is induced on the plate and the dipole moment of the plate is

$$\mu_{\text{ind}} = Q \cdot d = CV \cdot d = \varepsilon_0 \frac{A}{l} V \cdot d = \varepsilon_0 A d \cdot F$$

where Ad is the volume of the plate, and $F = V/l$ is the electric field strength.
Then the polarizability of the metallic plate is

$$\alpha = \frac{\mu_{\text{ind}}}{F} = \varepsilon_0 \cdot Ad = \varepsilon_0 \cdot \text{volume}$$

PROBLEM 10.3

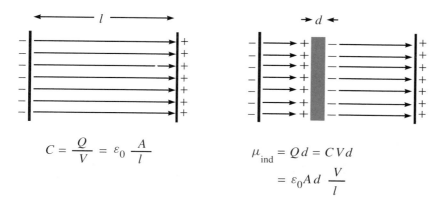

Figure 10.17 Polarization of a metallic plate inside a capacitor.

and we estimate for the polarizability α of a sphere with the same volume ($V = (4\pi/3) \cdot r^3$)

$$\alpha = \varepsilon_0 \cdot \frac{4\pi}{3} r^3$$

Actually, α of a conducting sphere is larger by a factor of 3:

$$\alpha = \varepsilon_0 \cdot 4\pi r^3 = 3\varepsilon_0 \cdot \text{volume}$$

11 Thermal Motion of Molecules

What is temperature, and how is this observable quantity related to molecular processes? What is heat? Why does heat flow from a hot body to a cold body? Why does the reverse process not happen?

Let us look at a gas. A container filled with a hot gas is brought into contact with a second container filled with a cold gas. We will show in this chapter that the hot gas is comprised of fast moving independent molecules, while the cold gas consists of slow molecules. Bringing the containers into contact induces an equilibrium through collisions of the molecules with the common wall. The fast molecules become slower, whereas the slow molecules become faster. The energy flowing from the container with the fast molecules to the container with the slow molecules constitutes the heat flowing from the hot body to the cold one.

Temperature measures the average translational energy of the molecules. We consider effects showing directly the motion of molecules in relation to temperature and heat.

11.1 Kinetic Gas Theory and Temperature

We first consider the simplest form of a molecular assembly, namely a gas, the particles of which are so far away from each other that the mutual attractive forces discussed in Chapter 10 are negligible. In an exact description the motion of these particles follows the rules of quantum mechanics. However, under certain conditions the quantum mechanical treatment can be replaced by a classical one (see Chapter 1).

Let us discuss an example. Hydrogen at room temperature is a gas. The gas exerts a pressure on the walls of a container, which depends on the number of particles and on the temperature. In the classical picture the pressure is the result of the collisions of the molecules. The mass of an hydrogen molecule is $m = 3 \times 10^{-27}$ kg, and its speed v is on the order of 10^3 m s^{-1} at room temperature (see Section 11.2). In the quantum mechanical treatment we would assign the particles a de Broglie wavelength

$$\Lambda = \frac{h}{mv} = 2 \times 10^{-10} \text{ m} \tag{11.1}$$

11.1 KINETIC GAS THEORY AND TEMPERATURE

according to equation (1.10). This wavelength is small compared to the dimensions of a macroscopic container (length = 1 cm) and small compared to the path between subsequent collisions (10^{-7} m, see Section 11.3), so the molecules can be well described as particles that obey the laws of classical mechanics.

11.1.1 Thermal Motion and Pressure

The molecules of a gas in a container are bouncing on the walls continuously. What is the pressure exerted on the walls?

Calculation of pressure

First we calculate the pressure the gas molecules exert on the walls of the container in which they are enclosed. For simplicity we neglect intermolecular forces; we attribute the same speed v to all molecules and assume elastic collisions of the molecules with the walls of the container. We consider N molecules in a cubic container of side edge L (Figure 11.1(a)); in our model $\frac{1}{6}N$ molecules are moving with speed v in the x direction, perpendicularly toward one of the six walls of the cube (area L^2) exerting a force f at their impact with the wall. The pressure P on a container wall is then

$$P = \frac{\text{force}}{\text{area}} = \frac{f}{L^2} \tag{11.2}$$

We use Newton's law for the calculation of f

$$\text{force} = \text{mass} \times \text{acceleration} = m_{\text{gas}} \frac{\text{velocity change}}{\Delta t} \tag{11.3}$$

m_{gas} is the mass of all the particles that collide with the wall in time Δt. According to the laws of elastic collisions a particle moves with the same speed, but into the opposite direction after its collision with the wall (Figure 11.2), thus the change in velocity is $2v$. Since the force exerted on the wall is opposite to the force exerted on the particles,

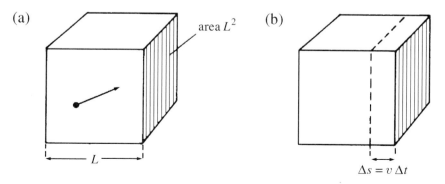

Figure 11.1 (a) Gas particles in a cubic container of length L. (b) one–sixth of the gas particles in partial volume $\Delta s \cdot L^2$ reach the wall in time Δt.

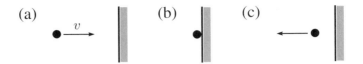

Figure 11.2 Elastic collisions of a particle with the wall. (a) before the collision, (b) at the moment of the collision, and (c) after the collision.

we obtain

$$f = m_{\text{gas}} \frac{2v}{\Delta t} \tag{11.4}$$

If ΔN gas particles reach the wall in time Δt, then

$$m_{\text{gas}} = m \cdot \Delta N \tag{11.5}$$

where m is the mass of one gas particle. In time Δt, only those particles which move toward the wall will reach it from a distance smaller than $\Delta s = v\Delta t$ (Figure 11.1(b)); these are one-sixth of all the particles in a volume

$$\Delta V = L^2 \Delta s = L^2 v \Delta t \tag{11.6}$$

This yields

$$\Delta N = \frac{1}{6} N \frac{\Delta V}{V} = \frac{1}{6} \frac{N}{V} L^2 v \Delta t \tag{11.7}$$

and finally

$$P = \frac{f}{L^2} = m \cdot \Delta N \cdot \frac{2v}{\Delta t} \frac{1}{L^2} = \frac{1}{3} \frac{N}{V} mv^2 \tag{11.8}$$

Since

$$E_{\text{trans}} = \tfrac{1}{2} mv^2 \tag{11.9}$$

is the kinetic energy of the translational motion of a gas particle, we can rewrite equation (11.8):

$$P = \frac{2N}{3V} \cdot \frac{mv^2}{2} = \frac{2}{3} \frac{N}{V} E_{\text{trans}} \tag{11.10}$$

or

$$\boxed{PV = \tfrac{2}{3} N \cdot E_{\text{trans}}} \tag{11.11}$$

This means that the product PV is a constant at a given translational energy of each gas molecule.

The same result is obtained when we take into account that the particles are moving in different directions (Box 11.1). As we will show in Chapter 12 the molecules in a gas do not have the same speed. For this reason v^2 and E_{trans} have to be replaced by their mean values $\overline{v^2}$ and $\overline{E}_{\text{trans}} = \tfrac{1}{2} m \overline{v^2}$.

11.1 KINETIC GAS THEORY AND TEMPERATURE

Box 11.1 – Pressure on the Wall of a Container

In the simplified model we assume that one-sixth of all particles are moving perpendicular to the wall. Actually, the particles are moving with arbitrary angles ϑ relative to the wall (Figure 11.3). Thus the force on the wall is proportional to its component perpendicular to the wall: v in equation (11.4) has to be replaced by $v \cdot \cos \vartheta$. To calculate the number of particles reaching the wall in time Δt, we have to replace $v/6$ in equation (11.7) by $v \cdot \cos \vartheta$. Then the pressure $P(\vartheta)$ exerted on the wall if all particles are moving under an angle ϑ toward the wall is

$$P(\vartheta) = 2\frac{N}{V}mv^2 \cos^2 \vartheta$$

The probability of finding a particle moving between ϑ and $\vartheta + d\vartheta$ is (see Figure 11.3(b))

$$\frac{2\pi \sin \vartheta \cdot d\vartheta}{4\pi} = \frac{1}{2} \sin \vartheta \cdot d\vartheta$$

and we obtain the total pressure P by integration

$$P = \int_0^{\pi/2} 2\frac{N}{V}mv^2 \cos^2 \vartheta \cdot \frac{1}{2} \sin \vartheta \cdot d\vartheta = \frac{1}{3}\frac{N}{V}mv^2$$

This is the same result as obtained in the simplified model (equation (11.8)).

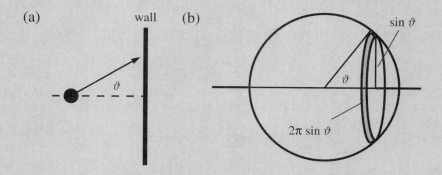

Figure 11.3 A particle reaching a wall at an angle ϑ.

Furthermore, the number N of particles can be expressed by the amount of substance n and Avogadro's constant N_A (see Box 11.2)

$$N = N_A \cdot n \tag{11.12}$$

Then we can rewrite equation (11.11) as

$$\boxed{PV = \tfrac{2}{3}N \cdot \overline{E}_{\text{trans}} = \tfrac{2}{3}nN_A \cdot \overline{E}_{\text{trans}}} \tag{11.13}$$

Box 11.2 – Molar Quantities

The number N of particles in a molecular assembly can be expressed by the amount of substance n. The amount of substance $n = 1$ mol is defined as an assembly of atoms or molecules that contains the same number of atoms or molecules as 12 g of the carbon isotope ^{12}C.

For example, from this definition we obtain the amount of substance in a reservoir of hydrogen gas when we examine the chemical reaction

$$2H_2 + C \longrightarrow CH_4$$

It is found experimentally that 12 g of the carbon isotope ^{12}C react with 4.032 g H_2. Since for each ^{12}C atom two H_2 molecules are consumed, the amount of substance $n_{H_2} = 1$ mol corresponds to 2.016 g H_2.

The quantity

$$M_m = \frac{\text{mass of the molecular assembly}}{\text{amount of substance}} = \frac{N \cdot m}{n}$$

is called the *molar mass*; for ^{12}C we get $M_m = 12$ g/mol, for H_2 we get $M_m = 2.016$ g/mol (N = number of molecules, m = mass of one molecule). For other molecules M_m is calculated in a corresponding manner (e.g., see table on the cover). The number of particles N is, on the other hand, proportional to the amount of substance n:

$$N = N_A \cdot n.$$

The proportionality factor N_A is called the *Avogadro constant*. N_A is obtained by measuring the lattice distances of crystalline carbon (isotope ^{12}C); in this way the number of lattice sites in 12 g of carbon can be counted. The value obtained for N_A is

$$N_A = 6.0221367 \times 10^{23} \text{ mol}^{-1}$$

Thus the molar mass M_m is related to its molecular mass m by

$$M_m = N_A \cdot m.$$

For example, for ^{12}C we have $M_m = 12$ g mol^{-1} and $m = (12 \text{ g mol}^{-1})/(6.022 \times 10^{23} \text{ mol}^{-1}) = 1.99 \times 10^{-26}$ kg.

Furthermore,

$$c = \frac{n}{V} \quad \text{and} \quad V_m = \frac{V}{n} = \frac{1}{c}$$

are the concentration and the molar volume, respectively.

11.1 KINETIC GAS THEORY AND TEMPERATURE

This equation relates the pressure P of a gas ($N = nN_A$ molecules) on the wall of a container of volume V to the average translational energy $\overline{E}_{trans}$ of each molecule.

Comparison with experiment

We consider a cylinder filled with 1 mol of hydrogen gas (i.e., with 2.016 g H$_2$) at a pressure $P_1 = 1 \text{ bar} = 1 \times 10^5 \text{ Pa} = 1 \times 10^5 \text{ N m}^{-2}$. (One bar is approximately the atmospheric pressure; 1 N is approximately the weight of a mass of 0.1 kg.) The cylinder is placed in a bath with melting ice. In the experiment we find that the volume of the gas is $V_1 = 22.712$ L. A pressure change to $P_2 = 2$ bar or $P_3 = \frac{1}{2}$ bar causes a change in the enclosed gas volume to $V_2 = \frac{1}{2}V_1$ or $V_3 = 2V_1$, respectively (Figure 11.4).

More generally we find for 1 mol of a gas in a container placed in melting ice

$$PV = 22.712 \text{ bar L} \quad \text{(1 mol gas, melting ice)} \tag{11.14}$$

Thus the volume V of the gas is inversely proportional to its pressure P (Figure 11.5): this is *Boyle's law* (1664; also known as Mariotte's law). Bernoulli (Daniel Bernoulli, 1738) demonstrated that this early finding can be explained by the given considerations (kinetic theory of gases).

From the measured data we can calculate the translational energy of a hydrogen molecule in a gas container placed in melting ice according to equation (11.13).

$$\overline{E}_{trans} = \frac{PV}{\frac{2}{3}nN_A} = \frac{22.712 \text{ bar L}}{\frac{2}{3} \text{ mol } 6.022 \times 10^{23} \text{ mol}^{-1}} = 5.66 \times 10^{-21} \text{ J} \tag{11.15}$$

(with 1 bar L = 10^5 N m^{-2} L = 100 N m = 100 J).

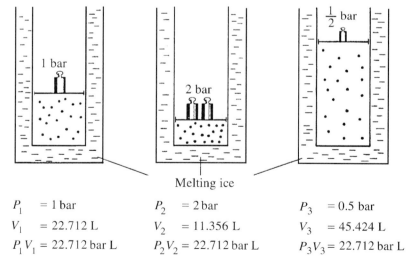

Figure 11.4 Compression and expansion of a gas which is in a container in melting ice ($n = 1$ mol); above the piston there is vacuum. More accurately, we consider an ice bath in an evacuated container such that the liquid water is in equilibrium with the ice and with the water vapor (triple point of water). In this case no air is present in the system.

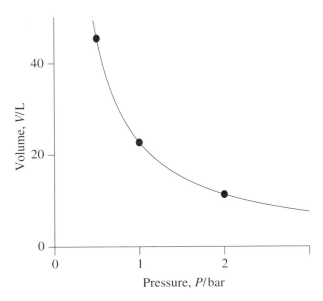

Figure 11.5 The volume of a gas which is in a container in melting ice as a function of pressure ($n = 1$ mol).

This means that, according to equation (11.9), the H_2 molecule (mass $m = 3.32 \times 10^{-27}$ kg) has a speed of

$$v = \sqrt{\frac{2\overline{E}_{\text{trans}}}{m}} = \sqrt{\frac{2 \times 5.66 \times 10^{-21}\,\text{J}}{3.32 \times 10^{-27}\,\text{kg}}} = 1850\,\text{m s}^{-1}$$

11.1.2 Avogadro's Law

Measuring the volume of 1 mol of gas at a pressure of 1 bar in a container in melting ice for different gases (e.g., N_2, CO_2, O_2, H_2) results in the same value: this is *Avogadro's law* (1776–1856). In other words, Avogadro's law and Boyle's law can be expressed by stating that equation (11.14), obtained from measurements with H_2, is true for all gases.

Can we understand this surprising result on the basis of the kinetic theory of gases? To do this we consider again a container with H_2 molecules (mass m_1, average kinetic energy $\frac{1}{2}m\overline{v_1^2}$) in a bath with melting ice. Then we add a CO_2 molecule (mass m_2) which is at rest in the beginning (Figure 11.6). According to the laws of elastic collisions, energy is transferred from the H_2 molecules to the CO_2 molecule, so its speed increases at the expense of the speed of the H_2 molecules. This procedure will continue until an equilibrium has been established: on a time average the same amount of energy will be transferred from the H_2 molecules to the CO_2 molecule as vice versa from the CO_2 molecule to the H_2 molecules. It can be shown (see Chapter 12) that this stage is reached for

$$\tfrac{1}{2}m_1\overline{v_1^2} = \tfrac{1}{2}m_2\overline{v_2^2} \qquad (11.16)$$

According to equation (11.16), particles with a small mass, on the average, move faster than particles with a larger mass. This can easily be shown for the collision of two particles

11.1 KINETIC GAS THEORY AND TEMPERATURE

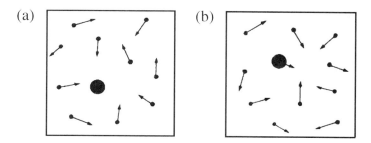

Figure 11.6 A CO_2 molecule in between H_2 gas molecules. (a) At the beginning the CO_2 molecule is at rest. (b) The motion of the CO_2 molecule after thermal equilibrium has been reached.

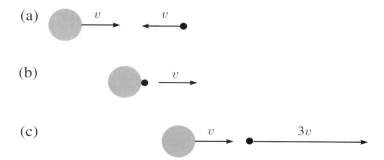

Figure 11.7 Central collision of two particles. The mass of the particle on the right is negligibly small compared to the mass of the particle on the left. (a) Before collision, (b) at the moment of collision, and (c) after collision.

with very different masses and equal speed (Figure 11.7). After the collision the speed of the heavy particle is practically unchanged, whereas the lighter particle moves with a speed three times higher than the heavy one.

According to the laws of mechanics, the mean translational energies of both kinds of molecules in dynamic equilibrium are equal. There is no change in the conclusion when we consider two separate containers with different kinds of molecules in a bath with melting ice (Figure 11.8). In this case, also the translational energies of both kinds of

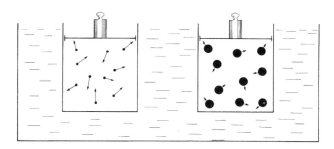

Figure 11.8 Two containers with different kinds of molecules in a bath of melting ice.

molecules must be equal. Thus with equation (11.11) we obtain

$$\frac{1}{2}m_1\overline{v_1^2} = \frac{1}{2}m_2\overline{v_2^2} = \frac{3}{2}\frac{P_1 V_1}{N_1} = \frac{3}{2}\frac{P_2 V_2}{N_2} \qquad (11.17)$$

In this way we can understand Avogadro's law on the basis of molecular motion. Two different gases with the same number of molecules ($N_1 = N_2$) under the same pressure ($P_2 = P_1$) in a container in melting ice have the same volume.

11.1.3 Temperature Equilibration and Heat

We consider in more detail what happens if we transfer a small container with gas molecules into a temperature bath. We imagine the temperature bath as a huge container filled with gas molecules; inside this container we place the small container. We assume that the molecules in the small container move slowly:

$$\overline{E}_{\text{trans, small container}} < \overline{E}_{\text{trans, temperature bath}}$$

The molecules in the temperature bath collide with the molecules of the wall of the small container, and energy is transferred through the wall. After some time the mean translation energy of the molecules in the small container equals the mean translation energy in the temperature bath:

$$\overline{E}_{\text{trans, small container}} = \overline{E}_{\text{trans, temperature bath}}$$

This theoretical statement is in agreement with the experimental fact that a hot body, brought in contact with a cold body, cools down and the cold body warms up until both bodies are equally warm (temperature equilibration). It suggests one can relate E_{trans} to the energy transferred from the hot to the cold body. This process will be discussed in more detail in Chapter 14. At this point we focus on the question: what is the relation between E_{trans} and temperature?

11.1.4 Ideal Gas Law: Definition of Absolute Temperature

Temperature measures the translational energy

When we transfer the cylinder filled with 1 mol of a gas at a pressure of $P = 1$ bar in a bath of melting ice (Figure 11.4) into a bath with boiling water the gas volume increases from 22.712 L to 31.026 L (Figure 11.9). This means that in the transition from an ice bath to a bath with boiling water the product PV of the gas increases:

$$(PV)_{\text{boiling water}} = (PV)_{\text{melting ice}} \cdot \frac{31.026}{22.712} \qquad (11.18)$$

A smaller value of PV corresponds to a lower temperature: the translational energy of the gas particles is smaller – that is, the particles move more slowly. The high temperature, however, corresponds to a high translational energy – that is, to a fast motion of the molecules (Figure 11.10).

From our considerations it seems obvious that the motion of molecules can be used to measure the temperature. We define, therefore, that the mean translational energy is

11.1 KINETIC GAS THEORY AND TEMPERATURE

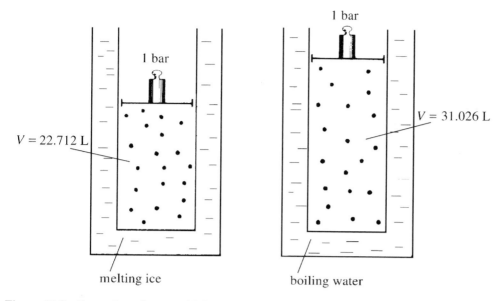

Figure 11.9 Expansion of a gas with increasing temperature.

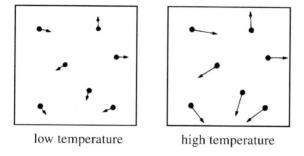

Figure 11.10 Thermal motion of gas molecules. Low speed at low temperature, high speed at high temperature.

proportional to the temperature T:

$$\overline{E}_{\text{trans}} = C \cdot T \tag{11.19}$$

where C is a constant. From equation (11.13) it follows

$$PV = \tfrac{2}{3} N \cdot \overline{E}_{\text{trans}} = \tfrac{2}{3} NC \cdot T = \tfrac{2}{3} n N_A \cdot C \cdot T \tag{11.20}$$

This enables us to measure the temperature T by measuring PV. We choose the temperature scale arbitrarily such that the temperature of an ice bath (more exactly: of a bath at the triple point of water) is $T = 273.160$ temperature units; we call this unit the kelvin (K). Then the temperature T is defined by

$$T = 273.160 \,\text{K} \cdot \frac{(PV)_T}{(PV)_{\text{triple point of water}}} \quad (11.21)$$

A triple point bath is an ice bath prepared from pure water without dissolved air; liquid, solid and gaseous water are in equilibrium at the melting point. T is proportional to PV (Figure 11.11); T is also called the *absolute temperature* because it is not conceivable that PV can become less than zero. The definition of T is chosen such that the temperature difference between a bath with melting ice (saturated with air at 1 atm = 1.013 bar, temperature 273.15 K) and boiling water (at 1 atm, temperature 273.160 K × 31.026/22.712 = 373.15 K) is 100 K. These two temperatures are the fixed points of the earlier defined Celsius scale (temperature t). Both scales are related by

$$\frac{t}{^\circ\text{C}} = \frac{T - 273.15\,\text{K}}{\text{K}}$$

Accordingly, 273.15 K corresponds to 0 °C and 373.15 K corresponds to 100 °C. The triple point of water (273.160 K) is then at 0.01 °C.

Relation (11.21) tells us how to measure the absolute temperature T: we compare the pressure of a gas in a container at constant volume in a temperature bath with its pressure when putting the container in a bath of the triple point temperature of water.

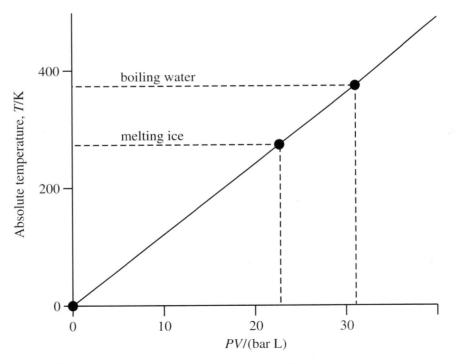

Figure 11.11 The absolute temperature T as a function of PV.

11.1 KINETIC GAS THEORY AND TEMPERATURE

Ideal gas law

Now we are able to specify the constant C in equation (11.19). Using $\overline{E}_{trans}$ from equation (11.15) we obtain

$$C = \frac{\overline{E}_{trans}}{T} = \frac{5.66 \times 10^{-23} \, \text{bar L}}{273.160 \, \text{K}} = 2.072 \times 10^{-25} \, \text{bar L K}^{-1}$$

It is customary to use the expression $\frac{3}{2}k$ instead of C in equation (11.19):

$$\boxed{\overline{E}_{trans} = \tfrac{3}{2}kT} \qquad (11.22)$$

We rewrite equation (11.20) as

$$\boxed{PV = NkT = nN_A kT = n\boldsymbol{R}T} \qquad (11.23)$$

The quantity k is called the *Boltzmann constant*:

$$k = \tfrac{2}{3} \cdot C = \tfrac{2}{3} \cdot 2.072 \times 10^{-25} \, \text{bar L K}^{-1} = 1.381 \times 10^{-25} \, \text{bar L K}^{-1}$$
$$= 1.381 \times 10^{-23} \, \text{J K}^{-1} \text{ (more accurately}: k = 1.380658 \times 10^{-23} \, \text{J K}^{-1}\text{)}.$$

The constant

$$\boldsymbol{R} = N_A k = 6.022 \times 10^{23} \cdot 1.381 \times 10^{-25} \, \text{bar L K}^{-1} \, \text{mol}^{-1}$$
$$= 8.314 \times 10^{-2} \, \text{bar L K}^{-1} \, \text{mol}^{-1} = 8.314 \, \text{J K}^{-1} \, \text{mol}^{-1}$$

is called the *gas constant* (more accurately: $\boldsymbol{R} = 8.314510 \, \text{J K}^{-1} \, \text{mol}^{-1}$).

Equation (11.23) is called the *ideal gas law* (Figure 11.12). This law can also be written in the form

$$\boxed{PV_m = \boldsymbol{R}T} \qquad (11.24)$$

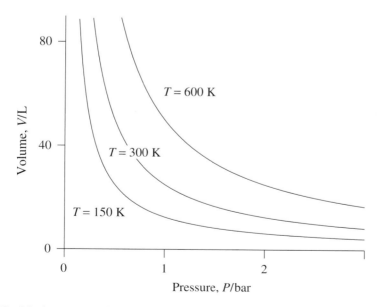

Figure 11.12 Ideal gas. V as a function of P (isotherms) for $n = 1$ mol and different temperatures T according to equation (11.23).

where

$$V_m = \frac{V}{n} \qquad (11.25)$$

is the molar volume.

Note: The volume V is an extensive variable, it is proportional to the amount of substance. In contrast, the molar volume V_m is an intensive variable. Molar quantities such as V_m are represented by the index "m" throughout this book.

Equations (11.22) and (11.23) are fundamental in physical chemistry. Equation (11.22) gives the mean translational energy E_{trans} of a molecule as a function of temperature T. Equation (11.23) connects pressure P, volume V, and temperature T of a gas.

Note that the temperature T is defined by equation (11.23). Equation (11.22) then follows from the fact that the pressure of a gas on the wall of a container is related to the translational energy E_{trans}.

Gas thermometer

At this point we have to describe more precisely our procedure of measuring the temperature. When plotting PV_m as a function of P according to equation (11.23), a line parallel to the P axis is expected. In reality, however, deviations occur at higher pressure (Figure 11.13); these result from the attraction forces between the gas molecules and from the fact that the volume of the molecules is not small enough to be neglected completely. At $P = 1$ bar the deviations from the ideal gas law are on the order of 1%. However, for $P \to 0$ all the curves in Figure 11.13 meet at the same point, independent of the nature of the gas. Following this observation we replace PV by the limit of PV for $P \to 0$ in the ideal gas law. Then instead of equation (11.23) we obtain

$$\lim_{P \to 0} PV = n\mathbf{R}T \qquad (11.26)$$

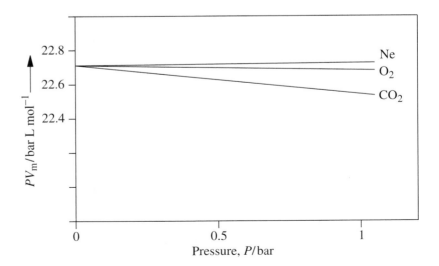

Figure 11.13 PV_m as a function of pressure P at $T = 273.16$ K.

11.2 Speed of Molecules in a Gas

Only this form enables us to define the temperature T exactly:

$$T = \frac{\lim_{P \to 0} PV}{\lim_{P \to 0} (PV)_{\text{triple point of water}}} \times 273.160 \text{ K} \tag{11.27}$$

(*gas thermometer*).

11.2 Speed of Molecules in a Gas

11.2.1 Mean Speed

From the translational energy we can easily calculate the speed of a gas molecule. According to equations (11.9) and (11.22) we have

$$\overline{v^2} = \frac{2\overline{E}_{\text{trans}}}{m} = \frac{3kT}{m} = \frac{3RT}{M_m} \tag{11.28}$$

Here the mass m of a molecule is expressed by the molar mass M_m and Avogadro's constant

$$M_m = m \cdot N_A \tag{11.29}$$

The speed is obtained by taking the square root of $\overline{v^2}$. Actually, it is more important to calculate $\bar{v}$, the mean speed. As will be shown in Chapter 12, we obtain

$$\bar{v} = \sqrt{\frac{8kT}{\pi m}} = \sqrt{\frac{8RT}{\pi M_m}} = 0.92 \cdot \sqrt{\overline{v^2}} \tag{11.30}$$

Example

For H_2 at $T = 300$ K with $M_m = 2.016$ g/mol we find

$$\bar{v} = \sqrt{\frac{8 \times 8.314 \text{ J K}^{-1} \text{ mol}^{-1} \times 300 \text{ K}}{\pi \times 2.016 \times 10^{-3} \text{ kg mol}^{-1}}} = \sqrt{3.15 \times 10^6 \text{ J kg}^{-1}}$$

$$= 1.77 \times 10^3 \text{ m s}^{-1} = 1.77 \text{ km/s}$$

The H_2 molecules move with a very high speed at room temperature. If we cool the gas down, the speed of the molecules becomes lower:

T/K	$\bar{v}$/(m s^{-1})
300	1770
100	1022
1	102
0.01	10

At absolute zero we have $\bar{v} = 0$, and any motion of the molecules stops. (As mentioned in Section 11.1, we neglect quantum effects leading to a zero-point energy.) Since

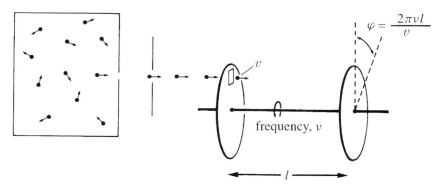

Figure 11.14 Apparatus for measuring the speed of molecules.

according to equation (11.28) $\overline{v^2}$ is inversely proportional to the mass of the molecules, heavy molecules move slower than light molecules. For example, the speed of Br_2 molecules ($M_m = 159\,\text{g mol}^{-1}$) at 300 K is $\overline{v} = 203\,\text{m s}^{-1}$, which is one-tenth of the speed of H_2 molecules.

The speed of molecules can be measured by firing a stream of particles (e.g., Na atoms) at two discs rotating on the same axis inside a vacuum chamber (Figure 11.14). The first of these discs has a slit through which the molecules can pass. They are deposited on the surface of the second disc and detected as a small spot. The molecules need a finite time to reach the second disc, so by the time the molecules collide with the second disc this disc has rotated by an additional angle φ. The speed can be calculated from the angle φ, the frequency ν of rotation, and the distance l between the two discs. The average speed found in this experiment is in good agreement with the calculated speed.

11.2.2 Effusion

A gas is in a container with a very small hole (diameter less than the mean free path, Section 11.3) connected to a vacuum line. The probability that a molecule enters the hole and escapes is proportional to its speed. Thus the rate of effusion is inversely proportional to the square root of the molecular mass (*Graham's law*). This effect allows us to separate isotopes. Light isotopes escape from the container faster than heavy isotopes (Problem 11.5).

11.3 Mean Free Path and Collision Frequency

Even though a molecule in a gas moves at a high speed, it can move forward only very slowly because of collisions with other molecules which deflect it from its original path (Figure 11.15).

It is of general interest to know the average path of a molecule between subsequent collisions (*mean free path*). The frequency of collisions related to the mean free path and to the speed $\overline{v}$ is of particular interest when considering the rate of chemical reactions (Chapter 21).

11.3 MEAN FREE PATH AND COLLISION FREQUENCY

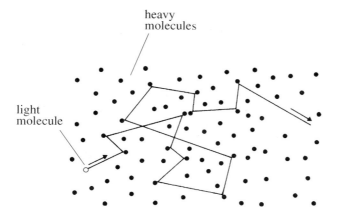

Figure 11.15 A zigzag path of a light molecule colliding with heavy molecules considered to be at rest.

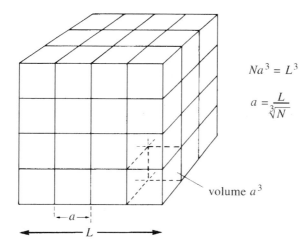

Figure 11.16 The average distance a of two molecules in a gas is estimated by assuming, for simplicity, a regular arrangement of gas molecules in a cube.

11.3.1 Mean Free Path

Before calculating the mean free path we ask for the average distance a between nearest neighbors. Let us imagine that the molecules are evenly distributed in a cube of length L (Figure 11.16); then $L^3 = N \cdot a^3$ where N is the total number of molecules in the cube and a is the average distance of two nearest neighbors. For a we obtain

$$a = \sqrt[3]{\frac{L^3}{N}} = \sqrt[3]{\frac{V}{N}} = \sqrt[3]{\frac{kT}{P}}$$

For $P = 1$ bar and $T = 273.16$ K the result for the average distance is $a = 3.3 \times 10^{-9}$ m = 3.3 nm.

However, the free path λ a molecule traverses between two collisions is much longer than the average distance a because the molecule passes between many neighboring molecules (Figure 11.15). To simplify the treatment, we assume that λ is the same in all collisions. To calculate λ we consider, as in Figure 11.15, a light particle moving among many heavy particles; for simplicity we imagine the heavy particles approximately at rest. We also approximate the particles as spheres. A collision occurs each time a light particle (radius r_1) touches a heavy particle (radius r_2) (Figure 11.17). We consider the path of the center of gravity of the light particle between two collisions as the axis of a cylinder of diameter $2(r_1 + r_2)$. All heavy particles which have their centers of gravity inside or on the surface of the considered cylinder collide with the light particle.

According to Figure 11.17, when the light particle has traveled a distance λ, one collision occurs on the average. This means that in a cylinder of volume

$$V_{cyl} = \pi(r_1 + r_2)^2 \lambda \tag{11.31}$$

only one heavy molecule can be found on the average. There are N heavy particles in the total volume V of the gas and one particle, on the average, is in volume V_{cyl}; thus

$$V_{cyl} = \frac{V}{N} \tag{11.32}$$

and from equations (11.31) and (11.32) we obtain

$$\lambda = \frac{V}{\pi(r_1 + r_2)^2 N} \tag{11.33}$$

Because of the distribution of the free path (see Foundation 11.1) we have to replace λ by the mean free path $\bar{\lambda}$. Furthermore, besides the collision diameter $r_1 + r_2$, we introduce the collision cross-section σ

$$\sigma = \pi(r_1 + r_2)^2 \tag{11.34}$$

Then we obtain

$$\boxed{\bar{\lambda} = \frac{V}{\sigma N}} \quad \text{(light particle among heavy particles)} \tag{11.35}$$

In our consideration we assumed that the heavy molecules are practically at rest. What happens when all molecules are moving with the same speed? Then, when two molecules are approaching each other between two collisions, the free path becomes smaller than calculated by equation (11.35). Let us assume that all molecules have the same mass

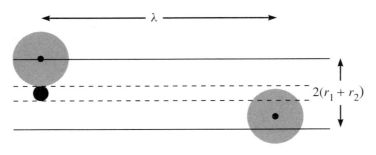

Figure 11.17 Calculation of the mean free path λ. A collision occurs when the two colliding molecules touch each other.

11.3 MEAN FREE PATH AND COLLISION FREQUENCY

(e.g., the motion of a selected H_2 molecule among a multitude of H_2 molecules); for all collisions where the speed of colliding molecule 1 is u and the speed of molecule 2 is $-u$ the mean free path is half the value calculated from equation (11.35). On the average, in this case we obtain:

$$\boxed{\bar{\lambda} = \frac{V}{\sqrt{2}\sigma N}} \quad \text{(particles of equal mass)} \tag{11.36}$$

In the general case (mean free path $\bar{\lambda}_1$ of a particle with mass m_1 moving among particles with mass m_2) we obtain

$$\bar{\lambda}_1 = \frac{V}{\sigma N} \cdot \frac{1}{\sqrt{1 + m_1/m_2}} \tag{11.37}$$

For $m_1 \ll m_2$ this leads to equation (11.35), and for $m_1 = m_2$ this leads to equation (11.36). In the case $m_1 \gg m_2$ (one very heavy particle moving among light particles) the mean free path approaches zero: many collisions with the light particles occur, while the heavy particle travels only a very small distance.

Example – H_2 molecule moving among Br_2 molecules

In this case a light particle is moving among heavy particles. With $(r_1 + r_2) = (138 + 312)\,\text{pm} = 450\,\text{pm}$, where the index 1 refers to the H_2 molecule and the index 2 refers to the Br_2 molecules, we obtain $\sigma = 6.36 \times 10^{-19}\,\text{m}^2$; at a pressure of $P = 1$ bar and a temperature of $T = 273.16$ K we have the number density $N/V = 6.022 \times 10^{23}/(22.711\,\text{L}) = 2.65 \times 10^{25}\,\text{m}^{-3}$; then with equation (11.35) we obtain

$$\bar{\lambda}_{H_2} = \frac{1}{6.36 \times 10^{-19}\,\text{m}^2 \cdot 2.65 \times 10^{25}\,\text{m}^{-3}} = 59\,\text{nm}$$

$\bar{\lambda}$ is then about 20 times larger than the average distance $a = 3.3$ nm between the gas molecules.

Example – H_2 molecule moving among H_2 molecules

With $(r_1 + r_2) = 2r_1 = 276$ pm we obtain $\sigma = 2.39 \times 10^{-19}\,\text{m}^2$; with $N/V = 2.65 \times 10^{25}\,\text{m}^{-3}$ we obtain from equation (11.36)

$$\bar{\lambda}_{H_2} = \frac{1}{\sqrt{2} \cdot 2.39 \times 10^{-19}\,\text{m}^2 \cdot 2.65 \times 10^{25}\,\text{m}^{-3}} = 112\,\text{nm}$$

Example – Br_2 molecule moving among N_2 molecules

In this case $r_1 + r_2 = (312 + 190)\,\text{pm} = 502$ pm, where the index 1 refers to the Br_2 molecule and the index 2 refers to the N_2 molecules. Then $\sigma = 7.92 \times 10^{-19}\,\text{m}^2$. With $N/V = 2.65 \times 10^{25}\,\text{m}^{-3}$ and $m_1/m_2 = 159.6/28.0 = 5.7$ we obtain from equation (11.37)

$$\bar{\lambda}_{Br_2} = \frac{1}{7.92 \times 10^{-19}\,\text{m}^2 \cdot 2.65 \times 10^{25}\,\text{m}^{-3}} \cdot \frac{1}{\sqrt{1 + 5.7}}$$
$$= 47 \times 0.39\,\text{nm} = 19\,\text{nm}$$

11.3.2 Collision Frequency

From $\bar{\lambda}$ and the speed $\bar{v}$ we can calculate the number z of collisions that occur to a molecule in a time interval Δt. The total traveled path in time Δt is

$$s = \bar{\lambda} z = \bar{v} \Delta t \tag{11.38}$$

Then

$$z = \frac{\bar{v} \Delta t}{\bar{\lambda}} \tag{11.39}$$

For the collision frequency ν we obtain

$$\nu = \frac{z}{\Delta t} = \frac{\bar{v}}{\bar{\lambda}} \tag{11.40}$$

Example

We consider H_2 molecules at $T = 300$ K and $P = 1$ bar. In this case the speed is $\bar{v} = 1.77 \times 10^3$ m s^{-1} and the mean free path is $\bar{\lambda} = 112$ nm. Then we obtain $\nu = 1.6 \times 10^{10}$ s^{-1}. Thus in every second a molecule collides about 10^{10} times with other molecules in the gas.

11.4 Diffusion

11.4.1 Mean Displacement

We ask how far a gas molecule can travel in the x direction from its initial position in a time Δt (diffusion path, see Figure 11.18). We consider a light molecule moving in a gas of heavy molecules. At time $t = 0$ the light molecule is at the origin of a x, y, z coordinate system. We assume that the molecule is moving in the x direction with speed v and that the direction of the particle, after a collision, can be either changed by 180° (central collision) or, with about equal probability, remain the same (grazing collision).

To simplify the treatment, we assume that the molecule collides after equal distances λ (i.e., we neglect the distribution of the free path). We consider two limiting cases.

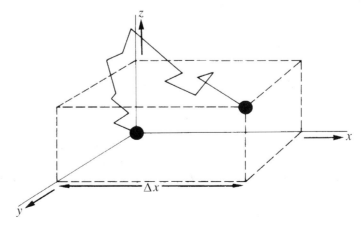

Figure 11.18 A zigzag path of a molecule. After time Δt the molecule traverses a distance Δx in the x direction.

11.4 DIFFUSION

(a) Many collisions occur in such a manner that after traveling the distance λ the molecule only bounces back and forth; then Δx cannot exceed λ:

$$\Delta x = \lambda \tag{11.41}$$

(b) After each collision the molecule moves by the distance λ in the x direction; then we find

$$\Delta x = j \cdot \lambda \tag{11.42}$$

where j is the number of collisions in time Δt.

We show in Box 11.3 and in Problem 11.9 that Δx^2, on the average, is given by the product of the distances reached in the two limiting cases:

$$\boxed{\overline{\Delta x^2} = j \cdot \lambda^2} \tag{11.43}$$

For example, after $j = 7$ collisions on the path, the traveled distance can be, by chance,

$$\Delta x = \lambda + \lambda + \lambda - \lambda + \lambda - \lambda + \lambda = 3\lambda$$

Box 11.3 – Averaging the Displacement Δx^2

We consider the displacement Δx of a particle moving randomly in the x direction (one-dimensional diffusion); at each collision, Δx can change by $+\lambda$ or by $-\lambda$. Let us assume that after j collisions the displacement is Δx_j. We can express Δx_j by the displacement Δx_{j-1} (displacement after $j-1$ collisions). If in the last collision the change in Δx was $+\lambda$, then

$$(\Delta x_j)^2 = (\Delta x_{j-1} + \lambda)^2 = (\Delta x_{j-1})^2 + 2\Delta x_{j-1}\lambda + \lambda^2$$

If in the last collision the change in Δx was $-\lambda$, then

$$(\Delta x_j)^2 = (\Delta x_{j-1} - \lambda)^2 = (\Delta x_{j-1})^2 - 2\Delta x_{j-1}\lambda + \lambda^2$$

Since the particle is moving randomly, both cases are equally probable; thus on the average we obtain

$$\overline{(\Delta x_j)^2} = \overline{(\Delta x_{j-1})^2} + \lambda^2$$

because, when averaging, the terms $+2\Delta x_{j-1}\lambda$ and $-2\Delta x_{j-1}\lambda$ occur statistically and cancel each other. In a corresponding way we obtain

$$\overline{(\Delta x_{j-1})^2} = \overline{(\Delta x_{j-2})^2} + \lambda^2$$

that is,

$$\overline{(\Delta x_j)^2} = \overline{(\Delta x_{j-2})^2} + 2\lambda^2 = \overline{(\Delta x_{j-3})^2} + 3\lambda^2 = \cdots = \overline{(\Delta x_1)^2} + (j-1) \cdot \lambda^2$$

Since $\overline{(\Delta x_1)^2} = \lambda^2$, we finally obtain $\overline{(\Delta x_j)^2} = j \cdot \lambda^2$.

(five times by λ in the $+x$ direction, and twice in the $-x$ direction). Then we obtain
$$\Delta x^2 = 9 \cdot \lambda^2$$
In a second case, the result of 7 collisions might be
$$\Delta x = -\lambda + \lambda - \lambda - \lambda + \lambda + \lambda - \lambda = -\lambda$$
$$\overline{\Delta x^2} = \lambda^2$$
On the average, according to equation (11.43), we expect
$$\overline{\Delta x^2} = 7 \cdot \lambda^2 \tag{11.44}$$

Note that the molecules move randomly, thus on the average a molecule will be found at time Δt with equal probability at a distance Δx to the right or to the left of its starting point; therefore, the average of Δx becomes zero. For this reason we proceeded by taking the average of the squares Δx^2.

For a three-dimensional motion of the molecules, only every third collision effectively leads to a displacement in the positive or negative x direction, so $\overline{\Delta x^2}$ is only one-third of the value calculated in equation (11.43). Furthermore, because of the distribution of the free path, we have to replace λ^2 by $\overline{\lambda^2}$
$$\boxed{\overline{\Delta x^2} = \tfrac{1}{3} j \cdot \overline{\lambda^2}} \tag{11.45}$$

Since $j\overline{\lambda}$ is the total path traversed by the molecule in time Δt
$$j\overline{\lambda} = \overline{v}\Delta t \tag{11.46}$$
it follows that
$$\overline{\Delta x^2} = \frac{1}{3} \cdot \frac{\overline{\lambda^2}}{\overline{\lambda}} \cdot \overline{v}\Delta t \tag{11.47}$$

$\overline{\lambda^2}$ is larger than $(\overline{\lambda})^2$, because in averaging the squares of large values of λ dominate. In Foundation 11.1 it is shown that
$$\overline{\lambda^2} = 2 \cdot (\overline{\lambda})^2 \tag{11.48}$$
Thus
$$\boxed{\overline{\Delta x^2} = \tfrac{2}{3}\overline{\lambda}\overline{v} \cdot \Delta t \quad \text{(light particle moving among heavy particles)}} \tag{11.49}$$

$\overline{\Delta x^2}$ increases proportionally to Δt. This means that the displacement increases proportional to the square root of time. For example, increasing the time by a factor of four results in an increase of the displacement by a factor of two only.

With the relations (11.30) and (11.35) we can rewrite equation (11.49) as
$$\overline{\Delta x^2} = \frac{2}{3} \cdot \frac{V}{\sigma \cdot N} \cdot \sqrt{\frac{8kT}{\pi m}} \cdot \Delta t \tag{11.50}$$

In the general case (particle of mass m_1 moving among particles of mass m_2) the mass m has to be replaced by the reduced mass μ
$$\mu = \frac{m_1 m_2}{m_1 + m_2} \tag{11.51}$$

11.4 DIFFUSION

For $m_1 \ll m_2$ this leads to equation (11.50), where m is the mass of the light particle moving among heavy particles.

For $m_1 \gg m_2$ (heavy particle moving among light particles) again we obtain equation (11.50) with m being the mass of the light particles.

How can we understand this surprising fact, that a heavy particle moving among light particles, in spite of its slow motion, reaches the same displacement as a light particle moving among heavy particles? When a heavy particle collides with a light particle it changes its direction only slightly. The persistence of its motion compensates for its low speed.

Example – displacement of Br_2 molecules

We consider a Br_2 molecule moving in air at atmospheric pressure and room temperature. The collision diameter is $r_1 + r_2 = (312 + 190)\,\text{pm} = 502\,\text{pm}$, if we treat air as nitrogen gas. Thus we obtain $\sigma = 7.92 \times 10^{-19}\,\text{m}^2$. Furthermore, the reduced mass is, with $m_{Br_2} = 26.4 \times 10^{-26}\,\text{kg}$ and $m_{N_2} = 4.65 \times 10^{-26}\,\text{kg}$,

$$\mu = \frac{26.4 \cdot 4.65}{26.4 + 4.65} \times 10^{-26}\,\text{kg} = 3.95 \times 10^{-26}\,\text{kg}$$

and we obtain $\sqrt{kT/\mu} = 324\,\text{m s}^{-1}$. Then the square of the displacement is (with the number density $N/V = 2.65 \times 10^{25}\,\text{m}^{-3}$ at 1 bar)

$$\overline{\Delta x^2} = 4.8 \times 10^{-8}\,\text{m} \cdot 324\,\text{m s}^{-1} \cdot \Delta t = 1.6 \times 10^{-5}\,\text{m}^2\,\text{s}^{-1} \Delta t$$

In $\Delta t = 15\,\text{s}$ the Br_2 molecule travels the distance $\sqrt{1.6 \times 10^{-5} \times 15\,\text{m}^2} = 1.6\,\text{cm}$ in the x direction; after $\Delta t = 150\,\text{s}$ the Br_2 molecule has advanced by 4.9 cm. The displacement does not increase proportional to Δt but, instead, proportional to $\sqrt{\Delta t}$.

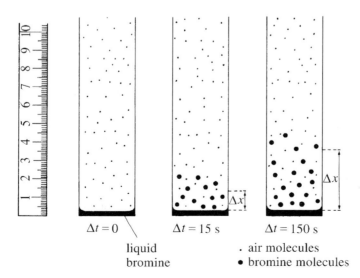

Figure 11.19 The advancing of the brown color of Br_2 gas in a cylinder filled with air.

This result can be checked experimentally by transferring a few drops of liquid bromine to the bottom of a cylinder filled with air and following the advancement of the brown-colored zone (Figure 11.19). When we plot the squares of the measured displacement versus the time Δt we obtain the straight line displayed in Figure 11.20: $\overline{\Delta x^2} = 1.6 \times 10^{-5}$ m^2 s$^{-1}\Delta t$, in good agreement with the theory.

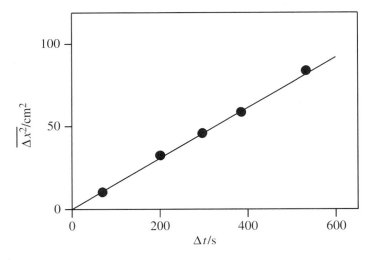

Figure 11.20 Mean square displacement $\overline{\Delta x^2}$ for the diffusion of Br$_2$ in air.

Example – displacement of nitrogen atoms adsorbed at a surface

It is most fascinating that single atoms adsorbed on a surface can be observed by using the scanning tunneling microscope (STM; see Chapter 23). Single nitrogen atoms adsorbed at a ruthenium single crystal have been observed. Each atom changes place after about 50 s. This can be seen in a sequence of STM pictures. The displacement Δx of each single atom was measured. In Figure 11.21 the average $\overline{\Delta x^2}$ is plotted versus time t. The straight line shows that our consideration of random motion applies even here and is not restricted to thermal motion and molecules in a gas. (see G. Ertl, Dynamik der Wechselwirkung von Molekülen mit Oberflächen, *Ber. Bunsenges. Phys. Chem.*, 1995, **99**, 1282).

11.4.2 Equation of Einstein and Smoluchowski

We have shown that the mean square displacement $\overline{\Delta x^2}$ is proportional to the diffusion time Δt. Usually this proportionality is written as

$$\overline{\Delta x^2} = 2D \cdot \Delta t \tag{11.52}$$

This is the *equation of Einstein and Smoluchowski*. D is called the *diffusion coefficient*.

11.4 DIFFUSION

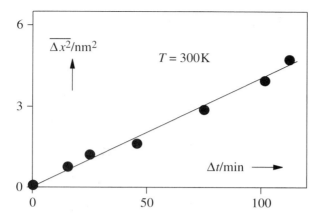

Figure 11.21 Nitrogen atoms adsorbed on a surface. Average of the square displacement $\overline{\Delta x^2}$ of selected nitrogen atoms versus time. For the slope we find $\overline{\Delta x^2}/\Delta t = 10^{-21}$ m² s^{-1}.

In the case of molecules diffusing in a gas we obtain for the diffusion coefficient from equations (11.50) and (11.51)

$$D = \frac{1}{2} \cdot \frac{V}{\sigma \cdot N} \cdot \sqrt{\frac{kT}{\mu}}, \quad \mu = \frac{m_1 m_2}{m_1 + m_2} \qquad (11.53)$$

Example – diffusion coefficient

For bromine molecules diffusing through a gas of nitrogen molecules we obtain with the data in Section 11.4.1 $D = \frac{1}{2} \cdot 1.6 \times 10^{-5}$ m² s^{-1} = 0.8×10^{-5} m² s^{-1}. Correspondingly, for the diffusion of H$_2$ molecules through a gas of bromine molecules we obtain $D = 3.5 \times 10^{-5}$ m² s^{-1}.

The equation of Einstein and Smoluchowski is of particular importance when considering Brownian motion of particles in solution (see Section 11.6.4). It allows one to calculate the time Δt required to travel a certain distance by diffusion.

11.4.3 Fick's Law

What amount of bromine vapor emerges through the opening of the cylinder with cross-section A and height H in Figure 11.22 in time Δt? We assume that there is a continuous flow of bromine vapor from the bottom of the cylinder (by vaporizing the liquid bromine) and that a bromine molecule reaching the opening will be trapped immediately (by a chemical reaction with a silver foil which covers the opening). Under these conditions a linear concentration gradient is established in the cylinder from the bottom to the top. Through any cross-section of the cylinder the same amount Δn of Br$_2$ diffuses in time Δt (Figure 11.22).

The concentration c_x at distance x from top (Figure 11.22(b)) is

$$c(x) = c_H \cdot \frac{x}{H} \qquad (11.54)$$

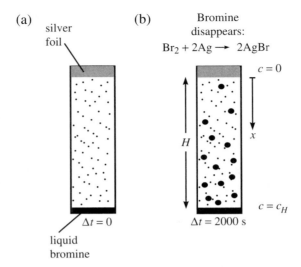

Figure 11.22 Diffusion of bromine vapor through a cylinder filled with air. Evaporation of liquid bromine on the bottom provides a constant flow of Br_2 gas; the bromine reaching the opening is trapped by a silver foil. A linear concentration gradient is formed inside the cylinder; the concentration decreases from c_H on the bottom to zero at the opening. A stationary state is reached after a time of 2000 s. This time is large compared to the time a bromine molecule needs to diffuse from the bottom to the top (see below).

where c_H is the concentration at the bottom ($x = H$). We consider the top layer between $x = 0$ and x. The amount Δn of Br_2 in this layer is

$$\Delta n = \frac{1}{2} c(x) \cdot x \cdot A = \frac{1}{2} c_H \cdot \frac{x^2}{H} \cdot A \qquad (11.55)$$

$\frac{1}{2} c(x)$ is the average concentration in this layer and $x \cdot A$ is the volume between $c = 0$ and $c = c(x)$. According to the equation of Einstein and Smoluchowski the mean square of the displacement x^2 can be expressed by the time Δt needed, on the average, for the bromine molecules in the top layer to be trapped:

$$\overline{x^2} = 2D \cdot \Delta t \qquad (11.56)$$

Then we obtain

$$\Delta n = c_H \cdot \frac{D \cdot \Delta t}{H} \cdot A \qquad (11.57)$$

Assuming that the amount Δn is trapped in time Δt we obtain

$$\boxed{\frac{\Delta n}{\Delta t} = AD \cdot \frac{c_H}{H}} \qquad (11.58)$$

The same amount passes each cross-section in time Δt. Using this relation we can calculate the amount of substance passing the opening of the cylinder in time Δt.

Example – bromine diffusing through a cylinder

With the diffusion coefficient $D = 0.8 \times 10^{-5}\,\text{m}^2\,\text{s}^{-1}$ calculated in Section 11.4.2 and for $H = 10\,\text{cm}$ we obtain

11.4 DIFFUSION

$$\Delta t = \frac{H^2}{2D} = \frac{100 \times 10^{-4}\,\text{m}^2}{2 \cdot 0.8 \times 10^{-5}\,\text{m}^2\,\text{s}^{-1}} = 625\,\text{s}$$

Thus the time needed for one bromine molecule to travel from the bottom to the top is about 10 minutes. The amount Δn of bromine molecules passing the opening with cross-section A during the time $\Delta t = 625$ s is

$$\Delta n = AD\frac{c_H}{H} \cdot \Delta t$$

The vapor pressure of bromine at room temperature is $P_{\text{vap}} = 100$ mbar and the corresponding concentration is, according to the ideal gas law $PV = n\boldsymbol{R}T$

$$c_H = \frac{n}{V} = \frac{P_{\text{vap}}}{\boldsymbol{R}T} = \frac{1 \times 10^4\,\text{N}\,\text{m}^{-2}}{8.314 \cdot 300\,\text{J}\,\text{mol}^{-1}} = 4.01 \times 10^{-3}\,\text{mol}\,\text{L}^{-1}$$

Then with $D = 0.8 \times 10^{-5}\,\text{m}^2\,\text{s}^{-1}$, $H = 10$ cm, $\Delta t = 625$ and $A = 1\,\text{cm}^2$ we obtain

$$\Delta n = 2.0 \times 10^{-5}\,\text{mol}$$

This corresponds to a mass of $\Delta m = \Delta n \cdot 159.8\,\text{g}\,\text{mol}^{-1} = 3.2$ mg. This means that in about 10 minutes only 3 mg of liquid bromine disappears; it takes 52 hours $\approx$ 2 days for the disappearance of 1 g of liquid bromine. This is observed in the experiment.

The quantity c_H/H in equation (11.58) is related to the concentration gradient of the diffusing particles inside the cylinder; for a sufficiently small diffusion path the concentration gradient can be written as dc/dx. Correspondingly, $\Delta n/\Delta t$ can be written as dn/dt. Then we obtain *Fick's law*:

$$\boxed{\frac{dn}{dt} = AD \cdot \frac{dc}{dx}} \qquad (11.59)$$

According to this relation the amount of substance passing any cross-sections of the cylinder is proportional to the time t, the cross-sectional area A, and the concentration gradient dc/dx. The proportionality constant D increases, obviously, with the average molecular speed $\bar{v}$ and the mean free path $\bar{\lambda}$.

11.4.4 Gravity Competing with Thermal Motion

In the above consideration the influence of gravity on the molecules has been neglected. If the molecules are under the influence of gravity then an atmospheric distribution is established in a cylinder filled with a gas. Each molecule is pulled down by a force $f = mg$ (m = mass of a molecule; g = gravitational acceleration) and diffuses up as a result of the thermal motion. At a certain height d the potential energy in the gravity field fd equals the thermal energy kT:

$$fd = mgd = kT, \quad d = \frac{kT}{mg}$$

For nitrogen gas ($m = 4.7 \times 10^{-26}$ kg), $g = 9.81$ m s^{-2}, and $T = 300$ K we obtain $d = 9$ km. The atmospheric distribution (see Box 11.4) is

$$c(x) = c_0 \cdot e^{-fx/(kT)} = c_0 \cdot e^{-mgx/(kT)} \tag{11.60}$$

where $c(x)$ is the concentration of the gas molecules at height x and c_0 is the concentration at $x = 0$.

Box 11.4 – Atmospheric Distribution

We consider a cylinder (cross-section A) filled with a gas (molecules with mass m) as displayed in Figure 11.23. We select a layer of thickness Δx (volume $V = A \cdot \Delta x$). The number N of molecules in this layer is related to the pressure P:

$$PV = NkT, \quad N = \frac{PV}{kT} = \frac{PA \cdot \Delta x}{kT}$$

The force of gravity acting on one molecule is $f = mg$ (g = gravitational acceleration). This means that there is a pressure change $\Delta P = P_{x+\Delta x} - P_x$ in the gas

$$\Delta P = -\frac{N \cdot mg}{A} = -\frac{P\Delta x \cdot mg}{kT}$$

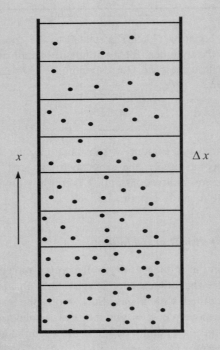

Figure 11.23 Layers of thickness Δx (atmospheric distribution).

Continued on page 383

> Continued from page 382
>
> By replacing the differences by differentials
>
> $$dP = -P\frac{mg}{kT} \cdot dx \quad \text{or} \quad \frac{dP}{P} = -\frac{mg}{kT} \cdot dx$$
>
> and integration we obtain the atmospheric distribution function
>
> $$P = P_0 \cdot e^{-mgx/(kT)} \quad \text{or} \quad c = c_0 \cdot e^{-mgx/(kT)}$$
>
> where P_0 is the pressure at $x = 0$ and c_0 is the concentration at $x = 0$.
>
> ---
>
> ### Example
>
> With $x = 8882$ m (this is the height of Mount Everest), $m = 28$ g mol^{-1}/$(6.022 \times 10^{23}$ mol$^{-1}) = 4.65 \times 10^{-26}$ kg (value for nitrogen) and $g = 9.81$ m/s^2 we obtain for $T = 273$ K (as an average of the temperature at $x = 0$ and $x = 8882$ m), $P = P_0 \cdot e^{-1.08} = 0.34 \cdot P_0$. This means that the atmospheric pressure on top of Mount Everest is only one-third of the atmospheric pressure on sea level. On the other hand, if we choose x on the order of laboratory dimensions ($x = 1$ m) we obtain $P = 0.99987 P_0$; this is a decrease by 1×10^{-4} only.

Thus d can be identified with the height x at which the concentration has dropped to c_0/e. This example shows that in a diffusion experiment in a laboratory ($d \approx 10$ cm) the influence of gravity is small enough to be neglected.

In an artificial gravity field in an ultracentrifuge (e.g., $g_{\text{centrifuge}} = 10^6$ m s^{-2}), d can be brought to laboratory dimensions ($d \approx 10$ cm) and the dependence of d on the mass of the molecules can be used to separate isotopes (see Problem 11.6).

11.5 Viscosity Arising from Collisions of Molecules

We consider a plate moving parallel to its surface in a gas. Then the gas molecules bouncing on the plate are accelerated in the direction of the motion of the plate and the plate is decelerated. Obviously the effect increases with increasing molecular speed, with increasing molecular mass, and with increasing free path (thus it decreases with increasing collision diameter). The measurement of this viscous response is an useful method to determine collision diameters.

11.5.1 Viscous Flow

We consider a square plate of area A which can be moved in a gas parallel to and between two large plates at rest, each at distance d from the moving plate (Figure 11.24). In order to move the plate at a constant speed u we have to apply a force f. If u is

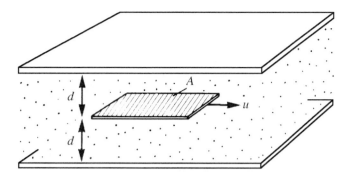

Figure 11.24 A plate moving with speed u in a gas between two resting plates.

sufficiently small (laminar flow) this force is proportional to u and to the surface area of the plate ($2A$) and inversely proportional to the distance d of the plate from the container wall:

$$f = \eta u \frac{2A}{d} \tag{11.61}$$

The proportionality factor η is called *viscosity*.

11.5.2 Calculating Viscosity

We can understand the occurrence of the force on the plate in the following way. A molecule that collides with the moving plate is for a short while captured on the surface of the plate; it assumes the velocity u in the direction of the moving plate. This causes an acceleration of the molecule and a corresponding deceleration of the plate unless the external force keeps the plate at a constant velocity u (Figure 11.25).

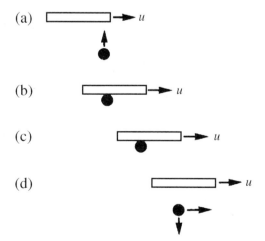

Figure 11.25 Processes responsible for the viscosity of a gas. A molecule bouncing back and forth between the moving plate and the stationary walls of the container is accelerated at the moving plate. After the collision (a) and (b) the particle is fixed for a short while on the surface of the plate and brought to velocity u (c); after leaving the plate (d) it keeps this velocity component.

11.5 VISCOSITY ARISING FROM COLLISIONS OF MOLECULES

Molecules approaching from the lower resting plate (velocity $u = 0$) are accelerated to the velocity u when reaching the moving plate. Then the force acting on the plate is

$$f = m_{\text{gas}} \cdot \frac{u}{\Delta t} \qquad (11.62)$$

where m_{gas} is the mass of all gas particles reaching the moving plate in time Δt.

When on its way from the moving plate toward the resting plate the molecules collide with other molecules, they transfer a part of their additional velocity component to these molecules. So those molecules which did not collide with the moving plate obtain an additional velocity in the direction of the moving plate (x direction). This additional velocity decreases linearly when proceeding from the moving plate to the resting plate (from u to zero, see Figure 11.26). Only molecules in a distance smaller than the mean free path $\bar{\lambda}$ can reach the moving plate without a collision with other gas molecules.

Then the velocity difference between the moving plate and the molecules hitting the plate is $u \cdot \bar{\lambda}/d$ instead of u. Then the force f is

$$f = m_{\text{gas}} \cdot \frac{u}{\Delta t} \cdot \frac{\bar{\lambda}}{d} \qquad (11.63)$$

As in the calculation of the gas pressure on the walls of a container (Section 11.1.1), we express m_{gas} by

$$m_{\text{gas}} = m \cdot \Delta N = m \cdot \frac{1}{6} \frac{N}{V} \cdot 2A \cdot \bar{v} \cdot \Delta t \qquad (11.64)$$

where $\bar{v}$ is the mean speed of the gas molecules. Then we obtain for the force f

$$f = m \cdot \frac{1}{6} \frac{N}{V} \cdot 2A \cdot \bar{v} \cdot \Delta t \cdot \frac{u}{\Delta t} \cdot \frac{\bar{\lambda}}{d} = \frac{1}{6} \frac{N}{V} m \bar{v} \bar{\lambda} \cdot u \cdot \frac{2A}{d} \qquad (11.65)$$

Comparison with equation (11.61) shows that the viscosity η of the gas is

$$\eta = \frac{1}{6} \frac{N}{V} m \cdot \bar{v} \bar{\lambda}$$

In this model the force acting on the plate is underestimated. From the distribution of λ and v it follows that the contribution of the molecules with larger λ or v is more effective than that estimated from the averages $\bar{\lambda}$ and $\bar{v}$. Also, molecules can reach more distant

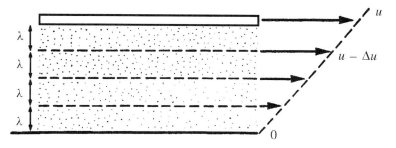

Figure 11.26 Moving plate at a distance $d = 4\lambda$ from the stationary container wall; the additional velocity in the different gas layers decreases linearly from u to zero.

layers after noncentral collisions. The rigorous calculation leads to a three times larger value of η:

$$\boxed{\eta = \frac{1}{2}\frac{N}{V}m \cdot \overline{v}\overline{\lambda}} \qquad (11.66)$$

11.5.3 Dependence of Viscosity on Pressure

Replacement of $\overline{\lambda}$ by equation (11.36) results in

$$\boxed{\eta = \frac{1}{2}\frac{m\overline{v}}{\sqrt{2}\cdot\sigma}} \qquad (11.67)$$

It seems surprising that η does not depend on the number density N/V or on the pressure

$$P = \frac{NkT}{V} = \frac{N}{V}\cdot kT$$

This is caused by a counteraction of P and $\overline{\lambda}$: when P decreases, the number of collisions with the moving plate decreases; however, the momentum transferred per collision increases because of the increasing mean free path $\overline{\lambda}$

$$\text{momentum} = m \cdot u\frac{\overline{\lambda}}{d} \quad (\overline{\lambda} < d)$$

so the momentum transferred per collision is accordingly larger.

At small pressure we have $\overline{\lambda} > d$; then the transfer of momentum in a collision of a gas particle with the moving plate is mu. In this case the viscosity decreases linearly with decreasing pressure P.

11.5.4 Increasing η with Increasing Temperature

There is another aspect in discussing equation (11.67). The viscosity of a gas increases with increasing temperature because the speed $\overline{v}$ of the molecules increases with temperature: in the same time interval more particles collide with the moving plate. This is in contrast to the behavior of a liquid, where η strongly decreases with increasing temperature (see Section 11.6.1). It was Maxwell who first predicted this unexpected behavior of gases on the basis of the kinetic theory; the experiments, carried out later, are in excellent agreement with his theory.

Following equation (11.67), the mean speed $\overline{v}$ and thus the viscosity η increase proportional to the square root of the temperature T. However, as shown in Foundation 11.2, the situation is more complicated because of the attraction forces acting between the molecules.

11.5.5 Calculating Collision Diameters

The viscosity of a gas can be experimentally determined by measuring the flow velocity of a gas flowing through a capillary (Figure 11.27). If $P_2 - P_1$ is the pressure difference

11.6 THERMAL MOTION IN LIQUIDS

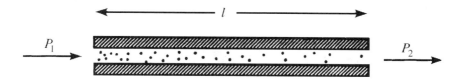

Figure 11.27 The flow of a gas through a capillary.

Table 11.1 Experimental viscosity η, calculated mean velocity $\bar{v}$, mean free path $\bar{\lambda}$ at $P = 1$ bar and collision diameter $2r$ for some gases (using equations (11.69)

	T/K	$\eta/10^{-6}$ g cm^{-1} s^{-1}	M/g mol^{-1}	$\bar{v}$/m s^{-1}	$\bar{\lambda}$/nm	$2r$/pm
H	273	69	1.008	2390	130	256
H$_2$	273	84	2.016	1690	112	276
CO$_2$	273	137	44.01	362	39	466
O$_2$	273	192	32.00	425	64.2	364
N$_2$	273	166	28.02	454	59.3	379
He	273	186	4.003	1200	176	220
Br$_2$	286	152	159.8	199	21.8	625

The viscosities of H atoms can be measured in gas discharge tubes filled with hydrogen gas

between the ends of the capillary of length l and with the inner radius R, then the volume ΔV of the gas flowing in time Δt into the capillary from left to right is

$$\Delta V = \frac{1}{\eta} \frac{\pi (P_2 - P_1)(P_1 + P_2) R^4}{16 l P_1} \Delta t \tag{11.68}$$

(Hagen–Poiseuille's law). The volume of the gas flowing out on the right-hand side of the capillary is larger because of the smaller pressure P_2. From the measured viscosity, the mean free path $\bar{\lambda}$ and the collision diameters of gas molecules can be calculated (Table 11.1) according to equations (11.66), (11.67), (11.36), (11.34), and (11.29):

$$\bar{\lambda} = 2 \frac{V}{N} \frac{1}{m\bar{v}} \eta, \quad \sigma = \frac{m\bar{v}}{2\sqrt{2}} \frac{1}{\eta}, \quad 2r = \sqrt{\frac{\sigma}{\pi}}, \quad \bar{v} = \sqrt{\frac{8kT}{\pi m}} \tag{11.69}$$

11.6 Thermal Motion in Liquids

11.6.1 Collisions in Liquids

Whereas atoms and molecules in a crystal have distinct positions, atoms and molecules in a fluid can change their positions as in a gas. At the same temperature, a molecule in a liquid has the same mean translational energy $\frac{3}{2}kT$ and the same mean speed as in the gas. However, the mean free path is much smaller, because the molecules are packed much closer in a liquid than in a gas.

Example – collision frequency

According to Section 11.3, the molecules in a gas at 1 bar and 273.16 K are spaced 3.3 nm apart, on the average. Accordingly, the average distance obtained for a liquid such as water is 0.3 nm when we suppose that in 1 cm^3 there are $\frac{1}{18}$ mol $\cdot N_A = 3.3 \times 10^{22}$ molecules. This distance is on the order of the diameter of molecules, so a particle cannot move freely between many molecules like in gases. The mean free path $\bar{\lambda}$ in liquids is therefore smaller than the average distance between the molecules; this means that $\bar{\lambda}$ is on the order of 0.1 nm. Thus the molecules in a liquid collide more frequently than the molecules in a gas. The collision frequency is then

$$\nu = \bar{v}/\bar{\lambda} = \frac{10^3 \text{ m s}^{-1}}{10^{-10} \text{ m}} = 10^{13} \text{ s}^{-1}$$

whereas the value calculated for gases (Section 11.3) is only $\nu = 10^{10}$ s^{-1}.

11.6.2 Diffusion Coefficient D of a Liquid

As a result of the fact that molecules in a liquid collide more frequently than the molecules in a gas, it takes much more time for a particle in a liquid to traverse a certain distance by diffusion. The particle collides many times with other particles in the liquid before it can change its position from (a) to (b) to (c) in Figure 11.28.

When changing its position, a particle travels the path a, where a is the mean distance between two molecules. If the position is changed after every ξth collision, then $\nu \Delta t/\xi$

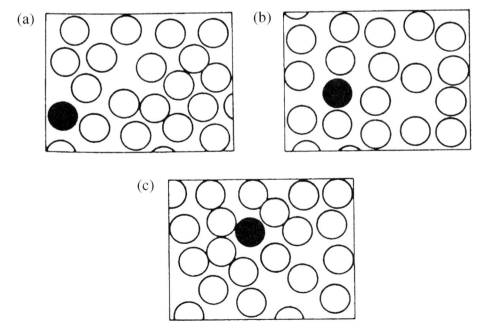

Figure 11.28 A change of position of a molecule which is moving forward inside a liquid.

is the number of position changes in time Δt, and the mean displacement square $\overline{\Delta x^2}$ is, according to equation (11.45),

$$\overline{\Delta x^2} = \frac{1}{3} j \cdot \overline{\lambda^2} = \frac{1}{3} \frac{v \Delta t}{\xi} \cdot a^2 = \frac{1}{3} \frac{v a^2}{\xi} \cdot \Delta t$$

By comparison with the equation of Einstein and Smoluchowski (equation (11.52)) we obtain

$$D = \frac{\overline{\Delta x^2}}{2 \Delta t} = \frac{1}{6} \frac{v a^2}{\xi} \qquad (11.70)$$

For a liquid such as water, $D = 1 \times 10^{-5}$ cm^2 s^{-1} and $a = 0.3$ nm, we calculate the number ξ of collisions:

$$\xi = \frac{1}{6} \frac{v a^2}{D} = \frac{1}{6} \frac{10^{13} \text{ s}^{-1} \cdot (0.3 \times 10^{-9} \text{ m})^2}{1 \times 10^{-5} \text{ cm}^2 \text{ s}^{-1}} = 150$$

About 150 collisions occur before a position change is taking place.

The frequency of position changes depends on the intermolecular forces which are determined by the specific surrounding of the molecule. The quantity D has, therefore, in contrast to gases, specific values for liquids.

11.6.3 Viscosity of a Liquid

It is also true that viscosity, which is much larger for liquids than for gases because of the close packing of the molecules, has specific values for liquids: at 300 K $\eta = 0.0089$ g s^{-1} cm^{-1} for water and $\eta = 9.54$ g s^{-1} cm^{-1} for glycerine. (1 poise = 1 g s^{-1} cm^{-1} = 10^{-1} kg s^{-1} m^{-1} is an unit for the viscosity frequently used in the literature.) In contrast to gases, the viscosity of liquids strongly decreases with temperature: at higher temperature it is easier for the molecules to overcome the energy barrier against positional changes depending on the strength of the intermolecular forces.

11.6.4 Stokes–Einstein Equation: Diffusion, Assembling, Interlocking

To estimate the diffusion coefficient D we consider the dissolved molecule as a sphere of radius r; we suppose that the force f acting on the molecule in order to pull it through the liquid with speed u is as large as would be expected according to hydrodynamics (*Stokes' law*):

$$f = 6 \pi \eta r u \qquad (11.71)$$

We consider the solvent as a continuum and neglect its molecular structure. In this approximation D is given by the Stokes–Einstein equation (see Box 11.5)

$$\boxed{D = \frac{kT}{6 \pi \eta r}} \qquad (11.72)$$

> **Box 11.5 – Derivation of the Stokes–Einstein Equation**
>
> Let us consider a dissolved particle in a force field, like the artificial gravity field of an ultracentrifuge. Each molecule is pulled down with the force f, and because of the thermal motion a distribution function is established. We can say that each particle is moving downwards under the action of the force f with a velocity
>
> $$u = \frac{f}{6\pi\eta r}$$
>
> (this is the Stokes equation; η is the viscosity of the liquid, and the dissolved particle is approximated by a sphere of radius r). The amount of substance moving downwards through a cross-section A at height x in time Δt is
>
> $$\Delta n_f = -c \cdot A \cdot u \Delta t = -cA \frac{f}{6\pi\eta r} \Delta t$$
>
> At the same time, however, the particle is diffusing upwards according to the concentration gradient dc/dx. The amount of substance moving upwards is, according to equation (11.59),
>
> $$\Delta n_D = -AD\frac{dc}{dx}\Delta t$$
>
> In the stationary state we have $\Delta n_f + \Delta n_D = 0$, and therefore
>
> $$D\frac{dc}{dx} = -\frac{fc}{6\pi\eta r} \quad \text{or} \quad \frac{dc}{c} = -\frac{f}{6\pi\eta r D} \cdot dx$$
>
> Integration yields $c(x) = c_0 \cdot e^{-fx/(6\pi\eta r D)}$.
> On the other hand, we can apply equation (11.60) to the same problem:
>
> $$c(x) = c_0 \cdot e^{-fx/(kT)}$$
>
> Equating the exponentials yields the Stokes–Einstein equation
>
> $$D = \frac{kT}{6\pi\eta r}$$

This equation is very useful for estimating diffusion coefficients of any particle in a liquid from the viscosity of the liquid and the average size of the particle. For example, the diffusion coefficient for a molecule with $r = 200$ pm diffusing in water ($\eta = 0.0089$ g s^{-1} cm^{-1}) becomes $D = 1.2 \times 10^{-5}$ cm^2 s^{-1}.

Equation (11.72) is justified for sufficiently large particles, because the solvent is treated as a continuum obeying hydrodynamics. Surprisingly, the equation can also be used to estimate D for molecules which are not much larger than the solvent molecules.

The Stokes–Einstein equation is very important for the estimation of motions in molecular systems. The assemblage of functional units, which is the basic process in

11.6 THERMAL MOTION IN LIQUIDS

molecular biology and supramolecular engineering, is strongly connected with the process in which molecules diffuse as a result of their thermal motion. The components that fit find each other and bind, whereas the unfitting ones separate, disperse, and are replaced by fit components. In this way, complex systems of molecules that interlock in a jigsaw-puzzle-like manner are formed (see Chapters 23 and 24).

Example – removal of excess thiosulfate from a photoplate by diffusion

How long does the watering process of a photoplate take, if the thickness of the emulsion is $d = 0.2$ mm? The emulsion is a gelatin gel, and the diffusion coefficient in the gel is the same as for water. For thiosulfate in water we obtain from equation (11.72) with $\eta = 0.0089$ g s^{-1} cm^{-1} and $r = 120$ pm the value $D = 2 \times 10^{-9}$ m^2 s^{-1}; then

$$\Delta t = \frac{\overline{\Delta x^2}}{2D} = \frac{(2 \times 10^{-4})^2 \, \text{m}^2}{2 \cdot 2 \times 10^{-9} \, \text{m}^2 \, \text{s}^{-1}} = 10 \, \text{s}$$

Example – Brownian Motion

By microscopic observation the x, y coordinates of a small sphere of diameter 2 μm in water ($\eta = 8.9 \times 10^{-4}$ kg s^{-1} m^{-1}) were determined at room temperature in the beginning, after 30 s, 60 s, and so on for 135 min = 8100 s. (Fig. 11.29). The diffusion coefficient D is, according to the Stokes–Einstein Equation (11.72), $D = 2.5 \times 10^{-13}$ m^2 s^{-1}. We compare the observed value of $(\Delta x)^2 = (40 \, \mu\text{m})^2 = 1.6 \times 10^{-9}$ m^2 with the average $\overline{\Delta x^2} = 2D\Delta t = 2 \cdot 2.5 \times 10^{-13} \cdot 8100 \, \text{m}^2 = 4 \times 10^{-9}$ m^2.

Example – bacteriophage docking at a bacterium

A bacteriophage attacks a bacterium (diameter 1 μm) in water. After 20 min the bacterium is destroyed, and the bacteriophage has produced 200 daughter phages. How long does it take before another bacterium at a distance of 5 μm is infected?

We approximate the bacteriophage as a sphere of radius $r = 100$ nm (Figure 11.30). Then, from equation (11.72) with $\eta = 0.0089$ g s^{-1} cm^{-1}, we get $D = 2.5 \times 10^{-8}$ cm^2 s^{-1} and

$$\Delta t = \frac{x^2}{2D} = \frac{(5 \times 10^{-6})^2 \, \text{m}^2}{2 \cdot 2.5 \times 10^{-8} \, \text{cm}^2 \, \text{s}^{-1}} = 5 \, \text{s}$$

This is, on the average, the time required to reach a distance of 5 μm. Since the bacterium covers an area of about 1 μm^2, the probability of finding a bacterium on the surface of a sphere with radius 5 μm is

$$\frac{1 \, \mu\text{m}^2}{4\pi \cdot (5 \, \mu\text{m})^2} = \frac{1}{300}$$

Thus about one of the 200 phages reaches the actual position of the bacterium in 5 s. This time is negligibly small compared to the time of 20 min needed to destroy the bacterium.

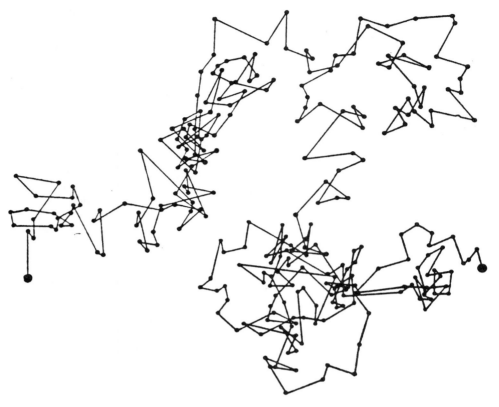

Figure 11.29 Brownian motion. The coordinates of a particle with diameter 2 μm, moving in water, is observed every 30 s for 135 min. The distance Δ_x between starting point and end point is 40 μm.

Figure 11.30 Bacteriophage approximated by a sphere.

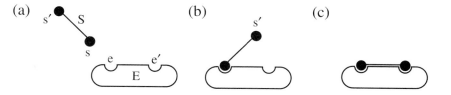

Figure 11.31 A substrate S approaching an enzyme E.

Example – enzyme assembling and interlocking with a substrate

An enzyme E reacts with a substrate S if the point s on the substrate has approached point e on the enzyme within about $d = 100$ pm (Figure 11.31). Then point s′ must find point e′; this process, however, requires a short time compared to the first process. The substrate is assumed to be in a concentration of 1 nmol L^{-1}, so we find one molecule of the substrate in a volume $V = (1\,\mu\text{m})^3$. We consider an enzyme molecule assumed to be fixed in the volume V. Let us divide up V into cells with an edge length of $d = 100$ pm. Point s of the substrate S is in one of these cells and diffuses from cell to cell until it reaches the cell with point e. We are looking for the average time $\bar{t}$ required for this process. According to the equation of Einstein–Smoluchowski the average time τ needed by S to reach the next cell is

$$\tau = \frac{d^2}{2D}$$

(D = diffusion coefficient of S). Thus

$$\bar{t} = \frac{V}{d^3}\tau = \frac{V}{2dD}$$

If we assume that the substrate molecule has a mean radius of $R = 1$ nm, then

$$D = \frac{kT}{6\pi\eta R} = 2.5 \times 10^{-6}\,\text{cm}^2\,\text{s}^{-1} \quad \text{for } \eta = 10^{-2}\,\text{g cm}^{-1}\,\text{s}^{-1}\ (\text{water})$$

and we find $\bar{t} = 20$ s. Note that this time is much larger than the time required for S to travel a distance of 1 μm by diffusion:

$$t' = \frac{(1\,\mu\text{m})^2}{2D} = 2\,\text{ms}$$

What is the reason for this difference? It is extremely improbable for the substrate molecule to meet the enzyme molecule on an arbitrary diffusion path of length 1 μm such that points s and e coincide. The substrate molecule must diffuse many times over a distance of 1 μm until it reaches the very particular location of the active site e on the enzyme (Figure 11.31).

11.7 Molecular Motion and Phases

We will shortly look at some exciting phenomena occurring when heating a substance. This helps in developing a lively picture of the effects created by thermal motion. The processes will be discussed in more detail in subsequent chapters.

What happens in heating a solid? We consider a piece of ice kept at a pressure of 1 bar. On heating, thermal fluctuations occur, thereby causing an expansion of the crystal lattice. At a distinct temperature (melting point: 273.16 K) the lattice becomes unstable. When the fluctuations reach a critical strength, the crystal structure breaks down into clusters of limited size: the piece of ice melts.

11.7.1 Melting Point

In general, a fluid is less compact than the ordered crystalline solid: the liquid has a smaller density than the coexisting solid. In the particular case of water, however, the structure of ice is very open. This is because of the tetrahedral arrangement of the water molecules connected by hydrogen bonds leaving an empty space in between (Figure 11.32). This structure is largely preserved within the clusters of liquid water, but the regions between the clusters are more dense: water has a higher density than ice at the melting point. The density maximum is reached at 4 °C. At higher temperatures the density decreases due to the increase in fluctuation.

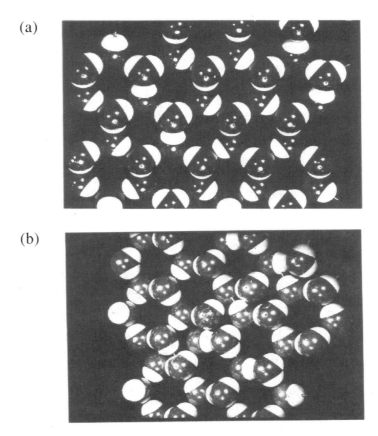

Figure 11.32 Structure of (a) an ice crystal and (b) liquid water. (Courtesy W. A. P. Luck).

11.7.2 Boiling Point

At a certain strength of fluctuation the liquid state becomes unstable (boiling point for water: 373.15 K) and the molecules evaporate: transition into the gaseous state. At lower external pressures the boiling point of the liquid is obviously reached at a lower temperature, since the dynamic equilibrium between evaporation and condensation is established at a lower temperature. On the other hand, the melting temperature is only slightly changed by decreasing the pressure. Then, at a distinct pressure (0.006 bar), the melting and boiling points coincide (at temperature 273.16 K). At this point (the triple point of water) we can see that solid, liquid, and gas coexist (Figure 11.33).

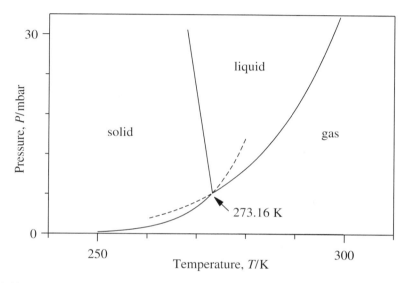

Figure 11.33 Phase diagram (solid, liquid, and gas). Pressure P versus temperature T. Solid lines: solid–gas, solid–liquid, and liquid–gas boundaries. Dashed lines: extrapolation of vapor pressure above solid (above the triple point, right) and above liquid (below the triple point, left). The triple point is at 273.16 K for water.

11.7.3 Critical Point

When increasing the temperature at a sufficiently high pressure (above 200 bar), the liquid water becomes a crowd of increasingly lively moving, more and more equally distributed molecules. A continuous transition into a highly compressed gas takes place. The situation is described more closely in Figure 11.34, where isotherms for water (pressure P versus molar volume V_m) are shown for different temperatures. When compressing the gas at 573 K condensation takes place when a pressure of 84.5 bar is reached (right-hand dot, $V_m = 0.39$ L mol^{-1}). When further decreasing the volume the pressure P remains constant, because liquid and gaseous water are co-existing: vapor pressure of water. The situation changes when decreasing the volume below $V_m = 0.025$ L mol^{-1} (left-hand dot): at this point all gaseous water has been transformed into liquid water; the liquid is only slightly compressible and therefore the pressure increases strongly.

With increasing temperature the difference in density between gas (saturated vapor) and coexisting liquid decreases: the horizontal parts (curves for 623 K and 633 K) become

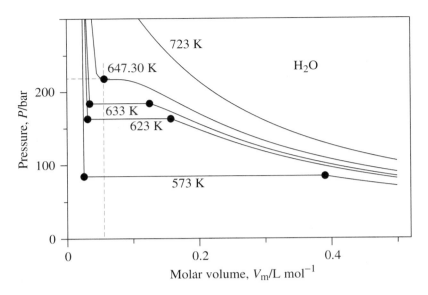

Figure 11.34 Isotherms (pressure P versus molar volume V_m) of water at different temperatures (573 K, 623 K, 633 K, 647.30 K, and 723 K). The dots indicate the limits where gaseous (right-hand side) and liquid (left-hand side) water are co-existing. The dashed lines indicate the critical point.

shorter. In the curve for 647.30 K this part shrinks to a point: we can no longer distinguish between vapor and liquid. This point is called the *critical point* ($V_m = 0.0563\,\text{L mol}^{-1}$, $P = 218$ bar). Above the critical point, when compressing the dilute gas (first behaving as an ideal gas) a continuous change into a liquid takes place (curve for 723 K). On cooling, this liquid will continuously develop its specific cluster structure.

11.7.4 Phases

Gas, liquid, and solid are called *phases*. A phase is a macroscopically homogeneous part of a system, uniform in chemical composition and physical state. Below the critical temperature the phase transition from gas to liquid is discontinuous, above the critical temperature it is continuous. The phase transitions from the solid to the gas (sublimation) and from the solid to the liquid (melting) are discontinuous. Also the phase transition from one crystal structure to another one is a discontinuous process: when compressing ice (say at 200 K) to 1900 bar the loose structure (ice 1) in Figure 11.32 becomes unstable, and a more compact structure (ice 2) is formed.

11.7.5 Clusters and Liquid Crystals

The kind of clusters formed in the liquid state strongly depends on the nature of the molecules. Cluster formation can be neglected when the bonding between the molecules is relatively unspecific (Section 11.6.1). With increasing tendency to aggregate and with more than one species, an unlimited number of possibilities to organize into specific clusters arises. Even with only one species, arrays of macroscopic orientational order can occur (liquid crystals). For further details see Chapters 22 and 23.

Problems

Problem 11.1 – Elastic Collisions of Spheres

Two spheres (masses m_1 and m_2; initial velocities v_1 and v_2) collide (central collision). What happens in the cases (a) $m_1 = m_2 = m$; $v_1 = 0$, $v_2 = v$ and (b) $m_1 \gg m_2$; $v_1 = v$, $v_2 = -v$?

Solution. Case (a): the velocities after the collision are u_1 and u_2; according to the conservation of energy and the conservation of momentum we can write

$$\tfrac{1}{2}mv^2 = \tfrac{1}{2}mu_1^2 + \tfrac{1}{2}mu_2^2, \quad mv = mu_1 + mu_2$$

Thus

$$u_2^2 = v^2 - u_1^2 \quad \text{and} \quad u_2^2 = (v - u_1)^2 = v^2 - 2vu_1 + u_1^2$$

By elimination of u_2 we obtain

$$u_1(v - u_1) = 0$$

with the solution $u_1 = v$ and $u_2 = 0$. This means that particle 2 has transferred its energy completely to particle 1.

Case (b): again we apply the laws of conservation of energy and momentum:

$$\tfrac{1}{2}m_1 v^2 + \tfrac{1}{2}m_2 v^2 = \tfrac{1}{2}m_1 u_1^2 + \tfrac{1}{2}m_2 u_2^2 \quad \text{and} \quad m_1 v - m_2 v = m_1 u_1 + m_2 u_2$$

Therefore

$$v^2 \cdot (m_1 + m_2) = m_1 u_1^2 + m_2 u_2^2 \quad \text{and} \quad v \cdot (m_1 - m_2) = m_1 u_1 + m_2 u_2$$

We solve for u_1^2:

$$u_1^2 = v^2 \left(1 + \frac{m_2}{m_1}\right) - u_2^2 \frac{m_2}{m_1}$$

$$u_1^2 = v^2 \left(1 - \frac{m_2}{m_1}\right)^2 - 2vu_2 \left(1 - \frac{m_2}{m_1}\right) \frac{m_2}{m_1} + \left(\frac{m_2}{m_1}\right)^2 u_2^2$$

We eliminate u_1^2:

$$u_2^2 \left(1 + \frac{m_2}{m_1}\right) - 2vu_2 \left(1 - \frac{m_2}{m_1}\right) = v^2 \left(3 - \frac{m_2}{m_1}\right)$$

Then for $m_1 \gg m_2$ we obtain

$$u_2^2 - 2vu_2 = 3v^2$$

with the solution

$$u_2 = 3v, \quad u_1 = v$$

The heavy particle moves with the same velocity as before the collision, and the light particle moves with three times the velocity of the heavy particle (see Figure 11.7).

This result can be easily understood. The light particle has a velocity of $-2v$ relative to the heavy particle, thus it will be reflected with a velocity $+2v$ relative to the heavy particle. Since the heavy particle is moving at a velocity v, the velocity of the light particle becomes $+3v$.

Problem 11.2 – Gas Bubbles Rising from the Bottom of a Sea

On the bottom of a sea (depth 10 m) a bubble of methane gas is formed (diameter $d_{bottom} = 1$ cm). What is the diameter of the bubble when it reaches the sea surface? The sea water temperature rises from 5 °C on the bottom to 25 °C on the surface.

Solution. On the bottom we have $P = 2$ bar (if we approximate the water pressure by 1 bar) and $T = 278$ K; on the surface we have $P = 1$ bar and $T = 298$ K. According to the ideal gas law the volume is proportional to T/P and the diameter is proportional to $V^{1/3}$, thus for the diameter $d_{surface}$ we obtain

$$d_{surface} = d_{bottom} \cdot \sqrt[3]{\frac{2 \text{ bar}}{1 \text{ bar}} \cdot \frac{298 \text{ K}}{278 \text{ K}}} = 1.3 \text{ cm}$$

Problem 11.3 – Pressure Change on Cooling in a Refrigerator

The air inside a refrigerator (volume 100 L) is chilled from 20 °C to 0 °C after closing the door. What is the force needed to open the refrigerator door (area 1 m²)? The refrigerator is supposed to be totally airtight.

Solution. P_1 and P_2 are the pressures at $T_1 = 293$ °C and at $T_2 = 273$ °C, respectively. Then

$$P_2 = P_1 \frac{T_2}{T_1} = 1 \text{ bar} \frac{273 \text{ K}}{293 \text{ K}} = 0.93 \text{ bar}$$

The force acting on the door is $f = (P_1 - P_2) \cdot 1 \text{ m}^2 = 7000 \text{ N} = 714 \, kp$.

Problem 11.4 – Gas Thermometer

The following molar volumes V_m are measured for SO_2 when the gas is examined at different pressures:

P/bar	V_m/(L mol^{-1})
0.253	89.13
0.506	44.30
0.660	33.83
1.013	21.89

What is the temperature of the gas?

Solution. We calculate the product PV_m and plot PV_m as a function of P (Figure 11.35). From the figure we obtain

$$\lim_{P \to 0} PV_m = 22.67 \text{ bar L mol}^{-1}$$

Thus the temperature is

$$T = \frac{\lim_{P \to 0} PV_m}{R} = \frac{22.67 \text{ bar L mol}^{-1}}{0.08314 \text{ bar L K}^{-1} \text{ mol}^{-1}} = 272.7 \text{ K}$$

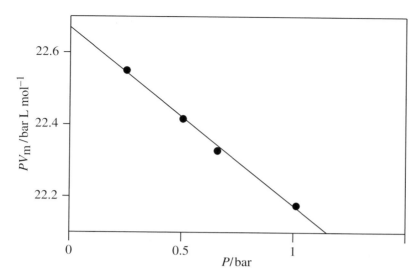

Figure 11.35 Product PV_m as a function of P: extrapolation of PV_m for $P \to 0$.

Problem 11.5 – Separation of Isotopes by Effusion

A mixture of $^{235}UF_6$ (0.1%) and $^{238}UF_6$ (99.9%) is allowed to flow from a container through a small hole into vacuum for a short time. By how much is the lighter isotope enriched as a consequence of its larger velocity?

Solution. According to equation (11.28) we have $v_1/v_2 = \sqrt{m_2/m_1} = 1.004$, if we attribute v_1, m_1 to the compound $^{235}UF_6$ and v_2, m_2 to the compound $^{238}UF_6$. The lighter molecules are faster by 0.4%, thus there is an enrichment from 0.1% to 0.1004%.

The process must be repeated many times to obtain an appreciable separation.

Problem 11.6 – Separation of Isotopes by Using an Ultracentrifuge

We consider a tube with the mixture of $^{235}UF_6$ and $^{238}UF_6$ in an ultracentrifuge rotating with a frequency $\nu = 500\,\text{s}^{-1}$ at a radius $r = 10\,\text{cm}$. What will be the concentration ratio of both isotopes at different distances x from the bottom of the tube?

Solution. The forces f_1 and f_2 acting in the centrifuge on the molecules with masses m_1 and m_2 are $f_1 = m_1(2\pi\nu)^2 r$ and $f_2 = m_2(2\pi\nu)^2 r$. We assume that the radius r of the centrifuge is much larger than the length of the tube with the gas molecules, that is, everywhere in the tube the forces are the same. According to equation (11.60) the concentrations c_1 and c_2 of the two gases at height x in the tube are $c_1 = a_1 e^{-f_1 x/(kT)}$ and $c_2 = a_2 e^{-f_2 x/(kT)}$ where a_1 and a_2 are the concentrations at $x = 0$. Thus for the ratio c_1/c_2 we obtain

$$\frac{c_1}{c_2} = \frac{a_1}{a_2} \cdot e^{-(f_1 - f_2)x/(kT)}$$

To obtain a_1/a_2 we note that the total amount of substance n_1 for gas 1 is 0.1% of the amount of substance n_2 for gas 2. We calculate n_1 as

$$n_1 = \int_0^\infty c_1 \cdot dV = a_1 A \int_0^\infty e^{-f_1 x/(kT)} \cdot dx$$

where A is the cross-section of the tube. With the substitution

$$z = \frac{f_1 x}{kT}$$

we obtain

$$n_1 = a_1 A \frac{kT}{f_1} \int_0^\infty e^{-z} \cdot dz = a_1 A \frac{kT}{f_1}$$

A corresponding expression is obtained for n_2. Then we have

$$\frac{n_1}{n_2} = \frac{a_1}{a_2} \cdot \frac{f_2}{f_1} = \frac{a_1}{a_2} \cdot \frac{m_2}{m_1} = 0.1\%$$

Thus

$$\frac{a_1}{a_2} = 0.1\% \cdot 0.99$$

For $T = 298\,\text{K}$ we obtain

$$\frac{f_1 - f_2}{kT} = \frac{(2\pi v)^2 r}{kT}(m_1 - m_2) = -0.012\,\text{cm}^{-1}$$

and thus

$$\frac{c_1}{c_2} = 0.1\% \cdot 0.99 \cdot e^{0.012 \cdot x/\text{cm}} = 0.1\% \cdot 0.99 \cdot (1 + 0.012 \cdot x/\text{cm})$$

At the bottom of the tube ($x = 0$) we get $c_1/c_2 = 0.99 \cdot 0.1\%$, at $x = 1\,\text{cm}$ we get $c_1/c_2 = 0.1\%$, and at $x = 2\,\text{cm}$ we get $c_1/c_2 = 1.01 \cdot 0.1\% = 0.101\%$.

The enrichment of $^{235}\text{UF}_6$ is better by a factor of two than that calculated in Problem 11.4.

Problem 11.7 – Diffusion Path

In Figures 11.36(a)–(c) three arbitrary choosen diffusion paths of a particle moving randomly in space (for a total of 200 collisions) are displayed (projection of the path on the x,y plane). The paths were obtained in a computer simulation, where the direction and the free path of the particle were calculated by using pseudorandom numbers. In Figure 11.36(d) the end points of the particle for 200 simulation experiments are displayed. Compare the mean square displacement $\overline{\Delta x^2} = 3.96\,\mu\text{m}^2$ taken from Figure 11.36(d) to what is expected from equation (11.45) on the basis of the data given in the figure.

Solution. From equation (11.45) we expect with the data given in the figure

$$\overline{\Delta x^2} = \tfrac{1}{3} j \overline{\lambda^2} = \tfrac{2}{3} j (\overline{\lambda})^2 = \tfrac{2}{3} \cdot 200 \cdot (0.18\,\mu\text{m})^2 = 4.3\,\mu\text{m}^2$$

in good agreement with the simulation.

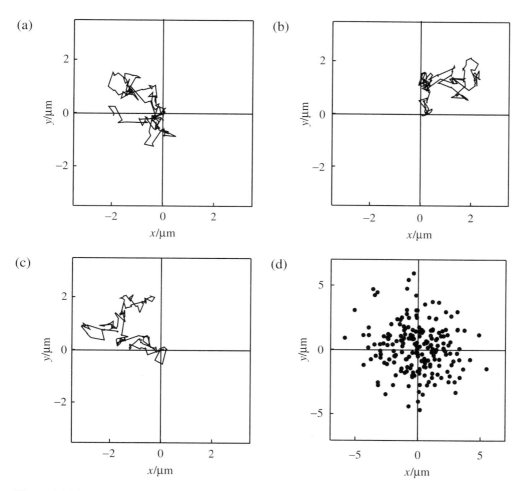

Figure 11.36 Simulated diffusion paths. (a)–(c) Arbitrary paths for $j = 200$ collisions, collision diameter 200 pm, number density $N/V = 2.65 \times 10^{25}$ m^{-3}. The resulting mean free path is $\bar{\lambda} = 0.18\,\mu$m. (d) Display of the end points of a diffusing particle for 200 simulations with the same parameters as in parts (a)–(c). The mean square displacement for all the points in (d) is $\overline{\Delta x^2} = 3.96\,\mu\text{m}^2$.

Problem 11.8 – Distribution Function for Diffusion Path

For a one-dimensional diffusion path (steps $+\lambda$, $-\lambda$) calculate the distribution function after eight diffusion steps. Extend the consideration to the distribution function in space.

Solution. The different stages of diffusion are shown in Table 11.2. We start at $x = 0$. After one step the particle can be at $x = -\lambda$ or at $x = +\lambda$ with equal probability. Then the mean square of the displacement after this step is $\overline{x^2} = \frac{1}{2}(\lambda^2 + \lambda^2) = \lambda^2$. After two steps the particle can be found at $x = -2\lambda$, at $x = 0$ (two possibilities), or at $x = +2\lambda$, and we obtain $\overline{x^2} = \frac{1}{4}(4\lambda^2 + 0 + 4\lambda^2) = 2\lambda^2$. The distributions for the following steps are shown in the scheme in Table 11.2. The mean square displacements after subsequent steps are collected in Table 11.3.

Table 11.2 Number N of possibilities to find a particle after taking subsequent steps in x and $-x$ direction

Step	Displacement in units of λ																
	To the left									To the right							
	8	7	6	5	4	3	2	1	0	1	2	3	4	5	6	7	8
0									1								
1								1		1							
2							1		2		1						
3						1		3		3		1					
4					1		4		6		4		1				
5				1		5		10		10		5		1			
6			1		6		15		20		15		6		1		
7		1		7		21		35		35		21		7		1	
8	1		8		28		56		70		56		28		8		1

Table 11.3 Mean square displacements after subsequent steps

Step	$\dfrac{\sum x^2}{\lambda^2}$	Number N of possibilities	$\dfrac{\overline{x^2}}{\lambda^2}$
1	2	2	1
2	8	4	2
3	24	8	3
4	64	16	4
5	160	32	5
6	384	64	6
7	896	128	7
8	2048	256	8

From Table 11.3 it can be seen that $\overline{x^2} = j \cdot \lambda^2$, where j is the number of steps. This is the same result as obtained in Box 11.3. According to Table 11.2, the number of possibilities N are shown as a function of displacement x in Figure 11.37 (dots) for $j = 8$. This distribution can be considered as a Gaussian distribution

$$\frac{dN}{dx} = a e^{-bx^2}$$

(solid line in Figure 11.37, with $a = 70/\lambda$ and $b = 0.063/\lambda^2$). The mean displacement is then

$$\overline{x^2} = \frac{\int_0^\infty x^2 \cdot a e^{-bx^2} \cdot dx}{\int_0^\infty a e^{-bx^2} \cdot dx} = \frac{1}{b} \frac{\int_0^\infty z^2 e^{-z^2} \cdot dz}{\int_0^\infty e^{-z^2} \cdot dz} = \frac{1}{2b} = 7.9 \cdot \lambda^2$$

This is very close to the theoretical value $\overline{x^2} = j \cdot \lambda^2$ with $j = 8$, and it shows that the Gaussian curve is a good approximation even for such a small value as $j = 8$. The Gaussian curve with $1/(2b) = j \cdot \lambda^2$ is an accurate description for large values of j.

PROBLEM 11.8

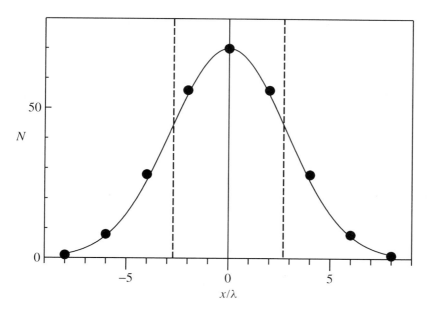

Figure 11.37 Distribution of one-dimensional diffusion path. Number N of possibilities to find a diffusing particle at displacement x after $j = 8$ diffusion steps of size λ. Dots: data according to Table 11.3. Solid line: Gaussian curve $dN/dx = ae^{-bx^2}$ with $a = 70/\lambda$ and $b = 0.063/\lambda^2$. Dashed lines: mean square displacement $\sqrt{\overline{x^2}}$.

In the case of the diffusion in space (see Problem 11.7) the relation for $\overline{x^2}$ is

$$\overline{x^2} = \frac{1}{3}\lambda^2 \cdot j = \frac{1}{2b}$$

and the probability of finding the particle in volume element $dx\,dy\,dz$ is

$$P(x, y, z) = const \cdot e^{-bx^2} \cdot e^{-by^2} \cdot e^{-bz^2} \cdot dx\,dy\,dz = const \cdot e^{-br^2} \cdot dx\,dy\,dz$$

where $b = 3/(2j\lambda^2)$ and $r = \sqrt{x^2 + y^2 + z^2}$ (distance to the original position of the particle). The probability of finding the particle in a distance between r and $r + dr$ (in a spherical shell of volume $4\pi r^2 \cdot dr$) is

$$P(r) \cdot dr = const \cdot 4\pi r^2 \cdot \exp\left(-\frac{3r^2}{2j\lambda^2}\right)$$

where the constant is determined by the condition

$$\int_0^\infty P(r) \cdot dr = 1$$

Problem 11.9 – Evaporation of Water at Room Temperature

We consider a measuring cylinder (inner diameter 3 cm, height 15 cm) with a layer (height 3 cm) of water on the bottom. Above the water surface gaseous water is in equilibrium with liquid water (vapor pressure of 30 mbar at room temperature). The water molecules escape from the cylinder by diffusion. How long does it take until 1 mL of liquid water has disappeared?

Solution. We apply Fick's law (equations 11.58, 11.59) in the form

$$\frac{\Delta n}{\Delta t} = AD \cdot \frac{c_H}{H}$$

(see Figure 11.22b). The concentration c_H of the water vapor on the bottom of the cylinder is

$$c_H = \frac{n}{V} = \frac{P_{vap}}{RT} = \frac{30\,\text{mbar}}{8.314 \cdot 300\,\text{J mol}^{-1}} = 1.2 \times 10^{-3}\,\text{mol L}^{-1}$$

We calculate the diffusion coefficient with

$$\sigma = \pi(r_{H_2O} + r_{N_2})^2 = \pi(231 + 190)^2 \times 10^{-24}\,\text{m}^2 = 5.6 \times 10^{-19}\,\text{m}^2$$

(we treat air as a nitrogen gas) and the number density $N/V = 2.65 \times 10^{25}\,\text{m}^{-3}$ of the nitrogen molecules. The reduced mass is

$$\mu = \frac{18 \times 28}{18 + 28} \cdot \frac{1}{6.02 \times 10^{23}}\,\text{g} = 1.8 \times 10^{-26}\,\text{kg}$$

Then the diffusion coefficient is

$$D = \frac{1}{2} \cdot \frac{1}{\sigma \cdot N/V} \cdot \sqrt{\frac{kT}{\mu}} = 1.6 \times 10^{-5}\,\text{m}^2\,\text{s}^{-1}$$

with $H = (15 - 3)\,\text{cm} = 12\,\text{cm}$ and $A = \pi \cdot 1.5^2\,\text{cm}^2 = 7.07\,\text{cm}^2$ we calculate the time Δt needed for the evaporation of 1 mL of liquid water (this corresponds to an amount of substance of $(1/18)\,\text{mol} = 0.056\,\text{mol}$) as

$$\Delta t = \Delta n \cdot \frac{H}{ADc_H} = 4.9 \times 10^5\,\text{s} = 5.7\,\text{days}$$

Thus it takes approximately one week for 1 mL of liquid water to evaporate.

Foundation 11.1 – Averaging the Free Path λ

We consider N_0 particles which can collide with other particles. In a range of the free path between λ and $\lambda + d\lambda$ the number of particles without any collision changes by $dN = -aN \cdot d\lambda$ (a = constant). By integration we obtain the number N of particles traveling a distance λ without any collision after a path λ: $N = N_0 e^{-a\lambda}$.

N decreases exponentially with λ. Thus the number of particles colliding in a range between λ and $\lambda + d\lambda$ is $dN = aN_0 e^{-a\lambda} \cdot d\lambda$. This is the distribution function of the free path. We calculate the averages $\overline{\lambda}$ and $\overline{\lambda^2}$.

$$\overline{\lambda} = \frac{1}{N_0} \int_0^\infty \lambda \cdot dN = \frac{1}{N_0} aN_0 \int_0^\infty \lambda e^{-a\lambda} \cdot d\lambda$$

$$= a \frac{1}{a^2} \int_0^\infty z \cdot e^{-z} \cdot dz = \frac{1}{a}$$

$$\overline{\lambda^2} = \frac{1}{N_0} \int_0^\infty \lambda^2 \cdot dN = \frac{1}{N_0} aN_0 \int_0^\infty \lambda^2 e^{-a\lambda} \cdot d\lambda$$

$$= a \frac{1}{a^3} \int_0^\infty z^2 \cdot e^{-z} \cdot dz = 2 \frac{1}{a^2}$$

Thus we obtain $\overline{\lambda^2} = 2 \cdot (\overline{\lambda})^2$.

Foundation 11.2 – Intermolecular Forces Affecting the Mean Free Path

The square root dependence of η on temperature no longer holds strictly for low temperatures. This is because intermolecular forces affect the mean free path of the molecules: $\overline{\lambda}$ becomes smaller than expected from equation (11.36):

$$\overline{\lambda} = \frac{V}{\sqrt{2}\sigma N} \cdot \frac{1}{1 + C/T}$$

The constant C (*Sutherland constant*) depends on the intermolecular forces. We can understand this effect on a qualitative level: two molecules, which would pass each other at a small distance in the absence of intermolecular forces, can collide because of their mutual attraction. Thus the number of collisions becomes somewhat larger than originally calculated. At higher velocity

Continued on page 406

Continued from page 405

(higher temperature) the molecules are for a shorter time in the region where they interact. Thus the effect is more pronounced at low temperature.

To derive an expression for C we assume that a constant force f is acting between the molecules 1 and 2 (Figure 11.38) as soon as molecule 1 enters the shaded area; outside this area the force is zero. Then during the time

$$t = \frac{d_0}{\bar{v}}, \quad \bar{v} = \sqrt{\frac{8kT}{\pi m}}$$

(where d_0 is the collision diameter, $d_0 = r_1 + r_2$) the molecule is accelerated by the constant force f and deflected in the y-direction by y. According to Newton's law

$$\frac{d^2 y}{dt^2} = \frac{f}{m}, \quad \frac{dy}{dt} = \frac{f}{m}t, \quad y = \frac{1}{2}\frac{f}{m}t^2$$

we obtain

$$y = \frac{1}{2}\frac{f}{m}t^2 = \frac{1}{2}\frac{f}{m}\frac{d_0^2}{\bar{v}^2} = \frac{1}{2}f d_0^2 \frac{\pi}{8kT}$$

Thus an increased collision diameter

$$d = d_0 + y = d_0 \cdot \left(1 + \frac{1}{2}f d_0 \frac{\pi}{8kT}\right)$$

is obtained, resulting in a modified expression for the mean free path:

$$\bar{\lambda} = \frac{V}{\sqrt{2}\sigma N} \cdot \frac{1}{\left(1 + \frac{1}{2}f d_0 \frac{\pi}{8kT}\right)^2} \approx \frac{V}{\sqrt{2}\sigma N} \cdot \frac{1}{1 + \frac{\pi}{8}\frac{f d_0}{kT}}$$

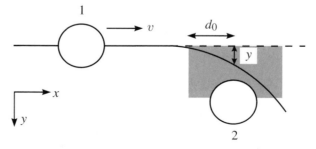

Figure 11.38 Forces acting between molecules cause a deflection from the straight path.

Continued on page 407

FOUNDATION 11.2

Continued from page 406

Thus we can identify the Sutherland constant as

$$C = \frac{\pi}{8} \frac{f d_0}{k}$$

Example:

For N_2 ($d_0 = 300$ pm) the dispersion force (see Chapter 10) at a distance d_0 is on the order of 10^{-11} N and we obtain $C = 85$ K, in reasonable agreement with the experimental value $C = 103$ K.

12 Energy Distribution in Molecular Assemblies

What happens inside a molecule when it is hit by a neighboring molecule? After a collision with another molecule, a molecule can reach an excited quantum state by taking up translational, rotational, vibrational, or electronic energy. Or it can transfer its energy to the collision partner. At higher temperature, higher internal energy states are reached. The Boltzmann distribution discussed in this chapter describes the population probability P_i of quantum state i of energy E_i as a function of temperature:

$$P_i \sim e^{-E_i/kT}$$

The Maxwell–Boltzmann distribution describes the distribution of the translational speed as a function of temperature:

$$\frac{dN}{dv} \sim v^2 \cdot e^{-mv^2/2kT}$$

This distribution follows when treating the molecules in a gas as classical particles, just as we did in Chapter 11. The relation also follows when determining the translational quantum states of the molecules in a gas sample and asking for the population of these states. This is a good chance to illustrate the correspondence of classical and quantum mechanical considerations, as discussed in Section 3.8.4.

The essence of the Boltzmann equation is easy to grasp, but its formal proof needs deeper reflections given in Section 12.6 and in Foundation 12.3 addressed to the more advanced student.

12.1 The Boltzmann Distribution Law

12.1.1 System Consisting of Two Quantum States

We consider sodium gas in a bath of temperature T; the Na atoms can exist either in the ground state (energy E_1) or in an electronically excited state (energy E_2), if we neglect higher excited states (Figure 12.1). If there are altogether N atoms in the Na vapor, then N_1 atoms can be in quantum state 1 and N_2 in quantum state 2. Obviously, at low enough temperature, all atoms are in the lower atomic quantum state and we have

$$N_2 = 0, \quad N_1 = N \quad \text{(low temperature)}$$

12.1 THE BOLTZMANN DISTRIBUTION LAW

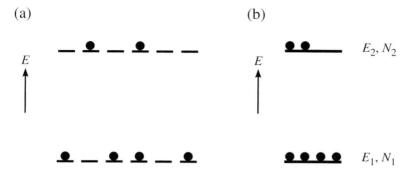

Figure 12.1 Energies of Na atoms (E_1: ground state, E_2: electronically excited state). (a) Four out of six atoms are in the ground state, two atoms are in the excited state. (b) Population of atomic quantum states 1 and 2.

With increasing temperature the population number N_2 will increase. When the mean translational energy of the Na atoms reaches the energy difference between both states

$$E_2 - E_1 = \tfrac{3}{2}kT \quad \text{(medium temperature)}$$

a considerable fraction of the molecules will be in quantum state 2.

At very high temperature it is equally probable that a sodium atom in the ground state is excited in a collision or that a sodium atom in the excited state is de-excited. This means that N_2 can never exceed N_1:

$$N_2 = N_1 = \tfrac{1}{2}N \quad \text{(high temperature)}$$

These three cases are demonstrated in Fig. 12.2.

As we shall prove in Section 12.6, for the two quantum states of the Na atoms we obtain

$$N_1 = C \cdot e^{-E_1/kT}, \quad N_2 = C \cdot e^{-E_2/kT} \tag{12.1}$$

and with $\Delta E = E_2 - E_1$

$$\frac{N_2}{N_1} = e^{-\Delta E/kT} \tag{12.2}$$

On the other hand we have to regard the condition

$$N_1 + N_2 = N \tag{12.3}$$

and it follows from equations (12.2) and (12.3) that

$$N_1 = N \cdot \frac{1}{1 + e^{-\Delta E/kT}}, \quad N_2 = N \cdot \frac{1}{1 + e^{+\Delta E/kT}} \tag{12.4}$$

From this relation we obtain $N_1 = N$ and $N_2 = 0$ at $T = 0$: all the atoms are in the ground state. The result for $T \to \infty$ is $N_1 = N_2 = N/2$. This means that each of the two quantum states is occupied by the same number of atoms. When $\Delta E = kT$, $N_1 = 0.73 \cdot N$ and $N_2 = 0.27 \cdot N$ (Figure 12.2).

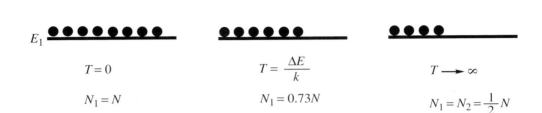

Figure 12.2 Population of two quantum states at different temperatures ($\Delta E = E_2 - E_1$).

According to Section 3.1, the electronic excitation energy of Na is $\Delta E = 3.37 \times 10^{-19}$ J. With this value we obtain the following fractions of excited Na atoms:

T/K	N_2/N_1
300	3×10^{-36}
1000	3×10^{-11}
2000	3×10^{-6}
20 000	0.23

2000 K is the temperature of the hottest part of the flame of a Bunsen burner; when a small piece of sodium chloride is heated to this temperature, the flame becomes intensely yellow. This results from the Na evaporating from the crystal lattice; the Na atoms are electronically excited, and by the emission of light (sodium D line) they return to the ground state. In Figure 12.3, N_1 and N_2 are illustrated as a function of T.

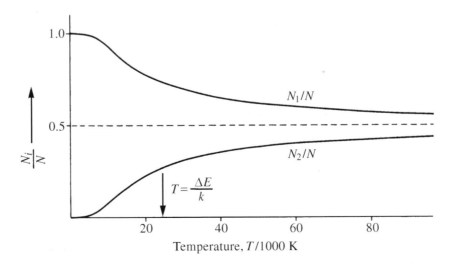

Figure 12.3 N_1 and N_2 as a function of T according to equation (12.4) with $\Delta E = 3.37 \times 10^{-19}$ J (excitation energy of a Na atom: yellow Na line).

12.1 THE BOLTZMANN DISTRIBUTION LAW

How can we understand why at high temperature both states are almost equally occupied? As a simple model, we imagine a container with a bottom made of two steps (Figure 12.4). On the bottom we place steel spheres. As long as the container is at rest, all the spheres will be on the lower step at lowest potential energy (Figure 12.4(a)). Now we shake the container, thus providing the steel spheres with kinetic energy. Weak shaking corresponds to low kinetic energy–that is, to low temperature. On a time average, most of the spheres are below and only few are up (Figure 12.4(b)). The stronger we shake, the more spheres will reach the upper step, but the spheres will not accumulate on the upper step. When strongly shaking the container, the spheres do not recognize any difference between the two parts of the container, and therefore they will be equally distributed (Fig. 12.4(c)).

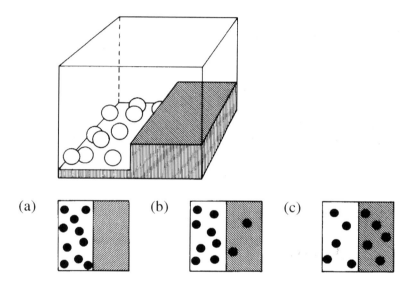

Figure 12.4 Model experiment for the demonstration of the distribution of molecules on two energy levels. (a) Box at rest; (b) box weakly shaken; (c) box vigorously shaken.

12.1.2 Systems Consisting of Many Quantum States

Now we proceed from a system with two quantum states to systems with many quantum states. As we will prove in Section 12.6, the population number N_i of a certain quantum state i is given by the Boltzmann distribution

$$N_i = C \cdot e^{-E_i/kT}, \quad i = 1, 2, 3, \ldots \quad (12.5)$$

where C is a constant. N_i decreases when the energy E_i increases (Figure 12.5). Then, using the condition that the sum over all population numbers N_i equals the total number of particles

$$N_1 + N_2 + N_3 + \cdots = N \quad (12.6)$$

and inserting in equation 12.5 we obtain

$$N = \sum_{i=1}^{\infty} N_i = \sum_{i=1}^{\infty} C \cdot e^{-E_i/kT} = C \cdot \sum_{i=1}^{\infty} e^{-E_i/kT} = C \cdot Z \quad (12.7)$$

The quantity

$$Z = \sum_{i=1}^{\infty} e^{-E_i/kT} \qquad (12.8)$$

is called the *partition function*. C is then

$$C = \frac{N}{Z} \qquad (12.9)$$

With this expression we rewrite the Boltzmann distribution as

$$N_i = \frac{N}{Z} \cdot e^{-E_i/kT} \qquad (12.10)$$

This distribution is demonstrated in Fig. 12.5.

For the probability P_i of finding a particle in quantum state i of energy E_i at temperature T we obtain

$$P_i = \frac{N_i}{N} = \frac{1}{Z} \cdot e^{-E_i/kT} \qquad (12.11)$$

These are the basic equations to determine the distribution of N molecules over all given quantum states in equilibrium with a temperature bath of temperature T.

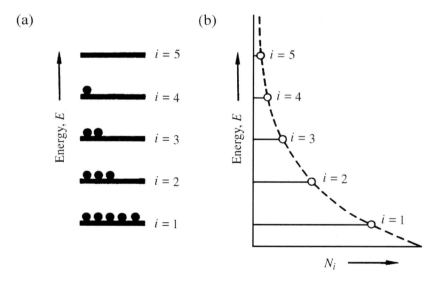

Figure 12.5 N_i as a function of E_i at constant temperature. (a) Population scheme; (b) illustration of the Boltzmann distribution (12.10). The population scheme is shown for the distribution of vibrational energy of a diatomic molecule (equal distances of energy levels).

12.2 Distribution of Vibrational Energy

We consider a diatomic molecule in vapor phase. The molecule can be in the vibrational ground state or in a vibrational excited state. We ask for the number N_n of molecules in a vibrational state with quantum number n.

12.2 DISTRIBUTION OF VIBRATIONAL ENERGY

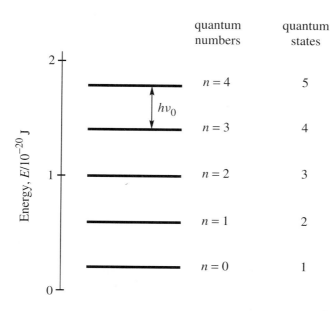

Figure 12.6 Quantum states of an oscillator with $h\nu_0 = 4.3 \times 10^{-21}$ J (data for I_2).

According to Chapter 9, the vibrational energy E_n of a diatomic molecule (nuclear masses m_1 and m_2) with a quantum number n is

$$E_n = \left(\tfrac{1}{2} + n\right) \cdot h\nu_0, \quad n = 0, 1, 2, 3, \ldots \tag{12.12}$$

where

$$\nu_0 = \frac{1}{2\pi}\sqrt{\frac{k_f}{\mu}} \quad \text{with} \quad \mu = \frac{m_1 \cdot m_2}{m_1 + m_2} \tag{12.13}$$

μ is the reduced mass and k_f is the force constant.

We consider I_2 molecules as an example. With $k_f = 173\,\text{N}\,\text{m}^{-1}$ and $\mu = 1.05 \times 10^{-25}$ kg we obtain $h\nu_0 = 4.27 \times 10^{-21}$ J. The corresponding quantum states are shown in Figure 12.6.

12.2.1 Population Number N_n

The population number N_n of molecules in quantum state with quantum number n is

$$N_n = \frac{N}{Z} e^{-E_n/kT}, \quad n = 0, 1, 2, 3, \ldots \tag{12.14}$$

Note that the index n (quantum number, $n = 0, 1, 2, 3, \ldots$) is to be distinguished from the index i (number of quantum state, $i = 1, 2, 3, \ldots$).

The partition function Z

$$Z = \sum_{n=0}^{\infty} e^{-E_n/kT} \tag{12.15}$$

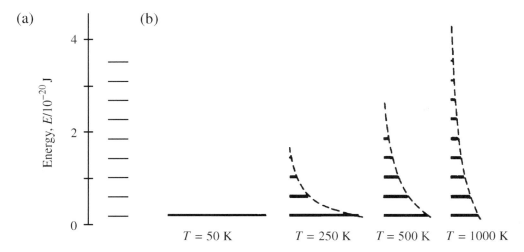

Figure 12.7 I_2 molecule: Population of vibration levels at various temperatures. (a) Energy levels, (b) population of the different quantum states; the lengths of the lines in the different energy levels are proportional to the number of particles. The population numbers were calculated numerically according to equations (12.14) and (12.15).

can be easily calculated using the relations (12.12) and (12.13); the calculation can be performed numerically or analytically (see Foundation 12.1). In this way we obtain the population numbers for I_2 shown in Figure 12.7 for different temperatures.

At low enough temperature, all molecules are in the lowest quantum state. With increasing temperature, more and more molecules occupy higher energy states. At room temperature, 64% of all I_2 molecules are in the ground state, and 36% are distributed over vibrationally excited states.

12.2.2 Total Vibrational Energy U_{vib}

By adding up the vibrational energies E_n of the N_n molecules in the distinct quantum states n we obtain the total vibrational energy U_{vib}

$$U_{vib} = \sum_{n=0}^{\infty} N_n E_n = \frac{N}{Z} \sum_{n=0}^{\infty} e^{-E_n/kT} \cdot E_n \qquad (12.16)$$

Again, the summation is carried out for different temperatures, either numerically or analytically (see Foundation 12.1). In Figure 12.8(a) $U_{vib,m}$ of I_2 is shown as a function of temperature. At low enough temperature, all N molecules are in the lowest energy state, thus

$$U_{vib} = N \cdot \tfrac{1}{2} h\nu_0 \quad \text{(for } T = 0\text{)} \qquad (12.17)$$

As in the two-level system discussed in Section 12.1.1, a minimum temperature is required for the molecules to overcome the energy difference $\Delta E = h\nu_0$ from the lowest to the first excited energy state, that is, the thermal energy kT must be on the order of ΔE

$$kT \approx h\nu_0 \qquad (12.18)$$

12.2 DISTRIBUTION OF VIBRATIONAL ENERGY

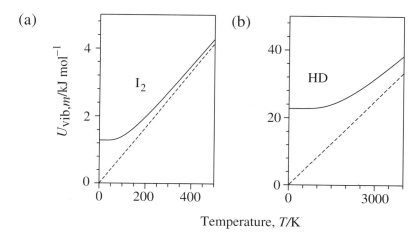

Figure 12.8 Molar vibrational energy $U_{\text{vib},m}$ as a function of the temperature T (solid lines). (a) I_2 gas, (b) HD gas. The dashed lines correspond to the classically expected curve ($U_{\text{vib},m} = N_A kT$). U_{vib} was calculated numerically according to equation (12.16).

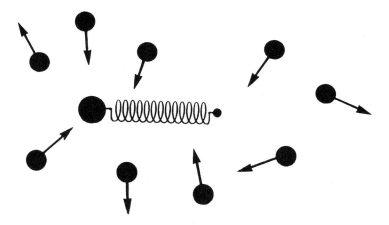

Figure 12.9 Oscillator in a temperature equilibrium with gas molecules.

For I_2, this condition is fulfilled at $T = h\nu_0/k = 4.27 \times 10^{-21}$ J$/(1.34 \times 10^{-23}$ J K$^{-1}) = 309$ K, in agreement with Figure 12.8. Contrary to the two-level system, U_{vib} is continuously increasing, without reaching a limiting value. This is, because with increasing temperature more and more higher energy levels can be reached. At high enough temperature U_{vib} increases linearly with T, approaching NkT, as derived in Foundation 12.1 (dashed line in Figure 12.8(a)):

$$U_{\text{vib}} = NkT \quad \text{(for large enough } T\text{)} \tag{12.19}$$

This expression is identical with the expression obtained when the oscillator is considered as a classical oscillator in equilibrium with a temperature bath of temperature T (*equipartition principle*, Figure 12.9, Box 12.1).

> **Box 12.1 – Equipartition Principle in the Classical Limit**
>
> For simplicity we consider a molecule like HCl in which the center of gravity coincides with the heavy nucleus (Figure 12.9). In the classical picture the light atom will have, on the average, the same kinetic energy $\frac{3}{2}kT$ as the gas molecules that hit it. The energy is equally distributed in the direction of the bond between the atoms and in the two directions perpendicular to it. Then, on the average, $\frac{1}{2}kT$ is stored as the kinetic energy of vibration. Additionally, the oscillator stores potential energy (stretching of the spring), which, on the average, equals the kinetic energy. Then
>
> $$U_{\text{vib,classic}} = U_{\text{vib,kin}} + U_{\text{vib,pot}} = N \cdot 2 \cdot \tfrac{1}{2}kT = NkT$$
>
> At high enough temperature the quantum mechanical consideration approaches the classical one; according to Figure 12.8 this happens for I_2 at temperatures much higher than 500 K and for HD at temperatures much higher than 5000 K. At lower temperatures the deviations from classical behavior are considerable.
>
> In polyatomic molecules with N_{atom} atoms several vibration modes are possible (Section 9.4.2):
>
> - ($3N_{\text{atom}} - 5$) vibration modes (linear molecules)
> - ($3N_{\text{atom}} - 6$) vibration modes (nonlinear molecules)
>
> In the classical model each of these vibration modes contributes kT to the internal energy.
>
> The translation contributes with three terms of $\frac{1}{2}kT$ (x, y, and z component of kinetic energy). The rotation of a linear molecule contributes with two terms of $\frac{1}{2}kT$ corresponding to two possible axes of rotation. The rotation of a nonlinear molecule contributes with three terms of $\frac{1}{2}kT$ corresponding to three possible axes of rotation.
>
> In the classical limit each of these terms contributes $\frac{1}{2}kT$ to the expression for the energy of a molecule. This statement is called the *equipartition principle*.

The temperature behavior of U_{vib} depends strongly on the force constant and the mass of the oscillator.

Example

For HD, $h\nu_0 = 75 \times 10^{-21}$ J; then the temperature T required to overcome the energy difference between the ground state and the first excited state is $T = h\nu_0/k = 5430$ K. At room temperature, practically all the molecules occupy the lowest vibrational state (Figure 12.8(b)).

12.3 Distribution of Rotational Energy

In our consideration, we have evaluated the expressions for the population numbers N_n and for the total energy U_{vib} numerically as well as analytically. Particularly, a general relation between the total energy U_{vib} and the partition function Z can be derived (see Foundation 12.1)

$$U_{vib} = NkT^2 \cdot \frac{d \ln Z}{dT} \tag{12.20}$$

12.3 Distribution of Rotational Energy

According to Chapter 9, the rotational energy E_n of a diatomic molecule is characterized by the quantum number n

$$E_n = B \cdot n(n+1), \quad n = 0, 1, 2, \ldots \tag{12.21}$$

with

$$B = \frac{h^2}{8\pi^2 \mu d^2} \tag{12.22}$$

where B is called the *rotational constant*. In contrast to the oscillator where only one wavefunction belongs to each quantum number, the energy states of a rotator are

$$g_n = 2n + 1 \quad \text{-fold degenerate} \tag{12.23}$$

We consider HD as an example. With $\mu = 1.12 \times 10^{-27}$ kg and $d = 74$ pm we obtain $B = 9.07 \times 10^{-22}$ J. The corresponding quantum states are shown in Figure 12.10.

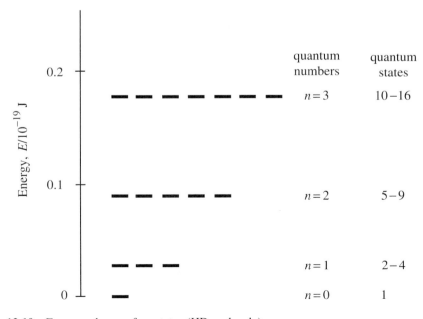

Figure 12.10 Energy scheme of a rotator (HD molecule).

12.3.1 Population Numbers N_n

The population numbers N_n of particles in the energy levels with quantum number n are

$$N_n = (2n+1) \cdot \frac{N}{Z} \cdot e^{-E_n/kT}, \quad n = 0, 1, 2, \ldots \quad (12.24)$$

because $2n+1$ quantum states belong to each quantum number n.

In this case the partition function is

$$Z = \sum_{n=0}^{\infty} (2n+1) \cdot e^{-E_n/kT} \quad (12.25)$$

On the one hand, N_n decreases exponentially with E_n; on the other hand, N_n increases with $2n+1$.

The partition function Z and the population numbers N_n for HD are calculated numerically. In this case, an analytical treatment, in contrast to the case of vibration, is not possible. The population numbers N_n are displayed in Figure 12.11 for $T = 100$ K. The population increases, reaching a maximum at $n = 1$; then it decreases with increasing n.

12.3.2 Total Rotational Energy U_{rot}

To calculate the energy U_{rot} of N rotators we proceed as in the case of the oscillator

$$U_{\text{rot}} = \sum_{n=0}^{\infty} N_n E_n = \frac{N}{Z} \sum_{n=0}^{\infty} (2n+1) \cdot e^{-E_n/kT} \quad (12.26)$$

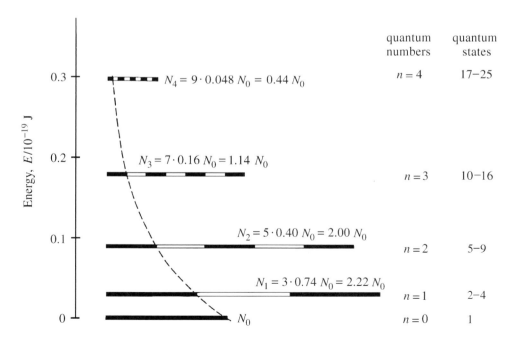

Figure 12.11 Population scheme of a rotator (HD) at $T = 100$ K. The lengths of the horizontal lines on the left from the dashed curve correspond to the number of particles per quantum state. N_0 is the number of particles in the lowest energy state ($n = 0$).

12.3 DISTRIBUTION OF ROTATIONAL ENERGY

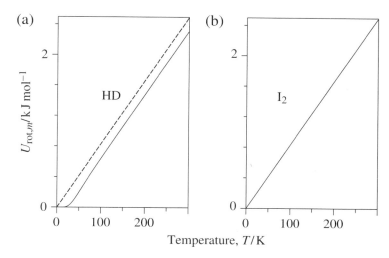

Figure 12.12 Molar rotational energy $U_{\text{rot},m}$ as a function of temperature (solid lines). (a) HD gas; (b) I_2 gas. The dashed line corresponds to the classically expected curve ($U_m = N_A kT$).

This expression is evaluated numerically (see Foundation 12.2). In Figure 12.12(a) $U_{\text{rot},m}$ of HD is shown as a function of temperature.

At $T = 0$ we obtain $U_{\text{rot}} = 0$ (all molecules occupy the lowest energy state with $E_0 = 0$); at high enough temperatures, U_{rot} approaches the value

$$U_{\text{rot}} = NkT \tag{12.27}$$

as expected from the equipartition principle (Box 12.1). The temperature T required for the rotator to overcome the energy difference between ground state and first excited state is $T = \Delta E/k = 2B/k$. In the case of HD ($B = 9.07 \times 10^{-22}$ J), we find $T = 132$ K.

At this temperature U_{rot} reaches 80% of the classical value; 92% are reached at 300 K (Figure 12.12(a)). For heavier molecules the classical value is achieved at much lower temperature (Figure 12.12(b)).

Example

For I_2, with $\mu = 1.05 \times 10^{-25}$ kg and $d = 267$ pm we obtain $B = 7.4 \times 10^{-25}$ J, thus we calculate $T = \Delta E/k = 2B/k = 0.1$ K. This shows that the equipartition principle is fulfilled already at extremely low temperature (Figure 12.12(b)).

Then the rotational energy of gas molecules at room temperature can be well described by the classical relation (12.27). This is because the spacing between the first and the second rotational levels is smaller by a factor of about 10 compared to the spacing of vibrational levels. For example, $\Delta E_{\text{vib},0\to 1} = 75 \times 10^{-21}$ J and $\Delta E_{\text{rot},0\to 1} = 1.8 \times 10^{-21}$ J for HD.

12.4 Translational Energy

In Chapter 11 we assumed that the translational motion of a molecule can be described by classical mechanics. The de Broglie wavelength of an H_2 molecule with speed $v = 10^3$ m s^{-1} is $\Lambda = 200$ pm, thus Λ is small compared to the free path. The quantum mechanical description should, therefore, lead to the same result as the classic consideration.

Is this really the case? According to equation (11.22) the mean translational energy of a molecule at temperature T is

$$\overline{E}_{\text{trans}} = \tfrac{3}{2}kT \tag{12.28}$$

Can we reproduce this classical result by quantum mechanical considerations?

12.4.1 Average Translational Energy According to Quantum Mechanics

According to quantum theory (Chapter 2) the energy of a particle of mass m in a cubic box of length L is

$$E_{n_x,n_y,n_z} = \frac{h^2}{8mL^2}\left(n_x^2 + n_y^2 + n_z^2\right), \quad n_x, n_y, n_z = 1, 2, 3, \ldots \tag{12.29}$$

(In Chapter 2 we used the same relation with the mass m_e of an electron, now m is the mass of the molecule.)

The probability P_{n_x,n_y,n_z} of finding a molecule in a state characterized by the quantum numbers n_x, n_y, n_z is, according to the Boltzmann distribution (12.11),

$$P_{n_x,n_y,n_z} = \frac{1}{Z} \cdot e^{-E_{n_x,n_y,n_z}/kT} \tag{12.30}$$

Then we find for the mean translational energy $\overline{E}_{\text{trans}}$ of the molecule

$$\overline{E}_{\text{trans}} = \sum_{n_x=1}^{\infty}\sum_{n_y=1}^{\infty}\sum_{n_z=1}^{\infty} E_{n_x,n_y,n_z} \cdot P_{n_x,n_y,n_z} \tag{12.31}$$

We obtain (Box 12.2)

$$\overline{E}_{\text{trans}} = \tfrac{3}{2}kT \quad \text{or} \quad U_{\text{trans}} = \tfrac{3}{2}NkT \tag{12.32}$$

This is the same result as in the classical consideration (Chapter 11).

Box 12.2 – Calculation of Translational Energy

According to Foundation 12.1, the total energy of translation is

$$U_{\text{trans}} = NkT^2 \cdot \frac{d \ln Z}{dT}$$

Continued on page 421

12.4 TRANSLATIONAL ENERGY

Continued from page 420

According to equation (12.15) the partition function Z for translation is

$$Z = \sum_{n_x, n_y, n_z} e^{-A(n_x^2 + n_y^2 + n_z^2)} \quad \text{with} \quad A = \frac{h^2}{8mL^2} \cdot \frac{1}{kT}$$

Then

$$Z = \sum_{n_x} e^{-An_x^2} \cdot \sum_{n_y} e^{-An_y^2} \cdot \sum_{n_z} e^{-An_z^2} = \left(\sum_{n_x} e^{-An_x^2} \right)^3$$

(the expressions for x, y, and z are identical).

To calculate the remaining sum, we approximate the summation by an integration. This means that we replace the area of the rectangles in Figure 12.13 (representing the sum) by the area under the curve (representing the integral). In this case n_x is treated as a continuous variable not restricted to integer values.

For any appreciable temperature many quantum states are in the range of kT, thus $A \ll 1$. For example, for He gas enclosed in a container of volume $V = 1$ L (corresponding to $L = 10$ cm) at $T = 10$ K we obtain (with $m = 6.6 \times 10^{-27}$ kg) $A = 6.2 \times 10^{-18}$. Then the two areas in Figure 12.13 are practically identical in cases of interest.

$$\sum_{n_x=1}^{\infty} e^{-An_x^2} = \int_0^{\infty} e^{-An_x^2} \cdot dn_x$$

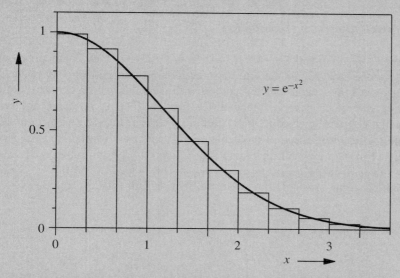

Figure 12.13 Summation replaced by an integration, demonstrated by the function $y = e^{-x^2}$. The sum is represented by the areas of the rectangles. The integral corresponds to the area below the curve.

Continued on page 422

> Continued from page 421
>
> Substituting
> $$v^2 = An_x^2, \quad dv = \sqrt{A} \cdot dn_x$$
>
> we obtain
> $$\int_0^\infty e^{-An_x^2} \cdot dn_x = \frac{1}{\sqrt{A}} \int_0^\infty e^{-v^2} \cdot dv = \frac{1}{\sqrt{A}} \cdot \frac{1}{2}\sqrt{\pi}$$
>
> Then
> $$Z = \frac{1}{8} C \cdot T^{3/2} \quad \text{with} \quad C = \frac{\pi m L^2 k}{h^2}$$
>
> Then we obtain
> $$\frac{d \ln Z}{dT} = \frac{1}{Z} \cdot \frac{dZ}{dT} = \frac{8}{CT^{3/2}} \cdot \frac{1}{8} C \cdot \frac{3}{2} T^{1/2} = \frac{3}{2} \frac{1}{T}$$
>
> and
> $$U_\text{trans} = NkT^2 \cdot \frac{d \ln Z}{dT} = NkT^2 \cdot \frac{3}{2} \frac{1}{T} = \frac{3}{2} NkT$$

12.4.2 Maxwell–Boltzmann Distribution

In Chapter 11 we assumed for simplicity that the molecules in a gas are moving with the same speed. This cannot be true. Let us assume that actually all molecules have accidentally the same speed v. They collide with each other. In a centered collision the speeds of equal molecules are exchanged (Chapter 11). This is, however, not the case for a noncentered collision (Figure 12.14). After the collision (Figure 12.14(c)), molecule 2 is at rest, and the resulting speed of molecule 1 is $v_1' = \sqrt{2} \cdot v$. Noncentered collisions occur frequently. Then, starting with the same speed, many of the molecules will have speeds higher or lower than the average speed after a while (speed distribution, Figure 12.15, dots). From classical considerations, Maxwell derived for this distribution

$$dN = N \cdot 4\pi \cdot \left(\frac{m}{2\pi kT}\right)^{3/2} \cdot v^2 \cdot e^{-mv^2/2kT} \cdot dv \tag{12.33}$$

(solid line in Figure 12.15). This is called the *Maxwell–Boltzmann distribution of speed*.

This distribution can be checked experimentally. Small speeds are rare; the most probable speed is

$$v_\text{P} = \sqrt{\frac{2kT}{m}} \tag{12.34}$$

12.4 TRANSLATIONAL ENERGY

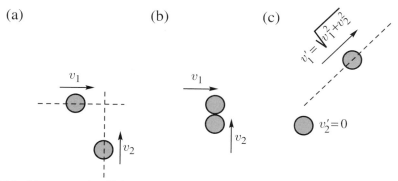

Figure 12.14 Noncentered collision of two identical molecules moving in directions perpendicular to each other. (a) Velocities v_1 and v_2 shortly before the collision, (b) velocities at collision; (c) velocities $v'_1 = \sqrt{v_1^2 + v_2^2}$ and $v'_2 = 0$ shortly after the collision. In the special case $v_1 = v_2 = v$ we obtain $v'_1 = \sqrt{2}v$.

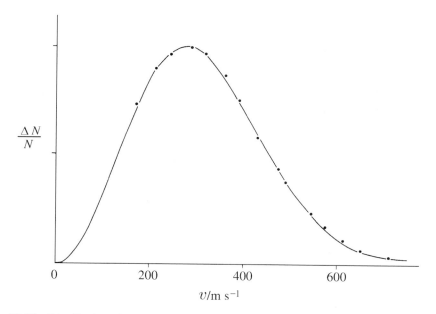

Figure 12.15 Distribution of speeds of thallium atoms in the gas state at 944 K; the ordinate scale is arbitrarily chosen. Dots: experiment, (R.C. Miller, P. Kusch, Velocity distributions in potassium and thallium atomic beams, *Phys. Rev.* 99, 1314 (1955)) solid line: calculation according to equation (12.33).

(Problem 12.2); the population decreases gradually at higher speeds. The average speed (Problem 12.2)

$$\bar{v} = \sqrt{\frac{8kT}{\pi m}} \qquad (12.35)$$

is higher than the most probable speed v_P because speeds $v > v_P$ occur more frequently than speeds $v < v_P$, thus they contribute more to the average.

The average translational energy

$$\overline{E}_{\text{trans}} = \tfrac{1}{2} m \overline{v^2} \qquad (12.36)$$

should be $\tfrac{3}{2}kT$, according to equation (12.32); this follows from equation (12.33) – see Problem 12.2.

As an example, the Maxwell–Boltzmann distribution for a gas of helium atoms is shown in Figure 12.16 for two different temperatures. The maximum of the distribution is considerably shifted to higher speeds when increasing the temperature of the gas from 100 K to 600 K.

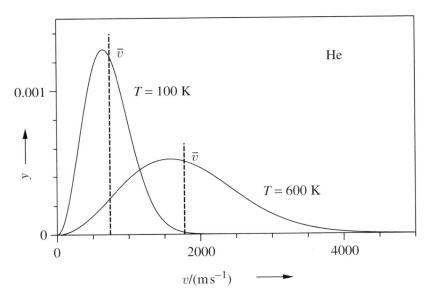

Figure 12.16 Distribution $y = \dfrac{dN}{N \cdot dv}$ of speeds for He atoms at 100 K and 600 K. The mean speed is obtained by averaging the speeds.

12.4.3 Deriving the Maxwell–Boltzmann Distribution

Can we understand the Maxwell–Boltzmann equation (12.33) from a quantum mechanical consideration?

We consider N molecules of mass m in a cubic box of side length L in a temperature bath of temperature T. The energy levels calculated from equation (12.29) are populated according to the Boltzmann distribution (equation (12.10)).

The probability of finding a molecule in a quantum state between E and $E + \Delta E$ is proportional to Δg, the number of quantum states between E and $E + \Delta E$. This number increases with increasing energy E. This can be easily seen in Figure 12.17(a) where the lowest 139 energy levels obtained from equation (12.29) are shown. In Figure 12.17(b) Δg is plotted as a function of the energy E (dots) and we can guess that Δg increases proportional to $\sqrt{E}$.

12.4 TRANSLATIONAL ENERGY

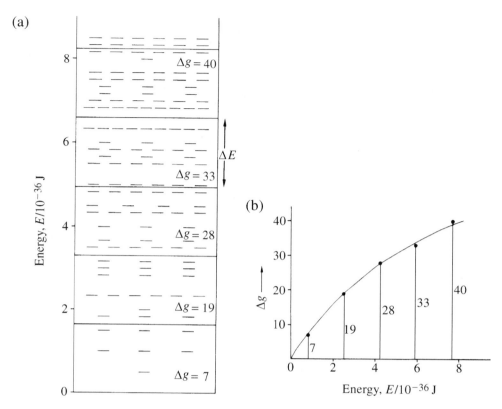

Figure 12.17 Number Δg of quantum states in the interval ΔE with increasing energy (lowest quantum state of H_2 in a cube of volume $V = 1 \, cm^3$, energies calculated according to equation (12.29). (a) Energy scheme (corresponding to Figure 3.5(b)); (b) Δg as a function of E (for $\Delta E = 1.66 \times 10^{-36}$ J).

In Box 12.3 it is shown that the number of quantum states below an energy E is

$$g = \frac{\pi}{6} \cdot \left(\frac{8mL^2}{h^2}\right)^{3/2} \cdot E^{3/2} \tag{12.37}$$

There is a large number of quantum states in the interval kT at any appreciable temperature. Therefore we consider g as a continuous function of E. We obtain by differentiation

$$\frac{dg}{dE} = \frac{\pi}{4} \cdot \left(\frac{8mL^2}{h^2}\right)^{3/2} \sqrt{E} \tag{12.38}$$

for sufficiently large values of E. If we replace the differentials dg and dE by the differences Δg and ΔE, we obtain the solid line in Figure 12.17(b).

Thus the probability of finding a molecule in quantum states between E and $E + dE$ is proportional to the Boltzmann factor and proportional to dg

$$dP_{E, E+dE} = C \cdot e^{-E/kT} \sqrt{E} \cdot dE \tag{12.39}$$

(constants condensed in C).

Box 12.3 – Number of Quantum States in a Gas

For the derivation of equation (12.37) we assume that the quantum numbers n_x, n_y, n_z are plotted on a coordinate system in the x, y, and z directions (Figure 12.13). To each quantum state of the molecule corresponds a distinct set of these quantum numbers, which again corresponds to a lattice point in the coordinate system in Figure 12.18.

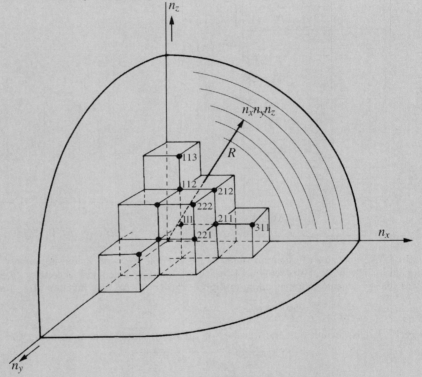

Figure 12.18 Counting the number g of quantum states.

g is the number of lattice points belonging to the set of n_x, n_y and n_z with energies between 0 and E; for example, the combinations 1,1,1; 1,2,1; 2,1,1; 1,1,2 belong to states for which $n_x^2 + n_y^2 + n_z^2 \leq 6$ (see also Figure 12.17).

The number of lattice points in Figure 12.18 is the same as the number of the unit cells, and we can describe g by the number of these unit cells. All the cubes belonging to the sets of n_x, n_y, n_z with energies between

$$E = 0 \quad \text{and} \quad E = \frac{h^2}{8mL^2}(n_x^2 + n_y^2 + n_z^2)$$

Continued on page 427

12.4 TRANSLATIONAL ENERGY

> *Continued from page 426*
>
> are inside one-eighth of a sphere with radius
>
> $$R = \sqrt{n_x^2 + n_y^2 + n_z^2}$$
>
> Then we obtain:
>
> $$g = \frac{\text{volume of } \tfrac{1}{8} \text{ of a sphere}}{\text{volume of a unit cell}} = \frac{1}{8} \cdot \frac{4}{3}\pi R^3 = \frac{1}{8} \cdot \frac{4}{3}\pi \cdot \left(n_x^2 + n_y^2 + n_z^2\right)^{3/2}$$
>
> We express the quantum numbers by the energy E
>
> $$n_x^2 + n_y^2 + n_z^2 = \frac{8mL^2}{h^2} \cdot E$$
>
> Then
>
> $$g = \frac{\pi}{6}\left(\frac{8mL^2}{h^2} \cdot E\right)^{3/2} = \frac{\pi}{6}\left(\frac{8mL^2}{h^2}\right)^{3/2} \cdot E^{3/2}$$

This is the distribution function of the translational energies in a gas. To obtain the distribution function of speeds, we replace the energy E by the speed v

$$E = \tfrac{1}{2}mv^2, \quad dE = mv \cdot dv \tag{12.40}$$

This results in

$$dP_{v,v+dv} = C' \cdot v^2 \cdot e^{-mv^2/2kT} \cdot dv \tag{12.41}$$

(constants condensed in C'). With $dN = N \cdot dP$, and evaluating C' by the condition

$$\int_{v=0}^{\infty} dP = 1 \tag{12.42}$$

(Problem 12.2) we obtain the Maxwell–Boltzmann distribution (12.33).

This agreement is remarkable since the quantum mechanical justification is so very different from the classical treatment.

12.4.4 Number of Translational Quantum States Available per Molecule

In the previous section we assumed that the number of available quantum states is large compared to the number of molecules and that the particles can be regarded as independent of each other. In all practical cases this really is the case.

As an example, we consider a gas of He atoms ($m = 7 \times 10^{-27}$ kg) in a box of $1000 \, \text{cm}^3$ ($L = 10 \, \text{cm}$) at room temperature; in this case the mean translational energy is $\overline{E}_{\text{trans}} = \frac{3}{2}kT \approx 6 \times 10^{-21}$ J. From equation (12.37) the number of available quantum states is obtained: $g = 10^{28}$ for $E = \overline{E}_{\text{trans}}$. If the helium gas is at a pressure of 1 bar the number of molecules in 1000 cm^3 is $N = 3 \times 10^{22}$. This means that there are more than 10^5 quantum states available for each molecule.

The occupation probability in the lowest quantum state is only 4.5 times higher than on the average (Problem 12.4).

At low temperatures, however, the assumption is no longer true that the number of quantum states is much higher than the number of molecules. From equation (12.37) we find $g = N$ for He at 0.03 K and 1 bar. This means that our present description does not hold for gases at temperatures near to the zero point of 0 K.

12.5 Distribution of Electronic Energy

The distances between energy levels of electronic states are so large that the thermal energy at room temperature is not sufficient to excite higher energy electronic states. Thus the molecules are in the electronic ground state (see Na atoms, Fig. 12.3). However, in exceptional cases, the population of electronically excited levels plays a role.

An example is the NO molecule. It has one unpaired electron with its spin oriented in the magnetic field of the nuclei. Therefore the lowest energy state splits into two levels which are very close together. The lower level is occupied at low temperature, and the upper level becomes gradually populated as the temperature increases. We neglect the fact that higher electronically excited states come into play at higher temperatures. The internal electronic energy U_{electr} is

$$U_{\text{electr}} = N_1 E_1 + N_2 E_2 \tag{12.43}$$

where N_1 and N_2 are given by equation (12.5). If we set $E_1 = 0$, then

$$U_{\text{electr}} = N_2 E_2 = N \cdot \frac{E_2}{1 + e^{E_2/kT}} \tag{12.44}$$

For a gas of NO molecules, U_{electr} is displayed in Figure 12.19 as a function of temperature with $E_2 = 2.4 \times 10^{-21}$ J. At absolute zero, $U_{\text{electr}} = 0$ (since $N_1 = N$, $N_2 = 0$); at high enough temperature $U_{\text{electr}} = \frac{1}{2} N E_2$ (since $N_1 = N_2 = N/2$).

In contrast to rotation and vibration (where many higher levels are available), U_{electrf} approaches a limit and it does not further increase with temperature.

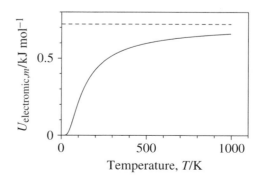

Figure 12.19 $U_{\text{electr},m}$ calculated for $E_1 = 0$, $E_2 = 2.4 \times 10^{-21}$ J (NO molecule). The limit for $T \to \infty$ is $0.5 \times \mathbf{N}_A E_2 = 0.72$ kJ mol^{-1} (dashed line).

12.6 Proving the Boltzmann Distribution

For simplicity we first ask how many possibilities exist to distribute a distinct energy over six particles, each of which can occupy three equally spaced energy states (spacing a, Figure 12.20). We have to consider that altogether only six particles are available, which means that

$$N_1 + N_2 + N_3 = 6 \qquad (12.45)$$

We suppose arbitrarily that the internal energy $4a$ is supplied to the system. Then

$$N_1 E_1 + N_2 E_2 + N_3 E_3 = 4a \qquad (12.46)$$

Thus only the three distributions A, B, and C shown in Figure 12.21 are possible. Which of these distributions is the most probable one?

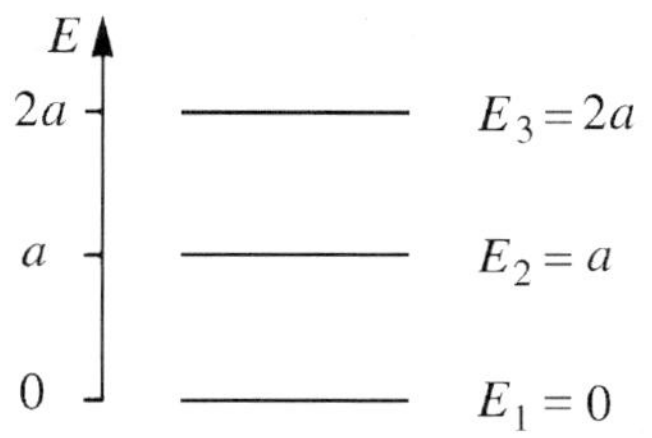

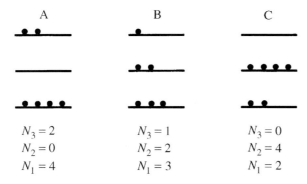

Figure 12.20 Energy scheme with three quantum states.

Figure 12.21 Possible populations of the quantum states in Figure 12.20 considering the conditions (12.45) and (12.46); six particles, total energy $4a$.

12.6.1 Distinguishable Particles

We assume that we can distinguish between the six particles. For example, in a crystal we can assign the different particles site numbers and distinguish between molecules that occupy different lattice sites. We distinguish the particles by numbers from 1 to 6. As demonstrated in Figure 12.22, there are 15 possibilities to realize distribution A by exchanging particles between the energy levels 1 and 3.

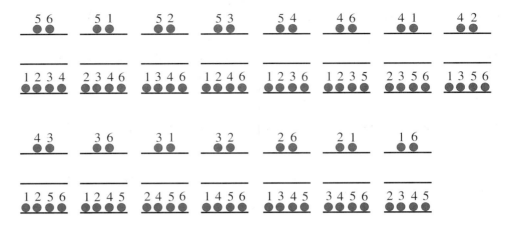

Figure 12.22 Possibilities of distributing particles 1 to 6 over quantum states 1 and 3 (type A in Figure 12.21), $\omega_A = 15$.

Number ω of configurations

How can we calculate the number ω of these possibilities (number of configurations)? In type A we imagine 6 digits (1 to 6) arranged in groups of 4 (corresponding to the population in level 1) and 2 digits (corresponding to the population in level 3). We ask how often we can exchange digits belonging to these different groups. This number is the total number of exchanges of the 6 digits ($= 6!$) divided by the number of exchanges in each group (4! and 2!, respectively). Then the number of configurations ω_A in type A can be represented by

$$\omega_A = \frac{6!}{4!2!} = 15 \quad \text{(type A)} \tag{12.47}$$

In general we obtain

$$\omega = \frac{N!}{N_1!N_2!N_3!\ldots} \tag{12.48}$$

Then for types B and C we have

$$\text{type B:} \quad \omega_B = \frac{6!}{3!2!1!} = 60$$

$$\text{type C:} \quad \omega_C = \frac{6!}{2!4!} = 15$$

Altogether we obtain

$$\Omega = \omega_A + \omega_B + \omega_C = 15 + 60 + 15 = 90$$

different configurations. Each of these configurations has equal probability. The probability P that type A is realized is therefore $15/90 = \frac{1}{6}$; the probabilities for types B and C are $\frac{2}{3}$ and $\frac{1}{6}$, respectively (Figure 12.23). We conclude that the six particles are predominantly distributed according to type B; but types A and C can also occur.

12.6 PROVING THE BOLTZMANN DISTRIBUTION

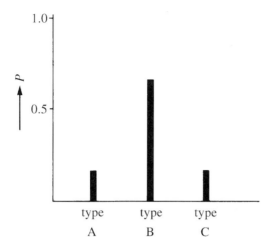

Figure 12.23 Probability P that a distribution A, B, or C in Figure 12.21 is realized.

When we repeat this calculation for 100 particles possessing an energy of $40a$ we find 21 different distribution types (Table 12.1). In calculating the factorials of large numbers it is reasonable to using Stirling's formula (see Box 12.4).

Table 12.1 Distribution types for a three level system according to Figure 12.21 with $N = 100$ particles possessing the energy $U = 40a$

	A	B	C	D	E	F	G	H	I	J	K
N_3	20	19	18	17	16	15	14	13	12	11	10
N_2	0	2	4	6	8	10	12	14	16	18	20
N_1	80	79	78	77	76	75	74	73	72	71	70
$\omega/10^{32}$	0	0	0	0	0	0	0.1	0.4	1.5	4.3	8.8

	L	M	N	O	P	Q	R	S	T	U
N_3	9	8	7	6	5	4	3	2	1	0
N_2	22	24	26	28	30	32	34	36	38	40
N_1	69	68	67	66	65	64	63	62	61	60
$\omega/10^{32}$	13.4	15.0	12.6	7.8	3.6	1.2	0.3	0	0	0

Box 12.4 – Stirling's Formula: $\ln N! = N \cdot \ln N - N$

Calculating $N!$ in the case when N is a large number is a difficult problem. However, in most cases it is appropriate to evaluate $N!$ by using a simple approximation (Stirling's formula).

Continued on page 432

> *Continued from page 431*
>
> From
> $$N! = 1 \cdot 2 \cdot 3 \cdot \ldots \cdot N$$
>
> we obtain
> $$\ln N! = \ln 1 + \ln 2 + \cdots + \ln N = \sum_{i=1}^{N} \ln i$$
>
> We approximate the summation by an integral (see Box 12.2).
>
> $$\ln N! = \int_0^N \ln i \cdot di = |i \ln i - i|_0^N = N \ln N - N$$
>
> For $N!$ we obtain
> $$N! = e^{N \ln N - N} = \left(e^{\ln N}\right)^N \cdot e^{-N} = N^N \cdot e^{-N}$$
>
> (*Stirling's formula*).
>
> ---
>
> **Example**
>
> | N | $N!$ | $\ln N!$ | $N^N \cdot e^{-N}$ | $N \ln N - N$ |
> |---|---|---|---|---|
> | 10 | 3.36×10^6 | 15.1 | 4.54×10^5 | 13.0 |
> | 60 | 8.32×10^{81} | 189 | 4.45×10^{80} | 186 |

Altogether there are

$$\begin{aligned}\Omega &= \omega_A + \omega_B + \cdots + \omega_U \\ &= \begin{pmatrix} 0 + 0 + 0 + 0 + 0 + 0 + 0.1 + 0.4 + 1.5 + 4.3 + 8.8 \\ + 13.4 + 15.0 + 12.6 + 7.8 + 3.6 + 1.2 + 0.3 + 0 + 0 \end{pmatrix} \times 10^{32} \\ &= 69.0 \times 10^{32}\end{aligned}$$

configurations.

80% of all configurations belong to types K to O, 60% to types L, M, and N and 20% belong to type M (Figure 12.24). The larger the number of particles under consideration, the closer concentrated are the distributions around that with the maximal value of ω.

In the present case (energy $40a$ distributed among 100 particles), according to distribution M, it is most probable that $N_1 = 68$ particles are in the lowest energy state, $N_2 = 24$ particles are in state 2, and $N_3 = 8$ particles are in the highest energy state. According to distributions L and N, slightly different population numbers are obtained.

12.6 PROVING THE BOLTZMANN DISTRIBUTION

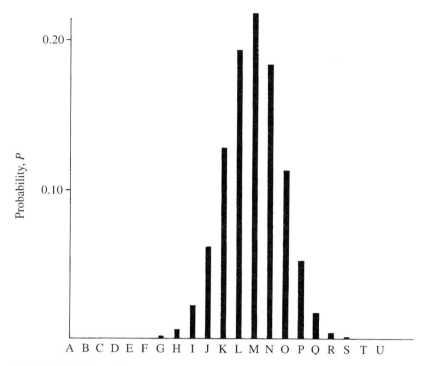

Figure 12.24 Probability P that a distribution A, B, C, ..., U in Table 12.1 is realized.

Calculating population numbers N_i

In Figure 12.25 the population numbers N_i corresponding to type M (Table 12.1) are plotted as a function of the energy E_i–that is, for $E_1 = 0$, $E_2 = a$, and $E_3 = 2a$. The points can be well described by the function

$$N_i = C \cdot e^{-\beta E_i} \qquad (12.49)$$

with $C = 68$ and $\beta = 1.04/a$ (solid line in Figure 12.25).

It can be shown that for a large number of particles this expression is valid for any number of quantum states available to the particles, and for any spacing of the energy levels (see Foundation 12.3).

If we decrease the internal energy in our example from $40a$ to $20a$, then the corresponding values will be $C = 82$ and $\beta = 1.76/a$ (Problem 12.5). The constant β decreases with increasing energy of the system, i.e., with increasing temperature. The quantity β is therefore a measure for the temperature of the system. At higher temperatures (i.e., with increasing energy uptake), the higher levels become more occupied and β becomes correspondingly smaller.

Relation between β and T

What is the relation between β and our existing measure of temperature T? According to Chapter 11, the mean translational energy $\overline{E}_{\text{trans}}$ is related to the temperature T:

$$\overline{E}_{\text{trans}} = \tfrac{3}{2}kT \qquad (12.50)$$

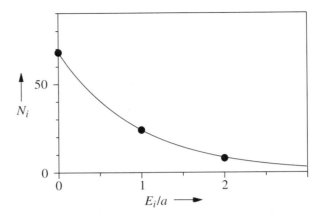

Figure 12.25 Population of quantum states i for the distribution M in Table 12.1 as a function of E_i; the quantity a is the spacing of the energy levels (Fig. 12.20).

In Section 12.4.1 we calculated $\overline{E}_{\text{trans}}$ as the average

$$\overline{E}_{\text{trans}} = \sum_{n_x,n_y,n_z} E_{n_x,n_y,n_z} \frac{1}{Z} \cdot e^{-E_{n_x,n_y,n_z}/kT}$$

Replacing the expression $e^{-E_{n_x,n_y,n_z}/kT}$ by $e^{-E_{n_x,n_y,n_z}\cdot\beta}$, we obtain for the mean translational energy

$$\overline{E}_{\text{trans}} = \frac{3}{2} \cdot \frac{1}{\beta} \tag{12.51}$$

Therefore we conclude that

$$\beta = \frac{1}{kT} \tag{12.52}$$

This leads to the Boltzmann distribution (12.1), (12.10), and (12.11).

12.6.2 Indistinguishable Particles

In counting the number of configurations, we have assumed that the particles are distinguishable. This assumption was reasonable for particles occupying different lattice sites which can be distinguished by using different site numbers.

The assumption is not correct when considering identical particles moving freely in a gas. It can be shown, however, that the Boltzmann distribution holds for indistinguishable particles as well (Foundation 12.3).

The reason for this fact is closely connected with the large number of quantum states available for the particles in a gas at room temperature. In Section 12.4.4 we calculated that the number of quantum states at room temperature exceeds the number of particles by a factor of 10^5. Thus most of the quantum states are unoccupied, a few are occupied with one particle, and there are essentially no quantum states occupied by more than one particle.

Each configuration in a system of indistinguishable particles corresponds to $N!$ configurations of distinguishable particles, because $N!$ is the number of possibilities to exchange

PROBLEM 12.1

N particles. Thus the values of Ω calculated for distinguishable particles have to be divided by $N!$; a division by a constant, however, does not change the distribution function. However, we shall see that the lack of distinguishability plays a very important role in other properties of gases (see, for example, Section 14.4).

Problems

Problem 12.1 – Population of Quantum States of a Rotator at Temperature T

Consider at what temperature the excitation energy $\Delta E_{0\to 1}$ equals kT for HD, HCl, CO, and CO_2; that is, when does the quantum mechanical description change into the classical one?

Solution. From

$$\Delta E_{0\to 1} = \frac{h^2}{4\pi^2 \mu d^2} = kT$$

we calculate the corresponding temperature

$$T = \frac{\Delta E_{0\to 1}}{k}$$

Then we obtain

	$\mu/10^{-27}$ kg	d/pm	$\Delta E_{0\to 1}/10^{-22}$ J	T/K
HD	1.12	74	18.2	132
HCl	1.63	127	4.2	30
CO	11.38	113	0.7	5
CO_2	53.1	116	0.16	1

Problem 12.2 – Maxwell–Boltzmann Distribution: v_P, $\bar{v}$, $\overline{v^2}$, and $\bar{E}_{trans}$

Calculate the constant C' in equation (12.41), the most probable speed v_P in equation (12.34), the mean speed $\bar{v}$ in equation (12.35), and the mean translational energy $\bar{E}_{trans}$ in equation (12.36).

Solution. Because of the condition

$$\int_0^\infty dP = 1$$

we obtain from equation (12.41)

$$\int_0^\infty dP = C' \int_0^\infty v^2 e^{-mv^2/2kT} \cdot dv = C' \left(\frac{2kT}{m}\right)^{3/2} \int_0^\infty x^2 e^{-x^2} \cdot dx$$

$$= C' \left(\frac{2kT}{m}\right)^{3/2} \cdot \frac{1}{4}\sqrt{\pi} = 1$$

Then

$$C' = \frac{4}{\sqrt{\pi}} \left(\frac{m}{2kT}\right)^{3/2} = 4\pi \cdot \left(\frac{m}{2\pi kT}\right)^{3/2}$$

and
$$dP_{v,v+dv} = 4\pi \cdot \left(\frac{m}{2\pi kT}\right)^{3/2} \cdot v^2 \cdot e^{-mv^2/2kT} \cdot dv$$

The most probable speed v_P corresponds to the maximum of the function
$$y = v^2 \cdot e^{-mv^2/2kT}$$
$$\frac{dy}{dv} = 2v \cdot e^{-mv^2/2kT} + v^2 \cdot \frac{-mv}{kT} \cdot e^{-mv^2/2kT} = ve^{-mv^2/2kT} \cdot \left[2 - \frac{mv^2}{kT}\right]$$

By setting this derivative equal to zero we obtain
$$v_P = \sqrt{\frac{2kT}{m}}$$

The mean speed is calculated by
$$\bar{v} = \int_0^\infty v \cdot dP = \int_0^\infty v \cdot C'v^2 \cdot e^{-mv^2/2kT} \cdot dv$$

We substitute
$$x = \frac{mv^2}{2kT}, \quad dx = \frac{mv}{kT} \cdot dv$$

Then we obtain
$$\bar{v} = C' \frac{1}{2} \left(\frac{2kT}{m}\right)^2 \cdot \int_0^\infty x \cdot e^{-x} \cdot dx = 4\pi \cdot \left(\frac{m}{2\pi kT}\right)^{3/2} \cdot \left(\frac{2kT}{m}\right)^2 \cdot \frac{1}{2} = \sqrt{\frac{8kT}{\pi m}}$$

Correspondingly, the mean square speed is calculated by
$$\overline{v^2} = \int_0^\infty v^2 \cdot dP = C' \left(\frac{2kT}{m}\right)^{5/2} \cdot \int_0^\infty x^4 \cdot e^{-x^2} \cdot dx$$
$$= 4\pi \cdot \left(\frac{m}{2\pi kT}\right)^{3/2} \cdot \left(\frac{2kT}{m}\right)^{5/2} \cdot \frac{3}{8}\sqrt{\pi} = \frac{3kT}{m}$$

Then the mean translational energy is
$$\overline{E}_{trans} = \tfrac{1}{2}m\overline{v^2} = \tfrac{3}{2}kT$$

Problem 12.3 – Number of Quantum States in a One-dimensional Gas

What is the number of quantum states for a one-dimensional model gas? Compare with the corresponding relation for a three-dimensional gas.

PROBLEM 12.4

Solution. The energy levels for a one-dimensional gas are

$$E = \frac{h^2}{8mL^2} \cdot n^2$$

In this case the number g of quantum states between $E = 0$ and $E = E_n$ equals the quantum number n:

$$g = n = \sqrt{\frac{8mL^2}{h^2}} \cdot \sqrt{E}$$

For a three-dimensional gas the number of quantum states is, according to equation (12.37),

$$g = \frac{\pi}{6} \left(\frac{8mL^2}{h^2} \right)^{3/2} \cdot E^{3/2}$$

This means that approximately $g_{\text{three–dimensional gas}} = \left(g_{\text{one–dimensional gas}} \right)^3$.

Problem 12.4 – Population Probability

Calculate the population probability of the lowest translational quantum state of a gas.

Solution. According to the Boltzmann distribution, the probability of finding a molecule in a translational state of energy $E_i = \frac{3}{2}kT$ equals

$$P_i = C \cdot e^{-E_i/kT} = C \cdot e^{-3/2} = C \cdot 0.22$$

For the lowest quantum state ($E_1 \ll kT$) the corresponding probability is $P_1 = C$. Then the occupation probability of the lowest quantum state is larger by only a factor of 4.5 compared with a quantum state corresponding to the mean energy. This means that, according to Section 12.4.4, $P_1 = 4.5 \times 10^{-5}$.

Problem 12.5 – Boltzmann Distribution

Calculate C and β in equation (12.49) for the case $U = 20a$.

Solution. For $N = 100$ and $U = 20a$ the following distributions on three energy levels with spacing a are possible:

	A	B	C	D	E	F	G	H	I	J	K
N_3	10	9	8	7	6	5	4	3	2	1	0
N_2	0	2	4	6	8	10	12	14	16	18	20
N_1	90	89	88	87	86	85	84	83	82	81	80
$\omega/10^{21}$	0	0	0	0	0.1	0.8	2.4	4.5	4.7	2.5	0.5

The maximum of ω is realized in the distribution I. Similarly to Figure 12.25 we plot N_i as a function of E_i (Figure 12.26). The points can be well described by the function. $N_i = C \cdot e^{\beta E_i}$ with $C = 82$ and $\beta = 1.76/a$.

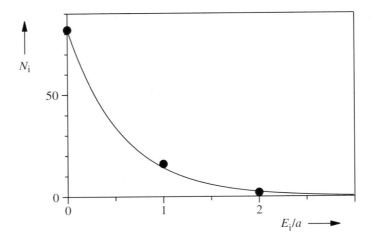

Figure 12.26 Evaluation of C and β by plotting N_i as a function of E_i/a. Dots: values from table, column 1. Solid line: calculated with $C = 80$ and $\beta = 1.76/a$.

Problems 12.6 and 12.7 – Superhelix as a Frozen Boltzmann Distribution

Problem 12.6

DNA double helices can be closed to a ring by an enzyme connecting the ends of the strands (Figure 12.27, $i = 0$). If the strand is twisted during the connection process, then the resulting ring will also be twisted (superhelix with $i = 1$, $i = 2$, $i = -1$, $i = -2$, and so on). A certain energy E_i has to be supplied for twisting the strand; this energy increases with the number i of twists in the superhelix. For N equal DNA molecules, there is a certain number N_0 of molecules forming a regular ring, as well as $N_{+1}, N_{+2}, \ldots N_{-1}$, $N_{-2} \ldots$ molecules with $1, 2, \ldots$ twists in one or other direction. After the closure of the ring, a frozen Boltzmann distribution is obtained. The various ring types can be separated by applying an electric field where they move with different speeds. For a DNA made up of $Z_{\text{basepair}} = 9850$ basepairs the distribution shown in Figure 12.28 is obtained.

Estimate the energies E_i and try to understand the observed distribution.

Hint. For estimating the energy E_i to twist the strand we assume that the DNA strand can be compared in its elastic properties with a laboratory rubber tube. By measuring the moment applied for twisting the tube the elastic energy can be determined and related to the energy of the twisted DNA.

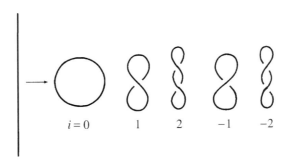

Figure 12.27 Twisting a ring consisting of DNA (superhelix).

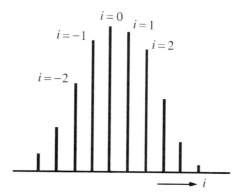

Figure 12.28 Probability of finding superhelices with $i = 0, 1, 2, \ldots$.

Solution. First we consider a macroscopic model. We imagine a rubber tube (diameter d, length L) twisted around its axis. The moment M connected with the twisting is proportional to the twisting angle φ

$$M = a \cdot \varphi, \quad \text{where} \quad a = \alpha \frac{d^4}{L}$$

To twist the tube i times around its axis the energy

$$E_i = \int_0^{i2\pi} a \cdot \varphi \cdot d\varphi = \frac{a}{2} 4\pi^2 \cdot i^2 = 2\pi^2 \cdot \alpha \frac{d^4}{L} \cdot i^2, \quad i = 0, \pm 1, \pm 2, \ldots$$

has to be supplied.

In a simple experiment with a laboratory rubber tube closed to a ring (length $L = 40$ cm, diameter $d = 1$ cm) which is cut at one point, twisted once, and closed, we find $M = 2 \times 10^{-2}$ N m; thus

$$\alpha = 0.02 \, \text{N m} \cdot \frac{40 \, \text{cm}}{1 \, \text{cm}^4 \cdot 2\pi} = 10^5 \, \text{N m}^{-2}$$

For a fair comparison with DNA we should consider a rubber cylinder rather than a tube, so α should be about $10^6 \, \text{N m}^{-2}$.

Then we obtain for DNA with $L = Z_{\text{basepair}} \cdot b$, where b is the contour length of one base pair, that is the extension in the direction of the helical axis ($b = 0.3$ nm), and with diameter $d = 3$ nm

$$E_i = 2\pi^2 \alpha \frac{d^4}{L} \cdot i^2 = \frac{A}{Z_{\text{basepair}}} \cdot i^2$$

with

$$A = 2\pi^2 \cdot 1 \times 10^6 \cdot \frac{(3 \times 10^{-9})^4}{0.3 \times 10^{-9}} \, \text{J} = 5 \times 10^{-18} \, \text{J} = 1200 kT$$

(for $T = 300$ K).

According to equation (12.11), the probability P_i of finding an i-fold twisted ring is

$$P_i \sim e^{-E_i/kT} = \exp\left(\frac{-A}{Z_{\text{basepair}} \cdot kT} \cdot i^2\right)$$

Then

$$\ln \frac{P_i}{P_0} = -\frac{A}{Z_{basepair} \cdot kT} \cdot i^2 \quad \text{or} \quad \frac{P_i}{P_0} = \exp\left(-\frac{A}{Z_{basepair} \cdot kT} \cdot i^2\right)$$

When plotting the experimental $\ln P_i/P_0$ values versus i^2 it follows from the slope of the straight line that $A/Z_{basepair} = 0.1kT$ and $A = 985kT$ for $Z_{basepair} = 9850$. With these data, P_i/P_0 is calculated and compared with the experiment in Figure 12.29. A fair agreement is obtained.

Experiments with DNA of different chain lengths reveal that the constant A is nearly independent of $Z_{basepair}$. The agreement of experiment and theory indicates that our very simple model is essentially correct.

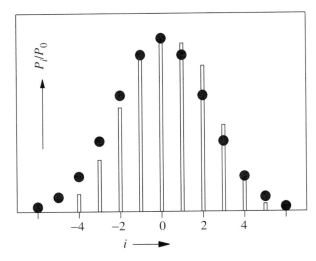

Figure 12.29 Comparison of the experimental probabilities P_i (bars), to the values calculated from the Boltzmann equation (dots). Experimental data: R.E. Depew, J.C. Wang, Conformational functuation of DNA helix, *Proc. Nat. Acad. Sci.* USA 72, 4275 (1975) J.C. Wang, Helical repeat of DNA in solution, *Proc. Nat. Acad. Sci.* USA 76, 200 (1979)).

Problem 12.7

In the example above ($Z_{basepair} = 9850$) the two ends can be linked without stress. What happens when we consider DNA made of additional five basepairs ($Z_{basepair} = 9855$)? Note that ten basepairs (one turn) allow, again, linking without stress. With five additional pairs, turning by $180°$ in either direction is needed for linking the two ends with lowest possible stress.

Solution. The least elastic energy to link the two ends is

$$E_0 = \int_0^\pi a\varphi \cdot d\varphi = \frac{a}{2}\pi^2$$

Generally,

$$E_i = \frac{a}{2}4\pi^2 \cdot i^2, \quad i = \pm\frac{1}{2}, \pm\frac{3}{2}, \pm\frac{5}{2}, \ldots$$

In contrast to example 1 where a maximum peak appears, we find two maximum peaks of equal height ($i = \pm\frac{1}{2}$).

Problem 12.8 – Internal Energy for Particles Rotating on a Circle

Calculate U_{rot} for rotators rotating on a circle (Figure 9.5).

Solution. In analogy to the three-dimensional rotator (Foundation 12.2) we approximate the partition function by

$$Z = 1 + 2 \cdot \sum_{n=1}^{\infty} e^{-Bn^2/kT} \approx 1 + 2 \cdot \int_1^{\infty} e^{-Bn^2/kT} \cdot dn$$

$$\approx 1 + 2 \cdot \sqrt{\frac{kT}{B}} \cdot \int_0^{\infty} e^{-x^2} \cdot dx = 1 + 2 \cdot \sqrt{\frac{kT}{B}} \cdot \frac{1}{2}\sqrt{\pi}$$

(the factor of two is because of the twofold degeneracy of the energy levels for $n \geq 1$). Then for sufficiently high temperature we have

$$\frac{d \ln Z}{dT} = \frac{1}{2} \cdot \frac{1}{T} \quad \text{and} \quad U_{rot} = \frac{1}{2} NkT$$

The result of the numerical calculation of U_{rot} for the full temperature range is shown in Figure 12.30 with the same numerical data as for HD (see Foundation 12.2).

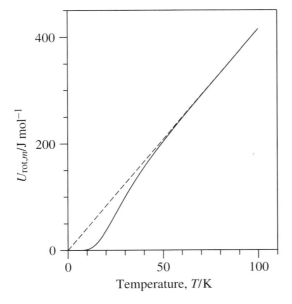

Figure 12.30 $U_{rot,m}$ for a particle rotating on a circle as a function of temperature (solid line, calculated by numerical integration); dashed line: classical limit. The numerical data are the same as for HD (see Figure 12.12).

Foundation 12.1 – Oscillator: Population Numbers N_n and Total Energy U_{vib}

Partition function for vibration

From equations (12.12) and (12.15) we obtain with $x = h\nu_0/kT$

$$Z = \sum_{n=0}^{\infty} e^{-(1/2+n)x} = e^{-x/2} \cdot \sum_{n=0}^{\infty} e^{-nx} = e^{-x/2} \cdot \sum_{n=0}^{\infty} \left(e^{-x}\right)^n = e^{-x/2} \cdot \sum_{n=0}^{\infty} y^n$$

where $y = e^{-x}$ is a number smaller than 1. Then

$$\sum_{n=0}^{\infty} y^n = 1 + y + y^2 + y^3 + \cdots = \frac{1}{1-y} = \frac{1}{1-e^{-x}}$$

and

$$Z = e^{-x/2} \cdot \frac{1}{1-e^{-x}} = e^{-h\nu_0/2kT} \cdot \frac{1}{1-e^{-h\nu_0/kT}}$$

Total energy U_{vib}

According to equation (12.16) we have

$$U_{vib} = \sum_{n=0}^{\infty} N_n E_n = \frac{N}{Z} \cdot \sum_{n=0}^{\infty} E_n \cdot e^{-E_n/kT} = N \cdot \frac{A}{Z}$$

with

$$Z = \sum_{n=0}^{\infty} e^{-E_n/kT}, \quad A = \sum_{n=0}^{\infty} E_n e^{-E_n/kT}$$

From

$$\frac{dZ}{dT} = \sum_{n=0}^{\infty} e^{-E_n/kT} \cdot \left(\frac{E_n}{kT^2}\right) = \frac{1}{kT^2} \cdot A$$

we obtain

$$\boxed{U_{vib} = NkT^2 \cdot \frac{dZ/dT}{Z} = NkT^2 \cdot \frac{d \ln Z}{dT}}$$

This is a useful equation relating the energy U to the partition function Z. Note that this relation is general and applies for U_{rot}, U_{trans} and $U_{electronic}$ as well.

Continued on page 443

FOUNDATION 12.1

Continued from page 442

Evaluating the expression for U_{vib}

From

$$Z = e^{-x/2} \cdot \frac{1}{1-e^{-x}} \quad \text{with} \quad x = \frac{h\nu_0}{kT}$$

we calculate

$$\ln Z = -\frac{x}{2} - \ln\left(1 - e^{-x}\right)$$

and

$$\frac{d\ln Z}{dT} = -\frac{1}{2} \cdot \frac{dx}{dT} - \frac{e^{-x}}{1-e^{-x}} \cdot \frac{dx}{dT}$$

Because of

$$\frac{dx}{dT} = -\frac{h\nu_0}{kT^2}$$

we have

$$\frac{d\ln Z}{dT} = \frac{h\nu_0}{kT^2} \cdot \left(\frac{1}{2} + \frac{1}{e^{h\nu_0/kT} - 1}\right)$$

and

$$U_{vib} = Nh\nu_0 \cdot \left(\frac{1}{2} + \frac{1}{e^{h\nu_0/kT} - 1}\right)$$

High temperature limit

For high temperature we can approximate U_{vib} by

$$e^{h\nu_0/kT} - 1 = e^x - 1 = 1 + x + \tfrac{1}{2}x^2 + \cdots - 1 \approx x + \tfrac{1}{2}x^2$$

Then

$$U_{vib} = Nh\nu_0 \cdot \left(\frac{1}{2} + \frac{1}{x + \tfrac{1}{2}x^2}\right) = Nh\nu_0 \cdot \left(\frac{1}{2} + \frac{1}{x} \cdot \frac{1}{1 + \tfrac{1}{2}x}\right)$$

$$= Nh\nu_0 \cdot \left(\frac{1}{2} + \frac{1}{x} \cdot \left[1 - \frac{1}{2}x\right]\right) = Nh\nu_0 \cdot \left(\frac{1}{2} + \frac{1}{x} - \frac{1}{2}\right)$$

$$= Nh\nu_0 \cdot \frac{kT}{h\nu_0} = NkT$$

Population numbers for I_2

In the following table the population of the vibrational levels of I_2 (expressed by N_n/N) and the internal energy U_{vib} of the vibration is shown for different temperatures (see Figures 12.7(b) and 12.8(a)). Data used for the evaluation:

Continued on page 444

Continued from page 443

$h\nu_0 = 4.27 \times 10^{-21}$ J; $E_n = 4.27 \times 10^{-21}$ J $\cdot (\frac{1}{2} + n)$. $U_{\text{vib},m}$ is the molar energy of vibration.

n	$E_n/10^{-20}$ J	N_n/N		
		125 K $Z = 0.32$	250 K $Z = 0.76$	500 K $Z = 1.59$
0	0.21	0.92	0.71	0.46
1	0.64	0.08	0.21	0.25
2	1.07	0.01	0.06	0.13
3	1.49		0.02	0.07
4	1.92		0.01	0.04
5	2.35			0.02
		$U_{\text{vib},m}$/kJ mol^{-1}		
		1.52	2.34	4.28

Foundation 12.2 – Rotator: Population Number N_n and Total Energy U_{rot}

Numerical calculation

The partition function Z for the rotator cannot be evaluated analytically as for the oscillator. Therefore we calculate Z numerically. The result for HD is displayed in the following table using the numerical data: $\mu = 1.116 \times 10^{-27}$ kg, $d = 74$ pm, $E_n = B \cdot n(n+1)$, $B = 0.00910 \times 10^{-19}$ J. $U_{\text{rot},m}$ is the molar energy of rotation. See also Figure 12.12(a).

The partition function Z is

$$Z = e^{-E_0/kT} + 3 \cdot e^{-E_1/kT} + 5 \cdot e^{-E_2/kT} + \cdots$$

With

$$\frac{E_n}{kT} = \frac{h^2}{8\pi^2 \mu d^2 \cdot kT} \cdot n(n+1) = x \cdot n(n+1)$$

we obtain

$$Z = 1 + 3 \cdot e^{-2x} + 5 \cdot e^{-6x} + 7 \cdot e^{-12x} + 9 \cdot e^{-20x} + \cdots$$

With $x = 6.67$ for $T = 10$ K,

$$Z = 1 + 3 \cdot e^{-13.3} + 5 \cdot e^{-40} + \cdots = 1.00$$

Continued on page 445

FOUNDATION 12.2

Continued from page 444

For $T = 100$ K,

$$Z = 1 + 3 \cdot e^{-1.33} + 5 \cdot e^{-4.00} + 7 \cdot e^{-8.00} + \cdots = 1.89$$

Correspondingly, the remaining values in the following table are obtained. The population numbers N_n follow from

$$N_n = \frac{N}{Z} \cdot e^{-E_n/kT}$$

n	$E_n/10^{-19}$ J	N_n/N			
		10 K $Z = 1.00$	50 K $Z = 1.22$	100 K $Z = 1.89$	500 K $Z = 7.91$
0	0	1.00	0.82	0.53	0.13
1	0.0182	0.00	0.18	0.42	0.29
2	0.0547		0.00	0.05	0.29
3	0.109			0.00	0.18
4	0.182				0.08
5	0.274				0.03
6	0.383				0.00
		$U_{\text{rot},m}$/kJ mol^{-1}			
		0.00	0.20	0.66	3.83

Energy at high temperature

At high temperature many rotational quantum states are occupied. Then we can replace the summation

$$Z = \sum_{n=0}^{\infty}(2n + 1) \cdot e^{-Bn(n+1)/kT}$$

by an integration (see Box 12.2):

$$Z = \int_0^{\infty}(2n + 1) \cdot e^{-Bn(n+1)/kT} \cdot dn$$

With the substitution

$$x = \frac{Bn(n + 1)}{kT} = \frac{B}{kT} \cdot (n^2 + n), \quad dx = \frac{B}{kT} \cdot (2n + 1) \cdot dn$$

we obtain

$$Z = \frac{kT}{B} \int_0^{\infty}(2n + 1) \cdot e^{-x} \cdot \frac{1}{2n + 1} \cdot dx = \frac{kT}{B} \int_0^{\infty} e^{-x} \cdot dx = \frac{kT}{B}$$

Continued on page 446

Continued from page 445

Then, in analogy to the oscillator (Foundation 12.1)

$$U_{\text{rot}} = NkT^2 \cdot \frac{\mathrm{d}\ln Z}{\mathrm{d}T} = NkT^2 \cdot \frac{B}{kT} \cdot \frac{k}{B} = NkT$$

This evaluation of the partition function is, of course, only valid for high enough temperature.

Foundation 12.3 – Boltzmann Distribution

How to find the maximum of $\ln \omega$

We ask for the maximum of the number ω of configurations. In practice, it is more convenient to search for the maximum of $\ln \omega$ instead of the maximum of ω. ω is a function depending on very many variables $N_1, N_2, \ldots$. Then for the maximum of $\ln \omega$ the condition

$$\delta \ln \omega = \left(\frac{\partial \ln \omega}{\partial N_1}\right)_{N_2, N_3, \ldots} \cdot \mathrm{d}N_1 + \left(\frac{\partial \ln \omega}{\partial N_2}\right)_{N_1, N_3, \ldots} \cdot \mathrm{d}N_2$$

$$+ \left(\frac{\partial \ln \omega}{\partial N_3}\right)_{N_1, N_2, \ldots} \cdot \mathrm{d}N_3 + \cdots = 0$$

must be fulfilled. Additionally, the conditions

$$N_1 + N_2 + N_3 + \cdots = N \quad \text{or} \quad \sum \mathrm{d}N_i = 0 \qquad \text{(condition 1)}$$

$$E_1 N_1 + E_2 N_2 + E_3 N_3 + \cdots = U \quad \text{or} \quad \sum E_i \cdot \mathrm{d}N_i = 0 \qquad \text{(condition 2)}$$

must be regarded. This means the particle numbers $N_1, N_2, \ldots$ are not independent of each other, and the maximum of $\ln \omega$ is not simply obtained by setting the partial derivatives equal to zero. The trick to continue is to include the conditions 1 and 2 in the expression for $\mathrm{d}\ln \omega$ (method of Lagrange's multipliers). Therefore, we multiply the expressions in conditions 1 and 2 with arbitrary constants α and β

$$\alpha \cdot \sum \mathrm{d}N_i = 0 \quad \text{and} \quad \beta \cdot \sum E_i \cdot \mathrm{d}N_i = 0$$

Continued on page 447

FOUNDATION 12.3

Continued from page 446

and subtract both expressions from $d\ln\omega$ (we are allowed to do this, because effectively we subtract zero):

$$\delta\ln\omega = \left[\left(\frac{\partial\ln\omega}{\partial N_1}\right) - \alpha - \beta E_1\right]\cdot dN_1 + \left[\left(\frac{\partial\ln\omega}{\partial N_2}\right) - \alpha - \beta E_2\right]\cdot dN_2$$
$$+ \left[\left(\frac{\partial\ln\omega}{\partial N_3}\right) - \alpha - \beta E_3\right]\cdot dN_3 + \cdots = 0$$

Because the constants α and β are arbitrary, we can choose any value for them we like. Arbitrarily we define these constants by setting

$$\left(\frac{\partial\ln\omega}{\partial N_1}\right) - \alpha - \beta E_1 = 0 \quad \text{and} \quad \left(\frac{\partial\ln\omega}{\partial N_2}\right) - \alpha - \beta E_2 = 0$$

Then we obtain for $\delta\ln\omega$

$$\delta\ln\omega = \left[\left(\frac{\partial\ln\omega}{\partial N_3}\right) - \alpha - \beta E_3\right]\cdot dN_3 + \left[\left(\frac{\partial\ln\omega}{\partial N_4}\right) - \alpha - \beta E_4\right]\cdot dN_4 + \cdots = 0$$

In this expression the remaining variables $N_3, N_4, N_5, \ldots$ are independent of each another. Then $\delta\ln\omega$ is zero for

$$\left(\frac{\partial\ln\omega}{\partial N_i}\right) - \alpha - \beta E_i = 0, \quad (i = 1, 2, 3, \ldots)$$

(condition for the maximum of $\ln\omega$)

How to calculate the population numbers N_i

Distinguishable particles. For distinguishable particles we calculate ω according to equation (12.48)

$$\omega = \frac{N!}{N_1!N_2!N_3!\ldots} \quad \text{(distinguishable particles)}$$

According to Stirling's formula (Box 12.4) we have $\ln N! = N\cdot\ln N - N$.

$$\ln\omega = N\cdot\ln N - N - N_1\cdot\ln N_1 + N_1 - \cdots$$

Then

$$\frac{\partial\ln\omega}{\partial N_1} = -\ln N_1 - N_1\frac{1}{N_1} + 1 = -\ln N_1$$

Inserting into the condition for the maximum of ω we obtain

$$-\ln N_1 - \alpha - \beta E_1 = 0.$$

The solution for N_1 is $N_1 = e^{-\alpha-\beta E_1} = e^{-\alpha}\cdot e^{-\beta E_1}$. Accordingly, we can proceed for arbitrary N_i

$$\boxed{N_i = e^{-\alpha}\cdot e^{-\beta E_i}}$$

Continued on page 448

Continued from page 447

Indistinguishable particles (high temperature). For indistinguishable particles we restrict our discussion to a gas at such a temperature that the number g_i of quantum states in an energy interval ΔE is much larger than the number of particles. Then quantum states can be occupied by one particle, or they are unoccupied, and we obtain the number of configurations ω_i by counting the possibilities to exchange empty and singly occupied quantum states.

If g_i quantum states are available for one particle, we obtain $\omega_i = g_i^1$; for n_i particles in the energy interval, $\omega_i = g_i^{n_i}$. However, this calculation includes all the possibilities where two particles are interchanged: these are $n_i!$ possibilities. Therefore

$$\omega_i = \frac{g_i^{n_i}}{n_i!}$$

The total number of quantum states ω is given by the product

$$\omega = \frac{g_1^{n_1}}{n_1!} \cdot \frac{g_2^{n_2}}{n_2!} \cdot \frac{g_3^{n_3}}{n_3!} \cdots \quad \text{(indistinguishable particles)}$$

since the number of arrangements at energy E_i is independent of the number of arrangements at other energies.

Then, proceeding as in the case of distinguishable particles, we obtain

$$\boxed{n_i = g_i \cdot e^{-\alpha} \cdot e^{-\beta E_i}}$$

The number N_i of particles in a given quantum state at energy E_i is

$$\boxed{N_i = \frac{n_i}{g_i} = e^{-\alpha} \cdot e^{-\beta E_i}}$$

Indistinguishable particles (low temperature). At low enough temperature our assumption that the number of quantum states in an energy interval ΔE is much larger than the number n_i of particles is no longer correct. In this case (number of quantum states comparable with the number of particles) the number of distinguishable configurations of the n_i indistinguishable particles is

$$\omega_i = \frac{(n_i + g_i - 1)!}{n_i!(g_i - 1)!}$$

Example

We consider the case $n_i = 3$ and $g_i = 3$. Then

$$\omega_i = \frac{5!}{3!2!} = 10$$

Continued on page 449

FOUNDATION 12.3

Continued from page 448

These configurations are:

•	•	•
• •	•	
• •		•
•	• •	
	• •	•
•		• •
	•	• •
• • •		
	• • •	
		• • •

The total number of configurations is the product

$$\omega = \frac{(n_1 + g_1 - 1)!}{n_1!(g_1 - 1)!} \cdot \frac{(n_2 + g_2 - 1)!}{n_2!(g_2 - 1)!} \cdot \frac{(n_3 + g_3 - 1)!}{n_3!(g_3 - 1)!} \cdots$$

Then proceeding as in the cases considered above we obtain

$$\boxed{n_i = \frac{g_i}{e^\alpha \cdot e^{\beta E_i} - 1}}$$

(*Bose distribution*). At high temperature this expression goes over in the Boltzmann distribution. This can also be seen for the expression for the number of configurations:

$$\lim_{g_i \gg n_i} \omega_i = \frac{(g_i + n_i - 1)!}{n_i!(g_i - 1)!} = \frac{g_i! \cdot (g_i + 1) \cdot (g_i + 2) \cdots (g_i + n_i)}{n_i! g_i!}$$

$$= \frac{g_i! \cdot g_i^{n_i}}{n_i! g_i!} = \frac{g_i^{n_i}}{n_i!}$$

13 Internal Energy *U*, Heat *q*, and Work *w*

We consider concepts and quantities describing macroscopic properties of matter: system and surroundings, state of system, heat, work, and internal energy. The thermodynamic system is a part of the universe of interest, to be distinguished from the surroundings, that is, the universe outside the boundary of the system. The system is considered in a given thermodynamic state; that is, it has a given composition, a given volume, and it is at a given temperature and under a given pressure.

The energy content of the system (the internal energy U) increases on heating: heat q is transferred from the temperature bath into the system, and work w is performed on the system:

$$\Delta U = q + w$$

At the end, the system reaches a new thermodynamic state. Thermodynamic features are related to heat and work. The internal energy is intimately related to the molecular quantum phenomena discussed in the previous chapters. The heat capacities C_V and C_P are determined by the energetic arrangement of quantum states. It is remarkable that simply heating a body gives information on the quantum nature of matter.

The internal energy U is a state variable: the change is exclusively determined by the properties of the thermodynamic states at the beginning and the end of a change of state. On the other hand, the heat q and the work w do not constitute changes of state variables: q and w depend on how the change of state has been carried out.

13.1 Change of State at Constant Volume

13.1.1 Change of Internal Energy ΔU; Heat q

Let us look at a gas consisting of hydrogen molecules in a closed container in a bath of temperature T (Figure 13.1). The energy of the gas is the sum of the electronic, translational, rotational, and vibrational energies of the H_2 molecules (Figure 13.2):

$$U = U_{\text{electr}} + U_{\text{trans}} + U_{\text{rot}} + U_{\text{vib}} \quad (13.1)$$

U is called the *internal energy*.

13.1 CHANGE OF STATE AT CONSTANT VOLUME

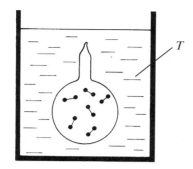

Figure 13.1 Hydrogen gas in a closed container in a bath of temperature T.

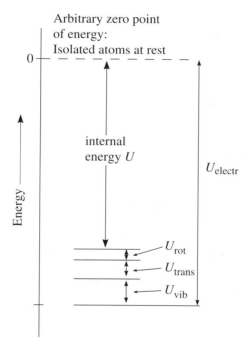

Figure 13.2 Contributions to the internal energy U of H_2. The electronic energy is strongly negative, the contributions of vibration, rotation, and translation are slightly positive; thus the internal energy U is still strongly negative.

What happens if we change the temperature, the volume, or the pressure of the gas considered in Figure 13.1? Then the gas is transformed from its original state into a different state. Such processes are called *changes of state*.

Let us first consider, instead of H_2, N atoms of He ($U_{\text{rot}} = U_{\text{vib}} = 0$). We increase the temperature of the gas from T_1 to T_2 at constant volume (change of state at constant volume, Figure 13.3). After thermal equilibration the He atoms take up translational energy upon heating, and the change of the internal energy U is

$$\Delta U = \Delta U_{\text{trans}} = U_{\text{trans},2} - U_{\text{trans},1} = \tfrac{3}{2}NkT_2 - \tfrac{3}{2}NkT_1 = \tfrac{3}{2}Nk \cdot (T_2 - T_1) \quad (13.2)$$

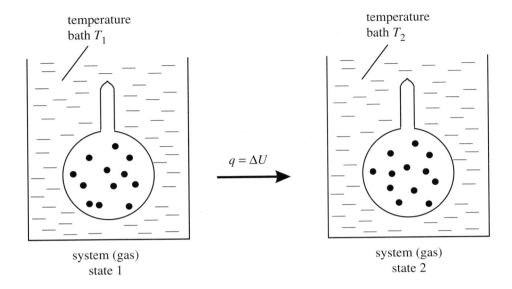

Figure 13.3 Heating of a gas of He atoms from T_1 to T_2 at constant volume.

At the same time, heat q is taken from the bath of temperature T_2 and supplied to the gas. The energy corresponding to the heat q is stored in the gas molecules, thus increasing the internal energy U of the gas. Then

$$\boxed{\Delta U = q \quad \text{(heating at constant volume)}} \tag{13.3}$$

When we use hydrogen molecules instead of helium atoms, we must include the contributions of rotation and vibration to the internal energy. According to Chapter 12, the H_2 molecules remain in the lowest vibrational state at moderate temperature; thus the contribution U_{vib} does not change on heating. On the other hand, the rotational energy can be considered as the classical limit

$$U_{\text{rot}} = NkT$$

Then instead of equation (13.2) we find

$$\Delta U = \Delta U_{\text{trans}} + \Delta U_{\text{rot}} = \tfrac{5}{2} Nk \cdot (T_2 - T_1) \tag{13.4}$$

In this case a larger amount of heat q is transferred from the temperature bath to the gas than for the He gas for the same temperature rise.

13.1.2 Heat Capacity C_V

Now we transfer the container with the gas from a bath of temperature T into a bath of temperature $T + \Delta T$, keeping the volume constant. Heat Δq flows from the bath into the container. For small ΔT the heat Δq is proportional to ΔT. The ratio $\Delta q / \Delta T$ in the limit $\Delta T \to 0$ is called the *heat capacity* C_V at constant volume. Thus the heat capacity C_V

13.2 TEMPERATURE DEPENDENCE OF C_V

is defined by

$$C_V = \lim_{\Delta T \to 0} \left(\frac{\Delta q}{\Delta T}\right)_{V=\text{const}} = \left(\frac{\partial q}{\partial T}\right)_V \tag{13.5}$$

The symbol ∂ denotes a partial differential. In the present case, the heat q depends on T and V, and the differentiation with respect to T has to occur such that the volume V is kept constant. C_V is an important measure for the ability of a substance to absorb heat at a given temperature.

According to equation (13.3) we have

$$dq = dU \quad \text{(constant volume)} \tag{13.6}$$

Thus we can express C_V by the change of internal energy instead of heat:

$$C_V = \left(\frac{\partial U}{\partial T}\right)_V \tag{13.7}$$

From equations (13.2) and (13.4), for a temperature change dT of a helium gas we obtain

$$dU = \tfrac{3}{2} Nk \cdot dT \quad \text{and} \quad C_V = \tfrac{3}{2} Nk \tag{13.8}$$

Correspondingly, for the hydrogen gas we obtain

$$dU = \tfrac{5}{2} Nk \cdot dT \quad \text{and} \quad C_V = \tfrac{5}{2} Nk \tag{13.9}$$

The molar heat capacity at constant volume is defined by

$$C_{V,m} = \frac{C_V}{n} \tag{13.10}$$

where n is the amount of substance. Because of

$$Nk = nN_A k = nR \tag{13.11}$$

the molar heat capacities for He and for H_2 are

$$C_{V,m}(\text{He}) = \tfrac{3}{2} R, \quad C_{V,m}(H_2) = \tfrac{5}{2} R$$

13.2 Temperature Dependence of C_V

13.2.1 Rotational and Vibrational Contribution to C_V

The molar heat capacity $C_{V,m}$ measured at different temperatures is shown in Figure 13.4 for He (atomic gas), HD (diatomic gas), and CO_2 (polyatomic gas). (We use HD instead of H_2 as an example because in H_2 a particular quantum effect leads to a more complicated behavior – see Section 13.2.1.)

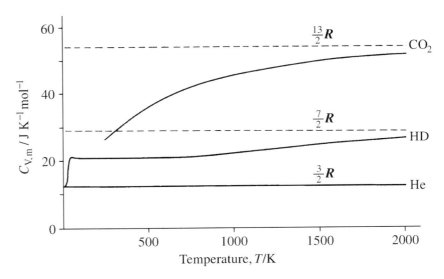

Figure 13.4 Experimental heat capacity $C_{V,m}$ for several gases as a function of temperature.

C_V curve of He. The experimental value $C_{V,m} = \frac{3}{2}R$ for He is, according to equation (13.8), due to the translational motion of the He atoms.

C_V curve of HD. The experimental value $C_{V,m} = \frac{5}{2}R$ for HD in the temperature range from 200 to 600 K is due to the contributions of the translational and rotational motion. At a low enough temperature, however, the contribution of the rotational motion becomes zero. This is in accordance with the theoretical $C_{V,m}$ curve for HD in Figure 13.5(b), which is obtained by calculating the slope of the theoretical curve for U (Figure 13.5(a)). It can be seen from Figure 13.5 that the contribution of the rotation is zero up to 10 K; above 10 K it is strongly increasing with temperature, leading to a constant value above 200 K (after passing a maximum at 50 K).

Above 600 K, $C_{V,m}$ is further increasing (Figure 13.4) due to the contribution of the vibrational motion (theoretical curve in Figure 13.6(b), calculated from the theoretical curve for the internal energy U in Figure 13.6(a), see Box 13.1). Because of the large vibrational frequency of HD, the classical limit

$$C_{V,m} = \tfrac{7}{2}R$$

is not reached below 2000 K (see Chapter 12).

C_V curve of CO_2. For CO_2, because of the much larger moment of inertia compared with HD, we expect that the classical limit of $\frac{5}{2}R$ for the contributions of translation and rotation is reached already at a temperature below 20 K. Because of the four different vibrational modes (Chapter 9), the classical limit for the contribution of vibration is

$$C_{V,\text{vib},m} = 4R \qquad (13.12)$$

The calculated $C_{V,m}$ curves (Figure 13.6(b)) are in excellent agreement with experiment. A closer inspection of the experimental and theoretical data reveals that at high temperatures the experimental values are slightly larger (by 5% for HD and by 1% for CO_2 at 2000 K). This is because the harmonic oscillator model used in the calculation is not strictly valid for higher quantum states.

13.2 TEMPERATURE DEPENDENCE OF C_V

When decreasing the temperature, $C_{V,m}$ does not decrease below $\frac{3}{2}R$ (that is the contribution of translation); this holds even at the lowest temperature at which gases can be examined experimentally. This is because the spacing between the two lowest translational quantum states is smaller by many orders of magnitude than those for rotation and vibration.

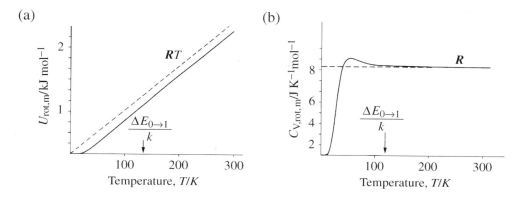

Figure 13.5 Calculated rotational contribution to U and C_V for HD versus temperature (solid lines). (a) $U_{rot,m}$ taken from Fig. 12.12(a), (b) $C_{V,rot,m}$ obtained from $U_{rot,m}$ by numerical differentiation. Dashed lines: classical limit.

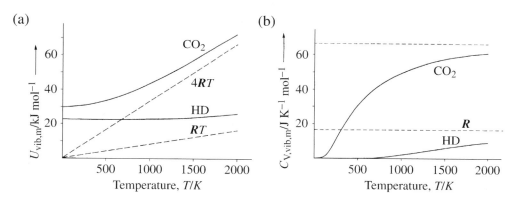

Figure 13.6 Calculated vibrational contribution to U and C_V for HD and CO_2 versus temperature (solid lines). For data see Box 13.1. Dashed lines: classical limit.

Box 13.1 – Calculation of $C_{V,vib}$ for HD and CO_2

The calculation is performed analytically. According to Foundation 12.1 the internal energy U_{vib} is

$$U_{vib} = Nh\nu_0 \cdot \left[\frac{1}{2} + \frac{1}{e^{a/T} - 1} \right]$$

Continued on page 456

> *Continued from page 455*
>
> where $a = h\nu_0/k$. Then we obtain for the heat capacity
>
> $$C_V = \frac{dU_{vib}}{dT} = Nh\nu_0 \cdot \frac{-1}{(e^{a/T}-1)^2} \cdot e^{a/T} \cdot \left(\frac{-a}{T^2}\right) = Nh\nu_0 \cdot \frac{a \cdot e^{a/T}}{T^2 \cdot (e^{a/T}-1)^2}$$
>
> The following data are used for the calculation:
>
> - HD: $\nu_0 = 1.14 \times 10^{14}\,\text{s}^{-1}$.
> - CO_2: $\nu_{bending} = 0.20 \times 10^{14}\,\text{s}^{-1}$; $\nu_{symmetric} = 0.42 \times 10^{14}\,\text{s}^{-1}$; $\nu_{antisymmetric} = 0.71 \times 10^{14}\,\text{s}^{-1}$. The bending mode is twofold degenerate, and therefore it contributes twice.
>
> In the following table the molar heat capacity $C_{V,m}$ (in $\text{J\,K}^{-1}\,\text{mol}^{-1}$) is listed.
>
> | | | Molar heat capacity, $C_{V,m}$/J K^{-1} mol^{-1} | | | |
> |---|---|---|---|---|---|
> | T/K | HD | CO_2 | | | |
> | | | bending | symmetric | antisymmetric | total |
> | 200 | 0.0 | 3.2 | 0.0 | 0.0 | 3.2 |
> | 300 | 0.0 | 7.5 | 0.5 | 0.0 | 8.0 |
> | 400 | 0.0 | 10.5 | 1.4 | 0.1 | 12.0 |
> | 600 | 0.1 | 13.5 | 3.6 | 1.0 | 18.1 |
> | 800 | 0.4 | 14.8 | 5.1 | 2.2 | 22.1 |
> | 1000 | 1.1 | 15.4 | 6.0 | 3.4 | 24.8 |
> | 1500 | 3.0 | 16.1 | 7.2 | 5.5 | 28.8 |
> | 2000 | 4.6 | 16.3 | 7.7 | 6.6 | 30.6 |

13.2.2 ortho- and para-H_2: Fascinating Quantum Effects on C_V

In Figure 13.7 the heat capacity C_V for H_2 is displayed. The theoretical curve calculated in the corresponding way as for HD (curve 1) is in strong disagreement with the experiment. How can we understand this discrepancy?

As outlined in Chapter 9, H_2 is a 3 : 1 mixture of ortho- and para-H_2. para-H_2 can only exist in rotational states with even quantum numbers ($n = 0, 2, 4, \ldots$), and ortho-H_2 in states with odd quantum numbers ($n = 1, 3, 5, \ldots$). Thus the excitation energy from the lowest to the next rotational state

$$\Delta E_{0 \to 2} = B \cdot [2(2+1) - 0] = 6B \quad \text{(para-H}_2\text{)} \tag{13.13}$$

$$\Delta E_{1 \to 3} = B \cdot [3(3+1) - 1(1+1)] = 10B \quad \text{(ortho-H}_2\text{)} \tag{13.14}$$

(where B is the rotational constant) is about two times larger for ortho- than for para-H_2. As a consequence, when warming up the gases from $T = 0$ para-H_2 takes up heat at lower temperature than ortho-H_2 (curves 2 and 3). At higher temperature, C_V of para-H_2 exhibits a maximum at $T = 168$ K, whereas C_V of ortho-H_2 approaches monotonically the classical

13.2 TEMPERATURE DEPENDENCE OF C_V

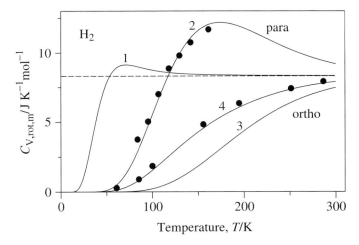

Figure 13.7 Rotational contribution to C_V for H_2 versus temperature. Solid lines: calculation. Curve 1: all rotational levels assumed to be accessible (as for HD). Curve 2: only even rotational levels accessible ($n = 0, 2, \ldots$, para-H_2). Curve 3: only odd rotational levels accessible ($n = 1, 3, \ldots$, ortho-H_2). Curve 4: natural mixture of ortho- and para-H_2 (3:1). Dots: experimental data for curve 2 and curve 4; pure ortho-H_2 cannot be isolated, thus experimental data for curve 3 are not available.

value Nk. The calculated C_V curve of H_2 (a 3:1 mixture of ortho-H_2 and para-H_2) is in excellent agreement with the experiment (Figure 13.7, curve 4). This is a fascinating macroscopic manifestation of a most astonishing quantum feature: the indistinguishability of protons.

13.2.3 Electronic Contribution to C_V ($C_{V,electr}$)

In Chapter 12 we calculated the electronic energy U_{electr} for the two-level system corresponding to the NO molecule (energy levels E_1 and E_2). From equation 12.44 we obtain C_V by differentiation:

$$U_{electr} = N \cdot \frac{E_2}{1 + e^{E_2/kT}}$$

$$C_{V,electr} = \left(\frac{\partial U_{electr}}{\partial T}\right)_V = NE_2 \frac{E_2 \cdot e^{E_2/kT}}{kT^2 \cdot [1 + e^{E_2/kT}]^2} \quad (13.15)$$

$C_{V,electr}$ for NO calculated from equation (13.15) and the corresponding molar quantity $C_{V,electr,m} = C_{V,electr}/n$ is compared with experiment in Figure 13.8.

13.2.4 C_V of Solids

The atoms in a solid vibrate in space (x, y, and z directions), but they do not translate or rotate. For simplicity we assume that the atoms are vibrating independently with the same frequency ν_0 (*Einstein's model of a solid*, see Figure 13.9). Then we expect the

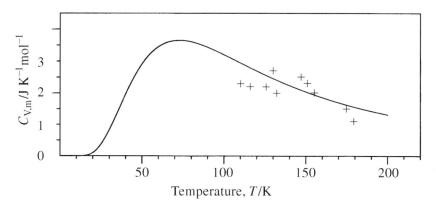

Figure 13.8 Molar heat capacity $C_{V,m}$ of NO (electronic contribution) versus temperature. Solid line: theory; crosses: experiment. No experimental data are available below the boiling point of NO.

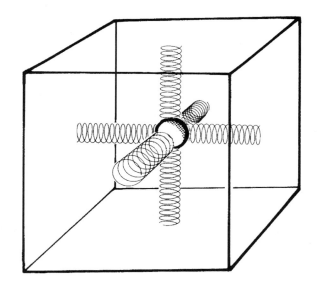

Figure 13.9 Three-dimensional oscillator.

same curve for $C_{V,m}$ as a function of temperature as in Figure 13.6, but $C_{V,m}$ should be three times larger than for a linear oscillator at high enough temperature:

$$\lim_{T \to \infty} C_{V,m} = 3R \tag{13.16}$$

This is in good agreement with the experimental data displayed in Figure 13.10. In this way we can understand the *rule of Dulong and Petit* that the molar heat capacity of metals at room temperature is about $3R$.

13.2 TEMPERATURE DEPENDENCE OF C_V

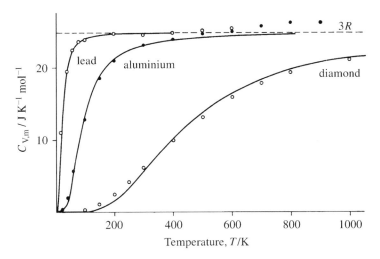

Figure 13.10 Heat capacity C_V of solids versus temperature. Dots and circles: experiment; solid lines: calculation; the vibrational frequencies are adjusted to the measured points: $\nu_0 = 1.33 \times 10^{12}\,\text{s}^{-1}$ (Pb), $\nu_0 = 6.00 \times 10^{12}\,\text{s}^{-1}$ (Al), and $\nu_0 = 2.83 \times 10^{13}\,\text{s}^{-1}$ (diamond). Dashed line: classical limit. At high enough temperature, the experimental values for Pb and Al exceed the classical limit; this means that the vibrations do not occur strictly harmonic.

13.2.5 Characteristic Temperature

The disappearance of C_V at low temperatures is a direct result of the quantum nature of matter. At $T = 0$ all molecules are in the lowest quantum state (Figure 13.11). The temperature θ when kT equals $E_2 - E_1$,

$$\theta = \frac{E_2 - E_1}{k} \tag{13.17}$$

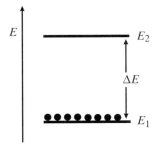

Figure 13.11 Population at the absolute zero point.

is called the *characteristic temperature*. For rotational and vibrational motions the characteristic temperatures are

$$\theta_{\text{rot}} = \frac{2h^2}{8\pi^2 \mu d^2 k}, \quad \Theta_{\text{vib}} = \frac{h\nu_0}{k} = \frac{h}{2\pi k} \cdot \sqrt{\frac{k_f}{\mu}} \tag{13.18}$$

For HD we obtain $\theta_{\text{rot}} = 131\,\text{K}$ and $\Theta_{\text{vib}} = 5480\,\text{K}$. For CO_2 we obtain $\theta_{\text{rot}} = 1.1\,\text{K}$ and $\theta_{\text{vib}} = 960\,\text{K}$ (corresponding to the bending mode). Among the solids considered above the vibration frequency ν_0 is much larger for diamond ($\Theta = 1350$ K) than for lead ($\Theta = 64$ K) because of the smaller nuclear mass and the larger force constant.

It is most remarkable that Θ reflects, very directly, the wave–particle duality of atomic nuclei. Only distinct energies of rotation and vibration are allowed, corresponding to distinct standing de Broglie waves representing these energy states. A body, when heated from absolute zero, can reach higher energy states only if the temperature exceeds a critical value.

The de Broglie wavelength of a nucleus decreases with increasing mass; accordingly, the characteristic temperature θ depends on the nuclear mass.

13.3 Change of State at Constant Pressure

13.3.1 Change of Internal Energy ΔU; Heat q and work w

Now let us consider a change of state at constant pressure (isobaric state change) as displayed in Figure 13.12.

We heat a gas at constant pressure from T_1 to T_2. Because the internal energy U of an ideal gas depends only on the temperature T, but not on the volume, the change ΔU is the same as for heating the gas at constant volume. However, more heat is taken from the bath with temperature T_2 because during the expansion of the gas the weight on the piston in Figure 13.12 has to be lifted. The work done by the gas against the piston to lift the weight (force f) by a distance ds can be calculated from the force f exerted on the piston

$$f \cdot ds = (P \cdot \text{area of piston}) \cdot \frac{dV}{\text{area of piston}} = P \cdot dV \qquad (13.19)$$

where dV is the volume change and P is the pressure of the gas. Then the total work done to lift the piston under constant pressure condition is

$$\int_{V_1}^{V_2} P \cdot dV = P \int_{V_1}^{V_2} dV = P(V_2 - V_1) = P\Delta V \qquad (13.20)$$

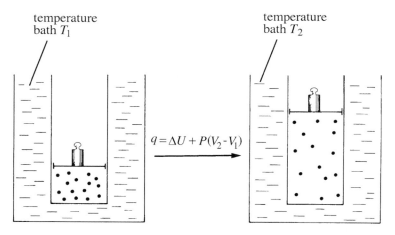

Figure 13.12 Heating a gas from T_1 to T_2 at constant pressure.

13.3 CHANGE OF STATE AT CONSTANT PRESSURE

The heat q necessary to perform the change of state is

$$q = \Delta U + P \Delta V \tag{13.21}$$

where $P \Delta V$ is the work performed on the surroundings; then

$$w = -P \Delta V \tag{13.22}$$

is the work done on the system (i.e., to the gas) and thus

$$\boxed{\Delta U = q - P \Delta V \quad \text{(heating at constant pressure)}} \tag{13.23}$$

Note: we attribute the positive sign to heat flowing to the system and to work done on the system. In the present case q is positive and w is negative.

Equation (13.23) means, as in the process at constant volume, that the energy performed on the gas (work as well as heat) is stored by the molecules as internal energy. In the above example, ΔU is less than q, since w is negative: work is done by the system.

Equations (13.3) and (13.23) are special cases for the *law of conservation of energy*.

Example – heat supplied for the isobaric heating of He gas

1 mol He gas is heated at $P = 1$ bar from state 1 ($T = 300$ K) to state 2 ($T = 400$ K). Then

$$\Delta U = \tfrac{3}{2} Nk\Delta T = \tfrac{3}{2} 6.02 \times 10^{23} \cdot 1.38 \times 10^{-23} \times 100 \, \text{J} = 1.25 \, \text{kJ}$$

Because of $PV_1 = NkT_1$ and $PV_2 = NkT_2$, we obtain

$$V_1 = \frac{NkT_1}{P} = 24.9 \, \text{L}, \quad V_2 = \frac{NkT_2}{P} = 33.2 \, \text{L}$$

Then $P\Delta V = 1 \, \text{bar} \cdot 8.3 \, \text{L} = 0.83 \, \text{kJ}$ and $q = (1.25 + 0.83) \, \text{kJ} = 2.08 \, \text{kJ}$. The heat supplied to the gas (2.08 kJ) is used to increase the internal energy of the gas (1.25 kJ) and to perform work on the piston (0.83 kJ).

13.3.2 Heat Capacity C_P

The heat capacity C_P (heat capacity at constant pressure) is defined by

$$C_P = \left(\frac{\partial q}{\partial T}\right)_P \tag{13.24}$$

According to equation (13.21) we have

$$dq = dU + P \cdot dV \quad \text{(constant pressure)} \tag{13.25}$$

and

$$C_P = \left(\frac{\partial U}{\partial T}\right)_P + P \left(\frac{\partial V}{\partial T}\right)_P \tag{13.26}$$

For an ideal gas, U does not depend on pressure or volume:

$$\left(\frac{\partial U}{\partial T}\right)_P = \left(\frac{\partial U}{\partial T}\right)_V = C_V \qquad (13.27)$$

Because of the ideal gas law we have

$$V = \frac{NkT}{P}, \quad \left(\frac{\partial V}{\partial T}\right)_P = \frac{Nk}{P}, \quad P\left(\frac{\partial V}{\partial T}\right)_P = Nk \qquad (13.28)$$

and

$$C_P = C_V + Nk \quad \text{(ideal gas)} \qquad (13.29)$$

C_P is larger than C_V because of the additional heat which has to be supplied to lift the weight on the piston. In other cases (e.g., for solids or for gases at high pressure) the relation between C_P and C_V is more complicated.

13.4 System and Surroundings; State Variables

We have already used the terms "system" and "state". Now we want to define these terms more precisely.

13.4.1 Defining System and Surroundings

In the macroscopic description of the behavior of molecular assemblies we must clearly distinguish between the "system" and the "surroundings." A system is a well-defined part of the universe (e.g., the oxygen molecules enclosed in the container in Figure 13.13). The part of the universe outside the boundary of the system is called *surroundings* (e.g., container, temperature bath and atmosphere above the system in Figure 13.13(a) or the container, the temperature bath and the weight on the piston in Figure 13.13(b)).

Generally the work w done on the system and the heat q flowing into the system increase the internal energy U of the system (law of energy conservation):

$$\boxed{\Delta U = q + w} \qquad (13.30)$$

Internal energy is stored in the form of electronic, vibrational, rotational, and translational energy of the molecular assembly.

13.4.2 Defining State and State Variables

The system displayed in Figure 13.13 can be described by the following data: $n_{O_2} = 1$ mol, $T = 273.15$ K, $P = 1$ bar, and $V = 22.71$ L. We can also say that the system is in a certain state, where the state is defined by macroscopic data (composition, temperature, pressure, volume, internal energy). Of course, this definition of the thermodynamic state must be distinguished from the definition of the quantum state of a single atom or molecule. Data such as temperature, pressure, volume, and internal energy are called *state variables*. State variables define the state. They are independent of how the state has been reached.

13.4 SYSTEM AND SURROUNDINGS; STATE VARIABLES

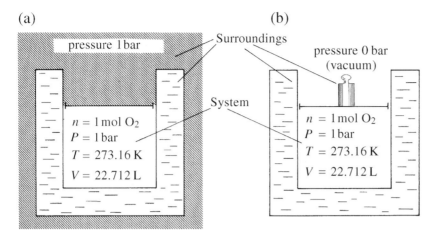

Figure 13.13 A system and its surroundings. Description of a state by composition, pressure, temperature, and volume. (a) Pressure (1 bar) exerted by the atmosphere; (b) pressure (1 bar) exerted by a weight.

A complete description of the state of the O_2 gas in Figure 13.13 is possible in different ways by a set of state variables – for example, by n, T, and P or n, T, and V, or n, V, and U. All other state variables are then uniquely defined by a state equation – for example, by the gas equation (13.28).

Whereas the change of a state variable is defined by the initial and the final state, for example,

$$\Delta T = T_2 - T_1, \quad \Delta P = P_2 - P_1, \quad \Delta U = U_2 - U_1 \qquad (13.31)$$

there are other quantities which depend upon how the change of state is performed. Such a quantity is the heat supplied to a system during a change of state (Box 13.2).

Box 13.2 – ΔU, q, and w in a Change of State

We consider a change from state 1 to state 2 carried out in two different ways (Figure 13.14). First, we evaporate water in a cylinder connected to a piston (Figure 13.14(a)) from state 1 ($T = 373$ K, $P = 1$ bar, $V = 0.018$ L; the water is a liquid) to state 2 ($T = 373$ K, $P = 1$ bar, $V = 31.2$ L; the water is a saturated vapor). After opening the valve, the water evaporates and lifts the weight on the piston. Then

$$w_a = -P\Delta V = -1 \text{ bar} \cdot (31.2 - 0.018) \text{ L} = -3.12 \text{ kJ}$$

In a second experiment, we evaporate the water into a vacuum (Figure 13.14(b)). No weight is lifted, and therefore $w_b = 0$. The heat supplied

Continued on page 464

Continued from page 463

to the system in case (a) is larger than that in case (b), because part of the heat is converted into work. On the other hand, the change ΔU of the internal energy is the same in both cases:

$$\Delta U = U_2 - U_1 = q_a + w_a = q_b + w_b$$

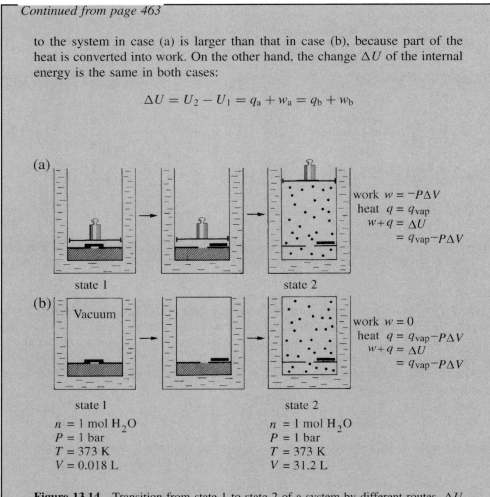

Figure 13.14 Transition from state 1 to state 2 of a system by different routes. ΔU is the same in both cases (U is a state function). The heat supplied in case (a) is larger than in case (b), because part of the energy is used for external work.

13.4.3 Closed, Isolated, and Open Systems

The systems we have examined above exchange energy with the surroundings (exchange of work or heat, see Figure 13.15(a)). There is no transfer of matter to and from the surroundings. They are called *closed systems*. Systems where matter is exchanged are called *open systems*.

Often it is useful to examine systems which are completely isolated from the surroundings: no exchange of heat, work, or matter. Such systems are called *isolated systems* (Figure 13.15(b)).

Since any change of state in an isolated system occurs without exchange of heat and work with the surroundings ($q = 0$, $w = 0$), we can say

13.4 SYSTEM AND SURROUNDINGS; STATE VARIABLES

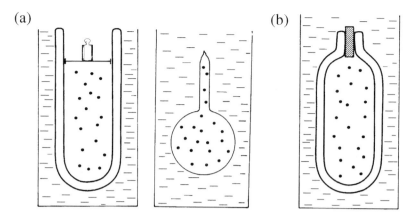

Figure 13.15 Closed and isolated systems. (a) Closed system: no exchange of matter (closed container), but work can be done on the system (left) or heat can be transferred (right). (b) Isolated system: heat cannot be transferred (heat isolation), matter cannot be transferred (closed container), and no work is done on the system.

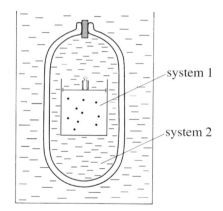

Figure 13.16 System 2, composed of system 1 and surroundings, constitutes an isolated system.

$$\boxed{\Delta U = 0 \quad \text{(isolated system)}} \qquad (13.32)$$

This is a very general statement, because any system together with devices for the storage of energy (e.g., temperature baths, springs, weights) can be considered as a new system, and thus as an isolated system (Figure 13.16).

13.4.4 Cyclic Processes

A change of state where the system is in the same state at the end as in the beginning is called a *cyclic process* (Figure 13.17). Then

$$\boxed{\Delta U = 0 \quad \text{(cyclic process)}} \qquad (13.33)$$

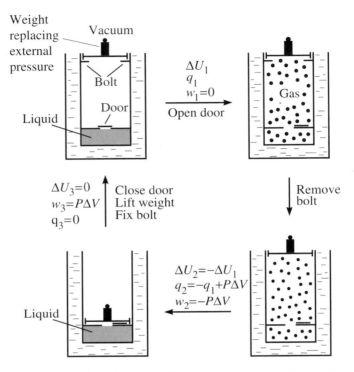

Figure 13.17 Evaporation and condensation of water as an example of a cyclic process.

Note that in the cyclic process displayed in Figure 13.17, $\Delta U = 0$, but $q \neq 0$ and $w \neq 0$; because of the law of conservation of energy (equation (13.30))

$$\Delta U = q + w$$

and equation (13.33) we obtain

$$w = -q \quad \text{(cyclic process)} \tag{13.34}$$

The work done on the system (w) equals the heat flowing from the system to the surroundings ($-q$) if the system goes through a cyclic process.

Problems

Problem 13.1 – Isothermal Expansion of an Ideal Gas

An ideal gas ($T = 300$ K, $P_1 = 2$ bar, $V_1 = 1$ L) expands isothermally (Figure 13.18(a)) by lifting a weight on a piston corresponding to the pressure $P_2 = 1$ bar. What is the volume V_2 of the gas after thermal equilibration, and what is the work w done on the gas and the heat q flowing in the gas during the process?

PROBLEM 13.1

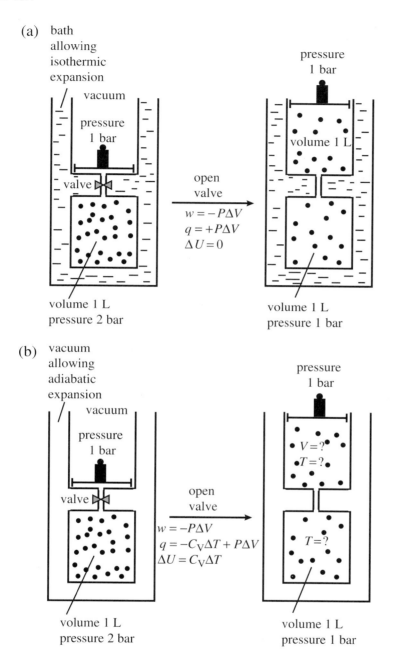

Figure 13.18 Expansion of a gas. (a) Isothermal (at constant temperature), (b) adiabatic (system thermally isolated from its surroundings).

Solution. Because of $P_1 V_1 = P_2 V_2$ we obtain

$$V_2 = V_1 \cdot \frac{P_1}{P_2} = 2\,\text{L}$$

The work $w = -P_2 \Delta V = -P_2 \cdot (V_2 - V_1) = -1\,\text{bar} \cdot 1\,\text{L} = -100\,\text{J}$. Because of $\Delta U = 0$ the heat is $q = -w = 100\,\text{J}$.

Problem 13.2 – Adiabatic Expansion of an Ideal Gas

An ideal gas ($T_1 = 300\,\text{K}$, $P_1 = 2\,\text{bar}$, $V_1 = 1\,\text{L}$; Figure 13.18(b)) expands adiabatically to the pressure $P_2 = 1\,\text{bar}$ (experimental setup as in Problem 13.1, but the gas is thermally isolated from the surroundings). We wait for thermal equilibration within the system (temperature T_2 in both volumes). Calculate T_2, V_2, and w.

Solution. In the adiabatic process we have $q = 0$, thus we obtain $\Delta U = w$. Because of $\Delta U = C_V \cdot (T_2 - T_1)$, $w = -P_2 \cdot (V_2 - V_1)$, and $P_1 V_1 = NkT_1$, $P_2 V_2 = NkT_2$, we obtain

$$C_V \cdot (T_2 - T_1) = -P_2 \cdot \left(\frac{NkT_2}{P_2} - \frac{NkT_1}{P_1} \right) = Nk \cdot \left(-T_2 + \frac{P_2}{P_1} \cdot T_1 \right)$$

Then

$$T_2 = T_1 \cdot \frac{P_2/P_1 + C_V/Nk}{1 + C_V/Nk}$$

With the data given above and $C_V = \frac{3}{2} Nk$ for an atomic gas it follows

$$T_2 = 300\,\text{K} \cdot 0.8 = 240\,\text{K}$$

$$V_2 = V_1 \cdot \frac{P_1 T_2}{P_2 T_1} = 1.6\,\text{L}$$

$$w = -P_2 \Delta V = -1\,\text{bar} \cdot 0.6\,\text{L} = -60\,\text{J}$$

14 Principle of Entropy Increase

A thermodynamic state of a system consisting of a large number of molecules is defined by its macroscopic properties: temperature, pressure, and composition. How does it happen that the system can transform from a certain state to another state, but the reverse process never occurs? The reason is not that the reverse process is impossible, but that it is extraordinarily improbable.

A system in a given thermodynamic state has a distinct number of quantum states representing the thermodynamic state, called the *number of configurations* Ω. Let us consider an isolated system. We ask if a change from some initial state 1 (Ω_1 configurations) to some final state 2 (Ω_2 configurations) can occur spontaneously (perhaps after a trigger). This is the case for $\Omega_2 > \Omega_1$.

The reverse process is not impossible, but the probability $P = \Omega_1/\Omega_2$ is extremely small. Spontaneously occurring processes are called (*thermodynamically*) *irreversible*. A reversion of the process can only occur when $\Omega_1 \approx \Omega_2$ – that is, when the number of configurations of the isolated system is practically unchanged; such a process is called a (*thermodynamically*) *reversible* process.

It is of fundamental importance to know how Ω changes in proceeding from one state to another, in order to judge whether the process can occur spontaneously. The number Ω is calculated for simple cases in this chapter.

It is useful to define a quantity called the *entropy* of a system in a given thermodynamic state: $S = k \cdot \ln \Omega$, where k is the Boltzmann constant. For processes in an isolated system S always increases since the number of configurations Ω increases. The significance of the quantity S defined here will become clear in the context of Chapter 15, where we will relate S to the calorimetric properties of matter.

14.1 Irreversible and Reversible Changes of State

14.1.1 Irreversible Changes

In Chapter 11 we considered thermal equilibration: a body of high temperature is brought in contact with a body of low temperature. Heat flows from the first body to the second body until both bodies have the same temperature.

The reverse process does not occur; that is, two bodies with the same temperature do not evolve to a state where they have different temperatures by means of spontaneous heat flow from one body to the other. Thus we can say:

Heat does not flow from a colder to a warmer body, even though this would be possible according to the law of energy conservation.

This statement, established by experience, is based on our considerations of large numbers of molecules.

The fast molecules in the hot body exchange energy with the slow molecules in the cold body. The reverse process, bringing two bodies of equal temperature into contact and one body gets warm and the other one gets cold, is extremely improbable. This is because it is extremely improbable that all molecules with high kinetic energy are observed in one body, and those with low kinetic energy in the other body.

The greater the number of particles in the two containers, the more improbable it is that the molecules in the upper container take on a greater energy than the molecules in the lower container. So with more particles, it becomes increasingly improbable that the temperature in the upper container exceeds the temperature in the lower container with no external interference. In practical cases we always have such large numbers of molecules that the reverse process of temperature equilibration becomes so improbable that we can view it as practically impossible. This consideration shows the statistical character of our statement.

Processes which occur by themselves only in a certain direction are called *irreversible*, since they cannot be reversed without external interaction. Examples of irreversible processes are temperature equilibration, mixing of two gases, expansion of a gas into a vacuum, and evaporation in vacuum (Figure 14.1).

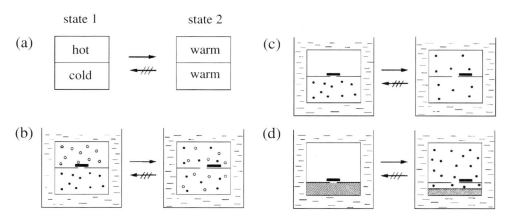

Figure 14.1 Examples of irreversible processes. (a) Temperature equilibration. (b) Mixing of two gases. (c) Expansion of a gas into a vacuum. (d) Evaporation into a vacuum.

14.1.2 Reversible Changes

The changes of state in Figure 14.1 can also be carried out indirectly, such that at each instant the system is practically in equilibrium.

14.1 IRREVERSIBLE AND REVERSIBLE CHANGES OF STATE

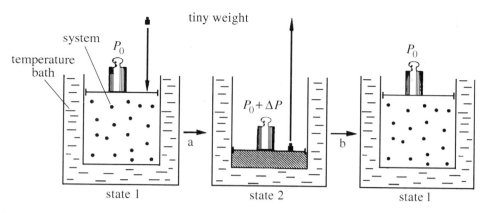

Figure 14.2 Condensation and evaporation of a liquid in a reversible path (e.g., water at 373 K, $P_0 = 1$ bar). State 1: liquid completely evaporated under the equilibrium vapor pressure P_0. State 2: liquid completely condensed under the pressure $P_0 + \Delta P$. Process a (states 1 → 2): pressure minimally increased by adding a tiny weight. Process b (states 2 → 1): pressure minimally decreased by taking off a tiny weight. Note that the temperature bath (symbolized by a basin) is large (heat exchange with the system does not change the temperature of the bath).

Reversible evaporation

For example, consider a vapor with equilibrium vapor pressure P_0 in a temperature bath as shown in Figure 14.2 (state 1). If we apply an external pressure which is greater than the vapor pressure by a very small amount ΔP, then the vapor will condense; this process occurs such that thermal equilibrium is maintained during the entire process (Figure 14.2, step a). If we remove the tiny additional weight, then the liquid again will slowly evaporate completely (Figure 14.2, step b). This process can be reversed at any time by adding the tiny additional weight. Because the additional weight can be arbitrarily small, a negligibly small increment of work from the outside is expended in reversing the evaporation process.

Processes which occur in the manner described above are called *reversible*, because they can be reversed at each moment through an arbitrarily small change of a state variable (e.g., pressure). In contrast, there are irreversible processes which can be run backward only through massive intervention; these occur in only one direction by themselves.

Reversible expansion of a gas

How can the expansion of a gas (Figure 14.1(c)) be carried out on a reversible path? Consider the reversible isothermal (i.e., occurring at constant temperature) expansion of a gas from pressure P_1 to pressure P_2. We imagine the pressure decreased in small steps, in which we remove small weights one after the other (Figure 14.3). We do this very slowly to allow heat exchange between system and temperature bath in each step; that is, the temperature of the system is kept practically constant during the process. Then for the work done on the system in each step we can write

$$dw = -P \cdot dV \qquad (14.1)$$

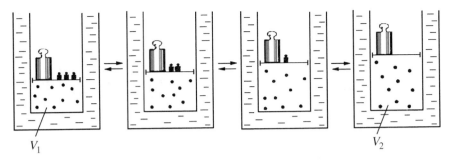

Figure 14.3 Reversible isothermal expansion of a gas (decrease of the pressure by stepwise removal of tiny weights) from volume V_1 to volume V_2.

The total work in the pressure change is then

$$w = \int_{V_1}^{V_2} dw = -\int_{V_1}^{V_2} P \cdot dV = -\int_{V_1}^{V_2} nRT \cdot \frac{dV}{V} \tag{14.2}$$

If we make sure that the temperature of the gas remains constant during the change of state (isothermal change of state), then we can take nRT outside the integral:

$$\boxed{w = -nRT \int_{V_1}^{V_2} \frac{dV}{V} = -nRT \ln \frac{V_2}{V_1}} \tag{14.3}$$

This integral corresponds to the shaded area in Figure 14.4. Because in this case $\Delta U = 0$ (ideal gas), the heat

$$\boxed{q = -w = nRT \cdot \ln \frac{V_2}{V_1}} \tag{14.4}$$

is transferred from the temperature bath to the gas. Through the addition or removal of tiny weight increments (Figure 14.3), we can at any time make the process run from left to right or vice versa.

Equation (14.4) is a key equation in physical chemistry occurring in many derivations in the following chapters. The energy (heat q) transferred from the temperature bath to the system when expanding a gas is converted into mechanical energy (work done $-w$ by the system).

Example – conversion of heat into mechanical work

We consider the reversible expansion of $n = 1$ mol of a gas at $T = 300$ K from a volume V_1 to a volume $V_2 = 2V_1$. In this case we obtain $q = -w = 1$ mol $RT \cdot \ln 2 = 1.73$ kJ. This is the energy required to lift a weight of 1 kg (force 10 N) by 173 m.

14.2 Distribution Possibilities

In attempting to find criteria for irreversibility we first consider the change in the distribution possibilities when changing the state of a system. In this section we show that

14.2 DISTRIBUTION POSSIBILITIES

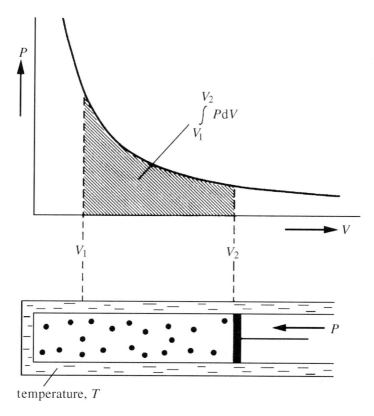

Figure 14.4 Reversible isothermal expansion of an ideal gas: calculation of reversible work.

the number of distribution possibilities increases in an irreversible process: the occasional reversion of the process is extremely improbable. In the limiting case of a reversible process the number of distribution possibilities remains unchanged. We illustrate this by considering two examples where the increase of distribution possibilities is obvious: the mixing of two gases (Figure 14.1(b)) and the expansion of a gas (Figure 14.1(c)).

14.2.1 Mixing of Two Gases

Consider the change of state illustrated in Figure 14.1(b). In State 1 all the molecules of the first type (open circles) are concentrated in the upper chamber, and all the molecules of the second type (filled circles) are concentrated in the lower chamber. When the partition is opened, the molecules distribute themselves over the whole volume, until State 2 (fully mixed) is reached. In this process the number of distribution possibilities increases.

We can easily understand this for a simple example. We consider two particles (1 and 2) of the first type, and two particles (3 and 4) of the second type. In the beginning, particles 1 and 2 are in the upper chamber, and particles 3 and 4 are in the lower. We denote this case symbolically as:

$$1\ 2\ |\ 3\ 4$$

At the end the following 16 possibilities for distribution between upper and lower chambers can be distinguished:

```
12 | 34     13 | 24     14 | 23     23 | 14
24 | 13     34 | 12      1 | 234     2 | 134
 3 | 124     4 | 123    123 | 4     124 | 3
134 | 2    234 | 1     1234 |          | 1234
```

The number of distribution possibilities has increased from 1 to 16 when proceeding from state 1 (molecules unmixed) to state 2 (molecules mixed). The number of distribution possibilities also increases on the whole in all other cases shown in Figure 14.1, as we will examine more closely in the following sections.

14.2.2 Expansion of a Gas

We examine the expansion of a gas (Figure 14.5). In state 1 the smaller volume V_1 and in state 2 the larger volume V_2 is available to the molecules. The larger the volume in which the molecules are distributed, the greater the number of distribution possibilities. We have seen that the process can not be run in the opposite direction by a minimal change of the conditions.

Certainly this statement is of a statistical nature: it is not absolutely impossible that the molecules transform from state 2 to state 1 by themselves, it is just extraordinarily improbable. We can easily calculate the probability P that all molecules appear in partial volume V_1 at the same time by chance. We assume that the gas particles are independent of each other. We consider the simple case

$$V_1 = \tfrac{1}{2} V_2 \tag{14.5}$$

and imagine that only one gas particle is found in the container. In this case it happens equally frequently that the particle stays in the upper or in the lower part of the container (Figure 14.6).

There are two possibilities for the particle to stay in V_2 which are a priori equally likely, and one possibility to stay in V_1. So the probability P that the particle is in the

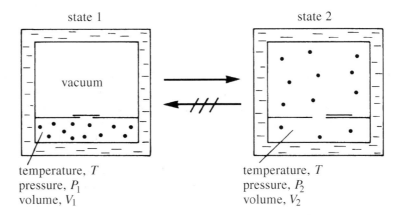

Figure 14.5 Expansion of an ideal gas into a vacuum.

14.2 DISTRIBUTION POSSIBILITIES

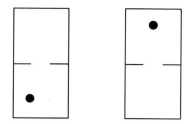

Figure 14.6 A particle can be found in the upper or the lower partial volume.

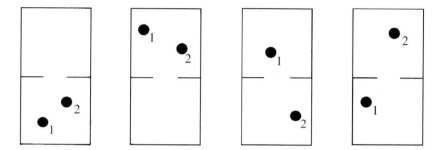

Figure 14.7 Configurations for the distribution of two particles in two partial volumes.

lower partial volume is

$$P = \tfrac{1}{2} \tag{14.6}$$

If we consider two particles, then there are four a priori equally probable possibilities; only in one case are both particles in the lower partial volume (Figure 14.7). So the probability of finding both particles in the lower partial volume is

$$P = \left(\tfrac{1}{2}\right)^2 = \tfrac{1}{4} \tag{14.7}$$

With three particles there are eight possibilities, so the probability that all three particles will be found in the lower partial volume is

$$P = \left(\tfrac{1}{2}\right)^3 = \tfrac{1}{8} \tag{14.8}$$

If we consider N particles, then we get

$$P = \left(\tfrac{1}{2}\right)^N \tag{14.9}$$

In cases of practical interest, N is extraordinarily large. If we imagine one mole of a gas in the container, then N is on the order of 10^{23}. The probability that all the molecules are in the lower half of the container is then

$$P = \left(\frac{1}{2}\right)^{10^{23}} = \frac{1}{2^{(10^{23})}} \approx \frac{1}{10^{(10^{22})}} = \frac{1}{10^{10000000000000000000000}}$$

This probability is inconceivably small. We can compare this value with the probability that every human on earth ($N \approx 10^{10}$) dies of a heart attack within 1 second, purely by chance. The life expectancy of a human is about 10^2 years $= 10^9$ seconds, and heart attacks are the cause of death for about a tenth of the population. So the probability that a person chosen at random will suffer a fatal heart attack within a certain given time interval of 1 second equals $0.1 \times 10^{-9} = 10^{-10}$. The probability for everyone to have a heart attack within the same second is then

$$P = (10^{-10})^N = (10^{-10})^{10^{10}} = \frac{1}{(10^{10})^{(10^{10})}} = \frac{1}{10^{(10^{11})}}$$

This probability is greater by an enormous factor than the probability that 10^{23} molecules will be found by chance in the volume V_1. It is about as large as the probability that all molecules out of $N = 10^{12}$ ($\simeq 10^{-11}$ g H_2) will gather in the partial volume V_1.

Our system thus changes from state 1 to state 2 (Figure 14.5), which has a much greater number of distribution possibilities. If it has once reached this state, then the probability P that it returns to state 1 through chance fluctuations is very small. It is thus practically impossible for the process in Figure 14.5 to run in reverse by itself.

In Box 14.1 it is shown that in the general case (V_1/V_2 arbitrary) the relationship

$$P = \left(\frac{V_1}{V_2}\right)^N \qquad (14.10)$$

Box 14.1 – Derivation of Equation (14.10)

First we consider the example $V_1 = \frac{2}{3}V_2$. In this case we divide the volume V_2 into three partial volumes (Figure 14.8). There are three possibilities of finding a particle in one of these three partial volumes; then the probability P of finding the particle in the partial volume V_1 is $P = \frac{2}{3}$. In the general case we divide the volume V_2 into partial volumes according to the ratio V_1/V_2 (e.g., 11 partial volumes for $V_1 = \frac{7}{11}V_2$). So we get $P = V_1/V_2$ for one particle. In analogy to the case $V_1 = \frac{1}{2}V_2$ we obtain equation (14.10) for N particles.

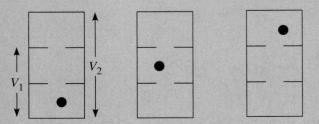

Figure 14.8 A particle can be found in one of three partial volumes.

is obtained in place of equation (14.9). So the smaller the value of V_1/V_2, the smaller that of P. The probability P thus decreases as a gas expands further and further in an irreversible state change.

14.3 Counting the Number of Configurations Ω

We have considered a large number of particles freely moving in a volume V_2. We have seen that the probability P to find all particles, by chance, in the partial volume V_1 is extremely small. At first glance it looks like a stupid question to ask for P but we will see that this is not the case.

14.3 Counting the Number of Configurations Ω

In the derivation of the Boltzmann distribution (Section 12.6) we have seen that the number Ω of configurations is given by the number of quantum states which can be occupied at a given temperature. So Ω is unambiguously determined if each of these quantum states is described by equal a priori occupation probability.

14.3.1 Probability P of Reversion of an Irreversible Process

The probability P discussed in Section 14.2 can be expressed by the number of configurations Ω of the system before and after an irreversible process, Ω_1 in state 1 and Ω_2 in state 2:

$$P = \frac{\Omega_1}{\Omega_2} \qquad (14.11)$$

In the consideration in Section 14.2 we could fix this ratio unambiguously, but not Ω_1 or Ω_2 by itself. In the following we determine P in simple cases by counting the number of configurations Ω_1 and Ω_2.

14.3.2 Particles Each in One of Three Energy States

In Section 12.6 we considered a system consisting of $N = 100$ particles which can occupy three quantum states with energies 0, a, and $2a$. The total energy of this system is assumed to be $U = 40a$. According to Table 12.1, the total number of configurations is $\Omega = 69.0 \times 10^{32}$. If all the particles were in the lowest quantum state (this would be the case at absolute zero: $T = 0$), then we would have $\Omega = 1$. With increasing temperature (thus increasing total energy U), the number of configurations Ω increases.

As outlined in Section 12.6.1, the case $U = 40a$ for $N = 100$ is realized when the system is in a temperature bath of temperature T where

$$kT = \frac{1}{1.04} a \approx a$$

14.3.3 Number of Configurations Ω of an Atomic Gas

Number of quantum states

Now let us discuss a more realistic example. We calculate Ω for a gas of N particles in the corresponding manner. We imagine a gas of N particles in a cube of volume $V = L^3$ at the temperature T. The average kinetic energy of a particle is

$$\overline{E}_{\text{trans}} = \tfrac{3}{2}kT \qquad (14.12)$$

Most of the particles are distributed over quantum states between $E = 0$ and an energy which is about twice the value of the average translational energy ($E = 2\overline{E}_{\text{trans}} = 3kT$). According to equation (12.37), the number of quantum states between $E = 0$ and $E = 3kT$ is

$$g = \frac{\pi}{6}\left[\frac{8mL^2}{h^2}\right]^{3/2}(3kT)^{3/2} \qquad (14.13)$$

Distinguishable particles

To form a clear picture, we will start with a major approximation. We will assume for simplicity that these g quantum states are occupied with the same probability and that all states with higher energy are unoccupied (in reality, the occupation probability decreases, according to the Boltzmann distribution, exponentially with $E/(kT)$).

For one particle, $\Omega = g$. In the case of two particles, there are g configurations of particle 2 for each configuration of particle 1, so $\Omega = g^2$. For N particles,

$$\Omega = g^N.$$

For example, for $g = 3$ and $N = 2$, there are altogether $\Omega = 3^2 = 9$ configurations (Figure 14.9).

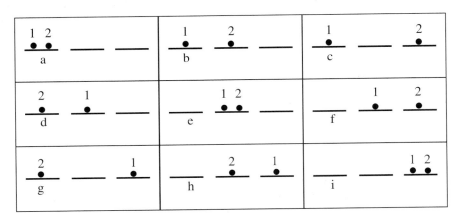

Figure 14.9 Possible configurations (a–i) for two particles occupying three quantum states ($N = 2$, $g = 3$).

Indistinguishable particles

We must now recall, however, that our N particles are indistinguishable (Chapter 12), so all possibilities which are formed through exchange of particles must not be included in the calculation of Ω.

In our numerical example (Figure 14.9), three configurations are formed through exchange of particles (d from b, g from c, and h from f), so because of indistinguishability, $\Omega = 6$ instead of $\Omega = 9$.

In the case of a gas at high enough temperature, g is much greater than N (Section 12.4.4), and for this case we can easily calculate Ω. Most quantum states are then

14.3 COUNTING THE NUMBER OF CONFIGURATIONS Ω

unoccupied, a few are singly occupied, and multiply occupied states play practically no role. Ω is then obtained easily by dividing the number of configurations for distinguishable particles by the number of possibilities to interchange N particles with each other. This number is $N!$, also called the *number of permutations*.

For example, for $N = 3$, there are $3! = 6$ permutations: 123, 132, 213, 231, 312, 321. Thus we find

$$\Omega = \frac{g^N}{N!} \quad \text{for } g \gg N \tag{14.14}$$

If we approximate $N!$ by Stirling's formula (see Box 12.4)

$$N! \approx N^N \cdot e^{-N} = \left(N \cdot \frac{1}{e}\right)^N \tag{14.15}$$

then with $V = L^3$ we obtain

$$\Omega = \left[\frac{g \cdot e}{N}\right]^N = \left[\frac{4}{3}\pi 6^{3/2} e \cdot \frac{(km)^{3/2}}{h^3} \frac{V}{N} T^{3/2}\right]^N \tag{14.16}$$

or

$$\ln \Omega = N \cdot \ln \frac{g \cdot e}{N} = N \cdot \ln \left[\frac{4}{3}\pi 6^{3/2} e \cdot \frac{(km)^{3/2}}{h^3} \frac{V}{N} T^{3/2}\right] \tag{14.17}$$

The rigorous calculation leads to the expression

$$\ln \Omega = N \cdot \ln \left[e^{5/2}(2\pi)^{3/2} \frac{(km)^{3/2}}{h^3} \frac{V}{N} T^{3/2}\right] \tag{14.18}$$

which differs only in the numerical factor 191.9 instead of $\frac{4}{3}\pi 6^{3/2} e = 167.3$. So Ω increases with the volume of the gas. That implies that the distance between neighboring energy levels decreases as the side length L and thus the volume V increase:

$$E = \frac{h^2}{8mL^2}\left(n_x^2 + n_y^2 + n_z^2\right), \quad L = \sqrt[3]{V},$$

according to equation (3.29). Increasing the volume produces many more quantum states available to the molecules and thus increases Ω (Figure 14.10).

We test whether the same value is obtained for the probability P of finding the gas in state 1 as in the classical calculation for the expansion of a gas. The result is

$$P = \frac{\Omega_1}{\Omega_2} = \frac{\left[e^{5/2}(2\pi)^{3/2}\frac{(km)^{3/2}}{h^3}\frac{1}{N}T^{3/2}\right]^N}{\left[e^{5/2}(2\pi)^{3/2}\frac{(km)^{3/2}}{h^3}\frac{1}{N}T^{3/2}\right]^N} \cdot \frac{V_1^N}{V_2^N} = \left(\frac{V_1}{V_2}\right)^N \tag{14.19}$$

This agrees with the result (equation (14.10)) from our previous approach.

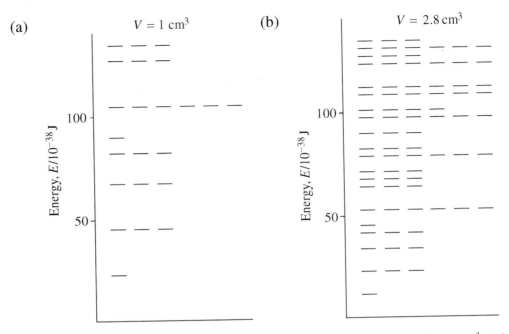

Figure 14.10 Translational quantum states for He in a cube of volume (a) $V = 1\,\text{cm}^3$ and (b) $V = 2.8\,\text{cm}^3$. Energies calculated from equation (3.29), $E = [h^2/(8mL^2)]\,(n_x^2 + n_y^2 + n_z^2)$ with $L = \sqrt[3]{V}$.

14.4 Entropy of a System: $S = k \cdot \ln \Omega$

In practice it is convenient to use the quantity

$$S = k \cdot \ln \Omega \tag{14.20}$$

rather than the number of configurations Ω for a system in a given state. S is called the *entropy* of the system in the state of interest. We will describe the advantages of S over Ω in more detail in the next chapter. Since Ω depends only on the state of the system and not on the way in which the state was reached, S is a state variable, like the internal energy U.

14.4.1 Entropy of Atomic Gases

For example, for the entropy of an atom gas from equation (14.18) we obtain

$$S = Nk \cdot \ln\left[e^{5/2}(2\pi)^{3/2}\frac{(km)^{3/2}}{h^3}\frac{V}{N}T^{3/2}\right] \tag{14.21}$$

(*Sackur–Tetrode equation*). Note that this equation becomes invalid at very low temperature, because in the derivation we assumed that the number of available quantum states is much larger than the number of particles.

14.4 ENTROPY OF A SYSTEM: $S = k \cdot \ln \Omega$

Example – entropy of rare gases

From equation (14.21) we can easily calculate the entropy of an atomic gas. For example, for argon gas ($m = 6.63 \times 10^{-26}$ kg) at $T = 298.15$ K and pressure 1 bar (number density $N/V = 2.43 \times 10^{25}$ m^{-3}) we obtain for the molar entropy $S_m = S/n$ with $n = N/N_A$

$$S_{m,298} = R \cdot \ln \left[e^{5/2} (2\pi)^{3/2} \frac{(km)^{3/2}}{h^3} \frac{V}{N} T^{3/2} \right] = 154.8 \text{ J K}^{-1} \text{ mol}^{-1}$$

Because of its smaller mass, helium ($m = 0.644 \times 10^{-26}$ kg) has a lower entropy ($S_{m,298} = 126.0$ J K^{-1} mol^{-1}) than argon; accordingly, the entropy of xenon ($m = 21.8 \times 10^{-26}$ kg) is larger ($S_{m,298} = 169.5$ J K^{-1} mol^{-1}) than that of argon.

Note that the differences in entropy of the rare gases are a direct consequence of the different spacing of the energy levels of the translational quantum states. Increasing the mass m results in a decrease of the spacing of the energy levels, thus more energy levels are available for the atoms and thus the number of configurations Ω increases.

14.4.2 Entropy of Diatomic Gases

In diatomic gases (e.g., HD, HCl, CO) the entropy S results from contributions of translational, rotational, and vibrational motion:

$$S = S_{\text{trans}} + S_{\text{rot}} + S_{\text{vib}} \tag{14.22}$$

S_{trans} is given by equation (14.21); S_{rot} is, according to Foundation 14.1,

$$S_{\text{rot}} = Nk \cdot \left[\ln \frac{8\pi^2 \mu d^2 kT}{h^2} + 1 \right] \tag{14.23}$$

The contribution S_{rot} increases with decreasing spacing of the energy levels – that is, with increasing moment of inertia μd^2.

Accordingly, the contribution S_{vib} is

$$S_{\text{vib}} = Nk \cdot \left[-\ln(1 - e^{-h\nu_0/kT}) + \frac{h\nu_0}{kT} \cdot \frac{1}{e^{h\nu_0/kT} - 1} \right] \tag{14.24}$$

Note that the expression for S_{rot} is an approximation for high enough temperature, whereas the expression (14.24) for S_{vib} follows from an exact analytical treatment. In the limit of high temperature ($h\nu_0/kT \ll 1$) we can approximate equation (14.24) by

$$S_{\text{vib}} = Nk \cdot \left[\ln \frac{kT}{h\nu_0} + 1 \right] \tag{14.25}$$

In this case it can be easily seen that the entropy also increases with decreasing spacing of the energy levels – that is, with decreasing vibration frequency ν_0.

Numerical data of the entropy for different diatomic gases are discussed in Chapter 15.

14.5 Entropy Change ΔS

Because the entropy S is a state variable, the entropy change in a transformation from a state 1 to a state 2,

$$\Delta S = S_2 - S_1 \tag{14.26}$$

is independent of the path. From this it follows that

$$\boxed{\Delta S \text{ (cyclic process)} = 0} \tag{14.27}$$

The entropy change for an arbitrary process is

$$\Delta S = k \ln \Omega_2 - k \ln \Omega_1 = k \cdot \ln \frac{\Omega_2}{\Omega_1} \tag{14.28}$$

Then for the expansion of a gas at constant temperature, using equation (14.21), we find

$$\Delta S = k \cdot \ln \left(\frac{V_2}{V_1}\right)^N = Nk \cdot \ln \frac{V_2}{V_1} \tag{14.29}$$

S increases with increasing volume ($V_2 > V_1$), reflecting the increasing number of distribution possibilities.

For the heating of an atomic gas from temperature T_1 to T_2 at constant volume, using equation (14.21), we obtain

$$\Delta S = k \cdot \ln \left(\frac{T_2^{3/2}}{T_1^{3/2}}\right)^N = \frac{3}{2} Nk \cdot \ln \frac{T_2}{T_1} \tag{14.30}$$

In this case S increases with increasing temperature ($T_2 > T_1$). This is because of the increasing number of quantum states available to the molecules at higher temperature.

14.5.1 Temperature Equilibration

The entropy change in temperature equilibration in the case of Figure 14.1(a) (N atoms of an atomic gas in the upper container at temperature T_2 and N atoms of an atomic gas in the lower container at temperature T_1) is

$$\Delta S = \Delta S_{\text{upper container}} + \Delta S_{\text{lower container}} \tag{14.31}$$

$$= \frac{3}{2} Nk \cdot \ln \frac{T}{T_2} + \frac{3}{2} Nk \cdot \ln \frac{T}{T_1} = \frac{3}{2} Nk \cdot \ln \frac{T^2}{T_1 T_2} \tag{14.32}$$

where

$$T = \frac{T_1 + T_2}{2} \tag{14.33}$$

is the average temperature after equilibration. For $T_1 < T_2$, the entropy in the upper container is decreasing while the entropy in the lower container is increasing.

For example, if we choose $T_1 = 200$ K and $T_2 = 600$ K, then $T = 400$ K, and we obtain

$$\Delta S = \frac{3}{2} Nk \cdot \ln \frac{400^2}{200 \times 600} = +\frac{3}{2} Nk \times 0.29$$

14.5 ENTROPY CHANGE ΔS

Thus the entropy increases, as expected: Ω, the number of configurations, increases for an irreversible process.

14.5.2 Mixing of Two Gases

We consider two separated containers A and B (each of volume V) with different ideal gases A and B at pressure P and temperature T. By connecting the containers the gases mix (process shown in Figure 14.1(b)). We calculate the entropy change in this irreversible process by counting the number of configurations before and after mixing (Problem 14.2):

$$\Delta S = (N_A + N_B)k \cdot \ln 2 \qquad (14.34)$$

ΔS is positive, thus the entropy increases for this irreversible process. This relation can also easily be obtained by calculating the probability P that the mixture separates into gases A and B (see Section 14.2.2).

$$P = \left(\tfrac{1}{2}\right)^{N_A} \cdot \left(\tfrac{1}{2}\right)^{N_B} = \left(\tfrac{1}{2}\right)^{N_A + N_B}$$

Then we obtain for the change of the entropy

$$\Delta S = k \ln \frac{\Omega_2}{\Omega_1} = k \ln \frac{1}{P} = k(N_A + N_B) \cdot \ln 2$$

14.5.3 Entropy Increase in an Irreversible Process in an Isolated System

In an isolated system, changes of state occur in which the system transforms from an initial state 1 to a final state 2 that has a greater number of configurations. We find

$$\boxed{\Omega_2 > \Omega_1 \text{ (isolated system)}} \qquad (14.35)$$

The probability that an isolated system transforms from state 2 to state 1 is thus negligibly small for a system with a large number of particles. An isolated system tends to go over into a state with the maximum possible value of Ω. The system is then said to be in an equilibrium state. Thermal equilibrium states differ from static equilibrium states in mechanical systems in that molecular fluctuations still occur, but they lead to no overall macroscopic process (dynamic equilibrium). From equations (14.28) and (14.35) we have

$$\boxed{\Delta S \geq 0 \text{ (isolated system)}} \qquad (14.36)$$

The equal sign is valid for the limiting case of a reversible process.

Law of Entropy Increase. *In an isolated system the entropy cannot decrease*

This law is in contrast to the law of conservation of energy:

$$\boxed{\Delta U = 0 \text{ (isolated system)}} \qquad (14.37)$$

Both expressions are completely general, so long as an isolated system is chosen.

14.5.4 Entropy of Subsystems

The entropy S has an interesting property:

> *The entropy of a system with a large number of molecules is the sum of the entropies of its parts.*

Distinguishable particles

We consider two subsystems separated by a wall (Figure 14.11). We assume that the particles in subsystem I are distinguishable from the particles in subsystem II (e.g., the molecules in a solid). So for each configuration in subsystem I there are Ω_{II} configurations for subsystem II. Thus the total system has

$$\Omega = \Omega_I \cdot \Omega_{II} \tag{14.38}$$

configurations. According to equation (14.20), we have

$$S_I = k \cdot \ln \Omega_I \quad \text{and} \quad S_{II} = k \cdot \ln \Omega_{II} \tag{14.39}$$

and we find

$$\boxed{S = S_I + S_{II}} \tag{14.40}$$

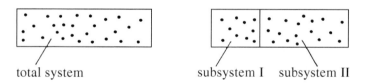

Figure 14.11 Entropy of subsystems: $S_{\text{total}} = S_I + S_{II}$.

Indistinguishable particles

Now we ask if equation (14.40) can also be applied to the case when the particles in subsystem I are indistinguishable from the particles in subsystem II (e.g., the molecules in an atomic gas). This is a more complicated situation.

We consider an atomic gas (N particles) in a container with volume V; we divide the container into two subsystems, each with volume $\frac{1}{2}V$ (so $\frac{1}{2}N$ particles are in each subsystem), by inserting a wall. This produces a completely new situation: the spacing of the energy levels in the smaller partial containers is larger than the spacing of the energy levels in the whole container. This means that the number of configurations in both subsystems together is smaller than that of the original system. However, it can be shown that equation (14.40) still holds even in this case, provided that N is large enough (see Box 14.2).

14.5 ENTROPY CHANGE ΔS

Box 14.2 – Indistinguishable Particles: Change of ΔS when Dividing a System into Subsystems

To elucidate the problem we take the simple case of $N = 2$ indistinguishable particles which are distributed over 4 quantum states. Here we have $\Omega_1 = 10$ configurations (Figure 14.12(a), configurations 1–10). After division into two equal subsystems I and II, we have $\Omega_2 = 4$ (Figure 14.12(b), configurations 1–4).

In general, we consider g quantum states available to two particles. Then we have g configurations with doubly occupied quantum states and $\frac{1}{2}(g^2 - g)$ configurations with singly occupied quantum states; so we have a total of

$$\Omega_1 = g + \tfrac{1}{2}(g^2 - g) = \tfrac{1}{2}g(g+1)$$

indistinguishable configurations (e.g., $\Omega_1 = \frac{1}{2}4(4+1) = 10$ for $g = 4$). After division into the subsystems we have

$$\Omega_2 = \Omega_I \cdot \Omega_{II} = \left(\frac{g}{2}\right) \cdot \left(\frac{g}{2}\right) = \left(\frac{g}{2}\right)^2$$

where Ω_I and Ω_{II} are the numbers of configurations of the subsystems. Then the ratio

$$\frac{\Omega_2}{\Omega_1} = \frac{\tfrac{1}{4}g^2}{\tfrac{1}{2}g(g+1)} = \frac{g}{2(g+1)}$$

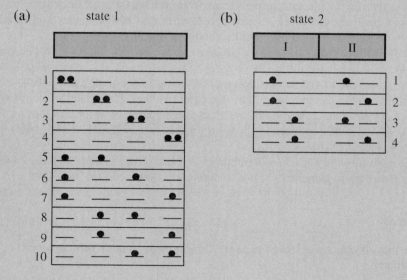

Figure 14.12 Illustration of the configurations for the case $N = 2$ particles and $g = 4$ quantum states. (a) Four quantum states available to both particles. (b) Only two quantum states available to each particle.

Continued on page 486

> *Continued from page 485*
>
> cannot become greater than 0.5 with increasing number of quantum states. That means that the number of configurations in the two partial systems is much smaller than the number of configurations in the whole system.
>
> However, if we compare the logarithms of the numbers of configurations
>
> $$\frac{\ln \Omega_2}{\ln \Omega_1} = \frac{\ln\left(\frac{1}{4}g^2\right)}{\ln\left(\frac{1}{2}g(g+1)\right)} = \frac{\ln \frac{1}{4} + 2\ln g}{\ln \frac{1}{2} + \ln g + \ln(g+1)}$$
>
> we see that this ratio converges to 1 at large g. Thus for a large number of quantum states we have $\ln \Omega_1 = \ln \Omega_2 = \ln \Omega_I \Omega_{II} = \ln \Omega_I + \ln \Omega_{II}$ and $S_1 = S_I + S_{II}$.

14.5.5 Entropy Change in Non-isolated Systems

Figures 14.1 and 14.2 exemplify the condition (14.36)

$$\Delta S \geq 0 \text{ (isolated system)}$$

In Figure 14.1(a) the system is isolated.

However, in Figures 14.1(b)–(d) and 14.2 the system is not isolated from the temperature bath. Then the condition (14.36) cannot be applied to the entropy change in the system: the entropy of the system can decrease on the expense of an entropy increase of the temperature bath. However, system and temperature bath together constitute a new system which is isolated and thus fulfills condition (14.36).

Figure 14.2 illustrates, for a reversible process, the entropy increase in one part of an isolated system and it illustrates the entropy decrease in another part. In process a the molecules, distributed over the gas volume, condense to a liquid: the number of configurations in the originally chosen system decreases. However, the heat of condensation is transferred to the temperature bath, and thus the number of configurations in the bath increases. The sum of configurations (of original system and temperature bath) and thus the entropy remain unchanged.

In process b the number of configurations in the original system increases and that in the temperature bath decreases (heat of evaporation flows into original system).

Problems

Problem 14.1 – Increase in the Number of Configurations with Temperature Equilibration

Figure 14.13 illustrates a simple example of the process of temperature equilibration. Calculate the probability P that, by chance, the molecules are distributed as in State 1.

Solution. As illustrated in Figure 14.13, only two energy levels are available to each molecule participating in the process. In the cold body, all four molecules occupy the lower energy level. In the hot body, two molecules occupy the lower energy level and

PROBLEM 14.1

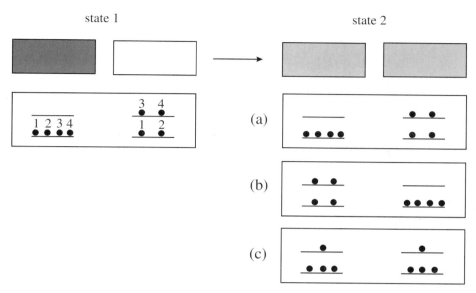

Figure 14.13 Description of temperature equilibration in a model system with only two quantum states available to each molecule. State 1: cold body (left) and hot body (right) thermally isolated. State 2: both bodies in heat contact, (a), (b), and (c) are different macro states: (a) 6 configurations; (b) 6 configurations; (c) 16 configurations.

two molecules occupy the upper energy level. Then according to equation (12.48) with $N = 4$ and $N_1 = 4$, we have

$$\Omega_{\text{cold}} = \frac{4!}{4!} = 1$$

Correspondingly, in the hot body we have $N = 4$, $N_1 = 2$, and $N_2 = 2$, so we obtain

$$\Omega_{\text{hot}} = \frac{4!}{2!2!} = 6$$

A first configuration of the hot body is: particles 1 and 2 in the lower level, particles 3 and 4 in the upper level (12|34). The five remaining configurations are (13|24), (14|23), (23|14), (24|13), and (34|12).

The number Ω_1 of configurations in state 1 (cold/hot) is the product of Ω_{cold} and Ω_{hot}:

$$\Omega_1 = \Omega_{\text{cold}} \cdot \Omega_{\text{hot}} = 6$$

We bring the bodies in contact and wait for thermal equilibration (state 2). The internal energy U must remain constant; then there are the following possibilities for state 2:

(a) left body: $N_1 = 4$, $N_2 = 0$; right body: $N_1 = 2$, $N_2 = 2$
(b) left body: $N_1 = 2$, $N_2 = 2$; right body: $N_1 = 4$, $N_2 = 0$
(c) left and right body: $N_1 = 3$, $N_2 = 1$

In all cases we have a total of six molecules in level 1 and two molecules in level 2. Then

$$\Omega_2 = \frac{8!}{6! \cdot 2!} = 28$$

The same value is obtained by adding the numbers of configurations of macro states (a) (6 configurations), (b) (6 configurations), and (c) (16 configurations). So Ω_2 is greater than Ω_1.

In our numerical example, the probability P that after equilibration the system will be found accidentally in state 1 is

$$P = \frac{6}{28} = 0.21$$

still relatively large.

If we increase the number of molecules, then P will decrease drastically. For example, for $N = 40$,

$$\Omega_1 = \frac{40!}{20! \cdot 20!} = 1.38 \times 10^{11}, \quad \Omega_2 = \frac{80!}{60! \cdot 20!} = 3.6 \times 10^{18}$$

so that we obtain

$$P = \frac{1.38 \times 10^{11}}{3.6 \times 10^{18}} = 4 \times 10^{-8}$$

For arbitrary N we obtain

$$\Omega_1 = \frac{N!}{\left(\frac{N}{2}\right)! \cdot \left(\frac{N}{2}\right)!} \quad \text{and} \quad \Omega_2 = \frac{(2N)!}{\left(\frac{3N}{2}\right)! \cdot \left(\frac{N}{2}\right)!}$$

If we approximate $N!$ by Stirling's formula (see Box 12.4) $N! \approx N^N \cdot e^N$ and set $N = 10^{23}$, then we can calculate

$$P = 10^{-10^{22}}$$

So we get a similarly small value of P as for the expansion of a gas in vacuum as considered in Section 14.2.2.

Problem 14.2 – Mixing Entropy

Calculate, by using the Sackur–Tetrode equation, the entropy change of a system of two separated gases, when the gases are mixed (see Figure 14.1(b)).

Solution. From equation (14.21) we obtain for the entropy of a system of two different gases A and B, each of them separated in volume V,

$$S_1 = N(A)k \cdot \ln\left(C\frac{V}{N(A)}\right) + N(B)k \cdot \ln\left(C\frac{V}{N(B)}\right)$$

with

$$C = e^{5/2}(2\pi)^{3/2}\frac{(km_A)^{3/2}}{h^3}T^{3/2}$$

PROBLEM 14.2

After mixing, both gases occupy a volume $2V$, and thus the entropy is

$$S_2 = N(A)k \cdot \ln\left(C\frac{2V}{N(A)}\right) + N(B)k \cdot \ln\left(C\frac{2V}{N(B)}\right)$$

For the entropy change we obtain

$$\Delta S = S_2 - S_1 = N(A)k \cdot \ln 2 + N(B)k \cdot \ln 2 = (N(A) + N(B))k \cdot \ln 2$$

This means that the entropy increases after mixing.

If the molecules A and B cannot be distinguished (i.e., A and B are identical particles, $N(A)$ particles in upper volume in Figure 14.1(b), $N(A)$ particles in lower volume) then

$$S_1 = 2N(A)k \cdot \ln\left(C\frac{V}{N(A)}\right), \quad S_2 = 2N(A)k \cdot \ln\left(C\frac{2V}{2N(A)}\right)$$

and $\Delta S = 0$. This means the entropy remains unchanged.

Note that a tiny difference between the two kinds of gas molecules (e.g., the tiny difference in mass between the isotopes $^{235}UF_6$ and $^{237}UF_6$ molecules) is sufficient to generate an entropy effect on mixing. It does not matter how much the gas molecules differ from each other.

Foundation 14.1 – Entropy for Rotation and Vibration

In the case of rotation and vibration, the centers of gravity of the molecules remain unchanged. Thus we can attribute site numbers to any molecule. Rotators and oscillators are distinguishable, in contrast to the description of the translational movement of the molecules. Then, according to Section 12.6, we have

$$\omega = \frac{N!}{N_1!N_2!\ldots}$$

and

$$\Omega = \omega_1 + \omega_2 + \omega_3 + \cdots$$

where $\omega_1, \omega_2, \omega_3 \ldots$ are the numbers of configurations belonging to different macro states. We compare numerical data obtained for Ω and $\omega_{\max}$, where $\omega_{\max}$ corresponds to the macro state with the maximum number of configurations.

As an example we refer to the situation described in Section 12.6 (Table 12.1), where $N = 6$ or $N = 100$ particles are distributed over three energy levels; we extend the consideration to $N = 1000$ and $N = 10\,000$:

N	Ω	$\ln \Omega$	$\omega_{\max}$	$\ln \omega_{\max}$	$\omega_{\max}/\Omega$	$\ln \omega_{\max}/\ln \Omega$
6	90	4.5	60	4.09	0.67	0.90
100	69×10^{32}	77.9	15×10^{32}	76.4	0.22	0.98
1000	16×10^{349}	806.5	1×10^{349}	803.7	0.063	0.997
10 000	164×10^{3503}	8072.5	3.6×10^{3503}	8068.7	0.022	0.9995

Apparently, the ratio $\omega_{\max}/\Omega$ becomes smaller with increasing N; this means that there are more and more distributions close to the distribution with maximum ω. In contrast the ratio $\ln \omega_{\max}/\ln \Omega$ converges to 1 with increasing N. Consequently, at high enough particle number we can set

$$\ln \Omega = \ln \omega_{\max} \quad \text{and} \quad S = k \cdot \ln \omega_{\max}$$

We calculate $\ln \omega_{\max}$ by approximating

$$\ln \omega_{\max} = \ln N! - \ln N_1! - \ln N_2! - \cdots$$

by Stirling's formula:

$$\ln \omega_{\max} = N \cdot \ln N - N - N_1 \cdot \ln N_1 + N_1 - N_2 \cdot \ln N_2 + N_2 - \cdots$$

$$= N \cdot \ln N - \sum_{i=1}^{\infty} N_i \ln N_i$$

Continued on page 491

FOUNDATION 14.1

Continued from page 490

We express the particle numbers N_i by the Boltzmann distribution

$$N_i = \frac{N}{Z} \cdot e^{-E_i/(kT)}$$

Then we obtain

$$\ln \omega_{\max} = N \cdot \ln N - \frac{N}{Z} \sum_{i=1}^{\infty} \left\{ e^{-E_i/kT} \cdot \left(\ln \frac{N}{Z} - \frac{E_i}{kT} \right) \right\}$$

$$= N \cdot \ln N - \frac{N}{Z} \cdot \left(\sum_{i=1}^{\infty} e^{-E_i/(kT)} \right) \cdot \ln \frac{N}{Z} + \frac{1}{kT} \sum_{i=1}^{\infty} \left[E_i \frac{N}{Z} e^{-E_i/(kT)} \right]$$

The first sum is identical with the partition function Z and the second sum is identical with the internal energy U. In this way we obtain

$$\ln \omega_{\max} = N \cdot \ln N - N \cdot \ln \frac{N}{Z} + \frac{U}{kT} = N \cdot \ln Z + \frac{U}{kT}$$

Then

$$S = k \cdot \ln \omega_{\max} = Nk \cdot \ln Z + \frac{U}{T}$$

According to Foundations 12.1 and 12.2 the partition functions for an oscillator and a rotator are

$$Z_{\text{vib}} = \frac{e^{-h\nu_0/(2kT)}}{1 - e^{-h\nu_0/(kT)}} \quad \text{and} \quad Z_{\text{rot}} = \frac{kT}{B}$$

The corresponding internal energies are

$$U_{\text{vib}} = Nh\nu_0 \cdot \left(\frac{1}{2} + \frac{1}{e^{h\nu_0/kT} - 1} \right), \quad U_{\text{rot}} = NkT$$

Then for the entropies we obtain

$$S_{\text{vib}} = Nk \cdot \left[-\frac{h\nu_0}{2kT} - \ln(1 - e^{-h\nu_0/(kT)}) \right] + \frac{Nh\nu_0}{2T} + \frac{1}{T} \frac{Nh\nu_0}{e^{h\nu_0/(kT)} - 1}$$

$$= Nk \cdot \left[-\ln(1 - e^{-h\nu_0/(kT)}) + \frac{h\nu_0}{kT} \frac{1}{e^{h\nu_0/(kT)} - 1} \right]$$

$$S_{\text{rot}} = Nk \cdot \ln \frac{kT}{B} + \frac{NkT}{T} = Nk \cdot \left[\ln \frac{kT}{B} + 1 \right]$$

Note that the expression for S_{rot} is valid only for high enough temperature.

15 Entropy S and Heat q_{rev}

When heating a body, the number of configurations Ω of the system (and thus the entropy $S = k \cdot \ln \Omega$ of the system) increases. Can we find a relationship between the entropy change $\Delta S_{a \to b}$ of the system when bringing it from state a to state b and the heat flowing from the surroundings into the system during this process? Such a relation would be useful since calorimetric measurements are easy to perform whereas counting the number of configurations is only possible in simple cases. We show that the entropy change $\Delta S_{a \to b}$ is related to the heat flowing into the system when bringing it on a reversible path from state a to state b:

$$\Delta S_{a \to b} = \sum_{a}^{b} \frac{q_{\text{rev},i}}{T_i}$$

where $q_{\text{rev},i}$ is the heat transferred in the i-th step, and T_i is the corresponding temperature.

First, we consider simple cases to demonstrate this relation between the entropy $S = k \cdot \ln \Omega$ and heat. Then we show, generally, that

$$\sum_{a}^{b} \frac{q_{\text{rev},i}}{T_i}$$

has the properties of $\Delta S_{a \to b}$ and thus defines $\Delta S_{a \to b}$. S is a state variable, because it has definite values for given states a and b of the system. Furthermore, we show that

$$\sum_{a}^{b} \frac{q_{\text{rev},i}}{T_i} \geq 0$$

for any process in an isolated system.

We demonstrate this by referring to the experimental fact that heat cannot flow from a temperature bath of low temperature to a temperature bath of high temperature without external interaction.

An interesting example is given illustrating the relationship between the number Ω of configurations and heat in an atomic gas (argon), where Ω can be easily calculated using the astonishing fact that the translation of the molecules in a gas is quantized. The value of S thus obtained agrees with the value obtained by summing up $q_{\mathrm{rev},i}/T_i$ when slowly heating argon, starting with the solid at $T = 0$ and ending up with the gas at the given temperature. This agreement is fascinating, showing the importance of quantum theory for an understanding of the macroscopic properties of matter.

15.1 Heat and Change of Entropy in Processes with Ideal Gases

We consider an irreversible process from an initial state 1 to a final state 2 of a system. On the one hand, we look at the change in entropy calculated from the number of configurations Ω. On the other hand, we consider a reversible path leading from state 1 to state 2 and look at the heat flow from the temperature bath into the system. Our examples are the expansion of a gas and the temperature equalization.

15.1.1 Expansion at Constant Temperature

Entropy increase

We consider the isothermal expansion of an ideal atomic gas (Figure 15.1(a)). The change of entropy in this process is, according to eqn. (14.28),

$$\Delta S = k \cdot \ln \frac{\Omega_2}{\Omega_1} \tag{15.1}$$

where Ω_1 and Ω_2 are the numbers of configurations before and after the expansion. ΔS can be expressed by equation (14.29)

$$\Delta S = Nk \cdot \ln \frac{V_2}{V_1} = n\boldsymbol{R} \cdot \ln \frac{V_2}{V_1} \tag{15.2}$$

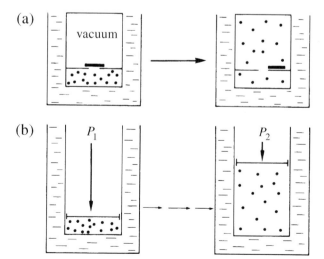

Figure 15.1 Isothermal expansion of an ideal gas. (a) Irreversible (expansion into a vacuum); (b) reversible (the pressure increases in small steps from the starting value P_1 to the final value P_2).

Heat flow q_{rev} from temperature bath into system

Now we ask if there are any relations between ΔS and calorimetric quantities. If the isothermal expansion is carried out reversibly (Figure 15.1(b)), then the heat supplied to the system is, according to equation (14.4),

$$q_{rev} = nRT \cdot \ln \frac{V_2}{V_1} \qquad (15.3)$$

From equations (15.2) and (15.3) we conclude that the entropy change ΔS can be expressed by

$$\boxed{\Delta S = \frac{q_{rev}}{T} \quad \text{(isothermal)}} \qquad (15.4)$$

if the change of state occurs at constant temperature.

15.1.2 Thermal Equilibration

Entropy increase at constant volume

In Section 14.5 we considered the temperature equilibration between gas molecules in a upper container at high temperature T_2 and a lower container at low temperature T_1 (see Figures 14.1(a) and 15.2). The change ΔS of the entropy in both containers after temperature equilibration is

$$\Delta S = \Delta S_{\text{upper container}} + \Delta S_{\text{lower container}} \qquad (15.5)$$

$\Delta S_{\text{upper container}}$ is the entropy change by cooling the gas in the upper container from T_2 to the equilibrium temperature

$$T = \frac{T_1 + T_2}{2} \qquad (15.6)$$

We obtained for an atomic gas (see Section 14.5.1)

$$\Delta S_{\text{upper container}} = \frac{3}{2} Nk \cdot \ln \frac{T}{T_2} \qquad (15.7)$$

Accordingly, the entropy change $\Delta S_{\text{lower container}}$ by heating the gas in the lower container is

$$\Delta S_{\text{lower container}} = \frac{3}{2} Nk \cdot \ln \frac{T}{T_1} \qquad (15.8)$$

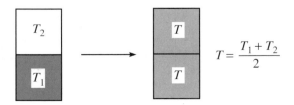

Figure 15.2 Temperature equilibration between two bodies (isolated system).

15.1 HEAT AND CHANGE OF ENTROPY IN PROCESSES WITH IDEAL GASES

Thus for ΔS we obtain

$$\Delta S = \frac{3}{2} Nk \cdot \ln \frac{T^2}{T_1 T_2} \tag{15.9}$$

The entropy change ΔS is positive, i.e., the entropy of the system increases.

Reversible heat q_{rev} for the process at constant volume

For heating the gas in the lower container reversibly we contact the gas in small steps with a series of temperature baths. Thus in each step the temperature of the bath is only slightly different from the temperature of the gas (Figure 15.3).

In the i-th step the heat

$$q_{rev,i} = C_V \cdot \Delta T \tag{15.10}$$

is supplied to the gas. Then

$$\frac{q_{rev,i}}{T_i} = \frac{C_V \cdot \Delta T}{T_i} \tag{15.11}$$

For the total process of heating, we sum up the contributions in each step:

$$\sum_{T_1}^{T} \frac{q_{rev,i}}{T_i} = \sum_{T_1}^{T} \frac{C_V \cdot \Delta T}{T_i} \tag{15.12}$$

To evaluate this expression we replace the summation by an integration

$$\int_{T_1}^{T} \frac{dq_{rev}}{T} = \int_{T_1}^{T} \frac{C_V \cdot dT}{T} \tag{15.13}$$

The heat capacity C_V for an atomic gas is

$$C_V = \tfrac{3}{2} Nk \tag{15.14}$$

Then we obtain

$$\sum_{T_1}^{T} \frac{q_{rev,i}}{T_i} = \frac{3}{2} Nk \int_{T_1}^{T} \frac{dT}{T} = \frac{3}{2} Nk \cdot \ln \frac{T}{T_1} \tag{15.15}$$

Comparison with equation (15.8) gives

$$\Delta S_{\text{lower container}} = \sum_{T_1}^{T} \frac{q_{rev,i}}{T_i} \tag{15.16}$$

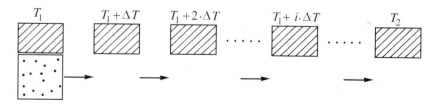

Figure 15.3 Reversible heating of an ideal gas at constant volume in small steps from the starting temperature T_1 to the final temperature T_2. The figure shows the system connected with the starting temperature T_1.

Correspondingly, for the upper container we obtain

$$\sum_{T_2}^{T} \frac{q_{\text{rev},i}}{T_i} = \frac{3}{2}Nk \int_{T_2}^{T} \frac{dT}{T} = \frac{3}{2}Nk \cdot \ln \frac{T}{T_2}$$

and, by comparison with equation (15.7),

$$\Delta S_{\text{upper container}} = \sum_{T_2}^{T} \frac{q_{\text{rev},i}}{T_i} \qquad (15.17)$$

In conclusion we can state:

$$\Delta S = \sum_{T_1}^{T} \frac{q_{\text{rev},i}}{T_i} + \sum_{T_2}^{T} \frac{q_{\text{rev},i}}{T_i} \quad \text{or} \quad \Delta S = \sum_{i} \frac{q_{\text{rev},i}}{T_i} \qquad (15.18)$$

This relation is valid for any change of state of an atomic gas, because we proved its validity for a volume change at constant temperature and for a temperature change at constant volume; any change of state can be realized by a combination of these two changes.

Further examples are discussed in Boxes 15.1 and 15.2.

Box 15.1 – ΔS for the evaporation of a Liquid

According to Figure 15.4, for evaporation of a liquid we obtain

$$w_{\text{rev}} = -P_{\text{vap}} \cdot (V_{\text{vap}} - V_{\text{liq}})$$

and

$$q_{\text{rev}} = \Delta U - w_{\text{rev}} = \Delta U + P_{\text{vap}} \cdot (V_{\text{vap}} - V_{\text{liq}})$$

(P_{vap}, V_{vap}: pressure and volume of the vapor; V_{liq}: volume of the liquid). Because of $V_{\text{vap}} \gg V_{\text{liq}}$ and $P_{\text{vap}} \cdot V_{\text{vap}} = nRT$ (vapor treated as an ideal gas), for ΔS we obtain

$$\Delta S = \frac{q_{\text{rev}}}{T} = \frac{\Delta U}{T} + nR$$

In this example, the change ΔU of the internal energy is a result of the strong attraction forces acting in liquids, but not in ideal gases. Energy has to be supplied to separate the molecules during evaporation. ΔU is then positive, and therefore q_{rev} and ΔS are positive. On the other hand, the heat q_{rev} is taken from the temperature bath, thus

$$(\Delta S)_{\text{temperature bath}} = -\frac{q_{\text{rev}}}{T}$$

Continued on page 497

Continued from page 496

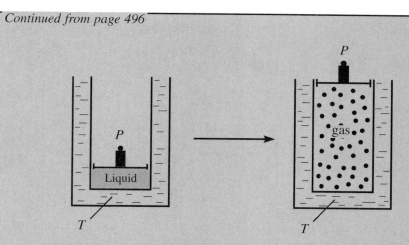

Figure 15.4 Evaporation of a liquid at constant pressure.

The entropy of the bath decreases in the same manner as the entropy of the system increases. When we consider the system and the temperature bath as a new system, which is isolated from the surroundings, then

$$\Delta S_{\text{isolated system}} = \Delta S_{\text{system}} + \Delta S_{\text{temperature bath}} = 0$$

in accordance with equation (14.36) for a reversible process.

Box 15.2 – ΔS for the Expansion of a Gas into Vacuum

When an ideal gas in an isolated container is expanded into vacuum (Figure 15.5), no work is done by the gas in this process, and the heat q is zero. Then, according to the law of energy conservation, its internal energy U does not change and its temperature T remains constant (after temperature equilibration between both partial volumes). To calculate ΔS we need the heat q_{rev} reversibly supplied to the gas for the same change of state. We imagine the change of state carried out by a reversible isothermal expansion of the gas through contact with a temperature bath (Figure 14.3). Then, according to equation (14.29) we have

$$\Delta S = \frac{q_{\text{rev}}}{T} = nR \cdot \ln \frac{V_2}{V_1}$$

Continued on page 498

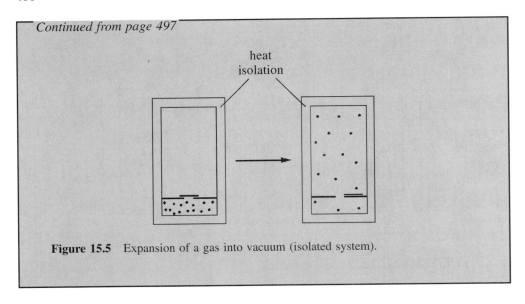

Figure 15.5 Expansion of a gas into vacuum (isolated system).

15.1.3 Cyclic Processes and Processes in Isolated Systems

For any cyclic process, according to Section 14.5, we obtain

$$\Delta S = 0 \quad \text{(cyclic process)} \tag{15.19}$$

because S is a state variable. Therefore for a cyclic process with an atomic gas we obtain

$$\sum_i \frac{q_{\text{rev},i}}{T_i} = 0 \quad \text{(cyclic process)} \tag{15.20}$$

Correspondingly, for a process in an isolated system we obtain, according to equation (14.36)

$$\Delta S \geq 0 \quad \text{(isolated system)} \tag{15.21}$$

and therefore

$$\boxed{\sum_i \frac{q_{\text{rev},i}}{T_i} \geq 0 \quad \text{(process in an isolated system)}} \tag{15.22}$$

15.1.4 Reversible Heat Engine with Ideal Gases (Carnot Cycle)

We consider the simplest case of a reversible cyclic process, namely, a cyclic process with an ideal gas connected to two temperature baths (*Carnot cycle*). The gas is first in contact with bath T_2 and is expanded (isothermal expansion) from volume V_a to volume V_b (Figure 15.6). In this process, heat $q_{\text{rev},2}$ is transferred from the bath to the gas. Then the gas is removed from the bath and expanded adiabatically (no heat exchange with the bath) until it is cooled down to temperature T_1 (volume V_c). Then it is contacted with bath T_1 and compressed from volume V_c to volume V_d. Heat $-q_{\text{rev},1}$ is transferred to bath T_1. Finally, the gas is removed from the bath and further compressed adiabatically (no

15.1 HEAT AND CHANGE OF ENTROPY IN PROCESSES WITH IDEAL GASES

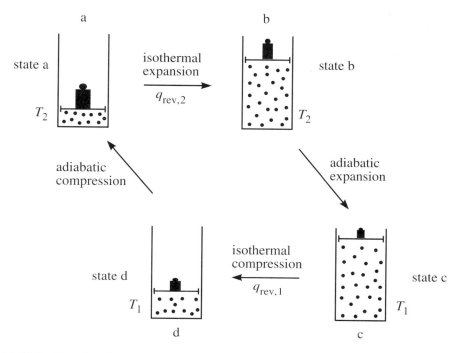

Figure 15.6 Carnot cycle.

heat exchange with the bath) until it is heated up to temperature T_2 and has reached its initial state (volume V_a). Then according to equation (15.20), for a cyclic process with an ideal gas we obtain

$$\frac{q_{rev,1}}{T_1} + \frac{q_{rev,2}}{T_2} = 0 \qquad (15.23)$$

Furthermore, $\Delta U = 0$ for a cyclic process and because of the law of energy conservation the work $-w_{rev}$ done by the gas in one cycle (Figure 15.7(a)) is

$$-w_{rev} = q_{rev,2} - (-q_{rev,1}) \qquad (15.24)$$

Then the efficiency of the Carnot engine (ratio of work done by the gas and heat removed from hot bath) is

$$\eta = \frac{-w_{rev}}{q_{rev,2}} = \frac{q_{rev,2} + q_{rev,1}}{q_{rev,2}} = 1 + \frac{q_{rev,1}}{q_{rev,2}} \qquad (15.25)$$

Because of equation (15.23)

$$\frac{q_{rev,1}}{q_{rev,2}} = -\frac{T_1}{T_2}$$

we obtain for the efficiency η

$$\boxed{\eta = 1 - \frac{T_1}{T_2}} \qquad (15.26)$$

Example – efficiency of heat engine and heat pump

For $T_2 = 373$ K and $T_1 = 273$ K we obtain

$$\eta = 1 - \frac{T_1}{T_2} = 0.27$$

This means that only 27% of the energy supplied by the warmer temperature bath can be transferred into mechanical work. On the other hand, if the cyclic process is carried out in the reverse direction, mechanical work is supplied to the system, and heat is transferred from the colder bath into the warmer bath. Then only 27% of the transferred heat is taken from the mechanical energy, and 73% is taken from the colder bath (principle of a heat pump, Figure 15.7(b)).

In order to perform a cyclic process, volume V_d cannot be arbitrary. Condition (15.23) leads to

$$q_{\text{rev},1} = -\frac{T_1}{T_2} \cdot q_{\text{rev},2} \tag{15.27}$$

and with

$$q_{\text{rev},1} = nRT_1 \cdot \ln \frac{V_d}{V_c} \quad q_{\text{rev},2} = nRT_2 \cdot \ln \frac{V_b}{V_a} \tag{15.28}$$

we obtain from equation (15.27)

$$nRT_1 \cdot \ln \frac{V_d}{V_c} = -\frac{T_1}{T_2} \cdot nRT_2 \cdot \ln \frac{V_b}{V_a} \tag{15.29}$$

Then it follows that

$$V_d = V_c \cdot \frac{V_a}{V_b} \tag{15.30}$$

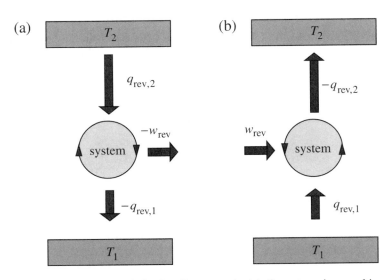

Figure 15.7 Flow of heat and work in the Carnot cycle (a) Carnot engine working as a heat engine; (b) Carnot engine working as a heat pump.

15.2 Heat and Change of Entropy in Arbitrary Processes

Can we generalize the remarkable relation (15.18)

$$\Delta S = \sum_i \frac{q_{\text{rev},i}}{T_i}$$

between ΔS and $\sum_i q_{\text{rev},i}/T_i$? We will see that this is indeed the case. This is most important, because this relation is of great practical importance.

However, we cannot prove this directly by counting the number of configurations Ω_1 and Ω_2 for any change of state, because the calculation of Ω is practically impossible for complex systems. Therefore, we proceed by showing that the quantity

$$\sum_i \frac{q_{\text{rev},i}}{T_i} \tag{15.31}$$

has the same properties as the entropy change:

$$\Delta S = 0 \quad \text{(cyclic process)}, \quad \Delta S \geq 0 \quad \text{(isolated system)}$$

Thus we have to prove proposals 1 and 2:

Proposal 1

The quantity

$$\sum_i \frac{q_{\text{rev},i}}{T_i}$$

is independent of the path leading from an initial state to a final state and therefore constitutes the change of a state variable. Thus

$$\sum_i \frac{q_{\text{rev},i}}{T_i} = 0 \quad \text{(cyclic process)} \tag{15.32}$$

Proposal 2

$$\sum_i \frac{q_{\text{rev},i}}{T_i} \geq 0 \quad \text{(isolated system)} \tag{15.33}$$

A general proof of proposals 1 and 2 is given in Box 15.3.

In conclusion we have shown that the relation

$$\Delta S = \sum_i \frac{q_{\text{rev},i}}{T_i} \tag{15.34}$$

is valid for an arbitrary process. Thus for heating a body at constant volume we have

$$\Delta S = \int_{T_1}^{T_2} \frac{C_V}{T} \cdot dT \quad \text{or} \quad \left(\frac{\partial S}{\partial T}\right)_V = \frac{C_V}{T} \tag{15.35}$$

and for heating at constant pressure we obtain

$$\Delta S = \int_{T_1}^{T_2} \frac{C_P}{T} \cdot dT \quad \text{or} \quad \left(\frac{\partial S}{\partial T}\right)_P = \frac{C_P}{T} \tag{15.36}$$

For a process at constant temperature T, such as the evaporation of a liquid, we have

$$\Delta S = \frac{q_{\text{rev}}}{T} = \frac{q_{\text{vap}}}{T} \tag{15.37}$$

where q_{vap} is the heat of evaporation. Correspondingly, this relation holds for any other phase transition (e.g., melting).

Box 15.3 – Proof of Proposals 1 and 2

Arbitrary Reversible Heat Engine (Two Temperature Baths Involved)

Proving proposal 1

How can we prove proposal 1? We consider an arbitrary reversible heat engine and show that its efficiency must be the same as the efficiency η of the Carnot machine. This means that the relation

$$\frac{q_{\text{rev},1}}{T_1} + \frac{q_{\text{rev},2}}{T_2} = 0 \quad \text{(cyclic process)}$$

must hold for any arbitrary reversible heat engine.

What is the efficiency η of an arbitrary reversible heat engine? We imagine, besides the Carnot engine, a second reversible heat engine (a fictional engine) with an efficiency

$$\eta^* = \frac{-w^*_{\text{rev}}}{q^*_{\text{rev},2}} > \eta$$

This means that the efficiency η^* is larger than the efficiency η for the Carnot engine. We couple this engine with a Carnot engine by using the Carnot engine as a heat pump.

The two engines are coupled such that the mechanical energy supplied by the fictional engine is used up by the Carnot engine to pump heat (Figure 15.8). Then the total work done on the surroundings in a cycle is zero.

However, because of $\eta^* > \eta$, more heat is pumped by the Carnot engine into the warmer bath than taken away from it by the fictional engine. All together, in a cycle (i.e., when both engines are in the same stage as in the beginning) heat is transferred from the colder bath to the warmer bath without any external work being done. Such a result, however, contradicts the statement that heat can flow only from a warmer to a colder body (see Sections 14.1.1 and 11.1.3).

Continued on page 503

15.2 HEAT AND CHANGE OF ENTROPY IN ARBITRARY PROCESSES

Continued from page 502

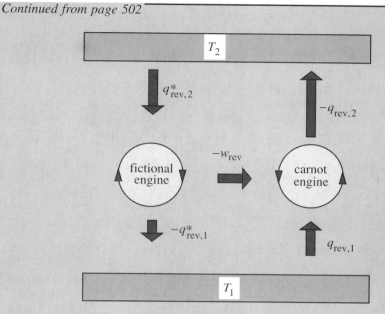

Figure 15.8 Carnot engine coupled to a reversibly working fictional engine with $\eta^* > \eta$.

What happens if $\eta^* < \eta$ and the fictional engine is working reversibly? Then again we can transfer heat from a colder body to a warmer body without supplying mechanical energy by using the fictional engine as heat pump and the Carnot machine as heat engine. In this case the fictional engine pumps more heat into the warmer bath than is removed by the Carnot engine.

From these considerations it follows that no reversibly working heat engine can exist with an efficiency different from the Carnot efficiency. In this way we have proved proposal 1 for a system connected to two temperature baths.

Arbitrary Heat Engine (Two Temperature Baths Involved)

Any practically realizable heat engine works nonreversibly. This means that such an engine yields less work than a Carnot engine when consuming the same amount of heat from bath T_2. Used as a heat pump, it needs more external mechanical energy to transfer a given amount of heat to bath T_2 than a Carnot machine (Figure 15.9). Thus, the relation

$$\eta_{\text{realistic}} < \eta$$

or

$$\frac{q_1}{T_1} + \frac{q_2}{T_2} < 0 \quad \text{(cyclic process, realistic heat engine)}$$

Continued on page 504

Continued from page 503

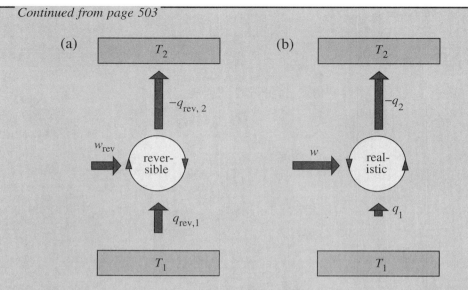

Figure 15.9 Comparison of a reversibly working heat pump with a realistic one ($\eta_{\text{realistic}} < \eta$).

must generally hold. The heat flowing into the system is less than in the reversible limit because more work is done on the system than in the reversible limit.

Arbitrary Process in an Isolated System (Two Temperature Baths Involved)

Proving proposal 2

How can we prove proposal 2? Let us consider the isolated system in Figure 15.10 consisting of a realistic heat engine, two temperature baths, and a device for exchanging mechanical energy between the engine and the surroundings. The baths are so large that the addition or removal of heat does not change their temperature. This process is totally independent of whether the engine itself works reversibly or not.

In a cycle heat $-q_2$ is pumped into bath T_2 and heat q_1 is removed from bath T_1. Thus

$$-q_2 = q_{\text{rev,bath2}} \quad \text{and} \quad -q_1 = q_{\text{rev,bath1}} \tag{15.38}$$

We are interested in the quantity

$$\sum_i \frac{q_{\text{rev},i}}{T_i}$$

for the isolated system consisting of temperature baths and heat engine when a cyclic process is carried out by the engine. The sum extends over engine, bath T_1, and bath T_2.

Continued on page 505

15.2 HEAT AND CHANGE OF ENTROPY IN ARBITRARY PROCESSES

Continued from page 504

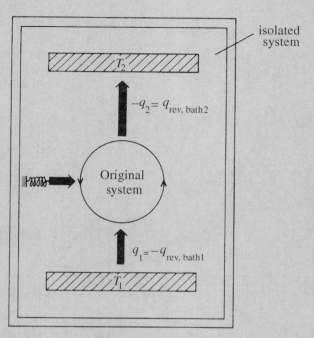

Figure 15.10 Heat engine and temperature baths T_1 and T_2 considered as an isolated system. Work is done on the original system (symbolized by a relaxing spring) and heat is transferred to the high temperature bath.

The contribution of the engine is zero (cyclic process); the contribution of the temperature baths is

$$-\frac{q_1}{T_1} \quad \text{for bath 1} \quad \text{and} \quad -\frac{q_2}{T_2} \quad \text{for bath 2}$$

Thus, with equation (15.36)

$$\sum_i \frac{q_{\text{rev},i}}{T_i} = \frac{q_{\text{rev,bath1}}}{T_1} + \frac{q_{\text{rev,bath2}}}{T_2} = -\frac{q_1}{T_1} - \frac{q_2}{T_2} \geq 0 \quad \text{(isolated system)}$$

where the equals sign corresponds to a reversible process. In this way we have proved proposal 2 for processes involving two temperature baths.

Arbitrary Process (Any Number of Temperature Baths Involved)

Our considerations can be easily extended to processes in which the changes of state occur at any chosen number of temperatures. We consider a cyclic process involving three temperatures (Figure 15.11). We imagine this process

Continued on page 506

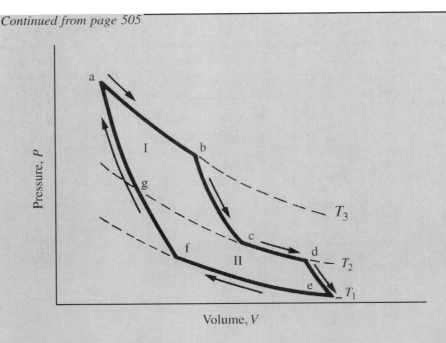

Figure 15.11 Cyclic process involving three temperatures.

as composed of the two partial processes I ($a \to b \to c \to g \to a$) and II ($g \to d \to e \to f \to g$). The sum of the heats supplied to the system in the partial isothermal steps is equal to that of the total system. The steps $a \to b \to c \to d$ and $e \to f$ occur in the partial processes as well as in the total process and the contributions of steps $c \to g$ (of I) and $g \to c$ (of II) cancel each other. Then for the partial processes we obtain

$$\frac{q_{\text{rev}}(a \to b)}{T_3} + \frac{q_{\text{rev}}(c \to g)}{T_2} = 0 \quad \text{(2 temperature baths involved)}$$

$$\frac{q_{\text{rev}}(e \to f)}{T_1} + \frac{q_{\text{rev}}(g \to d)}{T_2} = 0 \quad \text{(2 temperature baths involved)}.$$

From

$$q_{\text{rev}}(g \to c) + q_{\text{rev}}(c \to d) = q_{\text{rev}}(g \to d)$$

$$q_{\text{rev}}(g \to c) = -q_{\text{rev}}(c \to g)$$

it follows

$$q_{\text{rev}}(c \to g) + q_{\text{rev}}(g \to d) = q_{\text{rev}}(c \to d)$$

Continued on page 507

> Continued from page 506
>
> For the total process we find
>
> $$\frac{q_{\text{rev}}(a \to b)}{T_3} + \frac{q_{\text{rev}}(c \to d)}{T_2} + \frac{q_{\text{rev}}(e \to f)}{T_1} = 0$$
>
> (3 temperature baths involved)
>
> This consideration can be easily extended to cyclic processes with 4, 5, ... or any number of temperature baths. Then
>
> $$\sum_i \frac{q_{\text{rev},i}}{T_i} = 0 \quad \text{(cyclic process)}$$
>
> and
>
> $$\sum_i \frac{q_{\text{rev},i}}{T_i} \geq 0 \quad \text{(process in an isolated system)}$$
>
> Thus we have proved proposals 1 and 2.

15.3 Entropies of Substances

The relation

$$\Delta S \geq 0 \text{ (isolated system)}$$

is the fundamental condition for spontaneous chemical reaction, and therefore it is of great practical importance to know the entropies of reacting species in order to calculate entropy changes. In simple cases the entropy of a substance is obtained from quantum mechanics (see Section 14.4). Practically, this is not possible in all cases; however, entropies can be obtained as well by measuring the quantity

$$\sum_{T=0}^{T} \frac{q_{\text{rev},i}}{T_i}$$

In the following example this is demonstrated for argon. This case is particularly exciting because it allows us to compare the entropy obtained from quantum mechanics (Section 14.4) with the calorimetric value.

Example – entropy of argon from calorimetric data

We imagine solid Ar at absolute zero under a pressure of $P = 1$ bar. We heat the sample and measure the heat q_{rev} supplied in small steps. Then for the molar entropy at 298 K we obatain

$$S_{m,298} = \frac{1}{n} \sum_{T=0}^{298\,K} \frac{q_{rev,i}}{T_i} + S_{m,0}$$

where $S_{m,0}$ is the molar entropy at absolute zero at which argon is a solid. At this temperature all the atoms are in the vibrational ground state, thus $S_{m,0} = 0$.

We heat the sample at constant pressure $P = 1$ bar. First, we heat up to the melting point ($T_1 = 83.9$ K) and take into account the entropy increase during the melting process (molar melting heat $q_{m,1} = 1.18$ kJ mol^{-1}). Then we heat up to the boiling point ($T_2 = 87.3$ K) and consider the entropy increase during boiling (molar heat of evaporation $q_{m,2} = 6.52$ kJ mol^{-1}). Finally we heat up to $T_3 = 298.15$ K. Replacing the summation by an integration we obtain

$$S_{m,298} = \int_0^{T_1} \frac{C_{P,m} \cdot dT}{T} + \frac{q_{m,1}}{T_1} + \int_{T_1}^{T_2} \frac{C_{P,m} \cdot dT}{T} + \frac{q_{m,2}}{T_2} + \int_{T_2}^{T_3} \frac{C_{P,m} \cdot dT}{T}$$

In the last integral $C_{P,m}$ is the molar heat capacity of argon gas at constant pressure $\left(C_{P,m} = \frac{5}{2}R\right)$, thus

$$\int_{T_2}^{T_3} \frac{C_{P,m} \cdot dT}{T} = \frac{5}{2}R \int_{T_2}^{T_3} \frac{dT}{T} = \frac{5}{2}R \cdot \ln \frac{T_3}{T_2} = 25.5 \, J\,K^{-1}\,mol^{-1}$$

Furthermore,

$$\frac{q_{m,1}}{T_1} + \frac{q_{m,2}}{T_2} = 88.7 \, J\,K^{-1}\,mol^{-1}$$

The remaining two integrals are calculated numerically by plotting $C_{P,m}/T$ versus T (Table 15.1, Figure 15.12):

$$\int_0^{T_1} \frac{C_{P,m} \cdot dT}{T} + \int_{T_1}^{T_2} \frac{C_{P,m} \cdot dT}{T} = 39.9 \, J\,K^{-1}\,mol^{-1}$$

Then the result for the molar entropy of Ar at 298.15 K and $P = 1$ bar is $S_{m,298} = 154.1 \, J\,K^{-1}\,mol^{-1}$.

This value is in excellent agreement with the value $S_{m,298} = 154.8 \, J\,K^{-1}\,mol^{-1}$ calculated in Section 14.4 by using the Sackur–Tetrode equation. The agreement becomes even better by regarding that Ar behaves not strictly as an ideal gas at atmospheric pressure. For a comparison with the quantummechanical calculation Ar should be considered as ideal gas, and then the entropy is larger by $0.5 \, J\,K^{-1}\,mol^{-1}$ than in the real state, thus $S_{m,298} = 154.6 \, J\,K^{-1}\,mol^{-1}$.

Note that for the calculation based on the Sackur–Tetrode equation we used only data about the final state of the argon gas; we did not need any information about the behavior of Ar at low temperatures, neither do we need the enthalphy changes for the phase transitions. The distribution of the translational quantum states of the gas particles is decisive for the evaluation of the entropy. This result reveals that even in the case of the translational motion the quantum nature of matter is very important to explain the macroscopic behavior.

15.3 ENTROPIES OF SUBSTANCES

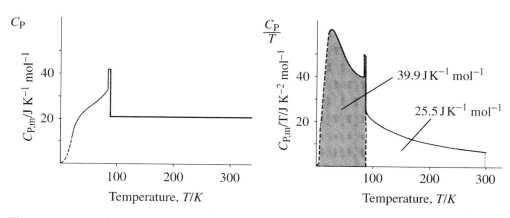

Figure 15.12 Evaluation of entropy by plotting C_P and C_P/T versus temperature (for data see Table 15.1).

Table 15.1 $C_{P,m}$ and $C_{P,m}/T$ for argon as a function of temperature

T/K	$C_{P,m}$/J K^{-1} mol^{-1}	$C_{P,m}/T$/J K^{-2} mol^{-1}	
20	11.76	0.588	Solid
40	22.09	0.552	Solid
60	26.59	0.443	Solid
80	32.13	0.401	Solid
83.85	33.26	0.397	Solid
83.85	42.05	0.501	Liquid
87.29	42.05	0.482	Liquid
87.29	20.79	0.238	Gas
100	20.79	0.208	Gas
300	20.79	0.069	Gas

Example – entropy of HD and HCl from calorimetric data

In analogy to the case of argon, the entropy of diatomic gases such as HD and HCl can be obtained from calorimetric data (Table 15.2, column 7). We find an excellent agreement with the quantummechanical values (Table 15.2, column 5).

Example – entropy of CO from calorimetric data

According to Table 15.2, the calorimetric value of S for CO is too small by about 4 J K^{-1} mol^{-1}. This discrepancy can be easily explained when discussing the dipole moment of CO. This dipole moment is very small, and, as a consequence, the molecules cannot arrange in the most favorable position when forming a crystal at the freezing point of 68 K: the energy difference between the two possible orientations is too small compared with kT. Once the crystal is formed, a CO molecule can no longer rotate, and the disordered state remains frozen.

What is the number of configurations for this state? When one CO molecule can exist in two configurations (CO or OC), then for N molecules $\Omega = 2^N$ (Figure 15.13). Then the molar entropy of CO at absolute zero is

$$S_\mathrm{m} = \frac{k}{n} \cdot \ln \Omega = \frac{k}{n} \cdot \ln(2^N) = \frac{Nk}{n} \cdot \ln 2 = R \cdot \ln 2 = 5.6\,\mathrm{J\,K^{-1}\,mol^{-1}}$$

in good agreement with the difference between calorimetric and quantummechanical value, $4.2\,\mathrm{J\,K^{-1}\,mol^{-1}}$ (Table 15.2). In contrast, the dipole moment of HCl is large, and in forming a crystal the orientations of all molecules are the same.

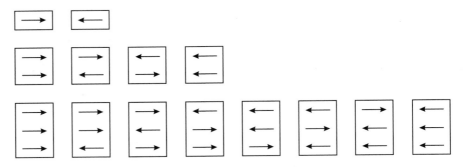

Figure 15.13 Distribution possibilities of CO molecules in a crystal. The directions of the arrows symbolize the orientations CO and OC, respectively. The number of configurations is $\Omega = 2$ for $N = 1$, $\Omega = 2^2$ for $N = 2$, and $\Omega = 2^3$ for $N = 3$.

Table 15.2 Comparison of entropies (in $\mathrm{J\,K^{-1}\,mol^{-1}}$) calculated quantummechanically (see Section 14.4) and obtained experimentally from $C_{P,m}$ data

	S_m from quantum mechanics				S_m from $C_{P,m}$ data		Difference
	trans	rot	vib	total	real	ideal	
He	126.0			126.0		124.7	−1.3
Ar	154.8			154.8	154.1	154.6	−0.1
Xe	169.5			169.5	169.2	169.8	+0.3
HD	122.5	20.8	2×10^{-6}	143.3	143.8	144.2	+0.9
HCl	153.6	33.1	7×10^{-5}	186.7	185.7	186.2	−0.5
CO	150.3	47.2	3×10^{-3}	197.5	192.4	193.3	−4.2

	$m/10^{-26}\,\mathrm{kg}$	$\mu/10^{-26}\,\mathrm{kg}$	$h\nu_0/10^{-20}\,\mathrm{J}$	d/pm
He	0.644			
Ar	6.63			
Xe	21.8			
HD	0.501	0.111	7.58	74.1
HCl	6.05	0.163	5.93	127.5
CO	4.65	1.139	4.31	112.8

The values correspond to room temperature (298.15 K) and atmospheric pressure (1.013 bar). The real gas correction takes account of the fact that the gases at atmospheric pressure do not behave completely ideally. m = molecular mass, μ = reduced mass, d = bond distance.

15.4 Thermodynamic Temperature Scale and Cooling

In Section 15.1.4 we used a heat engine as a heat pump to transfer heat from a cold bath to a hot bath. For a cycle of the heat pump according to equation (15.23) we have

$$\boxed{\frac{T_2}{T_1} = -\frac{q_{\text{rev},2}}{q_{\text{rev},1}}} \qquad (15.39)$$

This relation holds independently of the type of the working medium of the engine; that is, it is valid not only for changes of state of ideal gases. For this reason we can treat equation (15.39) as a "recipe" to measure the temperature. If we define temperature T_1 arbitrarily (e.g., the triple point of water), then we can determine T_2 by measuring $q_{\text{rev},1}$ and $q_{\text{rev},2}$ with any reversible heat engine (*thermodynamic temperature scale*).

For ideal gases, equation (15.45) reduces to the corresponding relation (11.27) of the gas thermometer. The "recipe" in equation (15.45) to measure the temperature can still be used in temperature regions where working with ideal gases is impossible.

Temperatures below 1 K can be reached by using a substance S containing atoms with unpaired electron spins (e.g., gadolinium sulfate) or with large nuclear spins (e.g., copper). Body S in a low temperature bath is exposed to a strong magnetic field orientating the spins. Heat is produced and transferred to the low temperature bath, similarly to what occurs in the isothermal compression of a gas where heat flows from the gas to the bath. Then S is isolated from the bath and the magnetic field is slowly reduced to zero (*adiabatic demagnetization*). S cools down similarly to what occurs in the adiabatic expansion of a gas (Problem 15.1). In the next step a strong magnetic field is applied to a section S_1 of S. The residual part of S constitutes the low temperature bath allowing essentially isothermal magnetization of S_1. Then S_1 is separated from the residual part of S and the magnetic field is reduced to zero, so further cooling of S_1 is reached. By using orientation of electron spins, temperatures of 2×10^{-3} K can be reached, and temperatures of 10^{-5} K by using nuclear spins. Thus temperatures lower than anywhere in nature are reached. For comparison, the lowest temperature in the universe is 3 K.

Lower and lower temperatures can be reached by repeating the process of cooling. To reach absolute zero, an infinite number of steps would be necessary.

15.5 Laws of Thermodynamics

In the previous sections we have introduced U and S as useful quantities describing the global properties of matter. Historically U and S were introduced without reference to the atomistic structure of matter when attempting to understand heat engines (classical thermodynamics). The bases of classical thermodynamics are the following statements introduced as postulates (laws of thermodynamics):

- *Zeroth Law*

    A state function exists, the temperature T, which has the same value in all systems which are in thermal equilibrium.

- *First Law*

    In an isolated system, $\Delta U = 0$ (law of conservation of energy).

- *Second Law*

    It is impossible to transfer heat from a colder to a warmer system without the occurrence of other simultaneous changes in the two systems or in the environment.
    In an isolated system

    $$\sum_i \frac{q_{\text{rev},i}}{T_i} \geq 0$$

    Since ΔS is defined as

    $$\Delta S = \sum_i \frac{q_{\text{rev},i}}{T_i}$$

    this statement is equivalent to

    $$\Delta S \geq 0 \quad \text{for an isolated system}$$

    (law of increasing entropy).

- *Third Law*

    The entropy of each pure substance in a perfect crystalline form is zero at the absolute zero point of temperature. In this context, "pure" means not only chemically pure, but also isotopically pure.
    A consequence of this statement is the fact that $T = 0$ cannot be reached. This statement is an alternative version of the Third Law.

In the present approach the laws of thermodynamics result from considerations on molecular assemblies; they are consequences, not postulates (statistical mechanics approach). We started by introducing temperature as a measure for the mean kinetic energy of molecules in a molecular aggregate at thermodynamic equilibrium. Then we considered the establishment of temperature equilibrium when two systems with different temperatures come into contact. Because of statistical reasons the molecules of the system with the higher temperature transfer energy to the molecules of the system with the lower temperature (Figure 14.1(a)). Temperature appears as a state variable (it has the same value for systems which are in thermal equilibrium), and heat appears as a form of energy. Thus, the statement that energy is being conserved (this is valid for mechanical systems) can immediately be used for processes in which mechanical energy is converted to heat, and vice versa.

We defined the entropy by

$$S = k \cdot \ln \Omega$$

The law of increasing entropy results immediately from the change in the number of configurations. Only later we found that the quantity

$$\sum_i \frac{q_{\text{rev},i}}{T_i}$$

has all properties of the change of entropy ΔS where S is defined as $k \cdot \ln \Omega$; thus $\sum_i q_{\text{rev},i}/T_i$ can be identified with ΔS.

From the fact that in a perfect crystal (no lattice disorder) only one realization is possible at zero point temperature ($\Omega = 1$), we have concluded that its entropy must be zero.

Problems

Problem 15.1 – Reversible Adiabatic Expansion

How much mechanical energy w must be transferred to the system when an ideal gas is expanded from volume V_1 (at temperature T_1) to volume V_2 (at temperature T_2) in an insulated cylinder (thus adiabatically) by a reversible path?

Solution. For an adiabatic change of state, $q = 0$ and $\Delta U = w$. If the temperature changes by dT, the change of the internal energy is

$$dU = C_V \cdot dT$$

The work dw supplied to the system is

$$dw = -P \cdot dV = -\frac{NkT}{V} \cdot dV$$

Then, because of $dU = dw$,

$$C_V \cdot dT = -\frac{NkT}{V} \cdot dV \quad \text{or} \quad \frac{dT}{T} = -\frac{Nk}{C_V} \cdot \frac{dV}{V}$$

and by integration we obtain

$$\ln \frac{T_1}{T_2} = -\frac{Nk}{C_V} \cdot \ln \frac{V_1}{V_2} = \ln \left(\frac{V_1}{V_2}\right)^{-Nk/C_V}$$

and

$$T_2 = T_1 \cdot \left(\frac{V_1}{V_2}\right)^{Nk/C_V} \quad \text{or} \quad \Delta T = T_2 - T_1 = T_1 \left[\left(\frac{V_1}{V_2}\right)^{Nk/C_V} - 1\right]$$

This equation relates the change in volume from V_1 to V_2 to the change in temperature from T_1 to T_2.

We calculate the work w by

$$w = \Delta U = C_V \cdot \Delta T = C_V T_1 \left[\left(\frac{V_1}{V_2}\right)^{Nk/C_V} - 1\right]$$

Example

For $V_2 = 2V_1$, $T_1 = 300$ K, and $C_V = \frac{3}{2} Nk$ (atomic gas)

$$\Delta T = T_1 \cdot \left[\left(\tfrac{1}{2}\right)^{2/3} - 1\right] = -T_1 \cdot 0.370 = -111 \text{ K}$$

and

$$w = -\tfrac{3}{2} Nk \cdot 111 \text{ K} = -N \cdot 2.3 \times 10^{-21} \text{ J}$$

w is smaller than that for an isothermal expansion:

$$w = -NkT_1 \cdot \ln \frac{V_2}{V_1} = -N \cdot 2.9 \times 10^{-21} \text{ J}$$

because after cooling the final pressure P_2 of the gas is smaller than for constant temperature.

Problem 15.2 – Entropy of Mixing

Two equal containers (each of volume V) contain two different ideal gases A and B (pressure P and temperature T). By connecting the containers the gases are mixed. What is the mixing entropy and what is the reversible work necessary to separate the gases?

Solution. According to Section 14.6.2, the entropy of mixing is

$$\Delta S = S_2 - S_1 = (N(\text{A}) + N(\text{B}))k \cdot \ln 2$$

The reversible work w_{rev} and the reversible heat q_{rev} to separate the two gases (Figure 15.14) are

$$w_{\text{rev}} = -N(\text{A})kT \cdot \ln \frac{V}{2V} - N(\text{B})kT \cdot \ln \frac{V}{2V} = (N(\text{A}) + N(\text{B}))kT \cdot \ln 2$$

$$q_{\text{rev}} = -w_{\text{rev}} = -(N(\text{A}) + N(\text{B}))kT \cdot \ln 2$$

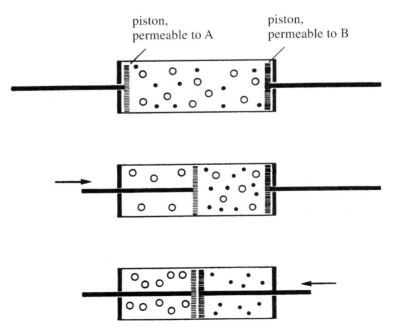

Figure 15.14 Separation of a mixture of two different gases A and B.

Problem 15.3 – Efficiency of Different Heat Engines

Compare the efficiency of (a) a low-pressure steam engine ($T_2 = 400$ K, vapor pressure about 5 bar); (b) a high-pressure steam engine ($T_2 = 600$ K, vapor pressure 100 bar); and (c) a gasoline engine ($T_2 = 1500$ K). Assume that they are reversible working engines with $T_1 = 300$ K.

Solution. From

$$\eta = \frac{T_2 - T_1}{T_2}$$

the following efficiencies are obtained:

	(a)	(b)	(c)
T_1/K	300	300	300
T_2/K	400	600	1500
η	25%	50%	80%

Problem 15.4 – Efficiency of a Power Station

How much heat is supplied on one day to the cooling towers of a power station with a power of 1000 MW, if the steam turbine is used with $T_2 = 600$ K and $T_1 = 300$ K?

Solution. The efficiency of the power station is, according to Problem 15.3, 50%. Then 50% of the heat supply is converted into electric power, and the heat supplied to the cooling towers is $q = 1000$ MW $\cdot$ 24 h $= 24 \times 10^6$ kW h. In the cooling towers, this heat is used to evaporate water. The amount of substance n of water needed for this process is $n = q/q_{m,\text{vap}}$ where $q_{m,\text{vap}} = 41.1$ kJ mol^{-1} is the molar heat of evaporation of water. Then

$$n = \frac{24 \times 10^6 \text{ kW h}}{41.1 \text{ kJ}} \text{ mol} = 2 \times 10^9 \text{ mol}$$

This corresponds to an amount of 36 000 m^3 of liquid water. In reality, the efficiency of the power station is smaller, and even more heat is supplied to the cooling towers.

Problem 15.5 – Heat Pump

A room has to be heated to $+20\,°$C when the outdoor temperature is $-20\,°$C. What is the electric power required when the heat pump is equivalent to a 2 kW heater?

Solution. With $T_2 =$ indoor temperature, $T_1 =$ outdoor temperature, and $q_2 =$ heat to be supplied at T_2 we obtain

$$w_{\text{rev}} = -\frac{T_2 - T_1}{T_2} \cdot q_2 = \frac{40 \text{ K}}{293 \text{ K}} \cdot q_2 = 0.14 \cdot q_2$$

Then the required electric power is 0.14×2 kW $= 280$ W, this is only 14% of the energy needed when using an electric heater.

Problem 15.6 – Air Conditioning

A heat pump is working as a refrigerator cooling a room down to $T_1 = 25\,°$C, while the outdoor temperature is $T_2 = 30\,°$C. Because of the leakage of the heat insulation, heat is permanently entering the room at a rate $dq/dt = A(T_2 - T_1)$ with $A = 200$ W K^{-1}. What is the electrical power P needed to maintain the temperature gradient $T_2 - T_1$?

Solution. In this case the heat q_1 must be removed from the room at the temperature T_1, and the heat $-q_2$ must be pumped to the surroundings at the temperature T_2. Then with $w_{rev} + q_1 = -q_2$ we obtain

$$w_{rev} = \frac{T_2 - T_1}{T_1} q_1$$

and

$$\text{Power} = \frac{dw_{rev}}{dt} = \frac{dq_1}{dt}\frac{T_2 - T_1}{T_1} = A(T_2 - T_1)\frac{T_2 - T_1}{T_1} = A\frac{(T_2 - T_1)^2}{T_1}$$

$$= 200 \text{ W K}^{-1} \frac{25 \text{ K}^2}{298 \text{ K}} = 16 \text{ W}$$

Note that the power increases with the square of the temperature gradient; for example, for $\Delta T = 10$ K the power is four times the value for $\Delta T = 5$ K. To maintain the same temperature gradient at a much lower temperature T_1 requires a much larger power; for example, to keep liquid helium at $T_1 = 4$ K when the temperature of the surroundings is $T_2 = 303$ K, then the power is 14 kW. Of course, this value can be considerably decreased by decreasing A (better heat insulation). However, A cannot be decreased to zero; this means that it is impossible to maintain a body at the temperature of absolute zero.

Problem 15.7 – Entropy and Configurations of Water

(a) $S_{273} = 3.9$ kJ K^{-1} for 1 kg of water at atmospheric pressure. What is the number of configurations Ω_{273} for water at $0\,°$C and atmospheric pressure?

(b) A kilogram of liquid water at $100\,°$C and atmospheric pressure has the entropy $S_{373} = S_{273} + 1300$ J K^{-1}. Calculate the number of configurations Ω_{373} for the water at $100\,°$C and compare to S_{273}.

Solution.

(a) $S_{273} = k \cdot \ln \Omega$, then $\Omega_{273} = e^{S_{273}/k} = e^{2.8 \times 10^{26}} = 10^{1.2 \times 10^{26}}$

(b) $\Omega_{373} = e^{S_{373}/k} = e^{3.7 \times 10^{26}} = 10^{1.6 \times 10^{26}} = \Omega_{273} \times 10^{4 \times 10^{25}}$

Note that a small increase in the exponent (from 1.2×10^{26} to 1.6×10^{26}) corresponds to a very big increase of Ω.

16 Criteria for Chemical Reactions

It is of practical interest to know the heat produced in a chemical reaction (such as the combustion of coal), as well as the temperatures which can be reached as a result of chemical reactions. It is useful to introduce the enthalpy, besides the internal energy U,

$$H = U + PV$$

The change of the enthalpy equals the heat transferred in a reaction at constant pressure. Enthalpy changes of reactions can be calculated from measured enthalpies of formation of all the species involved in the reaction. Why can a reaction such as the combustion of coal occur spontaneously (perhaps after a trigger), while other reactions, such as the reaction $CO_2 \rightarrow C + O_2$, are not possible for given conditions? Can we answer this question on the basis of the general statements made in Chapter 15?

It is useful to introduce the Gibbs energy $G = H - TS$, besides the enthalpy H and the entropy S, we show that a chemical reaction at constant pressure is spontaneous only if $\Delta G \leq 0$. H and G are state variables of the system. We consider $\Delta H - T\Delta S$ for reactions when all species involved in the reaction are separated in the beginning and at the end (e.g., combustion of carbon). Then ΔS is obtained from the entropies of each species. Enthalpies of formation and entropies are listed in tables for many compounds for the standard state (at pressure 1 bar). ΔG can be calculated for any desired reaction between these compounds, and it can be judged if the criterion $\Delta G < 0$ is fulfilled or not.

A simple example is "burning" (actually heating) limestone:

$$CaCO_3 \longrightarrow CaO + CO_2$$

In this case, at a certain temperature and a certain CO_2 pressure P_0, we have $\Delta G = 0$. Thus the reaction is spontaneous at $P_{CO_2} < P_0$, and it is impossible at $P_{CO_2} > P_0$. P_0 increases strongly with temperature reaching 1 bar at $T = 1145$ K. This is the minimum temperature for "burning limestone" at 1 bar.

16.1 Heat Exchange

16.1.1 Reaction at Constant Volume: q = ΔU

We consider a closed container filled with a stoichiometric mixture of hydrogen and oxygen in a bath at a temperature T (Figure 16.1). Ignition of the mixture, e.g., by a spark, causes a chemical reaction in the container (detonating gas reaction):

$$2H_2 + O_2 \longrightarrow 2H_2O \tag{16.1}$$

We wait until the temperature equilibration with the temperature bath has been established. A large enough bath will have the same temperature at the end as in the beginning. This process is called a *reaction at constant temperature*, even though immediately after the ignition the temperature in the container is much higher.

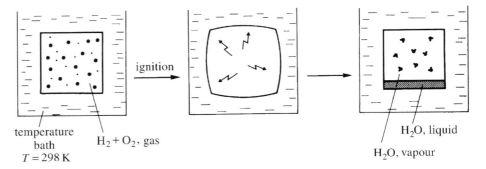

Figure 16.1 Detonating gas reaction performed at T = const and V = const. System: $H_2 + O_2$, gas, H_2 and O_2 in a stoichiometric mixture.

In this reaction new bonds are formed. The water molecules formed as products possess a lower internal energy than the starting materials: the change of internal energy is $\Delta U < 0$. Since the reaction is performed at constant volume, no work is done on the system, and therefore

$$\boxed{q = \Delta U \quad \text{(reaction at constant volume)}} \tag{16.2}$$

In this case the heat q supplied to the system (that is, to the gas mixture) is negative. This means that heat is transferred to the bath (*exothermic reaction*). This heat can be measured with a proper device (adiabatic bomb calorimeter, Figure 16.2, in which there is no heat exchange with the external surroundings of the bath). Then ΔU can be determined according to equation (16.2):

$$\Delta U = -564.4 \, \text{kJ mol}^{-1} \cdot n_{O_2} \quad (T = 298 \, \text{K})$$

where n_{O_2} is the amount of reacting O_2.

Only a few chemical reactions can be directly investigated in a calorimeter. But it is possible to study most reactions indirectly. If a reaction $1 \rightarrow 2$ has to be investigated, two indirect reactions $1 \rightarrow 3$ and $2 \rightarrow 3$ can be used:

16.1 HEAT EXCHANGE

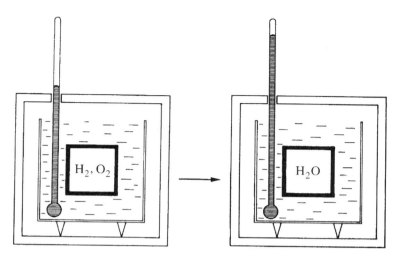

Figure 16.2 Adiabatic bomb calorimeter to measure the heat transferred into the temperature bath. The small increase of the temperature T of the bath is measured using a very sensitive thermometer, then the heat can be calculated from ΔT and the heat capacity of the calorimeter.

Then we can calculate $\Delta U_{1\to 2}$ indirectly

$$\Delta U_{1\to 2} = \Delta U_{1\to 3} + \Delta U_{3\to 2} = \Delta U_{1\to 3} - \Delta U_{2\to 3} \tag{16.3}$$

(*Hess's law*) because the sequence $1 \to 2 \to 3 \to 1$ constitutes a cyclic process with $\Delta U = 0$.

As an example we consider the hydrogenation of ethene

$$1 \to 2: \ C_2H_4 + H_2 \longrightarrow C_2H_6 \tag{16.4}$$

$\Delta U_{1\to 2}$ can be measured directly; the value

$$\Delta U_{1\to 2} = -134.1 \ \text{kJ mol}^{-1} \cdot n_{C_2H_4}$$

is found. $\Delta U_{1\to 2}$ can also be obtained indirectly by measuring the heat of oxidation of C_2H_4, H_2, and of C_2H_6. These reactions are rapid and clean if a large excess of pure oxygen is applied.

$$1 \to 3: \ C_2H_4 + H_2 + \tfrac{7}{2}O_2 \longrightarrow 2CO_2 + 3H_2O \tag{16.5}$$

$$2 \to 3: \ C_2H_6 + \tfrac{7}{2}O_2 \longrightarrow 2CO_2 + 3H_2O \tag{16.6}$$

These indirect reactions are chosen such that the products are the same. The difference of $1 \to 3$ and $2 \to 3$ yields just reaction $1 \to 2$. With

$$\Delta U_{1\to 3} = -1688.1 \ \text{kJ mol}^{-1} \cdot n_{C_2H_4}, \quad \Delta U_{2\to 3} = -1554.0 \ \text{kJ mol}^{-1} \cdot n_{C_2H_6}$$

at $T = 298$ K we obtain with $n_{C_2H_4} = n_{C_2H_6}$ at $T = 298$ K

$$\Delta U_{1\to 2} = (-1688.1 + 1554.0) \ \text{kJ mol}^{-1} \cdot n_{C_2H_4} = -134.1 \ \text{kJ mol}^{-1} \cdot n_{C_2H_4}$$

$\Delta U_{1\to 3}$ is actually obtained by adding the measured heats of the reactions

$$C_2H_4 + 3O_2 \longrightarrow 2CO_2 + 2H_2O \quad \text{and} \quad H_2 + \tfrac{1}{2}O_2 \longrightarrow H_2O$$

16.1.2 Reaction at Constant Pressure: $q = \Delta H$

Usually, chemical reactions are not carried out in closed containers, but in open vessels (e.g., test tubes, Erlenmeyer flasks) in which the reactants are under external pressure. Then the pressure remains constant during the reaction, whereas the volume of the reaction mixture changes (*reaction at constant pressure*). This is illustrated in Figure 16.3 for the detonating gas reaction, where the external air pressure is represented by a weight on the piston. In this reaction liquid water is formed from gaseous H_2 and O_2. The volume of the reactants decreases during the reaction: the weight sinks, and work is done on the system. Therefore the change of internal energy is

$$\Delta U = q + w = q - P\Delta V \quad \text{(reaction at constant pressure)} \tag{16.7}$$

The heat q transferred into the system is

$$q = \Delta U + P\Delta V = \Delta U + P(V_{H_2O} - V_{H_2,O_2}) \tag{16.8}$$

In our example q is smaller than ΔU, because ΔV is negative (the volume of liquid water is much smaller than the volume of H_2 and O_2). Whereas in a reaction at constant volume the heat q equals the change ΔU of the internal energy, the heat transferred at constant pressure is $\Delta U + P\Delta V$. Since most of the reactions are performed at constant pressure, it is useful to define the function

$$\boxed{H = U + PV} \tag{16.9}$$

which is called the *enthalpy*. H is a state variable, since it is composed of state variables – that is, variables which are independent of the way by which the system reaches this state. In a chemical reaction from state 1 to state 2 the change ΔH is given by

$$\Delta H = H_2 - H_1 = U_2 - U_1 + P_2 V_2 - P_1 V_1 \tag{16.10}$$

For reactions at constant pressure P this expression reduces to

$$\Delta H = \Delta U + P\Delta V \tag{16.11}$$

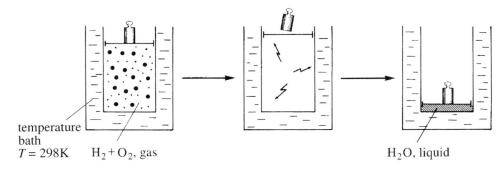

Figure 16.3 Detonating gas reaction performed at $T = $ const and $P = $ const. System: $H_2 + O_2$, gas, H_2 and O_2 in a stoichiometric mixture.

16.1 HEAT EXCHANGE

With equation (16.8) for a reaction at constant pressure we obtain

$$\boxed{q = \Delta U + P\Delta V = \Delta H} \quad \text{(reaction at constant pressure)} \quad (16.12)$$

The change ΔH is equal to the heat transferred into the system at constant pressure. The heat supplied to the system is converted to internal energy of the system and to work done by the system against the external pressure.

In our example of the detonating gas reaction, the volume of the liquid water is much smaller than the volume of H_2 and O_2. Neglecting the volume of the liquid and treating H_2 and O_2 as ideal gases leads to

$$P\Delta V = -RT \cdot (n_{H_2} + n_{O_2}) = -RT \cdot (2n_{O_2} + n_{O_2}) = -3RT \cdot n_{O_2} \quad (16.13)$$

leading to (at $T = 298$ K)

$$\Delta H = \Delta U - 3RT \cdot n_{O_2} = (-564.4 - 7.4)\,\text{kJ mol}^{-1} \cdot n_{O_2}$$
$$= -571.8\,\text{kJ mol}^{-1} \cdot n_{O_2}$$

Note that $n_{H_2} = n_{H_2O} = 2n_{O_2}$.

Example – heat needed for warming up water

The heat produced in burning 1 mol H_2 (285.9 kJ) is used to heat liquid water from room temperature (20 °C) to the boiling point (100 °C, $C_{P,m} = 75.3\,\text{J K}^{-1}\,\text{mol}^{-1}$, see Table 16.1). For heating 1 mol (18 g) water we need 75.3×80 J $= 6.02$ kJ. Thus $285.9/6.02 = 47.5$ mol water (corresponding to 855 g) can be heated from 20 °C to 100 °C.

In a more careful treatment we have to pay attention to the fact that the final states of the reactions at constant volume (Figure 16.1) and at constant pressure (Figure 16.3) are slightly different (Box 16.1).

Box 16.1 – $\Delta U_{P=\text{const}}$ from Measured $\Delta U_{V=\text{const}}$

Let us discuss the changes of state in Figure 16.4. States 1 and 3 correspond to the initial and final states in Figure 16.3 (reaction at $P = $ const); states 1 and 2 correspond to the initial and final states in Figure 16.1 (reaction at $V = $ const). Then

$$\Delta U_{V=\text{const}} = \Delta U_{1 \to 2}$$
$$\Delta U_{P=\text{const}} = \Delta U_{1 \to 3} = \Delta U_{1 \to 2} + \Delta U_{2 \to 3}$$

Thus ΔU is different in both processes.

Continued on page 522

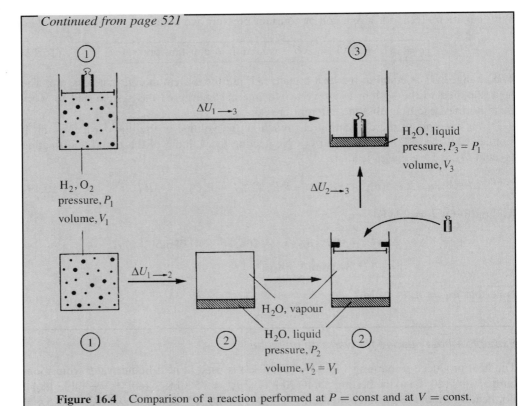

Figure 16.4 Comparison of a reaction performed at $P = \text{const}$ and at $V = \text{const}$.

In our example, $\Delta U_{2\rightarrow 3}$ corresponds to the following change: the water vapor which is in equilibrium with liquid water at temperature T is condensed to liquid water, and the liquid which was under the vapor pressure P_2 ($P_2 =$ 20 mbar at $T = 298$ K) is compressed. In state 2, however, only a very small fraction of the water is in the form of vapor, and liquid water is only slightly compressible, thus $\Delta U_{2\rightarrow 3}$ is extremely small (provided that the pressure of the H_2/O_2 mixture present in the beginning is on the order of 1 bar). Then the difference between $\Delta U_{V=\text{const}}$ and $\Delta U_{P=\text{const}}$ can be neglected.

16.2 Change of Internal Energy and Enthalpy

16.2.1 Temperature Dependence of ΔU and ΔH

The energies of bonding, translation, rotation, and vibration contribute to the internal energy U. Increasing the temperature causes an increase in U. According to equation (13.7), for a change of state at constant volume we obtain

$$\left(\frac{\partial U}{\partial T}\right)_V = C_V \tag{16.14}$$

16.2 CHANGE OF INTERNAL ENERGY AND ENTHALPY

Then

$$U_{T_2} = U_{T_1} + \int_{T_1}^{T_2} C_V \cdot dT \tag{16.15}$$

Now we consider a chemical reaction performed at temperature T_1 (change ΔU_1 of the internal energy). What will the change ΔU_2 be if the same reaction is performed at a temperature T_2?

To answer this question we imagine the reaction carried out at temperature T_1; after that the products are heated from T_1 to T_2. Then the change of internal energy is

$$\Delta U_{T_1} + \int_{T_1}^{T_2} C_{V,\text{products}} \cdot dT$$

On the other hand, if the reactants are first heated from T_1 to T_2 and the reaction is carried out at temperature T_2, the change of internal energy is

$$\Delta U_{T_2} + \int_{T_1}^{T_2} C_{V,\text{reactants}} \cdot dT$$

Both changes of internal energy are the same, thus we obtain

$$\Delta U_{T_1} + \int_{T_1}^{T_2} C_{V,\text{products}} \cdot dT = \Delta U_{T_2} + \int_{T_1}^{T_2} C_{V,\text{reactants}} \cdot dT \tag{16.16}$$

or

$$\boxed{\Delta U_{T_2} = \Delta U_{T_1} + \int_{T_1}^{T_2} \Delta C_V \cdot dT} \tag{16.17}$$

where

$$\Delta C_V = C_{V,\text{products}} - C_{V,\text{reactants}} \tag{16.18}$$

Correspondingly, for a chemical reaction performed at constant pressure we obtain

$$H_{T_2} = H_{T_1} + \int_{T_1}^{T_2} C_P \cdot dT \tag{16.19}$$

and

$$\boxed{\Delta H_{T_2} = \Delta H_{T_1} + \int_{T_1}^{T_2} \Delta C_P \cdot dT} \tag{16.20}$$

with

$$\Delta C_P = C_{P,\text{products}} - C_{P,\text{reactants}} \tag{16.21}$$

This treatment implies that no phase transitions (melting, evaporation) occur during heating. Otherwise, as shown in the examples in Foundations 16.2, the heat of the phase transition must be added.

Example – calculation of ΔH at 350 K for the detonating gas reaction

$$2H_2 + O_2 \longrightarrow 2H_2O$$

The molar heat capacities $C_{P,m} = C_P/n$ of H_2, O_2, and liquid H_2O at 298 K and 350 K are shown in Table 16.1.

Table 16.1 Molar heat capacity $C_{P,m}$ for different temperatures at a pressure of 1 bar

Compound	$C_{P,m}$/J K^{-1} mol^{-1}		
	at 298 K	at 350 K	average
H_2O (liquid)	75.3	75.6	75.5
H_2 (gas)	28.8	29.0	28.9
O_2 (gas)	29.4	29.7	29.6

In this case the heat capacities are only slightly changing with temperature, and we replace $C_{P,m}(T)$ by the average $\overline{C}_{P,m}$. Then equation (16.20) can be simplified:

$$\Delta H_{T_2} = \Delta H_{T_1} + \Delta \overline{C}_P \cdot (T_2 - T_1)$$

With

$$\Delta \overline{C}_P = n_{H_2O} \cdot \overline{C}_{P,m}(H_2O) - n_{H_2} \cdot \overline{C}_{P,m}(H_2) - n_{O_2} \cdot \overline{C}_{P,m}(O_2)$$
$$= n_{O_2} \cdot \left[2 \cdot \overline{C}_{P,m}(H_2O) - 2 \cdot \overline{C}_{P,m}(H_2) - \overline{C}_{P,m}(O_2) \right]$$
$$= n_{O_2} \cdot 63.6 \, \text{J K}^{-1} \, \text{mol}^{-1}$$

we obtain

$$\Delta H_{350} = (-571.8 \times 10^3 + 63.6 \cdot 52) \, \text{J mol}^{-1} \cdot n_{O_2} = -568.6 \, \text{kJ mol}^{-1} \cdot n_{O_2}$$

The heat produced when burning H_2 at 350 K is by 1% smaller than the heat produced when burning H_2 at 298 K. This is because the heat capacity of liquid water exceeds the heat capacities of H_2 and O_2. The high heat capacity of liquid water is caused by breaking hydrogen bonds by heating.

For larger changes of the temperature, the integral in equation (16.20) must be evaluated explicitly (see Foundation 16.1).

16.2.2 Molar Enthalpies of Formation from Elements $\Delta_f H^\ominus$

The enthalpy change for the formation of a compound from its elements under standard conditions (compounds separated and under standard pressure $P^\ominus = 1$ bar) is useful for obtaining enthalpy changes in arbitrary reactions. Figure 16.5 illustrates this process for the formation of H_2O from H_2 and O_2:

$$2 \, H_2 + O_2 \longrightarrow 2 \, H_2O$$

In this case ($T = 298$ K)

$$\Delta H = -571.8 \, \text{kJ mol}^{-1} \cdot n_{O_2} = -285.9 \, \text{kJ mol}^{-1} \cdot n_{H_2O}$$

according to Section 16.1.2. The quantity

$$\Delta_f H^\ominus(H_2O) = \frac{\Delta H \text{ (formation from elements, separated, 1 bar)}}{n_{H_2O}}$$

$$= -285.9 \, \text{kJ mol}^{-1} \quad \text{(for 298 K corresponding to 25 °C)}$$

16.2 CHANGE OF INTERNAL ENERGY AND ENTHALPY

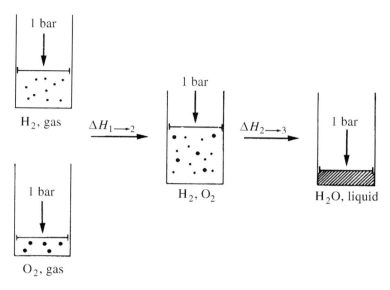

Figure 16.5 Formation of H$_2$O from the separated elements H$_2$ and O$_2$ at $P^\ominus = 1$ bar and $T = 298$ K. First the elements are mixed, then the reaction is started.

is the *molar enthalpy of formation under standard conditions*, i.e., the enthalpy of formation of 1 mol of H$_2$O from elements under standard conditions. The symbol $\ominus$ indicates standard conditions (reactants separated, pressure 1 bar). A subscript at the symbol Δ designates a molar quantity; additionally, subscript f indicates formation of a substance from elements; other subscripts will be used below to indicate chemical reactions (subscript r), vaporization (subscript vap), melting (subscript fus), sublimation (subscript sub).

For selected compounds the molar enthalpies of formation $\Delta_f H^\ominus_{298\,K}$ are listed in Appendix 1.

16.2.3 Molar Enthalpy of Reaction $\Delta_r H^\ominus$

The change in enthalpy under standard conditions (species separated, standard pressure $P^\ominus = 1$ bar) at temperature T can be obtained for any reaction between species A, B,... forming C, D,... from the enthalpies of formation of A, B,... and C, D,....

For example, for the reaction

$$CO + \tfrac{1}{2}O_2 \longrightarrow CO_2$$

we define the molar enthalpy of reaction under standard conditions as the enthalpy change for the reaction of 1 mol of CO and 0.5 mol of O$_2$ to 1 mol of CO$_2$ (species separated, pressure 1 bar):

$$\Delta_r H^\ominus = \Delta_f H^\ominus(CO_2) - \Delta_f H^\ominus(CO) - \tfrac{1}{2}\Delta_f H^\ominus(O_2)$$
$$= \left(-393.5 + 110.5 + \tfrac{1}{2} \cdot 0\right) \text{ kJ mol}^{-1} = -283.0 \text{ kJ mol}^{-1} \text{ (for 298 K)}$$

If we write, for the same reaction,

$$2CO + O_2 \longrightarrow 2CO_2$$

then $\Delta_r H^\ominus$ refers to 2 mol CO reacting with 1 mol O_2, thus

$$\Delta_r H^\ominus = 2 \cdot (-283.0 \, \text{kJ mol}^{-1}) = -566.0 \, \text{kJ mol}^{-1} \quad \text{(for 298 K)}$$

Generalization

This result can be generalized for the reaction

$$\alpha A + \beta B + \cdots \longrightarrow \gamma C + \delta D + \cdots \quad (16.22)$$

Then we obtain $\Delta_r H^\ominus$, the molar enthalpy change for the reaction of α mol A, β mol B... to γ mol C and δ mol D..., under standard conditions:

$$\Delta_r H^\ominus = \gamma \cdot \Delta_f H^\ominus(C) + \delta \cdot \Delta_f H^\ominus(D) - \alpha \cdot \Delta_f H^\ominus(A) - \beta \cdot \Delta_f H^\ominus(B) \quad (16.23)$$

Extensive and intensive quantities

We have considered the *extensive* quantities ΔU, ΔH, and ΔV related to the given systems. In contrast, $\Delta_f H^\ominus$ and $\Delta_r H^\ominus$ are molar quantities; these quantities and the pressure P do not depend on the size of the system. They are called *intensive* quantities.

Box 16.2 – Temperature of a Flame

What is the temperature of the flame of a Bunsen burner (Figure 16.6) fed with a mixture of hydrogen and air? In the flame the detonating gas reaction

$$2H_2 + O_2 \longrightarrow 2H_2O$$

takes place. Contrary to our previous considerations, the reaction occurs adiabatically: the reaction heat is used to warm up the reaction products from the initial temperature T_1 to the final temperature T_2. Then, neglecting the temperature dependence of the heat capacities and starting at $T_1 = 373$ K, where H_2O is in the gaseous state, we obtain

$$\Delta_r H^\ominus = -C_{P,m}^\ominus(\text{products}) \cdot \Delta T$$

and

$$\Delta T = \frac{-\Delta_r H^\ominus}{C_{P,m}^\ominus(\text{products})}$$

Because the Bunsen burner operates with air (4:1 mixture of N_2 and O_2), nitrogen must be warmed up additionally to the gaseous H_2O. Then

$$C_{P,m}^\ominus(\text{products}) = 2 \cdot C_{P,m}^\ominus(H_2O) + 4 \cdot C_{P,m}^\ominus(N_2) \quad \text{at} \quad T = 373 \, \text{K}$$

Continued on page 527

16.2 CHANGE OF INTERNAL ENERGY AND ENTHALPY

Continued from page 526

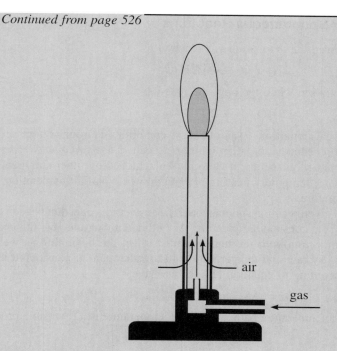

Figure 16.6 Scheme of a Bunsen burner.

From Foundation 16.1 we obtain

$$\Delta_r H^\ominus = (-571.8 + 4.72 + 87.8)\,\text{kJ mol}^{-1} = -479\,\text{kJ mol}^{-1} \quad \text{at} \quad 373\,\text{K}$$

and

$$C_{P,m}^\ominus(\text{products}) = (2 \times 34 + 4 \times 29.2)\,\text{J K}^{-1}\,\text{mol}^{-1} = 185\,\text{J K}^{-1}\,\text{mol}^{-1}$$

Then

$$\Delta T = \frac{479 \times 10^3}{185}\,\text{K} = 2590\,\text{K}$$

Actually, the heat capacities are larger at higher temperature; using the mean values 40 and 32 J K^{-1} mol^{-1} for H$_2$O and N$_2$ (see Foundation 16.1, Table 16.2), we obtain $\Delta T = 2360\,\text{K}$. Of course, in reality the process is not completely adiabatic, because it cannot be avoided that the air surrounding the flame is heated as well. (Our assumption that the reaction is complete at this temperature will be justified in Chapter 17.) In practice, temperatures around 1800 K can be reached with a Bunsen burner.

16.3 Conditions for Spontaneous Reactions

In Section 16.1 we discussed the detonating gas reaction
$$2H_2 + O_2 \longrightarrow 2H_2O$$
which takes place after ignition while the opposite reaction
$$2H_2O \longrightarrow 2H_2 + O_2$$
does not occur under usual conditions (say standard conditions at room temperature). A reaction that occurs immediately or after ignition is called a *spontaneous reaction*. The opposite reaction is not spontaneous, it can only be achieved by other simultaneous changes (in the present case: temporary heating of H_2O allowing partial dissociation into H_2 and O_2 at high temperature).

At what conditions is a given reaction spontaneous? To answer this question we consider an arbitrary change of state at constant temperature. As outlined in Section 16.1 this means that the system is in equilibrium with the temperature bath in the beginning as well as at the end of the process. System and temperature bath together can be considered as an isolated system. Then according to equation (14.36) we have

$$\Delta S_{\text{isolated system}} = \Delta S_{\text{system}} + \Delta S_{\text{temperature bath}} \geq 0 \tag{16.24}$$

The transfer of heat into a temperature bath at constant temperature is a reversible process (Section 15.2.4); then

$$\Delta S_{\text{temperature bath}} = -\frac{q}{T} \tag{16.25}$$

and with $\Delta S = \Delta S_{\text{system}}$ we obtain the key condition for the possibility of a spontaneous reaction:

$$\boxed{\Delta S - \frac{q}{T} \geq 0} \quad \text{(reaction at } T = \text{const)} \tag{16.26}$$

This expression can be written in the form

$$q - T\Delta S \leq 0 \quad \text{(reaction at } T = \text{const)} \tag{16.27}$$

In the case of a reaction at $V = \text{const}$ we have
$$q = \Delta U$$
and thus

$$\boxed{\Delta U - T\Delta S \leq 0 \quad (T = \text{const}, V = \text{const})} \tag{16.28}$$

In the case of a reaction at $P = \text{const}$ we have
$$q = \Delta H$$
and thus

$$\boxed{\Delta H - T\Delta S \leq 0 \quad (T = \text{const}, P = \text{const})} \tag{16.29}$$

These are the fundamental criteria for finding the conditions where a reaction can occur spontaneously.

16.3.1 Helmhoitz Energy and Gibbs Energy

It is useful to define the Helmholtz energy A

$$\boxed{A = U - TS} \tag{16.30}$$

16.3 CONDITIONS FOR SPONTANEOUS REACTIONS

A is a state variable since U, T, and S are state variables. Then we obtain for a reaction at constant temperature

$$\Delta A = \Delta U - T\Delta S \quad (T = \text{const}) \tag{16.31}$$

Thus we can write the condition (16.28) in the form

$$\Delta A \leq 0 \quad (T = \text{const}, V = \text{const}) \tag{16.32}$$

Correspondingly, we define the Gibbs energy G

$$G = H - TS \tag{16.33}$$

G is a state variable since H, T, and S are state variables. Then we obtain for a reaction at constant temperature

$$\Delta G = \Delta H - T\Delta S \quad (T = \text{const}) \tag{16.34}$$

Thus we can write the condition (16.29) in the form

$$\Delta G \leq 0 \quad (T = \text{const}, P = \text{const}) \tag{16.35}$$

The relations (16.32) and (16.35) are fundamental conditions which must be fulfilled for a reaction to proceed in a given direction. However, whether or not the reaction occurs can depend upon kinetic factors. In many cases the reaction has to be started by an initial ignition or by adding a catalyst or an enzyme (compounds which accelerate the reaction, but remain unchanged at the end of the reaction).

16.3.2 Reversible Work w_{rev}

What is the reversible work $-w_{\text{rev}}$ done by the system when proceeding from the initial state to the final state on a reversible path? Conservation of energy requires

$$\Delta U = q_{\text{rev}} + w_{\text{rev}} = T\Delta S + w_{\text{rev}} \tag{16.36}$$

Then

$$w_{\text{rev}} = \Delta U - T\Delta S \tag{16.37}$$

Therefore for a reaction at $T = \text{const}$ we obtain

$$\Delta A = w_{\text{rev}} \quad (T = \text{const}) \tag{16.38}$$

and for a reaction at $T = \text{const}$ and $P = \text{const}$ we obtain

$$\Delta G = w_{\text{rev}} + P\Delta V \quad (T = \text{const}, P = \text{const}) \tag{16.39}$$

The reversible work done on the system in a change of state at constant volume (no PV work occurs) must be $w_{\text{rev}} \leq 0$; the system must be able to do work on a reversible path.

In a change of state at constant pressure $-w_{\text{rev}}$ must be larger than the PV work done by the system in the actual reaction:

$$-w_{\text{rev}} \geq P\Delta V \quad \text{or} \quad w_{\text{rev}} + P\Delta V \leq 0 \tag{16.40}$$

Chemical reactions are usually carried out "at constant pressure and constant temperature". Therefore equations (16.34) and (16.35) are of particular interest. They will be used throughout this book. Equation (16.34), in the form

$$T\Delta S + \Delta G = \Delta H$$

expresses conservation of energy in the case of a reversible path:

| heat supplied to the system: $q_{\text{rev}} = T\Delta S$ | + | work done on the system minus PV work done on the system: $w_{\text{rev}} + P\Delta V = \Delta G$ | = | energy increase of the system minus PV work done on the system: $\Delta U + P\Delta V = \Delta H$ |

16.4 Change of Gibbs Energy

16.4.1 Molar Gibbs Energy of Formation from Elements $\Delta_f G^\ominus$

Generally, according to equation (16.34), we have

$$\Delta G = \Delta H - T\Delta S \quad (T = \text{const})$$

Then in the case of the formation of a substance from elements in molar quantity under standard conditions the Gibbs energy of formation is

$$\Delta_f G^\ominus = \Delta_f H^\ominus - T\Delta_f S^\ominus \quad (T = \text{const}) \tag{16.41}$$

where $\Delta_f S^\ominus$ is the change in entropy when forming 1 mol of substance from elements under standard pressure $P^\ominus = 1$ bar. $\Delta_f G^\ominus$ is the molar Gibbs energy of formation from elements under standard pressure $P^\ominus = 1$ bar. Values for this quantity at 298 K are included in Appendix 1.

Example – formation of H_2O (1 mol, 298 K, 1 bar) from separated elements

$\Delta_f G^\ominus$ for the formation of H_2O (which is liquid at 298 K and 1 bar) is obtained in the following way:

$$\Delta_f H^\ominus = -285.9 \text{ kJ mol}^{-1}$$
$$\Delta_f S^\ominus = S_m^\ominus(H_2O) - S_m^\ominus(H_2) - \tfrac{1}{2}S_m^\ominus(O_2)$$
$$= (69.9 - 130.6 - \tfrac{1}{2} \cdot 205.0) \text{ J K}^{-1}\text{ mol}^{-1}$$
$$= -163.2 \text{ J K}^{-1}\text{ mol}^{-1}$$

Then

$$\Delta_f G^\ominus = (-285.9 \times 10^3 - 298 \cdot (-163.2))\text{ J mol}^{-1} = -237.2 \text{ kJ mol}^{-1}$$

16.4 CHANGE OF GIBBS ENERGY

16.4.2 Molar Gibbs Energy of Reaction $\Delta_r G^\ominus$

$\Delta_r G^\ominus$, the change of the Gibbs energy G per mol under standard conditions (species separated, under standard pressure $P^\ominus = 1$ bar), at temperature T, is calculated from

$$\Delta_r G^\ominus = \Delta_r H^\ominus - T \cdot \Delta_r S^\ominus \tag{16.42}$$

$\Delta_r H^\ominus$ is obtained from the enthalpies of formation (Appendix 1), and $\Delta_r S^\ominus$ from the molar entropies $S_m^\ominus$ of the species involved in the reaction. The entropies $S_m^\ominus$ are calculated or derived from calorimetric data following the procedure described in Section 15.3; values for selected compounds are listed in Appendix 1.

The relation (16.42) is of great importance. It allows us to calculate the Gibbs energy for any suggested reaction provided the data for the reactants and products can be found in a table.

Example – detonating gas reaction

For the reaction

$$2H_2 + O_2 \longrightarrow 2H_2O$$

we calculate from Section 16.4.1

$$\Delta_r G^\ominus = 2 \cdot (-237.2)\,\text{kJ mol}^{-1} = -474.4\,\text{kJ mol}^{-1} \text{ at } 298\,\text{K}$$

Thus $\Delta_r G^\ominus_{298}$ is strongly negative, and the formation of H_2O can occur spontaneously under standard conditions.

Example – oxidation of CO at 298 K and 1 bar

We consider the reaction

$$2CO + O_2 \longrightarrow 2CO_2$$

According to Section 16.2.3 we obtain $\Delta_r H^\ominus = -566.0\,\text{kJ mol}^{-1}$, and from Appendix 1

$$\Delta_r S^\ominus = 2S_m^\ominus(CO_2) - 2S_m^\ominus(CO) - S_m^\ominus(O_2) = -173.6\,\text{J mol}^{-1}$$

Then

$$\Delta_r G^\ominus = -566.0\,\text{kJ mol}^{-1} - 298 \cdot (-173.6\,\text{J mol}^{-1}) = -514.3\,\text{kJ mol}^{-1}$$

16.4.3 Temperature Dependence of $\Delta G^\ominus$

By using the data in Appendix 1 we are able to calculate $\Delta G^\ominus_{298}$, the change of the Gibbs energy under standard conditions and at 298 K. What is $\Delta G^\ominus$ at any other temperature T?

We calculate $\Delta G^\ominus_T$ by

$$\Delta G^\ominus_T = \Delta H^\ominus_T - T \cdot \Delta S^\ominus_T \tag{16.43}$$

where, according to equation (16.20) and for standard conditions,

$$\Delta H_T^\ominus = \Delta H_{298}^\ominus + \int_{298}^{T} \Delta C_P^\ominus \cdot dT \tag{16.44}$$

For a change of state at constant pressure the change of entropy is

$$dS = \frac{dq_{rev}}{T} = \frac{C_P \cdot dT}{T}$$

Then in analogy to the derivation of equation (16.20) from equation (16.19), we obtain

$$\Delta S_T^\ominus = \Delta S_{298}^\ominus + \int_{298}^{T} \frac{\Delta C_P^\ominus}{T} \cdot dT \tag{16.45}$$

This treatment implies that no phase transitions (melting, evaporation) occur during heating. Otherwise, as shown in the examples in Foundation 16.2, the heat of the phase transition (e.g. $\Delta_{vap}H$ in the case of evaporation) must be added to $\Delta H_T^\ominus$ and $\Delta_{vap}H/T$ must be added to $\Delta S_T^\ominus$.

Example – burning limestone

We consider the reaction (Figure 16.7)

$$CaCO_3 \longrightarrow CaO + CO_2 \tag{16.46}$$

From Appendix 1 we derive for $T = 298$ K $\Delta_r H^\ominus = 177.4$ kJ mol^{-1}, $\Delta_r S^\ominus = 160.4$ J K^{-1} mol^{-1}, and $\Delta_r G^\ominus = (-394.4 - 604.0 + 1128.0)$ kJ mol^{-1} = 129.6 kJ mol^{-1}. $\Delta_r G^\ominus$ is positive at 298 K, thus reaction (16.46) cannot proceed spontaneously under standard conditions.

What is $\Delta_r G^\ominus$ at $T = 1200$ K? As shown in Foundations 16.1 and 16.2 from equations (16.43) to (16.45) we obtain $\Delta_r G^\ominus = -8.0$ kJ mol^{-1} at $T = 1200$ K. Note that $\Delta_r G^\ominus$ is negative at 1200 K; this means that reaction (16.46) can proceed spontaneously at 1200 K under standard conditions. This is the well-known process of burning limestone,

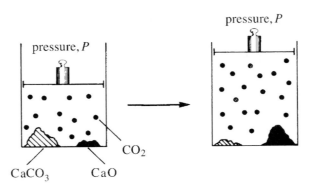

Figure 16.7 Formation of CO_2 and CaO from $CaCO_3$ (burning limestone).

16.4 CHANGE OF GIBBS ENERGY

which needs a minimum temperature to proceed at a CO_2 pressure of 1 bar overcoming the atmospheric pressure. This minimum temperature is determined by

$$\Delta G^\ominus = \Delta H^\ominus - T \cdot \Delta S^\ominus = 0$$

By calculating $\Delta G^\ominus$ for different temperatures (Figure 16.8) we find $T = 1145\,\text{K}$ for $\Delta_r G^\ominus = 0$. This is in agreement with experimental data (the measured temperature to burn limestone).

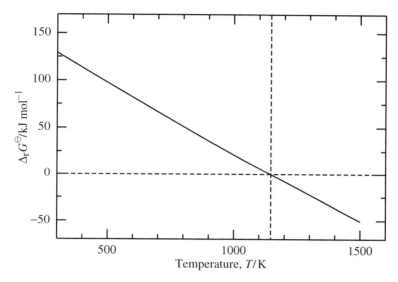

Figure 16.8 $\Delta_r G^\ominus$ for the reaction $CaCO_3 \to CaO + CO_2$ (burning limestone) as a function of temperature. At $T = 1145\,\text{K}$ we obtain $\Delta_r G^\ominus = 0$ (dashed lines).

16.4.4 Pressure Dependence of ΔG

How does ΔG change if we change the external pressure from the standard pressure $P^\ominus$ to an arbitrary pressure P? As for the temperature dependence we have to consider the changes in ΔH and ΔS.

As outlined in Box 16.1, we can usually neglect the change of ΔH with pressure (U and thus H of an ideal gas are independent of pressure; the same is true for solids and liquids, which are treated as incompressible). Furthermore, because of their incompressibility, the entropies of solids and liquids, essentially do not depend on the pressure. However, the volume of an ideal gas decreases with increasing pressure P; according to equation (14.29) we have, because of $P_1 V_1 = P_2 V_2$,

$$\boxed{S_T = S_T^\ominus - nR \cdot \ln \frac{P}{P^\ominus}} \qquad (16.47)$$

where S_T is the entropy at temperature T and pressure P, and $S_T^\ominus$ is the entropy at temperature T and standard pressure $P^\ominus = 1$ bar. Because of

$$\Delta G = \Delta H - T \Delta S \qquad (16.48)$$

we obtain the following for a chemical reaction with one gaseous compound as reaction product

$$\Delta G = \Delta G^\ominus + nRT \cdot \ln \frac{P}{P^\ominus} \qquad (16.49)$$

When the reaction is just in equilibrium, the limiting case is determined by the condition

$$\Delta G = 0 \qquad (16.50)$$

Then

$$\Delta G^\ominus = -nRT \cdot \ln \frac{P_{eq}}{P^\ominus} \quad \text{or} \quad \Delta_r G^\ominus = -RT \cdot \ln \frac{P_{eq}}{P^\ominus} \qquad (16.51)$$

where P_{eq} refers to the equilibrium pressure of the gaseous product.

Example – burning limestone (Figure 16.7)

In Section 16.4.3, for the reaction

$$CaCO_3 \longrightarrow CaO + CO_2$$

we calculated $\Delta_r G^\ominus = -8.0\,\text{kJ}\,\text{mol}^{-1}$ for $T = 1200\,\text{K}$ for the change of the Gibbs energy when 1 mol of $CaCO_3$ forms 1 mol of CaO and 1 mol of CO_2 at 1 bar and 1200 K. Then

$$\Delta_r G = \Delta_r G^\ominus + RT \cdot \ln \frac{P_{CO_2}}{P^\ominus}$$

$\Delta_r G$ increases with increasing pressure; finally $\Delta_r G$ switches from a negative value to a positive value (Figure 16.9). With $\Delta_r G = 0$ we obtain the following for the equilibrium

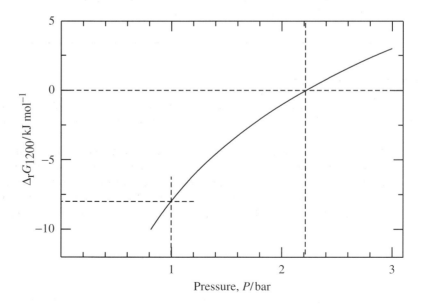

Figure 16.9 $\Delta_r G_{1200}$ for the reaction $CaCO_3 \rightarrow CaCO + CO_2$ (burning limestone) as a function of external pressure P. At $P = 1$ bar the change in Gibbs energy is negative, at $P = 2.2$ bar we obtain $\Delta_r G^\ominus = 0$ (dashed lines).

PROBLEM 16.1

pressure of CO_2:

$$\ln \frac{(P_{CO_2})_{eq}}{P^{\ominus}} = -\frac{\Delta_r G^{\ominus}}{RT} = \frac{8.0 \text{ kJ mol}^{-1}}{R \cdot 1200 \text{ K}} = 0.80$$

$$(P_{CO_2})_{eq} = 2.2 \text{ bar} \quad \text{at} \quad T = 1200 \text{ K}$$

This means that burning limestone at 1200 K is spontaneous up to a CO_2 pressure of 2.2 bar.

Problems

Problem 16.1 – Bunsen Burner Fed with Methane

What is the flame temperature of a Bunsen burner fed with methane instead of hydrogen?
Solution. We consider the reaction

$$CH_4 + 2O_2 \longrightarrow CO_2 + 2H_2O$$

As in Box 16.2 we start at $T_1 = 373$ K. Then we obtain

$$\Delta_r H^{\ominus}_{373} = \Delta_r H^{\ominus}_{298} + \int_{298}^{373} \Delta C^{\ominus}_{P,m} \cdot dT = (-802 + 3.0) \text{ kJ mol}^{-1}$$

$$= -799 \text{ kJ mol}^{-1}$$

$$C^{\ominus}_{P,m}(\text{products}) = C^{\ominus}_{P,m}(CO_2) + 2C^{\ominus}_{P,m}(H_2O) + 8C^{\ominus}_{P,m}(N_2)$$

$$= (40.3 + 2 \cdot 34.0 + 8 \times 29.2) \text{ J K}^{-1} \text{ mol}^{-1}$$

$$= 342 \text{ J K}^{-1} \text{ mol}^{-1}$$

Then in analogy to the procedure in Box 16.2 we obtain

$$\Delta T = \frac{799 \times 10^3}{342} \text{ K} = 2340 \text{ K}$$

Remember that $\Delta_r H^{\ominus}$ and $C^{\ominus}_{P,m}$ (products) refer to the reaction of 1 mol CH_4 and 2 mol O_2 to 1 mol CO_2 and 2 mol H_2O in the presence of 8 mol N_2.

Problem 16.2 – $\Delta_r H$ from Standard Enthalpies of Formation

Calculate $\Delta_r H^{\ominus}$ for some selected reactions from $\Delta_f H^{\ominus}$ at $T = 298$ K.
Solution.

Reaction	$\Delta_r H^{\ominus}$/kJ mol^{-1} at $T = 298$ K
$C_{\text{diamond}} + O_2 \longrightarrow CO_2$	-395.4
$C_{\text{graphite}} + O_2 \longrightarrow CO_2$	-393.5
$C_{\text{graphite}} \longrightarrow C_{\text{diamond}}$	$+1.9$
$2Ag + Cl_2 \longrightarrow 2AgCl$	-256.1
benzene $+ 3H_2 \longrightarrow$ cyclohexane	-206
cyclohexene $+ 3H_2 \longrightarrow$ cyclohexane	-116

Problem 16.3 – Temperature Regulation in the Human Body

A daily intake of 450 g glucose is sufficient to cover the human energy demand when no work is done by the human. Then all processes in the body finally end with the production of heat. This heat is transferred to the surroundings to keep the body at a temperature of 37°C. We assume that one-half of the heat is released by evaporating water and the second half is conducted by convection through clothing.

(a) What is the heat produced by the oxidation of 450 g glucose to CO_2 and H_2O?
(b) What would be the increase of temperature in the body in one day in the case of complete heat isolation?
(c) The temperature of the body is kept constant by evaporation of water; what amount of water is necessary?
(d) The temperature of the body ($T_2 = 37\,°C = 310\,K$) is kept constant by thermal conduction to the surroundings (temperature T_1) through clothing. We model the clothing by a bolster of air of thickness 1 cm (thermal conductivity $\kappa = 25 \times 10^{-5}\,J\,K^{-1}\,cm^{-1}\,s^{-1}$). Calculate T_1 and calculate the demand of glucose for $T_1 = -40\,°C = 233\,K$.
(e) What is the power of a continuously burning electrical bulb corresponding to a consumption of 450 g of glucose per day?

Solutions.

(a) Because of the reaction

$$C_6H_{12}O_6 \text{ (glucose)} + 6O_2 \longrightarrow 6H_2O + 6CO_2$$

we find $\Delta_r H^\ominus = -2802\,kJ\,mol^{-1}$ at 298 K.
450 g glucose correspond to 2.5 mol, thus the daily heat production is 7000 kJ.

(b) The increase of temperature is $\Delta T = q/C_P$. We approximate C_P by the heat capacity of water (the body consists mainly of water). For a mass of 80 kg we obtain $n = 4444$ mol of H_2O, thus $C_P = 75.3 \cdot 4444\,J\,K^{-1} = 334\,kJ\,K^{-1}$ and $\Delta T = 21\,K$.

(c) The heat of evaporation of water is $\Delta_{vap}H = 44\,kJ\,mol^{-1}$. Then the amount of water needed for cooling the body for one-half of the glucose intake is

$$n_{H_2O} = \frac{3500\,kJ}{44\,kJ\,mol^{-1}} = 80\,mol$$

This corresponds to 1.4 L of water.

(d) The surface of the human body is about $A = 1\,m^2 = 10^4\,cm^2$. The heat q passing the air bolster in one day is

$$q = \kappa \cdot \frac{A}{\text{thickness}} \cdot (T_2 - T_1) \cdot t$$

$$= 25 \times 10^{-5}\,J\,K^{-1}\,cm^{-1}\,s^{-1} \cdot \frac{10^4\,cm^2}{1\,cm} \cdot (T_2 - T_1) \times 10^5\,s$$

$$= 25 \times 10^4 \cdot (T_2 - T_1)\,J\,K^{-1}$$

This heat is compensated by 3500 kJ due to the combustion of 225 g glucose. Then

$$T_2 - T_1 = \frac{3500 \text{ kJ}}{25 \times 10^4 \text{ J}} \cdot \text{K} = 14 \text{ K}$$

and $T_1 = T_2 - 14 \text{ K} = 37°\text{C} - 14°\text{C} = 23°\text{C}$. On the other hand, for $T_2 - T_1 = 37°\text{C} - (-40°\text{C}) = 77 \text{ K}$ the daily intake is $225 \text{ g} \times 77/14 = 1240 \text{ g}$ glucose.

(e) We assume no loss in producing electrical energy. Then, according to Appendix 1, for the power we obtain

$$\text{Power} = \frac{\Delta G^\ominus}{24 \text{ h}} = \frac{2.5 \text{ mol} \cdot 2802 \text{ kJ mol}^{-1}}{86400 \text{ s}} = 81 \text{ W}$$

Problem 16.4 – Spontaneous Reactions

Do the following reactions occur spontaneously under standard conditions (reactants and products in their standard states)?

Reaction
1 $Fe + S \longrightarrow FeS$
2 $2AgI + Pb \longrightarrow PbI_2 + 2Ag$
3 $2Al + Fe_2O_3 \longrightarrow Al_2O_3 + 2Fe$
4 $C_2H_4 \longrightarrow$ polyethene
5 $CaCO_3 \longrightarrow CaO + CO_2$
6 H_2O (liquid) $\longrightarrow H_2O$ (vapor)
7 $2Ag + PbI_2 \longrightarrow 2AgI + Pb$

Solution. According to Appendix 1 the following values of $\Delta_r G^\ominus$ are obtained at 298 K:

Reaction	$\Delta_r G^\ominus$/kJ mol^{-1}
1	−97.5
2	−40.5
3	−883.4
4	−54.0
5	+129.7
6	+8.7
7	+40.8

Reactions 1–4 can occur spontaneously and do occur after initiation. Reactions 5–7 cannot occur.

Problem 16.5 – Burning Limestone on Mount Everest

What is the minimum temperature needed to burn limestone on the top of Mount Everest (atmospheric pressure 0.34 bar)?

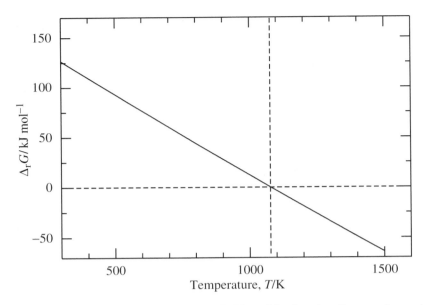

Figure 16.10 $\Delta_r G$ for the reaction $CaCO_3 \rightarrow CaCO + CO_2$ (burning limestone) as a function of temperature at an external pressure of $P = 0.34$ bar. At this pressure, we obtain $\Delta_r G^{\ominus} = 0$ at 1079 K (dashed lines); this is a lower temperature than in the case $P = P^{\ominus} = 1$ bar (Figure 16.8).

Solution. The pressure of CO_2 must overcome the atmospheric pressure of 0.34 bar. In Figure 16.10 $\Delta_r G$ is plotted versus temperature for $P_{CO_2} = 0.34$ bar

$$\Delta_r G = \Delta_r G^{\ominus} + RT \cdot \ln 0.34$$

From this plot we conclude that $\Delta_r G$ changes its sign at $T = 1079$ K. This is 66 K less than for atmospheric pressure at sea level.

Foundation 16.1 – How to Calculate $\int_{T_1}^{T_2} \Delta C_{P,m}^{\ominus} \cdot dT$

Over a large range of temperature the heat capacity $C_{P,m}^{\ominus}$ of a substance can change considerably (Figure 16.11), and it is no longer correct to replace $C_{P,m}^{\ominus}$ by its mean value. In most cases it is possible to approximate $C_{P,m}^{\ominus}$ by a power series (Table 16.2):

$$C_{P,m}^{\ominus} = c_1 + c_2 T + c_3 T^2$$

Then for reaction

$$\alpha A + \beta B + \cdots \longrightarrow \gamma C + \delta D + \cdots$$

we obtain

$$\int_{T_1}^{T_2} \Delta C_{P,m}^{\ominus} \cdot dT = \int_{T_1}^{T_2} (\Delta c_1 + \Delta c_2 \cdot T + \Delta c_3 \cdot T^2) \cdot dT$$
$$= \Delta c_1 \cdot (T_2 - T_1) + \Delta c_2 \cdot \tfrac{1}{2}(T_2^2 - T_1^2) + \Delta c_3 \cdot \tfrac{1}{3}(T_2^3 - T_1^3)$$

with

$$\Delta c_1 = \gamma c_1(C) + \delta c_1(D) + \cdots - \alpha c_1(A) - \beta c_1(B) - \cdots$$

(Δc_2, Δc_3 correspondingly).

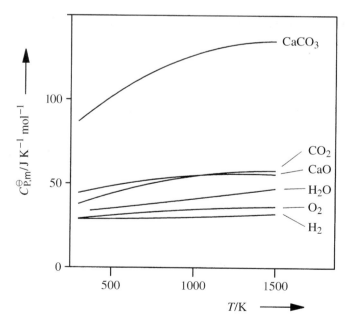

Figure 16.11 Molar heat capacity $C_{P,m}^{\ominus}$ of selected compounds versus temperature T.

Continued on page 540

Continued from page 539

Table 16.2 Coefficients c_1, c_2, and c_3 in the power series for the heat capacity $C_{P,m}^{\ominus}$ at $P = 1$ bar

Compound	$c_1/\text{J mol}^{-1}\text{ K}^{-1}$	$c_2/\text{J mol}^{-1}\text{ K}^{-2}$	$c_3/10^{-5}\text{ J mol}^{-1}\text{ K}^{-3}$
H_2	29.04	−0.0008	0.197
O_2	25.61	0.0132	−0.397
N_2	27.20	0.00535	−0.0121
NH_3	24.15	0.0392	−0.738
CO_2	26.47	0.0425	−1.440
HI	27.91	0.00314	0.119
I_2 (gas)	36.39	0.00233	−0.077
I	20.87	−0.00033	0.025
H_2O (gas)	30.15	0.0100	0.091
H_2O (liquid)	90.00	−0.0957	15.56
CaO	36.58	0.0296	−1.128
$CaCO_3$	60.38	0.0991	−3.305
CH_4	21.5	0.048	

The values are valid in the range 298 to 1500 K (for H_2O (liquid) 298 to 373 K, for H_2O(gas) 373 K to 1500 K).

Example – detonating gas reaction (temperature range 298 to 350 K)

$$2H_2 + O_2 \longrightarrow 2H_2O$$

In the considered range, H_2O is in the liquid state. Using the data in Table 16.2

$$\Delta c_1 = (2 \cdot 90.00 - 2 \cdot 29.04 - 25.61)\,\text{J mol}^{-1}\text{ K}^{-1}$$
$$= 96.3\,\text{J mol}^{-1}\text{ K}^{-1}$$

Δc_2 and Δc_3 are calculated correspondingly. Then we obtain

$$\int_{298}^{350} \Delta C_{P,m}^{\ominus} \cdot dT = \left[96.3 \cdot (350 - 298) - \frac{1}{2} 0.204 \cdot (350^2 - 298^2)\right.$$
$$\left. + \frac{1}{3} 31 \times 10^{-5} \cdot (350^3 - 298^3)\right]\,\text{J mol}^{-1}$$
$$= (5010 - 3437 + 1696)\,\text{J mol}^{-1} = 3.27\,\text{kJ mol}^{-1}$$

This is practically the same value as we obtain by using the averages $\overline{C}_{P,m}^{\ominus}$ in Table 16.1:

$$\int_{298}^{350} \Delta C_{P,m}^{\ominus} \cdot dT \approx \Delta \overline{C}_{P,m}^{\ominus} \cdot (T_2 - T_1)$$
$$= 63.6\,\text{J K}^{-1}\text{ mol}^{-1} \cdot 52\,\text{K} = 3.31\,\text{J mol}^{-1}$$

Continued on page 541

FOUNDATION 16.1

Continued from page 540

Example – detonating gas reaction (temperature range 298 to 1500 K)

Now we extend the temperature range to 1500 K. In this case we have to consider that water evaporates at 373 K at an applied pressure of 1 bar. To calculate $\Delta C_{P,m}^{\ominus}$ we use the data for liquid water up to 373 K and the data for gaseous water above 373 K. Then

$$\int_{298}^{1500} \Delta C_{P,m}^{\ominus} \cdot dT = \int_{298}^{373} \Delta C_{P,m}^{\ominus} \cdot dT + \int_{373}^{1500} \Delta C_{P,m}^{\ominus} \cdot dT$$

$$= (4.72 - 15.5)\, \text{kJ mol}^{-1}$$

Note that the second term is negative. This is because the heat capacity of gaseous water is much smaller than that of liquid water.

To calculate $\Delta_r H_{1500}^{\ominus}$ from $\Delta_r H_{298}^{\ominus}$ we have to account for the enthalpy change for the evaporation of two moles of liquid water at 373 K ($\Delta_{vap} H^{\ominus} = 40.95\, \text{kJ mol}^{-1}$):

$$\Delta_r H_{1500}^{\ominus} = \Delta_r H_{298}^{\ominus} + \int_{298}^{373} \Delta C_{P,m}^{\ominus} \cdot dT + 2 \times \Delta_{vap} H^{\ominus} + \int_{373}^{1500} \Delta C_{P,m}^{\ominus} \cdot dT$$

$$= (-571.8 + 4.72 + 81.9 - 15.5)\, \text{kJ mol}^{-1}$$

$$= -500.6\, \text{kJ mol}^{-1}$$

For the reverse reaction

$$2H_2O \longrightarrow 2H_2 + O_2$$

we obtain $\Delta_r H_{1500}^{\ominus} = +500.6\, \text{kJ mol}^{-1}$.

Example – burning limestone

For the reaction

$$CaCO_3 \longrightarrow CaO + CO_2$$

we obtain from Table 16.2

$$\int_{298}^{1200} \Delta C_{P,m}^{\ominus} \cdot dT = 2.67 \times 902\, \text{J mol}^{-1} + \left(-\frac{1}{2} 0.027 \cdot 1.35 \times 10^6\right) \text{J mol}^{-1}$$

$$+ \frac{1}{3} 0.74 \times 10^{-5} \cdot 1.70 \times 10^9\, \text{J mol}^{-1}$$

$$= -11.7\, \text{kJ mol}^{-1}$$

Continued on page 542

Continued from page 541

Then
$$\Delta_r H^\ominus_{1200} = (177.4 - 11.7)\,\text{kJ mol}^{-1} = 165.7\,\text{kJ mol}^{-1}$$

$\Delta_r H^\ominus_{1200}$ is smaller than $\Delta_r H^\ominus_{298}$ because the heat capacity of $CaCO_3$ is much larger than the sum of the heat capacities of CaO and CO_2 (see Figure 16.11).

Foundation 16.2 – How to Calculate ΔG_{T_2} from ΔG_{T_1}

We calculate ΔG according to
$$\Delta G = \Delta H - T \cdot \Delta S$$

In Foundation 16.1 we demonstrated how ΔH_{T_2} can be calculated from ΔH_{T_1}. In analogy we calculate
$$\Delta S_{T_2} = \Delta S_{T_1} + \int_{T_1}^{T_2} \frac{\Delta C_P}{T} \cdot dT$$

As in Foundation 16.1 we approximate $C^\ominus_{P,m}$ by a power series
$$C^\ominus_{P,m} = c_1 + c_2 T + c_3 T^2$$

Then for reaction
$$\alpha A + \beta B + \cdots \longrightarrow \gamma C + \delta D + \cdots$$

we obtain
$$\int_{T_1}^{T_2} \frac{\Delta C^\ominus_{P,m}}{T} \cdot dT = \int_{T_1}^{T_2} \left(\frac{\Delta c_1}{T} + \Delta c_2 + \Delta c_3 \cdot T\right) \cdot dT$$
$$= \Delta c_1 \cdot \ln\frac{T_2}{T_1} + \Delta c_2 \cdot (T_2 - T_1) + \frac{1}{2}\Delta c_3 \cdot (T_2^2 - T_1^2)$$

with
$$\Delta c_1 = \gamma c_1(C) + \delta c_1(D) + \cdots - \alpha c_1(A) - \beta c_1(B) - \cdots$$

(Δc_2, Δc_3 correspondingly).

Example – detonating gas reaction

$$2H_2 + O_2 \longrightarrow 2H_2O$$

Continued on page 543

FOUNDATION 16.2

Continued from page 542

From Table 16.2 we calculate $\Delta c_1 = 96.3 \, \text{J mol}^{-1} \, \text{K}^{-1}$, $\Delta c_2 = -0.204 \, \text{J mol}^{-1} \, \text{K}^{-2}$ and $\Delta c_3 = 31 \times 10^{-5} \, \text{J mol}^{-1} \, \text{K}^{-3}$. Then we obtain

$$\int_{298}^{350} \frac{\Delta C_{P,m}^{\ominus}}{T} \cdot dT = 10.1 \, \text{J K}^{-1} \, \text{mol}^{-1}$$

and

$$\Delta_r S_{350}^{\ominus} = (-326.4 + 10.1) \, \text{J K}^{-1} \, \text{mol}^{-1} = -316.3 \, \text{J K}^{-1} \, \text{mol}^{-1}$$

Above 373 K H$_2$O is in the gaseous state. Then

$$\Delta_r S_{1500}^{\ominus} = \Delta_r S_{298}^{\ominus} + \int_{298}^{373} \frac{\Delta C_{P,m}^{\ominus}(\text{liquid})}{T} \cdot dT$$

$$+ \frac{2 \times \Delta_{vap} H^{\ominus}}{T_{vap}} + \int_{373}^{1500} \frac{\Delta C_{P,m}^{\ominus}(\text{gas})}{T} \cdot dT$$

$$= (-326.4 + 14.1 + 219.6 - 21.8) \, \text{J K}^{-1} \, \text{mol}^{-1}$$

$$= -114.5 \, \text{J K}^{-1} \, \text{mol}^{-1}$$

(with $\Delta_{vap} H^{\ominus} = 40.95 \, \text{kJ mol}^{-1}$ at $T = 373 \, \text{K}$). Then with $\Delta_r H_{1500}^{\ominus}$ from Foundation 16.1 we obtain

$$\Delta_r G_{1500}^{\ominus} = (-500.6 \times 10^3 + 1500 \cdot 114.5) \, \text{J mol}^{-1} = -328.9 \, \text{kJ mol}^{-1}.$$

For the reverse reaction

$$2H_2O \longrightarrow 2H_2 + O_2$$

we obtain $\Delta_r G_{1500}^{\ominus} = +328.9 \, \text{kJ mol}^{-1}$.

Example – burning limestone

$$CaCO_3 \longrightarrow CaO + CO_2$$

$$\Delta_r S_{1200}^{\ominus} = \Delta_r S_{298}^{\ominus} + \int_{298}^{1200} \frac{\Delta C_{P,m}^{\ominus}}{T} \cdot dT$$

where $\Delta_r S_{m,298}^{\ominus} = 160.4 \, \text{J K}^{-1} \, \text{mol}^{-1}$ and from Table 16.2 we calculate $\Delta c_1 = 2.67 \, \text{J mol}^{-1} \, \text{K}^{-1}$, $\Delta c_2 = -0.027 \, \text{J mol}^{-1} \, \text{K}^{-2}$ and $\Delta c_3 = 0.74 \times 10^{-5} \, \text{J mol}^{-1} \, \text{K}^{-3}$. Then we obtain

$$\int_{298}^{1200} \frac{\Delta C_{P,m}^{\ominus}}{T} \cdot dT = \left[2.67 \cdot \ln \frac{1200}{298} + (-0.027 \cdot 902) \right.$$

$$\left. + \frac{1}{2} 0.74 \times 10^{-5} \cdot 1.35 \times 10^6 \right] \, \text{J K}^{-1} \, \text{mol}^{-1}$$

$$= -15.7 \, \text{J K}^{-1} \, \text{mol}^{-1}$$

Continued on page 544

Continued from page 543

Then

$$\Delta_r S^\ominus_{1200} = (160.4 - 15.7)\,\text{J K}^{-1}\,\text{mol}^{-1} = 144.7\,\text{J K}^{-1}\,\text{mol}^{-1}.$$

With $\Delta_r H^\ominus_{1200}$ from Foundation 16.1 we obtain

$$\Delta_r G^\ominus_{1200} = (165.7 \times 10^3 - 1200 \cdot 144.7)\,\text{J mol}^{-1} = -8.0\,\text{kJ mol}^{-1}.$$

17 Chemical Equilibrium

A chemical reaction at constant pressure can only be spontaneous if the condition $\Delta G < 0$ or, for a chemical reaction in molar terms, $\Delta_r G < 0$ is fulfilled. In the limiting case $\Delta G = 0$ (or $\Delta_r G = 0$) an equilibrium is established. In Chapter 16 we considered reactions where the reacting species were separated in the beginning and at the end. Now we extend this consideration to reactions in gas mixtures or dilute solutions. Again we consider the establishment of an equilibrium. As an example, we discuss the dissociation of water vapor into H_2 and O_2 at high temperature

$$H_2O \longrightarrow H_2 + \tfrac{1}{2}O_2$$

The partial pressures in the container are P_{H_2O}, P_{H_2}, and P_{O_2}. What is the condition for the above reaction to be spontaneous (i.e., to take place immediately or after ignition)?

What is the mixing ratio of H_2O, H_2, and O_2 in the equilibrium state? How does this ratio change with temperature and pressure? These questions can be answered by the fundamental equilibrium relation

$$\Delta_r G^\ominus = -RT \cdot \ln K$$

K is the equilibrium constant

$$K = \left(\frac{\widehat{P}_{H_2} \cdot \widehat{P}_{O_2}^{1/2}}{\widehat{P}_{H_2O}} \right)_{eq}$$

It gives the relation between P_{H_2O}, P_{H_2}, and P_{O_2} in the case of equilibrium. In this relation $\widehat{P}_{H_2} = P_{H_2}/P^\ominus$ is the numerical value of the partial pressure, measured in units of the standard pressure $P^\ominus = 1$ bar, in the equilibrium mixture; P_{O_2} and P_{H_2O} are defined correspondingly.

$\Delta_r G^\ominus$ must be distinguished from $\Delta_r G$. $\Delta_r G$ is the change in Gibbs energy when 1 mol of H_2O dissociates into H_2 and O_2 in a large container with a surplus of H_2O, H_2, and O_2 at any given partial pressures P_{H_2O}, P_{H_2}, and P_{O_2}. In contrast, $\Delta_r G^\ominus$ is the change in Gibbs energy when the dissociation of 1 mol of H_2O takes place under standard conditions (initial state: 1 mol of water vapor at pressure $P^\ominus = 1$ bar; final state: 1 mol of H_2 and 0.5 mol of O_2, separated each at pressure $P^\ominus = 1$ bar).

These results can be immediately extended to dilute solutions. In this case the particles of the solute (molecules or ions) do not interact with each other, like molecules in an ideal gas. In the expression for the equilibrium constant the gas pressure P has to be replaced by the osmotic pressure ^{osm}P (the pressure on a piston between solution and solvent which is permeable to the solvent and impermeable to the solute).

17.1 ΔG for Reactions in Gas Mixtures

17.1.1 Mass Action Law and Equilibrium Constant K

Reaction in a closed container (T = const) at constant volume V.

We consider the reaction

$$H_2O \longrightarrow H_2 + \tfrac{1}{2}O_2 \qquad (17.1)$$

under conditions where H_2O is in the gaseous state. We assume that H_2, O_2, and gaseous H_2O are enclosed in a container, and the partial pressures of the gases are P_{H_2}, P_{O_2}, and P_{H_2O}. We imagine that a small amount of H_2O (n_{H_2O}) reacts to form H_2 and O_2, according to equation (17.1). Can this process occur spontaneously, provided that conditions are such that the reaction is not hindered kinetically?

To answer this question we imagine the same change of state performed reversibly and we determine the reversible work w_{rev}. A spontaneous reaction in the container is only possible if work is done by the reacting system on the surroundings in this reversible change of state, i.e., if w_{rev} is negative (or zero in the limiting case) according to equations (16.32) and (16.38):

$$w_{rev} = \Delta A \leq 0 \qquad (17.2)$$

(reaction at $T = $ const and $V = $ const). To perform reaction (17.1) reversibly we consider a second container where H_2, O_2, and H_2O are at the same temperature as in the first container, but under the equilibrium partial pressures $P_{H_2,eq}$, $P_{O_2,eq}$, and $P_{H_2O,eq}$ (Figure 17.1).

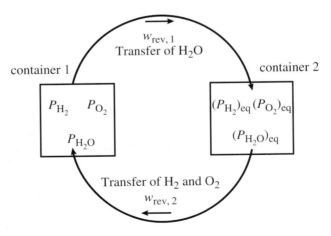

Figure 17.1 Performing the detonating gas reaction (17.1) reversibly at $T = $ const and $V = $ const. Reacting species under arbitrary partial pressures are in container 1 and those under equilibrium pressures are in container 2.

17.1 ΔG FOR REACTIONS IN GAS MIXTURES

Then we take a small amount of H_2O from container 1 and transfer it reversibly into container 2; correspondingly we transfer the products H_2 and O_2 from container 2 to container 1. In order to transfer H_2O from container 1 to container 2 reversibly, the partial pressure of H_2O must be changed reversibly from P_{H_2O} to $P_{H_2O,eq}$. In the ideal gas approximation the reversible work needed for this process is

$$w_{rev,H_2O} = n_{H_2O} RT \cdot \ln \frac{P_{H_2O,eq}}{P_{H_2O}} \tag{17.3}$$

Then w_{rev} for the whole process is

$$w_{rev} = n_{H_2O} RT \cdot \ln \frac{P_{H_2O,eq}}{P_{H_2O}} + n_{H_2} RT \cdot \ln \frac{P_{H_2}}{P_{H_2,eq}} + n_{O_2} RT \cdot \ln \frac{P_{O_2}}{P_{O_2,eq}} \tag{17.4}$$

(H_2 must be taken out from container 1 separately in a reversible manner. To overcome this problem we imagine a filter which is permeable to H_2 molecules, but not to O_2 and H_2O (e.g., palladium foil). In principle it should be possible to separate any gas with its particular semipermeable membrane. Practically this cannot be realized. In our considerations, which are typical for a mental exercise, we are interested in the realization possibility as a matter of principle, and not in the practical realization. Mental exercises play an important role in physical chemistry.)

The amount of H_2 and the amount of O_2 transferred from container 2 to container 1 is related to the amount of H_2O transferred from container 1 to container 2 by the stoichiometry of reaction (17.1):

$$n = n_{H_2O} = n_{H_2} = 2n_{O_2} \tag{17.5}$$

We use the relations $\ln a + \ln b = \ln ab$, and $x \ln b = \ln b^x$, and define the dimensionless quantities

$$\boxed{\widehat{P} = \frac{P}{P^{\ominus}} \quad \text{and} \quad \widehat{P}_{eq} = \frac{P_{eq}}{P^{\ominus}}} \tag{17.6}$$

($P^{\ominus}$ = standard pressure). Then we obtain from equation (17.4):

$$\boxed{w_{rev} = -nRT \cdot \ln K + nRT \cdot \ln Q} \tag{17.7}$$

In this expression

$$K = \left(\frac{\widehat{P}_{H_2} \cdot \widehat{P}_{O_2}^{1/2}}{\widehat{P}_{H_2O}} \right)_{eq} \tag{17.8}$$

is the *equilibrium constant* (involving the equilibrium pressures $P_{H_2,eq}$, $P_{O_2,eq}$, and $P_{H_2O,eq}$) and

$$Q = \frac{\widehat{P}_{H_2} \cdot \widehat{P}_{O_2}^{1/2}}{\widehat{P}_{H_2O}} \tag{17.9}$$

is the *reaction quotient* (involving the arbitrary pressures P_{H_2}, P_{O_2}, and P_{H_2O} in container 1).

We distinguish three different cases:

- *Case 1.* $Q < K$, then $w_{rev} < 0$ (spontaneous reaction). In this case the partial pressure of H_2O is higher than in equilibrium; thus the reaction will proceed toward the products H_2 and O_2 to reach equilibrium.
- *Case 2.* $Q > K$, then $w_{rev} > 0$. In this case the partial pressure of H_2O is lower than in equilibrium; thus the reaction will proceed towards the reactant H_2O to reach equilibrium: the reverse reaction to reaction (17.1) occurs spontaneously.
- *Case 3.* $Q = K$, then $w_{rev} = 0$ (equilibrium).

The equilibrium case can be realized in various ways. Whenever the partial pressures in the container fulfill relation (17.8), the reacting species are in equilibrium regardless of the special choice of P_{H_2}, P_{O_2}, or P_{H_2O} (*mass action law*). For example, P_{O_2} does not change if we arbitrarily increase the partial pressures of H_2O and of H_2 by the same factor.

Reaction (T = const) at constant pressure P

We have assumed that the reacting gases in container 1 (Figure 17.1) are at constant volume. Now we consider the same reaction when the reactants are in a container under a constant pressure P. This is useful to get familiar with the concepts introduced in Chapter 16. The volume changes when the reaction occurs and we obtain an additional contribution $-P\Delta V$ to w_{rev}:

$$w_{rev} = -n\boldsymbol{R}T \cdot \ln K + n\boldsymbol{R}T \cdot \ln Q - P\Delta V \tag{17.10}$$

Remember that ΔG is the reversible work dimished by the contribution of w_{rev} done on the system by the external pressure P. Then according to equation (16.39)

$$\boxed{\Delta G = w_{rev} + P\Delta V = -n\boldsymbol{R}T \cdot \ln K + n\boldsymbol{R}T \cdot \ln Q} \tag{17.11}$$

In our example, assuming ideal gas behavior, the volume increases by

$$\Delta V = n_{O_2}\frac{\boldsymbol{R}T}{P} + n_{H_2}\frac{\boldsymbol{R}T}{P} - n_{H_2O}\frac{\boldsymbol{R}T}{P} \tag{17.12}$$

and we obtain, because of equation (17.5),

$$P\Delta V = \boldsymbol{R}T \cdot \left(\tfrac{1}{2} + 1 - 1\right)n = \tfrac{1}{2}n\boldsymbol{R}T \tag{17.13}$$

Again, we can consider three different cases:

- case 1. $Q < K$, $\Delta G < 0$, reaction occurs spontaneously.
- case 2. $Q > K$, $\Delta G > 0$, reverse reaction occurs spontaneously.
- case 3. $Q = K$, $\Delta G = 0$, equilibrium.

It is useful to write equation (17.11) in the form

$$\Delta_r G = -\boldsymbol{R}T \cdot \ln K + \boldsymbol{R}T \cdot \ln Q \tag{17.14}$$

where $\Delta_r G$ is the change in Gibbs energy per mol dissociating water vapor under the given conditions.

17.1 ΔG FOR REACTIONS IN GAS MIXTURES

Generalization

The result can be easily generalized for the reaction

$$\alpha A + \beta B + \cdots \longrightarrow \gamma C + \delta D + \cdots \qquad (17.15)$$

carried out in a large container at constant pressure P with species A, B, C, and D at any given partial pressures P_A, P_B, P_C, and P_D.

Then, as a generalization of equation (17.14), we obtain

$$\boxed{\Delta_r G = -RT \cdot \ln K + RT \cdot \ln Q} \qquad (17.16)$$

where K is the equilibrium constant:

$$\boxed{K = \left(\frac{\widehat{P}_C^\gamma \cdot \widehat{P}_D^\delta \cdots}{\widehat{P}_A^\alpha \cdot \widehat{P}_B^\beta \cdots}\right)_{eq}} \qquad (17.17)$$

and Q is the reaction quotient:

$$\boxed{Q = \frac{\widehat{P}_C^\gamma \cdot \widehat{P}_D^\delta \cdots}{\widehat{P}_A^\alpha \cdot \widehat{P}_B^\beta \cdots}} \qquad (17.18)$$

$\Delta_r G$ is the change of Gibbs energy when α mol of species A and β mol of species B react forming γ mol of species C and δ mol of species D.

17.1.2 Equilibrium Constant K from $\Delta_r G^\ominus = -RT \cdot \ln K$

In the previous section we discussed the reaction

$$\alpha A + \beta B + \cdots \longrightarrow \gamma C + \delta D + \cdots \qquad (17.19)$$

with the reacting species mixed in one container. Now we consider the same reaction, but with the species A, B, ... and the species C, D, ... in separate reservoirs under standard pressure $P^\ominus = 1$ bar (Figure 17.2).

Amounts n_A, n_B, ... of species A, B, ... are reversibly taken from their reservoirs at $P^\ominus = 1$ bar and reversibly transferred into the equilibrium container. After reaction, the amounts n_C, n_D, ... of species C, D, ... are reversibly transferred into the corresponding reservoirs (reaction at constant temperature and constant pressure).

According to the consideration in Section 17.1.1, we obtain instead of equation (17.10)

$$w_{rev} = -nRT \cdot \ln K - P\Delta V \qquad (17.20)$$

where

$$n = \frac{1}{\alpha} n_A = \frac{1}{\beta} n_B = \frac{1}{\gamma} n_C = \frac{1}{\delta} n_D$$

The term $nRT \cdot \ln Q$ in equation (17.7) does not occur in this expression because the pressures in each container are 1 bar, thus $\ln Q = 0$. The term $-P\Delta V$ originates from the reversible work to take the species A, B, ... out of their reservoirs (work done on the

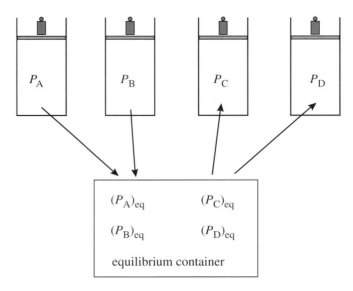

Figure 17.2 Performing reaction (17.19) reversibly at $T = $ const and $P = $ const. Reacting species under standard pressure $P^\ominus = 1$ bar in separate reservoirs.

system by the external pressure of 1 bar) and to bring the species C, D, ... into their reservoirs (work done by the system against the external pressure of 1 bar), correspondingly.

Because of $\Delta G = w_{\text{rev}} + P\Delta V$ ($T = $ const, $P = $ const), equation (17.20) can be written as

$$\Delta G^\ominus = -n\boldsymbol{R}T \cdot \ln K$$

or in the form

$$\boxed{\Delta_r G^\ominus = -\boldsymbol{R}T \cdot \ln K} \tag{17.21}$$

According to Section 16.4 the symbol $\ominus$ indicates that the reactants and products are under standard conditions (separated and at pressure $p^\ominus = 1$ bar). Equation (17.21) enables us to obtain the equilibrium constant from calorimetric data.

In the general case (any pressures P_A, P_B, P_C, and P_D in reservoirs A, B, C, and D) we obtain

$$\boxed{\Delta_r G = \Delta_r G^\ominus + \boldsymbol{R}T \ln Q} \tag{17.22}$$

where Q is given by equation (17.18) with the pressures P_A, P_B, P_C, and P_D in the reservoirs A, B, C, and D. Equation (17.22) enables us to decide if the reaction is spontaneous ($Q < K$; $\Delta G < 0$), in equilibrium ($Q = K$; $\Delta G = 0$), or if the reverse reaction is spontaneous ($Q > K$; $\Delta G > 0$).

$\Delta_r G^\ominus$ is obtained from the standard enthalpies of formation $\Delta_f H^\ominus$ and standard entropies $S^\ominus$ of the reacting species. As outlined in Chapter 16, $\Delta_f G^\ominus$ is tabulated for $T = 298$ K. These data must be converted to the desired temperature T.

Example – dissociation of water vapor

For the reaction

$$H_2O \longrightarrow H_2 + \tfrac{1}{2}O_2$$

17.1 ΔG FOR REACTIONS IN GAS MIXTURES

we find at 1500 K (Foundation 16.2), that $\Delta_r G^\ominus_{1500} = +164.4\,\text{kJ mol}^{-1}$. Then

$$\ln K = -\frac{164.4\,\text{kJ mol}^{-1}}{RT} = -13.2, \quad K = \left(\frac{\widehat{P}_{H_2} \cdot \widehat{P}^{1/2}_{O_2}}{\widehat{P}_{H_2O}}\right)_{eq} = 2 \times 10^{-6}$$

This shows that for $\widehat{P}_{H_2O} = 1$ the equilibrium is completely shifted toward the reactant H_2O. On the other hand, if the reaction is carried out at $T = 4000\,\text{K}$, $\Delta_r G^\ominus$ becomes much smaller, and the equilibrium is shifted towards the products: $\Delta_r G^\ominus_{4000} = +19\,\text{kJ mol}^{-1}$ and $K_{4000} = 0.6$. In this case, 80% of the water (initial pressure 1 bar) is dissociated into H_2 and O_2.

In the following we will omit, for simplicity, the index "eq" in the expression for the equilibrium constant K, since it is obvious that equilibrium conditions are considered in connection with K.

Example – detonating gas reaction

For the reverse reaction

$$H_2 + \tfrac{1}{2}O_2 \longrightarrow H_2O$$

we find $\Delta_r G^\ominus_{1500} = -158\,\text{kJ mol}^{-1}$ and the corresponding equilibrium constant is

$$K = \frac{\widehat{P}_{H_2O}}{\widehat{P}_{H_2} \cdot \widehat{P}^{1/2}_{O_2}} = \frac{1}{2 \times 10^{-6}} = 0.5 \times 10^6$$

Example – reaction written in different form

If we write the reaction in the form

$$2H_2 + O_2 \longrightarrow 2H_2O$$

we find $\Delta_r G^\ominus_{1500} = -328.8\,\text{kJ mol}^{-1}$. Then

$$K = \frac{(\widehat{P}_{H_2O})^2}{(\widehat{P}_{H_2})^2 \cdot \widehat{P}_{O_2}} = \left(\frac{1}{2 \times 10^{-6}}\right)^2 = 2.5 \times 10^{11}$$

Note that the equilibrium constant for one and the same equilibrium depends on the way the reaction is written:

$$H_2 + \tfrac{1}{2}O_2 \longrightarrow H_2O \qquad K = 0.5 \times 10^6$$
$$2H_2 + O_2 \longrightarrow 2H_2O \qquad K = 2.5 \times 10^{11}$$
$$H_2O \longrightarrow H_2 + \tfrac{1}{2}O_2 \qquad K = 2.0 \times 10^{-6}$$

To avoid confusion we use the symbol "$\rightarrow$" independent of whether or not the equilibrium is shifted to the left or to the right. When using the familiar symbol "$\rightleftharpoons$" the equilibrium constant K, by definition, is related to the reaction from left to right.

Example – formation of NH$_3$ from the elements at 298 K and 700 K

For the reaction
$$3H_2 + N_2 \longrightarrow 2NH_3$$

we find (Problem 17.4) $\Delta_r G^\ominus_{298} = -33.2 \text{ kJ mol}^{-1}$. Then

$$\ln K = -\frac{-33.2 \text{ kJ mol}^{-1}}{RT} = 13.4 \quad \text{and} \quad K = \frac{\widehat{P}^2_{NH_3}}{\widehat{P}^3_{H_2} \widehat{P}_{N_2}} = 6.6 \times 10^5$$

At this temperature the equilibrium is completely shifted in favor of the product NH$_3$, if the total pressure is 1 bar. Thus, the reaction can occur spontaneously, according to this criterion. However, because of kinetic reasons, the reaction cannot take place without an appropriate catalyst. Such catalysts require temperatures above 700 K.

At $T = 700$ K, however, we find (Problem 17.4) $\Delta_r G^\ominus_{700} = +54 \text{ kJ mol}^{-1}$ and $K = 9.3 \times 10^{-5}$. Then only a very small fraction of H$_2$ and N$_2$ can be converted into NH$_3$ under the same conditions.

The fraction of H$_2$ converted into NH$_3$ depends strongly on the total pressure

$$P_{\text{total}} = P_{H_2} + P_{N_2} + P_{NH_3}$$

As shown in Box 17.1, increasing P_{total} to 300 bar increases the ratio $P_{NH_3} : P_{\text{total}}$ to 0.37 for a stoichiometric mixture of H$_2$ and N$_2$. These considerations are important for the industrial synthesis of NH$_3$ from the elements. Only at high temperatures is a catalyst available to accelerate the reaction. Thus the reaction must be performed at high temperature (700 K) and high pressure (400–600 bar): the *Haber–Bosch process*.

Box 17.1 – Synthesis of Ammonia from Its Elements

For the reaction
$$N_2 + 3H_2 \longrightarrow 2NH_3$$

the equilibrium constant at 700 K is ($\widehat{P} = P/P^\ominus$)

$$K = \frac{\widehat{P}^2_{NH_3}}{\widehat{P}_{N_2} \widehat{P}^3_{H_2}} = 9.3 \times 10^{-5}$$

What is the yield of NH$_3$ at different total pressures $P_{\text{total}} = P_{H_2} + P_{N_2} + P_{NH_3}$?

We restrict our discussion to the case where N$_2$ and H$_2$ are in the stoichiometric ratio 1 : 3. Then in the beginning of the reaction the pressures of N$_2$ and H$_2$ are P_{N_2} and $3P_{N_2}$, respectively. When the equilibrium has been established, the equilibrium pressures are $P_{N_2,\text{eq}} = P_{N_2} - x$, $P_{H_2,\text{eq}} = 3P_{N_2} - 3x$, $P_{NH_3,\text{eq}} = 2x$, and the total pressure is $P_{\text{total,eq}} = 4P_{N_2} - 2x$. Then with $\widehat{x} = x/P^\ominus$ we

Continued on page 553

Continued from page 552

obtain

$$K = \frac{4\widehat{x}^2}{27(\widehat{P}_{N_2} - \widehat{x})^4} \quad \text{and} \quad \frac{\widehat{x}}{(\widehat{P}_{N_2} - \widehat{x})^2} = \sqrt{\frac{27}{4}K}$$

resulting in

$$\widehat{x} = \widehat{P}_{N_2} + \frac{1}{2a} - \sqrt{\frac{\widehat{P}_{N_2}}{a} + \frac{1}{4a^2}} \quad \text{with} \quad a = \sqrt{\frac{27}{4}K}$$

For different pressures P_{N_2} the following equilibrium pressures are obtained:

$\widehat{P}_{N_2}$	$\widehat{x}$	$\widehat{P}_{total,eq}$	$\widehat{P}_{NH_3,eq}$	$\widehat{P}_{H_2,eq}$	$\widehat{P}_{N_2,eq}$	$\dfrac{\widehat{P}_{NH_3,eq}}{\widehat{P}_{total,eq}}$
1	0.023	3.95	0.05	2.93	0.98	0.013
30	9.95	100	19.9	60.2	20.1	0.20
100	53.5	293	107	140	47	0.36
200	130	540	260	210	70	0.48

The strong shift of the equilibrium with pressure favoring the product NH_3 is due to the fact that the number of molecules is decreasing during the reaction (three molecules of H_2 react with one molecule of N_2 to produce two molecules of NH_3). Then the pressure decreases at constant volume or the volume decreases at constant pressure (Le Chatelier's principle of minimum constraint). For instance, at $T = 700$ K and $P_{H_2} = P_{N_2} = 1$ bar we obtain

$$P_{NH_3} = \sqrt{9.3 \times 10^{-5}} \text{ bar} = 0.0096 \text{ bar}$$

and at $P_{H_2} = P_{N_2} = 100$ bar we obtain

$$P_{NH_3} = \sqrt{9.3 \times 10^{-5} \times (100)^4} \text{ bar} = 96 \text{ bar}$$

17.1.3 Reactions Involving Gases and Immiscible Condensed Species

We consider the combustion of benzene

$$C_6H_6 \text{ (liquid)} + \tfrac{15}{2}O_2 \longrightarrow 6CO_2 + 3H_2O \text{ (liquid)} \tag{17.23}$$

Benzene and water are assumed to be incompressible, and their miscibility with the other partners is neglected. Then the transfer of these species from the reservoir to the equilibrium container and vice versa (a simple shift) does not contribute to ΔG. Thus these species do not appear in the expression for K:

$$\Delta_r G^\ominus = -RT \cdot \ln K \tag{17.24}$$

$$K = \frac{(\widehat{P}_{CO_2})^6}{(\widehat{P}_{O_2})^{15/2}} \tag{17.25}$$

Burning limestone (see Chapter 16)

$$CaCO_3 \text{ (solid)} \longrightarrow CaO \text{ (solid)} + CO_2 \tag{17.26}$$

is a very simple example of this kind:

$$K = \widehat{P}_{CO_2} \tag{17.27}$$

17.1.4 Van't Hoff Equation

According to equation (17.21), the equilibrium constant K is determined by $\Delta_r G^\ominus$

$$\ln K = -\frac{\Delta_r G^\ominus}{RT} \tag{17.28}$$

Changing the temperature changes $\Delta_r G^\ominus$ and thus K (see Section 17.1.2). To determine the temperature dependence of K in a small temperature interval we derive from equation (17.28)

$$\frac{d \ln K}{dT} = -\frac{1}{R}\left(\frac{1}{T}\frac{d\Delta_r G^\ominus}{dT} - \Delta_r G^\ominus \frac{1}{T^2}\right) \quad \text{with} \quad \Delta_r G^\ominus = \Delta_r H^\ominus - T \cdot \Delta_r S^\ominus \tag{17.29}$$

In the general case $\Delta G = \Delta H - T \cdot \Delta S$, we find

$$\left(\frac{\partial \Delta G}{\partial T}\right)_P = \left(\frac{\partial \Delta H}{\partial T}\right)_P - T \cdot \left(\frac{\partial \Delta S}{\partial T}\right)_P - \Delta S \tag{17.30}$$

According to Sections 16.2.1 and 16.4.3, we obtain

$$\left(\frac{\partial \Delta H}{\partial T}\right)_P = \Delta C_P, \quad \left(\frac{\partial \Delta S}{\partial T}\right)_P = \frac{\Delta C_P}{T} \tag{17.31}$$

Then we obtain

$$\left(\frac{\partial \Delta G}{\partial T}\right)_P = \Delta C_P - T \cdot \frac{\Delta C_P}{T} - \Delta S = -\Delta S \tag{17.32}$$

and

$$\frac{d\Delta_r G^\ominus}{dT} = \left(\frac{\partial \Delta_r G}{\partial T}\right)_{P=P^\ominus} = -\Delta_r S^\ominus$$

Then from equation (17.29) we obtain

$$\frac{d \ln K}{dT} = -\frac{1}{R} \cdot \left(-\frac{\Delta_r S^\ominus}{T} - \frac{\Delta_r H^\ominus}{T^2} + \frac{\Delta_r S^\ominus}{T}\right)$$

The first term and the third term cancel each other, thus the final result is

$$\boxed{\frac{d \ln K}{dT} = \frac{\Delta_r H^\ominus}{RT^2}} \tag{17.33}$$

(*van't Hoff equation*). This equation enables us to judge if K is increasing or decreasing with temperature. In addition, it is useful to obtain the enthalpy change of a reaction by measuring the equilibrium constant at different temperatures.

17.1 ΔG FOR REACTIONS IN GAS MIXTURES

If $\Delta H^\ominus$ is positive (endothermic reaction), then $\mathrm{d}\ln K/\mathrm{d}T$ is positive: K increases with temperature, that is, the equilibrium is shifted toward the products. This result is in accordance with *Le Chatelier's principle* of minimum constraint:

When an independent variable of a system at equilibrium is changed, the equilibrium shifts in the direction that tends to reduce the effect of this change. For example, when pressure is increased, the equilibrium shifts in the direction to reduce the number of molecules. If the temperature is increased in an endothermic reaction, the equilibrium shifts in the direction to decrease the number of the reactants.

17.1.5 Statistical Interpretation of K

It may be surprising that an energy-rich species can be a main product of a reaction. However, this fact can be immediately grasped by considering the number of configurations of that species. Let us consider the isomerization reaction between i-butane and n-butane:

```
         H
       H C H
   H   |   H                    H   H   H   H
   H C—C—C H       →        H C—C—C—C H
   H   H   H                    H   H   H   H
      i-Butane                      n-Butane
```

In this case the equilibrium constant

$$K = \frac{P_{n\text{-butane}}}{P_{i\text{-butane}}} \quad (17.34)$$

depends strongly on temperature. At low temperature the equilibrium is completely on the left-hand side, and with increasing temperature it is shifted to the right-hand side. Can we understand this astonishing behavior on grounds of molecular considerations?

i-Butane is more rigid than n-butane; thus the vibration modes of i-butane occur at higher energies than those for n-butane. On the other hand, the electronic energy of i-butane is lower than that for n-butane, because of the stronger C–C bonds (Figure 17.3). At low enough temperature all molecules accumulate in the lowest energy level of i-butane: K is small, and the isomerization reaction strongly favors reactants. At high temperature, energy levels in i-butane and in n-butane can be populated according to the Boltzmann distribution law. Because of the smaller spacing of the n-butane levels, there are many more molecules in the n-butane than in the i-butane: K is increasing with temperature.

In the present case the electronic energy is higher for the product ($\Delta H > 0$); because of the increasing number Ω of configurations $\Delta S > 0$. Thus $K < 1$ at low temperature, and $K > 1$ at high temperature (Figure 17.4(a)).

Now imagine a process where $\Delta H > 0$, but $\Delta S < 0$ (Figure 17.4(b)); then K can never reach 1 at high temperature. Correspondingly, the cases $\Delta H < 0$, $\Delta S > 0$ and $\Delta H < 0$, $\Delta S < 0$ are displayed in Figures 17.4(c) and (d).

In simple cases the number Ω of configurations of a substance at a given temperature can be calculated on the basis of quantum mechanics (Chapter 14); then equilibrium constants can be obtained directly without using thermochemical data.

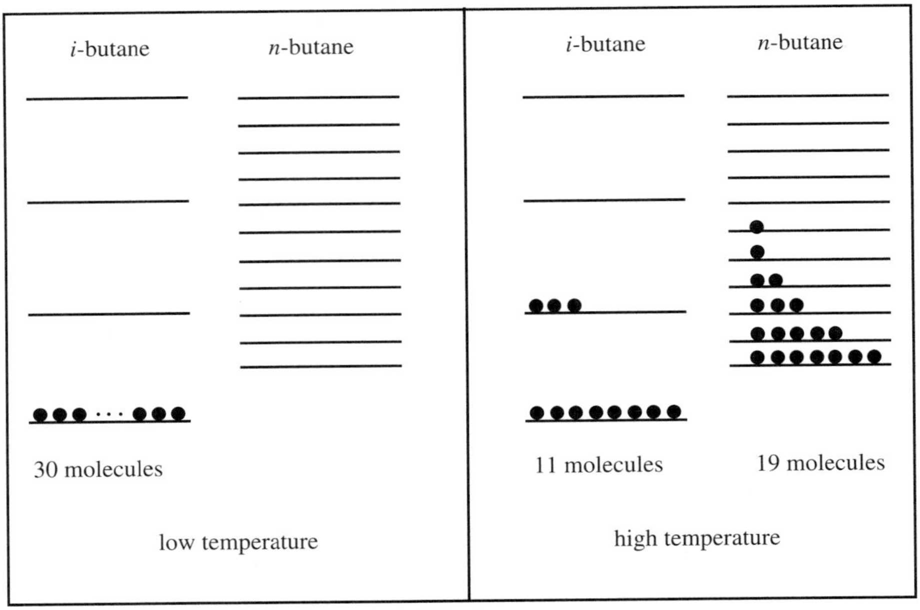

Figure 17.3 Electronic and vibronic energy of i-butane and n-butane. Population at low and high temperature. Low temperature: all molecules are in the i-form. High temperature: most molecules are in the n-form.

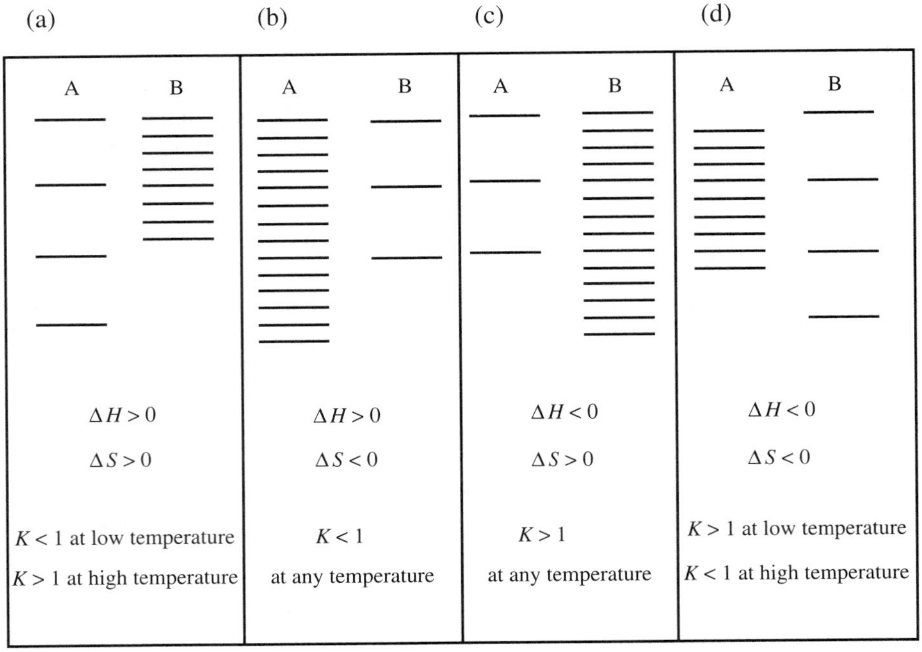

Figure 17.4 Equilibrium constant K of the equilibrium $A \rightleftharpoons B$ for different vibronic energy level systems for A and B.
(a) $\Delta H > 0$, $\Delta S > 0$; (b) $\Delta H > 0$, $\Delta S < 0$;
(c) $\Delta H < 0$, $\Delta S > 0$; (d) $\Delta H < 0$, $\Delta S < 0$.

17.1.6 Estimation of $K = f(T)$

In many cases it is useful to estimate K at a higher temperature, avoiding a rigorous treatment. According to equation (17.28) we obtain for a starting temperature T_1

$$\ln K_1 = -\frac{\Delta_r G_1^\ominus}{RT_1} = -\frac{\Delta_r H_1^\ominus}{RT_1} + \frac{\Delta_r S_1^\ominus}{R} \qquad (17.35)$$

If we neglect the temperature dependence of $\Delta H^\ominus$ and $\Delta S^\ominus$, then it follows for any other temperature T

$$\ln K = -\frac{\Delta_r H_1^\ominus}{RT} + \frac{\Delta_r S_1^\ominus}{R} \qquad (17.36)$$

and

$$\ln K = \ln K_1 - \frac{\Delta_r H_1^\ominus}{R}\left(\frac{1}{T} - \frac{1}{T_1}\right) \qquad (17.37)$$

Example – ammonia synthesis at 298 K and at 700 K

$$N_2 + 3H_2 \longrightarrow 2NH_3$$

$K_{298} = 6.6 \times 10^5$, $\ln K_{298} = 13.4$, $\Delta_r H_{298}^\ominus = -92.4\,\text{kJ mol}^{-1}$. Then from equation (17.37)

$$\ln K_{700} = 13.4 - \frac{-92.4\,\text{kJ mol}^{-1}}{8.314\,\text{J K}^{-1}\,\text{mol}^{-1}}\left(\frac{1}{700\,\text{K}} - \frac{1}{298\,\text{K}}\right) = -8.0$$

The resulting estimate $K_{700} = 3.4 \times 10^{-4}$ is in reasonable agreement with the value 9.3×10^{-5} obtained in the rigorous treatment (see Problem 17.4); we note that the change in K is about nine orders of magnitude. In Figure 17.5 $\ln K$ is displayed in the range $T = 300\,\text{K}$ to $T = 700\,\text{K}$.

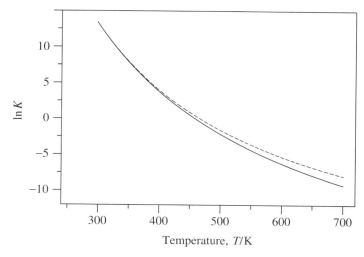

Figure 17.5 $\ln K$ for the formation of NH_3 from the elements H_2 and N_2 versus temperature. Solid line: exact calculation (from $\Delta_r G_T^\ominus$ as calculated in analogy to Foundation 16.2). Dashed line: $\ln K$ approximated by equation (17.37).

17.1.7 ΔH and ΔS from Measured K

In the approximation equation (17.37), $\ln K$ depends linearly on $1/T$. Thus plotting experimental $\ln K$ data versus $1/T$ should result in a straight line with slope $-\Delta_r H^\ominus/R$.

Example – iodine hydrogen reaction

Experimental data for K of the reaction

$$H_2 + I_2 \longrightarrow 2HI$$

are listed in Table 17.1. In Figure 17.6(a), $\ln K$ is plotted as a function of temperature, and in Figure 17.6(b) $\ln K$ is plotted versus $1/T$. From the slope of the straight line in Figure 17.6(b), $\Delta_r H^\ominus = -13.0\,\text{kJ mol}^{-1}$ is obtained. This is not far from $\Delta_r H^\ominus = -13.2\,\text{kJ mol}^{-1}$ calculated in Problem 17.1.

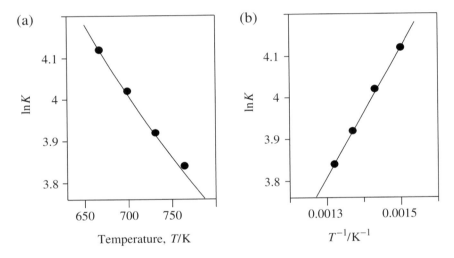

Figure 17.6 Evaluation of $\Delta_r H^\ominus$ for the reaction $I_2 + H_2 \rightarrow 2HI$ at $T = 700\,\text{K}$. For data see Table 17.1 (a) $\ln K$ vs T, (b) $\ln K$ vs 1/T.

Table 17.1 Experimental equilibrium constant $K = \widehat{P}_{HI}^2/(\widehat{P}_{H_2} \cdot \widehat{P}_{I_2})$ at Different Temperatures

T/K	K	$\ln K$	T^{-1}/K^{-1}
667	61.0	4.12	0.00150
699	55.2	4.02	0.00143
731	49.9	3.92	0.00137
764	45.9	3.84	0.00132

17.1.8 Vapor Pressure

In the equilibrium

$$H_2O \text{ (liquid)} \longrightarrow H_2O \text{ (vapor)} \tag{17.38}$$

the liquid is in condensed phase and therefore does not contribute to K. Then, according to Section 17.1.3,

$$K = \widehat{P}_{vap} \tag{17.39}$$

where P_{vap} is the vapor pressure. The liquid can be treated as incompressible, and the vapor can be treated as an ideal gas. In this case, ΔH does not depend on pressure (see Section 16.4.4) and we can replace $\Delta_r H^\ominus$ by $\Delta_{vap} H$, the enthalpy of evaporation (i.e., the heat to vaporize the liquid). Then from the van't Hoff equation (17.33) we obtain

$$\boxed{\frac{d \ln \widehat{P}_{vap}}{dT} = \frac{\Delta_{vap} H}{RT^2}} \tag{17.40}$$

This is the *Clausius and Clapeyron equation*.

Because of $\Delta_{vap} H > 0$ the vapor pressure is strongly increasing with temperature. Accordingly, from equation (17.37) we obtain

$$\ln \widehat{P}_{vap} = \ln \widehat{P}_{vap,1} - \frac{\Delta_{vap} H^\ominus}{R} \left(\frac{1}{T} - \frac{1}{T_1} \right) \tag{17.41}$$

As an example, the vapor pressure for water is displayed in Figure 17.7 (solid lines) as a function of temperature.

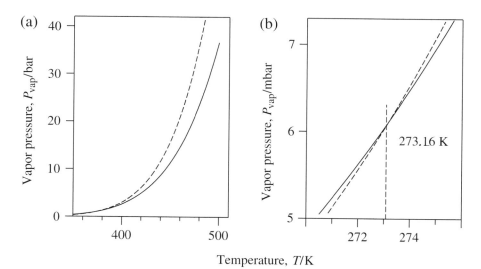

Figure 17.7 Vapor pressure P_{vap} of water as a function of temperature calculated from equation (17.41). Solid lines: vapor pressure above liquid water ($\Delta_{vap} H = 44.1 \text{ kJ mol}^{-1}$, $\widehat{P}_{vap,1} = 1$ at $T_1 = 373 \text{ K}$). Dashed lines: vapor pressure above solid water ($\Delta_{sub} H = (6.0 + 44.1) \text{ kJ mol}^{-1} = 50.1 \text{ kJ mol}^{-1}$, $\widehat{P}_{1,vap} = 6.11 \times 10^{-3}$ at the triple point $T_1 = 273.16 \text{ K}$) (a) Temperature range 350–500 K; (b) behavior at the triple point. Compare with Fig. 11.33.

Correspondingly, equations (17.40) and (17.41) can be applied to the vapor pressure above a solid – for example, in the equilibrium

$$H_2O \text{ (solid)} \longrightarrow H_2O \text{ (vapor)} \tag{17.42}$$

Then $\Delta_{vap}H$ has to be replaced by the molar enthalpy of sublimation:

$$\Delta_{sub}H = \Delta_{fus}H + \Delta_{vap}H \tag{17.43}$$

where $\Delta_{fus}H$ is the molar enthalpy of melting. Because of

$$\Delta_{sub}H > \Delta_{vap}H \tag{17.44}$$

the vapor pressure above a solid is more strongly increasing with temperature than the vapor pressure above the corresponding liquid (Figure 17.7, dashed lines).

17.2 ΔG for Reactions in Dilute Solution

17.2.1 Osmotic Pressure and Concentration

We consider an ideally diluted solution, where the dissolved molecules, similar to the particles of an ideal gas, do not interact with one another. This is the case for a dilute solution of sugar in water; the solution is in a container connected with a piston which allows water molecules to pass through, but not sugar molecules (Figure 17.8).

The dissolved molecules have a tendency to be diluted by the pure solvent above the piston, thus exerting a pressure on the piston. The influx of solvent is stopped by putting a corresponding weight on the piston. The pressure exerted on the piston is called *osmotic pressure* ^{osm}P. If we decrease the weight, the piston is pushed up, and the solution is diluted. This process takes place until the pressure exerted by the diluted solution corresponds to the smaller weight.

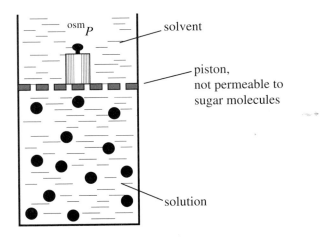

Figure 17.8 Osmotic pressure ^{osm}P of a solution of sugar in water. The semipermeable piston is a membrane with small pore size, permeable for water and inpermeable for sugar (e.g., a porous glass disk treated with copper(II) hexacyanoferrate(II)).

17.2 ΔG FOR REACTIONS IN DILUTE SOLUTION

When we increase the weight in Figure 17.8 slightly, the piston moves downward by a distance ds. In this process, reversible work

$$dw = -{}^{osm}P \cdot dV \tag{17.45}$$

is done on the system. If the system is in contact with a temperature bath of temperature T, the process is isothermal. Then

$$dw = dU - dq = dU - T \cdot dS \tag{17.46}$$

and therefore

$$^{osm}P = -\frac{dU}{dV} + T\frac{dS}{dV} \tag{17.47}$$

Because we have assumed that the sugar molecules do not interact with each other, the internal energy U does not change when changing the volume V. Then

$$^{osm}P = T\frac{dS}{dV} \quad (T = \text{const}) \tag{17.48}$$

When decreasing the volume V ($dV < 0$), the entropy S decreases, as we can see by considering the number of configurations Ω in the beginning and $\Omega + d\Omega$ at the end of the process. The treatment of gases in Section 14.2 can be applied to the present case, since the interactions with the solvent remain unchanged on diluting the solution. The number of configurations increases if the dissolved molecules occupy a larger volume. From equations (14.28) and (14.29) we obtain

$$dS = k \cdot \ln \frac{\Omega + d\Omega}{\Omega} = Nk \cdot \ln \frac{V + dV}{V} = Nk \cdot \ln\left(1 + \frac{dV}{V}\right) \tag{17.49}$$

where N is the number of sugar molecules. Because of

$$\ln(1 + x) = x \tag{17.50}$$

for small enough x the expression simplifies to

$$dS = Nk \cdot \frac{dV}{V}, \quad \frac{dS}{dV} = \frac{Nk}{V} \tag{17.51}$$

Then

$$^{osm}P = NkT\frac{1}{V} = RT\frac{n}{V} \tag{17.52}$$

or

$$^{osm}P = RT \cdot c \quad \text{with} \quad c = \frac{n}{V} \tag{17.53}$$

(c = concentration of the sugar solution). Thus the osmotic pressure is linearly increasing with the concentration c. Equation (17.53) agrees with the ideal gas equation (11.23) if the pressure P of the gas is replaced by the osmotic pressure ^{osm}P.

Example

For a sugar solution of concentration $c = 0.1 \, \text{mol L}^{-1}$ at 298 K with $R = 0.08314 \, \text{bar L K}^{-1} \, \text{mol}^{-1}$ we obtain $^{osm}P = 2.48 \, \text{bar}$. This is in agreement with the experimental value 2.66 bar within experimental error. At higher concentrations, however, the agreement becomes worse, because of the mutual interaction of the dissolved molecules.

17.2.2 Concentration and Molality

Instead of the concentration

$$c = \frac{n_{\text{solute}}}{V_{\text{solution}}} \qquad (17.54)$$

the molality

$$\underline{m} = \frac{n_{\text{solute}}}{m_{\text{solvent}}} \qquad (17.55)$$

is often used. Because of

$$\frac{m_{\text{solvent}} + m_{\text{solute}}}{V_{\text{solution}}} = \rho_{\text{solution}} \qquad (17.56)$$

(where ρ_{solution} is the density of the solution) we obtain

$$V_{\text{solution}} = \frac{m_{\text{solvent}} + m_{\text{solute}}}{\rho_{\text{solution}}} \qquad (17.57)$$

For highly diluted solutions ($m_{\text{solute}} \ll m_{\text{solvent}}$) with $\rho_{\text{solution}} \approx \rho_{\text{solvent}} = \rho$ we obtain

$$V_{\text{solution}} \approx \frac{m_{\text{solvent}}}{\rho} \qquad (17.58)$$

Then

$$c = \frac{n_{\text{solute}}}{m_{\text{solvent}}} \cdot \rho = \underline{m} \cdot \rho \qquad (17.59)$$

For diluted solutions in water ($\rho \approx 1\,\text{g cm}^{-3}$) both descriptions are equivalent, for example, the concentration $c = 1\,\text{mol L}^{-1}$ corresponds to the molality $\underline{m} = 1\,\text{mol kg}^{-1}$.

17.2.3 Depression of Vapor Pressure

We consider a dilute solution of a nonvolatile substance, say sugar in water. If a part of the solvent evaporates, then the dissolved molecules are confined to a smaller volume: the concentration increases during evaporation of the solvent. As a consequence, it is more difficult to evaporate the solution than the pure solvent: the vapor pressure P above the solution is lower than the vapor pressure P_s of the solvent.

How does the vapor pressure P depend on the concentration of the dissolved species? As derived in Box 17.2 we obtain

$$\frac{P_s - P}{P_s} = x \qquad (17.60)$$

or

$$\frac{P}{P_s} = 1 - x = x_s \quad \text{(Raoult's law)} \qquad (17.61)$$

where x is the mole fraction of the dissolved species and x_s is the mole fraction of the solvent

$$x = \frac{n}{n + n_s} \approx \frac{n}{n_s}, \quad x_s = \frac{n_s}{n + n_s} = 1 - x \qquad (17.62)$$

(n = amount of substance of the dissolved species, n_s = amount of substance of the solvent).

17.2 ΔG FOR REACTIONS IN DILUTE SOLUTION

Example – vapor pressure of sugar solution in water

The vapor pressure P_s of pure water at 298 K is 0.0031 bar. For a diluted solution of sugar in water (concentration $c = 1\,\mathrm{mol\,L^{-1}}$) we obtain

$$x = \frac{1\,\mathrm{mol}}{55.5\,\mathrm{mol}} = 0.018, \quad x_s = 0.982$$

Then $P = 0.982 \cdot P_s = 0.0030\,\mathrm{bar}$.

The elevation of the boiling point (ΔT_b) and the depression of the freezing point (ΔT_m) of the solution as compared to the pure solvent are immediate consequences of the depression of the vapor pressure (Figure 17.9). We presume that crystallization of the solvent takes place when decreasing the temperature of the solution. This is generally the case for the dilute solutions considered here.

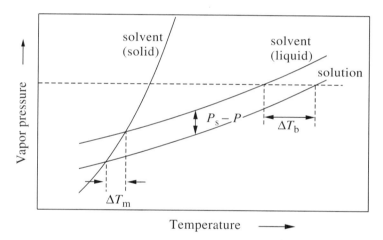

Figure 17.9 Depression ($P_s - P$) of vapor pressure, elevation ΔT_b of boiling point, and depression ΔT_m of freezing point. Relations for the calculation of ΔT_b and ΔT_m will be derived in Problems 17.7 and 17.8.

Box 17.2 – Raoult's Law

We imagine a container A with a dilute solution of concentration c and a container B with the pure solvent (Figure 17.10). We transfer a small amount of solvent (dn_s) reversibly at constant temperature T from A to B by pressing the solvent through a semipermeable membrane (see Figure 17.8). The reversible work $dw_{\mathrm{rev},1}$ for this process is

$$dw_{\mathrm{rev},1} = -{}^{\mathrm{osm}}P \cdot dV = {}^{\mathrm{osm}}P \cdot V_{s,m} \cdot dn_s$$

Continued on page 564

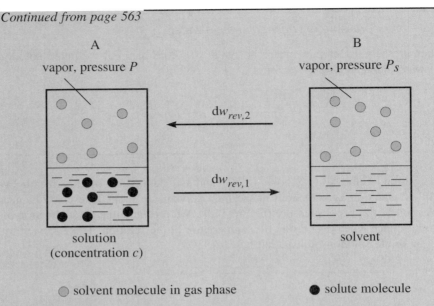

Figure 17.10 How to calculate the vapor pressure above a solution. Solvent from the solution (concentration c) in container A is reversibly transferred to the solvent in container B (step 1). Vapor from container B (vapor pressure P_s of the pure solvent) is transferred to container A, where the vapor is under pressure P (step 2).

($V_{s,m}$ = molar volume of the solvent). Then we transfer the same amount of vapor reversibly from B to A. The reversible work $dw_{rev,2}$ for this process is

$$dw_{rev,2} = dn_s \cdot RT \cdot \ln \frac{P}{P_s}$$

Both processes constitute a cyclic process, thus

$$^{osm}P \cdot V_{s,m} \cdot dn_s = -dn_s \cdot RT \cdot \ln \frac{P}{P_s}$$

Because of $^{osm}P = c \cdot RT$ and

$$\ln \frac{P}{P_s} = \ln\left(1 - \frac{P_s - P}{P_s}\right) \approx -\frac{P_s - P}{P_s}$$

we obtain

$$\frac{P_s - P}{P_s} = V_{s,m} \cdot c \quad \text{or} \quad \frac{P}{P_s} = 1 - V_{s,m} \cdot c$$

For dilute solutions the mole fraction x and the concentration c are

$$x = \frac{n}{n + n_s} \approx \frac{n}{n_s}, \quad c = \frac{n}{V_{s,m} \cdot n_s}$$

Continued on page 565

> Continued from page 564
>
> Then
> $$c = \frac{x}{V_{s,m}}$$
>
> and we obtain Raoult's law
> $$\frac{P_s - P}{P_s} = x, \quad \frac{P}{P_s} = 1 - x$$

17.2.4 Reversible Change of Concentration

By changing the pressure on the piston in the arrangement in Figure 17.8 we can concentrate or dilute the solution reversibly, in a similar way as for the compression or expansion of a gas. Therefore we can extend the considerations on gases to dilute solutions. What we have to do is to substitute the pressure of the gas by the osmotic pressure or the concentration of the dissolved species. Then the work w_{rev} to concentrate a sugar solution reversibly from a concentration c_1 to a concentration c_2 is

$$w_{\text{rev}} = nRT \cdot \ln \frac{^{\text{osm}}P_2}{^{\text{osm}}P_1} = nRT \cdot \ln \frac{c_2}{c_1} \tag{17.63}$$

The corresponding reversible heat transfer is, because of $\Delta U = 0$,

$$q_{\text{rev}} = -nRT \cdot \ln \frac{c_2}{c_1} \tag{17.64}$$

This results in a change of the entropy

$$\Delta S = \frac{q_{\text{rev}}}{T} = -nR \cdot \ln \frac{c_2}{c_1} \tag{17.65}$$

17.2.5 Mass Action Law: Solutions of Neutral Particles

In analogy to the gas reaction (17.19) we consider a chemical reaction

$$\alpha A + \beta B + \cdots \longrightarrow \gamma C + \delta D + \cdots \tag{17.66}$$

at constant temperature in aqueous solution. How can we perform this reaction reversibly?

To answer this question we assume that before and after the reaction the reacting species are in aqueous solution at standard concentrations $c^{\ominus} = 1$ mol L^{-1} in separate reservoirs. We take certain amounts $n_A, n_B, \ldots$ of species A, B, $\ldots$ of the starting materials from their reservoirs and transfer them reversibly to an equilibrium container (equilibrium concentrations $c_A, c_B, \ldots$). After reaction the products C, D, $\ldots$ are reversibly transferred to the

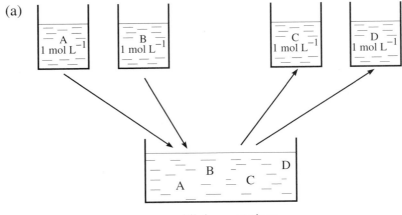

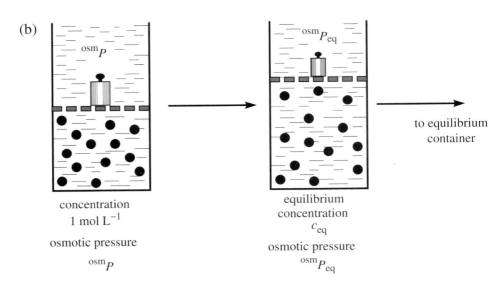

Figure 17.11 Performing reaction A + B → C + D in solution reversibly at T = const and P = const. (a) Reactants A and B under standard concentration $c^\ominus = 1\,\text{mol}\,\text{L}^{-1}$ in separate reservoirs. A and B are transferred into the equilibrium container. Then the products C and D are taken from the equilibrium container and transferred into separate reservoirs under standard concentration. (b) Reversible transfer of reactant A realized by reversibly changing the osmotic pressure of the solution of A from $^{\text{osm}}P$ to $^{\text{osm}}P_{\text{eq}}$.

corresponding reservoirs (Figure 17.11). In analogy to a gas reaction (equations (17.21), (17.22)) we obtain

$$\Delta_r G^\ominus = -RT \cdot \ln K, \quad \Delta_r G = \Delta_r G^\ominus + RT \cdot \ln Q \tag{17.67}$$

with

$$K = \left(\frac{\widehat{c}_C^\gamma \cdot \widehat{c}_D^\delta \cdots}{\widehat{c}_A^\alpha \cdot \widehat{c}_B^\beta \cdots}\right)_{\text{eq}}, \quad Q = \frac{\widehat{c}_C^\gamma \cdot \widehat{c}_D^\delta \cdots}{\widehat{c}_A^\alpha \cdot \widehat{c}_B^\beta \cdots} \tag{17.68}$$

17.2 ΔG FOR REACTIONS IN DILUTE SOLUTION

where $\widehat{c}$ is the dimensionless quantity

$$\widehat{c} = \frac{c}{c^\ominus}, \quad c^\ominus = 1 \text{ mol L}^{-1} \tag{17.69}$$

$\Delta_r G^\ominus$ is the change in Gibbs energy when α mol of species A and β mol of species B react to form γ mol of species C and δ mol of species D. All species in the beginning and at the end of the process are separated and at concentrations of $c^\ominus = 1 \text{ mol L}^{-1}$.

As in Section 17.2.1, these relations are only valid if the mutual interactions of the molecules can be neglected – that is, for small enough concentrations of the dissolved species. The treatment will be refined in Chapter 19 for concentrated solutions.

Example – complex formation in solution

Iodine forms a complex with dioxane in a solution of n-heptane.

$$I_2 + \text{dioxane} \longrightarrow [I_2 \cdot \text{dioxane}]$$

The equilibrium constant

$$K = \frac{\widehat{c}_{\text{complex}}}{\widehat{c}_{I_2} \cdot \widehat{c}_{\text{dioxane}}}$$

can be measured in a spectrophotometric experiment: $K = 9.3$ at $T = 298$ K. Then $\Delta_r G^\ominus_{298} = -5.5 \text{ kJ mol}^{-1}$.

17.2.6 Mass Action Law: Solutions of Charged Particles

Hitherto we have considered neutral molecules in solution. Now let us discuss the reaction

$$\text{HAc(aq)} + \text{NaCl(aq)} \longrightarrow \text{NaAc(aq)} + \text{HCl(aq)} \tag{17.70}$$

(HAc = acetic acid, NaAc = sodium acetate) where (aq) indicates that the corresponding species are dissolved in water. In this case NaCl, NaAc, and HCl dissociate into ions; since the concentrations of Na$^+$ and Cl$^-$ do not change in the reaction, we can write equation (17.70) as

$$\text{HAc} \longrightarrow \text{Ac}^- + \text{H}^+ \tag{17.71}$$

Can we still apply the mass action law,

$$K = \left(\frac{\widehat{c}_{H^+} \cdot \widehat{c}_{Ac^-}}{\widehat{c}_{HAc}} \right)_{\text{eq}} \tag{17.72}$$

to reactions of ions in solution? Because of the condition of electroneutrality, we cannot handle single ions.

Surprisingly, the mass action law can be applied, although a tricky operation is needed for the proof (Box 17.3).

Box 17.3 – How to Determine $\Delta_r G^\ominus$ for Reactions Involving Ions

As in Section 17.2.6 we determine $\Delta_r G^\ominus$ for reaction (17.70):

$$\text{HAc(aq)} + \text{NaCl(aq)} \longrightarrow \text{NaAc(aq)} + \text{HCl(aq)}$$

The reactants are in reservoirs at standard concentration $c^\ominus = 1\,\text{mol}\,\text{L}^{-1}$ (Figure 17.12). We transfer small amounts of HAc and NaCl reversibly into the equilibrium container and return NaAc and HCl reversibly into the corresponding reservoirs. Now the following problem arises: Na^+ and Cl^- have the same concentrations $c^\ominus$ in the reservoirs, but different concentrations c_{Na^+} and c_{Cl^-} in the equilibrium container. Because of the condition of electroneutrality, we cannot transfer the single ions Na^+ and Cl^- separately, but we have to transfer NaCl as a whole.

How can we overcome this difficulty? The trick is to transfer NaCl (amount n), in a first step, from the standard concentration to the concentration

$$c = \sqrt{c_{\text{Na}^+} \cdot c_{\text{Cl}^-}}$$

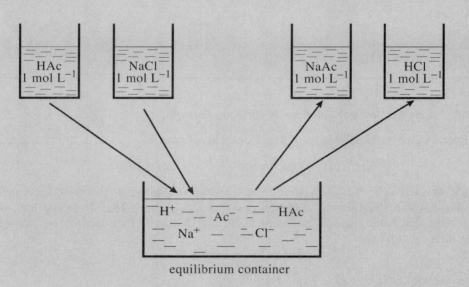

Figure 17.12 Performing the reaction $\text{HAc} \rightarrow \text{Ac}^- + \text{H}^+$ under standard conditions (reacting species in separate reservoirs at standard concentration $c^\ominus = 1\,\text{mol}\,\text{L}^{-1}$). The species are at equilibrium concentrations in the equilibrium container.

Continued on page 569

17.2 ΔG FOR REACTIONS IN DILUTE SOLUTION

Continued from page 568

The work needed in this step (transfer of an amount of n Na$^+$ and an amount of n Cl$^-$ is

$$w_{\text{rev},1} = nRT \cdot \ln \frac{c}{c^\ominus} + nRT \cdot \ln \frac{c}{c^\ominus}$$

$$= nRT \cdot \ln \frac{c^2}{(c^\ominus)^2} = nRT \cdot \ln \frac{c_{\text{Na}^+} \cdot c_{\text{Cl}^-}}{(c^\ominus)^2}$$

$$= nRT \cdot \ln \widehat{c}_{\text{Na}^+} + nRT \cdot \ln \widehat{c}_{\text{Cl}^-}$$

Then this solution is transferred into the equilibrium container by using a piston and a membrane permeable to Na$^+$ and Cl$^-$ (Figure 17.13). The work

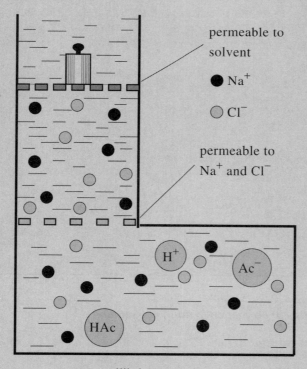

Figure 17.13 Transfer of NaCl (concentration c) into the equilibrium container (concentrations c_{Na^+}, c_{Cl^-}). The process is reversible, if $c = \sqrt{c_{\text{Na}^+} \cdot c_{\text{Cl}^-}}$.

Continued on page 570

> Continued from page 569
>
> needed for this step is
>
> $$w_{\text{rev},2} = nRT \cdot \ln \frac{c_{\text{Na}^+}}{c} + nRT \cdot \ln \frac{c_{\text{Cl}^-}}{c} - {}^{\text{osm}}P\Delta V$$
>
> $$= nRT \cdot \ln \frac{c_{\text{Na}^+} \cdot c_{\text{Cl}^-}}{c^2} - {}^{\text{osm}}P\Delta V$$
>
> Because of our special choice of $c = \sqrt{c_{\text{Na}^+} \cdot c_{\text{Cl}^-}}$ the first term in the equation for $w_{\text{rev},2}$ is zero and we obtain
>
> $$w_{\text{rev},2} = -{}^{\text{osm}}P\Delta V$$
>
> If, for example, $c_{\text{Na}^+} < c_{\text{Cl}^-}$, then the work to transfer Na$^+$ from c to c_{Na^+} is negative, whereas the work to transfer Cl$^-$ from c to c_{Cl^-} is positive; both contributions cancel each other. This means by supplying or removing a tiny additional weight on the piston the process can proceed in the forward or in the reverse direction. Then the total work required to transfer NaCl from its reservoir into the equilibrium container is
>
> $$w_{\text{rev}} = nRT \cdot \ln(\hat{c}_{\text{Na}^+} \cdot \hat{c}_{\text{Cl}^-}) - {}^{\text{osm}}P\Delta V$$
>
> The same consideration holds for NaAc and HCl, while HAc is neutral and needs no trick. Thus
>
> $$\Delta G^\ominus = nRT \cdot [\ln(\hat{c}_{\text{Na}^+} \cdot \hat{c}_{\text{Cl}^-}) + \ln \hat{c}_{\text{HAc}} - \ln(\hat{c}_{\text{Na}^+} \cdot \hat{c}_{\text{Ac}^-}) - \ln(\hat{c}_{\text{H}^+} \cdot \hat{c}_{\text{Cl}^-})]$$
>
> $$= -nRT \cdot \ln \frac{\hat{c}_{\text{H}^+} \cdot \hat{c}_{\text{Ac}^-}}{\hat{c}_{\text{HAc}}} = -nRT \cdot \ln K$$
>
> This shows that the mass action law equation (17.72) is indeed valid.

17.2.7 Gibbs Energy of Formation in Aqueous Solution

Solutions of neutral particles

In gas reactions, $\Delta_r G^\ominus$ is given by the Gibbs energies of formation of each species from elements, $\Delta_f G^\ominus$ (species separated, at $P^\ominus = 1$ bar). In the present case, $\Delta_r G^\ominus_{\text{aq}}$ is given by the Gibbs energies of formation in aqueous solution of each species, $\Delta_f G^\ominus_{\text{aq}}$ (species separated, in water at $c^\ominus = 1 \text{ mol L}^{-1}$).

$\Delta_f G^\ominus_{\text{aq}}$ is obtained from $\Delta_f G^\ominus$ by adding the molar Gibbs energy required to transfer the species from a reservoir with the pure substance into aqueous solution of concentration $c^\ominus = 1 \text{ mol L}^{-1}$ (Box 17.4). Values of $\Delta_f G^\ominus_{\text{aq}}$ are listed in Appendix 1 for selected compounds.

17.2 ΔG FOR REACTIONS IN DILUTE SOLUTION

Example

Calculate $\Delta G_{aq}^{\ominus}$ for the reaction

$$\text{NH}_2-\text{CH}_2-\overset{\overset{\text{O}}{\|}}{\text{C}}-\text{NH}-\underset{\underset{\text{CH}_3}{|}}{\text{CH}}-\text{COOH} + \text{H}_2\text{O} \quad \text{Glycylalanine}$$

$$\downarrow$$

$$\text{NH}_2-\text{CH}_2-\text{COOH} \quad + \quad \text{NH}_2-\underset{\underset{\text{CH}_3}{|}}{\text{CH}}-\text{COOH}$$

Glycine   Alanine

in aqueous solution. From Appendix 1 we obtain $\Delta_r G_{aq}^{\ominus} = (-371.3 - 379.9 + 733.9)\,\text{kJ}\,\text{mol}^{-1} = -17.3\,\text{kJ}\,\text{mol}^{-1}$. The hydrolysis of glycylalanine under standard conditions and at $T = 298\,\text{K}$ can occur spontaneously according to the present criterion ($K = 1.1 \times 10^3$). Actually, this reaction is kinetically hindered. This is of general importance: proteins, chains of amino acids linked by peptide bonds, would otherwise hydrolyze spontaneously.

Box 17.4 – How to Obtain $\Delta_f G_{aq}^{\ominus}$ from $\Delta_f G^{\ominus}$

To calculate the Gibbs energy for dissolving a given species in water we imagine a container filled with a saturated solution of that species (saturation concentration c, Figure 17.14). We add a certain amount of the pure species to this solution; since the solution is saturated, the additional amount remains as a precipitate on the bottom, and no work has to be done in this process.

Now we apply a semipermeable piston with a weight corresponding to the osmotic pressure of the saturated solution and fill up with solvent. Then the piston is slowly moved up until the precipitate has dissolved ($\Delta G = 0$). Finally, the saturated solution is reversibly diluted to the standard concentration $c^{\ominus}$. For this step we have

$$\Delta G = nRT \cdot \ln \frac{c^{\ominus}}{c} = -nRT \cdot \ln \hat{c}$$

if we neglect the interactions between the dissolved particles (n = amount of solute). Then the molar Gibbs energy of formation in aqueous solution is

Continued on page 572

$$\Delta_f G^\ominus_{aq} = \Delta_f G^\ominus - RT \cdot \ln \widehat{c}$$

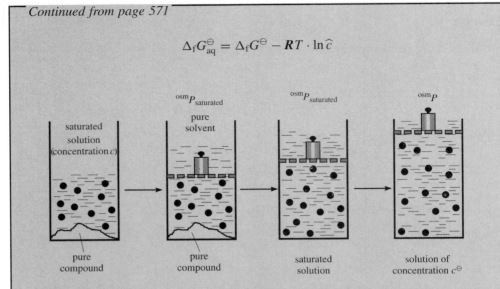

Figure 17.14 Isothermal reversible dissolution of a pure solid compound in a solvent. First, the pure compound is added to a saturated solution of concentration c. Finally, a solution of concentration $c^\ominus$ is obtained.

Example

For a saturated solution of glucose in water, $\widehat{c} = 3.1$ at $T = 298\,\text{K}$ and we obtain $\Delta_f G^\ominus_{aq} = -910.6\,\text{kJ}\,\text{mol}^{-1} - RT \cdot \ln 3.1 = -913.4\,\text{kJ}\,\text{mol}^{-1}$. This is only slightly different from the value $-914.5\,\text{kJ}\,\text{mol}^{-1}$ in Appendix 1, where the deviation from the behavior of a highly diluted solution is taken into account.

Solutions of charged particles

This case needs special attention, since a solution of isolated ions cannot be prepared. In spite of this we can define values of $\Delta_f G^\ominus_{aq}$ for each ionic species in the following sense.

Let us consider the solution of a salt such as NaCl. $\Delta_f G^\ominus_{aq}$ is obtained from $\Delta_f G^\ominus$ of crystalline NaCl as described in Box 17.3. $\Delta_f G^\ominus_{aq}$ can be considered as the sum of contributions of the corresponding ions:

$$\Delta_f G^\ominus_{aq}(\text{NaCl}) = \Delta_f G^\ominus_{aq}(\text{Na}^+) + \Delta_f G^\ominus_{aq}(\text{Cl}^-) \qquad (17.73)$$

$$\Delta_f G^\ominus_{aq}(\text{HCl}) = \Delta_f G^\ominus_{aq}(\text{H}^+) + \Delta_f G^\ominus_{aq}(\text{Cl}^-) \qquad (17.74)$$

It is useful to define Gibbs energies of formation of single ions as follows. Arbitrarily we set

$$\boxed{\Delta_f G^\ominus_{aq}(\text{H}^+) = 0} \qquad (17.75)$$

17.2 ΔG FOR REACTIONS IN DILUTE SOLUTION

Then we obtain

$$\Delta_f G_{aq}^{\ominus}(Cl^-) = \Delta_f G_{aq}^{\ominus}(HCl), \quad \Delta_f G_{aq}^{\ominus}(Na^+) = \Delta_f G_{aq}^{\ominus}(NaCl) - \Delta_f G_{aq}^{\ominus}(HCl) \quad (17.76)$$

Values for $\Delta_f G_{aq}^{\ominus}$ are listed in Appendix 1 for selected ions.

Example – $\Delta_r G_{aq}^{\ominus}$ and K from $\Delta_f G_{aq}^{\ominus}$ in a reaction involving ions

$$Fe^{2+} + Cu^{2+} \longrightarrow Fe^{3+} + Cu^+$$

From Appendix 1 we obtain $\Delta_r G_{aq}^{\ominus} = (50.2 - 10.5 + 84.9 - 65) \text{ kJ mol}^{-1} = 59.6 \text{ kJ mol}^{-1}$.

$$\ln K = -\frac{\Delta_r G_{aq}^{\ominus}}{RT} = -24.1, \quad K = \frac{\widehat{c}_{Fe^{3+}} \cdot \widehat{c}_{Cu^{1+}}}{\widehat{c}_{Fe^{2+}} \cdot \widehat{c}_{Cu^{2+}}} = 3.5 \times 10^{-11}$$

This means that in a solution of Fe^{2+} and Cu^{2+} a tiny amount of Fe^{3+} and Cu^{1+} are present, formed by electron exchange. For $c_{Fe^{2+}} = c_{Cu^{2+}} = 1 \text{ mol L}^{-1}$ we obtain

$$c_{Fe^{3+}} = c_{Cu^{1+}} = \sqrt{3.5 \times 10^{-11}} \text{ mol L}^{-1} = 6 \times 10^{-6} \text{ mol L}^{-1}$$

17.2.8 Part of Reactants or Products in Condensed or Gaseous State

We have seen in Section 17.1.3 that immiscible condensed species do not appear in the expression for the equilibrium constant for gas reactions. For the same reason condensed species do not appear in the expression for the equilibrium constant for reactions in solution.

Examples

$$AgCl \text{ (solid)} \longrightarrow Ag^+(aq) + Cl^-(aq) \quad K = \widehat{c}_{Ag^+} \cdot \widehat{c}_{Cl^-}$$

$$H_2O \text{ (liquid)} \longrightarrow H^+(aq) + OH^-(aq) \quad K = \widehat{c}_{H^+} \cdot \widehat{c}_{OH^-}$$

$$Zn \text{ (solid)} + Cu^{2+}(aq) \longrightarrow Zn^{2+}(aq) + Cu \text{ (solid)} \quad K = \frac{\widehat{c}_{Zn^{2+}}}{\widehat{c}_{Cu^{2+}}}$$

$$Hg \text{ (liquid)} + 2Ag^+(aq) \longrightarrow Hg_2^{2+}(aq) + 2Ag \text{ (solid)} \quad K = \frac{\widehat{c}_{Hg_2^{2+}}}{(\widehat{c}_{Ag^+})^2}$$

When in a reaction in solution a gaseous species is formed under standard pressure $P^{\ominus}$, then the gaseous species, as the condensed species, does not appear in the equilibrium constant. However, if the pressure P of the gas differs from the standard pressure, the work to transfer the gas reversibly from $P^{\ominus}$ to P must be added to ΔG.

Example

$$Zn \text{ (solid)} + 2H^+ \text{ (aq)} \longrightarrow Zn^{2+} \text{ (aq)} + H_2 \text{ (gas)}$$

In this case

$$\Delta_r G = \Delta_r G^\ominus + RT \cdot \ln \frac{\widehat{c}_{Zn^{2+}} \cdot \widehat{P}_{H_2}}{\widehat{c}_{H^+}^2}$$

Example – does Zn dissolve in acidic medium?

We consider the reaction

$$Zn + 2H^+ \longrightarrow Zn^{2+} + H_2$$

We have $\Delta_f G^\ominus(Zn) = \Delta_f G^\ominus_{aq}(H^+) = \Delta_f G^\ominus(H_2) = 0$ and, according to Appendix 1 at 298 K, $\Delta_f G^\ominus_{aq}(Zn^{2+}) = -147.2 \text{ kJ mol}^{-1}$. Then we obtain $\Delta_r G^\ominus_{aq}(Zn^{2+}) = \Delta_f G^\ominus_{aq}(Zn^{2+})$. $\Delta_r G^\ominus$ is negative under standard conditions; that is, Zn is dissolved in acids at a concentration $c^\ominus = 1 \text{ mol L}^{-1}$ and at an external pressure $P^\ominus = 1$ bar.

Example – autoprotolysis constant K_w of water

We consider the reaction

$$H_2O \text{ (liq)} \longrightarrow H^+ \text{ (aq)} + OH^- \text{ (aq)}$$

According to Appendix 1 at 298 K we have $\Delta_f G^\ominus(H_2O) = -237.2 \text{ kJ mol}^{-1}$ and $\Delta_f G^\ominus_{aq,m}(OH^-) = -157.3 \text{ kJ mol}^{-1}$. Then with $\Delta_f G^\ominus_m(H^+) = 0$ we obtain $\Delta_r G^\ominus = (-157.3 + 237.2) \text{ kJ mol}^{-1} = 79.9 \text{ kJ mol}^{-1}$ and

$$\ln K_w = -\frac{\Delta_r G^\ominus}{RT} = -32.2, \quad K_w = \widehat{c}_{H^+} \cdot \widehat{c}_{OH^-} = 1 \times 10^{-14}$$

In pure water H^+ and OH^- are present in concentrations $c_{H^+} = c_{OH^-} = 10^{-7} \text{ mol L}^{-1}$.

Example – complex formation

Zn^{2+} forms a complex with cyanide:

$$Zn^{2+} + 4CN^- \longrightarrow [Zn(CN)_4]^{2-}$$

With the data in Appendix 1 we obtain

$$\Delta_r G^\ominus = [+446.9 - (-147.2 + 4 \times 1654.7)] \text{ kJmol}^{-1} = -68.7 \text{ kJ mol}^{-1}$$

$$K = \frac{\widehat{c}_{complex}}{\widehat{c}_{Zn^{2+}} \cdot \widehat{c}_{CN^-}^4} = 1.1 \times 10^{12}$$

In a solution with Zn^{2+} and CN^-, both in a concentration of $1 \times 10^{-3} \text{ mol L}^{-1}$, the complex is present in a concentration of $c_{complex} = 1.1 \times 10^{12} \cdot (10^{-3})^5 \text{ mol L}^{-1} = 1.1 \times 10^{-3} \text{ mol L}^{-1}$.

PROBLEM 17.1

This chapter shows the relevance and importance of the mass action law in understanding chemical equilibria and it demonstrates how equilibrium constants can be obtained from Gibbs energy of formation.

Problems

Problem 17.1 – Equilibrium Constant Calculated from ΔH and ΔS

Calculate the equilibrium constants for the reactions (a) $I_2 \rightarrow 2I$ at 1273 K and (b) $H_2 + I_2 \rightarrow 2HI$ at 700 K.

Solution. Reaction (a):
From Appendix 1 we obtain at 298 K, $\Delta_r H^\ominus = 151.0\,\mathrm{kJ\,mol^{-1}}$ and $\Delta_r S^\ominus = 100.8\,\mathrm{J\,K^{-1}\,mol^{-1}}$ at 298 K if we consider gaseous I_2. Then using the data for c_1, c_2, and c_3 in Foundation 16.1 we calculate, according to Section 16.4.3, $\Delta_r G^\ominus = (154 - 1273 \times 0.1067)\,\mathrm{kJ\,mol^{-1}} = 18.9\,\mathrm{kJ\,mol^{-1}}$ at 1273 K. Then from equation (17.21) we obtain $K = 0.168$ at 1273 K.

Solution. Reaction (b):
From Appendix 1 we obtain $\Delta_r H^\ominus = -10.4\,\mathrm{kJ\,mol^{-1}}$ and $\Delta_r S^\ominus = 21.4\,\mathrm{J\,K^{-1}\,mol^{-1}}$ at 298 K if we use the values for gaseous I_2. Then in analogy to reaction (a) we have $\Delta_r G^\ominus = (-13.2 - 700 \times 0.0154)\,\mathrm{kJ\,mol^{-1}} = -23.9\,\mathrm{kJ\,mol^{-1}}$ at 700 K. Then from equation (17.21) we obtain $K = 61$. The directly measured value is $K = 55.2$.

Problem 17.2 – Dissociation of I_2

For the dissociation of iodine:

$$I_2 \longrightarrow 2I$$

we found in Problem 17.1 (above) $K = 0.168$ at $T = 1273$ K. What is the fraction of iodine atoms at total pressures of 0.1, 1.0, and 10 bar?

Solution.

$$K = \frac{\widehat{P}_I^2}{\widehat{P}_{I_2}}, \quad \widehat{P}_{\text{total}} = \widehat{P}_{I_2} + \widehat{P}_I$$

Then

$$\widehat{P}_{I_2} = \widehat{P}_{\text{total}} - \widehat{P}_I, \quad K = \frac{\widehat{P}_I^2}{\widehat{P}_{\text{total}} - \widehat{P}_I}$$

We solve for $\widehat{P}_I$:

$$\widehat{P}_I = -\tfrac{1}{2}K + \sqrt{K \cdot \widehat{P}_{\text{total}} + \tfrac{1}{4}K^2}$$

$\widehat{P}_{\text{total}}$	$\widehat{P}_{\text{I}}$	$\widehat{P}_{\text{I}_2}$	$\dfrac{\widehat{P}_{\text{I}}}{\widehat{P}_{\text{total}}}$
0.1	0.07	0.03	0.7
1.0	0.33	0.67	0.33
10	1.21	8.79	0.12

With increasing pressure the reaction is shifted to the left-hand side, thus the fraction of iodine atoms decreases in accordance with Le Chatelier's principle.

Problem 17.3 – Hydrogen–Iodine Equilibrium

For the equilibrium

$$H_2 + I_2 \longrightarrow 2HI$$

we find (see Problem 17.1 above) $K = 55.2$ at $T = 700$ K. What is the fraction of HI at a total pressure of 0.1, 1.0, and 10 bar?

Solution.

$$K = \frac{\widehat{P}_{\text{HI}}^2}{\widehat{P}_{\text{H}_2}\widehat{P}_{\text{I}_2}}, \quad \widehat{P}_{\text{total}} = \widehat{P}_{\text{H}_2} + \widehat{P}_{\text{I}_2} + \widehat{P}_{\text{HI}}$$

We restrict our discussion to a stoichiometric composition of the reaction mixture: $\widehat{P}_{\text{H}_2} = \widehat{P}_{\text{I}_2}$, $\widehat{P}_{\text{total}} = 2\widehat{P}_{\text{H}_2} + \widehat{P}_{\text{HI}}$.

$$K = \frac{\widehat{P}_{\text{HI}}^2}{\tfrac{1}{4}(\widehat{P}_{\text{total}} - \widehat{P}_{\text{HI}})^2} \quad \text{and} \quad \tfrac{1}{2}\sqrt{K} = \frac{\widehat{P}_{\text{HI}}}{(\widehat{P}_{\text{total}} - \widehat{P}_{\text{HI}})}$$

Then

$$\widehat{P}_{\text{HI}} = \frac{\widehat{P}_{\text{total}}}{1 + 2/\sqrt{K}} \quad \text{and} \quad \frac{\widehat{P}_{\text{HI}}}{\widehat{P}_{\text{total}}} = \frac{1}{1 + 2/\sqrt{K}} = 0.78$$

This fraction does not depend on the total pressure, contrary to the situation in Problem 17.2 (above). This is in accordance with Le Chatelier's principle, because the number of product species is the same as the number of the reacting species.

Problem 17.4 – Formation of NH$_3$ from Its Elements

For the reaction

$$3H_2 + N_2 \longrightarrow 2NH_3$$

we find from Appendix 1 $\Delta_r H^\ominus = -92.4$ kJ mol^{-1}, $\Delta_r S^\ominus = -198.6$ J K^{-1} mol^{-1}, and $\Delta G^\ominus = -33.2$ kJ mol^{-1} at 298 K. Calculate $\Delta_r G^\ominus$ and K for 700 K.

Solution. We proceed in analogy to Problem 17.1 (above):

$$\Delta_r H^\ominus_{700} = \Delta_r H^\ominus_{298} + \int_{298}^{700} \Delta C^\ominus_{P,m} \cdot dT$$

$$= (-92.4 - 13.6) \text{ kJ mol}^{-1} = -106.0 \text{ kJ mol}^{-1}$$

$$\Delta_r S_{700}^\ominus = \Delta_r S_{298}^\ominus + \int_{298}^{700} \frac{\Delta C_{P,m}^\ominus}{T} \cdot dT$$

$$= (-198.6 - 20.2)\,\text{J K}^{-1}\,\text{mol}^{-1} = -228.5\,\text{J K}^{-1}\,\text{mol}^{-1}$$

Then

$$\Delta_r G_{700}^\ominus = \Delta_r H_{700}^\ominus - T \cdot \Delta_r S_{700}^\ominus = +54\,\text{kJ mol}^{-1}$$

and

$$\ln K_{700} = -\frac{\Delta_r G_{700}^\ominus}{RT} = -9.28, \quad K_{700} = 9.3 \times 10^{-5}$$

Problem 17.5 – Boiling Point of Water on Mount Everest

What is the boiling point of water on the top of Mount Everest (8882 m above sea level)?

Solution. We start the calculation with equations (17.36) and (17.39), neglecting the temperature dependence of $\Delta H^\ominus$ and $\Delta S^\ominus$:

$$\ln \widehat{P}_{vap} = -\frac{\Delta_{vap} H^\ominus}{RT} + \frac{\Delta_{vap} S^\ominus}{R}$$

Then the boiling temperature is

$$T = \frac{-\Delta_{vap} H^\ominus}{R \cdot \ln \widehat{P}_{vap} - \Delta_{vap} S^\ominus}$$

The atmospheric pressure on the top of Mount Everest was calculated as 0.34 bar in Box 11.4. With $\Delta_{vap} H^\ominus = 44.1\,\text{kJ mol}^{-1}$ and $\Delta_{vap} S^\ominus = 118.8\,\text{J K}^{-1}\,\text{mol}^{-1}$ at 298 K we obtain

$$T = \frac{-44.1 \times 10^3}{8.314 \times \ln 0.34 - 118.8} = 345\,\text{K}$$

This is 28 K below the boiling point at sea level.

Problem 17.6 – Reverse Osmosis

We consider sea water with a salt concentration of $c_1 = 1\,\text{mol L}^{-1}$ below the piston in Figure 17.1(b). By exerting an external pressure on the piston, the sea water is concentrated, and pure water appears above the piston (reverse osmosis). What is the reversible work to extract 0.5 L of pure water from 1 L of sea water?

Solution. To extract 0.5 L of pure water, the concentration in the remaining sea water must be increased by a factor of 2. Thus with $n_{salt} = 1$ mol we obtain $w_{rev} = n_{salt} RT \cdot \ln 2 = 1.7\,\text{kJ}$. To purify the same amount of water by distillation, the heat of evaporation $\Delta H = n_{water} \cdot 44\,\text{kJ mol}^{-1} = 1200\,\text{kJ}$ is needed.

Problem 17.7 – Elevation of Boiling Point

From Figure 17.9, using Raoult' law (equation (17.60)) and the Clausius-Clapeyron equation (equation (17.40)) calculate the elevation of boiling point, ΔT_b, as a function of mole fraction x of the solute.

Solution. From Figure 17.9 we obtain

$$\frac{\Delta P}{\Delta T_b} = \frac{dP}{dT}$$

where $\Delta P = P_s - P$ and dP/dT is the slope of the vapor pressure/temperature curve above the liquid. According to Raoults law,

$$\Delta P = P_s \cdot x$$

According to the equation of Clausius and Clapeyron,

$$\frac{d\ln P}{dT} = \frac{\Delta_{vap}H}{RT^2} \quad \text{or} \quad \frac{dP}{dT} = P\frac{\Delta_{vap}H}{RT^2}$$

Then with $P_s = P$ we obtain

$$\Delta T_b = \frac{RT^2}{\Delta_{vap}H} \cdot x$$

Problem 17.8 – Depression of Melting Point

From Figure 17.9, using Raoult'law (equation (17.60)) and the Clausius-Clapeyron equation (equation (17.40)), applied to the melting and to the evaporation process, calculate the depression of melting point, ΔT_b, as a function of mole fraction x of the solute.

Solution. From Figure 17.9 we obtain

$$\frac{\Delta P}{\Delta T_b} = \left(\frac{dP}{dT}\right)_{solid} - \left(\frac{dP}{dT}\right)_{liquid}$$

where $\Delta P = P_s - P$ and $(dP/dT)_{solid}$ is the slope of the vapor pressure/temperature curve above the solid and $(dP/dT)_{liquid}$ is that above the liqid. According to Raoults law,

$$\Delta P = P_s \cdot x$$

According to the equation of Clausius and Clapeyron,

$$\left(\frac{d\ln P}{dT}\right)_{liquid} = \frac{\Delta_{vap}H}{RT^2} \quad \text{or} \quad \left(\frac{dP}{dT}\right)_{liquid} = P\frac{\Delta_{vap}H}{RT^2}$$

and

$$\left(\frac{d\ln P}{dT}\right)_{solid} = \frac{\Delta_{fus}H}{RT^2} \quad \text{or} \quad \left(\frac{dP}{dT}\right)_{solid} = P\frac{\Delta_{fus}H}{RT^2}$$

Then with $P_s = P$ we obtain

$$\Delta T_m = \frac{RT^2}{\Delta_{fus}H - \Delta_{vap}H} \cdot x$$

18 Reactions in Aqueous Solution and in Biosystems

A great number of compounds dissolve in water, where they interact in various ways. Important applications are based on this possibility. How can we understand reactions of acids and bases and of reductants and oxidants? These are fundamental questions in analytical chemistry and in biochemistry.

Energetic considerations are crucial in understanding the interplay of biomolecules. As in Chapter 17, we restrict our discussion to the important limiting case of dilute solutions.

18.1 Proton Transfer Reactions: Dissociation of Weak Acids

18.1.1 Henderson–Hasselbalch Equation

In the reaction

$$\text{HA} \longrightarrow \text{A}^- + \text{H}^+ \tag{18.1}$$

a proton is transferred from a weak acid HA to another molecule (H_2O in this case): the *proton transfer reaction*. The equilibrium constant is

$$K = \frac{\widehat{c}_{\text{A}^-} \cdot \widehat{c}_{\text{H}^+}}{\widehat{c}_{\text{HA}}} \tag{18.2}$$

(Note that $\widehat{c} = c/c^\ominus$, where $c^\ominus$ is the standard concentration.) For convenience we define the quantities

$$\boxed{pK = -\log K, \quad pH = -\log \widehat{c}_{\text{H}^+}} \tag{18.3}$$

(In practice, the definition of pH is based on the measurement of the voltage of an electrochemical cell with a hydrogen electrode, see Chapter 20.) Substituting these values in equation (18.2) yields

$$pK = -\log \frac{\widehat{c}_{A^-} \cdot \widehat{c}_{H^+}}{\widehat{c}_{HA}} = -\log \frac{\widehat{c}_{A^-}}{\widehat{c}_{HA}} + pH \qquad (18.4)$$

or

$$\boxed{pH = pK + \log \frac{c_{A^-}}{c_{HA}}} \qquad (18.5)$$

(the *Henderson–Hasselbalch equation*).

In the special case of a solution of a weak acid and its salt in a 1 : 1 ratio the concentration of A^- equals that of HA:

$$c_{A^-} = c_{HA} \qquad (18.6)$$

(the acid is almost undissociated and the salt is completely dissociated). Thus:

$$pH = pK \qquad (18.7)$$

Such solutions in sufficiently high concentration are called *buffer solutions*. The pH of a buffer solution depends only on the dissociation constant of the acid. It does not or only slightly change at the addition of small amounts of strong acids or bases (see Problem 18.2).

18.1.2 Degree of Dissociation

We dissolve a weak acid HA in water at a concentration c_{total}. What is the fraction of molecules forming A^- and H^+? According to equation (18.1)

$$c_{A^-} = c_{H^+}, \quad c_{HA} = c_{\text{total}} - c_{H^+}, \quad K = \frac{\widehat{c}_{H^+}^2}{\widehat{c}_{\text{total}} - \widehat{c}_{H^+}} \qquad (18.8)$$

Then the proton concentration is

$$\widehat{c}_{H^+} = -\tfrac{1}{2} K + \sqrt{\widehat{c}_{\text{total}} K + \tfrac{1}{4} K^2} \qquad (18.9)$$

The fraction of dissociated HA molecules is called the *degree of dissociation*, α:

$$\alpha = \frac{c_{A^-}}{c_{\text{total}}} = \frac{c_{H^+}}{c_{\text{total}}} = \frac{\widehat{c}_{H^+}}{\widehat{c}_{\text{total}}} \qquad (18.10)$$

With equation (18.9) we obtain:

$$\boxed{\alpha = \frac{-\tfrac{1}{2} K + \sqrt{\widehat{c}_{\text{total}} \cdot K + \tfrac{1}{4} K^2}}{\widehat{c}_{\text{total}}}} \qquad (18.11)$$

For large enough total concentration we obtain

$$\alpha = \sqrt{\frac{K}{\widehat{c}_{\text{total}}}}$$

18.1 PROTON TRANSFER REACTIONS: DISSOCIATION OF WEAK ACIDS

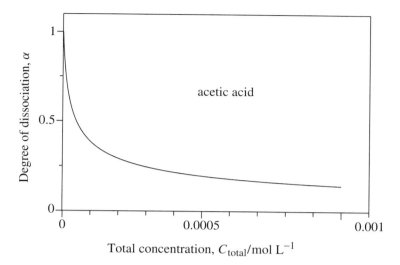

Figure 18.1 Degree of dissociation α for acetic acid versus the total concentration c_{total}. Calculated from equation (18.11) with $K = 2.4 \times 10^{-5}$.

Then α converges to zero with increasing $\widehat{c}_{\text{total}}$. For sufficiently small total concentration

$$\sqrt{\widehat{c}_{\text{total}} \cdot K + \frac{1}{4} K^2} = \frac{1}{2} K \cdot \sqrt{1 + \frac{4 \cdot \widehat{c}_{\text{total}}}{K}} \approx \frac{1}{2} K \cdot \left(1 + \frac{2 \cdot \widehat{c}_{\text{total}}}{K}\right)$$

and

$$\alpha = \frac{-\frac{1}{2} K + \frac{1}{2} K + \widehat{c}_{\text{total}}}{\widehat{c}_{\text{total}}} = 1$$

Then α converges to 1 (Figure 18.1).

18.1.3 Acid in a Buffer

A small amount of a weak acid is dissolved in a buffer solution with a given pH. Then, because of the Henderson–Hasselbalch equation (18.5),

$$\log \frac{\widehat{c}_{A^-}}{\widehat{c}_{HA}} = \text{pH} - \text{p}K \quad \text{or} \quad \frac{\widehat{c}_{A^-}}{\widehat{c}_{HA}} = 10^{\text{pH}-\text{p}K} \tag{18.12}$$

and from equation (18.10) it follows that

$$\alpha = \frac{c_{A^-}}{c_{\text{total}}} = \frac{c_{A^-}}{c_{A^-} + c_{HA}} = \frac{1}{1 + \frac{c_{HA}}{c_{A^-}}} = \frac{1}{1 + 10^{\text{p}K-\text{pH}}} \tag{18.13}$$

(Figure 18.2). For pH = pK we obtain $\alpha = \frac{1}{2}$.

How much of a strong acid can be dissolved in a buffer without appreciably changing the pH? This question is answered in Problem 18.2.

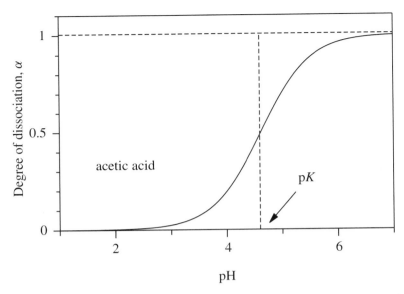

Figure 18.2 Degree of dissociation α for acetic acid versus $(\text{pH})_{\text{buffer}}$. Calculated from equation (18.13) with $pK = 4.62$.

18.1.4 Titration Curve of a Weak Acid

The addition of NaOH to a solution of a weak acid causes an increase of the pH. Protons are transferred from the acid to the OH⁻ ions forming water:

$$H^+ + OH^- \rightleftharpoons H_2O \qquad (18.14)$$

The dissociation constant K_w of water at 25 °C is

$$K_w = \widehat{c}_{H^+} \cdot \widehat{c}_{OH^-} = 10^{-14} \qquad (18.15)$$

What is the change of pH when more and more NaOH is added? To calculate the pH of the solution at different amounts of added NaOH we consider the dissociation of the acid according to equation (18.2)

$$K = \frac{\widehat{c}_{A^-} \cdot \widehat{c}_{H^+}}{\widehat{c}_{HA}}$$

Furthermore, we have to take into account the conditions of mass balance

$$c_{HA} + c_{A^-} = c_{\text{total}} \qquad (18.16)$$

and of electroneutrality

$$c_{H^+} + c_{Na^+} = c_{OH^-} + c_{A^-} \qquad (18.17)$$

Equations (18.15), (18.2), (18.16), and (18.17) constitute four equations for the four unknowns c_{H^+}, c_{OH^-}, c_{A^-}, and c_{HA}. We solve these equations for c_{H^+} and thus obtain pH versus amount n_{NaOH}, as shown in Foundation 18.1 and in Figure 18.3 for acetic acid as an example (*titration curve*).

18.1 PROTON TRANSFER REACTIONS: DISSOCIATION OF WEAK ACIDS

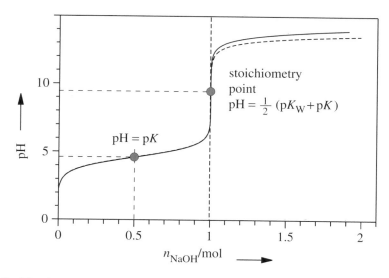

Figure 18.3 Titration curve for acetic acid (initial concentration $1\,\mathrm{mol\,L^{-1}}$, amount $n_{HA} = 1$ mol dissolved in $V = 1\,\mathrm{L}$) with NaOH. pH versus the added amount n_{NaOH} of NaOH. Solid line: volume of the solution assumed as constant; dashed line: volume V increasing with added solution of NaOH ($c_{NaOH,0} = 1\,\mathrm{mol\,L^{-1}}$). For numerical data see Foundation 18.1.

18.1.5 Stepwise Proton Transfer: Amino Acids

In the case of an amino acid, e.g., glycine, two different proton transfer reactions are possible, namely at the carboxyl group and at the amino group:

$$\underset{A}{H_3\overset{+}{N}-CH_2-\underset{\underset{O}{\|}}{C}-OH} \xrightarrow{K_1} \underset{B}{H_3\overset{+}{N}-CH_2-\underset{\underset{O}{\|}}{C}-O^-} + H^+$$

$$\underset{B}{H_3\overset{+}{N}-CH_2-\underset{\underset{O}{\|}}{C}-O^-} \xrightarrow{K_2} \underset{C}{H_2N-CH_2-\underset{\underset{O}{\|}}{C}-O^-} + H^+$$

Species A is stable in acidic solution, whereas species C is stable in alkaline medium. Species B (a zwitterion) exists at an intermediate pH (isoelectric point).

What is the pH of a solution of glycine in water? For the calculation we start with the equilibrium constants

$$K_1 = \frac{\widehat{c}_{H^+} \cdot \widehat{c}_B}{\widehat{c}_A}, \quad K_2 = \frac{\widehat{c}_{H^+} \cdot \widehat{c}_C}{\widehat{c}_B} \tag{18.18}$$

and include the conditions for mass balance and electroneutrality

$$c_A + c_B + c_C = c_{total}, \quad c_{H^+} = c_C - c_A \tag{18.19}$$

The formation of each C is accompanied by the formation of one H^+, and the formation of each A is accompanied by the loss of one H^+, if we neglect the dissociation of water.

Furthermore, only a small amount of B is converted into A or C, thus approximately

$$c_B = c_{total} \tag{18.20}$$

Then

$$K_1 = \frac{\widehat{c}_{H^+} \cdot \widehat{c}_{total}}{\widehat{c}_A}, \quad K_2 = \frac{\widehat{c}_{H^+} \cdot (\widehat{c}_{H^+} + \widehat{c}_A)}{\widehat{c}_{total}} \tag{18.21}$$

and we obtain

$$(\widehat{c}_{H^+})^2 = K_1 K_2 \cdot \frac{\widehat{c}_{total}}{\widehat{c}_{total} + K_1} \tag{18.22}$$

For sufficiently large $\widehat{c}_{total}$ this simplifies to

$$(\widehat{c}_{H^+})^2 = K_1 K_2, \quad pH = \tfrac{1}{2}(pK_1 + pK_2) \tag{18.23}$$

With $K_1 = 10^{-2.4}$ and $K_2 = 10^{-9.8}$ we obtain pH $= \tfrac{1}{2}(2.4 + 9.8) = 6.1$. As outlined in Problem 18.3, pH $= 5.4$ would be expected for $c_{total} = 0.1$, if the reaction from B to A would not take place; the pH would be 7.7, if the reaction from B to C could be neglected. This demonstrates the significance of the coupling of both equilibria.

18.2 Electron Transfer Reactions

18.2.1 Electron Transfer from Metal to Proton: Dissolution of Metals in Acid

When we insert a zinc rod into a beaker with dilute acid (Figure 18.4), the reaction

$$\tfrac{1}{2}Zn + H^+ \longrightarrow \tfrac{1}{2}Zn^{2+} + \tfrac{1}{2}H_2 \tag{18.24}$$

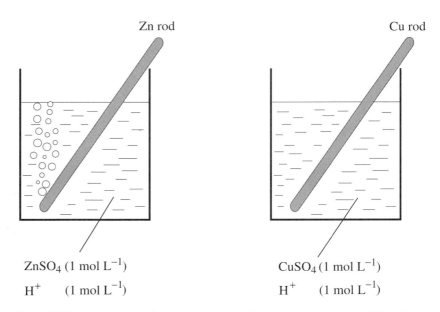

Figure 18.4 Different behavior of zinc and copper when immersed in an acidic solution of zinc sulfate and copper sulfate, respectively.

18.2 ELECTRON TRANSFER REACTIONS

occurs spontaneously under standard conditions ($\widehat{c}_{Zn^{2+}} = \widehat{c}_{H^+} = 1$, $\widehat{P}_{H_2} = 1$), as shown in Section 17.2.8 ($\Delta_r G^\ominus = -73.6\,\text{kJ}\,\text{mol}^{-1}$). In this reaction the metallic zinc transfers electrons to the protons in the solution:

$$\tfrac{1}{2}Zn \longrightarrow \tfrac{1}{2}Zn^{2+} + e^-, \quad e^- + H^+ \longrightarrow \tfrac{1}{2}H_2 \qquad (18.25)$$

Such reactions are called *electron transfer reactions* or *redox reactions*.

What happens if the concentrations and the pressure are different from the standard state values? According to Sections 17.1.2 and 17.2.8, the change of Gibbs energy is

$$\Delta_r G = \Delta_r G^\ominus + RT \cdot \ln \frac{(\widehat{c}_{Zn^{2+}})^{1/2} \cdot (\widehat{P}_{H_2})^{1/2}}{\widehat{c}_{H^+}} \qquad (18.26)$$

$\Delta_r G$ is the change in Gibbs energy when 0.5 mol Zn reacts with 1 mol H$^+$ (concentration c_{H^+}) forming 0.5 mol Zn^{2+} (concentration $c_{Zn^{2+}}$) and 0.5 mol H$_2$ (presure P_{H_2}).

Note: The value $\Delta_r G^\ominus = -73.6\,\text{kJ}\,\text{mol}^{-1}$ is half the value reported in Section 17.2.8 (see third example); there the value is related to the reaction of 1 mol Zn, whereas here we consider the reaction of 0.5 mol Zn.

Now let us immerse the zinc rod in a large volume of a solution containing H$^+$ and Zn^{2+} at a concentration of 1 mol L^{-1}; then

$$\widehat{c}_{Zn^{2+}} = \widehat{c}_{H^+} = 1 \quad \text{and} \quad \Delta_r G = \Delta_r G^\ominus + RT \cdot \ln(\widehat{P}_{H_2})^{1/2}$$

With increasing pressure of H$_2$, the change of Gibbs energy $\Delta_r G$ becomes larger and larger; $\Delta_r G$ switches at a certain pressure P_{H_2} from a negative value to a positive value. We calculate the equilibrium pressure $(\widehat{P}_{H_2})_{eq}$ by setting $\Delta_r G = 0$:

$$0 = \Delta_r G^\ominus + RT \cdot \ln(\widehat{P}_{H_2})_{eq}^{1/2}, \quad (\widehat{P}_{H_2})_{eq} = 6 \times 10^{25}$$

This means the reaction does cease proceeding spontaneously before H$_2$ reaches the incredible pressure of about 10^{26} bar.

On the contrary, for the reaction

$$\tfrac{1}{2}Cu + H^+ \longrightarrow \tfrac{1}{2}Cu^{2+} + \tfrac{1}{2}H_2 \qquad (18.27)$$

$\Delta_r G^\ominus = +32.5\,\text{kJ}\,\text{mol}^{-1}$. This means that copper does not dissolve spontaneously under standard conditions (in this case the pressure in equilibrium, $(P_{H_2})_{eq}$, would be 10^{-12} bar, see Figure 18.4).

18.2.2 Electron Transfer from Metal 1 to Metal 2 Ion: Coupled Redox Reactions

Let us consider a zinc rod immersed in a CuSO$_4$ solution. Then the reaction

$$\tfrac{1}{2}Zn + \tfrac{1}{2}Cu^{2+} \longrightarrow \tfrac{1}{2}Zn^{2+} + \tfrac{1}{2}Cu \qquad (18.28)$$

occurs. This reaction can be considered as the difference between the reactions (18.25) and (18.27). Then $\Delta_r G$ is the difference between

$$\Delta_r G_1 = \Delta_r G_1^\ominus + RT \cdot \ln \frac{(\widehat{c}_{Zn^{2+}})^{1/2} \cdot (\widehat{P}_{H_2})^{1/2}}{\widehat{c}_{H^+}} \tag{18.29}$$

and

$$\Delta_r G_2 = \Delta_r G_2^\ominus + RT \cdot \ln \frac{(\widehat{c}_{Cu^{2+}})^{1/2} \cdot (\widehat{P}_{H_2})^{1/2}}{\widehat{c}_{H^+}} \tag{18.30}$$

Thus

$$\Delta_r G = \Delta_r G_1 - \Delta_r G_2 = \Delta_r G_1^\ominus - \Delta_r G_2^\ominus + RT \cdot \frac{1}{2} \ln \frac{\widehat{c}_{Zn^{2+}}}{\widehat{c}_{Cu^{2+}}} \tag{18.31}$$

During the reaction Zn^{2+} is formed and Cu^{2+} is consumed. What are the equilibrium concentrations of both species? In equilibrium ($\Delta_r G = 0$) we obtain

$$\ln \left(\frac{\widehat{c}_{Zn^{2+}}}{\widehat{c}_{Cu^{2+}}} \right)_{eq} = -2 \frac{\Delta_r G_1^\ominus - \Delta_r G_2^\ominus}{RT} = 85.6 \quad \text{and} \quad \left(\frac{\widehat{c}_{Zn^{2+}}}{\widehat{c}_{Cu^{2+}}} \right)_{eq} = 10^{37}$$

The reaction continues until practically all the Cu^{2+} has been consumed.

18.2.3 Electron Transfer Coupled with Proton Transfer

We consider the reaction

$$\tfrac{1}{2} H_2Q + Fe^{3+} \longrightarrow \tfrac{1}{2} Q + Fe^{2+} + H^+ \tag{18.32}$$

where H_2Q is hydroquinone and Q is quinone

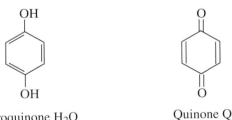

Hydroquinone H_2Q Quinone Q

This reaction can be treated as the difference of the reactions

$$\tfrac{1}{2} H_2Q \longrightarrow \tfrac{1}{2} Q + \tfrac{1}{2} H_2, \quad \Delta_r G_1^\ominus = 33.8 \, \text{kJ mol}^{-1} \tag{18.33}$$

and

$$Fe^{2+} + H^+ \longrightarrow Fe^{3+} + \tfrac{1}{2} H_2, \quad \Delta_r G_2^\ominus = 74.48 \, \text{kJ mol}^{-1} \tag{18.34}$$

Then

$$\Delta_r G = \Delta_r G_1^\ominus - \Delta_r G_2^\ominus + RT \cdot \ln \frac{(\widehat{c}_Q)^{1/2} \cdot \widehat{c}_{Fe^{2+}} \cdot \widehat{c}_{H^+}}{(\widehat{c}_{H_2Q})^{1/2} \cdot \widehat{c}_{Fe^{3+}}} \tag{18.35}$$

18.2 ELECTRON TRANSFER REACTIONS

At equilibrium ($\Delta_r G = 0$) we have

$$\ln\left(\frac{(\widehat{c}_Q)^{1/2} \cdot \widehat{c}_{Fe^{2+}} \cdot \widehat{c}_{H^+}}{(\widehat{c}_{H_2Q})^{1/2} \cdot \widehat{c}_{Fe^{3+}}}\right)_{eq} = -\frac{\Delta_r G_1^\ominus - \Delta_r G_2^\ominus}{RT} = 16.3, \quad K = e^{16.3} = 1.2 \times 10^7$$

In this case the ratio $c_{Fe^{2+}}/c_{Fe^{3+}}$ depends strongly on the acidity of the solution.

18.2.4 Electron Transfer to Proton at pH 7: $\Delta G^{\ominus\prime}$

As outlined in the previous sections, $\Delta G^\ominus$ values for reactions of the general scheme

$$\text{species} + H^+ \longrightarrow \text{product} + \tfrac{1}{2} H_2 \qquad (18.36)$$

are useful for determining ΔG of redox reactions between different species. Such values are listed in Appendix 2 for selected compounds.

A special case is reactions in buffered solution, i.e., the reaction takes place at a constant pH. Then c_{H^+} is a constant and it no longer appears in the expression for the equilibrium constant.

Most important in biochemistry are reactions taking place in a buffer solution with pH = 7. With the notation K' for the equilibrium constant in buffered solution at pH = 7 the corresponding molar change of the Gibbs energy is $\Delta_r G'$ (The prime indicates a reaction at pH = 7). For the reaction

$$\alpha A + \beta B + \cdots + H^+ \longrightarrow \gamma C + \delta D + \cdots + \tfrac{1}{2} H_2 \qquad (18.37)$$

we obtain

$$K' = \left(\frac{(\widehat{c}_C)^\gamma \cdot (\widehat{c}_D)^\delta \cdot \ldots \cdot \widehat{P}_{H_2}^{1/2}}{(\widehat{c}_A)^\alpha \cdot (\widehat{c}_B)^\beta \cdot \ldots}\right)_{eq} \qquad (18.38)$$

as compared to

$$K = \left(\frac{(\widehat{c}_C)^\gamma \cdot (\widehat{c}_D)^\delta \cdot \ldots \cdot \widehat{P}_{H_2}^{1/2}}{(\widehat{c}_A)^\alpha \cdot (\widehat{c}_B)^\beta \cdot \ldots \cdot \widehat{c}_{H^+}}\right)_{eq} = K' \cdot \frac{1}{\widehat{c}_{H^+}} \qquad (18.39)$$

Then with

$$\Delta_r G^{\ominus\prime} = -RT \cdot \ln K' \qquad (18.40)$$

and

$$\Delta_r G^\ominus = -RT \cdot \ln K \quad \text{with} \quad K = \frac{K'}{\widehat{c}_{H^+}}$$

we obtain

$$\boxed{\Delta_r G^{\ominus\prime} = \Delta_r G^\ominus - RT \cdot \ln \widehat{c}_{H^+}} \qquad (18.41)$$

For $c_{H^+} = 10^{-7}$ mol L^{-1} we obtain at $T = 298$ K

$$\Delta_r G^{\ominus\prime} = \Delta_r G^\ominus - RT \cdot \ln 10^{-7} = \Delta_r G^\ominus + 39.9 \text{ kJ mol}^{-1} \qquad (18.42)$$

Values of $\Delta_r G^{\ominus\prime}$ (electron transfer to the proton at pH7) for selected compounds are included in Appendix 2.

Example – $\Delta_r G^{\ominus\prime} > \Delta G^{\ominus}$: proton involved in removing electron

$$\tfrac{1}{2}Cu + H^+ \longrightarrow \tfrac{1}{2}Cu^{2+} + \tfrac{1}{2}H_2$$

$$K = \frac{(\widehat{c}_{Cu^{2+}})^{1/2} \cdot (\widehat{P}_{H_2})^{1/2}}{\widehat{c}_{H^+}}, \quad K' = (\widehat{c}_{Cu^{2+}})^{1/2} \cdot (\widehat{P}_{H_2})^{1/2}$$

$\Delta_r G^{\ominus} = 32.5\,\text{kJ mol}^{-1}$ (see Section 18.2.1), then $\Delta_r G^{\ominus\prime} = 72.5\,\text{kJ mol}^{-1}$ at 298 K.

Example – change of sign from $\Delta_r G^{\ominus}$ to $\Delta_r G^{\ominus\prime}$

$$\tfrac{1}{2}Cd + H^+ \longrightarrow \tfrac{1}{2}Cd^{2+} + \tfrac{1}{2}H_2$$

$\Delta_r G^{\ominus} = -38.8\,\text{kJ mol}^{-1}$, then $\Delta_r G^{\ominus\prime} = +1.2\,\text{kJ mol}^{-1}$ at 298 K. We observe a change of sign when proceeding from $\Delta_r G^{\ominus}$ to $\Delta_r G^{\ominus\prime}$: at $\widehat{c}_{Cd^{2+}} = 1$ and $\widehat{P}_{H_2} = 1$, cadmium dissolves in acid at $\widehat{c}_{H^+} = 1$ but not in water.

Example – no change from $\Delta_r G^{\ominus}$ to $\Delta_r G^{\ominus\prime}$

In the reaction

$$\tfrac{1}{2}\,\text{hydroquinone} \longrightarrow \tfrac{1}{2}\,\text{quinone} + \tfrac{1}{2}H_2$$

no H^+ is consumed in the reaction, thus $\Delta_r G^{\ominus}_{298} = \Delta_r G^{\ominus\prime}_{298}$.

18.3 Group Transfer Reactions in Biochemistry

Let us consider the reaction of adenosine triphosphate (ATP) to yield adenosine diphosphate (ADP) and phosphoric acid:

18.3 GROUP TRANSFER REACTIONS IN BIOCHEMISTRY

$$\text{ATP}^{4-} + \text{H}_2\text{O} \longrightarrow \text{ADP}^{3-} + \text{HPO}_4^{2-} + \text{H}^+ \tag{18.43}$$

We perform the reaction in a buffer solution at pH 7 and use the shorter notation

$$\text{ATP} \longrightarrow \text{ADP} + \text{P} \tag{18.44}$$

According to equation (18.40) the Gibbs energy $\Delta G^{\ominus'}$ is

$$\Delta_r G^{\ominus'} = -RT \cdot \ln K' \tag{18.45}$$

$$K' = \left(\frac{\widehat{c}_{\text{ADP}} \cdot \widehat{c}_{\text{P}}}{\widehat{c}_{\text{ATP}}} \right)_{\text{eq}} \tag{18.46}$$

$\Delta_r G^{\ominus'}$ is the molar change in Gibbs energy of the reaction at standard physiological conditions ($\widehat{c}_{\text{ATP}} = \widehat{c}_{\text{ADP}} = \widehat{c}_{\text{P}} = 1$; species separated at pH 7 in buffered solution). With $\Delta_r G^{\ominus'} = -30.5 \text{ kJ mol}^{-1}$ we obtain $K' = 2.22 \times 10^5$. The equilibrium (18.44) is completely shifted to the right for $\widehat{c}_{\text{P}} = 1$; therefore, ATP is called an energy-rich compound.

$\Delta_r G^{\ominus'}$ for the hydrolysis of selected substances important in biochemistry at pH 7 are listed in Appendix 3. Using these values, one can predict whether or not biochemical group transfer reactions can occur spontaneously.

Example

It is interesting to know whether a substance can be phosphorylated by ATP. Phosphorylated compounds are important intermediates in biochemical syntheses and degradation processes. Let us ask whether the reaction

$$\text{ATP} + \text{glucose} \longrightarrow \text{ADP} + \text{glucose-6-phosphate} \tag{18.47}$$

can occur spontaneously in aqueous solution at pH 7. To calculate $\Delta G^{\ominus'}$ for this reaction we combine reaction (18.44) with the reaction (P = phosphate)

$$\text{glucose-6-phosphate} + \text{H}_2\text{O} \longrightarrow \text{glucose} + \text{P} \tag{18.48}$$

($\Delta_r G^{\ominus'} = -13.8 \text{ kJ mol}^{-1}$). Therefore, for reaction (18.47) we have $\Delta_r G^{\ominus'} = (-30.5 + 13.8) \text{ kJ mol}^{-1} = -16.7 \text{ kJ mol}^{-1}$. This means that this reaction can occur spontaneously under standard physiological conditions (pH7) if the corresponding enzyme acting as a catalyst is present. The equilibrium constant is $K' = e^{-\Delta_r G^{\ominus'}/RT} = 845$. In this example a phosphate group is transferred from ATP to glucose; such reactions are called *group transfer reactions*.

18.3.1 Group Transfer Potential

Besides ATP there are other compounds that can transfer phosphate. The larger the value of $-\Delta_r G^{\ominus'}$ for the cleavage of the phosphate group, the higher the power of group transfer to other species. Therefore $-\Delta_r G^{\ominus'}$ is called the *group transfer potential*. The group transfer potential of ATP is $-\Delta_r G^{\ominus'} = 30.5 \text{ kJ mol}^{-1}$.

Example – group transfer potential

For the hydrolysis of acetylphosphate

$$\text{acetylphosphate} + H_2O \longrightarrow \text{acetic acid} + P$$

we find $\Delta_r G^{\ominus\prime} = -43.9\,\text{kJ}\,\text{mol}^{-1}$. Thus the group transfer potential of acetylphosphate is $43.9\,\text{kJ}\,\text{mol}^{-1}$. It is larger than that of ATP. Therefore, acetylphosphate can be used to convert ADP to ATP.

18.3.2 Coupling of Reactions by Enzymes

Enzymes can couple specific reactions for which $\Delta G > 0$ with other reactions for which $\Delta G < 0$. Let us consider the oxidation of 3-phosphoglycerolaldehyde R–CHO to phosphoglyceric acid R–COOH. This reaction plays an important role in the degradation of carbohydrates.

Direct burning of the aldehyde with O_2

$$R-CHO + \tfrac{1}{2}O_2 \longrightarrow R-COOH \tag{18.49}$$

would result in a strongly negative value of $\Delta_r G^{\ominus\prime}$ ($-263\,\text{kJ}\,\text{mol}^{-1}$). In the biological process, most of this Gibbs energy can be kept in the biological system. This is realized by coupling the oxidation process to a reduction process, namely the formation of NADH from NAD$^+$ ($\Delta_r G^{\ominus\prime} = 220\,\text{kJ}\,\text{mol}^{-1}$). At the same time, ATP is formed from ADP ($\Delta_r G^{\ominus\prime} = 30.5\,\text{kJ}\,\text{mol}^{-1}$):

Note: in analytical biochemistry the shift of the UV absorption to longer wavelengths when proceeding from NAD$^+$ to NADH (from 260 to 340 nm) is important (see Problem 18.4).

$$NAD^+ + H_2O \longrightarrow NADH + H^+ + \tfrac{1}{2}O_2 \qquad (18.50)$$

$$ADP + P \longrightarrow ATP \qquad (18.51)$$

The enzymatic reaction is the sum of reactions (18.49), (18.50), and (18.51):

$$R-CHO + NAD^+ + H_2O + ADP + P \longrightarrow R-COOH + NADH + ATP + H^+ \qquad (18.52)$$

Then $\Delta_r G^{\ominus\prime}$ for reaction (18.52) is $\Delta_r G^{\ominus\prime} = (-263+220+30.5)\,\text{kJ mol}^{-1} = -13\,\text{kJ mol}^{-1}$. The reaction proceeds such that with the help of an enzyme the aldehyde is oxidized to the acid, which, in turn, is phosphorylated. A second enzyme transfers the 1-phosphate group to ADP, and ATP is formed.

The Gibbs energy still decreases but much less than in reaction (18.49). The products NADH and ATP serve the organism as a reducing agent and energy carriers by taking part in many other reactions. All the processes that define the biochemical energy balance are based on such couplings with reactions in which the Gibbs energy decreases only slightly in every single reaction step.

18.4 Bioenergetics

In all biological systems three types of polymers (proteins, nucleic acids, and polysaccharides) as well as lipids (for membrane structure) must be synthesized. The formation of biopolymers from monomers is accompanied by an increase in the Gibbs energy by 15 to 20 kJ mol^{-1} per bond. For example, for the formation of the peptide bond between alanine and glycine, $\Delta_r G^{\ominus\prime} = 17.3\,\text{kJ mol}^{-1}$ (Section 17.2.7). Also the synthesis of lipids from sugar requires Gibbs energy.

The primary compounds that biosystems can use for biosynthesis are mostly CO_2, H_2O, N_2, and inorganic phosphate. The Gibbs energy suppliers in biosystems are NADPH formed from NADP$^+$, and ATP formed from ADP and P, with the help of sunlight. NADPH is used for the reduction of CO_2 (and of N_2 in some bacteria), whereby ATP is needed as an additional supplier of energy. The energy is stored in the form of glucose. It can be used on demand by formation of ATP from ADP.

18.4.1 Synthesis of Glucose

Energy supplied by NADPH and ATP

We consider the synthesis of glucose from CO_2. For the direct reaction

$$6CO_2 + 6H_2O \longrightarrow C_6H_{12}O_6 + 6O_2 \qquad (18.53)$$

$\Delta_r G^{\ominus\prime}$ is positive ($\Delta_r G^{\ominus\prime} = 2872\,\text{kJ mol}^{-1}$): the reaction cannot occur spontaneously.

In nature this process occurs in many small steps, by coupling reaction (18.53) to the reaction

$$12NADPH + 12H^+ + 6O_2 \longrightarrow 12NADP^+ + 12H_2O \qquad (18.54)$$

($\Delta_r G^{\ominus\prime} = -2640\,\text{kJ mol}^{-1}$, see Appendix 3).

For the reaction

$$6CO_2 + 12NADPH + 12H^+ \longrightarrow C_6H_{12}O_6 + 12NADP^+ + 6H_2O \qquad (18.55)$$

we find $\Delta_r G^{\ominus\prime} = (2872 - 2640)\,\text{kJ mol}^{-1} = 232\,\text{kJ mol}^{-1}$. This is still positive, so that the reaction cannot occur spontaneously under standard conditions. Indeed, another 18ATP molecules are converted in 18 enzymatic reactions according to reaction (18.44) to ADP and P transferring their energy to intermediates. According to Section 18.3, $\Delta_r G^{\ominus\prime}$ for this reaction sequence is $\Delta_r G^{\ominus\prime} = 18 \cdot (-30.5)\,\text{kJ mol}^{-1} = -549\,\text{kJ mol}^{-1}$. For the total reaction

$$6CO_2 + 12NADPH + 18ATP + 12H^+ \qquad (18.56)$$
$$\longrightarrow C_6H_{12}O_6 + 12NADP^+ + 18ADP + 18P + 6H_2O$$

we obtain $\Delta_r G^{\ominus\prime} = (232 - 549)\,\text{kJ mol}^{-1} = -317\,\text{kJ mol}^{-1}$. This reaction can occur spontaneously.

Sunlight – induced NADPH and ATP production

The synthesis of ATP and NADPH takes place in the photosynthetic system of plants. For the formation of 18ATP and of 12NADPH molecules ($\Delta_r G^{\ominus\prime} = (549 + 2640)\,\text{kJ mol}^{-1} = 3189\,\text{kJ mol}^{-1}$) about 50 light quanta have to be absorbed. Each light quantum has an energy of about 2.8×10^{-19} J (corresponding to the absorption maximum of chlorophyll at 700 nm); then 50 light quanta supply about 140×10^{-19} J per glucose molecule or $8500\,\text{kJ mol}^{-1}$. Only one-third of this amount is used to increase the Gibbs energy of the system. The low energetical yield in the primary process of the photosynthesis arises from the difficulty of keeping the electrical charges separated (see Chapter 23).

18.4.2 Combustion of Glucose

The combustion of glucose with O_2 to CO_2 and H_2O does not proceed in one step ($\Delta_r G^{\ominus\prime} = -2872\,\text{kJ mol}^{-1}$), but in many small steps which are coupled, each with a small decrease in $\Delta_r G$. We have considered one of these steps in the previous section (the oxidation of phosphoglycerolaldehyde to 3-phosphoglyceric acid (reaction 18.52)). Through enzymatic coupling with reactions in which ATP is synthesized from ADP and P, about half of the Gibbs energy is won back: 40% of the Gibbs energy initially available from the glucose ($2872\,\text{kJ mol}^{-1}$) would be sufficient for the formation of 94 molecules of ATP from ADP and P per molecule of glucose; however, only 38 molecules are formed.

18.4.3 Energy Balance of Formation and Degradation of Glucose

A schematic illustration of these energy transfer processes is shown in Figure 18.5. Each step is an oxidation process of one compound – that is, the removal of one electron, which is transferred to an oxidizing agent (e.g., $NADP^+$) which then becomes a reducing agent, conducting the electron to a stronger oxidizing agent. Finally an O_2 molecule is

18.4 BIOENERGETICS

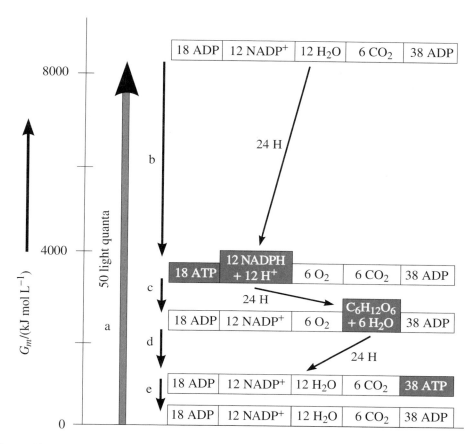

Figure 18.5 Biosynthesis of glucose and its degradation. Transfer of the absorbed light energy (step a) in chemical energy (ATP, NADPH, step b). Formation of glucose as an energy reservoir (step c). Degradation of glucose, forming ATP (step d). ATP (supplier of energy for the formation of biopolymers and maintenance of life processes) is converted into ADP (step e). Shaded areas: high-energy compounds occurring in the photo reaction.

formed which reacts with H^+ ions that have been formed initially from H_2O by electron removal

$$O_2 + 4H^+ + 4e^- \longrightarrow 2H_2O$$

Thus H_2O is restored.

Through enzymatic coupling of the various steps with the phosphorylation of ADP the Gibbs energy decreases only slightly in each step. One of the steps in this chain of coupled reactions is the transfer of an electron from cytochrome b [cyt $b(Fe^{2+})$] to the oxidized cytochrome c_1 [cyt c_1 (Fe^{3+})]. For the reaction

$$\text{cyt } c_1(Fe^{2+}) + H^+ \longrightarrow \text{cyt } c_1(Fe^{3+}) + \tfrac{1}{2}H_2 \qquad (18.57)$$

we find $\Delta_r G^{\ominus\prime} = 21 \text{ kJ mol}^{-1}$. For the reaction

$$\text{cyt } b(Fe^{2+}) + H^+ \longrightarrow \text{cyt } b(Fe^{3+}) + \tfrac{1}{2}H_2 \qquad (18.58)$$

we find $\Delta_r G^{\ominus\prime} = 4\,\text{kJ}\,\text{mol}^{-1}$. Then for the considered reaction

$$\text{cyt } b(\text{Fe}^{2+}) + \text{cyt } c_1(\text{Fe}^{3+}) \longrightarrow \text{cyt } b(\text{Fe}^{3+}) + \text{cyt } c_1(\text{Fe}^{2+}) \quad (18.59)$$

we obtain $\Delta_r G^{\ominus\prime} = (4 - 21)\,\text{kJ}\,\text{mol}^{-1} = -17\,\text{kJ}\,\text{mol}^{-1}$.

In biological systems (respiratory chain), ATP is formed by coupling this reaction with the synthesis of ATP from ADP and P. Three molecules of cyt $b(\text{Fe}^{2+})$ and three molecules of cyt $c_1(\text{Fe}^{3+})$ are used to form one molecule of ATP. Thus $\Delta_r G^{\ominus\prime} = (30.5 - 3 \times 17)\,\text{kJ}\,\text{mol}^{-1} = -20.5\,\text{kJ}\,\text{mol}^{-1}$.

Problems

Problem 18.1 – pH of Weak Acid for Different Total Concentrations

What is the pH of acetic acid at different total concentrations?

Solution. We calculate $\widehat{c}_{H^+}$ according to equation (18.9)

$$\widehat{c}_{H^+} = -\tfrac{1}{2}K + \sqrt{\widehat{c}_{\text{total}} \cdot K + \tfrac{1}{4}K^2}$$

with $K = 2.4 \times 10^{-5}$. With this relation a difficulty arises for $\widehat{c}_{\text{total}} = 0$: then we obtain $\widehat{c}_{H^+} = 0$, a clearly wrong result. The reason is that we neglected the dissociation of water in the derivation of equation (18.9). This neglect affects $\widehat{c}_{H^+}$ at very low acid concentration. We account for the dissociation of water by setting $\widehat{c}_{H^+} = 1.0 \times 10^{-7}$ for $\widehat{c}_{\text{total}} = 0$ and starting with the evaluation of equation (18.9) at $c_{\text{total}} = 3 \times 10^{-6}$ (for this value we calculate $\widehat{c}_{H^+} = 30 \times 10^{-7}$, and the dissociation of water can be neglected). The resulting $\text{pH} = -\log(\widehat{c}_{H^+})$ is displayed in Figure 18.6.

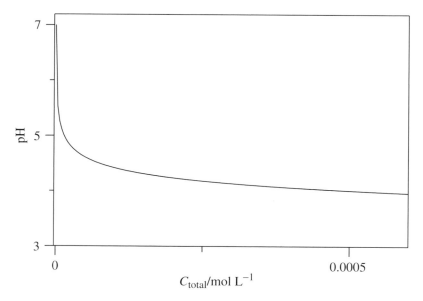

Figure 18.6 pH of acetic acid versus the total concentration of the acid.

Problem 18.2 – Buffer Solutions

1 mL of a solution of HCl (concentration 1 mol L^{-1}) is added to 100 mL of water (case 1) and to 100 mL of a buffer solution (case 2); thus the initial concentration c_a of HCl immediately after mixing is $c_a = 0.01$ mol L^{-1} in both cases. The buffer is a 1 : 1 mixture of acetic acid and sodium acetate, each of concentration $c_0 = 0.1$ mol L^{-1}. What is the change of pH in both cases?

Solution. We consider sodium acetate and hydrochloric acid as completely dissociated.

Case 1. The concentration of H$^+$ is equal to $c_a = 10^{-2}$ mol L^{-1}, thus pH = 2 and ΔpH = 2 − 7 = −5. This is a drastic change of pH.

Case 2. To calculate the change of pH in the buffer solution we start out with the buffer and add HCl:

$$K = \frac{\widehat{c}_{H^+} \cdot \widehat{c}_{Ac^-}}{\widehat{c}_{HAc}}$$

Because of stoichiometry and electroneutrality we have $\widehat{c}_{HAc} + \widehat{c}_{Ac^-} = 2\widehat{c}_0$, $\widehat{c}_{H^+} + \widehat{c}_{Na^+} = \widehat{c}_{OH^-} + \widehat{c}_{Cl^-} + \widehat{c}_{Ac^-}$. Furthermore, $\widehat{c}_{Na^+} = \widehat{c}_0$, $\widehat{c}_{Cl^-} = \widehat{c}_a$. Neglecting $\widehat{c}_{H^+}$ and $\widehat{c}_{OH^-}$ relative to $\widehat{c}_0$ and $\widehat{c}_a$, we obtain $\widehat{c}_{Ac^-} = \widehat{c}_0 - \widehat{c}_a$, $\widehat{c}_{HAc} = 2\widehat{c}_0 - \widehat{c}_0 + \widehat{c}_a = \widehat{c}_0 + \widehat{c}_a$. Then

$$\widehat{c}_{H^+} = K \frac{\widehat{c}_{HAc}}{\widehat{c}_{Ac^-}} = K \cdot \frac{\widehat{c}_0 + \widehat{c}_a}{\widehat{c}_0 - \widehat{c}_a}, \quad pH = pK - \log \frac{\widehat{c}_0 + \widehat{c}_a}{\widehat{c}_0 - \widehat{c}_a}$$

For $\widehat{c}_a = 0$ (i.e., for the pure buffer) we have pH = pK = 4.62.

For $\widehat{c}_a = 1 \times 10^{-2}$ and $\widehat{c}_0 = 0.1$ we have pH = 4.62 − log(0.11/0.09) = 4.53, thus ΔpH = −0.09. We see that the pH of the buffer solution remains nearly unchanged. In Figure 18.7, pH is plotted versus c_a for different buffer concentrations c_0.

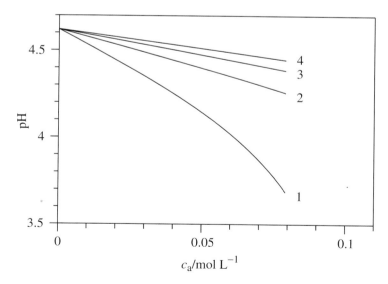

Figure 18.7 pH of a buffer solution (acetic acid–acetate buffer, pK = 4.62, concentration c_0) versus concentration c_a of added strong acid: $c_0 = 0.1$ mol L^{-1} (curve 1); $c_0 = 0.2$ mol L^{-1} (curve 2); $c_0 = 0.3$ mol L^{-1} (curve 3); $c_0 = 0.4$ mol L^{-1} (curve 4).

Problem 18.3 – pH of Amino Acids

In Section 18.1.5 the pH of amino acids was calculated.
(a) Check if the dissociation of water and K_1 besides c_{total} in equation (18.22)

$$(\widehat{c}_{H^+})^2 = K_1 K_2 \cdot \frac{\widehat{c}_{\text{total}}}{\widehat{c}_{\text{total}} + K_1}$$

can be neglected.
(b) Calculate the pH assuming that the reactions from B to A or from B to C (as in Section 18.1.5) do not take place.

Solution. (a) With the calculated pH = 6.1 we obtain from equation (18.21)

$$\widehat{c}_A = \frac{\widehat{c}_{H^+} \cdot \widehat{c}_{\text{total}}}{K_1} = 10^{-4.8}$$

for $\widehat{c}_{\text{total}} = 0.1$. Accordingly, from equation (18.18)

$$\widehat{c}_C = K_2 \frac{\widehat{c}_B}{\widehat{c}_{H^+}} \approx K_2 \frac{\widehat{c}_{\text{total}}}{\widehat{c}_{H^+}} = 10^{-4.7}$$

Then $\widehat{c}_{OH^-} = 10^{-7.9}$ given by the dissociation equilibrium of water can be neglected. K_1 can be neglected in the denominator of equation (18.22) as long as $\widehat{c}_{\text{total}} > 0.01$.

Solution. (b) Neglecting equilibrium 1: then $\widehat{c}_{H^+}$ is calculated from equation (18.9): $\widehat{c}_{H^+} = 3.9 \times 10^{-6} = 10^{-5.4}$; pH = 5.4 for $\widehat{c}_{\text{total}} = 0.1$ and pH = 4.9 for $\widehat{c}_{\text{total}} = 1$.

Neglecting equilibrium 2: in this case $\widehat{c}_A = \widehat{c}_{OH^-}$, because for each H$^+$ taken from the water one OH$^-$ is being formed.
Then

$$K_1 = \frac{\widehat{c}_{H^+} \cdot \widehat{c}_B}{\widehat{c}_{OH^-}} = \frac{\widehat{c}_{H^+} \cdot \widehat{c}_{\text{total}}}{K_w / \widehat{c}_{H^+}} \quad \text{and} \quad \widehat{c}_{H^+} = \sqrt{\frac{K_1 K_w}{\widehat{c}_{\text{total}}}}$$

Therefore $\widehat{c}_{H^+} = 10^{-7.7}$ (pH = 7.7) for $\widehat{c}_{\text{total}} = 0.1$; correspondingly, $\widehat{c}_{H^+} = 10^{-8.2}$ (pH = 8.2) for $\widehat{c}_{\text{total}} = 1$.

Problem 18.4 – Absorption Maximum of NADH

Explain the shift of the absorption maximum when proceeding from NAD$^+$ to NADH (see Section 18.3.2).

Solution: The absorption band of NADH is due to the fact that the molecular chain between N and O with the resonating structures

$$\text{N}-\text{C}=\text{C}-\text{C}=\text{O} \quad \text{and} \quad \overset{+}{\text{N}}=\text{C}-\text{C}=\text{C}-\overset{-}{\text{O}}$$

constitutes the chromophore with 6 π electrons; in this case from Table 8.2 we obtain $\lambda_{\text{max}} = 332$ nm. This is in contrast to NAD$^+$.

Foundation 18.1 – Titration of Acetic Acid by NaOH

Combining equations (18.2), (18.15), (18.16), and (18.17) results in

$$K = \widehat{c}_{H^+} \cdot \frac{\widehat{c}_{A^-}}{\widehat{c}_{total} - \widehat{c}_{A^-}} = \widehat{c}_{H^+} \cdot \left[\frac{\widehat{c}_{total}}{\widehat{c}_{A^-}} - 1\right]^{-1}$$

$$= \widehat{c}_{H^+} \cdot \left[\frac{\widehat{c}_{total}}{\widehat{c}_{H^+} + \widehat{c}_{Na^+} - K_w/\widehat{c}_{H^+}} - 1\right]^{-1}$$

Unfortunately, it is not possible to solve this equation analytically for $\widehat{c}_{H^+}$. Therefore we proceed by solving for $\widehat{c}_{Na^+}$:

$$\widehat{c}_{Na^+} = \frac{K \cdot \widehat{c}_{total}}{\widehat{c}_{H^+} + K} - \widehat{c}_{H^+} + \frac{K_w}{\widehat{c}_{H^+}}$$

n_{NaOH}, the amount of added NaOH, equals the amount of Na^+ ions in the solution:

$$n_{NaOH} = n_{Na^+} = V c_{Na^+} = V c^\ominus \widehat{c}_{Na^+}$$

with V = volume of the solution and $c^\ominus = 1$ mol L^{-1} (standard concentration). Then with $n_{HA,0} = c_{total} V$ ($n_{HA,0}$ = amount of acid at the start of the titration)

$$n_{NaOH} = \left[\frac{K \cdot n_{HA,0}}{\widehat{c}_{H^+} + K} - V c^\ominus \left(\widehat{c}_{H^+} - \frac{K_w}{\widehat{c}_{H^+}}\right)\right] = [A - VB]$$

with

$$A = \frac{K \cdot n_{HA,0}}{\widehat{c}_{H^+} + K} \quad B = \left(\widehat{c}_{H^+} - \frac{K_w}{\widehat{c}_{H^+}}\right) c^\ominus$$

In Table 18.1 n_{NaOH} is calculated for different values of $\widehat{c}_{H^+}$ for acetic acid as an example (dissociation constant $K = 2.4 \times 10^{-5}$, initial amount of acetic acid $n_{HA,0} = 1$ mol). It is assumed that the volume V ($V = 1$ L) of the solution does not change during the titration (this is possible by using NaOH in a high enough concentration). By plotting pH versus n_{NaOH} (using the data in Table 18.1) the titration curve in Figure 18.3 (solid line) is obtained.

At the beginning of the titration ($n_{NaOH} = 0$), $\widehat{c}_{H^+}$ is much larger than $K_w/\widehat{c}_{H^+}$, and $\widehat{c}_{H^+}$ is much larger than K:

$$n_{NaOH} \approx \frac{K \cdot n_{HA,0}}{\widehat{c}_{H^+}} - V c^\ominus \widehat{c}_{H^+} = 0$$

Then $\widehat{c}_{H^+} = \sqrt{K \widehat{c}_{total}}$ and pH = 2.31 for $\widehat{c}_{total} = 1$.

Continued on page 598

Continued from page 597

Table 18.1 Titration of acetic acid ($\widehat{c}_{total} = 1$, $n_{HA,0} = 1$ mol, $V = 1$ L) with NaOH

pH	$\widehat{c}_{H^+}$	A/mol	B/mol L^{-1}	n_{NaOH}/mol
2.31	4.9×10^{-3}	0.005	4.9×10^{-3}	0.000
4	1.0×10^{-4}	0.193	1.0×10^{-4}	0.193
4.62	2.3×10^{-5}	0.500	2.4×10^{-5}	0.500
5	1.0×10^{-5}	0.705	1.0×10^{-5}	0.705
9	1.0×10^{-9}	1.000	-1.0×10^{-5}	1.000
11	1.0×10^{-11}	1.000	-1.0×10^{-3}	1.001
12	1.0×10^{-12}	1.000	-1.0×10^{-2}	1.010
13	1.0×10^{-13}	1.000	-1.0×10^{-1}	1.100
13.75	1.8×10^{-14}	1.000	-0.555	1.555
14	1.0×10^{-14}	1.000	-1.000	2.000

Calculation of added amounts n_{NaOH} for various values of pH using $K = 2.4 \times 10^{-5}$ and $K_w = 10^{-14}$. n_{NaOH} in column 5 is calculated for constant volume.

For $n_{NaOH} = 0.5$ mol (one-half of the acid has been titrated) we can neglect the term VB; then

$$\frac{n_{NaOH}}{n_{HA,0}} \approx \frac{K}{\widehat{c}_{H^+} + K} = \frac{1}{2} \quad \text{and} \quad \widehat{c}_{H^+} = K, \text{pH} = 4.62$$

At the stoichiometry point ($n_{NaOH} = 1.0$ mol) with $n_{HA,0} = 1$ mol, $V = 1$ L we obtain

$$1 = \frac{K}{\widehat{c}_{H^+} + K} - \widehat{c}_{H^+} + \frac{K_W}{\widehat{c}_{H^+}}$$

$$\frac{\widehat{c}_{H^+}}{\widehat{c}_{H^+} + K} = -\widehat{c}_{H^+} + \frac{K_W}{\widehat{c}_{H^+}}$$

With the approximation $\widehat{c}_{H^+} \ll K$ and $\widehat{c}_{H^+} \ll K_W/\widehat{c}_{H^+}$ we calculate pH= $\frac{1}{2}(pK_W + pK) = 9.31$.

Finally, for $n_{NaOH} = 2.0$ mol there is an excess of NaOH by 1 mol L^{-1}, thus pH = 14.

Effect of increasing volume

Actually, the volume V of the solution increases during the titration:

$$V = V_0 + n_{NaOH} \cdot \frac{1}{c_{NaOH,0}}$$

Continued on page 599

FOUNDATION 18.1

Continued from page 598

(V_0 = starting volume, $c_{\text{NaOH},0}$ = contration of added NaOH solution) and we obtain

$$n_{\text{NaOH}} = A - VB = A - \left(V_0 + n_{\text{NaOH}} \cdot \frac{1}{c_{\text{NaOH},0}}\right) B$$

and

$$n_{\text{NaOH}} = \frac{A - V_0 B}{1 + B/c_{\text{NaOH},0}}$$

The corresponding values of n_{NaOH} are plotted in Figure 18.3 (dashed line); the effect of dilution is a less steep increase of pH at the end of the titration.

19 Chemical Reactions in Electrochemical Cells

What happens in an electrochemical cell? A chemical reaction takes place in the cell producing or consuming electrical energy. The cell has a distinct voltage E. When closing the electrical circuit by an external resistor, a chemical reaction occurs in the cell such that the condition

$$\Delta G < 0$$

is fulfilled. If we apply an external voltage $\phi > E$ the reverse reaction takes place. In the limiting case when ϕ is only slightly different from E the reaction is reversible. We obtain the change in Gibbs energy

$$\Delta G = -E \cdot I \cdot t$$

$I \cdot t$ (current × time) is the charge that has passed through the circuit in time t. ΔG for chemical reactions can be easily and accurately obtained from electrochemical measurements by using this relation. The conversion of electrical energy into chemical energy (electrolysis) and the conversion of chemical energy into electrical energy (batteries, fuel cells) are important applications.

19.1 ΔG and Potential E of an Electrochemical Cell Reaction

Figure 19.1 shows an electrochemical cell in which the reaction

$$AgI + \tfrac{1}{2}Pb \longrightarrow Ag + \tfrac{1}{2}PbI_2 \tag{19.1}$$

can occur essentially reversibly. A lead electrode and a silver electrode are immersed in a beaker containing a KI solution and solid PbI_2 and AgI. Pb and PbI_2 are in one part of the beaker, and Ag as well as AgI are in the other part, such that reaction (19.1) cannot occur directly. However, at the silver electrode the reaction

$$Ag^+ + e^- \longrightarrow Ag \tag{19.2}$$

19.1 ΔG AND POTENTIAL E OF AN ELECTROCHEMICAL CELL REACTION

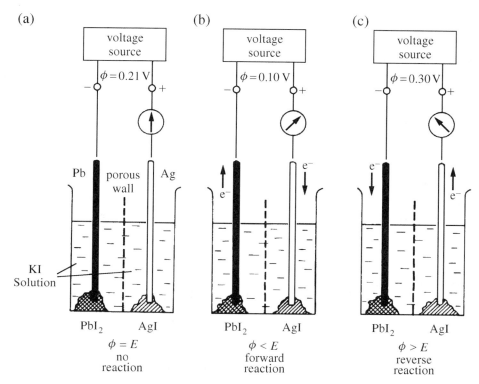

Figure 19.1 Electrochemical cell in which reaction (19.1) can occur reversibly at atmospheric pressure and $T = 298$ K. Dashed line: porous separating wall (sintered glass frit) preventing the mixing of the solutions in the half-cells by convection. The diffusion of Ag^+ and Pb^{2+} through the pores is neglected. The external voltage ϕ is (a) $\phi = E$; (b) $\phi < E$ (charging the cell), (c) $\phi > E$ (discharging the cell). KI is in excess, therefore K^+ and I^- are the carriers of the current through the frit.

and at the lead electrode the reaction

$$\tfrac{1}{2}Pb \longrightarrow \tfrac{1}{2}Pb^{2+} + e^- \tag{19.3}$$

take place. Thus the silver electrode is positively charged and the lead electrode is negatively charged.

The Pb^{2+} ions which are formed additionally according to reaction (19.3) react with I^- ions to give solid PbI_2.

$$\tfrac{1}{2}Pb^{2+} + I^- \longrightarrow \tfrac{1}{2}PbI_2 \tag{19.4}$$

Accordingly, solid AgI dissolves at the silver electrode:

$$AgI \longrightarrow Ag^+ + I^- \tag{19.5}$$

All four reactions together result in reaction (19.1). According to Appendix 1, $\Delta_r G^\ominus$ of this reaction is negative at 298 K: under standard conditions and at 298 K the reaction occurs spontaneously.

However, with the external circuit open, reaction (19.2) cannot proceed when the silver electrode is sufficiently positively charged; the same problem occurs with reaction (19.3) when the lead electrode is sufficiently negatively charged. This situation arises when the potential across the electrodes is $E = 0.210$ V. The situation remains unchanged when the electrodes are connected to an external voltage source with a voltage $\phi = 0.210\,\text{V}$ (Figure 19.1(a)): the ammeter shows zero current.

If we decrease ϕ below E (e.g., $\phi = 0.1\,\text{V}$), then electrons can move through the external circuit from the lead electrode to the silver electrode (Figure 19.1(b)). Reaction (19.1) takes place; the system (the solution and the electrodes) supplies electrical work to the surroundings (the external voltage source and the temperature bath).

On the other hand, if we increase ϕ above E (e.g., $\phi = 0.3\,\text{V}$), the reverse of reaction (19.1) takes place (Figure 19.1(c)); the surroundings supply electrical work to the system.

Minute changes of the external voltage ϕ enable us to reverse the direction of the chemical reaction; thus this reaction can be performed reversibly. Then it should be possible to relate the cell potential E to the change ΔG in Gibbs energy of the reaction.

What is the relation between ΔG and E? We consider the case where the external voltage ϕ is only slightly different from E, such that reaction (19.1) can just occur (the current I is extremely small). Then electrical energy is supplied by the system:

$$\text{electrical energy supplied by the system} = E \cdot I \cdot t \tag{19.6}$$

when the current I flows during time t. Since the cell is under atmospheric pressure, the volume of the solution in the cell can change by ΔV, and additional work $P\Delta V$ is supplied by the system. Then the total reversible work $-w_{\text{rev}}$ done by the system is

$$-w_{\text{rev}} = (\text{electrical energy supplied by the system}) + P\Delta V \tag{19.7}$$
$$= E \cdot I \cdot t + P\Delta V$$

For a change of state at constant temperature and constant pressure the change of Gibbs energy is, according to equation (16.39),

$$\Delta G = w_{\text{rev}} + P\Delta V \tag{19.8}$$

Therefore,

$$\Delta G = -E \cdot I \cdot t \tag{19.9}$$

During time t an amount n of electrons passes through the external circuit. Then the charge corresponding to I and t is

$$\text{charge} = I \cdot t = nN_A \cdot e \tag{19.10}$$

and we obtain

$$\boxed{\Delta G = -E \cdot nN_A \cdot e = -n\mathbf{F} \cdot E} \tag{19.11}$$

or

$$\boxed{\Delta_r G = \frac{\Delta G}{n} = -\mathbf{F} \cdot E} \tag{19.12}$$

where

$$\mathbf{F} = N_A \cdot e = 96485.309\,\text{C}\,\text{mol}^{-1} \tag{19.13}$$

is called the *Faraday constant*.

From equation (19.12) we can calculate $\Delta_r G$ for reaction (19.1) using the measured cell potential $E = 0.210\,\text{V}$ at 298 K: $\Delta_r G = -20.3\,\text{kJ}\,\text{mol}^{-1}$. (Note that $1\,\text{C}\,\text{V} = 1\,\text{J}$). This is just what we obtain by calculating $\Delta_r G$ from $\Delta_f G^{\ominus}$ for each species entering reaction (19.1) (separated solids at 1 bar), as shown in Problem 19.2.

19.1 ΔG AND POTENTIAL E OF AN ELECTROCHEMICAL CELL REACTION

Note that equation (19.11) is valid only if we write the reaction (19.1) such that just one electron passes the external circuit (see reactions (19.2) and (19.3)). $\Delta_r G$ is the change of Gibbs energy when 1 mol of electrons is involved.

By measuring the cell potential E at various temperatures, we obtain the change $\Delta_r S$ of entropy during the reaction. According to equation (17.32) we have

$$\left(\frac{\partial \Delta_r G}{\partial T}\right)_P = -\Delta_r S \qquad (19.14)$$

and by substituting $\Delta_r G$ by $-E\boldsymbol{F}$ we obtain

$$\Delta_r S = \boldsymbol{F} \cdot \left(\frac{\partial E}{\partial T}\right)_P \qquad (19.15)$$

Finally, from $\Delta_r G$ and $\Delta_r S$ we obtain $\Delta_r H$.

Example – evaluating $\Delta_r S$ and $\Delta_r H$ for reaction (19.1)

We find the following from Figure 19.2

$$\Delta_r S = \boldsymbol{F} \cdot \left(\frac{\partial E}{\partial T}\right)_P = \boldsymbol{F} \cdot \frac{-0.020 \text{ V}}{120 \text{ K}} = -16 \text{ J K}^{-1} \text{ mol}^{-1}$$

$$\Delta_r H = \Delta_r G + T \Delta_r S = (-20.3 \times 10^3 - 298 \times 16) \text{ J mol}^{-1}$$
$$= -25.1 \text{ kJ mol}^{-1}$$

The reaction takes place at a pressure of 1 bar and the species are separated at the beginning and at the end of the reaction, so we can identify $\Delta_r H$ and $\Delta_r S$ with $\Delta_r H^\ominus$ and $\Delta_r S^\ominus$. The values obtained from Appendix 1 are (Problem 19.2): $\Delta_r H^\ominus = -25.2 \text{ kJ mol}^{-1}$ and $\Delta_r S^\ominus = -16.3 \text{ J mol}^{-1} \text{ K}^{-1}$.

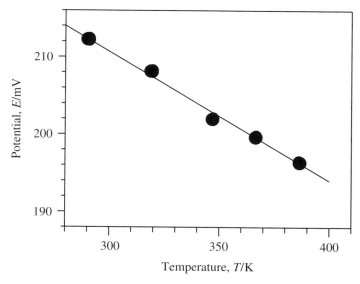

Figure 19.2 Potential E of the electrochemical cell in Figure 19.1 as a function of temperature T.

The negative sign of ΔS calculated in the above example means that heat $-q_{\text{rev}} = -T\Delta S > 0$ flows from the cell to the surroundings even when the reaction is carried out reversibly.

When the current flow is reversed (by slightly increasing ϕ) the reverse reaction takes place: heat flows from the surroundings into the cell (isothermal process), or the cell cools down when it is not connected with the heat bath (adiabatic process).

19.2 Concentration Cells

We consider, for simplicity, very dilute solutions of electrolytes and use concentrations instead of activities. To generalize we can always replace concentrations for activities, as outlined in Chapter 20.

19.2.1 Metal Electrodes

An electrochemical cell as displayed in Figure 19.3 is called a *concentration cell*. Two silver electrodes are immersed in $AgNO_3$ solutions of concentrations c_1 and c_2. Both solutions are separated by a sintered glass frit.

If $c_1 > c_2$, then the reaction

$$Ag^+(c_1) \longrightarrow Ag^+(c_2) \tag{19.16}$$

can occur spontaneously. We assume that the direct process (diffusion of Ag^+ ions through the sintered glass frit from left to right) is sufficiently slow (practically no change of concentration during measurement) and can thus be neglected. However, the process can take place indirectly by the electrode reactions

$$Ag^+(c_1) + e^- \longrightarrow Ag \text{ (left)} \tag{19.17}$$

$$Ag \longrightarrow Ag^+(c_2) + e^- \text{(right)} \tag{19.18}$$

Then the silver electrode on the right-hand side becomes negatively charged relative to the silver electrode on the left-hand side.

By applying a voltage ϕ in the external circuit which is slightly lower than the cell potential E we can achieve a practically reversible electron transport between the two electrodes from right to left. What will happen in the two electrode compartments? Ag^+ ions will be consumed (forming solid Ag according to reaction (19.17)) in the left compartment, and Ag^+ ions will be formed from solid Ag in the right compartment. Thus the overall reaction (19.16) takes place.

However, there is an additional process taking place at the same time. Because the solution is part of the electrical circuit, Ag^+ ions pass the sintered glass frit from right to left (from dilute to concentrated solution), and NO_3^- ions pass from left to right (from concentrated to dilute solution, Figure 19.3a). Then an additional change of the concentrations in the two compartments occurs. To avoid this difficulty we add an indifferent salt (e.g., KNO_3), to the solution such that its concentration is equal in both compartments and much larger than c_1 or c_2. Then the charge transport is mainly determined by this indifferent salt, and no work for the transport has to be supplied (Figure 19.3b). (In the arrangement of Figure 19.1, the KI solution has the same purpose.)

19.2 CONCENTRATION CELLS

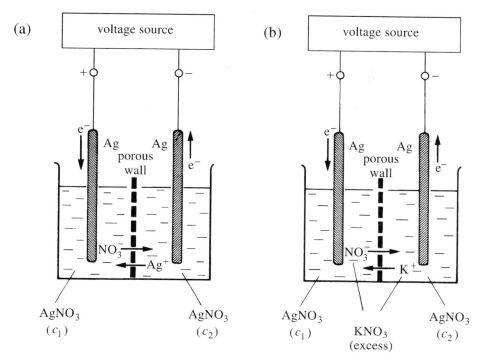

Figure 19.3 Concentration cell. The polarity of the cell is for $c_1 > c_2$. The direction of the current is such that reaction (19.16) takes place. (a) Transport of charge inside the cell by Ag^+ and NO_3^- ions. (b) Transport of charge by K^+ and NO_3^- ions in excess and in equal concentration in both compartments.

If we deal with ideally dilute solutions, then the change of Gibbs energy for reaction (19.16) is, according to equation (17.63),

$$\Delta_r G = RT \cdot \ln \frac{c_2}{c_1} \tag{19.19}$$

and

$$E = -\frac{\Delta_r G}{F} = -\frac{RT}{F} \cdot \ln \frac{c_2}{c_1} \tag{19.20}$$

Thus the cell potential E is determined by the ratio of the two concentrations. At $T = 298$ K

$$\frac{RT}{F} = 25.7 \times 10^{-3} \, \text{J C}^{-1} = 25.7 \, \text{mV}$$

and

$$E = -25.7 \, \text{mV} \cdot \ln \frac{c_2}{c_1} = -59.1 \, \text{mV} \cdot \log \frac{c_2}{c_1} \tag{19.21}$$

Example – Ag^+ ions in equilibrium with AgCl

Let us add some KCl into the right-hand compartment of the concentration cell (Figure 19.4). Then Ag^+ ions precipitate as AgCl. For an excess of KCl (concentration of Cl^- ions $0.1 \, \text{mol L}^{-1}$ after the AgCl has precipitated) the measured cell potential is

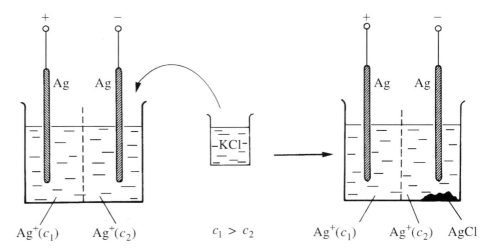

Figure 19.4 Measuring the Ag$^+$ concentration in the concentration cell in Figure 19.3 after adding a KCl solution to the right compartment. Transport of charge inside the cell by K$^+$ and NO$_3^-$ ions in excess.

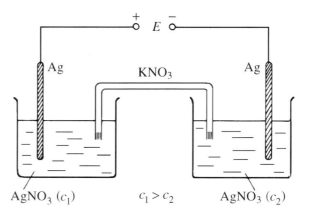

Figure 19.5 Concentration cell with salt bridge. The salt bridge is filled with a concentrated solution of KNO$_3$. It has the same effect as adding KNO$_3$ in excess in both compartments.

$E = 0.461$ V, when $c_1 = 0.1$ mol L^{-1}. Then the concentration c_2 of Ag$^+$ ions in equilibrium with Cl$^-$ and AgCl is $c_2 = c_1 \times 10^{-E/59.1\,\text{mV}} = 1.6 \times 10^{-9}$ mol L^{-1}. Even though practically all the Ag$^+$ ions have precipitated as AgCl, the tiny amount of still dissolved Ag$^+$ can be measured very accurately. In the present case we can determine the solubility product of AgCl as $K_{\text{sol}} = \widehat{c}_{\text{Ag}^+} \cdot \widehat{c}_{\text{Cl}^-} = 1.6 \times 10^{-10}$.

To avoid complications by the transport of charge through the sintered glass frit we added an indifferent salt in excess. Another possibility is to use two separate containers for the reaction solutions and to connect them by a salt bridge (Figure 19.5). A salt bridge is a glass tube filled with a concentrated solution of an electrolyte, the ions of which move practically with the same velocity and do not react with the ions of the solution at the electrodes. The potentials E measured with such an arrangement usually differ only by a

19.2 CONCENTRATION CELLS

few millivolts from the values expected according to equation (19.11); such an error can be neglected in most cases.

19.2.2 Gas Electrodes

We consider the electrochemical cell in Figure 19.6. Platinum electrodes in a bubble flow of hydrogen (hydrogen electrodes) are immersed in solutions with H^+ concentrations c_1 (left, HCl solution) and c_2 (right, NaOH solution). Again $c_1 > c_2$, and the reaction

$$H^+(c_1) \longrightarrow H^+(c_2) \tag{19.22}$$

can occur spontaneously. At the surface of the platinum electrodes the reactions

$$H^+(c_1) + e^- \longrightarrow \tfrac{1}{2}H_2(Pt) \quad \text{(left)} \tag{19.23}$$

$$\tfrac{1}{2}H_2(Pt) \longrightarrow H^+(c_2) + e^- \quad \text{(right)} \tag{19.24}$$

take place, and the right electrode becomes negatively charged relative to the left one.

What processes occur at the platinum electrodes? The solution in the neighborhood of the electrode is saturated with H_2. The H_2 molecules diffuse to the platinum surface and transfer electrons to the metal: conversion into H^+ ions. The H^+ ions disappear from the surface by diffusing into the solution. Also, the reverse process, the conversion of H^+ ions into H_2 molecules, takes place at the same time. Both processes are in a fast dynamic equilibrium. Although platinum is practically inert, it works as a catalyst for electron exchange.

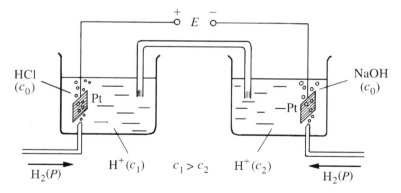

Figure 19.6 Concentration cell with gas electrodes (hydrogen electrodes).

Example – ionic product of water

The cell potential measured with the HCl/NaOH hydrogen cell is $E = 0.590$ V for $c_{HCl} = c_{NaOH} = 0.01 \text{ mol L}^{-1}$. Then, from equation (19.21)

$$\frac{c_2}{c_1} = 10^{-E/59.1\,\text{mV}} = 1.04 \times 10^{-10}$$

The HCl solution is completely dissociated into H^+ and Cl^- ions (strong acid), thus $c_1 = 0.01 \text{ mol L}^{-1}$, and for the concentration of H^+ ions in 0.01 molar NaOH we obtain

$c_2 = 0.01 \times 1.04 \times 10^{-10}$ mol L^{-1} = 1.04×10^{-12} mol L^{-1}. This example shows clearly that extremely small ion concentrations can be determined by measuring cell potentials. In the present case the concentration of the OH$^-$ ions in the NaOH solution is 0.01 mol L^{-1}, thus we can calculate the ionic product of water as $K_w = \widehat{c}_{H^+} \cdot \widehat{c}_{OH^-} = 0.01 \times 1.04 \times 10^{-12} = 1.04 \times 10^{-14}$.

In this case we used different H$^+$ concentrations in the two containers, but the hydrogen is under the same pressure P. Alternatively, we can choose equal H$^+$ concentrations, but different H$_2$ pressures P_1 and P_2 (Figure 19.7). This can be achieved by bubbling H$_2$ at atmospheric pressure (P_1) on the left side and, say, a 1:9 mixture of H$_2$ and argon as an inert gas on the right side; then $P_1/P_2 = 10$.

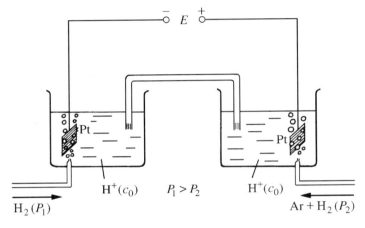

Figure 19.7 Concentration cell with hydrogen electrodes with different partial gas pressures.

When the circuit is closed the reaction

$$H_2(P_1) \longrightarrow H_2(P_2) \tag{19.25}$$

occurs spontaneously, and the reactions

$$H^+(c) + e^- \longrightarrow \tfrac{1}{2} H_2(P_2) \quad \text{(right)} \tag{19.26}$$

$$\tfrac{1}{2} H_2(P_1) \longrightarrow H^+(c) + e^- \quad \text{(left)} \tag{19.27}$$

take place at the electrodes. Then the right electrode becomes positively charged relative to the left one. From

$$\Delta_r G = RT \cdot \ln \frac{(\widehat{P}_2)^{1/2}}{(\widehat{P}_1)^{1/2}} \tag{19.28}$$

we obtain the cell potential

$$E = -\frac{\Delta_r G}{F} = -\frac{RT}{F} \cdot \ln \frac{(\widehat{P}_2)^{1/2}}{(\widehat{P}_1)^{1/2}} = -\frac{1}{2} \times 59.1 \text{ mV} \cdot \log \frac{\widehat{P}_2}{\widehat{P}_1} \tag{19.29}$$

19.3 Standard Potential $E^\ominus$

19.3.1 Nernst Equation

Next we discuss a cell with a hydrogen electrode connected to a copper electrode immersed in a Cu^{2+} solution (Figure 19.8). The cell reaction

$$\tfrac{1}{2}Cu^{2+} + \tfrac{1}{2}H_2 \longrightarrow \tfrac{1}{2}Cu + H^+ \qquad (19.30)$$

occurs spontaneously under standard conditions (Cu^{2+} is reduced to Cu by H_2). Then the copper electrode becomes positively charged relative to the hydrogen electrode. The change of Gibbs energy for this reaction is, according to equation (17.67),

$$\Delta_r G = \Delta_r G^\ominus + RT \cdot \ln Q \quad \text{with} \quad Q = \frac{\widehat{c}_{H^+}}{(\widehat{c}_{Cu^{2+}})^{1/2} \cdot (\widehat{P}_{H_2})^{1/2}} \qquad (19.31)$$

where Q is the reaction quotient. The potential E is

$$E = -\frac{\Delta_r G}{F} = -\frac{\Delta_r G^\ominus}{F} - \frac{RT}{F} \cdot \ln Q \qquad (19.32)$$

With

$$E^\ominus = -\frac{\Delta_r G^\ominus}{F} \qquad (19.33)$$

where $E^\ominus$ is called the *standard electrode potential*, we rewrite this equation as

$$\boxed{E = E^\ominus - \frac{RT}{F} \cdot \ln Q} \qquad (19.34)$$

This relation is called the *Nernst equation*. Note that $\Delta_r G$ is the change of Gibbs energy when one mol of electrons is involved in the reaction. Thus equation (19.34) is valid only if reaction (19.30) is written appropriately.

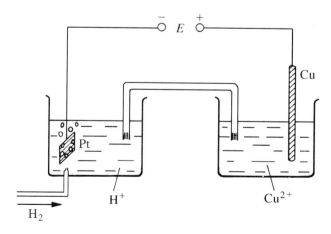

Figure 19.8 Potential of a Cu electrode in Cu^{2+} solution measured relative to a hydrogen electrode.

Such standard potentials have been measured for many reduction processes at $T = 298$ K; for some selected systems they are listed in Appendix 2.

For strongly oxidizing agents (that tend to accept electrons), $E^\ominus$ is strongly positive (the electrode is positive relative to the hydrogen electrode if all reactants are in their standard states). Accordingly, $E^\ominus$ is strongly negative for strongly reducing agents.

$$Ag^+ + e^- \longrightarrow Ag \qquad E^\ominus_{Ag^+/Ag} = +800 \, mV$$

$$\tfrac{1}{3}Fe^{3+} + e^- \longrightarrow \tfrac{1}{3}Fe \qquad E^\ominus_{Fe^{3+}/Fe} = -36 \, mV$$

$$\tfrac{1}{2}Zn^{2+} + e^- \longrightarrow \tfrac{1}{2}Zn \qquad E^\ominus_{Zn^{2+}/Zn} = -763 \, mV$$

Example – standard potential of Cu^{2+}/Cu

The potential of the cell in Figure 19.8 at $T = 298$ K is $E = 425$ mV for $c_{Cu^{2+}} = 0.001$ mol L^{-1}, $P_{H_2} = 1$ bar and $c_{H^+} = 0.001$ mol L^{-1}. What is the standard potential $E^\ominus$? From equation (19.34) we obtain

$$E^\ominus = E + \frac{RT}{F} \cdot \ln Q$$

We calculate Q as

$$Q = \frac{\widehat{c}_{H^+}}{(\widehat{c}_{Cu^{2+}})^{1/2} \cdot (\widehat{P}_{H_2})^{1/2}} = \frac{0.001}{\sqrt{0.001} \times \sqrt{1}} = 0.0316$$

Then $E^\ominus = (425 + 25.7 \cdot \ln 0.0316)$ mV $= 336$ mV.

Note that the standard potential $E^\ominus$ is defined for a reaction in a cell where one half-cell is the hydrogen electrode. The cell reaction is written such that H_2 appears on the left-hand side and H^+ on the right-hand side of the equation. This means that in the half-cell under consideration a reduction process takes place:

$$\tfrac{1}{2}Cu^{2+} + e^- \longrightarrow \tfrac{1}{2}Cu, \quad E^\ominus_{Cu^{2+}/Cu} = 336 \, mV$$

19.3.2 Practical Determination of $E^\ominus$

Generally the standard potentials $E^\ominus$ can be obtained from the measurement of the cell potential for any concentrations in both half-cells, as demonstrated above. The measurement can be simplified by using a hydrogen electrode with $P_{H_2} = 1$ bar and $c_{H^+} = 1$ mol L^{-1}. Then formally $\widehat{P}_{H_2}$ and $\widehat{c}_{H^+}$ are equal to one, and these quantities can be omitted in writing the Nernst equation.

This is not completely correct, because $c_{H^+} = 1$ mol L^{-1} is far from an ideally dilute solution. However, the experiment shows that the potential of a cell with a hydrogen electrode immersed in a HCl solution of $c_{HCl} = 1.184$ mol L^{-1} and under a H_2 pressure of 1 bar is just the same as formally expected for $\widehat{P}_{H_2} = 1$ and $\widehat{c}_{H^+} = 1$ assuming an

19.4 REDOX REACTIONS

ideally dilute solution. Such an electrode is commonly used as a reference electrode. It is called a *standard hydrogen electrode* (SHE).

Example

$$\tfrac{1}{2}Cu^{2+} + \tfrac{1}{2}H_2 \longrightarrow Cu + H^+ \qquad (19.35)$$

$$E = E^\ominus - \frac{RT}{F} \cdot \ln \frac{\widehat{c}_{H^+}}{(\widehat{c}_{Cu^{2+}})^{1/2} \cdot (\widehat{P}_{H_2})^{1/2}} \qquad (19.36)$$

$$E_{SHE} = E^\ominus - \frac{RT}{F} \cdot \ln \frac{1}{(\widehat{c}_{Cu^{2+}})^{1/2}} \qquad (19.37)$$

where E_{SHE} is the cell potential measured against a standard hydrogen electrode ($c_{HCl} = 1.184$ mol L^{-1}, $P_{H_2} = 1$ bar).

19.4 Redox Reactions

19.4.1 Fe^{3+}/Fe^{2+} Electrode

Let us consider a half-cell with a platinum electrode immersed in a solution containing a mixture of Fe^{2+} and Fe^{3+} connected to a standard hydrogen electrode. The platinum electrode can, as in the case of H$_2$ and H$^+$, exchange electrons with Fe^{2+} and Fe^{3+}:

$$Fe^{3+} + e^- \rightleftharpoons Fe^{2+} \qquad (19.38)$$

According to the cell reaction

$$Fe^{3+} + \tfrac{1}{2}H_2 \longrightarrow Fe^{2+} + H^+ \qquad (19.39)$$

the cell potential is

$$E_{SHE} = E^\ominus - \frac{RT}{F} \cdot \ln \frac{c_{Fe^{2+}}}{c_{Fe^{3+}}} \qquad (19.40)$$

For $c_{Fe^{2+}} = c_{Fe^{3+}}$ the cell potential at 298 K is $E_{SHE} = -441$ mV, thus $E^\ominus = -441$ mV.

Reactions where the active species exist in different oxidation states in solution (such as Fe^{2+}, Fe^{3+}) are called *redox reactions*.

19.4.2 Quinone/Hydroquinone Electrode

Another example for a redox system is the quinhydrone electrode (Figure 19.9), where the half-cell contains a mixture of quinone (Q) and hydroquinone (H$_2$Q). The half-cell reaction is

$$\tfrac{1}{2}Q + H^+ + e^- \rightleftharpoons \tfrac{1}{2}H_2Q \qquad (19.41)$$

Note that in this case the exchange of charge at the platinum electrode is accompanied by the production or consumption of a proton in the half-cell. According to the cell reaction

$$\tfrac{1}{2}Q + H^+ + \tfrac{1}{2}H_2(SHE) \longrightarrow \tfrac{1}{2}H_2Q + H^+(SHE) \qquad (19.42)$$

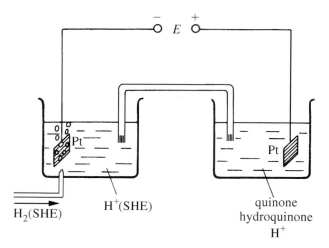

Figure 19.9 Quinone/hydroquinone electrode measured against a standard hydrogen electrode (SHE).

the cell potential is

$$E_{SHE} = E^\ominus - \frac{RT}{F} \cdot \ln \frac{(\widehat{c}_{H_2Q})^{1/2}}{(\widehat{c}_Q)^{1/2} \cdot \widehat{c}_{H^+}} \tag{19.43}$$

where c_{H^+} is the H$^+$ concentration in the Q/H$_2$Q half-cell. The proton is consumed in the half-cell and produced at the hydrogen electrode.

For $c_Q = c_{H_2Q}$ and $c_{H^+} = 1 \times 10^{-5}\,\text{mol L}^{-1}$ (pH = 5, acetic acid/acetate buffer) the measured cell potential is $E_{SHE} = +404\,\text{mV}$. Then $E^\ominus = (404 + 5 \times 59.1)\,\text{mV} = +700\,\text{mV}$.

19.4.3 NAD$^+$/NADH Electrode

As for the quinone/hydroquinone reaction the cell potential for the reaction

$$\tfrac{1}{2}\text{NAD}^+ + \tfrac{1}{2}\text{H}^+ + \tfrac{1}{2}\text{H}_2(\text{SHE}) \longrightarrow \tfrac{1}{2}\text{NADH} + \text{H}^+(\text{SHE}) \tag{19.44}$$

$$E_{SHE} = E^\ominus - \frac{RT}{F} \cdot \ln \frac{(\widehat{c}_{NADH})^{1/2}}{(\widehat{c}_{NAD^+})^{1/2} \cdot (\widehat{c}_{H^+})^{1/2}} \tag{19.45}$$

depends on pH.

In the three redox reactions discussed above, the dependence on pH is different, as we can see best by choosing 1:1 ratios for the redox partners ($c_{Fe^{2+}} = c_{Fe^{3+}}$; $c_Q = c_{H_2Q}$; $c_{NAD^+} = c_{NADH}$):

$$\text{Fe}^{3+}/\text{Fe}^{2+}: \qquad E_{SHE} = E^\ominus$$

$$\text{Q/H}_2\text{Q}: \qquad E_{SHE} = E^\ominus + \frac{RT}{F} \cdot \ln \widehat{c}_{H^+}$$

$$\text{NAD}^+/\text{NADH}: \qquad E_{SHE} = E^\ominus + \frac{1}{2}\frac{RT}{F} \cdot \ln \widehat{c}_{H^+}$$

19.4.4 Oxygen Electrode

We couple an oxygen electrode (oxygen bubbling over a platinum foil) to a standard hydrogen electrode. The cell potential for the reaction

$$\tfrac{1}{4}O_2 + H^+ + \tfrac{1}{2}H_2(\text{SHE}) \longrightarrow \tfrac{1}{2}H_2O + H^+(\text{SHE}) \tag{19.46}$$

is

$$E_{\text{SHE}} = E^\ominus - \frac{RT}{F} \cdot \ln \frac{1}{\widehat{c}_{H^+} \cdot (\widehat{P}_{O_2})^{1/4}} \tag{19.47}$$

with

$$E^\ominus = -\frac{\Delta_r G^\ominus}{F} \tag{19.48}$$

$\Delta_r G^\ominus$ is the change in Gibbs energy for the formation of 0.5 mol H_2O from 0.5 mol H_2 and 0.25 mol O_2 under standard conditions (species separated at $P = P^\ominus = 1$ bar). From Appendix 1 we obtain $\Delta_r G^\ominus = \tfrac{1}{2}\Delta_f G^\ominus(H_2O) = -118.6 \text{ kJ mol}^{-1}$ at 298 K. Then

$$E^\ominus = \frac{118.6 \text{ kJ mol}^{-1}}{F} = 1.23 \text{ V}$$

In contrast to the hydrogen electrode, the oxygen electrode does not work reversibly, because the electrode reaction is kinetically hindered. The reverse reaction, the splitting of water into H_2 and O_2

$$\tfrac{1}{2}H_2O \longrightarrow \tfrac{1}{2}H_2 + \tfrac{1}{4}O_2$$

does not take place at 1.23 V, a much higher voltage must be applied (overvoltage of 0.3 V).

19.5 Applications of Electrochemical Cells

19.5.1 Reference Electrodes

For the determination of $E^\ominus$ we used electrochemical cells with a hydrogen electrode. Since working with such electrodes is tedious, it is customary to use other reference electrodes which are easier to handle. Figure 19.10 shows a calomel electrode (Hg_2Cl_2) and a silver chloride electrode (AgCl). The half-cell reaction in the calomel electrode is

$$\tfrac{1}{2}Hg_2^{2+} + e^- \longrightarrow Hg \tag{19.49}$$

and the potential relative to the standard hydrogen electrode is

$$E_{\text{calomel}} = E^\ominus - \frac{RT}{F} \cdot \ln \frac{1}{(\widehat{c}_{Hg_2^{2+}})^{1/2}} \tag{19.50}$$

Since the electrode contains an excess of chloride ions, the concentration of Hg_2^{2+} is determined by the equilibrium (K_s = solubility product)

$$Hg_2^{2+} + 2Cl^- \rightleftharpoons Hg_2Cl_2 \tag{19.51}$$

$$\widehat{c}_{Hg_2^{2+}} \cdot (\widehat{c}_{Cl^-})^2 = K_s \tag{19.52}$$

and we obtain

$$E_{\text{calomel}} = E^\ominus - \frac{RT}{F} \cdot \ln \frac{\widehat{c}_{\text{Cl}^-}}{\sqrt{K_s}} \tag{19.53}$$

With $E^\ominus = 0.796\,\text{V}$ and $K_s = 2.038 \times 10^{-18}$ we obtain $E_{\text{calomel}} = 0.273\,\text{V}$ (for $c_{\text{Cl}^-} = 1\,\text{mol L}^{-1}$ at 298 K). In a similar way, for the silver chloride electrode we find

$$E_{\text{silver chloride}} = E^\ominus - \frac{RT}{F} \cdot \ln \frac{\widehat{c}_{\text{Cl}^-}}{K_{\text{AgCl}}} \tag{19.54}$$

With $E^\ominus = 0.800\,\text{V}$ and $K_{\text{AgCl}} = 1.58 \times 10^{-10}$ we obtain $E_{\text{silver chloride}} = 0.221\,\text{V}$ (for $c_{\text{Cl}^-} = 1\,\text{mol L}^{-1}$ at 298 K).

The experimental values are 0.281 V for the calomel electrode and 0.222 V for the silver chloride electrode (Figure 19.10); the small differences result from the fact that the present theory is valid only for ideally dilute solutions.

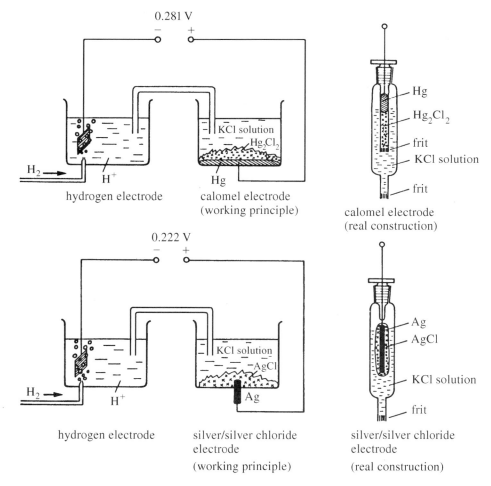

Figure 19.10 Calomel and silver chloride electrodes (reference electrodes). The voltages refer to $c_{\text{KCl}} = 1\,\text{mol L}^{-1}$.

19.5.2 Glass Electrodes

Two half-cells filled with solutions containing H^+ ions (one of the known concentration c_1, the second one with an unknown concentration c_2) are connected by a membrane permeable only to H^+ ions. Such membranes can be made from properly treated glass. In each half-cell a AgCl electrode is immersed (Figure 19.11). We consider the case $c_1 < c_2$. Since both silver chloride electrodes are identical, for the cell potential we obtain

$$E = \frac{RT}{F} \cdot \ln \frac{c_2}{c_1} = \frac{RT}{2.303F} \cdot \log \frac{c_2}{c_1} \qquad (19.55)$$

In the technical realization (Figure 19.11(b)) a glass ball permeable to H^+ ions is filled with a buffer solution of known pH. This solution is then connected to a silver chloride electrode. A second silver chloride electrode is inserted into the same glass tube. If c_2 is the concentration of the solution to be examined, then according to equation (19.55) we have, because of $pH = -\log \hat{c}_2$ and $pH_{buffer} = -\log \hat{c}_1$

$$\boxed{pH = pH_{buffer} - \frac{F}{2.303 \cdot RT} \cdot E} \qquad (19.56)$$

The pH value is then a linear function of the cell voltage. Equation (19.56) is the practical definition of pH; usually a KH_2PO_4/Na_2HPO_4 buffer with $0.1\ mol\ L^{-1}$ of NaCl is used as a standard (pH = 6.865 at $T = 298$ K). See Covington, A.K.; Bates, R.G.; Durst, R.A., Definition of pH Scales, Standard Reference Values, Measurement of pH and Related Terminology, *Pure Appl. Chem.*, 1985, **57**, 531–542.

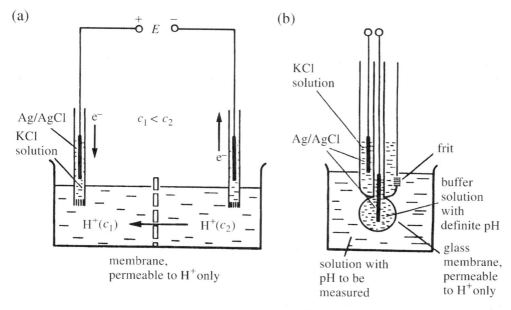

Figure 19.11 Glass electrode. (a) Fundamental setup; (b) technical realization.

19.5.3 Galvanic Elements

So far we have used electrochemical cells to measure thermodynamic quantities. In batteries and accumulators electrochemical processes occur in which chemical energy is converted into electrical energy. These devices can be used as energy sources or energy storages (Figure 19.12).

Leclanché element (Zn/graphite)

The electrode reactions are (Figure 19.12(a))

$$\tfrac{1}{2}Zn \longrightarrow \tfrac{1}{2}Zn^{2+} + e^-, \quad H^+ + e^- \longrightarrow \tfrac{1}{2}H_2 \qquad (19.57)$$

The reaction product Zn^{2+} is complexed by NH_4Cl into $\left[Zn(NH_3)_4\right]^{2+}$. A graphite rod surrounded by MnO_2 (mixed with graphite for electrical conductivity) is used as a counterelectrode; MnO_2 oxidizes H_2:

$$2MnO_2 + H_2 \longrightarrow 2MnO(OH) \qquad (19.58)$$

The cell voltage is 1.6 V. Solid zinc is used in the form of a beaker containing NH_4Cl and the counterelectrode (Figure 19.12(a)).

Alkaline manganese cell

The Leclanché cell works in acidic medium. An important modification is the alkaline manganese dioxide cell, where NH_4Cl is replaced by KOH. The electrode reactions are (Figure 19.12(b))

$$\tfrac{1}{2}Zn + OH^- \longrightarrow \tfrac{1}{2}Zn(OH)_2 + e^-, \quad H^+ + e^- \longrightarrow \tfrac{1}{2}H_2 \qquad (19.59)$$

The hydrogen is oxidized by MnO_2 in the same way as in the Leclanché cell. In this case, zinc is used in the form of powder in the inner part of the cell; the zinc powder is surrounded by alkaline MnO_2 placed in a beaker of stainless steel (Figure 19.12(b)). This is the most common cell used nowadays.

Lithium cell

Lithium cells (Figure 19.12(c)) are becoming more and more important because of their high cell voltage of 3 V, long lifetime, high energy content, and a broad temperature range. The reaction at the lithium electrode is

$$Li \longrightarrow Li^+ + e^- \qquad (19.60)$$

A layer of MnO_2 is used to remove electrons from the counterelectrode (stainless steel):

$$MnO_2 + e^- \longrightarrow [MnO_2]^- \qquad (19.61)$$

Because Li reacts directly with water, a mixture of propylene carbonate and dimethoxyethane is used as a solvent for the electrolyte (LiBr, $LiAlCl_4$).

19.5 APPLICATIONS OF ELECTROCHEMICAL CELLS

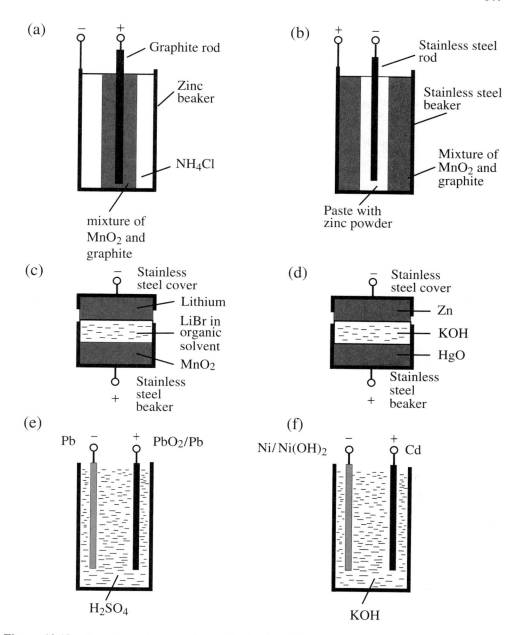

Figure 19.12 Batteries and accumulators. (a) Leclanché cell. (b) Alkaline manganese cell. (c) Lithium cell. (d) Mercury cell. (e) Lead accumulator. (f) Ni/Cd accumulator.

Besides MnO_2, SO_2 (dissolved in acetonitrile) or $SOCl_2$ can be used as electron-removing agents:

$$SO_2 + e^- \longrightarrow SO_2^-, \quad 2SO_2^- \longrightarrow S_2O_4^{2-} \qquad (19.62)$$

$$2SOCl_2 + 4e^- \longrightarrow S + SO_2 + 4Cl^-$$

Mercury cell (Mallory cell)

The overall electrode reaction is

$$Zn + HgO \longrightarrow ZnO + Hg \qquad (19.63)$$

resulting in a cell voltage of 1.35 V (Figure 19.12(d)).

Lead accumulator

Electrical energy can be stored in lead accumulators (Figure 19.12(e)). In loading a lead accumulator, Pb^{2+} (in the form of solid $PbSO_4$) is reduced to Pb at the negative electrode and oxidized to PbO_2 at the positive electrode:

$$PbSO_4 + H_2O \longrightarrow \tfrac{1}{2}Pb + \tfrac{1}{2}PbO_2 + H_2SO_4 \qquad (19.64)$$

This process requires an external voltage of 2 V. On the other hand, water can be split into H_2 and O_2 when applying a voltage above 1.23 V (Section 19.4.4). Why does this process not occur in the lead accumulator? This is because the production of H_2 at the negative electrode and of O_2 at the positive electrode is kinetically hindered. The overvoltage of H_2 at the lead electrode is 1.67 V, whereas there is no overvoltage for H_2 at a platinum electrode.

The fact that there is an overvoltage for H_2 at a lead electrode is crucial for the realization of a lead accumulator using an aqueous electrolyte. Without that overvoltage the lead accumulator could not work at all. In practice, traces of platinum (e.g., from analytical grade sulfuric acid produced in platinum vessels) are sufficient to catalyze the decomposition of water in the accumulator.

Ni/Cd accumulator

In loading an Ni/Cd accumulator (Figure 19.12(f)), Ni^{3+} is reduced to Ni and Ni^{2+}, and Cd is oxidized to Cd^{2+}:

$$2NiO(OH) + H_2O + e^- \longrightarrow Ni + Ni(OH)_2 + OH^-$$

$$\tfrac{1}{2}Cd + OH^- \longrightarrow \tfrac{1}{2}Cd(OH)_2 + e^-$$

19.5.4 Fuel Cells

In Chapter 15 we have proved by considering the Carnot process that it is impossible to convert heat completely into mechanical work. Thus it is not possible to convert the chemical energy stored in a fuel (e.g., coal or oil) completely into work by means of a heat engine.

According to equation (15.26) the work done by an idealized heat engine is

$$w_{\text{heat engine}} = \eta \cdot q_{\text{rev},2} = \left(1 - \frac{T_1}{T_2}\right) \cdot q_{\text{rev},2} \qquad (19.65)$$

where T_1 and T_2 are the working temperatures of the heat engine, and $q_{\text{rev},2}$ is the heat supplied by the combustion of the fuel at temperature T_2.

19.5 APPLICATIONS OF ELECTROCHEMICAL CELLS

Let us consider the combustion of carbon

$$C + O_2 \longrightarrow CO_2 \tag{19.66}$$

in a heat engine operating at the temperatures $T_1 = 300\,\text{K}$ and $T_2 = 800\,\text{K}$. Then $q_{\text{rev},2} = -\Delta H^{\ominus}_{800} = 394\,\text{kJ mol}^{-1} \cdot n_C$ and the work done by the engine is

$$\left(1 - \frac{300}{800}\right) \times 394\,\text{kJ mol}^{-1} \cdot n_C = 246\,\text{kJ mol}^{-1} \cdot n_C$$

n_C is the amount of carbon being oxidized. In principle, the same process can be carried out in an electrochemical cell (*fuel cell*). The work done by an idealized electrochemical cell equals $-\Delta G$. Then

$$w_{\text{fuel cell}} = -\Delta G^{\ominus} = -(\Delta H^{\ominus} - T\Delta S^{\ominus})$$
$$= (394 \times 10^3 + 298 \times 2.9)\,\text{J mol}^{-1} \cdot n_C$$
$$= 395\,\text{kJ mol}^{-1} \cdot n_C \text{ at } 298\,\text{K}$$

This is 60% more than can be obtained by burning coal in a heat engine.

Unfortunately, up to now it was not possible to construct fuel cells operating satisfactory with coal or similar fuels of technical importance. Fuel cells operating with the reaction

$$H_2 + \tfrac{1}{2}O_2 \longrightarrow H_2O$$

have been developed. In such a fuel cell, porous nickel electrodes are used, allowing a good contact of the gases with the electrolyte. However, the cell voltage (0.73 V) is only 60% of the theoretical value of 1.23 V. This is a somewhat higher efficiency than for a real heat engine.

19.5.5 Electrolysis

Instead of converting chemical energy into electrical energy or storing electrical energy in chemical compounds, we can perform chemical reactions by supplying electrical energy. As an example, we examine the electrolysis of a HCl solution in which hydrogen and chlorine are formed. When the reaction

$$HCl(aq) \longrightarrow \tfrac{1}{2}H_2 + \tfrac{1}{2}Cl_2 \tag{19.67}$$

takes place in an electrochemical cell (hydrogen electrode connected to a chlorine electrode in an HCl solution, Figure 19.13(a)), we expect, according to Appendix 2, a voltage of 1.36 V. Applying a somewhat higher voltage enables us to force the reaction (19.67) to occur: H_2 and Cl_2 are formed at atmospheric pressure. The formation of the gases can also be observed obviously in the electrolysis cell (Figure 19.13(b)).

By plotting the electrical current in the electrolysis cell as a function of the applied voltage ϕ, we obtain the curves displayed in Figure 19.14. We consider the curve for HCl. Above 1.36 V the current increases linearly with ϕ, because the migration velocities of H^+ and Cl^- increase linearly with ϕ; that is, in the same time more ions are discharged at the electrodes. Below 1.36 V, H^+ ions are discharged at the negative electrode and Cl^- ions at the positive electrode, until equilibrium concentrations of dissolved H_2 and

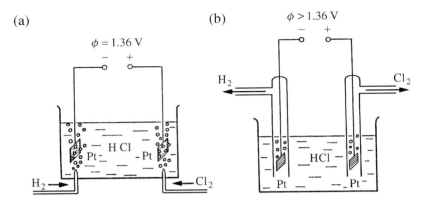

Figure 19.13 Electrochemical cell for reaction (19.67). (a) Cell working as a voltage source, (b) electrolysis cell.

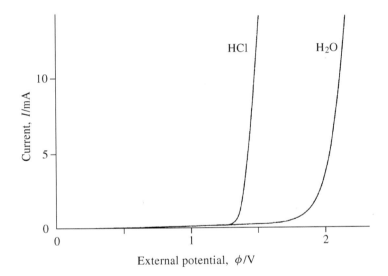

Figure 19.14 Electrolysis current I versus the external potential ϕ. The HCl electrolysis is carried out in an HCl solution (concentration $1\,\text{mol}\,\text{L}^{-1}$), and the H_2O electrolysis is carried out in an H_2SO_4 solution (concentration $1\,\text{mol}\,\text{L}^{-1}$), using platinum wire electrodes (diameter 1 mm, length 20 mm) in a distance of 40 cm (temperature 25°C).

Cl_2 at the electrodes have been established. Then we expect that after equilibration the current becomes zero. However, H_2 and Cl_2 molecules slowly disappear from the electrode surface by diffusion; in this way their concentrations become smaller than in equilibrium, and a small current is maintained to balance this diffusion effect.

19.6 Conductivity of Electrolyte Solutions

19.6.1 Mobility of Ions

We consider a solution containing positive and negative ions with charges ze. In an electric field the ions are accelerated, thus a current flows through the solution. First we restrict

19.6 CONDUCTIVITY OF ELECTROLYTE SOLUTIONS

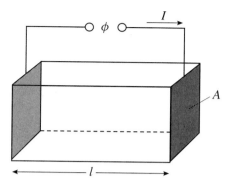

Figure 19.15 Conductivity cell.

our consideration to the case that only one kind of ion contributes to the current (we assume that the counterions are immobilized).

Two electrodes are immersed in the solution (Figure 19.15). If the potential across the electrodes is ϕ, the electrical field strength between the electrodes is

$$F = \frac{\phi}{l} \quad (19.68)$$

where l is the distance between the electrodes. The electrical field exerts a force f_{el} on the ions

$$f_{el} = ze \cdot \frac{\phi}{l} \quad (19.69)$$

Thus the positive ions are accelerated toward the negative electrode, and the negative ions are accelerated toward the positive electrode: the speed of the ions increases. On the other hand, the ions are decelerated by the force of friction f_{fric}. If the ions are treated as spheres moving in a viscous fluid, we can apply the law of Stokes

$$f_{fric} = -6\pi\eta r v \quad (19.70)$$

where η is the viscosity of the solvent, r is the radius of the sphere, and v is the speed of the ions. The speed v becomes constant when the resulting force on the ion becomes zero:

$$f_{el} + f_{fric} = 0 \quad (19.71)$$

Then

$$v = \frac{ze}{6\pi\eta r} \cdot \frac{\phi}{l} = u \cdot \frac{\phi}{l} \quad (19.72)$$

where u is the *electric mobility* of the ion. As a consequence of the movement of the ions, a current I can be measured in the external circuit.

$$I = \frac{dQ}{dt} \quad (19.73)$$

where dQ is the charge of the ions passing the cross-section A of the container in time dt.

During time dt the ions travel the distance

$$dx = v \cdot dt \quad (19.74)$$

Then the number of ions passing the cross-section A in time dt is

$$dN = c \cdot N_A A \cdot dx = c \cdot N_A A v \cdot dt \qquad (19.75)$$

where c is the concentration of the ions and N_A is Avogadro's constant. Then we obtain

$$I = \frac{dQ}{dt} = ze \cdot \frac{dN}{dt} = ze \cdot c \cdot N_A A v \qquad (19.76)$$

With

$$v = u\frac{\phi}{l}, \quad \mathbf{F} = eN_A \qquad (19.77)$$

(u = electric mobility), we obtain

$$I = \mathbf{F} czu\frac{A}{l} \cdot \phi \qquad (19.78)$$

This expression can be simplified by introducing the *conductivity* κ

$$\boxed{I = \kappa \cdot \frac{A}{l}\phi, \quad \kappa = \mathbf{F} czu} \qquad (19.79)$$

19.6.2 Generalization

In this consideration we restricted our discussion to the case where only one kind of ion is responsible for the current I; this is the case when the mobility of the counterions is sufficiently small. Generally, if cations (charge z_+e, concentration c_+, mobility u_+) and anions (charge $-z_-e$, concentration c_-, mobility u_-) contribute to the current, we obtain

$$\kappa = \mathbf{F}(c_+z_+u_+ + c_-z_-u_-) \qquad (19.80)$$

Because of electroneutrality,

$$c_+z_+ = c_-z_- \qquad (19.81)$$

and thus

$$\kappa = \mathbf{F} c_+z_+ \cdot (u_+ + u_-) \qquad (19.82)$$

In our consideration we assume independent migration of the ions. This is only appropriate for high dilution. In general, the measured conductivity $I \cdot l/(A \cdot \phi)$ depends on the electrostatic interaction between the ions and the conductivity of the pure solvent. κ, defined as conductivity of electrolyte minus conductivity of the pure solvent, is given by (19.82) only at high dilution. It is useful to introduce the *molar conductivity*

$$\Lambda = \frac{\kappa}{c} \qquad (19.83)$$

where c is the concentration of the electrolyte. Thus

$$\lim_{c \to 0} \Lambda = \lim_{c \to 0} \frac{\kappa}{c} = \mathbf{F} \cdot \frac{c_+z_+}{c} \cdot (u_+ + u_-) \qquad (19.84)$$

For example, for NaCl, CuCl$_2$, CuSO$_4$ we obtain $c_+/c = 1$; accordingly, $c_+/c = 2$ for H$_2$SO$_4$.

19.6 CONDUCTIVITY OF ELECTROLYTE SOLUTIONS

Example – speed of ions

What is the speed of an Na$^+$ ion moving in water between two electrodes at distance $l = 1$ cm (potential $\phi = 1$ V) at 25 °C and at 100 °C?
According to equation (19.72) we have

$$v = \frac{e}{6\pi\eta r} \cdot F = \frac{e}{6\pi\eta r} \cdot \frac{\phi}{l}$$

The viscosity of water at 25 °C is $\eta = 0.89 \times 10^{-2}$ g s^{-1} cm^{-1} and with $r = 200$ pm (regarding the hydration sphere of the ion) we obtain

$$v = \frac{1.60 \times 10^{-19} \text{ C}}{6\pi \times 0.89 \times 10^{-3} \text{ kg s}^{-1} \text{ m}^{-1} \cdot 200 \times 10^{-12} \text{ m}} \cdot \frac{1 \text{ V}}{10^{-2} \text{ m}} = 4.8 \times 10^{-3} \text{ mm s}^{-1}$$

Then the mobility Na$^+$ at 25°C is $u = vl/\phi = 4.8 \times 10^{-8}$ m^2 V^{-1}, in agreement with the measured value (see Table 19.1).

The viscosity of water at 100 °C is $\eta = 0.28 \times 10^{-2}$ g s^{-1} cm^{-1}; then $v = 15 \times 10^{-3}$ mm s^{-1}.

Thus the speed of the ions is remarkably low, in reasonable accordance with experimental data. The speed increases significantly with temperature because of the decreasing viscosity of the solvent.

The approximate agreement of the calculated values with experiment is remarkable, showing that the concepts of hydrodynamics considering the liquid as a continuum can be used even in the present case where the size of the moving ions is on the same order as the size of the particles in the liquid.

Using equation (19.72), we can calculate radii r of the ions from the mobility u (Table 19.1). The radii r are larger than the ionic radii discussed in Chapter 10, because of the solvation shell formed by the solvent molecules. The solvation shell increases with increasing field strength at the surface of the considered ion. Therefore Li$^+$ has the lowest mobility, and Rb$^+$ has a much higher mobility.

Table 19.1 Electric mobility u and radius r for some ions in aqueous solution at 25°C

Ion	$u/(10^{-8}$ m^2 s^{-1} V$^{-1})$	r/pm
Li$^+$	4.01	238
Na$^+$	5.19	184
Rb$^+$	7.92	120
Cs$^+$	7.85	122
F$^-$	5.70	168
Cl$^-$	7.91	121
I$^-$	7.96	120
H$^+$	36.2	26
OH$^-$	20.6	46

Example – mobility of H⁺

The mobility of an H^+ ion is about seven times higher than that for Na^+. Because of its hydration sphere, the radius r cannot be significantly smaller than 200 pm. What is then the reason for the high mobility of an H^+?

Water molecules are connected by hydrogen bonds. If an H^+ is attached to an H_2O molecule, an additional bond is formed:

$$\begin{array}{ccccccc}
 & H & & H & & H & & H \\
 & | & & | & & | & & | \\
H-O-H & --- & O-H & --- & O-H & --- & O-H \\
\oplus & & & & & &
\end{array}$$

Merely by exchanging one OH bond for a hydrogen bond and vice versa

$$\begin{array}{ccccccc}
 & H & & H & & H & & H \\
 & | & & | & & | & & | \\
H-O & --- & H-O-H & --- & O-H & --- & O-H \\
 & & \oplus & & & &
\end{array}$$

the positive charge can move to the adjacent water molecule. This process continues along the chain of hydrogen-bonded water molecules. It is much faster than the migration of a proton (*Grotthus mechanism*).

Problems

Problem 19.1 – Acceleration of Na⁺ Ions

What is the time required for Na^+ ions to reach a constant speed after switching on an electrical field strength of $1\,V\,cm^{-1}$?

Solution. According to Newton's law and equations (19.69) and (19.70), the acceleration of the ion is

$$\frac{dv}{dt} = \frac{1}{m} \cdot (f_{el} + f_{fric}) = \frac{ze}{m} \cdot F - \frac{6\pi\eta r}{m} \cdot v$$

or

$$\frac{dv}{dt} + av = b$$

with

$$a = \frac{6\pi\eta r}{m} \quad \text{and} \quad b = \frac{ze}{m} \cdot F$$

The solution of this inhomogeneous differential equation with the initial condition $v = 0$ at $t = 0$ is

$$v = \frac{b}{a} \cdot (1 - e^{-at})$$

With $\eta = 0.89 \times 10^{-2}\,g\,s^{-1}\,cm^{-1}$, $r = 200\,pm$, $m = 4 \times 10^{-26}\,kg$, and $F = 1\,V\,cm^{-1}$ we obtain $a = 8.4 \times 10^{13}\,s^{-1}$ and $b = 4.0 \times 10^8\,m\,s^{-2}$.

PROBLEM 19.2

We estimate the half life of this process:

$$t_{1/2} = \frac{1}{a} = \frac{m}{6\pi\eta r} = 10^{-14}\,\text{s}$$

Problem 19.2 – Thermodynamic and Electrochemical Data

Prove that ΔH, ΔS, and ΔG calculated from the cell potential of the electrochemical cell in Figure 19.1 are the same as calculated from $\Delta H^\ominus$, $\Delta S^\ominus$, and $\Delta G^\ominus$ in Appendix 1 for 298 K.

Solution. For the reaction

$$\text{AgI} + \tfrac{1}{2}\text{Pb} \longrightarrow \text{Ag} + \tfrac{1}{2}\text{PbI}_2$$

we find from Appendix 1

$$\Delta_r H^\ominus = \tfrac{1}{2}\Delta_f H^\ominus(\text{PbI}_2) - \Delta_f H^\ominus(\text{AgI})$$
$$= -\tfrac{1}{2}175.1\,\text{kJ mol}^{-1} - (-62.4\,\text{kJ mol}^{-1}) = -25.2\,\text{kJ mol}^{-1}$$

$$\Delta_r S^\ominus = \tfrac{1}{2}S_m^\ominus(\text{PbI}_2) + S_m^\ominus(\text{Ag}) - S_m^\ominus(\text{AgI}) - \tfrac{1}{2}S_m^\ominus(\text{Pb})$$
$$= \left(\tfrac{1}{2}175.2 + 42.7\right)\,\text{J K}^{-1}\,\text{mol}^{-1} - \left(114.2 + \tfrac{1}{2}64.8\right)\,\text{J K}^{-1}\,\text{mol}^{-1}$$
$$= -16.3\,\text{J K}^{-1}\,\text{mol}^{-1}$$

Then $\Delta_r G^\ominus = \Delta_r H^\ominus - T\Delta_r S^\ominus = -20.3\,\text{kJ mol}^{-1}$ at 298 K. This result is in agreement with the corresponding values calculated from the measured cell potential.

Problem 19.3 – Complex Formation

What will the change in E in Figure 19.2 be when KI is added to the AgNO$_3$ solution instead of KCl? What will E be when a KCl solution (0.1 mol L^{-1}) is added to the left compartment and a KCN solution to the right compartment? The solubility product of AgI is $K_s = 1.6 \times 10^{-16}$ and the complex formation constant of $[\text{Ag(CN)}_2]^-$ is $K = 10^{21}$. In all cases the concentration of excess reagent is 0.1 mol L^{-1}.

Solution. Precipitate of AgI: $\widehat{c}_{\text{Ag}^+} = 1.6 \times 10^{-16}/0.1 = 1.6 \times 10^{-15}$. Then

$$E = 59.1\,\text{mV} \cdot \log \frac{0.1}{1.6 \times 10^{-15}} = 815\,\text{mV}$$

In the right compartment the concentration of Ag$^+$ in equilibrium with the cyano complex is, because of

$$K = \frac{\widehat{c}_{[\text{Ag(CN)}_2]^-}}{\widehat{c}_{\text{Ag}^+} \cdot \widehat{c}_{\text{CN}^-}} = 10^{21}$$

$$\widehat{c}_{\text{Ag}^+} = \frac{0.1}{10^{21} \times (0.1)^2} = 10^{-20}$$

In the left compartment the concentration is $\widehat{c}_{Ag^+} = 1.6 \times 10^{-9}$. Then

$$E = 59.1\,\text{mV} \cdot \log \frac{10^{-20}}{1.6 \times 10^{-9}} = -662\,\text{mV}$$

Problem 19.4 – Calculation of $\Delta G^{\ominus\prime}$ and $E^{\ominus\prime}$

Calculate $\Delta G^{\ominus\prime}$ of the following reactions from the values of $E^{\ominus\prime}$ in Appendix 2.

1. $NAD^+ + H_2 \rightarrow NADH + H^+$
2. cyt $c_1 Fe(III) + H^+ \rightarrow$ cyt $c_1 Fe(III) + \frac{1}{2}H_2$

Solution

1. $\Delta G^{\ominus\prime} = 18.1\,\text{kJ mol}^{-1}$
2. $\Delta G^{\ominus\prime} = 61.2\,\text{kJ mol}^{-1}$

20 Real Systems

What are the effects of intermolecular forces on chemical properties? How much are chemical equilibria affected? Do we find new important effects due to these interactions, which we neglected so far?

We consider phase equilibria – for example, the dependence of the melting temperature on pressure. We investigate the behavior of gases at high pressure. Simple models (e.g., the van der Waals equation) are used to analyze changes of state in real systems. The principles behind important applications (e.g., liquefaction of gases) are discussed.

In real solutions the forces between the dissolved species are particularly strong in the case of ions. The effect of Coulomb interaction is treated by a simple model (Debye–Hückel theory).

20.1 Phase Equilibria

In Chapter 17 we determined the vapor pressure of a liquid (P_l) and of a solid (P_s). The melting point is the temperature where $P_l = P_s$. We neglected the volume of the condensed phase and treated the vapor as an ideal gas (Clausius–Clapeyron equation). By this simplification we could not account for the fact that the melting point depends on the pressure. For example, ice under a pressure of 200 bar melts at $-1.5\,°C$. Usually the melting temperature of a substance increases with pressure; in this sense water is an exception: ice is less dense than the coexisting water and, as a consequence, the melting point decreases with pressure. This can be immediately understood from Le Chatelier's principle.

In Box 20.1 we show that for the equilibrium

$$\text{solid} \longrightarrow \text{liquid} \tag{20.1}$$

we can apply the relation

$$\boxed{\frac{dP}{dT} = \frac{\Delta S}{\Delta V} \quad \text{or} \quad \frac{dP}{dT} = \frac{\Delta_{\text{fus}} S}{\Delta_{\text{fus}} V}} \tag{20.2}$$

where $\Delta S = S_{\text{liquid}} - S_{\text{solid}}$, $\Delta V = V_{\text{liquid}} - V_{\text{solid}}$ and $\Delta_{\text{fus}} S$ and $\Delta_{\text{fus}} V$ are the corresponding molar quantities. Equation (20.2) is called the *Clapeyron equation*. We grasp its significance by looking at an example.

Example – freezing point of ice

We consider the equilibrium

$$\text{water (solid)} \longrightarrow \text{water (liquid)}$$

What is the change in the temperature of the ice in equilibrium with the liquid water if we apply an external pressure of 200 bar?

We apply the Clapeyron equation (20.2)

$$dT = \frac{\Delta_{\text{fus}} V}{\Delta_{\text{fus}} S} \cdot dP$$

For water we have

$$\Delta_{\text{fus}} V = -1.62\,\text{cm}^3\,\text{mol}^{-1}, \quad \Delta_{\text{fus}} S = \frac{\Delta_{\text{fus}} H}{T_{\text{fus}}} = 22.0\,\text{J}\,\text{K}^{-1}\,\text{mol}^{-1}$$

Then for $dP = 200$ bar we obtain $dT = -1.5$ K. This means that the freezing point of ice at a pressure of 200 bar is 1.5 K below the freezing point at atmospheric pressure. For comparison, the pressure exerted on the ice by a skater is on the order of 100 bar if the weight of the skater is 800 N and the area of the skate in contact with the ice is 1 cm^2.

In the particular case of water $\Delta_{\text{fus}} V$ is negative, the density of the liquid at the melting point is higher than the density of ice (see Section 11.7.1). Therefore, an increase in the pressure extends the range of existence of liquid water and lowers the melting point in accordance with the principle of Le Chatelier of minimum constraint.

Example – Clausius–Clapeyron equation

Equation (20.2) can correspondingly be applied to the evaporation of a liquid:

$$\text{liquid} \longrightarrow \text{vapor}$$

We express $\Delta_{\text{vap}} S$ by

$$\Delta_{\text{vap}} S = \frac{\Delta_{\text{vap}} H}{T}$$

where $\Delta_{\text{vap}} H$ is the molar change in enthalpy of vaporization (1 mol of liquid is transformed into 1 mol of gas). Furthermore, we treat the vapor as an ideal gas neglecting the volume of the liquid:

$$\Delta_{\text{vap}} V = \frac{V_{\text{vapor}}}{n} = \frac{RT}{P}$$

Thus from the Clapeyron equation we obtain

$$\frac{dP}{dT} = P \cdot \frac{\Delta_{\text{vap}} H}{RT^2}$$

or

$$\boxed{\frac{d\ln P}{dT} = \frac{\Delta_{\text{vap}} H}{RT^2}}$$

This is the Clausius–Clapeyron equation, which we derived already in Chapter 17 from the van't Hoff equation.

> **Box 20.1 – Clapeyron Equation**
>
> We consider a phase transition, e.g. the fusion of a solid
>
> $$\text{solid} \longrightarrow \text{liquid}$$
>
> For equilibrium we have
>
> $$\Delta G = G_{\text{liquid}} - G_{\text{solid}} = 0, \quad \text{thus} \quad G_{\text{liquid}} = G_{\text{solid}}$$
>
> For differential changes maintaining equilibrium (e.g., for differential changes of pressure and temperature)
>
> $$dG_{\text{liquid}} = dG_{\text{solid}}$$
>
> The change dG when changing P and T follows from (16.33)
>
> $$G = H - TS, \quad dG = dH - d(TS), \quad dH = dU + d(PV)$$
>
> Then
>
> $$dG = dU + PdV + VdP - SdT - TdS$$
>
> With
>
> $$dU = dw + dq = -PdV + dq \quad \text{and} \quad dS = \frac{dq}{T}$$
>
> we obtain
>
> $$dG = VdP - SdT$$
>
> Then for the equilibrium of the solid with the liquid we obtain
>
> $$V_{\text{liquid}} \cdot dP - S_{\text{liquid}} \cdot dT = V_{\text{solid}} \cdot dP - S_{\text{solid}} \cdot dT$$
>
> or
>
> $$\frac{dP}{dT} = \frac{\Delta S}{\Delta V}$$
>
> with $\Delta S = S_{\text{liquid}} - S_{\text{solid}}$, $\Delta V = V_{\text{liquid}} - V_{\text{solid}}$.

20.2 Equation of State for Real Gases

In the preceding chapters we have described the behavior of gases by using the ideal gas law. In that treatment the forces acting between the molecules are neglected. In general, this simplifying assumption cannot be valid for real gases. At sufficiently small pressure, however, the molecules are far apart and the intermolecular forces do not appreciably affect the macroscopic properties. Then the ideal gas theory can be compared with experimental data extrapolated to the limit $P \to 0$.

20.2.1 van der Waals Equation

Our aim is to construct a model for real gases considering the attractive forces between the molecules and the effect of the molecular volume. For an ideal gas the equation of state is

$$PV = nRT \qquad (20.3)$$

To account for the effect of the molecular volume V_0 of the gas molecules we replace V in equation (20.3) by $V - V_0$. $V - V_0$ is the volume available for the molecules to move freely. The effect of V_0 is an increase of the pressure of the real gas compared with the pressure of an ideal gas.

On the other hand, the attractive forces between the molecules decrease the pressure P by an amount P_0. Thus the pressure P in equation (20.3) has to be replaced by $(P + P_0)$. Then

$$(P + P_0) \cdot (V - V_0) = nRT \qquad (20.4)$$

or, by using molar quantities,

$$(P + P_0) \cdot (V_m - b) = RT \qquad (20.5)$$

where $V_m = V/n$ is the molar volume, and the constant b is

$$b = \frac{V_0}{n} \qquad (20.6)$$

The volume V_0 and thus the constant b can be estimated (Box 20.2) as

$$V_0 = 4.2 \cdot V_{molecule} N_A \cdot n, \quad b = 4.2 \cdot V_{molecule} N_A \qquad (20.7)$$

where $V_{molecule}$ is the volume of one molecule.

According to equation (20.7), b/N_A should be about four times the volume $V_{molecule}$ of a molecule. Indeed, this is the case for noble gases – for example, for Ar: $b = 0.032 \, \text{L mol}^{-1}$, $r = 0.14 \, \text{nm}$, and $V_{molecule} = 0.012 \times 10^{-27} \, \text{m}^3$. Then

$$\frac{b}{N_A} = 0.053 \times 10^{-27} \, \text{m}^3 = 4.4 \cdot V_{molecule}$$

On the other hand, it must be admitted that for CO_2 ($r = 0.23$ nm) this ratio is $1.5 \cdot V_{molecule}$.

Box 20.2 – Estimation of V_0

To approximate the increase in pressure by the volume V_0 of the molecules we consider Figure 20.1, where two colliding molecules are displayed. Due to the size of the molecules (diameter $2r$), both molecules move ahead by an additional distance $2r \cdot \cos \alpha$ – that is, by a distance r, on the average. In each collision they make up for this additional distance. Thus, the molecules

Continued on page 631

20.2 EQUATION OF STATE FOR REAL GASES

> Continued from page 630
>
>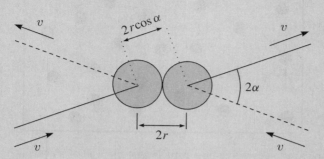
>
> **Figure 20.1** Collision frequency increases with increasing diameter of the molecules.
>
> are faster by a factor
>
> $$\frac{\bar{\lambda} + r}{\bar{\lambda}} = 1 + \frac{r}{\bar{\lambda}}$$
>
> where $\bar{\lambda}$ is the mean free path. Consequently, the frequency of collisions is increased by this factor. Then instead of equation (20.3) we obtain
>
> $$P = \left(1 + \frac{r}{\bar{\lambda}}\right) \cdot \frac{nRT}{V} = \frac{nRT}{(1 - r/\bar{\lambda}) \cdot V} = \frac{nRT}{V - rV/\bar{\lambda}} = \frac{nRT}{V - V_0}$$
>
> if $r \ll \bar{\lambda}$. Therefore, with equations (11.34) and (11.36) we obtain
>
> $$V_0 = \frac{r \cdot V}{\bar{\lambda}} = r \cdot \sqrt{2} \cdot \pi (2r)^2 \cdot N = 4\sqrt{2}\pi r^3 \cdot N$$
>
> The volume of a molecule, approximated as a sphere, is $V_{\text{molecule}} = \frac{4}{3}\pi r^3$. Then $V_0 = 3\sqrt{2} \cdot V_{\text{molecule}} \cdot N = 4.2 V_{\text{molecule}} \cdot N = 4.2 V_{\text{molecule}} \cdot N_A \cdot n$.

The attractive forces between the molecules (Figure 20.2) are accounted by

$$P_0 = \frac{a}{V_m^2} = a \cdot \frac{n^2}{V^2} \tag{20.8}$$

The constant a depends on the nature of the gas. The attraction of a selected molecule by its neighbors diminishes the momentum transferred to the wall of the container when the molecule hits the wall. This force is considered to be proportional to the number of molecules in the close neighborhood of the selected molecule, i.e. to the number density and thus proportional to $1/V_m$. Additionally, the number of molecules hitting the wall is proportional to $1/V_m$ as well. Then P_0 is proportional to $1/V_m^2$.

Thus equation (20.5) can be written as (*van der Waals equation*)

$$\boxed{\left(P + \frac{a}{V_m^2}\right) \cdot (V_m - b) = RT \quad \text{or} \quad P = \frac{RT}{V_m - b} - \frac{a}{V_m^2}} \tag{20.9}$$

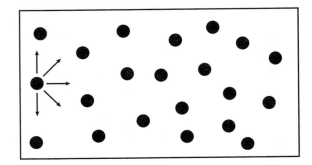

Figure 20.2 Attractive forces acting on a particle near the wall of a container.

Example – isotherms of CO_2

The constants a and b for CO_2 are (see below) $a = 3.66 \, \text{bar} \, \text{L}^2 \, \text{mol}^{-2}$, $b = 0.043 \, \text{L} \, \text{mol}^{-1}$. The PV curves calculated from equation (20.9) with these constants are shown in Figure 20.3(a) and compared with the corresponding curve for the ideal gas at 323 K. For comparison, experimental data are shown in Figure 20.3(b) for $T = 273 \, \text{K}$ and $T = 323 \, \text{K}$.

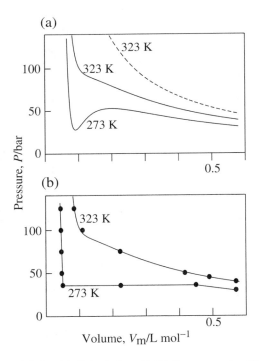

Figure 20.3 Isotherms of CO_2. (a) Calculation. Solid lines: van der Waals equation with $a = 3.66 \, \text{L}^2 \, \text{bar} \, \text{mol}^{-2}$ and $b = 0.043 \, \text{L} \, \text{mol}^{-1}$ for $T = 273 \, \text{K}$ and $T = 323 \, \text{K}$. Dashed line: ideal gas law for 323 K. (b) Experiment for 273 and 323 K (dots).

20.2 EQUATION OF STATE FOR REAL GASES

At high enough temperature and large enough volume we find reasonable agreement with the ideal gas law. At temperatures high enough above the critical temperature (see Section 11.7) the experimental data are well reproduced by equation (20.9). On the other hand, at temperatures below the critical temperature the van der Waals model fails because it does not account for the phase transition into the liquid state.

20.2.2 Critical Point and van der Waals Constants

The critical point (see Section 11.7) is the point of inflection with zero slope in the PV diagram; it is determined by the conditions

$$\left(\frac{\partial P}{\partial V}\right)_T = 0, \quad \left(\frac{\partial^2 P}{\partial V^2}\right)_T = 0 \tag{20.10}$$

Then we obtain (Box 20.3)

$$T_c = \frac{8 \cdot a}{27 \cdot R \cdot b}, \quad P_c = \frac{a}{27 \cdot b^2}, \quad V_{m,c} = 3 \cdot b \tag{20.11}$$

where T_c, P_c, and $V_{m,c}$ are the temperature, the pressure, and the molar volume at the critical point.

Box 20.3 – Critical Constants T_c, P_c, and V_c

By differentiation of equation (20.9) we obtain

$$\left(\frac{\partial P}{\partial V_m}\right)_T = -\frac{RT}{(V_m - b)^2} + \frac{2a}{V_m^3}$$

$$\left(\frac{\partial^2 P}{\partial V_m^2}\right)_T = \frac{2RT}{(V_m - b)^3} - \frac{6a}{V_m^4}$$

Because of the conditions (20.10) we obtain

$$\frac{RT_c}{(V_{m,c} - b)^2} = \frac{2a}{V_{m,c}^3}, \quad \frac{RT_c}{(V_{m,c} - b)^3} = \frac{3a}{V_{m,c}^4}$$

We divide the left-hand equation by the right-hand one and obtain $V_{m,c} = 3b$. Inserting this result into the left-hand equation, we obtain

$$\frac{RT_c}{(2b)^2} = \frac{2a}{(3b)^3} \quad \text{and} \quad T_c = \frac{8a}{27Rb}$$

Finally we insert $V_{m,c}$ and T_c in equation (20.9)

$$P_c = \frac{RT_c}{2b} - \frac{a}{(3b)^2} = \frac{R \cdot 8a}{2b \cdot 27Rb} - \frac{a}{(3b)^2} = \frac{a}{27b^2}$$

Example – evaluation of a and b from the critical data for CO_2

For CO_2 the experimental values for the critical data are $T_c = 304$ K, $P_c = 73.6$ bar, and $V_{m,c} = 0.095$ L mol^{-1}. Then from the relations for T_c and P_c in equation (20.11) we obtain $a = 3.66$ bar L^2 mol^{-2}, $b = 0.043$ L mol^{-1}. $V_{m,c} = 3 \cdot 0.043$ L mol^{-1} = 0.129 L mol^{-1} is only slightly larger than the measured $V_{m,c} = 0.094$ L mol^{-1}; thus the van der Waals equation leads to a reasonable description of the real gas at the critical point.

The situation below the critical temperature is not correctly described by equation (20.9), which predicts a maximum and a minimum of P in the PV diagram, contrary to the experiment. The condensation of a gas is a cooperative phenomenon which was not considered in the derivation of the van der Waals equation.

From equation (20.11) we find

$$\frac{V_{m,c} P_c}{T_c} = \frac{3 \cdot R}{8} = 3.13 \text{ J K}^{-1}$$

This means that the ratio $V_{m,c} P_c / T_c$ should be independent of the nature of the gas (*principle of corresponding states*). In the experiment, values between 2.28 and 2.54 are found. This agreement is surprising in view of the simple assumptions on which the van der Waals equation is based.

20.2.3 Virial Coefficients

What is the deviation of PV_m from the value RT at small pressure (close to ideal gas conditions)?

We multiply the van der Waals equation (20.9) by $(V_m - b)$

$$P \cdot (V_m - b) = RT - \frac{a \cdot (V_m - b)}{V_m^2} \quad (20.12)$$

and rearrange into

$$PV_m = RT + \left(Pb - \frac{a}{V_m} + \frac{ab}{V_m^2}\right) \quad (20.13)$$

For sufficiently large molar volume V_m the last term can be neglected and V_m can be approximated by the ideal gas law:

$$\boxed{PV_m = RT + \left(b - \frac{a}{RT}\right) \cdot P} \quad (20.14)$$

From this relation we expect that PV_m increases or decreases linearly with P depending on the particular values of a and b.

20.2 EQUATION OF STATE FOR REAL GASES

Example – slope of PV curve

For CO_2 at $T = 273$ K we calculate from equation (20.14)

$$b - \frac{a}{RT} = \left(0.043 - \frac{3.66}{0.0831 \cdot 273}\right) \text{L mol}^{-1} = -0.12 \text{L mol}^{-1}$$

This is just the same as the slope of the experimental PV curve (Figure 11.13).

For Ne at 273 K ($a = 0.210 \text{ bar L}^2 \text{ mol}^{-2}$ and $b = 0.017 \text{ L mol}^{-1}$) we obtain

$$b - \frac{a}{RT} = +0.007 \text{ L mol}^{-1}$$

In this case the PV curve is slightly increasing with P, again in agreement with the experiment (Figure 11.13).

At higher pressure, the dependence of PV on pressure becomes more complicated. For example, in the case of CO_2, PV decreases at low pressure and increases at high pressure; at $P = 600$ bar, PV is even larger than expected for an ideal gas. At high pressure the term ab/V_m^2 in equation (20.13) becomes important (Figure 20.4).

A formal description of such complicated behavior is given by the equation of state

$$PV_m = RT + c_1 P + c_2 P^2 + \cdots \qquad (20.15)$$

where the coefficients $c_1, c_2, \ldots$ are called *virial coefficients*. Comparing equation (20.15) with the van der Waals equation (20.14), we can identify the first virial coefficient as

$$c_1 = b - \frac{a}{RT} \qquad (20.16)$$

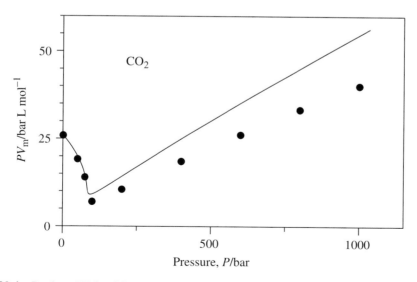

Figure 20.4 Product PV for CO_2 versus pressure P at 313 K. Dots: experiment. Solid line: calculated from the van der Waals equation (20.9) with $a = 3.66 \text{ L}^2 \text{ bar mol}^{-2}$ and $b = 0.043 \text{ L mol}^{-1}$. In detail, P is calculated for different values of V_m using equation (20.9); from the P/V_m data the product $P \cdot V_m$ is calculated and $P \cdot V_m$ is plotted versus P.

20.3 Change of State of Real Gases

20.3.1 Isothermal Compression of a van der Waals Gas

The reversible work w_{rev} for an isothermal compression of a gas from volume V_1 to V_2 is

$$w_{\text{rev}} = -\int_{V_1}^{V_2} P \cdot dV \qquad (20.17)$$

Inserting P from the van der Waals equation (20.9) we obtain (n = amount of the gas)

$$w_{\text{rev}} = -nRT \cdot \ln \frac{V_2 - nb}{V_1 - nb} + n^2 a \cdot \left(\frac{1}{V_1} - \frac{1}{V_2} \right) \qquad (20.18)$$

Box 20.4 – Isothermal Compression of CO_2

For the compression of 1 mol of an ideal gas at $T = 313\,\text{K}$ from $V_1 = 26\,\text{L}$ to $V_2 = 0.26\,\text{L}$ we obtain

$$w_{\text{rev,ideal}} = -nRT \cdot \ln \frac{0.26}{26} = 12.0\,\text{kJ}$$

For CO_2 with $a = 3.66\,\text{bar}\,\text{L}^2\,\text{mol}^{-2}$ and $b = 0.043\,\text{L}\,\text{mol}^{-1}$ we calculate the following from equation (20.18)

$$w_{\text{rev,real}} = nRT \cdot \ln \frac{0.217}{26} - 1.4\,\text{kJ} = 11.1\,\text{kJ}$$

For the ideal gas the final pressure is 100 bar, and for the real gas we calculate the following from equation (20.9)

$$P_2 = \frac{RT}{(0.26 - 0.043)\,\text{L}\,\text{mol}^{-1}} - \frac{3.66}{0.26^2}\,\text{bar} = 66\,\text{bar}$$

For an ideal gas the reversible work would be 12.0 kJ and the final pressure would be 100 bar. In spite of the great difference in the final pressure, there is only a small difference in w_{rev}.

20.3.2 Fugacity and Equilibrium Constant

In practical applications it is useful to operate with ΔG instead of w_{rev}. According to Chapter 16 we have for a change of state at constant temperature

$$\Delta G = \Delta H - T\Delta S = (\Delta U - T\Delta S) + \Delta(PV) = w_{\text{rev}} + \Delta(PV)$$

20.3 CHANGE OF STATE OF REAL GASES

Then ΔG for the isothermal compression of a gas from volume V_1 to volume V_2 is given by

$$\Delta G = w_{\text{rev}} + \Delta(PV) = w_{\text{rev}} + P_2 V_2 - P_1 V_1 \tag{20.19}$$

For an ideal gas we have

$$P_2 V_2 = P_1 V_1 \tag{20.20}$$

and thus

$$\Delta G = w_{\text{rev}} = -\int_{P_1}^{P_2} P \cdot dV = nRT \cdot \ln \frac{P_2}{P_1} \quad \text{(ideal gas)} \tag{20.21}$$

For a real gas this is no longer the case; it is shown in Box 20.5 that ΔG can be obtained as

$$\boxed{\Delta G = \int_{P_1}^{P_2} V \cdot dP} \tag{20.22}$$

Box 20.5 – Comparison of $\int V \cdot dP$ and $-\int P \cdot dV$

According to Equation (20.19) the change ΔG in Gibbs energy is related to the reversible work w_{rev} by

$$\Delta G = w_{\text{rev}} + P_2 V_2 - P_1 P_1$$

To compare ΔG with w_{rev} we consider an isothermal compression of a real gas from a large volume V_1 to a small volume V_2. PV diagrams for such a gas are shown in Figure 20.5 (a–c). In Figure 20.5(a) the shaded area corresponds to the reversible work

$$w_{\text{rev}} = -\int_{V_1}^{V_2} P \cdot dV$$

w_{rev} is positive (work is done on the system). The quantities $P_1 V_1$ and $P_2 V_2$ are given by the horizontal and the vertical rectangle, respectively, in Figure 20.5(b). Therefore $\Delta G = w_{\text{rev}} + P_2 V_2 - P_1 V_1$ corresponds to the shaded area in Figure 20.5(c). This area can be expressed as $\int_{P_1}^{P_2} V \cdot dP$, thus

$$\Delta G = \int_{P_1}^{P_2} V \cdot dP$$

In the special case of an ideal gas the areas of the two rectangles in Figure 20.5(b) are equal because of the Ideal Gas Law:

$$P_1 V_1 = P_2 V_2 = nRT$$

and we obtain

$$\Delta G = w_{\text{rev}}$$

Continued on page 638

Continued from page 637

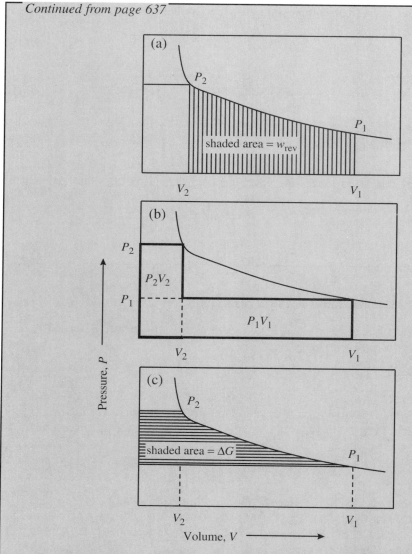

Figure 20.5 *PV* diagram of a real gas; comparison of ΔG and w_{rev}. (a) Shaded area $= -\int P \cdot dV = w_{\text{rev}}$. (b) Rectangles of area $P_2 V_2$ and $P_1 V_1$, respectively. (c) Shaded area $= -\int V \cdot dP = \Delta G$.

This can be also seen when inserting the Ideal Gas Law in the expression for ΔG:

$$\Delta G_{\text{ideal gas}} = \int_{P_1}^{P_2} V \cdot dP = n\boldsymbol{R}T \cdot \int_{P_1}^{P_2} \frac{1}{P} \cdot dP = n\boldsymbol{R}T \cdot \ln \frac{P_2}{P_1}$$

$$w_{\text{rev,ideal gas}} = -\int_{P_1}^{P_2} P \cdot dV = n\boldsymbol{R}T \int_{P_1}^{P_2} \frac{dP}{P} = n\boldsymbol{R}T \cdot \ln \frac{P_2}{P_1}$$

20.3 CHANGE OF STATE OF REAL GASES

Table 20.1 Fugacity f and fugacity coefficient ϕ for CO_2 at 313 K

P/bar	V_m/L mol^{-1}	ϕ	f/bar
25	0.92	0.9	22.5
50	0.38	0.8	40.1
75	0.18	0.69	51.5
100	0.069	0.58	58
1000	0.0040	0.284	284

For a real gas the volume V depends in a complicated way on P. Therefore it is useful to proceed as follows. First we consider the gas at a small enough pressure P_0, such that the ideal gas law is fulfilled. Then for the compression of the gas from pressure P_0 to pressure P we set formally

$$\Delta G = nRT \cdot \ln \frac{P \cdot \phi}{P_0} \qquad (20.23)$$

ϕ is a measure of the deviation of the behavior of the real gas from the behavior of an ideal gas. The quantity

$$f = \phi \cdot P \qquad (20.24)$$

is called *fugacity*, and ϕ is the fugacity coefficient. ϕ can be calculated from measured PV data (Problem 20.2). For CO_2 at 313 K the values in Table 20.1 are obtained.

Then for an isothermal compression from any arbitrary pressure P_1 to P_2 we can write

$$\Delta G = nRT \cdot \ln \frac{f_2}{f_1} \qquad (20.25)$$

In this way the thermodynamic relations for ΔG remain unchanged if we replace the pressures by fugacities.

Example – compression of CO_2

What is ΔG for the compression of 1 mol CO_2 at 313 K from 1 bar ($V_1 = 26$ L) to 66 bar ($V_2 = 0.26$ L)? With $\phi_1 = 1$ and $\phi_2 = 0.73$ (interpolation in Table 20.1) we obtain $f_1 = 1$ bar, $f_2 = 0.73 \times 66$ bar $= 48$ bar, and

$$\Delta G = nRT \cdot \ln \frac{f_2}{f_1} = 10 \text{ kJ for } n = 1 \text{ mol}$$

This is essentially the same as ΔG calculated from w_{rev} using the van der Waals model (see Box 20.4):

$$\Delta G = w_{\text{rev}} + P_2 V_2 - P_1 V_1$$
$$= 11.1 \text{ kJ} + (66 \cdot 0.26 - 1 \cdot 26) \text{ bar L} = 10.2 \text{ kJ}$$

Example – ammonia equilibrium

$\Delta_r G^\ominus$ for the synthesis of NH_3 from its elements

$$N_2 + 3H_2 \longrightarrow 2NH_3$$

Table 20.2 Calculation of K for the synthesis of NH_3 from its elements at $T = 723\,\text{K}$

$\widehat{P}_{\text{total}}$	$\widehat{P}_{NH_3}$	$\widehat{P}_{N_2}$	$\widehat{P}_{H_2}$	$K_{\text{ideal}}/10^{-5}$	$K/10^{-5}$
4.03	0.0337	1.0	3.0	4.20	4.20
10	0.207	2.47	7.45	4.20	4.10
51	4.64	11.4	34.5	4.62	4.10
101	16.56	21.2	63.5	5.04	3.97
304	107.9	49.0	147	7.48	3.48
608	326	70.4	211	16.13	4.10
1013	703	77.5	232	51.07	9.41

is related to the equilibrium constant by $\Delta_r G^{\ominus} = -RT \cdot \ln K$ with the equilibrium constant

$$K = \frac{(\widehat{P}\phi)^2_{NH_3}}{(\widehat{P}\phi)_{N_2}(\widehat{P}\phi)^3_{H_2}} = \frac{\widehat{P}^2_{NH_3}}{\widehat{P}_{N_2} \cdot \widehat{P}^3_{H_2}} \cdot \frac{\phi^2_{NH_3}}{\phi_{N_2} \cdot \phi^3_{H_2}}$$

In Table 20.2, K is calculated for different total pressures up to 1013 bar.

The quantity

$$K_{\text{ideal}} = \frac{\widehat{P}^2_{NH_3}}{\widehat{P}_{N_2} \cdot \widehat{P}^3_{H_2}}$$

is constant at low pressures; serious deviations occur above 100 bar. On the other hand, the equilibrium constant K (partial pressures replaced by fugacities) is constant up to 600 bar. The remaining deviation at 1013 bar is due to interactions between different gases in the mixture; the fugacities can not account for this effect because they are determined for pure gases.

20.3.3 Adiabatic Expansion into Vacuum

We consider an adiabatic expansion of a gas into vacuum (Figure 20.6(a)). Because the system is isolated, no heat or work is supplied to the system:

$$q = 0, \quad w = 0 \tag{20.26}$$

Then for the change of internal energy we obtain

$$\Delta U = w + q = 0 \tag{20.27}$$

The internal energy of an ideal gas depends on temperature, but not on the volume. Then it follows that the temperature of an ideal gas does not change in the experiment (after temperature equilibration).

In a real gas the molecules attract each other. After expansion, they are at larger distance. During the process of expansion they are moving against the attractive forces. As a consequence, their speed is slowed down: after temperature equilibration, the temperature of the gas has decreased.

What is the decrease in temperature? In a thought experiment we split the process in Figure 20.6(a) into two partial processes (Figure 20.6(b)). First we expand the gas in an isothermal process (temperature T_1); in this process the internal energy changes by ΔU_a

20.3 CHANGE OF STATE OF REAL GASES

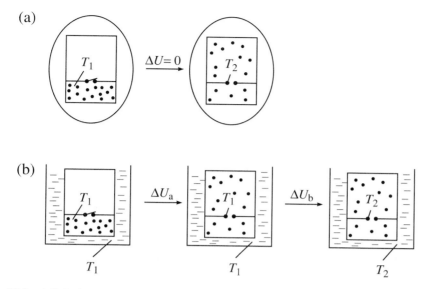

Figure 20.6 Adiabatic expansion of a real gas into vacuum. (a) Direct process: the temperature decreases from T_1 to T_2. (b) Indirect process: isothermal expansion at T_1 (ΔU_a) followed by cooling down to T_2 at constant volume.

(gas at the same temperature, but molecules at larger distance). In a second step we cool the gas down to the temperature T_2 at constant volume. In this process the internal energy changes by ΔU_b (molecules at the same distance, but at lower temperature). Because the initial and the final states are the same as in the adiabatic process, we obtain

$$\Delta U = \Delta U_a + \Delta U_b = 0 \tag{20.28}$$

Step b constitutes a cooling at constant volume, thus

$$\Delta U_b = C_V \cdot (T_2 - T_1) \tag{20.29}$$

ΔU_a corresponds to the energy to separate the molecules against the attractive forces reducing the pressure of the gas, according to equation (20.4), by P_0; then, using the van der Waals model, because of equation (20.8) we obtain

$$\Delta U_a = \int_{V_1}^{V_2} P_0 \cdot dV = \int_{V_1}^{V_2} an^2 \cdot \frac{1}{V^2} \cdot dV = -an^2 \cdot \left(\frac{1}{V_2} - \frac{1}{V_1} \right) \tag{20.30}$$

and thus

$$T_2 - T_1 = \frac{\Delta U_b}{C_V} = -\frac{\Delta U_a}{C_V} = \frac{a}{C_{V,m}} \cdot \left(\frac{1}{V_{m,2}} - \frac{1}{V_{m,1}} \right) \tag{20.31}$$

Example – adiabatic expansion of CO_2

We expand 1 mol of CO_2 adiabatically from volume $V_1 = 2.27\,\text{L}$ to $V_2 = 22.7\,\text{L}$. With $a = 3.66\,\text{bar}\,\text{L}^2\,\text{mol}^{-2}$ and $C_{V,m} = 2.5R$ (neglecting the vibrational contribution to $C_{V,m}$)

we obtain

$$T_2 - T_1 = \frac{3.66 \text{ bar L}^2 \text{ mol}^{-2}}{2.5 \cdot 0.08314 \text{ bar L K}^{-1} \text{ mol}^{-1}} \cdot \left(\frac{1}{22.7} - \frac{1}{2.27}\right) \text{ mol L}^{-1} = -7 \text{ K}$$

The temperature of the gas decreases by 7 K.

20.3.4 Joule–Thomson Effect

In this experiment (which is the basic effect used in cooling and liquefying gases) the real gas, passing a throttle valve, is adiabatically expanded from pressure P_1 (volume V_1) to pressure P_2 (volume V_2), as shown in Figure 20.7. The work done in this process is

$$w = -P_2 V_2 + P_1 V_1 \tag{20.32}$$

and the heat q is zero (adiabatic process). Thus the change ΔU in internal energy is

$$\Delta U = w = -P_2 V_2 + P_1 V_1.$$

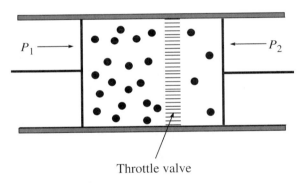

Figure 20.7 Joule–Thomson experiment. Adiabatic expansion of a gas from pressure P_1 to pressure P_2 using a throttle valve.

Then, using the van der Waals model according to equation (20.14), we obtain

$$\Delta U = -n\mathbf{R} \cdot (T_2 - T_1) - n \cdot \left(b - \frac{a}{\mathbf{R}T_2}\right) \cdot P_2 + n \cdot \left(b - \frac{a}{\mathbf{R}T_1}\right) \cdot P_1 \tag{20.33}$$

As in the preceding section, the process can be split into two partial processes a and b:

$$\Delta U = \Delta U_a + \Delta U_b \tag{20.34}$$

where ΔU_a and ΔU_b are, as in the case of the adiabatic expansion into vacuum,

$$\Delta U_a = -an^2 \cdot \left(\frac{1}{V_2} - \frac{1}{V_1}\right) \tag{20.35}$$

$$\Delta U_b = C_V \cdot (T_2 - T_1) \tag{20.36}$$

20.3 CHANGE OF STATE OF REAL GASES

In this case ΔU is nonzero, contrary to the process of expansion into vacuum. As shown in Box 20.6, we obtain for the change in temperature

$$T_2 - T_1 = \frac{1}{C_{P,m}} \cdot \left(\frac{2a}{RT_1} - b\right) \cdot (P_2 - P_1) \tag{20.37}$$

Note that the sign of the temperature change depends on the values of a, b, and T_1: if $2a/RT_1$ is larger than b, the temperature decreases; otherwise the temperature increases. The quantity

$$\mu_{JT} = \frac{\Delta T}{\Delta P} = \frac{1}{C_{P,m}} \cdot \left(\frac{2a}{RT_1} - b\right) \tag{20.38}$$

is called the *Joule–Thomson coefficient*.

Box 20.6 – Joule–Thomson Effect: Calculation of ΔT

From equations (20.34) and (20.36) we obtain

$$C_V \cdot (T_2 - T_1) = \Delta U - \Delta U_a$$

ΔU is given by equation (20.33); in the expression (20.35) for ΔU_a we approximate V_1 and V_2 by the ideal gas law: (we are allowed to do this because V_1 and V_2 appear only in a correction term of the van der Waals equation.) Then we obtain

$$\Delta U = -nR \cdot (T_2 - T_1) - n\left(b - \frac{a}{RT_2}\right) \cdot P_2 + n\left(b - \frac{a}{RT_1}\right) P_1$$

$$\Delta U_a = -n\frac{a}{RT_2} \cdot P_2 + n\frac{a}{RT_1} \cdot P_1$$

and

$$C_V \cdot (T_2 - T_1) = -nRT \cdot (T_2 - T_1) - n\left(b - \frac{2a}{RT_2}\right) \cdot P_2 + n\left(b - \frac{2a}{RT_1}\right) \cdot P_1$$

Because of $C_V + nRT = C_P$ we obtain the following expression for $T_2 - T_1$

$$C_P \cdot (T_2 - T_1) = -n\left(b - \frac{2a}{RT_2}\right) \cdot P_2 + n\left(b - \frac{2a}{RT_1}\right) \cdot P_1$$

If the temperatures T_1 and T_2 differ only slightly (e.g., like in the expansion experiment in the previous section, where $\Delta T/T_1 \approx 2\%$) we approximate T_2

Continued on page 644

> *Continued from page 643*
>
> on the right-hand side by T_1. Then it follows that
>
> $$C_P \cdot (T_2 - T_1) = -n \left(b - \frac{2a}{RT_1}\right) \cdot (P_2 - P_1)$$
>
> or
>
> $$T_2 - T_1 = \frac{1}{C_{P,m}} \cdot \left(\frac{2a}{RT_1} - b\right) \cdot (P_2 - P_1)$$

Example – cooling of CO_2 in the Joule–Thomson experiment

For CO_2 ($C_{P,m} = \frac{7}{2}R$, $a = 3.66\,\text{bar}\,\text{L}^2\,\text{mol}^{-2}$, $b = 0.043\,\text{L}\,\text{mol}^{-1}$) we obtain the following for $P_2 - P_1 = -10$ bar and $T_1 = 300$ K:

$$T_2 - T_1 = \frac{2}{7 \cdot 0.08314} \cdot \left(\frac{2 \cdot 3.66}{0.08314 \cdot 300} - 0.043\right) \cdot (-10)\,\text{K} = -8.6\,\text{K}$$

The temperature decreases by about 9 K, in accordance with the experimental value of 10 K.

With increasing starting temperature T_1 the term $2a/(RT_1)$ in equation (20.37) becomes smaller. If

$$\frac{2a}{RT_1} = b \qquad (20.39)$$

then no temperature effect occurs in the process. The corresponding temperature

$$T_i = \frac{2a}{b \cdot R} \qquad (20.40)$$

is called the *inversion temperature*.

Example – inversion temperature

For CO_2 we obtain the following from equation (20.40)

$$T_i = \frac{2 \cdot 3.66}{0.043 \cdot 0.08314}\,\text{K} = 2040\,\text{K} \text{ (experiment: 1500 K)}$$

For He ($a = 0.035\,\text{bar}\,\text{L}^2\,\text{mol}^{-2}$, $b = 0.024\,\text{L}\,\text{mol}^{-1}$) we obtain $T_i = 35$ K (experiment: 40 K), and for Ar ($a = 1.345\,\text{bar}\,\text{L}^2\,\text{mol}^{-2}$, $b = 0.032\,\text{L}\,\text{mol}^{-1}$) $T_i = 1010$ K (experiment: 723 K).

20.3 CHANGE OF STATE OF REAL GASES

To cool a gas down by means of the Joule–Thomson effect it is important that the starting temperature is below the inversion temperature of the gas. For example, to liquefy helium, the helium gas must be cooled down below 40 K before starting the Joule–Thomson process.

Liquid N_2 (boiling point 80 K) and liquid He (boiling point 4 K) are produced in industry by using the Linde process based on the Joule–Thomson effect. With He, evaporated at reduced pressure, temperatures of about 1 K can be reached. Lower temperatures are available by adiabatic demagnetization (see Section 15.4).

According to equation (20.40), the inversion temperature should be independent of pressure. Experimentally, however, T_i decreases with increasing pressure. This effect can also be seen in the van der Waals model, if we use the full van der Waals equation (20.13) instead of the approximation (20.14), see Box 20.7.

Box 20.7 – Joule–Thomson Effect: Inversion Temperature

By inserting the full van der Waals equation (20.13) in equation (20.32) and again approximating the molar volume V_m by RT/P, we obtain the following instead of equations (20.37) and (20.38), using the relation

$$P_2^2 - P_1^2 = (P_2 - P_1) \cdot (P_2 + P_1),$$

$$\Delta T = \frac{1}{C_{P,m}} \cdot \left[\left(\frac{2a}{RT_1} - b \right) \cdot (P_2 - P_1) - \frac{ab}{R^2 T_1^2} \cdot (P_2 - P_1) \cdot (P_2 + P_1) \right]$$

$$\mu_{JT} = \frac{1}{C_{P,m}} \cdot \left[\frac{2a}{RT_1} - b - \frac{ab \cdot (P_2 + P_1)}{R^2 T_1^2} \right]$$

This expression differs from equation (20.38) by the term

$$\frac{ab \cdot (P_2 + P_1)}{R^2 T_1^2}$$

At low pressure P_1 or at high temperature T_1 we can neglect this term. At high pressure or at low temperature, however, this term becomes important. This means that even at temperatures below the inversion temperature given by equation (20.40) it is possible that a heating effect results instead of a cooling effect, provided that the pressure is high enough or the temperature is low enough.

Then the condition for the inversion temperature T_i is

$$\frac{2a}{RT_i} - b - \frac{ab \cdot (P_2 + P_1)}{R^2 T_i^2} = 0$$

Continued on page 646

Continued from page 645

and we can resolve for T_i:

$$T_i = \frac{1}{b\mathbf{R}} \cdot \left[a \pm \sqrt{a^2 - ab^2(P_1 + P_2)} \right]$$

In Figure 20.8 the inversion temperature of N_2 is plotted against the pressure $P_2 + P_1$. Inside the shaded area we have $\Delta T < 0$. The main features of the experimental curve are well reproduced by the van der Waals model (area with $\Delta T < 0$ enclosed by dashed line). However, the van der Waals model cannot account quantitatively for the observed effect.

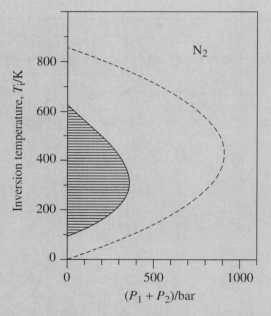

Figure 20.8 Inversion temperature T_i plotted against $(P_2 + P_1)$. Solid line: experiment for N_2. Inside the shaded area a cooling effect occurs, and outside the shaded area the gas is warmed up. Dashed line: calculated for N_2 from the van der Waals model with $a = 1.390 \, \text{L}^2 \, \text{bar} \, \text{mol}^{-2}$ and $b = 0.0391 \, \text{L} \, \text{mol}^{-1}$ Experimental results from (J. R. Roebuck and H. Osterberg, *Phys. Rev.*, 1935, **48**, 450).

20.4 Change of State of Real Solutions

20.4.1 Partial Molar Volume

For an ideal solution, (*note*: do not confuse an ideal solution with an ideally diluted solution!), state variables such as V, U, H, and G are obtained by adding the corresponding

20.4 CHANGE OF STATE OF REAL SOLUTIONS

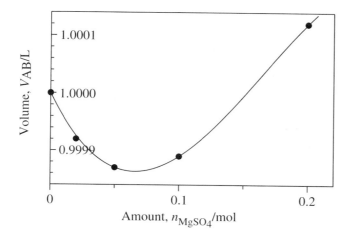

Figure 20.9 Volume V_{AB} of a solution of $MgSO_4$ in 1000 mL water as a function of the amount n_{MgSO_4} at 298 K.

contributions of solvent (A) and solute (B)

$$V_{AB} = n_A V_A + n_B V_B \tag{20.41}$$

This simple relation is not fulfilled in real solutions: V_{AB} depends on n_A and n_B in a very complicated manner. For instance, a dilute solution of $MgSO_4$ in water has a smaller volume than the water used to prepare the solution (Figure 20.9). This is because the water is compressed in the Coulomb field of the Mg^{2+} ions.

How can we reasonably describe the volume V_{AB} of a real solution? We consider a solution of a given composition n_B/n_A; we add a small amount dn_B of solute and a small amount dn_A of solvent. Then the change dV_{AB} of the volume V_{AB} is

$$dV_{AB} = \left(\frac{\partial V}{\partial n_A}\right)_{n_B} \cdot dn_A + \left(\frac{\partial V}{\partial n_B}\right)_{n_A} \cdot dn_B \tag{20.42}$$

The partial derivatives are called *partial molar volumes* of solvent (v_A) and solute (v_B). Then

$$dV_{AB} = v_A \cdot dn_A + v_B \cdot dn_B \tag{20.43}$$

$$v_A = \left(\frac{\partial V}{\partial n_A}\right)_{n_B}, \quad v_B = \left(\frac{\partial V}{\partial n_B}\right)_{n_A}$$

Now we prepare a solution by adding alternately in many small steps small amounts of solvent and solute in the fixed ratio n_A/n_B. Then all changes of volume are equal and we obtain

$$V_{AB} = v_A \cdot n_A + v_B \cdot n_B \tag{20.44}$$

This equation corresponds to equation (20.41), only the molar volumes $V_{m,A}$ and $V_{m,B}$ are replaced by the partial molar volumes v_A and v_B.

The partial molar volumes v depend on the composition of the solution. For example, for a solution of $MgSO_4$ in water, v_B is negative at small concentrations and positive at large concentrations.

Comparing the differential

$$dV_{AB} = d(n_A v_A + n_B v_B) = dn_A \cdot v_A + dv_A \cdot n_A + dn_B \cdot v_B + dv_B \cdot n_B \tag{20.45}$$

with equation (20.43) reveals

$$n_A \cdot dv_A + n_B \cdot dv_B = 0 \tag{20.46}$$

Example – v_B of MgSO$_4$

From the slope of the curve in Figure 20.9 (volume V_{AB} versus the amount n_B of MgSO$_4$) we calculate

$$v_B = \left(\frac{\partial V}{\partial n_B}\right)_{n_A} = -0.67 \times 10^{-3} \, \text{L mol}^{-1} \quad \text{for } n_B = 0.05 \, \text{mol}$$

and $v_B = +2.24 \times 10^{-3} \, \text{L mol}^{-1}$ for $n_B = 0.15$ mol.

20.4.2 Chemical Potential

In analogy, similar relations hold for other state variables. The change dG_{AB} of the Gibbs energy at constant temperature and constant pressure can be described as

$$dG_{AB} = \mu_A \cdot dn_A + \mu_B \cdot dn_B \tag{20.47}$$

where

$$\mu_A = \left(\frac{\partial G}{\partial n_A}\right)_{n_B}, \quad \mu_B = \left(\frac{\partial G}{\partial n_B}\right)_{n_A} \tag{20.48}$$

are partial molar Gibbs energies, in analogy to the partial molar volumes discussed in Section 20.4.1. They are called *chemical potentials*. Similarly to equations (20.44) and (20.46) we find

$$G_{AB} = \mu_A \cdot n_A + \mu_B \cdot n_B \tag{20.49}$$

$$n_A \cdot d\mu_A + n_B \cdot d\mu_B = 0 \tag{20.50}$$

This relation is called the *Gibbs Duhem equation* (in the special case of a two component system at constant temperature).

Example – mass action law derived from chemical potentials

We consider the special case of an ideally diluted solution as we did in Chapter 16. The change in Gibbs energy when transferring a dissolved species from a container with standard concentration $c^\ominus$ into a reaction vessel with the concentration c is $\Delta G = nRT \cdot \ln \widehat{c}$. Thus the Gibbs energy of the species in the reaction vessel is $G = G^\ominus + nRT \cdot \ln \widehat{c}$. Therefore, its chemical potential is

$$\mu = \left(\frac{\partial G}{\partial n}\right) = \frac{G}{n} = \frac{G^\ominus}{n} + RT \cdot \ln \widehat{c}$$

20.4 CHANGE OF STATE OF REAL SOLUTIONS

Then

$$\mu = \mu^\ominus + RT \cdot \ln \widehat{c}$$

where

$$\mu^\ominus = \frac{G^\ominus}{n}$$

Now we consider a chemical reaction

$$\alpha A + \beta B \longrightarrow \gamma C + \delta D$$

We imagine the reactants reversibly transferred from the containers with the standard concentrations into the equilibrium container; correspondingly, the products are transferred from their equilibrium concentrations to standard concentration. Then $\Delta G = 0$ for this process (equilibrium condition). Using equation (20.49) we calculate ΔG from the Gibbs energies of the different species.

$$\Delta G = G_C + G_D - G_A - G_B$$
$$= \mu_C \cdot n_C + \mu_D \cdot n_D - \mu_A \cdot n_A - \mu_B \cdot n_B$$

or in molar terms

$$\Delta_r G = \mu_C \cdot \gamma + \mu_D \cdot \delta - \mu_A \cdot \alpha - \mu_B \cdot \beta$$

$\Delta_r G$ is the change in Gibbs energy when removing α mol A from container A and β mol B from container B and bringing γ mol C into container C and δ mol D into container D. With $\mu = \mu^\ominus + RT \cdot \ln \widehat{c}$ we obtain

$$\Delta_r G = (\mu_C^\ominus \cdot \gamma + \mu_D^\ominus \cdot \delta - \mu_A^\ominus \cdot \alpha - \mu_B^\ominus \cdot \beta) + RT \cdot \ln \frac{\widehat{c}_C^\gamma \cdot \widehat{c}_D^\delta}{\widehat{c}_A^\alpha \cdot \widehat{c}_B^\beta}$$

$$= \Delta_r G^\ominus + RT \cdot \ln K$$

Then it follows because of the equilibrium condition $\Delta_r G = 0$

$$\boxed{\Delta_r G^\ominus = -RT \cdot \ln K}$$

This is the same result as obtained in Section 17.1.2 by considering the reversible work w_{rev}.

20.4.3 Activities and Equilibrium Constants in Solutions with Ions

In real gases we have replaced pressures by fugacities. We use the same approach to real solutions. This is especially important for solutions of ions. Deviations from the condition for ideally dilute solutions occur at concentrations as low as $0.01 \, \text{mol L}^{-1}$ due to the far reaching interactions between ions by Coulomb forces.

Activity Coefficient

To account for interactions in real solutions we replace the concentration c by the activity a

$$a = c \cdot \gamma \tag{20.51}$$

(γ = activity coefficient). We transfer reversibly a small amount n of a selected kind of ion from an ideally diluted solution ($\gamma = 1$, concentration c_0) to a solution of the same ion with concentration c. Then the change ΔG in Gibbs energy is

$$\Delta G = n\boldsymbol{R}T \cdot \ln \frac{\gamma \cdot c}{c_0} = n\boldsymbol{R}T \cdot \ln \frac{a}{c_0} \tag{20.52}$$

This equation defines the activity coefficient γ in a manner similar to the way that equation (20.23) defines the fugacity coefficient ϕ. Then, instead of equation (17.68) we can write

$$K = \left(\frac{\hat{a}_C^\gamma \cdot \hat{a}_D^\delta \cdots}{\hat{a}_A^\alpha \cdot \hat{a}_B^\beta \cdots}\right)_{\text{eq}}, \quad Q = \frac{\hat{a}_C^\gamma \cdot \hat{a}_D^\delta \cdots}{\hat{a}_A^\alpha \cdot \hat{a}_B^\beta \cdots}$$

Debye–Hückel Theory

How can we estimate the activity coefficient γ? First we imagine what happens when the electrostatic interaction between the ions is neglected. Then the reversible work to transform the ions from c_0 to c is

$$w_{\text{rev,ideal}} = n\boldsymbol{R}T \cdot \ln \frac{c}{c_0} \tag{20.53}$$

The difference between w_{rev} and $w_{\text{rev,ideal}}$ is due to the fact that a selected ion (charge e in the case of a singly charged ion) in the concentrated solution is preferentially surrounded by counterions. In contrast, in the ideally dilute solution the ions are statistically distributed. Because of the Coulomb attraction forces, w_{rev} is decreased when compared to $w_{\text{rev,ideal}}$.

According to the Debye–Hückel theory the additional reversible work is

$$w_{\text{rev}} - w_{\text{rev,ideal}} = -\frac{1}{2} \cdot \frac{e^2}{4\pi\varepsilon_0\varepsilon \cdot r_D} \cdot N \tag{20.54}$$

when the counterions are at an average distance r_D from the selected ion; ε is the dielectric constant of the solvent, ε_0 is permittivity of the vacuum, and N is the number of ions (see Box 20.8).

Box 20.8 – Electrical Work to Transfer Ions into Concentrated Solution

To calculate the electrical work in equation (20.54) we have to transfer an ion reversibly from an ideally diluted solution to a concentrated solution. In a thought experiment we first remove the charge from the ion and transfer the neutral particle into the concentrated solution. Then we charge the particle from $Q = 0$ to $Q = e$. On charging the ion to Q the counterions in the concentrated solution will arrange in the neighborhood of the ion such that an

Continued on page 651

20.4 CHANGE OF STATE OF REAL SOLUTIONS

> *Continued from page 650*
>
> opposite charge $-Q$ at distance r_D from the ion will result. The work to add a small charge dQ to Q is then
>
> $$-\frac{1}{4\pi\varepsilon_0\varepsilon} \cdot \frac{Q \cdot dQ}{r_D}$$
>
> Then the total work to charge the ion from $Q = 0$ to $Q = e$ is
>
> $$-\frac{1}{4\pi\varepsilon_0\varepsilon \cdot r_D} \cdot \int_0^e Q \cdot dQ = -\frac{1}{2} \cdot \frac{e^2}{4\pi\varepsilon_0\varepsilon \cdot r_D}$$
>
> Multiplication with the number N of ions leads to the expression in equation (20.54).

Let us approximate ΔG in our model by w_{rev} (neglecting the volume change of the solution). Then from equations (20.52), (20.53), and (20.54) we obtain

$$w_{rev} = n\mathbf{R}T \cdot \ln \frac{\gamma \cdot c}{c_0}, \quad w_{rev,ideal} = n\mathbf{R}T \cdot \ln \frac{c}{c_0}$$

and

$$w_{rev} - w_{rev,ideal} = n\mathbf{R}T \cdot \ln \gamma = -\frac{1}{2} \cdot \frac{e^2}{4\pi\varepsilon_0\varepsilon \cdot r_D} \cdot N \quad (20.55)$$

or (with $n\mathbf{R} = Nk$)

$$\ln \gamma = \frac{-e^2}{2kT \cdot 4\pi\varepsilon_0\varepsilon \cdot r_D} \quad (20.56)$$

This relation can easily be generalized to ions with arbitrary charge ze. Then

$$\boxed{\ln \gamma = \frac{-(ze)^2}{2kT \cdot 4\pi\varepsilon_0\varepsilon \cdot r_D}} \quad (20.57)$$

According to equation (20.57), the activity coefficient γ decreases with decreasing distance r_D (i.e., with increasing concentration of the solution). The average distance r_D is obtained from the theory of Gouy and Chapman (Foundation 20.1):

$$\frac{1}{r_D} = \frac{\kappa}{1 + \kappa R} \quad (20.58)$$

$$\kappa = \sqrt{\frac{2e^2 N_A}{\varepsilon_0 \varepsilon kT}} \cdot \sqrt{I} \quad (20.59)$$

where

$$I = \frac{1}{2} \cdot \sum_i c_i z_i^2 \qquad (20.60)$$

is the ionic strength, c_i and $z_i e$ are the concentration and the charge of ion species i and R is the smallest possible distance between ion and counterion. Then from equation (20.57) we obtain the *Debye–Hückel law*

$$\boxed{\ln \gamma = -\frac{A \cdot z^2 \sqrt{I}}{1 + B \cdot R \sqrt{I}}} \qquad (20.61)$$

with the abbreviations

$$A = \frac{e^3}{(4\pi\varepsilon_0)^{3/2}} \cdot \sqrt{\frac{2\pi N_A}{k^3}} \cdot \frac{1}{(\varepsilon T)^{3/2}} \qquad (20.62)$$

$$B = \frac{2e}{(4\pi\varepsilon_0)^{1/2}} \cdot \sqrt{\frac{2\pi N_A}{k}} \cdot \frac{1}{(\varepsilon T)^{1/2}} \qquad (20.63)$$

In the case of water ($\varepsilon = 80$) at 298 K we find $A = 1.175 \, \text{L}^{1/2} \, \text{mol}^{-1/2}$ and $B = 0.330 \times 10^{10} \, \text{m}^{-1} \text{L}^{1/2} \, \text{mol}^{-1/2}$.

For small enough ionic strength I equation (20.61) simplifies to

$$\ln \gamma = -A \cdot z^2 \cdot \sqrt{I} \qquad (20.64)$$

This equation is called the *Debye–Hückel limiting law*.

Example – activity coefficient in NaCl solution

What is the activity coefficient γ of Na^+ ions in a $0.01 \, \text{mol L}^{-1}$ solution of NaCl in water? First we calculate the ionic strength I, $I = \frac{1}{2} \cdot (c_{Na^+} + c_{Cl^-}) = 0.01 \, \text{mol L}^{-1}$. Assuming $R = 0$, from equation (20.61) we obtain $\ln \gamma = -1.175 \cdot \sqrt{0.01} = -0.1175$ and $\gamma = 0.89$. Considering the influence of R the electrostatic interaction is decreased, thus γ increases. Then with $R = 500 \, \text{pm}$ (regarding the hydration sphere of an ion) we obtain $1 + BR\sqrt{I} = 1 + 0.330 \times 10^{10} \cdot 500 \times 10^{-12} \cdot \sqrt{0.01} = 1 + 0.17 = 1.17$ and $\ln \gamma = -0.1175/1.17 = -0.100$ and $\gamma = 0.90$.

Example – activity coefficient in MgSO$_4$ solution

For 0.01 molar MgSO$_4$ we have $I = \frac{1}{2} \cdot (0.01 \cdot 4 + 0.01 \cdot 4) \, \text{mol L}^{-1} = 0.04 \, \text{mol L}^{-1}$. Then with $R = 0$ we obtain $\ln \gamma = -1.175 \cdot \sqrt{0.04} \cdot 4 = -0.94$ and $\gamma = 0.39$ and with $R = 500 \, \text{pm}$ we have $1 + BR\sqrt{I} = 1.34$, $\ln \gamma = -0.70$, and $\gamma = 0.49$. Note that the Coulomb interaction increases strongly with the ionic charge!

Activity coefficients for cations (γ_+) and for anions (γ_-) cannot be measured separately. Then the Debye–Hückel theory can only be checked indirectly by measuring the mean

PROBLEM 20.1

activity coefficient $\gamma_\pm$ defined by

$$\gamma_\pm = \left(\gamma_+^{z_+} \cdot \gamma_-^{z_-}\right)^{1/(z_+ + z_-)} \tag{20.65}$$

where $z_+ e$ is the charge of the cation and $-z_- e$ is the charge of the anion. For example, in the case of HCl dissociating into H$^+$ and Cl$^-$ ions we have $z_+ = z_- = 1$ and $\gamma_\pm = (\gamma_+ \cdot \gamma_-)^{1/2} = \sqrt{\gamma_+ \cdot \gamma_-}$ (see Problem 20.4).

Problems

Problem 20.1 – Freezing Point of Mercury

Imagine a vertical tube filled with mercury (height $h = 10$ m). What is the freezing point of the mercury at the bottom when the freezing point at the top is 234.3 K?

Solution. The mercury exerts a pressure of $P = hg\rho = 10\,\text{m} \times 9.81\,\text{m}\,\text{s}^{-2} \times 13.6\,\text{g}\,\text{cm}^{-3} = 13.3\,\text{bar}$. From the Clapeyron equation (20.2) we obtain with $\Delta_{\text{fus}} H = 2.29\,\text{kJ}\,\text{mol}^{-1}$ and $\Delta_{\text{fus}} V = 0.517\,\text{cm}^3\,\text{mol}^{-1}$

$$dT = \frac{234.3 \times 0.517 \times 10^{-6}}{2.29 \times 10^3} \times 13.3 \times 10^5 \,\text{K} = +0.07\,\text{K}$$

The freezing point at the bottom is about 0.1 K higher than that at the top; this means freezing of the mercury starts at the bottom of the tube, contrary to the freezing of water.

Problem 20.2 – Determination of Fugacity Coefficient ϕ

Determine ϕ for a real gas from experimental PV data for CO_2.

Solution. From equations (20.22) and (20.23) we obtain

$$\int_{P_0}^{P} V_m \cdot dP = RT \cdot \ln \frac{P}{P_0} + RT \cdot \ln \phi$$

$$= \int_{P_0}^{P} V_{m,\text{ideal}} \cdot dP + RT \cdot \ln \phi$$

where P_0 is a small enough pressure such that $\phi_0 = 1$. Then

$$\ln \phi = \frac{1}{RT} \int_{P_0}^{P} (V_m - V_{m,\text{ideal}}) \cdot dP$$

where $V_{m,\text{ideal}} = RT/P$. The molar volume V_m is obtained from experimental PV data (Table 20.3, for $n_{CO_2} = 1$ mol and $T = 313$ K).

In Figure 20.10, $V_m - V_{m,\text{ideal}}$ is plotted versus the pressure P. The shaded area in the figure corresponds to

$$\int_{0\,\text{bar}}^{100\,\text{bar}} (V_m - V_{m,\text{ideal}}) \cdot dP = -14\,\text{bar L}$$

Then

$$\ln \phi = -\frac{1}{RT} \cdot 14\,\text{bar L} = -0.54, \quad \phi = 0.58$$

Table 20.3 Calculation of the fugacity coefficient ϕ of CO_2 ($T = 313$ K) from experimental PV data

P/bar	V_m	$V_{m,ideal}$ /L mol^{-1}	$(V_m - V_{m,ideal})$
0	–	–	0
1.013	25.60	25.67	−0.070
20	1.190	1.310	−0.120
28.6	0.786	0.909	−0.123
42.4	0.469	0.613	−0.127
50.6	0.380	0.513	−0.133
75.9	0.184	0.343	−0.159
101	0.069	0.257	−0.188

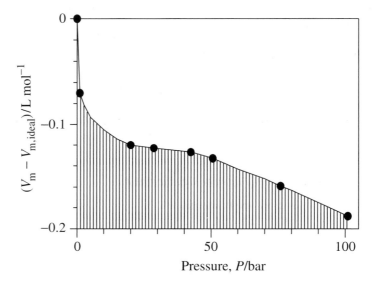

Figure 20.10 $V_m - V_{m,ideal}$ versus pressure P for the data in Table 20.3. The shaded area is -14 bar L.

Problem 20.3 – Charge Distribution ρ(r)

Prove that the total charge of the cloud of the counterions in the treatment of a spherical distribution equals $-e$.

Solution.

$$\text{Total charge} = \int_0^\infty \rho(r) \cdot 4\pi r^2 \cdot dr$$

$$= -2n \cdot \frac{e^2}{kT} \cdot \frac{e}{4\pi\varepsilon_0\varepsilon} \cdot 4\pi \cdot \int_0^\infty r \cdot e^{-\kappa r} \cdot dr$$

$$= -2n \cdot \frac{e^2}{kT} \cdot \frac{e}{4\pi\varepsilon_0\varepsilon} \cdot 4\pi \cdot \frac{1}{\kappa^2} = -e$$

Problem 20.4 – Activity Coefficient $\gamma_\pm$ for HCl

The cell potential E of the cell

$$\text{Pt}|\text{H}_2(\text{gas})|\text{HCl}(\text{aq})|\text{AgCl}(\text{solid})|\text{Ag}$$

is measured for different concentrations c of HCl at 298 K (columns 1 and 2 in Table 20.4). Calculate $\gamma_\pm$.

Solution. The reaction in the cell is

$$\tfrac{1}{2}\text{H}_2 + \text{Ag}^+ \longrightarrow \text{H}^+ + \text{Ag}$$

and the resulting cell potential

$$E = E^\ominus_{\text{Ag}^+/\text{Ag}} - \frac{RT}{F} \cdot \ln \frac{\widehat{a}_{\text{H}^+}}{\widehat{a}_{\text{Ag}^+}}$$

Because of $\widehat{a}_{\text{Ag}^+} \cdot \widehat{a}_{\text{Cl}^-} = K_s$ we obtain

$$\frac{\widehat{a}_{\text{H}^+}}{\widehat{a}_{\text{Ag}^+}} = \frac{\widehat{a}_{\text{H}^+} \cdot \widehat{a}_{\text{Cl}^-}}{K_s} = \frac{\widehat{c}_{\text{H}^+} \cdot \widehat{c}_{\text{Cl}^-} \cdot \gamma_\pm}{K_s} = \frac{\widehat{c}^2 \cdot \gamma_\pm^2}{K_s}$$

Then

$$E = E^\ominus_{\text{Ag}^+/\text{Ag}} + \frac{RT}{F} \cdot \ln K_s - 2\frac{RT}{F} \cdot \ln \widehat{c} - 2\frac{RT}{F} \cdot \ln \gamma_\pm$$

Table 20.4 Deriving the activity coefficient $\gamma_\pm$ of HCl solutions at 308 K by measuring the cell potential E. The experimental values of $\gamma_\pm$ (column 4) are compared with the result of the Debye–Hückel law (equation (20.61) with $R = 500$ pm, column 5) and with that of the Debye–Hückel limiting law (equation (20.64), column 6). According to equation (20.65), in this case the mean activity coefficient is $\gamma_\pm = \gamma$. Experimental data from: H. S. Harned, R. W. Ehlers, The thermodynamics of aqueous hydrochloric acid solutions from electromotive force measurements, J. Amer. Chem. Soc. 55, 2179 (1933)

c	E	$\left(E + 2\dfrac{RT}{F} \cdot \ln \widehat{c}\right)$	$\gamma_\pm$ (exp.)	$\gamma_\pm$ (calc.)	
(mol L^{-1})	(mV)	(mV)		(equ. 20.61)	(equ. 20.64)
0.00573	494.0	220.1	0.93	0.93	0.92
0.00675	485.5	220.1	0.93	0.93	0.91
0.00707	483.2	220.4	0.93	0.93	0.91
0.00820	475.9	220.9	0.92	0.92	0.90
0.00914	470.0	220.8	0.92	0.92	0.90
0.01045	463.4	221.3	0.91	0.91	0.89
0.01249	454.6	222.0	0.90	0.91	0.88
0.01492	445.3	222.1	0.90	0.90	0.87
0.01617	441.5	222.6	0.89	0.90	0.87
0.01814	435.5	222.7	0.89	0.90	0.86
0.02558	418.2	223.6	0.87	0.88	0.84
0.05254	382.3	225.9	0.84	0.85	0.77
0.1091	345.8	228.2	0.80	0.82	0.69
0.3964	280.2	231.1	0.76	0.77	0.49
0.9862	228.7	227.9	0.81	0.74	0.33

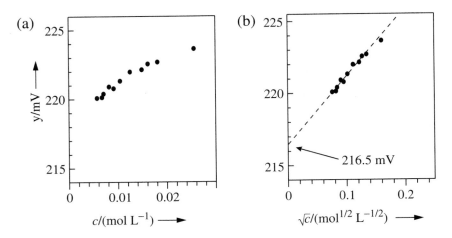

Figure 20.11 Evaluating E' by extrapolating the expression $y = E + 2\frac{RT}{F} \cdot \ln \widehat{c}$ for $\widehat{c} \to 0$, where c is the concentration of HCl. Dots: experiment; dashed line: extrapolation. (a) y plotted versus c, (b) y plotted versus $\sqrt{c}$.

With

$$E' = E^{\ominus}_{Ag^+/Ag} + \frac{RT}{F} \cdot \ln K_s$$

we obtain the following for the activity coefficient

$$\ln \gamma_{\pm} = \frac{F}{2RT}(E' - E) - \ln \widehat{c}$$

To determine E' we rearrange this expression:

$$y = E + 2\frac{RT}{F} \ln \widehat{c} = E' - 2\frac{RT}{F} \ln \gamma_{\pm}$$

For small enough concentration the activity coefficient reaches the limiting value $\gamma_{\pm} = 1$, i.e.

$$\lim_{\widehat{c} \to 0} y = \lim_{\widehat{c} \to 0} \left(E + 2\frac{RT}{F} \ln \widehat{c} \right) = E'$$

E' can be obtained by plotting $y = (E + 2RT/F \cdot \ln \widehat{c})$ versus c and extrapolating for $c \to 0$. However, y is not a linear function of c (Figure 20.11(a)) and therefore it is difficult to perform the extrapolation. On the other hand, according to the Debye–Hückel limiting law, $\ln \gamma_{\pm}$ decreases proportional to $\sqrt{c}$; thus plotting y versus $\sqrt{c}$ results in a straight line (Figure 20.11(b)) and from the intercept with the ordinate we obtain $E' = 0.2165$ V. Using this value we calculate $\gamma_{\pm}$ (Table 20.4, column 4).

From the Debye–Hückel law (equation (20.61)) with $R = 500$ pm we obtain the data in Table 20.4, column 5, which are in excellent agreement with the experiment up to a concentration of 0.1 mol L^{-1}. In column 6 the result of the Debye–Hückel limiting law (equation (20.64)) is shown; these values are satisfactory up to a concentration of 0.01 mol L^{-1}.

Foundation 20.1 – Distribution of Ions at a Charged Plate

Charge distribution between two plates

First we consider a simplified model. We imagine two parallel metal plates at a distance d in a solution with ions (charges $+e$ and $-e$) that do not react with the plate. The plates are biased by a voltage ϕ_0 (Figure 20.12). Then the negative ions tend to accumulate at the positive plate, and the positive ions accumulate at the negative plate; on the other hand, the thermal motion of the ions counteracts this effect. What is the resulting distribution of the ions in the proximity of the plate?

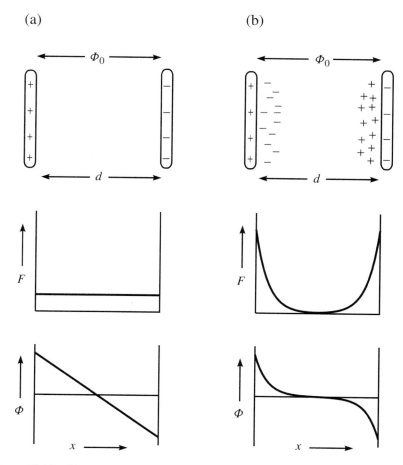

Figure 20.12 Plate capacitor biased by a voltage ϕ_0. Electrical field strength F and potential ϕ. (a) Plates in vacuum; (b) plates in a solution containing ions with charges $+e$ and $-e$.

Continued on page 658

Continued from page 657

The charges of the ions (charge density $\rho(x)$) compensate the charges on the plates (Figure 20.12(b)). The electric field strength F is strongly decreased in the central region within the plates. Each ion finds itself in a potential determined by the charge distribution of the ions in its neighborhood.

We focus on the region next to the positive plate on the left. Following the assumption of Gouy and Chapman we consider a selected ion in the average potential $\phi(x)$, where x is the distance from the plate. Then the potential energies of a negative and a positive ion are

$$E_- = -e\phi, \quad E_+ = +e\phi$$

The negative ion is attracted to the plate, the positive ion is repelled; the zero point of potential energy is at the center between the two plates.

How can we obtain $\phi(x)$? How large are the number densities n_+ and n_- of the positive and of the negative ions? According to the Boltzmann law the number densities

$$n_+ = \frac{N_+}{V} \quad \text{and} \quad n_- = \frac{N_-}{V}$$

are

$$n_- = n \cdot e^{e\phi/kT} \quad \text{and} \quad n_+ = n \cdot e^{-e\phi/kT}$$

where n is the number density of the bulk (or the number density in the absence of the electric field). Then for the charge density $\rho(x)$ we obtain

$$\rho = \frac{\text{charge}}{\text{volume}} = n_+ e - n_- e = (n_+ - n_-) \cdot e$$

and, inserting the expressions for n_+ and n_-, we obtain

$$\rho = ne \cdot (e^{-b} - e^{b}) \quad \text{with} \quad b = \frac{e\phi}{kT}$$

At low enough potential ϕ and high enough temperature we have $b \ll 1$ (e.g., for $\phi = 1$ mV and $T = 298$ K: $b = 0.039$) and we can approximate the exponential by the two first terms of the power series:

$$\rho = ne \cdot (1 - b - 1 - b) = -ne \cdot 2b = -2n \frac{e^2}{kT} \cdot \phi$$

In order to find $\phi(x)$ we consider an additional relation between ρ and ϕ from electrostatics:

$$\frac{d^2 \phi}{dx^2} = -\frac{\rho}{\varepsilon_0 \varepsilon}$$

(*Poisson equation*, see below).

Continued on page 659

Continued from page 658

With this equation it follows that

$$\frac{d^2\phi}{dx^2} = 2n\frac{e^2}{\varepsilon_0\varepsilon kT}\cdot\phi$$

The solution of this differential equation is

$$\phi = \frac{\phi_0}{2}\cdot e^{-\kappa x} \quad \text{with} \quad \kappa = \sqrt{\frac{2ne^2}{\varepsilon_0\varepsilon kT}}$$

and for the charge density ρ we obtain

$$\rho = -2n\frac{e^2}{kT}\cdot\frac{\phi_0}{2}\cdot e^{-\kappa x}$$

where $\phi_0/2$ is the potential change from the left plate to the center between the plates.

Then the mean distance $\bar{x}$ of the center of gravity of the negative charge cloud from the left plate is

$$\bar{x} = \frac{\int_0^\infty x\rho\cdot dx}{\int_0^\infty \rho\cdot dx} = \frac{\int_0^\infty xe^{-\kappa x}\cdot dx}{\int_0^\infty e^{-\kappa x}\cdot dx} = \frac{1}{\kappa}$$

$\bar{x}$ corresponds to r_D in equation (20.57).

Charge distribution around an ion

We approximate the ion as a point charge and ask for the charge distribution $\rho(r)$. We start with the Poisson equation (see below)

$$2r\frac{d\phi}{dr} + r^2\frac{d^2\phi}{dr^2} = -\frac{1}{\varepsilon_0\varepsilon}\cdot\rho\cdot r^2$$

and with

$$\rho = -2n\cdot\frac{e^2}{kT}\cdot\phi$$

Then

$$2r\frac{d\phi}{dr} + r^2\frac{d^2\phi}{dr^2} = \frac{1}{\varepsilon_0\varepsilon}\cdot 2n\cdot\frac{e^2}{kT}\cdot\phi\cdot r^2$$

The solution for the potential ϕ is (as can be verified by inserting the solution into the differential equation)

$$\phi = A\cdot\frac{1}{r}\cdot e^{-\kappa r} \quad \text{with} \quad \kappa = \sqrt{\frac{2ne^2}{\varepsilon_0\varepsilon kT}}$$

Continued on page 660

Continued from page 659

The constant A follows from the condition that in the limit $\kappa \to 0$ the potential ϕ is identical with the potential of a point charge:

$$\lim_{\kappa \to 0} \phi = A \cdot \frac{1}{r} = \frac{e^2}{4\pi\varepsilon_0\varepsilon} \cdot \frac{1}{r}$$

Then

$$\phi = \frac{e}{4\pi\varepsilon_0\varepsilon} \cdot \frac{1}{r} \cdot e^{-\kappa r}$$

Inserting ϕ in the equation for ρ we obtain:

$$\rho = -2n \cdot \frac{e^2}{kT} \cdot \frac{e}{4\pi\varepsilon_0\varepsilon} \cdot \frac{1}{r} \cdot e^{-\kappa r}$$

We calculate $1/r_D$ as the average of $1/r$

$$\overline{(1/r)} = \frac{\int_R^\infty \frac{1}{r}\rho \cdot 4\pi r^2 \cdot dx}{\int_R^\infty \rho \cdot 4\pi r^2 \cdot dx} = \frac{\kappa}{1+\kappa R}$$

R is the minimum distance between the selected ion and the counterions. By identifying $1/r_D$ with the average of $1/r$ we obtain equation (20.58). For small enough κ (i.e., for small enough number density n) the relation for r_D simplifies to $r_D = 1/\kappa$.

So far we have considered singly charged ions described by the number density $n = N/V$ (N cations, N anions in volume V). In the general case we have different kinds of ions (number density n_i, charge $z_i \cdot e$ of ion species i). Then, in the expression for κ the term n has to be replaced by $\frac{1}{2}\sum_i n_i \cdot z_i^2$. Usually, the number density is expressed by the concentration:

$$n_i = \frac{N_i}{V} = c_i \cdot N_A$$

Then

$$\kappa = \sqrt{\frac{2e^2}{\varepsilon_0\varepsilon kT}} \cdot \sqrt{\frac{1}{2}\sum_i n_i \cdot z_i^2} = \sqrt{\frac{2e^2 N_A}{\varepsilon_0\varepsilon kT}} \cdot \sqrt{\frac{1}{2}\sum_i c_i \cdot z_i^2}$$

or

$$\kappa = \sqrt{\frac{2e^2 N_A}{\varepsilon_0\varepsilon kT}} \cdot \sqrt{I} \quad \text{with} \quad I = \sqrt{\frac{1}{2}\sum_i c_i \cdot z_i^2}$$

I is called the *ionic strength*.

Continued on page 661

FOUNDATION 20.1

Continued from page 660

Poisson equation

We consider a spherical charge distribution $\rho(r)$ and ask for the electric field strength $F(r_1)$ at distance r_1 from the center of the sphere (Figure 20.13). The charges outside the sphere of radius r_1 do not contribute to F, because the field strength inside a spherical cavity is zero. The charges inside the sphere of radius r_1

$$Q_1 = \int_0^{r_1} \rho(r) \cdot 4\pi r^2 \cdot dr$$

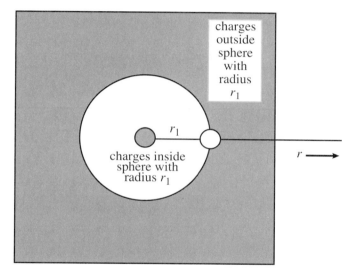

Figure 20.13 Spherical distribution of charges. Electrical field inside and outside a sphere with radius r_1.

act as a point charge located in the center of the sphere. Thus

$$F(r_1) = \frac{Q_1}{4\pi\varepsilon_0\varepsilon \cdot r_1^2} = \frac{1}{4\pi\varepsilon_0\varepsilon \cdot r_1^2} \cdot \int_0^{r_1} \rho(r) \cdot 4\pi r^2 \cdot dr$$

$F(r_1)$ is connected with the potential by

$$F(r_1) = -\left(\frac{d\phi}{dr}\right)_{r_1}$$

Therefore

$$r_1^2 \cdot \left(\frac{d\phi}{dr}\right)_{r_1} = -\frac{1}{\varepsilon_0\varepsilon} \cdot \int_0^{r_1} \rho(r) \cdot r^2 \cdot dr$$

Continued on page 662

Continued from page 661

By differentiation and omitting the index 1 it follows

$$\frac{d}{dr}\left(r^2 \cdot \frac{d\phi}{dr}\right) = -\frac{1}{\varepsilon_0 \varepsilon} \cdot \rho(r) \cdot r^2$$

$$\left[2r\frac{d\phi}{dr} + r^2 \frac{d^2\phi}{dr^2}\right] = -\frac{1}{\varepsilon_0 \varepsilon} \cdot \rho \cdot r^2$$

This equation is called the *Poisson equation*. For large enough r we can neglect the first term in the brackets:

$$\frac{d^2\phi}{dr^2} = -\frac{1}{\varepsilon_0 \varepsilon} \cdot \rho \qquad (20.66)$$

This is the form of the Poisson equation applied above for the potential inside two plates.

21 Kinetics of Chemical Reactions

In the previous chapters we discussed the conditions for a chemical reaction to occur spontaneously in a chosen direction. We did not ask, however, how much time is taken until the equilibrium has been established. In most chemical reactions, this question is of great importance.

For instance, the condition $\Delta G \leq 0$ is fulfilled for the oxidation of organic compounds by the oxygen in air. However, this reaction is usually extremely slow, and most organic compounds are stable in air at room temperature.

In chemical synthesis it is crucial that the attempted reaction be fast and that side reactions or consecutive reactions be sufficiently slow. This can be achieved by an appropriate choice of the conditions (temperature, solvent, catalyst).

Molecules can react only if they are in contact with each another. If the reaction occurs immediately upon contact, diffusion to the right place is the rate-determining step (a diffusion-controlled reaction). In the majority of cases, however, mere contact is not sufficient. Then it is crucial that the kinetic energy of the colliding molecules exceeds a certain limit (activation energy). In this case the rate-determining step is the accumulation of sufficient kinetic energy.

This approach (collision theory) is appropriate for reactions of simple molecules. If the reactants are more complex, the energy distribution in the activated complex is important to open a reaction channel. Then it is more appropriate to treat the activated complex as being in equilibrium with the reactants; the reaction products are formed by dissociation of the activated complex (transition state theory).

Investigating reaction sequences and coupled reactions is a fascinating subject. Chain reactions, autocatalytic reactions, and reactions oscillating in time and space are examples of strongly growing fields in reaction kinetics.

21.1 Collision Theory for Gas Reactions

21.1.1 Counting the Number of Collisions

In Chapter 11 we considered the mean speed of a molecule in a gas and its mean path determining the frequency of collisions.

We imagine one molecule A moving in a gas of $N(B)$ molecules of type B. For simplicity we assume that molecules B have much larger mass than molecules A; then we can consider molecules B to be practically at rest. According to equation (11.40), the number z of collisions of one molecule A with molecules B in time Δt is

$$z = \frac{\bar{v} \cdot \Delta t}{\bar{\lambda}} \tag{21.1}$$

where $\bar{v}$ is the mean speed and $\bar{\lambda}$ is the mean free path of molecule A. Then the total number Z of collisions of molecules A with molecules B in a gas of $N(A)$ molecules A and $N(B)$ molecules B is

$$Z = z \cdot N(A) = \frac{\bar{v} \cdot \Delta t}{\bar{\lambda}} \cdot N(A) \tag{21.2}$$

According to equations (11.30) and (11.33) we have

$$\bar{v} = \sqrt{\frac{8kT}{\pi m_A}}, \quad \bar{\lambda} = \frac{V}{\pi (r_A + r_B)^2 N(B)} \tag{21.3}$$

where V is the volume of the gas. Then we obtain

$$Z = \sqrt{\frac{8kT}{\pi m_A}} \cdot \pi (r_A + r_B)^2 \frac{1}{V} \cdot N(A)N(B) \cdot \Delta t \tag{21.4}$$

In the general case when the masses m_A and m_B are arbitrary, $\bar{\lambda}$ has to be calculated from equation (11.37); then we obtain

$$Z = \sqrt{\frac{8kT}{\pi \mu}} \cdot (r_A + r_B)^2 \frac{1}{V} \cdot N(A)N(B) \cdot \Delta t \quad \text{with} \quad \mu = \frac{m_A m_B}{m_A + m_B} \tag{21.5}$$

Example – hydrogenation of ethene

$$H-H + H_2C=CH_2 \longrightarrow H_3C-CH_3$$

In this case $m_A = 3.25 \times 10^{-27}$ kg and $m_B = 4.66 \times 10^{-26}$ kg (thus $\mu = 3.04 \times 10^{-27}$ kg), and $r_A + r_B = 300$ pm. For $V = 1\,\text{cm}^3$ and $N(B) = 2.43 \times 10^{19}$ (this is the number of particles in a gas of volume $V = 1\,\text{cm}^3$ at 298 K and 1 bar) the number of collisions Z is obtained as follows:

$$Z = 1.86 \times 10^3 \,\text{m s}^{-1} \cdot 2.83 \times 10^{-19}\,\text{m}^2 \cdot \frac{2.43 \times 10^{19}}{10^{-6}\,\text{m}^3} \cdot N(A) \cdot \Delta t$$

$$= 1.3 \times 10^{10}\,\text{s}^{-1} \cdot N(A) \cdot \Delta t$$

This means that one molecule A undergoes 1.3×10^{10} collisions with molecules B in one second.

On the other hand, if we imagine the number of molecules B reduced to $N(B) = 1$ (one molecule A colliding with one molecule B in a volume of $1\,\text{cm}^3$), the number of collisions is only 5×10^{-10} in one second; this means that it takes $\frac{1}{5} 10^{10}$ s $= 64$ years for one collision to occur. It takes 10^{19} times longer than in the first case until the A molecule collides. This gives us a lively picture of the smallness of the molecules.

21.1.2 Activation

We have seen in Chapter 12 that the speeds of the molecules in a gas are distributed over a considerable range (Maxwell–Boltzmann distribution). High speeds are rare but occasionally a molecule may accumulate, by many favorable collisions, sufficient energy to be able to react in an appropriate collision with an appropriate reaction partner. We first consider a very simple case to grasp the essentials.

As an example we choose the reaction between a deuterium atom and a hydrogen molecule

$$D + H_2 \longrightarrow HD + H \qquad (21.6)$$

This reaction proceeds as follows. The D atom approaches an H_2 molecule until the repulsion of the electrons becomes effective (Chapter 4). On approaching, further energy

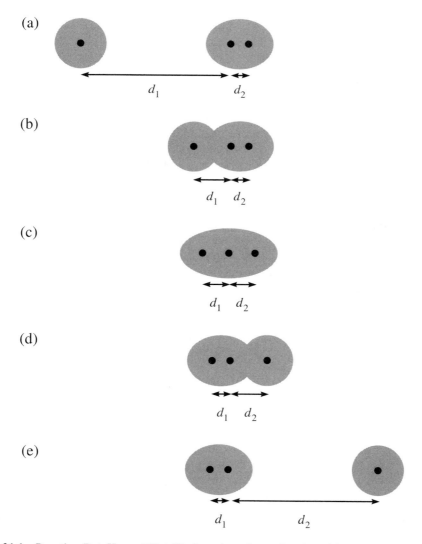

Figure 21.1 Reaction $D + H_2 \rightarrow HD + H$ via activated complex (transition state, stage c).

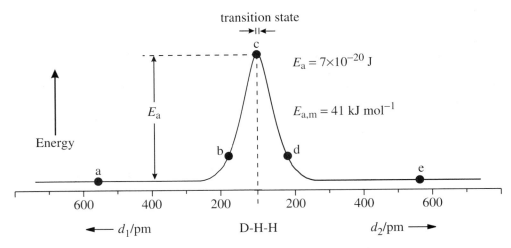

Figure 21.2 Reaction $D + H_2 \rightarrow HD + H$. Energy of the system in the various stages of the reaction. The curve was drawn from numerical data in: B. Liu, Ab initio potential surface for linear H_2, J. Chem. Phys. 58, 1925 (1973), Table 10.

must be supplied to deform the electron clouds of the reacting particles (Figure 21.1(a) and (b)). This can be achieved by means of the kinetic energy of the colliding particles (translational energy in the direction of the collision). Then a configuration can be reached with equal bond lengths (Figure 21.1(c)). Finally, a transition to the final state occurs (Figure 21.1(d) and (e)). The various stages of the reaction are characterized by the distances d_1 and d_2 between adjacent nuclei.

The energy E of the system is displayed in Figure 21.2 for different distances d_1 and d_2 corresponding to the stages (a) to (e) in Figure 21.1. The maximum of the curve in Figure 21.2 (at $d_1 = d_2 = 100$ pm) corresponds to the activation energy $E_a = 7 \times 10^{-20}$ J of the reaction (activated complex of transition state); then the molar activation energy is $E_{a,m} = E_a \cdot N_A = 41$ kJ mol^{-1}. This value is much smaller than the energy required for breaking the H–H bond (Section 4.3: 7.1×10^{-19} J $\cdot$ 6.02×10^{23} mol^{-1} = 427 kJ mol^{-1}).

In this special example the energy of the reactants is essentially the same as the energy of the products. In the general case the energy of the products can be lower (exothermic reaction) or higher (endothermic reaction) than the energy of the reactants. As an example, the energy in the course of the exothermic reaction

$$Cl + HI \longrightarrow HCl + I \tag{21.7}$$

is displayed in Figure 21.3.

21.1.3 Accumulation of Kinetic Energy

For many reactions the activation energy $E_{a,m}$ is on the order of 50 kJ mol^{-1}, whereas the average kinetic energy at room temperature is, according to Chapter 11, only $\frac{3}{2}RT = 3.7$ kJ mol^{-1}. Of course, this is not enough to surmount the activation barrier. In Chapter 11 we have shown that in a noncentered collision the speed of a selected molecule can increase by a factor of $\sqrt{2}$; then the energy of this molecule increases by a factor of two.

21.1 COLLISION THEORY FOR GAS REACTIONS

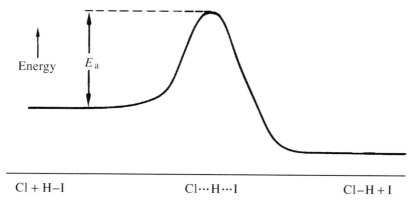

Figure 21.3 Energy diagram for the reaction $Cl + HI \rightarrow HCl + I$ ($\Delta_r H = -133 \, \text{kJ mol}^{-1}$).

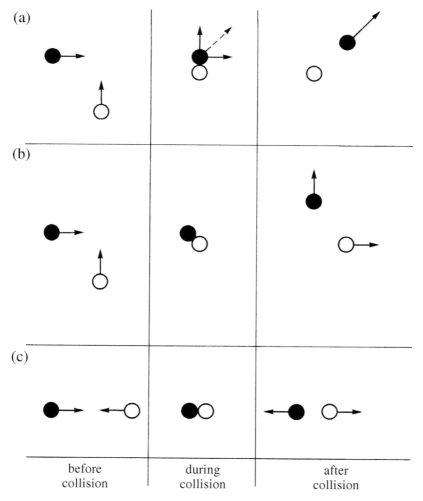

Figure 21.4 Representative possibilities for two particles to collide with each other. (a) Transfer of energy from the white to the black particle; (b) and (c) no energy transfer.

This means that the energy taken up in one collision is on the order of kT. Such a process can occur, in principle, many times, resulting in an energy of the molecule much higher than the average energy and sufficient to surmount the activation barrier.

The probability for the accumulation of so much energy is very low. Look at Figure 21.4, where three representative possibilities of how collisions can occur are displayed. There is only one out of these three possibilities (case a) where energy is transferred from one collision partner to another. Therefore we estimate the probability of transferring the energy kT to a selected molecule as $\frac{1}{3}$. Then the probability to transfer the energy $2kT$ is $\frac{1}{3} \cdot \frac{1}{3} = \frac{1}{9}$. Consequently, to accumulate the activation energy E_a, the probability P is

$$P = \left(\tfrac{1}{3}\right)^{E_a/kT} = 3^{-E_a/kT} = 3^{-E_{a,m}/RT} \tag{21.8}$$

This estimate is essentially in agreement with the result of the rigorous treatment (Box 21.1)

$$P = e^{-E_a/kT} = e^{-E_{a,m}/RT} \tag{21.9}$$

Box 21.1 – Activation Factor

For simplicity we consider a light particle A (mass m) colliding with a much heavier particle B. According to Section 11.4, the probability dP_s of finding a light particle with a velocity component, in a given direction s (Figure 21.5), between v_s and $v_s + dv_s$ is

$$dP_s = \mathrm{const} \cdot e^{-mv_s^2/(2kT)} \cdot dv_s = f(v_s) \cdot dv_s$$

The probability that this particle hits a heavy particle is proportional to

$$v_s \cdot dP_s = v_s f(v_s) \cdot dv_s$$

(fast particles approach the heavy particle in a shorter time than slow particles). Therefore the fraction of collisions involving a kinetic energy

$$\frac{m}{2} v_s^2 \geq E_a$$

between light and heavy particle is

$$P = \frac{\int_{v_{s,a}}^{\infty} v_s f(v_s) \cdot dv_s}{\int_{0}^{\infty} v_s f(v_s) \cdot dv_s} = \frac{\int_{v_{s,a}}^{\infty} v_s \cdot e^{-mv_s^2/(2kT)} \cdot dv_s}{\int_{0}^{\infty} v_s \cdot e^{-mv_s^2/(2kT)} \cdot dv_s}$$

where $\frac{1}{2} m v_{s,a}^2 = E_a$. The integrals are solved by the substitution

$$x = \frac{mv_s^2}{2kT}, \quad dx = \frac{m}{kT} v_s \cdot dv_s$$

Continued on page 669

21.2 RATE EQUATION FOR GAS REACTIONS

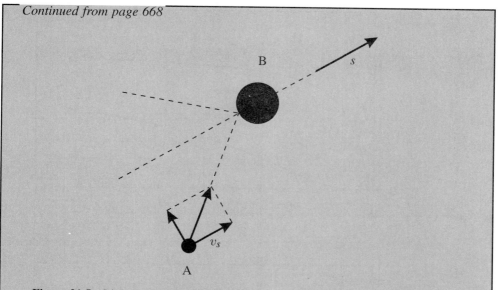

Figure 21.5 Light particle A colliding with heavy particle B. v_s is the velocity component of particle A in the s-direction (in direction of the center of mass of particle A).

Then

$$\int v_s \cdot e^{-mv_s^2/(2kT)} \cdot dv_s = \frac{kT}{m} \cdot \int e^{-x} \cdot dx = -\frac{kT}{m} \cdot e^{-x}$$

and with $x_a = E_a/(kT)$ we obtain

$$P = \frac{|e^{-x}|_{x_a}^{\infty}}{|e^{-x}|_0^{\infty}} = \exp(-x_a) = \exp\left(-\frac{E_a}{KT}\right)$$

Note that the probability P is strongly increasing with temperature (with increasing temperature, fewer collisions are necessary to accumulate the activation energy). In Figure 21.6, P is displayed as a function of temperature.

To account for the correct orientation of the molecules in the collision initiating the chemical reaction, we introduce the steric factor α. Then the number of successful collisions is

$$Z_{\text{successful}} = Z \cdot e^{-E_{a,m}/RT} \cdot \alpha \tag{21.10}$$

21.2 Rate Equation for Gas Reactions

21.2.1 Rate Constant and Frequency Factor

In the chemical reaction

$$A + B \longrightarrow C \tag{21.11}$$

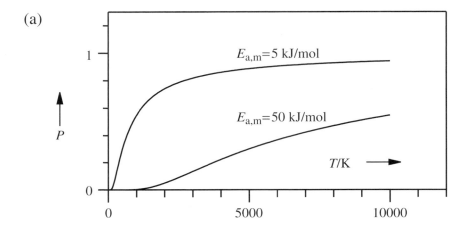

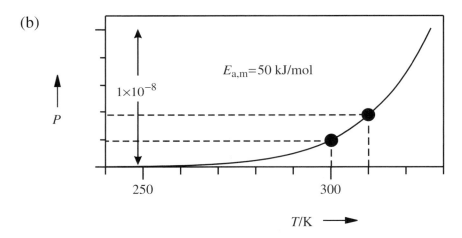

Figure 21.6 Probability P for a particle to surmount the activation barrier at temperature T. (a) Temperature range 0–10 000 K, $E_{a,m} = 5\,\text{kJ}\,\text{mol}^{-1}$ and $E_{a,m} = 50\,\text{kJ}\,\text{mol}^{-1}$; (b) temperature range 240–330 K, $E_{a,m} = 50\,\text{kJ}\,\text{mol}^{-1}$. Increasing the temperature from 300 K to 310 K results in an increase of P from 2.0×10^{-9} to 3.8×10^{-9}.

one molecule A and one molecule B disappear after every successful collision. Then the change $\Delta N(\text{A})$ in the number of molecules A is

$$\Delta N(\text{A}) = -Z_{\text{successful}} = -Z \cdot e^{-E_{a,m}/RT} \cdot \alpha \qquad (21.12)$$

with equations (21.4) and (21.5)

$$Z = \sqrt{\frac{8kT}{\pi\mu}} \cdot \pi\,(r_\text{A} + r_\text{B})^2\,\frac{1}{V} \cdot N(\text{A})N(\text{B}) \cdot \Delta t$$

21.2 RATE EQUATION FOR GAS REACTIONS

We convert the particle numbers $N(A)$ and $N(B)$ into concentrations c_A and c_B, and convert $\Delta N(A)$ into Δc_A:

$$c_A = \frac{N(A)}{V \cdot N_A}, \quad c_B = \frac{N(B)}{V \cdot N_A}, \quad \Delta c_A = \frac{\Delta N(A)}{V \cdot N_A} \quad (21.13)$$

and replace the differences by differentials; then we obtain

$$\boxed{\frac{dc_A}{dt} = -k_r \cdot c_A \cdot c_B} \quad (21.14)$$

with

$$\boxed{k_r = A_f \cdot e^{-E_{a,m}/RT}} \quad (21.15)$$

$$\boxed{A_f = \sqrt{\frac{8kT}{\pi\mu}} \cdot \pi \cdot (r_A + r_B)^2 \cdot N_A \cdot \alpha} \quad (21.16)$$

(gas reaction through activated complex).

The constant k_r is called the *rate constant*. Equation (21.15) was first suggested by Arrhenius as an empirical equation; therefore this equation is called the *Arrhenius equation*. The constant A_f depends on the frequency of collisions, therefore it is called the *frequency factor*.

The term $\sqrt{8kT/(\pi\mu)}$ is related to the speed, $\pi(r_A + r_B)^2$ is related to the mean free path; these two factors determine the frequency of collisions. The factor $e^{-E_{a,m}/RT} \cdot \alpha$ is the probability that a collision is successful (that is, the energy is sufficient to surmount the activation barrier and the collision is sterically appropriate).

In special cases, when $E_a = 0$, every collision leads to a reaction; then

$$\boxed{k_r = \sqrt{\frac{8kT}{\pi\mu}} \cdot \pi(r_A + r_B)^2 \cdot N_A \cdot \alpha} \quad (21.17)$$

(gas reaction, diffusion-controlled).

Example – frequency factor and rate constant

For reaction (21.6)

$$D + H_2 \longrightarrow HD + H$$

the activation energy is, according to Figure 21.2, $E_{a,m} = 41$ kJ mol^{-1}. Then the activation factor is $e^{-E_{a,m}/RT} = 4.9 \times 10^{-7}$. With $m_D = 2m_H = 3.3 \times 10^{-27}$ kg ($\mu = 1.6 \times 10^{-27}$ kg), $r_D = 128$ pm, and $r_{H_2} = 138$ pm we calculate for $T = 298$ K

$$A_f = 2.6 \times 10^3 \, \text{m s}^{-1} \cdot \pi \cdot 7.1 \times 10^{-24} \, \text{m}^2 \cdot 6.02 \times 10^{23} \, \text{mol}^{-1} \cdot \alpha$$
$$= 3.5 \times 10^{11} \, \text{L mol}^{-1} \, \text{s}^{-1} \cdot \alpha$$
$$k_r = 3.5 \times 10^{11} \, \text{L mol}^{-1} \, \text{s}^{-1} \cdot \alpha \cdot 4.9 \times 10^{-7}$$
$$= 1.7 \times 10^5 \cdot \alpha \, \text{L mol}^{-1} \, \text{s}^{-1}$$

The experimental value is $2 \times 10^5 \,\mathrm{L\,mol^{-1}\,s^{-1}}$. Thus the steric factor is approximately $\alpha = 1$.

21.2.2 Reactants A and B are Identical Molecules

We discuss the case where the reactants in equation (21.11) are identical molecules (e.g., dimerization, disproportionation reactions):

$$2A \longrightarrow C \quad \text{or} \quad 2A \longrightarrow C + D \tag{21.18}$$

In this case, two molecules A disappear in one successful collision

$$\frac{dc_C}{dt} = -\frac{1}{2}\frac{dc_A}{dt} = k_r c_A^2 \quad \text{or} \quad \frac{dc_A}{dt} = -2k_r c_A^2 \tag{21.19}$$

This relation, which is in accordance with the rules of IUPAC, is not trivial. On the one hand, $-\Delta N(A)$ is twice the number of successful collisions; on the other hand, applying the consideration in Section 21.1.1, each collision is counted twice, thus $Z_{\text{successful}}$ in equation (21.10) must be divided by two. Thus k_r has to be identified with one-half of the rate constant calculated from the expressions (21.15) and (21.16).

21.3 Rate Equation for Reactions in Solution

21.3.1 Reaction Through Activated Complex

Let us consider a reaction of particles A and B in a liquid, like iodide and chloromethane dissolved in methanol reacting to chloride and iodomethane.

$$I^- + ClCH_3 \longrightarrow Cl^- + ICH_3 \tag{21.20}$$

In contrast to a gas reaction, A collides with particles B as well as with solvent molecules.

The mean free path is much shorter in a liquid than in a gas, thus it takes longer for a particle to diffuse around and to collide with a reaction partner. However, when the particle has reached a favorable position at the reaction partner it can collide many times; thus its chance to collide successfully is much higher than in a gas. Surprisingly, these two effects largely compensate and the reaction rates are similar in a reaction in solution and in a gas reaction, as will be shown in the following.

If ν is the collision frequency of particle A, then

$$z' = \nu \cdot \Delta t \tag{21.21}$$

is the total number of collisions in time Δt. What is then the number z of collisions of A with B? To answer this question we calculate the probability P to find particle A at the site of a particle B. For simplicity we assume that a solvent molecule as well as a molecule A can be described by a sphere of radius r_A. After a certain number of collisions, particle A exchanges its place with a neighboring solvent molecule (Figure 21.7). Then

$$P = \frac{\text{number of favorable sites}}{\text{total number of sites}} = \frac{\text{volume of favorable sites}}{\text{volume of solution}} \tag{21.22}$$

$$= \frac{N(B) \cdot \frac{4}{3}\pi r_A^3}{V}$$

21.3 RATE EQUATION FOR REACTIONS IN SOLUTION

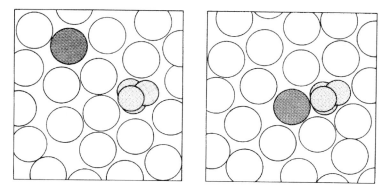

Figure 21.7 Reaction in solution (methanol): $I^- + CH_3Cl \rightarrow Cl^- + CH_3I$. (a) Reactants separated by solvent molecules; (b) favorable position of I^- to react with CH_3Cl.

and the number of collisions of a molecule A with molecules B is

$$z = z' \cdot P = \frac{4}{3}\pi r_A^3 \cdot \frac{N(B)}{V} \cdot v \cdot \Delta t \tag{21.23}$$

Following the same arguments as in the derivation of the relations (21.14), (21.15), and (21.16) for reactions in the gas phase we obtain

$$\boxed{\frac{dc_A}{dt} = -k_r \cdot c_A \cdot c_B} \tag{21.24}$$

with

$$\boxed{k_r = A_f \cdot e^{-E_{a,m}/RT}} \tag{21.25}$$

$$\boxed{A_f = v \cdot \tfrac{4}{3}\pi r_A^3 \cdot N_A \cdot \alpha} \tag{21.26}$$

(reaction in solution through activated complex).

What is the difference between equation (21.26) and the corresponding expression (21.16) for a gas reaction? We can write the frequency as

$$v = \frac{\bar{v}}{\bar{\lambda}_{\text{solution}}} \tag{21.27}$$

where the mean speed $\bar{v}$ is given by equation (21.3); the mean free path $\bar{\lambda}_{\text{solution}}$ is smaller than the diameter of a solvent molecule. With the estimate

$$\bar{\lambda}_{\text{solution}} = \tfrac{1}{3} r_A \tag{21.28}$$

we find

$$v = \frac{3\bar{v}}{r_A} \tag{21.29}$$

and

$$A_f = \frac{3\bar{v}}{r_A} \cdot \frac{4}{3}\pi r_A^3 \cdot N_A \cdot \alpha = \bar{v} \cdot \pi \cdot 4r_A^2 \cdot N_A \cdot \alpha \tag{21.30}$$

This corresponds to equation (21.16) for $r_A = r_B$.

Although the mean free path is much smaller in a solution than in a gas and thus the collision frequency v is accordingly higher, the number of successful collisions is comparable in both cases. This is because very many collisions with solvent molecules have to take place in solution until the reacting particle A has reached a favorable location at molecule B.

21.3.2 Diffusion Controlled Reaction

Many reactions (e.g., electron transfer reactions) occur whenever the molecules A and B touch each other or reach a critical distance. Then the number $Z_{\text{successful}}$ of the successful collisions equals the number of complexes AB formed in time Δt. The rate-determining step then is the diffusion of A to the location of B. We consider a molecule A exchanging its location in the liquid in time τ; that is, it is moving along a distance

$$a = 2 \cdot r_A \tag{21.31}$$

where a is the distance between two adjacent locations (solvent molecule and molecule A described by a sphere of radius r_A). When we treat the molecular movement as a diffusion process, then according to the Einstein–Smoluchowski relation (11.52) we have

$$\tau = \frac{x^2}{2D} = \frac{a^2}{2D} = \frac{4r_A^2}{2D} \tag{21.32}$$

Let us first assume that molecule B reacts with molecule A when A is just at the right site. In our present case the reaction occurs whenever the molecules A and B touch each other. If both molecules are considered as spheres of radius r_A in closest packing (molecule B surrounded by 12 molecules A) there are 12 right sites. Then the probability P of finding a molecule A at a right site at molecule B is

$$P = \frac{\text{volume of all molecules B}}{\text{volume of solution}} \times 12 = \frac{N(B)\frac{4\pi}{3}r_A^3}{V} \times 12$$

Then the number of successful collisions of $N(A)$ molecules A in time Δt is

$$Z_{\text{successful}} = N(A) \cdot \frac{\Delta t}{\tau} \cdot P \tag{21.33}$$

$N(A) \cdot \Delta t / \tau$ is the number of exchanges of location of molecules A taking place during time Δt. In this way we obtain

$$Z_{\text{successful}} = \frac{N(A)N(B)}{V} \cdot 4\pi \cdot 2r_A \cdot D \cdot \Delta t \tag{21.34}$$

Additionally, we have to consider that the molecules B are not at rest, but change their locations by diffusion. Thus D must be replaced by $D_A + D_B$ (diffusion coefficients of molecules A and B). We obtain (in agreement with the rigorous treatment)

$$Z_{\text{successful}} = \frac{N(A)N(B)}{V} \cdot 4\pi \cdot (r_A + r_B) \cdot (D_A + D_B) \cdot \Delta t \tag{21.35}$$

Then

$$\frac{dc_A}{dt} = -k_r c_A c_B \qquad (21.36)$$

with

$$k_r = 4\pi \cdot (r_A + r_B) \cdot (D_A + D_B) \cdot N_A \qquad (21.37)$$

(diffusion-controlled reaction in solution).

The reaction rate depends on the speed of diffusion of the reaction partners (diffusion coefficients $D_A + D_B$) and on the distance $(r_A + r_B)$ at which the reactants touch and are able to react.

21.4 Transition State Theory

21.4.1 Eyring Equation for Bimolecular Reaction

The collision theory does not take into account that the activated complex may have many different realization possibilities leading to a forward or a backward reaction. This point of view becomes important with increasing complexity of the reactants.

To account for this situation in a simplified way we assume an equilibrium between the reactants and the transition complex, without considering the collisions explicitly. The transition complex M can dissociate, forming either reaction product C or the reactants A and B:

$$A + B \xrightleftharpoons{K^\ddagger} M \xrightarrow{k^\ddagger} C \qquad (21.38)$$

Then

$$\frac{dc_C}{dt} = c_M \cdot k^\ddagger \qquad (21.39)$$

Assuming a fast equilibrium we obtain

$$\frac{\widehat{c}_M}{\widehat{c}_A \cdot \widehat{c}_B} = K^\ddagger \qquad (21.40)$$

Thus

$$c_M = \widehat{c}_M \cdot c^\ominus = K^\ddagger \cdot \widehat{c}_A \cdot \widehat{c}_B \cdot c^\ominus = K^\ddagger \cdot c_A c_B \cdot \frac{1}{c^\ominus} \qquad (21.41)$$

where $c^\ominus = 1 \text{ mol L}^{-1}$ is the standard concentration. Then we obtain

$$\frac{dc_C}{dt} = k_r \cdot c_A c_B \quad \text{with} \quad k_r = k^\ddagger \cdot K^\ddagger \cdot \frac{1}{c^\ominus} \qquad (21.42)$$

To estimate $k^\ddagger$ we consider the activated complex extending a certain distance along the coordinate in Figure 21.2 (*reaction coordinate*). The kinetic energy for this translational motion is

$$\tfrac{1}{2}m\bar{v}^2 = \tfrac{1}{2}kT \qquad (21.43)$$

where m is the effective mass of the transition complex. The corresponding de Broglie wavelength is

$$\Lambda = \frac{h}{m\bar{v}} \qquad (21.44)$$

The transition state can be considered as extending over a range of $\Lambda/2$: in the model of a particle in a box this corresponds to the smallest possible energy state; we neglect higher energy states because they require further activation.

Then the estimated lifetime τ of the activated complex is the time needed by the complex to travel the distance $\Lambda/2$:

$$\tau = \frac{\frac{1}{2}\Lambda}{\bar{v}} = \frac{1}{2}\frac{h}{m\bar{v}^2} = \frac{1}{2}\frac{h}{kT} \tag{21.45}$$

(neglecting the difference between $\overline{v^2}$ and $\bar{v}^2$). The activated complex decomposes either into the reactants or into the products. Then the change in concentration of product C formed during the lifetime τ of the activated complex is $\frac{1}{2}c_M$, and the rate of formation of C is

$$\frac{dc_C}{dt} \approx \frac{1}{2}\frac{c_M}{\tau} = \frac{1}{2}c_M \cdot \frac{2kT}{h} = c_M \cdot \frac{kT}{h} \tag{21.46}$$

Thus, by comparison with equation (21.39) we obtain

$$k^{\ddagger} = \frac{kT}{h} \tag{21.47}$$

and

$$\boxed{k_r = \frac{1}{c^{\ominus}} \cdot \frac{kT}{h} \cdot K^{\ddagger}} \tag{21.48}$$

This relation is called the *Eyring equation*. Note that $c^{\ominus}$ is the standard concentration.

21.4.2 Activation Enthalpy and Activation Entropy

In our consideration we have assumed an equilibrium between reactants and the transition complex. Therefore we can use the relations

$$\Delta_r G^{\ominus} = \Delta_r H^{\ominus} - T\Delta_r S^{\ominus}$$

(see Chapter 16) and

$$\Delta_r G^{\ominus} = -RT \cdot \ln K$$

(see Chapter 17) to determine the equilibrium constant K. In our case we replace K, $\Delta_r H^{\ominus}$ and $\Delta_r S^{\ominus}$ by $K^{\ddagger}$, $\Delta_r H^{\ddagger}$ (*activation enthalpy*) and $\Delta_r S^{\ddagger}$ (*activation entropy*), respectively.

$$\Delta_r G^{\ddagger} = \Delta_r H^{\ddagger} - T\Delta_r S^{\ddagger} = -RT \cdot \ln K^{\ddagger} \tag{21.49}$$

Then we obtain

$$K^{\ddagger} = e^{-\Delta_r H^{\ddagger}/RT} \cdot e^{\Delta_r S^{\ddagger}/R} \tag{21.50}$$

Then for the rate constant k_r we obtain

$$\boxed{k_r = A_f \cdot e^{-\Delta_r H^{\ddagger}/RT}} \quad \text{with} \quad \boxed{A_f = \frac{kT}{hc^{\ominus}} \cdot e^{\Delta_r S^{\ddagger}/R}} \tag{21.51}$$

Compared with the result of the collision theory, the dependence of k_r on temperature is the same, if we neglect the small difference in the temperature dependence of A_f and

21.5 TREATMENT OF EXPERIMENTAL DATA

replace $E_{a,m}$ by $\Delta_r H^{\ddagger}$. For reactions in gas phase we have to consider the change of volume. Then

$$\Delta_r H^{\ddagger} = E_{a,m} + P\Delta V_m = E_{a,m} - RT$$

Usually the difference between $\Delta_r H^{\ddagger}$ and $E_{a,m}$ is small and we will neglect it in the following.

As outlined in Chapters 13, 14, and 16 the quantities $\Delta_r H^{\ddagger}$ and $\Delta_r S^{\ddagger}$ are related to the distribution of the energy levels of the particles A, B, and M; this distribution is determined by Boltzmann's law.

The simplest case for the application of the Eyring equation (21.48) is the reaction of two atoms forming a diatomic transition complex:

$$A + B \rightleftharpoons (AB)^{\ddagger} \tag{21.52}$$

Then A and B occupy only translational energy levels, and $(AB)^{\ddagger}$ occupies translational and rotational levels.

21.4.3 Decay Reaction

In equation (21.38) we considered a bimolecular reaction leading to the activated complex. In the case of first-order decay or rearrangement reactions

$$A \underset{}{\overset{K^{\ddagger}}{\rightleftharpoons}} M^{\ddagger} \longrightarrow \text{product} \tag{21.53}$$

the equilibrium constant is

$$K = \frac{\widehat{c}_{M^{\ddagger}}}{\widehat{c}_A} = \frac{c_{M^{\ddagger}}}{c_A} \tag{21.54}$$

and instead of equations (21.57) we obtain

$$\boxed{k_r = A_f \cdot e^{-\Delta_r H^{\ddagger}/RT}} \quad \text{with} \quad \boxed{A_f = \frac{kT}{h} \cdot e^{\Delta_r S^{\ddagger}/R}} \tag{21.55}$$

21.5 Treatment of Experimental Data

21.5.1 Rate Constants

How can we deduce rate constants, activation energies, and steric factors from experimental data? Let us consider the reaction

$$A + B \longrightarrow C \tag{21.56}$$

A and B in equal initial concentration

First we assume that the initial concentrations of A and B are the same

$$c_{A,0} = c_{B,0} = a \tag{21.57}$$

Then, because one molecule A reacts with one molecule B, $c_A = c_B$ at any time, and we obtain from equation (21.14)

$$\boxed{\frac{dc_A}{dt} = -k_r \cdot c_A^2} \qquad (21.58)$$

By integration we obtain

$$\int \frac{dc_A}{c_A^2} = -k_r \cdot \int dt \quad \text{and} \quad -\frac{1}{c_A} = -k_r t + C \qquad (21.59)$$

where C is a constant. C is determined by the initial condition (21.57); for $t = 0$ and $c_A = a$ we have

$$C = -\frac{1}{a} \qquad (21.60)$$

Then we obtain

$$\boxed{\frac{1}{c_A} = \frac{1}{a} + k_r t} \quad \text{or} \quad \boxed{c_A = \frac{1}{k_r t + 1/a}} \qquad (21.61)$$

In Figure 21.8(a), c_A is displayed as a function of time. To evaluate the rate constant from experimental data we plot $1/c_A$ versus t. Then a straight line with slope k_r is obtained (Figure 21.8(b)). The time when c_A decreases to one-half of its original value is called the *half-life* $t_{1/2}$:

$$\frac{1}{2}a = \frac{1}{k_r t_{1/2} + 1/a}, \quad t_{1/2} = \frac{1}{a \cdot k_r} \qquad (21.62)$$

In this case the half-life depends on the initial concentration a.

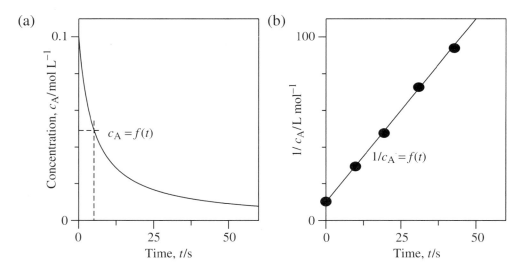

Figure 21.8 Reaction $A + B \rightarrow C$ with initial conditions $a = c_{A,0} = 0.1 \, \text{mol L}^{-1}$ and $b = c_{B,0} = 0.1 \, \text{mol L}^{-1}$. (a) Concentration c_A versus time t; (b) $1/c_A$ versus time t (dots). From the slope of the straight line ($2.0 \, \text{L mol}^{-1} \, \text{s}^{-1}$) we obtain $k_r = 2.0 \, \text{L mol}^{-1} \, \text{s}^{-1}$; the half-life is $t_{1/2} = 1/(0.1 \times 2 \, \text{s}^{-1}) = 5 \, \text{s}$.

21.5 TREATMENT OF EXPERIMENTAL DATA

B in large excess over A

A second simple case for reaction (21.56) can be realized by applying B in large excess over A. Then c_B is practically constant during the reaction:

$$\frac{dc_A}{dt} = -(k_r c_B) \cdot c_A = -k'_r \cdot c_A \tag{21.63}$$

By integration we obtain

$$\int \frac{dc_A}{c_A} = -k'_r \cdot \int dt, \quad \ln \frac{c_A}{a} = -k'_r \cdot t \tag{21.64}$$

and

$$\boxed{c_A = a \cdot e^{-k'_r t}} \tag{21.65}$$

In Figure 21.9(a), c_A is displayed as a function of time. The rate constant can be evaluated from experimental data by plotting $\ln(c_A/a)$ versus time (Figure 21.9(b)); then the slope of the straight line is $-k'_r$. The half-life follows from

$$\frac{1}{2}a = a \cdot e^{-k'_r t_{1/2}}, \quad t_{1/2} = \frac{\ln 2}{k'_r} \tag{21.66}$$

In this case, $t_{1/2}$ is independent of the initial concentration a.

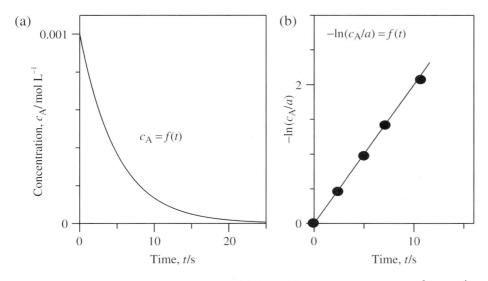

Figure 21.9 Reaction $A + B \longrightarrow C$ with initial conditions $a = c_{A,0} = 1 \times 10^{-3}$ mol L^{-1} and $b = c_{B,0} = 0.1$ mol L^{-1}. (a) Concentration c_A versus time t; (b) $-\ln(c_A/a)$ versus time t (dots). From the slope of the straight line (0.2 s^{-1}) we obtain $k_r = 0.2$ s$^{-1}/(0.1$ mol L$^{-1}) = 2$ L mol^{-1} s^{-1}; the half life is $t_{1/2} = \ln 2/(0.2$ s$^{-1}) = 3.5$ s.

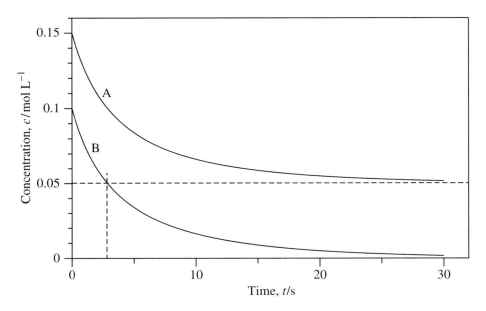

Figure 21.10 Reaction $A + B \rightarrow C$ (general case) with initial conditions $a = c_{A,0} = 0.15 \text{ mol L}^{-1}$ and $b = c_{B,0} = 0.1 \text{ mol L}^{-1}$ and rate constant $k_r = 2 \text{ L mol}^{-1}\text{ s}^{-1}$.

General case

The general case for reaction (21.56) is treated in Problem 21.1. The result for c_A (Figure 21.10) is

$$c_A = \frac{b - a}{(b/a) \cdot \exp[k_r \cdot (b - a) \cdot t] - 1} \qquad (21.67)$$

where a and b are the initial concentrations of the species A and B, respectively.

Equations (21.61) and (21.65) are special cases of equation (21.67). In the first case the reaction rate is proportional to c_A^2, and in the second case the rate is proportional to c_A. Therefore the corresponding reactions are called *second-order* and *first-order* reactions.

Example – first-order reaction

21.5 TREATMENT OF EXPERIMENTAL DATA

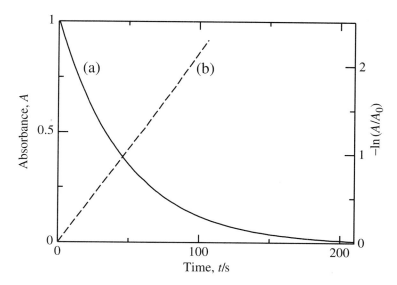

Figure 21.11 First-order reaction: reaction (21.68) with $c_{OH^-} = 0.2\,\mathrm{mol\,L^{-1}}$ and initial concentration $c_{R^+,0} = 5 \times 10^{-5}\,\mathrm{mol\,L^{-1}}$ at 296 K. The reaction is followed by measuring the absorbance A at 590 nm (optical path length 3 cm). (a) Solid line: absorbance A versus time t (left hand scale). (b) Dashed line: $-\ln(c/c_0) = -\ln(A/A_0)$ versus time t (right hand scale).

We consider the bleaching of the triphenylmethan dye cation R^+

$$R^+ + OH^- \longrightarrow ROH \tag{21.68}$$

at 296 K with OH^- in excess ($c_{OH^-} = 0.2\,\mathrm{mol\,L^{-1}}$, $c_R = 5 \times 10^{-5}\,\mathrm{mol\,L^{-1}}$). The dye cation R^+ is trigonal planar, and the positive charge is uniformly distributed over the molecular skeleton; therefore, R^+ behaves similar as a cyanine cation, absorbing light at about 600 nm. The structure of the reaction product ROH is tetrahedral, and the π electrons are confined to the isolated benzene rings, thus ROH does not absorb light in the visible (bleaching of de cation R^+). The concentration c_{R^+} (measured by following the absorbance of the dye at 590 nm) is displayed as a function of time in Figure 21.11 (curve a). From the logarithmic plot (curve b) we obtain $k_r' = 0.0214\,\mathrm{s^{-1}}$ and $k_r = k_r'/c_{OH^-} = 0.11\,\mathrm{L\,mol^{-1}\,s^{-1}}$.

Example – second-order reaction

We consider the disproportionation of $HBrO_2$ at 293 K (initial concentration $2 \times 10^{-5}\,\mathrm{mol\,L^{-1}}$) in acidic solution

$$2HBrO_2 \longrightarrow HOBr + HBrO_3 \tag{21.69}$$

measured by the absorbance of $HBrO_2$ at 240 nm (Figure 21.12, curve a). In this case two identical molecules react and, according to equation (21.19)

$$\frac{dc}{dt} = -2k_r \cdot c^2$$

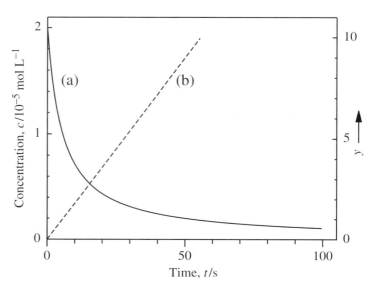

Figure 21.12 Second-order reaction: reaction (21.69) followed at 20 °C by measuring the absorbance A at 240 nm (optical path length 10 cm). (a) Solid line: concentration $c = c_{HBrO_2}$ versus time t. (b) Dashed line: $y = (c_0/c - 1)$ versus time t, where $c_0 = 2 \times 10^{-5}$ mol L^{-1}.

we obtain in analogy to equation (21.61)

$$\frac{1}{c} = \frac{1}{c_0} + 2k_r t \quad \text{or} \quad \frac{c_0}{c} - 1 = 2k_r c_0 t$$

From a plot of $y = c_0/c - 1$ versus time (curve b) we obtain $k_r = 4500$ L mol^{-1} s^{-1}.

This reaction is an important step in the Belousov–Zhabotinsky reaction discussed in Section 21.6.10. Note that the half-life of this reaction is 5 s, in spite of the high value of the rate constant. This is due to the low initial concentration $c_0 = 2 \times 10^{-5}$ mol L^{-1} used in this experiment. If we increase c_0 to 2×10^{-2} mol L^{-1}, the half-life would be only 5 ms.

21.5.2 Activation Energy and Frequency Factor

According to equation (21.15)

$$k_r = A_f \cdot e^{-E_{a,m}/RT} \tag{21.70}$$

the rate constant depends strongly on temperature. By measuring k_r at different temperatures we can evaluate $E_{a,m}$. From equation (21.70) we obtain

$$\boxed{\ln \frac{k_r}{k_{r,1}} = -\frac{E_{a,m}}{R}\left(\frac{1}{T} - \frac{1}{T_1}\right)} \tag{21.71}$$

where $k_{r,1}$ is the rate constant at temperature T_1. By plotting $\ln(k_r/k_{r,1})$ versus $1/T$ we obtain $E_{a,m}$ from the slope of the resulting straight line. Then the frequency factor A_f follows by inserting $E_{a,m}$ in equation (21.70).

21.5 TREATMENT OF EXPERIMENTAL DATA

The chemist's rule of thumb

For many reactions in solution which are important for the organic chemist the activation energy is on the order of $50\,\text{kJ}\,\text{mol}^{-1}$. By increasing the temperature, these reactions can be considerably accelerated. According to equation (21.70) an increase of temperature by ΔT results in a change of the rate constant from $k_\text{r}(T)$ to $k_\text{r}(T + \Delta T)$:

$$\frac{k_\text{r}(T + \Delta T)}{k_\text{r}(T)} = \frac{e^{-E_{\text{a,m}}/R(T+\Delta T)}}{e^{-E_{\text{a,m}}/RT}} \approx \exp\left(\frac{E_{\text{a,m}}\Delta T}{RT^2}\right)$$

if we neglect the small temperature dependence of the number of collisions. Then changing the temperature from 300 K to 310 K (this is a change by 3%) and assuming $E_{\text{a,m}} = 50\,\text{kJ}\,\text{mol}^{-1}$ we obtain $k_\text{r}(310\,\text{K}) = 1.95 \cdot k_\text{r}(300\,\text{K})$. This is in accordance with the chemist's rule of thumb that an increase of temperature by 10 K increases the reaction rate by a factor of two.

Example – Arrhenius plot

The bleaching of the triphenylmethane dye cation R^+ (reaction (21.68)) was measured at different temperatures (see Table 21.1).

In Figure 21.13(a), $k'_\text{r} = k_\text{r} \cdot c_{\text{OH}^-}$ is plotted versus T. In Figure 21.13(b) $\ln(k'_\text{r}/k'_{\text{r},1})$ is plotted versus $1/T$; from the slope $-7640\,\text{K}$ of the straight line we obtain $E_{\text{a,m}} = R \cdot 7640\,\text{K} = 63.5\,\text{kJ}\,\text{mol}^{-1}$. The frequency factor A_f is calculated from equation (21.70): $A_\text{f} = 1.8 \times 10^{10}\,\text{L}\,\text{mol}^{-1}\,\text{s}^{-1}$.

From equation (21.30) we can estimate A_f by using $r_\text{A} = r_{\text{OH}^-} = 250\,\text{pm}$, and $\mu \approx m_{\text{OH}^-} = 3 \times 10^{-26}\,\text{kg}$. Then $\bar{v} = 609\,\text{m}\,\text{s}^{-1}$ at 298 K and $A_\text{f} = 3 \times 10^{11}\,\text{L}\,\text{mol}^{-1}\,\text{s}^{-1} \cdot \alpha$. By comparison with the experiment we find $\alpha = 0.06$.

Table 21.1 Rate constants $k'_\text{r} = k_\text{r} \cdot c_{\text{OH}^-}$ for reaction (21.68) at different temperatures T; $c_{\text{OH}^-} = 0.2\,\text{mol}\,\text{L}^{-1}$

Experiment	T/K	$k'_\text{r}/\text{s}^{-1}$	$\ln(k'_\text{r}/k'_{\text{r},1})$	$1/(T/\text{K})$
1	293	0.0170	0.000	3.413×10^{-3}
2	296	0.0214	0.230	3.378×10^{-3}
3	299	0.0272	0.473	3.345×10^{-3}
4	301	0.0325	0.648	3.322×10^{-3}
5	303	0.0389	0.828	3.299×10^{-3}
6	307	0.0540	1.156	3.257×10^{-3}
7	311	0.0780	1.524	3.215×10^{-3}

Example – hydrogenation of ethene

For the reaction

$$\text{H}_2 + \text{H}_2\text{C}=\text{CH}_2 \longrightarrow \text{H}_3\text{C}-\text{CH}_3$$

the activation energy is $E_{\text{a,m}} = 180\,\text{kJ}\,\text{mol}^{-1}$. Then at $T = 1000\,\text{K}$

$$P = e^{-E_{\text{a,m}}/RT} = 4 \times 10^{-10}$$

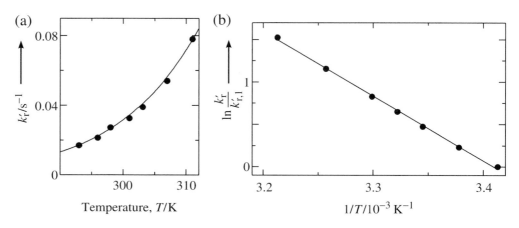

Figure 21.13 Reaction (21.68) measured at different temperatures. (a) k'_r versus temperature T; (b) $\ln(k'_r/k'_{r,1})$ versus $1/T$. For data see Table 21.1.

About 10^{10} collisions have to occur until two molecules react with one another. Using the data in Section 21.1.1 for the rate constant at 1000 K we obtain

$$k_r = 2.4 \, \text{L mol}^{-1} \, \text{s}^{-1} \cdot \alpha$$

The experimental value is $k_r = 0.24 \, \text{L mol}^{-1} \, \text{s}^{-1}$, thus the steric factor is $\alpha = 0.1$. The half life of the reaction at 1000 K for the reactants at a pressure of 1 bar (concentration $0.012 \, \text{mol L}^{-1}$) is $t_{1/2} = 35 \, \text{s}$.

Example – recombination of ethyl radicals

The reaction

$$^\bullet CH_2-CH_3 + {}^\bullet CH_2-CH_3 \longrightarrow CH_3CH_2CH_2CH_3$$

is diffusion controlled ($E_a = 0$). With $\mu = 24 \times 10^{-27}$ kg, $(r_A + r_B) = 500$ pm from equation (21.17) we obtain the rate constant for $T = 298$ K

$$k_r = 3 \times 10^{11} \, \text{L mol}^{-1} \, \text{s}^{-1} \cdot \alpha$$

The experimental value is $k_r = 1.6 \times 10^{11} \, \text{L mol}^{-1} \, \text{s}^{-1}$, corresponding to $\alpha = 0.5$. The half life of the reaction at 298 K for the reactants at a concentration of $1 \times 10^{-5} \, \text{mol L}^{-1}$ is $t_{1/2} = 0.6 \, \mu s$.

Example – harpoon reaction

The reaction

$$K + Br_2 \longrightarrow KBr + Br$$

is diffusion controlled ($E_a = 0$). The steric factor is $\alpha = 5$. How can we explain that α is larger than one? When a K atom approaches a Br_2 molecule and a critical distance R is

21.5 TREATMENT OF EXPERIMENTAL DATA

reached, an electron can be transferred from K to Br_2:

$$K + Br_2 \longrightarrow K^+ + Br_2^- \longrightarrow KBr + Br$$

Then the Coulomb attraction initiates a fast reaction of both ions, and the effective collision cross section is πR^2 instead of $\pi(r_A + r_B)^2$. This means that we can identify the steric factor as

$$\alpha = \frac{R^2}{(r_A + r_B)^2}$$

With $r_A + r_B = 400$ pm and $R = 800$ pm we estimate $\alpha = 4$ (K.T. Gillen, A.M. Rulis, R.B. Bernstein, *J. Chem. Phys.* 54, 2831 (1971)).

Example – electron transfer reaction

Anthracene molecules dissolved in acetonitrile are excited by light (the asterisk symbolizes the excited state). They react with diethylaniline molecules. With $(r_A + r_B) = 700$ pm and $(D_A + D_B) = 3.8 \times 10^{-5}$ cm^2 s^{-1} we obtain the rate constant from equation (21.37): $k_r = 2.02 \times 10^{10}$ L mol^{-1} s^{-1}. This value agrees exactly with the experimental value 2.0×10^{10} L mol^{-1} s^{-1} (D. Rehm, A. Weller, *Ber Bunsenges. Phys. Chem.* 73, 834 (1969)).

Example – neutralization reaction: $H^+ + OH^- \longrightarrow H_2O$

Because of the hydration spheres the radii r_A and r_B are on the order of 250 pm. The diffusion coefficients are $D_{H^+} = 9 \times 10^{-5}$ cm^2 s^{-1} and $D_{OH^-} = 5 \times 10^{-5}$ cm^2 s^{-1}. With these values we calculate $k_r = 5 \times 10^{10}$ L mol^{-1} s^{-1}. The experimental value is $k_r = 15 \times 10^{10}$ L mol^{-1} s^{-1} (H. Strehlow, W. Knoche, *Fundamentals of Chemical Relaxation*, Verlag Chemie, Weinheim 1977).

21.5.3 Activation Enthalpy and Activation Entropy

In the preceding examples the steric factor α was found in the range 0.1 to 1 (with the exception of the special case of the harpoon reaction). However, many reactions are extremely slow, and the resulting steric factor is 10^{-5} or even less, or it can be larger than one; then the collision theory becomes unreasonable. In both cases it is advisable to discuss the reaction in terms of the transition state theory.

Example – dimerization of cyclopentadiene

The rate constant is $k_r = 16.8 \times 10^{-6}$ L mol^{-1} s^{-1} at $T = 343$ K, and the activation energy is $E_{a,m} = 62$ kJ mol^{-1}. Then the frequency factor is

$$A_f = \frac{k_r}{e^{-E_{a,m}/RT}} = 4.7 \times 10^4 \text{ L mol}^{-1} \text{ s}^{-1}$$

From the transition state theory for gas-phase reactions we obtain the following from equation (21.51)

$$\Delta_r H^{\ddagger} \approx E_{a,m} = 62 \text{ kJ mol}^{-1}$$

and

$$\Delta_r S^{\ddagger} = R \cdot \ln \frac{A_f h c^{\ominus}}{kT} = -157 \text{ J K}^{-1} \text{ mol}^{-1}$$

$\Delta_r S^{\ddagger}$ is strongly negative, indicating that the number of configurations of the activated complex is much smaller than that for the reactants.

For comparison, in section 21.2.1 we found $A_f = 3.5 \times 10^{11}$ L mol^{-1} s^{-1} at 298 K for the reaction

$$D + H_2 \longrightarrow HD + H$$

This results in $\Delta_r S^{\ddagger} = -32$ J K^{-1} mol^{-1} (or $\alpha = 1$ in terms of the collision theory).

Example – hardboiling of eggs

Usually 10 minutes is required to hardboil an egg at 100 °C (boiling point of water at standard pressure). On the top of a mountain 2500 m above sea level the atmospheric pressure is 780 mbar at T = 300 K (see Box 11.4); then the boiling point of water is 364 K = 91 °C (in analogy to Problem 17.5) and it takes 12 hours to hardboil the egg. This dramatic change in rate with temperature is far beyond the chemist's rule of thumb. From these data we estimate the rate constants

$$k_r(373) = \frac{1}{10 \text{ min}} = 0.1 \text{ min}^{-1} \qquad k_r(364) = \frac{1}{12 \text{ hours}} = 1.4 \times 10^{-3} \text{ min}^{-1}$$

Then from equation (21.71) we obtain

$$E_{a,m} = -\frac{R}{1/(373 \text{ K}) - 1/(364 \text{ K})} \cdot \ln \frac{0.1}{1.4 \times 10^{-3}} = 535 \text{ kJ mol}^{-1}$$

It is surprising that the activation energy is about ten times the value for ordinary organic reactions. The frequency factor is calculated as

$$A_f = \frac{k_r}{e^{-E_{a,m}/RT}} = 1 \times 10^{72} \text{ s}^{-1}$$

21.5 TREATMENT OF EXPERIMENTAL DATA

From the transition state theory (see equation (21.55) for first-order reactions) we obtain $\Delta H^{\ddagger} = E_{a,m} = 535 \text{ kJ mol}^{-1}$ and

$$\Delta_r S^{\ddagger} = R \cdot \ln \frac{A_f h}{kT} = +1100 \text{ J K}^{-1} \text{ mol}^{-1}$$

In the boiling process native protein is denatured. Whereas the native protein is a highly organized structure (low entropy), the denatured protein has random structure (high entropy). Thus the frequency factor A_f is extremely large. On the other hand, the factor $\exp(-\Delta_r H^{\ddagger}/RT)$ is extremely small due to the large activation enthalpy. As a consequence, the reaction rate is in an easily measurable range.

21.5.4 Tunneling of Proton and Deuteron

Particles can tunnel through an energy barrier instead of surmounting the barrier. Usually this effect can be neglected, but in special cases it is the rate determining process. In chemical kinetics tunneling happens in electron and proton transfer reactions. Then these reactions are much faster than can be expected from transition state theory.

Example – tunneling of proton and deuteron

When excited by light, molecule 1 (with $R=CH_3$) transforms into molecule 2. Then molecule 2 rearranges by shifting a proton by a distance of $r = 200 \text{ pm}$ (formation of molecule 3). At high temperatures the rate constant k_r for the proton shift follows the Arrhenius equation ($E_{a,m} = 50 \text{ kJ mol}^{-1}$, thus $E_a = 8.3 \times 10^{-20} \text{ J}$); the result is the same if we replace the proton by a deuteron.

At low temperatures, however, k_r becomes constant. By extrapolation for $T \to 0$ it is found that k_r is 10^8 times larger for a shift of H than for a shift of D. This tremendous difference can be understood by considering tunneling of the proton and of the deuteron through the activation barrier. The rate constant k_r is proportional to the probability density ψ^2 of the proton and of the deuteron, respectively, at distance r

$$k_r = a\psi^2$$

We describe the proton as a particle (mass m, energy E) in a potential trough of depth V_0. According to Box 3.1 and Figure 3.2 the wavefunction outside the trough is

$$\psi \sim e^{-br} \quad \text{with} \quad b = \frac{2\pi\sqrt{2m}}{h} \cdot \sqrt{V_0 - E}$$

$V_0 - E$ corresponds to the activation barrier E_a, and we obtain

$$\psi^2 \sim \exp(-C\sqrt{m}), \quad C = \frac{4\pi}{h}\sqrt{2E_a} \cdot r = 1.55 \times 10^{15} \text{ kg}^{-1/2} \quad (\text{for } r = 200 \text{ pm})$$

Then the ratio of the rate constants k_H/k_D is

$$\frac{k_H}{k_D} = \frac{\exp(-C\sqrt{m_H})}{\exp(-C\sqrt{m_D})} = \exp(-C \cdot [\sqrt{m_H} - \sqrt{m_D}])$$

$$= \exp(-C\sqrt{m_H} \cdot [1 - \sqrt{2}])$$

With $C \cdot \sqrt{m_H} = 1.55 \times 10^{15} \cdot 4.09 \times 10^{-14} = 63.4$, we obtain

$$\frac{k_H}{k_D} = \exp(-63.4 \times [1 - \sqrt{2}]) = 2 \times 10^{11}$$

This estimate shows that tunneling of the H nucleus is by many orders of magnitude more probable than tunneling of the heavier D nucleus. This is in agreement with the experimental ratio $k_H/k_D = 10^8$ (K.H. Grellmann, U. Schmitt, H. Weller, Tunnel Effect on the Kinetics of Sigmatropic Proton Shift. *Chem. Phys. Lett.* **88**, 40 (1982); G. Bartelt, A. Eychmueller, K.H. Grellmann, *Chem. Phys. Lett.* **118**, 568 (1985); U. Kensy, M.M. Gonzalez, K.H. Grellmann, *Chem. Phys.* **170**, 381 (1993)).

Example – tunneling in triplet state

The enol compound E was investigated in a hydrocarbon glass at low temperature

When exciting by light, E goes over into the excited singlet state of the keto form K, then into the triplet state of K, into the triplet state of E and finally back into the ground state of E:

$$E \longrightarrow {}^1K^* \longrightarrow {}^3K^* \longrightarrow {}^3E^* \longrightarrow E$$

The rate of reaction ${}^3K^* \longrightarrow {}^3E^*$ (migration of proton from nitrogen to oxygen), below 25 K, is independent of temperature. When replacing the proton by a deuteron the rate of this reaction is 2000 times slower. We estimate the ratio k_H/k_D by using the values given in Section 10.4 for the hydrogen bond in water: $E_{a,m} = 0.5 \times 10^{-19}$ J $\times N_A = 30$ kJ mol^{-1}, $r = (180-97)$ pm $\cong 80$ pm. Following the procedure in the preceding example, we obtain

$$C\sqrt{m_H} = 4.8 \times 10^{14} \times 4.09 \times 10^{-14} = 19.6$$

and

$$\frac{k_H}{k_D} = \exp(-19.6 \times [1-\sqrt{2}]) \cong 3000$$

in good agreement with the experiment. (H. Eisenberger, B. Nickel, A.A. Ruth, W. Al-Soufi, K.H. Grellmann, *J. Phys. Chem.* 95, 10510 (1991); W. Al-Soufi, K.H. Grellman, B. Nickel, *J. Phys. Chem.* 95, 10503 (1991)).

21.6 Complex Reactions

In the cases considered so far, chemical reactions have been described by simple first- and second-order rate equations. However, in most processes the situation is much more complicated. For example, back reactions have to be considered (establishment of equilibrium); additionally, consecutive or side reactions can occur. In special cases the reaction path can be strongly affected by products occurring during the reaction (autocatalysis, oscillations).

21.6.1 Reactions Leading to Equilibrium

In the discussion of chemical equilibria we have seen that, in general, reactions lead to an equilibrium between reactants and products. In terms of reaction kinetics this means that the products react with each other, forming reactants:

$$A + B \xrightarrow{k_1} C + D \tag{21.72}$$

$$C + D \xrightarrow{k_{-1}} A + B \tag{21.73}$$

Then the rate of formation of molecules A is

$$\frac{dc_A}{dt} = -k_1 \cdot c_A c_B + k_{-1} \cdot c_C c_D \tag{21.74}$$

Let us assume, for simplicity, that at time $t=0$ the concentrations of molecules A and B are equal ($c_{A,0} = c_{B,0} = a$) and that no molecules C and D are present. Then at time t: $c_C = c_D = a - c_A$ and thus

$$\frac{dc_A}{dt} = -k_1 \cdot c_A^2 + k_{-1} \cdot (a - c_A)^2 \tag{21.75}$$

In the special case $k_1 = k_{-1} = k_r$ this differential equation can be easily integrated (Problem 21.2), and we obtain c_A as a function of time:

$$\boxed{c_A = c_B = \tfrac{1}{2} a \cdot (1 + e^{-2k_r a t})} \tag{21.76}$$

Correspondingly

$$\boxed{c_C = c_D = \tfrac{1}{2} a \cdot (1 - e^{-2k_r a t})} \tag{21.77}$$

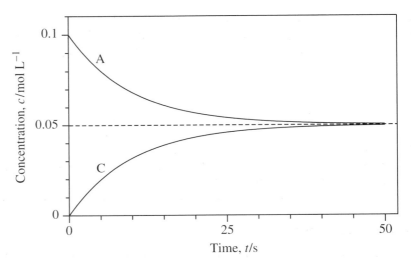

Figure 21.14 Chemical reaction leading to equilibrium (reactions (21.72) and (21.73)): $A + B \rightleftharpoons C + D$. Rate constants $k_1 = k_{-1} = 0.5 \, \text{L mol}^{-1} \, \text{s}^{-1}$; initial concentrations $a = c_{A,0} = b = c_{B,0} = 0.1 \, \text{mol L}^{-1}$, $c_{C,0} = c_{D,0} = 0$.

The concentrations c_A and c_C are displayed in Figure 21.14 as a function of time. After infinite time the equilibrium concentrations

$$c_{A,\infty} = c_{C,\infty} = \tfrac{1}{2}a \tag{21.78}$$

have been established.

In the general case (arbitrary k_1 and k_{-1}, different initial concentrations of $c_{A,0}$ and $c_{B,0}$) the integration of the differential equation (21.74) is no longer so easy. Therefore, we restrict consideration to the calculation of the equilibrium concentrations. After an infinite time the concentrations of the reactants and of the products do not change, thus $dc_A/dt = 0$ and

$$-k_1 \cdot c_A c_B + k_{-1} \cdot c_C c_D = 0 \tag{21.79}$$

leading to

$$\frac{c_C \cdot c_D}{c_A \cdot c_B} = \frac{k_1}{k_{-1}} \tag{21.80}$$

Because the expression on the left-hand side represents the equilibrium constant K

$$K = \frac{\widehat{c}_C \cdot \widehat{c}_D}{\widehat{c}_A \cdot \widehat{c}_B} = \frac{c_C \cdot c_D}{c_A \cdot c_B} \tag{21.81}$$

we can relate K to the ratio of the rate constants k_1 and k_{-1}:

$$\boxed{K = \frac{k_1}{k_{-1}}} \tag{21.82}$$

21.6 COMPLEX REACTIONS

Example – iodine hydrogen equilibrium

For the equilibrium

$$H_2 + I_2 \longrightarrow 2HI$$

the equilibrium constants determined from the rate constants k_1 and k_{-1} of the forward and reverse reaction are in good agreement with the constants obtained from equilibrium measurements (Table 21.2).

Table 21.2 Rate constants k_1, k_{-1} and equilibrium constant K (from measurement of equilibrium) for the $H_2 + I_2 \rightleftharpoons 2HI$ Reaction

T/K	$k_1/\text{L mol}^{-1}\text{s}^{-1}$	$k_{-1}/\text{L mol}^{-1}\text{s}^{-1}$	k_1/k_{-1}	K
667	1.56×10^{-2}	2.59×10^{-4}	60.4	61.0
699	6.70×10^{-2}	12.4×10^{-4}	54.0	55.2

How to interpret kinetic data

By this example we learn how to interpret kinetic data. A reaction model involving a bimolecular reaction leads to a second-order rate law. However, vice versa, when a reaction follows a second-order rate law, we cannot conclude that the basic process is a simple bimolecular process.

As an example, because of the observed second-order kinetics, it was believed for a long time that the iodine hydrogen reaction is based on direct collisions between H_2 and I_2 molecules. Meanwhile it was found that this reaction is initiated by a thermal dissociation of iodine:

$$I_2 \rightleftharpoons 2I$$

which can be described by the dissociation constant

$$K' = \frac{\widehat{c}_I^2}{\widehat{c}_{I_2}}$$

Then I atoms react with H_2

$$H_2 + 2I \xrightarrow{k_3} 2HI$$

and we obtain

$$\frac{dc_{HI}}{dt} = k_3 \cdot c_{H_2} \cdot c_I^2 = k_3 \cdot c_{H_2} \cdot K' \cdot \widehat{c}_{I_2} = k_3 K' \frac{1}{c^\ominus} \cdot c_{H_2} \cdot c_{I_2}$$

In spite of the complicated kinetics, the rate equation for the forward reaction is formally second-order with $k_1 = k_3 K'/c^\ominus$.

21.6.2 Parallel Reactions

Same reaction order in both reactions

We consider a main reaction between molecule A and molecule B to yield C:

$$A + B \xrightarrow{k_1} C \tag{21.83}$$

At the same time a parallel reaction occurs, forming a side product D:

$$A + B \xrightarrow{k_2} D \tag{21.84}$$

Then

$$\frac{dc_A}{dt} = -k_1 \cdot c_A c_B - k_2 \cdot c_A c_B = -(k_1 + k_2) \cdot c_A c_B \tag{21.85}$$

In this case, A decreases according to a second-order reaction, but the effective rate constant is $k_1 + k_2$.

Different reaction orders in both reactions

The situation is more complicated when the side reaction follows a different reaction order than the main reaction:

$$A + B \longrightarrow C \text{ (main reaction)} \tag{21.86}$$

$$2A \longrightarrow D \text{ (side reaction)}$$

Let us assume, for simplicity, that B is in large excess over A; then the main reaction is first-order, and the side reaction is second-order.

Example – yield of main reaction

We consider the formation of 2-formyl-methyl-norbornene (C) from cyclopentadiene (A) and crotonaldehyde (B)

with the dimerization of cyclopentadiene as a side reaction

21.6 COMPLEX REACTIONS

As calculated in Problem 21.3 we find for the concentration of C at the end of the reaction ($t \to \infty$)

$$\frac{c_{C,\infty}}{c_{A,0}} = \frac{\ln(1+x)}{x} \quad \text{with} \quad x = \frac{2c_{A,0}k_2}{k_1 c_{B,0}}$$

What must be the excess of B over A to obtain a yield of 80% when $k_1 = 6.6 \times 10^{-6}\,\text{L mol}^{-1}\,\text{s}^{-1}$ and $k_2 = 16.8 \times 10^{-6}\,\text{L mol}^{-1}\,\text{s}^{-1}$ at 70 °C? For $c_{C,\infty}/c_{A,0} = 0.8$ we obtain $x = 0.54$. Then

$$\frac{c_{B,0}}{c_{A,0}} = \frac{2k_2}{k_1} \cdot \frac{1}{x} = \frac{33.6}{6.6} \cdot \frac{1}{0.54} = 9.4$$

This means that to obtain 80% of the main product, reactant B must be in an about tenfold excess over A.

21.6.3 Consecutive Reactions

General case

We imagine molecules A reacting with B to form an intermediate C; in a consecutive reaction, the product D is formed from C and B:

$$A + B \xrightarrow{k_1} C, \quad C + B \xrightarrow{k_2} D \tag{21.87}$$

We restrict our consideration to the case where B is in large excess. Then the rate equations for this sequence of reactions are with $k'_1 = k_1 c_{B,0}$ and $k'_2 = k_2 c_{B,0}$:

$$\frac{dc_A}{dt} = -k'_1 c_A \quad \text{(disappearance of A)} \tag{21.88}$$

$$\frac{dc_C}{dt} = k'_1 c_A - k'_2 c_C \quad \text{(formation of C)} \tag{21.89}$$

$$\frac{dc_D}{dt} = k'_2 c_C \quad \text{(formation of D)} \tag{21.90}$$

Equation (21.88) can be easily integrated as (a = initial concentration of A)

$$\boxed{c_A = a \cdot e^{-k'_1 t}} \tag{21.91}$$

Then

$$\frac{dc_C}{dt} = k'_1 a \cdot e^{-k'_1 t} - k'_2 c_C \tag{21.92}$$

This differential equation is solved in Problem 21.4:

$$\boxed{c_C = a \frac{k'_1}{k'_2 - k'_1} \left(e^{-k'_1 t} - e^{-k'_2 t} \right)} \tag{21.93}$$

and for the concentration of D we obtain

$$\boxed{c_D = a \frac{1}{k'_2 - k'_1} \left[k'_1 \left(e^{-k'_2 t} - 1 \right) - k'_2 \left(e^{-k'_1 t} - 1 \right) \right]} \tag{21.94}$$

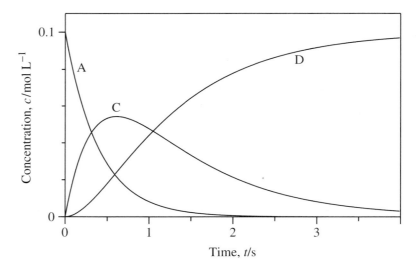

Figure 21.15 Consecutive reaction (21.87) with B in large excess: $A + B \rightarrow C + D$, $C + B \rightarrow D$. Concentrations c_A, c_C, and c_D versus time t. Initial concentrations $a = c_{A,0} = 0.1 \text{ mol L}^{-1}$, $c_{C,0} = c_{D,0} = 0$. Rate constants $k'_1 = 2.5 \text{ s}^{-1}$, $k'_2 = 1 \text{ s}^{-1}$.

This solution is illustrated in Figure 21.15. As expected, the concentration of the intermediate C increases initially (its formation is faster than its decomposition); it reaches a maximum and finally it decreases. The maximum is reached (see Problem 21.4) at

$$t_{\max} = \frac{\ln \frac{k'_2}{k'_1}}{k'_2 - k'_1} \tag{21.95}$$

Example – accumulation of intermediate

The synthesis of the laundry detergent paraffinsulfochloride (B) from paraffin (A), sulfur dioxide, and chlorine is a technically important process:

$$\underset{A}{\overset{R^1}{\underset{R^2}{|}}{HCH}} + SO_2 + Cl_2 \xrightarrow{k_1} \underset{B}{\overset{R^1}{\underset{R^2}{|}}{HC-SO_2-Cl}} + HCl$$

However, B can further react in a consecutive reaction:

$$\underset{B}{\overset{R^1}{\underset{R^2}{|}}{HC-SO_2-Cl}} + SO_2 + Cl_2 \xrightarrow{k_2} \underset{C}{\overset{R^1}{\underset{R^2}{|}}{Cl-SO_2-C-SO_2-Cl}} + HCl$$

21.6 COMPLEX REACTIONS

In this case the rate constants k_1 and k_2 are equal, thus $k_1' = k_2'$. Then from equation (21.93) we obtain with $\varepsilon = k_2' - k_1'$, applying l'Hospital's rule,

$$\lim_{\varepsilon \to 0} c_B = \lim_{\varepsilon \to 0} ak_1' \left[\frac{e^{-k_1' t}(1 - e^{-\varepsilon t})}{\varepsilon} \right] = \lim_{\varepsilon \to 0} ak_1' \cdot e^{-k_1' t} \cdot t \cdot e^{-\varepsilon t} = ak_t' \cdot e^{-k_1' t} \cdot t$$

The maximum of c_B is

$$c_{B,\max} = ae^{-1} = \frac{a}{2.718} = 0.37a$$

This means that in this process the yield of the main product B cannot exceed 37%.

Approximation by rate-determining step

In many consecutive reactions one reaction step is so slow that it can be considered as rate determining. Then the treatment can be simplified.

Example – reaction of bromide with bromate

In the Belousov–Zhabotinsky reaction (see Section 21.6.10) the inhibitor bromide is removed by the reaction

$$5Br^- + BrO_3^- + 6H^+ \longrightarrow 3Br_2 + 3H_2O \tag{21.96}$$

where BrO_3^- is in large excess over bromide.

The rate equation for this reaction is

$$\frac{dc_{Br^-}}{dt} = -k_r \cdot c_{BrO_3^-} \cdot c_{Br^-} \cdot c_{H^+}^2 \tag{21.97}$$

This cannot mean that reaction (21.96) is an elementary reaction, because it is practically impossible for five bromide ions to collide simultaneously with bromate to form the reaction products. Actually, the reaction proceeds in three consecutive steps:

$$Br^- + BrO_3^- + 2H^+ \xrightarrow{k_1} HOBr + HBrO_2 \tag{21.98}$$

$$Br^- + HBrO_2 + H^+ \xrightarrow{k_2} 2HOBr \tag{21.99}$$

$$Br^- + HOBr + H^+ \xrightarrow{k_3} Br_2 + H_2O \tag{21.100}$$

Reaction (21.98) is rate determining ($k_1 \ll k_2, k_3$). When one bromide reacts with bromate, then four additional bromide ions react with the products HOBr and $HBrO_2$; then

$$\frac{dc_{Br^-}}{dt} = -5k_1 \cdot c_{BrO_3^-} \cdot c_{Br^-} \cdot c_{H^+}^2 \tag{21.101}$$

and we can identify k_r with $5k_1$.

Steady-state approximation

In many cases one intermediate in a consecutive reaction appears only in small concentration. Consequently, the change of this concentration is also small. Approximately, we can treat this change as zero:

$$\frac{dc_{\text{intermediate}}}{dt} = 0$$

In this way the treatment of the reaction can be simplified. This approximation is called the *steady-state approximation*.

Example – bromination of malonic acid

In the Belousov–Zhybotinsky reaction (see Section 21.6.10) malonic acid $HOOC-CH_2-COOH$ (MA) is brominated by dissolved bromine. This reaction proceeds via the enol form (MAE) of the acid, $HOOC-CH=C(OH)_2$, which is in equilibrium with the acid in its keto form

$$MA \underset{k_{-1}}{\overset{k_1}{\rightleftharpoons}} MAE \qquad (21.102)$$

$$HOOC-CH=C(OH)_2 + Br_2 \xrightarrow{k_2} HOOC-CHBr-COOH + HBr \qquad (21.103)$$

Then

$$\frac{dc_{Br_2}}{dt} = -k_2 \cdot c_{MAE} \cdot c_{Br_2}$$

$$\frac{dc_{MAE}}{dt} = k_1 c_{MA} - k_{-1} \cdot c_{MAE} - k_2 \cdot c_{MAE} \cdot c_{Br_2}$$

The enol form MAE is formed as an intermediate in a small concentration. Applying the steady-state approximation

$$\frac{dc_{MAE}}{dt} = 0$$

and assuming MA in great excess over bromine ($c_{MA} = c_{MA,0}$) we obtain for the enol concentration

$$c_{MAE} = \frac{k_1 c_{MA,0}}{k_{-1} + k_2 c_{Br_2}}$$

Then the rate equation for bromine can be written as

$$\frac{dc_{Br_2}}{dt} = -\frac{k_1 k_2 \cdot c_{MA,0} \cdot c_{Br_2}}{k_{-1} + k_2 \cdot c_{Br_2}}$$

We discuss two limiting cases.

Case 1. For high enough Br_2 concentration we can neglect k_{-1} in the denominator. Then

$$\frac{dc_{Br_2}}{dt} = -k_1 \cdot c_{MA,0} \quad \text{and} \quad c_{Br_2} = c_{Br_2,0} - k_1 \cdot c_{MA,0} \cdot t \qquad (21.104)$$

This means that Br_2 disappears linearly with time (zero-order kinetics, Figure 21.16(a)).

21.6 COMPLEX REACTIONS

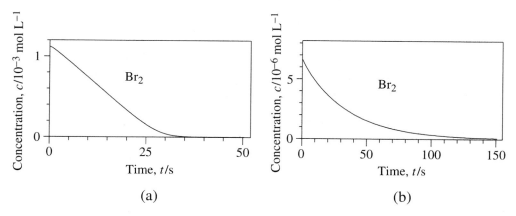

Figure 21.16 Bromination of malonic acid via enol form at 20 °C in acidic medium (c_{H^+} = 1.29 mol L^{-1}): reactions (21.103) and (21.102). Bromine is injected into a malonic acid solution and the concentration c of bromine is measured. (a) High bromine concentration ($c_{\text{malonic acid},0}$ = 0.016 mol L^{-1}, $c_{\text{Br}_2,0}$ = 1.1 × 10^{-3} mol L^{-1}): linear decay of Br$_2$. (b) Low bromine concentration ($c_{\text{malonic acid},0}$ = 0.001 mol L^{-1}, $c_{\text{Br}_2,0}$ = 6.5 × 10^{-6} mol L^{-1}): exponential decay of Br$_2$ (A. Sirimungkala, H. D. Försterling, V. Dlask, R. J. Field, Bromination Reactions Important in the Mechanism of the Belousov-Zhabotinsky System, J. Phys. Chem. A, 103, 1038 (1999)).

Case 2. For low enough Br$_2$ concentration, $k_2 c_{\text{Br}_2}$ can be neglected in the denominator. Then

$$\frac{dc_{\text{Br}_2}}{dt} = -\frac{k_1 k_2}{k_{-1}} \cdot c_{\text{MA},0} \cdot c_{\text{Br}_2} \quad \text{and} \quad c_{\text{Br}_2} = c_{\text{Br}_2,0} \cdot e^{-k'_r t} \quad (21.105)$$

with $k'_r = k_1 k_2 c_{\text{MA},0} / k_{-1}$. This means that Br$_2$ disappears exponentially with time (Figure 21.16(b)).

21.6.4 Chain Reactions

In the iodine hydrogen reaction, iodine radicals are important as intermediates (Section 21.6.1). In the bromine hydrogen reaction

$$\text{Br}_2 + \text{H}_2 \longrightarrow 2\text{HBr} \quad (21.106)$$

hydrogen radicals are formed besides bromine radicals because of the higher reactivity of the bromine molecules. This proceeds in the following steps:

- (a) Production of radicals

$$\text{Br}_2 \xrightarrow{k_1} 2\text{Br} \quad (21.107)$$

- (b) Conservation of radicals

$$\text{Br} + \text{H}_2 \underset{k_{-2}}{\overset{k_2}{\rightleftharpoons}} \text{HBr} + \text{H} \quad (21.108)$$

$$\text{H} + \text{Br}_2 \xrightarrow{k_3} \text{HBr} + \text{Br} \quad (21.109)$$

- (c) Removal of radicals

$$2\text{Br} \xrightarrow{k_4} \text{Br}_2 \tag{21.110}$$

Such a mechanism is called a *chain reaction*. The H and Br radicals are called *chain carriers*. Br is first formed in the chain-initiating step (a). Br and H are conserved in the *chain-propagation* steps (b), until they are removed by a *chain-termination* or *chain-breaking* step (c).

From these reactions we obtain the following rate equation for the overall process (see Box 21.2)

$$\frac{dc_{\text{HBr}}}{dt} = \frac{A}{c_{\text{Br}_2} + B \cdot c_{\text{HBr}}} \cdot (c_{\text{Br}_2})^{3/2} \cdot c_{\text{H}_2} \tag{21.111}$$

with

$$A = 2k_2 \sqrt{\frac{k_1}{k_4}} \quad \text{and} \quad B = \frac{k_{-2}}{k_3}$$

This result is in agreement with the empirical rate law.

Box 21.2 – Rate Law for Chain Reaction

To solve the rate equations

$$\frac{dc_{\text{HBr}}}{dt} = k_2 \cdot c_{\text{Br}} \cdot c_{\text{H}_2} - k_{-2} \cdot c_{\text{HBr}} \cdot c_{\text{H}} + k_3 \cdot c_{\text{H}} \cdot c_{\text{Br}_2}$$

$$\frac{dc_{\text{Br}}}{dt} = 2k_1 \cdot c_{\text{Br}_2} - k_2 \cdot c_{\text{Br}} \cdot c_{\text{H}_2} + k_{-2} \cdot c_{\text{HBr}} \cdot c_{\text{H}} + k_3 \cdot c_{\text{H}} \cdot c_{\text{Br}_2} - 2k_4 \cdot c_{\text{Br}}^2$$

$$\frac{dc_{\text{H}}}{dt} = k_2 \cdot c_{\text{Br}} \cdot c_{\text{H}_2} - k_{-2} \cdot c_{\text{HBr}} \cdot c_{\text{H}} - k_3 \cdot c_{\text{H}} \cdot c_{\text{Br}_2}$$

we apply the steady-state approximation for the Br and H radicals:

$$\frac{dc_{\text{Br}}}{dt} = \frac{dc_{\text{H}}}{dt} = 0$$

Then

$$2k_1 c_{\text{Br}_2} - k_2 c_{\text{Br}} c_{\text{H}_2} + k_{-2} c_{\text{HBr}} c_{\text{H}} + k_3 c_{\text{H}} c_{\text{Br}_2} - 2k_4 c_{\text{Br}}^2 = 0$$

$$k_2 c_{\text{Br}} c_{\text{H}_2} - k_{-2} c_{\text{HBr}} c_{\text{H}} - k_3 c_{\text{H}} c_{\text{Br}_2} = 0$$

By adding these equations we obtain $2k_1 c_{\text{Br}_2} - 2k_4 c_{\text{Br}}^2 = 0$ and

$$c_{\text{Br}} = \sqrt{\frac{k_1}{k_4}} \cdot \sqrt{c_{\text{Br}_2}}$$

Continued on page 699

> *Continued from page 698*
>
> We insert this expression in the third equation and solve for c_H
>
> $$c_H = \frac{k_2 c_{H_2}}{k_{-2} c_{HBr} + k_3 c_{Br_2}} \cdot \sqrt{\frac{k_1}{k_4}} \cdot \sqrt{c_{Br_2}}$$
>
> Furthermore, from the second equation we obtain $k_2 c_{Br} c_{H_2} - k_{-2} c_{HBr} c_H = k_3 c_H c_{Br_2}$. By inserting this relation in the first equation we obtain
>
> $$\frac{dc_{HBr}}{dt} = 2 k_3 c_H c_{Br_2}$$
>
> $$= \frac{2 k_2 \sqrt{k_1/k_4}}{c_{Br_2} + (k_{-2}/k_3) c_{HBr}} \cdot \sqrt{c_{Br_2}^3} \cdot c_{H_2}$$

21.6.5 Branching Chain Reactions

In the chain-propagating steps more radicals can be formed than consumed. Then the reaction is called a *branched chain reaction*. An example is the detonating gas reaction

$$2H_2 + O_2 \longrightarrow 2H_2O \tag{21.112}$$

with the following steps:

$$\text{Initiation: } H_2 + O_2 \longrightarrow 2OH \tag{21.113}$$
$$\text{Propagation: } OH + H_2 \longrightarrow H_2O + H \tag{21.114}$$
$$\text{Branching: } H + O_2 \longrightarrow OH + O \tag{21.115}$$
$$\text{Branching: } O + H_2 \longrightarrow OH + H \tag{21.116}$$
$$\text{Termination: } H \longrightarrow \text{wall} \tag{21.117}$$
$$\text{Termination: } H + O_2 + M \longrightarrow HO_2 + M^* \tag{21.118}$$

where M is a third body acting as an inert collision partner. Then the overall reaction in the propagation and branching steps is

$$2H_2 + O_2 + OH \longrightarrow H_2O + H + 2OH \tag{21.119}$$

This means that the number of H and OH radicals increases exponentially with time. The result is an explosion. At low pressure the reaction is terminated by collisions with the wall: reaction (21.117), *lower explosion limit*. At high pressure the reaction is terminated by formation of unreactive hydroperoxide radicals HO_2, which diffuse to the wall: reaction (21.118), *upper explosion limit*. At very high pressures the gas mixture becomes again explosive: HO_2 radicals react with H_2 and form OH radicals and H_2O before reaching the wall.

21.6.6 Enzyme Reactions (Michaelis–Menten Mechanism)

Enzymes act as very effective catalysts, reducing the activation energy of a reaction significantly. Thus the reaction rate can be increased by many orders of magnitude. As an example we consider the hydrolysis of urea by using the enzyme urease

$$O=C(NH_2)_2 + 2H_2O \xrightarrow{\text{enzyme urease}} (NH_4)_2CO_3 \tag{21.120}$$

In a first step a complex UE between urea (U) and the enzyme (E) is formed:

$$U + E \underset{k_{-1}}{\overset{k_1}{\rightleftharpoons}} UE \tag{21.121}$$

In a second step the complex hydrolyzes into the product (P) and the enzyme E:

$$UE \xrightarrow{k_2} P + E \tag{21.122}$$

Then the rate equations for U, E, UE, and P are

$$\frac{dc_U}{dt} = -k_1 \cdot c_U \cdot c_E + k_{-1} \cdot c_{UE} \tag{21.123}$$

$$\frac{dc_{UE}}{dt} = k_1 \cdot c_U \cdot c_E - k_{-1} \cdot c_{UE} - k_2 \cdot c_{UE} \tag{21.124}$$

$$\frac{dc_E}{dt} = -\frac{dc_{UE}}{dt} \tag{21.125}$$

$$\frac{dc_P}{dt} = k_2 c_{UE} \tag{21.126}$$

These rate equations cannot be integrated rigorously. Therefore we apply the steady-state approximation for the complex UE (UE is present in a very low concentration):

$$\frac{dc_{UE}}{dt} = k_1 \cdot c_U \cdot c_E - k_{-1} \cdot c_{UE} - k_2 \cdot c_{UE} = 0 \tag{21.127}$$

Then with

$$c_E + c_{UE} = c_{E,0} \tag{21.128}$$

where $c_{E,0}$ is the initial concentration of the enzyme, we obtain

$$c_{UE} = \frac{k_1 \cdot c_U \cdot c_{E,0}}{k_1 c_U + k_{-1} + k_2} \tag{21.129}$$

and

$$\boxed{\frac{dc_P}{dt} = k_2 \cdot c_{UE} = \frac{k_2 \cdot c_U}{c_U + K_M} \cdot c_{E,0}} \tag{21.130}$$

with

$$K_M = \frac{k_{-1} + k_2}{k_1}$$

21.6 COMPLEX REACTIONS

This rate equation is called the *Michaelis–Menten equation*. The initial reaction rate

$$v_0 = \left(\frac{dc_P}{dt}\right)_{t=0}$$

increases linearly with the concentration of the enzyme. At low initial concentrations of urea we can neglect c_U in the denominator of equation (21.130):

$$v_0 = \frac{k_2}{K_M} \cdot c_{U,0} \cdot c_{E,0} \quad (c_U \ll K_M) \qquad (21.131)$$

Then v_0 increases linearly with the initial urea concentration.

At high initial concentrations of urea we can neglect K_M in the denominator of equation (21.130):

$$v_0 = k_2 \cdot c_{E,0} \quad (c_U \gg K_M) \qquad (21.132)$$

Then v_0 is independent of c_U; this is because the equilibrium (21.121) is completely shifted to the right-hand side (the maximum amount of UE has been formed). Thus by plotting v_0 versus $c_{U,0}$ (Figure 21.17) we can evaluate k_2/K_M from the initial slope and k_2 from the final value. Alternatively, we can rearrange equation (21.130):

$$\frac{1}{v_0} = \frac{c_{U,0} + K_M}{k_2 \cdot c_{U,0} \cdot c_{E,0}} = \frac{1}{k_2 \cdot c_{E,0}} + \frac{K_M}{k_2 \cdot c_{E,0}} \cdot \frac{1}{c_{U,0}} \qquad (21.133)$$

By plotting $1/v_0$ versus $1/c_{U,0}$ (Figure 21.18) we obtain K_M/k_2 from the slope and k_2 from the intercept with the ordinate (*Lineweaver–Burk plot*).

The close fit between enzyme and substrate is crucial for the specificity of enzyme reactions. Inhibition changing the enzymatic activity play a fundamental role in triggering

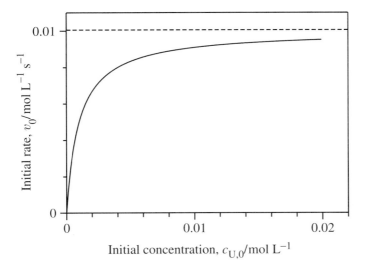

Figure 21.17 Enzyme reaction: hydrolysis of urea by enzyme urease. Initial rate v_0 versus initial concentration $c_{U,0}$ of urea ($c_{E,0} = 1 \times 10^{-5}$ mol L^{-1}, $K_M = 1 \times 10^{-3}$ mol L^{-1}, $k_2 = 1000$ s^{-1}).

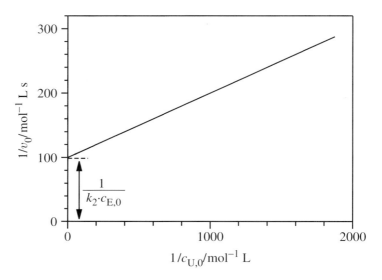

Figure 21.18 Enzyme reaction: Lineweaver–Burk plot. Data as in Figure 21.17. Ordinate intercept $= 1/(k_2 \cdot c_{E,0})$, slope $= K_M/(k_2 \cdot c_{E,0})$.

the interplay of enzymes determining their functional network. Several kinds of inhibitors are known. We consider *competitive inhibitors* which are most important. A molecule that resembles the substrate can bind reversibly to the active site of the enzyme and therefore compete with the substrate. The inhibition can be reversed by diluting the inhibitor or by increasing the concentration c_S of the substrate. The initial rate v_0 of the reaction is diminished by the inhibitor, but its influence is less at high concentration $c_{S,0}$ when the active site is readily occupied by the substrate. In addition to the Michaelis Menten scheme

$$S + E \underset{k_{-1}}{\overset{k_1}{\rightleftharpoons}} SE \overset{k_2}{\to} P + E$$

we have to introduce the relations

$$I + E \underset{k'_{-1}}{\overset{k'_1}{\rightleftharpoons}} IE \quad \text{and} \quad c_E + c_{SE} + c_{IE} = c_{E,0}$$

where I is the inhibitor and IE is the inhibitor-substrate complex. We assume equilibrium between I, E, and IE:

$$K_I = \frac{k'_{-1}}{k'_1} = \frac{c_I \cdot c_E}{c_{IE}}$$

Then instead of equations (21.129) and (21.130) we obtain

$$c_{SE} = \frac{k_1 \cdot c_S \cdot c_{E,0}}{k_1 c_S + (k_{-1} + k_2)(1 + c_I/K_I)}$$

$$\frac{dc_P}{dt} = \frac{k_2 \cdot c_S}{c_S + K'_M} \cdot c_{E,0} \quad \text{with} \quad K'_M = K_M \cdot \left(1 + \frac{c_I}{K_I}\right)$$

21.6 COMPLEX REACTIONS

Example

The enzyme succinic acid dehydrogenase catalyses the oxidation of succinic acid forming fumaric acid:

$$\text{HOOC-CH}_2\text{-CH}_2\text{-COOH} \longrightarrow \text{HOOC-CH=CH-COOH}$$

Malonic acid

$$\text{HOOC-CH}_2\text{-COOH}$$

acts as a competitive inhibitor since its structure resembles that of succinic acid. Thus it can bind to the active site of the enzyme, but it cannot act as an alternative substrate. In this case the equilibrium constant is $K_I = 10^{-5}\,\text{mol}\,\text{L}^{-1}$, i.e. at the concentration $c_I = 10^{-5}\,\text{mol}\,\text{L}^{-1}$ and small concentration c_S, the initial rate v_0 is reduced to one-half of the value without inhibitor.

In many cases a reaction product acts as a competitive inhibitor. This is an important form of regulation: the enzymatic reaction is turned off when sufficient product for further requirements is present.

21.6.7 Autocatalytic Reactions

In many processes (e.g., oscillating chemical reactions) autocatalysis plays an important role. In an autocatalytic reaction a reactant appears as a product of the reaction:

$$\text{A} + \text{B} \xrightarrow{k_r} \text{C} + 2\text{B} \tag{21.134}$$

The reaction can be started with a very low concentration of B. Then the reaction is very slow in the beginning. With increasing concentration of B the reaction becomes faster and faster. Because of the consumption of reactant A, however, the reaction rate decreases after passing a maximum. The rate equation for B is

$$\frac{dc_B}{dt} = k_r \cdot c_A \cdot c_B \tag{21.135}$$

When one molecule of B has been formed, one molecule of A has disappeared. Thus, the concentration c_A can be written as

$$c_A = a - (c_B - b) \tag{21.136}$$

where a and b are the initial concentrations of A and B. Thus

$$\frac{dc_B}{dt} = k_r \cdot c_B \cdot (a + b - c_B) \tag{21.137}$$

The solution of this differential equation is (Problem 21.5):

$$c_B = \frac{a+b}{1 + (a/b) \cdot e^{-(a+b)k_r t}} \tag{21.138}$$

Correspondingly,

$$c_A = (a+b) - c_B = \frac{a+b}{1 + (b/a) \cdot e^{+(a+b)k_r t}} \quad (21.139)$$

The concentrations c_A and c_B are displayed in Figure 21.19 (S-shaped curves are typical for autocatalytic behavior).

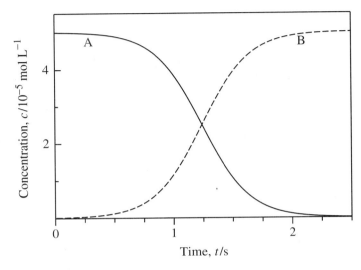

Figure 21.19 Autocatalytic reaction (21.134). Concentrations c_A and c_B calculated from equations (21.138) and (21.139). $k_r = 1 \times 10^5 \, \text{L mol}^{-1} \, \text{s}^{-1}$. Initial conditions: $a = 5 \times 10^{-5} \, \text{mol L}^{-1}$, $b = 1 \times 10^{-7} \, \text{mol L}^{-1}$.

Example – formation of the enzyme trypsin

The reaction of trypsinogen to trypsin is catalyzed by the product trypsin (Figure 21.20).

Example – autocatalytic oxidation of Ru^{2+}

In most cases, autocatalytic reactions proceed in several steps. One important example is the oxidation of Ru^{2+} by acidic bromate. Ru^{2+} is used in the form of its trisbipyridyl complex: $[Ru(bipy)_3]^{2+}$. This reaction is one of the key reactions in the Belousov–Zhabotinsky reaction (see Section 21.6.10).

We consider an aqueous solution of Ru^{2+} and acidic bromate containing traces ($10^{-7} \, \text{mol L}^{-1}$) of $HBrO_2$; bromate and H^+ are in large excess over Ru^{2+}. BrO_3^- reacts with $HBrO_2$ forming BrO_2 which is immediately removed by reaction with Ru^{2+}:

$$BrO_3^- + HBrO_2 + H^+ \xrightarrow[\text{slow}]{k_2} 2BrO_2 + H_2O \quad (21.140)$$

$$BrO_2 + Ru^{2+} + H^+ \xrightarrow[\text{fast}]{k_1} HBrO_2 + Ru^{3+} \quad (21.141)$$

21.6 COMPLEX REACTIONS

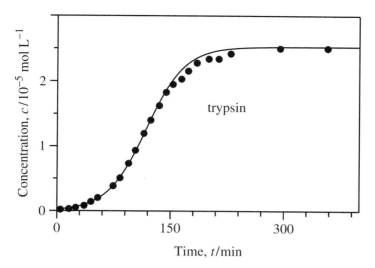

Figure 21.20 Autocatalytic formation of trypsin from trypsinogen (at pH 8, 25 °C in 0.1 M Tris/HCl buffer + 0.05 M CaCl$_2$). The concentration c of trypsin is displayed. Initial concentrations: trypsin: 2.5×10^{-7} M, trypsinogen: 2.5×10^{-5} M. Dots: experiment (H. Lachmann, private communication); solid line: calculated from equation (21.138) with rate constant $k_r = 1.55 \times 10^3$ L min^{-1} mol^{-1}.

By investing one HBrO$_2$ in the first reaction, two HBrO$_2$ and two Ru^{3+} are produced in the second reaction. The HBrO$_2$ molecules thus formed enter into the first reaction, leading to the overall reaction

$$2\text{Ru}^{2+} + \text{BrO}_3^- + \text{HBrO}_2 + 3\text{H}^+ \rightarrow 2\text{Ru}^{3+} + 2\text{HBrO}_2 + \text{H}_2\text{O} \tag{21.142}$$

In this way an exponential increase in the amount of HBrO$_2$ and Ru^{3+} takes place as long as Ru^{2+} is available; then the reaction stops.

The corresponding rate equations are (with $k_1' = k_1 c_{\text{H}^+}$ and $k_2' = k_2 c_{\text{BrO}_3^-} \cdot c_{\text{H}^+}$)

$$\frac{dc_{\text{Ru}^{2+}}}{dt} = -k_1' \cdot c_{\text{Ru}^{2+}} \cdot c_{\text{BrO}_2} \tag{21.143}$$

$$\frac{dc_{\text{BrO}_2}}{dt} = -k_1' \cdot c_{\text{Ru}^{2+}} \cdot c_{\text{BrO}_2} + 2k_2' \cdot c_{\text{HBrO}_2} \tag{21.144}$$

$$\frac{dc_{\text{HBrO}_2}}{dt} = +k_1' \cdot c_{\text{Ru}^{2+}} \cdot c_{\text{BrO}_2} - k_2' \cdot c_{\text{HBrO}_2} \tag{21.145}$$

It is not possible to integrate these rate equations rigorously in analytical form. An approximative treatment is given in Box 21.3. The corresponding decay of Ru^{2+} is shown in Figure 21.21 (curve 1) together with the experimental data (dots). This decay is slightly faster than in the experiment; this is because reaction (21.141) is reversible and HBrO$_2$ disproportionates according to

$$2\text{HBrO}_2 \longrightarrow \text{HOBr} + \text{BrO}_3^- + \text{H}^+ \tag{21.146}$$

Including these reactions and solving the differential equations numerically we obtain curve 2 in Figure 21.21, which is in agreement with the experiment.

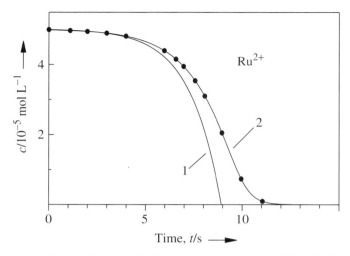

Figure 21.21 Autocatalytic oxidation of Ru^{2+} by acidic bromate at 20 °C. Initial concentrations: $c_{BrO_3^-} = 0.01\,mol\,L^{-1}$, $c_{Ru^{2+}} = 5 \times 10^{-5}\,mol\,L^{-1}$, $c_{H^+} = 1.29\,mol\,L^{-1}$. Dots: experiment. Solid lines: calculation; curve 1: calculated from equations (21.143), (21.144), and (21.145) with the approximation in Box 21.3; curve 2: calculated by including reversibility of reaction (21.141) and of disproportionation of $HBrO_2$ (21.146). Initial concentration of $HBrO_2$ in the calculation: $1 \times 10^{-7}\,mol\,L^{-1}$. For rate constants used in the calculation see Box 21.5 (Table 21.3, reactions 1–5 from top). Experimental data from Y. Gao, H.D. Försterling, *J. Phys. Chem.* 99, 8638 (1995).

Box 21.3 – Approximate Treatment of Autocatalytic Oxidation of Ru^{2+}

We consider equations (21.140) and (21.141). In the present case, reaction (21.140) is much faster than reaction (21.141), thus BrO_2 is in a low concentration and we can apply the steady-state approximation

$$\frac{dc_{BrO_2}}{dt} = 0$$

Then from equation (21.144) we obtain

$$c_{BrO_2} = \frac{2k_2' \cdot c_{HBrO_2}}{k_1' \cdot c_{Ru^{2+}}}$$

and from equation (21.145) we have

$$\frac{dc_{HBrO_2}}{dt} = +k_2' \cdot c_{HBrO_2}$$

This means that $HBrO_2$ grows exponentially, as long as Ru^{2+} is present:

$$c_{HBrO_2} = c_{HBrO_2,0} \cdot e^{k_2' t}$$

For Ru^{2+} we obtain from equation (21.143)

$$\frac{dc_{Ru^{2+}}}{dt} = -2k_2' \cdot c_{HBrO_2} = -2c_{HBrO_2,0} \cdot k_2' \cdot e^{k_2' t}$$

Continued on page 707

21.6 COMPLEX REACTIONS

> *Continued from page 706*
>
> and by integration
>
> $$c_{Ru^{2+}} = c_{Ru^{2+},0} - 2c_{HBrO_2,0} \cdot \left(e^{k_2't} - 1\right)$$
>
> The reaction stops when all Ru^{2+} has been consumed. (see Figure 21.21, curve 1, calculated with $c_{Ru^{2+},0} = 5 \times 10^{-5}$ mol L^{-1}, $c_{HBrO_2,0} = 1 \times 10^{-7}$ mol L^{-1}, $c_{BrO_3^-} = 0.01$ mol L^{-1}, $c_{H^+} = 1.29$ mol L^{-1}, $k_2 = 48$ mol^{-1} L s^{-1}, $k_2' = 0.619$ s^{-1}).

21.6.8 Inhibition of Autocatalysis

Reactants removing intermediates of the autocatalytic process inhibit the onset of the autocatalysis. Let us add Br$^-$ to the aqueous solution of Ru^{2+} and acidic bromate considered in the example in Section 21.6.7. In this example we have seen that HBrO$_2$ (initially present as a trace amount) is produced in an autocatalytic process. Its amount is growing exponentially. However, with added bromide the intermediate HBrO$_2$ is removed in a fast process:

$$HBrO_2 + Br^- + H^+ \xrightarrow{k_3} 2HOBr \qquad (21.147)$$

Then the rate of step (21.141) in the oxidation of Ru^{2+} is slowed down: only a small fraction of HBrO$_2$ can be oxidized to BrO$_2$. This situation is displayed in Figure 21.22, where bromide in a small concentration (2% of the initial Ru^{2+} concentration) is added. The bromide concentration decreases because of reaction (21.147) and of direct reaction

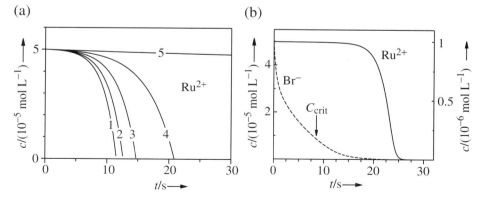

Figure 21.22 Inhibition of autocatalytic oxidation of Ru^{2+} by addition of bromide. Conditions as in Figure 21.21. (a) Concentration of Ru^{2+} versus time curves for constant bromide concentration, calculated as described in Box 21.4. Bromide concentrations for curves 1–5: 0, 0.2, 0.5, 1.0, 2.0 × 10^{-7} mol L^{-1}. (b) Concentrations of Ru^{2+} (left hand scale) and of Br$^-$ (right hand scale) for initial bromide concentration of 1 × 10^{-6} mol L^{-1} (bromide reacts with bromate, thus its concentration decreases with time). For rate constants used in the calculation see Box 21.5 (Table 21.3, reactions 1–7 from top).

with bromate (see Section 21.6.3). As soon as the bromide concentration falls below a critical limit (c_{crit} in Figure 21.22) the autocatalytic process starts. As shown in Box 21.4, the critical bromide concentration is

$$c_{crit} = \frac{k'_2}{k'_3} \tag{21.148}$$

With $k'_2 = 48 \times 0.01\,\text{s}^{-1} = 0.48\,\text{s}^{-1}$ and $k'_3 = 2.5 \times 10^6\,\text{L mol}^{-1}\,\text{s}^{-1}$ we obtain $c_{crit} = 1.92 \times 10^{-7}\,\text{mol L}^{-1}$.

During the inhibitory phase of the reaction the Ru^{2+} concentration is practically constant; then it decreases abruptly. This behavior can be considered as a *clock reaction*. A well-known clock reaction is the *Landolt reaction*, where excess iodate is reduced to iodide by sulfite. After the sulfite has been consumed, iodide reacts with iodate and forms iodine, and the brown color of iodine appears abruptly.

Box 21.4 – Critical Bromide Concentration

To estimate c_{crit} in the inhibition of the autocatalytic oxidation of Ru^{2+} we consider the rate equations (21.143) to (21.145) for the autocatalytic process and include the inhibition process (21.147). Then with $k'_3 = k_3 \cdot c_{H^+}$

$$\frac{dc_{HBrO_2}}{dt} = k'_1 \cdot c_{Ru^{2+}} \cdot c_{BrO_2} - k'_2 \cdot c_{HBrO_2} - k'_3 \cdot c_{HBrO_2} \cdot c_{Br^-}$$

Using the steady-state approximation for c_{BrO_2} we find

$$\frac{dc_{HBrO_2}}{dt} = c_{HBrO_2} \cdot \left(k'_2 - k'_3 \cdot c_{Br^-}\right)$$

and

$$c_{HBrO_2} = c_{HBrO_2,0} \cdot e^{(k'_2 - k'_3 \cdot c_{Br^-})t}$$

For the concentration of Ru^{2+}, in analogy to the consideration in Box 21.3, we obtain

$$c_{Ru^{2+}} = c_{Ru^{2+},0} - \frac{2k'_2 \cdot c_{HBrO_2,0}}{k'_2 - k'_3 \cdot c_{Br^-}} \cdot \left[e^{(k'_2 - k'_3 \cdot c_{Br^-})t} - 1\right]$$

The concentration of Ru^{2+} is shown in Figure 21.22(a) for $k'_2 = 48 \times 0.01\,\text{s}^{-1} = 0.48\,\text{s}^{-1}$, $k'_3 = 2.5 \times 10^6\,\text{L mol}^{-1}\,\text{s}^{-1}$, $c_{HBrO_2,0} = 1 \times 10^{-7}\,\text{mol L}^{-1}$, $c_{Ru^{2+},0} = 5 \times 10^{-5}\,\text{mol L}^{-1}$, and for different bromide concentrations:

$$c_{Br^-} = 0,\, 0.2 \times 10^{-7},\, 0.5 \times 10^{-7},\, 1 \times 10^{-7},\, \text{and } 2 \times 10^{-7}\,\text{mol L}^{-1}.$$

For $k'_2 > k'_3 \cdot c_{Br^-}$ the exponent is positive: exponential growth of $HBrO_2$. For $k'_2 < k'_3 \cdot c_{Br^-}$ we expect an exponential decay. Then for the critical bromide concentration we obtain

$$k'_2 - k'_3 \cdot c_{crit} = 0, \quad c_{crit} = \frac{k'_2}{k'_3}$$

21.6 COMPLEX REACTIONS

21.6.9 Bistability

When an autocatalytic reaction is performed in a flow reactor instead of in a batch reactor, instabilities can occur. Different steady states can be established, depending on the flow conditions. This behavior is called *bistability*. A flow reactor (Figure 21.23a) consists of a well-stirred reaction vessel with a continuous inlet of reactants and a continuous outlet of products. Thus steady-state concentrations inside the reactor are established. Let us assume that we feed the reactor with Ru^{2+} and bromate. Then certain steady-state concentrations of Ru^{2+} and Ru^{3+} are established, depending on the flow rate. At low flow rate there is enough time for the autocatalytic reaction (discussed in the example in Section 21.6.7) to convert all Ru^{2+} into Ru^{3+}. At high flow rate Ru^{3+} disappears fast through the outlet of the reactor, and the steady-state concentration of Ru^{3+} is small.

Now let us consider a medium flow rate; additionally we add bromide continuously in inlet 3. That is, we consider a situation similar to that discussed in Section 21.6.8, but now we have a steady state: Ru^{3+} is removed and Ru^{2+} is supplied. We start with $c_{Br^-} = 0$; then the steady-state concentration of Ru^{2+} corresponds to point 1 in Figure 21.23b (steady state I; the concentration of Ru^{2+} is relatively low because the autocatalytical formation of Ru^{3+} with $HBrO_2$ as an intermediate is effective). If we increase c_{Br^-} in the inlet flow the Ru^{2+} concentration changes only slightly.

At point 2 a critical bromide concentration is reached to inhibit the autocatalytic reaction by removal of $HBrO_2$ according to reaction (21.147). The Ru^{3+} present in the solution disappears by the outlet flow; at the same time the Ru^{2+} concentration increases by the inlet flow. The system switches from steady state I to steady state II (point 3).

Note that a high concentration of bromide is necessary for the inhibition at point 2 because the rate of formation of $HBrO_2$ by the autocatalytic process is high (corresponding to the strong decay of Ru^{2+} in the final part of curves 1 and 2 in Figure 21.21).

If we decrease c_{Br^-} slightly, the system will not switch back to steady state I, because the autocatalytic process cannot start. This is because the rate of formation of $HBrO_2$ by

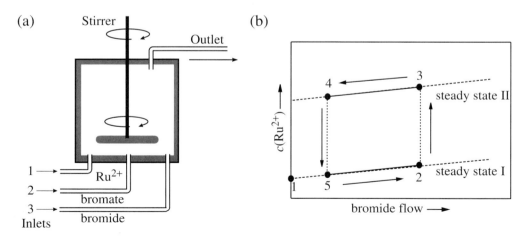

Figure 21.23 Reaction in a flow reactor. (a) Flow reactor with three inlets for reactants and one outlet for products. Inlet 1, Ru^{2+}; inlet 2, bromate; inlet 3, bromide. (b) Bistability in the Ru^{2+}-bromate-bromide system. Steady-state Ru^{2+} concentration $c_{Ru^{2+}}$ at different inlet concentrations of bromide. The starting point is 1 (no bromide flow); then the bromide concentration in the inlet flow is increased up to point 2 and decreased again from point 3 to point 4.

the autocatalytic process is low (corresponding to the very slow decay of Ru^{2+} at the starting point of curves 1 and 2 in Figure 21.21).

When we decrease c_{Br^-} further, the rate of formation of $HBrO_2$ can compete with the rate of consumption by bromide. Then the autocatalytic reaction starts again, and the system switches back to steady state I (point 5).

The essence of bistability is that in steady state I Ru^{2+} is rapidly oxidized to Ru^{3+} in the autocatalytic sequence (21.140), (21.141); then $HBrO_2$ appears immediately in such an amount that its removal by reaction (21.147) is negligible. In steady state II the inhibitor Br^- removes the intermediate $HBrO_2$ by reaction (21.147) before the autocatalytic sequence (21.140), (21.141) can be getting effective. How can we reach steady state I at a given moderate Br^- flow? We have to increase the flow, starting from a value where autocatalysis is not suppressed (e.g., from point 1 in Figure 21.23(b)). Accordingly, if we start from a value where autocatalysis is suppressed (from top right in Figure 21.23(b)), steady state II is reached.

21.6.10 Oscillating Reactions (Belousov–Zhabotinsky Reaction)

Periodic changes in population dynamics are well known, e.g. the famous case of the two kinds of fishes: large fishes (predators) and small fishes (prey). The small fishes live in an unlimited supply of food, and the large fishes eat the small ones. The number of predator fishes, initially small, increases by eating more and more prey fishes. As a consequence, the number of prey fishes decreases; accordingly, because of lack of food, the number of predator fishes decreases also. Then, the population of the prey fishes can increase again. This cycle is repeated periodically. The dynamics of this problem can be described by the *Lotka-Volterra mechanism*, assuming autocatalytic growth of species A and B, if we denote A = prey fish, B = predator fish:

$$A \xrightarrow{k_1} 2A$$
$$A + B \xrightarrow{k_2} 2B$$
$$B \xrightarrow{k_3} C$$

The last process describes the natural death of B. The corresponding differential equations are:

$$\frac{dN(A)}{dt} = k_1 \cdot N(A) - k_2 \cdot N(A) \cdot N(B)$$
$$\frac{dN(B)}{dt} = k_2 \cdot N(A) \cdot N(B) - k_3 \cdot N(B)$$

where $N(A)$ and $N(B)$ are the population numbers of A and B. A numerical solution of these differential equations is shown in Figure 21.24. Can such a periodic behavior be realized by means of chemical reactions?

According to Section 21.6.9 the oxidation of Ru^{2+} to Ru^{3+} in the presence of excess acidic bromate is an autocatalytic process showing bistability in a flow reactor. By adding an agent which is able (1) to reduce Ru^{3+} to Ru^{2+} and (2) to generate a product inhibiting the autocatalytic reaction, it is possible to reach bistability in a batch reactor. The system switches periodically from steady state I to steady state II and vice versa.

21.6 COMPLEX REACTIONS

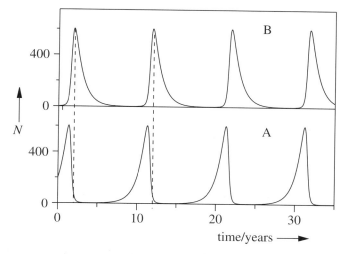

Figure 21.24 Lotka Volterra model. Populations of prey fishes A and of predator fishes B versus time. Initial conditions: $N(A)_0 = 200$, $N(B)_0 = 2$. The rate constants are $k_1 = 1$ year^{-1}, $k_2 = 0.01$ year^{-1}, $k_3 = 1$ year^{-1}.

This happens in the Belousov–Zhabotinsky reaction, where bromomalonic acid (BrMA; COOH–CHBr–COOH) is initially present besides Ru^{2+}, BrO_3^-, and traces of BrO_2. First Ru^{2+} is oxidized to Ru^{3+} according to equations (21.140) and (21.141):

$$Ru^{2+} + BrO_2 + H^+ \longrightarrow Ru^{3+} + HBrO_2$$

$$HBrO_2 + BrO_3^- + H^+ \longrightarrow 2BrO_2 + H_2O$$

Ru^{3+} reacts with bromomalonic acid to form bromomalonyl radicals (BrMA•; COOH–$\overset{\bullet}{C}$Br–COOH)

$$Ru^{3+} + BrMA \longrightarrow Ru^{2+} + BrMA\bullet + H^+ \qquad (21.149)$$

These radicals recombine, forming the inhibitor bromide and an organic product P_1

$$2BrMA\bullet \longrightarrow Br^- + P_1 + H^+ \qquad (21.150)$$

(Figure 21.25).

At the same time, the bromomalonyl radicals remove BrO_2 from the autocatalytic process (second inhibition reaction):

$$BrMA\bullet + BrO_2 \longrightarrow P_2 \qquad (21.151)$$

P_2 reacts further:

$$P_2 \longrightarrow Br^- + P_3$$

thereby forming the inhibitor bromide.

The oscillatory behavior of this system is displayed in Figure 21.26. The reaction is started by an injection of Ru^{2+} into an acidic mixture of bromate and bromomalonic acid.

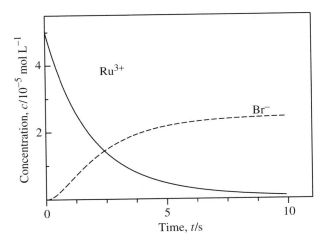

Figure 21.25 Reaction of Ru^{3+} with excess bromomalonic acid. Initial conditions: $c_{BrMA} = 0.01$ mol L^{-1}, $c_{Ru^{3+}} = 5 \times 10^{-5}$ mol L^{-1}. Decay of Ru^{3+} (solid line) and formation of bromide (dashed line).

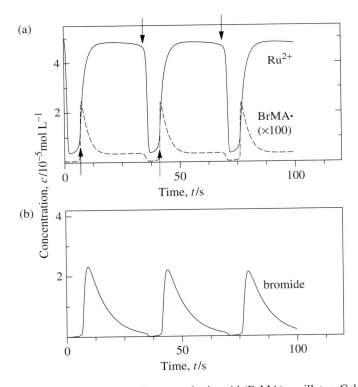

Figure 21.26 Ruthenium–bromate–bromomalonic acid (BrMA) oscillator. Calculated concentrations c of Ru^{2+} and BrMA• (bromomalonyl radical) (a) and of bromide (b). Initial conditions: $c_{bromate} = 0.1$ mol L^{-1}, $c_{BrMA} = 0.01$ mol L^{-1}, $c_{Ru^{2+}} = 5 \times 10^{-5}$ mol L^{-1}, $c_{H^+} = 1.29$ mol L^{-1}, $c_{HBrO_2} = 1 \times 10^{-6}$ mol L^{-1}. Arrows on bottom of part (a): inhibition of autocatalytic reaction by bromide and bromomalonyl radicals. Arrows on top of part (a): onset of autocatalytic reaction, when all bromide has been consumed. For reactions and rate constants used in the calculation see Box 21.5 (Tables 21.3 and 21.4).

21.6 COMPLEX REACTIONS

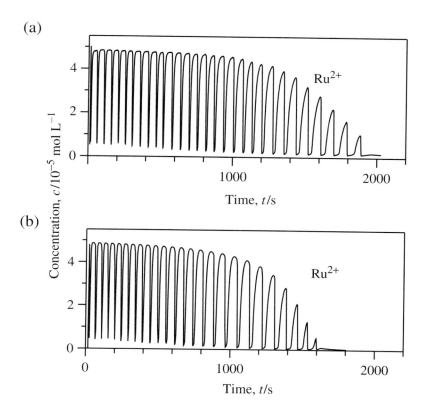

Figure 21.27 Oscillatory system as in Figure 21.26, but the concentration c of Ru^{2+} is followed for 2000 s. (a) Experiment; Ru^{2+} is measured by its absorbance at 450 nm. (b) Theory; For rate constants used in the calculation see Box 21.5 (Table 21.3). Experimental data from Y. Gao, H.D. Försterling, *J. Phys. Chem.* **99**, 8638 (1995).

Since no inhibitor is present, the autocatalytic reaction starts immediately, thus Ru^{2+} is consumed and Ru^{3+} is formed.

Then because of reactions (21.150) and (21.151), bromide and bromomalonyl radicals BrMA• are formed, thereby inhibiting the autocatalysis (arrows on bottom of Figure 21.26(a)). As a consequence, Ru^{3+} is reduced to Ru^{2+}; the bromide formed during this process disappears by its reaction with bromate. When the bromide concentration is below a critical value the autocatalysis starts again (arrows on top of Figure 21.26(a)). This process continues periodically until the bromomalonic acid has been consumed (Figure 21.27).

Oscillations can also be obtained by starting the reaction with malonic acid instead of bromomalonic acid. In this case Ru^{3+} reacts with malonic acid instead of bromomalonic acid, and no production of bromide is possible. However, HOBr, the final product of the autocatalytic reaction, brominates malonic acid. After a short induction time enough bromomalonic acid has accumulated to start oscillations (Figure 21.28(a)). The amplitude of these oscillations increases with time due to the increasing concentration of bromomalonic acid. After a longer time the BrMA concentration is kept on a nearly constant level and undamped oscillations occur (Figure 21.28(b)).

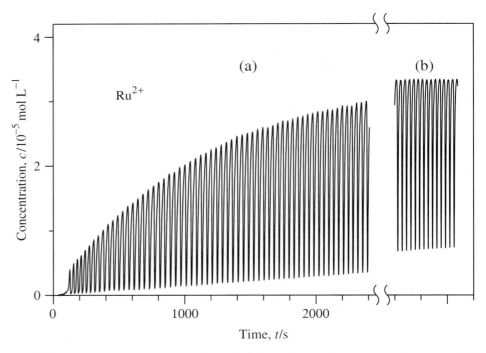

Figure 21.28 Oscillatory system as in Figure 21.27, but the reaction was started with an addition of malonic acid ($c = 0.1 \, \text{mol L}^{-1}$) into a mixture of bromate (0.1 M) and Ru^{2+} (5×10^{-5} M). (a) $t = 0\text{--}2400$ s. (b) $t = 4000\text{--}4500$ s. At $t = 0$ all ruthenium salt is in the oxidized form (Ru^{3+}).

21.6.11 Chemical Waves

Circular Waves

First, let us consider a prairie fire, as an analogon of a chemical wave. We assume a large area covered with dry grass. Thus the grass is in an excitable state: a single spark is enough to start burning at a certain spot. A circular firefront propagates, in analogy to a water wave. The fire is fed by dry grass at its front leaving burned grass behind it. This means the medium behind the firefront is no more in an excitable stae, and a second firefront cannot start from the same spot, in contrast to a water wave. Finally, when the firefront runs against a solid wall it is extinguished, in contrast to a water wave showing reflection at the wall. These stages are shown in Figure 21.29a. Two colliding firefronts extinguish each other, also in contrast to two water waves showing interference phenomena. After a recovery time, new grass growths, dries out in the sun and the dry grass is ready to generate a new firefront. This means the phenomenon shows periodicity in time and space.

Now let us consider a chemical wave. As in the example of the prairie fire we need an excitable chemical medium. Such a medium can be found in the Belousov-Zhabotinsky reaction discussed in Section 21.6.10. We prepare a thin layer of this medium (mixture of bromomalonic acid, bromate, and ruthenium complex in sulfuric acid solution) in a Petri dish. Such a system is usually selfoscillatory. However, under appropriate conditions, the inhibitor bromide can be kept above its critical value; then oscillations cannot start

21.6 COMPLEX REACTIONS

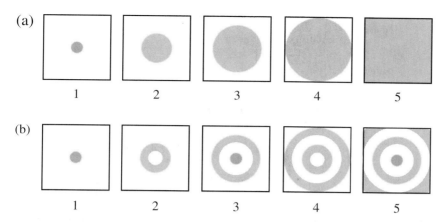

Figure 21.29 Propagation of waves in an excitable medium. (a) Prairiefire. Dry grass (white area) is burned by a spark at a small spot (1). The firefront propagates from inside to outside, leaving burned grass behind (grey area, 2–4). Finally, the firefront reaches solid walls and is extinguished; the full area, originally covered with dry grass, is burned out (5). (b) Chemical wave. An excitable chemical medium (Ru^{2+}, bromate, and bromomalonic acid, white area) is excited at a small spot by touching with a silver wire (1). The wavefront (Ru^{3+} at the front) propagates from the inside to the outside, leaving the medium in an non-excitable state (grey area, no Ru^{2+} available to start the autocatalytic reaction). The Ru^{3+} behind the wavefront is reduced by bromomalonic acid, and after a short while the medium is again in an excitable state. A new wave is induced by touching again with a silver wire (3). Thus a series of waves propagates simultaneously (3–5). Finally, the first wave reaches the solid walls and is extinguished.

spontaneously. Decreasing the bromide concentration at a small spot for a short while (e.g. by touching with a silver wire, thus producing Ag^+ ions removing Br^-) can start the autocatalytic process. Once the reaction has been started a circular chemical wavefront propagates, just as in the case of the firefront.

In the beginning, the thin layer is characterized by an uniformly ditributed concentration of Ru^{2+} (orange color). When the chemical wave propagates a circular ring of Ru^{3+} is formed, running from the inside to the outside, easily detectably by eye because of the color change from orange to colorless. Compared with the firefront, the system inside the circular wavefront recovers much faster: Ru^{3+} is reduced by excess bromomalonic acid and the layer is ready for the start of a second wavefront before the first wavefront has reached the borders of the excitable medium (Figure 21.29b). As a consequence, the concentration profile of Ru^{3+} in the wave is steep in its front region (due to the fast autocatalytic process) and it is flat behind the front (due to the slow process of reduction of Ru^{3+} by bromomalonic acid).

In the actual experiment a high bromide concentration in the medium is established by illumination of the layer containing the Belousov–Zhabotinsky (BZ) medium with visible light. Ru^{2+} in the excited state is capable to oxidize bromomalonic acid; thus an additional amount of bromide is produced preventing the onset of the autocatalytic process. A small part in the center of the layer is kept in the dark (Figure 21.30). Thus a wavefront can start spontaneously in the central part of the Petri dish, reach the illuminated part and start a wavefront there. In Figure 21.31 a sequence of several wavefronts is shown following each other with a period of 30 s.

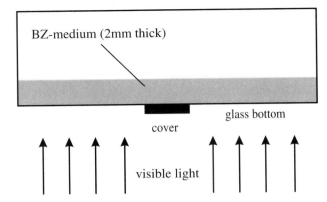

Figure 21.30 Experimental setup for the observation of self-induced chemical waves. On the bottom of a Petri dish a thin layer of excitable BZ medium (bromate 0.8 mol L^{-1}, bromomalonic acid 0.12 mol L^{-1}, Ru^{2+} 6×10^{-4} mol L^{-1}, dissolved in H_2SO_4 (0.8 mol L^{-1})) is placed. The layer is kept in an excitable, but not selfoscillatory state by shining visible light. The center of the layer is covered such that it cannot be reached by the light; thus this part of the medium is self-oscillatory and waves are propagating spontaneously. (T.Akagi, T.Matsumura-Inoue, Dynamic Photochemical Response of Ruthenium(II)/(III) Redox Chemical Waves in Belousov-Zhabotinsky Reaction, Chemistry Letters 1996, 557).

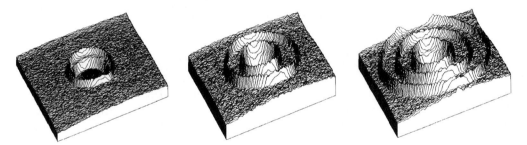

Figure 21.31 Propagation of Ru^{3+} waves in a thin layer of BZ medium (see Figure 21.30). The Ru^{3+} concentration is measured by recording the transmitted light intensity using a video camera. The transmittance is converted into Ru^{3+} concentration which is shown as a 3D representation. The three pictures are taken every 30 s.

Chemical Pinwheel

We consider an excitable BZ medium confined to a ring-shaped area, and a circular chemical wave propagating in this medium (Figure 21.32(a,b). White zone: excitable medium; grey inner circle: no excitable medium). In the wavefront the catalyst (e.g. the ruthenium complex) is in the oxidized state, due to the fast running autocatalytic process. After reaching the outer and inner bounderies of the excitable medium the upper and the lower part of the wavefront are extinguished. Thus the circular wave splits into two arms propagating clockwise and counterclockwise around the ring (Figure 21.32(c)). We expect that, after a rotation by 180^0 they meet at the opposite side of the ring and extinguish each other. This can be avoided by dying out one of the two arms before they meet:

21.6 COMPLEX REACTIONS   717

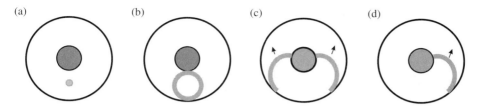

Figure 21.32 Formation of a pinwheel in a ring-shaped excitable medium (white area). The boundaries of the outer circle and of the inner circle (grey zone: no excitable medium) act as solid walls, extinguishing the waves. (a) Starting a circular wave. (b) The circular wave reaches the boundaries, (c) leading to two pinwheels rotating in opposite directions. (d) Remaining one-armed pinwheel rotating counterclockwise.

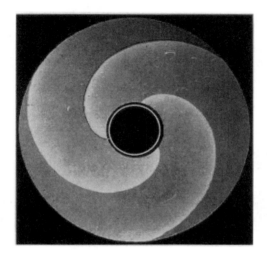

Figure 21.33 Three-armed chemical pinwheel rotating counterclockwise. The time needed for one rotation is on the order of one hour. Composition of the excitable medium: bromate 0.03 mol L^{-1}, bromomalonic acid 0.03 mol L^{-1}, malonic acid 0.08 mol L^{-1}; ferroin (trisphenanthroline complex of Fe^{2+}) is added as a catalyst instead of the ruthenium complex used in the experiment in Figure 21.31. The reaction takes place on the surface of a circular polysulfone membrane (pore size 0.45 μm, outer diameter 40 mm) with a hole cut in the center (diameter 10 mm). The membrane is in contact with a solution containing bromate, bromomalonic acid and malonic acid; the ferroin is adsorbed at the surface of the membrane. Thus the excitable areas on the membrane are red (strong absorption of ferroin), and at the locations of the wavefronts (which are colorless because of the weak absorption of ferriin), the white color of the membrane appears. (A.Lazar, H. D. Försterling, H. Farkas, P. Simon, A. Volford, Z. Noszticzius, Waves of excitation on nonuniform membrane rings, caustics, and reverse involutes, Chaos 7, 731 (1997)).

touching the wavefront with an iron wire (e.g. an iron paperclip) is sufficient to bring the catalyst back in the reduced state. Then, the remaining arm rotates permanently around the ring, like the arm of a pinwheel (Figure 21.32(d)). In the same manner, a second and a third arm can be introduced, which, after equilibration, rotate at equal distances from each other (Figure 21.33).

Box 21.5 – Rate Constants of Belousov–Zhabotinsky Reaction

In the numerical simulations of the Belousov–Zhabotinsky reaction displayed in Figures 21.21, 21.22, and 21.26–21.28 the rate constants in Tables 21.3 and 21.4 are used. Note that Ru^{2+} denotes the trisbipyridyl complex of ruthenium(II).

Note that the formation of BrO_2 occurs via the precursor Br_2O_4 and that the disproportionation of bromous acid occurs via a protonated form. The intermediate P_2 disappears by two parallel reactions.

The abbreviations are: BrMA, bromomalonic acid; BrMA•, bromomalonyl radical; BrTA, bromotartronic acid; MOA, mesoxalic acid; BrMAE, enol form of bromomalonic acid. P_1 can be identified as bromoethenetricarboxylic acid, P_3 has not yet been identified.

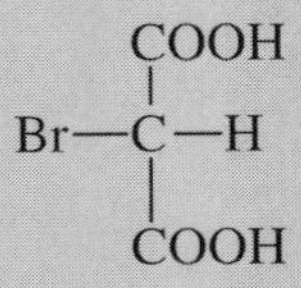

bromomalonic acid (BrMA)

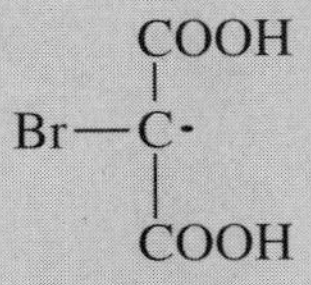

bromomalonyl radical (BrMA•)

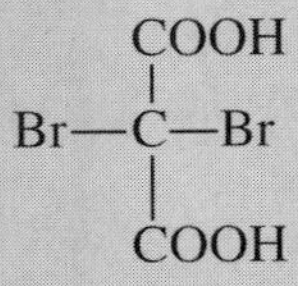

dibromomalonic acid (Br_2MA)

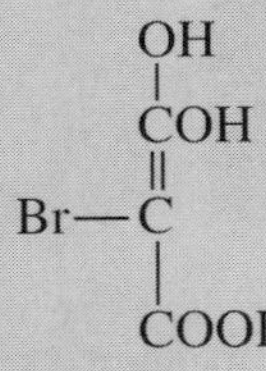

enol form of bromomalonic acid (BrMAE)

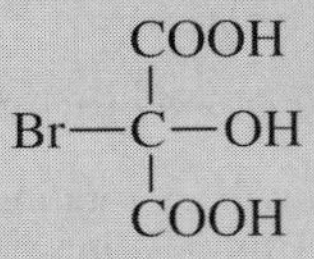

bromotartronic acid (BrTA)

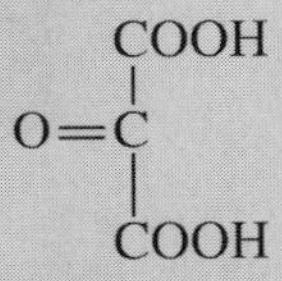

mesoxalic acid (MOA)

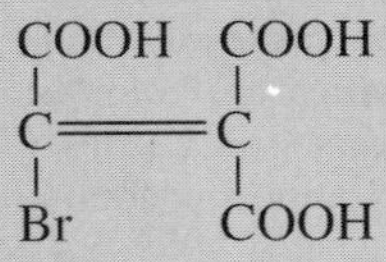

bromoethenetricarboxylic acid (P_1)

Continued on page 719

Continued from page 718

Table 21.3 Rate constants of reactions which are important for the autocatalysis and for the inhibition by bromide (temperature 20 °C) (Y. Gao, H. D. Försterling, Oscillations in the Bromomalonic Acid/Bromate System Catalyzed by $[Ru(bipy)_3]^{2+}$, J.Phys.Chem. 99, 8638 (1995))

Reaction	$k_{forward}$	$k_{reverse}$
$Ru^{2+} + BrO_2 + H^+ \longrightarrow Ru^{3+} + HBrO_2$	4×10^6	0
$HBrO_2 + HBrO_3 \longrightarrow Br_2O_4 + H_2O$	4.8×10^1	3.2×10^3
$Br_2O_4 \longrightarrow 2BrO_2$	7.5×10^4	1.4×10^9
$HBrO_2 + H^+ \longrightarrow H_2BrO_2^+$	2×10^6	1×10^8
$H_2BrO_2^+ + HBrO_2 \longrightarrow HOBr + HBrO_3$	1.7×10^5	0
$Br^- + HBrO_2 + H^+ \longrightarrow 2HOBr$	2.5×10^6	2×10^{-5}
$Br^- + HOBr + H^+ \longrightarrow Br_2 + H_2O$	8×10^9	80
$Br^- + HBrO_3 + H^+ \longrightarrow HOBr + HBrO_2$	0.12×10^1	3.2

The units of the rate constants are s^{-1}, if one reactant appears in the chemical reaction, $L\,mol^{-1}\,s^{-1}$ for two reactants, and $L^2\,mol^{-2}\,s^{-1}$ for three reactants. The solvent water is included in the rate constants.

Table 21.4 Rate constants of reactions with bromomalonic acid BrMA (temperature 20 °C) Y. Gao, H. D .Försterling, Oscillations in the Bromomalonic Acid/Bromate System Catalyzed by $[Ru(bipy)_3]^{2+}$, J.Phys.Chem. 99, 8638 (1995)

Reaction	$k_{forward}$	$k_{reverse}$
$Ru^{3+} + BrMA \longrightarrow Ru^{2+} + BrMA\bullet + H^+$	5.5×10^1	1.3×10^5
$2BrMA\bullet \longrightarrow Br^- + P_1 + H^+ + CO_2$	1×10^8	0
$BrMA\bullet + BrO_2 \longrightarrow P_2$	4×10^9	0
$P_2 \longrightarrow Br^- + P_3$	6.2×10^{-1}	0
$P_2 \longrightarrow BrTA + HBrO_2$	4.6×10^{-1}	0
$BrTA \longrightarrow Br^- + MOA$	0.15×10^1	0
$BrMA \longrightarrow BrMAE$	1.2×10^{-2}	800
$BrMAE + Br_2 \longrightarrow Br_2MA + Br^- + H^+$	3.5×10^6	0
$BrMAE + HOBr \longrightarrow Br_2MA + H_2O$	6.6×10^4	0

Units of the rate constants see Table 21.3.

21.7 Experimental Methods

Only for very slow reactions can the concentrations of the reactants be analyzed by taking samples from the reaction mixture. In most kinetic investigations the concentration is measured by selecting a physical quantity proportional to the concentration. To follow a reaction, mixing methods can be applied when the mixing time is smaller than the half-life of the reaction. Fast reactions can be investigated by using relaxation methods.

21.7.1 Flow Methods

In the stopped flow method, reactants A and B are in two cylinders. By means of two pistons they are rapidly transferred through a mixing chamber and a measuring cell into a third cylinder (Figure 21.34). When the piston in that third cylinder reaches the stop face, the flow is stopped abruptly. Then the concentration in the measuring cell is followed by selecting an appropriate physical property (absorbance, electrical conductivity, optical rotation, fluorescence). With this method, reactions with a half-life of about 1 ms can be investigated. Instead of stopping the flow, we can apply a continuous flow (velocity v) and measure the steady-state concentration along a flow tube (Figure 21.35). Then the reactants at point $x = vt$ along the tube have concentrations corresponding to the time t after mixing.

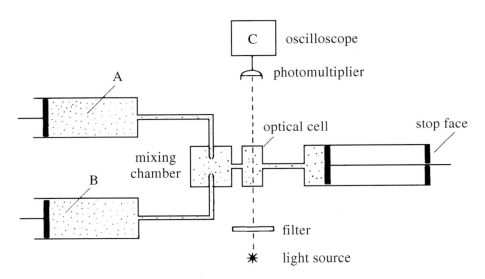

Figure 21.34 Stopped flow apparatus.

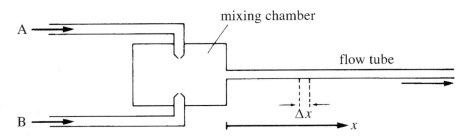

Figure 21.35 Flow method.

21.7.2 Flash Photolysis

To study the recombination of I radicals

$$2I \longrightarrow I_2 \qquad (21.152)$$

a flash light pulse is applied to a reaction vessel containing I_2 molecules; thus I_2 dissociates into I atoms (Figure 21.36). Then the recombination of the atoms can be followed spectroscopically (Figure 21.37). This method can be applied if the half-life of the reaction is longer than the decay time of the flash pulse. Typical values are $1-50\,\mu s$ for a flash from a xenon discharge tube. When using a nitrogen or excimer laser, this time is on the

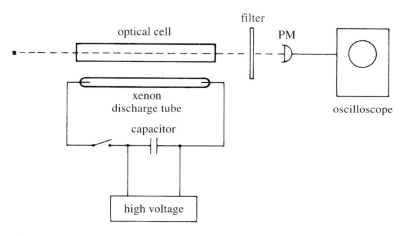

Figure 21.36 Flash light apparatus.

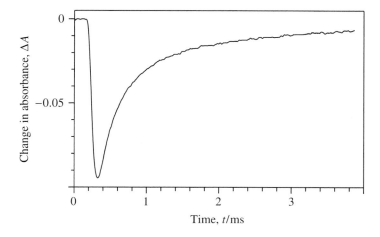

Figure 21.37 Change ΔA of the absorbance of I_2 molecules after flash photolysis of gaseous I_2 at $25\,°C$ ($\lambda = 540\,nm$, optical path length $d = 15\,cm$). Immediately after the flash ΔA is negative because I_2 molecules disappear. Then ΔA increases because of the recombination of I atoms.

order of 1 ns. By applying advanced laser techniques the time scale can be extended into the femtosecond (10^{-15} s) range.

21.7.3 Relaxation Method

Instead of removing one reactant from the reaction mixture, it is sufficient to induce a fast change of the equilibrium concentrations. This can be realized by changing temperature, pressure, or electric field strength (*temperature, pressure,* or *field jump method*, see Figure 21.38). We consider the equilibrium

$$A + B \underset{k_{-1}}{\overset{k_1}{\rightleftharpoons}} C \tag{21.153}$$

According to the van't Hoff equation (17.40)

$$\frac{d \ln K}{dT} = \frac{\Delta_r H^\ominus}{RT^2} \tag{21.154}$$

we can estimate the relative change of the equilibrium constant K for a change of temperature. Because of $d \ln K = dK/K$ we obtain

$$\frac{dK}{K} = \frac{\Delta_r H^\ominus}{RT^2} \cdot dT \tag{21.155}$$

Thus for $\Delta_r H^\ominus = 10\,\text{kJ}\,\text{mol}^{-1}$, $T = 298\,\text{K}$, and $dT = 5\,\text{K}$ we obtain a relative change of K by 7%. This is only a small change, but it is enough to follow the relaxation of the system into the new equilibrium by using sensitive detection methods (Figure 21.39).

Let us denote the concentrations in the new equilibrium with $c_{A,\infty}$, $c_{B,\infty}$, and $c_{C,\infty}$. Then

$$\frac{dc_C}{dt} = k_1 c_{A,\infty} c_{B,\infty} - k_{-1} c_{C,\infty} = 0 \tag{21.156}$$

After the temperature jump the concentrations are $c_A = c_{A,\infty} + x$, $c_B = c_{B,\infty} + x$, $c_C = c_{C,\infty} - x$, where x is a measure of the deviation of the system from the new equilibrium.

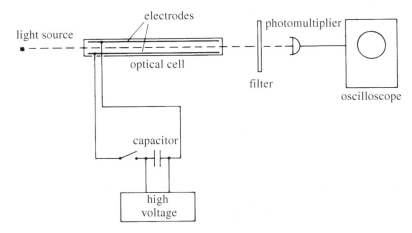

Figure 21.38 Temperature jump apparatus. The capacitor is discharged through the solution with the reactants. The equilibrium is shifted by the resulting temperature jump caused by the discharge.

21.7 EXPERIMENTAL METHODS

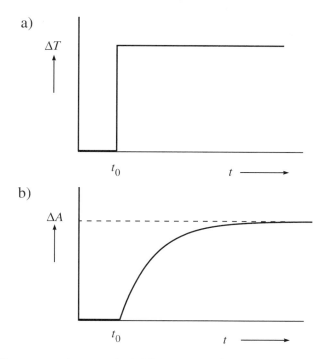

Figure 21.39 Temperature jump method. Time course of temperature change ΔT (a) and of absorbance change ΔA (b), which is proportional to x, the deviation from equilibrium.

Then the rate of concentration change of molecules C is

$$\frac{dc_C}{dt} = k_1(c_{A,\infty} + x) \cdot (c_{B,\infty} + x) - k_{-1}(c_{C,\infty} - x) \quad (21.157)$$

$$= k_1 c_{A,\infty} c_B - k_{-1} c_{C,\infty} + k_1 x (c_{A,\infty} + c_{B,\infty}) + k_{-1} x + k_1 x^2$$

By inserting the equilibrium condition (21.156), neglecting the term with x^2 and using the equation

$$\frac{dc_C}{dt} = -\frac{dx}{dt} \quad (21.158)$$

we obtain

$$\frac{dx}{dt} = -k_r \cdot x \quad \text{with} \quad k_r = k_1 \cdot (c_{A,\infty} + c_{B,\infty}) + k_{-1} \quad (21.159)$$

Then

$$x = x_0 \cdot e^{-k_r t} \quad (21.160)$$

x decreases exponentially with time, where $x_0 = c_{A,0} - c_{A,\infty} = c_{C,\infty} - c_{C,0}$ ($c_{A,0}$ and $c_{C,0}$ are the equilibrium concentrations of A and C before the temperature jump). The change ΔA of the absorbance is proportional to the change of the concentration of reactant C:

$$\Delta A \sim \Delta c_C = c_C - c_{C,0} = (c_{C,\infty} - x) - (c_{C,\infty} - x_0) = x_0 \cdot (1 - e^{-k_r t})$$

By measuring ΔA as a function of time (Figure 21.39(b)) we obtain k_r. To evaluate k_1 and k_2, the experiment is repeated for different concentrations $c_{A,\infty}$ and $c_{B,\infty}$. Then k_r is plotted versus $(c_{A,\infty} + c_{B,\infty})$. According to (21.159), k_1 is obtained from the slope and k_{-1} from the intercept of the straight line.

Another possibility is to make use of the equilibrium constant

$$K_1 = \frac{\widehat{c}_{C,\infty}}{\widehat{c}_{A,\infty} \cdot \widehat{c}_{B,\infty}} = \frac{k_1}{k_{-1}} \cdot c^\ominus \qquad (21.161)$$

Then we obtain

$$k_1 = \frac{k_r}{(c_{A,\infty} + c_{B,\infty}) + c^\ominus/K_1}, \quad k_{-1} = \frac{k_1}{K_1} \cdot c^\ominus \qquad (21.162)$$

Example – deprotonation of indicator dye

We consider the reaction of the protonated form of the indicator dye alizarin yellow R (A, yellow) with OH^- to the deprotonated form (C, orange-red)

$$A + OH^- \underset{k_2}{\overset{k_1}{\rightleftharpoons}} C + H_2O \qquad (21.163)$$

In this case, in contrast to the indicator methylorange (see Figure 8.16), the alkaline form absorbs at longer wavelengths compared with the acidic form, in accordance with the considerations in Chapter 8. The following values of k_r are measured at 298 K (temperature after the temperature jump) with total indicator concentration 0.98×10^{-4} mol L^{-1} at different pH (C_A and c_{OH^-} refer to the equilibrium values):

pH	$(c_A + c_{OH^-})/$mol L^{-1}	k_r/s^{-1}
10.30	3.26×10^{-4}	4.43
10.49	4.57×10^{-4}	4.60
10.60	5.71×10^{-4}	5.46
10.70	7.02×10^{-4}	5.92

These values are plotted in Figure 21.40. From the slope of the straight line we obtain $k_1 = 4.0 \times 10^7$ L mol^{-1} s^{-1} and from the intercept we obtain $k_{-1} = 3.2 \times 10^4$ s^{-1}.

21.7 EXPERIMENTAL METHODS

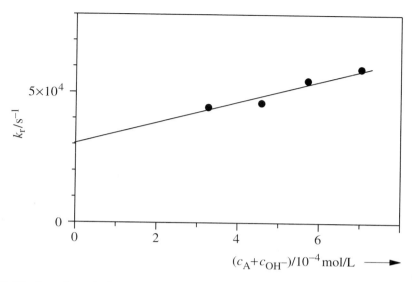

Figure 21.40 Reaction of alizarin yellow (A) with OH^-. Rate constant k_r versus $(c_A + c_{OH^-})$. M. Callaghan Rose, J. Struehr, Kinetics of proton transfer reactions in aqueous solution. III. Rates of internally hydrogen-bonded systems. *J. Am. Chem. Soc.* **90**, 7205 (1968).

Example – dissociation of water

For the reaction

$$H^+ + OH^- \underset{k_{-1}}{\overset{k_1}{\rightleftharpoons}} H_2O$$

the value $k_r = 2.9 \times 10^4 \, s^{-1}$ is obtained from a pressure jump experiment with $c_{H^+} = c_{OH^-} = 10^{-7} \, mol \, L^{-1}$ at $T = 298 \, K$. The equilibrium constant is $K_w = \widehat{c}_{H^+} \cdot \widehat{c}_{OH^-} = 1.0 \times 10^{-14}$. Because of the equilibrium condition $k_1 c_{H^+} \cdot c_{OH^-} - k_{-1} c_{H_2O} = 0$ we obtain

$$\frac{k_1}{k_{-1}} = \frac{c_{H_2O}}{c_{H^+} \cdot c_{OH^-}} = \frac{c_{H_2O}}{K_w \cdot (c^\ominus)^2}$$

Then from equation (21.162) we obtain

$$k_1 = \frac{2.9 \times 10^4}{2 \times 10^{-7} + 1/(55.5 \times 10^{14})} \, L \, mol^{-1} \, s^{-1}$$

$$= 1.5 \times 10^{11} \, L \, mol^{-1} \, s^{-1}$$

$$k_{-1} = 3 \times 10^{-5} \, s^{-1}$$

The forward reaction is diffusion controlled. However, the reverse reaction is extremely slow: it takes $\tau = 1/k_{-1} = 3 \times 10^4 \, s = 8 \, h$ for one H_2O molecule to dissociate into H^+ and

OH$^-$. (H. Strehlow, W. Knoche, *Fundamentals of Chemical Relaxation*, Verlag Chemie, Weinheim 1977).

Example – bromine hydrolysis

For the reaction

$$\text{Br}^- + \text{HOBr} + \text{H}^+ \underset{k_{-1}}{\overset{k_1}{\rightleftharpoons}} \text{Br}_2 + \text{H}_2\text{O}$$

the values $k_1 = 1.61 \times 10^{10}\,\text{L}^2\,\text{mol}^{-2}\,\text{s}^{-1}$ and $k_{-1} = 110\,\text{s}^{-1}$ are obtained from a temperature jump experiment at 20 °C. M. Eigen, K. Kustin, The kinetics of halogen hydrolysis, *J. Am. Chem. Soc.* **84**, 1355 (1962)

Problems

Problem 21.1 – Second-order Reaction

Solve the rate equation for reaction (21.56) in the general case that the initial concentrations of the reactants A and B (a and b respectively), are orbitrary. Check the solution for the special cases $a = b$ and $b \gg a$.

Solution. The rate equation is

$$\frac{dc_A}{dt} = -k_r c_A c_B$$

For one reacting molecule A, one molecule B disappears. Then

$$c_A = a - x, \quad c_B = b - x, \quad \frac{dc_A}{dt} = -\frac{dx}{dt}$$

and

$$\frac{dx}{dt} = k_r \cdot (a-x) \cdot (b-x)$$

We separate the variables and integrate by partial fractions:

$$\int \frac{dx}{(a-x)\cdot(b-x)} = k_r \int dt$$

Then the integrated rate equation is

$$\frac{1}{b-a} \cdot \ln\frac{b-x}{a-x} = k_r t + C$$

Initial condition: $x = 0$ at $t = 0$, thus $C = 1/(b-a) \cdot \ln(b/a)$; then

$$\ln\frac{b-x}{a-x} = (b-a) \cdot k_r t + \ln\frac{a}{b} \quad \text{and} \quad \frac{b-x}{a-x} = \frac{b}{a} \cdot e^{(b-a)k_r t}$$

Then

$$c_A = \frac{b-a}{\dfrac{b}{a} \cdot e^{(b-a)k_r t} - 1}$$

Limiting case $a = b$:
We set $b = a + z$. Then, by using l'Hospital's rule

$$\lim_{z \to 0} c_A = \lim_{z \to 0} \frac{z}{\left(1 + \dfrac{z}{a}\right) \cdot e^{zk_r t} - 1}$$

$$= \lim_{z \to 0} \frac{1}{e^{zk_r t} \left[\dfrac{1}{a} + k_r t + \dfrac{z}{a k_r t}\right]} = \frac{1}{\dfrac{1}{a} + k_r t}$$

Limiting case $b \gg a$:
In this case $b - a \approx b$ and we can neglect the term -1 in the denominator. Thus

$$c_A = \frac{b}{\dfrac{b}{a} \cdot e^{bk_r t}} = a \cdot e^{-bk_r t}$$

Problem 21.2 – Reaction Leading to Equilibrium

Integrate the differential equation (21.75)

$$\frac{dc_A}{dt} = -k_1 \cdot c_A^2 + k_{-1} \cdot (a - c_A)^2$$

in the special case $k_1 = k_{-1}$.

Solution. By inserting $k_1 = k_{-1} = k_r$ and separating the variables, we obtain

$$\frac{dc_A}{2c_A - a} = -k_r a \cdot dt$$

Then

$$\int \frac{dc_A}{2c_A - a} = \frac{1}{2} \cdot \ln(2c_A - a) = -k_r a t + A$$

where A is an integration constant. At $t = 0$ the concentration c_A equals a, thus $A = \frac{1}{2} \ln a$ and

$$\ln \frac{2c_A - a}{a} = -2k_r a t \quad \text{or} \quad \frac{2c_A - a}{a} = e^{-2k_r a t}$$

Then $c_A = \frac{1}{2} a \cdot (1 + e^{-2k_r a t})$. For $t \to \infty$ (equilibrium) we obtain $c_A = \frac{1}{2} a$.

Problem 21.3 – Yield of Main Reaction

For two parallel reactions $A + B \to C$ and $2A \to D$ (with the reactant B in great excess over A) calculate the yield of C.

Solution. The rate equation for A is

$$\frac{dc_A}{dt} = -k_1' \cdot c_A - 2k_2 \cdot c_A^2$$

with $k_1' = k_1 c_{B,0}$ (B in great excess). By integration we find

$$c_A = c_{A,0} \cdot \frac{k_1' \cdot e^{-k_1' t}}{k_1' + 2k_2 c_{A,0} \cdot [1 - \exp(-k_1' t)]}$$

Furthermore, the rate equation for C is

$$\frac{dc_C}{dt} = k_1' \cdot c_A = c_{A,0} k_1' \cdot \frac{k_1' \cdot e^{-k_1' t}}{k_1' + 2k_2 c_{A,0} \cdot [1 - \exp(-k_1' t)]}$$

We integrate this rate equation by using the substitution $z = \exp(-k_1' t)$, $dz = -k_1' \cdot \exp(-k_1' t) \cdot dt$. Then we obtain

$$c_C = \frac{k_1'}{2k_2} \cdot \ln\left[1 + 2c_{A,0} \cdot \frac{k_2}{k_1'} \cdot [1 - \exp(-k_1' t)]\right]$$

The yield of C at the end of the reaction ($t \to \infty$) is

$$\frac{c_{C,\infty}}{c_{A,0}} = \frac{k_1'}{2k_2 c_{A,0}} \cdot \ln\left(1 + \frac{2c_{A,0} k_2}{k_1'}\right) = \frac{\ln(1+x)}{x}$$

with

$$x = \frac{2c_{A,0} k_2}{k_1'} = \frac{2c_{A,0} k_2}{k_1 c_{B,0}}$$

Problem 21.4 – Consecutive Reaction

Solve the differential equation (21.92)

$$\frac{dc_C}{dt} + k_2' c_C = k_1' a \cdot e^{-k_1' t}$$

Solution. We set $c_C = y \cdot f$ where y is the solution of the corresponding homogeneous differential equation

$$\frac{dy}{dt} + k_2' \cdot y = 0$$

Then

$$y = \alpha \cdot e^{-k_2' t} \quad \text{and} \quad c_C = \alpha \cdot e^{-k_2' t} \cdot f$$

(α = integration constant). Insertion into the original differential equation results in a differential equation for f:

$$\frac{df}{dt} = \frac{k_1'}{\alpha} a \cdot e^{-(k_1' - k_2') t}$$

By integration we obtain (β = integration constant)

$$f = -\frac{k_1'}{\alpha} \cdot \frac{a}{k_1' - k_2'} \cdot e^{-(k_1' - k_2') t} + \beta$$

Then

$$c_C = -\frac{k_1' a}{k_1' - k_2'} \cdot e^{-k_1' t} + \alpha \beta \cdot e^{-k_2' t}$$

PROBLEM 21.5

At time $t = 0$ the concentration c_C equals zero, then

$$\alpha\beta = \frac{k_1' a}{k_1' - k_2'}$$

and

$$c_C = -\frac{k_1' a}{k_1' - k_2'} \cdot \left(e^{-k_1' t} - e^{-k_2' t}\right)$$

What is the time t_{max} required to reach the maximum concentration of c_C? To answer this question we differentiate c_C with respect to t and set the differential quotient equal to zero:

$$-k_1' \cdot e^{-k_1' t} + k_2' \cdot e^{-k_2' t} = 0$$

Then

$$e^{-(k_1' - k_2')t} = \frac{k_2'}{k_1'}$$

and

$$t_{max} = \frac{\ln(k_2' - k_1')}{k_1' - k_2'}$$

Problem 21.5 – Autocatalytic Reaction

Solve the differential equation (21.137)

$$\frac{dc_B}{dt} = k_r \cdot c_B \cdot (a + b - c_B)$$

for the autocatalytic reaction.

Solution. We separate the variables

$$\frac{dc_B}{c_B \cdot (a + b - c_B)} = k_r \cdot dt$$

and integrate. The left integral is solved by using the method of partial fractions; then

$$\frac{1}{a+b} \cdot \ln \frac{c_B}{a+b-c_B} = k_r t + C$$

where C is an integration constant. At $t = 0$ the concentration of B is b, thus

$$C = \frac{1}{a+b} \cdot \ln \frac{b}{a}$$

and we obtain

$$\ln \frac{a c_B}{b(a+b-c_B)} = (a+b)k_r t \quad \text{and} \quad c_B = \frac{a+b}{1 + \frac{a}{b} \cdot e^{-(a+b)k_r t}}$$

Problem 21.6 – Unimolecular Reactions

We consider the isomerizatiom of 3-methyl-cyclobutene into 4-methyl-butadiene:

$$\text{A} \longrightarrow CH_2=CH-CH=CH-CH_3$$

(A: 3-methyl-cyclobutene; B: 4-methyl-butadiene)

Following the Lindemann mechanism of unimolecular reactions the first step in this reaction is a collision between two molecules A, leading to one molecule A in a vibrationally excited state, A^*

$$A + A \underset{k_{-1}}{\overset{k_1}{\rightarrow}} A^* + A$$

In A^* the weakest bond breaks, forming the product B after reorganization

$$A^* \overset{k_2}{\rightarrow} B$$

We have to take into account that collisions between A and B can produce A^* as well

$$A + B \underset{k_{-3}}{\overset{k_3}{\rightarrow}} A^* + B$$

Derive the rate equation for the formation of the product B. Show that the reaction is first order. How does the first order rate constant k' depend on the initial concentration $c_{A,0}$? Compare with the following experimental data (H. M. Frey, D. C. Marshall, Thermal Unimolecular Isomerization of Cyclobutenes, Trans. Faraday Soc. 61, 1715 (1965)) measured at 123.5°C:

$p_{A,0}$/Torr	$c_{A,0}/10^{-6}$ mol L^{-1}	$k'/10^{-4}$ s^{-1}
0.010	4.03	0.48
0.020	8.06	0.57
0.054	21.8	0.79
0.102	41.1	0.92
0.195	78.6	1.03
0.261	105	1.10
0.504	203	1.17

Solution. The rate equations for B and for A^* are

$$\frac{dc_B}{dt} = k_2 \cdot c_{A^*}$$

$$\frac{dc_{A^*}}{dt} = k_1 \cdot c_A^2 - k_{-1} \cdot c_A \cdot c_{A^*} - k_2 \cdot c_{A^*} + k_3 \cdot c_A c_B - k_{-3} \cdot c_B \cdot c_{A^*}$$

Because A and B are of comparable complexity and of equal mass, the rate constants k_1, k_3 and k_{-1}, k_{-3} are assumed to be equal. The intermediate A^* occurs in a small concentration, thus we can apply the steady state approximation $dc_{A^*}/dt = 0$. Thus

$$c_{A^*} = \frac{k_1 \cdot c_A \cdot (c_A + c_B)}{k_{-1} \cdot (c_A + c_B) + k_2}$$

and with $c_A + c_B = c_{A,0}$ and $c_A = c_{A,0} - c_B$ we obtain

$$\frac{dc_B}{dt} = k_2 \cdot c_{A^*} = \frac{k_1 k_2 \cdot c_{A,0}}{k_{-1} \cdot c_{A,0} + k_2} \cdot (c_{A,0} - c_B)$$
$$= k' \cdot (c_{A,0} - c_B)$$

Thus the product B is formed according to a first order kinetics, the first order rate constant k'

$$k' = \frac{k_1 k_2 \cdot c_{A,0}}{k_{-1} \cdot c_{A,0} + k_2} = \frac{(k_1 k_2/k_{-1}) \cdot c_{A,0}}{c_{A,0} + k_2/k_{-1}}$$

depending on the initial concentration $c_{A,0}$. At low $c_{A,0}$ we expect $k' = k_1 \cdot c_{A,0}$, i.e. k' increases linearly with $c_{A,0}$. At high $c_{A,0}$ we expect $k = k_1 k_2/k_{-1}$, i.e. k' becomes independent of the initial concentration $c_{A,0}$.

This can be easily understood: At low $c_{A,0}$ only few collisions between A and B occur, thus the formation of A^* is rate determining. At high $c_{A,0}$ the time between two collisions between A^* and A or B becomes shorter than the lifetime of A^*, and the isomerization becomes rate determining.

The prediction of the theory is compared with the experimental data in Figure 21.41. The experimental points are fairly reproduced. However, measurements at higher initial concentrations reveal that the theory needs improvement. Essentially, it has to be taken into account that the unimolecular reaction step depends on the specific structure of A^*: such improvements were introduced by Hinshelwood, Slater, Rice-Ramsberger-Kassel (RRK-Theorie), and Rice-Ramsberger-Kassel-Marcus (RRKM-Theorie).

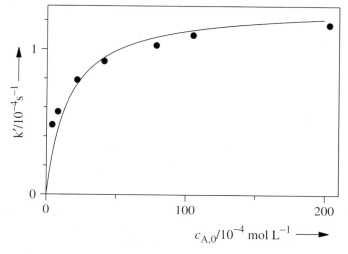

Figure 21.41 Experimental data for the isomerization of 3-methyl-cyclobutene (dots) compared with the prediction of the Lindemann mechanism (solid line). The calculation was carried out with the values $k'_\infty = k_1 k_2/k_{-1} = 1.3 \times 10^{-4}\,\text{s}^{-1}$ and $k_1 = 0.084\,\text{M}^{-1}\,\text{s}^{-1}$.

22 Organized Molecular Assemblies

We have considered both simple and increasingly complex systems of molecules. Now we focus on systems where the specific molecular arrangement is crucial for the pertinent properties: the molecular arrangement at surfaces, in membranes, and in polymers. These assemblies can be called organized since the specific arrangement of molecules gives rise to very specific properties that can be achieved by planning the molecular arrangement. Organized assemblies of molecules have an enormous richness in properties resulting from the multitude of interactions. The interplay between the different kinds of molecules leads to exciting phenomena.

Basic bioreactions and technical processes occur at interfaces; they are bound to membranes, vesicles, monolayers, and micelles. Liquid crystals are important functional components in display systems. The huge variety of properties of artificial polymers and biopolymers is related to the particular interaction of the molecular chains. The principles of self-organizing molecular structures are basic in understanding the occurrence of complexity in natural and artificial supramolecular systems.

22.1 Interfaces

22.1.1 Surface Tension and Interfacial Tension

A molecule in the bulk of a liquid is bound to the neighbor molecules surrounding it. For a molecule at the surface the neighbors on top are missing (Figure 22.1). Such a molecule, therefore, has a higher energy than the others. Thus a liquid becomes spherical when forces changing its shape are absent (a drop of oil in water or a liquid in free space without gravitational forces): the sphere, for a given volume, has the smallest surface. When increasing the surface of a liquid, the number of molecules in energy-rich interfacial positions increases. The work to be done at constant pressure and temperature to increase the area of the surface by a small amount dA in a reversible way is proportional to dA

$$dG = \gamma \cdot dA \qquad (22.1)$$

22.1 INTERFACES

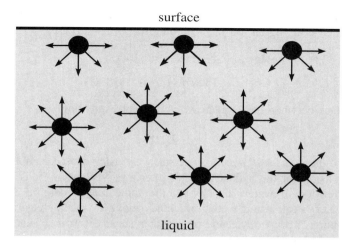

Figure 22.1 Higher energy of a molecule at the surface than in the bulk of a liquid.

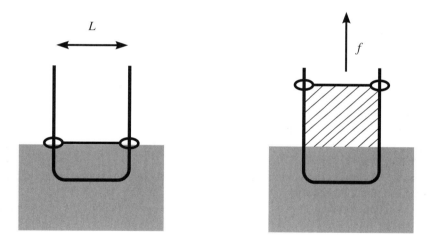

Figure 22.2 Withdrawing a lamella from a soap solution. Force f equals $\gamma 2L$.

γ is called the *surface tension*. The meaning of γ is seen if we withdraw a lamella from a soap solution (Figure 22.2). Work must be done to pull out the frame; lifting it by dh requires a force f. In the reversible limit the work is (considering both sides of the lamella)

$$\mathrm{d}w = f \cdot \mathrm{d}h = \gamma \cdot \mathrm{d}A = \gamma \cdot 2L \cdot \mathrm{d}h \tag{22.2}$$

where L is the length of the movable frame. Thus

$$\gamma = \frac{f}{2L} \tag{22.3}$$

γ is between 18 and 30 mN m^{-1} for most organic liquids. For clean water at 20 °C it is as large as 72.8 mN m^{-1} because hydrogen bonds are broken when increasing A. It is much smaller for a soap solution than for clean water because soap molecules proceed to the surface. For molten metals or molten salts γ is large because the Coulomb attraction must be overcome when bringing an ion from the interior to the surface:

$$\text{Hg: } \gamma = 470\,\text{mN m}^{-1}\ (20\,°\text{C})$$

$$\text{Cu: } \gamma = 1300\,\text{m Nm}^{-1}\ (1130\,°\text{C})$$

For the interface between two liquids we obtain correspondingly

$$dG = \gamma \cdot dA \qquad (22.4)$$

where γ is called the *interfacial tension*. The surface tension of n-decane is $23.9\,\text{mN m}^{-1}$; the interfacial tension for the n-decane/water interface, $51.2\,\text{mN m}^{-1}$, is almost as large as the surface tension of water. This is for the same reason (Figure 22.3): hydrogen bonds in the water break when water molecules move to the increasing decane/water interface. Therefore, a drop of decane on a water surface forms a lens. The geometry of the lens is such that the sum of the forces acting in the direction tangential to the surface and perpendicular to the element dl on the line between the three phases is zero (Figure 22.4(a)), if gravitational forces are neglected. For the horizontal components of the force f we obtain:

$$\gamma_{\text{decane/air}} \cdot \cos\alpha \cdot dl + \gamma_{\text{decane/water}} \cdot \cos\beta \cdot dl = \gamma_{\text{water/air}} \cdot dl \qquad (22.5)$$

and for the vertical components:

$$\gamma_{\text{decane/air}} \cdot \sin\alpha = \gamma_{\text{decane/water}} \cdot \sin\beta \qquad (22.6)$$

A quite different situation is observed when replacing n-decane by an aliphatic alcohol, e.g. n-octyl alcohol. On the one hand, $\gamma_{\text{octyl alcohol/air}} = 28\,\text{mN m}^{-1}$ is only moderately larger than $\gamma_{\text{decane/air}}$ because the hydrocarbon tails of the molecules at the surface face the air in both cases. On the other hand, however, $\gamma_{\text{octyl alcohol/water}}$ is only $8.5\,\text{mN m}^{-1}$ because an octyl alcohol molecule at the interface can form a hydrogen bond with a water molecule, diminishing the work to bring it to the surface.

As a consequence, when increasing the area of the n-octyl alcohol by dA, then the area of the water/air interface decreases correspondingly and we obtain

$$dG = \gamma_{\text{water/air}} \cdot (-dA) + \gamma_{\text{water/octyl alcohol}} dA + \gamma_{\text{octyl alcohol/air}} \cdot dA$$

$$= dA \cdot (-72.8 + 8.5 + 28)\,\text{mN m}^{-1} = -36\,\text{mN m}^{-1} \cdot dA$$

dG is negative, thus octyl alcohol spreads on water (Figure 22.4(b)).

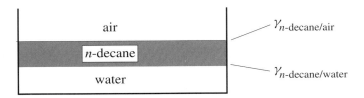

Figure 22.3 Interfacial tension. Hydrogen bonds break when water molecules move to the n-decane/water interface. Therefore $\gamma_{n-\text{decane/water}}$ ($51.2\,\text{mN m}^{-1}$) is much higher than the surface tension of n-decane ($23.9\,\text{mN m}^{-1}$).

22.1 INTERFACES

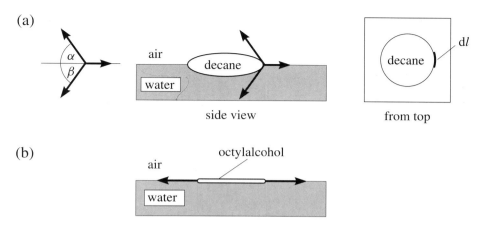

Figure 22.4 Two liquids in contact: (a) *n*-decane and water, formation of a lens; (b) *n*-octyl alcohol and water, spreading.

What happens when dipping a bar into a liquid? A molecule at the interface is attracted to the bar and toward the bulk of the liquid. Depending on which forces predominate we can distinguish between two cases. (1) The forces attracting the molecule to the bar predominate. When the surface of the bar in Figure 22.5(a) is covered by a film of the liquid, the bar gets wet. An example is water on clean glass; the surface of the glass is called *hydrophilic*. (2) When the forces between the molecules in the liquid facing the solid and the molecules in the bulk of the liquid predominate, the bar in Figure 22.5(b) keeps dry. Examples are mercury at glass and water at paraffin; the surface of paraffin is called *hydrophobic*.

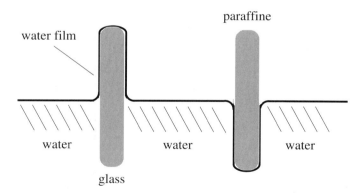

Figure 22.5 Dipping a solid (glass and paraffin) into a liquid (water): (a) wetting the surface of glass; (b) not wetting the surface of paraffin.

When dipping a glass capillary into water, the water/air interface grows by forming a liquid film on the wall of the capillary. The interfacial area can be diminished by lifting liquid in the capillary. When lifting the liquid by dh the water/air interface is diminished by $2\pi r \mathrm{d}h$, the change in Gibbs energy is

$$\mathrm{d}G_1 = \gamma \cdot \mathrm{d}A = -\gamma \cdot 2\pi r \cdot \mathrm{d}h \tag{22.7}$$

(r is the inner radius of the capillary, Figure 22.6(a)). The liquid is lifted until the energy decrease by diminishing the water surface equals the energy increase dG_2 by lifting the liquid in the capillary against gravity:

$$dG_2 = \pi r^2 hg\rho \cdot dh \tag{22.8}$$

(g is the acceleration due to gravity, and ρ is the density of the liquid). The maximum height h_{max} is reached for

$$dG_1 + dG_2 = 0 \tag{22.9}$$

Then

$$h_{max} = \frac{2\gamma}{g\rho r} \tag{22.10}$$

The meniscus in the capillary is not planar, but roughly a half-sphere (Figure 22.6(b)). The reason is seen when proceeding from Figure 22.6(a) to Figure 22.6(b). The surface of the half-sphere is

$$A_{meniscus} = 2\pi r^2 \tag{22.11}$$

while the corresponding surface in the case of a planar meniscus is

$$A_{meniscus} = \pi r^2 + 2\pi r \cdot r = 3\pi r^2 \tag{22.12}$$

Thus the work to decrease the surface in proceeding from Figure 22.6(a) to Figure 22.6(b) is

$$\Delta G = \gamma 2\pi r^2 - \gamma 3\pi r^2 = -\gamma \pi r^2 \tag{22.13}$$

The energy $\gamma \pi r^2$ is used to lift the liquid against gravity beyond h. In reality the meniscus is not a half-sphere but has the shape as in Figure 22.6(c). In this case, G has reached its minimum.

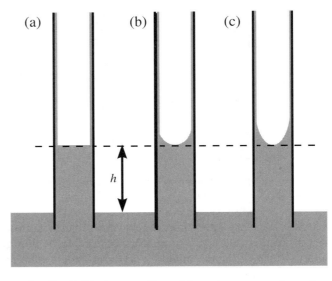

Figure 22.6 A wetting liquid lifts in a capillary: (a) meniscus approximated as being flat; (b) as being half-spherical; (c) real shape of meniscus.

22.1.2 Surface Films

We consider a solution of a surfactant in water (e.g., Na-oleate). When increasing the concentration of the surfactant, the surface excess Γ of the solute (i.e., the amount of surfactant per unit area of surface) increases and the surface tension decreases. For a dilute solution the relation

$$\Gamma = \frac{n_{\text{surfactant}}}{A} = -\frac{c}{RT}\frac{d\gamma}{dc} \qquad (22.14)$$

holds (*Gibbs adsorption equation*, Box 22.1).

Γ increases with the concentration c and with the tendency of the solute to decrease the surface tension $-d\gamma/dc$; furthermore, it decreases with increasing thermal motion counteracting the separation. For checking the relation Γ can be measured by removing the surface film by stripping it off with a fast moving razor blade.

Box 22.1 – Surface Film; Gibbs Adsorption Equation

We transfer a small amount n of solute from solution a (concentration c) to solution b (concentration $c + \Delta c$) in two reversible ways (Figure 22.7).

(1) Osmotically (Chapter 17):

$$\Delta G_{\text{osm}} = nRT \cdot \ln \frac{c + \Delta c}{c} = nRT \frac{\Delta c}{c}$$

(2) We remove the amount n by drawing a lamella (area A) from solution a and supply this amount reversibly to solution b. The work in withdrawing the lamella is $\Delta G_1 = \gamma A$.

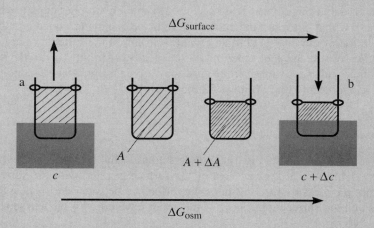

Figure 22.7 Reversible transfer of surfactant from solution of concentration c (left) to solution of concentration $c + \Delta c$ (right) as a surface film on a lamella (withdrawing lamella from solution on the left) (a), diminishing area, transfer lamella to solution on the right (b).

Continued on page 738

> Continued from page 737
>
> Now we slightly diminish the area of the lamella to reach the surface excess $\Gamma + \Delta\Gamma$ as in solution b. The corresponding work is $\Delta G_2 = \gamma \Delta A$. Finally we transfer the lamella reversibly to solution b:
>
> $$\Delta G_3 = -(\gamma + \Delta\gamma) \cdot (A + \Delta A) = -\gamma A - \gamma \Delta A - \Delta\gamma A$$
>
> (neglecting the term $\Delta\gamma\Delta A$). Thus
>
> $$\Delta G_{\text{surface}} = \Delta G_1 + \Delta G_2 + \Delta G_3 = -\Delta\gamma A$$
>
> By equalizing the work for both processes we obtain with $\Gamma = n/A$:
>
> $$nRT\frac{\Delta c}{c} = -\Delta\gamma A = -\Delta\gamma\frac{n}{\Gamma}$$
>
> and
>
> $$\boxed{\Gamma = -\frac{c}{RT} \cdot \frac{d\gamma}{dc}}$$
>
> (differences replaced by differentials).

22.1.3 Insoluble Monolayers

We consider a surfactant which is insoluble in water, for example, a fatty acid molecule that consists of a hydrophilic part (carboxyl group) and a hydrophobic part (a long hydrocarbon chain) which prevents the molecule from entering the aqueous subphase. We spread a solution of the fatty acid in an organic solvent on the water surface. After evaporation of the solvent the molecules, first statistically distributed over the surface, are pushed together with a movable barrier until they are densely packed forming a compact monolayer one molecule thick (Figure 22.8).

Surface pressure

A force $f = \Pi L$ (weight W, Figure 22.8) is acting on the movable barrier of length L to establish equilibrium at a given area of the surface covered with the surfactant. Π is called the *surface pressure*. It constitutes the difference between the surface tension γ of clean water and the surface tension γ_s of the water covered with the surfactant:

$$\Pi = \gamma - \gamma_s \tag{22.15}$$

The force γL shifting the barrier towards the clean surface is only partly counteracted by the force $\gamma_s L$. Then the additional force

22.1 INTERFACES

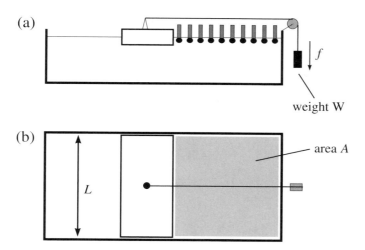

Figure 22.8 Insoluble monolayer. Fatty acid molecules are spread on a water surface and compressed by a movable barrier. (a) Side view; (b) view from top. Surface pressure $\Pi = f/L$.

$$f = \gamma L - \gamma_s L = \Pi L \qquad (22.16)$$

is required to keep equilibrium.

Π is almost zero as long as the surfactant molecules are apart. Π increases strongly when a compact monolayer is reached (Figure 22.9). When the force f reaches a critical value, sections of the monolayer begin to overturn. The film collapses and becomes irreversibly destroyed. When using fatty acid as surfactant, the steep increase of Π with decreasing area occurs at $A = N \times 0.2\,\text{nm}^2$, where N is the number of molecules in the layer. This value agrees well with the cross-section of a fatty acid chain obtained from bond lengths and van der Waals radii.

Monolayer transfer

A monolayer of densely packed fatty acid molecules can be transferred to the surface of a solid support, for example, a glass slide that has been dipped in the water before producing the film. When taking the glass slide out of the water the slide is covered by the film. The carboxyl groups of the fatty acid molecules bind to the surface of the glass slide, and the ends of the hydrocarbon chains are now facing the air (Figure 22.10, step 1). In this way the surface of the slide, which is hydrophilic initially, becomes hydrophobic.

When dipping in the slide again, it is covered by a second layer of fatty acid. The ends of the hydrocarbon chains of the fatty acids on the glass slide and those at the water surface bind to each other (Figure 22.10, step 2). The carboxyl groups are facing to the water. When taking out the slide a third monolayer of fatty acid is attached (Figure 22.10, step 3). This technique to produce multilayers of fatty acid (the *Langmuir–Blodgett method*) can be extended and applied to many other amphiphilic molecules (see Chapter 23).

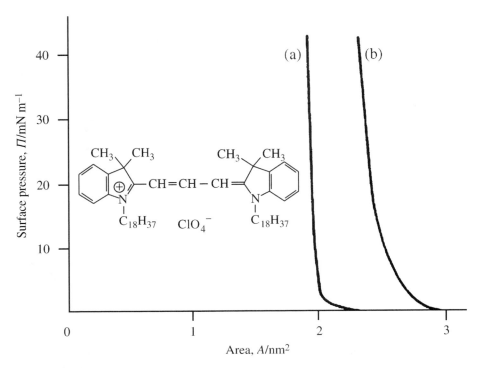

Figure 22.9 Insoluble monolayer. Surface pressure Π versus area A per 10 arachidic acid molecules. Compact monolayer is reached when Π is strongly increasing with decreasing A. (a) Fatty acid (cadmium salt of arachidic acid $C_{19}H_{39}$–COOH). The area per molecule is $0.2\,\mathrm{nm}^2$ (cross-section of the hydocarbon chain), i.e. $2\,\mathrm{nm}^2$ per 10 arachidic acid molecules. (b) Fatty acid and dye in molar ratio 10:1. Additional area of $0.4\,\mathrm{nm}^2$ due to the two hydrocarbon chains per dye molecule incorporated in the fatty acid matrix. The chromophores below matrix are not crowded and, therefore, do not effectively contribute to the surface pressure. (H. Bücher, H. Kuhn, B. Mann, D. Möbius, L. v.Szentpály, P. Tillmann, Photographic Science and Engineering 11, 233 (1967)).

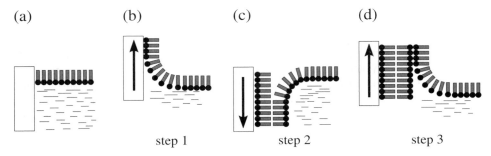

Figure 22.10 Transfer of fatty acid monolayers to a solid support (glass slide): Langmuir–Blodgett method. (a) Glass slide in contact with monolayer on water surface. (b) Glass slide moved upwards: monolayer with hydrophilic groups attached on the glass surface. (c) Glass slide moved downwards: second monolayer with hydrophobic groups attached. (d) Glass slide moved upwards: third monolayer attached. (B. Blodgett, J. Amer. Chem. Soc. 57, 1007 (1935); I. Langmuir, J. Franklin Inst. 218, 143 (1934); Proc. Roy. Soc. London, Ser. A 170, 15 (1939)).

22.1.4 Solid Surfaces

The surface of a solid, in general, is not well defined. However, when monolayers have been transferred to a glass slide, the surface is relatively well defined. This is also the case when the glass slide is covered with a monolayer by adsorption of appropriate molecules interlocking into a well-ordered array (*self-assembled monolayers*).

Example – monolayers on glass

A clean glass slide is dipped into a solution of fatty acid in hexane. The head groups bind to the hydrophilic glass surface and a compact monolayer of well-ordered hydrocarbon chains is formed. In this case the monolayer is bound to the surface by electrostatic forces.

When using chlorinated silane compounds, such as dimethyldichlorosilane, the Cl atoms react with the OH groups bound to Si at the glass surface and a well ordered monolayer is bound to the glass surface by covalent bonds (hydrophobic surface).

When putting a drop of a liquid on a solid surface, a contact angle Θ is established between the solid surface and the liquid (Figure 22.11(a)). Θ is determined by the balance of cohesive forces measured by γ_{lg}, γ_{sl}, and γ_{sg} (energy to create a unit area of liquid/gas, solid/liquid and solid/gas interface, respectively):

$$\gamma_{sg} = \gamma_{sl} + \gamma_{lg} \cdot \cos \Theta \tag{22.17}$$

Thus

$$\cos \Theta = \frac{\gamma_{sg} - \gamma_{sl}}{\gamma_{lg}} \tag{22.18}$$

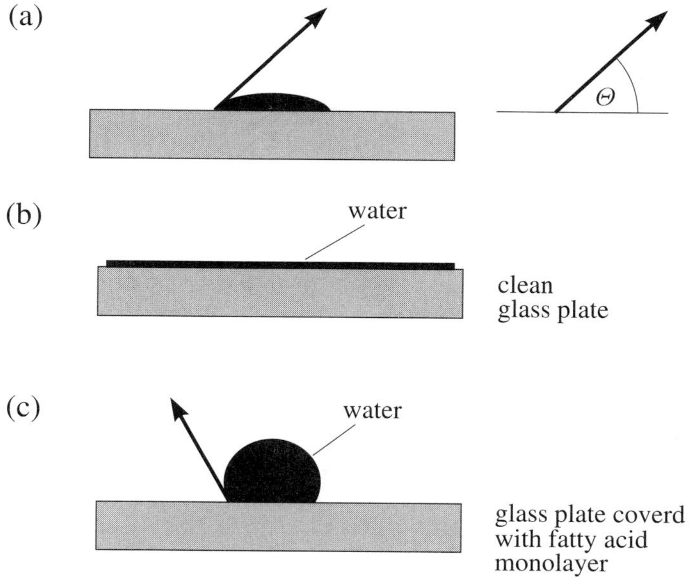

Figure 22.11 Drop of liquid on solid. (a) Definition of contact angle Θ; (b) complete wetting for water on a clean glass plate ($\Theta = 0°$); (c) water on a fatty acid monolayer ($\Theta = 109°$).

When $\gamma_{sl} = \gamma_{sg}$ the contact angle becomes $\Theta = 90°$. An example is water on a solid surface consisting of long-chain alcohols, e.g. $CH_3-(CH_2)_{20}-OH$.

For water on a nickel surface we find $\Theta = 27°$, and $\Theta = 105°$ is obtained for water on a paraffin surface. The contact angle for water on a clean glass surface is zero: complete wetting (Figure 22.11(b)). (In this case equation (22.18) is no longer valid because the equilibrium cannot be established.) The situation changes dramatically when a monolayer of a fatty acid is bound to the glass surface: $\Theta = 109°$ (Figure 22.11(c)). By depositing an appropriate monolayer on top of the glass slide, any contact angle can be purposely achieved.

Example – contact angle on different monolayers

A well-ordered mixed monolayer of $CH_3(CH_2)_{11}SH$ (A) and $HO(CH_2)_{11}SH$ (B) is obtained by self-assembling on a gold surface and reacting the SH group with the gold. The sample is immersed in octane and the contact angle of a drop of water on the surface is measured. Figure 22.12(a) shows the result for a monolayer of A (left), A and B in a 1 : 1 molar mixing ratio (center) and B (right). The contact angle Θ changes significantly when octane is replaced by air (Figure 22.12(b)).

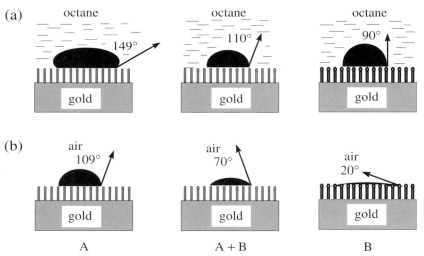

Figure 22.12 Drop of water on differently organized monolayers. Left, monolayer A = $CH_3(CH_2)_{11}SH$ (CH_3 groups at surface), right, monolayer B = $HO(CH_2)_{11}SH$ (OH groups at surface), and center, mixed monolayer of A and B in a 1 : 1 molar ratio (A. Ulman, Ultrathin Organic Films, Academic Press, San Diego, 1991; S. O. Evans, P. Sanassy, A. Ulman, Thin Solid Films 243, 325 (1994)).

22.1.5 Micelles

When a soluble surfactant dissolved in water reaches a certain concentration (e.g., $0.2\, mol\, L^{-1}$ for sodium dodecyl sulfate), the molecules begin to associate (*critical micelle*

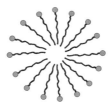

Figure 22.13 Formation of a spherical micelle above the critical micelle concentration.

concentration). The hydrocarbon chains, when forming the aggregate, keep together and the hydrophilic heads face the water. The water molecules located around the tail of the surfactant molecule before aggregating are now in the bulk, forming additional hydrogen bonds (Figure 22.13). This results in a negative contribution to the enthalpy change in micelle formation.

On the other hand, the charged heads become close to each other and their repulsion leads to a positive contribution to ΔH. The two effects essentially compensate, and thus the enthalpy of micelle formation is small. The temperature dependence of the critical micelle concentration must be small; this can be observed. Therefore, the formation of micelles taking place above the critical micelle concentration must be accompanied by an increase in entropy to allow ΔG to be negative: $\Delta S = 140 \, \text{J K}^{-1} \, \text{mol}^{-1}$ is measured at 300 K.

This is a surprising result. We have to keep in mind that during micelle formation, two species are separated from their mixture. Such a process corresponds to a decrease in entropy. What is going on? The water shell surrounding the surfactant molecule before taking part in aggregation has an ordered ice-like structure which breaks down when the water molecules in the shell become part of plain water. This gives rise to an entropy production exceeding the entropy consumption in the separation process.

Micelles fulfilling the given requirements can be very differently shaped depending on the nature of the surfactant and its concentration. Spherical micelles are predominantly formed by surfactants with bulky or charged head groups slightly above the critical micelle concentration. With increasing concentration there is a competition of the micelles for water molecules forming their solvation shell; this favors cylindrical structures which become arranged in ordered patterns, as indicated in Figure 22.14.

22.2 Liquid Crystals

Molecules of specific shape (sufficiently long or discus-like) aggregate in a specific way. A substance consisting of such molecules remains in an ordered state above the melting point. Although the three-dimensional lattice breaks down on heating above the melting point, the molecules are not disordered on a macroscopic scale as in an ordinary liquid, but still have a macroscopic orientational order. Only when the temperature is increased above a critical value (*clearing point*) does the thermal motion dominate. Transition into the liquid state occurs. These aggregates existing in the temperature range between melting point and clearing point are called *liquid crystals*.

Figure 22.14 Rod-like micelles (courtesy E. Sackmann).

Let us consider typical examples. The compound 4-methoxybenzyliden-4′-butylaniline (MBBA)

$$CH_3-\ddot{\underset{..}{O}}-\bigcirc-CH=\ddot{N}-\bigcirc-C_4H_9$$

has a melting point of 22 °C and a clearing point of 47 °C; pentylcyanobiphenyl (5CB)

$$C_5H_{11}-\bigcirc-\bigcirc-C\equiv N\!:$$

has a melting point of 23 °C and a clearing point of 35 °C. The MBBA molecule is 2500 pm long and 500 pm in diameter. Then in a good approximation these molecules can be treated as rods. The rods are oriented in an exact parallel manner in the crystalline state (Figure 22.15, left), and approximately parallel above the melting point (Figure 22.15, center: nematic state). At sufficiently high temperature the thermal motion destroys the macroscopic order (Figure 22.15, right: transition into an ordinary liquid).

An even stronger orientation is achieved in the smectic state (Figure 22.16), where the molecules are arranged in distinct layers. If the molecules are oriented perpendicular to the layer planes, the arrangement is called *smectic A*; if the molecules are somehow tilted, the arrangement is called *smectic C*.

22.2 LIQUID CRYSTALS

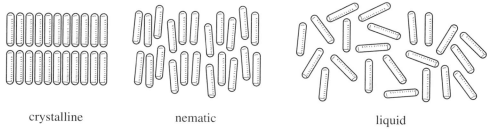

Figure 22.15 Phase transitions in a nematic liquid crystal.

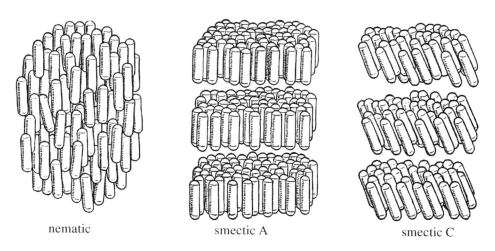

Figure 22.16 Comparison of nematic and smectic liquid crystals (G. Solladié, R. G. Zimmermann, *Angew. Chem. Int. Ed. Engl.* **23**, 348 (1984)).

If the molecules are chiral, then adjacent layers can form a helix like structure: *cholesteric state* (Figure 22.17). An example is cholesteryl benzoate

with a melting point of 145 °C and a clearing point of 179 °C.

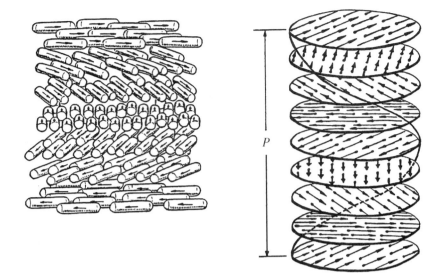

Figure 22.17 Arrangement of molecules in a cholesteric liquid crystal (left); schematic representation emphasizing helical structure with period length P (right). Reference see Figure 22.16.

22.2.1 Birefringence

Because of the parallel orientation of the individual molecules, the index of refraction is different for light polarized parallel and perpendicular to the orientation of the molecules. This can be shown in an experiment where a prism is filled with a nematic liquid crystal (Figure 22.18). The surfaces of the glass plates of the prism have been slightly scratched parallel to the edge of the prism such that all liquid crystal molecules are aligned parallel to that direction. As a consequence of the different indices of refraction ($\hat{n}_\parallel$ and $\hat{n}_\perp$), a light beam passing the prism is split into two beams with different polarization (Figure 22.18). When warming up the liquid crystal in the prism the difference in the indices of refraction becomes smaller and smaller, and thus the two beams come closer together; they coincide at the clearing point. The temperature dependence of $\hat{n}_\parallel$ and $\hat{n}_\perp$ is displayed in Figure 22.19 in the case of *p*-azoxyanisole (PAA)

22.2.2 Selective Reflection

Because adjacent layers in a cholesteric liquid crystal are twisted, the plane of polarized light passing through the liquid is changed. The effect depends on the height h_{helix} of one period of the helix-like structure. If the wavelength λ of the incident light in the medium equals h_{helix}

$$\lambda = h_{helix} \tag{22.19}$$

22.2 LIQUID CRYSTALS

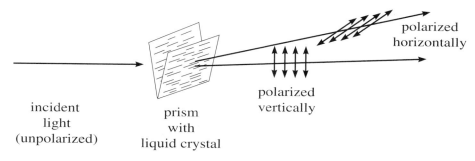

Figure 22.18 Birefringence of a nematic liquid crystal measured with a prism.

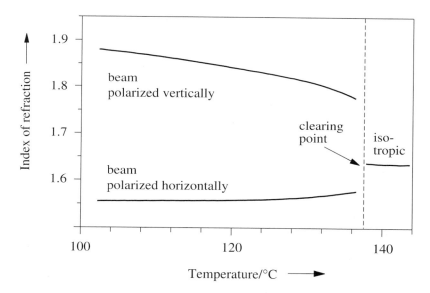

Figure 22.19 Indices of refraction $\hat{n}_\parallel$ (beam polarized vertically) and $\hat{n}_\perp$ (beam polarized horizontally) for a nematic liquid crystal versus temperature (PAA, $\lambda = 589\,\mathrm{nm}$). Above the clearing point the liquid is isotropic (G. Heppke, Ch. Bahr in: Bergmann–Schaefer, Lehrbuch des Experimentalphysik, Vol. 5, de Gruyter, Berlin 1992, p. 397).

then the liquid crystal acts like a periodic structure for the passing light. This leads to the occurrence of interference colors. Because h_{helix} changes with temperature, the observed color changes as well. This effect can be used to indicate the surface temperature of a sample.

22.2.3 Electro-optical Effects

A nematic liquid crystal is placed between two parallel glass plates at a distance of about 1 μm. The glass plates are scratched, as in the case of the prism experiment. Thus the molecules can be aligned parallel to the scratching direction. Then the glass

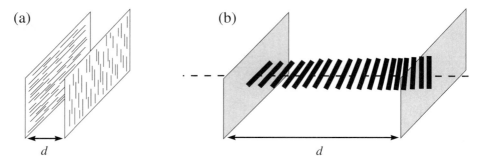

Figure 22.20 Twisting liquid crystal molecules. (a) Arrangement of two glass plates at distance d scratched perpendicular to each other; (b) helical arrangement of liquid crystal molecules placed between both plates.

plates are twisted relative to each other at an angle of 90°; this results in a corresponding twisting of the molecules in the thin layer (helical arrangement, Figure 22.20). As in the case of a cholesteric crystal, the plane of light passing the arrangement is changed.

A polarizer is attached on the left of the arrangement, and an analyzer rotated by 90° is attached on the right. Then, in the absence of the twisted nematic liquid crystal, light coming from the left cannot pass the arrangement. In the presence of the liquid crystal, however, the direction of polarization of the light wave is rotated 90^0 and light can pass the analyzer.

The situation changes drastically when an electric field is applied between the glass plates. Such a field can be established by coating the surface of the glass plates with a thin layer of a conducting, but still transparent, material (e.g., indium–tin dioxide) and applying a voltage. By the electric field strength the long axes of the molecules are turned in the direction of the field: there is a strong polarizability in the direction of the long axis caused by the mobility of the π electrons. As a consequence, the liquid crystal (which is under a strain by its being twisted), changes into a liquid-like state by an extremely small perturbation. A field of $10^6 \, \text{V m}^{-1}$ is sufficient to trigger the phase transition. Then the light can no longer pass the corresponding polarizer.

An important application of this effect is an optical display (Figure 22.21). Usually, in such a display the light passes the arrangement from the top and is reflected at a mirror located at the bottom. Then, without the electric field the display appears white, and with applied field it becomes black. To avoid any chemical change in the liquid crystal, an alternating voltage of 2–3 V with a frequency of 1000 Hz is applied (Schadt–Helfrich display or TN (twisted nematics) display). A current of $1 \, \mu\text{A cm}^{-2}$ is observed which can be related to the turning of the liquid crystal molecules triggered by the applied alternating field. This current is much smaller than the leakage current of a typical battery.

Typical response times for switching a TN display are on the order of 10 ms (Figure 22.22), and the minimum voltage needed to switch between both states is on the order of some volts (Figure 22.23).

22.2 LIQUID CRYSTALS

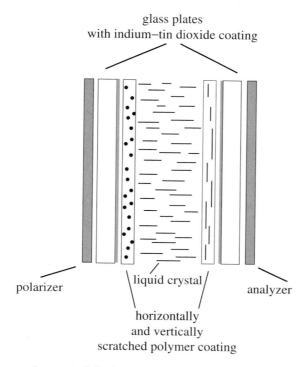

Figure 22.21 Scheme of an optical display using a twisted liquid crystal (Schadt–Helfrich effect). Polarizer and analyzer are twisted by 90°. Thus without applied external voltage light passes the cell (the display is transparent). With a voltage applied between both glass plates light can no longer pass (the display becomes dark).

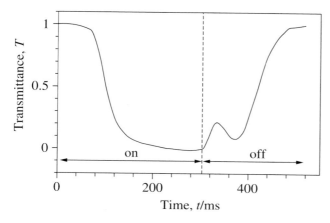

Figure 22.22 Time response of the transmittance T of a liquid crystal display when switching the external voltage $V = 8$ V on and off (M. Schadt, *Mol. Cryst. Liq. Cryst.*, 1988, **165**, 405).

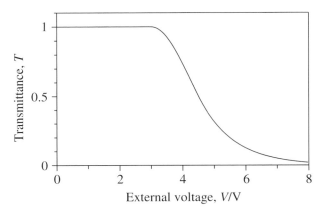

Figure 22.23 Response of the transmittance T of a liquid crystal display when gradually increasing the external voltage V (M. Schadt, W. Helfrich, *Appl. Phys. Lett.*, 1971, **18**, 127).

22.3 Membranes

22.3.1 Soap Lamella

A soap lamella can be made of a solution of the amphiphile sodium dodecyl sulfate (an amphiphile is a molecule with a hydrophilic and an oleophilic part)

in water. The lamella is stable when being kept in air saturated with the solvent vapor to avoid evaporation of liquid from the lamella. The lamella becomes increasingly thinner by solvent flow into the bulk of the solution and reaches a constant thickness of almost molecular dimension in a spontaneous process. This can be seen by the disappearance of the interference colors of the lamella well known from soap bubbles (occurrence of a *black lamella*). The hydrophobic ends of the hydrocarbon chains of the surfactant are facing the air and the hydrophilic groups are facing the water layer bearing the counterions of the surfactant. In this way an equilibrium is reached governed by the Coulomb attraction of the negatively charged surfactant monolayers with the positive counterions and the Coulomb repulsion of the two monolayers (Figure 22.24).

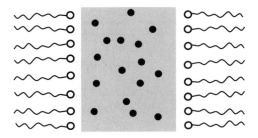

Figure 22.24 Soap lamella, equilibrium reached at 3 nm thickness of water layer. Open circles: negative charges of head groups; full circles: positive charges of counterions.

22.3.2 Black Lipid Membranes

A hole of about 1 mm diameter in a Teflon substrate in contact with water is covered with a tiny amount of the amphiphile phosphatidylcholine in an organic solvent (*n*-decane).

A film forms which becomes increasingly thinner, and finally a double layer is present. The hydrophilic groups of the surfactant are facing the water solutions on the left and on the right (Figure 22.25) and the hydrocarbon chains face each other. Similar to soap lamellas, the formation of the double layer membrane can be seen by a sudden appearance of a black spot that is gradually extending over the entire hole by extruding the organic solvent. The bilayer, like the black lamella of soap, is reached in equilibration.

The presence of the membrane can be easily checked by measuring the electric resistance which drops from $10^8 \, \Omega$ to $1 \, \Omega$ when the membrane collapses. The structure can be checked by measuring the capacity of the film (applying an alternating voltage between the two solutions). The capacitance C of a dielectric layer of thickness d and area A is

$$C = \varepsilon_r \varepsilon_0 \frac{A}{d} \tag{22.20}$$

For $A = 1 \, \text{cm}^2$ a capacitance of $0.5 \, \mu\text{F}$ and a resistance of $10^8 \, \Omega$ was measured; then with $\varepsilon_r = 2.5$ we obtain a distance $d = 4.4 \, \text{nm}$, which is in reasonable agreement with the assumption that the membrane consists of an ordered double layer.

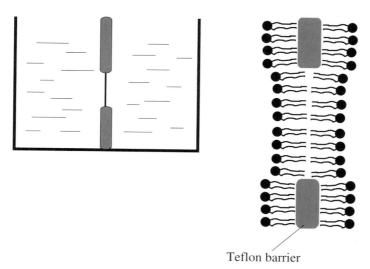

Figure 22.25 Black lipid membrane (phosphatidylcholine). (M. Montal, P. Mueller, Formation of bimolecular Membranes from Lipid Monolayers, Proc. Natl. Acad. Sci USA 69, 3561 (1972)).

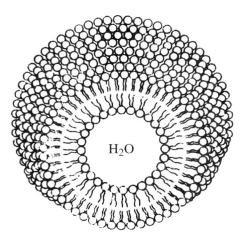

Figure 22.26 Liposome. Spherical particle: Aqueous medium inside and outside. Cross-section showing double layer structure. Same reference as for Figure 22.14.

22.3.3 Liposomes

When a micellar solution of phosphatidylcholine is exposed to ultrasound the micelles are disrupted and spherical particles consisting of a bilayered membrane (called *lipid vesicles* or *liposomes*) with diameter of about 20 nm are formed (Figure 22.26). Larger liposomes of 100 nm diameter are obtained by pressing the micellar solution through a nucleopore filter. Liposomes are useful models mimicking biomembranes (see Chapter 23).

22.3.4 Biomembranes

Biomembranes have many similarities with black lipid membranes. Resistance and capacity, thickness, and permeation of water are similar. Essentially, biomembranes can be considered as lipid bilayers with proteins and other compounds incorporated or attached. The proteins have a tremendous richness in specific properties. Proteins and complexes of interlocking proteins are embedded in the lipid matrix forming distinct domains with complex functionality (see examples in Chapter 23).

Small molecules can easily diffuse laterally in the membrane, indicating that the lipid matrix behaves liquid-like. This is important for an understanding of the variety of membrane-bound reactions in biosystems. The mobility of molecules in the lipid matrix can be measured by incorporating fluorescent molecules in membranes and bleaching a membrane section. From the time dependence of the borderline broadening, the diffusion coefficient can be evaluated.

Example – diffusion coefficient from borderline broadening

In an erythrocyte membrane the border width B of incorporated fluorescent molecules increases with the square root of time. After 1 minute the border width increases to $10\,\mu m$.

22.3 MEMBRANES

Using the Einstein–Smoluchowski relation with $\overline{x^2} = B^2$:

$$B^2 = 2Dt$$

we obtain

$$D = \frac{B^2}{2t} = \frac{(10^{-3}\,\text{cm})^2}{2 \times 60\,\text{s}} = 8 \times 10^{-9}\,\text{cm}^2\,\text{s}^{-1}$$

Diffusion coefficients on the order of 10^{-7} to $10^{-9}\,\text{cm}^2\,\text{s}^{-1}$ have been measured. (E. Sackmann, Physical Foundations of the Molecular Organization and Dynamics of Membranes in: W. Hoppe, W. Lohmann (Eds.): Biophysics, Springer, Heidelberg 1983)

In biomembranes, ions are transferred through specific channels. This is demonstrated by contacting a capillary with the membrane (Figure 22.27). The tip of the capillary (1 to 2 μm inner diameter) covers an area of the membrane with only one channel (*patch clamp*

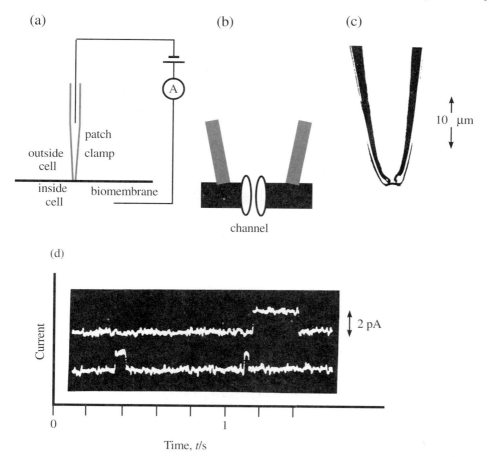

Figure 22.27 Biomembrane: single-channel recording. (a) Electric circuit using the patch clamp technique; (b) enlargement of the channel in the membrane; (c) photograph of the tip of a capillary; (d) two traces showing the electric current measured on opening and closing a single channel (E. Neher, B. Sakmann, J. H. Steinbach, *Pflügers Archiv*, 1978, **375**, 219).

technique). To avoid leakage the capillary must be appropriately prepared (polishing by heat treatment) and contact between capillary and membrane must be carefully controlled. A voltage up to 100 mV is applied between the capillary and the inside of the cell. Then the current flow through the membrane is measured (Figure 22.27). The signal shifts back and forth in steps of equal height, indicating opening and closing of the single channel in the range of the capillary. The channel is open for about 10^{-4} s; then a current of about 10^{-12} A is observed corresponding to a number of

$$\frac{10^{-4} \times 10^{-12}}{1.6 \times 10^{-19}} \approx 500$$

ions passing through the channel. This is a powerful method to investigate the properties of the channels and how they can be affected by an external influence – for example, by drugs.

The transport of macromolecules through membranes requires a sophisticated machinery and is not well understood. Only for some small proteins does the penetration takes place without a translocation machinery; the mechanism has been elucidated by investigating a number of mutants with various patterns of charged amino acids and studying their ability to penetrate the membrane.

Figure 22.28 illustrates the translocation of the newly synthesized coat protein of a bacteriophage through the bacterial membrane. The protein can be considered to consist

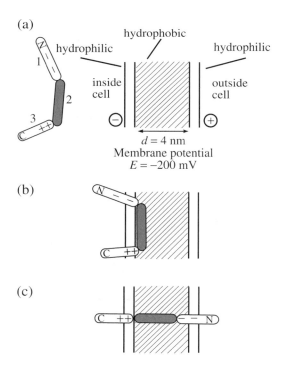

Figure 22.28 Small protein penetrating biomembrane. (a) Protein and bacterial membrane. (b)–(c) Stages of penetrating the membrane. Segments 1 and 3 are hydrophilic, segment 2 is hydrophobic. The membrane potential is needed to stabilize stage c relative to stage b. (D. Kiefer, X. Hu, R. Dalbey, A. Kuhn, EMBO Journal 16, 2197 (1997)).

of three α-helical segments 1, 2, and 3 consisting of 18, 18, and 14 amino acids, respectively (Figure 22.28(a)). Segments 1 and 3 are hydrophilic, and segment 2 is strongly hydrophobic. The protein inserts into the membrane (a to b), and the region with the N-terminal is moved from the inside of the bacterial membrane to the outside (b to c).

The Gibbs energy to transfer a given segment of the protein from water into the interior of the membrane is approximately given by the sum of empirical values of the Gibbs energies of each amino acid in the segment. The contribution of a hydrophobic amino acid such as leucine is negative ($\Delta_r G^\ominus_{Leu} = -11.7 \, kJ \, mol^{-1}$) while it is strongly positive for charged amino acids such as aspartic acid ($\Delta_r G^\ominus_{Asp} = +38.5 \, kJ \, mol^{-1}$) (D.M. Engelman, T.A. Steitz, A. Goldman, *Ann. Rev. Biophys. Biochem.* 15, 321 (1998)); an estimate (see Problem 22.1) shows that this value is plausible.

An evaluation shows that the transfer of Section 2 into the membrane (step a to b in Figure 22.28) is accompanied by a strong decrease in Gibbs energy. Section 1 can then be transferred to the outside of the membrane (step b to c) in contrast to Section 3. Stage c is stabilized due to the membrane potential $E = -0.2 \, V$. The membrane potential is found to be necessary for the transfer which is accompanied by a decrease of Gibbs energy (shift of negative charge; $\Delta_r G^\ominus_{b \to c}$ is on the order of $EF \approx -20 \, kJ \, mol^{-1}$).

In contrast, a transfer of segment 3 instead of segment 1 from inside to outside (shift of positive charge) would increase the Gibbs energy correspondingly.

22.4 Macromolecules

22.4.1 Random Coil: A Chain of Statistical Chain Elements

We first consider a polymer, a macromolecule consisting of equal monomers in a chain such as polyethylene, polystyrene, or polymethylcellulose. The molar mass of the polymer is

$$M_m = Z \cdot M_{m,monomer} \tag{22.21}$$

where $M_{m,monomer}$ is the molar mass of the monomer and Z is the number of monomers in the chain. The shape of the molecule in dilute solution in an appropriate solvent and the properties depending on the shape (such as viscosity, flow birefringence, light scattering) are subject to statistical considerations. A molecule such as butane can exist in three different conformations by rotating about C—C bonds (one *trans* and two *gauche* conformations, Figure 22.29).

The number of conformations of equal or nearly equal energy increases with the number Z of monomers. Neglecting the energy difference between *trans* and *gauche* and steric restrictions there are 3^{Z-1} equally probable conformations of a hydrocarbon chain. There is only one fully extended conformation and an immense number of coiled conformations. The molecule changes its conformation continuously by internal diffusion processes.

The complexity and diversity of interfacial interactions do not allow more than a qualitative description (see S.H. White, W.C. Windey, *Biochim. Biophys. Acta* 1376, 339 (1998).

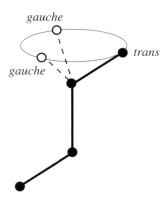

Figure 22.29 Conformations of butane – one *trans* and two *gauche* conformations. The butane molecule can be considered as composed of two monomer units $-CH_2-CH_2-$. Then the number of conformations is $3^{Z-1} = 3^1 = 3$.

We ask for the average size and for the distribution in size and shape of the polymer in a simplified manner. Two quantities are important to approach this task – the contour length L (the length of the line along the thread) and the length l of each section of the thread where the memory on the direction of the contour line is practically lost (Figure 22.30).

l depends on the nature of the monomer. In the case of a hydrocarbon, assuming *trans* and *gauche* conformations equally probable, we expect $l = 0.4$ nm; in reality, the *trans* conformation is about three times more probable than each of the *gauche* conformations,

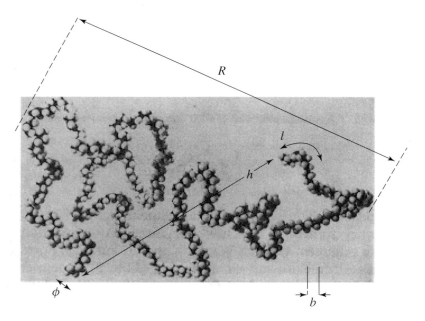

Figure 22.30 Parameters of a coil. R is the coil diameter, h is the distance between the chain ends, b is the length of a monomer unit, and ϕ is the cross-sectional diameter of the chain; l is the length of the statistical chain element. The length L of the chain is $L = bZ$, and the molar mass of the polymer is $M_m = ZM_{m,monomer}$, where Z is the number of monomer units and $M_{m,monomer}$ is the molar mass of the monomer.

22.4 MACROMOLECULES

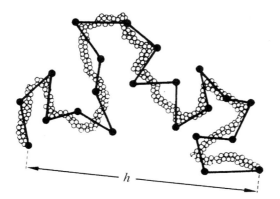

Figure 22.31 Polymer coil (total chain length L) replaced by chain of $N_{stat} = L/l$ statistical chain elements of length l. Each element is directed at random. l depends on the nature of the monomer. For a hydrocarbon, $l = 1$ nm (W. Kuhn and H. Kuhn, Helv. Chim. Acta 26, 1394 (1943)).

then $l = 1.0$ nm. In cellulose acetate we find $l = 11$ nm, and in polystyrene we find $l = 3.5$ nm.

An important simplification to describe the essential features of polymers is obtained by replacing the molecular chain by a chain of straight elements of length l (Figure 22.31). These chain elements of length l are called *statistical chain elements*. Each of these elements is directed at random, i.e., each direction in space has an a priori equal probability.

In this approximation the shape of the molecule can be compared with a diffusion path. In terms of diffusion theory (Chapter 11) the total length L of the coil corresponds to the length of the diffusion path $vt = j \cdot \lambda$, the number of statistical chain elements, N_{stat}, corresponds to j, and the length l of a chain element corresponds to the mean free path $\overline{\lambda}$. Thus the considerations of Chapter 11 can be immediately applied to the present case.

According to Problem 11.8 the probability $P(r) \cdot dr$ of finding the particle at time t at a distance between r and $r + dr$ from its original position is

$$P(r) \cdot dr = const \cdot 4\pi r^2 \cdot \exp\left(-\frac{3r^2}{2j \cdot \overline{\lambda^2}}\right) \cdot dr$$

where

$$\overline{r^2} = j \cdot \overline{\lambda^2}$$

is the mean square of the displacement of the particle after j collisions. Then, in the present case, the probability $P(h) \cdot dh$ of finding the distance between the chain ends between h and $h + dh$ equals

$$P(h) \cdot dh = const \cdot 4\pi h^2 \cdot \exp\left(-\frac{3h^2}{2N_{stat} \cdot l^2}\right) \cdot dh \qquad (22.22)$$

and the average of h^2 is

$$\overline{h^2} = N_{stat} \cdot l^2 = L \cdot l \qquad (22.23)$$

since the number of statistical chain elements is

$$N_{stat} = \frac{L}{l} \qquad (22.24)$$

Now the length l of the statistical chain element can be defined explicitly: $\overline{h^2}$ in the model must be equal to $\overline{h^2}$ of the true chain. L is given by the number Z of monomers in the chain and the contour length b per monomer:

$$L = b \cdot Z \tag{22.25}$$

Thus we obtain

$$l = \frac{\overline{h^2}}{L} = \frac{\overline{h^2}}{b \cdot Z} \tag{22.26}$$

b and Z are given by the molecular structure and $\overline{h^2}$ is obtained from optical and hydrodynamic properties of polymer solutions, as will be shown below.

The picture of a random coil can be extended to cases such as DNA double helices where the molecular strand is bent by thermal fluctuation due to its elasticity, not due to conformational changes. As outlined in Box 22.2, a bending of 90° occurs over a section of 100 nm along the strand at room temperature, on the average. That is, in this section the memory on the direction of the contour line is effectively lost, and DNA strands are coils consisting of statistical chain elements of about 100 nm length.

The result can be checked for DNA ($L = 50$ μm) carrying a fluorescent dye as a tracer. Then the molecule can be directly observed in an optical microscope. R, the coil diameter (see Figure 22.30), can be measured and the probability distribution of R is obtained (Figure 22.32, dots). The solid line in Figure 22.32 gives the theoretical distribution function $P(R)$ for $l = 100$ nm. The value agrees with the value of l estimated in Box 22.2.

When heating a DNA double strand, the hydrogen bonds in the base pairs break in a narrow range of temperatures. This cooperative process ("melting") leads to

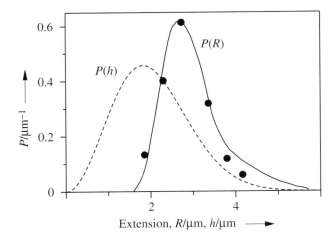

Figure 22.32 DNA double helix with $Z = 1.5 \times 10^5$ basepairs ($l = 100$ nm, $L = 50$ μm). Distributions $P(R)$ of maximum extension R (solid line) and $P(h)$ of distance h of chain ends (dashed line). The distribution of R is measured on fluorescence-labeled strands observed in the microscope (dots) and compared with the theoretical distribution. $P(R) \cdot dR$ is the probability of finding the maximum extension of the coil between R and $R + dR$. On the average, R is $1.4 \cdot \sqrt{L \cdot l}$, this is somewhat larger than $\sqrt{\overline{h^2}}$ (Experiment: M. Yanagida, Y. Hiraoka, I. Katsura, Cold Springs Harbor Symposia on Quantitative Biology 47, 177 (1982). Theory: S. Hoops, H. Kuhn, W. Huber, L. Eckert, J. Polymer Science 20, 101, (1956)).

22.4 MACROMOLECULES

single-stranded DNA forming a random coil strongly decreased in size due to many different conformations available at higher temperature.

Actin filaments (molecules of the protein actin arranged like the row of beads in a necklace with a cross-section of 7 nm diameter) form random coils, directly observed by labeling with fluorescent dye molecules. The observed length of the statistical chain element is $l = 3.4\,\mu\text{m}$ in the case of freshly polymerized material (many defects in the sample), in agreement with estimates assuming similar elastic properties as for a laboratory rubber tube (Box 22.2). The statistical chain element becomes longer by a factor of ten by anealing ($l \approx 30\,\mu\text{m}$). Nets of actin filaments play a central role for the mechanical stabilization of cells.

Box 22.2 – Bending of DNA Strand and Actin Filament

In Problem 12.6 we used a laboratory rubber tube or a rubber cylinder as a model to estimate the energy required to twist a DNA double strand. Now we estimate the energy for bending a DNA double strand by fixing a rubber cylinder (length l, diameter $2r$) at one end and bending it by applying a force f at the other end perpendicular to the cylinder axis, producing a shift s of the end. Then

$$f = A_{\text{Young}} \frac{3\pi r^4 s}{4l^3}$$

where A_{Young} is Young's modulus for bending. The elastic energy is

$$E = \int_0^s f \cdot ds = A_{\text{Young}} \frac{3\pi r^4}{4l^3} \cdot \frac{s^2}{2}$$

We can write s as

$$s = l\frac{\alpha}{2}$$

where α is the bending angle. Then

$$E = A_{\text{Young}} \frac{3\pi}{32} \cdot \frac{r^4}{l} \alpha^2$$

For $\alpha = \pi/2$ (bending by 90°) we obtain

$$E_{\alpha=\pi/2} = A_{\text{Young}} \frac{3\pi^3}{128} \cdot \frac{r^4}{l}$$

In an experiment we measure the force to bend a rubber cylinder of length $l = 27\,\text{cm}$ and diameter $2r = 1\,\text{cm}$ by 90°; this force is $f = 0.6\,\text{N}$ and the elastic energy is about $E = f \cdot l/4 \approx 0.1\,\text{J}$. In this way we estimate

$$A_{\text{Young}} \frac{3\pi^3}{128} = E\frac{l}{r^4} = 4 \times 10^7\,\text{N}\,\text{m}^{-2}$$

Continued on page 760

> *Continued from page 759*
>
> Now let us consider a DNA double helix of length $l = 100$ nm and radius $r = 1.5$ nm; for the energy to bend the helix by $90°$ we obtain
>
> $$E_{\alpha=\pi/2} = 4 \times 10^7 \, \text{N m}^{-2} \cdot \frac{r^4}{l} = 2 \times 10^{-21} \, \text{J}$$
>
> This is on the order of kT at room temperature. We conclude that the length of the statistical chain element is about 100 nm at room temperature.
>
> Actin filaments are threads of radius $r = 3.5$ nm. Assuming again the value
>
> $$A_{\text{Young}} \frac{3\pi^3}{128} = 4 \times 10^7 \, \text{N m}^{-1}$$
>
> the estimated length of the statistical chain element at $T = 300$ K is
>
> $$l = A_{\text{Young}} \frac{3\pi^3}{128} \cdot \frac{r^4}{kT} = 4 \, \mu\text{m}$$
>
> This agrees with the real situation.

22.4.2 Length of Statistical Chain Element from Light Scattering

Solutions of macromolecules have a certain turbidity which allows conclusions concerning the molecular size. Let us illuminate a dilute solution of macromolecules with light polarized in the z direction (Figure 22.33).

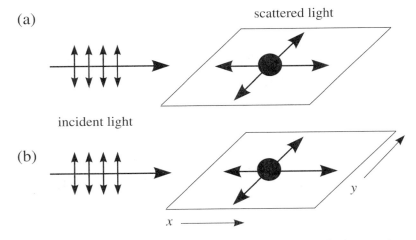

Figure 22.33 Light scattering of coil in x,y plane. The coil is described as a sphere of radius $r_{\text{coil}} = \frac{1}{2}\sqrt{\overline{h^2}}$. The incident light is polarized in the z-direction. (a) $r_{\text{coil}} \ll \lambda$: equal intensity of forward and back scattering. (b) $r_{\text{coil}} = \lambda/8$: intensity of forward scattering is twice that of back scattering. This is due to interference effects, see Figure 22.34(a) and Problem 22.3.

22.4 MACROMOLECULES

We first consider molecules that are small compared to the wavelength λ of light (Figure 22.33(a)). Then each molecule can be considered as a dipole oscillating with the frequency of the incident light; thus it emits light of the same intensity in all directions in the xy plane. No light is emitted in the z direction (Hertz antenna equation see Foundation 8.5). The amplitude of the oscillating dipole depends on the difference in refractive index of the solution (refractive index $\hat{n}$) relative to that of the solvent (refractive index $\hat{n}_0$). Then for the intensity $I_{\text{scattered}}$ of the scattered light in the xy plane we obtain (see Problem 22.2)

$$I_{\text{scattered}} = I_{\text{incident}} \cdot 4\pi^2 \left[\frac{(\hat{n} - \hat{n}_0) \cdot \hat{n}}{m/V} \right]^2 \frac{mM_m}{d^2 \lambda^4 N_A} \quad (22.27)$$

where d is the distance between the sample and the observer; m is the dissolved mass in volume V of the sample. The strong dependence of $I_{\text{scattered}}$ with λ is due to the increasing speed of the oscillating electron when increasing its frequency.

This method allows us to determine the molar mass M_m from the scattering intensity. In the present case this method does not give information on the size of the particles.

As soon as the particle is no longer small compared to the wavelength of the incident light, the intensity of the light scattered in the forward direction is higher than the intensity in the backward direction (Figure 22.33(b)); conclusions on the size of the particle can then be drawn.

The effect is easily grasped by modeling the particle by two scattering centers A and B at distance $a = r_{\text{coil}}$ (Figure 22.34(a)). Looking at light scattered in the direction of the incident light, the delay of the wave emitted in center B relative to center A is compensated by the shorter path, thus both beams arrive at the observer (point P_2) with identical phase. This is not the case for light scattered in the opposite direction; the amplitude is accordingly smaller (point P_1). As shown in Problem 22.3 the intensity ratio is

$$\frac{I_{\text{backward}}}{I_{\text{forward}}} = 1 - \left(\frac{2\pi a}{\lambda} \right)^2 \quad (22.28)$$

This ratio is 1 for $a \ll \lambda$ and 0.4 for $a = \lambda/8$. By measuring this ratio the diameter of the coil, $\sqrt{\overline{h^2}} \approx 2a$ can be obtained, thereby allowing us to determine l according to equation (22.26).

For a coil of larger size the model used (two scattering centers) becomes unrealistic. The summation over the contributions of each statistical chain element to the scattered wave gives the expression (P. Debye, *J. Phys. Chem.*, 1951, **51**, 18; H. Kuhn, *Helv. Chim. Acta* **29**, 432 (1946))

$$\frac{I_\varphi}{I_{\text{forward}}} = \frac{2}{\alpha^2}(e^{-\alpha} - 1 + \alpha) \quad \text{with} \quad \alpha = \frac{8\pi^2}{3} \frac{\overline{h^2}}{\lambda^2} \sin^2 \frac{\varphi}{2} \quad (22.29)$$

The incident light is polarized in z direction. I_φ is the intensity of the light scattered in the xy plane in a direction forming an angle φ with the direction of the incident beam ($I_{\varphi=0} = I_{\text{forward}}$). For α smaller than about 1 we obtain practically the same expression as in equation (22.28).

Experimentally, it is not easy to measure light scattering at 0 and 180°, but the light scattering can be measured well at 45° and 135°. The ratio $I_{135°}/I_{45°}$ obtained from equation (22.29) is displayed in Figure 22.34(b) as a function of $\sqrt{\overline{h^2}}/\lambda$.

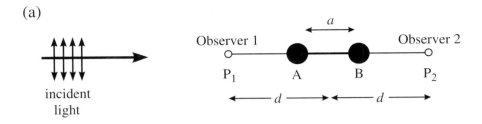

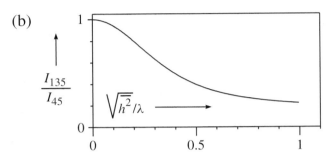

Figure 22.34 Light scattering on macromolecules. (a) Light scattered from points A and B reach point P_2 with equal phase. In contrast, point P_1 is reached by light scattered at points A and B with phase shift $2a/\lambda$, where $a = \frac{1}{2}\sqrt{\overline{h^2}} = r_{\text{coil}}$. (b) Scattering function I_{135}/I_{45} versus $\sqrt{\overline{h^2}}/\lambda$, according to equation (22.29).

22.4.3 Length of Statistical Chain Element from Hydrodynamic Properties

Diffusion

A coil consisting of many statistical chain elements can be viewed, in a first approximation, as a sphere of radius

$$r_{\text{coil}} = \tfrac{1}{2}\sqrt{Ll} = \tfrac{1}{2}\sqrt{\overline{h^2}} \tag{22.30}$$

including solvent in large excess over the polymer material. The coil is considered as undrained when moving through the solvent. The diffusion coefficient (Chapter 11) is given by the Stokes–Einstein equation

$$D = \frac{kT}{6\pi\eta r} \quad \text{with} \quad r = r_{\text{coil}} \tag{22.31}$$

Because of

$$L = Zb \quad (b = \text{length of a monomer unit}) \tag{22.32}$$

we obtain

$$2r_{\text{coil}} = \sqrt{L \cdot l} = \sqrt{Zbl} \tag{22.33}$$

22.4 MACROMOLECULES

and thus

$$D = \frac{kT}{3\pi\eta} \cdot \frac{1}{\sqrt{bl}} \frac{1}{\sqrt{Z}} \qquad (22.34)$$

Sedimentation

In the artificial gravity field of an ultracentrifuge, a polymer molecule is driven with a force f. In free space this force would be

$$f = \frac{M_m}{N_A} \cdot g_{\text{centrifuge}} \qquad (22.35)$$

where M_m/N_A is the mass of the polymer molecule and $g_{\text{centrifuge}}$ is the acceleration in the ultracentrifuge:

$$g_{\text{centrifuge}} = \omega^2 r_{\text{axis}} \qquad (22.36)$$

(ω = angular speed, r_{axis} = distance from axis of revolution). Typically, $g_{\text{centrifuge}}$ is on the order of 10^5 m s^{-2}, that is 10 000 times the acceleration of free fall. In the solvent (density ρ) we obtain

$$f = \frac{M_m}{N_A} g_{\text{centrifuge}} \cdot (1 - V_{\text{part}}\rho) \qquad (22.37)$$

V_{part} is the partial specific volume of solute:

$$V_{\text{part}} = \frac{V_{\text{solution}} - V_{\text{solvent}}}{m_{\text{solute}}} \qquad (22.38)$$

If the volume of the solution can be considered as volume of the solvent plus volume of the solute we obtain

$$V_{\text{part}} = \frac{1}{\rho_{\text{solute}}} \quad \text{or} \quad V_{\text{part}} \cdot \rho = \frac{\rho}{\rho_{\text{solute}}}$$

For $\rho_{\text{solute}} > \rho$ the macromolecule moves away from the center of the axis of the centrifuge.

The molecule is accelerated in the centrifuge until equilibrium with the friction is reached at a certain speed v. The friction is given by Stokes's law. Then

$$f = 6\pi\eta \cdot r_{\text{coil}} \cdot v = \frac{M_m}{N_A}(1 - V_{\text{part}}\rho) \cdot g_{\text{centrifuge}} \qquad (22.39)$$

The ratio

$$s = \frac{v}{g_{\text{centrifuge}}} = \frac{M_m \cdot (1 - V_{\text{part}}\rho)}{N_A \cdot 6\pi\eta r_{\text{coil}}} \qquad (22.40)$$

is called the *sedimentation constant*. Substituting $6\pi\eta r_{\text{coil}}$ by using the Stokes–Einstein equation (22.31) we obtain

$$s = \frac{M_m D}{RT} \cdot (1 - V_{\text{part}} \cdot \rho) \qquad (22.41)$$

We obtain M_m from D and s. Equation (22.41) is not restricted to statistical coils, but holds for any small particle (Problem 22.5). In typical cases, the sedimentation speed v is very small: about 1 cm per day for $g_{\text{centrifuge}} = 10^5$ m s^{-2}.

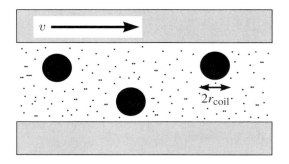

Figure 22.35 Increasing viscosity by suspending spheres (full circles) in a solvent (indicated by small dots).

Viscosity

When spheres are suspended in a liquid, the viscosity is increased (Figure 22.35). At high dilution, when the interaction between the spheres can be neglected, the viscosity increases proportional to the volume fraction x_{volume} where

$$x_{\text{volume}} = \frac{\text{volume of spheres in suspension}}{\text{volume of suspension}} = \frac{V_s}{V} \qquad (22.42)$$

Then

$$\eta = \eta_0 \cdot (1 + a x_{\text{volume}}) \qquad (22.43)$$

where η is the viscosity of the suspension, and η_0 is the viscosity of the pure solvent (for definition of viscosity see Chapter 11). According to Einstein (Ann. Phys. 1906, **19**, 289.) the constant a is $a = 2.5$.

The coil including solvent is essentially undrained when applying a flow gradient. It behaves roughly like a sphere of radius r_{coil}. Then the volume of the spheres is

$$V_s = N \tfrac{4}{3} \pi r_{\text{coil}}^3 \qquad (22.44)$$

where N is the number of coils. Then from equations (22.42) to (22.44) we obtain

$$x_{\text{volume}} = \frac{N}{V} \tfrac{4}{3} \pi r_{\text{coil}}^3 = \frac{1}{a}\left(\frac{\eta}{\eta_0} - 1\right) \qquad (22.45)$$

and

$$\left(\frac{\eta}{\eta_0} - 1\right) = a \frac{N}{V} \tfrac{4}{3} \pi r_{\text{coil}}^3 \qquad (22.46)$$

The number N of coils can be expressed by

$$N = n N_A = \frac{m}{M_m} N_A \qquad (22.47)$$

where m is the mass of the polymer dissolved in volume V and M_m is its molar mass. Furthermore, because of

$$2 r_{\text{coil}} = \sqrt{h^2} \qquad (22.48)$$

22.4 MACROMOLECULES

and with $a = 2.5$ we obtain:

$$[\eta] = \underbrace{\left(\frac{\frac{\eta}{\eta_0} - 1}{m/V}\right)}_{\text{high dilution}} = N_A \cdot 2.5 \cdot \frac{\pi}{6} \cdot (\overline{h^2})^{3/2} \cdot \frac{1}{M_m} \qquad (22.49)$$

$[\eta]$ is called the *Staudinger index*; it is a measure of the relative viscosity increase by the dissolved polymer in the limit of high dilution.

22.4.4 Refined Theory: Macroscopic Models

In reality, our assumption that the coil is a sphere and that it is undrained when moving it through the solvent or exposing it to a flow gradient is by far not strictly correct. Treating the solvent as a homogeneous liquid of macroscopic viscosity is a reasonable simplification, but finding the resistance of a random coil is still a formidable problem. It can be bypassed by performing experiments with macroscopic models of coils.

The models are made by bending wires in a random fashion to form statistical coils in the range from $N_{\text{stat}} = 1$ to 100. This is achieved by dividing the contour length L into L/l_{segment} equal segments of length l_{segment} and by bending each segment randomly by an angle between 0° and 180°. This kind of model of random coils (Figure 22.36a) fits better to the contour line of the actual molecule than the model made of stretched segments (Figure 22.31). It can be demonstrated that the length l_{segment} of a statistical segment must be made by a factor 1/1.53 shorter than the length l of a statistical chain element in order to correspond to the same value of $\overline{h^2}$. The models are dropped in a viscous liquid and the speed of fall is measured to obtain a refined description of sedimentation and diffusion of the corresponding macromolecules; furthermore, they are rotated about their axes of gyration to search for the behavior of macromolecules in a flow gradient.

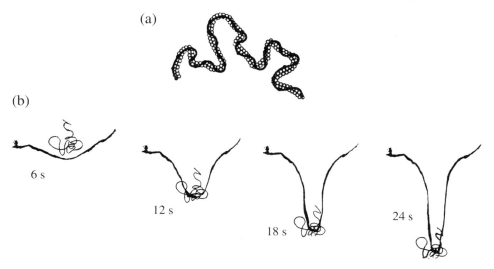

Figure 22.36 Macroscopic model of random coil. (a) Model made of wire by bending it accordingly. (b) Model falling through a viscous liquid. A horizontal strip of ink in the viscous liquid is carried along with the model showing that the coil is essentially undrained. (H. Kuhn, F. Moning, W. Kuhn, Helv. Chim. Acta 36, 731 (1953)).

Figure 22.36b shows such a model during its fall through a viscous liquid. The limited draining is seen by the deformation of the mark – a horizontal strip of ink in the viscous liquid. The model is drained at the border and essentially undrained in the center.

From experiments with many models a statistical average can be obtained. In the considered translation experiments on macroscopic models the force f can be represented by

$$f = C \cdot \eta \cdot \overline{h^2} \cdot v = 2C \cdot \eta \cdot r_{coil} \cdot v$$

For $N_{stat} \gg 1$ we obtain $2C = 15$; this value is close to the value $6\pi = 18.8$ expected from Stokes Law, which we used in our simplified model of an undrained sphere.

For $N_{stat} = 1$ we obtain $2C = 30/[2 + \ln(l/\phi)]$, where ϕ is the diameter of the chain (Figure 22.30). That is, $2C = 6\pi \cdot 0.36$ for $l = 10\phi$ and $2C = 6\pi \cdot 0.24$ for $l = 100\phi$: The measured resistance for a bar of length l and diameter ϕ is smaller than for a sphere of corresponding volume; the resistance becomes smaller when decreasing the diameter ϕ. From many model experiments carried out in a broad range of values for l, N_{stat}, and ϕ the constant C (*form factor*) can be represented by

$$C = \frac{7.5}{1 + 0.5 \cdot \ln(l/\phi)/\sqrt{N_{stat}}} = \frac{7.5}{1 + 0.5\sqrt{l/L} \cdot \ln(l/\phi)} \qquad (22.50)$$

This expression includes the two limiting cases considered above.

With $L = bZ$ we can rewrite the form factor as

$$C = \frac{3\pi \cdot 0.8}{1 + \alpha/\sqrt{Z}} \quad \text{with} \quad \alpha = 0.5\sqrt{\frac{l}{b}} \cdot \ln\frac{l}{\phi}$$

Diffusion

Introducing the form factor C we obtain for the diffusion constant from equation (22.34):

$$D = \frac{1}{0.8} \cdot \frac{kT}{3\pi\eta} \cdot \frac{1}{\sqrt{bl}} \frac{1}{\sqrt{Z}} \left(1 + \alpha \frac{1}{\sqrt{Z}}\right) \qquad (22.51)$$

or

$$D \cdot Z = \frac{1}{0.8} \cdot \frac{kT}{3\pi\eta} \cdot \frac{1}{\sqrt{bl}} (\sqrt{Z} + \alpha)$$

Sedimentation

Accordingly, from equation (22.41) we obtain the sedimentation constant with $M_m = Z \cdot M_{m,monomer}$:

$$s = A \cdot (\sqrt{Z} + \alpha), \quad A = \frac{M_{m,monomer} \cdot (1 - V_{part} \cdot \rho)}{0.8 \cdot N_A \cdot 3\pi\eta \cdot \sqrt{bl}} \qquad (22.52)$$

22.4 MACROMOLECULES

Viscosity

According to rotation experiments with models made of wire the factor $2.5\pi/6 = 1.3$ in equation (22.49) has to be replaced by

$$\frac{0.45}{1+\sqrt{\frac{l}{L}}\cdot\left(\ln\frac{l}{\phi}-1.6\right)}$$

For $L \gg l$ we obtain 0.45 instead of 1.3: the draining is more effective than in the simplified model assuming undrained spheres. For short chains with $L = l$ and $l = 10 \cdot \phi$ we obtain the factor 0.26.

Then instead of equation (22.49) we obtain

$$[\eta] = N_A \cdot \frac{0.45}{1+\sqrt{\frac{l}{L}}\cdot\left(\ln\frac{l}{\phi}-1.6\right)} \cdot (\overline{h^2})^{3/2} \cdot \frac{1}{M_m} \qquad (22.53)$$

For $L \gg l$ this equation reduces to

$$[\eta] = A \cdot (\overline{h^2})^{3/2}\frac{1}{M_m}, \quad A = 2.6 \times 10^{23} \,\text{mol}^{-1} \qquad (22.54)$$

This theoretical equation agrees exactly with an empirical equation called the *Flory equation*.

With $\overline{h^2} = lL = lbZ$ and $M_m = M_{m,\text{monomer}} \cdot Z$, we can rewrite equation (22.53) as

$$[\eta] = N_A \cdot 0.45 \cdot (lb)^{3/2} \cdot \frac{1}{M_{m,\text{monomer}}}\frac{Z}{\sqrt{Z}+\beta} \quad \text{with} \quad \beta = \sqrt{\frac{l}{b}}\cdot\left(\ln\frac{l}{\phi}-1.6\right) \qquad (22.55)$$

Example – diffusion of methyl cellulose in water

In Figure 22.37, DZ is plotted versus $\sqrt{Z}$. As expected from equation (22.51), the experimental points are on a straight line with y intercept $1.1 \times 10^{-5}\,\text{cm}^2\,\text{s}^{-1}$ and slope $0.26 \times 10^{-5}\,\text{cm}^2\,\text{s}^{-1}$:

$$D \cdot Z = \frac{1}{0.8} \cdot \frac{kT}{3\pi\eta} \cdot \frac{1}{\sqrt{bl}}(\sqrt{Z}+\alpha) = (0.26\sqrt{Z}+1.1) \times 10^{-5}\,\text{cm}^2\,\text{s}^{-1}$$

With $\eta = 1.0 \times 10^{-2}$ poise $= 10^{-3}\,\text{N m}^{-2}\,\text{s}$ (viscosity of water), $b = 0.515$ nm (contour length of monomer from X-ray data), and $T = 300$ K, we obtain $l = 9$ nm, and $\phi = 0.8$ nm. The actual thickness of the chain is about 1.3 nm, indicating that the hydrodynamic approximation is appropriate.

Example – sedimentation of cellulose acetate in acetone

According to Figure 22.38 and equation (22.52), the expected linear dependence when plotting the sedimentation constant s versus $\sqrt{Z}$ is observed:

$$s = \frac{1 - V_{\text{part}} \cdot \rho}{\eta} \cdot \frac{M_{m,\text{monomer}}}{0.8 \cdot N_A \cdot 3\pi \cdot \sqrt{bl}}(\sqrt{Z}+\alpha) = (0.39\sqrt{Z}+2.0) \times 10^{-13}\,\text{s}$$

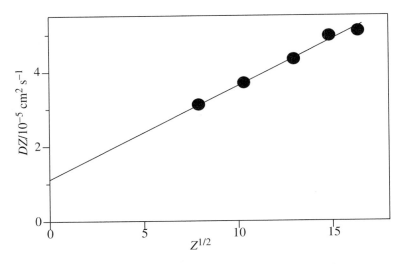

Figure 22.37 Diffusion of methyl cellulose in water; DZ versus $\sqrt{Z}$. (A. Polson, Kolloid Zeitschrift 83, 173 (1938)).

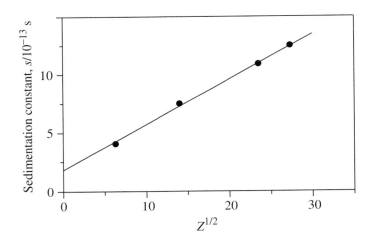

Figure 22.38 Sedimentation of cellulose acetate in acetone; sedimentation constant s versus $\sqrt{Z}$. (J. Singer, On the Concentration Dependence of the Rates of Sedimentation and Diffusion of Macromolecules in Solution., J. Chem. Phys. 15, 341 (1947)).

With $M_{m,monomer} = 260\,\text{g mol}^{-1}$, $b = 0.515\,\text{nm}$, $\rho = 0.79\,\text{g cm}^{-3}$, $\eta = 0.3 \times 10^{-2}$ poise $= 0.3 \times 10^{-3}\,\text{N m}^{-2}\,\text{s}$ (viscosity of acetone), and $V_{part} = 0.68\,\text{cm}^3\,\text{g}^{-1}$, we obtain $l = 10\,\text{nm}$ and $\phi = 1.0\,\text{nm}$ from slope and y intercept. Note that s is extremely small.

Example – viscosity of cellulose acetate in acetone

When introducing the data given in the second example, we obtain the following from equation (22.55)

22.4 MACROMOLECULES

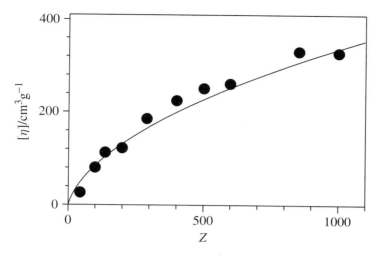

Figure 22.39 Viscosity of cellulose acetate in acetone. $[\eta]$ versus Z (A. Sookme and M. Harris, Ind. Eng. Chem. 37, 475 (1945)).

$$[\eta] = N_A \cdot 0.45 \cdot (lb)^{3/2} \cdot \frac{1}{M_{m,monomer}} \cdot \frac{Z}{\sqrt{Z} + \beta} = \frac{12Z}{\sqrt{Z} + 3.1} \text{ cm}^3 \text{ g}^{-1}$$

The experimental points (Figure 22.39) agree with the theoretical expectation.

22.4.5 Uncoiling Coil

Fully unraveling coil in flowing medium

We consider a coil fixed at one end (Figure 22.40(a)). In a weak current the coil is slightly deformed, and when increasing the current the coil begins to unroll at the fixed end where the strain is strongest because the entire chain, resisting against current, pulls on the end. At sufficiently large current the molecule becomes fully uncoiled. What minimum velocity v_c of the flow is needed to unravel the coil?

In the fully unraveled coil the energy to orient a given statistical chain element by the stream surrounding it must be greater than the energy of thermal fluctuations. This must be the case even for the statistical element constituting the open end (Figure 22.40(b)). The energy E to turn the element oriented in the flow by 180° (to move the center by the distance l against the current) must be greater than kT. We approximate the friction of the element by the friction of a sphere of radius $l/4$. According to Stokes' law (see Chapter 11) with $E = f \cdot l$ (f is the force of friction) we obtain for the speed required for fully unraveling the coil

$$E = f \cdot l = 6\pi\eta \frac{l}{4} v \cdot l \geq kT \tag{22.56}$$

or

$$v \geq v_c, \quad v_c = \frac{kT}{\frac{3}{2}\pi\eta l^2} \tag{22.57}$$

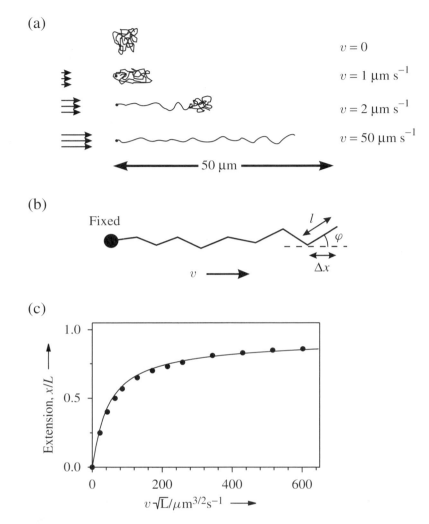

Figure 22.40 (a) DNA double helix fixed at an end. Uncoiling by a current v larger than $v_c = 50\,\mu m\,s^{-1}$; $L = 50\,\mu m$; $l = 100\,nm$; $N_{stat} = 500$. The complete and partial unraveling of a DNA duplex strand carrying a fluorescent dye can be seen in an optical microscope. The molecule is in slightly moving water and one end is fixed (M. Yanagida, Y. Hiraoka, I. Katsura, Cold Springs Harbor Symposia on Quantitative Biology 47, 177 (1982); F. Brochard, DNA: Observing Stems and Flowers, Physics World, Dec. 1995, page 24; A. Buguin, F. Brochard-Wyart, Macromolecules 29, 4937 (1996)).
(b) Critical current for uncoiling. The molecule is described as a chain of statistical chain elements. The frictional force of flowing solvent counteracts thermal fluctuations. $\Delta x = l \cdot \cos\varphi$ is the extension of the statistical chain element in the x-direction.
(c) DNA double helix. The molecular chain end is bound to a small bead of 0.3 μm diameter. The bead is pulled by a laser beam and moved through the solvent with constant speed. The force pulling the bed is due to the polarization of the bead in the electric field of light; the bead is driven toward the point of maximum field. Plot of fractional extension x/L versus $v\sqrt{L}$ for samples with $L = 50$ μm. Dots: Experiment (T.T. Perkins, D.E. Smith, R.G. Larson, S. Chu, Stretching of a Single Tethered Polymer in a Uniform Flow, Science 268, 83 (1995). The theoretical curve (solid line) is described by equation (22.59), with $l = 100\,nm$, $\eta = 1 \times 10^{-3}$ N m^{-2} s, and $T = 300$ K.

Example – flow to unravel a coil

For a DNA strand in water we have $l = 100$ nm and $\eta = 0.01$ poise $= 1 \times 10^{-3}$ N m^{-2} s; then from equation (22.57) we obtain $v_c = 100\,\mu\text{m s}^{-1}$. Thus a tiny current is sufficient, and unraveling a coil is a very sensitive sensor for currents.

What is the force acting on the fixed end at speed v_c?

Each statistical chain element exerts a resistance force $6\pi\eta\frac{1}{4}lv_c$ and the force acting on the fixed end is the sum of the resistance forces of all N_{stat} statistical elements

$$f = 6\pi\eta\tfrac{1}{4}lv_c \cdot N_{\text{stat}} = \tfrac{3}{2}\pi\eta L v_c \qquad (22.58)$$

Then with the expression for v_c in equation (22.57) we obtain

$$f \approx \frac{kT}{l} N_{\text{stat}} = 2 \times 10^{-11}\,\text{N} = 20\,\text{pN}$$

for $N_{\text{stat}} = 500$, $l = 100$ nm, and $T = 300$ K.

When attaching a magnetic bead to the end of the chain of a DNA double helix dissolved in water ($L = 40\,\mu\text{m}$) and pulling the bead with a magnetic field, the force f and the speed v can be measured. When applying a force of 30 pN a speed of $100\,\mu\text{m s}^{-1}$ is observed (fully uncoiled chain, D. Wirtz, Phys. Rev. Letters 75, 2436 (1995)). Thus we obtain

$$\frac{f}{v} = 0.3\,\text{pN}\,\mu\text{m}^{-1}\,\text{s}$$

On the other hand, according to the simple model discussed in this section (equation 22.58)

$$\frac{f}{v} = \frac{3\pi}{2}\eta L = \frac{3\pi}{2} \cdot 10^{-3} \cdot 40 \cdot 10^{-6}\,\text{N m}^{-1}\,\text{s} = 0.2\,\text{pN}\,\mu\text{m}^{-1}\,\text{s}$$

in excellent agreement with the measurement.

Partially unraveling coil in flowing medium

The complete and partial unrolling of a DNA strand carrying a fluorescent dye can be seen in an optical microscope (Figure 22.40(a)). The molecule is in slightly moving water, and one end is fixed. What speed v keeps a section unraveled or loosely coiled? We consider a statistical chain element number n (numbering from the fixed end). The chain between element n and the fixed end exerts a force f (see Problem 22.6):

$$f = \frac{kT}{2l}\left[\frac{1}{(1 - \overline{\Delta x_n}/l)^2} - 1 + \frac{4\overline{\Delta x_n}}{l}\right]$$

$\overline{\Delta x_n}$ is the average extension of chain element n in the x-direction (Figure 22.40(b)).

On the other hand, the chain between element n and the open end acts on the element with a hydrodynamic force f'. For simplicity we consider this section as undrained, thus

$$f' = 6\pi\eta \cdot l \cdot \tfrac{1}{2}\sqrt{N_{\text{stat}} - n} \cdot v$$

We equalize f and f', thus obtaining the average contribution $\overline{\Delta x}$ of the element n to the extension x of the molecule in the direction of the current.

$$\frac{kT}{2l}\left[\frac{1}{(1-\overline{\Delta x_n}/l)^2} - 1 + \frac{4\overline{\Delta x_n}}{l}\right] = 6\pi\eta \cdot l \cdot \tfrac{1}{2}\sqrt{N_{\text{stat}} - n} \cdot v \tag{22.59}$$

In Problem 22.7 it is shown how the extension x and the fractional extension x/L can be obtained from this relation.

The fractional extension x/L is plotted versus $v\sqrt{L}$ in Figure 22.40(c) (solid line); the experiment (dots in Figure 22.40(c)) is in good agreement with the theory. It is easy to estimate the extension for small speed when the coil is deformed but not unraveled. Then we assume that the hydrodynamic force f'

$$f' = 6\pi\eta \cdot r_{\text{coil}} \cdot v$$

acts on the open end which is attracted toward the fixed end at distance x by a force f, given on the average, by the equation

$$f = \frac{3xkT}{\overline{h^2}} \quad \text{for} \quad x \ll L$$

This force is due to the thermal collisions of the solvent molecules with each element of the filament considered to be fixed at the ends. Obviously, f increases with T and with x. The proof of this relation is given below in Box 22.4.

By equalizing the two forces we obtain with $r_{\text{coil}} = \tfrac{1}{2}\sqrt{\overline{h^2}}$

$$\frac{3x}{\overline{h^2}} \cdot kT = 6\pi\eta \cdot \tfrac{1}{2}\sqrt{\overline{h^2}} \cdot v$$

or, with $\overline{h^2} = l \cdot L$ and $L/l = N_{\text{stat}}$

$$\frac{x}{L} = \frac{3\pi\eta l^{3/2}}{3kT} \cdot \sqrt{L} \cdot v = \frac{2}{3}\frac{v}{v_c} \cdot \sqrt{N_{\text{stat}}} \tag{22.60}$$

(initial part of dashed curve in Figure 22.40(c)).

With $l = 100$ nm, $\eta = 1$ cP $= 1 \times 10^{-3}$ N m^{-2} s, and $T = 300$ K for the initial slope we calculate

$$\frac{x/L}{v\sqrt{L}} = \frac{3\pi\eta l^{3/2}}{3kT} = \frac{1}{41 \, \mu\text{m}^{3/2} \, \text{s}^{-1}}$$

The measured slope depends slightly on L:

$$\frac{x/L}{v\sqrt{L}} = \frac{1}{58 \, \mu\text{m}^{3/2} \, \text{s}^{-1}} \quad \text{(for } L = 22 \, \mu\text{m)}, \qquad \frac{x/L}{v\sqrt{L}} = \frac{1}{55 \, \mu\text{m}^{3/2} \, \text{s}^{-1}} \quad \text{(for } L = 84 \, \mu\text{m)}$$

22.4 MACROMOLECULES

In the refined theory the factor 3π in equation (22.60) is replaced by the form factor C (see equation (22.50)). The form factor is $C = 6.75$ for $L = 22\,\mu\text{m}$ and $C = 7.08$ for $L = 84\,\mu\text{m}$). The ratio of the two slopes should then be

$$\frac{C\,(L = 22\,\mu\text{m})}{C\,(L = 84\,\mu\text{m})} = \frac{6.75}{7.09} = 0.95$$

in accordance with the experimental ratio $55/58 = 0.95$.

Unraveling Coil by Force Applied at Chain Ends

The force to unravel a coil can be measured by fixing one end, attaching a magnetic bead (iron grain in latex particles of 200 nm diameter) to the other end and using a magnetic field instead of a current to unravel the coil. A force of 100 pN unravels a DNA double helix ($L = 50\,\mu\text{m}$). For comparison, 1000 pN is the force to stretch a C–C bond by 1% (see Chapter 6). The breaking point of the DNA chain is reached when stretching it with a force of 500 pN. This corresponds to a force per area of cross-section of $10^8\,\text{N m}^{-2}$, i.e. the breaking limit is similar to that of steel.

At small forces f, when the chain is largely coiled (distance h between chain ends small compared to contour length L) we can treat f as proportional to the distance h between chain ends. The ratio $f/h = 4 \times 10^{-9}\,\text{N m}^{-1}$ was measured for a DNA double strand in water ($L = 33\,\mu\text{m}$) at 300 K. By comparison with the relation (see Box 22.4)

$$f = \frac{3kT}{l \cdot L} h, \quad l = \frac{h}{f} \cdot \frac{3kT}{L}$$

we obtain the length l of the statistical chain element: $l = 100$ nm. This is in agreement with the value found in Section 22.4.1.

At large forces, when the chain is uncoiled, it is still found to be slightly extending proportional with the increasing additional force. For a DNA double helix ($L = 33\,\mu\text{m}$ an extension of $\Delta h = 1\,\mu\text{m}$ is found at a force of $\Delta f = 15$ pN. Thus the chain is elastically stretched and we obtain for the Young's modulus by stretching

$$M_m = \frac{\Delta f \cdot L}{\pi r^2 \Delta h} = 10^8\,\text{N m}^{-2}$$

with $r = 1.5$ nm. For comparison, $M_m = 10^{11}\,\text{N m}^{-2}$ for steel and $M_m = 10^6\,\text{N m}^{-2}$ for the rubber cylinder which we used as a good model to mimic the twisting and bending of double stranded DNA (see Problem 12.6 and Box 22.2). In this comparison, the DNA chain is less easy to stretch than the rubber cylinder.

Example – polyethylenglycol

We consider a single chain ($L = 160$ nm) of polyethylenglycol $HO\!\!-\!\!(CH_2\!-\!CH_2\!-\!O)_z\!H$. The chain is spanned between tip and surface of an atomic force microscope. The force is measured as a function of the elongation h (Figure 22.41, AFM, see Chapter 23). At small forces f the ratio $f/h = 1 \times 10^{-4}\,\text{N m}^{-1}$ was measured; thus the length of the statistical chain element is

$$l = \frac{h}{f} \cdot \frac{3kT}{L} = \frac{1}{1 \times 10^{-4}\,\text{N m}^{-1}} \cdot \frac{3 \cdot 4.2 \times 10^{-21}\,\text{Nm}}{160 \times 10^{-9}\,\text{m}} = 1\,\text{nm},$$

in agreement with the value obtained from viscosity measurements.

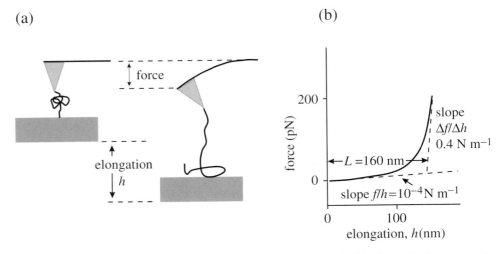

Figure 22.41 Polyethylene-glycol in hexane. Individual chain stretched by force f using an atomic force microscope. h is the elongation of the chain. (a) Experimental setup. One end of the polymer chain is fixed to a tip connected to a spring; the second end of the polymer is fixed to a movable plate. When the plate is moved downward by an elongation h, a force acts on the polymer chain which is measured by the bending of the spring. (b) Force acting on the polymer chain versus elongation h. (F. Oesterhelt, M. Rief, H.E. Gaub, New Journal of Physics 1, 6.1–6.6 (1999)).

At a large force f an increase Δf causes an additional elongation Δh; according to Figure 22.41 we obtain

$$\Delta f = \Delta h \cdot 0.4 \, \text{N m}^{-1}$$

With $L = 160$ nm we rewrite Δf as

$$\Delta f = \Delta h \cdot \frac{160 \, \text{nm}}{L} \cdot 0.4 \, \text{N m}^{-1} = \frac{\Delta h}{L} \cdot 160 \, \text{nm} \cdot 0.4 \, \text{N m}^{-1} = \frac{\Delta h}{L} \cdot 4 \times 10^{-8} \, \text{N}$$

This can be compared with the force required to extend a hydrocarbon zigzag chain by C–C–C bond angle distortion:

$$\Delta f = \frac{\Delta h}{L} k'_f d_{\text{eq}} \frac{2 \sin \alpha/2}{\cos^2 \alpha/2} = \frac{\Delta h}{L} \cdot 3 \times 10^{-8} \, \text{N}$$

where $k'_f = 36 \, \text{N m}^{-1}$, $d_{\text{eq}} = 154$ pm, $\alpha = 109°$ (see Section 5.2.4)

22.4.6 Restoring Coil

Now let us stop the flow that has completely unrolled the coil and see in what time the coil is restored (Figure 22.42(a)). An initial coil develops at the open end and moves with speed v toward the fixed end including gradually the uncoiled portion (length x). As

22.4 MACROMOLECULES

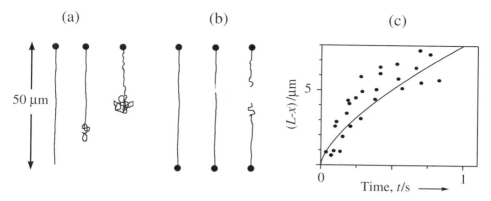

Figure 22.42 Restoring a coil from a fully uncoiled chain. (a) Shape of coil fixed at one end at different times t after stopping the flow. (b) Coils fixed at both ends and cut by UV radiation. (c) Theory (solid line) according to equation (22.61) compared with experiment (dots) observing single fluorescence-labeled DNA molecules (M. Yanagida, Y. Hiraoka, I. Katsura: Cold Spring Harbor Symposia on Quantitative Biology 47(2), 177 (1982)).

shown in Box 22.3, x decreases with time t proportional to $t^{2/3}$:

$$x = L - \left(\frac{kT}{2\pi\eta}\right)^{2/3} \frac{1}{l} \cdot t^{2/3} \tag{22.61}$$

The extended strand of a DNA double helix carrying a fluorescent dye can be fixed at both ends. When illuminating, the strand occasionally breaks (some light quantum initiates a photochemical reaction) and the growth of the two coils and their motion to the fixed-points can be observed (Figure 22.42(b)).

The measured decrease of length x with time (experimental points in Figure 22.42(c)) follows the theoretical relation (22.61) with $\eta = 1 \times 10^{-2}$ poise $= 10^{-3}\,\mathrm{N\,m^{-2}\,s}$, $l = 100$ nm, and $T = 300$ K:

$$L - x = 8\,\mathrm{s}^{-2/3}\,\mu\mathrm{m} \cdot t^{2/3}$$

In another experiment the chain is bound to a small bead. The bead is pulled by a laser beam until the chain is fully uncoiled; then the bead is stopped and the evolution of the shape of the molecule is observed.

According to (22.61) the time to restore a coil of contour length $L-x$ is independent of the total chain length L; this is justified by the experiment. However, when the fractional extension becomes small, (22.61) is no longer valid. Then, similar to equation (22.60), the speed dx/dt is given by

$$\frac{dx}{dt} = -\frac{3kT}{3\pi\eta \cdot (lL)^{3/2}} \cdot x$$

Thus

$$x \sim e^{-t/\tau} \quad \text{with} \quad \tau = \frac{\pi\eta(lL)^{3/2}}{kT}$$

In this case τ is expected to be proportional to $L^{3/2}$, in agreement with the experiment. (T.S. Perkins, S.R. Quake, D.E. Smith, S. Chu, Relaxation of a Single DNA Molecule Observed by Optical Microscopy, Science 264, 822 (1994)).

> **Box 22.3 – Time to Restore Random Coil from Unraveled Chain**
>
> After stopping the current, coiling begins at the open end where the tension along the chain is smallest. After a while a coil has formed, including a part of the chain of contour length $L - x$. This coil moves with speed v. We approximate the coil as a sphere of radius
>
> $$r_{\text{coil}} = \tfrac{1}{2}\sqrt{(L-x)l}$$
>
> (hydrodynamic resistance $f = 6\pi\eta r_{\text{coil}} \cdot v$). The force acting on the statistical chain element connecting the coiled with the uncoiled portion is given by
>
> $$f = 6\pi\eta \tfrac{1}{2}\sqrt{(L-x)l} \cdot v$$
>
> This force counteracts the force f' caused by thermal collisions between the molecular chain and the molecules of the solvent. This effect tends to a random orientation of the statistical chain element. For keeping the chain element well oriented (probability for $\varphi = 0°$ considerably larger than for $\varphi = 180°$, see Figure 22.41) the energy $f' \cdot l$ to turn the element must be about kT, according to the Boltzmann law:
>
> $$f' = \frac{kT}{l}$$
>
> Therefore
>
> $$v = \frac{kT}{3\pi\eta\sqrt{L-x} \cdot l^{3/2}} = -\frac{dx}{dt}$$
>
> We rewrite this as
>
> $$-\sqrt{L-x} \cdot dx = \frac{kT}{3\pi\eta l^{3/2}} \cdot dt$$
>
> and by integration we obtain
>
> $$\tfrac{2}{3}(L-x)^{3/2} = \frac{kT}{3\pi\eta l^{3/2}} \cdot t$$
>
> Thus
>
> $$x = L - \left(\frac{kT}{2\pi\eta}\right)^{2/3} \frac{1}{l} \cdot t^{2/3}$$

22.4.7 Motion Through Entangled Polymer Chains

Draining solvent through meshwork

The sedimentation in a centrifuge of a polymer of polystyrene with $Z = 2300$ is about ten times as fast as that of a polymer with $Z = 26$, because the sedimentation constant s

22.4 MACROMOLECULES

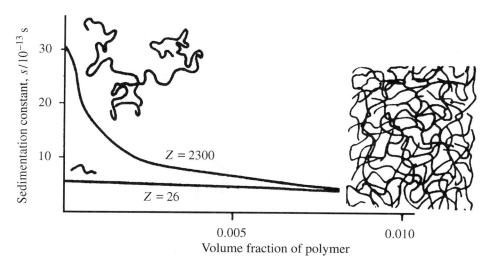

Figure 22.43 Sedimentation constant of isolated molecules and entanglements (H. Mosimann, Helv. Chim. Acta 26, 61 (1943); P. Baertschi, Helv. Chim. Acta 34, 1324 (1951)).

is essentially proportional to $\sqrt{Z}$, as discussed in Section 22.4.4. This is true for isolated polymer molecules that do not interact – that is, at sufficiently high dilution.

However, already at a dilution of 1 volume % of polymer material the chains penetrate each other, and the sedimentation speed is equal in both cases (Figure 22.43). The sedimentation speed is determined by the resistance against the draining of solvent through the entanglement. The process can be simulated with macroscopic models (by sucking a liquid through a cotton plug mimicking the molecular meshwork). The simulation shows that it is reasonable to treat the solvent as a liquid of macroscopic viscosity.

Moving unraveled chain through entangled polymer chains

Let us consider a DNA double helix (labeled with a fluorescent dye) bound to a bead at one end in a solution of other unlabeled DNA chains (Figure 22.44). The bead is pulled by a laser beam and moved along a curved path sufficiently fast to uncoil the chain bound to the bead. The whole chain follows the trajectory of the bead because it stays in the same tube. As mentioned in Figure 22.40(c), the force pulling the bead is due to the polarization of the bead in the electric field of the light; the bead is driven toward the point of maximum field.

Moving random coil by winding through meshwork: gel electrophoresis of DNA fragments

Gels can be considered as entanglements of chain molecules in a surplus of solvent. Bonds between the chains fix the structure. The motion of a macromolecule in a gel when applying a moderate electric field (about $1\,\mathrm{V\,cm^{-1}}$) is an important method for separating proteins and nucleic acids.

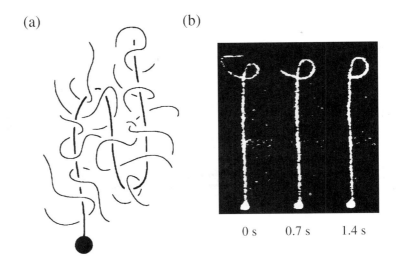

Figure 22.44 Fluorescently stained DNA strand driven through entangled polymer chains T. T. Perkins, D. E. Smith, S. Chu, *Science*, 1994, **264**, 819. (a) DNA strand bound to a bead in the entanglement. (b) Images of the fluorescent DNA strand driven by an optical tweezer. The images are taken every 0.7 s.

Example – electrophoresis of DNA fragments

A mixture of double-stranded DNA fragments was separated by electrophoresis in an polyacrylamid gel (5%). After electrophoresis for 180 min, the gel was immersed in a staining solution containing a fluorescent dye (ethidium) and then photographed under UV light.

As shown in Figure 22.45, the DNA solution was placed in a small hole in a layer of the gel (meshwork with pore size $\delta < 100$ nm) and a field of $5\,\mathrm{V\,cm^{-1}}$ was applied (voltage 100 V between platinum wires at distance 20 cm). In Figure 22.46 the traveling distance x is plotted versus Z, the number of the base pairs.

In the following we consider two limiting cases: length L of the DNA strand small compared to δ and L large compared to δ.

L small compared to δ. We first consider a short strand (L smaller than the mesh size δ). It should move with roughly the same speed as in free solvent. This speed v_0 is given by equalizing the force f due to the applied field F with the hydrodynamic resistance, according to equation (22.58):

$$f = -QF = 6\pi\eta \tfrac{1}{4} l v_0 \cdot N_{\text{stat}}$$

Each monomer carries the charge $-2e$ (phosphate groups in both strands of the double helix). Then the charge is $Q = -2e \cdot Z$ where Z is the number of monomers (basepairs) in the strand. With

$$Z = \frac{L}{b} \quad \text{and} \quad N_{\text{stat}} = \frac{L}{l}$$

22.4 MACROMOLECULES

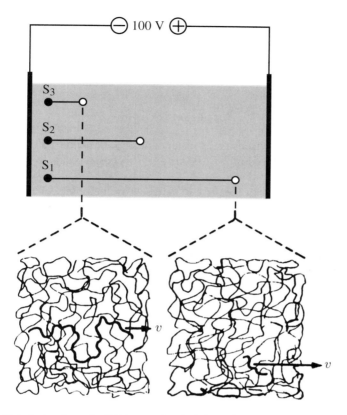

Figure 22.45 Gel electrophoresis. DNA strands wind through meshwork toward the anode. s_1, s_2, and s_3 are the starting points for three DNA strands with different chain lengths L.

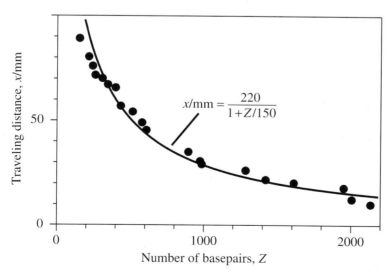

$$x/\text{mm} = \frac{220}{1+Z/150}$$

Figure 22.46 DNA electrophoresis. Displacement x of the separated nucleic acids versus the number Z of basepairs. Dots: experiment; solid line: theory according to equation (22.66) (courtesy Andreas Kuhn).

we obtain

$$f = 2eF \cdot \frac{L}{b} = 6\pi\eta\frac{1}{4}v_0 \cdot L \tag{22.62}$$

Remember that b is the contour length per monomer. Thus we obtain

$$v_0 = \frac{4}{3\pi}\frac{eF}{b\eta}, \quad (L < \delta) \tag{22.63}$$

With $\eta = 10^{-3}\,\text{N m}^{-2}\,\text{s}$ (viscosity of water) and $b = 0.34\,\text{nm}$ we obtain for a field of $5\,\text{V cm}^{-1} = 500\,\text{V m}^{-1}$ a speed of $1 \times 10^{-4}\,\text{m s}^{-1}$; this corresponds to a traveling distance $x_0 = v_0 \cdot t = 1080\,\text{mm}$ in $t = 180\,\text{min}$.

There are polymer chains in the spaces between the netpoints of the meshwork, thus the viscosity η is effectively larger than the viscosity of water and the traveled distance is accordingly smaller. The measured traveling distance in 180 min (Figure 22.46), extrapolated to sufficiently small values of Z, is $x_0 = 220\,\text{mm}$.

L large compared to δ. The traveling distance x decreases with increasing chain length. The strands have to wind through the meshwork when L is longer than the pore size δ. The speed v of the molecule in the meshwork in the direction of the applied force (x direction), according to Foundation 22.1, is

$$v = \frac{l}{3\pi\eta L^2}f$$

and therefore, according to equation (22.62),

$$v = \frac{l}{3\pi\eta L^2} \cdot 2eF\frac{L}{b} = v_0\frac{l}{2L} \quad (L > \delta) \tag{22.64}$$

What is the essence of this relation? Effectively, the chain is driven by the force acting on the statistical chain element at the end heading toward the anode. The forces acting on the charges along the residual chain are compensated by the opposing forces of the network in which the chain is entangled: effectively, the force

$$2eF\frac{l}{b}$$

is drawing the entire strand, provided that the statistical chain element at the end can stick through the meshwork toward the anode. This force is acting on this statistical chain element (the engine is in front and pulls the train). It equalizes with the hydrodynamic resistance of the chain. The molecule is free to move in this way in about one third of the cases. Otherwise the statistical chain element at the end is blocked in the meshwork and thus the entire molecule is blocked. Then we obtain

$$2eF\frac{l}{b} \cdot \frac{1}{3} = \pi\eta vL \quad \text{or} \quad v = \frac{2eFl}{b \cdot 3\pi\eta L}$$

This relation is identical with equation (22.64); for the factor π see Foundation 22.1.

General Case. The general relation

$$v = v_0\frac{l/2}{L + l/2} \tag{22.65}$$

22.4 MACROMOLECULES

reduces to the limiting cases with $L \gg \delta$ and $L \ll \delta$. Then the traveling distance x is given by

$$x = x_0 \frac{1}{1 + 2L/l} \quad (22.66)$$

In the considered case ($x_0 = 220$ mm, $L = Zb$, $b = 0.34$ nm, $l = 100$ nm) we obtain

$$x = 220 \text{ mm} \cdot \frac{1}{1 + Z/150} \quad (22.67)$$

Figure 22.46 shows that x calculated from this relation agrees with the experimental result. (Note that the value $x_0 = 220$ mm is reasonable, but has to be considered as an adjustable parameter).

In this consideration we have neglected diffusion. This is appropriate. You can easily estimate that the traveling distance by diffusion through the meshwork in 180 min is 1 mm (see Foundation 22.1); this is small compared to x.

22.4.8 Rubber Elasticity

In order to see the essence of rubber elasticity we consider a chain fixed in both ends (Figure 22.47). It exerts an attractive force on the ends due to its Brownian motion; the force increases with distance h and with temperature. The force has statistical reasons: the number of conformations of the chain decreases with increasing distance h. On the average the force f is

$$f = kT \frac{3h}{Ll} \quad (22.68)$$

as shown in Box 22.4. Equation (22.68) is valid for $h \ll L$. The force f is infinite for $h = L$, if L is considered to be constant. f is zero for $h = 0$; in this case, the chain, by its thermal motion, acts on point P_2 (Figure 22.47) equally often in the direction to the left and to the right, thus the force is zero on the average.

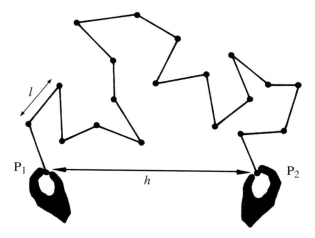

Figure 22.47 Random coil held tight at chain ends. Contracting force due to thermal fluctuation.

Box 22.4 – Force Contracting a Chain Held at Chain Ends

We consider a chain in solution fixed at chain end P_1. The second chain end P_2 is free to diffuse but is bound to P_1 with a force (k_f = force constant) $f = -k_f \cdot h$. According to Boltzmann's Law we have

$$P(h) \cdot dh = \text{const} \cdot 4\pi h^2 \exp\left(\frac{-E}{kT}\right) \cdot dh$$

Then the probability to find the chain ends in distances between h and $h+dh$ is

$$P(h) \cdot dh = \text{const} \cdot 4\pi h^2 \cdot \exp\left(-\frac{k_f h^2}{2kT}\right) \cdot dh$$

The average

$$\overline{h^2} = \frac{\int h^2 P(h) \cdot dh}{\int P(h) \cdot dh} = \frac{3}{2}\frac{2kT}{k_f}$$

On the other hand, we have according to equation (22.23),

$$\overline{h^2} = Ll$$

Thus

$$k_f = \frac{3}{2}\frac{2kT}{Ll} = \frac{3kT}{Ll}$$

and

$$f = -\frac{3kT}{Ll} \cdot h$$

f is the attraction force on the ends of a polymer molecule caused by the thermal motion of the chain elements. Then we obtain

$$P(h) \cdot dh = \text{const} \cdot 4\pi h^2 \exp\left(-\frac{3kT}{Ll}\frac{h^2}{2kT}\right) \cdot dh \qquad (22.69)$$

$$= \text{const} \cdot 4\pi h^2 \exp\left(-\frac{3}{2}\frac{h^2}{Ll}\right) \cdot dh \qquad (22.70)$$

This confirms the distribution given by equation (22.22).

h is considered to be small ($h \ll L$). For $h = L$ the force f must be infinite. Generally

$$f = \frac{kT}{2l}\left[\frac{1}{(1 - h/L)^2} - 1 + \frac{4h}{L}\right]$$

Continued on page 783

22.4 MACROMOLECULES

> *Continued from page 782*
>
> (C. Bustamante, Entropy Elasticity of λ-Phaga DNA, Science 265, 1599 (1994)). This relation reduces to equation (22.68) for $h \ll L$:
>
> $$f \approx \frac{kT}{2l}\left[\frac{1}{1-2h/L} - 1 + \frac{4h}{L}\right] \approx \frac{kT}{2l}\left[1 + 2\frac{h}{L} - 1 + \frac{4h}{L}\right] = kT \cdot \frac{3h}{L \cdot l}$$
>
> The force f becomes infinite for $h = L$. The force f can be related to the average contribution of a statistical chain element, $\overline{\Delta h}$,
>
> $$\overline{\Delta h} = \frac{h}{N_{\text{stat}}} \quad \text{or} \quad \frac{\overline{\Delta h}}{l} = \frac{h}{N_{\text{stat}} \cdot l} = \frac{h}{L}$$
>
> Then
>
> $$f = \frac{kT}{2l}\left[\frac{1}{(1-\overline{\Delta h}/l)^2} - 1 + \frac{4\overline{\Delta h}}{l}\right]$$

Rubber consists of molecular chains connected by bridges made of sulfur atoms (net points, Figure 22.48). A force f_{total} is to be applied to stretch a band of rubber. This is due to the fact that chains connecting net points are stretched and a counteracting force

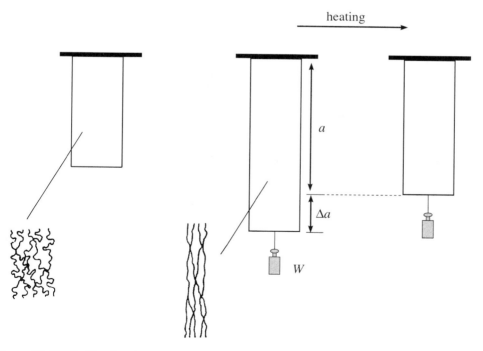

Figure 22.48 Rubber band extended by weight W. Contraction when heating: increased thermal fluctuations.

according to equation (22.68) occurs. In the case of small extensions the force f_{total} to stretch the band from its original length a to $a + \Delta a$ is given by

$$f_{\text{total}} = \frac{\Delta a}{a^2} \cdot 3NkT \quad \text{for} \quad \Delta a \ll a \tag{22.71}$$

where N is the number of chains connecting net points in the sample (Box 22.5).

Box 22.5 – Stretching a Chain

For simplicity we assume that the lines connecting the end points of the chains (net points) have the direction of the x, y, and z axis, $N/3$ in each axis (Figure 22.49). Furthermore, we assume, before stretching, that h has the average value $h_1 = \sqrt{Ll}$. After stretching in the z direction we obtain,

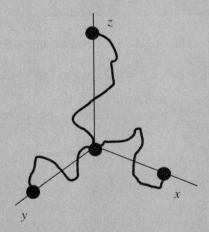

Figure 22.49 Rubber idealized by molecular chains, $N/3$ oriented in the x direction, and $N/3$ each in the y and z directions. Dots: net points.

$$h_2 = \sqrt{Ll} \cdot \left(1 + \frac{\Delta a}{a}\right)$$

for the z directed chains ($\frac{1}{3}N$ chains). The corresponding work is

$$w_z = \frac{N}{3} \cdot \int_{h_1}^{h_2} k_{\text{f}} h \cdot dh = \frac{3kT}{2Ll}(h_2^2 - h_1^2) \cdot \frac{N}{3}$$

$$= \frac{3}{2}kT \cdot \left[\left(1 + \frac{\Delta a}{a}\right)^2 - 1\right] \cdot \frac{N}{3}$$

Continued on page 785

22.4 MACROMOLECULES

> Continued from page 784
>
> The volume is unchanged when stretching the sample. This means that the cross-sectional diameter is diminished by a factor α where
>
> $$\left(1 + \frac{\Delta a}{a}\right)\alpha^2 = 1 \quad \text{and} \quad \alpha = \frac{1}{\sqrt{1 + \Delta a/a}}$$
>
> Then
>
> $$w_x = w_y = \frac{3}{2}kT \cdot \left(\frac{1}{1 + \frac{\Delta a}{a}} - 1\right) \cdot \frac{N}{3}$$
>
> and
>
> $$w = w_x + w_y + w_z = \frac{3}{2}kT \cdot \left[\left(1 + \frac{\Delta a}{a}\right)^2 + \frac{2}{1 + \frac{\Delta a}{a}} - 3\right] \cdot \frac{N}{3}$$
>
> For $\Delta a/a \ll 1$ we obtain
>
> $$w = kT\frac{N}{2}\left[\left(\frac{\Delta a}{a}\right)^2 + 2\left(\frac{\Delta a}{a}\right)^2\right] = kT\frac{3}{2}N \cdot \left(\frac{\Delta a}{a}\right)^2$$
>
> Then the force is
>
> $$f_{\text{total}} = \frac{dw}{d\Delta a} = 3kTN \cdot \frac{\Delta a}{a^2}$$

An interesting aspect of rubber elasticity is due to the fact that the number of conformations decreases in stretching. All conformations have essentially equal energy, and the enthalpy change can be neglected. The work supplied by stretching the sample is transferred as heat into the thermal bath in an isothermal process, or the sample is heated up in an adiabatic process and cools down when releasing a stretched sample.

A sample stretched with a given weight contracts on heating:

$$\Delta a = f_{\text{total}} \cdot \frac{a^2}{3N_{\text{thread}} \cdot k} \cdot \frac{1}{T} \tag{22.72}$$

This behavior is in contrast to the normal case of a body extending on increasing the temperature.

Even rubber with missing sulfur bridges behaves as an elastic material when the body is subject to a fast change in shape, giving no time for disentanglement (Figure 22.50(a)). In this case, points of entangled chains act like net points. In slow processes, however, the material behaves like a viscous liquid (Figure 22.50(b)).

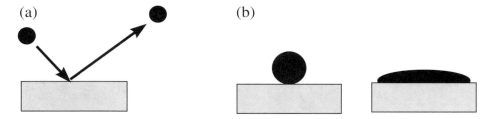

Figure 22.50 Behavior of a silicon rubber with missing sulfur bridges. (a) Elastic response on a fast change of shape. (b) Liquid-like response on slow changes.

22.5 Supramolecular Structures

The formation of supramolecular structures is a fundamental process in building biosystems and furnishes the guiding principle in constructing supramolecular systems. We give an example each for a natural and for a designed self-assembled structure, illustrating the principles. Other examples showing the use of the concepts in constructing supramolecular machines are discussed in Chapter 23.

Figure 22.51 shows a fascinating example of a self-assembling system. A protein binds to an RNA chain, and by distinct interlocking of the protein molecules a cylindrical aggregate with an RNA spiral inside is formed: the tobacco mosaic virus. The process can be reversed, whereby the virus disintegrates.

Assembling and disintegration are processes taking place near equilibrium, and small changes of conditions reverse the processes. This is crucial for building the correct structure. Weak noncovalent interactions allow an easy correction of mismatches. A molecule which is added to the aggregate and correctly interlocks is stabilized by additional interlocking with the next molecule being added: aggregation is a cooperative process.

This mutual interlocking cannot take place in the case of a defect; then the arrangement is unstable and is immediately corrected when the interactions are weak. This condition

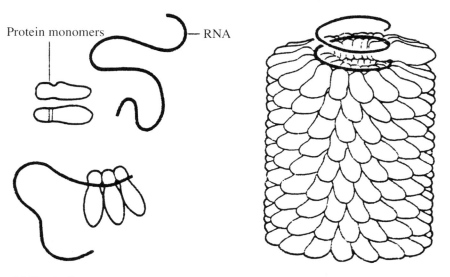

Figure 22.51 Self-assembly of tobacco mosaic virus. (courtesy H. Tschesche).

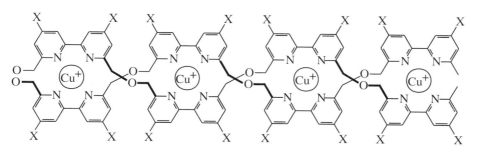

Figure 22.52 Self-assembly by interlinking bipyridine units. (J.M. Lehn, Supramolecular Chemistry, Concepts and Perspectives, p. 147, VCH Verlag, Weinheim 1995).

is important for eliminating a defect; otherwise the defect would be frozen. The precise formation of the cooperative system depends on the mutual stabilization of many weakly bonded mutually interlocking components.

These principles must be considered in designing supramolecular structures. Figure 22.52 shows the self-organization of a double helix of an oligobipyridine and Cu(I) ions used for interlinking the bipyridine units.

Problems

Problem 22.1 – Contact angles

Calculate the contact angles α and β in Figure 22.4 from equations (22.5) and (22.6), using the data given in the text.

Solution. With the notations $a = \gamma_{n-\text{decane/air}} = 23.9 \text{ mN m}^{-1}$, $b = \gamma_{n-\text{decane/water}} = 51.2 \text{ mN m}^{-1}$, and $c = \gamma_{\text{water/air}} = 72.8 \text{ mN m}^{-1}$ from equations (22.5) and (22.6) we obtain

$$a \cdot \cos \alpha + b \cdot \cos \beta = c$$

$$a \cdot \sin \alpha = b \cdot \sin \beta$$

We replace $\sin \alpha$ by $\sqrt{1 - \cos^2 \alpha}$ in the second equation, eliminate $\cos \alpha$ from both equations and solve for $\cos \beta$:

$$\cos \beta = \frac{1 - (b/a)^2 - (c/a)^2}{-2(bc/a^2)} = 0.98, \quad \beta = 11^0$$

Inserting in the second equation we obtain

$$\sin \alpha = \frac{b}{a} \cdot \sin \beta = 0.41, \quad \alpha = 24^0$$

Problem 22.2 – Transfer of Charge e from Water into a Membrane

Calculate the work w for this process.

Solution. The charge (the COO^- side group of amino acid Asp including a solvation shell) is replaced by a sphere of radius $r = 0.5$ nm.

We calculate the energy w to transfer the sphere with charge e from dielectric medium with relative permittivity ε_1 (in this case water, $\varepsilon_1 = 80$) into a dielectric medium with relative permittivity ε_2 (in this case hydrocarbon portion, $\varepsilon_2 = 2.5$). The energy dw to transfer a charge dQ is

$$dw = -\frac{1}{4\pi\varepsilon_0}\frac{e-Q}{\varepsilon_1 r}\cdot dQ + \frac{1}{4\pi\varepsilon_0}\frac{Q}{\varepsilon_2 r}\cdot dQ$$

Thus by integration we obtain

$$w = \int_0^e dw = -\frac{1}{4\pi\varepsilon_0 r}\int_0^e \left(\frac{e-Q}{\varepsilon_1} - \frac{Q}{\varepsilon_2}\right)\cdot dQ$$

$$= -\frac{1}{4\pi\varepsilon_0 r}\left(\frac{e^2}{\varepsilon_1} - \frac{1}{2}e^2 - \frac{1}{2}\frac{e^2}{\varepsilon_2}\right)$$

$$= -\frac{e^2}{4\pi\varepsilon_0 2r}\left(\frac{1}{\varepsilon_1} - \frac{1}{\varepsilon_2}\right)$$

$$= -1.44 \text{ eV}\left(\frac{1}{80} - \frac{1}{2.5}\right) = 0.56\,\text{eV} = 8.9\times 10^{-20}\,\text{J}$$

(corresponding to $54\,\text{kJ mol}^{-1}$).

Problem 22.3 – Light Scattering (diameter of molecules $\ll \lambda$)

Derive equation (22.27) from the Hertz antenna equation (see Foundation 8.5).

Solution. The amplitude F_0 of the electric field at large distance d of an oscillating dipole in the xy plane ($d \gg \lambda$) is given by

$$F_0 = \frac{1}{4\pi\varepsilon_0}\frac{Q\xi_0}{c_0^2 d}\omega^2$$

where Q is the charge oscillating in the z direction. The amplitude ξ_0 is given by the amplitude of the incident light: $Q\xi_0 = \alpha F_{0,\text{incident}}$. Then the intensity of the light scattered by the sample containing N particles is

$$\frac{I_{\text{scattered}}}{I_{\text{incident}}} = \frac{F_{0,\text{scattered}}^2}{F_{0,\text{incident}}^2} = \left(\frac{1}{4\pi\varepsilon_0}\right)^2 \alpha^2 \frac{1}{c_0^4 d^2}\omega^4 N$$

α is related to the polarization P of the given dilute solution (refractive index $\widehat{n}$, volume V) additional to the polarization of the solvent (refractive index $\widehat{n}_0$, volume V)

$$\frac{P}{F_{\text{incident}}} = \varepsilon_0(\widehat{n}^2 - \widehat{n}_0^2)V = \alpha N$$

Because of $\widehat{n}^2 - \widehat{n}_0^2 = (\widehat{n} - \widehat{n}_0)(\widehat{n} + \widehat{n}_0) \approx (\widehat{n} - \widehat{n}_0)2\widehat{n}$, we obtain

$$\alpha = \frac{\varepsilon_0(\widehat{n} - \widehat{n}_0)2\widehat{n}V}{N}$$

With

$$N = \frac{m}{M_m}N_A \quad \text{and} \quad \omega = 2\pi\frac{c_0}{\lambda}$$

this results in equation (22.27).

Problem 22.4 – Light scattering (diameter of molecules $<\lambda/4$

What is the ratio of the intensities of scattered light arriving at points P_1 and P_2 (Figure 22.34), respectively?

PROBLEM 22.5

Solution. The waves scattered from centers A and B arrive at P_2 with the same phase:

$$F(P_2) = 2F_0 \cos\left[\omega\left(t - \frac{d}{c_0}\right)\right]$$

At point P_1 it arrives with a phase shift $2a/c_0$ (c_0 = speed of light):

$$F(P_1) = F_0 \cos\left[\omega\left(t - \frac{d-a}{c_0}\right)\right] + F_0 \cos\left[\omega\left(t - \frac{d+a}{c_0}\right)\right]$$

$$= 2F_0 \cos\left[\omega\left(t - \frac{d}{c_0}\right)\right]\cos\frac{\omega a}{c_0}$$

where $\omega/c_0 = 2\pi/\lambda$. This gives an intensity ratio for the light arriving in P_1 and P_2

$$\frac{I(P_1)}{I(P_2)} = \frac{F^2(P_1)}{F^2(P_2)} = \cos^2\frac{2\pi a}{\lambda}$$

Because of

$$\cos x = 1 - \tfrac{1}{2}x^2 + \cdots$$

for sufficiently small x we have

$$\cos^2 x = \left(1 - \tfrac{1}{2}x^2\right)^2 = 1 - x^2$$

Then we obtain

$$\frac{I(P_1)}{I(P_2)} \approx 1 - \left(\frac{2\pi a}{\lambda}\right)^2$$

Note that in our model $2a = \sqrt{\overline{h^2}}$ where a is the distance between points P_1 and P_2.

Problem 22.5 – Special Case of Scattering Relation (22.29)

Consider relation (22.29) for backward scattering at $\alpha < 1$.
  Solution:

$$e^{-\alpha} - 1 + \alpha \approx 1 - \alpha + \tfrac{1}{2}\alpha^2 - \tfrac{1}{6}\alpha^3 - 1 + \alpha = \alpha^2\left(\tfrac{1}{2} - \tfrac{1}{6}\alpha\right)$$

Then for $\varphi = 180°$ (backward scattering) we obtain

$$\alpha(180°) = \frac{8\pi^2}{3}\frac{\overline{h^2}}{\lambda^2}$$

and

$$\frac{I_{\text{backward}}}{I_{\text{forward}}} \approx 1 - \frac{\alpha(180°)}{3} = 1 - \frac{8\pi^2}{9}\frac{\overline{h^2}}{\lambda^2}$$

Problem 22.6 – Molar Mass from Diffusion and Sedimentation

The following data have been obtained for ribosomes: sedimentation constant $s = 82.6 \times 10^{13}$ s, diffusion coefficient $D = 1.52 \times 10^{-7}$ cm^2 s^{-1}, partial molar volume $V_{\text{partial}} = 0.61$ cm^3 g^{-1}, density $\rho = 1$ g cm^{-3} at 20 °C. Calculate the molar mass M_m of the ribosomes.

Solution. According to equation (22.41) we have

$$M_m = \frac{RT \cdot s}{D \cdot (1 - V_{partial} \cdot \rho)} = 3.4 \times 10^6 \text{ g mol}^{-1}$$

Problem 22.7 – Extension of an unraveled coil

We consider a chain fixed at one end in a current of speed v. The extension x is less than the contour length L, due to the thermal collisions of the chain elements. Estimate the average extension x.

Solution: We consider a statistical chain element number n (numbering from the fixed end). According to Box 22.4, the average contribution $\overline{\Delta x_n}$ of the element n to the extension x is related to f:

$$f = \frac{kT}{2l} \cdot \left[\frac{1}{\left(1 - \overline{\Delta x_n}/l\right)^2} - 1 + \frac{4\overline{\Delta x_n}}{l} \right]$$

Equalizing f and the hydrodynamic force f'

$$f' = 3\pi\eta l \sqrt{N_{stat} - n} \cdot v$$

gives

$$\frac{kT}{2l} \cdot \left[\frac{1}{\left(1 - \overline{\Delta x_n}/l\right)^2} - 1 + \frac{4\overline{\Delta x_n}}{l} \right] = 3\pi\eta l \sqrt{N_{stat} - n} \cdot v$$

or

$$\frac{1}{\left(1 - \overline{\Delta x_n}/l\right)^2} - 1 + \frac{4\overline{\Delta x_n}}{l} = \frac{4v}{v_c} \sqrt{N_{stat} - n}$$

The total extension x is obtained by summing up over all elements n

$$x = \sum_{n=1}^{N_{stat}} \overline{\Delta x_n}$$

For example, we consider the case $L = 50\,\mu\text{m}$, $l = 100\,\text{nm}$ (thus $N_{stat} = L/l = 500$) and $v = 14\,\mu\text{m s}^{-1}$. Then with $\eta = 1 \times 10^{-3}\,\text{N m}^{-2}$ and $T = 300\,\text{K}$

$$v_c = \frac{2kT}{3\pi\eta l^2} = 87.9\,\mu\text{m s}^{-1}$$

Inserting different values for n, we obtain the following data for $\overline{\Delta x_n}$:

n	$\overline{\Delta x_n}/\text{nm}$	$\overline{\Delta x_n}/l$
1	71	0.71
100	70	0.70
200	68	0.68
300	64	0.64
400	56	0.56
500	0	0.00

PROBLEM 22.7

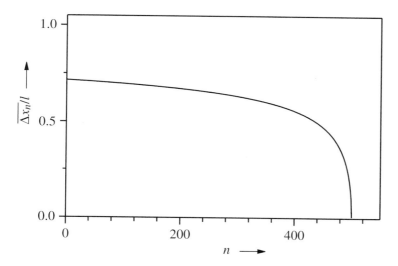

Figure 22.53 $\overline{\Delta x_n}/l$ versus n for $L = 50\,\mu\text{m}$, $l = 100\,\text{nm}$, $v = 14\,\mu\text{m}\,\text{s}^{-1}$, $\eta = 1 \times 10^{-3}\,\text{N}\,\text{m}^{-2}$, and $T = 300\,\text{K}$.

These data are illustrated in Figure 22.53.

By numerically summing up the values $\overline{\Delta x_n}$ for n in the range from 1 to 500 the following results are obtained for different values of the flow speed v:

$v/\mu\text{m}\,\text{s}^{-1}$	$v\sqrt{L}/\mu\text{m}^{3/2}\,\text{s}^{-1}$	x/L
0	0	0
14	100	0.62
28	200	0.74
56	400	0.82
85	600	0.85

The fractional extension x/L is plotted in Figure 22.40(c) as a function of $v\sqrt{L}$.

Foundation 22.1 – Mobility of DNA in a Meshwork

Diffusion

We first consider the diffusion of the strand in the absence of an electric field. We assume that the pore size is of the order of l and that $N_{stat} \gg 1$. The molecule, by its Brownian motion, winds itself through the meshwork. At time $t = 0$ the chain is fixed in the meshwork and the molecule needs a certain time τ to leave this cage – that is, until chain end 1, essentially, reaches the original position of chain end 2, or, with equal probability, 2 reaches the original position of 1. This time τ, on the average, is given by

$$L^2 = 2D_L\tau \quad \text{or} \quad \tau = \frac{L^2}{2D_L}$$

The chain considered to be enclosed in a narrow channel diffuses along this channel in the direction of the curved contour line of length L. The diffusion coefficient D_L depends on the hydrodynamic resistance f when moving the chain along its axis. This is easier than moving it perpendicularly to the axis: $f_\parallel$, for a given speed, is about one half of $f_\perp$. On the average, we obtain

$$f = \frac{1}{3}(f_\parallel + 2f_\perp) = \frac{5}{3}f_\parallel$$

Then, with our estimate in equation (22.62), we obtain

$$f = \frac{3\pi}{2}\eta L v$$

So we obtain

$$f_\parallel = \frac{3}{5} \cdot \frac{3\pi}{2}\eta L v \approx \pi \eta L v$$

Thus, corresponding to the Stokes–Einstein equation (see Chapter 11), for the diffusion coefficient D_L for a chain with length L we have

$$D_L = \frac{kT}{\pi \eta L}$$

Thus

$$\tau = \frac{L^2}{2D_L} = \frac{L^2}{2}\frac{\pi \eta L}{kT} = \frac{\pi \eta L^3}{2kT}$$

Now we consider the global diffusion in space of the coil in the entangled network of the gel. In time τ the shift of the center of the coiled molecule in an arbitrary direction in space is about

$$2r_{coil} = \sqrt{Ll}$$

Continued on page 793

FOUNDATION 22.1

Continued from page 792

In a time $t \gg \tau$ the coil as an entity makes a random walk of t/τ steps. Remember that in the relation $\overline{\Delta x^2} = \lambda^2 \cdot j$ derived in Chapter 11, the square λ^2 of the free path is replaced by $(2r_{\text{coil}})^2 = L \cdot l$ and j is replaced by $\frac{1}{3} t/\tau$. Then the shift Δx of the coil in the x direction is given by

$$\overline{\Delta x^2} = \tfrac{1}{3} L l \frac{t}{\tau}$$

This random walk can be described by introducing a diffusion coefficient D to be distinguished from D_L (that described the micro diffusion of the strand along its channel). Again, according to Einstein and Smoluchowski, we have

$$\overline{\Delta x^2} = 2Dt = \frac{1}{3} L l \frac{t}{\tau} = \frac{1}{3} L l \frac{2kT}{\pi \eta L^3} \cdot t$$

Thus

$$D = \frac{kT}{3\pi \eta L^2 / l}$$

Now we consider the coil as being exposed to a force acting in the x direction driving the coil to move with speed v. Generally the diffusion coefficient is related to f/v by

$$D = \frac{kT}{f/v}$$

(in the case of a sphere, as discussed in Chapter 11, we have $f/v = 6\pi \eta r$). In the present case we obtain

$$\frac{f}{v} = \frac{3\pi \eta L^2}{l} \quad \text{or} \quad v = \frac{l}{3\pi \eta L^2} \cdot f$$

23 Supramolecular Machines

Supramolecular machines are organized systems of cooperating molecules, where the individual molecules constitute the functional components. Such systems have a great potential as sensors and actuators, information processing devices, memories with high storage capacity, tailored catalysts for complex chemical reactions, and energy conversion systems. The most intricate devices of this kind are living systems. Today's artificial supramolecular machines are very simple, but the methods to fabricate them pave the way to construct much more complex systems. The limits of supramolecular engineering are human imagination and creativity. Modern chemistry finds itself at the outset of a fascinating development based on a new paradigm, the planned design of organized entities.

Simple prototypes of supramolecular machines are obtained by designing and synthesizing molecules that interlock like the parts of a jigsaw puzzle. They change position by diffusing and interlock in the intended way whenever opportunities arise (*lock-and-key principle*). The interlocking is further supported by programmed external manipulations and by appropriate changes of the environmental conditions (*programmed environmental change principle*). Possibilities of interlocking identical molecules were discussed in Chapter 22. These can be extended to be used for this new aim.

We illustrate this by discussing some typical examples: devices to study energy transfer and electron transfer between molecules, switching devices based on photoinduced isomerization, and devices for amplifying signals by inducing cooperative processes. We consider molecular recognition, replica formation, and signal transfer through molecular wires.

23.1 Energy Transfer Illustrating the Idea of a Supramolecular Machine

Tools and machines are functional units, systems where different components interact to perform tasks that cannot be done by any component. Scissors, constituting an instrument based on the cooperation of two blades, are a simple example (Figure 23.1(a)). The idea of a supramolecular machine, of an entity where single molecules constitute the cooperating functional components can be illustrated by considering two interacting dye molecules D and A. D is a donor molecule absorbing blue light and emitting green light, and A is

23.2 PROGRAMMED INTERLOCKING MOLECULES

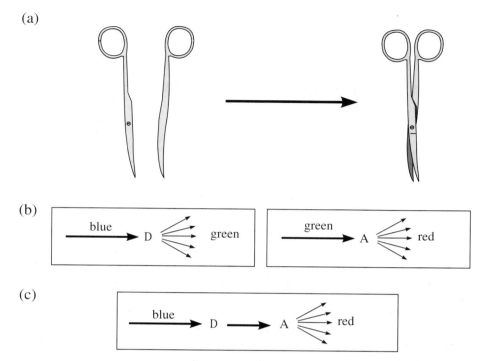

Figure 23.1 Functional units. (a) Two blades combined to give a pair of scissors. (b) Isolated dyes D and A emitting green and red light after excitation. (c) Dyes D and A combined to give an energy transducing device.

an acceptor molecule absorbing green light and emitting red light (Figure 23.1(b)). A is considered to be separated from D by a small but fixed distance; if D is excited by blue light, the excitation energy is transferred to A and A emits a quantum of red light (Figure 23.1(c)). If, however, the distance between D and A is slightly increased, the interaction of the two molecules becomes negligible and then, instead, D itself emits a quantum of green light. The pair D and A, when close together, forms a functional unit, a simple kind of supramolecular machine. The solid parts of the supramolecular machine are the dye molecules, while the photon is the moving functional element. The entity is an energy transducing device.

Appropriate dye molecules can be switched by light or protons or electrons from an active to an inactive form and vice versa. Then the arrangement is a supramolecular switching element emitting, for example, red light when D and A are in the active form, green light when D is active and A inactive, and no light when D is inactive and A is active or inactive. Functional units of increasing complexity and intricacy can be constructed by appropriately arranging more and more molecules serving as functional components.

The crucial question is how to fabricate the assembly of dye molecules such that the relative positions of the constituent molecules will be fixed in a purposefully planned manner.

23.2 Programmed Interlocking Molecules

In biosystems, most intricate supramolecular machines—for example, machines transforming light energy into chemical energy by electron or proton pumping—are present.

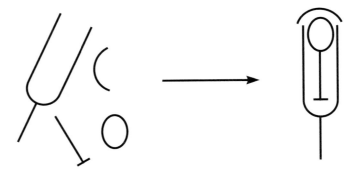

Figure 23.2 Forming complicated structures by interlocking components.

They emerge by self-assembly from molecular components of complicated structure able to interlock in distinct manner. This principle (Figure 23.2) can be used to construct artificial supramolecular machines from chemically synthesized complex molecules with regions that are complementary to each other, recognize each other, assemble, and interlock (*lock-and-key principle*).

The attempt to construct supramolecular machines by self-assembly of molecules in solution is one of the main attempts in modern preparative chemistry (*supramolecular chemistry*). However, this goal requires considerable design ingenuity. In this section we focus on an approach demanding less preparative effort: to assist interlocking by exposing the molecules to programmed environmental changes (*supramolecular engineering*). In this way we can construct complex assemblies by using relatively simple interlocking molecules. Thus basic fabricational concepts are the lock-and-key principle and the *programmed environmental change principle*.

A possibility to facilitate the self-assembly process is the confinement of carefully designed molecules to a surface (e.g., a water surface). The molecules, in Brownian motion, meet each other occasionally in a way that promotes mutual interlocking and binding, as planned. This process is further assisted by pushing the molecules together in order to form a compact monolayer film (Figure 23.3). The complexity of the organized

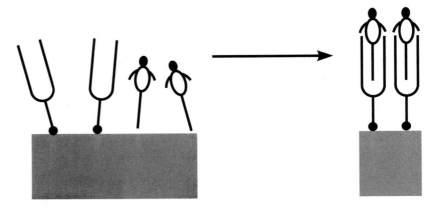

Figure 23.3 Interlocking on a surface (organized monolayer).

23.2 PROGRAMMED INTERLOCKING MOLECULES

Box 23.1 – Manipulation of Monolayer Assemblies

In Figure 23.4, different manipulation processes with monolayers are shown. The precision of the procedures can be checked by using energy transfer as a molecular ruler.

(a) Cleavage at hydrophobic interface

Upper diagram: Monolayer assembly of Donor D and acceptor A on a glass plate is covered with a polyvinyl alcohol (PVA) film. Cleavage at the hydrophobic interface between distinct monolayers is induced by lifting the PVA film. The PVA film serves as a transfer vehicle.

Before cleavage: when shining a light on the assembly, energy is absorbed by D and transferred to A, resulting in a fluorescence of A. After cleavage: shining the light on the glass slide results in a fluorescence of D; shining the light on the PVA film results in a fluorescence of A. This shows molecular precision in cleavage.

Lower diagram: Monolayer A carried by the PVA film is allowed to contact a monolayer of different dye molecules D' on a glass plate.

(b) Cleavage at hydrophilic interface

We consider an assembly of two monolayers A and B on glass slide 1. The assembly is immersed in a water trough, where the water surface is covered with a fatty acid monolayer. A cleavage at the hydrophobic interface between monolayers A and B occurs, allowing monolayer B to float on the water surface. The fatty acid monolayer holds the floating B-monolayer together.

Subsequently, slide 2, with a hydrophobic surface, is allowed to contact the floating monolayer, B. Both slides (slide 1 with A and slide 2 with B) are then coated with a layer of fatty acid by producing a fatty acid layer on the surface of the water and lifting the slides upward.

(c) Dry transfer from glass to PVA film

Dry transfer of a monolayer assembly from a glass tube to a PVA film covered with a monolayer. The transfer is implemented by rolling the tube on the PVA film.

(d) Transfer of organized structure

The multilayer assembly made by sequentially depositing monolayers on a glass slide is overlaid with a PVA thin film. The assembly and the PVA film

Continued on page 798

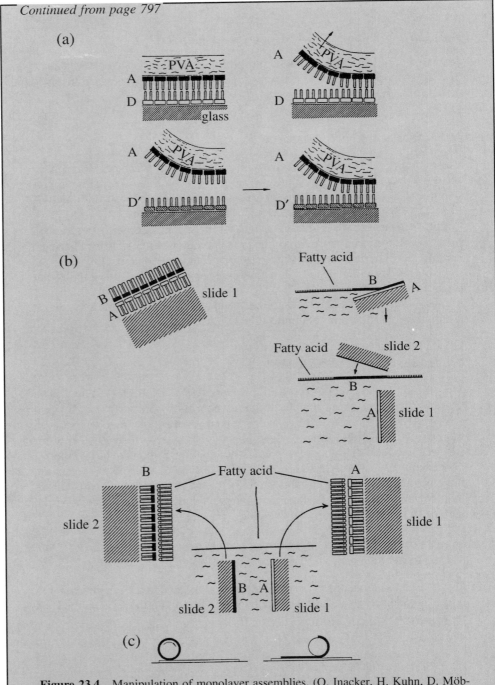

Figure 23.4 Manipulation of monolayer assemblies. (O. Inacker, H. Kuhn, D. Möbius, G. Debuch, Z. Phys. Chem. Neue Folge, 101, 337 (1976); D. Möbius, Acc. Chem. Res. 14, 63 (1981); H. Kuhn, Pure and Appl. Chem. 53, 2105 (1981)).

23.2 PROGRAMMED INTERLOCKING MOLECULES

Continued from page 798

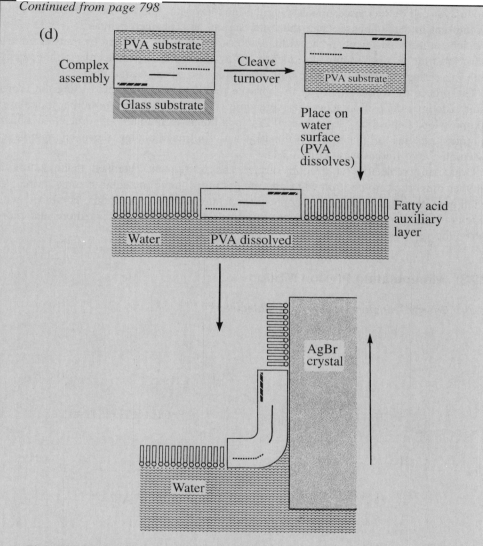

Figure 23.4 (*Continued*).

are stripped off the glass slide, turned upside down, and placed on the water surface. The PVA film, which serves as a vehicle for this transfer, is then dissolved, leaving the complex assembly on the water surface. A fatty acid monolayer surrounding the assembly aids further manipulation, shown here for deposition on a AgBr crystal. The AgBr crystal, while being lifted upward, is then coated with the monolayer assembly. This method allows the dye to be positioned at various distinct distances from the surface of the AgBr crystal.

monolayer can be increased by allowing the monolayer to bind additional molecules that are present in the subphase under the water surface in a planned manner.

Such carefully designed and accordingly constructed layers can then be transferred to a solid substrate in a programmed sequence. Thus, organized supramolecular arrangements with increasing complexity can be built according to a given design plan.

The procedure follows the way to fabricate multilayers of fatty acid using the Langmuir–Blodgett deposition sequence (spreading solution, solvent evaporation, monolayer compression, deposition, see Chapter 22); additional techniques are crucial to obtain complex systems of planned functionality and for manipulating designed monolayer assemblies in different ways (Box 23.1).

Organizing molecules at a solid surface and subsequent chemical manipulation is another important route following the programmed environmental change principle. Intelligent chemical synthesis of sophisticated interlocking compounds, combined with ingenious programmed manipulations, should be the way to fabricate more and more intriguing supramolecular machines.

23.3 Manipulating Photon Motion

23.3.1 Energy Transfer Between Dye Molecules

How can we fabricate a simple supramolecular machine? Let us look on Figure 23.5. A monolayer assembly for studying energy transfer from dye D to dye A is obtained by first transferring a fatty acid monolayer from water to a glass slide (upward trip, layer 1); then the slide is covered with a mixed monolayer of dye D and fatty acid in great excess (downward trip, layer 2). The intended assembly is obtained in superimposing additional layers in a planned sequence – for example, two layers of fatty acid (upward and downward trip, layers 3 and 4) and a mixed monolayer of dye A in a surplus of fatty acid (layer 5). It should be mentioned that the molecules in these layers are in a compact and well-defined arrangement.

Dye D absorbs blue light and emits green light; dye A absorbs green light and emits red light (absorption and fluorescence spectra in Figure 23.6).

Then, dye D is excited by blue light. The intensities I of the fluorescence of D (green) and A (red) are measured at various distances d between chromophores D and A; d is

23.3 MANIPULATING PHOTON MOTION

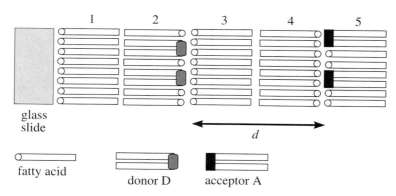

Figure 23.5 Monolayer assembly to study energy transfer from donor D to acceptor A in distance d.

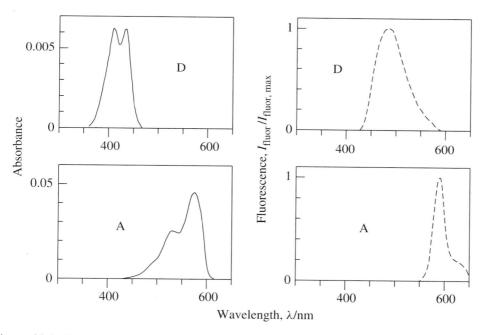

Figure 23.6 Dyes D and A: absorption spectra (solid lines) and fluorescence spectra (relative fluorescence intensity $I_{\text{fluor}}/I_{\text{fluor,max}}$, dashed lines). The spectra were measured in mixed monolayers of dye and fatty acid (molar ratio 1 : 20 for D, 1 : 1 for A) on a glass plate.

controlled by the number of fatty acid spacer layers (Figure 23.7(a)). The experimental data are compared with the theoretical relations derived in Foundation 23.1:

$$\left(\frac{I_d}{I_\infty}\right)_D = \frac{1}{1 + \left(\frac{d_0}{d}\right)^4}, \qquad \left(\frac{I_d}{I_0}\right)_A = \frac{1}{1 + \left(\frac{d}{d_0}\right)^4} \qquad (23.1)$$

$I_{d,D}$ is the intensity of the green fluorescence of D at distance d; $I_{\infty,D}$ is the intensity in the case when the acceptor is at a great distance. $I_{d,A}$ is the intensity of the red fluorescence

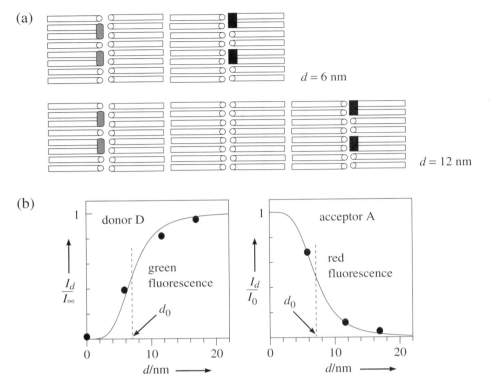

Figure 23.7 Energy transfer from donor D to acceptor A. (a) Monolayer assemblies with different numberes of fatty acid spacer layers. (b) I_d/I_∞ for donor fluorescence and I_d/I_0 for acceptor fluorescence versus distance d. Dots, experiment; solid lines, calculated from equation (23.1). D. Möbius, Z. Naturforsch. 24a, 251 (1969).

of A at distance d; $I_{0,A}$ is the intensity in the case when acceptor and donor are close together. d_0 is the distance where

$$\left(\frac{I_d}{I_\infty}\right)_D = \left(\frac{I_d}{I_0}\right)_A = \frac{1}{2}$$

$(I_d/I_\infty)_D$ and $(I_d/I_0)_A$ are plotted versus distance d in Figure 23.7(b).

An interesting possibility to make use of the energy transfer between D and A is shown in Figure 23.8. D and A are linked to two different antibodies which bind to the same

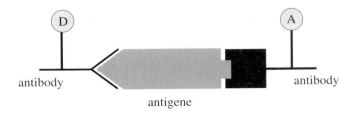

Figure 23.8 Donor D and acceptor A linked to different antibodies interlock with the corresponding antigen. (J. M. Lehn, 'Supramolecular Chemistry', VCH Weinheim, 1995).

23.3 MANIPULATING PHOTON MOTION

antigen at two different positions. As a result, D and A become sufficiently close to allow energy transfer. The arrangement is used as immunoassay.

23.3.2 Functional Unit by Coupling Dye Molecules

We consider dye molecules 1 and 2 in the arrangement shown in Figure 23.9(a). The interaction of the two chromophores results in a change of the absorption spectrum (Figures 23.9(b) and (c)). This is due to exactly the same reason as the intramolecular coupling effect discussed in Chapter 8. Each dye molecule can be replaced by an oscillator exposed to the oscillating field of the incident light and to the field of the oscillator corresponding to the other dye molecule (Figure 23.10(a)). The absorption band of the dye

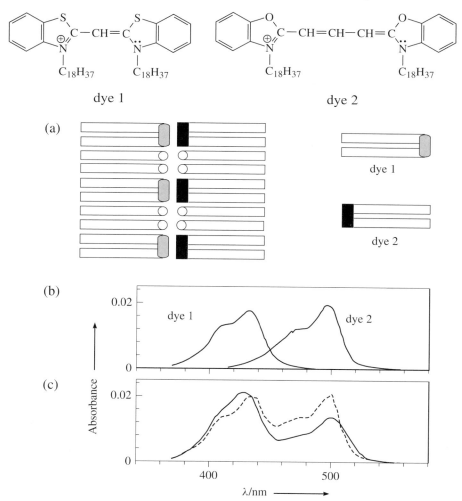

Figure 23.9 Change of absorption spectra of dyes 1 and 2 when both dyes are in a functional unit. (a) Functional unit, arrangement with distance $d = 0$. (b) Absorption spectra of isolated dyes. (c) Absorption spectra of dyes in functional unit (solid line); superposition of the spectra of the isolated dyes (dashed line). (V. Czikkely, G. Dreizler, H.D. Försterling, H. Kuhn, J. Sondermann, P. Tillmann, H. Wiegand, Z. Naturforsch. 24a, 1823 (1969)).

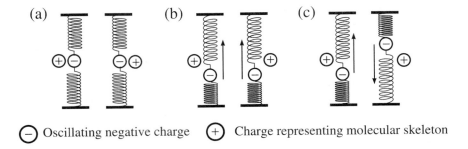

Figure 23.10 Shift of absorption spectrum in Figure 23.9 explained by the oscillator model. (a) Two coupled oscillators in rest. (b) Oscillators vibrating in phase (strong absorption band). (c) Oscillators vibrating out of phase (weak absorption band).

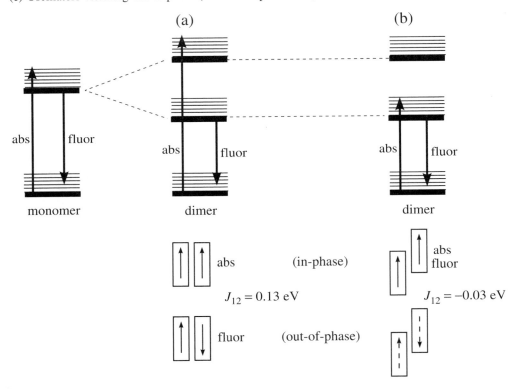

Figure 23.11 Coupling dyes in a dimer arrangement. Shift of absorption and of fluorescence bands in forming the dimer depends on geometry; interaction described by two-electron coupling integral J_{12}, see Foundation 8.3. Arrangement (a): shift of absorption band to shorter waves ($J_{12} > 0$); fluorescence after thermal de-excitation. The lifetime of the dimer is increased compared with the monomer because the transition moments compensate (transition dipoles not to be confused with permanent dipoles). This leads to a stronger competition with nonradiative processes resulting in a weaker fluorescence. Arrangement (b): shift of absorption and fluorescence bands to longer waves ($J_{12} < 0$), strong fluorescence.

23.3 MANIPULATING PHOTON MOTION

molecule absorbing at short wavelengths is enhanced and shifted to shorter wavelengths (*in-phase oscillation*, Figure 23.10(b)), the band of the dye absorbing at long wavelengths is diminished and shifted to longer wavelengths (*out-of-phase oscillation*, Figure 23.10(c)). The effect can be calculated as shown in Chapter 8 in agreement with experiment. The arrangement is a molecular functional unit, and the entity behaves quite differently from each part.

The effect is easily understood in the classical picture. In this picture the repulsion of the oscillating dipoles in the in-phase oscillation leads to an increased frequency (shorter wavelength). The attraction of the dipoles in the out-of-phase oscillation decreases the frequency (longer wavelength).

If molecules 1 and 2 are identical, particularly large changes of the absorption and of the fluorescence appear. For instance, the green fluorescence of the monomer of dye 1 changes to a yellow fluorescence when forming the sandwich dimer.

This is due to the large Stokes shift of the dimer (shift between absorption and fluorescence maximum, see Figure 23.11(a)): the high-energy excited state is reached in absorption (in-phase oscillation), but the fluorescence originates from the low-energy excited state (out-of-phase oscillation).

The coupling effect is strongly dependent on the geometry of the dimer. In the geometry of Figure 23.11(b), the absorption corresponding to the in-phase oscillation (strong band) is at long wavelengths (attraction of the dipoles). The out-of-phase oscillation is at short wavelengths (repulsion of dipoles). The transition dipole moments of the two molecules (dashed arrows) compensate for each other (no absorption).

23.3.3 Dye Aggregate as Energy Harvesting Device

The primary energy source of life is solar energy collected by an antenna system trapped in the reaction center of plants and bacteria, where the solar energy is converted into chemical energy. We consider an artificial energy channeling system.

A 1:1 mixture of dye D and octadecane is spread at a water surface and the layer is compressed. A well organized monolayer is formed. The chromophores are tightly

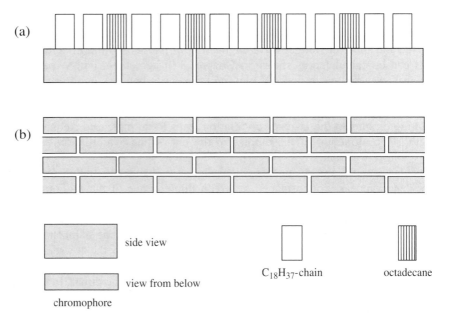

Figure 23.12 Dye D in a brickstone work arrangement. (a) Side view; the spaces between adjacent dye molecules are filled by octadecane molecules. The chromophores are perpendicular to the layer plane. (b) View from below.

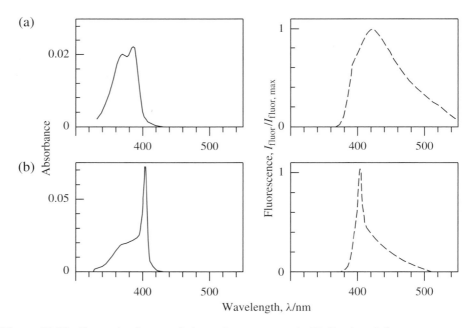

Figure 23.13 Dramatic change of absorption spectrum (solid lines) and fluorescence spectrum (dashed lines) of dye D in a monolayer when the monomers are not ordered (a), and ordered in a brickstone work arrangement (b). The spectra were measured in a mixed monolayer of the dye (with arachidic acid in case a, and with octadecane in case b) in a molar ratio 1 : 1 on a glass plate. (H. Bücher, H. Kuhn, *Chem. Phys. Lett.*, 1970, **6**, 183, page 185).

23.3 MANIPULATING PHOTON MOTION

packed in a brickstone work arrangement, with the octadecane chains fitting the holes left between the hydrocarbon substituents of the dye molecules (Figure 23.12). A dramatic effect occurs in the absorption and fluorescence (Figure 23.13). A very narrow and intense absorption band and an equally narrow and almost coinciding fluorescence band appear. The absorption band is shifted to long wavelengths relative to the absorption of the monomer in a surplus of fatty acid.

Dye A can be incorporated in the aggregate monolayer as an energy acceptor (Figure 23.14). Each A molecule matches exactly into the space of a D molecule

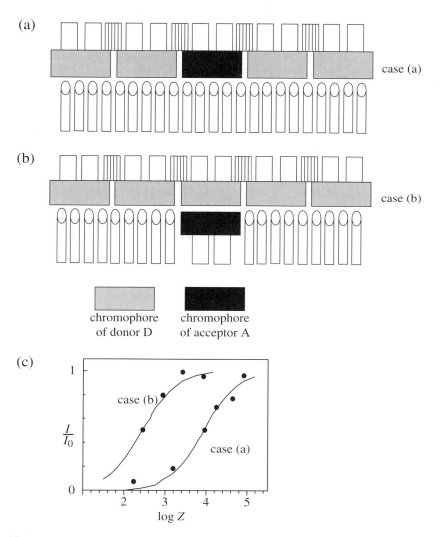

Figure 23.14 Energy transfer from exciton (donor) to acceptor. (a) and (b): acceptor molecules A incorporated in an aggregate monolayer of donor molecules D. Case (a): acceptor A substituting one donor molecule. ($Z = 10\,000$ for $I/I_0 = 0.5$). Case (b): acceptor A in contact with donor molecule, but incorporated in the fatty acid layer ($Z = 300$ for $I/I_0 = 0.5$). (c) Relative fluorescence intensity I/I_0 of donor versus $\log Z$ (Z = ratio of number of donor molecules to acceptor molecules). D. Möbius, H. Kuhn, J. Appl. Phys. 64, 5138 (1988).

(Figure 23.14(a)). A surprising phenomenon is observed even at a dilution of one molecule of A per 10 000 molecules of D ($Z = 10\,000$). The blue fluorescence of the host is quenched and the green fluorescence of the energy acceptor A appears: the energy absorbed by D moves over a large distance and is trapped by A. How can we understand these phenomena?

Formation of excited domain (exciton)

We first consider the pure aggregate consisting of molecules D. In the aggregate, due to the tight packing, the oscillators replacing each dye molecule are strongly coupled. This means that when a light quantum is absorbed the excitation is distributed over a certain domain and can be described as an in-phase oscillation of all oscillators replacing the dye molecules in the domain. Nearest-neighbor oscillators are coupled according to Figure 23.11(b), thus a shift of the absorption maximum to longer wavelengths occurs. Restricting the treatment to nearest-neighbor interactions (4 neighbors), the shift is $\Delta E = 8J_{12} = -0.16$ eV, where $J_{12} = -1.25 \times 10^{-19}$ J $= -0.02$ eV is the two-electron repulsion integral considered in Foundation 8.3. A more detailed approximation (including all neighbors) gives $\Delta E = -0.24$ eV. This corresponds to a shift of $\Delta \lambda = 30$ nm, in good agreement with the measured shift of 23 nm (Figure 23.13).

The number N of molecules contributing to the excited domain is determined by a compromise between the attracting forces of the in-phase oscillating dipoles constituting the excited portion and the thermal motion tending to break the cooperation. The energy shift is given by the sum of bond energies between the N coherent dipole oscillators decreasing the energy of the excited state. Thus the bond energy in a pair of oscillators at the border of the excited domain is $-\Delta E/N$. This energy, keeping the oscillator at the border in phase with the excited domain, must agree with the energy kT knocking this oscillator out of phase. Therefore

$$\frac{-\Delta E}{N} = kT \tag{23.2}$$

or

$$N = -\frac{\Delta E}{kT} = \frac{0.24 \text{ eV}}{k} \cdot \frac{1}{T} = 3000 \text{ K} \cdot \frac{1}{T} \tag{23.3}$$

and for room temperature we obtain

$$N = \frac{3000 \text{ K}}{300 \text{ K}} = 10$$

This excited domain is called an *exciton*.

Lifetime of fluorescence of exciton

According to Foundation 8.5 the fluorescence lifetime of a monomer is, in terms of the classical oscillator model,

$$\tau_0 = \frac{3m\varepsilon_0 c_0^3}{2Q^2 \pi v_0^2 \hat{n}} \tag{23.4}$$

(m, Q, v_0 = mass, charge, and frequency of the oscillator, ε_0 = permittivity of the vacuum, c_0 = speed of light in vacuum, $\hat{n}$ = index of refraction). In the aggregate N

oscillators are oscillating in phase. Then we obtain the lifetime by replacing m by Nm and Q by NQ:

$$\tau_{agg} = \frac{3(Nm)\varepsilon_0 c_0^3}{2(NQ)^2 \pi v_0^2 \widehat{n}} = \frac{1}{N}\tau_0 \tag{23.5}$$

Furthermore, with equation (23.3) it follows

$$\tau_{agg} = \frac{T}{3000\,\text{K}}\tau_0 \tag{23.6}$$

From equation (23.4) we calculate $\tau_0 = 5 \times 10^{-9}$ s for $m = m_e$, $Q = e$, $\widehat{n} = 1.5$, and $v_0 = 0.75 \times 10^{15}\,\text{s}^{-1}$. This value is in agreement with the experiment. Then at room temperature we have

$$\tau_{agg} = \frac{300\,\text{K}}{3000\,\text{K}} \times 5 \times 10^{-9}\,\text{s} = 5 \times 10^{-10}\,\text{s}$$

In fact, an increased emission rate (called *superradiance*) and a proportionality of τ_{agg} with temperature T are observed in experiments performed in the temperature range from 90 K to 300 K. At low temperature, τ_{agg} becomes constant. In this case the lifetime of the fluorescent excited state is short; therefore the time to reach the fluorescent excited state is rate-determining (it takes about 100 collisions to dissipate the vibrational energy).

Speed of exciton

The exciton moves over the monolayer. To estimate its speed v we consider the area A covered by the exciton during its life time τ_{agg}

$$A = L_{exciton} \cdot v \cdot \tau_{agg} \tag{23.7}$$

where $L_{exciton}$ is the width of the exciton and $v \cdot \tau_{agg}$ is the distance traveled by the exciton. In order to reach an acceptor molecule with 50% probability, A must be equal to the aggregate area in which just one acceptor molecule is present, namely to Za where a is the area covered by a donor molecule. Remember that Z is the number of donor molecules per acceptor molecule. Then Za is the area of the aggregate where you find one acceptor molecule, on the average. Then $A = Z \cdot a$ and

$$v = \frac{A}{L_{exciton} \cdot \tau_{agg}} = \frac{Za}{L_{exciton} \cdot \tau_{agg}} \tag{23.8}$$

With $Z = 10\,000$, $a = 1500 \times 400\,\text{pm}^2$, $L_{exciton} = 5$ nm (at 300 K) and $\tau_{agg} = 5 \times 10^{-10}$ s we obtain $v = 2.4\,\text{km s}^{-1}$. Such a high speed must be assumed to understand the observation that only a tiny amount of acceptor molecules is required to quench the donor fluorescence.

How can we understand the fast motion of the exciton? When the light quantum has been absorbed, the excited domain is compressed, due to the cohesive forces between the coherent oscillators. This is best seen in the classical picture of in-phase oscillating dipoles replacing the excited domain (Figure 23.15). The charges change sign simultaneously, thus nearest-neighbor charges are always opposite; the attraction of charges 1a and 2b, 1b and 3a, 2b and 5a, etc. leads to a compression of the excited domain.

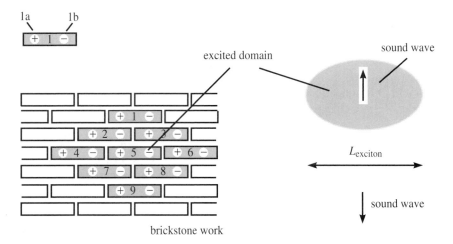

Figure 23.15 Formation of an exciton. Left: excited domain in a brickstone work arrangement of coupled oscillating dipoles. Right: exciton (carried by sound wave) moving upward, and sound wave moving downward. (H. Kuhn, C. Kuhn, in: J-Aggregates, Ed. T. Kobayashi, World Scientific, Singapore, 1996).

Thus a sound wave is produced moving in the upward direction along with a corresponding sound wave moving in the downward direction. One wave carries the exciton (symmetry breaking). The exciton holds the sound wave together, due to the cohesion within the excited domain. So the excited domain constitutes a *soliton*–that is, a confined excited domain of constant shape moving without energy loss until its annihilation by emitting a fluorescent light quantum or by exciting an acceptor molecule.

Efficiency of energy transfer

In the arrangement of case (a) in Figure 23.14(a), one molecule of A per 10 000 donor molecules D ($Z = 10\,000$) is sufficient for a 50% quenching of the fluorescence of the donor at room temperature. This is because of the high probability of energy transfer in this arrangement (acceptor A incorporated in the layer of D molecules).

In the arrangement of case (b) in Figure 23.14(b), the D aggregate monolayer is superimposed on a monolayer of fatty acid containing traces of acceptor A. Then each A molecule is still in direct contact with D molecules at essentially equal distance from its nearest neighbors. There is a tremendous difference between the two cases because in case (a) the acceptor is exposed to a strong field of the adjacent oscillating dipoles, while in case (b) the oscillator fields partially compensate each other.

Thus the probability of energy transfer is reduced in case (b), and the effect of A as an exciton trap is considerably smaller. The fluorescence of D is quenched for one molecule of A per 300 donor molecules ($Z = 300$), in contrast to $Z = 10\,000$ in case (a) (Figure 23.14(c); see Foundation 23.2).

23.3.4 Solar Energy Harvesting in Biosystems

Energy funneling in nature is less efficient than in the device of Figure 23.14(a) discussed above: only about 300 dye molecules are present collecting solar energy and funneling it

23.3 MANIPULATING PHOTON MOTION

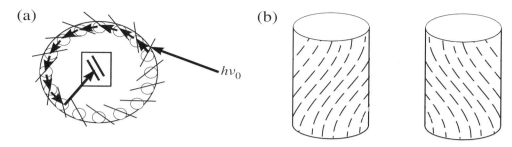

Figure 23.16 Energy funneling in nature. (a) Antenna system in purple bacterium guiding the light to the reaction center (special pair). (W. Kühlbrandt, Many Wheels Make Light Work, Nature 374, 497 (1995); S. Karrasch, P. Bullough, R. Ghosh, EMBO J. 14, 631 (1995)) (b) Antenna system of green bacterium. Tube of chlorophyll c molecules in a brickstone work arrangement (left-handed and right-handed species supposed to be in equal amounts).

to a reaction center (instead of $Z = 10\,000$). However, a broad spectral range of sunlight is utilized, while in the previous case light is absorbed in the very narrow spectral range of the aggregate band. The requirement of the funneling device in biosystems is achieved by different types of collector molecules and different arrangements. In the antenna system in purple bacteria a first group of 18 dye molecules (bacteriochlorophyll) is arranged in a ring around the reaction center (Figure 23.16(a)). The chromophores are shifted against each other. We have seen that this particular kind of arrangement of chromophores leads to a strong bathochromic shift, i.e. a shift of the absorption maximum to longer wavelengths (Section 23.3.2, Figure 23.11(b)).

Another group of bacteriochlorophyll molecules consists of molecules that are separated from each other in a distinct way by a protein moiety. Accordingly, these groups of molecules that are isolated from each other absorb at shorter wave lengths than the first group.

Furthermore, carotene molecules are incorporated in a well-organized way. They absorb at even shorter wavelengths, allowing energy transfer to the second group of bacteriochlorophyll.

The edge-to-edge distances between the three groups of antenna molecules are small compared to the critical distance r_0 for energy transfer (Foundation 23.1). This allows efficient energy transfer from the third to the second and from the second to the first group absorbing at 875 nm and from there to the special pair in the reaction center (see Section 23.4.5). The special pair of bacteriochlorophyll molecules absorbs at 965 nm, as required for optimal energy transfer from the antenna molecules to the reaction center. The molecules in this pair, again, are arranged according to Figure 23.11(b). They are tightly fixed in this particular geometry, leading to this large shift to 965 nm.

Both in the first group of antenna molecules and in the special pair, the long wavelength absorption is reached by the very particular arrangement of bacteriochlorophyll according to Figure 23.11(b). It is fascinating that this arrangement allowing the specific coupling of chromophores is present to use the broad spectral range of solar energy for transformation into chemical energy.

This sophisticated arrangement of the chromophores in the purple bacteria antenna is achieved by fixing them in a protein matrix. In the antenna complex of green bacteria, chlorophyll c molecules are in a brickstone work arrangement forming tubes (chlorosomes, Figure 23.16(b)). The tubes do not contain proteins. They result from a self-assembly process where the chlorophyll c molecules are linked by C=O···H–O···Mg bridges.

(A. R. Holzwarth, K. Schaffner, *Photosynthesis Research*, 1994, **41**, 225). We explain the circular and linear dichroism spectra by assuming that left-handed and right-handed helical tubes are present in equal amounts (Problem 23.5). These tubes absorb at 750 nm, and energy is transferred via bacteriochlorophyll a (absorbing at 800 nm) to the reaction center (absorbing at 866 nm). The bacteriochlorophyll is localized in a membrane between tubes and reaction center, so again the relevant distance is small compared to r_0, allowing efficient energy transfer.

It is most remarkable to see that the trick to arrange the chromophores in a brickstone work allowing coupling according to Figure 23.11(b) or Figure 23.12 has been selected by nature repeatedly to achieve the required bathochromic shifts.

23.3.5 Manipulating Luminescence Lifetime by Programming Echo Radiation Field

We have seen that the lifetime of luminescence can be strongly decreased by tightly packing the molecules. They couple to each other, absorption and emission are cooperative processes. Another fascinating phenomenon is coupling a luminescent molecule to its own echo radiation field, which depends on details of the environment producing the echo.

Let us look at a luminescent molecule in front of a metal layer serving as a mirror (Figure 23.17(a)). The monolayer assembly technique allows us to position the luminescent molecules at any defined distance d from the mirror. The luminescent molecules can be excited by a light flash and the number of quanta of luminescent light emitted during small time intervals after the flash can be counted. Thus the average lifetime τ of the excited state is obtained. For the europium complex

we consider the ratio τ/τ_0, where τ_0 is the average lifetime of luminescence of the sample without mirror. τ/τ_0 is displayed in Figure 23.17(b) as a function of distance d between the europium complex and the mirror: surprisingly, τ/τ_0 depends strongly on distance d. Why can this be the case? Why should the molecule know all about what is present in the remote neighborhood when it is going to emit a quantum of light? How can the lifetime of the excited state of a molecule depend on objects far apart?

23.3 MANIPULATING PHOTON MOTION

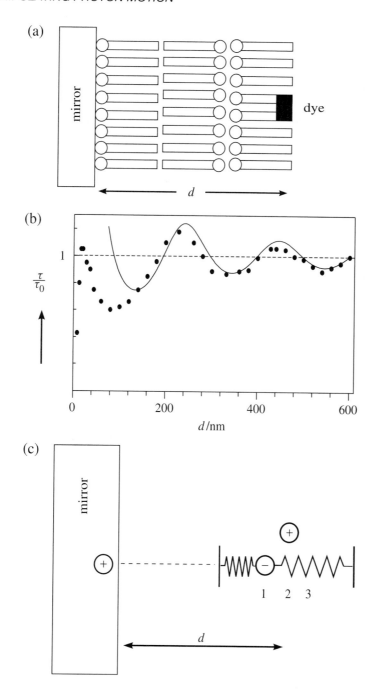

Figure 23.17 Radiation echo field. (a) Luminescent dye molecule (europium complex, $\lambda_0 = 612$ nm) in distance d in front of a mirror. (b) τ/τ_0 ($\tau =$ lifetime of the europium complex) versus distance d. Dots, experiment (K. H. Drexhage, J. Luminescence 1, 345 (1970)); solid line, theory according to Foundation 23.3. (c) Illustration of the phase shift between the electron (assumed to oscillate between points 1, 2, and 3) and the induced positive charge on the mirror; the fixed positive charge in point 2 represents the molecular skeleton (as in Figure 23.10).

These questions can be answered by considering the wave–particle duality. The strangeness of this duality is illustrated dramatically by the echo effect. According to Chapter 8 the excited molecule behaves in a way like a classical oscillator emitting light waves that are reflected at the mirror and then interact with the oscillator.

This echo field interacts with the oscillator replacing the excited molecule. Depending on the phase of the echo field relative to the phase of the oscillator an acceleration or retardation of the oscillator occurs (Foundation 23.3). Thus, the lifetime becomes longer or shorter depending on distance d between oscillator and mirror.

Essentially, the light echo arrives at the oscillator with a delay time of

$$\Delta t = \frac{2d}{c} \tag{23.9}$$

(c = speed of light in the medium).

We compare Δt with T, the period time of the oscillation. First we consider the case $\Delta t = \frac{1}{4}T$ (that is for $d = \frac{1}{8}cT$). Then the oscillator, when it is at point 1 (negative charge in Figure 23.17(c)), polarizes the metal maximally. The field of this polarized charge arrives at the oscillator with a delay, when the oscillator is just passing point 2, i.e., when it has its maximum speed. In this case the oscillator is decelerated and the lifetime τ is shortened.

Correspondingly, for $\Delta t = \frac{3}{4}T$ (that is for $d = \frac{3}{8}cT$) the oscillator is accelerated and the lifetime τ becomes longer.

For $\Delta t = T$ and $\Delta t = \frac{1}{2}T$ the maximum field of the echo arrives when the oscillator speed is zero; in this case the lifetime τ remains unchanged.

The distance d corresponding to $\Delta t = \frac{1}{4}T$ is

$$d = \frac{\Delta t \cdot c}{2} = \frac{T \cdot c}{8} = \frac{\lambda}{8} = \frac{\lambda_0}{8\widehat{n}} \tag{23.10}$$

where λ is the wavelength in the medium and λ_0 is the wavelength in the vacuum. With $\lambda_0 = 612$ nm and $\widehat{n} = 1.5$ we obtain

$d = 51$ nm   for $\Delta t = \frac{1}{4}T$ (τ shortened)

$d = 102$ nm   for $\Delta t = \frac{1}{2}T$ (τ unchanged)

$d = 153$ nm   for $\Delta t = \frac{3}{4}T$ (τ longer)

$d = 204$ nm   for $\Delta t = T$ (τ unchanged)

This simple picture gives the maxima and minima in Figure 23.17(b) at correct distances, but shifted by 70 nm. This discrepancy disappears in the refined approach discussed in Foundation 23.2. However, a difference between theory and experiment remains because silver is not an ideal (i.e. non-absorbing) reflector. This is of particular relevance at small distances d. Thus silver acts as an energy acceptor like the acceptor dye in Figure 23.5, quenching the luminescence (Figure 23.17(b), small distances: the experimental points are below the theoretical curve).

The fluorescence lifetime has been measured for single dye molecules near a metal surface (a carbocyanine dye in a polymer film) and also near the film/air interface acting as a mirror. The above results were confirmed. The experiments illustrate very directly, at the single molecule level, how the emission can be manipulated by distinct changes of the remote neighborhood of the molecule. (See X. S. Xie, R. C. Dunn, *Science*, 1994,

265, 361; W. P. Ambrose, P. M. Goodwin, J. C. Martin, R. A. Keller, *Science*, 1994, **265**, 364; J. K. Trautman, *Proc. of Welsch Foundation*, 39th Conference (Nanophase Chemistry, p. 173.))

23.3.6 Nonlinear Optical Phenomena

We consider a molecule in a static electric field F parallel to its axis; for simplicity, the molecule is assumed to be rotationally symmetric. The field induces a dipole moment μ_{ind}

$$\mu_{ind} = \alpha F + \beta F^2 + \gamma F^3 \tag{23.11}$$

The coefficients α, β, and γ are the polarizability and the first and second hyperpolarizability.

We consider a polyene molecule with 10 double bonds. In this case, by symmetry, μ_{ind} changes its sign when inverting the direction of the electric field; thus $\beta = 0$. Following the considerations in Chapter 7 (search for self-consistency between bond lengths and

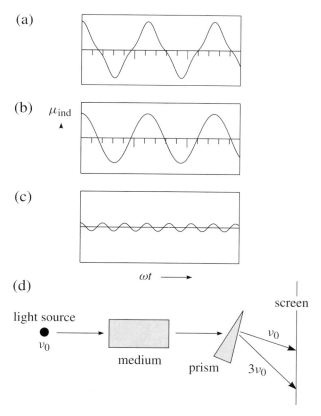

Figure 23.18 Third harmonic generation in a nonlinear optical material with $\alpha = 2.3 \times 10^{-38}$ C m^2 V^{-1} and $\gamma = 1.2 \times 10^{-57}$ C m^4 V^{-3}. Dipole moment μ_{ind} induced by oscillating field strength F (frequency ν_0, amplitude 50×10^8 V m^{-1}). (a) μ_{ind} versus ωt; (b) and (c) contributions of frequencies ν and 3ν. Full-scale range of ordinate: ± 100 debye. (d) Optical arrangement to detect the high frequency contribution.

electron density) the values $\alpha = 2.3 \times 10^{-38}$ C m^2 V^{-1} and $\gamma = 1.2 \times 10^{-57}$ C m^4 V^{-3} are obtained (Problem 23.3); these values are in agreement with experimental data. The positive sign of γ (μ_{ind} increases more than proportionally to F) is due to the fact that the π electrons, with increasing field, are first largely confined to the double bonds and can more freely move at higher field strengths.

Now we consider the polyene in the oscillating field $F = F_0 \cdot \cos \omega t$ of a light wave polarized in the direction of the chain ($\omega = 2\pi \nu$). If the frequency ν of the light wave is lower than the resonance frequency ν_0 (maximum of the absorption band), then the molecule, in the classical picture, oscillates in phase. In a weak radiation field μ_{ind} is proportional to F:

$$\mu_{ind} = \alpha F = \alpha F_0 \cdot \cos \omega t \tag{23.12}$$

Then the oscillating induced dipole emits a wave of frequency ν.

In a strong field the term γF^3 cannot be neglected. Then μ_{ind} can be represented by a superposition of cosine functions of frequencies ν and 3ν

$$\mu_{ind} = \alpha F + \gamma F^3 = \alpha F_0 \cdot \cos \omega t + \gamma F_0^3 \cdot \cos^3 \omega t \tag{23.13}$$

Because of

$$\cos^3 \alpha = \tfrac{1}{4}(\cos 3\alpha + 3 \cos \alpha) \tag{23.14}$$

we obtain

$$\mu_{ind} = \left(\alpha F_0 + \tfrac{3}{4} \gamma F_0^3\right) \cdot \cos \omega t + \tfrac{1}{4} \gamma F_0^3 \cdot \cos 3\omega t \tag{23.15}$$

Therefore, light of frequency 3ν, along with light of frequency ν, is emitted by the oscillating induced dipole as shown in Figures 23.18(a)–(c) (third harmonic generation).

This effect can be used to produce high-frequency radiation from low-frequency laser radiation. A layer of a nonlinear optical material is illuminated by laser light (Figure 23.18(d)). The oscillating dipoles replacing each dye molecule are coherently coupled to the incident light wave. The light waves emitted by each oscillator superimpose. The superposition of the waves of frequency ν constitutes the refracted and reflected light beams. Similarly, waves of frequency 3ν superimpose to the tripled frequency light beam.

When using a polyene with a donor at one end (left-hand side) and an acceptor at the other end (a *push–pull polyene*)

a

the polarization is strong when an external field is directed such that a negative charge can accumulate at the nitrogen atom according to

b

23.3 MANIPULATING PHOTON MOTION

When the field is directed in the reverse direction the polarization is weak. In the first case the resonance structure (b), in chemist's language, becomes dominant. Then the first hyperpolarizability β is no longer zero, and the induced dipole moment is

$$\mu_{ind} = \alpha F + \beta F^2 + \gamma F^3 \tag{23.16}$$
$$= \alpha F_0 \cdot \cos \omega t + \beta F_0^2 \cdot \cos^2 \omega t + \gamma F_0^3 \cdot \cos^3 \omega t$$
$$= \tfrac{1}{2}\beta F_0^2 + \left(\alpha F_0 + \tfrac{3}{4}\gamma F_0^3\right) \cdot \cos \omega t + \tfrac{1}{2}\beta F_0^2 \cdot \cos 2\omega t + \tfrac{1}{4}\gamma F_0^3 \cdot \cos 3\omega t$$

Then light of frequency 2ν, in addition, is emitted (*second harmonic generation*).

Purposely constructed organized molecular systems are important to optimize nonlinear optical properties. The volume polarization P of the material

$$P = \frac{\text{induced dipole moment}}{\text{volume}} = \chi_e^{(1)} F + \chi_e^{(2)} F^2 + \chi_e^{(3)} F^3 \tag{23.17}$$

depends on the nature of the molecules in the material and on their arrangement. This can be seen in the aggregate displayed in Figure 23.19(a), which has a particularly large value of $\chi_e^{(3)}$. The polarization of a molecule in the aggregate enhances the polarization of the neighbor molecules, thus supporting the applied field F enormously.

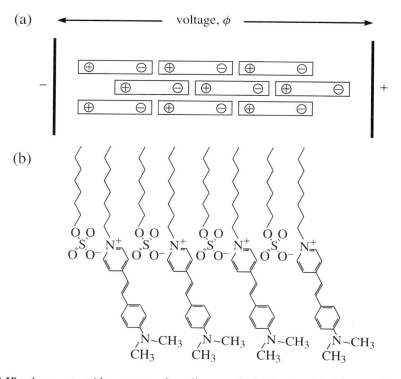

Figure 23.19 Aggregate with pronounced nonlinear optical behavior. (a) Scheme of an arrangement serving as $\chi_e^{(3)}$ material. (b) Control of packing of hemicyanine with an amphiphilic counterion (octadecyl sulfate): large $\chi_e^{(2)}$ by parallel chromophore chains. G. Ashwell, R. Hargreaves, C. Balwin, G. Balvia, C. Brown, *Nature*, 1992, **357**, 393).

In order to obtain large values of $\chi_e^{(2)}$, it is important to interlock the appropriate chromophores in a strictly parallel orientation. This is achieved by constructing monolayers where amphiphilic anions (like octadecyl sulfate) act as spacers between cationic hemicyanine (Figure 23.19(b)). Multilayers of this compact and well-ordered structure can be made. Replacing the anion octadecyl sulfate by bromide ions reduces $\chi_e^{(2)}$ by one order of magnitude, showing the importance of exactly planned interlocking.

23.4 Manipulating Electron Motion

23.4.1 Photoinduced Electron Transfer in Designed Monolayer Assemblies

In Figure 23.20(a) an electron acceptor A is positioned in the vicinity of the electron donor dye D.

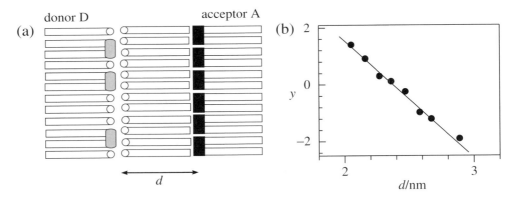

D and A are separated by a fatty acid spacer layer with a thickness d. This thickness can be varied by changing the hydrocarbon chain length of the spacer molecule.

Figure 23.20 Electron transfer from excited donor D to electron acceptor A. (a) Arrangement of layer with D and layer with A separated by fatty acid molecules. (b) $y = \ln(I_\infty/I - 1)$ for the donor fluorescence versus length d of the fatty acid chain. (H. Kuhn, D. Möbius, Investigations of Surfaces and Interfaces, Part B, Eds. B. Rossiter, R.C. Baetzold, Physical Methods of Chemistry Series, 2nd edn., Vol. IXB, Wiley, New York 1993, page 465).

23.4 MANIPULATING ELECTRON MOTION

The donor is excited by irradiation with light at 366 nm. In this case, electron transfer

$$D^* + A \longrightarrow D^+ + A^-$$

is competing with de-excitation of the dye D and its emission of a quantum of fluorescence light. Electron transfer thus proceeds at the expense of fluorescence. Therefore, the rate of electron transfer can be measured by the amount of fluorescence quenching–that is, the suppression of fluorescence.

I, the fluorescence intensity of D, depends on k_{fluor} (the rate constant of the decay of the fluorescence) and $k_{transfer}$ (the rate constant for electron transfer from D to A):

$$I = \frac{k_{fluor}}{k_{fluor} + k_{transfer}} I_\infty \qquad (23.18)$$

I_∞ is the fluorescence intensity for $d \to \infty$ (no electron transfer). Then

$$\frac{I_\infty}{I} = 1 + \frac{1}{k_{fluor}} \cdot k_{transfer} = 1 + \tau \cdot k_{transfer} \qquad (23.19)$$

where τ is the fluorescence lifetime, and

$$\frac{I_\infty}{I} - 1 = \tau \cdot k_{transfer} \qquad (23.20)$$

As outlined in Foundation 4.4 and in Foundation 23.4, electron transfer is a typical tunneling process, where

$$k_{transfer} \sim e^{-const \cdot d} \qquad (23.21)$$

Therefore we expect that $\ln(I_\infty/I - 1)$ depends linearly on the distance d between D and A. Actually, such a behavior is found in the experiment (Figure 23.20(b)).

This arrangement is a simple supramolecular machine that can be coupled with other functional components to form more complex entities that are useful in investigating photophysical processes.

23.4.2 Switching by Photoinduced Electron Transfer

We consider a device consisting of a monolayer arrangement of the dye cations A, B, and C.

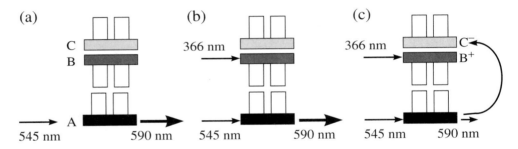

The device is irradiated at 545 nm (absorption maximum of dye A), and the fluorescence of dye A is measured at 590 nm (Figure 23.21(a)). Then an additional irradiation at 366 nm is switched on inducing electron transfer from dye B to dye C (Figure 23.21(b)). Transient radicals B^+ and C^- are formed in this way. Whereas C absorbs in the UV, the absorbance of C^- is at 600 nm. Thus the radical C^- acts as an energy acceptor for molecule A, quenching its fluorescence at 590 nm (Figure 23.21(c)). After switching

Figure 23.21 Photoinduced electron transfer. Monolayer arrangement consisting of dyes A, B, and C. (a) Irradiation of A with 545 nm: fluorescence of A. (b) Additional irradiation of B at 366 nm: electron transfer from B to C. (c) Molecule C^- formed in process (b) acts as energy acceptor for A: quenching of fluorescence. (D. Möbius, Ber. Bunsenges. Phys. Chem. 82, 848 (1978)).

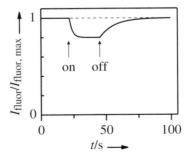

Figure 23.22 Fluorescence intensity I_{fluor} of dye A in the arrangement in Figure 23.21. I_{fluor} is switched on and off by irradiation with light of 366 nm.

off the 366 nm irradiation, the radicals C⁻ disappear by reaction with oxygen, and the fluorescence intensity increases to the original value (Figure 23.22).

This arrangement demonstrates the programmed interplay between three different single molecules in a distinct geometry. It can be considered as a switch with optical readout. It is turned on and off by UV radiation absorbed by molecule B, which then triggers molecule C to change into a molecule C⁻ that interplays with molecule A, thereby quenching its fluorescence.

23.4.3 Monolayer Assemblies for Elucidating the Nature of Photographic Sensitization

The energy transfer discussed in Section 23.3.1 has a longer range than electron transfer. Thus we can discriminate between energy transfer or electron transfer processes. For example, the method can be applied to the elucidation of the nature of the photographic sensitization process.

The surface of a silver bromide grain of a photographic plate is covered with dye molecules acting as sensitizer for the photographic process. When exciting the dye molecules with visible light, electrons are generated moving around in the AgBr grain; thereby a latent image is produced—that is, a cluster of silver atoms. The latent image is later developed. The developer is a reducing agent. The latent image acts as nucleation center initiating the reduction process by this agent.

We ask the question: Does the electron moving around in the AgBr grain originate (a) from electron transfer (injection of an electron from the excited dye into the AgBr grain) or (b) from energy transfer from the excited dye to energy acceptors (Ag atoms near the surface of the AgBr grain that accept the energy from the excited dye and then release electrons into the AgBr grain)?

By the manipulation method described in Box 23.1 (Figure 23.4(d)) the sensitizing action of the dye can be studied for many different geometries. The sensitizing dye molecule D

can be in direct contact (Figure 23.23, case 1) or at a distance of 5 nm (Figure 23.23, case 2). By using very pure AgBr crystals it was found that the sensitization in case 2 was only 1% of that in case 1. This shows that the photographic process is mainly caused by electron transfer; additionally, a small but distinct contribution of energy transfer can be observed. In contrast, when purposely positioning energy acceptors (silver atoms) near the surface of AgBr, the photographic process is strongly sensitized also in case 2: the

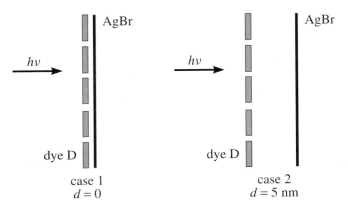

Figure 23.23 Photographic sensitization. Dye D in distance d from a AgBr surface. Case 1: $d = 0$ (direct contact). Case 2: $d = 5$ nm. (R. Steiger, H. Hediger, P. Junod, H. Kuhn, D. Möbius, Photogr. Sci. Eng. 24, 185 (1980)).

sensitization of the photographic process is governed by energy transfer from the excited dye to those acceptors in the AgBr grain near the surface.

The photographic sensitization process can be manipulated in many ways by positioning an additional dye with a planned geometrical relation. For example, the dye A

can be used for this purpose. Desired functionalities can be achieved by using the additional dye either as an amplifier or as a competitor of AgBr (Figure 23.24).

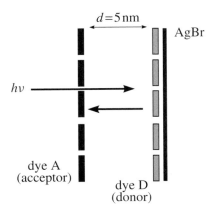

Figure 23.24 Dye A as a competitor for AgBr in the sensitizing process. Reference see Figure 23.23.

23.4 MANIPULATING ELECTRON MOTION

23.4.4 Conducting Molecular Wires

Conduction through short wires

We considered the electron tunneling arrangement in Figure 23.20 with a fatty acid as spacer layer. This layer can be replaced by a mixed layer of fatty acid and of molecules with a π electron system W (quinthiophene molecules)

connecting electron donor D and acceptor A. Then the electron transfer from D to A is strongly enhanced (Figure 23.25): no electron transfer at $d = 5$ nm with fatty acid spacers, but strong electron transfer with incorporated W molecules (quenching of the fluorescence of D). In this case the rate constant of electron transfer is about 10^9 s^{-1}.

Figure 23.25 Arrangement of dyes A and D as in Figure 23.20, but the fatty acid spacer layer is replaced by two mixed layers of fatty acid and quinthiophene molecules (wire W). (H. Kuhn, D. Möbius, Investigations of surfaces and interfaces, Part B, eds. B. W. Rossiter and R. C. Baetzold, 'Physical Methods of Chemistry Series', 2nd edn., Vol IX B, Wiley, 1993, page 467).

Figure 23.26 shows an arrangement demonstrating the trans-membrane electron transfer from a reducing agent D (sodium dithionite) outside a phospholipid vesicle to an oxidizing agent A (potassium ferricyanide) inside. The π electron system incorporated in the bilayer membrane accepts the electron from D and donates it to A. The π electron system used in these experiments is a zwitterionic caroviologene

Only 150 active molecules among 150 000 phospholipid molecules per vesicle are sufficient to accelerate the electron transfer by a factor of 8.

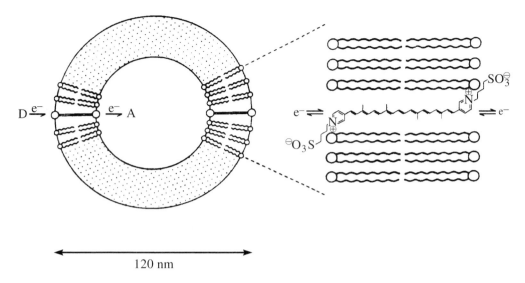

Figure 23.26 Electron transfer through a phospholipid vesicle from donor D (sodium dithionite) to acceptor A ($K_3[Fe(CN)_6]$), through a π electron system (caroviologene). (S. Kugimiya, T. Lazrak, M. Blanchard-Desce, J. M. Lehn, *J. Chem. Soc., Chem. Commun.*, 1991, 1179).

Conduction along DNA double strand

Intercalating ruthenium and rhodium complexes in a nucleic acid double helix offers the possibility of studying the photoinduced electron conduction from Ru^{2+} to Rh^{3+} through the chain of stacking base pairs. The dipyridylphenazin ligand of Ru^{2+} and the phenan-threnchinondiimin ligand of Rh^{3+} are intercalated, separated by 11 base pairs (distance 4 nm). The Ru^{2+} complex is excited, the electron moves to the intercalated ligand, through the stack of base pairs and is captured by the Ru^{3+} acceptor. The luminescence of the Ru^{2+} is completely quenched by the Rh^{3+} complex, indicating a rate constant for electron transfer larger than $3 \times 10^9 \, s^{-1}$. (See J. K. Barton, R. E. Holmlin, P. J. Dandliker, *Angew. Chem., Int. Ed. Engl.*, 1997, **36**, 2715.) We have seen that $10^9 \, s^{-1}$ is the approximate rate of electron transfer from an excited dye through a π electron system to an acceptor at a distance of 5 nm (Figure 23.25).

In contrast, when acridine is intercalated in a DNA double helix at various distances d from a guanine cytosine base pair, the fluorescence of the acridine is increasingly quenched with decreasing d, and the evaluation of the experimental data, as outlined in Section 23.4.1, gives a rate constant $k_{transfer} = 10^6 \, s^{-1}$ for $d = 1$ nm, and $k_{transfer} = 10^8 \, s^{-1}$ for $d = 0.68$ nm. (See K. Fukui, K. Tanaka, *Angew. Chem. Int. Ed. Engl.*, 1998, **110**, 167, A. Harriman, *Angew. Chem. Int. Ed. Engl.*, 1999, **38**, 945).

In this case the electron transfer from the guanine to the excited acridine takes place at the energy level of the ground state and is much slower than in the case of the Ru^{2+} to Rh^{3+} electron transfer. When exciting the acridine an electron hole is left, an electron from the adjacent base in the stack follows; in this way the hole moves to the distant guanine which is oxidized.

It can be expected that electron transfer through much longer chains of π electron systems is possible; designed molecular conductors have a great impact as information carriers.

23.4 MANIPULATING ELECTRON MOTION

Soliton conduction through long wires

We discuss the possibility that an electron can be transferred essentially loss-free along a long π electron chain. How should this work? In Chapter 7 the electron density distribution and the bond lengths in a π electron system are obtained by searching for self-consistency between π electron density and bond length. In a polyene there are alternating single and double bonds. If, in a computer simulation, an electron is added to a long-chain polyene, self-consistency is reached for charges extending over a confined region of about 15 carbon atoms with essentially equal bond lengths. This distinct charged region constitutes a *soliton*. It is assumed that the charge carriers in conducting polymers such as polyacetylene are solitons.

Soliton switching of a long cyanine

We have seen in Chapter 7 that a cyanine dye such as

$$(CH_3)_2\ddot{N}-(CH=CH)_n-CH=\overset{\oplus}{N}(CH_3)_2 \qquad (A)$$

$$(CH_3)_2\overset{\oplus}{N}=(CH-CH)_n=CH-\ddot{N}(CH_3)_2 \qquad (B)$$

has equal bond lengths. A and B are limiting structures describing the resonance hybrid. An interesting phenomenon is expected above some critical value of the number n of conjugated double bonds: A and B describe two equivalent tautomeric forms. The molecule either exists in state A or in state B. The theoretical method described in Section 7.4.2 shows that this is the case for $n \geq 15$. Thus, a longer cyanine forms a push–pull polyene which can be considered as a switching element (Figure 23.27). The calculation shows that an electric field of about 10^8 V m^{-1} is required to push the molecule from structure A to structure B and vice versa.

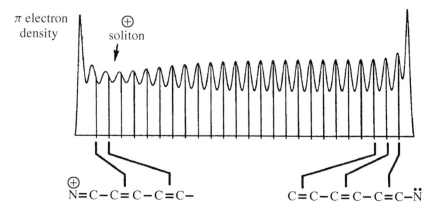

Figure 23.27 A long cyanine (26 double bonds) in an asymmetric state (soliton on the left-hand side). By applying an external electric field the soliton can be shifted to the right-hand side. (Ch. Kuhn, *Synthetic Metals*, 1991, **41**, 3681).

The soliton is expected to move with a speed of 10 km s^{-1}. This is faster than the speed of the soliton in the aggregates in Section 23.3.3. This fast speed is due to the coupling of the electron motion with the lattice deformation, and therefore it is related to the fast propagation of the sound wave along the molecular chain. The result indicates the possibility of molecular signal switching and transfer.

23.4.5 Solar Energy Conversion: The Electron Pump of Plants and Bacteria

The conversion of solar energy into chemical energy in plants is a process necessary to maintain life on earth. Electrons are excited by light, charges separate, and reduction and oxidation processes take place leading to Gibbs energy-rich products (Chapter 18). We emphasize the physicochemical aspect: the search is on for the principles behind the bioprocesses, and for possibilities to develop artificial systems on that basis.

Basic mechanism of electron pumping

Let us consider the electron transfer process from a donor dye D to an acceptor dye A. As a result of the excitation of D, an electron is transferred from D to A: charge separation into D^+ and A^-. In order to prevent this pair of separated charges from subsequently recombining, the distance between D^+ and A^- must be sufficiently large to avoid recombining by quantum mechanical electron tunneling (see Section 23.4.1).

A basic problem is how the electron can be transferred from D^* (photoexcited D) over a large distance to A fast enough to compete with de-excitation of D^*. A π electron system acting as a molecular wire (W) is required to shift the electron from D^* to A over a long distance. A secondary electron donor D' must be present to restore (reduce) the oxidized dye D^+.

But how can the low-energy electron be transferred from D' to D^+? You need a sufficiently high barrier to be sure that the electron, after excitation, moves through the wire. On the other hand, the barrier must allow an electron of D' to pass the barrier later on. How can this be achieved? The high barrier must be sufficiently narrow (1 nm thickness) to allow tunneling from D' to D^+ in due time.

Bacterial electron pump

This idealized arrangement is actually present in the bacterial reaction center (Figure 23.28). The molecular wire is represented by two π electron systems W_1 (monomer bacteriochlorophyll) and W_2 (bacteriopheophytin), and the barrier between D and D' is represented by a protein portion of 1 nm thickness, as proposed.

The initial step of electron transfer in the reaction center is particularly critical in avoiding back transfer (charge recombination). It is essential to bring the excited electron to W_2 via W_1 and, at the same time, to lose as little energy as possible. The process takes a few picoseconds, and the loss of energy is about 0.15 eV. The level of the excited state of D and the level of W_1 coincide (as a result of evolutionary improvement), allowing fastest removal of the electron from D^*. The next step, further charge separation by shifting the electron from W_2 to A, can be somewhat slower but is essentially completed in 230 ps.

23.4 MANIPULATING ELECTRON MOTION

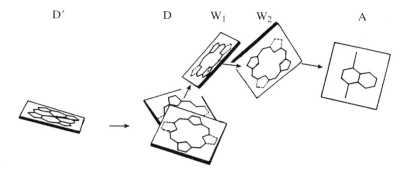

Figure 23.28 Scheme of the reaction center of purple bacterium. Electron transfer from excited "special pair" of bacteriochlorophyll molecules, D, via molecular wires W_1 (monomer bacteriochlorophyll) and W_2 (bacteriopheophytin) to acceptor A. Finally, an electron is transferred from D' to D^+. (Arrangement of chromophores in protein matrix according to J. Deisenhofer, O. Epp, K. Miki, R. Huber, H. Michel, J. Mol. Biol. 180, 385 (1984)).

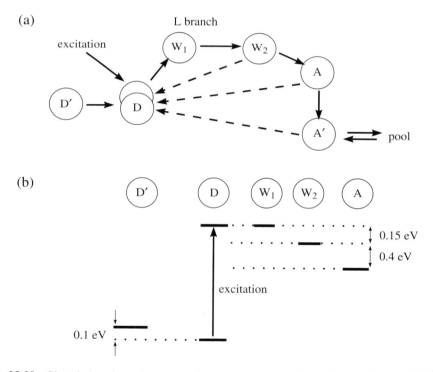

Figure 23.29 Photoinduced electron pump in reaction center of purple bacterium. (a) Photoinduced electron transfer to produce high-energy compounds (solid lines). The dashed lines indicate possibilities for charge recombination. (b) Balance of energy levels for optimal pumping efficiency.

In order to avoid losses, the precise geometrical arrangement and the matching of energy levels of each chromophore component are crucial. The probability of charge recombination by electron tunneling from W_1^- to D^+ or W_2^- to D^+ or A^- to D^+ must be kept as small as possible. Therefore, the tunneling path (dashed lines in Figure 23.29(a)) must be confined to the protein moiety to keep the tunneling probability as small as possible. The tunneling barrier is then much higher than for a path through the π electron portion. This explains why the chromophores are arranged in a banana-shaped structure.

Optimum yield. Compromise between minimizing energy loss and minimizing recombination rate

The precise geometry is needed to avoid recombination by tunneling. W_2^- and A^- are at lower energy levels than D^* and W_1^- to avoid thermally activated recombination. The lifetime of the vibronically relaxed state of W_2^- (230 ps) and the lifetime of A^- (1 ms) result from the precisely matched energy levels and matched geometry. By this matching, charge recombination by tunneling and by thermally assisted recombination is avoided during the lifetimes of W_2^- and A^-.

The vibronically relaxed state of W_2^- is 0.15 eV below D^*, and A^- is about $(0.4 + 0.15)$ eV below D^* (Figure 23.29(b)). This allows W_2^- and A^- to exist for 230 ps and 1 ms, respectively, as required. This requirement causes an unavoidable energy loss of such a photoinduced electron pump to keep an optimal yield. Shifting W_2^- or A^- to a higher energy level leads to a higher recombination rate reducing the yield.

The requirement to keep A^- for 1 ms is related to the further steps. The electron is transferred from A^- to an acceptor A' acting as a shuttle carrying electrons to a pool. A' is a quinone that becomes a hydroquinone after accepting two electrons; then it becomes unbound and diffuses into the pool delivering the two electrons. Afterwards it is trapped again in the reaction center at its original location eventually reached on a random walk.

The nearly optimal arrangement of the chromophores and of the matched energy levels is the result of a continuous selection in the course of evolution. The electron pump in plants (photosynthesis) is closely related to the bacterial pump.

Tunneling processes in the bacterial reaction center

The optimal arrangement of the chromophores can be estimated by using a simple relationship between the electron transfer rate constant k_r and the edge-to-edge distance d between the donor and acceptor orbitals (see Foundation 23.4):

$$k_r = a \cdot e^{-bd} \tag{23.22}$$

with $a = 10^{15}\,\text{s}^{-1}$ and $b = 14\,\text{nm}^{-1}$. This relationship is based on a tunneling barrier of about 2 eV appropriate for the electron transferred from D^* to W_1, W_2, A, A' and back to D^+ from A^- and $(A')^-$. It describes the experimental data astonishingly well (Table 23.1).

We can grasp the meaning of equation (23.22) by considering the excited electron at location a in Figure 23.30: this location corresponds to the antinode of the wavefunction at the edge. The probability of finding the electron at antinode a is $P_a = c_a^2 = 0.1$ in the case of bacteriochlorophyll, since the wave of the electron in the excited state or in the reduced state has 10 essentially equal antinodes, as outlined in Section 8.8.4. The electron behaves as if it would hit the wall with a frequency

$$\nu = \frac{v}{L} \tag{23.23}$$

23.4 MANIPULATING ELECTRON MOTION

Table 23.1 Calculated and experimental electron transfer rate constants for processes in the reaction center of the purple bacterium. Calculation based on equation (23.22) with $a = 10^{15}\,\text{s}^{-1}$ and $b = 14\,\text{nm}^{-1}$

Process	d/nm	k_r/s^{-1}	
		Calculated	Experiment
$D^* \to W_1$	0.55	4×10^{11}	4×10^{11}
$W_1^- \to W_2$	0.48	1×10^{12}	1×10^{12}
$W_2^- \to A$	0.96	1×10^{9}	4×10^{9}
$A^- \to A'$	1.35	1×10^{7}	4×10^{7}
$A^- \to D^+$	2.24	2×10^{1}	6×10^{1}
$(A')^- \to D^+$	2.34	6×10^{0}	2×10^{0}

d is the edge-to-edge distance between the donor and acceptor orbitals. See C. C. Moser, J. M. Keske, K. Warncke, R. S. Farid, P. L. Dutton, *Nature*, 1992, **355**, 796; W. Zinth, private communication.

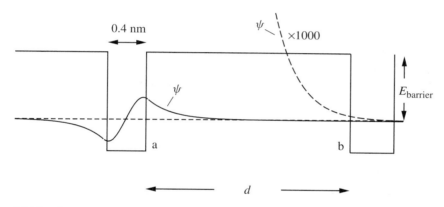

Figure 23.30 Tunneling of electron from location a (antinode of electron wavefunction in donor) to location b (acceptor) in distance $d = 2\,\text{nm}$ through a rectangular potential barrier of height $E_\text{barrier} = 1.3 \times 10^{-19}\,\text{J} = 0.7\,\text{eV}$. Solid and dashed lines: wavefunction ψ of tunneling electron. We use $E_\text{barrier} = 0.7\,\text{eV}$ instead of 2 eV, because for 2 eV the initial exponential decrease of the wavefunction is too steep for illustrating.

where L is the extension of the p orbital and the speed v is given by the de Broglie relation.

$$v = \frac{h}{m_e \Lambda} \quad \text{with} \quad \Lambda = L \tag{23.24}$$

Then the frequency ν is

$$\nu = \frac{h}{m_e L^2} \tag{23.25}$$

According to Box 3.1 we obtain

$$P = e^{-2bd} \quad \text{with} \quad b = \frac{2\pi}{h}\sqrt{2 m_e E_\text{barrier}} \tag{23.26}$$

where P is the probability of the electron to tunnel through a barrier in one hit, and where E_barrier and d are the height and width of this barrier.

Then the tunneling rate constant k_r is

$$k_r = v \cdot c_a^2 \cdot P = \frac{h}{m_e L^2} \cdot c_a^2 \cdot e^{-2bd} \qquad (23.27)$$

With $L = 0.4$ nm, $c_a^2 = 0.1$, and $E_{\text{barrier}} = 2$ eV (Foundation 23.4) we obtain

$$k_r = 0.5 \times 10^{15} \cdot \exp(-14\,\text{nm}^{-1} \cdot d)\,\text{s}^{-1}$$

23.4.6 Artificial Photoinduced Electron Pumping

An electron pump based on the mechanism discussed in Section 23.4.5 can be realized by appropriately assembling monolayers. Another possibility is to incorporate the functional components D′, D, W, and A in one and the same molecule (Figure 23.31(a)). Dye D is excited and charge is transferred from D via W to A and subsequently from D′ to D$^+$. In this case an intermediate energy transfer step takes place (from b to c in Table 23.2). The charges remain separated for as long as 55 μs. Thus the basic requirement for an electron pump can be achieved: a sufficiently long-living charge separation state to allow chemical oxidation and reduction processes.

An intermediate energy transfer step is avoided in the arrangement in Figure 23.31(b); see Table 23.3. The charges remain separated for 280 ns. The quantum yield of formation of state d is 80%; the residual processes are de-excitation (b → a) and early recombination (c → a).

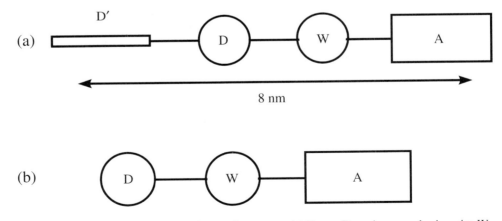

Figure 23.31 Artificial photosynthetic reaction center. (a) Donor D = zinc–porphyrin, wire W = free base porphyrin, acceptor A = naphthoquinone–benzoquinone derivative, donor D′ = β-carotene. (D. Gust, T.A. Moore, A.L. Moore, *IEEE Engineering in Medicine and Biology*, 1994, **13**, 58; *Acc. Chem. Res.*, 1993, **26**, 198.) (b) Donor D = zinc–diporphyrin, wire W = free base porphyrin, acceptor A = pyromellitimid. (A. Osuka, S. Nakajima, T. Okada, S. Taniguchi, K. Nozaki, T. Ohno, I. Yamazaki, Y. Nishimura, N. Mataga, *Angew. Chem. Int. Ed. Engl.*, 1996, **35**, 92).

23.4 MANIPULATING ELECTRON MOTION

Table 23.2 Photoinduced electron transfer from D′ to A in the artificial reaction center displayed in Figure 23.31(a)

	State				Process leading to this state	k_r/s^{-1}
a	D′	D	W	A		
b	D′	D*	W	A	Excitation of D	
c	D′	D	W*	A	Energy transfer to W	2×10^{10}
d	D′	D	W$^+$	A$^-$	Electron transfer from W* to A	7×10^{8}
e	D′	D$^+$	W	A$^-$	Electron transfer from D to W$^+$	
f	(D′)$^+$	D	W	A$^-$	Electron transfer from D′ to D$^+$	

Table 23.3 Photoinduced electron transfer from D to A in the artificial reaction center displayed in Figure 23.31(b)

	State			Process leading to this state	k_r/s^{-1}
a	D	W	A		
b	D*	W	A	Excitation of D	
c	D$^+$	W$^-$	A	Electron transfer to W	$> 5 \times 10^{10}$
d	D$^+$	W	A$^-$	Electron transfer from W$^-$ to A	4×10^{9}

23.4.7 Tunneling Current Through Monolayer

A monolayer of a fatty acid is deposited on a glass slide covered with a metal layer (aluminium) and a second metal layer is produced on top by condensation from metal vapor. A voltage V is applied between the metal layers (area A) and the current I is measured (Figure 23.32(a)). I depends on the number of carbon atoms in the fatty acid chain (determining the thickness d of the fatty acid layer). The current decreases exponentially with d:

$$I = V \cdot \frac{A}{d} \cdot a \cdot e^{-bd}$$

with $a = 10^3 \, \text{A V}^{-1} \, \text{cm}^{-1}$ and $b = 15.5 \, \text{nm}^{-1}$. This behavior is typical for electron tunneling: the probability of tunneling decreases exponentially with distance d. The constant b is somewhat larger than in equation (23.22). This is because the tunneling electron is at a lower level and the potential barrier is higher.

In the present case the hydrocarbon chains of length l are perpendicular to the layer plane ($d = l$) and $2r = 0.475 \, \text{nm}$ is the distance between the axes of adjacent chains (Figure 23.32(a)). What is expected to happen when the hydrocarbon chains are inclined and the distance d becomes $d = l \cos \varphi$, where φ is the tilt angle (Figure 23.32(b))?

Essentially, tunneling along the highly polarizable region of the C–C bonds takes place, the current per chain is unchanged in spite of the smaller distance d (*through-bond tunneling*).

$$I_{\text{through-bond}} = V \cdot \frac{A}{d} \cdot a \cdot e^{-bl}$$

In addition, the electron might reach the positive electrode by tunneling to the neighboring chain. Then, on the one hand, the tunneling path along the chain of the highly polarizable C–C bonds is decreased by the distance $l_3 = 2r \tan \varphi$ (Figure 23.32(c)); then the tunneling distance is

$$l_1 + l_2 = l - l_3 = l - 2r \tan \varphi$$

On the other hand, the electron has an additional tunneling path through the vacuum-like region between the highly polarizable cores of the chains: *through-space tunneling*. In

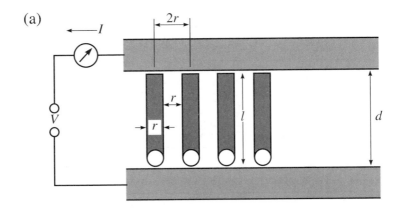

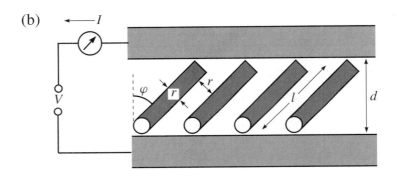

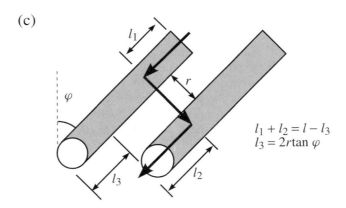

Figure 23.32 Tunneling current I through monolayer of fatty acid molecules between two parallel metal electrodes in distance d. (a) Fatty acid molecules arranged perpendicularly to the electrodes. Distance between the axes of two molecules $2r$. Distance d equal to the chain length l. (b) Fatty acid molecules tilted by an angle φ. Distance d equal to $l \cdot \cos\varphi$. (c) Tunneling path including *through-bond* tunneling (tunneling distance $l_1 + l_2$) and *through-space* tunneling (tunneling distance r between the highly polarizable cores of adjacent molecules): this core is indicated in the figure. This path is competing with through-bond tunneling along the chain (tunneling distance l).

this case the estimated tunneling distance is r.

$$I_{\text{through-space}} = V \cdot \frac{A}{d} \cdot a \cdot \frac{1}{4\pi} e^{-b(l-2r\tan\varphi)} \cdot e^{-b_{\text{vacuum}}r}$$

Thus the sum of both effects is

$$I = I_{\text{through-bond}} + I_{\text{through-space}}$$

From the tunneling barrier height of the vacuum (about 4 eV) we obtain $b_{\text{vacuum}} = 22$ nm^{-1} (see Foundation 23.4). The factor $1/4\pi$ in the second term is the estimated fraction of the surface of the chain that is in contact with the neighboring chain and where tunneling to the neighboring chain takes place. The second term increases with increasing tilt angle; at $\varphi = 45^0$ it is about equal to the first term.

This effect is observed in the case of an arrangement similar to the arrangement considered here but with an electrolyte instead of a second metal electrode and an alkanethiol monolayer instead of a fatty acid monolayer. The metal electrode is a mercury drop. It is covered by the monolayer. When extending the drop a continuous increase of the tilt angle φ is achieved and a corresponding increase in current is measured. (K. Slowinski, R.V. Chamberlain, C.J. Miller, M. Majda, *J. Am. Chem. Soc.* 119, 11910 (1997)).

23.4.8 Electron Transfer Through Proteins

Proteins can be modified by binding a luminescent group at exactly defined amino acids*. For instance, a ruthenium complex is bound in cyt c to amino acids His 39 or His 72, and the electron transfer rate from the excited Ru complex to the iron porphyrin complex is measured by fluorescence quenching of the Ru complex. The rate constant for the electron transfer is similar in both cases, $k_r = 10^6$ s^{-1} and $k_r = 3 \times 10^5$ s^{-1}, although the edge-to-edge distance between the π electron systems is different (1230 and 840 pm, respectively).

According to equation (23.22) we expect $k_r = 10^7$ s^{-1} for $d = 1230$ pm and $k_r = 10^{10}$ s^{-1} for $d = 840$ pm. This is essentially in agreement with the first case; but what is the reason for the discrepancy in the second case? From His 72 the electron has to pass an empty space of 600 pm, whereas in the derivation of equation (23.22) it was assumed that the electron tunnels through a dielectric medium. In a vacuum the wavefunction of the electron decays faster than in the dielectric medium, and the rate of electron transfer is much smaller.

The electron transfer rate in proteins can be treated by summing up the decay contributions of each bond along the tunneling pathway. Considering the fact that the electron tunnels along the easiest path, the pathway is defined as the route of least decay. This would be the edge-to-edge connecting line for a uniform medium. In general, it deviates from this line since polarizable portions are preferred, and empty portions are avoided.

23.5 Manipulating Nuclear Motion

23.5.1 Light-induced Change of Monolayer Properties

Periodic change of monolayer area

Azobenzene can be converted from the *cis* form to the *trans* form and vice versa by illuminating at 365 nm and 436 nm, respectively.

---

*(D.R. Casimiro, J.H. Richards, J.R. Winkler, H.B. Gray, *J. Phys. Chem.*, 1993, **97**, 13073; D.S. Wuttke, M.J. Bjerrum, J.R. Winkler, H.B. Gray, *Science*, 1992, **256**, 1007; D.N. Beratan, J.N. Betts, J.N. Onuchic, *Science*, 1991, **252**, 1285).

cis ⇌ trans (365 nm / 436 nm)

$R_1 = -(CH_2)_5 - CH_3$

$R_2 = -(CH_2)_{10} - C\begin{smallmatrix}O\\OH\end{smallmatrix}$

This effect can be used for switching. A *cis*-azobenzene derivative is bound to a polymer (esterification with polyvinyl alcohol) and spread on a water surface; the film is kept at constant surface pressure (8 mN m^{-1}). The film expands to three times its area when illuminated with 365 nm light and contracts again to the original area when illuminated with 436 nm light (Figure 23.33).

When such a monolayer, kept at constant area, is illuminated with light of 436 nm (365 nm), the surface pressure and surface potential are increased (decreased). This change in surface pressure is due to the fact that the *cis* form is more bulky, and the surface potential changes because *trans*-azobenzene has no dipole moment, while the dipole moment of *cis*-azobenzene is 3 debye.

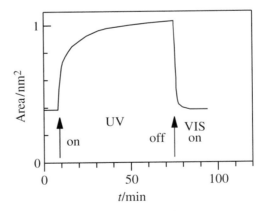

Figure 23.33 Light-induced expansion and contraction of monolayers of azobenzene induced by illumination with UV light (366 nm) and VIS light (436 nm). (T. Seki, T. Tamaki, *Chem. Lett.*, 1993, 1739).

Monolayer commanding liquid crystal orientation

A monolayer of *trans*-azobenzene on a solid surface is covered by a nematic liquid crystal (Figure 23.34). The molecules in the liquid crystal are aligned perpendicular to the solid surface. When illuminated with 436 nm light, the *trans–cis* isomerization of azobenzene induces an alignment parallel to the surface and the reverse effect is induced by 365 nm

23.5 MANIPULATING NUCLEAR MOTION

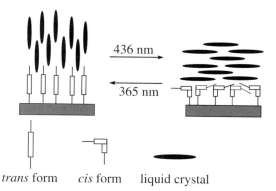

Figure 23.34 Command surface effect. Azo dyes are switched by light from the trans into the cis form and back; as a consequence, a liquid crystal changes its orientation (K. Ichimura, Y. Suzuki, T. Seki, A. Hosoki, K. Aoki, *Langmuir* **4**, 1214 (1988)).

light. The photoinduced *trans–cis* and *cis–trans* isomerization is greatly amplified by the changing alignment of the liquid crystal.

High-density data storage

To reach the goal of high-density data storage, the switching molecules must be densely packed; at the same time, unhindered *cis–trans* and *trans–cis* isomerization must be allowed. This is achieved by constructing a mixed monolayer of an azobenzene derivative with positive head groups acting as the switching element and an amphiphile with negative head groups acting as spacers (DMPA = dimyristoyl phosphatidic acid) (J. Maak, R. Ahuja, D. Möbius, H. Tashibana, H. Matsumoto, *Thin Solid Films* **242**, 122 (1994)):

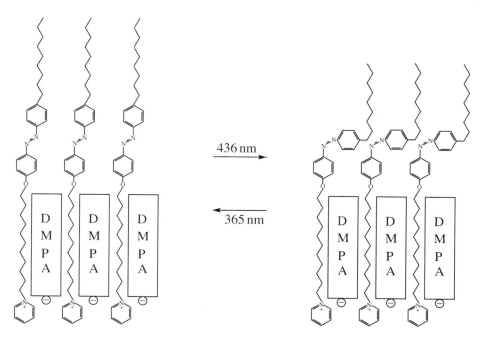

The amphiphile DMPA is dimyristoyl phosphatidic acid. In this way a well-organized assembly is achieved. This system can be switched as the system in Figure 23.34.

Another approach to high-density chromophore packing and unhindered *cis–trans* isomerization is to use bacteriorhodopsin, a protein bound to a chromophore (for chromophore structure and structural change see Section 23.5.2).

A thin film of isolated bacteriorhodopsin can be used as a highly resolving optical storage system. Bleaching the purple color (formation of deprotonated chromophore) and restoring the purple color in the dark can be varied by altering conditions and by using different mutants. Switching times between 100 µs and a minute can be reached. (Ch. Bräuchle, N. Hampp, D. Oesterhelt, in: Molecular Electronics. Properties, Dynamics and Applications. Ed. G. Mahler, V. May, M. Schreiber, Marcel Dekker, New York 1996, p. 359–381).

23.5.2 Solar Energy Conversion in Halobacteria

In Section 23.4.5 we have discussed the physicochemical aspects of solar energy conversion basic to life on earth. Another mechanism to convert solar energy into chemical energy used by halobacteria is equally interesting in its supramolecular engineering features. When illuminated, protons are transferred from the inside of the cell to the outside by a membrane-bound device discussed below. A membrane potential is built up and a proton back-transfer drives the formation of energy-rich compounds (especially

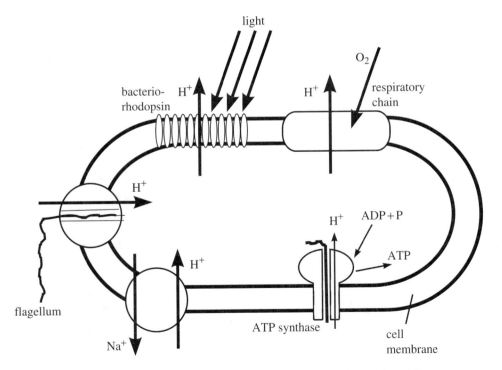

Figure 23.35 Halobacterial cell: scheme of the cell. Transport of H^+ by a light driven proton pump (bacteriorhodopsin) produces an H^+ gradient between inside and outside the cell. The gradient ΔpH and the membrane potential $\Delta\varphi$ drive the ADP $\to$ ATP reaction: $-\Delta_r G = \boldsymbol{R}\boldsymbol{T} \cdot 2.303 \cdot \Delta pH - \boldsymbol{F} \cdot \Delta\varphi$.

23.5 MANIPULATING NUCLEAR MOTION

ATP) and is the energy source of a motor that drives the rotation of the flagella which propel the bacterial motion (see Figure 23.35). The motion is triggered by a sensory device. The flagella are fastened to the membrane through a molecular bearing. The rotational motion of the flagellum can be visualized by fastening it on a flat surface. Then the cell body rotates, which is observed in the light microscope. In the case of the mitochondrial ATP-synthase it has been shown that the synthase is fastened to the membrane and its axis is rotating in a proton gradient. ATP is produced in a highly cooperative process. In a surplus of ATP the process can be reversed and the rotation of the axis can be visualized by binding an actin filament to the axis and observing its fluorescence in the light microscope.

*Photoinduced proton pump**

Trans to cis isomerization and proton translocation
Retinal in halobacteria is attached to a protein by a protonated Schiff base linkage to the side chain of the amino acid lysine.

This complex is called *bacteriorhodopsin*. Bacteriorhodopsin is incorporated in the cell membrane (Figure 23.36, left). Light isomerizes the all-*trans* chromophore into the 13-*cis* form. In a cycle the all-*trans* form is reestablished by thermal isomerization. In each cycle a proton is pumped from inside the cell to the outside. In bright light about 50 protons are pumped per second by each bacteriorhodopsin molecule.

The chromophore divides the channel between inside and outside into one part connected to the inside of the cell (called *cytoplasmic channel*) and another part connected to the outside (called *extracellular channel*). It is assumed that the *trans* isomer can exchange protons only with the outside part; the *cis* isomer, correspondingly, is connected only to the inside part (Figure 23.37).

There is an aspartic acid in the extracellular channel (D85, pK = 3.5) acting as a proton acceptor of the protonated Schiff base of the excited chromophore (Figure 23.37, step a to b). Therefore, the Schiff base in the excited chromophore must have a pK lower than 3.5.

---

*D. Oesterhelt, W. Stoeckinius, *Proc. Natl. Acad. Sci. USA* **70**, 2853 (1973); U. Haupts, J. Titor, D. Oesterhelt, General Aspects for Ion Translocation by Halobacterial Retinal Proteins: The Isomerization/Switch/Transfer (IST) Model, Biochemistry 36, 2 (1997); J.K. Lanyi, *Biochem. Biophys. Acta* **1183**, 241 (1993); D. Oesterhelt, *Israel J. Chem.* 35, 475 (1995).

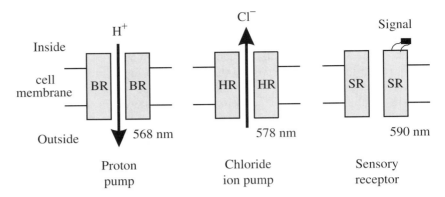

Figure 23.36 Halobacterial cell: unit of rhodopsin consisting of seven transmembrane α-helices forming a channel. Left: Transport of H^+ (proton pump, BR = bacteriorhodopsin). Center: Transport of Cl^- (chloride ion pump, HR = halorhodopsin). Right: Sensory receptor (SR = sensorrhodopsin).

On the other hand, there is an aspartic acid in the cytoplasmic channel (D96, pK = 10) acting as a proton donor of the Schiff base of the chromophore in the ground state which must then have a pK higher than 10 (actually, it is 12). The Schiff base accepts the proton which is then translocated from the cytoplasmic into the extracellular channel due to the thermal *cis* to *trans* reisomerization of the chromophore (Figure 23.37, step c to d).

Empty D96 accepts a proton from the inside of the cell and protonated D85 releases the proton to the outside. Thus, effectively, a proton is pumped from the inside (pH 7) to the outside (pH 4).

Schiff base: photoinduced change from pK > 10 to pK < 3.5

How can this extreme pK change of the Schiff base (from pK > 10 to pK < 3.5) be achieved when exciting the chromophore? It is important to bear in mind that the chromophore is much more polarizable after excitation than in its ground state. The chromophore has alternating single and double bonds and the electron in the HOMO is largely confined to the double bonds. In the LUMO the electron is more mobile. Therefore, when the *trans*-isomer (protonated Schiff base) is excited, the π electron cloud is polarized in the electrostatic field of the negatively charged side groups of two amino acids located close to the Schiff base in the extracellular channel (D85 and D212, see Figure 23.37).

The π electron cloud is repelled by these charges. The Coulomb field acting at the π electrons is approximately 4×10^9 V m^{-1} and restricted to the section of the chromophore chain between atoms 7 to 14 (see Problem 23.4).

The π electron density distribution of the chromophore in this electrostatic field can be calculated according to Chapter 7. In the calculation it is found that the π electron charge at the N atom decreases by 0.2e when exciting the chromophore. Accordingly, the bonding of the proton to the Schiff base nitrogen is weakened by 0.5 eV when the *trans*-isomer is excited. We describe the equilibrium

by the equilibrium constant K. K_1 is the equilibrium constant before excitation, and K_2

23.5 MANIPULATING NUCLEAR MOTION

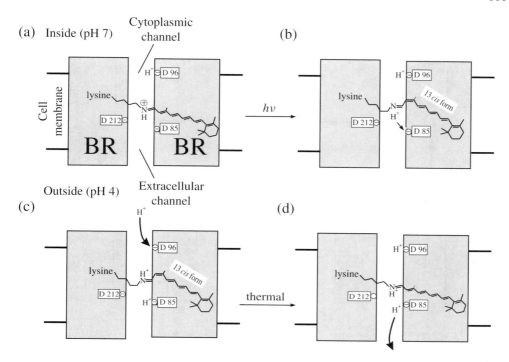

Figure 23.37 Proton pump in halobacterial cell. H$^+$ ions are pumped from inside to outside by means of light induced *trans* → *cis* and thermal *cis* → *trans* isomerization of retinal bound to protein (called *bacteriorhodpsin*). Acidification from pH 7 to pH 4 is achieved in an unbuffered saline suspension of cells.

is the equilibrium constant after excitation:

$$\Delta_r G_1^\ominus = -RT \cdot \ln K_1 \quad \text{and} \quad \Delta_r G_2^\ominus = -RT \cdot \ln K_2 \tag{23.28}$$

Then

$$\Delta pK = -\frac{1}{2.303} \cdot (\ln K_2 - \ln K_1) = -\frac{1}{2.303} \frac{\Delta_r G_2^\ominus - \Delta_r G_1^\ominus}{RT} \tag{23.29}$$

With $\Delta_r G_2^\ominus - \Delta_r G_1^\ominus = N_A \times 0.5\,\text{eV} = 48\,\text{kJ mol}^{-1}$ we obtain $\Delta pK = -8.3$, slightly larger than the difference between the experimentally determined pK values of D85 and D96 (3.5 and 10).

Trans to cis isomerization, proton left behind in extracellular channel
We consider the scheme in Figure 23.38. The excited *trans*-form (pK < 3.5), when changing into the *cis*-form, is assumed to leave the proton behind in the extracellular channel (steps 1 → 2 and 2 → 3). There are water molecules in this channel*, and the proton is assumed to be accepted by the nearby water molecule. Later on, in about 70 µs, the proton is accepted by D85 (step 3 → 4), and finally, in about 4 ms, a proton leaves the channel to the outside (step 6 → 1).

---

*(E. Pebay-Peyroula, G. Rummel, J.P. Rosenbusch, E.M. Landau, *Science* **277**, 1676 (1997); J. LeCoutre, J. Tittor, D. Oesterhelt, K. Gerwert, *Proc. Natl. Acad. Sci. USA* **92**, 4962 (1995))

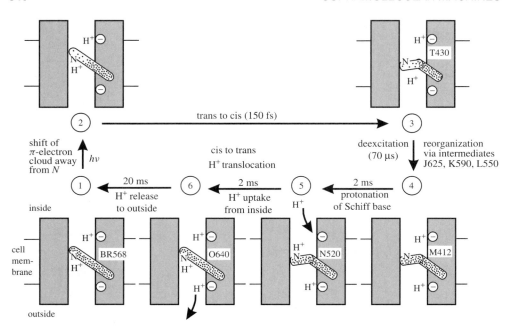

Figure 23.38 Photoinduced proton pump. Overall mechanism. π electron cloud of chromophore indicated by dots. 1 → 2: Excitation of bacteriorhodopsin BR568 by a light pulse. Shift of π-electron cloud away from N atom (causing a pK change of protonated Shiff base from 12 to 3). 2 → 3: *Trans* to *cis* isomerization to the excited transient T430. 3 → 4: Deexcitation of T430. Shift of π-electron cloud back to N atom (pK change from 3 to 12). Occurrence of deprotonated Schiff base M412. 4 → 5: Protonation of Schiff base, occurrence of N520. 5 → 6: *cis* to *trans* isomerization, translocation of H^+ from cytoplasmic to extracellular channel, uptake of proton from inside. 6 → 1: Release of proton to outside.

Note: Several transient stages are reached before M412 (deprotonated Schiff base absorbing at 412 nm) shows up in about 70 µs after flash illumination (step 3 → 4 in Figure 23.38): J625 (in 500 fs), K590 (in 3 ps), L550 (in 1 ns). The absorption of these intermediates is evidence for the protonated Schiff base. Therefore, in this picture, the Schiff base accepts a proton from a water molecule present in the cytoplasmic channel immediately after *trans* to *cis* isomerization and deexcitation. This protonation must happen within 500 fs* (occurrence of J625). The OH^- ion left behind reacts again with the proton of the Schiff base: Occurrence of M412.

The postulated initially occurring deprotonated *cis* form has not yet been observed directly as a spectral intermediate. Before the occurrence of J625, transient absorption at 430 nm shows up in 150 fs after flash which must be assigned to an excited form (T430 in Figure 23.38, step 2 → 3), since a corresponding induced emission has been observed*.

In the proposed model this should be the excited deprotonated *cis* form. Can this be the case? Following the theoretical treatment in Section 8.8, the excited deprotonated form should have a strong absorption peak at 470 nm. This is in agreement with the measured peak at 430 nm, supporting the proposal.

In the usually considered model the strong pK change of the Schiff base is related to a conformation change in the protein which occurs at a distinct stage in the cycle induced

---

*(J. Dobler, W. Zinth, W. Kaiser, D. Oesterhelt, *Chem. Phys. Letters* **144**, 215 (1988))

23.5 MANIPULATING NUCLEAR MOTION

by the photoinduced *trans-cis* isomerization, while in the present model for the proton and the chloride ion pump and the sensory receptor the strong pK change occurs in the initial step.

Translocation of proton from cytoplasmic to extracellular channel

In the steps considered, the chromophore is excited, a proton moves from inside into the cytoplasmic channel, and a proton moves from the extracellular channel to the outside. To complete the cycle, a proton must move from the cytoplasmic to the extracellular channel. This takes place in the thermally activated re-isomerization of the *cis* form into the *trans* form (Figure 23.38, step 5 → 6). The electron remains in the HOMO, the pK remains at about 12. Therefore, the proton remains bound to the nitrogen atom in the thermal *cis* to *trans* reisomerization. This is the process in the cycle where a proton is translocated from the cytoplasmic into the extracellular channel. Then the original situation in the dark is present again.

Driving force of proton pump

The driving force of the proton pump is given by the work done by the system when one mole of protons is transferred on a reversible path from the outside of the cell (pH$_{out}$) to the inside (pH$_{in}$). This work is done by carrying 1 mol of protons from a concentrated into a dilute solution:

$$-\Delta_r G = -F \cdot \Delta\varphi + 2.303 \cdot RT \cdot \Delta\text{pH}, \qquad (23.30)$$

where

$$\Delta\text{pH} = \text{pH}_{in} - \text{pH}_{out}, \Delta\varphi = \varphi_{in} - \varphi_{out}$$

and $\Delta\varphi = 280\,\text{mV}$ is the membrane potential. For this contribution a value of $F \cdot 280\,\text{mV} = 27\,\text{kJ}\,\text{mol}^{-1}$ was experimentally determined (H. Michel, D. Oesterhelt, FEBS Lett. 65, 175 (1976)).
Absorption shift of chromophore by protein. As discussed in Chapter 8, the absorption peaks of the protonated and unprotonated Schiff base in solution are at 440 nm and 359 nm, respectively, in reasonable agreement with the theoretical expectation. In bacteriorhodopsin the absorption of the chromophore changes during the cycle. The peak assigned to the protonated Schiff base is observed between 550 nm and 625 nm depending on the protein environment; the peak of the unprotonated Schiff base is observed at 412 nm. Why is the absorption band of the protonated and unprotonated chromophore in the pumping device at much longer wavelengths than in solution?

As outlined above, the electrostatic field of the negative charges close to the Schiff base polarizes the π electron cloud in the excited state much more strongly than in the ground state. Therefore, the LUMO is lowered more than the HOMO, and the peak is expected to be shifted to 570 nm for the protonated form (Figure 23.39) and to 413 nm for the unprotonated form. This is in agreement with the experiment.

In summary, the essence of the pumping mechanism is the photoinduced π electron shift due to the increase in mobility of the electron in the excited state. This shift causes the proton to separate from the chromophore and to remain behind in the extracellular

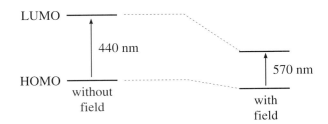

Figure 23.39 Shift of the absorption band of the protonated Schiff base of retinal by the external field exerted by the negative charges of D85 and D212 (Figure 23.37).

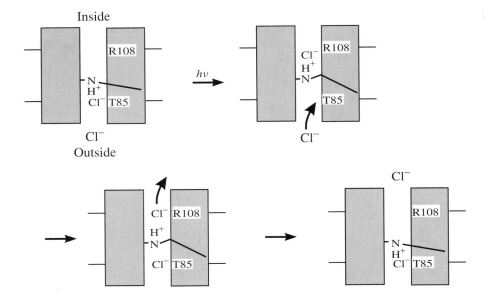

Figure 23.40 Chloride ion pump in halobacterium.

channel. In the thermal reisomerization step a new proton that has been linked to the chromophore, is shifted from the cytoplasmic into the extracellular channel.

Chloride ion pump

Halobacteria live in an almost saturated NaCl solution (4.5 mol L^{-1}) (J.K. Lanyi, Physiol. Rev. 70, 319, (1990)). When the cell grows its volume must be doubled, the inside must be filled with a concentrated KCl solution (4.5 mol L^{-1}). The required active transport of Cl$^-$ from outside to inside has been reached by slightly changing the proton pump, then acting as a Cl$^-$ pump (Figure 23.36). The proton donor D96 is missing. The proton acceptor D85 in the extracellular channel is replaced by the amino acid threonine; threonine forms, together with a conserved arginine, the chloride binding site in halorhodopsin (T85 in Figure 23.40).

Then we expect that the proton, in the photoinduced *trans–cis* isomerization, is not removed from the Schiff base since the proton acceptor D85 is absent. The Cl$^-$ present at position T85 is assumed to be picked up by the proton at the Schiff base and it is captured by the highly polarizable excited chromophore which is polarized by the charge of the Cl^- ion and therefore attracts Cl^-.

In this way, the Cl^- is considered to be dragged along and transferred from the extracellular to the cytoplasmic channel. After deexcitation, when the chromophore is no longer strongly polarizable, the Cl^- is no longer bound. It is released into the cytoplasmic space, assisted by Cl^- acceptor R108 in the cytoplasmic channel. Again, the change in polarizability of the chromophore in exciting and de-exciting is considered to be essential for pumping: Cl^- is captured by the chromophore after excitation, and, after *trans–cis* isomerization and deexcitation, it is released.

*Sensory receptors**

Halobacteria possess highly efficient sensory receptors (Figure 23.36) constructed in a manner very similar to the proton pump, but the proton donor D96 is missing, while the proton acceptor D85 is present. In the one receptor (called SRI) the thermally stable protonated trans form absorbs at 587 nm. When photoexcited, a form with a deprotonated Schiff base (absorption band at 373 nm) occurs after 0.5 ms; it is protonated and thermally reisomerized forming the stable trans form in 1 s (Figure 23.41). Thus a proton is shifted from the cytoplasmic into the extracellular channel. The unprotonated intermediate corresponds, in the proton pump, to the form absorbing at 412 nm occurring 70 μs after photoisomerization and disappearing in 2 ms. The 500 times longer lifetime of the unprotonated Schiff base in the sensory receptor is due to the absence of the proton donor D96.

The presence of the form with the 373 nm peak suppresses directional reorientation of the bacterium (swimming reversal), thus favoring cell migration into regions with higher intensity of orange-red light. This is beneficial for photosynthesis.

When the intermediate with the 373 nm peak is photoexcited, a form with a protonated Schiff base is obtained which is converted into the stable form in 100 ms (Figure 23.41). This photoinduced process triggers the bacterium to reversal of swimming direction and thus to escape from the region with high UV light intensity. In this way, damaging of the cell is avoided. The process is induced by a transducer protein which is activated by the photoinduced isomerization or by the subsequent proton translocation. The process is a dramatic single photon event.

The close relation between proton pump, chloride ion pump and sensory receptor is fascinating. By mutations in the proteines and variation in the color of the illuminating light the functionality in all three cases can be varied in many ways. For example, the proton acceptor D85 in bacteriorhodopsin can be replaced by an amino acid with a neutral residue. Then the proton, in the photoinduced *trans* to *cis* isomerization, is not anymore kept in the extracellular channel but it is shifted into the cytoplasmic channel. It is shifted back again during the thermal *cis–trans* re-isomerization, thus there is no net transport. However, when the unprotonated *cis* isomer formed during the cycle is excited by blue light we obtain the unprotonated *trans* form. Therefore, the net transport corresponds to the shift of a proton from the extracellular into the cytoplasmic channel. When replacing the proton-acceptor D85 of bacteriorhodopsin by a chloride ion acceptor (threonine) the

---

*(D. Oesterhelt, W. Marwan, Signal transduction in halobacteria, Chapter 5 in: *The Biochemistry of Archaae*, Edt.: M. Kates *et al.*, Elsevier 1993; J.L. Spudich, *Cell* **79**, 747 (1994))

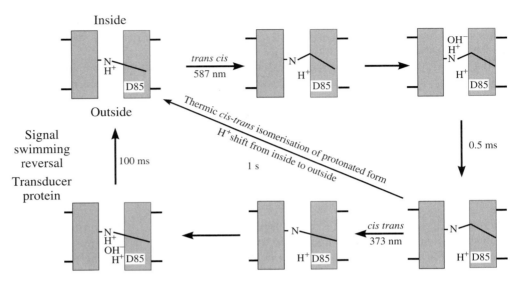

Figure 23.41 Sensory receptor in halobacterium.

proton pump can be transformed into a chloride ion pump. These kinds of experiments support the model considerations.

23.5.3 Photoinduced Sequence of Amplification Steps: The Visual System

Human rod cells contain retinal which, in the dark, is in the 11-*cis* form attached to a protein (opsin) by a protonated Schiff base linkage to a lysine side chain. The complex is called *rhodopsin*. The initial reaction in the vision process is the photoinduced *cis–trans* isomerization (see Section 8.8):

11-*cis* form (500 nm)

all-*trans* form (543 nm)

The absorption peak shifts from 500 nm to 543 nm within 1 ps (the *trans*-form absorbs at longer wavelengths than the *cis*-form). The change in shape causes a conformation change of the surrounding protein portion, resulting in a shift of the absorption peak from 543 nm

23.5 MANIPULATING NUCLEAR MOTION

to 497 nm within nanoseconds, along with a further shift to 480 nm within microseconds. A protein (transducin) is activated within milliseconds. About 500 molecules of activated transducin are formed in a domino effect that activates a protein cascade which results in blocking of about a million specific membrane channels for Na^+; this change in the membrane constitutes the event of sensory transduction and ultimately leads to an actively propagated nervous impulse. Thus a single light quantum absorbed in a human rod cell can trigger a nervous impulse in a tremendously efficient chemical amplification process.

23.5.4 Mechanical Switching Devices

Supramolecular chemistry offers a fascinating possibility to design machines based on photoinduced electron transfer triggering distinct shifts of molecular parts. Bistability of a molecule is an essential property in viewing the possibility of molecular information storage.

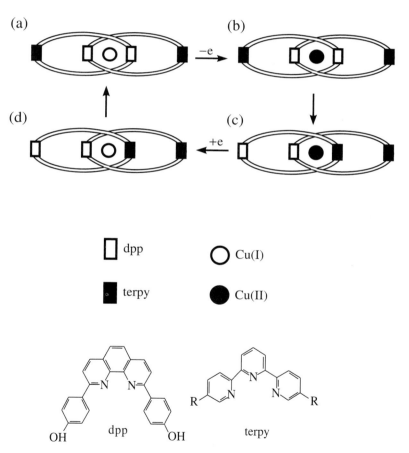

Figure 23.42 Catenane ring rotating upon oxidation and reduction process ($Cu^+ \rightarrow Cu^{2+} \rightarrow Cu^+$). dpp: 2,9-diphenyl-1,10-phenanthroline; terpy: 2,2′,6′,2″-terpyridine. The coordination number of Cu(I) is 4, and of Cu(II) is 5. Then Cu(II) prefers a five-fold coordinated complex with a dpp and a terpy unit, whereas Cu(I) prefers a four-fold coordinated complex with two dpp units. (A. Livoreil, C.O. Diedrich-Buchecker, J.P. Sauvage, *J. Am. Chem. Soc.*, 1994, **116**, 9399).

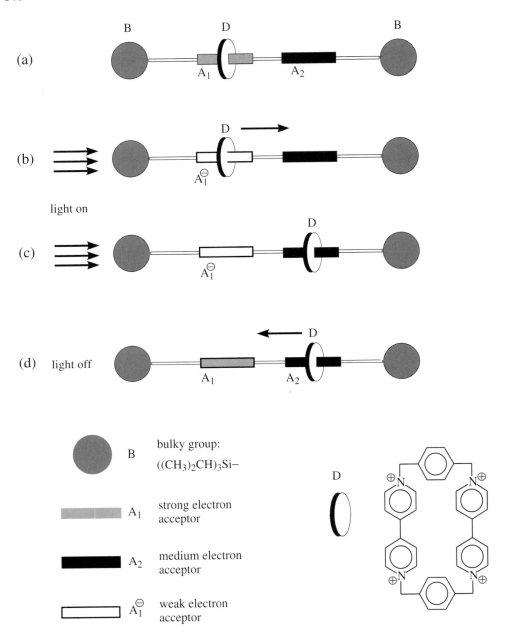

Figure 23.43 Supramolecular device (rotaxane) photoinduced movement of ring D (electron donor) on a strand with electron acceptors A_1 and A_2. B is a bulky group preventing unthreading of the strand. (a) Because A_1 is a strong acceptor and A_2 is a medium acceptor only, the ring is bonded to A_1. (b) By irradiation with light, photosensitive molecules in the solution (not shown in the figure) reduce A_1 to the weak electron acceptor $A_1^{\ominus}$. (c) Consequently, the ring D moves to the stronger electron acceptor A_2. (d) By switching the light off, the $D-A_1$ bond is re-formed. (R. Ballardini, V. Balzani, M. Teresa Gandolfi, L. Prodi, M. Venturi, D. Philp, H.G. Ricketts, J.F. Stoddart, *Angew. Chem. Int. Ed. Engl.*, 1993, **32**, 1301; D. Philp, J.F. Stoddart, *Angew. Chem. Int. Ed. Engl.*, 1996, **35**, 1154; P.R. Ashton, P. T. Glink, J. F. Stoddart, P. A. Tasker, A. I. White, D.J. Williams, *Chem. Eur. J.*, 1996, **2**, 729).

23.6 MOLECULAR RECOGNITION AND REPLICA FORMATION

Figure 23.42 shows catenane, consisting of two interlocked rings. The rings are connected by Cu^+ ions which form tetrahedral bonds. Cu^+ is oxidized to Cu^{2+} (step a → b). Then the complex becomes unstable because doubly charged Cu^{2+} tends to form a solvation shell and the link between the two rings is opened up. The compound undergoes a complete reorganization process and the system stabilizes again forming the stable 5-coordinate Cu^{2+} complex (step b → c). Upon reduction, a conformational change takes place which regenerates the starting complex (steps c → d and d → a).

Figure 23.43 shows a molecule (called *rotaxane*) synthesized by threading a strand with electron acceptors A_1 and A_2 through a ring with electron donor D (like a needle, because of chemical interaction of donor and acceptors) and tying bulky groups B to the strand ends. The structure forms automatically, under the appropriate conditions, in a mixture containing the three ingredients: the rings, the strands, and the plugs that prevent the strands from becoming unthreaded. A_1 is the stronger acceptor, thus normally the ring is bonded to A_1 (Figure 23.43(a)). By irradiation with light, photosensitive molecules in the solution transfer an electron to A_1, thus reducing A_1 and weakening the A_1–D bond. By switching light on and off, the ring moves from A_1 to A_2 and vice versa (Figures 23.43(b)–(d)).

23.6 Molecular Recognition and Replica Formation

Molecular engineering is based on the self-assembly or induced assembly and interlocking of complementary molecular components. This is also the basis of molecular recognition: copying at the molecular level, designing specific catalysts, chemical sensors, and biosensors.

23.6.1 Multisite Recognition of a Molecule at a Surface Layer

A mixed monolayer of the three components 1, 2 and 3

is at the air/water interface (The rectangles symbolize fatty acid groups). Flavin adenine dinucleotide (FAD), an important coenzyme, is highly diluted in the aqueous subphase. It binds to the monolayer very specifically (Y. Oishi, Y. Torii, M. Kumamori, K. Suehiro, K. Ariga, K. Taguchi, A. Kamino, T. Kunitake, Two-Dimensional Molecular Patterning Through Molecular Recognition, Chem.Lett. 1998, 411). One FAD molecule is linked to one molecule of 1, two molecules of 2 and one molecule of 3 by hydrogen bonds indicated by arrows. The flavin is bound to 1, each of the two phosphates is bound to one molecule of 2 and the adenine is bound to 3:

The unique molecular ordering is crucial for juxtaposing these units and inter-locking the five molecules.

23.6.2 Catalytic Reaction in Solution

An example of a purposely constructed catalytic system is indicated in Figure 23.44(a). A concave receptor with two receptor sites permits the interlinkage of guest molecules. An acetylphosphate molecule reacts with a nitrogen of the receptor. It is cleaved, the phosphorus atom is linked to the nitrogen and the acetic acid molecule leaves. A phosphate ion is linked to positively charged nitrogen atoms in the neighborhood. Because of the close contact of the two phosphate groups, pyrophosphate is formed and the receptor is restored.

Figure 23.44(b) shows a self-replicating complex. The idea is that an adenine moiety, bound to a cleverly designed adenine acceptor moiety, can recognize a molecule constituting the other moiety. The complex can thus be used as a template in the autocatalytic synthesis of additional identical molecules from a mixture of the two components. This system constitutes a very exciting case of an autocatalytic biomimetic process.

23.6.3 Catalytic Reaction in Organized Media

Figure 23.45(a) shows a monolayer assembly made by chemisorption of a silane at a glass plate, surface chemical modification (oxidation of a $-CH=CH_2$ end group into $-COOH$),

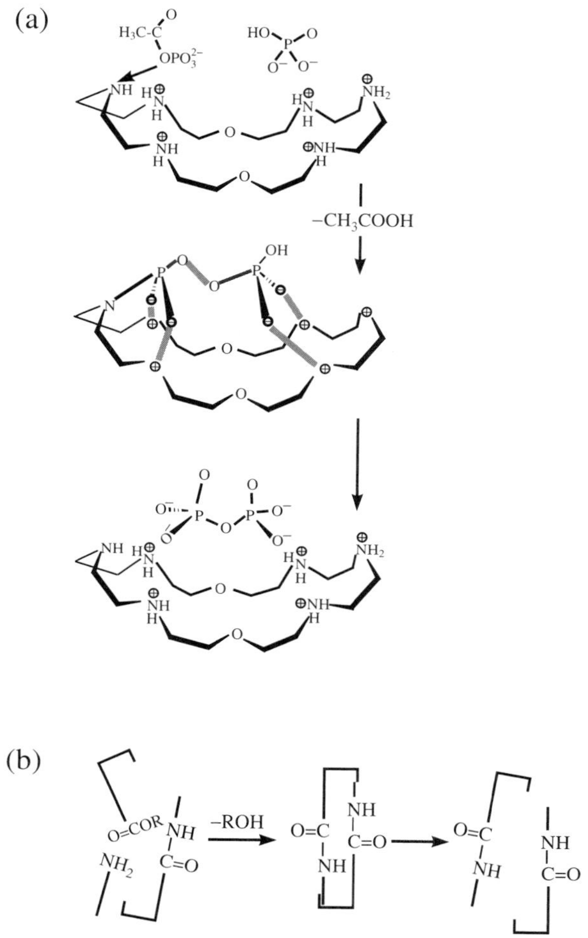

23.6 MOLECULAR RECOGNITION AND REPLICA FORMATION

Figure 23.44 Catalytic reaction in solution. (a) Catalytic formation of pyrophosphate from acetylphosphate and an phosphate ion. (b) Self-replicating complex: autocatalytic synthesis of identical molecules from a mixture of the two components. (Lehn, *Angew. Chem. Int. Ed. Engl.*, 1988, **27**, 89; J. Rebek, *Angew. Chem. Int. Ed. Engl.*, 1990, **29**, 245, scheme 17, p. 267).

and adsorption of a second layer of silane. The layers are connected by hydrogen bonds. The protons in the hydrogen bonds can be exchanged by copper ions (Figure 23.45(b)). The intercalation proceeds through lateral transport.

A glass slide covered with the monolayer assembly is touched at an edge with a drop of a solution containing Cu^{2+} ions. The Cu^{2+} ions replace H^+ ions connecting the two monolayers. The replacement proceeds with a speed as high as 1 cm s^{-1} indicating high cooperativity in the proton flux to the outside compensating for the flux of Cu^{2+}.

Instead of Cu^{2+}, molecules such as aniline can be intercalated. Also the silane can be intercalated; it reacts forming a double layer identical with the original layer. The layer

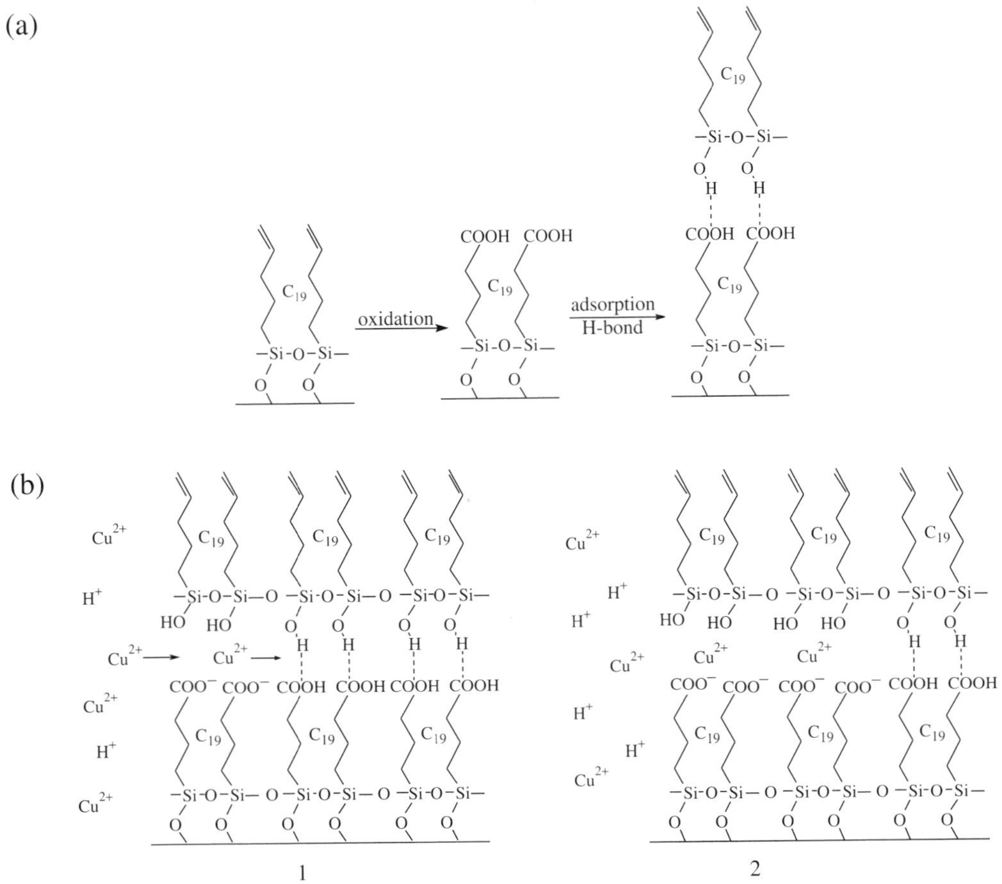

Figure 23.45 Monolayer assembly kept together by hydrogen bonds. (a) Formation of silane layers connected by hydrogen bonds. (b) The protons in the hydrogen bonds can be exchanged by Cu^{2+} ions. (R. Maoz, S. Matlis, J. Sagiv, Nature 384, 150 (1996); R. Maoz, R. Yam, G. Berkovic, J. Sagiv, in: *Organic Thin Films and Surfaces: Directions for the Nineties*, (A. Ulman, ed.), Academic Press, San Diego, pp. 41–68, (1995).).

assembly then serves as a template for the growth of its copy. This process of replication forming an intercalated stacked bilayer can be continued. It constitutes a most remarkable autocatalytic process.

23.6.4 Molecular Replica Formation

This is the basis of genetic information storage and copying in biosystems (replication of DNA, transcription by forming messenger RNA). A simple experiment demonstrating the formation of a molecular replica is shown in Figure 23.46.

23.6 MOLECULAR RECOGNITION AND REPLICA FORMATION

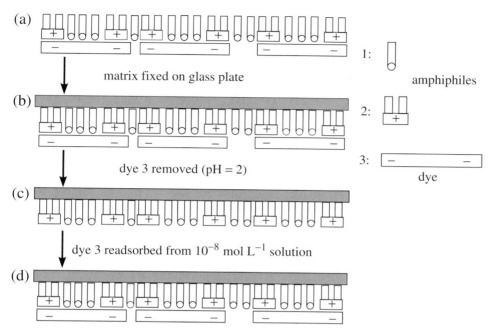

Figure 23.46 Molecular replica formation. D. Möbius, Z. Phys. Chem. Neue Folge 154, 121 (1987); M.T. Martin, D. Möbius, This Solid Films 284–285, 663 (1996).

A monolayer is obtained by co-spreading, on a water surface, neutral amphiphile 1, positively charged amphiphile 2, and negatively charged porphyrin 3 in molar ratio 16:4:1. Amphiphiles 1 and 2 form a compact monolayer; the dye is bound to the monolayer on the side facing the water subphase. It is attracted by ionic forces. The counterions in the matrix (amphiphile 2) are arranged in the pattern given by the charges of the adsorbent dye lying flat under the monolayer (Figure 23.46(a)).

The pattern can be frozen by transfer of the monolayer from the water surface to the surface of a solid using the Langmuir–Blodgett technique (Figure 22.10). A glass plate with a hydrophobic surface is dipped in which is then covered with the monolayer assembly, the dye facing the water (Figure 23.46(b)).

The charge pattern stays fixed when removing the dye by changing the pH of the subphase from 5 to 2 (Figure 23.46(c)). The glass plate is transferred to a 10^{-8} molar solution of dye 3 at pH 5 (Figure 23.46(d)). The absorption at 420 nm (absorption maximum of the dye) shows that the dye is fully re-adsorbed. In contrast, for a monolayer with statistically distributed charges the amount adsorbed is less than 10% of the amount adsorbed on the organized matrix under the same conditions.

In principle, it is possible to copy information at the molecular level and to separate the copy from the original. This can be demonstrated by incorporating dye A molecules in a monolayer and freezing their position and orientation. On this layer a layer with dye B molecules is deposited. The dye B molecules are selected such that they can form sandwich pairs with dye A molecules (see Section 23.3.2). In this way, information (about the arrangement of molecules A) is transferred to layer B. Then the arrangement in layer B is fixed by freezing, and both layers are separated by using the method discussed in Box 23.1(b).

23.6.5 Biosensors

Figure 23.47 shows as an example, a biosensor for glucose based on the enzyme glucose oxidase (GOD) as recognizing molecule. A platinum electrode is covered with a monolayer of GOD and polypyrrole (conducting polymer similar to polyacetylene, Section 23.4.4). Glucose reacts with GOD

$$\text{glucose} \xrightarrow{\text{GOD}} \text{gluconolactone} + e^-$$

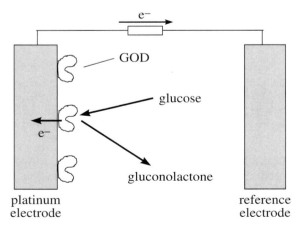

Figure 23.47 Biosensor for glucose. (M. Aizawa, *IEEE Engineering in Medicine and Biology* **13**, 94 (1994)).

23.7 Addressing and Positioning Molecules

The electron produced in the reaction is transferred to the platinum electrode by polypyrrole. Thus the reaction can be followed by measuring the current I in the circuit.

23.7 Addressing and Positioning Molecules

We have considered large populations of identical supramolecular machines which are exposed to the same stimulus (photons, electrons, protons) and respond equally. In contrast, processing information at the molecular level requires addressing individual molecules arranged in a functional network and receiving signals from individual molecules. Experimental methods pointing into this goal are scanning tunneling microscopy (STM), atomic force microscopy (AFM), scanning near-field optical microscopy (SNOM), besides single molecule excitation spectroscopy (discussed in Sections 8.2.2 and 8.9.2).

23.7.1 STM

A metallic tip of atomic dimensions (e.g., from tungsten) is kept very closely above a conducting surface (distance $d < 1$ nm). A small voltage (of about 100 mV) is applied between surface and tip, resulting in a tunneling current of about 2 nA corresponding to 2×10^{-9} A$/1.6 \times 10^{-19}$ C $\approx 10^{10}$ electrons per second (Figure 23.48).

This tunneling current can be estimated from the tunneling rate discussed in Section 23.4. From equation (23.22) we obtain, with $d = 800$ pm, $k_r = 1.4 \times 10^{10}$ s^{-1}. Thus an electron needs 10^{-10} s for tunneling; this means that 10^{10} electrons tunnel per second, in accordance with the experimental value. An increase of d by 100 pm reduces k_r by a factor of $e^{1.4} = 4$. This sensitivity on d allows one to image the structure of the surface by scanning it with the tiny tungsten tip. The tip is moved vertically during scanning such that the distance d is kept constant, and a constant current is measured. Then the vertical movement measures the surface structure of the sample.

In Figure 23.49 the STM picture of a conducting oligomer is displayed. The oligomer consists of an octathiophene with hydrocarbon chains acting as spacer layers. The chains are stacked, forming well-ordered supramolecular wires.

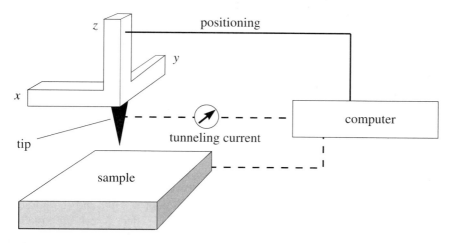

Figure 23.48 Scheme of a scanning tunneling microscope.

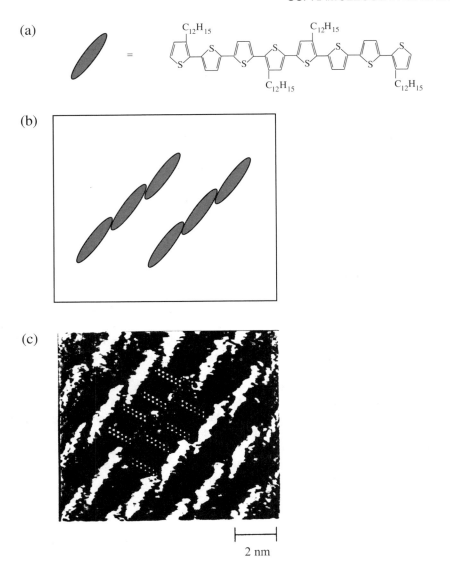

Figure 23.49 Arrangement of chains of octathiophene molecules (monolayers adsorbed on graphite surface). (a) Structural formula; (b) scheme of the structure; (c) STM picture: two molecules in the center of the picture are covered by space-filling models to indicate the four hydrocarbon substituents of each molecule; the space between the octathiophene chains is closely packed by hydrocarbon chains. (P. Bäuerle, T. Fischer, B. Bidlingmeier, A. Stabel, J. Rabe, *Angew. Chem. Int. Ed. Engl.*, 1995, **34**, 303).

23.7.2 AFM

In AFM, similarly to STM, a tip is kept at the surface of a solid and the attractive or repulsive force is measured by the deflection of a spring. The forces are smaller than 10^{-8} N (usually the forces are on the order of 10^{-11} N), thus the tip does not damage the surface. (For comparison, 10^{-8} N is the force needed to compress a C—C bond by 10%

23.7 ADDRESSING AND POSITIONING MOLECULES

of its length, and 10^{-9} N is the force to shift a C atom by 30 pm when decreasing the C–C–C bond angle – see Chapter 5.) The force between surface and tip can be measured as a function of distance (attraction followed by repulsion when diminishing the distance between tip and surface). By moving the tip over the surface and recording the change in deflection of the spring the surface structure can be investigated. It is possible to shift the tip in such an accurate way that single atoms can be picked up by the tip and set down at a different location on the surface.

As an example, we consider a surface with copper atoms (Cu(1,1,1)-plane on a single crystal). By shifting iron atoms along the surface it is possible to arrange them along a circle. The result can be observed in a tunneling microscope (Figure 23.50). Each of the 48 iron atoms in the circle can be seen. Tunneling electrons produce standing waves in the cage formed by the circle. The density maxima corresponding to these waves are seen. The wavelength is 3 nm; this is the wavelength of the electrons injected by the tip.

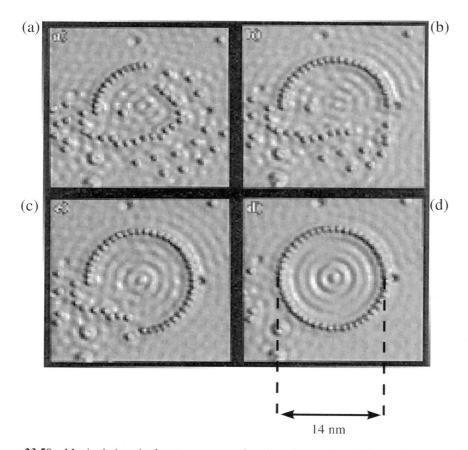

Figure 23.50 Manipulating single atoms on a surface by using an atomic force microscope. Iron atoms are arranged in a circle of 14 nm diameter on a Cu(111) surface (sequence a to d). After completion of the circle, standing electron waves inside the circular potential trough are observed. (C.M. Lieber, J. Liu, P.E. Sheehan, *Angew. Chem. Int. Ed. Engl.*, 1996, **35**, 68; M.F. Crommie, C.P. Lutz, D.M. Eigler, *Science*, 1993, **262**, 218).

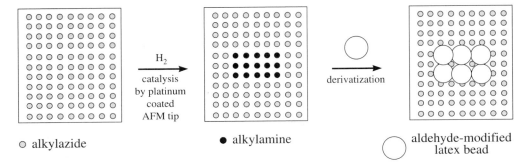

Figure 23.51 Chemical synthesis of nanostructures. (W.T. Müller, D.L. Klein, T. Lee, J. Clarke, P.L. McEuen, P.G. Schultz, *Science*, 1995, **268**, 272).

The standing waves are like the waves of an electron in a box discussed in Chapter 2, only in the present case you have a pill box instead of a rectangular box. It is fascinating to see that standing waves do not appear before the circle of iron atoms has been completed.

An interesting possibility is using the AFM tip for catalyzing a surface reaction, aiming at a chemical synthesis of nanostructures. A monolayer containing terminal azide groups is touched with a tip coated with platinum. In the presence of H_2 the azide groups are reduced to amino groups. Nanostructures are formed by binding aldehyde-modified latex beads to the amino groups (Figure 23.51).

23.7.3 SNOM

In Section 23.3.1 we considered the energy transfer from energy donor D to energy acceptor A. D was treated as a light source of molecular size, and A was treated as an absorbing body in the near field of the light source, with the light source considered as an oscillating point dipole. The near field is the electric field of the light source at a distance r smaller than about one-half of the wavelength (Coulomb field of the dipole). Then the near field increases proportional to $1/r^3$ when decreasing r.

Energy transfer allows us to monitor a molecular assembly by visible light with a resolution determined by the characteristic length of energy transfer ($r_0 \approx 10$ nm), far below the wavelength of visible light. Let us consider an optical fiber pulled to a small tip (diameter ≥ 20 nm) and illuminated by a laser beam (Figure 23.52). The tip corresponds to the fluorescent dye molecule acting as a light source, and the surface corresponds to the acceptor layer interacting with the near field of the light source. This arrangement is a useful tool to address a molecular assembly with visible light (scanning near-field optical microscope, SNOM).

Using this device it is possible to visualize individual dye molecules fixed in a polymer film. Figure 23.53 shows the fluorescence of 23 individual molecules when scanning an area of 1 μm². The directional dependence of the fluorescence of each molecule indicates its orientation. Bleaching after a longer illumination period does not lead to a continuous weakening of all signals, but to a sudden disappearance of individual signals in statistical sequence. This confirms that the signals belong to individual dye molecules. This effect has great future prospects as a possibility to address molecular devices.

23.7 ADDRESSING AND POSITIONING MOLECULES

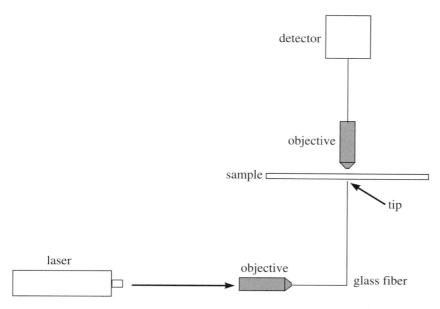

Figure 23.52 Scheme of a scanning near-field optical microscope (SNOM). Laser light is coupled to the surface of the sample by using a glass fiber with a very small tip (diameter ≥ 20 nm). The sample can be moved in x and y directions; in this way the surface can be scanned.

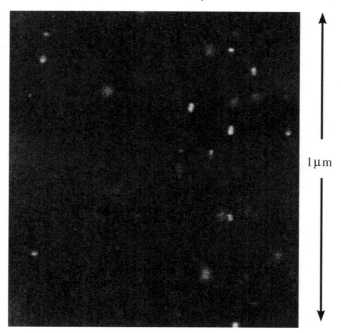

Figure 23.53 Visualization of the fluorescence of single molecules by using SNOM. Lipophile cyanine molecules are on the surface of a thin film of polymethylmethacrylate in a dilution of about 20 molecules per μm². (T. Basché, *Angew. Chem. Int. Ed. Engl.*, 1994, **33**, 1723; W.E. Moerner, T. Basché, *Angew. Chem. Int. Ed. Engl.*, 1993, **32**, 457; E. Betzig, R.J. Chichester, *Science*, 1993, **262**, 1422).

Problems

Problem 23.1 – Derivation of Equation for ε in Foundation 23.4

Derive the equation for ε in Foundation 23.4.

Solution. According to Foundation 4.4, for the oscillation of an electron between two protons we obtain

$$2\varepsilon = \varepsilon_2 - \varepsilon_1 = 2(H_{aa}S_{ab} - H_{ab})$$

In that case, for reasons of symmetry we have $H_{ab} = H_{ba}$ and $H_{aa} = H_{bb}$. In our present case, however, $H_{ab} \neq H_{ba}$ and $H_{aa} \neq H_{bb}$. Then

$$2\varepsilon = H_{aa}S_{ab} - H_{ab} + H_{bb}S_{ab} - H_{ba}$$

$$H_{ab} = \int \psi_a \left[-\frac{h^2}{8\pi^2 m} \left(\frac{\partial^2}{\partial x^2} + \frac{\partial^2}{\partial y^2} + \frac{\partial^2}{\partial z^2} \right) + V_a \right] \psi_b \cdot d\tau$$

$$= \int \psi_a \mathcal{H}_b \psi_b \cdot d\tau + \int \psi_a V_a \psi_b \cdot d\tau = E_b S_{ab} + \int \psi_a V_a \psi_b \cdot d\tau$$

Correspondingly,

$$H_{ba} = E_a S_{ab} + \int \psi_a V_b \psi_b \cdot d\tau, \quad H_{aa} = E_a, \quad H_{bb} = E_b$$

Then it follows

$$2\varepsilon = E_a S_{ab} - E_b S_{ab} - \int \psi_a V_a \psi_b \cdot d\tau + E_b S_{ab} - E_a S_{ab} - \int \psi_a V_b \psi_b \cdot d\tau$$

$$= -\int \psi_a (V_a + V_b) \psi_b \cdot d\tau$$

Problem 23.2 – Dielectric Medium as a Tunneling Barrier

How much below the vacuum is the height of the barrier formed by a dielectric medium? To answer this question we ask for the energy to transfer an electron from the vacuum into a dielectric medium of relative permittivity ε_r? The medium is polarized; we can treat this effect by replacing the electron by a charged sphere of radius $r = 0.2$ nm (Fermi radius).

Solution. We calculate the energy w to transfer the sphere with charge e from vacuum into the dielectric medium with relative permittivity ε_r. The energy dw to transfer the charge dQ is

$$dw = -\frac{1}{4\pi\varepsilon_0} \frac{e-Q}{r} \cdot dQ + \frac{1}{4\pi\varepsilon_0} \frac{1}{\varepsilon_r} \frac{Q}{r} \cdot dQ$$

Thus by integration we obtain

$$w = \int_0^e dw = -\frac{1}{4\pi\varepsilon_0 r} \left(-e^2 + \frac{1}{2}e^2 + \frac{1}{\varepsilon_r} \cdot \frac{1}{2}e^2 \right) = \frac{-1}{4\pi\varepsilon_0} \frac{e^2}{2r} \left(1 - \frac{1}{\varepsilon_r} \right)$$

$$= -3.6 \text{ eV} \left(1 - \frac{1}{\varepsilon_r} \right)$$

In the case $\varepsilon_r = 2.5$ (hydrocarbon portion, $w = -2.2\,\text{eV}$) the energy barrier is 2.2 eV below vacuum.

Problem 23.3 – Induced Dipole Moment

Table 23.4 gives the induced dipole moment μ_{ind} at different external field strengths F for a polyene with 10 double bonds. These data are calculated according to the method described in Section 7.2. Calculate the polarizability α and the second hyperpolarizability γ.

Solution. According to the table, the induced dipole moment is linearly increasing up to $F = 10^8\,\text{V m}^{-1}$ with a slope 0.7×10^{-8} debye $\text{V}^{-1}\,\text{m} = 2.3 \times 10^{-38}\,\text{C m}^2\,\text{V}^{-1}$, thus μ_{ind} can be described as $\mu_{\text{ind}} = \alpha F$ with $\alpha = 2.3 \times 10^{-38}\,\text{C m}^2\,\text{V}^{-1}$. At higher field strengths μ_{ind} is increasing much stronger than linearly (Figure 23.54). Then μ_{ind} is described by $\mu_{\text{ind}} = \alpha F + \gamma F^3$. The constant γ is obtained from the measured data as

$$\gamma = \frac{\mu_{\text{ind}} - \alpha F}{F^3}.$$

This expression is evaluated in Table 23.4 (columns 3 and 4) leading to $\gamma = 3.6 \times 10^{-28}$ debye $\text{m}^3\,\text{V}^{-3} = 1.2 \times 10^{-57}\,\text{C m}^4\,\text{V}^{-3}$.

Table 23.4 Induced dipole moment μ_{ind} calculated for a polyene with ten double bonds at different field strengths F. Evaluation of the second hyperpolarizability γ

$F/10^8\,\text{V m}^{-1}$	μ_{ind}/debye	$\mu_{\text{ind}} - \alpha F$/debye	$(\mu_{\text{ind}} - \alpha F)/F^3$/debye m^3V^{-3}
0.0	0.00	–	–
0.1	0.07	–	–
0.5	0.35	–	–
1.0	0.69	–	–
10	7.3	0.4	4.0×10^{-28}
20	16.7	2.9	3.6×10^{-28}
30	30.4	9.7	3.6×10^{-28}
40	50.6	23	3.6×10^{-28}
50	79.5	45	3.6×10^{-28}

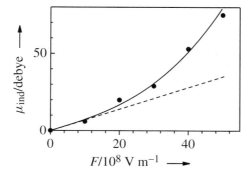

Figure 23.54 Induced dipole moment μ_{ind} for a polyene with 10 double bonds versus field strength F. Dots: data from quantum mechanical treatment. Solid line: description by $\mu_{\text{ind}} = \alpha F + \gamma F^3$ (data from Table 23.4). Dashed line: contribution of the linear term αF to μ_{ind}.

Problem 23.4 – Proton Pump: Field of Charged Amino Acids at Chromophore

The negatively charged amino acids D212 and D85 (Figure 23.37) are assumed to be at distance $d = 0.4\,\text{nm}$ from the N atom and from C atom 14, respectively (Figure 23.55). What is their field F at the chromophore?

Solution. The potential energy of a π electron at the N atom in the field of charges 1 and 2 is

$$E_\text{P} = \frac{e^2}{4\pi\varepsilon_0}\left(\frac{1}{d} + \frac{1}{\sqrt{d^2+a^2}}\right) = 1.1\times 10^{-18}\,\text{J} = 6.7\,\text{eV}$$

Figure 23.56 shows E_P at N, C_{14}, C_{12}, C_{10}, C_8, and C_6. The field between C_{14} and C_8 with $a = 0.24\,\text{nm}$, is then

$$F = \frac{1}{e}\cdot\frac{(6.7-3.1)\,\text{eV}}{3a} = 5\times 10^9\,\text{V m}^{-1}$$

Note: The H_2O molecules in the extracellular channel are assumed to be fixed during the *trans–cis* isomerization, so a relative permittivity not much larger than the assumed value $\varepsilon_\text{r} = 1$ should be appropriate.

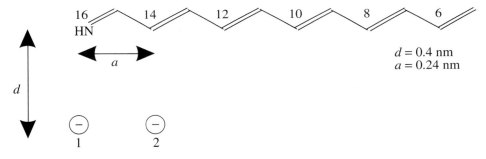

Figure 23.55 Bacteriorhodopsin. Chromophore in the field of carboxyl groups of amino acids D85 and D212.

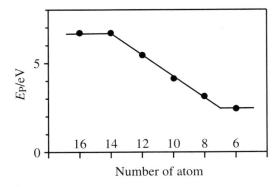

Figure 23.56 Potential energy E_P of a π electron at atoms 16, 14, ..., 6 of a retinal molecule in the field of the charges displayed in Figure 23.55.

Problem 23.5 – Circular Dichroism and Structural Features of Chlorosomes

The circular dichroism of chlorosomes changes sign within the absorption band at 740 nm (Figure 23.57), similar to the example in Section 8.11. However, the linear dichroism of stretched samples is constant within the absorption band* – in contrast to the situation in Section 8.11 where the coupled oscillations are polarized perpendicular to each other. Try to rationalize this surprising effect and the values

$$\left(\frac{\Delta\varepsilon}{\varepsilon}\right)_{750\,nm} = -10^{-2}, \quad \left(\frac{\Delta\varepsilon}{\varepsilon}\right)_{720\,nm} = +10^{-2}$$

taken from Figure 23.57 (H. Tamiaki, M. Amakawa, Y. Shimono, R. Tanikaga, A. R. Holzwarth, K. Schaffner, *Photochemistry and Photobiology*, 1996, **63**, 92).

The exciton is assumed to have the same size as the exciton in the brickstone work aggregate in Section 23.3.3 (distributed over about 10 molecules at 300 K). For simplicity we consider the exciton as confined to molecules 1 and 2 (Figure 23.57(a), that is, we replace the excited portion by two coupled oscillators (Figure 23.57(b), extracted from Figure 23.57(a)).

Solution. Using the values given in Figure 23.57(b) and relation (8.69, case a, λ_1), we find $g = -10^{-2}$ for the right-handed helix (Figure 23.57(a); correspondingly, from relation (8.69, case b, λ_1) we find $g = +10^{-2}$ for the corresponding left-handed helix.

These g values are in agreement with the corresponding measured ratios $\Delta\varepsilon/\varepsilon$. The result indicates that cylinders of left- and right-handed helices are present in a 1 : 1 ratio.

However, what is the reason for the splitting of the circular dichroism band? Such a splitting should not occur for mirror images. Indeed, the two forms are not exactly mirror images. The chlorophyll molecule is chiral, and for that reason the excitation energy for the left-handed helix should be somewhat different from that for the right-handed helix.

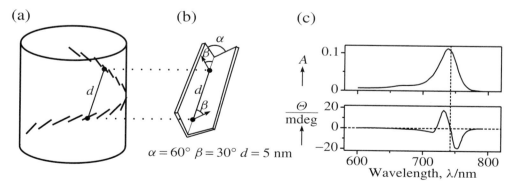

Figure 23.57 Circular dichroism of chlorosomes. (a) Geometrical arrangement; (b) replacement of excited portions by coupled oscillators. (c) Corresponding absorption spectrum (top) and CD spectrum (bottom).

---

*F. van Mouvik, K. Griebenow, B. van Haeringen, A.R. Holzwarth, R. van Grondelle, in: M. Baltscheffsky (Ed.), *Current Research in Photosynthesis*, Vol. 11, 141, Kluver Academic Publishers 1990

Foundation 23.1 – Energy Transfer

We consider an excited dye molecule (energy donor D) and an energy acceptor molecule (A) at a distance r. The transition moments of the fluorescence of D and of the absorption of A are in the direction of r. We begin with the classical approach, which is formally identical to the quantum mechanical treatment as outlined in Foundation 8.2 for related cases. We replace molecule D by a classical oscillator: frequency ν_0, charge Q, mass m, amplitude ξ_0, and energy $\bar{E} = \frac{1}{2}k_f \xi_0^2$. The electric field acting on acceptor A is simply the Coulomb field of the oscillator replacing D (see equation (10.31):

$$F = F_0 \cos 2\pi \nu_0 t, \quad F_0 = \frac{2Q\xi_0}{4\pi\varepsilon_0 r^3}$$

where ε_0 is the permittivity of vacuum and $Q\xi_0$ is the dipole moment μ of the oscillator. It is assumed that r is small compared to wavelength $\lambda = c_0/\nu_0$, but large compared to the dimension of D (then the point dipole approximation is sufficiently accurate). The mean power $\overline{dE/dt}$ absorbed by the acceptor A is proportional to ε_A (molar extinction coefficient of A at the wavelength of fluorescence of the donor) and proportional to the square of the field F_0 of the donor at the position of the acceptor (Foundation 8.1).

$$\overline{\left(\frac{dE}{dt}\right)}_{\text{energy transfer}} = g\varepsilon_A F_0^2 \quad \text{with} \quad g = 3 \cdot \frac{2.303 \times c_0 \varepsilon_0}{2N_A}$$

The factor 3 in the present case takes into account the fact that the transition moment of the acceptor molecule A is parallel to field F. In contrast, the molar decadic absorption coefficient ε_A of A as defined in Foundation 8.1 refers to a statistical spatial distribution of the transition moments of the absorbing molecules (or to an isotropic oscillator). Then

$$\overline{\left(\frac{dE}{dt}\right)}_{\text{energy transfer}} = g\varepsilon_A \left(\frac{2Q\xi_0}{4\pi\varepsilon_0 r^3}\right)^2 = g\varepsilon_A \frac{4Q^2 \xi_0^2}{16\pi^2 \varepsilon_0^2} \cdot \frac{1}{r^6}$$

The power emitted by the oscillator D (Hertz antenna equation) is, according to Foundation 8.5 (with $\omega = 2\pi\nu_0$),

$$\overline{\left(\frac{dE}{dt}\right)}_{\text{fluorescence of D}} = \frac{1}{4\pi\varepsilon_0} \cdot \frac{Q^2 \omega^4}{3c_0^3} \xi_0^2 = \frac{1}{4\pi\varepsilon_0} \cdot \frac{16\pi^4 Q^2 \nu_0^4}{3c_0^3} \xi_0^2$$

Then

$$\overline{\left(\frac{dE}{dt}\right)}_{\text{fluorescence of D}} = \frac{4Q^2 \pi^3 \nu_0^4}{3\varepsilon_0 c_0^3} \xi_0^2$$

Continued on page 863

FOUNDATION 23.1

Continued from page 862

At a certain distance $r = r_0$ the emission is quenched to half of the value without acceptor A. Therefore

$$\overline{\left(\frac{dE}{dt}\right)}_{\text{energy transfer}} = \overline{\left(\frac{dE}{dt}\right)}_{\text{fluorescence of D}}$$

or:

$$g\varepsilon_A \frac{4Q^2}{16\pi^2\varepsilon_0^2}\xi_0^2 \cdot \frac{1}{r_0^6} = \frac{4Q^2\pi^3\nu_0^4}{3\varepsilon_0 c_0^3}\xi_0^2$$

Therefore we obtain

$$r_0 = \left(\frac{9\varepsilon_A \times 2.303 \times c_0^4}{128\pi^5 N_A \nu_0^4} \cdot 4\right)^{1/6}$$

This relation holds for quantum yields of fluorescence in the absence of the acceptor, $\phi = 1$ (energy emitted exclusively by radiation), for refractive index of the medium $\hat{n} = 1$ (vacuum), and for orientation of the transition moments parallel to r. In the general case, the factor 4 in the brackets has to be replaced by $\kappa^2\phi/\hat{n}^4$, where κ is a measure of the orientation of the transition moments (Figure 23.58):

$$r_0 = \left(\frac{9\varepsilon_A \times 2.303 \times c_0^4}{128\pi^5 N_A \nu_0^4} \cdot \frac{\kappa^2\phi}{\hat{n}^4}\right)^{1/6}$$

In terms of quantum theory, the probability of energy transfer is

$$P = \frac{\overline{(dE/dt)}_{\text{energy transfer}}}{\overline{(dE/dt)}_{\text{fluorescence}} + \overline{(dE/dt)}_{\text{energy transfer}}} = \frac{1}{1+C}$$

with

$$C = \frac{\overline{(dE/dt)}_{\text{fluorescence}}}{\overline{(dE/dt)}_{\text{energy transfer}}} = \left(\frac{r}{r_0}\right)^6$$

Then we obtain

$$P = \frac{1}{1+(r/r_0)^6}$$

Apparently, r_0 is the distance where $P = \frac{1}{2}$. Correspondingly, the probability of emission of light is

$$1 - P = \frac{1}{1+(r_0/r)^6}$$

(*Förster equation*).

If the donor molecule D is spherically surrounded by a layer of acceptor molecules A, then dE/dt is proportional to the number of molecules A in

Continued on page 864

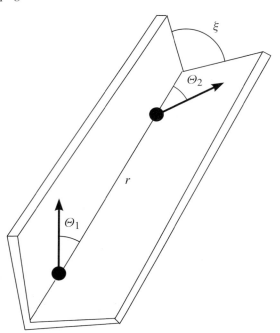

Figure 23.58 Orientation of the transition moments of molecules 1 and 2. The orientation factor κ in the expression for r_0 is $\kappa = 2\cos\Theta_1\cos\Theta_2 - \sin\Theta_1\sin\Theta_2\cos\xi$

the layer; this number increases with r^2 changing the $1/r^6$ dependence into a $1/r^4$ dependence. There is no essential change in the result if the acceptor molecules are in a plane in distance d from the donor:

$$P = \frac{1}{1+(d/d_0)^4}, \quad \text{with} \quad d_0 = \frac{1}{4\pi}\left(\frac{9}{4}\right)^{1/4}\frac{\lambda}{\tilde{n}}(\phi\alpha_A)^{1/4} \quad (23.31)$$

if the transition moments are statistically distributed in the layer plane and $\alpha_A = 1 - T_A$ is the absorbance factor of the layer (T_A = transmittance of the layer).

The probability of fluorescence of the donor is $(1 - P)$, thus

$$\left(\frac{I_d}{I_\infty}\right)_{\text{donor}} = 1 - P = \frac{1}{1+(d_0/d)^4}$$

where I_d is the fluorescence intensity of the donor at distance d (Figure 23.7). Correspondingly, for the fluorescence of the acceptor we obtain

$$\left(\frac{I_d}{I_0}\right)_{\text{acceptor}} = P = \frac{1}{1+(d/d_0)^4}$$

Foundation 23.2 – Energy Transfer from Exciton to Acceptor

We consider the exciton during the short time when it is surrounding the acceptor molecule. According to Foundation 23.1, the mean power of energy transfer for a fast moving exciton is

$$\overline{\left(\frac{dE}{dt}\right)}_{\text{energy transfer}} = g\varepsilon_A F_0^2 \quad \text{with} \quad g = \frac{3}{2} \cdot \frac{2.303 \times c_0 \varepsilon_0}{N_A}$$

ε_A is the molar decadic absorption coefficient of the acceptor at the wavelength of fluorescence of donor, ε_0 is the permittivity of vacuum, and F_0 is the amplitude of the electric field acting on the acceptor molecule in the direction of the transition moment of A.

F_0 is given by the sum of contributions from the oscillating charges representing the exciton.

Case (a). Arrangement in Figure 23.14(a)

For simplicity we first restrict our discussion to nearest-neighbor molecules and use the point dipole model as discussed in Box 10.3. Then, taking into account the four neighbors in Figure 23.59, we obtain

$$F_0 = 4 \times \frac{Q\xi_0}{4\pi\varepsilon_0 r^3} \cdot (3\cos^2\beta - 1)$$

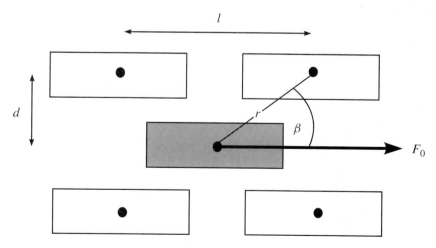

Figure 23.59 Contribution of the neighbor molecules to the electric field strength F_0.

Continued on page 866

Continued from page 865

Because of $r = d/\sin\beta$ we obtain

$$F_0 = \frac{Q\xi_0}{4\pi\varepsilon_0 d^3} \cdot \gamma \quad \text{with} \quad \gamma = 4 \times (3\cos^2\beta - 1) \cdot \sin^3\beta$$

where β is related to d and l:

$$\tan\beta = \frac{d}{l/2}$$

With $l = 1500$ pm and $d = 400$ pm we obtain $\beta = 28°$ and $\gamma = 4 \times 1.34 \times 0.10 = 0.54$.

Thus

$$\left(\frac{dE}{dt}\right)_{\text{energy transfer}} = g\varepsilon_A \cdot \left(\frac{Q\xi_0}{4\pi\varepsilon_0 d^3} \cdot \gamma\right)^2$$

Using this relation we estimate the time t_{abs} required by the acceptor molecule to absorb the exciton. The power of energy transfer can be expressed by

$$\left(\frac{dE}{dt}\right)_{\text{energy transfer}} = \frac{E}{t_{\text{abs}}}$$

where E is the energy of the N classical oscillators constituting the exciton:

$$E = \tfrac{1}{2} k_f \cdot \xi_0^2 N \quad \text{with} \quad k_f = 4\pi^2 \nu_0^2 m$$

Thus

$$t_{\text{abs}} = \frac{E}{(dE/dt)_{\text{energy transfer}}} = 8\pi^2 \varepsilon_0^2 \cdot \frac{k_f N d^6}{g\varepsilon_A Q^2 \gamma^2}$$

This relation holds for vacuum (index of refraction $\hat{n} = 1$). In the general case a factor $\hat{n}^3$ has to be applied. With $\hat{n} = 1.5$, $d = 400$ pm, $m = m_e$, $Q = e$, $N = 10$, $\varepsilon_A = 0.16 \times 10^5$ L mol^{-1} cm^{-1}, $\nu_0 = 7.5 \times 10^{14}$ s^{-1}, and $\gamma = 0.54$ we obtain $t_{\text{abs}} = 0.1$ ps. This time is considerably shorter than the time t_{pass} the exciton needs to pass the acceptor:

$$t_{\text{pass}} = \frac{\text{width of exciton}}{\text{speed}} = \frac{1000 \text{ pm}}{2000 \text{ m s}^{-1}} = 0.5 \text{ ps}$$

Therefore, the exciton will be absorbed by the acceptor as soon as it reaches the acceptor molecule.

Case (b). Arrangement in Figure 23.14(b)

In this case the D molecule adjacent to A contributes mostly to F_0. Neglecting the contributions of the next neighbors and again using the point dipole model,

Continued on page 867

Continued from page 866

F_0 is given by

$$F_0 = \frac{Q\xi_0}{4\pi\varepsilon_0 r^3}$$

where r is the distance between the adjacent molecules in both layers; r can be estimated as $r = 1.8 \times d$. Then

$$F_0 = \frac{Q\xi_0}{4\pi\varepsilon_0 d^3} \cdot (\gamma)' \quad \text{with} \quad (\gamma)' = \left(\frac{1}{1.8}\right)^3 = 0.17$$

Considering the statistical distribution of the transition moments of A in the layer plane gives $(\gamma)' = 0.06$. Then it follows that

$$\frac{t_{\text{abs}}(\text{case b})}{t_{\text{abs}}(\text{case a})} = \frac{\gamma^2}{(\gamma')^2} = \frac{0.54^2}{0.06^2} = 81$$

and

$$t_{\text{abs}}(\text{case b}) = 81 \times 0.1 \text{ ps} = 16 \cdot t_{\text{pass}}$$

(with $t_{\text{pass}} = 0.5$ ps). Thus a 50% fluorescence quenching should be reached at about $Z = 10\,000/16 = 600$ in case b instead of $Z = 10\,000$ in case a. This is indeed observed (Figure 23.14(c)).

The calculation explains the enormous quenching effect in case a, and the much smaller effect in case b.

Foundation 23.3 – Radiation Echo Field

Excited dye molecule without mirror

We treat the excited dye molecule as a classical oscillator with frequency v_0, charge Q, mass m, amplitude ξ_0, and energy $E = \frac{1}{2}k_f\xi_0^2$. The equation of the motion of the oscillator is

$$m\frac{d^2\xi}{dt^2} = -k_f\xi - \rho\frac{d\xi}{dt}$$

where ρ is the damping constant. ρ is so small that damping takes many oscillations. The lifetime τ_0 given by

$$\frac{dE}{dt} = -\frac{E}{\tau_0}$$

Continued on page 868

Continued from page 867

increases with decreasing damping constant and increasing mass m. We find (see below)
$$\tau_0 = \frac{m}{\rho}$$

In our case the damping is purely due to the emission of light (quantum yield of luminescence $\phi = 1$). Then, according to Foundation 8.5, we obtain
$$\tau_0 = \frac{6\pi m \varepsilon_0 c_0^3}{Q^2 \omega^2}$$

with $\omega = 2\pi \nu_0$. As shown below, the electric field strength F on the axis of the oscillating dipole at distance r is (Hertz equation)
$$F = A\xi + B\frac{d\xi}{dt}$$

with
$$A = \frac{3m\omega}{\tau_0 Q}\left[\frac{\sin \gamma}{\gamma^2} + \frac{\cos \gamma}{\gamma^3}\right]$$

$$B = \frac{3m}{\tau_0 Q}\left[\frac{\cos \gamma}{\gamma^2} - \frac{\sin \gamma}{\gamma^3}\right]$$

and
$$\gamma = \frac{\omega}{c_0} r$$

Excited dye molecule in the field F of a mirror

If the dye molecule is in a distance $d = r/2$ from a mirror, then an additional force QF (F = electric field of the light wave emitted from the dipole and reflected on the mirror) is acting on the corresponding dipole:
$$m\frac{d^2\xi}{dt^2} = -k_f \xi - \rho\frac{d\xi}{dt} + QF$$
$$= -(k_f - QA)\cdot\xi - (\rho - QB)\cdot\frac{d\xi}{dt}$$

Then the lifetime τ is
$$\tau = \frac{m}{\rho - QB}$$

instead of
$$\tau_0 = \frac{m}{\rho}$$

Continued on page 869

FOUNDATION 23.3

Continued from page 868

in the absence of the mirror. For the ratio τ_0/τ we obtain

$$\frac{\tau_0}{\tau} = \frac{\rho - QB}{\rho} = 1 - \frac{QB}{\rho} = 1 - \frac{QB}{m}\tau_0$$

Then

$$\frac{\tau_0}{\tau} = 1 - 3 \cdot \left[\frac{\cos\gamma}{\gamma^2} - \frac{\sin\gamma}{\gamma^3}\right] \quad \text{(oscillator perpendicular to mirror)}$$

Correspondingly, in the case of an oscillating dipole parallel to the mirror we find

$$\frac{\tau_0}{\tau} = 1 - \frac{3}{2} \cdot \left[\frac{\cos\gamma}{\gamma^2} - \frac{\sin\gamma}{\gamma^3} + \frac{\sin\gamma}{\gamma}\right] \quad \text{(oscillator parallel to mirror)}$$

In Figure 23.17 τ/τ_0 (oscillator parallel to mirror) is displayed as a function of distance d; γ is evaluated as

$$\gamma = \frac{\omega r}{c_0} = \frac{\omega 2d}{c_0} = \frac{4\pi d}{\lambda}$$

This expression is valid for an oscillator in vacuum. In the case of Figure 23.17, the oscillator is in a medium with refractive index $\hat{n} = 1.5$. Then

$$\gamma = \frac{4\pi\hat{n}d}{\lambda}$$

The agreement with measured data (Figure 23.17) shows that the simplification (mirror considered as perfect conductor) is reasonable to explain this strange echo effect.

Note 1: lifetime of oscillator

According to Foundation 8.1 the energy E and the power $-\overline{dE/dt}$ emitted by a weakly damped oscillator with damping constant ρ, force constant k_f, elongation ξ, and amplitude ξ_0 are

$$E = \tfrac{1}{2}k_f\xi_0^2 \quad \text{and} \quad -\overline{\left(\frac{dE}{dt}\right)} = \rho\overline{\left(\frac{d\xi}{dt}\right)^2}$$

With

$$\xi = \xi_0 \cos\omega t, \quad \frac{d\xi}{dt} = -\xi_0\omega \sin\omega t, \quad \text{and} \quad \omega = \sqrt{\frac{k_f}{m}}$$

we obtain

$$-\overline{\left(\frac{dE}{dt}\right)} = \rho\xi_0^2 \cdot \omega^2 \overline{\sin^2\omega t} = \frac{1}{2}\rho\xi_0^2 \cdot \frac{k_f}{m}$$

and it follows

$$\tau_0 = -\frac{E}{(dE/dt)} = \frac{\tfrac{1}{2}k_f\xi_0^2}{\tfrac{1}{2}\rho\xi_0^2 k_f/m} = \frac{m}{\rho}$$

Continued on page 870

Continued from page 869

Note 2: Conversion of the Hertz equation

In Foundation 8.5 we have calculated the electric field strength F (components F_a and F_b, see Figure 8.49(a)) at point P in distance r from an oscillating dipole (Hertz equation). The displacement of the oscillator is $\xi = \xi_0 \cos(\omega t + \gamma)$, where γ is the phase shift

$$\gamma = \frac{\omega r}{c_0} = \frac{2\pi r}{\lambda}$$

We restrict our consideration to the case that the angle between the directions of the dipole and r is $0°$ ($\vartheta = 0°$, see Figure 8.49(a)). Then $F_b = 0$ and $F = F_a$, and we can write the expression for the field F in the form

$$F = \frac{1}{4\pi\varepsilon_0} \cdot 2Q\xi_0 \frac{\omega^3}{c_0^3} \left[-\frac{\sin(\omega t - \gamma)}{\gamma^2} + \frac{\cos(\omega t - \gamma)}{\gamma^3} \right]$$

We express the first factor by the lifetime τ_0 (see Foundation 8.5)

$$\tau_0 = \frac{6\pi m \varepsilon_0 c_0^3}{Q^2 \omega^2}$$

$$F = \frac{3m\omega}{\tau_0 Q} \xi_0 \cdot \left[-\frac{\sin(\omega t - \gamma)}{\gamma^2} + \frac{\cos(\omega t - \gamma)}{\gamma^3} \right]$$

By using the relations

$$\sin(\alpha - \beta) = \sin\alpha \cos\beta - \cos\alpha \sin\beta$$
$$\cos(\alpha - \beta) = \cos\alpha \cos\beta + \sin\alpha \sin\beta$$

this expression is converted into

$$F = \frac{3m\omega}{\tau_0 Q} \xi_0 \cdot \left[\sin\omega t \left(-\frac{\cos\gamma}{\gamma^2} + \frac{\sin\gamma}{\gamma^3} \right) + \cos\omega t \left(\frac{\sin\gamma}{\gamma^2} + \frac{\cos\gamma}{\gamma^3} \right) \right]$$

Because of

$$\xi = \xi_0 \cos\omega t \quad \text{and} \quad \frac{d\xi}{dt} = -\xi_0 \omega \sin\omega t$$

we obtain the Hertz equation in the form

$$F = A\xi + B\frac{d\xi}{dt}$$

ξ_0 is the actual amplitude of the oscillator at time t; because of the damping term, ξ_0 decreases exponentially with time.

Foundation 23.4 – Electron Transfer Between π-Electron Systems

In Foundation 4.4 we have seen that an electron, first at proton a, oscillates between protons a and b in distance d. The probability P_b of finding the electron at proton b is

$$P_b = \sin^2 \frac{\varepsilon_2 - \varepsilon_1}{2\hbar} t = \sin^2 \frac{\varepsilon}{\hbar} t \quad \text{with} \quad \varepsilon = \frac{1}{2}(\varepsilon_2 - \varepsilon_1)$$

and ε_1 and ε_2 are the energies of the system in the ground state and the first excited state, respectively.

We generalize this consideration for an electron oscillating between two different molecules a and b. Then we find (Problem 23.1)

$$\varepsilon = -\int \psi_a \frac{V_a + V_b}{2} \psi_b \cdot dx\, dy\, dz$$

ψ_a and ψ_b are the wavefunctions and V_a and V_b are the potential energies of the electron in the field of molecules a and b, respectively; ε is called the *transfer matrix element*. As in Foundation 4.4, ε depends exponentially on the distance d between the two molecules:

$$\varepsilon \sim e^{-\alpha d} \quad \text{with} \quad \alpha = \frac{2\pi}{h}\sqrt{2m_e E_{\text{barrier}}}$$

where E_{barrier} is the energy barrier to be surmounted by the electron.

In this derivation of the transfer probability P_b, the energies E_a and E_b of the electron in molecules a and b have been assumed to be equal. For $\Delta E = E_b - E_a \neq 0$ the result of the corresponding consideration is

$$P_b = \frac{1}{1 + \left(\frac{\Delta E}{2\varepsilon}\right)^2} \cdot \sin^2\left(\sqrt{1 + \left(\frac{\Delta E}{2\varepsilon}\right)^2} \cdot \frac{\varepsilon}{\hbar} t\right)$$

This means that P_b decreases strongly with increasing ΔE. For $\Delta E = 0$ a maximum of P_b is achieved. The situation is similar to two coupled penduli when the first one is excited at $t = 0$. The excitation is transferred back and forth between both penduli only in the case of resonance (when both penduli are identical).

Rate of electron transfer

As an example, we consider the electron transfer from bacteriopheophytin to the quinone in the reaction center of the photosynthetic apparatus:

Continued on page 872

Continued from page 871

[Structure: porphyrin-like macrocycle (a) connected by distance d to a quinone (b)]

a   b

The electron transfer is accompanied by a decrease in Gibbs energy $G(\Delta_r G^\ominus = -45\,\text{kJ mol}^{-1})$. As mentioned above, the maximum electron transfer rate is achieved when the energy of the electron is the same in both systems (resonance). This condition can be realized by a transition from the original (vibrationless) state (wavefunction ψ_a, energy E_a) to a vibronically excited state of the same energy (wavefunction ψ_b, energy $E_b = E_a$). E_a and E_b are fluctuating in a fluctuating medium (protein), thus ΔE is fluctuating as well. The energetic match required for electron transfer is only occasionally reached for a short time t_c (time between fluctuations; t_c is about 100 fs).

This time is short compared with the oscillation period of the electron:

$$P_b = \sin^2 \frac{\varepsilon}{\hbar} t$$

Then it follows:

$$\frac{\varepsilon}{\hbar} = \frac{\pi}{T} \quad \text{or} \quad T = \frac{\pi \hbar}{\varepsilon}$$

A typical value for ε (see below) is $\varepsilon = 1 \times 10^{-4}\,\text{eV} = 2 \times 10^{-23}\,\text{J}$; then $T = 10\,\text{ps}$ and we can simplify the expression for P_b:

$$P_b = \sin^2 \frac{\varepsilon}{\hbar} t \approx \left(\frac{\varepsilon}{\hbar} t\right)^2$$

The probability of energy match is given by the ratio between uncertainty in energy, given by the Heisenberg relation (Chapter 2)

$$\Delta E = \frac{h}{t_c}$$

and the range of fluctuations, E_{fluct},

$$\frac{\Delta E}{E_{\text{fluct}}} = \frac{h}{t_c E_{\text{fluct}}}$$

E_{fluct} is on the order of $0.1\,\text{eV} = 0.16 \times 10^{-19}\,\text{J}$. Then the rate constant k_r for electron transfer is

$$k_r = \begin{array}{c}\text{number}\\\text{of trials}\\\text{per unit time}\end{array} \times \begin{array}{c}\text{probability}\\\text{of energy}\\\text{match}\end{array} \times \begin{array}{c}\text{probability}\\\text{of electron transfer}\\\text{during match}\end{array}$$

Continued on page 873

FOUNDATION 23.4

Continued from page 872

$$k_r = \frac{1}{t_c} \times \frac{h}{t_c E_{\text{fluct}}} \times \left(\frac{\varepsilon}{\hbar} t_c\right)^2 = \frac{2\pi}{\hbar} \varepsilon^2 \frac{1}{E_{\text{fluct}}}$$

Evaluation of ε

According to Foundation 9.2, the wavefunctions of molecules a and b can be treated as products of the electronic part ψ_{el} (depending on the electron coordinates x, y, z) and the vibronic part ψ_{vib} (depending on the nuclear coordinates ξ): $\psi_a = \psi_{a,el} \cdot \psi_{a,vib}$, $\psi_b = \psi_{b,el} \cdot \psi_{b,vib}$ (Born–Oppenheimer approximation). Then for ε we obtain

$$\varepsilon = -\int \psi_{a,el} \cdot \psi_{a,vib} \frac{V_a + V_b}{2} \psi_{b,el} \cdot \psi_{b,vib} \cdot dx\, dy\, dz \cdot d\xi$$

Following the same procedure as for the calculation of transition moments in Foundation 9.2 we obtain

$$\varepsilon = \varepsilon_{\text{electronic}} \cdot \left(\frac{\beta^n}{n!} e^{-\beta}\right)^{1/2}$$

with

$$\varepsilon_{\text{electronic}} = \int \psi_{a,el} \frac{V_a + V_b}{2} \psi_{b,el} \cdot dx\, dy\, dz \quad \text{and} \quad \beta = \frac{E^*}{h\nu_0}$$

n is the quantum state of vibration in molecule b reached after electron transfer; E^* is the vibration energy of all vibrating nuclei.

Example – contribution of vibration to ε

During the electron transfer from bacteriopheophytin to quinone the quinone is reduced to the semiquinone:

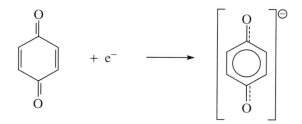

Thus alternating bond lengths are transformed into equal bond lengths, giving rise to bond vibrations indicated by arrows:

Continued on page 874

Continued from page 873

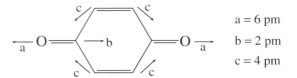

a = 6 pm
b = 2 pm
c = 4 pm

The bond lengths in bacteriopheophytin, due to the large extension of the π system, remain nearly unchanged. We describe all vibrations in the quinone by one frequency ν_0 (C–C valence frequency, $\nu_0 = 4 \times 10^{13}$ s^{-1}) with elongations a, b, and c.

E^* is expressed by the force constants k_f and the elongations ξ_i of all vibrating nuclei (assuming the same mass $m = m_C$)

$$E^* = \sum \frac{1}{2} k_f \xi_i^2 = 2\pi^2 \nu_0^2 m \sum \xi_i^2$$

(k_f expressed by the frequency ν_0). Then

$$\beta = \frac{2\pi^2 \nu_0 m}{h} \sum \xi_i^2$$

With

$$\sum \xi_i^2 = 2\xi_O^2 + 4\xi_{C,1}^2 + 2\xi_{C,2}^2 = 2a + 4c + 2b = 144 \, \text{pm}^2$$

we obtain $\beta = 3.4$; including the donor in the calculation results in $\beta = 4$. The quantum number $n = 3$ is estimated as the next integer to $-\Delta G^\ominus/(h\nu_0) = 2.8$. Then we obtain $\varepsilon = \varepsilon_{\text{electronic}} \times \sqrt{0.5}$.

Evaluation of $\varepsilon_{\text{electronic}}$

In evaluating the expression

$$\varepsilon_{\text{electronic}} = \int \psi_{a,\text{el}} \frac{V_a + V_b}{2} \psi_{b,\text{el}} \cdot dx\,dy\,dz$$

we restrict to the contribution of the shaded antinodes, where the edges of donor and acceptor are closest together, to $\psi_{a,\text{el}}$ and $\psi_{b,\text{el}}$:

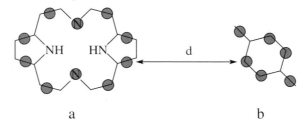

a

b

Continued on page 875

FOUNDATION 23.4

Continued from page 874

We describe the overlap of the wavefunctions by the coefficients c_a and c_b, where c_a^2 and c_b^2 are the probabilities of finding the electron in antinode a and b, respectively (before and after transfer).

Furthermore, the overlap depends on the geometry, which can be described by a geometrical factor κ. Then using the relation derived in Foundation 4.4 for electron tunneling from proton a to proton b, we obtain

$$\varepsilon_{\text{electronic}} = E_{\text{barrier}} \cdot c_a c_b \kappa \cdot e^{-\alpha d} \quad \text{with} \quad \alpha = \frac{2\pi}{h}\sqrt{2m_e E_{\text{barrier}}}$$

In the considered case of electron transfer from bacteriopheophytin to quinone $c_a = 0.303$ and $c_b = 0.28$ (calculated from the π electron densities), $\kappa = 0.5$ (from the geometry available from X-ray experiments) and $d = 0.96$ nm. E_{barrier} is the energy to transfer the electron from the donor into the dielectric medium. This energy is less than the energy E_{ion} required to transfer the electron into the vacuum. In Problem 23.2 we derive

$$E_{\text{barrier}} = E_{\text{ion}} - 3.6\,\text{eV} \cdot \left(1 - \frac{1}{\varepsilon_r}\right)$$

where $\varepsilon_r = 2.5$ is the relative permittivity of a medium of σ electrons (as discussed in Chapter 8). This approximation is appropriate for the rigid portion of the protein matrix between the chromophores. E_{ion} can be evaluated from the reduction potential of the bacteriopheophytin: $E_{\text{ion}} = 3.9\,\text{eV}$. Then we obtain $E_{\text{barrier}} = 2\,\text{eV}$ and $\alpha = 7\,\text{nm}^{-1}$. Thus the result for $\varepsilon_{\text{electronic}}$ is

$$\varepsilon_{\text{electronic}} = 2\,\text{eV} \times 0.303 \times 0.28 \times 0.5 \cdot e^{-7\,\text{nm}^{-1}\cdot d} = 0.085\,\text{eV} \cdot e^{-7\,\text{nm}^{-1}\cdot d}$$

Evaluation of Rate Constant k_r

In summary, we obtain for the rate constant k_r of electron transfer

$$k_r = \frac{2\pi}{\hbar}\varepsilon^2 \cdot \frac{1}{E_{\text{fluct}}} = \frac{2\pi}{\hbar}\varepsilon_{\text{electronic}}^2 \cdot \frac{\beta^n}{n!}e^{-\beta} \cdot \frac{1}{E_{\text{fluct}}}$$

$$= \frac{2\pi}{\hbar} \cdot [E_{\text{barrier}}^2 \cdot (c_a c_b \kappa)^2 \cdot e^{-2\alpha d}] \cdot \frac{\beta^n}{n!}e^{-\beta} \cdot \frac{1}{E_{\text{fluct}}}$$

or

$$k_r = A \cdot e^{-\text{const}\cdot d}$$

with

$$A = \frac{2\pi}{\hbar} \cdot [E_{\text{barrier}}^2 \cdot (c_a c_b \kappa)^2] \cdot \frac{\beta^n}{n!}e^{-\beta} \cdot \frac{1}{E_{\text{fluct}}}, \quad \text{const} = 2\alpha$$

where

$$\alpha = \frac{2\pi}{h}\sqrt{2m_e E_{\text{barrier}}}$$

Continued on page 876

Example

In the present case, with $E_{\text{barrier}} = 2\,\text{eV}$, $c_a c_b \kappa = 0.042$, $\beta = 4$, $n = 3$, and $E_{\text{fluct}} = 0.1\,\text{eV}$ we obtain

$$A = \frac{2\pi}{\hbar}[7.1 \times 10^{-3}\,\text{eV}^2] \cdot \frac{64}{6} \cdot e^{-4} \cdot \frac{1}{0.1\,\text{eV}} = 1.3 \times 10^{14}\,\text{s}^{-1}$$

and

$$\text{const} = 2\frac{2\pi}{h}\sqrt{2m_e E_{\text{barrier}}} = 1.4 \times 10^{10}\,\text{m}^{-1} = 14\,\text{nm}^{-1}$$

Then for $d = 0.96\,\text{nm}$ we obtain

$$k_r = 1.3 \times 10^{14}\,\text{s}^{-1} \cdot e^{-14 \cdot 0.96} = 2 \times 10^8\,\text{s}^{-1}$$

in reasonable agreement with experiment ($k_r = 4 \times 10^9\,\text{s}^{-1}$).

The natural quinone in the reaction center can be replaced by many other quinones. What is the consequence for the corresponding rate constant k_r? By changing the quinone we change $\Delta_r G^\ominus$ for the electron transfer process, thus we change n, the next integer to $\Delta_r G^\ominus/(N_A h\nu_0)$. n affects k_r in the contribution $\beta^n/n!$. With $\beta = 4$ and 5, respectively, and $h\nu_0 = 0.17\,\text{eV}$ we obtain the following values:

$-\Delta_r G^\ominus/\text{kJ mol}^{-1}$	$-\Delta_r G^\ominus/(N_A h\nu_0)$	n	$\dfrac{\beta^n}{n!}$ for $\beta = 4$	for $\beta = 5$
0	0.0	0	1	1
10	0.6	1	4	5
30	1.8	2	8	21
40	2.4	3	11	26
60	3.5	4	11	26
80	4.7	5	5	22

If we consider $\varepsilon_{\text{electronic}}$, β, and E_{fluct} as unchanged, we expect that k_r is increasing and, after a maximum at $n = 3$, slowly decreasing with increasing $-\Delta_r G^\ominus$. Such a behavior is observed in experiment (Figure 23.60). The experimental points are widely spread, essentially because $\varepsilon_{\text{electronic}}$ depends on structural details at the quinones and d is not strictly constant.

From this consideration we conclude that the dependence of k_r on $\Delta_r G^\ominus$ is not dramatic. Thus the electron tunneling rate in rigid σ electron media can be well described by $k_r = 10^{15}\,\text{s}^{-1} \cdot e^{-14\,\text{nm}^{-1} \cdot d}$.

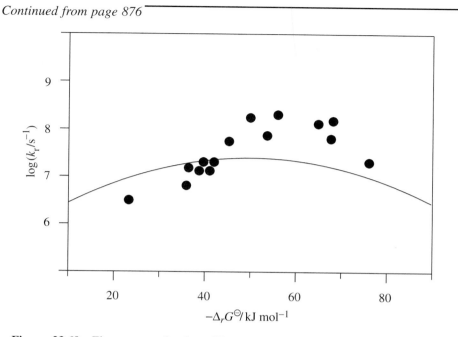

Figure 23.60 Electron transfer from W_2^- to A for various artificial acceptors A differing in their reduction potential. Rate constant k_r versus $-\Delta_r G^\ominus$. Rigid medium: moderate dependence on $-\Delta_r G^\ominus$. Individual properties of acceptors prevail. Dots: experiment (same reference as in Table 23.1). Solid line: theory:

$$\log k_r = \text{const} + \log \frac{\beta^n}{n!}.$$

Foundation 23.5 – Electron Transfer in Soft Medium

In Foundation 23.4 we described the different electron transfer steps in the reaction center by assuming a rigid, just fluctuating medium. This assumption is not always fulfilled.

For example, $\Delta_r G^\ominus$ for the electron transfer from D' to D$^+$ (Figure 23.29) can be varied by changing the environment of D (using mutants with changed amino acids in the surroundings of D). Then the measured rate constants k_r increase much more strongly with increasing $-\Delta_r G^\ominus$ than they would according to Figure 23.60 in Foundation 23.4. This indicates that the assumption of a rigid medium of D' is inadequate in this case.

Actually, the surroundings of D' must be soft. Then reorganization of the medium takes place and the Franck–Condon factor

Continued from page 877

$$\frac{\beta^n}{n!} \cdot e^{-\beta} \cdot \frac{1}{E_{\text{fluct}}}$$

used in the rigid medium model must be replaced by

$$\frac{1}{\sqrt{4\pi kT\lambda}} \cdot \exp\left(\frac{-(\Delta_r G^\ominus/N_A + \lambda)^2}{4\lambda kT}\right)$$

where λ is the reorganization energy. Then we obtain the *Marcus equation*

$$k_r = \frac{2\pi}{\hbar}\varepsilon_{\text{electronic}}^2 \cdot \frac{1}{\sqrt{4\pi kT\lambda}} \cdot \exp\left(\frac{-(\Delta_r G^\ominus/N_A + \lambda)^2}{4\lambda kT}\right)$$

In this equation, λ and $\varepsilon_{\text{electronic}}$ are treated as adjustable parameters; the measured dependence of k_r on $\Delta_r G^\ominus$ is well described with $\lambda = 0.5\,\text{eV}$ and $\varepsilon_{\text{el}} = 2 \times 10^{-5}\,\text{eV}$ (Figure 23.61).

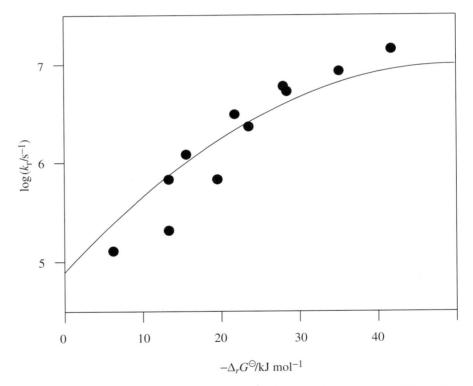

Figure 23.61 Electron transfer from D' to D^+ for various mutants differing in $-\Delta_r G^\ominus$. Rate constant k_r versus $-\Delta_r G^\ominus$. Soft medium. Dots: experiment (W. Lubitz, private communication). Solid line: Marcus equation, with $\epsilon_{\text{electronic}} = 21 \times 10^{-5}\,\text{eV}$, $\lambda = 0.5\,\text{eV}$.

Continued on page 879

Continued from page 878

It is reasonable to assume that D, W_1, W_2, and A are in a rigid environment and D′ is in a soft environment. This idea is supported by the fact that D′ is not fixed; instead, it can diffuse between its location in the reaction center (where it is oxidized) and a pool (where the reduced form is restored by accepting an electron).

24 Origin of Life

Life is the most fascinating form of matter. It is a fundamental problem to understand the origin of living forms and the processes leading to the first manifestations of life.

In the preceding chapters we have considered the thinking scheme of the physical chemist and his or her struggle to understand the behavior of material systems. What is the particularity of living systems in physicochemical terms, how was it possible that life emerged, and what was the leading mechanism? Our discussion of supramolecular devices in Chapter 23 should be helpful in attempting to answer these questions. Living organisms are certainly the most intricate supramolecular entities, and reflections on the fabrication of supramolecular devices should be shedding light on the mechanism leading to the origin of life. The underlying design principles in fabricating supramolecular devices are the *lock-and-key concept* and the *programmed environmental change concept*. Again, the two concepts appear as the guiding principles. The "program" driving the origin of life, is considered to be given by chance at some particular point. A huge variety of situations is given on the early earth and the particular situation in this particular location is assumed to lead to the emergence of life.

Modeling engineering aspects on the origin of life is intimately linked to the particular kind of modeling in physical chemistry, where it is crucial to search for extremely simplified descriptions of given situations in order to grasp the principles. We search for the basic mechanism behind the processes of the origin of life. We do not attempt to trace the historic path but to stimulate thinking in physicochemical terms: discussing a simple model path of consistent steps that can lead to a machine similar to the genetic apparatus of biosystems which should stimulate attempts to find artificial self-organizing complex systems. Ideas on the mechanism of the origin of life are helpful in tracing limits in thinking in physicochemical terms.

24.1 Investigation of Complex Systems

24.1.1 Need for Simplifying Models

Rigorous predictions on the basis of quantum mechanics encounter insurmountable problems, even in very simple cases. In order to proceed, simplifications have to be made. The search for ingenious models is a central task in physical chemistry. These models must be as simple as possible but sufficient for a reasonable description of molecular systems.

The essential idea of the physicochemical endeavor is to reveal unexplored realities and to plan new ones. Most fascinating is the question whether the appearance of life can be understood on the basis of physical chemistry.

24.1.2 Increasing Simplification with Increasing Stages of Complexity

The treatment of increasingly complex systems is subject to further and further simplifications in order not to lose the overview. This problem arose already in the first chapters in the discussion of simple molecules and assemblies where quantum mechanical considerations were replaced by deterministic thinking schemes. For example, describing molecules by their chemical structural formula plays an important role. This extremely useful tool becomes too complicated in describing the network of biochemical metabolism. The description again becomes too complicated when attempting to grasp general aspects on the origin of life. All irrelevant details must be abstracted in order to condense the problem to the fundamentals.

In the course of investigations of processes in biological systems, no behavior has been observed which contradicts the laws of physical chemistry. This observation supports the hypothesis that all material processes in living matter are of a physicochemical nature. But can the appearance of living forms also be understood as a physicochemical phenomenon?

24.2 Can Life Emerge by Physicochemical Processes?

24.2.1 Bioevolution as Process of Learning

Multiplication, variation, selection

According to the findings of molecular biology, every living organism carries a prescription to build copies. The message is carried in the form of a very specific sequence of four types of building blocks (monomers) arranged to form a molecular chain. According to that message the organism is built by using a very particular information translation mechanism and a replication mechanism for the message to be copied (Figure 24.1).

Occasionally an error due to a mismatch during the copying process can occur which does not affect the power of reproduction; – that is, a wrong monomer is introduced at some point on the chain. This is an extremely rare event and the biological copying process is extremely exact. When, however, an error does occur in copying the genetic message, the individual built by translating the changed message will become somewhat different. In most cases this is disadvantageous because the organism is a complex functional structure and like in a clockwork, minute changes can have catastrophic consequences.

Consequently, the changed individual is usually unable to survive. This yields an important error filter: individuals with a changed genetic message are eliminated. Rarely, a copying error results in improved survival chances for the changed individual. Thus, individuals will survive which are better adjusted to the environment. This adaptation and adjustment process is continued through generations. The resulting biological evolution process is a process of learning that requires many generations; the ultimate goal of evolution is to acquire the message to survive in an increasingly complex environment. Further and further adaptation to the environment ensues.

With the appearance of the first system that carries a message, a property of matter is manifested which beforehand did not even appear allusively.

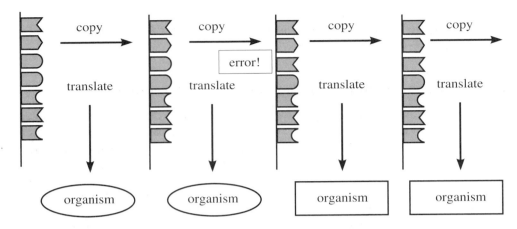

Figure 24.1 Bioevolution as a process of learning. The genetic message is copied and translated. The translation product is the organism which interacts with the environment. In the multiplication process a mismatch can lead to a copying error. The message is changed, and a correspondingly changed translation product is formed.

24.2.2 Model Case for the Learning Mechanism

Drawing structures adapting to a tub

For simplicity we consider drawn structures adapting to a tub symbolizing the environment. This should focus on the essence of the learning mechanism. In the beginning, drawn structures are obtained in the following way: we proceed from left to right in equal steps along a zigzag path, obtained by throwing a coin that determines at each step whether the next step should be up or down (Figure 24.2). Now half of the structures obtained in this manner are eliminated. Which structures remain is decided by chance; however, the better the structure is adapted to the tub, the greater, in our game, is its chance to survive. Therefore structure 4 in Figure 24.2 has a priori a better survival chance. In the next phase we form copies of all surviving structures, but the copies are not always correct. In each step there is a small probability that a copying error will occur – for example, a step up instead of a step down or vice versa (Figure 24.3) so that the altered chain is shifted up (in the first example) or down (in the second example). Or the chain occasionally becomes longer by a link (third example) or shorter. Afterwards, the selection continues according to the previous recipe. Doubling and selection of structures continues for many cycles. Drawing structures which are close to the given tub are selected because they disappear less frequently than others (Figure 24.4). If the shape of the tub is changed and the process of multiplication and selection continued, then after a sufficient number of generations we will find mainly structures which are closely adapted to the new form of the tub (Figure 24.5).

This example illustrates a mechanism in which structures are created by random decisions, and in a process of multiplication, variation, and selection the structures adapt to the tub. The system behaves as if it recognizes the shape of the tub and adapts to that shape. We call this process of creation of forms by variation and selection of fit forms "learning" – that is, learning in the sense of finding out what is fit, not in the sense of memorizing.

24.2 CAN LIFE EMERGE BY PHYSICOCHEMICAL PROCESSES?

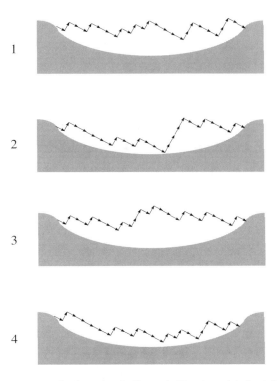

Figure 24.2 Various structures 1–4 randomly formed. Structure 4 is best fitted to the given form of a tub; thus its survival chances are increased in our game. The a priori probability for survival of structure 4 is higher than for structures 1–3.

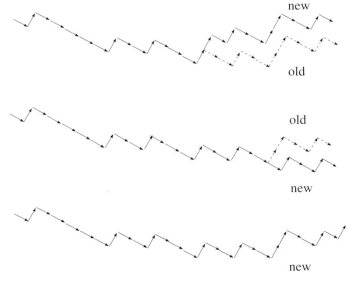

Figure 24.3 Various copying errors; they occur with low probability (e.g., one error in 30 copying steps).

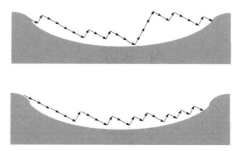

Figure 24.4 Adaptation of the structure to the given form of a tub after many generations.

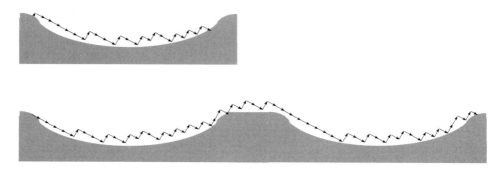

Figure 24.5 Adaptation to a changing form of the tub.

Optimum error frequency

From this example (Figures 24.2–24.5) it is clear that in an adapting (learning) system the frequency of errors should be neither too high nor too low, in order to allow adaptation. When errors occur too often the system forgets what it has learned before. When errors occur too seldom, the system is not able to adapt. It does not adjust fast enough to changes in the tub form. The optimum frequency in the given example is one error in about every 30th step in the copying process. Then there is a 97% probability in each step that no error will occur. Thus enough error-free copies are present in each cyle, so that the information is preserved in the population, but also enough errors to ensure adaptation. Accordingly, in longer chains the copying precision per chain element has to be better.

A simple quantitative relation of this sort between the amount of information and the optimal error frequency should be valid for any learning system. Therefore it can be applied to the process of evolution of living organisms as well. Evolution is a process of learning. The evolving system behaves as if its know-how to survive in an increasingly complex environment would gradually grow in the course of many generations. In this case the amount of transferred genetic information is very large; therefore, errors in one step are extremely seldom. Therefore, there are no principal reasons opposing the idea that sophisticated organisms were formed from first learning systems. On the one hand increasingly complex forms developed (higher forms of life), and on the other hand forms which are robust due to their relative simplicity (viruses and bacteria).

24.2 CAN LIFE EMERGE BY PHYSICOCHEMICAL PROCESSES?

24.2.3 Modeling the Emergence of the Genetic Apparatus

Basic questions

The crucial question concerning the origin of life is whether learning systems can emerge from dead matter (Section 24.2.1). The problem of how life originated has a chemical aspect: to understand how the building blocks were formed on the early planet and how their replication started. Also it has a biological aspect: how the evolutionary explosion of life took place after the first simple living cells were present.

The origin of life also has a molecular engineering aspect to be emphasized in the present context (Figure 24.6): to understand how the biomachinery developed from the given building blocks – that is:

- how it was possible for evolution to start and proceed;
- by what basic mechanism the genetic apparatus and the genetic code can be formed;
- how the particular molecular building blocks must be constructed independent of their specific chemical composition;
- how the environment must be designed to promote the process.

Prebiotic chemistry has given good reasons to assume that the building blocks of early life – the amino acids glycine, alanine, valine, and glutamic acid and the nucleic bases guanine, cytosine, adenine, uracil, and ribose – were present on a prebiotic earth and that short strands of nucleic acids or similar compounds were present at particular locations and were able to replicate. This assumption is not unreasonable in view of the recent progress in attempts to replicate short nucleic acid strands. The basic mechanisms of bioevolution are known and the phenomenon is well established. But there is no information on the stages between the prebiotic phase and the phase where simple living cells were present possessing the incredibly complex and intricate genetic apparatus (Box 24.1). There are no intermediates indicating how such systems might have evolved.

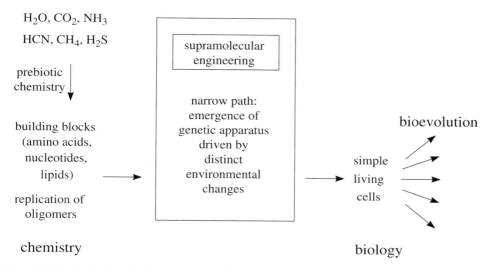

Figure 24.6 Prebiotic chemistry, supramolecular engineering (emergence of genetic apparatus), bioevolution.

Box 24.1 – Genetic Apparatus

The message coded in the DNA strand as a sequence of four kinds of monomers (Figure 24.1) is multiplied by replication based on complementary pairing; it is translated into proteins – that is, linear sequences of 20 kinds of amino acids. This step is accomplished by first transcribing the message from the

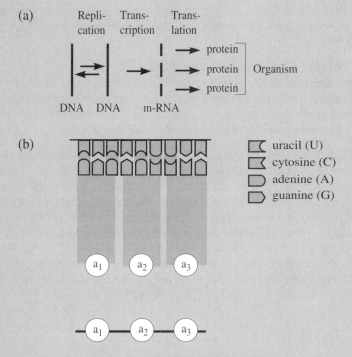

Figure 24.7 Simplified version of present-day protein synthesis. (a) Polypeptide bonds are formed from amino acids. The proteins form the organism and take part in the mechanism of replication, transcription, and translation. (b) The synthesis is guided by a messenger nucleic acid strand, to which adapter molecules carrying the amino acids a_1, a_2, and a_3 are attached.

DNA to the messenger ribonucleic acid (mRNA) (Figure 24.7). The translation into proteins takes place by means of *adapter* or *transfer ribonucleic acid* (tRNA) molecules. For every amino acid a_1, a_2, a_3, ..., at least one specific molecule of tRNA exists, to which the amino acid can become attached; the tRNA also carries a specific *anticodon* nucleotide triplet. Adapter molecules can, in turn, couple by complementary base pairing to a nucleic acid strand that contains the message for the protein in question (*messenger ribonucleic acid*). Attachment of the base triplet of an adapter molecule to a base triplet on a messenger nucleic acid strand can only occur if corresponding base pairs

Continued on page 887

> Continued from page 886
>
> are *complementary*. The four bases G (guanine), C (cytosine), A (adenine), and U (uracil) are pairwise complementary; bases G and C can readily form three hydrogen bonds with each other, while bases A and U can readily form two hydrogen bonds. Put more simply, if there is a C on the messenger strand, there must be a G in the corresponding place on the adapter molecule, and so on.
>
> For example, the first codon triplet UCU on the messenger strand (read in the 5'–3' direction) must correspond to an adapter molecule with the anticodon triplet AGA (read in the 3'–5' direction), which carries, for example, the amino acid a_1 (Figure 24.7). Amino acids are therefore linked into protein chains with an amino acid sequence that is determined by the sequence in the messenger. These proteins then form the functional structure of the organism. Proteins constitute essential components in all parts of the genetic apparatus.

The task of shedding light on how life emerged is quite different from the familiar tasks in physics, chemistry, and biology where the pertinent questions are approached by solving many special, well-defined problems, each being checked by comparison with facts.

The difficult situation in the origin of life is similar to the situation in attempting to understand the evolution of the universe. Here also the evolutionary product is known – our universe – but we have no direct evidence on how it emerged.

Evolution of the universe and evolution of life: the big bang and the tiny bang

In attempting to understand the evolution of the universe, a long sequence of events has been proposed, namely, a logic chain of hypothetical processes based on the present knowledge in particle physics. This long chain leads to a most astonishing result at the end of the thought experiment, a possible explanation of basic facts in astrophysics which cannot be rationalized otherwise, such as the cosmic background radiation and the relation between redshift and distance of stars.

Now, let us try to approach the problem of the origin of life in the same manner: to begin with some prebiotically reasonable conditions and reasonable properties of the building blocks and to see what kind of processes should take place according to what is expected from physics and chemistry. Can we rationalize the emergence of an intricate machinery of the kind of the genetic apparatus by inventing a long sequence of logically related physicochemical processes? Can we find a rational basis for the relation between amino acids and corresponding codes which appears at first quite ambiguous? The essence is not finding the detailed historic path but finding any reasonable path leading to a rationalization of that marvelous phenomenon.

The kind of theoretical modeling required here is familiar in physical chemistry: to start with reasonable simplifying assumptions and to investigate carefully the logical consequences. Thus, modeling basic mechanisms in the evolution of the genetic machinery should be considered as a useful exercise in physical chemistry, focussing on principles and stimulating future attempts in supramolecular engineering.

The crucial point is to find appropriate conditions leading to the evolution of a lifelike system at the end of a long sequence of consistent steps, each being simple. Einstein's advice must always be kept in mind: a theory should be as simple as possible but not simpler.

Important aspects are:

A spatial and temporal structure is present in the very special locations. This structure imposes a cycle of multiplication of simple strands, variation, and selection. Energy-rich building blocks are available. With this in mind, there are no difficulties from the point of view of thermodynamics to rationalize the self-organization of matter.

Microdiversity of the environment serves as an evolutionary gradient. Neighborhood regions with slightly different structural properties cannot be populated in the beginning but by casually occurring slightly improved forms. More and more complex forms evolve by colonizing further and further regions. In this way, you avoid difficulties in rationalizing the evolution of increasingly complex forms.

Thus, physical objects are viewed to come into being that behave as if they have a purpose, an intention, and an aim. This begins with the first system that evolves by multiplication, variation, and selection. Let us call physical objects with this fundamentally new property (not present in any ancestral form in a prebiotic universe) *living*. The appearance of such objects and the explosion of life is a tiny bang in the universe. This should emphasize the fundamental nature of the event leading to life.

Strategy: search for sequences of small steps

In our search for reasonable steps that lead to lifelike systems, we first focus on engineering aspects and then reflect on the chemical realization of the proposed constructional principles.

Probability considerations in the search for a meaningful model are illustrated for the first evolutionary step, the emergence of a replicating strand. We expect (see example, below) that a short strand serving as matrix for replication will appear among a reasonable number of spontaneously formed strands. In contrast, a longer replicable strand will not appear (the probability of its appearance is almost zero) even among a huge number of spontaneously formed strands. However, long strands can be easily obtained by linking small replicable strands when such strands have accumulated.

This shows that using a special model one is able to understand, in principle, the formation of a complex system in many small evolutionary steps under suitable plausible conditions, whereby each step occurs with a probability close to 1. With large steps there is only a minute probability that the same system can be formed.

Example – replicable strand made up of 60 monomers

Let us consider a solution of monomers that can interlink forming strands. A strand which is able of replicate is assumed to be formed only when the monomers are connected in distinct manner. The probability that a monomer will attach to a strand accordingly is, say, 1/100. The probability of finding a strand made of 60 monomers capable of replicating is $(1/100)^{60} = 10^{-120}$. Out of 10^{120} strands, only one is a correct sample. This probability is so small that if we could fill the whole universe with such strands (each strand would require $10^{-20}\,\text{cm}^3$), it would still be extremely improbable to have a correct sample among them.

24.2 CAN LIFE EMERGE BY PHYSICOCHEMICAL PROCESSES?

The probability of obtaining a correct strand of 10 monomers under the same conditions is $(1/100)^{10} = 10^{-20}$. In 1 millimole of substance (6×10^{20} strands) we would find, on the average, six correct strands. Then we should find at least one correct strand with a probability very close to 1. The corresponding situation applies for throwing dice, when we inquire about the probability of throwing a "6" 26 consecutive times. The probability is $(1/6)^{26} = 10^{-20}$. Playing with 6×10^{20} dice at the same time would lead to an average of six dice showing a "6" each time after throwing.

Now let the special strand with 10 monomers multiply by replication. Some strands fuse under plausible conditions, forming strands of 60 monomers that are able to replicate. In this way the result, which would have been reached only with minute probability spontaneously, is obtained with a probability close to 1.

Note that combinatorial chemistry, a fascinating new method of synthesis, is based on this principle of selecting a distinct combination out of a tremendous number of possibilities.

This example illustrates the importance of viewing small steps in proposing reasonable thought experiments.

Therefore, attempts to model the origin of life should start with short strands. It is crucial to consider many small steps in modeling a reasonable path; trying to reach the same goal in larger steps leads into basic difficulties. This indicates the strategy: search for a gapless sequence of small prebiotically reasonable physicochemical model steps and see to what the thought experiment leads when setting appropriate initial conditions and boundary conditions. Can the clearly defined goal be reached, a genetic apparatus?

Note that we are searching for basic principles; the actual processes in the origin of life must have been much more complicated.

Sketching the path

We can only sketch this path. We first focus on the logic frame of the process – that is, the search for the properties of the components and the conditions needed to drive self-assembly. The goal is finding conditions leading to a simplest possible path of obvious conceptual steps that suggest themselves.

The basic problem occurring throughout the evolutionary process is how to save the recipe to construct the machinery reached so far – that is, how to copy the genetic message with the required accuracy. The systems of increasing complexity and sophistication emerging in the course of evolution require increasing accuracy in the copying process. Again and again a new mechanism is needed that allows a better suppression of copying errors than the presently available mechanism. Obviously, the occurrence of the new mechanism leads to an evolutionary boom followed by a stagnation until a new and better solution of the error suppression problem appears.

Engineering features

Attempting to grasp the essence of this process is the purpose of a model of the origin of life. We try to illuminate this evolutionary searching procedure by starting with simple assumptions and investigating what changes in conditions might lead to further and further evolutionary steps; the underlying engineering principles are the lock-and-key concept and the programmed environmental change concept.

At first we consider simple monomers of one kind with the property of forming strands that copy under appropriate conditions by lateral bonding monomers to a template strand, and by linking adjacent monomers. The system has no evolutionary power since it cannot diversify. Then we ascribe an additional property to the monomers and to the environment. We investigate how far the evolutionary process can proceed. After that we add another property, and so on (Box 24.2, options a–g).

This analysis tells us that two kinds of monomers are needed to avoid an early evolutionary barrier. They must be complementary to each other and form strands that interlock in an antiparallel fashion. Only in this case can an aggregate serving as a translation device be constructed.

Box 24.2 – Search for Conditions Driving the Emergence of a Genetic Translation Device

Properties of monomers required for evolution Lock-and-key concept	Features of environment required for evolution Programmed environmental change concept	Result of evolution process
(a) One kind of monomer	Time: periodic (e.g. day/night) driving force space: compartmentalized (e.g. porous rock) to keep strands in a milieu appropriate for copying	Short strands form and multiply by copying
(b) See case (a)	Additionally to case (a): microheterogeneity Region with larger pores colonized by longer strands: extension of populated areas by increasing strand length	Longer strands formed by fusion evolve Length limited by occasional mismatch preventing links between monomers
(c) Two kinds of monomers: complementary	See case (b)	Folding Replication structures; most closely packed form (hairpin) only possible with monomers of type 2

Continued on page 891

24.2 CAN LIFE EMERGE BY PHYSICOCHEMICAL PROCESSES?

> Continued from page 890

(d) Additionally to case (c): hairpins formed from type 2 monomers bind each other weakly laterally	Additionally to cases (b) and (c): Larger porous region (colonized by aggregate-forming strands)	Aggregates of hairpins Device to preserve know-how to make hairpins (erroneous copies rejected in aggregate formation)
(e) Additionally to case (d): hairpins bind to assembler strand	Additionally to case (d): Larger porous regions (colonized by hairpins-assembler-devices (HA-devices))	HA-devices Aggregate compact and resistant (hairpins better interlocked by binding to assembler)
(f) Additionally to case (e): Monomers of a new variety (a-monomers). They bind to open end of hairpin and make links to each other	As in case (e) a-oligomers act as impediments allowing to colonize less restricted regions	HA-device catalyzes forming a-oligomers: potential of translation as by-product
(g) Additionally to case (f): Special a-oligomer intervenes in replication process and lowers probability of mismatch in replication	See case (f)	Translation device and special a-oligomer preserved Breakthrough of translation apparatus

The model is considered in Foundation 24.1 in more detail, still restricting our discussion to demands dictated by engineering considerations; that is, we do not take into account specific chemical aspects. We show that the emergence of a translation apparatus appears as a by-product in the evolution of an aggregate that emerged by its power to act as an error filter.

The step initiating this development is the formation of an aggregate of precisely interlocking molecules accompanied by the disappearance of molecules that do not match this condition. This step is of fundamental importance: a main difficulty in the search for reasonable model pathways is the huge variety of harmful side reactions accompanying the reaction in focus. They lead to a dead end. Besides the highly specific aggregate-filtering process, these side reactions are assumed to play no appreciable role. We neglect them, i.e., we apply Occams' razor (preferring the simplest of competing theories to the more complex ones).

Then, our main task in the search for a possible pathway leading to the emergence of a genetic apparatus, is to construct environmental situations that lead to the formation of increasingly sophisticated aggregates. The formation of aggregates of precisely interlocking molecules appears to be natures' trick to overcome side reactions.

The scheme in Figure 24.8 is thought to give the fundamental changes in the system structure and their logic connections leading to a pressure to colonize more and more different regions. This requires an increasing complexity ultimately connected with the increasing populated area.

Computer simulation of thought experiment

The thought experiment considered in Foundation 24.1 emphasizing this basic structure of the evolutionary process is supported by a computer simulation carried out as follows:

Storing fitness parameters. Many devices are taken into account as possibilities that may or may not be realized at some evolutionary level. Each device (defined by the particular sequences of the strands involved) is represented by values of parameters that describe its survival chance (*fitness parameters*); these parameters are stored in a dictionary relating fitness and sequence.

Applying an evolutionary search program. The following processes are simulated in the computer. Isolated strands are exposed to a periodically changing environment. Each cycle consists of three phases. (i) *Multiplication phase*: the strands are driven to replicate with occasional variation by copying-error due to mismatch. (ii) *Construction phase*: aggregates of strands are made by random choices and random trials to interlock the strands. (iii) *Selection phase*: the fitness parameters of all aggregates thus obtained are taken from the dictionary.

Then a random selection of devices biased by fitness parameters takes place. The surviving aggregates dissociate into isolated strands. Increasingly complex aggregates constituting devices of increasing survival capability evolve in the course of very many cycles.

It is fascinating to observe the computer automatically developing, in the course of 4000 cycles, a genetic apparatus and a genetic 4-letter code. This process, however, occurs only in a small fraction of runs. With other random seeds the evolution becomes stuck. This supports the idea that many starts of an evolutionary process were needed until the breakthrough leading to the explosion of life occurred. The computer simulation is helpful in overcoming the fundamental difficulty to see any possible and feasible mechanism of how the genetic machinery could have emerged, but the simplicity of the simulation, in contrast to the prebiotic reality, should be kept in mind.

Chemical features

Special chemical features are considered in Foundation 24.2. It is shown that the genetic code can be rationalized by introducing amino acids in the approximate order of their

24.2 CAN LIFE EMERGE BY PHYSICOCHEMICAL PROCESSES?

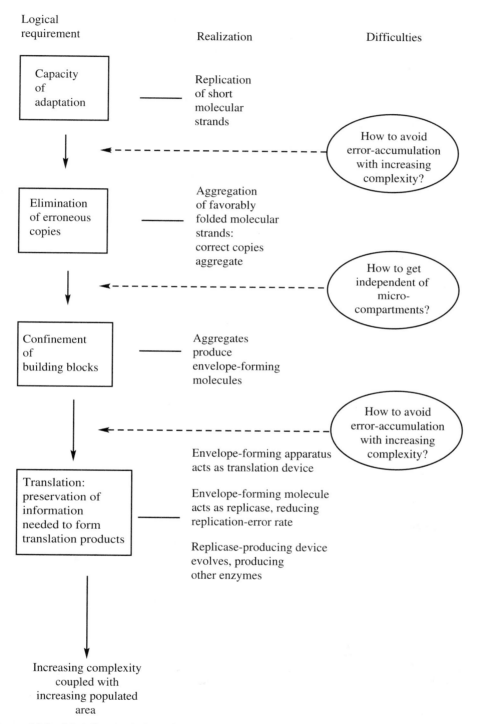

Figure 24.8 Modeling evolution of early life: logical requirements, their realization, and barriers to overcome.

availability. This reconstruction of the modern code supports the model approach considered. It remains a most challenging problem how the conditions resulting from these engineering considerations can be met either with the prebiotically available molecules or in future attempts in supramolecular engineering.

The basic structure of the genetic apparatus according to the present analysis had to evolve in three unavoidable steps. (1) RNA replication. (2) RNA → protein translation: a system unable to develop into systems of increasing complexity and sophistication. (3) DNA replication, DNA → RNA transcription, RNA → protein translation: a system able to form more and more complex systems by integrating more and more subsystems. At this point the confined evolutionary path opens up and a divergent evolution in many different directions takes place: the explosive evolution of life.

The proposed process is not tied to specific terrestrial realization. Important features do not depend on specific monomers, although, of course, quite definite requirements must be met by suitable monomers; among those important features are antiparallel template-driven replication, reading frame, and triplet attachment of hairpins to assembler strand. Basic is the fact that a huge variety of situations is given on the early planet, and at many locations a huge number of spontaneously formed oligomers is present. At some location the condition to replicate is fulfilled which may initiate the emergence of life.

24.2.4 Later Evolutionary Steps: Emergence of an Eye With Lens

Modeling of later evolutionary steps is a challenge to physicochemical ways of thinking. So far we have used a simple way of modeling the evolution of an intricate machine. Can this method be used to give some insight into later evolutionary processes?

The evolution of an eye with a lens producing an image on the retina is a particularly intriguing question. The given model starts with a flat batch of light-sensitive cells sandwiched between a transparent protective layer and a layer of a dark pigment (Figure 24.9, stage 1). It is assumed that the bending of the layer, its thickness and the local dependence of the refractive index of the transparent layer are slightly modified by

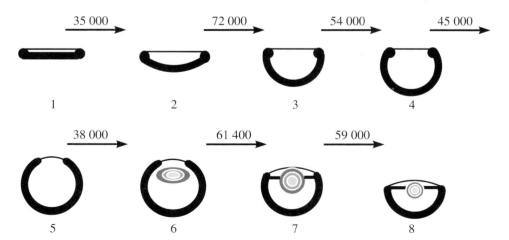

Figure 24.9 Computer simulation of the evolution of an eye with lens. Estimated number of generations for each step is indicated. (D. E. Nilson, S. Palger, Proc. Roy. Soc. Lond. B 256, 53 (1994)).

casual events. Individuals with the better sensitivity for light are assumed to have the better chance to survive. What happens, as a consequence of these assumptions, in the course of many generations?

Figure 24.9 indicates the development taking place according to a computer simulation. In stages 2 to 5 the photoreceptor layer and the pigment layer invaginate to form a hemisphere. The protective layer forms a vitreous body which fills the cavity. In stages 6 to 8 a graded-index lens appears by a local increase in refractive index. The lens shrinks and a flat iris gradually forms. In this way a sharply focused system develops.

The total number of generations to proceed from stage 1 to stage 8 is in the order of 400 000; this corresponds to a time period of less than 10^6 years, if we estimate one year per generation. This is small compared to the time actually available in bioevolution for the emergence of an eye with lens.

24.3 General Aspects of Life

24.3.1 Information and Knowledge

According to Section 24.2.3, the basic mechanism leading to lifelike processes is seen in the occurrence of entities which we may call "individuals" (which are at first simply replicating strands) that interact with the "environment". An individual either survives the "selection phase" or disappears; only surviving individuals have a chance to produce copies in the following "multiplication phase". The process is repeated many times, and new forms of individuals gradually occur due to copying errors in the replication process. Forms survive and evolve that are fit. In this way individuals of increasing intricacy are created. The genetic information of an individual (the message to produce copies in the given environment) and the individual's knowledge (to behave as if the individual would know how to survive and multiply in a given environment) increase accordingly. Genetic information and knowledge describe the quantity and quality of the genetic message, respectively.

24.3.2 Processing Information and Genesis of Information and Knowledge

Computer and biosystem

A computer, a device for processing information, can be described, in principle, as a system of switches; each switch is represented by a particle in a modulated potential, changing periodically, thus driving the switch through a switching phase, a storage phase, and a reset phase (Figure 24.10). In the switching phase the switch is set by the directing field of other switches that are in the storage phase. Accordingly, in its storage phase the switch directs other switches that are in the switching phase. In the reset phase the switch is made ready for its use in the next period. Each switch, in the storage phase, is said to carry one bit of information. (R. Landauer, Irreversibility and Heat Generation in the Computing Process, IBM J. Res. Dev. 5, 183 (1961).

It is fascinating to see that this mechanism is equally basic to biosystems where each switch is represented by a complementary pair of nucleotides. The replication process represents the switching phase. The switch is set by the directing field of the template strand. This field determines which of the two complementary nucleotides is incorporated

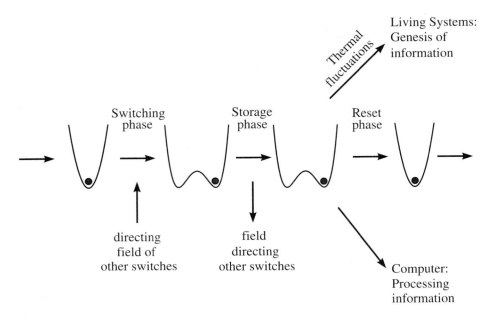

Figure 24.10 Switching processes in a computer and in biosystem.

in the growing daughter strand. In the storage phase the considered nucleotide, now incorporated in a template strand, directs further processing. The degradation of the strand into monomers represents the reset phase. (H. Kuhn, Origin of Life and Physics: Diversified Microstructure – Inducement to form Information Carrying and Knowledge-Accumulating Systems, IBM J. Res. Dev. 32, 37 (1988)).

In spite of this accordance, there is a fundamental difference between the computer and the living system. In the computer the probability of a copying error (particle jumping from one into another minimum by thermal noise) is so small that the probability of an error during the complete computation is still very small. In the living system the rate of errors (incorporation of a nucleotide that is not complementary to the nucleotide in the template) is small but finite: it cannot be too small or too large. The rate must allow the preservation of the accumulated information on the one hand and the production of variations by occasional copying errors on the other hand (see Section 24.2.2).

In the computer, information is handled in a deterministic process. In the biosystem, information is created by variation and the quality of the created information, the knowledge, grows by selection. Knowledge increases in each selection step in the course of evolution. Thus knowledge may be measured by the total information to be thrown away in all selection processes leading to a given evolutionary stage.

Minimum energy dissipation in information processing

A general question concerning information processing devices is: what is the minimum energy dissipation? This is best seen when considering biosystems. The replication process, in principle, can be performed reversibly. But entropy is ultimately produced in the reset process, the degradation of the strands into monomers. The information represented by the

24.3 GENERAL ASPECTS OF LIFE

distinct sequence of the monomers in the strand is destroyed. We consider, for simplicity, a strand with only two kinds of monomers, say, G and C. Then the minimum entropy produced is the entropy of mixing (Section 14.5.2):

$$\Delta S = k(N_G + N_C) \cdot \ln 2$$

In principle, a reversible degradation would be possible by binding the strand to a complementary strand and reversing the replication process. A computer based on this idea can be imagined that works in reversible steps throughout the computation; it stores the result and processes backward in reversible steps (reversible computing). (C. H. Bennet, Logical Reversibility in Computing, IBM J. Res. Dev. 17, 525 (1973) R. Landauer, Nature 335, 779 (1988)).

Creating information, in contrast to processing information in this manner, is ultimately irreversible.

Minimum energy dissipation in a measurement and the Maxwell demon

In any measurement the measuring instrument is exposed to some directing field produced by the object to be investigated. The measuring process, in principle, can be represented by a switch set by this directing field in the switching phase. The bit of information is processed in the storage phase, and in the reset phase the measuring instrument is made ready for the next measurement. Again, the measuring process can be reversible, but entropy ($k \cdot \ln 2$ per bit, when the two possible results are a priori equally probable) is ultimately produced in the reset phase. (C. H. Bennett, Maxwell Demon, Szilard Engine, and the Second Law, Scient. Amer. 255, 107 (1987)).

This consideration is important in understanding the famous Maxwell demon Paradox. The molecules in a gas, in the beginning, are equally distributed between two equal volumes separated by a wall with a door. The demon when he sees a molecule approaching from the left quickly opens the door and quickly closes it when the molecule has passed (Figure 24.11(a)). In this way more and more gas molecules accumulate in the volume on the right, in seeming disagreement with the principle of entropy increase.

The demon must perform a measurement. As mentioned, this can be done without entropy production. However, in the reset phase, when getting the measuring device ready for the next measurement, entropy is ultimately produced. It can be shown that the minimum entropy produced in this way is larger than the entropy decrease by accumulating the gas molecules on the right. No entropy decrease takes place in the isolated system including the demon.

This is illustrated in Figure 24.11(b). We consider ideal gases in two containers of volume V. Both containers are in contact with a temperature bath of temperature T. N_1 particles are in the left container, and $N_2 = 2N_1$ particles are in the right container, where N_1 is a large number.

Additionally, a small compartment of volume v is present which is connected with the two volumes by doors a and b. In the beginning door a is open and door b is closed. v is chosen such that it is much smaller than V/N_1 or V/N_2, the volume per molecule in the left or right container; then volume v is usually empty.

The Maxwell Demon (Figure 24.11(a)) is replaced by a device that closes door a and then performs a measurement deciding whether volume v is empty or not (Figure 22.11(b)). If volume v is empty door a is opened again. The process (opening and closing door a and performing a measurement) is repeated until v is found to be not empty. Then door b

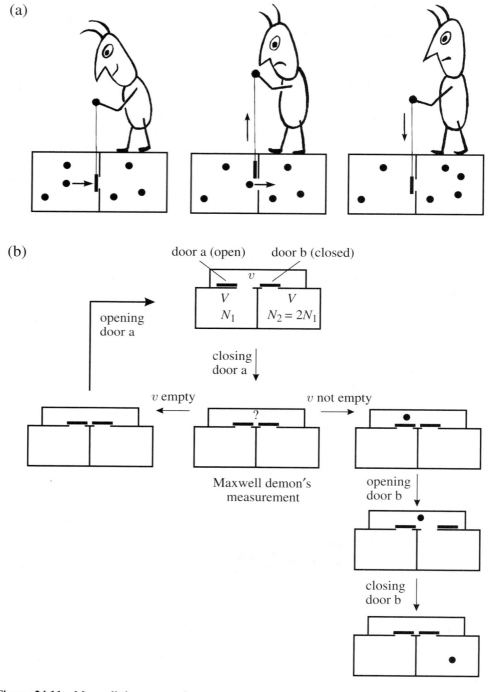

Figure 24.11 Maxwell demon paradox.

24.3 GENERAL ASPECTS OF LIFE

is opened and later closed again. Then in most cases the particle has entered the container on the right because of $v \ll V/N_2$. Door a is then opened and the cycle is repeated.

This means the demon (which is watching continuously until he sees a particle coming and then opens the door in Figure 24.11(a)) is replaced by a device that is testing continuously volume v for being empty or not empty. As soon as v is found to be not empty door b is opened and the particle enters the container on the right.

The decrease of the entropy of the system, when shifting a particle from the left to the right (corresponding to an isothermal compression of the ideal gas) is

$$-\Delta S_{\text{system}} = k \ln \frac{N_2}{N_1} = k \cdot \ln 2$$

We call n the number of measurements that must be taken, on the average, until a particle is shifted from the left to the right volume. Then the heat produced in the surroundings by n times resetting the measuring instrument is

$$q_{\text{surroundings}} = nkT \cdot \ln 2$$

In the ideal gas approximation no heat is produced when the particle moves from v to the right container. Therefore we obtain

$$\Delta S_{\text{suroundings}} = \frac{q_{\text{surroundings}}}{T} = nk \cdot \ln 2$$

The ratio

$$r = \frac{-\Delta S_{\text{system}}}{\Delta S_{\text{suroundings}}} = \frac{k \cdot \ln 2}{nk \cdot \ln 2} = \frac{1}{n}$$

increases with decreasing n. In the present case that v is much smaller than V/N_1, we obtain $n \gg 1$ and therefore $r \ll 1$. This means that the entropy decrease in the system is much smaller than the entropy increase in the surroundings.

In the general case that v is comparable with V/N_1, the ratio r is a function of v and V/N_1. In Foundation 24.3 it is shown that r reaches a maximum of $r = 0.187$ at $v/(V/N_1) = 0.183$ (see Figure 24.23).

24.3.3 Limits of Physicochemical Ways of Thinking

Questions concerning the limits of physicochemical ways of thinking lead to questions concerning processes in the central nervous system: can we identify a priori contradictions to the idea that all processes in the brain are in accord with the laws of physical chemistry?

In order to consider these limit questions, we imagine a simple physical model replacing the real complex system. We hope that the model will reproduce some essential features of the real system, in spite of its simplicity.

Device that develops an internal model of its "world"

The model to be considered here is an apparatus made of an observation device (that gives data that may be described by points in a plane), a device for storing these data (memory), and a device which handles the data by a process analogous to the biological evolution process (Figures 24.12 and 24.13): The device produces structures according to

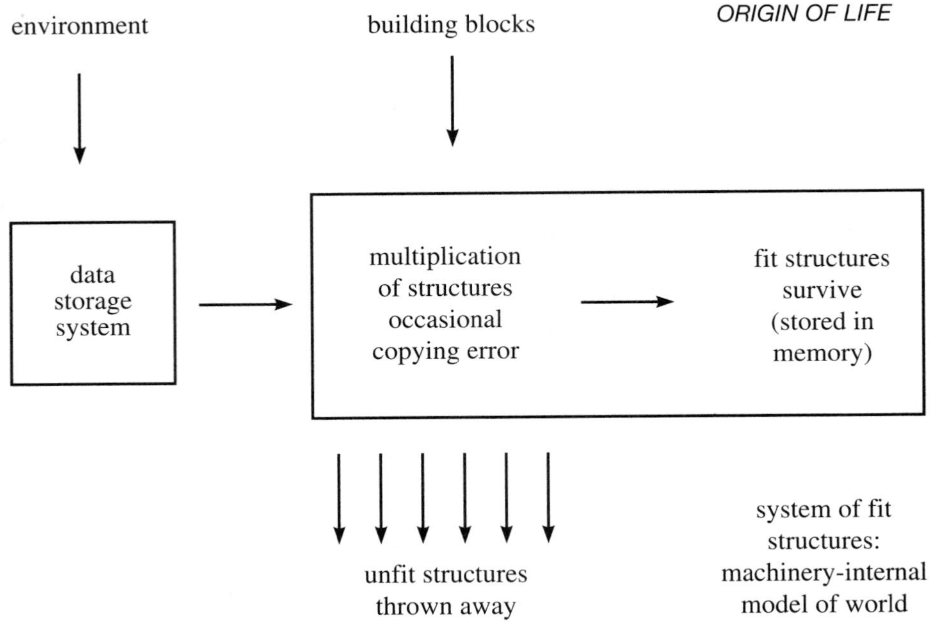

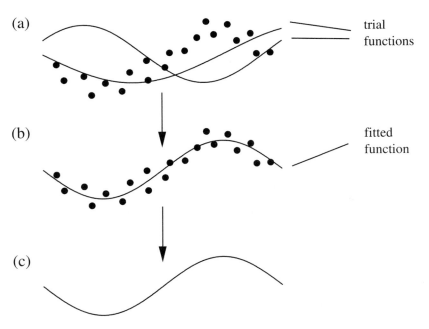

Figure 24.12 Model of a learning system.

Figure 24.13 Fitting of trial functions to stored observation data. (a) Observed data (dots) and arbitrary trial functions. (b) Fitted trial function. (c) Discharge of the memory by deleting observation data, but storing the fitted trial function.

24.3 GENERAL ASPECTS OF LIFE

a given recipe and selects the structure that happens to be reasonably accommodated to a multitude of stored data. Through multiplication of the structure, with occasional errors in the copying process, and selection, the adaptation becomes increasingly better. This is demonstrated in Figure 24.13(a). The circles are stored data, and the curves are trial functions. When an adapted function has been obtained in this manner (Figure 24.13(b)) the observation data are deleted in the memory (Figure 24.13(b)) so the memory (of limited size) is again ready to store further observation data.

According to the same mechanism a multitude of adapted functions which gradually accumulate is replaced by an overall structural scheme. This process is considered to occur in further stages and to lead to an increasingly sophisticated ordering scheme. In this way a scheme evolves which is accommodated to all observation data that the apparatus has stored in the course of time. This overall structural scheme of the apparatus may be called its internal model of the world.

Creating the laws of nature

Our thought experiment just described illustrates what basically happens in the process leading to the creation of a law of nature: the observation devices, our measuring instruments, give experimental data that are recorded. Structural schemes covering more and more recorded data gradually develop, i.e., theories of increasing generality are created by humans in a way corresponding to the way new forms emerge in bioevolution. (Note that the basic step in the measuring process is a yes-or-no decision, similar to the survival and multiplication or disappearance in the evolution process. The initiating stimulus is either amplified and recorded or there is no response.)

Mysterious new principles required to understand brain function?

Let us assume that the functioning of our brain is based on producing and selecting schemes accommodated to sensory data. Possible reactions of our bodies on environmental occurrences are tried within our internal model of the world before they are released as impulses for action. According to this view the knowledge of the external world is received through our senses, and the physical world cannot be more than the availability of this structural scheme. Our brain does not seem to supply us with any other information about the physical world. Thus the claim that our brain works similarly to an extremely sophisticated form of the device considered cannot be readily contradicted.

In Chapter 1 we pointed out an important step in the development of our model of reality. A sequence of impressions is seen as the perception of objects in space. The breakthrough of this scheme to order impressions is extremely important because it leads to an enormous simplification in the description of our experience and brings basic selection advantages. Therefore, it must be a basic step in the evolution of the central nervous system and the development of the baby. The interpretation of impressions as being based on processes in space and time is needed to get along in daily life. Our mind tends to perceive visual impressions as objects in space. This is illustrated in particular in cases where the impression allows two different interpretations and where we switch back and forth between the corresponding two perceptions; for example, in the case of the 12 lines in Figure 24.14(a) we see alternately a cube according to Figure 24.14(b), left and right, respectively. The ordering of impressions in daily life is limited, as can be seen in Chapter 1. Our familiar ordering scheme must be rejected when attempting to understand quantum effects.

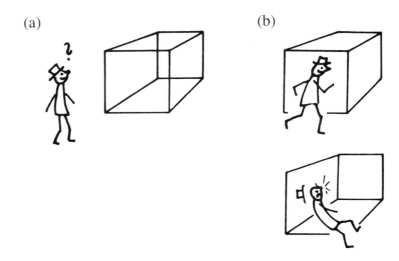

Figure 24.14 Visual impression (a) either perceived expecting that process (b), right above will take place or, alternatively process (b), right below.

Limitations in physical theory

The axioms of geometry are conventions using words like "straight line", which are useful in describing visual impressions. They are not statements on the nature of the physical space. The general laws of physics, as expressed by Einstein, are free inventions of the human mind which are selected on the basis of their usefulness. We cannot say whether these inventions are true or false, but merely whether they are useful schemes in describing facts. We have illustrated this in introducing the Schrödinger equation.

Similarly to the human mind, a sophisticated version of the model device considered above exposed to the physical world should create useful schemes to describe a multitude of facts, all based on trial, random changes, and selection, in a multitude of repetitions.

A remarkable aspect is that the device considered can be described within the framework of classical physics (quantum effects play no role in describing the process of knowledge accumulation). On the other hand, the accumulating knowledge (the evolving scheme ordering the input signals) depends on the received data and can be, for example, the scheme of quantum mechanics.

One of the greatest challenges in science is the search for a detailed knowledge of the processes in the brain. Let us assume for a moment that we possess this detailed knowledge and know the correlation of every facet of our consciousness (the feeling of happiness or sadness, the feeling of having a free will to act, seeing colors, listening to music) with the corresponding state of the cortical network in all physicochemical details. Then we would know all the correlations of our immediate feelings and impressions with processes in the brain, but we would not see behind the mystery of this correlation. The situation is not different in physics, where we know the general laws but do not understand why they are the way they are. Any question behind the scheme to describe nature cannot be answered and is therefore meaningless in the framework of the theory. In Chapters 1–3 we considered this problem. It is important to see this limitation in each physical theory.

Problems

Problem 24.1 – Aggregation of Folded Strands

A requirement for selection is that after the growth phase the folded strands must reach each other and precisely interlock in a time span short compared to the decay phase. We estimate a time $t = 10$ s required for aggregating two molecules. What is the volume V of a compartment including both molecules to allow the process in the given time?

Solution. A schematic view of the situation is depicted in Figure 24.15. Suppose that the aggregation of the folded strands A and B requires that the points a and b approach each other more closely than the critical distance d of about 0.1 nm. The strands must also have the correct mutual orientation. Assume that strand A remains stationary and that b is at the center of strand B; then point b participates only in the translational diffusion of B.

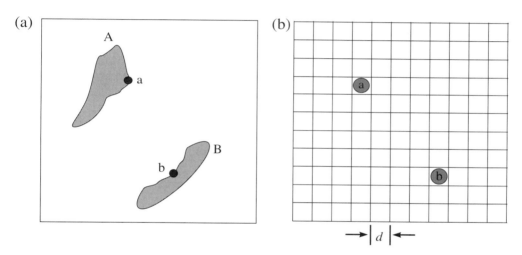

Figure 24.15 Aggregation of folded strands. (a) Strands A and B with active sites a and b. (b) Volume V divided in cubic cells of edge length d; sites a and b at different locations.

To estimate the average time it takes for point b to approach point a within a distance d, volume V (assumed to be cubic) is divided into cubic cells of edge length d. The average time τ it takes for point b to move from one cell to the next is then, according to the Einstein–Smoluchowski equation,

$$\tau = \frac{d^2}{2D}$$

where D is the diffusion constant of strand B. The average time t required for point b to reach the cell containing point a is

$$t = \tau \cdot \frac{V}{d^3} = \frac{d^2}{2D} \cdot \frac{V}{d^3} = \frac{V}{2D \cdot d}$$

Aggregation of strands A and B still requires that their mutual orientation be appropriate (steric factor α). With $D = 5 \times 10^{-11}$ m^2 s^{-1}, $d = 0.1$ nm, $t = 10$ s and $\alpha = 0.01$ we obtain

$$V = t \cdot 2D \cdot d \cdot \alpha = 1 \times 10^{-21} \text{ m}^3 = 10^6 \text{ nm}^3$$

This is the estimated pore volume (cube of an edge measuring roughly 100 nm); for comparison, the diameter of a bacterium is roughly 1000 nm = 1 μm.

Problem 24.2 – Replication Error Rate of a Bacterium and a Human

Estimate the replication error probability per base in a bacterium consisting of 10^3 proteins. What is different in a human with about 40×10^3 proteins?

Solution. Consider a protein consisting of about 300 amino acids, each requiring three bases for its establishment. Therefore, the DNA strand of a bacterium should consist of about $N_{\text{total}} = 3 \times 300 \times 10^3 = 9 \times 10^5 \approx 10^6$ bases; this is indeed the case. The information given by the sequence of bases in the DNA should be transferred to the next generation. Thus the replication error probability P per base should not be larger than 10^{-6} to allow for a sufficient amount of error-free copies. Furthermore, P should not be much smaller than 10^{-6} to allow for as many mutations as possible to allow adaptability. Thus we estimate $P \approx 10^{-6}$.

In the case of a human it must be taken into account that the information $N_{\text{total}} = 3 \times 300 \times 40 \times 10^3 = 36 \times 10^6 \approx 4 \times 10^7$ must be transferred by the germ cell lineage (about 100 cell divisions in a row) to the next generation. Thus, P should be about

$$P \approx \frac{1}{100} \cdot \frac{1}{N_{\text{total}}} \approx 2 \times 10^{-10}$$

This is actually the case. This accuracy is achieved by sophisticated repair mechanisms.

Problem 24.3 – Time Needed to Evolve a Bacterium

Evolution up to the period in which genetic machinery appeared must have occurred rapidly, because of the large error frequency that dominated these early stages. This time is small compared to the time required for the instruction of the approximately 1000 proteins of a bacterium. Estimate the time needed to evolve a bacterium.

Let us assume that one protein after another is inserted into the form existing at the time. This is achieved by lengthening of the DNA strand by $N = 10^3$ nucleotides for each new protein consisting of 300 amino acids.

For simplicity we assume that each protein is instructed in some hundred steps of optimization. Between each of these steps the situation must be awaited in which random distribution in the occupancies of the non-instructed locations of amino acids is reached that produces another step of optimization. This implies $1/P = 10^6$ generations per step, and $P = 10^{-6}$ is the replication error probability per base estimated for an early bacterium (see Problem 24.2).

Solution. Thus $10^6 \times 10^2 = 10^8$ generations are needed per protein and therefore a total of $10^8 \times 10^3 = 10^{11}$ generations to instruct the 1000 proteins. This takes 3×10^8 years assuming one day per generation. This time is small compared to the about 10^9 years available for the process in geological history. This is a pessimistic estimate since proteins develop largely by gene duplication which strongly decreases the time for instructing each additional protein.

Problem 24.4 – Maximum Genetic Information Carried by DNA

Estimate the maximum genetic information that can be stored in a human by DNA.

Solution. According to Problem 24.2, the replication error probability per base in a human to copy the genetic information is required to be about $P = 10^{-10}$. Another requirement is that the loss of information during time t between two subsequent cell divisions caused by thermal collisions is not larger than the loss by replication.

We consider the probability P' that a given base is exchanged by another base during time t due to an occasional strong collision. P' cannot be larger than about 10^{-10} in order to store the genetic information of a human or a higher organism. P' must be smaller than P: $P' < P$. We can estimate P' from the relation

$$P' = t \cdot \nu \cdot e^{-\Delta E/(kT)}$$

where ν is the collision frequency and ΔE is the activation energy. The collision frequency is in the order of 10^{13} s^{-1}, and we consider a time span $t = 0.1$ year $= 3 \times 10^6$ s for the time between cell divisions in the germ cell lineage. The activation energy is certainly smaller than the bond energy of a C–C bond (6×10^{-19} J); assuming that DNA is well protected the activation energy might be as high as half of the bond energy: $\Delta E = 3 \times 10^{-19}$ J. Then for $T = 300$ K

$$P' = 3 \times 10^6 \, \text{s} \times 10^{13} \, \text{s}^{-1} \times 3 \times 10^{-32} \approx 10^{-12}$$

Thus the condition $P' < P$ is fulfilled.

This estimate indicates that a barrier is reached in the genetic information to be stored in a human or in any higher organism. More information cannot be stored by DNA at room temperature.

How to overcome this barrier? Another mechanism is needed to transfer a larger information from one generation to the next. Artificial information storage (writing, computer) was the way to overcome this basic barrier in the evolution of life: the human society reached the critical stage to transfer a larger information from one generation to the next than can be transported by DNA.

Foundation 24.1 – The Emergence of a Simple Genetic Apparatus Viewed as a Supramolecular Engineering Problem. A Thought Experiment.

A short template marks the beginning

We consider a solution of two kinds of monomers capable of assembling covalent links into short polymer strands (Figures 24.16(a) and (b)). The two kinds of monomers are complementary: they weakly attract each other (e.g., by hydrogen bonding). The first short polymer strand serves as a template for further strands. Once the template strand has formed, monomers are attached because of their lock-and-key properties (Figure 24.16(c)): formation of a double strand (Figure 24.16(d)). The incident signaling the dawn of life is an unusual event that occurred, for instance, during drying and redissolving of a solution and leads to the formation of the first short template strand.

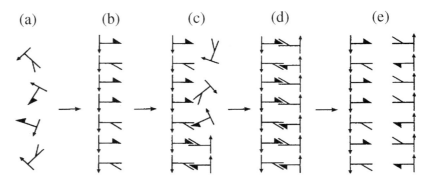

Figure 24.16 Monomers capable of assembling covalent links into short polymer strands. (a) Monomers. (b) Monomers assembled to a strand by chance. (c) and (d) Strand as matrix for the assemblage of a double strand at low temperature. (e) Splitting of the double strand into single strands at high temperature (replication).

Intricate cycles of temperature drive replication

Further essentials in our discussion are cyclical temperature changes that drive strand replication and spatial conditions providing compartmentation. These may be given by day-and-night and various shadow-casting cycles and pores in rocky material. A double strand which is formed at low temperature can be split into two single strands at high temperature (the weak hydrogen bonds break, but the strong covalent bonds persist); replication occurs when lowering the temperature (Figure 24.16(e)). The arrangement of the monomers in the copy is complementary to that in the template strand (with regard to their lock-and-key properties).

Continued on page 907

FOUNDATION 24.1

Continued from page 906

Strand evolution requires rare replication errors

The arrangement of the monomers in the first template strand is determined by chance, so all of the copies have the same or the complementary accidental structure. During the replication process, however, errors can occur. (Note that an error, in the present context, is the occurrence of a monomer unit in the daughter strand that is not complementary to the corresponding monomer unit in the template). Most of these errors are detrimental; by chance and very rarely, however, such an error imparts an advantage upon the replicate strand, improving its adaptation to environmental conditions (e.g., to the porous structure in rocky material, see Figure 24.17(a)), and its descendants eventually replace their less efficient fellow strands.

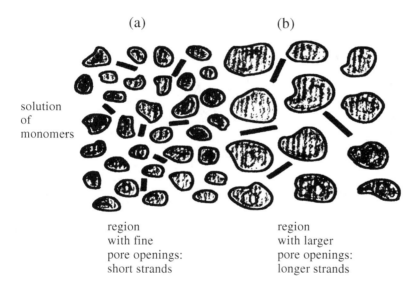

Figure 24.17 Adaptation of strands to environmental conditions (porous structure in rocky material): (a) region with fine pore structure; (b) region with larger openings.

Strand lengthening – colonization of larger porous regions

We consider short strands which are well adapted to the fine pore structure in Figure 24.17(a). By chance and under very special circumstances (e.g., partial drying and redissolution), two short strands can combine to make a longer one. What is important at this point is the presence of a neighboring region, suffused by monomers, with pores that offer almost – but not quite – adequate confinement for the existing short strands (Figure 24.17(b)). The longer strands are then better adapted to the large pore size region, because they are constrained from being lost by diffusion; the populated area extends by the colonization of pores with larger openings by longer strands. However, replication errors

Continued on page 908

Continued from page 907

set an upper limit for this process as well. It cannot be avoided that for longer strands the replication error probability is much larger than for the shorter ones, and we need mechanisms to eliminate error copies.

Strand folding: hairpins – most resistant structure

A long strand is exposed to chemical attack from the environment. How can the reactive groups along the strand be protected? A very effective possibility for the strand is to fold back onto itself to form weak bonds between complementary groups. In Figure 24.18(a), as an example, four adjacent groups are complementary. The most favorable conformation is a complete

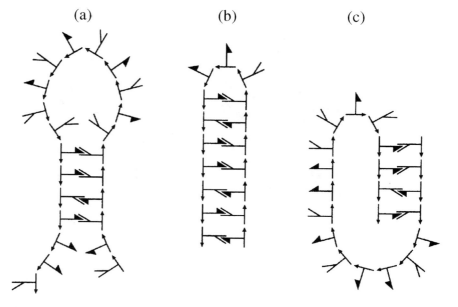

Figure 24.18 Protection of a strand by folding. (a) Partial hairpin; (b) perfect hairpin (antiparallel orientation); (c) no hairpins possible for monomers oriented parallel.

fitting of all groups along the strand (Figure 24.18(b)), forming a perfect hairpin. The hairpin conformation requires an antiparallel monomer orientation. With monomers oriented parallel, the formation of hairpins is impossible (Figure 24.18(c)).

The perfect replicate of a hairpin forms again a hairpin. The replicate differs, however, in one monomer in the middle of the loop [(+)-hairpin and (−)-hairpin; Figure 24.19]. We will see below that this is important for the establishment of what is viewed as the very first, primitive translation device.

Continued on page 909

FOUNDATION 24.1

Continued from page 908

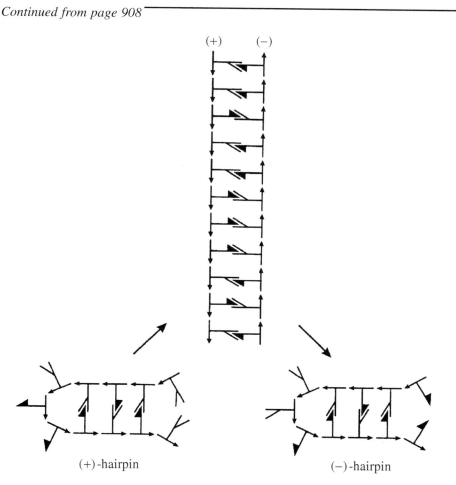

Figure 24.19 The replicate of a hairpin forms a hairpin again: (+)- and (−)-hairpins.

Aggregation of hairpins: an error filter

A strand folded into a hairpin is already well protected against attacks from outside. This protection can easily be improved by aggregation (Figure 24.20, picket-fence-like structure). The aggregates are much larger than single hairpins; hence the aggregates are able to colonize the larger pores, whereas single hairpins cannot. There is another advantage of aggregation: erroneous copies of hairpins, which do not fit into the aggregate, will be eliminated.

On the other hand, errors that do not affect incorporation into the aggregate are not lethal; this means that the further evolution of the strands can occur. Thus, aggregates of hairpins constitute a supramolecular functional unit. A machinery is present to preserve the know-how of making hairpins.

Continued on page 910

Continued from page 909

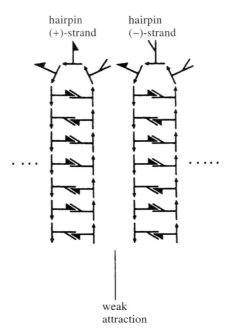

Figure 24.20 Aggregation of hairpins (picket-fence-like structure).

Assembler strand guides hairpin aggregation

The evolution toward picket-fence-like hairpin aggregates is a very important step to stabilize the strands. Nevertheless, the aggregation can still be improved. The weakly bonding groups in the hairpin loop are considered to be fixed to complementary groups on an open (i.e., unfolded) strand (Figure 24.21). Such a strand (assembler strand or A-strand) greatly speeds the formation of interlocking picket-fence-like aggregates: the three-dimensional diffusion of the hairpins is converted into a one-dimensional diffusion along the A-strand.

Open strands always exist, because of errors in the replication of hairpins. Usually they are lost by diffusion, but if, by chance, one of them has the appropriate sequence for serving as A-strand, then it becomes part of the hairpin-assembler (HA) device. Once in existence, such a strand-aided assemblage of hairpins is greatly favored by selection.

A new variety (a-monomers) appears: a-oligomers formed by HA device enable colonization of larger porous regions

Although the replicate of a hairpin differs in the middle of the loop only, there is the possibility that its open end is different. We can imagine that replication

Continued on page 911

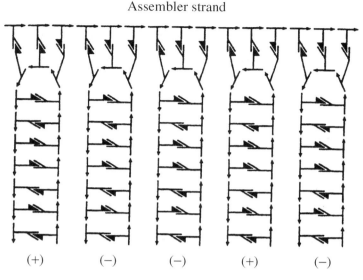

Figure 24.21 Hairpin aggregates fixed to complementary groups on an unfolded strand.

errors lead to an unmatched pair of monomers in the open end of the hairpin (Figures 24.19, 24.22). Then the open end is capable of binding a new variety of monomers (a-monomers). We assume that these new monomers line up along the hairpin ends and that covalent bond formation occurs between neighbors (formation of a-oligomers). These oligomers are assumed to be capable of agglutination, they assist the restraining function of pores by narrowing pore openings and making larger pores accessible to colonization, which provides an evolutionary gradient. We assume that variety a_1 is binding to the end of the (+)-strand and that variety a_2 is binding to the end of the (−)-strand (Figure 24.22).

A-strands carry the pattern, the prescription for the a-oligomers. In this way a translation apparatus has come into existence as a by-product of an error filter. However, this device is of no benefit at this stage: any special sequence of monomers in the A-strand that leads to an oligomer with an especially beneficial sequence of a_1 and a_2 is not preserved, because of copying errors.

Breakthrough of translational device

A special a-oligomer is assumed to improve the replication by intercalation. This way the prescription for the a-oligomers can be preserved if the replication

Continued from page 911

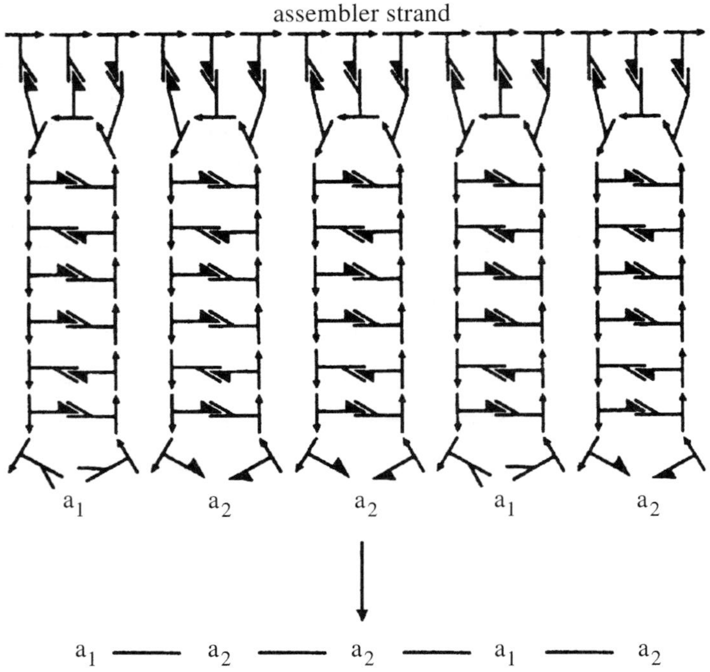

Figure 24.22 Open ends of hairpin aggregates bind a-monomers (formation of a-oligomers).

error probability is below a certain critical value. The required firm relationship between the monomer sequence of the A-strand and the sequence of the a-oligomer requires the linkage of three adjacent monomers in the hairpin loop to three complementary monomers along the A-strand, in addition to a weak lateral attraction between neighboring hairpins. The monomers of an A-strand are arrayed into adjoining triplets, and to each triplet is attached a complementary triplet in the loop of a hairpin.

A triplet provides the smallest number of complementary nucleotides for a firm and stable aggregate. Triplets offer also sufficient coding possibilities, and unless it is really required for coding needs, an attachment by more than three monomer pairs would represent an evolutionary burden rather than an advantage.

The stability of HA-devices, of course, depends on quite definite requirements to be met by suitable monomers.

Breakthrough of an integrative translation device

The cooperation of an increasing number of HA-devices, each providing a beneficial oligomer, is only possible by enclosure in an envelope and by overcoming

Continued on page 913

Continued from page 912

a dead end reached at a certain level of complexity, with the pollution by (−)-assembler strands producing nonsensical oligomers. By continuing the search for further small modeling steps in the way shown above for the earlier steps, it is found that the barrier, in the thought experiment, is overcome by a basic transformation in the genetic machinery. This is achieved by the appearance of a third monomer variety D that is a modified version of the original monomers of the first variety.

This fundamental change leads to a system in which the genetic message for constructing all HA-devices of the cooperative is carried by a strand made of D-monomers. This D-strand replicates, and in this way the genetic message of the cooperative (coded by the monomer sequence in the D-strand) is multiplied. The detailed model path leading to the machinery cannot be given here. The machinery has basically the same organizational structure as the genetic apparatus of the biosystems (Box 24.1), which is most remarkable.

Foundation 24.2 – Attempts to Model the Origin of Life

Can all the requirements of the thought experiment discussed in Foundation 24.1 be fulfilled by the molecules constituting the genetic apparatus in biosystems? Can such considerations contribute to the understanding of the origin of life on that basis?

The earliest phase

Among candidates for the monomers of the first variety are the nucleotides guanidine (G), cytosine (C), adenine (A), and uracil (U). We favor G and C for the beginning, because they basepair with three rather than two hydrogen bonds. Polynucleotide strands have the essential features required for HA-devices (forming hairpins, lateral attraction of hairpins by the intercalation of Ca^{2+} and Mg^{2+} ions). For the a monomers a_1 and a_2 we choose the most prevalent amino acids alanine and glycine. The earliest "code" in our model involves just these two nucleotides and two amino acids.

Breaking symmetry: chirality of nucleotides and its evolutionary benefit

Nucleotides are chiral because ribose, one of their components, is chiral, and nucleotides with D-ribose have opposite chirality from those containing the

Continued on page 914

Continued from page 913

L-isomer. Nucleotides with the same chirality, either D or L, are more appropriate for precise template-assisted replication than are strands containing monomers of mixed chirality. Monomers of incorrect chirality are rejected in this buildup of new daughter strands. They do not fit into the niche at the growth position created by the template strand and the already existing portion of the daughter strand (*lock-and-key principle*).

Another most significant aspect of strand chirality is that hairpins consisting of nucleotides with D-ribose have a helical twist that is opposite to that of hairpins made up of nucleotides with L-ribose. Hairpins of opposite twist do not interlock properly.

It is most likely that D-ribose and L-ribose were present in equal amounts on prebiotic earth, and this would therefore be true for nucleotides as well. In this view the fact that present-day RNA contains the D-isomer rather than the L-isomer of ribose is the result of a "frozen accident". Thus the chirality of the amino acids entering the genetic machinery (L-amino acids) is considered to be a consequence of the chirality of nucleotides (lock-and-key principle).

The process can have started at many locations on the early earth resulting, occasionally, in systems with L-ribose. They occasionally interacted with D-systems; thereby one chiral form became stronger at the expense of the other one and finally the D-forms survived the struggle.

The view of symmetry breaking by a frozen accident is supported by the recent important finding (M. Bolli, R. Micura, A. Eschenmoser, Pyranosyl-RNA: Chiroselective Self-assembly. Chemistry and Biology 4, 309 (1997); A. Eschenmoser: Chemical Ethiology of Nucleic Acid Structure, Science 284, 2118 (1999)) that pyranosyl-RNA (which is a stronger and more selective pairing system than the natural furanosyl-RNA) is obtained in pure L or pure D form by oligomerization of G—C—C—GcP tetramers (where cP means cyclophosphate). These tetramers consist of D and L nucleotides linked at random (mixed D and L tetramers besides pure D and pure L tetramers). Alone, pure D (or pure L) tetramers oligomerize by self-assembly:

$$\text{-------- G—C—C—G}_{cP}\ \text{G—C—C—G}_{cP}\ \text{G—C—C—G}_{cP}\text{----}$$

$$\text{-------- G}_{cP}\text{—C—C—G}\quad \text{G}_{cP}\text{—C—C—G}\quad \text{G}_{cP}\text{—C—C—G ----}$$

$$\Big\downarrow \text{some weeks}$$

$$\text{—G—C—C—G—G—C—C—G—G—C—C—G—}$$

$$\text{G—C—C—G—G—C—C—G—G—C—C—G}$$

Each tetramer carries cP that opens up by linking the phosphate group to the 4' end of the adjacent tetramer. This reaction occurs under mild conditions and only in the presence of the template.

Continued on page 915

FOUNDATION 24.2

Continued from page 914

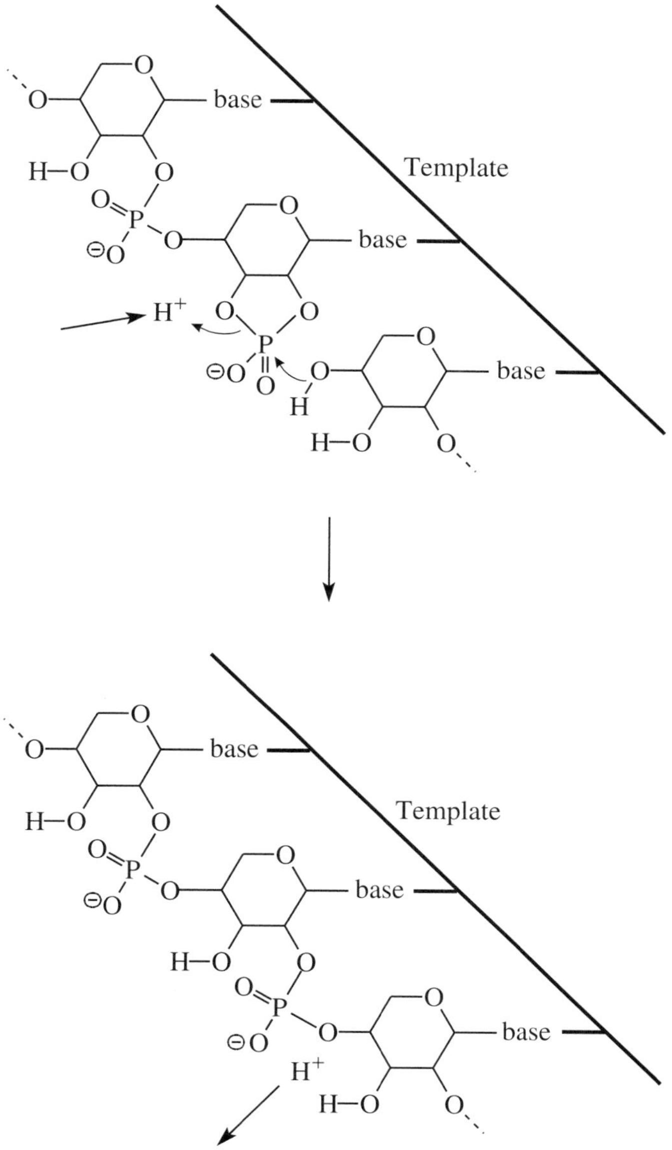

The complementary strands are strictly antiparallel, thus a crucial condition to form HA-devices is fulfilled. Hairpins can evolve by variations in the sequence. Replication and formation of a HA device remains to be demonstrated. The question arises: how can an evolution based on pyranosyl-RNAs go over into an evolution based on the natural furanosyl-RNAs? We may imagine

Continued on page 916

Continued from page 915

that pyranosyl-RNA based systems produce oligopeptides that serve as enzymes allowing replication of furanosyl-RNAs. An evolution of furanosyl-RNA based systems is then possible. These systems, at a certain evolutionary level, do not anymore depend on the pyranosyl-RNA systems which then disappear.

Recently, replication and exponential amplification of DNA analogues has been achieved (A. Luther, R. Brandsch, G. v. Kiedrowski, Nature 1998, **396**, 245) emphasizing the idea that cycles of particular environmental changes are the basic driving force in the origin of life (see Box 24.2). Templates have been exposed to a periodically changing environment, allowing, in succession, complementary binding of oligomers at template strands, chemical ligation, and liberation of new strands from templates. Original templates and new strands, after liberation, are immobilized to avoid formation of stable duplexes. Cyclic repetition allows an exponential increase in the amount of strands.

The earliest HA-device

The fact that complementary hairpins differ only in the middle of their loops is of crucial importance for the earliest HA-device considered as a thought experiment (Foundation 24.1). The earliest HA-device contained just the near-identical (+) and (−) versions of the "same" hairpin differing at the very end of their legs. They must be slightly different at their leg ends because (+) and (−) hairpins have to carry different amino acids; the complementarity requirement for monomers at the leg ends must therefore be relaxed. The stability of HA-devices, built from such near-identical hairpin neighbors, is, of course, favorably affected by their close fit.

For reasons detailed below, we make the assumption that hairpins with C in the loop were invested with glycine and that "G-hairpins" were invested with alanine. Therefore, looking at the middle nucleotide of a triplet along an A-strand, G would "code" for glycine and C would code for alanine. There is, however, another requirement to be met, because there is a need for recognizing the location of different code letters on the A-strand, a need for a "reading frame".

The reading frame requirement

In the early stages of evolution, there is the demand for a way of telling where a triplet along the A-strand begins and ends; that is, there has to be a reading frame. Specifically, our model requires that in all triplets along the A-strand, the first nucleotide must be G and the third must be C, assigning any coding function to the middle nucleotide. The reading frame has therefore the form 5′-GNC-3′, where N first stands for G or C and later for A or U as well. For a corresponding hairpin loop, the complementary sequence is 3′-CN′G-5′ or 5′-GN′C-3′, N′ being the nucleotide complementary to N.

Continued on page 917

FOUNDATION 24.2

Continued from page 916

However, as the loops of complementary (+) and (−) hairpins differ in the middle nucleotide only, they are both of the same type, namely, 3'-CN'G-5'. This implies that the same GNC reading frame applies to both (+) and (−) hairpins.

Stage-by-stage evolution of the code

The translation device becomes increasingly sophisticated by incorporation of increasingly complex translation products stabilizing the attachment of hairpins to the assembler strand. Strength in base pairing should be no longer beneficial. An original pyranosyl-RNA based system should be replaced by the more flexible and fast furanosyl-RNA system. The reading frame gradually loses in importance. Triplet position 1 is also pressed into coding service, and finally all three positions are used for this.

Each step of relaxation of constraints on the code by improvement of the translation machinery confers a selective advantage, increasing the number of amino acids that can be coded for. In this way the modern code can be reconstructed, when it is assumed: (1) that the order in which amino acids are introduced into the code is in the order of their availability; (2) that the code, first restricting to GNC (N = G,C,A,U) changes over to GNN (first position sufficient for docking), then the code changes to PuNN (Pu = G,A; a purine is sufficient for docking); later the third position is used for coding and the first position is opened for pyrimidines C and U; (3) that A in position 2 is associated with the most polar amino acid to be introduced, G, C, U are associated with amino acids in the order of decreasing polarity. This is a generalization of the initial codes GAC for Glu, Asp; GGC for Gly; GCC for Ala; GUC for Val.

The modern code results from these postulates when taking into consideration that important amino acids are provided with an additional codon at the stage when position 1 is opened for C and U (Table 24.1). Later, part of the codons for Leu, Gln, and Stop are used, instead, for special amino acids Phe, Tyr, His, and Cys. Finally, one of the codons for Ile and two of the codons for Stop are used for the latest amino acids Met, Trp, and Sec (Sec is the recently discovered amino acid number 21, selenocysteine). The insight thus afforded strongly suggests that the ancestral self-reproducing system resembled the HA-device.

It is, however, important to recognize that a basic change in the structure of the genetic machinery is required before the evolution of the genetic code can proceed – the breakthrough of an integrative translation device. In the special case of bioevolution the third monomer variety, D, is deoxyribonucleic acid. This leads to a device with the basic features of the genetic apparatus of biosystems (Box 24.1). The considerations in Boxes 24.2 and Foundation 24.1 indicate that the emergence of such a device should not be restricted to the special case of bioevolution based on nucleotides and amino acids.

Continued on page 918

Continued from page 917

Table 24.1 Hypothetical Stages in the Development of a Genetic Code

	A	→	B	→	C	→	D	→	E
Decreasing availability ↓	GNC		PuNN		Third position opens for coding		First position opens for Py		Modern genetic code
Gly	GGC		GGN						GGN
Ala	GCC		GCN						GCN
Asp	GAC		GAN		GAPy				GAPy
Glu	GAC		GAN		GAPu				GAPu
Val	GUC		GUN						GUN
Stop	UPuPu		UPuN						UGA, UAPu
Leu			AUN						
Ile			AUN						AUPy, AUA
Ser			AGN		AGPy				AGPy
Thr			ACN						ACN
Lys			AAN		AAPu				AAPu
Arg					from Ser AGPu				AGPu
Gln					from Lys AAPy				
Asn					from Lys AAPy				AAPy
Pro							CCN		CCN
							Arg CGN		CGN
							Gln CAN		CAPu
							Leu CUN		CUN
							Leu UUN		UUPu
							Ser UCN		UCN
Phe									from Leu UUPy
Tyr									from Stop UAPy
His									from Gln CAPy
Cys									from Stop UGPy
Met									from Ile AUG
Trp									from Stop UGG
Sec									from Stop UGA

Stage A: code GNC for the five most abundant amino acids Gly, Ala, Val, and Glu; Asp (not distinguished at this stage). Oligomerization is stopped by the best suppressing-binding triplet UPuPu considered as stop codon.

Stage B: code PuNN. Purine G or A in first position of triplet allows code for the ten most abundant amino acids (Asp; Glu and Leu; Ile not distinguished at this stage). Third position no longer used as reading frame.

Stage C: third position opens for coding: distinction between Asp and Glu; AGPu and AAPy open for Arg and Gln; Asn (not distinguished at this stage), using codes originally used for Ser and Lys.

Stage D: first position opens for Py: code for Pro and additional codes for particularly important amino acids. UPuN is reserved for stop at this stage.

Stage E: emergence of modern code: UUPy, UAPy, CAPy, UGPy open for special amino acids; finally AUG, UGG, UGA open for latest amino acids, using codes originally used for Leu, Stop, Gln and Ile.

Continued on page 919

Table 24.2 The 21 amino acids. Single letter codes, abbreviations and full names.

Code	Abbreviation	Full name	Code	Abbreviation	Full name
A	ala	alanine	N	asn	asparigine
C	cys	cysteine	P	pro	proline
D	asp	aspartic acid	Q	gln	glutamine
E	glu	glutamic acid	R	arg	arginine
F	phe	phenylalanine	S	ser	serine
G	gly	glycine	T	thr	threonine
H	his	histidine	V	val	valine
I	ile	isoleucine	W	trp	tryptophane
K	lys	lysine	Y	tyr	tyrosine
L	leu	leucine		sec	selenocysteine
M	met	metheonine			

Foundation 24.3 – Maxwell's Demon: Production of Entropy

We demonstrate that the production of entropy in the surroundings (resets of the measuring device) is larger than the gain in entropy in the system even under optimal conditions.

The probability P that the small volume v is empty after closing door a is

$$P = \left(\frac{V-v}{V}\right)^{N_1} = \left(1 - \frac{v}{V}\right)^{N_1}$$

The probability p to find one particle out of N_1 particles in volume v is

$$p = N_1 \frac{v}{V}$$

Then we obtain

$$P = \left(1 - \frac{p}{N_1}\right)^{N_1}$$

In our case N_1 is a large number; because of the identity

$$\lim_{x \to \infty} \left(1 - \frac{p}{x}\right)^x = e^{-p}$$

we can write P as

$$P = e^{-p}$$

Correspondingly, the probability that v is not empty is

$$1 - P = 1 - e^{-p}$$

Continued on page 920

Continued from page 919

On the average, $1/(1-P)$ measurements must be taken until the small volume v is found to be not empty. In this event there is a little more than one particle in v, on the average, namely,

$$\frac{N_1}{V} v \cdot \frac{1}{1-P} = p \cdot \frac{1}{1-P}$$

particles.

Now door b is opened, and these particles move into the right container. However, after equilibration and, subsequently, closing door b, there will be

$$\frac{N_2}{V} v$$

particles in volume v, on the average. Thus, effectively,

$$z = \frac{N_1}{V} v \cdot \frac{1}{1-P} - \frac{N_2}{V} v = \frac{N_1}{V} v \cdot \frac{1}{1-P} - \frac{2N_1}{V} v = p \cdot \frac{2P-1}{1-P}$$

particles move into volume 2. The number of measurements per moving particle is

$$n = \frac{1}{1-P} \cdot \frac{1}{z}$$

Therefore we obtain for the ratio r

$$r = \frac{-\Delta S_{\text{system}}}{\Delta S_{\text{surroundings}}} = \frac{k \cdot \ln 2}{n \cdot k \cdot \ln 2} = z \cdot (1-P) = p \cdot (2P-1) = p \cdot (2e^{-p} - 1)$$

The entropy production in resetting the measuring device is $k \cdot \ln 2$. This is the minimum entropy production in the reset phase when the two possible results are a priori equally probable. However, this is not the case in our example and this must be taken into account when asking for the maximum value of the ratio r. The measuring device, entering the reset phase, can be in the state corresponding to empty volume v (probability $P = \exp(-p)$) or in the state corresponding to not empty volume v (probability $Q = 1 - P$). Knowing that P is different from Q we can reset the measuring device by applying a biasing force favoring the state of a a priori higher probability. Then the erasure cost per bit is less than $kT \cdot \ln 2$ (the entropy production is less than $k \cdot \ln 2$). It can be shown that the produced entropy is

$$\Delta S_{\text{surroundings}} = k \cdot \left(P \cdot \ln \frac{1}{P} + Q \cdot \ln \frac{1}{Q} \right)$$

This expression gives $k \cdot \ln 2$ for $P = Q = \frac{1}{2}$:

$$k \cdot \left(\frac{1}{2} \cdot \ln 2 + \frac{1}{2} \cdot \ln 2 \right) = k \cdot \ln 2 = k \cdot 0.693$$

Continued on page 921

FOUNDATION 24.3

Continued from page 920

Then we obtain for the ratio r (biased resetting)

$$r = \frac{-\Delta S_{\text{system}}}{\Delta S_{\text{surroundings}}} = p \cdot (2e^{-p} - 1) \cdot \frac{\ln 2}{P \cdot \ln(1/P) + Q \cdot \ln(1/Q)}$$

Remember that $P = e^{-p}$ and $Q = 1 - P = 1 - e^{-p}$. This means that r is a function of p, with $p = vN_1/V$. This function is shown in Figure 24.23. It

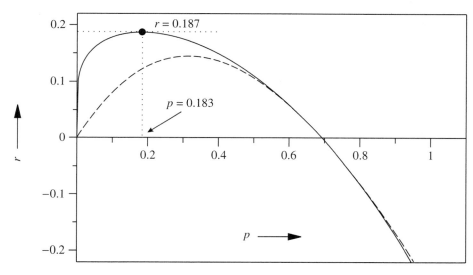

Figure 24.23 Ratio r as a function of the probability $p = vN_1/V$. Solid line: r calculated for biased resetting. Dashed line: r calculated without resetting.

can be seen that the ratio r has a maximum of $r = 0.187$ at $p = 0.183$. r becomes zero at $p = \ln 2 = 0.693$. This means that the entropy production in the surroundings is at least $1/0.187 = 5.35$ times the entropy decrease in the system.

Glossary

Fundamental Constants (the values for c_0, μ_0 and g_n are assigned by definition)

N_A	Avogadro constant	6.0221367×10^{23} mol^{-1}
m_p	proton rest mass	$1.6726231 \times 10^{-27}$ kg
m_e	electron rest mass	$9.1093897 \times 10^{-31}$ kg
e	elementary charge	$1.60217733 \times 10^{-19}$ C
$F = N_A e$	Faraday constant	96485.309 C mol^{-1}
h	Planck constant	$6.6260755 \times 10^{-34}$ J s
$\hbar = h/(2\pi)$		$1.05457266 \times 10^{-34}$ J s
c_0	speed of light in vacuum	2.99792458×10^8 m s^{-1}
k	Boltzmann constant	1.380658×10^{-23} J K^{-1}
$R = N_A k$	gas constant	8.314510 J K^{-1} mol^{-1}
μ_0	permeability of vacuum	$4\pi \times 10^{-7}$ H m^{-1}
$\varepsilon_0 = 1/(\mu_0 c_0^2)$	permittivity of vacuum	$8.854187816 \times 10^{-12}$ C^2 J^{-1} m^{-1}
$a_0 = \dfrac{\varepsilon_0 h^2}{\pi m_e e^2}$	Bohr radius	$0.529177249 \times 10^{-10}$ m
$\mu_B = \dfrac{eh}{4\pi m_e}$	Bohr magneton	$9.2740154 \times 10^{-24}$ J T^{-1}
$\mu_N = \dfrac{m_e}{m_p}\mu_B$	nuclear magneton	$5.0507866 \times 10^{-27}$ J T^{-1}
g_n	standard acceleration of free fall	9.80665 m s^{-2}

Useful Notations

1 eV corresponds to 96.48 kJ mol^{-1} = 23.05 kcal mol^{-1}, or to the energy of a quantum of light of wavelength 1240 nm or wavenumber 8065 cm^{-1}.

0°C corresponds to 273.15 K, 100°C corresponds to 373.15 K.

Molar volume of ideal gas at 1 bar, 0°C: 22.71108 L mol^{-1}.

ångstrom (Å) = 10^{-10} m; m s^{-1} = (millisecond)$^{-1}$; m s^{-1} = meters per second.

Data from IUPAC, Quantities, Units and Symbols in Physical Chemistry, Recommendations 1993, Blackwell Scientific Publication, 1995

SI Base Units

length	metre	m
mass	kilogram	kg
time	second	s
electric current	ampere	A
thermodynamic temperature	kelvin	K
amount of substance	mole	mol

SI Derived or Related Units

frequency	hertz	Hz	$= s^{-1}$
force	newton	N	$= m\ kg\ s^{-2}$
	kilogram-force	kgf	$= 9.80665\ N$
	dyne	dyn	$= 10^{-5}\ N$
energy	joule	J	$= Nm = kg\ m^2\ s^{-2}$
		L bar	$10^2\ J$
	erg	erg	$= 10^{-7}\ J$
	electron volt	eV	$= 1.60218 \times 10^{-19}\ J$
	calorie	cal	$= 4.184\ J$
power	watt	W	$= J\ s^{-1} = V\ A$
volume	litre	L	$= 10^{-3}\ m^3$
pressure	pascal	Pa	$= N\ m^{-2}$
	bar	bar	$= 10^5\ Pa$
	atmosphere	atm	$= 1.013\ bar$
	torr	Torr	$= 1/760\ atm$
	millimetre of mercury	mm Hg	$= Torr$
electric charge	coulomb	C	$= A\ s$
electric potential, electromotive force	volt	V	$= J\ C^{-1}$
electric resistance	ohm	Ω	$= V\ A^{-1}$
electric conductance	siemens	S	$= \Omega^{-1}$
electric capacitance	farad	F	$= C\ V^{-1}$
magnetic flux density	tesla	T	$= V\ s\ m^{-2}$
	gauss	G	$= 10^{-4}\ T$
inductance	henry	H	$= V\ A^{-1}\ s$
electric dipole moment	coulombmetre		$= C\ m$
	debye	D	$= 3.3356 \times 10^{-30}\ C\ m$
magnetic moment		$A\ m^2$	$= J\ T^{-1}$

Prefixes

a	f	p	n	μ	m	c	d
atto	femto	pico	nano	micro	milli	centi	deci
10^{-18}	10^{-15}	10^{-12}	10^{-9}	10^{-6}	10^{-3}	10^{-2}	10^{-1}

da	h	k	M	G	T	P	E
deka	hecto	kilo	mega	giga	tera	peta	exa
10^{1}	10^{2}	10^{3}	10^{6}	10^{9}	10^{12}	10^{15}	10^{18}

Greek Alphabet

Alpha	A	α	Iota	I	ι	Rho	P	ρ
Beta	B	β	Kappa	K	κ	Sigma	Σ	σ
Gamma	Γ	γ	Lambda	Λ	λ	Tau	T	τ
Delta	Δ	δ	Mu	M	μ	Upsilon	Υ	υ
Epsilon	E	ε	Nu	N	ν	Phi	Φ	ϕ
Zeta	Z	ς	Xi	Ξ	ξ	Chi	X	χ
Eta	H	η	Omnicron	O	o	Psi	Ψ	ψ
Theta	Θ	θ	Pi	Π	π	Omega	Ω	ω

Some Useful Physicochemical Relations

$$\boxed{\Lambda = \frac{h}{mv}} \quad \text{(page 12)}$$

$$\boxed{-\frac{h^2}{8\pi^2 m_e}\frac{d^2\psi}{dx^2} + V\psi = E\psi} \quad \text{(page 42)}$$

$$\boxed{E_{trans} = \tfrac{3}{2}kT, \quad PV = n\boldsymbol{R}T} \quad \text{(page 367)}$$

$$\boxed{\overline{\Delta x^2} = 2D \bullet \Delta t} \quad \text{(page 378)}$$

$$\boxed{N_i = \frac{N}{Z}e^{-E_i/(kT)}, \qquad Z = \sum_{i=1}^{\infty} e^{-E_i/(kT)}} \quad \text{(page 412)}$$

$$\boxed{S = k \bullet \ln \Omega} \quad \text{(page 480)}$$

$$\boxed{\Delta U = 0, \quad \Delta S \geq 0 \quad \text{(isolated system)}} \quad \text{(page 483)}$$

$$\boxed{\Delta S = \sum_i \frac{Q_{rev,i}}{T_i}} \quad \text{(page 501)}$$

$$\boxed{\Delta_r G^\ominus = \Delta_r H^\ominus - T\Delta_r S^\ominus = -\boldsymbol{R}T \bullet \ln K} \quad \text{(pages 531, 550)}$$

Some Useful Mathematical Relations

$$N! = 1 \times 2 \times 3 \times \cdots \times N$$

$$\ln N! = N \bullet \ln N - N, \qquad N! = N^N \bullet e^{-N} \quad \text{for} \quad N \gg 1$$

$$e = \lim_{n \to \infty} \left(1 + \frac{1}{n}\right)^n = 2.718282$$

$$\ln 10 = 2.302585$$

$$\ln x = \ln 10 \bullet \log x = 2.30285 \bullet \log x$$

$$\pi = 3.141593$$

$$\ln(1+x) \approx x \quad \text{for} \quad x \ll 1$$
$$\sqrt{1+x} \approx 1 + \tfrac{1}{2}x \quad \text{for} \quad x \ll 1$$
$$\frac{1}{1+x} \approx 1 - x \quad \text{for} \quad x \ll 1$$

Periodic Table and Molar Masses of the Stable Elements (in g mol^{-1})

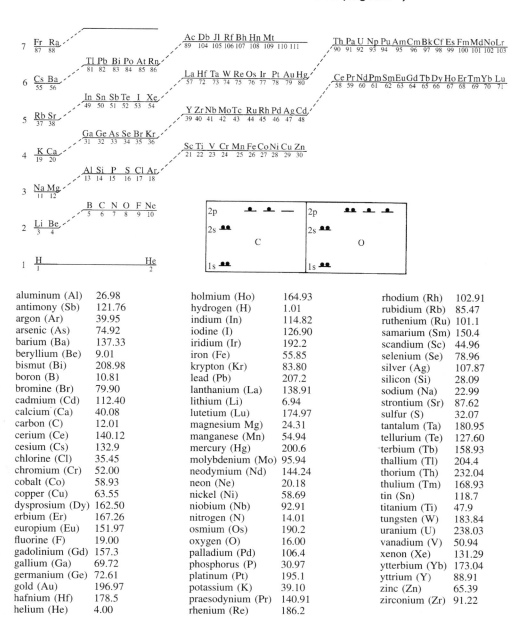

aluminum (Al)	26.98	holmium (Ho)	164.93	rhodium (Rh)	102.91
antimony (Sb)	121.76	hydrogen (H)	1.01	rubidium (Rb)	85.47
argon (Ar)	39.95	indium (In)	114.82	ruthenium (Ru)	101.1
arsenic (As)	74.92	iodine (I)	126.90	samarium (Sm)	150.4
barium (Ba)	137.33	iridium (Ir)	192.2	scandium (Sc)	44.96
beryllium (Be)	9.01	iron (Fe)	55.85	selenium (Se)	78.96
bismut (Bi)	208.98	krypton (Kr)	83.80	silver (Ag)	107.87
boron (B)	10.81	lead (Pb)	207.2	silicon (Si)	28.09
bromine (Br)	79.90	lanthanium (La)	138.91	sodium (Na)	22.99
cadmium (Cd)	112.40	lithium (Li)	6.94	strontium (Sr)	87.62
calcium (Ca)	40.08	lutetium (Lu)	174.97	sulfur (S)	32.07
carbon (C)	12.01	magnesium Mg)	24.31	tantalum (Ta)	180.95
cerium (Ce)	140.12	manganese (Mn)	54.94	tellurium (Te)	127.60
cesium (Cs)	132.9	mercury (Hg)	200.6	terbium (Tb)	158.93
chlorine (Cl)	35.45	molybdenium (Mo)	95.94	thallium (Tl)	204.4
chromium (Cr)	52.00	neodymium (Nd)	144.24	thorium (Th)	232.04
cobalt (Co)	58.93	neon (Ne)	20.18	thulium (Tm)	168.93
copper (Cu)	63.55	nickel (Ni)	58.69	tin (Sn)	118.7
dysprosium (Dy)	162.50	niobium (Nb)	92.91	titanium (Ti)	47.9
erbium (Er)	167.26	nitrogen (N)	14.01	tungsten (W)	183.84
europium (Eu)	151.97	osmium (Os)	190.2	uranium (U)	238.03
fluorine (F)	19.00	oxygen (O)	16.00	vanadium (V)	50.94
gadolinium (Gd)	157.3	palladium (Pd)	106.4	xenon (Xe)	131.29
gallium (Ga)	69.72	phosphorus (P)	30.97	ytterbium (Yb)	173.04
germanium (Ge)	72.61	platinum (Pt)	195.1	yttrium (Y)	88.91
gold (Au)	196.97	potassium (K)	39.10	zinc (Zn)	65.39
hafnium (Hf)	178.5	praesodynium (Pr)	140.91	zirconium (Zr)	91.22
helium (He)	4.00	rhenium (Re)	186.2		

List of Symbols

a	coefficient, distance	G	Gibbs energy
a_0	Bohr radius	G_m	molar Gibbs energy
b	coefficient	H	enthalpy
b	contour length per monomer unit	H_m	molar enthalpy
c	coefficient	H	Hamilton function
c_0	speed of light in vacuum	$\mathcal{H}$	Hamilton operator
d, d_0	distance, bond length	I	electric current
e	elementary charge	I	intensity of light
f	force	K	equilibrium constant
g	number of quantum states	L	length
h	Planck's constant	L	contour length of chain
h	distance between chain ends	M_m	molar mass
i, j	indices	$M_{monomer}$	molar mass of monomer
k	Boltzman constant	N	number of particles
k_f	force constant	N_A	Avogadro constant
k_r	rate constant	P	pressure
l	length	^{osm}P	osmotic pressure
$l_{segment}$	length of segment	P	bond order
l	length of statistical chain element	P	probability
		Q	electric charge
m_e	rest mass of electron	R	maximum extension of coil
m_p	rest mass of proton	$\boldsymbol{R}$	gas constant
n	amount of substance, quantum number	S	entropy
		S_m	molar entropy
q	heat	T	absolute temperature
r	distance	T	kinetic energy
r_{coil}	radius of coil	U	internal energy
s	coordinate	U_m	molar internal energy
t	time	V	volume
u, v	speed	V_m	molar volume
x, y, z	coordinates	V	electric voltage
z	collision number	V	potential energy
A	area, Helmholtz energy	Z	number of monomer units
B	magnetic flux density	Z	partition function
C	capacity	$P^{\ominus}$	standard pressure $= 1$ bar
C_V, C_P	heat capacity	$c^{\ominus}$	standard concentration $= 1$ mol L^{-1}
D	diffusion constant	$\widehat{P}$	$= P/P^{\ominus}$
E	energy	$\widehat{c}$	$= c/c^{\ominus}$
E_a	energy of activation	$U^{\ominus}, H^{\ominus}$	U, H under standard conditions
$E_{a,m}$	molar energy of activation	$S^{\ominus}$	S under standard conditions
E	electromotive force	$G^{\ominus}$	G under standard conditions
F	electric field strength	$C_V^{\ominus}, C_P^{\ominus}$	C_V, C_P under standard conditions
$\boldsymbol{F}$	Faraday constant	$\Delta_f H$	molar enthalpy of formation

LIST OF SYMBOLS

$\Delta_f G$	molar Gibbs energy of formation	λ	wavelength
$\Delta_r H$	molar change of H in reaction	λ	free path
$\Delta_r S$	molar change of S in reaction	μ	dipole moment
$\Delta_r G$	molar change of G in reaction	μ	chemical potential
$\Delta_{vap} H$	molar enthalpy of evaporization	μ	reduced mass
$\Delta_{fus} H$	molar enthalpy of melting	μ_0	permeability of the vacuum
$C_{V,m}$	molar heat capacity at constant volume	μ_B	Bohr magneton
$C_{P,m}$	molar heat capacity at constant pressure	ν	frequency
$\widehat{n}$	refractive index	ξ	coordinate
α	steric factor, angle	ρ	electron density
α	polarizability	σ	cross section
α	stoichiometric factor	σ	screening constant
α	Lagrange multiplicator	τ	lifetime
β	Lagrange multiplicator	ϕ	wavefunction
β	angle	ϕ	angle
γ	activity coefficient	φ	quantum efficiency
γ	stoichiometric factor	χ	electric susceptibility
δ	change in length	ψ	wavefunction
δ	chemical shift	ω	circular frequency
ε	energy	ω	number of configurations
ε	molar decadic absorption coefficient	Δ	change of quantity
ε	relative permittivity	Λ	deBroglie wavelength
ε_0	permittivity of the vacuum	Θ	angle
ζ	coordinate	Π	surface pressure
η	viscosity	Φ	wavefunction
θ	angle	Ψ	wavefunction
ϑ	angle	Ω	number of configurations
κ	electric conductivity		

Appendices

Appendix 1

Enthalpy of formation $\Delta_f H^\ominus_{298}$, entropy $S^\ominus_{298}$, Gibbs energy of formation $\Delta_f G^\ominus_{298}$, and heat capacity $C^\ominus_{P,298}$ at 298.15 K. Solids (s) and liquids (l) under $P = 1$ bar; gases (g) under $P = 1$ bar in ideal state; substances dissolved in water (aq) under $c = 1\,\text{mol L}^{-1}$ in ideal state. The data are corrected for an assumed ideal gas under $P = 1$ bar and an assumed ideally diluted solution under $c = 1\,\text{mol L}^{-1}$.

References: Landolt-Börnstein, *Zahlenwerte und Funktionen*, 2. Bd., 4. Teil, Springer, Berlin 1961

F.D. Rossini, D.D. Wagman, W.H. Evans, S. Levine, I. Jaffe, *Selected Values of Thermodynamic Properties*, Natl. Bur. Stand. Circ. 500 (1952)

D.D. Wagman, W.H. Evans, V.B. Parker, I. Halow, S.M. Bailey, R.H. Schumm, *Selected Values of Chemical Thermodynamic Properties*, Technincal Notes 270-3 and 270-4 (Natl. Bur. Stand., Washington 1968, 1969

I. Barin, O. Knacke, *Thermochemical Properties of Inorganic Substances*, Springer, Berlin 1973

F.D. Rossini, K.S. Pitzer, R.L. Arnett, R.M. Braun, G.C. Pimentel, *Selected Values of Physical and Thermodynamic Properties of Hydrocarbons and Related Compounds*, Carnegie Press, Pittsburgh 1953

D.R. Stull, H. Prophet, *JANAF Thermodynamical Tables*, Natl. Stand. Ref. Data Ser., Natl. Bur. Stand. 37 (1971)

H.D. Brow, *Biochemical Microcalorimetry*, Academic Press, New York 1969

R. Chang, *Physical Chemistry with Applications to Biological Systems*, Macmillan, London 1978

R.C. Weast, *Handbook of Physics and Chemistry*, CRC Press, Cleveland 1998

substance	state	$\Delta_f H^\ominus_{298}$ kJ mol^{-1}	$S^\ominus_{m,298}$ J K^{-1} mol^{-1}	$\Delta_f G^\ominus_{298}$ kJ mol^{-1}	$C^\ominus_{p,m,298}$ J K^{-1} mol^{-1}
Ag	s	0	42.7	0	25.5
Ag$^+$	aq	105.9		77.1	
AgCl	s	−126.1	96.1	−108.7	50.8
AgI	s	−62.4	114.2	−66.3	54.4

substance	state	$\Delta_f H^\ominus_{298}$ kJ mol^{-1}	$S^\ominus_{m,298}$ J K^{-1} mol^{-1}	$\Delta_f G^\ominus_{298}$ kJ mol^{-1}	$C^\ominus_{p,m,298}$ J K^{-1} mol^{-1}
Al	s	0	28.3	0	24.3
Al^{3+}	aq	−524.7		−481.2	
Al$_2$O$_3$	s	−1675.3	51.0	−1581.9	79.0
Ar	g	0	154.8	0	20.8
Au	s	0	47.4	0	25.2
Au^{3+}	aq			137.0	
AuCl$_4^-$	aq	−325.5		−235.1	
Br$_2$	l	0	152.1	0	55.7
Br$^-$	aq	−120.9		102.9	
C (graphite)	s	0	5.7	0	8.6
C (diamond)	s	1.9	2.4	2.9	6.1
CO	g	−110.5	197.9	−137.3	29.1
CO$_2$	g	−393.5	213.6	−394.4	37.1
CO$_3^-$	aq	−676.3		−528.1	
CN$^-$	aq	151.0		165.7	
(CN)$_2$	g	−308.9	242.3	−320.5	56.9
Ca	s	0	41.6	0	26.3
Ca^{2+}	aq	−543.0		−553.0	
CaO	s	−635.1	39.7	−604.0	42.8
CaCO$_3$ (calcite)	s	−1206	92.9	−1128.0	81.9
Cd	s	0	51.8	0	26.0
Cd^{2+}	aq	−72.4		−77.7	
Cl$_2$	g	0	223.0	0	33.9
Cl$^-$	aq	−167.5		−131.2	
HCl	g	−92.3	186.7	−95.3	29.1
HCl	aq	−167.5		−131.2	
Cu	s	0	33.1	0	24.5
Cu^{1+}	aq	51.9		50.2	
Cu^{2+}	aq	64.4		65.0	
[Cu(NH$_3$)$_4$]$^{2-}$	aq	−334.4		−256	
[Cu(CN)$_4$]$^{3-}$	aq			566.5	
F$_2$	g	0	202.7	0	31.5
F$^-$	aq	−329.1		−276.5	
Fe	s	0	27.3	0	25.2
Fe^{2+}	aq	−87.9		−84.90	
Fe^{3+}	aq	−47.7		−10.5	
Fe$_2$O$_3$	s	−825.5	87.4	−743.6	104.6
FeS	s	−95.1	67.4	−97.6	50.5
H	g	218.0	114.6	203.3	20.8
H$_2$	g	0	130.6	0	28.8
H$^+$	aq	0		0	

substance	state	$\Delta_f H^\ominus_{298}$ kJ mol^{-1}	$S^\ominus_{m,298}$ J K^{-1} mol^{-1}	$\Delta_f G^\ominus_{298}$ kJ mol^{-1}	$C^\ominus_{p,m,298}$ J K^{-1} mol^{-1}
Hg	l	0	76.0	0	27.8
Hg_2^{2+}	aq			153.6	
Hg^{2+}	aq			174.6	
Hg_2Cl_2	s	−264.9	192.5	−210.5	101.6
I	g	106.6	180.7		20.8
I_2	s	0	116.7	0	55.0
I^-	aq	55.9		51.7	
HI	g	25.9	206.3	1.3	29.2
K	s	0	64.7	0	29.5
K^+	aq	−251.2		−282.0	
Li	s	0	29.1	0	23.6
Li^+	aq	−278.5		−293.8	
N_2	g	0	191.5	0	29.1
NH_3	g	−46.2	192.5	−16.6	35.7
NH_4^+	aq	−132.8		−79.5	
NO	g	90.3	210.6	86.7	29.9
NO_2	g	33.1	240.0	51.2	37.9
NO_3^-	aq	−334.4		−256	
Na	s	0	51.4	0	28.4
Na^+	aq	−239.7		−261.9	
NaCl	s	−411.1	72.1	−384.0	49.7
NaCl	aq	−407.2			
NaOH	s	−732.9	84.5	−623.4	59.5
$Na_2SO_4 \cdot 10H_2O$	s	−4324	592.9	−3644	575.7
$Na_2SO_4 \cdot 10H_2O$	aq	−4245			
O	g	249.2	161.0	231.8	21.9
O_2	g	0	205.0	0	29.4
OH^-	aq	−229.9		−157.3	
H_2O	l	−285.9	69.9	−237.2	75.3
H_2O	g	−241.8	188.7	−228.6	33.6
Pb	s	0	64.8	0	26.8
Pb^{2+}	aq	1.6		−24.3	
PbI_2	s	−175.1	175.2	−173.4	81.2
PbO_2	s	−270.1	76.5	−212.4	64.4
S (rhombic)	s	0	31.9	0	22.6
Zn	s	0	41.6	0	25.1
Zn^{2+}	aq	−152.4		−147.2	
$[Zn(CN)_4]^{2-}$	aq	343.1		446.9	
acetate	aq	−485.6		−376.9	
acetic acid	aq	−485.3		−404.1	
acetic acid	l	−487.0	159.8	−392.5	123.4
alanine	aq	−554.9		−371.3	
alanine	s	−563.7		−372.0	
benzene	g	82.9	269.2	129.7	81.7
benzene	l	49.0	173.2		136.1
butadiene	g	111.9	278.7	152.4	79.5
cyclohexane	g	−123.1	298.2	31.7	106.3
cyclohexane	l	−156.2	204.1	26.8	156.5

substance	state	$\Delta_f H^\ominus_{298}$ kJ mol^{-1}	$S^\ominus_{298}$ J K^{-1} mol^{-1}	$\Delta_f G^\ominus_{298}$ kJ mol^{-1}	$C^\ominus_{p,298}$ J K^{-1} mol^{-1}
cyclohexane	g	−7.1	310.7	105.1	105.0
cyclohexane	l		216.2		140.2
ethane	g	−84.7	225.5	−32.9	52.7
ethanol	aq	−278.7		−182.0	
ethanol	l	−227.6	161.0	−174.8	112.0
ethene	g	52.3	219.5	68.1	43.6
ethyne	g	226.7	200.8	209.8	43.9
formic acid	l	−409.2	129.0	−346.0	99.0
formiate	aq	−410.0		−334.7	
glucose (α-D)	aq	−1263.1	264.0	−914.5	
glucose (α-D)	s	−1274.5	212.1	−910.6	218.9
glycine	aq	−523.1		−379.9	
glycine	s	−537.3		−377.8	
glycyl-alanine	aq			−733.9	
n-hexane	g	−167.2	386.8	0.3	146.7
n-hexane	l	−198.8	296.0	−4.3	195.0
hydroquinone	s	−363.0			142.0
methanol	aq	−245.9	−175.2		
methanol	l	−238.7	126.8	−166.3	81.6
phenol	s	−162.8	142.0	−47.5	134.7
polyethylene	s	−53.7	45.0	14.1	51.0
quinone	s	−186.8			132.0
saccharose	s	−2221	360	−1542	425

Appendix 2

Change of Gibbs energy $\Delta_r G^\ominus_{298}$ and $\Delta_r G^{\ominus'}_{298}$ (at pH7) and standard potentials $E^\ominus_{298}$ at 298.15 K for reactions in aqueous solution, according to the scheme: reactant + $\frac{1}{2}$H$_2$ → product + H$^+$.

References: D. Dobos, *Electrochemical Data*, Elsevier, Amsterdam 1975 I.G. Morris, *Physikalische Chemie für Biologen*, Verlag Chemie, Weinheim 1976

reaction		$\Delta_r G^\ominus_{298}$ kJ mol^{-1}	$E^\ominus_{298}$ mV	$\Delta_r G^{\ominus'}_{298}$ kJ mol^{-1}
Ag$^+$ + $\frac{1}{2}$H$_2$	→ Ag + J$^+$	−77.2	+800	117.2
$\frac{1}{3}$Al^{3+} + $\frac{1}{2}$H$_2$	→ $\frac{1}{3}$Al + H$^+$	+164.6	−1706	−124.6
$\frac{1}{3}$Au^{3+} + $\frac{1}{2}$H$_2$	→ $\frac{1}{3}$Au + H$^+$	−137	+1420	177.0
$\frac{1}{3}$[AuCl$_4$]$^-$ + $\frac{1}{2}$H$_2$	→ $\frac{1}{3}$Au + $\frac{4}{3}$Cl$^-$ + H$^+$	−95.5	+994	135.9
$\frac{1}{2}$Br$_2$ + $\frac{1}{2}$H$_2$	→ Br$^-$ + H$^+$	−104.9	1087	144.9
$\frac{1}{2}$Ca^{2+} + $\frac{1}{2}$H$_2$	→ $\frac{1}{2}$Ca + H$^+$	276.9	−2870	−236.9
$\frac{1}{2}$Cl$_2$ + $\frac{1}{2}$H$_2$	→ Cl$^-$ + H$^+$	−131.0	1358	171.0
Cu$^+$ + $\frac{1}{2}$H$_2$	→ Cu + H$^+$	−50.4	522	−90.4
$\frac{1}{2}$Cu^{2+} + $\frac{1}{2}$H$_2$	→ $\frac{1}{2}$Cu + H$^+$	−32.8	340	−72.8
$\frac{1}{2}$Cd^{2+} + $\frac{1}{2}$H$_2$	→ $\frac{1}{2}$Cd + H$^+$	38.9	−403	−1.1

reaction		$\Delta_r G^\ominus_{298}$ kJ mol$^-$	$E^\ominus_{298}$ mV	$\Delta_r G^{\ominus'}_{298}$ kJ mol^{-1}
$\frac{1}{2}F_2 + \frac{1}{2}H_2$	$\to F^- + H^+$	−276.9	2870	−316.9
$\frac{1}{2}Fe^{2+} + \frac{1}{2}H_2$	$\to \frac{1}{2}Fe + H^+$	42.5	−441	2.5
$\frac{1}{3}Fe^{3+} + \frac{1}{2}H_2$	$\to \frac{1}{3}Fe + H^+$	3.5	−36	−36.5
$Fe^{3+} + \frac{1}{2}H_2$	$\to Fe^{2+} + H^+$	−74.3	770	−114.3
$\frac{1}{2}Hg_2^{2+} + \frac{1}{2}H_2$	$\to Hg + H^+$	−76.8	796	−116.8
$\frac{1}{2}Hg^{2+} + \frac{1}{2}H_2$	$\to \frac{1}{2}Hg + H^+$	−87.3	905	−127.3
$\frac{1}{2}I_2 + \frac{1}{2}H_2$	$\to I^- + H^+$	−51.6	535	−91.6
$K^+ + \frac{1}{2}H_2$	$\to K + H^+$	282.1	−2924	242.1
$Li^+ + \frac{1}{2}H_2$	$\to Li + H^+$	293.8	−3045	253.88
$Na^+ + \frac{1}{2}H_2$	$\to Na + H^+$	261.6	−2711	221.6
$\frac{1}{2}Pb^{2+} + \frac{1}{2}H_2$	$\to \frac{1}{2}Pb + H^+$	12.2	−126	−27.8
$\frac{1}{2}Zn^{2+} + \frac{1}{2}H_2$	$\to \frac{1}{2}Zn + H^+$	73.6	−763	33.6
$\frac{1}{2}[Zn(CN)_4]^{2-} + \frac{1}{2}H_2$	$\to \frac{1}{2}Zn + 2CN^- + H^+$	108	−1120	68.0
$\frac{1}{2}$Quinone $+ \frac{1}{2}H_2$	$\to \frac{1}{2}$Hydroquinone $+ H^+$	−67.5	700	−67.5
$\frac{1}{2}$methylenblue $+ \frac{1}{2}H_2$	$\to \frac{1}{2}$leucomethylenblue $+ H^+$	−41.0	425	−41.0
$\frac{1}{2}NAD^+ + \frac{1}{2}H_2$	$\to \frac{1}{2}NADH + H^+$	21.9	227	−18.1
Cytb$(Fe^{3+}) + \frac{1}{2}H_2$	$\to$ Cytb$(Fe^{2+}) + H^+$	−3.9	40	−43.9
Cytc$_1$$(Fe^{3+}) + \frac{1}{2}H_2$	$\to$ Cytc$_1$$(Fe^{2+}) + H^+$	−21.2	220	−61.2

Appendix 3 Useful Mathematical Relations

Stirling's formula (as derived in Box 12.4):

$$\ln N! = N \cdot \ln N - N, \quad N! = N^N \cdot e^{-N}$$

A more sophisticated form of Stirling's formula is

$$\ln N! = N \cdot \ln N - N + \tfrac{1}{2} \cdot \ln(2\pi N), \quad N! = \sqrt{2\pi N} \cdot N^N \cdot e^{-N}$$

Useful numbers

$$e + \lim_{n\to\infty}\left(1 + \frac{1}{n}\right)^n = 2.718282, \quad \ln 10 = 2.302585, \quad \pi = 3.141593$$

Undefinite Integrals

$$\int \frac{1}{a-x} \cdot dx = -\ln(a-x)$$

$$\int \frac{1}{(a-x)(b-x)} \cdot dx = \frac{1}{b-a} \cdot \ln \frac{b-x}{a-x}$$

$$\int \frac{x}{(ax^2-b)^2} \cdot dx = -\frac{1}{2a} \cdot \frac{1}{ab^2-b}$$

APPENDIX 3 USEFUL MATHEMATICAL RELATIONS

$$\int \sqrt{1-x}\,dx = -\frac{2}{3}\sqrt{(1-x)^3}$$

$$\int \sqrt{a^2-x^2}\,dx = \frac{1}{2}a^2 \cdot \left(\arcsin\frac{x}{a} + \frac{x}{a}\sqrt{1-\frac{x^2}{a^2}}\right)$$

$$\int \frac{1}{\sqrt{a-x}}\,dx = -2\sqrt{a-x}$$

$$\int \frac{x}{\sqrt{1-x^2}}\,dx = -\sqrt{1-x^2}$$

$$\int x\cos x\,dx = x\sin x + \cos x$$

$$\int x^n \cos x\,dx = x^n \sin x - n\int x^{n-1}\sin x\,dx$$

$$\int \sin x \cos x\,dx = \frac{1}{2}\sin^2 x$$

$$\int \sin nx \cos kx\,dx = \frac{1}{2}\cdot\left[\frac{-\cos(n+k)x}{n+k} - \frac{\cos(n-k)x}{n-k}\right] \quad \text{for } k\neq n$$

$$\int \sin nx \sin kx\,dx = \frac{1}{2}\cdot\left[\frac{-\sin(n+k)x}{n+k} - \frac{\sin(n-k)x}{n-k}\right] \quad \text{for } k\neq n$$

$$\int \sin^2 x\,dx = \frac{1}{2}(x - \cos x \sin x) = \frac{1}{2}\left(x - \frac{1}{2}\sin 2x\right)$$

$$\int \cos^2 x\,dx = \frac{1}{2}(x + \cos x \sin x) = \frac{1}{2}\left(x + \frac{1}{2}\sin 2x\right)$$

$$\int \sin x \sin 2x\,dx = -\frac{2}{3}\sin x \cos 2x + \frac{1}{3}\cos x \sin 2x$$

$$\int \sin 2x \cos x\,dx = \frac{1}{3}\sin^3 x$$

$$\int \arcsin x\,dx = x\arcsin x + \sqrt{1-x^2}$$

$$\int x\cdot e^{-x}\,dx = -(x+1)\cdot e^{-x}$$

$$\int x^2\cdot e^{-x}\,dx = -(x^2+2x+2)\cdot e^{-x}$$

$$\int x^n\cdot e^{-x}\,dx = -x^n\cdot e^{-x} + n\int x^{n-1}\cdot e^{-x}\,dx$$

$$\int x\cdot e^{-x^2}\,dx = -\frac{1}{2}e^{-x^2}$$

$$\int \ln x\,dx = x\cdot(\ln x - 1)$$

$$\int x \cdot \ln x \cdot dx = \frac{1}{2}x^2(\ln x - \frac{1}{2})$$

$$\int \frac{\ln x}{x} \cdot dx = \frac{1}{2}(\ln x)^2$$

Definite Integrals

$$\int_0^\infty x^n \cdot e^{-x} \cdot dx = n!$$

$$\int_0^\infty \frac{x^n}{e^x - 1} \cdot dx = n! \cdot \sum_{k=1}^\infty \frac{1}{k^{n+1}}$$

$$\int_0^\infty e^{-x^2} \cdot dx \frac{1}{2}\sqrt{\pi}$$

$$\int_0^\infty x \cdot e^{-x^2} \cdot dx = \frac{1}{2}$$

$$\int_0^\infty x^2 \cdot e^{-x^2} \cdot dx = \frac{1}{2} \int_0^\infty e^{-x^2} \cdot dx = \frac{1}{4}\sqrt{\pi}$$

$$\int_0^\infty x^3 \cdot e^{-x^2} \cdot dx = \frac{1}{2}$$

$$\int_0^\infty x^4 \cdot e^{-x^2} \cdot dx = \frac{3}{2} \int_0^\infty x^2 \cdot e^{-x^2} \cdot dx \frac{3}{8}\sqrt{\pi}$$

$$\int_0^\infty x^{2n} \cdot e^{-x^2} \cdot dx = \frac{1 \times 2 \times 3 \ldots \times (2n-1)}{2^{n+1}}\sqrt{\pi}$$

Goniometric Functions

$$\sin \alpha = \cos\left(\frac{\pi}{2} - \alpha\right) = \sqrt{1 - \cos^2 \alpha}$$

$$\sin(\alpha + \beta) = \sin \alpha \cos \beta + \cos \alpha \sin \beta$$

$$\sin(\alpha - \beta) = \sin \alpha \cos \beta - \cos \alpha \sin \beta$$

$$\cos(\alpha + \beta) = \cos \alpha \cos \beta - \sin \alpha \sin \beta$$

$$\cos(\alpha - \beta) = \cos \alpha \cos \beta + \sin \alpha \sin \beta$$

$$\sin \alpha + \sin \beta = 2 \sin \frac{\alpha + \beta}{2} \cos \frac{\alpha - \beta}{2}$$

MATHEMATICAL APPENDIX

$$\sin\alpha - \sin\beta = 2\cos\frac{\alpha+\beta}{2}\cos\frac{\alpha-\beta}{2}$$

$$\cos\alpha + \cos\beta = 2\cos\frac{\alpha+\beta}{2}\cos\frac{\alpha-\beta}{2}$$

$$\cos\alpha - \cos\beta = -2\sin\frac{\alpha+\beta}{2}\cos\frac{\alpha-\beta}{2}$$

$$\sin\alpha + \cos\beta = 2\sin\left(\frac{\pi}{4} + \frac{\alpha-\beta}{2}\right)\cos\left(-\frac{\pi}{4} + \frac{\alpha+\beta}{2}\right)$$

$$\sin\alpha - \cos\beta = 2\cos\left(\frac{\pi}{4} + \frac{\alpha-\beta}{2}\right)\sin\left(-\frac{\pi}{4} + \frac{\alpha+\beta}{2}\right)$$

Power Series

$$e^x = 1 + x + \frac{1}{2!}x^2 + \frac{1}{3!}x^3 + \ldots$$

$$\ln(1+x) = x - \frac{1}{2}x^2 + \frac{1}{3}x^3 - \frac{1}{4}x^4 + \ldots \text{ for } x < 1$$

$$\ln(1-x) = -x - \frac{1}{2}x^2 - \frac{1}{3}x^3 - \frac{1}{4}x^4 + \ldots \text{ for } x < 1$$

$$\ln\frac{1+x}{1-x} = 2x + \frac{2}{3}x^3 + \frac{2}{5}x^5 + \ldots \text{ for } x < 1$$

$$\sin x = x - \frac{1}{3!}x^3 + \frac{1}{5!}x^5 + \ldots$$

$$\cos x = 1 - \frac{1}{2!}x^2 + \frac{1}{4!}x^4 + \ldots$$

$$(1+x)^n = 1 + \frac{n}{1!}x + \frac{n(n-1)}{2!}x^2 + \frac{n(n-1)(n-2)}{3!}x^3 + \ldots$$

$$\frac{1}{1+x} = 1 - x + x^2 - x^3 + \ldots$$

$$\sqrt{1+x} = 1 + \frac{1}{2}x - \frac{1}{8}x^2 + \frac{1}{16}x^3 - \ldots$$

Mathematical Appendix

We prove that

$$\cos\alpha - \sin(\alpha - \beta) = \sqrt{2} \cdot \sqrt{1 + \sin\beta} \cdot \cos\left(\alpha - \frac{\beta}{2} + \frac{\pi}{4}\right)$$

Solution. Using the relations

$$\sin\alpha = -\cos\left(\alpha + \frac{\pi}{2}\right)$$

$$\cos\alpha + \cos\beta = 2\cos\frac{\alpha+\beta}{2}\cos\frac{\alpha-\beta}{2}$$

we obtain

$$\cos\alpha - \sin(\alpha - \beta) = \cos\alpha + \cos\left(\alpha - \beta + \frac{\pi}{2}\right)$$
$$= 2\cos\frac{2\alpha - \beta + \pi/2}{2}\cos\frac{\beta - \pi/2}{2}$$
$$= 2\cos\left(\alpha - \frac{\beta}{2} + \frac{\pi}{4}\right)\cos\left(\frac{\beta}{2} - \frac{\pi}{4}\right)$$

Because of the relation

$$\cos(\alpha - \beta) = \cos\alpha\cos\beta + \sin\alpha\sin\beta$$

it follows that

$$\cos\left(\frac{\beta}{2} - \frac{\pi}{4}\right) = \cos\frac{\beta}{2}\cos\frac{\pi}{4} + \sin\frac{\beta}{2}\sin\frac{\pi}{4}$$
$$= \frac{1}{\sqrt{2}}\left(\cos\frac{\beta}{2} + \sin\frac{\beta}{2}\right)$$

Furthermore,

$$\left(\cos\frac{\beta}{2} + \sin\frac{\beta}{2}\right)^2 = \cos^2\frac{\beta}{2} + 2\cos\frac{\beta}{2}\sin\frac{\beta}{2} + \sin^2\frac{\beta}{2}$$
$$= 1 + 2\cos\frac{\beta}{2}\sin\frac{\beta}{2} = 1 + \sin\beta$$

(because of $\cos\alpha\sin\alpha = \frac{1}{2}\sin 2\alpha$). Thus

$$\cos\alpha - \sin(\alpha - \beta) = 2\cos\left(\alpha - \frac{\beta}{2} + \frac{\pi}{4}\right)\frac{1}{\sqrt{2}}\sqrt{1 + \sin\beta}$$

Further Reading

Textbooks in Physical Chemistry

E.A. Molwyn-Hughes, *Physical Chemistry*, Pergamon Press, London 1957
J. Rose, *Dynamical Physical Chemistry*, Pitman 1961
N.K. Adam, *Physical Chemistry*, Clarendon, Oxford 1962
G.H. Duffey, *Physical Chemistry*, McGraw Hill, New York 1962
W.J. Moore, *Physical Chemistry*, Prentice-Hall, 1962
J.C. Slater, *Introduction to Chemical Physics*, McGraw Hill, New York 1963
S.H. Maron, C.F. Prutton, Macmillan, New York 1965
W.F. Sheehan, *Physical Chemistry*, Allyn and Bacon, Boston 1970
G.W. Castellan, *Physical Chemistry*, Addison-Wesley, Menlo Park 1971
A.W. Adamson, *A Textbook of Physical Chemistry*, Academic Press, New York 1973
R.M. Rosenberg, *Principles of Physical Chemistry*, Oxford University Press, New York 1977
R. Chang, *Physical Chemistry with Applications to Biological Systems*, London 1978
B.A.W. Coller, I.R. McKinnon, I.R. Wilson, *Principles of Physical Chemistry*, Arnold, London 1978
R.S. Berry, S.A. Rice, J. Ross, Physical Chemistry, Wiley, New York 1980
J. Tinoco, K. Sauer, J.C. Wang, *Physical Chemistry, Principles and Applications in Biological Sciences*, Prentice-Hall 1985
R.A. Alberty and R.J. Silbey, *Physical Chemistry*, Wiley, New York 1992
G.M. Barrow, *Physical Chemistry*, McGraw Hill, New York 1996
D.A. Quarrie, J.D. Simon, *Physical Chemistry*, Univ. Science Books, Sansalito, 1997
C.E. Dykstra, *Physical Chemistry*, Prentice Hall, 1997
P.W. Atkins, *Physical Chemistry*, Oxford University Press, Oxford 1998

Chapter 1

A. Einstein, Über einen die Erzeugung und Verwandlung des Lichts betreffenden heuristischen Gesichtspunkt, *Ann. Phys.* (Leipzig) 17, 132 (1905)
A. Einstein, Über die vom Relativitätsprinzip geforderte Trägheit der Energie, *Ann. Phys.* (Leipzig) 23, 371 (1907)
R. Millikan, A Direct Photoelectric Determination of Planck's "h", *Phys. Rev.* 7, 355 (1916)

M.L. DeBroglie, Recherches sur la Theorie des Quanta, *Ann. Phys.* (Paris) 3, 22 (1925)

C. Davisson, L.H. Germer, Diffraction of Electrons by a Crystal of Nickel, *Phys. Rev.* 30, 705 (1927)

C. Jönsson, D. Brandt, S. Hirschi, Electron Diffraction at Multiple Slits, *Am. J. Phys.* 42, 4 (1974)

B. Cagnac, J.C. Pebay-Peyroula, *Modern Atomic Physics: Fundamental Principles*, Macmillan, London 1975

K.J. Laidler, *The World of Physical Chemistry*, Oxford Univ. Press, Oxford 1993

Chapter 2

R.L. Snow, J.L. Bills, The Pauli Principle and Electronic Repulsion in Helium, *J. Chem. Educ.* 51, 585 (1974)

J.W. Moore, W.G. Davies, Illustration of Some Consequences of the Indistinguishabilty of Electrons, *J. Chem. Educ.* 52, 312 (1975)

K.M. Jinks, A Particle in a Chemical Box, *J. Chem. Educ.* 52, 312 (1975)

F.O. Ellison, C.A. Hollingsworth, The Probability Equals Zero Problem in Quantum Mechanics, *J. Chem. Educ.* 53, 767 (1976)

A. Shimony, The Reality of the Quantum World, *Scient. Amer.* Jan. 1988, 36

C. Dykstra, *Quantum Chemistry and Molecular Spectroscopy*, Prentice Hall, Englewood Cliffs 1992

K. Volkamer, M.W. Lerom, More About the Particle-in-a-box System: the Confinement of Matter and the Wave-particle Dualism, *J. Chem. Educ.* 69, 100 (1992)

L.C. Venema, J.W.G. Wildöer, J.W. Janssen, S.J. Tans, H.L.J.T. Tuinstra, L.P. Kouwenhoven, C. Dekker, Imaging Electron Wave Functions of Quantized Energy Levels in Carbon Nanotubes, *Science* 283, 52 (1999)

Chapter 3

W. Heisenberg, *The Physical Principles of Quantum Theory*, University of Chicago Press, Chicago 1930

J.C. Slater, The Virial and Molecular Structure, *J. Chem. Phys.* 1, 687 (1935)

M. Jammer, *The Conceptual Development of Quantum Mechanics*, McGraw-Hill, New York 1966

F. Hund, *The History of Quantum Theory*, Harrap, London 1974

C.E. Dykstra, *Ab Initio Calculation of the Structures and Properties of Molecules*, Elsevier, Amsterdam 1988

W.J. Moore, Schrödinger: *Life and Thought*, Cambridge University Press, Cambridge 1989

A. Pais, *Niels Bohr's Times: in Physics, Philosophy and Polity*, Clarendon Press, Oxford 1991

D.C. Cassidy, Uncertainty: *The Life and Science of Werner Heisenberg*, W.H. Freeman, New York 1992

G.C. Schatz, M.A. Ratner, *Quantum Mechanics in Chemistry*, Prentice Hall, Englewood Cliffs, New Jersey 1993

M. Nauenburg, C. Stroud, J. Yeazell, The Classical Limit of an Atom, *Scient. Amer.* June 1994, 24

J.D. McGervey, *Quantum Mechanics*, Academic Press, New York 1995

P.L. Lang, Visualization of Wavefunctions Using Mathematica, *J. Chem. Educ.* 75, 506 (1998)

Chapter 4

L. Pauling, E.B. Wilson, *Introduction to Quantum Mechanics*, McGraw-Hill, New York 1935

J.O. Hirschfelder, J.F. Kincaid. Application of the Virialtheorem to Approximate Molecular and Metallic Eigenfunctions, *Phys. Rev.* 52, 658 (1937)

C.A. Coulson, The Energy and Screening Constants of the Hydrogen Molecule, *Trans. Faraday Soc.* 33, 1479 (1937)

H. Wind, Electron Energy for H_2^+ in the Ground State, *J. Chem. Phys.* 42, 2371 (1965)

Ch.L. Beckel, B.D. Hansen, Theoretical Study of H_2^+ Ground Electronic State Spectroscopic Properties, *J. Chem. Phys.* 53, 3681 (1970)

D.M. Bishop, L.M. Cheung, Accurate One- and Two-electron Diatomic Molecular Calculations, in Advances in Quantum Chemistry, Vol. 12, Ed. Per-Olov Löwdin, Academic, New York 1980

J.N. Murrell, S.F.A. Kettle, J.M. Tedder, *The Chemical Bond*, Wiley, New York 1985

B. Webster, *Chemical Bonding Theory*, Blackwell Scientific, Oxford 1990

L. Kevan, M.K. Bowman (Eds.), *Modern Pulsed and Continuous-Wave Electron Spin Resonance*, Wiley, New York 1990

N.M. Artherton, *Principles of Electron Spin Resonance*, Ellis Horwood, PTR Prentice Hall, New York 1993

R. Parson, Visualizing the Variation Principle: an Intuitive Approach to Interpreting the Theorem in Geometric Terms, *J. Chem. Educ.* 70, 115 (1993)

M.J. Winter, *Chemical Bonding*, Oxford University Press, Oxford 1993

J.A. Weil, J.R. Bolton, J.E. Wertz, *Electron Paramagnetic Resonance: Elementary Theory and Practical Applications*, Wiley, New York 1994

A. Hincliffe, *Modelling Molecular Structures*, Wiley, New York 1996

Chapter 5

J.C. Slater, Atomic Shielding Constants, *Phys. Rev.* 36, 57 (1930)

L. Pauling, *The Nature of the Chemical Bond, Cornell University Press, Ithaca*, New York 1940

J.C. Slater, One-electron Energies of Atoms, Molecules and Solids, *Phys. Rev.* 98, 1039 (1955)

A.C. Wahl, Molecular Orbital Densities: Pictorial Approach, *Science* 151, 961 (1966)

M. Karplus, R.N. Porter, *Atoms and Molecules*, Benjamin, New York 1970

G. Simons, M.E. Zandler, E.R. Talaty, Nonempirical Electronegativity Scale, *J. Am. Chem. Soc.* 98, 7869 (1976)

E.U. Condon, H. Odabasi, *Atomic Structure*, Cambridge University Press, Cambridge 1980

K.D. Sen, *Electronegativity*. In Structure and Bonding, 6., Springer, New York 1987

N. Agmon, Ionization Potentials for Isoelectronic Series, *J. Chem. Educ.* 65, 42 (1988)

R.D. Allendoerfer, Teaching the Shapes of the Hydrogenlike and Hybrid Atomic Orbitals, *J. Chem. Educ.* 67, 37 (1990)

R.T. Myers, The Periodicity of Electron Affinity, *J. Chem. Educ.* 67, 307 (1990)

N.C. Pyper, M. Berry, Ionization Energies Revisited, *Educ. in Chem.* 27, 135 (1990)

P.G. Nelson, Relative Energies of 3d and 4s Orbitals, *Educ. in Chem.* 29, 84 (1992)

N. Shenkuan, The Physical Basis of Hund's Rule: Orbital Contraction Effects, *J. Chem. Educ.* 69, 800 (1992)

J.E. Howell, The Heaviest Elements, *J. Chem. Educ.* 76, 295 (1999)

D.C. Hoffman, D.M. Lee, Chemistry of the Heaviest Elements – One Atom at a Time, *J. Chem. Educ.* 76, 332 (1999)

Chapter 6

L. Pauling, *The Nature of the Chemical Bond*, Cornell University Press, Ithaca, New York 1940

M. Kotani, Y. Mizuno, K. Kayama, E. Ischiguno, Electronic Structure of Simple Homonuclear Diatomic Molecules, *J. Phys. Soc. Jpn.* 12, 707 (1957)

Ch.Y. Hsu, M. Orchin, A Simple Method for Generating Sets of Orthonormal Hybrid Atomic Orbitals, *J. Chem. Educ.* 50, 114 (1973)

A.B. Sannigrahi, T. Kir, Molecular Orbital Theory of Bond Order and Valency, *J. Chem. Educ.* 65, 647 (1988)

C.A. Hollingsworth, Accidental Degeneracies of the Particle in a Box, *J. Chem. Educ.* 67, 999 (1990)

A.D. Buckingham, T.W. Rowlands, Can Addition of a Bonding Electron Weaken a Bond?, *J. Chem. Educ.* 68, 282 (1991)

G.J. Miller, J.G. Verkade, A Pictorial Approach to Molecular Orbital Bonding in Polymers: Non-Mathematical but Honest, *J. Chem. Educ.* 76, 428 (1999)

Chapter 7

Table 7.4: D.W.J. Cruickshank, R.A. Sparks, *Proc. Roy. Soc.* London A258, 270 (1960); F.R. Ahmed, J. Trotter, *Acta Cryst.* 16, 503 (1963); N. Norman, B. Post, *Acta Cryst.* 14, 503 (1961); J.M. Robertson, H.M.M. Ahearer, G.A. Sim, D.G. Watson, *Acta Cryst.* 15, 1 (1962)

E. Hückel, Quantentheoretische Beiträge zum Problem der aromatischen und ungesättigten Verbindungen, *Z. Phys.* 76, 628 (1932)

J. Slater, One-electron Energies of Atoms, Molecules and Solids, *Phys. Rev.* 98, 1039 (1955)

H. Kuhn, Physical Basis of the Free-electron Gas Model of Branched Molecules, *J. Chem. Phys.* 25, 293 (1956)

F.Bär, W. Huber, G. Handschig, H. Kuhn, Nature of the Free Electron Model. The Case of the Polyenes and Polyacetylenes, *J. Chem. Phys.* 32, 370 (1960)

A. Streitwieser, *Molecular Orbital Theory for Organic Chemists*, New York 1961

K. Ruedenberg, The Physical Nature of the Chemical Bond, *Rev. Mod. Phys.* 34, 326 (1962)

E. Heilbronner, P.A. Straub, Hückel Molecular Orbitals, Springer, Berlin 1966

C.P. Smyth, Determination of Dipole Moments, in A. Weissberger, B.W. Rossiter (Edts.), *Techniques of Chemistry* 4, 397, Wiley-Interscience, New York 1972

G. Maier, H.O. Kalinowski, K. Euler, Cyclobutadien: Mesomerie oder Valenzisomerie?, *Angew. Chem.* 94, 706 (1982)

A. Hincliffe, *Computational Quantum Chemistry*, Wiley, New York 1988

J.R. Dias, A Facile Hückel Molecular Orbital Solution of Buckminsterfullerene Using Chemical Graph Theory, *J. Chem. Educ.* 66, 1012 (1989)

Chapter 8

Tables 8.2, 8.3, 8.4: S.S. Malhorta, Mc. Whiting, *J. Chem. Soc.* (London) 1960, 3812; H. Kuhn The electron Gas Theroy of the Color of Natural and Artificial Dyes, in *Progress in the Chemistry of Organic Natural Products* (ed. L. Zechmeister) 16, 169 (1952), 17, 404 (1959); F. Sondheimer, D.A. Ben-Efraim, R. Wolovsky, *J. Am. Chem. Soc.* 83, 1675 (1961)

H. Kuhn, A Quantum-Mechanical Theory of Light Absorption of Organic Dyes and Similar Compounds, *J. Chem. Phys.* 17, 1198 (1949)

G. Herzberg, *Molecular Spectra and Molecular Structure III. Electronic Spectra of Polyatomic Molecules*, D. van Nostrand Reinhold, New York 1950

W. Kauzman, *Quantum Chemistry*, Academic Press, New York 1957

S.J. Strickler, R.A. Berg, Relationship between Absorption Intensity and Fluorescence Lifetimes of Molecules, *J. Chem. Phys.* 37, 818 (1962)

F.P. Schäfer, *Dye Lasers*, Springer, Berlin 1973

J.E. Crooks, *The Spectrum in Chemistry*, Acad. Press, London 1978

J.R. Lakowicz, *Principles of Fluorescence Spectroscopy*, Plenum Press, New York 1983

K. Nassau, *The Physics and Chemistry of Color*, Wiley, New York 1983

C.N. Banwell, *Fundamentals of Molecular Spectroscopy*, McGraw-Hill, 1983

J.I. Steinfeld, *Molecules and Radiation*, MIT Press, Cambridge 1985

D. Rendel, *Fluorescence and Phosphorescence*, Wiley, New York 1987

W. Demtröder, *Laser Spectroscopy, Springer Series in Chemical Physics*, Vol. 5, Springer, Berlin 1988

D.L. Matthews, M.D. Rosen, Soft-X-Ray Lasers, *Scient. Amer.* December 1988, 60

V.S. Letokhov, Detecting Individual Atoms and Molecules with Lasers, *Scient. Amer.* September 1988, 54

A.E. Siegman, *Lasers, University Science Books*, Mill Valley 1988

W.S. Struve, *Fundamentals of Molecular Spectroscopy*, Wiley, New York 1989

H.P. Freund, R.K. Parker, Free-Electron Lasers, *Scient. Amer.* April 1989, 56

J. Eichler, H.J. Eichler, Laser, Springer, Berlin 1990

A.H. Zewail, The Birth of Molecules, *Scient. Amer.* Dec. 1990, 40

D.L. Andrews, *Lasers in Chemistry*, Springer, Berlin 1990

S. Svanberg, *Atomic and Molecular Spectroscopy*, Springer, Berlin 1991

D.L. Andrews, *Applied Laser Spectroscopy: Techniques, Instrumentation, and Applications*, Wiley, New York 1992

H.H. Perkampus, *UV-VIS Atlas of Organic Compounds*, Wiley-VCH, Weinheim 1992

G.C. Schatz, M.A. Ratner, *Quantum Mechanics in Chemistry*, Prentice Hall, Englewood Cliffs 1993

C. Reichardt, Solvatochromic Dyes as Solvent Polarity Probes, *Chem. Rev.* 94, 2319 (1994)

P.F. Bernath, *Spectra of Atoms and Molecules*, Oxford University Press, New York 1995

T. Basché (Ed.), *Optical Probing of Single Molecules*, Wiley, New York 1996

T. Basché, W.E. Moerner, M. Orrit, U.P. Wild (eds), *Single Molecule Optical Detection, Imaging and Spectroscopy*, Wiley-VCH, Weinheim 1997

M. Hollas, *High Resolution Spectroscopy*, Wiley-VCH, Weinheim 1998

J.J.C. Mulder, The π-Electron System of Monocyclic Polyenes $C_{2n}H_{2n}$ with Alternating Single and Double Bonds, *J. Chem. Educ.* 75, 594 (1998)

P.L. Gourley, Nanolasers, *Scient. Amer.* March 1998, 40

M.S. Feld, K. An, The Single-Atom Laser, *Scient. Amer.* August 1998, 56

R.M. Hochstrasser, Ultrafast Spectroscopy of Protein Dynamics, *J. Chem. Educ.* 75, 559 (1998)

J.L. McHale, *Molecular Spectroscopy*, Prentice Hall, Upper Saddle River, 1999

I.N. Levine, *Molecular Spectroscopy*, Wiley, New York 1999

Chapter 9

J. Strong, Pure Rotation Spectrum of the HCl Flame, *Phys. Rev.* 45, 877 (1934)

G. Herzberg, *Molecular Spectra and Molecular Structure II. Infrared and Raman Spectra of Polyatomic Molecules*, D. van Nostrand Reinhold, New York 1945

G. Herzberg, *Molecular Spectra and Molecular Structure I. Spectra of Diatomic Molecules*, D. van Nostrand Reinhold, New York 1950

C.H. Townes, A.L. Schalow, *Microwave Spectroscopy*, McGraw-Hill, New York 1955

E.B. Wilson, J.C. Decius, P.C. Cross, *Molecular Vibrations, Dover Publications*, New York 1955

E.K. Plyler, E.D. Tidwell, *The Rotational Constants of Hydrogen Chloride, Z. Elektrochem.* 64, 717 (1960)

R.B. Sanderson, Measurement of Rotational Line Strengths in HCl by Asymptotic Fourier Transform Techniques, *Appl. Opt.* 6, 1527 (1967)

W.H. Wing, G.A. Ruff, W.E. Lamb, J.J. Spezeski, Observation of the Infrared Spectrum of the Hydrogen Molecular Ion HD^+, *Phys. Rev. Lett.* 36, 1488, (1976)

J.E. Crooks, *The Spectrum in Chemistry*, Acad. Press, London 1978

K.P. Huber, G. Herzberg, *Molecular Spectra and Molecular Structure. IV. Constants of Diatomic Molecules*, D. Van Nostrand Reinhold, New York 1979

O. Sonnich-Mortensen, S. Hassing, *Advances in Infrared and Raman Spectroscopy 6*, Heyden, New York 1980

C.P. Slichter, *Principles of Magnetic Resonance*, Springer, Berlin 1980

R.K. Harris, *Nuclear Magnetic Resonance Spectroscopy: A Physicochemical View*, Pitman, London 1983

R.K. Harris, *Nuclear Magnetic Resonance Spectroscopy*, Longman, London 1986

J.K.M. Sanders, B.K. Hunter, *Modern NMR Spectroscopy*, Oxford University Press, Oxford 1987

D. Shaw, *Fourier Transform NMR Spectroscopy*, Elsevier, Amsterdam 1987

A.E. Derome, *Modern NMR Techniques for Chemistry Research*, Pergamon Press, Oxford 1987

P.L. Goodfriend, Diatomic Vibrations Revisited, *J. Chem. Educ.* 64, 753 (1987)

A.R. Lacey, A Student Introduction to Molecular Vibrations, *J. Chem. Educ.* 64, 756 (1987)

J.G. Verkade, A Novel Pictorial Approach to Teaching Molecular Motions in Polyatomic Molecules, *J. Chem. Educ.* 64, 411 (1987)

F. Ahmed, A Good Example of the Franck-Condon Principle, *J. Chem. Educ.* 64, 427 (1987)

R.B. Snadden, The Iodine Spectrum Revisited, *J. Chem. Educ.* 64, 919 (1987)

R.R. Ernst, G. Bodenhausen, A. Wokaun, *Principles of Nuclear Magnetic Resonance in One and Two Dimensions*, Clarendon Press, Oxford 1988

R. Freyman, A Handbook of Nuclear Magnetic Resonance, Longman, *Sci. and Techn.*, 1988

F.A. Bovey, L. Jelinski, P.A. Mirau, *Nuclear Magnetic Resonance Spectroscopy*, Academic Press, New York 1988

H.O. Kalinowski, S. Berger, S. Braun. *Carbon-13 NMR Spectroscopy*, Wiley, New York 1988

B. Schrader, *Raman/Infrared Atlas of Organic Compounds*, Wiley-VCH, Weinheim 1989

E.I. von Nagi-Felsobuki, Hückel Theory and Photoelectron Spectroscopy, *J. Chem. Educ.* 66, 821 (1989).

E. Breitmeier, W. Voelter, *Carbon-13 NMR Spectroscopy: High-Resolution Methods and Applications in Organic Chemistry*, Wiley-VCH, Weinheim 1990

C.P. Slichter, *Principles of Magnetic Resonance*, Springer, Berlin 1992

H.F. Blanck, Introduction to a Quantum Mechanical Harmonic Oscillator Using a Modified Particle-in-a-box Problem, *J. Chem. Educ.* 69, 98 (1992).

M. Buback, H.P. Vögele, *FT-NIR Atlas*, Wiley-VCH, Weinheim 1993

J.M. Hollas, *Modern Spectroscopy*, Wiley, New York 1996

T. Buyana, *Molecular Physics*, World Scientific, Singapore 1997

S. Braun, H.O. Kalinowski, S. Berger, *150 and More Basic NMR Experiments*, Wiley-VCH, Weinheim 1998

H. Friebolin, *Basic One- and Two-Dimensional NMR-Spectroscopy*, Wiley-VCH, Weinheim 1998

R.S. Macomber, *A Complete Introduction to Modern NMR Spectroscopy*, Wiley-VCH, Weinheim 1998

H. Günzler, H.M. Heise, *IR Spectroscopy*, Wiley-VCH, Weinheim 1999

J.L. McHale, *Molecular Spectroscopy*, Prentice Hall, Upper Saddle River, 1999

Chapter 10

F. London, The General Theory of Molecular Forces, *Trans. Faraday Soc.* 33, 8 (1937)

E.L. Burrows, S.F.A. Kettle, Madelung Constants and Other Lattice Sums, *J. Chem. Educ.* 52, 58 (1975)

P. Schuster, G. Zundel, C. Sandorfy, *The Hydrogen Bond, Recent Developments in Theory and Experiments*, North-Holland, Amsterdam 1976

P. Schuster, Intermolecular Forces – an Example of Fruitful Cooperation of Theory and Experiment, *Angew. Chem. Int. Ed. Engl.* 20, 546 (1981)

G.C. Maitland, M. Rigby, E.B. Smith, W.A. Wakeham, *Intermolecular Forces: Their Origin and Determination*, Clarendon Press, Oxford 1981

J. Israelachvili, *Intermolecular and Surface Forces*, Academic Press, New York 1985

M. Rigby, E.B. Smith, W.A. Wakeham, G.C. Maitland, *The Forces between Molecules*, Oxford University Press, Oxford 1986

A.D. Buckingham, P.W. Fowler, J.M. Hutson, Theoretical Studies of van der Waals Molecules and Intermolecular Forces, *Chem. Rev.* 88, 963 (1988)

C.E. Dykstra, Electrical Polarization in Diatomic Molecules, *J. Chem. Educ.* 65, 198 (1988)

C.E. Dykstra, Intermolecular Electrical Interaction: A Key Ingredient in Hydrogen Bonding, *Acc. Chem. Res.* 21, 355 (1988)

A.L. Weisenhorn, P. Maivald, H.J. Butt, P.K. Hansma, Measuring Adhesion, Atraction, and Repulsion between Surfaces in Liquids with an Atomic-force Microscope, *Phys. Rev.* B45, 11226 (1992)

D. Hadzadi (Ed.), *Theoretical Treatments of Hydrogen Bonding*, Wiley, New York 1997

Chapter 11

J.C. Maxwell, Illustration of the Dynamical Theory of Gases, Part I. On the Motions and Collisions of Perfectly Elastic Spheres, *Philos. Mag.* 19, 19 (1860)

A. Einstein, Elementare Theorie der Brownschen Bewegung, *Z. Elektrochem.* 14, 235 (1908)

S. Chapman, The Kinetic Theory of a Gas Constituted of Spherrically Symmetrical Molecules, Philos. *Trans. R. Soc.* London Ser. A211, 433 (1912)

D.D. Present, *Kinetic Theory of Gases*, McGraw-Hill, New York 1958

W. Kauzmann, *Kinetic Theory of Gases*, Addison-Wesley, Reading 1966

T. Kenney, Graham's Law: Defining Gas Velocities, *J. Chem. Educ.* 67, 871 (1990)

G. Turrel, *Gas Dynamics: Theory and Applications*, Wiley, New York 1997

Chapter 12

R.C. Miller, P. Kusch, Velocity Distributions in Potassium and Thallium Atomic Beams. *Phys. Rev.* 99, 1314 (1955)

J.D. Fast, *Entropy, Philips' Technische Bibliothek*, Eindhoven 1960

N. Davidson, *Statistical Thermodynamics*, McGraw-Hill, New York 1962

D.A. McQuarrie, *Statistical Thermodynamics, Harper and Row*, New York 1985

T.L. Hill, *An Introduction to Statistical Thermodynamics, Dover*, New York 1986

D. Chandler, *Introduction to Modern Statistical Mechanics*, Oxford University Press, New York 1987

R.P.H. Gasser, W.G. Richards, *Entropy and Energy Levels*, Oxford University Press, Oxford 1988

O.V. Lounasmaa, G. Pickett, The ^{3}He Superfluids, *Scient. Amer.* June 1990, 64

C.E. Hecht, *Statistical Thermodynamics and Kinetic Theory*, W.H. Freeman, New York 1990

J.W. Whalen, *Molecular Thermodynamics*, Wiley, New York 1991

R.P. Bauman, *Modern Thermodynamics with Statistical Mechanics*, Macmillan, New York 1992

N.C. Craig, *Entropy Analysis: An Introduction to Chemical Thermodynamic Reviews*, Wiley, New York 1992

M. Plischke, B. Bergensen, *Equilibrium Statistical Physics*, World Scientific, Singapore 1994

R.P.H. Gasser, W.G. Richards, *An Introduction to Statistical Thermodynamics*, World Scientific, Singapore 1995

J. Goodisman, *Statistical Mechanics for Chemists*, Wiley, New York 1997

E.A. Cornell, C.E. Wiemann, The Bose-Einstein Condensate, *Scient. Amer.* March 1998, 26

Chapter 13

A. Einstein, Die Plancksche Theorie der Strahlung und die Theorie der spezifischen Wärme, *Ann. Phys.* (Leipzig) 22, 180 (1907)

P. Debye, Zur Theorie der spezifischen Wärmen, *Ann. Phys.* (Leipzig) 39, 789 (1912)
K.F. Bonhoeffer, P. Harteck, Über Para- und Orthowassertsoff, *Z. Phys. Chem.* B4, 113 (1929)
A. Eucken, K. Hiller, Der Nachweis einer Umwandlung des Orthowasserstoffes in Parawasserstoff durch Messung der spezifischen Wärme, *Z. Phys. Chem.* B4, 142 (1929)
J. Kestin, J.R. Dorfmann, *A Course in Statistical Thermodynamics*, Academic Press, New York (1971)
E.A. Cornell, C.E. Wiemann, The Bose-Einstein Condensate, *Scient. Amer.* March 1998, 26

Chapter 14

J.D. Fast, *Entropy, Philips' Technische Bibliothek*, Eindhoven 1960
E.F. Meyer, Thermodynamics of "mixing" of Ideal Gases: a Persistent Pitfall, *J. Chem. Educ.* 64, 676 (1987)
J.W. Whalen, *Molecular Thermodynamics. A Statistical Approach*, Wiley, New York 1991

Chapter 15

D. DeKleerk, M.J. Steenland, *Adiabatic Demagnetization, in Progress in Low Temperature Physics*, Vol. 1, Ed. C.J. Gorter, North-Holland, Amsterdam 1955
C.M. Herzfeld, *The Thermodynamic Temperature Scale, in Temperature. Its Measurement and Control*, Vol. 3, Pt. I, Ed. F.G. Brickwedde, Reinhold, New York 1962
J.B. Fenn, *Engines, Energy and Entropy*, Freeman, San Francisco 1982

Chapter 16

W.C. Gardiner, The Chemistry of Flames, *Scient. Am.* 246, 87 (Feb. 1982)
R.D. Freeman, Conversion of Standard (1 atm) Thermodynamic Data to the New Standard-State Pressure, 1 bar (10^5 Pa), *Bull. Chem. Thermodynamics* 25, 523 (1982)
W.E. Dasent, *Inorganic Energetics*, Cambridge University Press, Cambridge 1982
E.A. Gislason, N.C. Craig, General Definition of Work and Heat in Thermodynamic Processes, *J. Chem. Educ.* 64, 670 (1987)
E.B. Smith, *Basic Chemical Thermodynamics*, Oxford University Press, Oxford 1990
M. Binnewies, E. Milke, *Thermochemical Data of Elements and Compounds*, Wiley-VCH, Weinheim 1998

Chapter 17

H. Taylor, R.H. Christ, Rate and Equilibrium Studies on the Thermal Reaction of Hydrogen and Iodine, *J. Am. Chem. Soc.* 63, 1377 (1941)
J.R. Overton, *Determination of Osmotic Pressure*, in A. Weissberger, B.W. Rossiter (Edts.), Techniques of Chemistry 5, 309, Wiley, New York 1971
E.L. Skau, J.C. Arthur, Determination of Melting and Freezing Temperatures, in A. Weissberger, B.W. Rossiter (Edts.), *Techniques of Chemistry* 5, 105, Wiley, New York 1971

R. Holub, P. Vónka, *The Chemical Equilibrium of Gaseous Systems*, Reidel, Boston 1976

M.J. Blandamer, *Chemical Equilibria in Solution*, Ellis Horwood/Prentice Hall, Hemel Hempstead 1992

J. Gold, V. Gold, Le Chatelier's Principle and the Laws of van t'Hoff, *Educ. in Chem.* 22, 82 (1985)

R.T. Allsop, N.H. George, Le Chatelier – a Redundant Principle?, *Educ. in Chem.* 21, 54 (1984)

K. Denbigh, *The Principles of Chemical Equilibrium*, Cambridge University Press 1986

N.C. Craig, The Chemist's Delta, *J. Chem. Educ.* 64, 668 (1987)

H.F. Frantzen, The Freezing Point Depression Law in Physical Chemistry, *J. Chem. Educ.* 65, 1077 (1988)

J.J. MacDonald, Equilibria and $\Delta G^{\ominus}$, *J. Chem. Educ.* 67, 745 (1990)

B.L. Earl, The Direct Relation Between Altitude and Boiling Point, *J. Chem. Educ.* 67, 45 (1990)

E.R. Boyko, J.F. Belliveau, Simplification of Some Thermodynamic Equations, *J. Chem. Educ.* 67, 743 (1990)

T. Solomon, Standard Enthalpies of Ions in Solution, *J. Chem. Educ.* 68, 41 (1991)

I.M. Klotz, R.M. Rosenberg, *Chemical Thermodynamics*, Wiley, New York 1994

F.H. Chapple, The Temperature Dependence of $\Delta G^{\ominus}$ and the Equilibrium Constant, K_{eq}; Is there a Paradox?, *J. Chem. Educ.* 75, 342 (1998)

Chapter 18

I. Klotz, *Energy Changes in Biochemical Reactions*, Academic Press, New York 1967

R. Chang, *Physical Chemistry with Applications to Biological Systems*, Macmillan, London 1978

R.C. Bohinski, *Modern Concepts in Biochemistry*, Allyn and Bacon, Boston 1979

H.T. Witt, Energy Conversion in the Functional Membrane of Photosynthesis. Analysis by Light Pulse and Electric Pulse Methods, *Biochim. Biophys. Acta* 505, 355 (1979)

G.M. Barrow, *Physical Chemistry for the Life Sciences*, McGraw-Hill, New York 1981

J.T. Edsall, H. Guttfreund, *Biothermodynamics*, Wiley, New York 1983

F.M. Harold, *The Vital Force: A Study of Bioenergetics*, W.H. Freeman, New York 1986

R. deLevie, Explicit Expression of the General Form of the Titration Curve in Terms of Concentration: Writing a Single Closed-form Expression for the Titration Curve for a Variety of Titrations Without Using Approximations or Segmentation, *J. Chem Educ.* 70 209 (1993)

H. Rilbe, *pH and Buffer Theory – A New Approach*, Wiley, New York 1996

G.S. Patterson, A Simplified Method for Finding the pK_a of an Acid-Base Indicator by Spectrophotometry, *J. Chem. Educ.* 76, 395 (1999)

Chapter 19

R.A. Durst, Mechanism of the Glass Electrode Response, *J. Chem. Educ.* 44, 175 (1967)

A.J. Bard, L.R. Faulkner, *Electrochemical Techniques: Fundamentals and Applications*, Wiley-Interscience, New York 1979

A.J. Bard, L.R. Faulkner, *Electrochemical Methods: Fundamentals and Applications*, Wiley, New York 1980

D. Linden (Edt.), *Handbook of Batteries and Cells*, McGraw-Hill, New York 1984

A.K. Covington, R.G. Bates, D.A. Durst, Definition of pH Scales, Standard Reference Values, and Related Terminology, *Pure Appl. Chem.* 57, 531 (1985)

A.J. Bard, R. Parsons, J. Jordan, *Standard Potentials in Aqueous Solution*, Marcel Dekker, New York 1985

P. Reiger, *Electrochemistry*, Prentice Hall, Englewood Cliffs 1987

J. Goodisman, *Electrochemistry: Theoretical Foundations*, Wiley, New York 1987

H.G. Hert, B.M. Braun, K.J. Müller, R. Mauer, What is the Physical Significance of the Pictures Representing the Grotthus H^+ Conductance Mechanism?, *J. Chem. Educ.* 64, 777 (1987)

Y. Marcus, Ionic Radii in Aqueous Solutions, *Chem. Rev.* 88 1475 (1988)

D.R. Crow, *Principles and Applications of Electrochemistry*, Chapman and Hall, London 1988

C.D.S. Tuck, *Modern Battery Construction*, Ellis Horwood, New York 1991

J.O'M. Bockris, B.E. Conway, R.E. White (Edts.), *Modern Apsects of Electrochemistry*, Vol. 22, Plenum, New York 1992

D.B. Hibbert, *Introduction to Electrochemistry*, Macmillan, Basingstoke 1993

D.K. Gosser, *Cyclic Voltammetry, Simulation and Analysis of Reaction Mechanisms*, VCH, Weinheim 1994

M. Wakihara, O. Yamamoto (Edt.), *Lithium-Ion Batteries*, Wiley-VCH, Weinheim 1998

C.H. Hamann, A. Hamnett, W. Vielstich, *Electrochemistry*, Wiley-VCH, Weinheim 1998

P.J. Morgan, E. Gileadi, Alleviating the Common Confusion Caused by Polarity in Electrochemistry, *J. Chem. Educ.* 66, 912 (1989)

R.M. Wightman, D.O. Wipf, High-Speed Cyclic Voltammetry, *Acc. Chem. Res.* 23, 64 (1990)

R. Parsons, Electrical Double Layer: Recent Experimental and Theoretical Developments, *Chem. Rev.* 90, 813 (1990)

Chapter 20

P. Debye, E. Hückel, Zur Theorie der Elektrolyte, *Physik. Zeitschr.* 24, 185 (1923)

A. Reisman, *Phase Equilibria*, Academic Press, New York 1970

J.H. Dymond, E.B. Smith, *The Virial Coefficients of Pure Gases and Mixtures*, Oxford University Press, Oxford 1980

B.D. Wood, *Applications of Thermodynamics*, Addison-Wesley, New York 1982

P.A. Rock, *Chemical Thermodynamics*, University Science Books, Mill Valley 1983

J.M. Smith, H.C. Van Ness, *Introduction to Chemical Engineering Thermodynamics*, McGraw-Hill, New York 1987

S.J. Winn, The Fugacity of a van der Waals Gas, *J. Chem. Educ.* 65, 772 (1988)

J.G. Eberhard, The Many Faces of the van der Waals Equation of State, *J. Chem. Educ.* 66, 906 (1989)

H. Xijun, Y. Xiuping, Influences of Temperature and Pressure on Chemical Equilibrium in Nonideal Systems, *J. Chem. Educ.* 68, 295 (1991)

R. Battino, The Critical Point and the Number of Degrees of Freedom, *J. Chem. Educ.* 68, 276 (1991)

L.L. Combs, An Alternative View of Fugacity, *J. Chem. Educ.* 69, 218 (1992)

R.M. Noyes, Thermodynamics of a Process in a Rigid Container, *J. Chem. Educ.* 69, 470 (1992)

D'Alessio, On the Fugacity of a van der Waals Gas: An Approximate Expression that Separates Attractive and Repulsive Forces, *J. Chem. Educ.* 70, 96 (1993)

L.F. Loucks, Regression Methods to Extract Partial Molar Volume Values in the Method of Intercepts, *J. Chem. Educ.* 76, 425 (1999)

Chapter 21

H. Eyring, H. Gershinowitz, C.E. Sun, The Absolute Rate of Homogeneous Atomic Reactions, *J. Chem. Phys.* 3, 786 (1935)

R.J. Field, E. Körös, R.M. Noyes, Oscillations in Chemical Systems. Thorough Analysis of Temporal Oscillation in the Bromate-Cerium-Malonic Acid System, *J. Am. Chem. Soc.* 94, 8649 (1972)

B. Chance, E.K. Pye, A.K. Ghosh, B. Hess, *Biological and Biochemical Oscillators*, Acad. Press, New York 1973

W. Knoche, *Pressure-Jump Methods*, in G.G Hammes (Edt.), Techniques of Chemistry 6B, 187, Wiley-Interscience, New York 1974

H.O. Pritchard, Why Atoms Recombine More Slowly as the Temperature Goes up, *Acc. Chem. Res.* 9, 99 (1976)

J.M. Anderson, Computer Simulation in Chemical Kinetics, *J. Chem. Educ.* 53, 561 (1976)

G. Nicolis, I. Prigogine, *Self-Organization in Nonequilibrium Systems: From Dissipative Structures to Order through Fluctuations*, Wiley, New York 1977

H. Haken, *Synergetics, An Introduction*, Springer, Berlin 1977

H. Haken, *Chaos and Order in Nature*, Springer, Berlin 1981

L. Meites, *An Introduction to Chemical Equilibrium and Kinetics*, Pergamon Press, Oxford 1981

K.H. Ebert, P. Deuflhard, W. Jäger, *Modelling of Chemical Reaction Systems*, Springer, Berlin 1981

J.W. Moore, R.G. Pearson, *Kinetics and Mechanism*, Wiley, New York 1981

P. Pechukas, Recent Developments in Transition State Theory, *Ber. Bunsenges. Phys. Chem.* 86, 372 (1982)

R. Schmidt, V.N. Sapunov, *Non-formal Kinetics*, VCH, Weinheim 1982

C. Vidal, A. Pacault, *Non-Equilibrium Dynamics in Chemical Systems*, Springer, Berlin 1984

H. Maskill, *The Extent of Reaction and Chemical Kinetics, Educ. in Chem.* 21, 122 (1984)

L. Rensing, N.I. Jäger, *Temporal Order*, Springer, Berlin 1985

R.J. Field, M. Burger, *Oscillations and Traveling Waves in Chemical Systems*, Wiley, New York 1985

S.R. Logan, The Meaning and Significance of "the Activation Energy" of a Chemical Reaction, *Educ. in Chem.* 23, 148 (1986)

D.R. Herschbach, Molecular Dynamics of Elementary Chemical Reactions, *Chemica Scripta* 27, 327 (1987)

J.C. Polanyi, Some Concepts in Reaction Dynamics of Elementary Reactions, *Chemica Scripta* 27, 229 (1987)

K.J. Laidler, *Chemical Kinetics, Harper and Row*, New York 1987

F. Mata-Perez, J.F. Perez-Benito, The Kinetic Rate Law for Autocatalytic Reactions, *J. Chem. Educ.* 64, 925 (1987)

S.K. Scott, Oscillations in Simple Models of Chemical Systems, *Acc. Chem. Res.* 20, 186 (1987)

H.G. Schuster, *Deterministic Chaos*, VCH Weinheim 1988

K.J. Laidler, Rate-controlling Step: a Necessary or Useful Concept?, *J. Chem. Educ.* 65, 250 (1988)

K.J. Laidler, Just What is a Transition State?, *J. Chem. Educ.* 65, 540 (1988)

R.T. Raines, D.E. Hansen, An Intuitive Approach of Steady-State Kinetics, *J. Chem. Educ.* 65, 757 (1988)

C.E. Klots, The Reaction Coordinate and its Limitations: an Experimental Perspective, *Acc. Chem. Res.* 21, 16 (1988)

A.H. Zewail, Laser Femtochemistry, *Science* 242, 1645 (1988)

J.I. Steinfeld, J.S. Francisco, W.L. Hase, *Chemical Kinetics and Dynamics*, Prentice Hall, Englewood Cliffs, New Jersey 1989

I.R. Epstein, The Role of Flow Systems in Far-from-equilibrium Dynamics, *J. Chem. Educ.* 66, 191 (1989)

R.J. Field, The Language of Dynamics, *J. Chem. Educ.* 66, 188 (1989)

R.M. Noyes, Some Models of Chemical Oscillators, *J. Chem. Educ.* 66, 190 (1989)

C.E. Hecht, *Statistical Thermodynamics and Kinetic Theory*, Freeman, New York 1990

P. Gray, S.K. Scott, *Chemical Oscillations and Instabilities*, Clarendon Press, Oxford 1990

H. Maskill, The Arrhenius Equation, *Educ. in Chem.* 27, 111 (1990)

R.G. Gilbert, S.C. Smith, *Theory of Unimolecular and Recombination Reactions*, Blackwell Scientific, Oxford 1990

J.C. Reeve, Some Provocative Opinions on the Terminology of Chemical Kinetics, *J. Chem. Educ.* 68, 728 (1991)

W. Jahnke, A.R. Winfree, Recipes for Belousov-Zhabotinsky Reagents, *J. Chem. Educ.* 68, 320 (1991)

A.H. Zewail, Femtosecond Transition-State Dynamics, *Faraday Discuss. Chem. Soc.* 91, 1 (1991)

P.J. Ortoleva, *Nonlinear Chemical Waves*, Wiley, New York 1992

H. Strehlow, *Rapid Reactions*, Wiley-VCH, Weinheim 1992

J.N. Spencer, Competitive and Coupled Reactions, *J. Chem. Educ.* 69, 281 (1992)

R.J. Field, L. Györgyi, *Chaos in Chemistry and Biochemistry*, World Scientific, Singapore 1993

Ch.Y Ng, T. Baer, I. Powis, *Unimolecular Ion-Molecule Reaction Dynamics*, Wiley, New York 1994

S.H. Strogatz, *Nonlinear Dynamics and Chaos*, Addison-Wesley, Reading 1994

S.K. Scott, *Oscillations, Waves and Chaos in Chemical Kinetics*, Oxford Univ. Press, Oxford 1994

M.A. Marek, S.C. Müller, T. Yamaguchi, K. Yoshikawa, *Dynamics and Regulation in Nonlinear Chemical Systems*, North-Holland, 1995

R. Kapral, K. Showalter (Ed.), *Chemical Waves and Patterns*, Kluwer, Dordrecht 1995

P. Brumer, M. Shapiro, Laser Control of Chemical Reactions, *Scient. Amer.* March 1995, 34

G.D. Billing, K.V. Mikkelsen, *Introduction to Molecular Dynamics and Chemical Kinetics*, Wiley, New York 1996

K.A. Holbrook, M.J. Pilling, S.H. Robertson, *Unimolecular Reactions*, Wiley, New York 1996

I.R. Epstein, J.A. Pojman, *An Introduction to Nonlinear Chemical Dynamics*, Oxford University Press, Oxford 1998

A.L. Fradkov, A.Y. Pogromsky, *Introduction to Control of Oscillations and Chaos*, World Scientific, Singapore 1998

S.C.Müller, V.S.Zykov, *Spiral Waves in Active Media*, World Scientific, Singapore 1998

R.M. Hochstrasser, Ultrafast Spectroscopy of Protein Dynamics, *J. Chem. Educ.* 75, 559 (1998)

J.I. Steinfeld, J.S. Francisco, W.L. Hase, *Chemical Kinetics and Dynamics*, Prentice Hall, Upper Saddle River, 1999

Chapter 22

W. Kuhn, Über quantitative Deutung der Viskosität und Strömungsdoppelbrechung von Suspensionen, *Koll. Z.* 62, 269 (1933)

W. Kuhn, *Über die Gestalt fadenförmiger Moleküle in Lösungen, Kolloid Zeitschrift 68*, 2 (1934)

W. Kuhn, H. Kuhn, Rigidity of Chain Molecules and its Determination from Viscosity and Flow Birefringence in Dilute Solutions, *J. Colloid Science* 3, 11 (1948)

G.L. Gaines, Jr., *Insoluble Monolayers at Liquid-Gas Interfaces*, Wiley-Interscience, New York 1966

P. Flory, *Principles of Polymer Chemistry*, Cornell University Press, Ithaca and London 1971

P.G. deGennes, Reptation of a Polymer Chain in the Presence of Fixed Obstacles, *J. Chem. Phys.* 55, 572 (1971)

P.G. de Gennes, *The Physics of Liquid Crystals*, Clarendon Press, Oxford (1974)

B.J. Berne, R. Pecora, *Dynamic Light Scattering*, Wiley-Interscience, New York 1976

D. Freifelder, *Physical Biochemistry*, W.H. Freeman, New York 1978

E.G. Richards, *An Introduction to Physical Properties of Large Molecules in Solution*, Cambridge University Press, Cambridge 1980

H.R. Allcock, F.W. Lampe, *Contemporary Polymer Chemistry*, Prentice-Hall, Englewood Cliffs 1981

Single-Channel Recording, Edt. B. Sackmann, E. Neher, Plenum Press, New York and London, 1983

G.W. Gray, J.W. Goodby, *Smectic Liquid Crystals*, Leonhard Hill, Glasgow, 1984

F.W. Billmeyer, *Textbook of Polymer Science*, Wiley, New York 1984

L.H. Sperlin, *Physical Polymer Science*, Wiley-Interscience, New York 1986

J.P. Queslet, J.E. Mark, Advances in Rubber Elasticity and Characterization of Elastomeric Network, *J. Chem. Educ.* 64, 491 (1987)

P.G. de Gennes, Short Range Order Effects in the Isotropic Phase of Nematics and Cholesterics, *Mol. Cryst. Liquid Cryst.* 12, 193 (1971)

D. Demus, 100 Years Liquid Crystal Chemistry, *Mol. Cryst. Liq. Cryst.* 165, 45 (1988)

P.G. deGennes, *Scaling Concepts in Polymer Physics*, Cornel University Press, Ithaca and London 1988

M. Schadt, The Twisted Nematic Effect: Liquid Crystal Displays and Liquid Crystal Materials, *Mol. Cryst. Liq. Cryst.* 165, 405 (1988)

D. Langevin, Microemulsions, *Acc. Chem. Res.* 21, 255 (1988)

N.F. Zhou, The Availability of a Simple Form of Gibbs Adsorption Equation for Mixed Surfactants, *J. Chem. Educ.* 66, 137 (1989)

A.W. Anderson, *Physical Chemistry of Surfaces*, Wiley-Interscience, New York 1990

S.R. Morrison, *The Chemical Physics of Surfaces*, Plenum, New York 1990

P.J. Collings, *Liquid Crystals: Nature's Delicate Phase of Matter*, Wiley, New York 1991

D. Langevin, Micelles and Microemulsions, *Ann. Rev. Phys. Chem.* 43, 341 (1992)

Y. Moroi, *Micelles*, Plenum, New York 1992

H. Stegemeyer, Ed.: *Liquid Crystals*, Steinkopff, Darmstadt 1994

J. Käs, H. Strey, E. Sackmann, Direct Visualization of Reptation for Semiflexible Actin Filaments, *Nature* 368, 226 (1994)

A.W. Adamson, A.P. Gast, *Physical Chemistry of Surfaces*, Wiley, New York 1997

H.G. Elias, *An Introduction to Polymer Science*, Wiley-VCH, Weinheim 1997

H. Rief, F. Oesterhelt, B. Heymann, H.E. Gaub, Single Molecule Force Spectroscopy on Polysaccharides by Atomic Force Microscopy, *Science* 275, 1295 (1997)

M. Rief, M. Gautel, F. Oesterhelt, J. Fernandez, H.E. Gaub, Reversible Unfolding of Individual Titin Immunoglobulin Domains by AFM, *Science* 276, 1109 (1997)

D. Demus (Ed.), *Handbook of Liquid Crystals*, Wiley-VCH, Weinheim 1998

G.W. Gray (Ed.), *Physical Properties of Liquid Crystals*, Wiley-VCH, Weinheim 1999

Chapter 23

Th. Förster, Energiewanderung und Fluoreszenz, *Naturwissenschaften* 33, 166 (1946)

H. Bücher, K.H. Drexhage, M. Fleck, H. Kuhn, D. Möbius, F.P. Schäfer, J. Sondermann, W. Sperling, P. Tillmann, J. Wiegand, Controlled Transfer of Excitation Energy Through Thin Layers, *Mol. Cryst.* 2, 199 (1967)

H. Kuhn, Classical Aspects of Energy Transfer in Molecular Systems, *J. Chem. Phys.* 53, 101 (1970)

A. Warshel, Charge Stabilization Mechanism in the Visual and Purple Membrane Pigments, *Proc. Natl. Acad. Sci.* USA 75, 2558 (1978)

A. Warshel, Conversion of Light Energy to Electrostatic Energy in the Proton Pump of Halobacterium Halobium, *Photochem. Photobiol.* 30, 285 (1979)

H. Kuhn, Electron Transfer Mechanism in the Reaction Center of Photosynthetic Bacteria, *Phys. Rev.* A34, 3409 (1986)

K.V. Mikkelsen, M.A. Ratner, Electron Tunneling in Solid-state Electron-transfer Reactions, *Chem. Rev.* 87, 113 (1987)

H. Ringsdorf, B. Schlarb, J. Venzmer, Molecular Architecture and Function of Polymeric Oriented Systems: Models for the Study of Organization, Surface Recognition, and Dynamics of Biomembranes, *Angew. Chem. Int. Ed. Engl.* 27, 113 (1988)

H.K. Wickramasinghe, Scanned-Probe Microscopes, *Scient. Amer.* October 1989, 74

A.M. Wolsky, R.F. Giese, E.J. Daniels, The New Superconductors: Prospects for Applications, *Scient. Amer.* February 1989, 44

F.T. Hong (Ed.), *Molecular Electronics. Biosensors and Biocomputers*, Plenum Press, New York 1989

K.B. Mullis, The Unusual Origin of the Polymerase Chain Reaction, *Scient. Amer.* April 1990, 56

A. Ulman, *Ultrathin Organic Films, from Langmuir-Blodgett to Self-Assembly*, Academic Press, San Diego 1991

Ch. Bräuchle, N. Hampp, D. Oesterhelt, Optical Applications of Bacteriorhodopsin and its Mutated Variants, *Adv. Mater.* 3, 420 (1991)

G.A. Ashwell, R.C. Hargreaves, C.E. Baldwin, G.S. Bahra, C.R. Brown, Improved Second-harmonic Generation from Langmuir-Blodgett Films of Hemicyanine Dyes, *Nature* 357, 393 (1992)

G. Calzaferri, N. Gfeller, Thionine in the Cage of Zeolite L, *J. Phys. Chem.* 96, 3428 (1992)

G.H. Wagniére, *Linear and Nonlinear Optical Properties of Molecules*, Wiley, New York 1993

G. Wegner, Control of Molecular and Supramolecular Architecture of Polymers, Polymer Systems and Nanocomposites, *Mol. Cryst. Liq. Cryst.* 235, 1 (1993)

H. Kuhn, D. Möbius, Monolayer Assemblies, in *Investigations of Surfaces and Interfaces*-Part B, B.W. Rossiter, R.C. Baetzold (Eds.) *Physical Methods of Chemistry Series*, Vol. IXB, Wiley, New York 1993

Ch. Kuhn, W.F. van Gunsteren, Dynamics of Solitons in Polyacetylene in the Step-Potential Model, *Sol. State Comm.* 87, 203 (1993)

J.W. Lichtman, Confocal Microscopy, *Scient. Amer.* Aug. 1994, 30

J. Rebek, Synthetic Self-Replicating Molecules, *Scient. Amer.* July 1994, 34

A. Ulman (Ed.), *Organic This Films and Surfaces: Directions for the Nineties, Thin Films Vol. 20*, Academic Press, San Diego 1995

G.M. Whiteside, Selfassembling Materials, *Scient. Amer.*, Sept. 1995, 114

D.W. Urry, Elastic Biomolecular Machines, *Scient. Amer. Jan.* 1995, 44

G. McDermott, S.M. Prince, A.A. Freer, A.M. Hawthornthwalte-Lawless, M.Z. Papiz, R.J. Cogdell, Crystal Structure of an Integral Membrane Light-Harvesting Complex from Photosynthetic Bacteria, *Nature* 374, 517 (1995)

H. Tamiaki, M. Amakawa, Y. Shimono, R. Tanikaga, A.R. Holzwarth, K. Schaffner, Synthetic Zinc and Magnesium Chlorin Aggregates as Models for Supramolecular Antenna Complexes in Chlorosomes of Green Photosynthetic Bacteria, *Photochem. Photobiol.* 63, 92 (1996)

A. Ruaudel-Teixier, Supermolecular Architecture in Langmuir-Blodgett Films: Control and Chemistry, *Heterogeneous Chem. Rev.* 3, 1 (1996)

E. Sackmann, Supported Membranes: Scientific and Practical Applications, *Science* 271, 43 (1996)

H.T. Witt, Primary Reactions of Oxygenic Photosynthesis, *Ber. Bunsenges. Phys. Chem.* 100, 1923 (1996)

H. Sigl, G. Brink, M. Seufert, M. Schulze, G. Wegner, E. Sackmann, Assembly of Polymer/Lipid Composite Films on Solids Based on Hairy Rod LB-Films, *Eur. Biophys. J.* 25, 249 (1997)

H. Noji, R. Yasuda, M. Yoshida, K. Kinosita, Direct Observation of the Rotation of F_1-ATPase, *Nature* 386, 299 (1997)

S.M. Block, Real Engines of Creation, *Nature* 386, 217 (1997)

W.D. Schubert, O. Klukas, N. Krauss, W. Saenger, P. Fromme, H.T. Witt, Photosystem I of Synechococcus Elongatus at 4 Å Resolution: Comprehensive Structure Analysis, *J. Mol. Biology* 272, 741 (1997)

N. Bloembergen, Nonlinear Optical Instrumentation, *J. Chem. Educ.* 75, 555 (1998)

M.W. Berns, Laser Scissors and Tweezers, *Scient. Amer.* April 1998, 52

G. Kaim, P. Dimroth, Voltage-Generated Torque Drives the Motor of the ATP Synthase, *EMBO J.* 17, 5887 (1998)

A.R. Holzwarth, K. Schaffner, Self-Assembly of Synthetic Zinc Chlorins in Aqueous Microheterogeneous Media to an Artificial Supramolecular Light-Harvesting Device, *Helv. Chim. Acta* 82, 797 (1999)

L. Scheibler, P. Dumy, M. Boncheva, K. Leufgen, H.J. Mathieu, M. Mutter, H. Vogel, Functional Molecular Thin Films: Topological Templates for the Chemoselective Ligation of Antigene Peptides to Self-Assembled Monolayers, *Angew. Chem. Int. Ed.* 38, 696 (1999)

Chapter 24

M. Eigen, Self-Organization of Matter and the Evolution of Biological Macromolecules, *Naturwissenschaften* 58, 465 (1971)

H. Kuhn, Self-Organization of Molecular Systems and Evolution of the Genetic Apparatus, *Angew. Chem. Int. Edit. Engl.* 11, 798 (1972)

M. Eigen, P. Schuster, The Hypercycle, *Naturwissenschaften* 64, 541 (1977)

H. Kuhn, J. Waser, Molecular Self-Organization and the Origin of Life, *Angew. Chem. Int. Edit. Engl.* 20, 500 (1981)

W. Hoppe, W. Lohmann, H. Markl, H. Ziegler (eds) *Biophycis*, Springer Berlin Heidelberg 1983 (see chapter 17, Evolution)

M. Eigen, Stufen zum Leben, Piper Verlag, München 1987 [P. Wooley, Transl., Steps Toward Life, Oxford University Press, Oxford 1992]

Maxwell's Demon, Edt. H.S. Leff, A.F.R ex, *Princeton U.P.*, Princeton N.J. 1990

Ch. deDuve, *Blueprint for a Cell. The Nature and Origin of Life*, Neil Patterson, Burlington 1991

F.M. Richards, The Protein Folding Problem, *Scient. Amer.* Jan. 1991, 34

S.A. Kauffman, *The Origins of Order: Self-Organization and Selection in Evolution*, Oxford University Press, Oxford 1993

R.F. Gesteland, J.F. Atkins (Edts.), *The RNA World*, Cold Spring Harbor Laboratory Press, 1993

H. Kuhn, J. Waser, *A Model of the Origin of Life and Perspectives in Supramolecular Engineering*, Chapter 7 in: The Lock-and-Key Principle, pages 247–306, Edt. J.P. Behr, Wiley, New York 1994

R. Tjian, Molecular Machines That Control Genes, *Scient. Amer.* Feb. 1995, 38

A. Eschenmoser, Chemical Etiology of Nucleic Acid Structure, *Science* 284, 2118 (1999)

R. F. Gesteland, T. R. Cech, J. F. Atkins (Eds) *The RNA World*, second edition, Cold Spring Harbor Laboratory Press, 1999.

L. F. Landweber, Experimental RNA Evolution, *TREE* 14, 353 (1999).

Index

Absolute temperature 364–9
  as function of PV 366
Absorbance 204–7
Absorbed power 261
Absorption
  from excited states 240–1
  of light 204–77
Absorption bands 223, 238–9, 249
  polarization 213–16
  splitting 252
  strength 213–16
  vibrational structure 331–2
Absorption coefficient 205, 261
  calculation 261–2
Absorption lines 285
Absorption maxima
  of dyes 209–13
  of polyenes 221
Absorption shift of chromophore by protein 827
Absorption spectra 207, 215, 227, 791
  of dyes 803
Accumulators 617
Acetic acid, titration by NaOH 597–9
Acetone
  sedimentation of cellulose acetate in 767
  viscosity of cellulose acetate in 768
Activated complex 672–4, 676
Activation 665–6
Activation energy 666, 671, 682–5
  experimental data 677–89
Activation enthalpy 676–7, 685–7
Activation entropy 676–7, 685–7
Activation factor 668–9
Activity coefficient 649–50, 650
  in $MgSO_4$ solution 652
  in NaCl solution 652
Adenine 348
Adenosine diphosphate (ADP) 588–9, 836

Adenosine triphosphate (ATP) 588–9, 591–2, 820, 821
Adiabatic process 467–8, 642
Adiabatic bomb calorimeter 519
Adiabatic expansion
  into vacuum 640–2
  of CO_2 641–2
Adiabatic process 642
Aggregation 333
Alizarin yellow 724
Alkaline manganese cell 616, 617
Amidinium cation 187, 208
Amino acids 583–4, 892
Ammonia synthesis 552, 557, 576, 640
  equilibrium 639–40
  synthesis
    at 298 K and at 700 K 557
    from its elements 552–3
Amount of substance 360
  equilibrium 639–40
  synthesis
    at 298 K and at 700 K 557
    from its elements 552–3
Amphiphiles 750, 751, 851
Anharmonicity 302
Anisotropy factor 249, 252–3, 276–7
Antenna equation of Hertz 274–6, 748
Antenna molecules 811
Anthracene 240, 241, 250–3, 685
Antibonding orbitals 169, 183
Anti-Stokes lines 306
Antisymmetric function 98
Antisymmetric spatial function 102
Antisymmetric spin wavefunction 102
Antisymmetric stretching mode 300
Antisymmetric wavefunctions 99–103, 319
Antisymmetry postulate in H_2 320
Aqueous solutions, reactions in 579–99
Arrhenius equation 671
Arrhenius plot 683

Atmospheric distribution 382–3
Atom models 131
Atomic force scanning microscopy (AFM) 333, 772, 853–6
Atomic functions 92–3
Atomic number 122, 123
Atomic orbitals 97, 121, 140
   combination 91–6
   compressed 94–5
Atomic spectra 20
Attraction energy 352
Attractive forces 333–4
   between molecules 631
Aufbau principle and periodic table 120–4
Autocatalytic oxidation of Ru^{2+} 704–7
Autocatalytic reactions 703–7
Autoprotolysis constant of water 574
Average potential energy 57
Average distance 58, 72
   kinetic energy 57
Avidin 352
Avogadro's constant 261, 359
Avogadro's law 362–4
Azacyanine cations 208–9, 216–17, 254–5
Azobenzene 833–5
p-Azoxyanisole (PAA) 746

Bacterial electron pump 826–8
Bacterial reaction center, tunneling processes in 828–30
Bacteriochlorophyll 232–3, 811
Bacteriophage
   approximated by sphere 392
   docking at bacterium 391
Bacteriorhodopsin 837, 839, 840, 843, 844
Balmer series 61, 64, 65
Band broadening 210
Band gap 219–23
Bar 361
Barometric formula 383
Barrier, potential 104, 114, 829, 833, 875
Bathochromic shift 230
Batteries 617
BeH_2
   box model 154
   hybridization 148
   polymerization 154
Beer–Lambert law 205–6
Belousov–Zhabotinsky reaction 682, 695, 696, 710–14, 716
   rate constants 718–19
Bending force constants 136
Bending modes 298, 301, 302
Bending of DNA double strand 759–60
Benzene 163, 171, 173, 175–7, 185, 189, 223, 280–1, 322
   combustion 553

Bernoulli 361
Beryllium ion 34–5
Big bang 875–8
Bimolecular reaction, Eyring equation for 675–6
Bimolecular reaction 671
Biochemistry, group transfer reactions in 588–91
Bioenergetics 591–4
Bioevolution as process of learning 881–4
Biomembranes 752–4
Biomolecules, light absorption 228–33
Biosensors 852–3
Biosystems
   and computer 895–6
   reactions in 579–99
   solar energy harvesting 810–11
Biotin 352
Biological standard state 587
Birefringence 746
Bistability 709–10
Black lipid membranes 751
Bleaching 752
Bohr atomic model 69–70
Bohr correspondence principle 70–1, 215, 294
Bohr magneton 99–100
Bohr radius 29, 53, 70, 103
Boiling point 395, 563, 577
Bond
   dative 161
   hydrogen 348
   three-center 161
Boltzmann constant 367, 469
Boltzmann distribution 244, 408–12, 420, 429–35, 446, 449, 477, 478, 491, 658
Boltzmann equation 408
Boltzmann factor 427
Boltzmann Law 408–12, 658, 782
Bond alternation 187, 200–3, 256
   HOMO–LUMO gap by 219–23
Bond angle 126, 131–2, 136
Bond energy 165
Bond lengths 129–31, 165, 171, 184–7, 202–3, 220, 222, 223, 307
Bond models 124–6
Bond polarity 126–9
Bonding 85–116, 140–62
Bond order 185, 187, 192
   and electron density 89
   basic features 19–38
   nature of 91
   orbitals 154, 169
   π electrons 163–5
Borderline broadening 752–3
Born–Oppenheimer approximation 319, 332
Bose distribution 449

Boundary condition 41–2
Box model
 for oscillator 291–3
 H atom 27–31
 H_2^+ molecule 32–3
 butadiene 169
 ethene 166–7
 O_2 molecule 154–6
Box trial function 30
Box wavefunctions 86–9, 140, 154–5
Boyle's law 361
Br_2 molecule 373, 377
Branched molecules 177–81
Branched string, vibrations of 178
Branching chain reactions 699
Bromate, bromide reaction with 695
Bromide
 critical concentration 708
 reaction with bromate 695
Bromination of malonic acid 696–7
Bromine diffusion through a cylinder 380–1
Bromine hydrogen reaction 697
Bromine radicals 697
Bromomalonic acid 711
Bromomalonyl radicals 711
Brownian motion 391, 392, 767
Buffer solution 581, 595
Bunsen burner 526–7
Burning limestone 532–4, 541, 543, 554
Butadiene 163, 168, 169, 174, 183–5, 221
Butane, conformations 756
i-Butane 555
n-Butane 555
β-Carotene
 cis peak 258
 cis-$trans$ isomers 229

C—C 133–5, 164–7, 174, 755
C=C 165, 174
C≡C 165
C—H 217
C—O 133
Calomel electrode 614
Calorimeter 519
Capacitance 354, 751
Capillary rise 736
Carbon atoms 220
Carbon monoxide, residual entropy 509–10
Carnot cycle 498–500
Carnot process 618
β-Carotene 228–30, 811
Catalytic reaction
 in organized media 848
 in solution 848
Catenane 845
CD spectrum 249, 251

Cellulose acetate 767–9
Cell, electrochemical 600–20
Cell potential 602
Celsius scale 366
Center of gravity 127, 286, 291
Cesium electrode 1–3
CH_4 molecule 150
Chain-breaking 698
Chain carriers 698
Chain held at chain ends 782–3
Chain-propagation 698
Chain reactions 697
 rate law for 698–9
Chain stretching 774, 784–5
Chain-termination 698
Change of entropy. See Entropy
Change of state 463–4, 469–72, 521–2
 at constant pressure 460–2
 at constant volume 450–3
 distribution possibilities in 472–7
 of real gases 636–45
 of real solutions 646–53
Characteristic temperature 458–60
Charge density 173–5, 184–7
Charge distribution
 around an ion 659
 between two plates 657
CH_3CN molecules 346
Chemical equilibrium 545–78
Chemical pinwheel 716–17
Chemical potential 648–9
Chemical reactions 517–44
 at constant pressure 520–2
 at constant volume 518–20
 kinetics of 663
Chemical shift 321–2
Chemical waves 714–17
Chemist's rule of thumb 683
Chiral molecule 248
Chloride ion pump 842–3
 in halobacterium 842
Chlorine atom 278–9
Chlorophyll 232–3
Chlorophyll c molecules 811
Cholesteric state 745
Cholesteryl benzoate 745
Chromophores 811–2, 828
Circular dichroism (CD) 249–50
 of cyanine dye 253, 259
 of spirobisanthracene 250–3
Circular polarized light waves 250
Circular waves 714–15
cis–$trans$ isomerization 834–44
Classical oscillator 261–6, 289, 293–4
Clausius–Clapeyron equation 559, 627, 628–9
Clapeyron equation 627
Clearing point 743

Closed systems 464–5
Clusters 396
CO
Coexisting phases 395
Coil restoring 774
Coil uncoiling 769–73
Coil unraveling 773–4
   force applied at chain end 775
Collision cross-section 372
Colloid 756
Colour of dyes 213
   force constant from Raman spectrum 309
   rotational Raman spectrum 306
CO_2 299, 300, 362, 363
   adiabatic expansion 641–2
Collision diameters 351, 386–7, 406
Collision frequency 374, 388, 674
Collision of particles 405
Collision theory for gas reactions 663–9
Collisions
   counting the number of 663–4
   elastic 358
   in liquids 387–8
Combustion 517
   compression 639
Competitive inhibitors 702, 703
Complex formation 574
   in solution 567
Complex reactions 689–719
Complex systems 880–1
Computer and biosystem 895–6
Computer simulation of thought experiment 892
Concentration
   and molality 562
   reversible change 565
Concentration cells 604–8
Concentration gradient 381
Condensation 471
Condensed state, part of reactants or products 573–5
Conduction
   DNA double strand 824
   soliton 825
   through molecular wires 823–6
Conductivity cell 621
Configuration 477
Conformations of butane 756
Consecutive reactions 693–7
Construction phase 892
Contact angle on monolayers 742
Contour length 756
Cooling 511, 644
Correlation 97, 225–8, 267–9, 350
Correspondence principle 69, 70
Coulomb attraction 27, 29, 37, 91, 187, 333, 334, 650, 685, 750
   electron cloud 32

Coulomb energy 91, 341
Coulomb field 28, 647
Coulomb forces 29, 89, 103, 118, 649
   singlet–triplet splitting caused by 103
Coulomb repulsion 31, 91, 130, 237, 333, 750
Coulomb's law 36
Coupled oscillators 228, 267–9
   normal modes 270–4
   oscillator strength 270
   resonance frequencies 270
Coupled redox reactions 585–6
Coupling of reactions by enzymes 590–1
Coupling transitions with parallel transition moments 267–9
Covalent radii 129–31
Covalent bond 124–39
Critical constants 633
Critical micelle concentration 742–4
Critical point 395–6, 633–4
Crotonaldehyde 692–3
Cross-section, collision 372, 664, 671
Crystal 333–40
CsCl lattice 339
Cu-phthalocyanine 223–4
Cyanine dye cations 212, 218, 219, 234, 236
Cyclic π electron systems 171–3
Cyclic processes 465–6, 498
Cyclobutadiene 192
Cyclohexene 176, 177
Cyclopentadiene 692–3
   dimerization of 686
Cytoplasmic channel 837, 841
Cytosine 348

Dative bonds 152, 153
Davisson–Germer experiment 11
de Broglie relation 12, 41, 280
de Broglie wave 22, 163, 172, 204, 282, 293, 342
de Broglie wavelength 89, 178, 292, 333, 356, 675
Debye–Hückel law 652
Debye–Hückel theory 650
Decay reaction 677
Decay time, see Life time
Degeneracy 52, 159, 280
Degree of dissociation 580–1
Delocalized electrons 154–9
Demagnetization, adiabatic 511
Depression
   freezing point 563, 578
   of vapor pressure 562
Depression, freezing point 578
Detonating gas reaction 520, 523, 531, 540, 541, 543, 546, 551
Deuteron, tunneling 687–9

INDEX

Diatomic molecule   130, 135, 284, 294–7, 296–7, 312
   infrared spectra   294–7
   Raman spectra   307–8
   reduced mass   286–7, 290
   rotation   286
   rotational motion   287
   vibration of   291
   vibrational–rotational motion   296
   vibrational–rotational Raman spectra   307–8
Dichroism   215
Diethyne   164
Diffraction   3–7
   definition   3
   electrons   11, 14
   of light   14
   on double slit   6
Diffusion   374–83, 402
   of methyl cellulose in water   767–8
   removal of excess thiosulfate from photoplate   391
Diffusion coefficient   378, 379, 388–9, 674, 752–3, 763
Diffusion constant   766
Diffusion controlled reactions   674–5
Diffuse double layer   657
Dimerization of cyclopentadiene   686
Dimyristoyl phosphatidic acid (DMPA)   835–6
Dipole energy   345
Dipole forces   344–6
Dipole moment   127, 137, 187–9, 193, 283, 351
Dipole–dipole interaction   345
Dispersion energy   351, 352
Dispersion forces   349–52
Displacement   375
Dissociation
   degree   580–1
   of water   550–1, 574, 607–8
   of water vapor   545, 550–1
   of weak acids   579–84
Dissociation constant   582, 691
Dissolution of metals in acid   584–5
Distinguishable particles   429–34, 447, 478, 484
Distribution possibilities in change of state   472–7
DNA – RNA – protein translation   886, 894
DNA biosynthesis   231
DNA double helix   758, 770
DNA double strand   758
   bending of   759–60
   conduction   824
DNA fragments, gel electrophoresis   778
DNA mobility in meshwork   792–3
Double bonds   163, 176, 177

Double slit experiment, wavelength from   5–6
Doubly occupied orbitals   269
Double layer   657
Duality, wave-particle   1–17
Dulong and Petit rule   458
Dye aggregate as energy harvesting device   805–11
Dye cations   207–8, 236
Dye laser   243–7
Dye molecules   204–7, 210, 243, 245, 266, 805–6, 856, 857
   energy states   239
   energy transfer between   800–3
   functional unit by coupling   803–5
Dyes
   absorption maxima   209–13
   absorption spectra   803
   bleaching of   752, 856
   with cyclic electron cloud   223–4
   see also specific dyes

Ebullioscopy   578
Effective atomic number   123
Effective charge   152–4
Effusion   370, 399
Eggs, hardboiling   686–7
Eigenfunction   42
Einstein–Smoluchowski equation   378–81, 753–4
Einstein's model of a solid   457
Elastic collisions   358
Elasticity,
   rubber   781
   single molecular chain   773, 782–3
Electrical energy supplied by system   602
Electrochemical cell potential   603, 608
Electrochemical cell reaction   600–4
Electrochemical cells   600–26
   applications   613–20
Electrode reactions   604
Electrolysis   619–20
Electrolyte solution, conductivity   620–4
Electron(s)   10–13
   behavior   15
   between two walls   24–5, 41, 48
   comparison with photons   13
   diffraction   11, 14
   energy   24
   in a box   26, 45, 50
   in potential trough of finite depth   42, 77–8
   in rectangular box   49
   magnetic interactions   103
   most probable and average distance from nucleus   58
   outside a potential trough   43–4
   particle nature   10
   radial probability distribution   54
   wave nature   11

Electron acceptor   151, 818, 824–5, 845
Electron charge   174, 175, 181
Electron cloud   31, 89, 91, 130, 140, 167, 168, 179, 348
  Coulomb attraction   32
  ground state   27, 28
  hydrogen molecule   35–6
Electron configuration   120, 125, 147, 149
Electron density   126
  and chemical bonding   89
Electron diffraction   11, 14
Electron distribution   126
  heteroatoms as probes for   216–19
  π electron system   173
Electron donor   151, 818, 824–5, 846
Electron gas   342–4
Electron motion, manipulation   818–34
Electron pair bond   124–7, 140
  properties   151–4
Electron paramagnetic resonance (EPR)   101
Electron pumping   826
  artificial photoinduced   830
  optimum yield   827
Electron shell   31
Electron spin   102
Electron transfer
  coupled with proton transfer   586–7
  pi-electron systems   871–7
  from metal to metal   585–6
  from metal to proton   584–5
  in soft medium   877–9
  rate constant   875
  reactions   584–8, 674, 685
  through proteins   833
  to proton at pH 7   587–8
Electron tunneling   818, 823, 871–7
Electronegativity   126–9
Electronegativity scale   128–9
Electroneutrality   622
Electronic energy distribution   428
Electronic spectra, vibrational structure of   312–18
Electronic states, degeneracy   141–2
Electro-optical effects   747–8
Electrostatic interaction   650
Electron-electron repulsion   96, 118–20
Electronegativity   127–9
Electron spin resonance   101
Electrophoresis   777–81
Elevation, boiling point   563, 577
Ellipticity   248–9
Emission band narrowing   245–6
Emission of light   204–77
  spontaneous   233–42
  stimulated   242–7
Emission spectra, hydrogen atom   61–4
Energy curve   135

Energy diagram   24, 47
  hydrogen atom   64
Energy dissipation
  in information processing   896
  in measurement   897–8
Energy distribution in molecular assemblies   408–49
Energy funneling in nature   811
Energy harvesting device, dye aggregate as   805–10
Energy levels   434
  degeneracy   140–6
Energy splitting in magnetic field   99
Energy states   20–1
Energy transfer   667, 844, 862–4
  between dye molecules   800–3
  efficiency   810
  from exciton to acceptor   865–7
  monolayer assembly to study   800
  supramolecular machine   794–5
Enthalpy   517, 522–7, 676–7, 685–7
  definition   520
EPR   101, 105
ESR   101, 105
Entropy   480, 482–6, 492–517, 676–7, 685–7
  arbitrary processes   501–7
  argon from calorimetric data   507–9
  atomic gases   480–1
  CO from calorimetric data   509–10
  diatomic gases   481
  for rotation and vibration   490–1
  HD and HCl from calorimetric data   509
  ideal gas   493–500
  increase
    at constant volume   494
    in irreversible process in isolated system   483
    law of   483
  non-isolated systems   486
  of mixing   488, 514
  rare gases   481
  substances   507–10
  subsystems   484
  system   469–91
Enzymes
  assembling and interlocking with substrate   393
  coupling of reactions by   590–1
  reactions   700–3
Equation of Einstein and Smoluchowski   378–81, 753–4
Equation of state   629
Equilibrium, reactions leading to   689–91
Equilibrium condition   649, 737
Equilibrium constant   545–9, 551, 555, 636–40, 677, 724
Equipartition principle   415, 416

INDEX

Ethane 134
  stretching bond 134,
  hydrogenation of ethene 683–4
Ethanol 323–4
Ethene 163, 165–70, 184, 185
  hydrogenation 519, 664, 683–4
Ethyl radicals, recombination of 684
Ethyne 164, 166
Evaporation 471, 496–7
Evolutionary search program 880
Excess charge 188–9
Excimer laser 246–7
Excitation energy 242
Excitation spectroscopy, single molecule 841
Excited domain formation 808
Excited states 19
  absorption from 240–1
  helium 98
  hydrogen atom 58–64
  trial functions 68–9
Exciton
  formation 808
  lifetime of fluorescence 808–9
  speed 809
Expansion of a gas 365, 474–7
Extensive quantities 526
Extinction coefficient 205
Extracellular channel 837, 838–40
Eye with lens, emergence 894
Eyring equation for bimolecular reaction 675–6

Faraday constant 602
Fast Fourier transformation 321
F–Cl molecule 126
Fe^{3+}/Fe^{2+} electrode 611
FEMO 165, 212
Fermi resonance 303
Fermi energy 342, 343
Fick's law 379–81
Field strength 214
First law 511
First-order reaction 680–2
Fitness parameters 892
Flame temperature 526–7
Flash photolysis 720–1
Flavin adenine dinucleotide (FAD) 848
Flory equation 767
Flow methods 720
Fluorescence
  excitation spectra 236
  intensity 801–2, 807, 818, 820
  lifetime 274–6, 803
  quantum yield 242
  quenching 242, 817, 819
  relative to absorption 238–9
  spectrum 806

Fluorine
  atom 131
  molecule 125
Free energy see Helmholtz energy
Free enthalpy see Gibbs energy
Force constants 130, 165, 301
  of CO from Raman spectrum 309
  of H_2^+ 316–17
Form factor 766
Formal charge 152–4
2-Formyl-methyl-norbornene 692–3
Foerster equation 863
Franck–Condon principle 312, 313
Franck–Hertz experiment 20–1, 98
Free-electron model 165–81, 183, 184,
  195–7, 211, 237
Freezing point 563
  of ice 628
Frequency factor 669–72, 682–5
Fuel cells 618–19
Fugacity 636–40
Fugacity coefficient 639, 650
Fullerenes 187
Fulvene 189, 193–5
Fundamental equilibrium relation 545

Galvanic elements 616
Gas constant 367
Gas electrodes 607–8
Gas expansion 474–7
  reversible 471–2
  with increasing temperature 365
Gas mixtures, reactions in 473–4, 483, 546–60
Gas molecules, thermal motion of 365
Gas particles 357
Gas reactions
  collision theory for 663–9
  rate equation for 669–72
Gas thermometer 368–9, 398
Gaseous state, part of reactants or products 573–5
gauche conformation 743, 744
Gay Lussac's law 364
Gel electrophoresis of DNA fragments 778
Genetic apparatus 885–94, 906–13
Genetic code 885, 892, 917–8
Genetic translation device 890–1
Gibbs adsorption equation 737–8
Gibbs Duhem equation 648
Gibbs energy 517, 528–35, 545, 600, 648, 650, 712, 755, 756, 826
  pressure dependence 533–4
  reactions at constant pressure 548–9
  reactions at constant volume 546–8
  reactions in dilute solution 560–75
  reactions involving ions 568–70

Gibbs energy of formation 530
   in aqueous solution 571–2
      solutions of charged particles 572–3
      solutions of neutral particles 570–1
Glass
   electrodes 615
   monolayers on 727
Glucose
   biosensor 852
   combustion 592
   degradation 592–4
   energy balance of formation 592–4
   synthesis 591–2
Glucose oxidase (GOD) 852
Gravity competing with thermal motion 381–3
Glycylalanine 571
Gouy Chapman model 658
Graham's law 370
Ground state 19
   electron cloud 27, 28
   H_2^+ molecule ion 86
   helium 96–7
   hydrogen atom 27–31, 53–8
   hydrogen molecule ion 31–3
   population inversion 243
   trial function 66
Group transfer potential 589–90
Group transfer reactions in biochemistry 588–91
Grotthus mechanism 624
Guanidinium cation 178, 179
Guanidinium ion 180
Guanine 348

H atom 27–31
H_2^+ molecule 31
H_2O molecule 125, 126, 148–50, 348
H_2S molecule 148–50
HA-devices 891
Haber–Bosch process 552
Hagen–Poiseuille's law 387
Hairpins 891
Halobacteria 230, 836–44
   chloride ion pump in 842
   sensory receptor in 843–4
   solar energy conversion in 836–44
Halobacterial cell 836, 837
   proton pump in 837
Hamiltonian function 70–1
Hamiltonian operator 45, 71, 265
Hardboiling of eggs 686–7
Harmonic oscillator 291, 302
   classical 289, 293
   quantum theory 290–4
Harpoon reaction 684–5

HCl molecule 278, 284, 294, 349
   deflection of H atom 288
   equilibrium state 287
   rotational motion 279
   rotational motion of H nucleus 280, 281
Heat 450–68, 492–516
   arbitrary processes 501–7
   ideal gas 364, 493–500
   produced in burning 521
Heat capacity 539–40
   at constant pressure 461–2
   at constant volume 452–3
   electronic contribution to 457
   of solids 457–8
   quantum effects on 456–7
   rotational and vibrational contribution to 453–5
   temperature dependence of 453–60
Heat conversion into mechanical work 472
Heat engine
   arbitrary 502–4
   efficiency 500
   reversible 502–3
Heat exchange 518–22
Heat pump 504
   efficiency 500
Heat transfer 520
Heisenberg uncertainty principle 25–6
Helium
   excited states 98
   ground state 96–7
   ionization energy 118
Helium atom 34, 96–8, 350
   energy in ground state 111–12
Helium ion 34
Helium-like systems 34–5
Helium molecule, nonexistence of He_2 36
Helmholtz energy 528–9
Henderson–Hasselbalch equation 579–80
Hertz antenna equation 274–6, 748
Hertz equation 6
Hess's law 519
Heteroatoms as probes for electron distribution 216–19
High-density data storage 835
Highest occupied molecular orbital. *See* HOMO
H–N–H angle 126
H–O–H angle 126
HMO model 181–4, 198
HOMO 211, 212, 217–23, 223, 228–31, 838, 841
Hooke's Law 133
Hückel $4n + 2$ rule 177
Hückel molecular orbital (HMO) method 181–5
Hückel molecular orbital (HMO) model 181–4, 198–9

INDEX

Hund's rule 121, 122
Hybrid atomic orbitals 147–51
Hybrid functions 142–4, 154
Hydrodynamic properties, statistical chain element 762–5
Hydrogen 356
Hydrogen atom 33–4, 131, 278–9
  box model 28
  emission spectra 61–4
  energy 54
  energy diagram 64
  energy splitting in magnetic field 99
  excited states 58–64
  ground state 27–31, 53–8
  hybridization of functions 144–6
  ionization energy 30, 118
  lowest energy state of electron 55
  probability densities for electron in 56
  Schrödinger equation 80–1
  wavefunctions 53, 62
Hydrogen bond 333, 347–8, 850
  aggregate vibrations 347–8
Hydrogen cyanide 164
Hydrogen ion
  mobility 623–4
Hydrogen molecule ion 33–7, 85–95
  box model 32–3
  electron charge 86
  exact wavefunction and energy 106–8
  force constant of 316–17
  formation, from H atom and proton 32
  ground state 31–3
  mobility 623–4
Hydrogen molecule 362, 373
  and hydrogen ion 35–6
  antisymmetry postulate in 320
  electron cloud 35–6
Hydrogen radicals 697
Hydrogenation 176–7
  of ethene 519, 664, 683–4
Hydrophilic surface 712
Hydrophobic surface 712
Hyperpolarizability 815, 817

Ideal gas 493–500
  expansion at constant temperature 493–4
  expansion into vacuum 497–8
  reversible heat engine 498–500
  reversible heating at constant volume 495
  theory 629
Ideal gas law 364–9, 462, 634
Ideal solution 646–7
Ideal dilute solution 560–2
Immiscible condensed species 553
Indicator dyes 222
Indistinguishable particles 434–5, 448, 478–9, 484–6
Induced dipole moment 304, 305

Induction energy 346
Induction forces 348–9
Infrared spectra 297, 300
Information and knowledge 895–9
Information processing, energy dissipation 896
Inhibitor-substrate complex 702
In-phase oscillation 805, 806
Insoluble monolayers 738–40
Integrated absorption 261–6
  calculation 262–4
Intensive quantities 526
Interfaces 732–43
Interfacial tension 732–6
Interference
  computer simulation 14
  definition 4
Interlocking components 796
Intermolecular forces 333–55, 627
  affecting mean free path 405–7
Internal energy 450–68, 517, 522–7
Intersystem crossing 239, 240
Inversion of ground-state population 243
Inversion temperature 644–6
Iodine hydrogen
  equilibrium 691
  reaction 557, 697
Iodine molecule 129, 132
Iodine radicals 697
Ion distribution at charged plate 657–62
Ion mobility 620–2
Ion pair 334, 335
Ion transfer into concentrated solution 650–1
Ionic crystals 36–7
  forces in 333–40
Ionic Gibbs energy of formation 572–3
Ionic strength 660
Ionization energy 96, 351
  helium atom 118
  hydrogen atom 30, 118
  lithium atom 119
Ionization process 315
IR absorption of CO_2 303
IR spectrum 302
Irreversible changes of state 469–70
Irreversible process 469
  entropy increase in 483
  probability of reversion 477
Isolated system
  arbitrary process 504–5
  entropy increase in irreversible process in 483
Isolated systems 464–5, 498
Isothermal compression 636
Isothermal process 467–8
Isotope effect 326
Isotopes, separation 399

Joule–Thomson effect 642–6

Kelvin scale 366, 369, 511
Kinetic energy 20, 24, 30, 89, 341–2, 358, 477, 675
    accumulation 666–9
Kinetic gas theory and temperature 356–69
Kinetics of chemical reactions 663
Knowledge and information 845–9
Kuhn length (statistical chain element) 757

Lambert and Beer law 205, 206, 261
Landolt reaction 708
Langmuir–Blodgett deposition sequence 800
Langmuir–Blodgett method 739
Laplacian operator 45
Laser 243–7
    HeNe 6
    dye 243–7
    excimer 246–7
Lattice energy 341, 344
Lattice structure 335
Lattice types 334–40
Law of Entropy Increase 483
Law of Lambert and Beer 205, 206, 261
Laws of nature 901
LCAO 91–3, 181
    approximation 93
    integrals in H_2^+ 108–10
    model, O_2 molecule 155, 157, 158
    trial function 91–3, 159
    wavefunctions 155, 157
Lead accumulator 617, 618
Le Chatelier's principle 555, 627, 628
Leclanché element 616–7
Leclanché cell 617
Leclanché element 616
Life
    evolution 887–90
    general aspects 895
    origin of 880–905
    physicochemical processes 892–905
Lifetime 235, 274–6
Light
    absorption 204–77
    definition 1
    diffraction 5, 8, 14
    emission 204–77
    intensity 7
    particle nature 1–3
    quanta 3
    wave nature 3–7
Light scattering 302–4, 788
Light scattering, statistical chain element 760–2

Lindemann–Hinshelwood mechanism 729–31
Linear combination of atomic orbitals. See LCAO
Linear hybrid functions 144, 145
Linear hybrid orbitals 148, 164
Linear molecules, rotation of 282–4
Linear π electron systems 165–71
Linear triatomic molecule 300
Lineweaver–Burk plot 701
Lipid vesicles 752
Liposomes 752
Liquid crystals 396, 743–9
    orientation, monolayers in 834
    phase transitions 744–7
Liquid nitrogen 645
Liquids
    collisions in 387–8
    thermal motion in 387–93
Lithium
    atom 147, 344
        ionization energy 119
    cell 616–17
    crystal 342–4
    ions 34–7, 341, 344
    lattice 340
Localized electrons 147–51
Lock-and-key concept 794, 796, 880, 889, 906
Lone electron pair 253
Lotka–Volterra mechanism 710
Lower explosion limit 699
Lowest unoccupied molecular orbital. See LUMO
Luminescence lifetime 812–15
LUMO 211, 212, 217, 218, 219, 219–23, 223, 228–31, 838, 841
Lyman Series 64
Lysine 837

Macromolecules 755–86
Macroscopic models
    random coil, refined theory 765–9
Madelung constant 336–7, 339
Magnetic field
    current loop 99–100
    energy splitting in 99
Magnetic flux density 101
Magnetic interactions of electrons 103
Magnetic moment 100
Magneton
    Bohr 99
    nuclear 320
Mallory cell (mercury cell) 617, 618
Malonic acid, bromination of 696–7
Mariotte's law 361

Mass action law 546
  derived from chemical potentials 648
  solutions of charged particles 567
  solutions of neutral particles 565–7
Maxwell–Boltzmann distribution 408, 422–7, 665
Maxwell–Boltzmann equation 424
Maxwell demon 897–8
MBBA 744
Mean displacement 374–7
Mean energy 343–4
Mean free path 370–3, 672–4
  intermolecular forces affecting 405–7
Mean speed 358, 369
Melting point 394
Membranes 750–5
Membrane potential 615, 754, 841
Meniscus 736
Mercury cell (Mallory cell) 617, 618
Messenger ribonucleic acid (mRNA) 886
Metal electrodes 604–7
Metal lattice 340
4-Methoxybenzyliden-4′-butylaniline (MBBA) 744
Methyl cellulose, diffusion in water 767
Methylorange 222
Methane
  hybrid orbitals 150
  combustion
$MgSO_4$ 648
$MgSO_4$ solution, activity coefficient in 652
Micelles 742
Michaelis–Menten mechanism 700–3
Mixing of gases. *See* Gas mixtures
Mixing entropy 483
Mobility of DNA 792
Molality and concentration 562
Molar enthalpy
  of formation 524–5
  of reaction 525–6
Molar Gibbs energy
  of formation 530
  of reaction 531
Molar heat capacity 453, 458, 523–4, 539
Molar mass 360
Molar quantities 360
Molar rotation 247
Molar rotational energy 419
Molar vibrational energy 415
Molar volume 368, 630
Molecular assemblies
  energy distribution in 408–49
Molecular crystals 353
Molecular engineering 794
Molecular motion 393–6
Molecular orbitals 97, 131, 140, 154–9
Molecular recognition 847–53
Molecular replica formation 850–2

Molecular structure, simple molecules 124–36
Molecular volume 630–1
Molecular wires, conduction through 823–6
Molecules
  thermal motion of 356–407
  with π electron systems 163–203
Moment of inertia 287
Momentum transferred per collision 386
Monochromatic light 5
Monolayer assemblies
  elucidating photographic sensitization 821–2
  manipulation 797–9
  photoinduced electron transfer in 818–19
  switching 819–21
  to study energy transfer 800
Monolayers
  contact angle on 742
  light-induced change of properties 833–4
  liquid crystal orientation 834
  on glass 740
  transfer 739
  tunneling current through 831–3
Morse potential 134–5
Multiplication phase 892, 895

NaCl
  crystal 336
  lattice 337
  solution, activity coefficient in 652
$NAD^+/NADH$ electrode 612
NADPH 591–2, 592
NaOH, titration of acetic acid by 597–9
Natural line width 235
Nematic state 744
Nernst equation 609–10
Neutralization reaction 685
NH_3 formation from elements at 298 K and 700 K 552
Ni/Cd accumulator 617, 618
Nitrogen
  atoms 126, 377
  liquid 645
  molecule 373
NMR 320–4
NO molecules 428
Nonlinear optical phenomena 815–18
Normalization condition 92, 142
Normalization of wavefunctions 47–50
Normalized functions 142
Nuclear magnetic resonance (NMR) 320–4
  signals 323
  spectrum 323–4
Nuclear motion 278
  manipulation 833–47
  quantization 280

Nuclear spin   318–20
Nuclear wavefunctions   319
Nuclei, particle and wave properties   278–332
Nucleotides, chirality   914
Number of configurations   430, 449, 469, 477–9
   of atomic gas   477–9
Number of quantum states   477–8

Octatetraen   202–3
One-dimensional free-electron model of π electrons   170–1
One-electron coordinates   97
One-electron wavefunctions   97, 237
Open systems   464–5
Optical activity   247–60
Optical phenomena, nonlinear   815–18
Optical rotation   247
Optical rotatory dispersion (ORD)   247–53
Optimum error frequency   884
Orbital energy   121
Origin of life   880–921
   modeling   913–19
Orthogonality of wavefunctions   50–2
Orthohydrogen   318–20, 456–7
Oscillation
   electron between protons   112–16
   reactions   710–13
Oscillation period   115
Oscillator
   box model   291–3
   chemical   712
   classical   261–6, 289, 293–4
   harmonic   291, 302
   quantum mechanical   287–94
   ruthenium–bromate–bromomalonic acid (BrMA)   712
   three-dimensional   458
   see also Coupled oscillators
Oscillator strength   213–14, 227, 264
   coupled oscillators   270
   cyanine dyes   266–7
   quantum mechanical treatment   265–7
Osmotic pressure   560–1
Out-of-phase oscillation   805, 806
Overlap   35, 124–6, 131–3, 148–54, 165, 182
Overlap integral   182
Oxidation of CO   531
Oxygen electrode   613
Oxygen molecule
   box model   154–6
   charge cloud   157
   LCAO model   155, 157, 158

p-azoxyanisole (PAA)   746
PAA   746

Paraffinsulfochloride   694–5
Parahydrogen   318–20, 456–7
Parallel reactions   692–3
Partial molar volume   646–8
Partial specific volume of solute   763
Partial volume   475
Particle
   between parallel walls   22–5
   in a box   45–7
Particles each in one of three energy states   477
Partition function   412, 413, 418, 444
   for vibration   442
Patch clamp technique   753
Pauli exclusion principle   36–7, 102–3, 118
P branch   297
Pentylcyanobiphenyl (5CB)   744
Periodic perturbation   3
Periodic table   117–39
   and Aufbau principle   120–4
   basic principles   117–18
Permittivity of the vacuum   27
Peptide link   571
Phase equilibria   627–8
Phase transitions, liquid crystals   745
Phases   396
Phosphatidylcholine   751
Phosphorescence   233–6
   relative to absorption   238–9
Phosphoric acid   588–9
Photoelectric effect   1–3
Photoelectron spectrum   167, 314
Photoelectron spectroscopy   312–17
Photographic sensitization, monolayer assemblies for elucidating   821–2
Photoinduced electron transfer
   in monolayer assemblies   818–19
   switching   819–21
Photoinduced proton pump   837–41
Photoionization   315
Photons   3
   behavior   15
   comparison with electrons   13
   motion manipulation   800–18
Photosensitization   826–8, 836–8
Phthalocyanine   223–7, 233, 256
π electron cloud   838
π electron density   180, 838
π electron orbitals   155, 167, 168, 174
π electron pair   212
π electron systems   824
   electron transfer   871–7
   linear   165–71
π electrons   163–204
   bonding   163–5
   coupling   225–8
Planck's constant   9

INDEX

Poisson equation   659, 661, 662
Poiseuille's formula   387
Polarizability   348–51, 798
Polarization of absorption bands   215–16
Polyatomic molecules   130, 297–302, 317–18
  infrared spectra   297–9
  Raman spectra   311
Polyenes   187, 207–9, 219, 223, 317, 815–16
  absorption maxima   221
Polyethylenglycol   773
Polymer chains   776
  moving unraveled chain through   777
Polymers   755
Polystyrene, sedimentation   776
Population numbers   413–14, 418, 433
Porphyrin   225–8, 233
  coupling effect   272–4
Potential energy   24, 53, 87, 89, 217, 220, 291, 316
Potential trough   42, 77, 217, 220, 231
Potential of electrochemical cell   601
Prebiotic chemistry   885
Pressure
  and thermal motion   357–62
  calculation   357
  on wall of container   359
  osmotic   560–1
  vapor   381, 559
Pre-exponential factor   671
Principle of corresponding states   634
Probability   25, 48, 54, 72, 174, 412, 475, 668
Probes for electron distribution, heteroatoms as   216–19
Programmed environmental change concept   794, 796, 880, 889, 906
Programmed interlocking molecules   795–800
Programming echo radiation field   812–15
Proteins
  electron transfer through   833
  synthesis   886
  translation   894
Proton   320, 323
  electron transfer to   587–8
  tunneling   687–9
Proton pump
  driving force   841–2
  in halobacterial cell   839
Proton transfer
  electron transfer coupled with   586–7
  reactions   579–84
Proton translocation   837–41
  from cytoplasmic to extracellular channel   841
Pseudoisocyanine bromide   210
Pumping   243–4

Push–pull polyene   816, 825
PV curve   635

Quantum effects on heat capacity   456–7
Quantum mechanical oscillator   287–94
Quantum mechanical rotator   278–89
Quantum mechanical tunneling   104
Quantum mechanics   28, 102, 420, 880
Quantum numbers   61, 64, 82, 216, 284, 289, 298, 417, 427
Quantum states   408–13, 417, 424–30, 433, 434, 448, 477–8
Quantum yield of fluorescence   242
Quenching, fluorescence   242, 819
Quinone   586, 864
Quinone/hydroquinone electrode   611–12
Quinthiophene molecules   823

Radiation echo field   812–4, 867–70
Raman effect   302, 304
Raman spectra   302–12
  force constant of CO from   309
  of polyatomics   311
Random coil   755–9
  macroscopic models   765
  time to restore from unraveled chain   776
  winding through meshwork   777–8
Random walk   401–3
Raoult's Law   563–5
Rate constants   669–72
  Belousov–Zhabotinsky reaction   718–19
  electron transfer   875
  experimental data   677–89
Rate determining step   695
Rate equations   698
  gas reactions   669–72
  reactions in solution   672–5
Rate law for chain reaction   698–9
Rayleigh line   304
Rayleigh scattering   302–4
R branch   297
Reactants as identical molecules   672
Reaction coordinate   675
Reaction quotient   547, 549
Recombination of ethyl radicals   684
Redox reactions   585, 611–13
Reduced mass of diatomic molecules   290
Reduced mass
  diffusion   376
Reference electrodes   611, 613–14
Refractive indices   747, 760
Relaxation methods   719
Repelling forces   333–4
Replica formation   847–53
Replicable strand   888–9
Resonance energy   175–7

Resonance frequencies of coupled oscillators 270
Retinal 230–1
Reversible adiabatic expansion 513
Reversible changes of state 470–2
Reversible evaporation 471
Reversible expansion of a gas 471–2
Reversible heat 495
Reversible heat engine, ideal gas 498–500
Reverse osmosis 577
Reversible process 469
Reversible work 473, 529–30, 546, 548
Rhodamine 244
Rhodopsin 844
RNA replication 894
Rod-like micelles 744
Rotation
 in a plane 283–4
 of diatomic molecule 286
 of linear molecules 282–4
Rotational energy distribution 417–19
Rotational motion of diatomic molecule 287
Rotational Raman spectra 304–7
 of CO 306
Rotational spectra 282–7
Rotator, solution of the Schrödinger equation 330–1
Rotatory dispersion 247–8
RRK-theory 731
Ru^{2+}, autocatalytic oxidation 704–7
Rubber elasticity 781–6
Rule of Dulong and Petit 458
Ruthenium–bromate–bromomalonic acid (BrMA) oscillator 712
Rydberg constant 64

Sackur–Tetrode equation 480, 508
Scanning near-field optical microscopy (SNOM) 853, 856
Scanning tunneling microscopy (STM) 377, 853–4
Scattering light 303
Schadt–Helfrich effect 748, 749
Schiff base 230, 837–44
Schrödinger equation operator form 45, 39–64, 65, 85, 86, 279, 288, 293, 330–1
 hydrogen atom 80–1
 one-dimensional 42–4
 three-dimensional 44–7
 time-dependent 52, 79–80, 265
 time-independent 79–80
Second harmonic generation 817
Second law 512
Second-order reaction 680–2
Sedimentation 763
 cellulose acetate in acetone 767
 polystyrene 776

Sedimentation constant 763, 766, 767, 768
Selection rules 266, 295, 306
Selection phase 892
Selective reflection 746–7
Self-assembled monolayers 741
Self-consistency 200
Semipermeable membrane 560, 615
Sensory receptors 843–4
 in halobacterium 844
Shielding constant 123
σ electrons 167
σ orbitals 155
Silver chloride electrode 614
Single molecule
 absorption 210
 emission 235
Singlet state 236–8
Singlet–triplet splitting caused by Coulomb forces 103
Slater atomic functions 184
Slater atomic orbitals 181
Slater functions 121, 158
Slater wavefunctions 123–4
Smectic A 744
Smectic C 744
Smectic state 744
SNOM 853, 856
Soap lamella 750
Sodium chloride lattice 335, 337–9
Sodium dodecyl sulfate 750
Soft medium, electron transfer in 877–9
Solar energy 805
 conversion 826–30
 in halobacteria 836–44
 harvesting in biosystems 810–12
Solid surfaces 741–2
Soliton 810
 conduction 825
 switching 825
Solutions
 activities and equilibrium constants in 649
 rate equation for reactions in 672–5
Solvent draining through meshwork 776–7
STM, see scanning tunneling microscopy
Space-filling models 131, 132
Spectral hole burning 210
Speed of ions 623
Speed of molecules in a gas 369–70
Spin variables 102
Spirobisanthracene, circular dichroism of 250–3
Splitting of absorption band 252
Splitting
 of EPR signal 105
Spontaneous emission 233–42
Spontaneous reactions 527–30
Standard conditions 524, 549
Standard electrode potential 609–11

INDEX

Standard hydrogen electrode (SHE) 611, 613
Standard potential, practical determination 610–11
Standing waves 22–7, 40
  in a box 46
State and state variables 462–3
Stationary states 19
Statistical chain element 755–60, 771
  hydrodynamic properties 762–5
  light scattering 760–2
Staudinger index 765
Steady-state approximation 696, 698
Step potential model 201–2
Stepwise proton transfer 583–4
Steric factor 672
  experimental data 677–89
Stern–Volmer equation 242
Stimulated emission 242–7
Stirling's formula 431–2, 447, 479, 490
Stokes–Einstein equation 389, 390, 762
Stokes law 621
Stokes lines 306
Strand evolution 907
Strand folding 908
Strand lengthening 907–8
Stretching
  Hooke's law 133–4
  modes 298
  Morse function 134–5
Succinic acid dehydrogenase 703
Sulfur atom 148
Sunlight 592
Superhelix 438–40
Superradiance 809
Supramolecular chemistry 796
Supramolecular devices
  fabrication 880
  for photoinduced movement 846
Supramolecular engineering 794, 796
  genetic apparatus 885–94, 906–13
Supramolecular machines 794–879
  energy transfer 794–5
  fabrication 800
  prototypes 794
Supramolecular structures 786
Surface films 737–8
Surface pressure 738
Surface tension 732–36
Surfactant 737
Sutherland constant 405, 407
Symmetric cyanine dye 208, 211, 213, 223
Symmetric function 98
Symmetric spatial wavefunction 102
Symmetric spin function 102
Symmetric stretching mode 300, 302

Symmetric wavefunctions 319
System and surroundings 462

Temperature and kinetic gas theory 356–69
Temperature dependence of heat capacity 453–60
Temperature equilibration 482–3
  and heat 364
Temperature jump method 737–8
Terrylene molecule 236
Tetrahedral hybrid functions 146, 149
Tetrahedral hybrid orbitals 150, 151
Tetrahedral hybridization 154
Tetramethylsilane (TMS) 321–2
Thermal energy 381
Thermal equilibration 494–6
Thermal motion
  and pressure 357–62
  gravity competing with 381–3
  in liquids 387–93
  of gas molecules 365
  of molecules 356–407
Thermodynamic laws 511–12
Thermodynamic temperature scale 511
Third law 512
Three-dimensional box model 165–70
Three-dimensional oscillator 458
Thymine 348
Time-dependent probability density 114
Tiny bang 887–8
Titration curve of weak acid 582
Titration of acetic acid by NaOH 597–9
TN-display 748
Total rotational energy 418
Total vibrational energy 414–17
trans conformation 755, 756
Transfer ribonucleic acid (tRNA) 886
Transition moment 216, 282
Transition state theory 675–7
Translation device 912–13
Translational energy 364–8, 420–8, 433, 434
Transmittance 204–7
Traveling wave 9
Trial wavefunction 291
  excited states 68–9
  ground state 66
Trial function 30, 66–7, 73, 159, 291
Triatomic linear molecule 300
Trigonal planar hybrid functions 144, 146
Trigonal planar hybrid orbitals 164
Triple bonds 163
Triple point of water 361, 559
Triplet state 236–8
  tunneling 688–9
Triplet–triplet absorption 240, 241
Trypsin 704
Tuning 246

Tunneling 112–16, 823
Tunneling microscopy 853
  current through monolayer 831–3
  deuteron 687–9
  electron 823
  in bacterial reaction center 828–30
  probability 116
  process 819
  proton 687–9
  quantum mechanical 104
  triplet state 688–9
Twisted nematics 748
Two-electron system 238
Two-electron wavefunction 97
Twofold degeneracy 142

Ultracentrifuge 383, 399
  separation of isotopes 399
  sedimentation of macromolecule 763
Unimolecular reaction 729–31
Universe, evolution 875–8
Upper explosion limit 699

Valence electrons 154, 171
van der Waals
  constants 633–4
  equation 630–4, 636
  gas 636
  model 642
  radii 129–31
Van't Hoff equation 554–5
Vapor pressure 559–60
  depression 562–3
  sugar solution in water 563
Variation principle 19, 28–31, 65–9, 292
  justification 65–9
  proof 82–4
Vibrating string 40
  displacement 22
Vibration, selection rules 296–7
Vibrational energy distribution 412–17
Vibrational modes 297–8
  for CO_2 310
Vibrational Raman spectrum for CO_2 311
Vibrational-rotational Raman spectra of
  diatomic molecules 307–8
Vibrational-rotational spectra 294–302
Vibrational states of string 23
Vibrational structure
  electronic absorption bands 331–2
  electronic spectra 312–18
Vibrations

branched string 178
hydrogen bond aggregates 347–8
Virial coefficient 634–5
Virial theorem 57, 91, 93–4, 106–8
Viscosity
  arising from collisions of molecules 383–7
  calculation 384–6
  cellulose acetate in acetone 768
  of a gas, dependence on pressure 386
  of a gas, increasing with temperature 386
  of solution of macromolecules 764, 767
  liquid 389
Viscous flow 383–4
Visual system 844–5
Vitamin B_{12} 231–2

Water
  bond model 125, 132
  hydrogen bonding 347
  structure of ice crystal and liquid 394
  phase diagram 395
  critical behavior 396
  heat of formation 518
  formation in detonating gas reaction 543, 699
  thermal dissociation equilibrium 551
  boiling, dependence on pressure 577
  ionic dissociation constant 607
  proton mobility 624
  dissociation 550–1, 574, 607–8
Water vapor, dissociation of 545, 550–1
Wave equation 39–41
Wavefunctions 23, 26, 45, 86, 102, 180
  antisymmetry 99–103
  hydrogen atom 53, 62
  normalization 47–50
  orthogonality 50–2
Wave number 297, 310, 317
Wavelength from double slit experiment 5–6
  new knowledge 15
Wilson cloud chamber 10
Work function 3
Work 450–68
  reversible 529–30, 546, 548

Young's modulus for bending 759

Zero-point energy 414–15
Zinc dissolution in acidic medium 574
Zn-tetraphenyl-porphyrin 227
ZnS lattice 337, 338